T0299218

ORTHOGONAL POLYNOMIALS IN THE SPECTRAL ANALYSIS OF MARKOV PROCESSES

In pioneering work in the 1950s, S. Karlin and J. McGregor showed that the probabilistic aspects of certain Markov processes can be studied by analyzing the orthogonal eigenfunctions of associated operators. In the decades since, many authors have extended and deepened this surprising connection between orthogonal polynomials and stochastic processes.

This book gives a comprehensive analysis of the spectral representation of the most important one-dimensional Markov processes, namely discrete-time birth–death chains, birth–death processes and diffusion processes, and brings together all the main results from the extensive literature on the topic with detailed examples and applications. It also features an introduction to the basic theory of orthogonal polynomials and has a selection of exercises at the end of each chapter. The book is suitable for graduate students with a solid background in stochastic processes as well as researchers in orthogonal polynomials and special functions who want to learn about applications of their work to probability.

Encyclopedia of Mathematics and Its Applications

This series is devoted to significant topics or themes that have wide application in mathematics or mathematical science and for which a detailed development of the abstract theory is less important than a thorough and concrete exploration of the implications and applications.

Books in the **Encyclopedia of Mathematics and Its Applications** cover their subjects comprehensively. Less important results may be summarized as exercises at the ends of chapters. For technicalities, readers can be referred to the bibliography, which is expected to be comprehensive. As a result, volumes are encyclopedic references or manageable guides to major subjects.

ENCYCLOPEDIA OF MATHEMATICS AND ITS APPLICATIONS

All the titles listed below can be obtained from good booksellers or from Cambridge University Press.
For a complete series listing visit www.cambridge.org/mathematics.

131 H. Morimoto *Stochastic Control and Mathematical Modeling*
132 G. Schmidt *Relational Mathematics*
133 P. Kornerup and D. W. Matula *Finite Precision Number Systems and Arithmetic*
134 Y. Crama and P. L. Hammer (eds.) *Boolean Models and Methods in Mathematics, Computer Science, and Engineering*
135 V. Berthé and M. Rigo (eds.) *Combinatorics, Automata and Number Theory*
136 A. Kristály, V. D. Rădulescu and C. Varga *Variational Principles in Mathematical Physics, Geometry, and Economics*
137 J. Berstel and C. Reutenauer *Noncommutative Rational Series with Applications*
138 B. Courcelle and J. Engelfriet *Graph Structure and Monadic Second-Order Logic*
139 M. Fiedler *Matrices and Graphs in Geometry*
140 N. Vakil *Real Analysis through Modern Infinitesimals*
141 R. B. Paris *Hadamard Expansions and Hyperasymptotic Evaluation*
142 Y. Crama and P. L. Hammer *Boolean Functions*
143 A. Arapostathis, V. S. Borkar and M. K. Ghosh *Ergodic Control of Diffusion Processes*
144 N. Caspard, B. Leclerc and B. Monjardet *Finite Ordered Sets*
145 D. Z. Arov and H. Dym *Bitangential Direct and Inverse Problems for Systems of Integral and Differential Equations*
146 G. Dassios *Ellipsoidal Harmonics*
147 L. W. Beineke and R. J. Wilson (eds.) with O. R. Oellermann *Topics in Structural Graph Theory*
148 L. Berlyand, A. G. Kolpakov and A. Novikov *Introduction to the Network Approximation Method for Materials Modeling*
149 M. Baake and U. Grimm *Aperiodic Order I: A Mathematical Invitation*
150 J. Borwein et al. *Lattice Sums Then and Now*
151 R. Schneider *Convex Bodies: The Brunn–Minkowski Theory (Second Edition)*
152 G. Da Prato and J. Zabczyk *Stochastic Equations in Infinite Dimensions (Second Edition)*
153 D. Hofmann, G. J. Seal and W. Tholen (eds.) *Monoidal Topology*
154 M. Cabrera García and Á. Rodríguez Palacios *Non-Associative Normed Algebras I: The Vidav–Palmer and Gelfand–Naimark Theorems*
155 C. F. Dunkl and Y. Xu *Orthogonal Polynomials of Several Variables (Second Edition)*
156 L. W. Beineke and R. J. Wilson (eds.) with B. Toft *Topics in Chromatic Graph Theory*
157 T. Mora *Solving Polynomial Equation Systems III: Algebraic Solving*
158 T. Mora *Solving Polynomial Equation Systems IV: Buchberger Theory and Beyond*
159 V. Berthé and M. Rigo (eds.) *Combinatorics, Words and Symbolic Dynamics*
160 B. Rubin *Introduction to Radon Transforms: With Elements of Fractional Calculus and Harmonic Analysis*
161 M. Ghergu and S. D. Taliaferro *Isolated Singularities in Partial Differential Inequalities*
162 G. Molica Bisci, V. D. Radulescu and R. Servadei *Variational Methods for Nonlocal Fractional Problems*
163 S. Wagon *The Banach–Tarski Paradox (Second Edition)*
164 K. Broughan *Equivalents of the Riemann Hypothesis I: Arithmetic Equivalents*
165 K. Broughan *Equivalents of the Riemann Hypothesis II: Analytic Equivalents*
166 M. Baake and U. Grimm (eds.) *Aperiodic Order II: Crystallography and Almost Periodicity*
167 M. Cabrera García and Á. Rodríguez Palacios *Non-Associative Normed Algebras II: Representation Theory and the Zel'manov Approach*
168 A. Yu. Khrennikov, S. V. Kozyrev and W. A. Zúñiga-Galindo *Ultrametric Pseudodifferential Equations and Applications*
169 S. R. Finch *Mathematical Constants II*
170 J. Krajíček *Proof Complexity*
171 D. Bulacu, S. Caenepeel, F. Panaite and F. Van Oystaeyen *Quasi-Hopf Algebras*
172 P. McMullen *Geometric Regular Polytopes*
173 M. Aguiar and S. Mahajan *Bimonoids for Hyperplane Arrangements*
174 M. Barski and J. Zabczyk *Mathematics of the Bond Market: A Lévy Processes Approach*
175 T. R. Bielecki, J. Jakubowski and M. Niewęgłowski *Structured Dependence between Stochastic Processes*
176 A. A. Borovkov, V. V. Ulyanov and Mikhail Zhitlukhin *Asymptotic Analysis of Random Walks: Light-Tailed Distributions*
177 Y.-K. Chan *Foundations of Constructive Probability Theory*
178 L. W. Beineke, M. C. Golumbic and R. J. Wilson (eds.) *Topics in Algorithmic Graph Theory*
179 H.-L. Gau and P. Y. Wu *Numerical Ranges of Hilbert Space Operators*
180 P. A. Martin *Time-Domain Scattering*

Encyclopedia of Mathematics and Its Applications

Orthogonal Polynomials in the Spectral Analysis of Markov Processes

Birth–Death Models and Diffusion

MANUEL DOMÍNGUEZ DE LA IGLESIA
Universidad Nacional Autónoma de México

CAMBRIDGE
UNIVERSITY PRESS

University Printing House, Cambridge CB2 8BS, United Kingdom

One Liberty Plaza, 20th Floor, New York, NY 10006, USA

477 Williamstown Road, Port Melbourne, VIC 3207, Australia

314–321, 3rd Floor, Plot 3, Splendor Forum, Jasola District Centre, New Delhi – 110025, India

103 Penang Road, #05–06/07, Visioncrest Commercial, Singapore 238467

Cambridge University Press is part of the University of Cambridge.

It furthers the University's mission by disseminating knowledge in the pursuit of education, learning, and research at the highest international levels of excellence.

www.cambridge.org
Information on this title: www.cambridge.org/9781316516553
DOI: 10.1017/9781009030540

First published 2022

A catalogue record for this publication is available from the British Library.

Library of Congress Cataloging-in-Publication Data
Names: Iglesia, Manuel Domínguez de la, author.
Title: Orthogonal polynomials in the spectral analysis of Markov processes : birth-death models and diffusion / Manuel Domínguez de la Iglesia.
Description: Cambridge ; New York, NY : Cambridge University Press, 2022. |
Series: Encyclopedia of mathematics and its applications |
Includes bibliographical references and index.
Identifiers: LCCN 2021029856 (print) | LCCN 2021029857 (ebook) |
ISBN 9781316516553 (hardback) | ISBN 9781009030540 (epub)
Subjects: LCSH: Birth and death processes (Stochastic processes) | Orthogonal polynomials. |
Spectral theory (Mathematics) | BISAC: MATHEMATICS / Mathematical Analysis
Classification: LCC QA274.75 .I42 2022 (print) | LCC QA274.75 (ebook) |
DDC 519.2/33–dc23
LC record available at https://lccn.loc.gov/2021029856
LC ebook record available at https://lccn.loc.gov/2021029857

ISBN 978-1-316-51655-3 Hardback

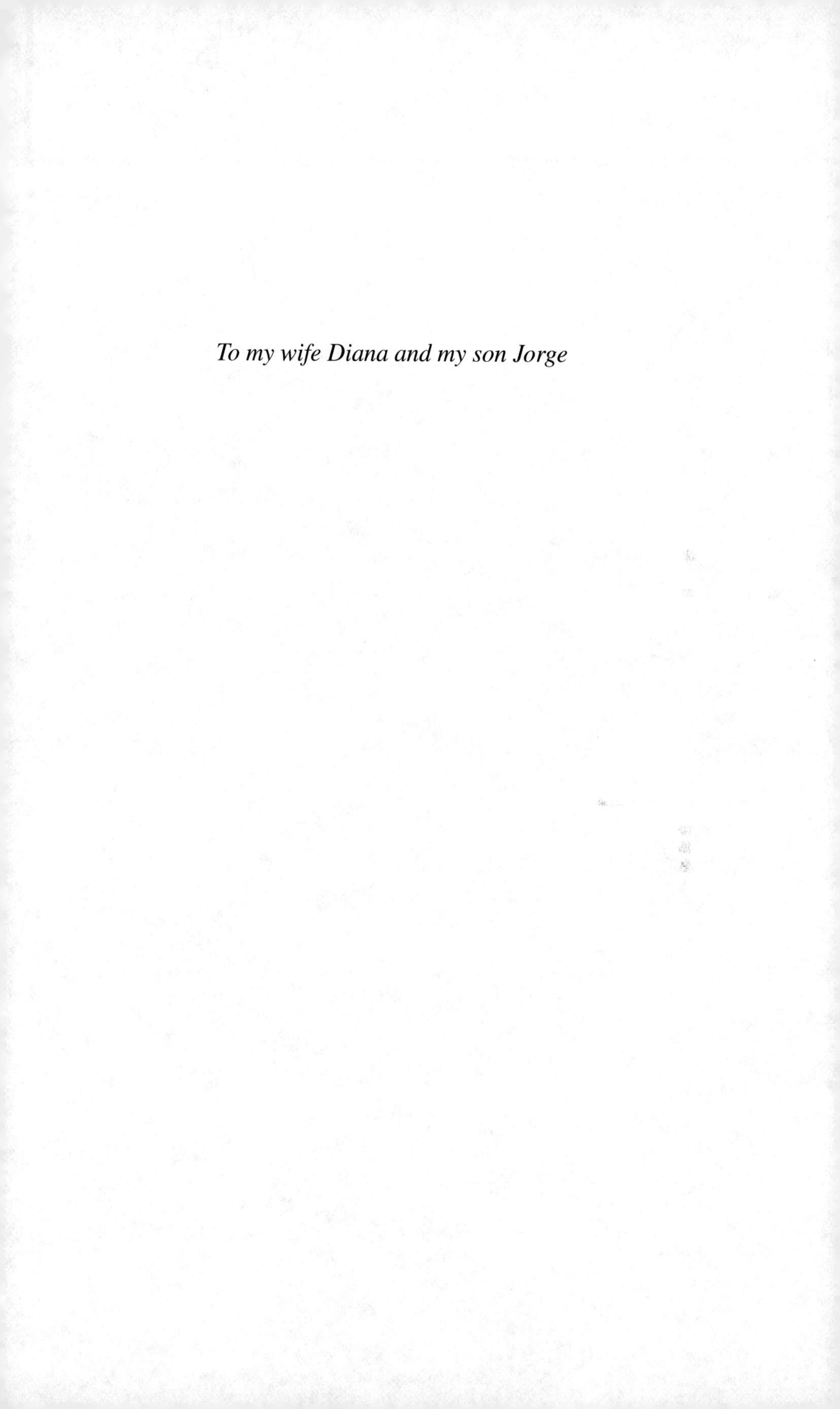

To my wife Diana and my son Jorge

Contents

Preface

The connection between stochastic processes, special functions and orthogonal polynomials has a long history. From the 1930s N. Wiener and later K. Itô knew about the connection between Hermite polynomials and integration theory with respect to Brownian motion. Around the 1950s many authors like M. Kac [80], W. Feller [53]–[56], E. Hille [71], W. Ledermann and G. E. Reuter [111], J. F. Barrett and D. G. Lampard [6], S. Karlin and J. McGregor [82]–[89], H. P. McKean [116] and D. G. Kendall [103] established an important connection between the transition probability functions of diffusion processes, continuous-time birth–death processes and discrete-time birth–death chains (in this order) by means of a spectral representation. This spectral representation is based on the spectral analysis of the infinitesimal operator associated with these special types of Markov processes and many probabilistic aspects can be analyzed in terms of the corresponding orthogonal eigenfunctions and eigenvalues. In the following years these relationships were developed further, finding connections with other stochastic processes like random matrices, Sheffer systems, Lévy processes, stochastic integration theory or Stein's method. For a brief account of all these relations see [129].

The main goal of this monograph is to give a comprehensive analysis of the main results related to the spectral representation of the most important one-dimensional Markov processes, namely discrete-time birth–death chains (also called random walks in some references, see [87]), birth–death processes and diffusion processes. Since the pioneering work of S. Karlin and J. McGregor in the 1950s, many authors have contributed to finding more applications of the spectral representation of the transition probability functions of these processes. This monograph tries to gather all the important results that appear in many publications over the last 60 years in one common text. The contents of this monograph served as a one-semester graduate advanced course taught at the Instituto de Matemáticas of the Universidad Nacional Autónoma de México in Fall 2018. The interested audience can be divided into two categories. On the one hand, it is intended for graduate students who have a solid background in the field of stochastic processes but are not so familiar with the theory of special functions and orthogonal polynomials. This monograph will give them alternative methods and ways of studying basic Markov processes by spectral methods. On the other hand, the book may also serve for students or researchers who are familiar with the theory of special functions and orthogonal polynomials but want to learn more about the connection between basic Markov processes and orthogonal polynomials.

In the experience of the author, graduate students are typically more familiar with probability theory and stochastic processes. This is the reason why an introduction to orthogonal polynomials is included in Chapter 1. This chapter also includes the concept of the *Stieltjes transform* and some of its properties, which will play a very important role in the spectral analysis of discrete-time birth–death chains and birth–death processes. A section about the spectral theorem for orthogonal polynomials (or *Favard's theorem*) will give insights about the relation between tridiagonal Jacobi matrices and spectral probability measures. We will focus then on the *classical families of orthogonal polynomials*, both of a continuous and a discrete variable. These families are characterized by the fact that they are eigenfunctions of a second-order differential or difference operator of hypergeometric type solving certain *Sturm–Liouville problems*. These classical families are part of the so-called *Askey scheme*.

In Chapter 2 we will perform the spectral analysis of discrete-time birth–death chains on $\mathbb{N}_0$, which are the most basic and important discrete-time Markov chains. These chains are characterized by a tridiagonal one-step transition probability matrix. We will obtain the so-called *Karlin–McGregor integral representation formula* of the n-step transition probability matrix in terms of orthogonal polynomials with respect to a probability measure ψ with support inside the interval $[-1,1]$. We will give an extensive collection of examples related to orthogonal polynomials, including gambler's ruin, the Ehrenfest model, the Bernoulli–Laplace model and the Jacobi urn model. The chapter ends with applications of the Karlin–McGregor formula to probabilistic aspects of discrete-time birth–death chains, such as recurrence, absorption, the strong ratio limit property and the limiting conditional distribution. Finally we will apply spectral methods to discrete-time birth–death chains on $\mathbb{Z}$, which are not so much studied in the literature.

In Chapter 3 we will perform the spectral analysis of birth–death processes on $\mathbb{N}_0$, which are the most basic and important continuous-time Markov chains. In this case, these processes will be characterized by an infinitesimal operator, which is a tridiagonal matrix whose spectrum is inside the interval $(-\infty,0]$. Again, we will obtain the *Karlin–McGregor integral representation formula* of the transition probability functions of the process in terms of orthogonal polynomials with respect to a probability measure ψ with support inside the interval $[0,\infty)$. Although many of the results are similar or equivalent to those of discrete-time birth–death chains, the methods and techniques are quite different. For instance, in this chapter, we will have to prove that the Karlin–McGregor representation formula is in fact a transition probability function of a birth–death process, something that was not necessary for the case of discrete-time birth–death chains. We will also provide an extensive collection of examples related to orthogonal polynomials, including the $M/M/k$ queue with $1 \leq k \leq \infty$ servers, the continuous-time Ehrenfest and Bernoulli–Laplace urn models, a genetics model of Moran and linear birth–death processes. As in the case of discrete-time birth–death chains, we will apply the Karlin–McGregor formula to probabilistic aspects of birth–death processes, such as processes with killing, recurrence, absorption, the strong ratio limit property, the limiting conditional distribution, the decay parameter, quasi-stationary distributions and bilateral birth–death processes on $\mathbb{Z}$.

In Chapter 4 we will perform the spectral analysis of one-dimensional diffusion processes, which are the most basic and important continuous-time Markov processes where now the state space is a continuous interval contained in $\mathbb{R}$. Diffusion processes are characterized by an infinitesimal operator, which is a second-order differential operator with a drift and a diffusion coefficient. We will obtain a spectral representation of the transition probability density of the process in terms of the orthogonal eigenfunctions of the corresponding infinitesimal operator, for which we will have to solve a *Sturm–Liouville problem* with certain boundary conditions. An analysis of the behavior of these boundary points will also be made. We will also give an extensive collection of examples related to special functions and orthogonal polynomials, including Brownian motion with drift and scaling, the Orstein–Uhlenbeck process, a population growth model, the Wright–Fisher model, the Jacobi diffusion model and the Bessel process, among others. Finally, we will study the concept of quasi-stationary distributions, for which the spectral representation will play an important role.

I would like to thank F. Alberto Grünbaum for introducing me to the fascinating connection between orthogonal polynomials and Markov processes. Back in 2009 I was visiting him as an undergraduate student at the University of California, Berkeley and we were studying one example of matrix-valued orthogonal polynomials coming from group representation theory which had a nice interpretation in terms of two-dimensional Markov chains. This was my first connection to the subject that brought me to write this monograph. I would also like to thank Eric A. van Doorn for reading the manuscript and providing an important list of corrections and additional material to include in the book. Unfortunately he tragically passed away before being able to read the final version of this book. In closing I would like to thank the staff at Cambridge University Press for their support and cooperation during the preparation of this book.

Manuel Domínguez de la Iglesia[1]
Ciudad de México

[1] Partially supported by PAPIIT-DGAPA-UNAM grants IA102617 and IN104219 (México) and CONACYT grant A1-S-16202 (México).

1

Orthogonal Polynomials

In this chapter we introduce some basic definitions and properties about the theory of special functions and orthogonal polynomials on the real line. In the first section we will introduce some basic special functions and the concept of the *Stieltjes transform*, which will be used frequently in the text. In Section 1.2 we will give some properties of the general theory of orthogonal polynomials. Section 1.3 is devoted to the *spectral theorem* and in particular applied to orthogonal polynomials, in which case it is usually called *Favard's theorem*. In Sections 1.4 and 1.5 we will focus on the so-called *classical orthogonal polynomials*, both of a continuous and a discrete variable. These special families, apart from being orthogonal, are characterized by the fact that they are eigenfunctions of a second-order differential operator (in the continuous variable) or a second-order difference operator (in the discrete variable) of the *Sturm–Liouville* type. Finally, in Section 1.6, we describe the *Askey scheme*, which is a way of organizing orthogonal polynomials of hypergeometric type into a hierarchy, where the classical orthogonal polynomials are included. This chapter is based on references [3, 9, 16, 74, 135, 137, 142].

1.1 Some Special Functions and the Stieltjes Transform

The *Gamma function* is a complex-valued function that extends the domain of the factorial function of a nonnegative integer $n!$. It was introduced by Euler in 1789 and it is defined by its integral representation

$$\Gamma(z) = \int_0^\infty e^{-t} t^{z-1} dt, \quad \operatorname{Re} z > 0. \tag{1.1}$$

Integrating by parts we obtain the functional equation

$$z\Gamma(z) = \Gamma(z+1), \quad \operatorname{Re} z > 0.$$

The formula above can also be written as

$$(z)_n \Gamma(z) = \Gamma(z+n), \quad n \geq 0,$$

where $(z)_n$ is the *Pochhammer symbol*

$$(z)_n = \begin{cases} 1, & \text{if } n = 0, \\ z(z+1)\cdots(z+n-1), & \text{if } n \geq 1. \end{cases} \tag{1.2}$$

From here we also observe that if n is a nonnegative integer, then $\Gamma(n+1) = n!$.

The *Beta function* is defined by the integral

$$\mathrm{B}(x,y) = \int_0^1 t^{x-1}(1-t)^{y-1}\,dt, \quad \mathrm{Re}\,x,\ \mathrm{Re}\,y > 0. \tag{1.3}$$

It is symmetric, i.e. $\mathrm{B}(x,y) = \mathrm{B}(y,x)$, and it is related to the Gamma function by the well-known formula

$$\mathrm{B}(x,y) = \frac{\Gamma(x)\Gamma(y)}{\Gamma(x+y)}.$$

A *hypergeometric series* $\sum_{n=0}^{\infty} c_n$ is a series for which $c_0 = 1$ and the ratio of consecutive terms is a rational function of the summation index n, i.e. one for which

$$\frac{c_{n+1}}{c_n} = \frac{P(n)}{Q(n)},$$

where $P(n)$ and $Q(n)$ are polynomials. In this case, c_n is called a hypergeometric term. If the polynomials are completely factored, the ratio of successive terms can be written as

$$\frac{c_{n+1}}{c_n} = \frac{P(n)}{Q(n)} = \frac{(n+a_1)(n+a_2)\cdots(n+a_p)}{(n+b_1)(n+b_2)\cdots(n+b_q)(n+1)},$$

where the factor $n+1$ in the denominator is present for historical reasons of notation. From here we define the *generalized hypergeometric function* as

$${}_pF_q\left(\begin{matrix} a_1,\ldots,a_p \\ b_1,\ldots,b_q \end{matrix};x\right) = \sum_{n=0}^{\infty} c_n x^n = \sum_{n=0}^{\infty} \frac{(a_1)_n \cdots (a_p)_n}{(b_1)_n \cdots (b_q)_n} \frac{x^n}{n!}. \tag{1.4}$$

We can also use the following notation for generalized hypergeometric functions:

$${}_pF_q(a_1,\ldots,a_p;b_1,\ldots,b_q;x).$$

This series is absolutely convergent for all x if $p \leq q$ and for $|x| < 1$ if $p = q+1$. It is divergent for all $x \neq 0$ if $p > q+1$, as long as the series is not finite. Observe that when one of the parameters of the numerator $a_i, i = 1,\ldots,p$, is a negative integer, then the generalized hypergeometric function is a polynomial.

Many of the known special functions can be represented in terms of generalized hypergeometric functions. For example, the simplest cases of ${}_0F_0$ and ${}_1F_0$ correspond to the exponential series and the binomial series, respectively. Indeed,

$${}_0F_0(-;-;x) = \sum_{n=0}^{\infty} \frac{x^n}{n!} = e^x,$$

$$ {}_1F_0(a;\ -;x) = \sum_{n=0}^{\infty} \frac{(a)_n x^n}{n!} = \sum_{n=0}^{\infty} \frac{\Gamma(z+n)}{\Gamma(a)\Gamma(n+1)} x^n = \sum_{n=0}^{\infty} \binom{a+n-1}{n} x^n = (1-x)^{-a}. $$

If $p = 2$ and $q = 1$, the function becomes what is called the *Gaussian hypergeometric function* ${}_2F_1(a,b;c;x)$ and it is related to the solutions of *Euler's hypergeometric differential equation*

$$ x(1-x)y''(x) + [c-(a+b+1)x]y'(x) - aby(x) = 0. \tag{1.5} $$

We will see later the relation of this equation with the Jacobi polynomials. All families of orthogonal polynomials in the Askey scheme admit a representation in terms of hypergeometric series, as we will see later. For more information about generalized hypergeometric functions see [3, Chapter 2].

The *Stieltjes transform* (also known as the Cauchy transform) of a measure ψ defined on $\mathbb{R}$ is defined as the complex-valued function

$$ B(z;\psi) = \int_{\mathbb{R}} \frac{d\psi(x)}{x-z}, \quad z \in \mathbb{C} \setminus \mathbb{R}. \tag{1.6} $$

This transform is related to the *generating function of the moments* of the measure ψ, since, formally

$$ B(z;\psi) = -\frac{1}{z}\int_{\mathbb{R}} \frac{1}{1-x/z} d\psi(x) = -\frac{1}{z}\sum_{n=0}^{\infty}\int_{\mathbb{R}} \frac{x^n}{z^n}\, d\psi(x) = -\sum_{n=0}^{\infty} \frac{\mu_n}{z^{n+1}}, \tag{1.7} $$

where $\mu_n = \int_{\mathbb{R}} x^n d\psi(x)$ are the *moments* of the measure. In the case where $\operatorname{supp}(\psi) \subseteq [-A,A]$, then $|\mu_n| \leq 2A^n$, implying that the series (1.7) is absolutely convergent for $|z| > A$. In this case, the Stieltjes transform is completely determined in terms of the moments of the measure ψ. In general, the expansion of the Stieltjes transform (1.6) has to be interpreted as an asymptotic expansion of the Stieltjes transform $B(z;\psi)$ as $|z| \to \infty$.

There is a formula which allows to calculate the measure ψ if we have information about the corresponding Stieltjes transform. This formula is known as the *Perron–Stieltjes inversion formula.* It has several versions, but the one we will use in this text is included in the following result.

Proposition 1.1 ([51, Theorem X.6.1]) *Let ψ be a probability measure with finite moments and $B(z;\psi)$ its Stieltjes transform* (1.6). *Then*

$$ \int_a^b d\psi(x) + \frac{1}{2}\psi(\{a\}) + \frac{1}{2}\psi(\{b\}) = \frac{1}{\pi}\lim_{\varepsilon\downarrow 0}\int_a^b \operatorname{Im}B(x+i\varepsilon;\psi)\, dx, \tag{1.8} $$

where $\psi(\{a\}) \geq 0$ is the magnitude or size of the mass at an isolated point a. If the measure is absolutely continuous at a then $\psi(\{a\}) = 0$.

Proof Observe that

$$2i\mathrm{Im}B(z;\psi) = B(z;\psi) - \overline{B(z;\psi)} = B(z;\psi) - B(\overline{z};\psi) = \int_{\mathbb{R}} \left[\frac{1}{x-z} - \frac{1}{x-\overline{z}}\right] d\psi(x)$$
$$= \int_{\mathbb{R}} \frac{z-\overline{z}}{|x-z|^2} d\psi(x) = 2i \int_{\mathbb{R}} \frac{\mathrm{Im}z}{|x-z|^2} d\psi(x).$$

Therefore

$$\mathrm{Im}B(x+i\varepsilon;\psi) = \int_{\mathbb{R}} \frac{\varepsilon}{|s-(x+i\varepsilon)|^2} d\psi(s) = \int_{\mathbb{R}} \frac{\varepsilon}{(s-x)^2+\varepsilon^2} d\psi(s).$$

Integrating and exchanging integrals (which is allowed since the integrand is positive) we have that

$$\int_a^b \mathrm{Im}B(x+i\varepsilon;\psi)\, dx = \int_{\mathbb{R}} \left[\int_a^b \frac{\varepsilon}{(s-x)^2+\varepsilon^2} dx\right] d\psi(s).$$

The internal integral can be calculated explicitly by making the change of variables $y = (x-s)/\varepsilon$:

$$\chi_\varepsilon(s) = \int_a^b \frac{\varepsilon}{(s-x)^2+\varepsilon^2}\, dx = \int_{(a-s)/\varepsilon}^{(b-s)/\varepsilon} \frac{1}{1+y^2}\, dy = \arctan y \Big|_{y=(a-s)/\varepsilon}^{y=(b-s)/\varepsilon}.$$

We have that $0 \le \chi_\varepsilon(s) \le \pi$ and when we take the limit (which is also allowed using the Lebesgue dominated convergence theorem since ψ is a probability measure and $\chi_\varepsilon(s)$ is bounded and positive) we have that

$$\lim_{\varepsilon\downarrow 0} \chi_\varepsilon(s) = \begin{cases} \pi, & \text{if } \quad a < s < b, \\ \frac{\pi}{2}, & \text{if } \quad s = a \text{ or } s = b. \end{cases}$$ □

As a consequence of the previous proposition we also have the formula

$$\int_a^b d\psi(x) = \frac{1}{\pi} \lim_{\varepsilon\downarrow 0} \lim_{\eta\downarrow 0} \int_{a+\eta}^{b-\eta} \mathrm{Im}B(x+i\varepsilon;\psi)\, dx. \tag{1.9}$$

When the measure is absolutely continuous with respect to the Lebesgue measure, i.e. $d\psi(x) = \psi(x)\, dx$ (abusing the notation), we have

$$\psi(x) = \frac{1}{\pi} \lim_{\varepsilon\downarrow 0} \mathrm{Im}B(x+i\varepsilon;\psi) = \lim_{\varepsilon\downarrow 0} \frac{B(x+i\varepsilon;\psi) - B(x-i\varepsilon;\psi)}{2\pi i}. \tag{1.10}$$

Finally, for measures that have an absolutely continuous part and a discrete part, there is a direct way to calculate the size of the jump. Indeed, assume that $\psi = \widehat{\psi} + \psi(\{a\})\delta_a$, where $\delta_a(x) = \delta(x-a)$ is the Dirac delta distribution which is defined, as usual, by $\int_{\mathbb{R}} f(x)\delta(x-a)\, dx = f(a)$. Then, since the Stieltjes transform is linear, we have

$$B(z;\psi) = B(z;\widehat{\psi}) + \frac{\psi(\{a\})}{a-z}.$$

Evaluating at $z = a + i\varepsilon$ and taking imaginary parts, we have

$$\mathrm{Im}B(a+i\varepsilon;\psi) = \mathrm{Im}B(a+i\varepsilon;\widehat{\psi}) + \mathrm{Im}\frac{\psi(\{a\})}{-i\varepsilon} = \mathrm{Im}B(a+i\varepsilon;\widehat{\psi}) + \frac{\psi(\{a\})}{\varepsilon}.$$

Therefore we get

$$\psi(\{a\}) = \varepsilon\mathrm{Im}B(a+i\varepsilon;\psi) - \varepsilon\mathrm{Im}B(a+i\varepsilon;\widehat{\psi}). \tag{1.11}$$

Taking limits as $\varepsilon \downarrow 0$ we observe that $B(a+i\varepsilon;\widehat{\psi})$ is bounded since $\widehat{\psi}$ is absolutely continuous. Therefore the meaningful isolated points (where $\psi(\{a\}) > 0$) must be those satisfying

$$\lim_{\varepsilon\downarrow 0}\mathrm{Im}B(a+i\varepsilon;\psi) = \infty,$$

while the size of the jump at $x = a$ is given by

$$\psi(\{a\}) = \lim_{\varepsilon\downarrow 0}\varepsilon\mathrm{Im}B(a+i\varepsilon;\psi) \geq 0. \tag{1.12}$$

Example 1.2 Let $B(z;\psi)$ be given by

$$B(z;\psi) = \frac{1}{1-z}, \quad z \in \mathbb{C}\setminus\{1\}.$$

According to (1.11) there will be a pole at $z = 1$, so it is a candidate for a singular part of the measure. Assume that $\psi = \widehat{\psi} + \psi(\{1\})\delta_1$, where $\widehat{\psi}$ is the absolutely continuous part. Then, by (1.10), we have

$$\widehat{\psi}(x) = \frac{1}{\pi}\lim_{\varepsilon\downarrow 0}\mathrm{Im}\frac{1}{1-x-i\varepsilon} = \frac{1}{\pi}\lim_{\varepsilon\downarrow 0}\mathrm{Im}\left(\frac{1-x+i\varepsilon}{(1-x)^2+\varepsilon^2}\right) = \frac{1}{\pi}\lim_{\varepsilon\downarrow 0}\frac{\varepsilon}{(1-x)^2+\varepsilon^2}.$$

We observe that if $x \neq 1$, then $\widehat{\psi}(x) = 0$. Therefore the measure ψ consists only of a singular part at $x = 1$. The value of $\psi(\{1\})$ is given by (1.12) and it is easy to see that

$$\psi(\{1\}) = \lim_{\varepsilon\downarrow 0}\varepsilon\mathrm{Im}B(1+i\varepsilon;\psi) = \lim_{\varepsilon\downarrow 0}\varepsilon\frac{\varepsilon}{\varepsilon^2} = 1.$$

Therefore $\psi(x) = \delta_1(x)$. ◊

Example 1.3 Consider the Stieltjes transform given by

$$B(z;\psi) = -2z + 2\sqrt{z^2-1}, \quad z \in \mathbb{C}\setminus[-1,1],$$

where the branch of the square root is determined by analytic continuation from positive values for real $z > 1$. We observe that there are no singular points, so the measure will consist only of an absolutely continuous part. From (1.10) we get

$$\begin{aligned}\psi(x) &= \frac{1}{\pi}\lim_{\varepsilon\downarrow 0}\mathrm{Im}B(x+i\varepsilon;\psi)\\ &= \frac{1}{\pi}\lim_{\varepsilon\downarrow 0}\left(-2\varepsilon + 2\mathrm{Im}\sqrt{x^2-\varepsilon^2+2ix\varepsilon-1}\right) = \frac{2}{\pi}\mathrm{Im}\sqrt{x^2-1}.\end{aligned}$$

The last part has only imaginary part when $|x| \leq 1$. Therefore

$$\psi(x) = \frac{2}{\pi}\sqrt{1-x^2}, \quad |x| < 1,$$

which is the *Wigner semicircle distribution.* ◊

In Chapters 2 and 3 we will see several examples of computation of measures using the Perron–Stieltjes inversion formula.

Remark 1.4 As we have seen in (1.7), the Stieltjes transform is related to the generating function of the moments of a probability measure ψ. This is not exactly the same as the usual *moment generating function*, which is defined as

$$M_X(t) = \mathbb{E}(e^{tX}) = \sum_{n=0}^{\infty} \mu_n \frac{t^n}{n!},$$

where X is the random variable associated with the probability measure ψ. This moment generating function is more related to the *Laplace transform.* Indeed, assume that the probability measure is absolutely continuous and supported on $[0,\infty)$. Then the Laplace transform is defined by

$$\mathcal{L}[\psi](s) = \int_0^{\infty} e^{-sx}\psi(x)\,dx.$$

Then we have $\mathcal{L}[\psi](-t) = M_X(t)$. The Stieltjes transform arises naturally as an iteration of the Laplace transform. Indeed, if we call $\phi(s) = \mathcal{L}[\psi](s)$ then, formally, we have

$$\begin{aligned}
\mathcal{L}[\phi](t) &= \int_0^{\infty} e^{-st}\phi(s)ds = \int_0^{\infty} e^{-st}\left(\int_0^{\infty} e^{-su}\psi(u)du\right)ds \\
&= \int_0^{\infty} \psi(u)\left(\int_0^{\infty} e^{-s(t+u)}ds\right)du \\
&= \int_0^{\infty} \psi(u)\left(-\frac{1}{t+u}e^{-s(t+u)}\Big|_{s=0}^{s=\infty}\right)du = \int_0^{\infty} \frac{\psi(u)}{t+u}du, \quad \mathrm{Re}(t) > 0.
\end{aligned}$$

Therefore $B(t;\psi) = \mathcal{L}^2[\psi](-t)$. A good reference about Stieltjes transforms in connection with the Laplace transform can be found in Chapter VIII of [146]. ◊

1.2 General Properties of Orthogonal Polynomials

Let ψ be a positive Borel measure on $\mathbb{R}$ with infinite support and let us assume that the moments

$$\mu_n = \int_{\mathbb{R}} x^n d\psi(x), \quad n \geq 0,$$

exist and are finite. We normalize the measure in such a way that $\mu_0 = 1$, so we have a probability measure. Following *Lebesgue's decomposition theorem* any Borel measure on the real line can be decomposed into three measures such that

$$\psi = \psi_c + \psi_d + \psi_{sc},$$

where ψ_c is absolutely continuous, ψ_d is discrete and ψ_{sc} is singular continuous. The absolutely continuous measure ψ_c is classified by the Radon–Nikodym theorem and can always be written (abusing the notation) as $d\psi_c(x) = \psi_c(x)dx$, with respect to the Lebesgue measure. The discrete measure ψ_d can always be written as

$$d\psi_d(x) = \sum_k \psi(\{x_k\})\delta(x - x_k)\, dx,$$

where k runs over a countable set, x_k are the mass points, $\psi(\{x_k\})$ are the sizes or magnitudes of these jumps and $\delta(x - a)$ is the Dirac delta distribution. Finally, the singular continuous measure ψ_{sc} is defined over a set of measure 0. The *Cantor measure* (the probability measure on the real line whose cumulative distribution function is the Cantor function) is an example of a singular continuous measure. In this text we consider positive Borel measures on $\mathbb{R}$ with either only an absolutely continuous part or only a discrete part (or a combination of both).

Associated with this measure ψ we can consider the Hilbert space L^2_ψ with the inner product

$$(f, g)_\psi = \int_{\mathbb{R}} f(x)g(x)\, d\psi(x), \tag{1.13}$$

of all measurable real functions f such that $(f,f)_\psi = \|f\|^2_\psi < \infty$. If the support of the measure is given by $\mathcal{S} \subseteq \mathbb{R}$, then this space will be written as $L^2_\psi(\mathcal{S})$. When $\mathcal{S}$ is a countable set, for example $\mathbb{N}_0 = \{0, 1, \ldots\}$, this space is usually denoted by $\ell^2_\psi(\mathbb{N}_0)$.

We say that $(p_n(x))_n$ is a *sequence of polynomials* if each element is a polynomial of degree exactly n in the real variable x. A sequence of polynomials is *monic* if the leading coefficient of each polynomial is exactly 1. A sequence of polynomials $(p_n)_n$ is *orthogonal* with respect to a Borel measure ψ if

$$(p_n, p_m)_\psi = \int_{\mathbb{R}} p_n(x)p_m(x)\, d\psi(x) = d_n^2 \delta_{nm},$$

where $d_n^2 = \|p_n\|^2_\psi > 0$. If the norm is always identically 1, we say that the polynomial sequence is *orthonormal* and we denote it by $(P_n)_n$. When we work with the sequence of monic orthogonal polynomials, we will use the notation $(\widehat{P}_n)_n$ and its norms will be denoted by $\|\widehat{P}_n\|^2_\psi = \zeta_n$.

Given a Borel measure ψ on $\mathbb{R}$ with infinite support and finite moments, it will always be possible to build a sequence of orthogonal polynomials. A direct way is through the *Gram–Schmidt orthogonalization process* applied to the set $\{1, x, x^2, \ldots\}$.

This method builds the polynomials one by one taking into account that all the previous ones have already been calculated. Specifically

$$\begin{aligned}
\widehat{P}_0(x) &= 1, \\
\widehat{P}_1(x) &= x - \frac{(\widehat{P}_0, x)_\psi}{(\widehat{P}_0, \widehat{P}_0)_\psi}\widehat{P}_0(x), \\
&\ \vdots \qquad\qquad \vdots \\
\widehat{P}_k(x) &= x^k - \sum_{j=0}^{k-1} \frac{(\widehat{P}_j, x^k)_\psi}{(\widehat{P}_j, \widehat{P}_j)_\psi}\widehat{P}_j(x).
\end{aligned}$$

Once they have been computed, the monic polynomials can be normalized by dividing them by $\|\widehat{P}_k\|_\psi = \sqrt{\zeta_k}$. Observe that the monic orthogonal polynomials have always real coefficients.

Another way to define orthogonal polynomials is through determinants associated with the moments. Consider the determinant

$$\Delta_n = \begin{vmatrix} \mu_0 & \mu_1 & \cdots & \mu_n \\ \mu_1 & \mu_2 & \cdots & \mu_{n+1} \\ \vdots & \vdots & \ddots & \vdots \\ \mu_n & \mu_{n+1} & \cdots & \mu_{2n} \end{vmatrix}, \quad \Delta_{-1} = 1.$$

The quadratic form associated with the matrix of the previous determinant, which we denote by (Δ_n), is always positive definite. Indeed, for any real vector $v = (v_0, v_1, \ldots, v_n)^T$, we have that

$$v^T(\Delta_n)v = \sum_{j,k=0}^{n} \mu_{j+k} v_j v_k = \int_{\mathbb{R}} \left[\sum_{j=0}^{n} v_j x^j \right]^2 d\psi(x),$$

which is clearly positive. Thus $\Delta_n > 0, n \geq 0$. $\Delta_n, n \geq 0$ are usually called *Hankel determinants*.

It is easy to see that the sequence of polynomials defined by

$$p_n(x) = \begin{vmatrix} \mu_0 & \mu_1 & \cdots & \mu_{n-1} & 1 \\ \mu_1 & \mu_2 & \cdots & \mu_n & x \\ \vdots & \vdots & \ddots & \vdots & \vdots \\ \mu_n & \mu_{n+1} & \cdots & \mu_{2n-1} & x^n \end{vmatrix}, \quad n \geq 0, \tag{1.14}$$

is orthogonal with respect to the measure ψ. To see that, we simply evaluate the inner product $(p_n, x^m)_\psi = 0, m = 0, 1, \ldots, n-1$ observing that we always have a repeated column, so the determinant is 0. Alternatively, we have $(p_n, x^n)_\psi = \Delta_n > 0$. Thus

$$p_n(x) = \Delta_{n-1}x^n + \text{lower degree terms},$$

and we have that

$$(p_n, p_n)_\psi = (p_n, \Delta_{n-1} x^n)_\psi = \Delta_{n-1}\Delta_n.$$

Therefore, the polynomials

$$P_n(x) = \frac{1}{\sqrt{\Delta_{n-1}\Delta_n}} p_n(x)$$

are orthonormal, and the leading coefficient is given by $h_n = \sqrt{\Delta_{n-1}/\Delta_n} = \zeta_n^{-1/2}$. The monic family can be written as

$$\widehat{P}_n(x) = \frac{1}{\Delta_{n-1}} p_n(x) = \sqrt{\frac{\Delta_n}{\Delta_{n-1}}} P_n(x).$$

Finally, let us see another way to generate the orthogonal polynomials recurrently. Assume that we have a sequence of orthogonal polynomials $(p_n)_n$. The polynomial $xp_n(x)$ has degree $n+1$ and can be expressed as a linear combination of the $n+1$ first polynomials, i.e.

$$xp_n(x) = \sum_{j=0}^{n+1} d_{n,j} p_j(x).$$

Now, multiplying by $p_k(x)$ and evaluating the inner product, it is easy to see, using the orthogonal relations, that the coefficients $d_{n,j} = 0, j = 0, 1, \dots, n-2$. Therefore, only the last three coefficients remain and every family of orthogonal polynomials satisfies a *three-term recurrence relation* of the form

$$xp_n(x) = a_n p_{n+1}(x) + b_n p_n(x) + c_n p_{n-1}(x), \quad n \geq 0, \quad p_{-1} = 0, \tag{1.15}$$

where

$$a_n = \frac{(xp_n, p_{n+1})_\psi}{(p_{n+1}, p_{n+1})_\psi}, \quad b_n = \frac{(xp_n, p_n)_\psi}{(p_n, p_n)_\psi}, \quad c_n = \frac{(xp_n, p_{n-1})_\psi}{(p_{n-1}, p_{n-1})_\psi}.$$

We observe that the coefficient b_n is always real. Moreover, for the orthonormal family $P_n(x)$ we have, comparing the coefficients of x^{n+1} in (1.15), that $a_n = h_n/h_{n+1} = \sqrt{\zeta_{n+1}/\zeta_n} > 0$, and that $c_n = (xP_n, P_{n-1})_\psi = (P_n, xP_{n-1})_\psi = a_{n-1}$. Therefore the sequence of orthonormal polynomials $(P_n)_n$ satisfies a three-term recurrence relation of the form

$$xP_n(x) = a_n P_{n+1}(x) + b_n P_n(x) + a_{n-1} P_{n-1}(x), \quad a_n > 0, \quad b_n \in \mathbb{R}. \tag{1.16}$$

For the monic family $(\widehat{P}_n)_n$ the three-term recurrence relation will be given by

$$x\widehat{P}_n(x) = \widehat{P}_{n+1}(x) + \alpha_n \widehat{P}_n(x) + \beta_n \widehat{P}_{n-1}(x), \quad \widehat{P}_0(x) = 1, \quad \widehat{P}_1(x) = x - \alpha_0, \tag{1.17}$$

where $\alpha_{n-1} \in \mathbb{R}, \beta_n > 0$ for $n \geq 1$. The relations between these coefficients and the coefficients of the orthonormal family are given by

$$a_n = \sqrt{\frac{\zeta_{n+1}}{\zeta_n}}, \quad \alpha_n = b_n, \quad \beta_n = \frac{\zeta_n}{\zeta_{n-1}}.$$

Observe that $\zeta_n = \beta_n \cdots \beta_1$.

Another way of writing this recurrence relation is in matrix form. Denoting the column vector of orthonormal polynomials by $P(x) = (P_0(x), P_1(x), \dots)^T$, we have that $xP(x) = JP(x)$, where J is the tridiagonal symmetric matrix

$$J = \begin{pmatrix} b_0 & a_0 & 0 & 0 & 0 & \cdots \\ a_0 & b_1 & a_1 & 0 & 0 & \cdots \\ 0 & a_1 & b_2 & a_2 & 0 & \cdots \\ 0 & 0 & a_2 & b_3 & a_3 & \cdots \\ \vdots & \vdots & \vdots & \ddots & \ddots & \ddots \end{pmatrix}. \tag{1.18}$$

This matrix plays a very important role and it is called a *Jacobi matrix*. In particular, we will find this kind of matrix in the one-step transition probability matrix of a one-dimensional discrete-time birth–death chain and in the infinitesimal operator of a birth–death process, as we will see in the next two chapters. The inverse result, i.e. for a family of polynomials defined by (1.16), where there exists a positive measure for which they are orthogonal, is known as Favard's theorem or the spectral theorem for orthogonal polynomials. We will see more details in Section 1.3.

The powers of J can be computed formally using orthogonality properties. Observe that the relation $xP(x) = JP(x)$ implies that $x^nP(x) = J^nP(x)$. Therefore, multiplying by $P^T(x)$, integrating with respect to the measure ψ and looking at the (i,j) entry, we obtain

$$\int_{\mathbb{R}} x^n P_i(x) P_j(x) d\psi(x) = \sum_{k\geq 0} \int_{\mathbb{R}} J^n_{ik} P_k(x) P_j(x) d\psi(x) = J^n_{ij}. \tag{1.19}$$

From here we observe that the moments $(\mu_n)_n$ of the measure ψ can be computed from J^n_{00}. In general, the diagonal coefficients J^n_{ii} are the moments of the measure $d\psi_i(x) = P_i^2(x) d\psi(x)$.

The identity (1.19) can be extended to any analytic function defined on supp(ψ) of the form $f(x) = \sum_{n\geq 0} c_n x^n$ as

$$\int_{\mathbb{R}} f(x) P_i(x) P_j(x)\, d\psi(x) = \sum_{n\geq 0} \int_{\mathbb{R}} c_n x^n P_k(x) P_j(x)\, d\psi(x) = \sum_{n\geq 0} c_n J^n_{ij} = f(J)_{ij}.$$

For instance, the function $f(x) = (1-zx)^{-1}$ with $z^{-1} \in \mathbb{C} \setminus \text{supp}(\psi)$ gives, formally, that

$$(I - zJ)^{-1}_{00} = \int_{\mathbb{R}} \frac{P_0^2(x)}{1-xz}\, d\psi(x) = \int_{\mathbb{R}} \frac{d\psi(x)}{1-xz} = \sum_{n\geq 0} \mu_n z^n, \tag{1.20}$$

i.e. the generating function of the moments of ψ. In terms of the Stieltjes transform $B(z;\psi)$ defined by (1.6), we have that

$$(I - zJ)^{-1}_{00} = -\frac{1}{z} B\left(\frac{1}{z};\psi\right). \tag{1.21}$$

Theorem 1.5 *Let J be the Jacobi matrix given by* (1.18) *and denote by $J^{(0)}$ the Jacobi matrix built from J by removing the first row and column. Then we have*

$$(I - zJ)^{-1}_{00} = \frac{1}{1 - b_0 z - a_0^2 z^2 (I - zJ^{(0)})^{-1}_{00}}.$$

Proof Write the Jacobi matrix J in (1.18) as

$$J = \left(\begin{array}{c|cc} b_0 & a_0 & \cdots \\ \hline a_0 & & \\ 0 & & J^{(0)} \\ \vdots & & \end{array}\right).$$

Using the well-known formula for the inverse of a 2×2 block matrix

$$\begin{pmatrix} A & B \\ C & D \end{pmatrix}^{-1} = \begin{pmatrix} (A - BD^{-1}C)^{-1} & * \\ * & * \end{pmatrix},$$

applied to the matrix $I - zJ$, we get

$$(I - zJ)^{-1}_{00} = \left[1 - zb_0 - a_0^2 z^2 e_0^T (I - zJ^{(0)})^{-1} e_0\right]^{-1} = \frac{1}{1 - zb_0 - a_0^2 z^2 (I - zJ^{(0)})^{-1}_{00}},$$

where e_0 is the canonical vector $e_0 = (1, 0, \ldots)^T$. □

Remark 1.6 If we assume that associated with the Jacobi matrices J and $J^{(0)}$ there exist positive measures ψ and $\psi^{(0)}$, respectively, then we have, using (1.20), that

$$\int_{-1}^{1} \frac{d\psi(x)}{1-xz} = \frac{1}{1 - b_0 z - a_0^2 z^2 \displaystyle\int_{-1}^{1} \frac{d\psi^{(0)}(x)}{1-xz}}.$$

This formula relates the generating functions of the moments of the measures ψ and $\psi^{(0)}$. In terms of Stieltjes transforms, relation (1.21) gives

$$B(z;\psi) = -\frac{1}{z - b_0 + a_0^2 B(z;\psi^{(0)})}. \tag{1.22}$$

Example 1.7 Consider the Jacobi matrix given by

$$J = \left(\begin{array}{c|ccccc} 0 & 1/2 & 0 & 0 & \cdots \\ \hline 1/2 & 0 & 1/2 & 0 & \cdots \\ 0 & 1/2 & 0 & 1/2 & \cdots \\ \vdots & \vdots & \ddots & \ddots & \ddots \end{array}\right).$$

Then we have that $J^{(0)} = J$ and consequently $\psi = \psi^{(0)}$. From (1.22) we obtain a quadratic equation for $B(z;\psi) = B(z)$, given by

$$B^2(z) + 4zB(z) + 4 = 0.$$

Therefore

$$B(z) = -2z \pm 2\sqrt{z^2 - 1}.$$

On the one hand, we can discard the negative solution of $B(z)$ since as $z \to \infty$ the Stieltjes transform should vanish. On the other hand, the function $B(z)$ is well defined as a single-valued function in the complex plane from which we have removed the interval $[-1, 1]$. If we approach the cut from above, $B(z)$ has a nontrivial imaginary part coming from the square root. This square root has positive values for $\operatorname{Re} z > 1$ and negative values for $\operatorname{Re} z > 1$. Therefore we have

$$B(z;\psi) = -2z + 2\sqrt{z^2 - 1}, \quad z \in \mathbb{C} \setminus [-1, 1],$$

and the branch of the square root is determined by analytic continuation from positive values of $\operatorname{Re} z > 1$. This example is precisely the one studied in Example 1.3 and the spectral measure is given by the Wigner semicircle distribution using the Perron–Stieltjes inversion formula. ◊

Multiplying (1.16) by $P_n(y)$ and (1.16) (for $x = y$) by $P_n(x)$, and subtracting both formulas, we get the telescopic relation

$$\begin{aligned}(x - y)P_n(x)P_n(y) = a_n[&P_{n+1}(x)P_n(y) - P_n(x)P_{n+1}(y)] \\ &- a_{n-1}[P_n(x)P_{n-1}(y) - P_{n-1}(x)P_n(y)],\end{aligned}$$

which iterating, adding and dividing by $x - y$ gives the *Christoffel–Darboux formula*

$$K_n(x, y) \doteq \sum_{j=0}^{n} P_j(x)P_j(y) = a_n \left[\frac{P_{n+1}(x)P_n(y) - P_n(x)P_{n+1}(y)}{x - y}\right]. \tag{1.23}$$

Taking $y \to x$ we get the *confluent Christoffel–Darboux formula*

$$\sum_{j=0}^{n} P_j^2(x) = a_n[P'_{n+1}(x)P_n(x) - P_{n+1}(x)P'_n(x)]. \tag{1.24}$$

The kernel $K_n(x,y)$ generated by the Christoffel–Darboux formula has the *reproducing kernel property*, i.e for every polynomial p of degree n, we have that

$$\int_{\mathbb{R}} p(x)K_n(x,y)d\psi(x) = p(y).$$

In terms of the monic family $(\widehat{P}_n)_n$ the Christoffel–Darboux formula can be written as

$$\sum_{j=0}^{n} \frac{\widehat{P}_j(x)\widehat{P}_j(y)}{\zeta_j} = \frac{\widehat{P}_{n+1}(x)\widehat{P}_n(y) - \widehat{P}_n(x)\widehat{P}_{n+1}(y)}{\zeta_n(x-y)} \tag{1.25}$$

and the confluent formula as

$$\sum_{j=0}^{n} \frac{\widehat{P}_j^2(x)}{\zeta_j} = \frac{\widehat{P}'_{n+1}(x)\widehat{P}_n(x) - \widehat{P}_{n+1}(x)\widehat{P}'_n(x)}{\zeta_n}. \tag{1.26}$$

Observe that the Christoffel–Darboux formula is a property that holds for every (monic) sequence of polynomials generated by the three-term recurrence relation (1.17), no matter if they are orthogonal or not with respect to some measure. The sequence ζ_n is generated by $\zeta_n = \beta_n \cdots \beta_1$. In the following result we will prove certain properties of the zeros of these polynomials.

Proposition 1.8 *The zeros or roots of the monic polynomials $\widehat{P}_n$ generated by the three-term recurrence relation* (1.16) *are all real and simple. Moreover the zeros of $\widehat{P}_{n+1}$ and $\widehat{P}_n$ interlace. If the polynomials are orthogonal with respect to some measure ψ, then these zeros lie in the smallest closed interval containing supp(ψ) for all $n \geq 1$.*

Proof Let u be a complex zero of $\widehat{P}_n$. Since the coefficients of $\widehat{P}_n$ are all real then $\bar{u}$ is also a complex zero of $\widehat{P}_n$. Taking $x = u$ and $y = \bar{u}$ in the Christoffel–Darboux formula (1.25) we get a contradiction since the right-hand part should be 0 while the left-hand side is > 1 (sum of positive absolute values of complex numbers). Then all zeros must be real. Alternatively, if we had a multiple zero then the confluent Christoffel–Darboux formula (1.26) will give the same contradiction.

If $\widehat{P}_{n+1}$ and $\widehat{P}_n$ have a zero in common, then by the recursion formula (1.17), it is also a zero of $\widehat{P}_{n-1}$. Following this reasoning this zero is a zero of $\widehat{P}_0 = 1$, but this is a contradiction. Regarding the interlacing property, for $n < 2$ there is nothing to prove. For $n \geq 2$, (1.26) implies that $\widehat{P}'_{n+1}(x)\widehat{P}_n(x) - \widehat{P}_{n+1}(x)\widehat{P}'_n(x) > 0$. Assume that $y_1 < y_2$ are two consecutive zeros of $\widehat{P}_{n+1}$. Then the above inequality implies that $\widehat{P}'_{n+1}(y_j)\widehat{P}_n(y_j) > 0, j = 1,2$. Since $\widehat{P}'_{n+1}(y_1)$ and $\widehat{P}'_{n+1}(y_2)$ must have different signs, since they are simple, it follows from the previous inequality that $\widehat{P}_n(y_j), j = 1, 2$, has different signs. Thus, $\widehat{P}_n$ has a zero in the range (y_1, y_2) by Bolzano's theorem.

Finally let $[a, b]$ the smallest closed interval containing $\operatorname{supp}(\psi)$ and $c_1, \ldots, c_j$ the zeros of $\widehat{P}_n$ contained in $[a, b]$. If $j < n$ then the orthogonality implies that $\int_{\mathbb{R}} \widehat{P}_n(x) \prod_{k=1}^{j}(x - c_k) d\psi(x) = 0$. But this a contradiction because the integrand does not change signs on $[a, b]$. Therefore $j = n$. □

For a fixed n, let $x_{n,j}$, $j = 1, \ldots, n$ denote the zeros of $\widehat{P}_n$ arranged in the following form:

$$x_{n,1} < x_{n,2} < \cdots < x_{n,n}. \tag{1.27}$$

The interlacing property says that each sequence $(x_{n,i})_n$ is monotone, therefore the limits exist. Define them as

$$\xi_i = \lim_{n \to \infty} x_{n,i} \quad \text{and} \quad \eta_j = \lim_{n \to \infty} x_{n,n-j+1}, \quad i, j \geq 1. \tag{1.28}$$

We have that

$$-\infty \leq \xi_i \leq \xi_{i+1} < \eta_{j+1} \leq \eta_j \leq \infty, \quad i, j \geq 1.$$

The interval $[\xi_1, \eta_1]$ is usually called the *true interval of orthogonality* and it is the smallest closed interval containing $\operatorname{supp}(\psi)$. Therefore $\xi_1 = \inf \operatorname{supp}(\psi)$ and $\eta_1 = \sup \operatorname{supp}(\psi)$. If we call

$$\sigma = \lim_{i \to \infty} \xi_i \quad \text{and} \quad \tau = \lim_{j \to \infty} \eta_j,$$

then we have

$$-\infty \leq \xi_i \leq \sigma \leq \tau \leq \eta_j \leq \infty, \quad i, j \geq 1.$$

Therefore, defining the (possible finite) sets

$$\Xi = \{\xi_1, \xi_2, \ldots\} \quad \text{and} \quad H = \{\eta_1, \eta_2, \ldots\},$$

we have that (see [16, II.4.2])

$$\operatorname{supp}(\psi) = \bar{\Xi} \cup S \cup \bar{H},$$

where the bar denotes closure and $S \subset (\sigma, \tau)$. Also σ is the smallest and τ is the largest limit point of $\operatorname{supp}(\psi)$.

Another way to prove Proposition 1.8 is to write the monic polynomials $\widehat{P}_n$ as the characteristic polynomial of the truncated Jacobi matrix of dimension $n \times n$ in the following form:

$$\widehat{P}_n(x) = \begin{vmatrix} x - \alpha_0 & -1 & 0 & \cdots & 0 & 0 & 0 \\ -\beta_1 & x - \alpha_1 & -1 & \cdots & 0 & 0 & 0 \\ \vdots & \ddots & \ddots & \ddots & \vdots & \vdots & \vdots \\ 0 & 0 & \cdots & & -\beta_{n-2} & x - \alpha_{n-2} & -1 \\ 0 & 0 & 0 & \cdots & 0 & -\beta_{n-1} & x - \alpha_{n-1} \end{vmatrix}. \tag{1.29}$$

The zeros of the orthogonal polynomials play an important role in the approximation of integrals of the form $\int_{\mathbb{R}} f(x)d\psi(x)$ by *Gaussian quadrature formulas*. For a fixed n, let $x_{n,j}, j = 1, \dots, n$ denote the zeros of $\widehat{P}_n$ arranged in the form (1.27).

Theorem 1.9 *For any n, there exist positive numbers $\lambda_1, \dots, \lambda_n$, such that*

$$\int_{\mathbb{R}} p(x)d\psi(x) = \sum_{k=1}^{n} \lambda_k p(x_{n,k}) \tag{1.30}$$

for all polynomials p of degree at most $2n - 1$. The values λ_k admit the following representation:

$$\lambda_k = \int_{\mathbb{R}} \frac{\widehat{P}_n(x)}{\widehat{P}'_n(x_{n,k})(x - x_{n,k})} d\psi(x) = \int_{\mathbb{R}} \left[\frac{\widehat{P}_n(x)}{\widehat{P}'_n(x_{n,k})(x - x_{n,k})} \right]^2 d\psi(x). \tag{1.31}$$

Proof A proof can be found in [74, Theorem 2.4.1]. □

The numbers $\lambda_1, \dots, \lambda_n$ are usually called *Christoffel numbers*. Observe from (1.31) that they are always positive.

Proposition 1.10 *The Christoffel numbers have the following properties:*

$$1 = \sum_{k=1}^{n} \lambda_k, \tag{1.32}$$

$$\lambda_k = -\frac{\zeta_n}{\widehat{P}_{n+1}(x_{n,k})\widehat{P}'_n(x_{n,k})}, \quad k = 1, \dots, n,$$

$$\frac{1}{\lambda_k} = K_n(x_{n,k}, x_{n,k}), \quad k = 1, \dots, n.$$

Proof A proof can be found in [74, Theorem 2.4.2]. □

The three-term recurrence relation (1.16) is a second-order difference equation and therefore it must have two linearly independent solutions. One is given by $P_n(x)$, and the other can be constructed using the initial conditions

$$P_0^{(0)}(x) = 0, \quad P_1^{(0)}(x) = 1/a_0,$$

which makes $P_n^{(0)}$ a polynomial of degree $n - 1$. These polynomials are called *associated polynomials*, 0th associated polynomials, numerator polynomials or polynomials of the second kind. Multiplying the recurrence relation (1.16) by $P_n^{(0)}$ and subtracting (1.16) (for $P_n^{(0)}$) multiplied by P_n, we can see that

$$a_{n-1}\left[P_n(x)P_{n-1}^{(0)}(x) - P_n^{(0)}(x)P_{n-1}(x)\right] = a_{n-1}\begin{vmatrix} P_n(x) & P_n^{(0)}(x) \\ P_{n-1}(x) & P_{n-1}^{(0)}(x) \end{vmatrix} = -1 \neq 0.$$

Then they are linearly independent. This relation (called a Casoratian determinant) also shows that the zeros of $P_n^{(0)}(x)$ are all real, simple and interlace with the zeros of $P_n(x)$. We also have the following integral representation:

$$P_n^{(0)}(x) = \int_{\mathbb{R}} \frac{P_n(x) - P_n(y)}{x - y} d\psi(y), \quad n \geq 0. \tag{1.33}$$

Indeed, let us use $R_n(x)$ to denote the right-hand side of (1.33). Then $R_0(x) = 0$ and $R_1(x) = 1/a_0$. For x not real and $n > 0$, we have, using (1.16), that

$$\begin{aligned} &a_n R_{n+1}(x) - (x - b_n) R_n(x) + a_{n-1} R_{n-1}(x) \\ &\quad = \int_{\mathbb{R}} \frac{-a_n P_{n+1}(y) + (x - b_n) P_n(y) - a_{n-1} P_{n-1}(y) + y P_n(y) - y P_n(y)}{x - y} d\psi(y) \\ &\quad = \int_{\mathbb{R}} \frac{(x - y) P_n(y)}{x - y} d\psi(y) = 0, \quad n > 0. \end{aligned}$$

Another way to generate the associated polynomials is by using the Jacobi matrix $J^{(0)}$ built from the Jacobi matrix J in (1.18) by removing the first row and column (see Theorem 1.5). The Stieltjes transforms of the spectral measures associated with both Jacobi matrices are related by the formula (1.22). For more information about how to compute the spectral measure associated with the associated polynomials see [61, 136] or more recently [31].

There is an important asymptotic result that relates these two solutions of the three-term recurrence relation with the Stieltjes transform of the corresponding measure ψ.

Theorem 1.11 (Markov's theorem) *Let ψ be a positive measure defined in a bounded interval $[a,b]$ and consider the corresponding orthonormal polynomials $P_n(x)$ and the associated polynomials $P_n^{(0)}(x)$. Then we have that*

$$\lim_{n \to \infty} \frac{P_n^{(0)}(z)}{P_n(z)} = \int_a^b \frac{d\psi(x)}{z - x}, \quad z \in \mathbb{C} \setminus [a,b],$$

and the convergence is uniform on compact subsets of $\mathbb{C} \setminus [a,b]$.

Proof Details of the proof can be found in [3, Section 5.5] or [34, Chapter 3]. □

There is a nice interpretation of the previous theorem in terms of *continued fractions* (see for instance [16, Chapter IV]), given by

$$\int_a^b \frac{d\psi(x)}{z - x} = \cfrac{1}{z - b_0 - \cfrac{a_0^2}{z - b_1 - \cfrac{a_1^2}{z - b_2 - \cfrac{a_2^2}{z - b_3 - \cdots}}}}, \quad z \in \mathbb{C} \setminus [a,b]. \tag{1.34}$$

This formula can be regarded as an alternative way of computing the Stieltjes transform of a probability measure ψ and eventually computing the measure by the Perron–Stieltjes inversion formula.

Finally we give without proof the necessary conditions for the completeness of a sequence of orthogonal polynomials in the space L^2_ψ.

Theorem 1.12 *Let ψ be an absolutely continuous positive Borel measure defined on an interval (a,b) and assume that for some $c > 0$, we have*

$$\int_a^b e^{c|x|}\psi(x)dx < \infty.$$

Let $(P_n)_n$ be a sequence of orthonormal polynomials with respect to ψ. Then, for any $f \in L^2_\psi(a,b)$, we have that

$$f(x) = \sum_{n=0}^{\infty}(f,P_n)_\psi P_n(x),$$

in the sense that the partial sums of the series converge in norm in the space $L^2_\psi(a,b)$. Moreover, we have Parseval's identity

$$\|f\|^2_\psi = \sum_{n=0}^{\infty}(f,P_n)^2_\psi. \tag{1.35}$$

Proof Details of the proof, using tools from Fourier analysis, can be found in [3, Section 6.5]. □

A measure ψ satisfying Parseval's identity (1.35) is usually called *extremal.*

1.3 The Spectral Theorem for Orthogonal Polynomials

In linear algebra, when we have a linear operator acting on $\mathbb{C}^n$, we may ask ourselves under what conditions a finite-dimensional square matrix associated with the operator can be diagonalized. In finite dimensions it is enough to analyze the spectrum or eigenvalues associated with this matrix. However, when we work with infinite-dimensional vector spaces, the situation is not as simple. The spectral theorem has a broader context in the theory of linear operators on Hilbert spaces equipped with an inner product. The spectral theorem identifies a class of linear operators that can be modeled by multiplication operators. In particular, self-adjoint operators will be of special interest. In general, given a self-adjoint linear operator A defined on a Hilbert space $\mathcal{H}$, we will always be able to find a measure ψ defined on a certain measurable space $\mathcal{S}$ and a unitary operator $U\colon \mathcal{H} \to L^2_\psi(\mathcal{S})$ such that

$$\left(UAU^{-1}f\right)(x) = F(x)f(x),$$

for a certain measurable and bounded real function F defined on $\mathcal{S}$. This is a generalization of the finite-dimensional case.

In the context of orthogonal polynomials, the operator A is identified with the symmetric tridiagonal Jacobi matrix J defined by (1.18), which is obviously self-adjoint (symmetric) in the Hilbert space $\mathcal{H} = \ell^2(\mathbb{N}_0)$. We want to find a measure ψ and a complete orthonormal basis in $L^2_\psi(\mathcal{S})$ with $\mathcal{S}$ a real interval (for example, the orthonormal polynomials) in such a way that $F(x) = x$. In the literature this result is usually called *Favard's theorem*, since J. Favard proved it in 1935. However, other authors, such as A. Wintner in 1929 or M. H. Stone in 1932, proved the same theorem some years before, or almost at the same time, such as J. A. Shohat in 1936.

There are several ways to prove the spectral theorem for orthogonal polynomials and in these notes we will see two different ones. The first one builds a distribution that is located at the zeros of the monic polynomials $\widehat{P}_n$ generated by a three-term recurrence relation and then we take $n \to \infty$. This version can be found in [74, Section 2.5] or more extensively in [16, Chapter II]. A second (more general) method uses functional analysis tools and spectral theory of self-adjoint operators (see [27, Chapter 2], [109] or [133]). We keep both proofs since the first one will be used in the spectral representation of birth–death processes in Chapter 3 while the second one will be in the spectral representation of one-dimensional discrete-time birth–death chains in Chapter 2 and bilateral birth–death processes in Chapter 3.

Other methods, which will not be discussed here, are related to the theory of positive linear functionals $\mathcal{L}$ defined by $\mathcal{L}(x^n) = \mu_n$ (see [16, p. 21]); or also solving the *moment problem*, which is divided into three depending on whether the support of the measure is finite (Hausdorff), semi-infinite (Stieltjes) or $\mathbb{R}$ (Hamburger). For more information, see [2, 133, 137]. There is a close connection between these problems and the theory of *continued fractions*.

Constructive Method Using the Zeros of the Polynomials

We will prove the spectral theorem for the monic family of polynomials $(\widehat{P}_n)_n$ defined by the three-term recurrence relation (1.17). From Proposition 1.8 we know that the zeros of $\widehat{P}_n$ are all real and simple and in addition those of $\widehat{P}_n$ and $\widehat{P}_{n-1}$ interlace. For a fixed n, let $x_{n,j}, j = 1, \dots, n$ be the zeros of $\widehat{P}_n$ arranged in the form (1.27). Since the polynomial is monic, we have $\widehat{P}_n(x) > 0$ for $x > x_{n,1}$ and therefore $(-1)^{j-1}\widehat{P}'_n(x_{n,j}) > 0$. Hence, using the confluent formula (1.24), we get that $(-1)^{j-1}\widehat{P}_{n-1}(x_{n,j}) > 0$. Therefore the sequence defined by

$$\rho(x_{n,j}) = \frac{\zeta_{n-1}}{\widehat{P}'_n(x_{n,j})\widehat{P}_{n-1}(x_{n,j})}, \quad 1 \le j \le n, \tag{1.36}$$

takes positive values. Using the Christoffel–Darboux formula (1.25) for monic polynomials with $\zeta_j = \beta_j \cdots \beta_1$ and the confluent formula (1.26), the expression (1.36) can be rewritten, taking $x = x_{n,r}$ and $y = x_{n,s}$, as

$$\rho(x_{n,r}) \sum_{k=0}^{n-1} \frac{\widehat{P}_k(x_{n,r})\widehat{P}_k(x_{n,s})}{\zeta_k} = \delta_{rs}. \tag{1.37}$$

The real matrix U defined by

$$U = (u_{r,k}), \quad 1 \leq r, k \leq n, \quad u_{r,k} = \sqrt{\rho(x_{n,r})}\frac{\widehat{P}_{k-1}(x_{n,r})}{\sqrt{\zeta_{k-1}}},$$

satisfies $UU^T = I$. Therefore $U^T U = I$ by the uniqueness of the inverse, i.e. U is a unitary matrix. From the (r,s) entry of $U^T U = I$ it follows that

$$\sum_{r=1}^{n} \rho(x_{n,r})\widehat{P}_k(x_{n,r})\widehat{P}_j(x_{n,r}) = \zeta_k \delta_{jk}, \quad j,k = 0,1,\ldots,n-1. \tag{1.38}$$

The previous identity shows that there exists some discrete orthogonality of the polynomials $\widehat{P}_n(x)$ when we restrict their support to the corresponding zeros. The values $\rho(x_{n,j}), j = 1,\ldots,n$ are the corresponding sizes of the jumps at those zeros. Note also that for $k = j = 0$, we have $\sum_{r=1}^{n} \rho(x_{n,r}) = 1$, and the total sum of all these quantities is exactly 1.

We now introduce a sequence of distribution functions $(\psi_n)_n$ defined by

$$\psi_n(-\infty) = 0, \quad \lim_{x \downarrow x_{n,j}} \psi_n(x) - \lim_{x \uparrow x_{n,j}} \psi_n(x) = \rho(x_{n,j}). \tag{1.39}$$

Theorem 1.13 ([74, Theorem 2.5.2]) *Given a sequence of polynomials $(\widehat{P}_n)_n$ generated by the three-term recurrence relation* (1.17) *with $\alpha_{n-1} \in \mathbb{R}$ and $\beta_n > 0$ for all $n \geq 1$, there exists a distribution function ψ such that*

$$\int_{\mathbb{R}} \widehat{P}_n(x)\widehat{P}_m(x)d\psi(x) = \zeta_n \delta_{nm}.$$

Proof From (1.38), we have that

$$1 = \zeta_0 = \int_{\mathbb{R}} d\psi_n(x) = \psi_n(\infty) - \psi_n(-\infty) = \sum_{r=1}^{n} \rho(x_{n,r}).$$

Therefore the functions ψ_n are uniformly bounded. Helly's selection principle (see [137, Introduction]) allows us to find a subsequence $(\psi_{n_k})_k$ of $(\psi_n)_n$ that converges to a distribution ψ, which is also non-decreasing and bounded. The same principle gives that if for all n, the moments of ψ_n exist, then the moments of ψ also exist and also the moments of the subsequence ψ_{n_k} converge to the moments of ψ. Since x^n can be written as a linear combination of the polynomials $\widehat{P}_j, j = 0,1,\ldots,n$, from (1.38) we see that the moments of ψ_n exist. Therefore, taking limits in (1.38) as $n \to \infty$, we obtain the result. □

The previous result shows existence but not uniqueness. In fact there exist families of measures having all the same moments (see [74, Example 2.5.3]). Let us now see that if the coefficients α_n and β_n are bounded, then the measure is unique.

Theorem 1.14 ([74, Theorem 2.5.5]) *If the sequences $(\alpha_n)_n$ and $(\beta_n)_n$ are bounded, then the orthogonality measure ψ of Theorem 1.13 is unique.*

Proof First, if the sequences $(\alpha_n)_n$ and $(\beta_n)_n$ are bounded, then the support of ψ is also bounded. This is due to the representation (1.29) of the polynomials in terms of the truncated matrix J_n of the Jacobi matrix J. The zeros of $\widehat{P}_n$ are the eigenvalues of J_n. Therefore $|x_{n,j}| < 3M$, where M is an upper bound of both sequences $(\alpha_n)_n$ and $(\beta_n)_n$ (see [74, Theorem 1.1.1]). Therefore the support of each ψ_n is contained in the interval $(-3M, 3M)$.

Let ν be another orthogonality measure that has the same moments as the measure ψ. For any $a > 0$, we have

$$\int_{|x|\geq a} d\nu(x) \leq a^{-2n} \int_{|x|\geq a} x^{2n} d\nu(x) \leq a^{-2n} \int_{\mathbb{R}} x^{2n} d\nu(x) = a^{-2n} \int_{\mathbb{R}} x^{2n} d\psi(x).$$

Applying the quadrature formula (1.30) for $p(x) = x^{2n}$ and using (1.32), we have that

$$\int_{|x|\geq a} d\nu(x) \leq a^{-2n} \sum_{k=1}^{n+1} \lambda_k (x_{n+1,k})^{2n} \leq (A/a)^{2n} \sum_{k=1}^{n+1} \lambda_k = (A/a)^{2n},$$

where $|x_{n,j}| \leq A$ for all $n \geq 1$ and $1 \leq j \leq n$. Since a is a free parameter, in particular, if $a > A$, then $\int_{|x|\geq a} d\nu(x) = 0$ by taking $n \to \infty$ in the previous inequality. Therefore the support of ν is contained in $[-A, A]$. Let us now see that $\nu = \psi$. For $|x| \geq 2A$, we have that $\sum_{k=0}^{n} t^k x^{-k-1}$ converges to $1/(x-t)$ for all $t \in [-A, A]$ since $|t/x| < 1$ as a consequence of $|x| \geq 2A > A \geq |t|$. Thus

$$\int_{\mathbb{R}} \frac{d\psi(t)}{x-t} = \int_{\mathbb{R}} \lim_{n\to\infty} \sum_{k=0}^{n} \frac{t^k}{x^{k+1}} d\psi(t) = \lim_{n\to\infty} \int_{\mathbb{R}} \sum_{k=0}^{n} \frac{t^k}{x^{k+1}} d\psi(t) = \lim_{n\to\infty} \sum_{k=0}^{n} \frac{\mu_k}{x^{k+1}},$$

using the dominated convergence theorem, since $|t/x| \leq 1/2$. The last limit only depends on the moments and is the same for all measures that have the same moments. Then $F(x) = \int_{\mathbb{R}} \frac{d\psi(t)}{x-t}$ is uniquely determined for any x outside the circle $|x| = 2A$. Since F is analytic in $x \in \mathbb{C} \setminus [-A, A]$, F is unique. Then the theorem is a consequence of the Perron–Stieltjes inversion formula (1.8). □

For the more general case see for instance [74].

Methods from Functional Analysis and Spectral Theory

This method is based on important results on functional analysis and spectral analysis of linear operators in Hilbert spaces, which will not be proved in these notes. For more information about these results, see [125, 127].

Let $\mathcal{H}$ be a Hilbert space with an inner product $(\cdot,\cdot)$ and denote by $\mathcal{B}(\mathcal{H})$ the set of all linear operators of $\mathcal{H}$ in $\mathcal{H}$. For an operator $T \in \mathcal{B}(\mathcal{H})$, the *resolvent operator* is defined by $R(z) = (T - z)^{-1}$. The values of $z \in \mathbb{C}$ for which $R(z)$ is a bounded linear operator are called *regular values* and are denoted by $\rho(T)$. The complement of the *resolvent set* $\rho(T)$ is called the *spectrum* of T and is denoted by $\sigma(T)$. For a bounded operator T the spectrum $\sigma(T)$ is a compact subset of the disk of radius $\|T\| = \inf_{u\in\mathcal{H}}(\|Tu\|/\|u\|)$. Moreover, if T is self-adjoint, i.e. $(Tu, v) = (u, Tv)$ for all $u, v \in \mathcal{H}$, then $\sigma(T) \subset \mathbb{R}$, so that $\sigma(T) \subset [-\|T\|, \|T\|]$.

A *resolution of the identity* E of the Hilbert space $\mathcal{H}$ is a map $E\colon \mathbb{R} \to \mathcal{B}(\mathcal{H})$ such that for any Borel sets $A, B \subseteq \mathbb{R}$ we have (i) $E(A)$ is a self-adjoint projection, i.e. $E(A)^2 = E(A)$, (ii) $E(A \cap B) = E(A)E(B)$, (iii) $E(\emptyset) = 0, E(\mathbb{R}) = I_{\mathcal{H}}$, (iv) $A \cap B = \emptyset$ implies $E(A \cup B) = E(A) + E(B)$ and (v) for all $u, v \in \mathcal{H}$ the map $A \mapsto E_{u,v}(A) = (E(A)u, v)$ is a complex Borel measure. The spectral measure for orthogonal polynomials will be constructed from the map $A \mapsto E_{e_0,e_0}(A) = (E(A)e_0, e_0)$, where e_0 is the first canonical vector of the space $\ell^2(\mathbb{N}_0)$, as we will see below.

Theorem 1.15 (Spectral theorem) *Let $T\colon \mathcal{H} \to \mathcal{H}$ be a bounded self-adjoint linear operator. Then there exists a unique resolution of the identity E of $\mathcal{H}$ such that $T = \int_{\mathbb{R}} t dE(t)$, i.e.*

$$(Tu, v) = \int_{\mathbb{R}} t dE_{u,v}(t).$$

Moreover, E is supported on the spectrum $\sigma(T)$ and any of the spectral projections $E(A), A \subset \mathbb{R}$, commutes with T.

Proof See [127, Section 12.22] or also [51, 125]. □

For any continuous function f defined on the spectrum $\sigma(T)$, we can define the operator $f(T) = \int_{\mathbb{R}} f(t)dE(t)$, i.e. $(f(T)u, v) = \int_{\mathbb{R}} f(t)dE_{u,v}(t)$. Then $f(T)$ is a bounded operator with norm $\|f(T)\| = \sup_{x\in\sigma(T)} |f(x)|$. In particular this can be applied to $f(x) = 1/(x-z), z \in \rho(T)$ and $f(T)(z) = R(z)$, the resolvent operator. The spectral measure E can be obtained from the resolvent operator using the Perron–Stieltjes inversion formula. Indeed, for an open interval $(a, b) \subset \mathbb{R}$, we have that

$$E_{u,v}((a,b)) = \lim_{\delta\downarrow 0}\lim_{\varepsilon\downarrow 0} \frac{1}{2\pi i} \int_{a+\delta}^{b-\delta} [(R(x+i\varepsilon)u, v) - (R(x-i\varepsilon)u, v)]\, dx.$$

Compare with the Perron–Stieltjes inversion formula in (1.9). For unbounded linear operators there is also a spectral theorem, but it is a little bit more technical than the bounded case (see [27, 109, 125]).

Let J be the symmetric tridiagonal Jacobi operator (1.18) with $b_n \in \mathbb{R}, a_n > 0$, $n \geq 0$ and assume that these coefficients are bounded. J is an operator defined on the Hilbert space $\ell^2(\mathbb{N}_0)$ given by

$$\ell^2(\mathbb{N}_0) = \left\{ (a_n)_{n\in\mathbb{N}_0} : \sum_{n=0}^{\infty} |a_n|^2 < \infty \right\}.$$

For the orthonormal canonical basis $(e_n)_n$ of $\ell^2(\mathbb{N}_0)$, this Jacobi operator is defined as

$$Je_n = \begin{cases} a_n e_{n+1} + b_n e_n + a_{n-1} e_{n-1}, & \text{if } n \geq 1, \\ a_0 e_1 + b_0 e_0, & \text{if } n = 0. \end{cases} \tag{1.40}$$

The operator J is extended (by continuity) to a self-adjoint linear bounded operator in $\ell^2(\mathbb{N}_0)$, since the canonical basis is dense in $\ell^2(\mathbb{N}_0)$, the coefficients of J are bounded, and also $(Ju, v) = (u, Jv)$, for all $u, v \in \ell^2(\mathbb{N}_0)$.

Lemma 1.16 *There exists a polynomial $P_n(x)$ of degree n such that $e_n = P_n(J)e_0$. In particular e_0 is a cyclic vector for the action of J, i.e. the linear subspace $\{J^n e_0 : n \in \mathbb{N}_0\}$ is dense in $\ell^2(\mathbb{N}_0)$.*

Proof The candidate for this polynomial is precisely the one generated by the three-term recurrence relation (1.16). We will prove it by induction on n. Let P_{n+1} be the polynomial of degree $n + 1$ generated by the recurrence formula

$$P_{n+1}(x) = \frac{1}{a_n}\left[(x - b_n)P_n(x) - a_{n-1}P_{n-1}(x)\right], \quad n \geq 0, \quad P_0(x) = 1, \quad P_{-1}(x) = 0.$$

For $n = 0$ is trivial. For $n = 1$ we have, using the previous formula and (1.40), that

$$P_1(J)e_0 = \frac{1}{a_0}[Je_0 - b_0 e_0] = e_1.$$

Assuming the result is true for n, for $n + 1$ we have, using again (1.40), that

$$\begin{aligned} P_{n+1}(J)e_0 &= \frac{1}{a_n}\left[(J - b_n I)P_n(J)e_0 - a_{n-1}P_{n-1}(J)e_0\right] \\ &= \frac{1}{a_n}\left[(J - b_n I)e_n - a_{n-1}e_{n-1}\right] = e_{n+1}. \end{aligned}$$

Since the monomials can be uniquely written as a linear combination of the polynomials P_n and vice versa, we have that the linear subspace $\{J^n e_0 : n \in \mathbb{N}_0\}$ is dense in $\ell^2(\mathbb{N}_0)$, since $(e_n)_n$ is dense in $\ell^2(\mathbb{N}_0)$. □

Now we are in the position of applying the spectral theorem to J, i.e. there exists a unique resolution of the identity E of $\ell^2(\mathbb{N}_0)$ such that

$$(Ju, v) = \int_{\mathbb{R}} t dE_{u,v}(t), \quad u, v \in \ell^2(\mathbb{N}_0).$$

From here we can define the positive Borel measure ψ by means of $\psi(A) = E_{e_0,e_0}(A) = (E(A)e_0, e_0)$. Since $E(A)$ is a projection then $\psi(A) = (E(A)^2 e_0, e_0) = (E(A)e_0, E(A)e_0) \geq 0$. Also $E(\mathbb{R}) = I_{\ell^2(\mathbb{N}_0)}$, implying that ψ is a probability measure. The support is contained in the interval $[-\|J\|, \|J\|]$, since J is bounded and in particular has finite moments. In addition, the spectral measure E is completely determined by ψ, since, using the previous lemma and that E commutes with J, we have that

$$\begin{aligned}(E(A)e_k, e_l) &= (E(A)P_k(J)e_0, P_l(J)e_0) = (P_k(J)P_l(J)E(A)e_0, e_0) \\ &= \int_A P_k(x)P_l(x)d\psi(x). \end{aligned} \tag{1.41}$$

With this measure we can define the space $L^2_\psi(\mathbb{R})$ given by

$$L^2_\psi(\mathbb{R}) = \left\{ f \text{ measurable} : \int_{\mathbb{R}} |f(x)|^2 d\psi(x) < \infty \right\}.$$

Let us now see the proof of Favard's theorem (or the spectral theorem for orthogonal polynomials).

Theorem 1.17 *Let J be a bounded Jacobi operator. Then there exists a unique probability measure ψ supported on a real compact interval such that for every polynomial P, the map $U: P(J)e_0 \mapsto P$ extends to a unitary operator $\ell^2(\mathbb{N}_0) \to L^2_\psi$ such that $UJ = MU$, where $M: L^2_\psi \to L^2_\psi$ is the multiplication operator $(Mf)(x) = xf(x)$. Moreover, the sequence $P_n = Ue_n$ is a set of orthonormal polynomials with respect to ψ. Therefore, the operator J can be diagonalized in the following way:*

$$(UJU^{-1}f)(x) = (Mf)(x) = xf(x), \quad f \in L^2_\psi.$$

This means that the following diagram is commutative:

$$\begin{array}{ccc} \ell^2(\mathbb{N}_0) & \xrightarrow{U} & L^2_\psi \\ \Big\downarrow J & & \Big\downarrow M \\ \ell^2(\mathbb{N}_0) & \xrightarrow{U} & L^2_\psi. \end{array}$$

Proof By Lemma 1.16, U sends dense subsets of $\ell^2(\mathbb{N}_0)$ in dense subsets of L^2_ψ, since the polynomials are dense in L^2_ψ, because the support of the measure ψ is bounded. Using (1.41) we have, for all pairs of polynomials P and Q, that

$$\begin{aligned}(P(J)e_0, Q(J)e_0) &= (\bar{Q}(J)P(J)e_0, e_0) = \int_{\mathbb{R}} \bar{Q}(x)P(x)d\psi(x) \\ &= (P, Q)_\psi = (UP(J)e_0, UQ(J)e_0)_\psi. \end{aligned}$$

Therefore U is a unitary operator and extends in a unique way to a unitary operator by the density on both spaces. To see that $UJ = MU$, we show it for the canonical base, i.e. $UJe_n = MUe_n$, for all n. Let $P_n = Ue_n$. Then $P_n \in L^2_\psi$ is a polynomial of degree n and they are also orthogonal with respect to ψ, since

$$\int_{\mathbb{R}} P_n(x)P_m(x)d\psi(x) = (Ue_n, Ue_m)_\psi = (e_n, e_m) = \delta_{nm}.$$

It remains to prove that the coefficients of the three-term recurrence relation for the polynomials P_n coincide with the coefficients of the operator J. But this is true from the relationship between these coefficients with the measure ψ, since, from (1.40) and the spectral theorem, we have that

$$a_n = (Je_n, e_{n+1}) = \int_{\mathbb{R}} x dE_{e_n, e_{n+1}}(x) = \int_{\mathbb{R}} x P_n(x) P_{n+1}(x) d\psi(x),$$

$$b_n = (Je_n, e_n) = \int_{\mathbb{R}} x dE_{e_n, e_n}(x) = \int_{\mathbb{R}} x (P_n(x))^2 d\psi(x).$$

The uniqueness follows from the fact that the moments of ψ are uniquely determined since P_n is a family of orthonormal polynomials in L^2_ψ. Since the measure has compact support and its Stieltjes transform is analytic in a neighborhood of ∞, the Perron–Stieltjes inversion formula gives a unique measure determined by its moments. □

For more details about spectral theory and special functions, as well as the unbounded case, see [27, 109, 125].

1.4 Classical Orthogonal Polynomials of a Continuous Variable

So far we have seen general properties for any family of orthogonal polynomials. In this section we will focus on special families called *classical*, which apart from being orthogonal on an interval $\mathcal{S} = (a,b)$ (and therefore eigenfunctions of a Jacobi operator) will also be eigenfunctions of a second-order differential operator in the continuous variable x of the form

$$\mathcal{D} = p(x)\frac{d^2}{dx^2} + q(x)\frac{d}{dx} + r(x), \tag{1.42}$$

where $p(x) > 0$ for all $x \in (a,b)$ and $q(x), r(x)$ are real functions. This kind of differential operator will correspond to the infinitesimal operators of diffusion processes, as we will see in Chapter 4.

We saw in the previous section that the Jacobi operator is equivalent to the multiplication operator in the space $\ell^2(\mathbb{N}_0)$. This operator is self-adjoint (or symmetric) and therefore we can apply the spectral theorem to that Hilbert space. In the same way, now for the continuous differential operator (1.42), we will be interested in self-adjoint differential operators in the space $L^2_\psi(a,b)$, where ψ is a probability measure (or *weight function*), with respect to the inner product (1.13), that is,

$$(\mathcal{D}f, g)_\psi = (f, \mathcal{D}g)_\psi$$

for any $f, g \in C^2(a,b)$, i.e. twice continuously differentiable functions on (a,b), that vanish outside of some closed subinterval of $\text{supp}(\psi) = (a,b)$ (finite or infinite). In this section we will assume that the measure ψ is absolutely continuous with respect to the Lebesgue measure, in which case we write, abusing the notation, $d\psi(x) = \psi(x)dx$.

Proposition 1.18 *The differential operator $\mathcal{D}$ defined by* (1.42) *is self-adjoint with respect to a measure ψ if and only if $q\psi = (p\psi)'$. Therefore, the operator $\mathcal{D}$ can be rewritten as*

$$\mathcal{D} = p\frac{d^2}{dx^2} + \frac{(p\psi)'}{\psi}\frac{d}{dx} + r = \frac{1}{\psi}\frac{d}{dx}\left(p\psi\frac{d}{dx}\right) + r. \tag{1.43}$$

Proof On the one hand, if the operator $\mathcal{D}$ is self-adjoint, we have, integrating by parts and taking into account the boundary behavior of f and g, that

$$0 = (\mathcal{D}f, g)_\psi - (f, \mathcal{D}g)_\psi = \int_a^b \left[p\left(f'g - fg'\right)' + q\left(f'g - fg'\right)\right]\psi(x)\,dx$$
$$= \int_a^b \left(f'g - fg'\right)\left[q\psi - (p\psi)'\right]dx.$$

In particular, for $f = \mathbf{1}_{(a,b)}$, where $\mathbf{1}_A$ is the indicator function, and $g \neq 0$, we have

$$0 = -\int_a^b g'\left[q\psi - (p\psi)'\right]dx = \int_a^b g\left[q\psi - (p\psi)'\right]'dx.$$

Therefore $q\psi - (p\psi)' = c$, where c is a constant. If $f = x$ and $g \neq 0$, we have

$$0 = c\int_a^b \left(g - xg'\right)dx = 2c\int_a^b g\,dx$$

for any g, so we get $c = 0$. Therefore the symmetry implies that $q\psi = (p\psi)'$. On the other hand, it is easy to see that the condition $q\psi = (p\psi)'$ implies symmetry of the differential operator. □

In this text, we will be interested in finding solutions of the following regular *Sturm–Liouville problem* with *separated* boundary conditions:

$$\begin{aligned}
\mathcal{D}\phi = \frac{1}{\psi(x)}\frac{d}{dx}\left(p(x)\psi(x)\frac{d\phi(x)}{dx}\right) + r(x)\phi(x) &= -\lambda\phi(x), \quad \text{on} \quad (a,b),\\
c_{1,1}\phi(a) + c_{1,2}\left(p(x)\psi(x)\phi'(x)\right)(a) &= 0,\\
c_{2,1}\phi(b) + c_{2,2}\left(p(x)\psi(x)\phi'(x)\right)(b) &= 0,
\end{aligned} \tag{1.44}$$

where $c_{i,j}, i,j = 1,2$ are real, $(c_{i,1}, c_{i,2}) \neq (0,0), i = 1,2$, and $\lambda \in \mathbb{C}$. A classical result about the behavior of this Sturm–Liouville problem is the following:

Theorem 1.19 *Consider the regular Sturm–Liouville problem with separated boundary conditions* (1.44) *such that $p, \psi > 0$ on (a,b). Then:*

1. *All eigenvalues are real and simple.*
2. *There is an infinite but countable number of eigenvalues $-\lambda_n, n \in \mathbb{N}_0$, bounded from below, and can be ordered in the following way:*

$$-\infty < \lambda_0 < \lambda_1 < \cdots < \lambda_n \to \infty.$$

3. *If ϕ_n is an eigenfunction of $\mathcal{D}$ then ϕ_n has exactly n zeros inside the open interval (a,b).*
4. *The sequence of eigenfunctions $(\phi_n)_n$ can be normalized in such a way that they form an orthonormal sequence in the space $L^2_\psi(a,b)$ which is also complete.*

Proof A proof can be found in [151, Theorem 4.6.2]. □

Remark 1.20 It is remarkable that the eigenvalues are bounded from above (even if the support is not bounded). Indeed, consider a normalized eigenfunction ϕ of $\mathcal{D}$ with eigenvalue $-\lambda$. Using integration by parts and the boundary values in (1.44), we have that

$$
\begin{aligned}
-\lambda = (\mathcal{D}\phi,\phi)_\pi &= \int_a^b \phi(x)[p(x)\psi(x)\phi'(x)]'dx + \int_a^b \phi(x)^2 r(x)\psi(x)\,dx \\
&= \phi(x)p(x)\psi(x)\phi'(x)\Big|_a^b - \int_a^b p(x)\psi(x)\phi'(x)^2 dx + \int_a^b \phi(x)^2 r(x)\psi(x)\,dx \\
&= \int_a^b [\phi(x)^2 r(x) - p(x)\phi'(x)^2]\psi(x)\,dx + \frac{c_{1,1}}{c_{1,2}}\phi(a)^2 - \frac{c_{2,1}}{c_{2,2}}\phi(b)^2.
\end{aligned}
$$

From here it is clear that $-\lambda$ is bounded from above by a real constant. If $c_{1,2} = 0$ then $\phi(b) = 0$ and the second term of the right-hand side drops out. Same if $c_{2,2} = 0$. The case where $\phi(a) = \phi(b) = 0$ gives that

$$
-\lambda = \int_a^b [\phi(x)^2 r(x) - p(x)\phi'(x)^2]\psi(x)dx \leq \max\{r(x), x \in [a,b]\}. \tag{1.45}
$$

Therefore we obtain a bound for all eigenvalues of $\mathcal{D}$. In particular, if $r(x) = 0$ then $\lambda \geq 0$. The same inequality holds is we assume that $p(a)\psi(a) = p(b)\psi(b) = 0$, as we will see below. ◊

Let us now look for *polynomial eigenfunctions* of $\mathcal{D}$, i.e. $\mathcal{D}p_n = -\lambda_n p_n$, with $\deg p_n = n$ and $-\lambda_n$ the sequence of eigenvalues independent of x. For this, we need to assume certain conditions on the boundary of the orthogonality support of ψ. Assume that the orthogonality support of ψ is given by a real interval (a,b) (which may or may not be bounded) and assume that ψ, ψ', p, p', q and r extend to continuous functions on the closed interval $[a,b]$. Assume further that $f,g \in C^2(a,b) \cap L^2_\psi(a,b)$ and that $f, f', g, g' \in C^0([a,b])$. Finally, assume that $\mathcal{D}$ is self-adjoint (then $q\psi = (p\psi)'$). From the proof of Proposition 1.18, we can deduce that

$$
(\mathcal{D}f,g)_\psi - (f,\mathcal{D}g)_\psi = p\psi(f'g - fg')\Big|_{x=a}^{x=b}. \tag{1.46}
$$

If $p\psi$ vanishes at both points of the boundary then we do not need additional conditions. Otherwise, conditions must be imposed on the functions f,g. In any case ψ must also have a finite integral in the interval (a,b). If we also force these eigenfunctions to be polynomials of all degrees, the differential operator $\mathcal{D}$ has to take polynomials of degree n into polynomials of degree n. Suppose that there polynomials of degrees 0, 1, 2 which are eigenfunctions of $\mathcal{D}$. If $p_0(x)=1$ then $\mathcal{D}p_0 = r$, where r must be constant, so we can always take $r=0$ (since the constant can always go to the eigenvalue part). If $p_1(x) = x$ then

$$\mathcal{D}p_1 = \frac{(p\psi)'}{\psi} = p' + p\frac{\psi'}{\psi} = q. \tag{1.47}$$

This expression must be a polynomial of degree at most 1. Finally, taking $p_2(x) = x^2/2$, we have that

$$\mathcal{D}p_2 = p + x\left(p' + p\frac{\psi'}{\psi}\right) = p + x\mathcal{D}p_1 = p + xq.$$

The above expression must be a polynomial of degree at most 2. This forces that p must necessarily be a polynomial of degree at most 2. Since we are interested in polynomial eigenfunctions of all degrees, the boundary conditions (1.46) force $p\psi$ to vanish at the boundary of the orthogonality support of ψ. In summary, we look for polynomial solutions $p_n(x)$ that are eigenfunctions of a differential operator of the form

$$\mathcal{D} = p\frac{d^2}{dx^2} + q\frac{d}{dx} = \frac{1}{\psi}\frac{d}{dx}\left(p\psi\frac{d}{dx}\right), \quad \deg p \leq 2, \quad \deg q \leq 1, \tag{1.48}$$

where $\mathcal{D}p_n = -\lambda_n p_n$ and

$$(p\psi)' = q\psi, \tag{1.49}$$

with the boundary conditions $p\psi \to 0$ as $x \to \{a,b\}$. The equation (1.49) is usually called the *Pearson equation*. From Theorem 1.19 and Remark 1.20 we have that all polynomial eigenfunctions $(p_n)_n$ of $\mathcal{D}$ are orthogonal with respect to ψ and $\lambda_n \geq 0$.

Now we want to characterize all families of orthogonal polynomials that are eigenfunctions of a differential operator of the Sturm–Liouville type (1.48). This problem was already solved by S. Bochner in [10], although E. Routh in [126] already raised and partially solved this question.

Since p is a polynomial of degree at most 2, after normalization (affine maps on the line or multiplication by constants), it can be reduced to five cases, depending on the degree and roots of the polynomial p.

Case I: p constant. In this case, from (1.47), we have that ψ'/ψ must be a polynomial of degree at the most 1. Solving the differential equation, we have that $\psi(x) = e^h$, where h is a real polynomial of degree at most 2. After normalization, it can be reduced to two cases, $\psi(x) = e^{-x}$ or $\psi(x) = e^{\pm x^2}$. In the first case, the boundary conditions say that $p\psi$ has to vanish at the boundary of some real interval, but e^{-x} never meets this criterion since it is always positive. The same applies for the case of e^{x^2}. Therefore the only case left is $\psi(x) = e^{-x^2}$, in which case the orthogonality interval must be $\mathcal{S} = (-\infty, \infty) = \mathbb{R}$. The polynomial $q = \psi'/\psi = -2x$ is a polynomial of degree exactly 1. Therefore the operator (1.48) is in this case

$$\mathcal{D} = \frac{d^2}{dx^2} - 2x\frac{d}{dx}, \quad \text{in} \quad L^2_\psi(\mathbb{R}), \quad \psi(x) = e^{-x^2}. \tag{1.50}$$

Identifying coefficients, we have $\lambda_n = 2n$. Therefore orthogonal polynomials with respect to the measure e^{-x^2}, which are called *Hermite polynomials* and are denoted by $H_n(x)$, satisfy that $\mathcal{D}H_n = -2nH_n$. This measure, once normalized, corresponds to the *normal* or *Gaussian distribution*. We will later give some structural formulas related to this family.

Case II: p linear. After normalization, we can always reduce p to $p(x) = x$. From (1.47), we have $\psi'/\psi = b + \alpha/x$. Solving the differential equation, we have that, up to constants, $\psi(x) = x^\alpha e^{bx}$. The boundary conditions $p\psi = x\psi = x^{\alpha+1}e^{bx}$ and the integrability of ψ, force the orthogonality interval to be $\mathcal{S} = (0, \infty) = \mathbb{R}^+$ and $b < 0, \alpha > -1$. Therefore, after normalizing, $\psi(x) = x^\alpha e^{-x}, \alpha > -1$. The polynomial $q = (x\psi)'/\psi = \alpha + 1 - x$ is a polynomial of degree exactly 1. Therefore the operator (1.48) is in this case

$$\mathcal{D} = x\frac{d^2}{dx^2} + (\alpha + 1 - x)\frac{d}{dx}, \quad \text{in} \quad L^2_\psi(\mathbb{R}^+), \quad \psi(x) = x^\alpha e^{-x}, \quad \alpha > -1. \tag{1.51}$$

Identifying coefficients, we have $\lambda_n = n$. Therefore orthogonal polynomials with respect to the measure $x^\alpha e^{-x}$, which are called *Laguerre polynomials* and are denoted by $L_n^{(\alpha)}(x)$, satisfy $\mathcal{D}L_n^{(\alpha)} = -nL_n^{(\alpha)}$. This measure, once normalized, corresponds to the *Gamma distribution* (see (1.1)) and as a special case, for $\alpha = 0$, the *exponential distribution*. We will later give some structural formulas related to this family.

Case III: p quadratic with different real roots. After normalization we can use the polynomial $p(x) = 1 - x^2$. From (1.47), we have $\psi'/\psi = \beta/(1+x) + \alpha/(1-x)$. Solving the differential equation, we have that, up to constants, $\psi(x) = (1-x)^\alpha (1+x)^\beta$. The boundary conditions $p\psi = (1-x^2)\psi = (1-x)^{\alpha+1}(1+x)^{\beta+1}$ and the integrability of ψ, force the orthogonality interval to be $\mathcal{S} = (-1, 1)$ and $\alpha, \beta > -1$. The polynomial $q = ((1-x^2)\psi)'/\psi = \beta - \alpha - x(\alpha + \beta + 2)$ is a polynomial of degree exactly 1. Therefore the operator (1.48) is in this case

$$\mathcal{D} = (1-x^2)\frac{d^2}{dx^2} + [\beta - \alpha - x(\alpha + \beta + 2)]\frac{d}{dx}, \tag{1.52}$$
$$\text{in} \quad L^2_\psi(-1, 1), \quad \psi(x) = (1-x)^\alpha(1+x)^\beta, \quad \alpha, \beta > -1.$$

Identifying coefficients, we have $\lambda_n = n(n+\alpha+\beta+1)$. Therefore orthogonal polynomials with respect to the measure $(1-x)^\alpha(1+x)^\beta$, which are called the *Jacobi polynomials* and are denoted by $P_n^{(\alpha,\beta)}(x)$, satisfy $\mathcal{D}P_n^{(\alpha,\beta)} = -n(n+\alpha+\beta+1)P_n^{(\alpha,\beta)}$. Jacobi polynomials can also be defined in any other bounded interval $[a,b]$, in which case $p(x) = (b-x)(x-a)$. A commonly used choice is for $a = 0$ and $b = 1$, i.e. $[0,1]$, in which case the measure, once normalized, corresponds to the *Beta distribution* (1.3). Taking $(1-x)/2$ as the new x-variable we obtain the weight function $x^\alpha(1-x)^\beta$ while the differential operator is given by

$$\mathcal{D} = x(1-x)\frac{d^2}{dx^2} + [\alpha+1-x(\alpha+\beta+2)]\frac{d}{dx},$$

which is the hypergeometric differential equation (1.5) for $a = \beta+1, b = 0, c = \alpha+1$ or $a = 0, b = \beta+1, c = \alpha+1$. We will later give some structural formulas related to these families.

Case IV: p quadratic with different complex roots. We can normalize to $p(x) = 1+x^2$. From (1.47), we have $\psi'/\psi = (-2ax+b)/(1+x^2)$. Solving the differential equation, we have, up to constants, that

$$\psi(x) = (1+x^2)^{-a}e^{b\arctan x}.$$

The boundary conditions and the positivity and integrability of ψ force the orthogonality interval to be $\mathcal{S} = (-\infty,\infty)$ and $a > 1/2, b \in \mathbb{R}$. The polynomial $q = ((1+x^2)\psi)'/\psi = 2x(1-a)+b$ is a polynomial of degree exactly 1. Therefore the operator (1.48) is in this case

$$\mathcal{D} = (1+x^2)\frac{d^2}{dx^2} + [2x(1-a)+b]\frac{d}{dx}, \tag{1.53}$$
$$\text{in } L^2_\psi(\mathbb{R}), \quad \psi(x) = (1+x^2)^{-a}e^{b\arctan x}, \quad a > 1/2, \quad b \in \mathbb{R}.$$

Identifying coefficients, we have $\lambda_n = n(2a-n-1)$. However, the range where a is defined breaks the possibility of the existence of a sequence of orthogonal polynomials of all degrees, since it is necessary that all moments of the measure are finite, and in this case there is only a finite number of finite moments, so there will be a finite number of orthogonal polynomials $p_n(x;a,b)$, called *Romanovski polynomials*. The number of polynomials that can be constructed is for $n < 2a-1$, which are the values for which λ_n is positive. This implies that the operator $\mathcal{D}$ will have a finite discrete spectrum in combination with an absolutely continuous spectrum. We will return to this example in Chapter 4 in relation to diffusion processes. Observe that for $b = 0$, ψ is reduced to the *t-student distribution*. For any other b, ψ is a generalization of this distribution. Finally, E. Routh in 1884 found an expression of these finite number of orthogonal polynomials in terms of the Jacobi polynomials [126]. More specifically,

$$p_n(x;a,b) = (-i)^n P_n^{(-a+ib/2,-a-ib/2)}(ix).$$

Case V: p quadratic with double root. We can normalize to $p(x) = x^2$. From (1.47), we have $\psi'/\psi = -a/x + b/x^2$. Solving the differential equation, we have, up to constants, that

$$\psi(x) = x^{-a}e^{-b/x}.$$

The boundary conditions, and the positivity and integrability of ψ, force the orthogonality interval to be $\mathcal{S} = (0,\infty) = \mathbb{R}^+$ and $a > 1, b \geq 0$. The polynomial $q = (x^2\psi)'/\psi = x(2-a) + b$ is a polynomial of degree exactly 1. Therefore the operator (1.48) is in this case

$$\mathcal{D} = x^2 \frac{d^2}{dx^2} + [x(2-a) + b]\frac{d}{dx},$$
$$\text{in} \quad L^2_\psi(\mathbb{R}^+), \quad \psi(x) = x^{-a}e^{-b/x}, \quad a > 1, b \geq 0. \tag{1.54}$$

Identifying coefficients, we have $\lambda_n = n(a-n-1)$. However, again, there is only a finite number of finite moments (for $n < a-1$). Therefore there will be a finite number of orthogonal polynomials $y_n(x;a,b)$. Also in this case the operator $\mathcal{D}$ will have a finite discrete spectrum in combination with an absolutely continuous spectrum. We will return to this example in Chapter 4 in relation to diffusion processes. There are two special cases of polynomial solutions, but they are not orthogonal with respect to any measure supported on the real line. On the one hand, when $b = 0$, we have that $y_n(x;a,0) = x^n$ is a solution of the differential equation $\mathcal{D}y_n(x;a,0) = -\lambda_n y_n(x;a,0)$, but the monomials are not orthogonal with respect to any measure in $\mathbb{R}$. On the other hand, for $b = 1$ and $a = 2-\alpha$, we have $y_n(x;2-\alpha,1) = B_n(x;\alpha,1)$, where $B_n(x;\alpha,\beta)$ are the so-called *Bessel polynomials*. However, these polynomials are orthogonal on the unit circle, but not on any real interval (for more information about Bessel polynomials, see [74]).

The families of Hermite, Laguerre and Jacobi polynomials are usually called the *classical families of orthogonal polynomials of a continuous variable*. Therefore, we have proved the following (somewhat informally stated) result:

Theorem 1.21 *Up to normalization, the classical orthogonal polynomials (Hermite, Laguerre and Jacobi) are the only families of orthogonal polynomials that are eigenfunctions of a second-order differential operator of the form* (1.43) *(i.e. self-adjoint with respect to a positive measure supported on the real line).*

The polynomial eigenfunctions of this kind of differential operator also have special representations in terms only of the corresponding polynomial p and the measure ψ. This is called *Rodrigues' formula* and will have important consequences in terms of formulas satisfied by these polynomials.

Proposition 1.22 (Rodrigues' formula) *The solutions of $\mathcal{D}p_n = -\lambda_n p_n$ where $\mathcal{D}$ is given by* (1.48) *can be written, up to normalization, as*

$$p_n(x) = \frac{1}{\psi(x)} \frac{d^n}{dx^n}\left[p^n(x)\psi(x)\right] = \psi^{-1}\left(p^n\psi\right)^{(n)}. \tag{1.55}$$

Proof A proof can be found in [9, Section 4.2]. □

It is possible to prove that the leading coefficient of the polynomial p_n in (1.55) is given by

$$k_n = \frac{(-1)^n}{n!} \prod_{m=0}^{n-1}\left[\lambda_n + mq' + \frac{1}{2}m(m-1)p''\right]. \tag{1.56}$$

Rodrigues' formula also gives an easy evaluation of integrals of the form

$$(f, p_n)_\psi = \int_a^b f(x)p_n(x)\psi(x)\,dx.$$

Since $p_n\psi = (p^n\psi)^{(n)}$, the function $p^n\psi$ vanishes quite fast at the endpoints of the interval (a,b). Therefore, if f is smooth enough, we can integrate by parts n times without obtaining terms at the boundary, concluding that

$$\int_a^b p_n(x)f(x)\psi(x)\,dx = (-1)^n \int_a^b p^n(x)f^{(n)}(x)\psi(x)\,dx.$$

If we apply this formula to the function $f = p_n$ and using (1.56), we get an explicit expression of the norms of p_n, i.e.

$$\begin{aligned}\|p_n\|_\psi^2 &= (-1)^n \int_a^b p_n^{(n)} p^n(x)\psi(x)\,dx \\ &= \prod_{m=0}^{n-1}\left[\lambda_n + mq' + \frac{1}{2}m(m-1)p''\right]\int_a^b p^n(x)\psi(x)\,dx. \end{aligned} \tag{1.57}$$

Another important consequence of Rodrigues' formula (1.55) is that we can obtain an integral representation of the orthogonal polynomials p_n using the well-known Cauchy integral formula for the nth derivative. Indeed,

$$p_n(x)\psi(x) = \frac{n!}{2\pi i}\oint_C \frac{p^n(z)}{(z-x)^{n+1}}\psi(z)\,dz, \tag{1.58}$$

where C is a certain complex contour enclosing the points $x \in (a,b)$. As a consequence, we have the following:

Proposition 1.23 *Let*

$$G(x,s) = \sum_{n=0}^{\infty} p_n(x)\frac{s^n}{n!}$$

be the generating function of the orthogonal polynomials p_n. If p_n satisfies (1.58), *then the generating function can be written in the following form:*

$$G(x,s) = \frac{\psi(\zeta)}{\psi(x)} \frac{1}{1 - sp'(\zeta)}, \quad \zeta - sp(\zeta) = x. \tag{1.59}$$

Proof Substituting (1.58) in the generating function, we have

$$G(x,s) = \frac{1}{2\pi i} \oint_C \sum_{n=0}^{\infty} \frac{s^n p^n(z)}{(z-x)^n} \frac{\psi(z)}{\psi(x)} \frac{dz}{z-x} = \frac{1}{2\pi i} \oint_C \frac{\psi(z)}{\psi(x)} \frac{dz}{z - x - sp(z)}.$$

We can assume that the contour C contains a solution $z = \zeta(x,s)$ of the equation $z - x = sp(z)$. Since the residue at that point is given by $\psi(\zeta)\psi^{-1}(x)\left[1 - sp'(\zeta)\right]^{-1}$, we get (1.59). □

Now, using all the considerations above, we will give a list of the most important and significant formulas for each of the classical families of orthogonal polynomials in the continuous variable.

1.4.1 Hermite Polynomials

The orthogonality interval $\mathcal{S}$, the measure ψ, and the coefficients p and q of (1.50) are given by

$$\mathcal{S} = \mathbb{R} = (-\infty, \infty), \quad \psi(x) = e^{-x^2}, \quad p(x) = 1, \quad q(x) = -2x.$$

The most general way to define the Hermite polynomials is through Rodrigues' formula

$$H_n(x) = (-1)^n e^{x^2} \left(e^{-x^2}\right)^{(n)}, \quad n \geq 0. \tag{1.60}$$

Another way to write the Hermite polynomials is in terms of the hypergeometric function (1.4). Indeed,

$$H_n(x) = (2x)^n {}_2F_0 \left(\begin{matrix} -n/2, -(n-1)/2 \\ - \end{matrix} ; -\frac{1}{x^2} \right), \quad n \geq 0.$$

The leading coefficient is 2^n and we have the symmetric relation $H_n(-x) = (-1)^n H_n(x)$. They satisfy the second-order differential equation

$$H_n''(x) - 2xH_n'(x) + 2nH_n(x) = 0, \quad n \geq 0. \tag{1.61}$$

From (1.57) we can compute the norms, using the well-known formula $\int_{-\infty}^{\infty} e^{-x^2} dx = \sqrt{\pi}$. Then

$$\|H_n\|_{\psi}^2 = 2^n n! \sqrt{\pi}, \quad n \geq 0. \tag{1.62}$$

Taking into account the initial factor $(-1)^n$, from (1.59) we obtain the generating function for the Hermite polynomials

$$G(x,s) = \sum_{n=0}^{\infty} H_n(x)\frac{s^n}{n!} = e^{2xs-s^2}. \tag{1.63}$$

Another interesting formula is the Poisson kernel:

$$\sum_{n=0}^{\infty} \frac{H_n(x)H_n(y)}{n!}\left(\frac{u}{2}\right)^n = \frac{1}{\sqrt{1-u^2}}\exp\left(\frac{2uxy}{1+u} - \frac{u^2(x-y)^2}{1-u^2}\right). \tag{1.64}$$

The three-term recurrence formula is given by

$$xH_n(x) = \frac{1}{2}H_{n+1}(x) + nH_{n-1}(x), \quad n \geq 0, \quad H_{-1} = 0, \quad H_0 = 1. \tag{1.65}$$

Therefore, the corresponding Jacobi matrix is given by

$$J = \begin{pmatrix} 0 & 1/2 & & & & \\ 1 & 0 & 1/2 & & & \\ & 2 & 0 & 1/2 & & \\ & & 3 & 0 & 1/2 & \\ & & & \ddots & \ddots & \ddots \end{pmatrix}.$$

1.4.2 Laguerre Polynomials

The orthogonality interval $\mathcal{S}$, the measure ψ and the coefficients p and q of (1.51) are given by

$$\mathcal{S} = \mathbb{R}_+ = (0,\infty), \quad \psi(x) = x^\alpha e^{-x}, \alpha > -1, \quad p(x) = x, \quad q(x) = \alpha + 1 - x.$$

The most general way to define the Laguerre polynomials is through Rodrigues' formula

$$L_n^{(\alpha)}(x) = \frac{1}{n!}x^{-\alpha}e^x\left(x^{n+\alpha}e^{-x}\right)^{(n)}, \quad n \geq 0. \tag{1.66}$$

Another way to write the Laguerre polynomials is in terms of the hypergeometric function (1.4). Indeed,

$$L_n^{(\alpha)}(x) = \frac{(\alpha+1)_n}{n!}\,{}_1F_1\left(\begin{matrix} -n \\ \alpha+1 \end{matrix}; x\right), \quad n \geq 0,$$

where $(z)_n$ is the Pochhammer symbol defined by (1.2). The leading coefficient is given by $(-1)^n/n!$. They satisfy the second-order differential equation

$$x[L_n^{(\alpha)}]''(x) + (\alpha + 1 - x)[L_n^{(\alpha)}]'(x) + nL_n^{(\alpha)}(x) = 0, \quad n \geq 0. \tag{1.67}$$

From (1.57) we can compute its norms, using (see (1.1)) that $\int_0^\infty x^\alpha e^{-x} dx = \Gamma(\alpha+1)$. Then

$$\|L_n^{(\alpha)}\|_\psi^2 = \frac{\Gamma(n+\alpha+1)}{n!}, \quad n \geq 0. \tag{1.68}$$

Taking into account the normalization (1.66), from (1.59) we obtain the generating function for the Laguerre polynomials

$$G(x,s) = \sum_{n=0}^{\infty} L_n^{(\alpha)}(x) \frac{s^n}{n!} = \frac{e^{-xs/(1+s)}}{(1-s)^{\alpha+1}}. \tag{1.69}$$

Another interesting generating function is

$$\sum_{n=0}^{\infty} L_n^{(\alpha)}(x) s^n = \frac{e^{xs/(s-1)}}{(1-s)^{\alpha+1}}, \tag{1.70}$$

while the Poisson kernel is given by

$$\sum_{n=0}^{\infty} \frac{n!}{\Gamma(n+\alpha+1)} L_n^{(\alpha)}(x) L_n^{(\alpha)}(y) u^n = \frac{\exp\left(-\frac{u(x+y)}{1-u}\right) I_\alpha\left(\frac{2\sqrt{xyu}}{1-u}\right)}{(xyu)^{\alpha/2}(1-u)}, \quad |u| < 1, \tag{1.71}$$

where $I_\alpha(z) = i^{-\alpha} J_\alpha(iz)$ and J_α is the standard *Bessel function* defined by

$$J_\alpha(z) = \sum_{n=0}^{\infty} \frac{(-1)^n}{n!\,\Gamma(n+\alpha+1)} \left(\frac{z}{2}\right)^{2n+\alpha}, \tag{1.72}$$

which is one solution of the Bessel differential equation

$$z^2 y''(z) + z y'(z) + (z^2 - \alpha^2) y(z) = 0. \tag{1.73}$$

The three-term recurrence formula is given by ($L_{-1}^{(\alpha)} = 0, L_0^{(\alpha)} = 1$)

$$-xL_n^{(\alpha)}(x) = (n+1)L_{n+1}^{(\alpha)}(x) - (2n+\alpha+1)L_n^{(\alpha)}(x) + (n+\alpha)L_{n-1}^{(\alpha)}(x), \quad n \geq 0. \tag{1.74}$$

Therefore the Jacobi matrix (with eigenvalue $-x$) is given by

$$J = \begin{pmatrix} -\alpha-1 & 1 & & & \\ \alpha+1 & -\alpha-3 & 2 & & \\ & \alpha+2 & -\alpha-3 & 3 & \\ & & \ddots & \ddots & \ddots \end{pmatrix}.$$

We also have the following connection formulas with the Hermite polynomials:

$$H_{2n}(x) = (-1)^n n!\, 2^{2n} L_n^{(-1/2)}(x^2), \quad H_{2n+1}(x) = (-1)^n n!\, 2^{2n+1} x L_n^{(1/2)}(x^2) \tag{1.75}$$

Finally, we have the following Laplace transformation type formula:

$$\int_0^\infty e^{-sx} x^\alpha L_n^{(\alpha)}(\lambda x) L_m^{(\alpha)}(\kappa x) dx = \frac{\Gamma(m+n+\alpha+1)}{m!\,n!} \frac{(s-\lambda)^n (s-\kappa)^m}{s^{m+n+\alpha+1}} \tag{1.76}$$

$$\times {}_2F_1\left(\begin{matrix} -m, -n \\ -m-n-\alpha \end{matrix} ; \frac{s(s-\lambda-\kappa)}{(s-\lambda)(s-\kappa)} \right), \quad \mathrm{Re}\, s > 0,$$

where ${}_2F_1$ is the Gaussian hypergeometric function (see (1.4)).

1.4.3 Jacobi Polynomials

The orthogonality interval $\mathcal{S}$, the measure ψ, and the coefficients p and q of (1.52) are given by

$$\mathcal{S} = (-1, 1), \quad \psi(x) = (1-x)^\alpha (1+x)^\beta, \alpha, \beta > -1,$$
$$p(x) = 1 - x^2, \quad q(x) = \beta - \alpha - (\alpha + \beta + 2)x.$$

The most general way to define the Jacobi polynomials is through Rodrigues' formula

$$P_n^{(\alpha,\beta)}(x) = \frac{(-1)^n}{n!\,2^n} (1-x)^{-\alpha} (1+x)^{-\beta} \left((1-x)^{\alpha+n} (1+x)^{\beta+n} \right)^{(n)}, \quad n \geq 0. \tag{1.77}$$

Another way to write the Jacobi polynomials is in terms of the hypergeometric function (1.4). Indeed,

$$P_n^{(\alpha,\beta)}(x) = \frac{(\alpha+1)_n}{n!} {}_2F_1\left(\begin{matrix} -n, n+\alpha+\beta+1 \\ \alpha+1 \end{matrix} ; \frac{1-x}{2} \right), \quad n \geq 0.$$

The leading coefficient is given by

$$\frac{(n+\alpha+\beta+1)_n}{2^n n!}, \quad n \geq 0.$$

They satisfy the second-order differential equation

$$(1-x^2)[P_n^{(\alpha,\beta)}]''(x) + [\beta - \alpha - (\alpha+\beta+2)x][P_n^{(\alpha,\beta)}]'(x)$$
$$+ n(n+\alpha+\beta+1) P_n^{(\alpha,\beta)}(x) = 0, \quad n \geq 0. \tag{1.78}$$

From (1.57) we can compute its norms, using that

$$\int_{-1}^1 (1-x)^\alpha (1+x)^\beta dx = 2^{\alpha+\beta+1} \int_0^1 s^\alpha (1-s)^\beta ds = 2^{\alpha+\beta+1} \mathrm{B}(\alpha+1, \beta+1),$$

where $\mathrm{B}(x,y)$ is the Beta function (1.3). Then

$$\|P_n^{(\alpha,\beta)}\|_\psi^2 = \frac{2^{\alpha+\beta+1}\Gamma(n+\alpha+1)\Gamma(n+\beta+1)}{n!\,(2n+\alpha+\beta+1)\Gamma(n+\alpha+\beta+1)}, \quad n \geq 0. \tag{1.79}$$

Taking into account the normalization (1.77), from (1.59) we see that s must be exchanged by $-s/2$. Therefore it is necessary to look for solutions of

$$\zeta = x - \frac{s}{2}\left(1-\zeta^2\right).$$

Therefore

$$\zeta = \frac{1}{s}\left[1-\sqrt{1-2xs+s^2}\right],$$

and we obtain the relations

$$\frac{1-\zeta}{1+x} = \frac{2}{1-s+\sqrt{1-2xs+s^2}}, \quad \frac{1+\zeta}{1+x} = \frac{2}{1+s+\sqrt{1-2xs+s^2}}.$$

Therefore we obtain the generating function for the Jacobi polynomials

$$G(x,s) = \sum_{n=0}^{\infty} P_n^{(\alpha,\beta)}(x)\frac{s^n}{n!} = \frac{2^{\alpha+\beta}}{R(1-s+R)^{\alpha}(1+s+R)^{\beta}}, \quad R = \sqrt{1-2xs+s^2}. \tag{1.80}$$

The three-term recurrence relation for $n \geq 0$ is given by $\left(P_{-1}^{(\alpha,\beta)} = 0, P_0^{(\alpha,\beta)} = 1\right)$

$$\begin{aligned} xP_n^{(\alpha,\beta)}(x) &= \frac{2(n+1)(n+\alpha+\beta+1)}{(2n+\alpha+\beta+1)(2n+\alpha+\beta+2)}P_{n+1}^{(\alpha,\beta)}(x) \\ &+ \frac{\beta^2-\alpha^2}{(2n+\alpha+\beta)(2n+\alpha+\beta+2)}P_n^{(\alpha,\beta)}(x) \\ &+ \frac{2(n+\alpha)(n+\beta)}{(2n+\alpha+\beta)(2n+\alpha+\beta+1)}P_{n-1}^{(\alpha,\beta)}(x). \end{aligned} \tag{1.81}$$

Therefore the Jacobi matrix is given by

$$J = \begin{pmatrix} \frac{\beta-\alpha}{\alpha+\beta+2} & \frac{2}{\alpha+\beta+2} & & & \\ \frac{2(1+\alpha)(1+\beta)}{(2+\alpha+\beta)(2+\alpha+\beta+1)} & \frac{\beta^2-\alpha^2}{(2+\alpha+\beta)(2+\alpha+\beta+2)} & \frac{4(1+\alpha+\beta+1)}{(2+\alpha+\beta+1)(2+\alpha+\beta+2)} & & \\ & \frac{2(2+\alpha)(2+\beta)}{(4+\alpha+\beta)(4+\alpha+\beta+1)} & \frac{\beta^2-\alpha^2}{(4+\alpha+\beta)(4+\alpha+\beta+2)} & \frac{6(2+\alpha+\beta+1)}{(4+\alpha+\beta+1)(4+\alpha+\beta+2)} & \\ & & \ddots & \ddots & \ddots \end{pmatrix}.$$

Using Leibniz's rule in (1.77) we get an explicit expression of the Jacobi polynomials

$$\begin{aligned} P_n^{(\alpha,\beta)}(x) &= \frac{(-1)^n}{2^n}\sum_{k=0}^{n}(-1)^k \\ &\times \frac{1}{k!\,(n-k)!}\frac{(\alpha+1)_n(\beta+1)_n}{(\alpha+1)_{n-k}(\beta+1)_k}(1-x)^{n-k}(1+x)^k, \quad n \geq 0. \end{aligned}$$

As a consequence we have the following properties:

$$P_n^{(\alpha,\beta)}(1) = \frac{(\alpha+1)_n}{n!}, \quad P_n^{(\alpha,\beta)}(-1) = (-1)^n \frac{(\beta+1)_n}{n!}, \quad n \geq 0. \tag{1.82}$$

In many cases it is useful to consider the Jacobi polynomials on the interval $[0,1]$, making the change of variables $y = \frac{1+x}{2}$ such that the endpoint $x = -1$ becomes $y = 0$ and the endpoint $x = 1$ becomes $y = 1$. It also changes the parameter α by β. In this case, all properties for the Jacobi polynomials can be extended in a natural way. In particular, for $n \geq 0$, they satisfy the second-order differential equation

$$\begin{aligned} y(1-y)[Q_n^{(\alpha,\beta)}]''(y) + [(1-y)(\alpha+1) - y(\beta+1)][Q_n^{(\alpha,\beta)}]'(y) \\ + n(n+\alpha+\beta+1)Q_n^{(\alpha,\beta)}(y) = 0, \end{aligned} \tag{1.83}$$

and the measure ψ is given by $\psi(y) = y^\alpha(1-y)^\beta, \alpha, \beta > -1$. The three-term recurrence relation satisfied by the polynomials $Q_n^{(\alpha,\beta)}(y)$ (normalized in such a way that $Q_n^{(\alpha,\beta)}(1) = 1$) is given by

$$yQ_n^{(\alpha,\beta)}(y) = a_n Q_{n+1}^{(\alpha,\beta)}(y) + b_n Q_n^{(\alpha,\beta)}(y) + c_n Q_{n-1}^{(\alpha,\beta)}(y), \quad n \geq 0,$$

where the coefficients a_n, b_n, c_n can be written as

$$\begin{aligned} a_n &= \frac{(n+\beta+1)(n+1+\alpha+\beta)}{(2n+\alpha+\beta+1)(2n+2+\alpha+\beta)}, \quad n \geq 0, \\ b_n &= \frac{(n+\beta+1)(n+1)}{(2n+\alpha+\beta+1)(2n+2+\alpha+\beta)} + \frac{(n+\alpha)(n+\alpha+\beta)}{(2n+\alpha+\beta+1)(2n+\alpha+\beta)}, \quad n \geq 0, \\ c_n &= \frac{n(n+\alpha)}{(2n+\alpha+\beta+1)(2n+\alpha+\beta)}, \quad n \geq 1. \end{aligned} \tag{1.84}$$

There are some special cases of the Jacobi polynomials for which all formulas simplify considerably, namely the Gegenbauer polynomials, Legendre polynomials and Chebychev polynomials. In fact some of these polynomials appeared chronologically before the Jacobi polynomials.

1. *Gegenbauer polynomials*. These are the Jacobi polynomials for the parameters $\alpha = \beta = \lambda - 1/2, \lambda \neq 0$. They are also called *ultraspherical polynomials*. They are usually denoted by $C_n^{(\lambda)}(x)$. These polynomials satisfy the second-order differential equation

$$(1-x^2)[C_n^{(\lambda)}]'' - (1+2\lambda)x[C_n^{(\lambda)}]' + n(n+2\lambda)C_n^{(\lambda)} = 0, \quad n \geq 0,$$

and the three-term recurrence relation ($C_{-1}^{(\lambda)} = 0, C_0^{(\lambda)} = 1$)

$$xC_n^{(\lambda)} = \frac{n+1}{2(n+\lambda)} C_{n+1}^{(\lambda)} + \frac{n+2\lambda-1}{2(n+\lambda)} C_{n-1}^{(\lambda)}, \quad n \geq 0.$$

Therefore the Jacobi matrix is given by

$$J = \begin{pmatrix} 0 & \frac{1}{2\lambda} & & & \\ \frac{2\lambda}{2(1+\lambda)} & 0 & \frac{2}{2(2+\lambda)} & & \\ & \frac{1+2\lambda}{2(2+\lambda)} & 0 & \frac{3}{2(3+\lambda)} & \\ & & \ddots & \ddots & \ddots \end{pmatrix}.$$

2. *Legendre polynomials.* These are the Jacobi polynomials for the parameters $\alpha = \beta = 0$. They are also called *spherical polynomials* and are orthogonal on the interval $[-1, 1]$ with respect to the constant weight 1. They are usually denoted by $P_n(x)$. They satisfy the second-order differential equation

$$(1 - x^2)P_n''(x) - 2xP_n'(x) + n(n + 1)P_n(x) = 0, \quad n \geq 0,$$

and the three-term recurrence relation ($P_{-1} = 0, P_0 = 1$)

$$xP_n(x) = \frac{n + 1}{2n + 1}P_{n+1}(x) + \frac{n}{2n + 1}P_{n-1}(x), \quad n \geq 0.$$

Therefore the Jacobi matrix is given by

$$J = \begin{pmatrix} 0 & 1 & & & \\ 1/3 & 0 & 2/3 & & \\ & 2/5 & 0 & 3/5 & \\ & & \ddots & \ddots & \ddots \end{pmatrix}.$$

3. *Chebychev polynomials.* There are two types, the *Chebychev polynomials of the first kind* (when $\alpha = \beta = -1/2$), which are usually denoted by $T_n(x)$ and the *Chebychev polynomials of the second kind* (when $\alpha = \beta = 1/2$), which are usually denoted by $U_n(x)$. They are related to *de Moivre's formula* and play an important role in approximation theory and many other areas of mathematics.

Chebychev polynomials of the first kind, orthogonal with respect to $\psi(x) = \frac{1}{\sqrt{1-x^2}}$, are defined by the three-term recurrence relation

$$T_0 = 1, \quad T_1(x) = x, \quad 2xT_n(x) = T_{n+1}(x) + T_{n-1}(x), \quad n \geq 1. \tag{1.85}$$

Therefore the Jacobi matrix is given by

$$J = \begin{pmatrix} 0 & 1 & & & \\ 1/2 & 0 & 1/2 & & \\ & 1/2 & 0 & 1/2 & \\ & & \ddots & \ddots & \ddots \end{pmatrix}.$$

They also satisfy the second-order differential equation

$$(1-x^2)T_n''(x) - xT_n'(x) + n^2 T_n(x) = 0, \quad n \geq 0.$$

Chebychev polynomials of the first kind can also be defined as the unique polynomials satisfying

$$T_n(\cos\theta) = \cos n\theta, \quad n \geq 0, \quad x = \cos\theta \in [-1,1]. \tag{1.86}$$

Therefore we have an explicit expression of the zeros of $T_n(x)$, given by

$$x_{n,k} = \cos\left(\frac{(2k-1)\pi}{2n}\right), \quad k = 1, \ldots, n.$$

Chebychev polynomials of the second kind, orthogonal with respect to $\psi(x) = \sqrt{1-x^2}$, are defined by the three-term recurrence formula

$$U_0 = 1, \quad U_1(x) = 2x, \quad 2xU_n(x) = U_{n+1}(x) + U_{n-1}(x), \quad n \geq 1. \tag{1.87}$$

Therefore the Jacobi matrix is given by

$$J = \begin{pmatrix} 0 & 1/2 & & & \\ 1/2 & 0 & 1/2 & & \\ & 1/2 & 0 & 1/2 & \\ & & \ddots & \ddots & \ddots \end{pmatrix}.$$

They also satisfy the second-order differential equation

$$(1-x^2)U_n''(x) - 3xU_n'(x) + n(n+2)U_n(x) = 0, \quad n \geq 0.$$

Chebychev polynomials of the second kind can also be defined in a trigonometric way as $U_{n-1}(\cos\theta)\sin\theta = \sin n\theta$ or by the Dirichlet kernel

$$U_n(\cos\theta) = \frac{\sin((n+1)\theta)}{\sin\theta}, \quad n \geq 0, \quad x = \cos\theta \in [-1,1]. \tag{1.88}$$

Therefore we have an explicit expression of the zeros of $U_n(x)$, given by

$$x_{n,k} = \cos\left(\frac{k\pi}{n+1}\right), \quad k = 1, \ldots, n. \tag{1.89}$$

Chebychev polynomials also satisfy a pair of mutual recurrence relations:

$$T_{n+1}(x) = xT_n(x) - (1-x^2)U_{n-1}(x), \quad U_n(x) = xU_{n-1}(x) + T_n(x), \tag{1.90}$$

and are also connected by the following relation:

$$T_n(x) = \frac{1}{2}(U_n(x) - U_{n-2}(x)). \tag{1.91}$$

1.5 Classical Orthogonal Polynomials of a Discrete Variable

In the previous section we studied orthogonal polynomials with respect to an absolutely continuous and positive function defined on a real interval. Now we will study discrete measures ψ with mass points at integer numbers $n \in \mathbb{Z}$ and jumps of magnitude $\psi_n = \psi(\{n\}) > 0$. Most of the results of the previous section can be naturally extended, but now the role of the differential operator will be replaced by a second-order difference operator acting on the (discrete) variable of the polynomials.

Let $\psi = (\psi_n)_{n\in\mathbb{Z}}$ be a sequence of nonnegative real numbers and consider the Hilbert space $\ell^2_\psi(\mathbb{Z})$ with the inner product

$$(f,g)_\psi = \sum_{n=-\infty}^{\infty} f(n)g(n)\psi_n, \tag{1.92}$$

of all sequences $(f(n))_{n\in\mathbb{Z}}$ such that $(f,f)_\psi = \|f\|^2_\psi < \infty$. If the sequence is defined in a subset $\mathcal{S}$ of $\mathbb{Z}$, then this space will be denoted by $\ell^2_\psi(\mathcal{S})$. Usually we normalize the discrete measure ψ in such a way that $\sum_{n=-\infty}^{\infty} \psi_n = 1$, i.e. it is a probability measure.

Although the inner product only depends on the values at the integer numbers, it will be convenient to consider the polynomials as if they were defined in the continuous variable. If we assume that all moments are finite, i.e. $\sum_{n\in\mathbb{Z}} |n|^m \psi_n < \infty$, $m \geq 0$, we can use all the definitions and properties already mentioned in Section 1.2. In particular, we can define the polynomials by means of the determinant (1.14) and these satisfy a three-term recurrence relation of the form (1.15). In the same way we can use the Christoffel–Darboux formulas, the spectral theorem and also an analogue of Theorem 1.12 about completeness.

Once we have defined orthogonal polynomials of a discrete variable we wonder which families, apart from being orthogonal, satisfy certain second-order difference equation acting on the variable x (which in this case it remains discrete). First we have to replace the concept of the derivative in a discrete sense, using the following notation:

$$\Delta f(x) = f(x+1) - f(x) \quad \text{and} \quad \nabla f(x) = f(x) - f(x-1).$$

Each operator takes polynomials of degree n into polynomials of degree $n-1$. Therefore, the combination of both

$$\Delta\nabla f(x) = \nabla\Delta f(x) = f(x+1) - 2f(x) + f(x-1)$$

decreases the degree of a polynomial by 2 and plays the role of the second derivative. It is convenient to write these operators as *shift operators*

$$\mathfrak{s}_m f(x) = f(x+m), \quad m \in \mathbb{Z},$$

in such a way that we have

$$\Delta = \mathfrak{s}_1 - \mathfrak{s}_0, \quad \nabla = \mathfrak{s}_0 - \mathfrak{s}_{-1}, \quad \Delta\nabla = \nabla\Delta = \mathfrak{s}_1 - 2\mathfrak{s}_0 + \mathfrak{s}_{-1} = \Delta - \nabla.$$

Therefore a generic second-order difference operator can be written as

$$\mathcal{D} = p_1\mathfrak{s}_1 + p_0\mathfrak{s}_0 + p_{-1}\mathfrak{s}_{-1}, \tag{1.93}$$

where p_1, p_{-1} and p_0 are real functions. Another way to write the difference operator, which appears in several references, is

$$\mathcal{D} = \sigma(x)\Delta\nabla + \tau(x)\Delta. \tag{1.94}$$

In this case we have the relations $p_1 = \sigma + \tau, p_{-1} = \sigma$ and $p_0 = -2\sigma - \tau$, or equivalently $\sigma = p_{-1}, \tau = p_1 - p_{-1}$ and $p_1 + p_{-1} = 2\sigma + \tau$.

Since we want $\mathcal{D}$ in (1.93) to be symmetric or self-adjoint with respect to $(\cdot,\cdot)_\psi$ in (1.92) we should have

$$\begin{aligned} 0 &= (\mathcal{D}f,g)_\psi - (f,\mathcal{D}g)_\psi = (p_1\mathfrak{s}_1 f,g)_\psi - (f,p_{-1}\mathfrak{s}_{-1}g)_\psi \\ &\quad + (p_{-1}\mathfrak{s}_{-1}f,g)_\psi - (f,p_1\mathfrak{s}_1 g)_\psi \\ &= \sum_{n=-\infty}^{\infty} [p_1\psi - \mathfrak{s}_1(p_{-1}\psi)](n)f(n+1)g(n) \\ &\quad + \sum_{n=-\infty}^{\infty} [p_{-1}\psi - \mathfrak{s}_{-1}(p_1\psi)](n)f(n-1)g(n), \end{aligned}$$

for every f and g that vanish at all but a finite number of integers. Choosing g such that it vanishes at all points except at a single point x and f to vanish at all points except at $x+1$ or at $x-1$, we conclude that the symmetry condition is equivalent to

$$\mathfrak{s}_{-1}(p_1\psi) = p_{-1}\psi, \quad \mathfrak{s}_1(p_{-1}\psi) = p_1\psi. \tag{1.95}$$

Since $\mathfrak{s}_1$ and $\mathfrak{s}_{-1}$ are inverse operators, these two conditions are mutually equivalent. In terms of the operator (1.94), these conditions are equivalent to $\Delta(\sigma\psi) = \tau\psi$. The symmetry implies, as in the continuous case, that if two eigenfunctions of $\mathcal{D}$ have different eigenvalues, then they are orthogonal.

We now wonder under what conditions we have a family of polynomial eigenfunctions such that $\mathcal{D}$ (with positive coefficients p_1 and p_{-1}) is symmetric with respect to some discrete weight ψ with finite moments. We assume that $\psi_x > 0$ for integers x in a certain interval that is either infinite or has $N+1$ points, and is zero elsewhere. There will be three interesting cases depending on how many points we are considering in the support:

1. $\psi_x > 0$ if and only if $0 \leq x \leq N$.
2. $\psi_x > 0$ if and only if $x \geq 0$.
3. $\psi_x > 0$ for all $x \in \mathbb{Z}$.

We will see that the third case is not possible and that the first and second cases give two families of polynomials respectively. The Pearson equations (1.95) force $p_{-1}(0) = 0$ in cases 1 and 2 and $p_1(N) = 0$ in case 1. With a change of notation in the term of order 0, the difference operator (1.93) can be rewritten as

$$\mathcal{D} = p_1\Delta - p_{-1}\nabla + r, \tag{1.96}$$

where $r = p_0 - p_1 - p_{-1}$. The difference operator $\mathcal{D}$ must take polynomials of degree n into polynomials of degree n. Suppose that there are polynomials of degrees 0, 1, 2 which are eigenfunctions of $\mathcal{D}$. Since $\mathcal{D}(1) = r$, then r must be a constant and we can always take $r = 0$. For degree 1, since $\Delta(x) = \nabla(x) = 1$, we have that $\mathcal{D}(x) = p_1 - p_{-1}$. Therefore $p_1 - p_{-1}$ must be a polynomial of degree 1. Finally, since $\Delta(x^2) = 2x+1$ and $\nabla(x^2) = 2x-1$, we have that $\mathcal{D}(x^2) = 2x(p_+ - p_{-1}) + p_1 + p_{-1}$, and $p_1 + p_{-1}$ must be a polynomial of degree at most 2. Therefore p_1 and p_{-1} are polynomials of degrees at most 2 and at least one has a positive degree. If both polynomials have degree 2, then they must have the same leading coefficient. A difference operator as in (1.96) with these conditions on its coefficients is usually called, as in the continuous case, of *hypergeometric* type.

The Pearson equation that we need to solve now is (from (1.95))

$$\psi_{x+1} = \frac{p_1(x)}{p_{-1}(x+1)}\psi_x = \frac{\sigma(x) + \tau(x)}{\sigma(x+1)}\psi_x = \varphi(x)\psi_x,$$

which can be solved recursively giving

$$\psi_x = \prod_{j=0}^{x-1} \varphi(j)\psi_0.$$

Let us see, as in the continuous case, the different cases that may appear.

Case I: One of $p_{\pm 1}$ has degree 0, so the other has degree 1. Then $\varphi(x)$ is linear in one direction or the other, i.e. $\varphi(x) = \mathcal{O}(|x|)$, which implies that case 3 is not possible, since otherwise the moments would not be finite. Since $p_{-1}(0) = 0$, this forces the constant polynomial to be p_1, so that case 1 is not possible neither since $p_1(N)$ should be 0. Therefore, the only possible case is 2. Normalizing by taking $p_1(x) = 1$, we have $p_{-1}(x) = x/a, a > 0$. Therefore $\varphi(x) = a/(x+1)$. Solving the Pearson equation we have that

$$\psi_x = \frac{a^x}{x!}\psi_0, \quad x = 0, 1, 2, \ldots.$$

Since $\sum_{x=0}^{\infty}\psi_x = \psi_0 e^a$, we can choose $\psi_0 = e^{-a}$ to normalize the measure. This distribution is known as the *Poisson distribution* and the corresponding orthogonal polynomials are called the *Charlier polynomials*, denoted by $C_n(x;a)$. We will see later some structural formulas related to this family.

Suppose now that both $p_{\pm 1}$ have degree 1. If the leading coefficients are not equal, then ψ_x is linear in one direction or the other and again case 3 is not possible, since the moments would not be finite. If the leading coefficients are equal, then, asymptotically $\varphi(x)-1 \sim b/x$ for some constant b, which implies that the products $\prod_{j=0}^{x}\varphi(j)$, and $\prod_{j=0}^{x}\varphi(j)^{-1}$ are both identically 1 (and $p_1 = p_{-1}$), or one grows with x and the other one decreases with $1/x$ as $x \to \infty$. This again excludes case 3. Then, if both have the same degree, we have cases 1 or 2 and $p_{-1}(0) = 0$. There will be two cases:

Case II: $p_{\pm 1}$ of degree 1 with infinite support. We normalize in such a way that $p_{-1}(x) = x$ and $p_1(x) = c(x+b)$ where $b, c > 0$ and $c < 1$. In this way we have that all moments are finite. Therefore $\varphi(x) = c(x+b)/(x+1)$ and solving the Pearson equation we have

$$\psi_x = \frac{(b)_x}{x!}c^x\psi_0, \quad x = 0, 1, 2, \ldots .$$

Since $\sum_{x=0}^{\infty}\psi_x = \psi_0(1-c)^{-b}$ (binomial series) we can choose $\psi_0 = (1-c)^b$ to normalize the measure. This distribution is known as the *Pascal distribution* and the corresponding orthogonal polynomials are the *Meixner polynomials*, denoted by $M_n(x;b,c)$. We will see later some structural formulas related to this family.

Case III: $p_{\pm 1}$ of degree 1 with finite support. Following the same steps as before, we can normalize in such a way that $p_1(x) = p(N-x)$ (since $p_1(N) = 0$) and $p_{-1}(x) = qx$, where $p, q > 0$, $p+q = 1$. Therefore solving the Pearson equation we have

$$\psi_x = \binom{N}{x}\left(\frac{p}{q}\right)^x \psi_0, \quad x = 0, 1, 2, \ldots, N.$$

Since $\sum_{x=0}^{N}\psi_x = \psi_0 q^{-N}$ we can choose $\psi_0 = q^N$ to normalize the measure. This distribution is known as the *binomial distribution* and the corresponding orthogonal polynomials are the *Krawtchouk polynomials*, denoted by $K_n(x;p,N)$. We will see later some structural formulas related to this family.

Case IV: one of $p_{\pm 1}$ has degree 2. Then, since $p_1 - p_{-1}$ has degree 1, we have that both $p_{\pm 1}$ has degree 2 and with the same leading coefficient. Then

$$\varphi(x) = \frac{p_1(x)}{p_{-1}(x+1)} = \frac{1+ax^{-1}+bx^{-2}}{1+cx^{-1}+dx^{-2}} = 1 + \frac{a-c}{x} + \mathcal{O}\left(\frac{1}{x^2}\right).$$

The condition on the moments exclude cases 2 and 3, so we only have case 1 with finite support. Since $p_{-1}(0) = 0 = p_1(N)$ there are two cases, which can be normalized by

$$p_{-1}(x) = x(N+1+\beta-x), \quad p_1(x) = (N-x)(x+\alpha+1), \quad \alpha,\beta > -1,$$

or

$$p_{-1}(x) = x(x-\beta-N-1), \quad p_1(x) = (N-x)(-x-\alpha-1), \quad \alpha,\beta < -N.$$

Therefore solving the Pearson equation (for the first case) we have that the measure is given by

$$\psi_x = \frac{(N-x+1)_x(\alpha+1)_x\psi_0}{x!\,(N+\beta+1-x)_x} = \binom{N}{x}\frac{(\alpha+1)_x(\beta+1)_{N-x}\psi_0}{(\beta+1)_N}.$$

Since $\sum_{x=0}^{N}\psi_x = \psi_0\frac{(\alpha+\beta+2)_N}{(\beta+1)_N}$ we can choose $\psi_0 = \frac{(\beta+1)_N}{(\alpha+\beta+2)_N}$ to normalize the measure. Therefore

$$\psi_x = \binom{N}{x}\frac{(\alpha+1)_x(\beta+1)_{N-x}}{(\alpha+\beta+2)_N} = \frac{\binom{\alpha+x}{x}\binom{\beta+N-x}{N-x}}{\binom{\alpha+\beta+N+1}{N}},$$

which is the *negative hypergeometric distribution* and the corresponding orthogonal polynomials are the *Hahn polynomials*, denoted by $Q_n(x;\alpha,\beta,N)$. We will see later some structural formulas related to this family. If we set $\alpha = -\tilde{\alpha}-1$ and $\beta = -\tilde{\beta}-1$ (so we are in the second case of coefficients p_1, p_{-1}), we arrive at

$$\psi_x = \frac{\binom{\tilde{\alpha}}{x}\binom{\tilde{\beta}}{N-x}}{\binom{\tilde{\alpha}+\tilde{\beta}}{N}},$$

which is the standard *hypergeometric distribution*.

The families of Charlier, Meixner, Krawtchouk and Hahn polynomials are usually called the *classical families of orthogonal polynomials of a discrete variable*. Therefore, we have proved the following (somewhat informally stated) result:

Theorem 1.24 *Up to normalization, the classical orthogonal polynomials (Charlier, Meixner, Krawtchouk and Hahn) are the only families of orthogonal polynomials that are eigenfunctions of a second-order difference operator of the form* (1.96) *(i.e. self-adjoint with respect to a positive discrete measure supported on the real line).*

Again we have Rodrigues' formulas for the classical discrete families.

Proposition 1.25 (Rodrigues' formula) *The solutions of $\mathcal{D}P_n = -\lambda_n P_n$ where $\mathcal{D}$ is given by* (1.96) *can be written, up to normalization, as*

$$P_n(x) = \frac{1}{\psi_x}\nabla^n\left(\psi^{(n)}(x)\right), \tag{1.97}$$

where

$$\psi^{(n)}(x) = \psi_x \prod_{k=0}^{n-1} p_1(x+k).$$

Proof A proof can be found in [9, Section 5.2]. □

Since $\nabla = \mathfrak{s}_0 - \mathfrak{s}_{-1}$, we can expand $\nabla^n = (\mathfrak{s}_0 - \mathfrak{s}_{-1})^n$ and rewrite (1.97) as

$$P_n = \frac{1}{\psi} \sum_{k=1}^{n} (-1)^{n-k} \binom{n}{k} \mathfrak{s}_{-k}(\psi^{(n)}).$$

As in the continuous case, we can obtain expressions for the norms of P_n. Observe that as long as the sums are convergent, we have that

$$\sum_k g(k)\nabla f(k) = -\sum_k f(k)\Delta g(k).$$

Therefore, denoting $P_n^{(k)} = (\Delta)^k P_n$, we have

$$\begin{aligned} -\lambda_n ||P_n||_\psi^2 &= (\mathcal{D}P_n, P_n)_\psi = \sum_k (\psi \mathcal{D} P_n)(k) P_n(k) \\ &= \sum_k [\nabla(\psi p_1 P_n^{(1)})](k) P_n(k) = -\sum_k [\psi p_1 P_n^{(1)}](k) P_n^{(1)}(k) = -||P_n^{(1)}||_{\psi^{(1)}}^2. \end{aligned}$$

In general, we have that

$$\lambda_n^{(k)} ||P_n^{(k)}||_{\psi^{(k)}}^2 = ||P_n^{(k+1)}||_{\psi^{(k+1)}}^2, \quad k = 0, 1, \ldots, n-1,$$

where

$$\lambda_n^{(k)} = \lambda_n + (\mathfrak{s}_0 + \mathfrak{s}_1 + \cdots + \mathfrak{s}_{k-1})\Delta p_1 - k\Delta p_{-1} = \lambda_n + (\mathfrak{s}_k - \mathfrak{s}_0)p_1 - k\Delta p_{-1}.$$

Iterating, we obtain

$$P_n^{(n)} ||P_n||_\psi^2 = (-1)^n ||P_n^{(n)}||_{\psi^{(n)}}^2 = (-1)^n [P_n^{(n)}]^2 ||1||_{\psi^{(n)}}^2,$$

and therefore

$$||P_n||_\psi^2 = (-1)^n P_n^{(n)} \sum_k \psi^{(n)}(k). \tag{1.98}$$

The constant value $P_n^{(n)}$ is given by

$$P_n^{(n)} = \prod_{k=0}^{n-1} [-\lambda_n - (\mathfrak{s}_k - \mathfrak{s}_0)p_1 + k\Delta p_{-1}] = \prod_{k=0}^{n-1} [-\lambda_n^{(k)}].$$

Now, using all the considerations above, we will give a list of the most important and significant formulas for each of the classical discrete families of orthogonal polynomials.

1.5.1 Charlier Polynomials

The support of the measure is $\mathbb{N}_0$. The measure and the coefficients $p_{\pm 1}$ are given by

$$p_1(x) = 1, \quad p_{-1}(x) = \frac{x}{a}, \quad \psi_x = e^{-a}\frac{a^x}{x!}, \quad a > 0.$$

Then $\psi^{(k)} = \psi$. The most common way to define the Charlier polynomials is using Rodrigues' formula

$$C_n(x;a) = \frac{x!}{a^x}\nabla^n\left[\frac{a^x}{x!}\right], \quad n \geq 0. \tag{1.99}$$

They can also be defined in terms of the hypergeometric function (1.4)

$$C_n(x;a) = {}_2F_0\left(\begin{matrix}-n, -x \\ -\end{matrix}; -\frac{1}{a}\right), \quad n \geq 0.$$

They satisfy the second-order difference equation

$$aC_n(x+1;a) - (x+a)C_n(x;a) + xC_n(x-1;a) = -nC_n(x;a), \quad n \geq 0. \tag{1.100}$$

The norms are given by

$$||C_n(x;a)||_\psi^2 = \frac{n!}{a^n}, \quad n \geq 0, \tag{1.101}$$

while the leading coefficient is given by $(-a)^{-n}$. The generating function for the Charlier polynomials is given by

$$G(x,t;a) = \sum_{n=0}^{\infty}\frac{C_n(x;a)}{n!}t^n = e^t\left(1 - \frac{t}{a}\right)^x. \tag{1.102}$$

The three-term recurrence relation is given by ($C_{-1}(x;a) = 0, \quad C_0(x;a) = 1$)

$$-xC_n(x;a) = aC_{n+1}(x;a) - (n+a)C_n(x;a) + nC_{n-1}(x;a), \quad n \geq 0. \tag{1.103}$$

Therefore, the corresponding Jacobi matrix is given by

$$J = \begin{pmatrix} -a & a & & & \\ 1 & -a-1 & a & & \\ & 2 & -a-2 & a & \\ & & \ddots & \ddots & \ddots \end{pmatrix},$$

which is the infinitesimal operator of an $M/M/\infty$ queue, as we will see in Chapter 3. Observe that we have the *duality relation*

$$C_n(x;a) = C_x(n;a), \quad x,n \in \mathbb{N}_0. \tag{1.104}$$

Then the recurrence formula (1.103) is equivalent to the second-order difference equation (1.100). Finally, the Charlier polynomials satisfy the following connection formula with the Laguerre polynomials:

$$C_n(x;a) = (-1)^n \frac{n!}{a^n} L_n^{(x-n)}(a), \quad n \geq 0.$$

1.5.2 Meixner Polynomials

The support of the measure is $\mathbb{N}_0$. The measure and the coefficients $p_{\pm 1}$ are given by

$$p_1(x) = c(x+b), \quad p_{-1}(x) = x, \quad \psi_x = (1-c)^b \frac{(b)_x}{x!} c^x, \quad b > 0, \quad 0 < c < 1.$$

The most common way to define the Meixner polynomials is using Rodrigues' formula

$$M_n(x;b,c) = \frac{x!}{(b)_x c^x} \nabla^n \left[\frac{(b+n)_x c^x}{x!} \right], \quad n \geq 0.$$

They can also be defined in terms of the hypergeometric function (1.4)

$$M_n(x;b,c) = {}_2F_1\left(\begin{matrix} -n, -x \\ b \end{matrix} ; 1 - \frac{1}{c} \right), \quad n \geq 0.$$

They satisfy the second-order difference equation

$$\begin{aligned} &c(x+b)M_n(x+1;b,c) - (x + c(x+b))M_n(x;b,c) + xM_n(x-1;b,c) \\ &= n(c-1)M_n(x;b,c), \quad n \geq 0. \end{aligned} \tag{1.105}$$

The norms are given by

$$||M_n(x;b,c)||_\psi^2 = \frac{n!}{(b)_n c^n}, \quad n \geq 0, \tag{1.106}$$

while the leading coefficient is given by $\frac{1}{(b)_n}\left(\frac{c-1}{x}\right)^n$. The generating function for the Meixner polynomials is given by

$$G(x,t;b,c) = \sum_{n=0}^{\infty} (b)_n \frac{M_n(x;b,c)}{n!} t^n = \left(1 - \frac{t}{c}\right)^x (1-t)^{-x-b}. \tag{1.107}$$

Another important generating function is given by $(x \geq y)$

$$\begin{aligned} \tilde{G}(x,y,t;b,c) &= \sum_{n=0}^{\infty} M_n(x;b,c) M_n(y;b,c) \frac{(b)_n}{n!} t^n \\ &= \frac{(1-t/c)^{x+y}}{(1-t)^{x+y+b}} \frac{y!}{(b)_y} \sum_{k=0}^{x} \binom{x}{k} (-1)^k \left(\frac{1-t/c^2}{1-t/c}\right)^k \left(\frac{1-t}{1-t/c}\right)^k \frac{(x+b)_{y-k}}{(y-k)!}. \end{aligned} \tag{1.108}$$

The three-term recurrence relation is given by ($M_{-1}(x;b,c)=0$, $M_0(x;b,c)=1$)

$$\begin{aligned} &c(n+b)M_{n+1}(x;b,c)-(n+c(n+b))M_n(x;b,c)+nM_{n-1}(x;b,c)\\ &\quad = x(c-1)M_n(x;b,c), \quad n\geq 0. \end{aligned} \tag{1.109}$$

Therefore the corresponding Jacobi matrix is given by

$$J=\frac{1}{1-c}\begin{pmatrix} -cb & cb & & & \\ 1 & -c(1+b)-1 & c(1+b) & & \\ & 2 & -c(2+b)-2 & c(2+b) & \\ & & \ddots & \ddots & \ddots \end{pmatrix}.$$

Observe that we have the duality relation

$$M_n(x;b,c)=M_x(n;b,c), \quad x,n\in\mathbb{N}_0. \tag{1.110}$$

Then the recurrence formula (1.109) is equivalent to the second-order difference equation (1.105). Finally, the Meixner polynomials satisfy the following connection formula with the Jacobi polynomials:

$$M_n(x;b,c)=\frac{n!}{(b)_n}P_n^{(b-1,-n-b-x)}\left(\frac{2-c}{c}\right).$$

1.5.3 Krawtchouk Polynomials

The support of the measure is $\{0,1,\ldots,N\}$. The measure and the coefficients $p_{\pm 1}$ are given by

$$p_1(x)=p(N-x), \quad p_{-1}(x)=qx, \quad \psi_x=\binom{N}{x}p^xq^{N-x}, \quad p,q>0, \quad p+q=1.$$

The most common way to define the Krawtchouk polynomials is using Rodrigues' formula

$$K_n(x;p,N)=\frac{q^x}{p^x\binom{N}{x}}\nabla^n\left[\binom{N-n}{x}\left(\frac{p}{q}\right)^x\right], \quad n=0,1,\ldots,N.$$

They can also be defined in terms of the hypergeometric function (1.4)

$$K_n(x;p,N)={}_2F_1\left(\begin{matrix}-n,\,-x\\ -N\end{matrix};\frac{1}{p}\right), \quad n=0,1,\ldots,N. \tag{1.111}$$

For $n=0,1,\ldots,N$ they satisfy the second-order difference equation

$$\begin{aligned} &p(N-x)K_n(x+1;p,N)-[p(N-x)+xq]K_n(x;p,N)\\ &\quad +xqK_n(x-1;p,N)=-nK_n(x;p,N). \end{aligned} \tag{1.112}$$

The norms are given by

$$||K_n(x;p,N)||_\psi^2=\frac{(-1)^n n!\,q^n}{(-N)_n p^n}, \quad n=0,1,\ldots,N, \tag{1.113}$$

while the leading coefficient is given by $\frac{1}{(-N)_n p^n}$. One generating function for the Krawtchouk polynomials is given by

$$G(x,t;p,N) = \sum_{n=0}^{N} \binom{N}{n} K_n(x;p,N) t^n = \left(1 - \frac{qt}{p}\right)^x (1+t)^{N-x}. \tag{1.114}$$

The three-term recurrence relation is given by $(K_{-1}(x;p,N) = 0, K_0(x;p,N) = 1)$

$$\begin{aligned} p(N-n)K_{n+1}(x;p,N) &- [p(N-n)+nq]K_n(x;p,N) \\ &+ nqK_{n-1}(x;p,N) = -xK_n(x;p,N). \end{aligned} \tag{1.115}$$

Therefore the corresponding Jacobi matrix is given by

$$J = \begin{pmatrix} -pN & pN & & & \\ q & -p(N-1)-q & p(N-1) & & \\ & 2q & -p(N-2)-2q & p(N-2) & \\ & & \ddots & \ddots & \ddots \\ & & & Nq & -Nq \end{pmatrix}.$$

This Jacobi matrix will be related to the Ehrenfest urn model (discrete and continuous-time) as we will see in Chapters 2 and 3. Observe that we have the dual relation

$$K_n(x;p,N) = K_x(n;p,N), \quad x,n \in \{0,1,\ldots,N\}. \tag{1.116}$$

Then the recurrence formula (1.115) is equivalent to the second-order difference equation (1.112).

1.5.4 Hahn Polynomials

The support of the measure is $\{0,1,\ldots,N\}$. The coefficients $p_{\pm 1}$ are given by

$$\begin{aligned} p_1(x) &= (N-x)(x+\alpha+1), \\ p_{-1}(x) &= x(N+\beta+1-x), \quad \alpha,\beta > -1 \quad \text{or} \quad \alpha,\beta < -N. \end{aligned}$$

The measure can be written as

$$\psi_x(\alpha,\beta,N) = \binom{\alpha+x}{x}\binom{\beta+N-x}{N-x}.$$

The most common way to define the Hahn polynomials is using Rodrigues' formula

$$\psi_x(\alpha,\beta,N)Q_n(x;\alpha,\beta,N) = \frac{(-1)^n(\beta+1)_n}{(-N)_n}\nabla^n\left[\psi_x(\alpha+n,\beta+n,N-n)\right].$$

They can also be defined in terms of the hypergeometric function (1.4)

$$Q_n(x;\alpha,\beta,N) = {}_3F_2\left(\begin{matrix} -n, n+\alpha+\beta+1, -x \\ \alpha+1, -N \end{matrix}; 1\right), \quad n = 0,1,\ldots,N. \tag{1.117}$$

They satisfy the second-order difference equation

$$B(x)Q_n(x+1;\alpha,\beta,N) - [B(x)+D(x)]Q_n(x;\alpha,\beta,N) + D(x)Q_n(x-1;\alpha,\beta,N)$$
$$= -n(n+\alpha+\beta+1)Q_n(x;\alpha,\beta,N), \tag{1.118}$$

where

$$B(x) = (N-x)(x+\alpha+1), \quad D(x) = x(\beta+N+1-x).$$

The norms are given by

$$||Q_n(x;\alpha,\beta,N)||_\psi^2 = \frac{(-1)^n(n+\alpha+\beta+1)_{N+1}(\beta+1)_n n!}{(2n+\alpha+\beta+1)(\alpha+1)_n(-N)_n N!}, \qquad n = 0,1,\ldots,N, \tag{1.119}$$

while the leading coefficient is given by $\frac{(n+\alpha+\beta+1)_n}{(-N)_n(\alpha+1)_n}$. The three-term recurrence relation is given by ($Q_{-1}(x;\alpha,\beta,N) = 0, Q_0(x;\alpha,\beta,N) = 1$)

$$A_nQ_{n+1}(x;\alpha,\beta,N) - (A_n+C_n)Q_n(x;\alpha,\beta,N)$$
$$+ C_nQ_{n-1}(x;\alpha,\beta,N) = -xQ_n(x;\alpha,\beta,N), \tag{1.120}$$

where

$$A_n = \frac{(n+\alpha+\beta+1)(n+\alpha+1)(N-n)}{(2n+\alpha+\beta+1)(2n+\alpha+\beta+2)}, \quad C_n = \frac{n(n+\alpha+\beta+N+1)(n+\alpha)}{(2n+\alpha+\beta)(2n+\alpha+\beta+1)}.$$

1.6 The Askey Scheme

Apart from the classical families of orthogonal polynomials, there are some other families which are also eigenfunctions of a second-order difference operator of the form

$$a(x)p_n(\lambda(x+1)) + b(x)p_n(\lambda(x)) + c(x)p_n(\lambda(x-1)) = -\lambda_n p_n(\lambda(x)), \tag{1.121}$$

where now $\lambda(x)$ is not necessarily a linear real function, so the support of the measure is not uniformly distributed, as in the case of the classical discrete families. For completeness, we will give the definitions of these remaining families, some of them will be used later in these notes.

Dual Hahn Polynomials

For the Hahn polynomials there is no self-duality as in the case of the Charlier, Meixner and Krawtchouk polynomials. Nevertheless, there exists a dual family. Consider

$$\lambda(x) = x(x+\gamma+\delta+1),$$

and the weight

$$\psi_x(\gamma,\delta,N) = \frac{(2x+\gamma+\delta+1)(\gamma+1)_x(-N)_xN!}{(-1)^x(x+\gamma+\delta+1)_{N+1}(\delta+1)_xx!}. \tag{1.122}$$

For $\gamma > -1$ and $\delta > -1$ or for $\gamma < -N$ and $\delta < -N$ dual Hahn polynomials, denoted by $R_n(\lambda(x);\gamma,\delta,N)$, are orthogonal with respect to the measure (1.122) and the norms are given by

$$\|R_n(\lambda(x);\gamma,\delta,N)\|_\psi^2 = \binom{\gamma+n}{n}^{-1}\binom{\delta+N-n}{N-n}^{-1}, \quad n=0,1,\ldots,N.$$

We have the following duality relation with Hahn polynomials:

$$R_n(\lambda(x);\gamma,\delta,N) = Q_x(n;\gamma,\delta,N).$$

Therefore dual Hahn polynomials satisfy difference equations and three-term recurrence relations which are "dual" with respect to Hahn polynomials formulas. Indeed, the second-order difference equation is given by

$$\begin{aligned} &B(x)R_n(\lambda(x+1);\gamma,\delta,N) - (B(x)+D(x))R_n(\lambda(x);\gamma,\delta,N) \\ &\quad + D(x)R_n(\lambda(x-1);\gamma,\delta,N) = -nR_n(\lambda(x);\gamma,\delta,N), \end{aligned}$$

where

$$B(x) = \frac{(x+\gamma+\delta+1)(x+\gamma+1)(N-x)}{(2x+\gamma+\delta+1)(2x+\gamma+\delta+2)}, \quad D(x) = \frac{x(x+\gamma+\delta+N+1)(x+\gamma)}{(2x+\gamma+\delta)(2x+\gamma+\delta+1)},$$

while the three-term recurrence formula is given by

$$\begin{aligned} &A_nR_{n+1}(\lambda(x);\gamma,\delta,N) - [A_n+C_n]R_n(\lambda(x);\gamma,\delta,N) + C_nR_{n-1}(\lambda(x);\gamma,\delta,N) \\ &\quad = -\lambda(x)R_n(\lambda(x);\gamma,\delta,N), \end{aligned} \tag{1.123}$$

where

$$A_n = (N-n)(n+\gamma+1), \quad C_n = n(\delta+N+1-n).$$

This family will be related to the Bernoulli–Laplace urn model (discrete and continuous time), as we will see in Chapters 2 and 3.

Racah Polynomials

The Racah polynomials are defined by

$$\begin{aligned} &\mathcal{R}_n(\lambda(x);\alpha,\beta,\gamma,\delta) \\ &\quad = {}_4F_3\left(\begin{matrix}-n, n+\alpha+\beta+1, -x, x+\gamma+\delta+1 \\ \alpha+1, \beta+\delta+1, \gamma+1\end{matrix};1\right), \quad n=0,1,\ldots,N, \end{aligned}$$

where $\lambda(x) = x(x+\gamma+\delta+1)$ and $\alpha+1$ or $\beta+\delta+1$ or $\gamma+1$ is equal to $-N$, for $N \in \mathbb{N}_0$. The Racah polynomial $\mathcal{R}_n$ is a polynomial of degree n in $\lambda(x)$. Racah polynomials are orthogonal with respect to weights ψ_x on the points $\lambda(x)$, where

$$\psi_x = \frac{(\alpha+1)_x(\beta+\delta+1)_x(\gamma+1)_x(\gamma+\delta+1)_x((\gamma+\delta+3)/2)_x}{x!\,(\gamma+\delta-\alpha+1)_x(\gamma-\beta+1)_x(\delta+1)_x((\gamma+\delta+1)/2)_x}.$$

From the definition, dual Racah polynomials are again Racah polynomials with α, β interchanged with γ, δ. For the coefficients of the recurrence relation (1.121) see [106, 121].

Wilson Polynomials

The Wilson polynomials are defined by

$$\frac{W_n(x^2;a,b,c,d)}{(a+b)_n(a+c)_n(a+d)_n} = {}_4F_3\left(\begin{matrix}-n, n+a+b+c+d-1, a+ix, a-ix \\ a+b, a+c, a+d\end{matrix}; 1\right).$$

The Wilson polynomial W_n is a polynomial of degree n in x^2. If $\mathrm{Re}(a,b,c,d) > 0$, Wilson polynomials are orthogonal with respect to the measure

$$\psi(x) = \left|\frac{\Gamma(a+ix)\Gamma(b+ix)\Gamma(c+ix)\Gamma(d+ix)}{\Gamma(2ix)}\right|^2, \quad x \in [0,\infty).$$

For the coefficients of the recurrence relation (1.121) see [106, 121].

Continuous Hahn and Continuous Dual Hahn Polynomials

The continuous dual Hahn polynomials are defined by

$$\frac{S_n(x^2;a,b,c)}{(a+b)_n(a+c)_n} = {}_3F_2\left(\begin{matrix}-n, a+ix, a-ix \\ a+b, a+c\end{matrix}; 1\right),$$

where a, b, c have positive real parts. If one of these parameters is not real then one of the other parameters is its complex conjugate. Continuous dual Hahn polynomials are orthogonal with respect to the measure

$$\psi(x) = \left|\frac{\Gamma(a+ix)\Gamma(b+ix)\Gamma(c+ix)}{\Gamma(2ix)}\right|^2, \quad x \in [0,\infty).$$

Continuous Hahn polynomials are defined by

$$\frac{n!\,p_n(x;a,b,c,d)}{i^n(a+c)_n(a+d)_n} = {}_3F_2\left(\begin{matrix}-n, n+a+b+c+d-1, a+ix \\ a+c, a+d\end{matrix}; 1\right).$$

If $\mathrm{Re}(a,b,c,d) > 0$, $c = \bar{a}$ and $d = \bar{b}$ continuous Hahn polynomials are orthogonal with respect to the measure $\psi(x) = |\Gamma(a+ix)\Gamma(b+ix)|^2$ on $\mathbb{R}$. For the coefficients of the recurrence relation (1.121) of both families see [106, 121].

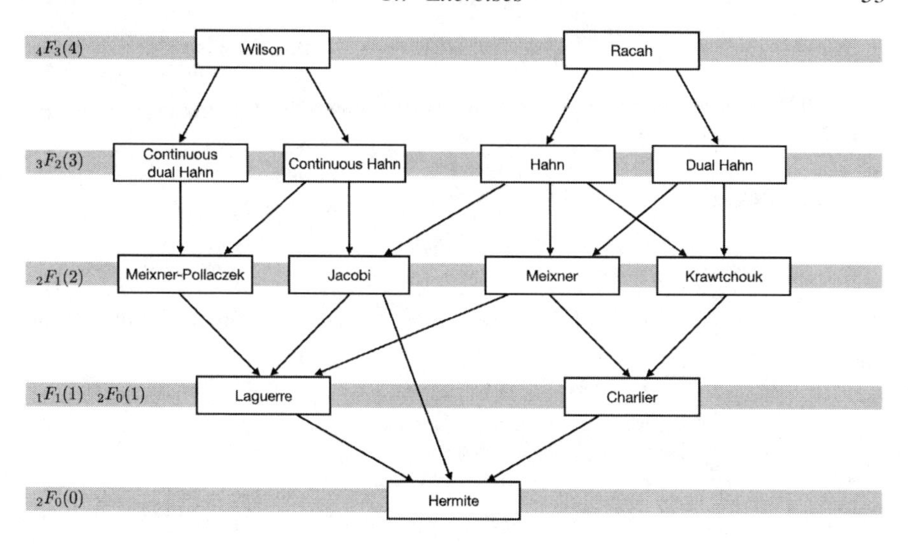

Figure 1.1 The Askey scheme

Meixner–Pollaczek Polynomials

The Meixner–Pollaczek polynomials are defined by

$$P_n^{(\lambda)}(x;\phi) = \frac{(2\lambda)_n}{n!} e^{in\phi} {}_2F_1\left(\begin{matrix} -n, \lambda + ix \\ 2\lambda \end{matrix} ; 1 - e^{-2i\phi} \right).$$

If $\lambda > 0$ and $0 < \phi < \pi$, Meixner–Pollaczek polynomials are orthogonal with respect to the measure $\psi(x) = e^{(2\phi-\pi)x}|\Gamma(\lambda + ix)|^2$ on $\mathbb{R}$. For the coefficients of the recurrence relation (1.121), see [106, 121].

The previous families and the classical families can be gathered in the so-called *Askey scheme* in Figure 1.1, where the arrows mean the limit transitions between them. The exact limit relations between all these polynomials can be found for instance in [106, 121].

1.7 Exercises

1.1 Prove, in a similar way to the proof of Proposition 1.1, the following extension of the Perron–Stieltjes inversion formula:

$$\int_a^b p(x)d\psi(x) + \frac{1}{2}p(a)\psi(\{a\}) + \frac{1}{2}p(b)\psi(\{b\})$$
$$= \frac{1}{\pi} \lim_{\varepsilon \downarrow 0} \int_a^b \operatorname{Im}\left[p(x+i\varepsilon)B(x+i\varepsilon;\psi)\right] dx,$$

where p is any polynomial with real coefficients.

1.2 Compute the probability measure ψ of the following Stieltjes transforms using the Perron–Stieltjes inversion formula:

(a) $B(z;\psi) = -\dfrac{1}{z+i}, \quad z \in \mathbb{C} \setminus \mathbb{R}.$

(b) $B(z;\psi) = \dfrac{z}{1-z^2}, \quad z \in \mathbb{C} \setminus \{-1,1\}.$

(c) $B(z;\psi) = \log\left(\dfrac{z-1}{z}\right), \quad z \in \mathbb{C} \setminus [0,1].$

1.3 (a) Expand the determinant of $\widehat{P}_{n+1}$ given by (1.29) to prove that they satisfy the three-term recurrence relation (1.17).

(b) Prove that the associated polynomials defined by (1.33) satisfy a three-term recurrence relation as in (1.16).

1.4 (a) Prove that

$$\int_{\mathbb{R}} \frac{P_n(x)}{x-z} d\psi(x) = P_n(z)B(z;\psi) + P_n^{(0)}(z),$$

where P_n are the orthonormal polynomials with respect to a measure ψ, $P_n^{(0)}$ are the associated polynomials and $B(z;\psi)$ is the Stieltjes transform.

(b) Prove, using (a), that

$$\sum_{n=0}^{\infty} |P_n(z)B(z;\psi) + P_n^{(0)}(z)|^2 \leq \int_{\mathbb{R}} \frac{d\psi(x)}{|x-z|^2} < \infty.$$

1.5 Check the validity of Theorem 1.11 for $n=0,1,2,3$ by writing the ratio $P_n^{(0)}(z)/P_n(z)$ as a truncated continued fraction of the form (1.34).

1.6 Let $\mathcal{D}$ be the second-order differential operator (1.42) and consider a solution u of the non-homogeneous equation $\mathcal{D}u = f$ for some function f. A *gauge transformation* is given by $u(x) = \varphi(x)v(x)$, where $\varphi(x) \neq 0$. Prove that $\mathcal{D}_\varphi v = f/\varphi$ where

$$\mathcal{D}_\varphi = p(x)\frac{d^2}{dx^2} + \left[2p(x)\frac{\varphi'(x)}{\varphi(x)} + q(x)\right]\frac{d}{dx} + p(x)\frac{\varphi''(x)}{\varphi(x)} + q(x)\frac{\varphi'(x)}{\varphi(x)} + r(x).$$

Using the gauge transformation solve the following questions:

(a) Find a gauge transformation such the coefficient of the first derivative of $\mathcal{D}_\varphi$ is 0 (Schrödinger form).

(b) Prove that there exists a unique (up to a constant) probability measure ψ such that $\mathcal{D}$ is symmetric with respect to ψ.

(c) Let ψ be a probability measure supported on (a,b). Prove that there exists a gauge transformation of the form $u = \phi v$ (in $\mathcal{D}u = f$) such that $\mathcal{D}_\phi$ is symmetric with respect to ψ.

1.7 Let $\mathcal{D}$ be the second-order differential operator (1.42) and consider a solution u of the non-homogeneous equation $\mathcal{D}u = f$ for some function f. A *Liouville transformation* is given by $u(x) = v(g(x))$ for some function g. Find the differential operator $\mathcal{D}_g$ associated with this transformation.

1.8 Let $\mathcal{D}_H, \mathcal{D}_L$ and $\mathcal{D}_J$ denote the Hermite, Laguerre and Jacobi second-order differential operators (see (1.50), (1.51) and (1.52), respectively). Find a gauge transformation of the corresponding eigenfunctions, i.e. $\phi_n = f(x)P_n(x)$ for some function $f(x)$, such that ϕ_n is a system of orthogonal functions in the spaces $L^2(\mathbb{R}), L^2(\mathbb{R}^+)$ and $L^2(-1,1)$, respectively, and compute the coefficients of the new second-order differential operators after this transformation.

1.9 Prove that the sequence of monomials $(x^n)_{n\in\mathbb{N}_0}$ is not orthogonal with respect to any measure supported on $\mathbb{R}$.

1.10 Consider $(p_n)_n$ the sequence of polynomials defined by Rodrigues' formula (1.55) and let us call $\varphi_n(x) = p_n(x)\psi(x)/n!$. Prove, using the integral formula (1.58), that the functions φ_n satisfy the three-term recurrence formula of the form

$$a_n\varphi_{n+1}(x) = b_n(x)\varphi_n(x) + c_n\varphi_{n-1}(x), \quad b_n(x) = b_{n0} + b_{n1}x,$$

and the difference-differential relation of the form

$$p(x)\varphi_n'(x) = a_n(x)\varphi_n(x) + b_n\varphi_{n-1}(x), \quad a_n(x) = a_{n0} + a_{n1}x,$$

where $\deg p(x) \le 2$. Compute all coefficients in the case of the Hermite, Laguerre and Jacobi polynomials. (For more details see [9, pp. 103–105].)

1.11 Find the generating functions for the Romanovski and Bessel polynomials using the corresponding Rodrigues' formulas.

1.12 Let $p_n = x^n + \sum_{k=0}^{n-1} a_{n,k}x^k$ be the polynomial solutions of the differential equation

$$\sigma(x)p_n''(x) + \tau(x)p_n'(x) = -\lambda_n p_n,\ \sigma(x) = \sigma_2 x^2 + \sigma_1 x + \sigma_0,\ \tau(x) = \tau_1 x + \tau_0.$$

Find a recurrent way of computing the coefficients $a_{n,k}, k = 0, 1, \ldots, n-1$, in terms of the coefficients of σ and τ.

1.13 Prove formulas (1.62)–(1.65) for the Hermite polynomials, (1.68)–(1.71) and (1.74)–(1.74) for the Laguerre polynomials and (1.79)–(1.82) for the Jacobi polynomials.

1.14 Prove that if $a^2 + b^2 = 1$, then

$$H_n(ax + by) = \sum_{k=0}^{n} \binom{n}{k} H_{n-k}(x)H_k(y)a^{n-k}b^k.$$

1.15 Let $u_n(x) = e^{-x^2/2}H_n(x)$, where H_n are the Hermite polynomials. These functions are called the *Hermite functions*.

(a) Show that $u_n(x)$ satisfies the second-order differential equation

$$u_n''(x) + (2n + 1 - x^2)u_n(x) = 0, \quad n \ge 0.$$

This equation is equivalent to the *Schrödinger equation* for the harmonic oscillator in quantum mechanics.

(b) Show that $u_n(x)$ are eigenfunctions of the *Fourier transform* with eigenvalues $(-i)^n$, i.e.

$$\frac{1}{\sqrt{2\pi}}\int_{-\infty}^{\infty} u_n(x)e^{-ixt}dx = (-i)^n u_n(t), \quad n \geq 0.$$

1.16 Prove the following addition formula for the Laguerre polynomials:

$$L_n^{(\alpha+\beta+1)}(x+y) = \sum_{k=0}^{n} L_{n-k}^{(\alpha)}(x)L_k^{(\beta)}(y).$$

1.17 Prove that

$$x^{\alpha/2}e^{-x/2}L_n^{(\alpha)}(x) = \frac{(-1)^n}{2}\int_0^{\infty} J_\alpha(\sqrt{xy})y^{\alpha/2}e^{-y/2}L_n^{(\alpha)}(y)dy,$$

for $\alpha > -1, n \geq 0$, where $L_n^{(\alpha)}(x)$ are the Laguerre polynomials and $J_\alpha(z)$ is the standard Bessel function defined by (1.72).

1.18 Let $\Pi_m(x) = (x-m+1)_m$, where $(z)_n$ is the Pochhammer symbol defined by (1.2). Show that $\Delta\Pi_m = m\Pi_{m-1}$ and conclude that $\Delta^k\Pi_m(0) = m!$ if $k = m$ and 0 otherwise. Use this to prove that if f is any polynomial, then it has a discrete Taylor expansion of the form

$$f(x+y) = \sum_{k\geq 0}\frac{\Delta^k f(x)}{k!}(y-k+1)_k.$$

1.19 Prove formulas (1.101)–(1.104) for the Charlier polynomials, (1.106)–(1.110) for the Meixner polynomials, (1.113)–(1.116) for the Krawtchouk polynomials and (1.119)–(1.120) for the Hahn polynomials.

1.20 Show that

$$C_n(x+y;a) = \sum_{k=0}^{n}\binom{n}{k}(-y)_{n+k}a^{-n-k}C_k(x;a),$$

where $C_n(x;a)$ are the Charlier polynomials defined by (1.99).

2

Spectral Representation of Discrete-Time Birth–Death Chains

In this chapter we describe the probabilistic and asymptotic aspects of any discrete-time birth–death chain on $\mathbb{N}_0$ in terms of the spectral representation of the corresponding one-step transition probability (tridiagonal) matrix P. These processes have also been studied under the name of *random walks* (see the seminal paper [87] or more recently [13, 22, 30, 46, 62, 69]), but we will use the term discrete-time birth–death chain, or simply birth–death chain, since random walks are usually understood as discrete-time birth–death chains with constant transition probabilities. After a first section of basic definitions and fundamental properties about discrete-time Markov chains, we will obtain in Section 2.2 the *Karlin–McGregor integral representation formula* of the n-step transition probabilities $P_{ij}^{(n)}$ in terms of orthogonal polynomials with respect to a probability measure ψ with support inside the interval $[-1,1]$. In Section 2.3 we will study some families of polynomials related to the original orthogonal polynomials, which will play a very important role in the next sections. After this, in Section 2.4, we will study a collection of examples related to orthogonal polynomials where we can apply the spectral representation already introduced in the previous sections. We will focus on examples with finite state space, examples where we can get an expression of the Stieltjes transform of the spectral measure and use the Perron–Stieltjes inversion formula and finally the birth–death chain generated by the Jacobi polynomials. Among these examples we will study gambler's ruin and some well-known urn models like Ehrenfest, Bernoulli–Laplace or Jacobi models. In Section 2.5 we will apply the Karlin–McGregor formula to study the probabilistic and asymptotic aspects of any birth–death chain. First we will see how to study *recurrence and absorption*. Second we will study the so-called *strong ratio limit property*, namely the behavior of the ratio $P_{ij}^{(n)}/P_{kl}^{(n)}$ as $n \to \infty$. And finally, we will look at the concept of *limiting conditional distribution*, which basically is a stationary distribution conditioned on the event that the absorption has not occurred yet. We will also give a short summary of other interesting results. To end the chapter, in Section 2.6 we will study the spectral representation of discrete-time birth–death chains on $\mathbb{Z}$, where there are not so many results as in the case of birth–death chains on $\mathbb{N}_0$.

2.1 Discrete-Time Markov Chains

We will use references [5, 57, 100, 120, 140] in this section. Let $(\Omega, \mathcal{F}, \mathbb{P})$ be a probability space and let us consider a Markov chain $\{X_n, n=0,1,\ldots\}$ with state space $\mathcal{S} \subseteq \mathbb{Z}$. The state space $\mathcal{S}$ can be a finite collection of discrete points, the space of nonnegative integers $\mathbb{N}_0$ or the whole set of integers $\mathbb{Z}$. In this chapter we will mainly work with $\mathcal{S} = \{0,1,2,\ldots\} = \mathbb{N}_0$, unless otherwise specified. Recall that Markov chains have the Markov property, i.e. for all $n \geq 0$ and any states $i_0, \ldots, i_{n-1}, i, j \in \mathcal{S}$, we have that

$$\begin{aligned}&\mathbb{P}(X_{n+1} = j | X_0 = i_0, X_1 = i_1, \ldots, X_{n-1} = i_{n-1}, X_n = i) \\ &= \mathbb{P}(X_{n+1} = j | X_n = i) = P_{ij}^{n,n+1}.\end{aligned}$$

If at time n the chain is in state $i \in \mathcal{S}$, the probabilities that at time $n+1$ the chain is in state $j \in \mathcal{S}$ are called *one-step transition probabilities*. In general, transition probabilities depend on the time n. However, in these notes we will assume that the Markov chain is *homogeneous* or stationary in time, so that $P_{ij}^{n,n+1}$ does not depend on n and will be denoted by P_{ij}. It is usual to represent these values P_{ij} as the following matrix:

$$P = \begin{pmatrix} P_{00} & P_{01} & P_{02} & \cdots \\ P_{10} & P_{11} & P_{12} & \cdots \\ P_{20} & P_{21} & P_{22} & \cdots \\ \vdots & \vdots & \vdots & \end{pmatrix}. \tag{2.1}$$

The matrix $P = (P_{ij})$ is called the *one-step transition probability matrix* of the Markov chain. This matrix is always a *stochastic matrix*, i.e. it satisfies the conditions

$$P_{ij} \geq 0, \quad i,j \in \mathcal{S}, \quad \sum_{j=0}^{\infty} P_{ij} = 1, \quad i \in \mathcal{S}.$$

In some cases there may exist a state i^* such that $\sum_{j=0}^{\infty} P_{i^*,j} < 1$ (substochastic) in which case it is interpreted in terms of an ignored state i^* which is a permanent absorbing state of the Markov chain. Then the probability that the Markov chain is absorbed at state i^* is given by $1 - \sum_{j=0}^{\infty} P_{i^*,j} > 0$.

The process is completely characterized once we know the transition probability matrix (2.1) and the initial distribution of X_0. We will define the probability of reaching the state j in n transitions of time starting at any other state i by

$$P_{ij}^{(n)} = \mathbb{P}(X_n = j | X_0 = i).$$

Again these probabilities form a (sub) stochastic matrix $P^{(n)}$, called the *n-step transition probability matrix*, which satisfies the following properties:

$$P^{(0)} = I, \quad P^{(1)} = P, \quad P^{(n+1)} = P^{(n)} P = P P^{(n)}.$$

The last relation is also known as the *Chapman–Kolmogorov equations*, which can be written in general as

$$P_{ij}^{(n)} = \sum_{k \in \mathcal{S}} P_{ik}^{(k)} P_{kj}^{(n-k)}, \quad n \geq 0.$$

It is clear that $P^{(n)} = P^n$, i.e. the nth power of the transition probability matrix P in (2.1). Another way to write $P_{ij}^{(n)}$ is

$$P_{ij}^{(n)} = \sum_{i_k \in \mathcal{S}} P_{i,i_0} P_{i_0,i_1} \cdots P_{i_{n-1},j},$$

where the sum takes values at all indices $i_k, k = 0, \ldots, n-1$.

Let us now study the classification of states. A state $j \in \mathcal{S}$ is said to be *accessible* from a state i, and is denoted by $i \to j$, if there exists an integer $n \geq 0$ such that $P_{ij}^{(n)} > 0$. If two states i and j are accessible to each other, they are said to be *communicated* and denoted by $i \longleftrightarrow j$. The concept of communication is an equivalence relation. We say that a Markov chain is *irreducible* if the equivalence relation induced by communication induces only a single class. In other words, it is irreducible if all states communicate with each other. We define the *period* of a state i, $d(i)$, as the greatest common divisor of all integers $n \geq 1$ such that $P_{ii}^{(n)} > 0$, that is, $d(i) = \gcd\{n \geq 1 \colon P_{ii}^{(n)} > 0\}$. If $P_{ii}^{(n)} = 0$ for all $n \geq 1$ we define $d(i)$ as $d(i) = 0$. If $i \longleftrightarrow j$ then $d(i) = d(j)$. A Markov chain in which each state has period 1 is called *aperiodic*. An example of Markov chain with period 2 is the symmetric discrete-time birth–death chain.

A vector $\pi = (\pi_i)_{i \in \mathcal{S}}$ is an *invariant vector* (or stationary vector) for the Markov chain if

$$\pi_i \geq 0, \quad i \in \mathcal{S} \quad \text{and} \quad \pi P = \pi. \tag{2.2}$$

If we can normalize this vector in such a way that we have a probability distribution (i.e. $\sum_{i \in \mathcal{S}} \pi_i = 1$) then it will be called an *invariant distribution* (or stationary distribution) for the Markov chain. Observe from (2.2) that $\pi P^n = \pi$ for all $n \geq 0$.

Let $i, j \in \mathcal{S}$. For any $n \geq 1$ we define the *first passage time probabilities* as

$$f_{ij}^{(n)} = \mathbb{P}(X_n = j, X_m \neq i, m = 1, 2, \ldots, n-1 | X_0 = i), \tag{2.3}$$

that is, the probability that, starting from i, the chain arrives *for the first time* to the state j in exactly n transitions of time. We obviously have $f_{ij}^{(n)} \leq P_{ij}^{(n)}$. An alternative way of writing these probabilities is the following:

$$f_{ij}^{(n)} = \sum_{i_k \neq j} P_{i,i_0} P_{i_0,i_1} \cdots P_{i_{n-1},j},$$

where the sum takes values at all indices $i_k, k = 0, \ldots, n-1$, such that $i_k \neq j$. This sum can also be written as the (i,j) entry of the matrix $\widetilde{P}^{n-1} P$, where $\widetilde{P} = P Q_j$ and

$Q_j = \sum_{i \neq j} E_{ii} = I - E_{jj}$. Here E_{ij} denotes the matrix with 1 at entry (i,j) and 0 otherwise. Observe that $\widetilde{P} = PQ_j$ avoids passing through the state j by erasing the jth column of P. The matrix Q_j is the orthogonal projection into $\mathcal{S} \setminus \{j\}$. Therefore

$$f_{ij}^{(n)} = \left(\widetilde{P}^{n-1}P\right)_{ij} = \left(\underbrace{PQ_jPQ_j\cdots PQ_j}_{n-1}P\right)_{ij}.$$

These probabilities can also be defined in terms of first passage times. Let us call

$$T_j = \min\{n \geq 1 \colon X_n = j\}, \tag{2.4}$$

the *first passage time* to the state j. Then $f_{ij}^{(n)} = \mathbb{P}(T_j = n | X_0 = i)$. These probabilities are related to the probabilities $P_{ij}^{(n)}$ by the formula

$$P_{ij}^{(n)} = \sum_{k=1}^{n} f_{ij}^{(k)} P_{jj}^{(n-k)}, \quad n \geq 1.$$

Observe that in the case where we have an absorbing state -1 only accesible from state 0 with probability, say q, we can define the first passage time probabilities $f_{i,-1}^{(n)}$ in the same form. The absorption occurs for the first time at the nth transition. We have the following straightforward relation:

$$f_{i,-1}^{(n)} = qP_{i,0}^{(n-1)}. \tag{2.5}$$

Let us define the generating functions associated with the sequences $P_{ij}^{(n)}$ and $f_{ij}^{(n)}$ as follows:

$$P_{ij}(s) = \sum_{n=0}^{\infty} P_{ij}^{(n)} s^n, \quad F_{ij}(s) = \sum_{n=0}^{\infty} f_{ij}^{(n)} s^n, \quad |s| < 1. \tag{2.6}$$

These generating functions are related by the formulas (see for instance [57])

$$P_{ij}(s) = F_{ij}(s)P_{jj}(s), \quad \text{if} \quad i \neq j, \tag{2.7}$$

$$P_{ii}(s) = 1 + F_{ii}(s)P_{ii}(s). \tag{2.8}$$

Definition 2.1 With the previous notation we have that:

- A state $i \in \mathcal{S}$ is *recurrent* if $\mathbb{P}(T_i < \infty | X_0 = i) = 1$. In terms of the probabilities $f_{ii}^{(n)}$, we must have that $\sum_{n=1}^{\infty} f_{ii}^{(n)} = 1$, or equivalently $\lim_{s \uparrow 1} F_{ii}(s) = 1$.
- A state $i \in \mathcal{S}$ is *transient* if it is not recurrent, i.e. $\mathbb{P}(T_i < \infty | X_0 = i) < 1$, or that $\sum_{n=1}^{\infty} f_{ii}^{(n)} < 1$, or equivalently $\lim_{s \uparrow 1} F_{ii}(s) < 1$. ◊

Recurrence is preserved by communication. From (2.8) we see that

$$P_{ii}(s) = \frac{1}{1 - F_{ii}(s)}.$$

Therefore it follows that a state i is recurrent if and only if

$$\sum_{n=1}^{\infty} P_{ii}^{(n)} = \infty.$$

The *expected return time* from state i to state j is given by

$$\tau_{ij} = \mathbb{E}(T_j | X_0 = i) = \sum_{n=1}^{\infty} n f_{ij}^{(n)} + \infty \left(1 - \lim_{s \uparrow 1} F_{ij}(s) \right). \tag{2.9}$$

It is clear that if a state is transient then $\tau_{ii} = \infty$. These quantities can also be written in terms of $F_{ij}(s)$. Indeed,

$$\tau_{ij} = \lim_{s \uparrow 1} \frac{d}{ds} F_{ij}(s).$$

In terms of $P_{ij}(s)$ we have, using (2.7) and (2.8), that

$$\tau_{ij} = \lim_{s \uparrow 1} \frac{d}{ds} \frac{P_{ij}(s)}{P_{jj}(s)}, \quad i \neq j, \quad \tau_{ii} = \lim_{s \uparrow 1} \frac{1}{(1-s) P_{ii}(s)}. \tag{2.10}$$

The expected return time will give an additional classification of recurrent states:

Definition 2.2 With the previous notation we have that:

- A state $i \in \mathcal{S}$ is *null recurrent* if it is recurrent and $\tau_{ii} = \infty$.
- A state $i \in \mathcal{S}$ is *positive recurrent* (or ergodic) if it is recurrent and $\tau_{ii} < \infty$. ◊

In the following result, which can be found in [120, Theorem 1.8.3], we display the relation between the expected return times τ_{ij} and the behavior of $P_{ij}^{(n)}$ as $n \to \infty$ for irreducible and aperiodic Markov chains.

Theorem 2.3 (Convergence theorem for Markov chains) *For any irreducible and aperiodic Markov chain we have, for any pair of states $i, j \in \mathcal{S}$, that*

$$\lim_{n \to \infty} P_{ij}^{(n)} = \frac{1}{\tau_{jj}}.$$

If additionally the Markov chain is positive recurrent, then the vector $\pi = (\pi_j)_{j \in \mathcal{S}}$ with components $\pi_j = 1/\tau_{jj}$ is an invariant distribution for the Markov chain.

If the state space is finite, i.e. $\mathcal{S} = \{0, 1, \ldots, N\}$, then we can derive a spectral representation of the n-step transition probability matrix $P_{ij}^{(n)}$ in terms of the eigenvalues of P. In particular this works well when P is assumed to be *weakly symmetric* or *reversible*, i.e. there exists a vector $m = (m_i)_{i=0}^{N}$ such that $m_i P_{ij} = m_j P_{ji}$ for all $i, j \in \mathcal{S}$. In that case, if we define the diagonal matrix $\Pi = \text{diag}\{\sqrt{m_0}, \sqrt{m_1}, \ldots \sqrt{m_N}\}$, we have that $S = \Pi P \Pi^{-1}$ is a symmetric matrix. Therefore all the eigenvalues of P are real and located inside the interval $[-1, 1]$ (applying the Perron–Frobenius theorem). Then we can write $S = MDM^T$, where M is an orthogonal real matrix and D a diagonal matrix with the eigenvalues of P located in the main diagonal. Therefore we obtain

$$P^{(n)} = \Pi^{-1} M D^n M^T \Pi, \quad n \geq 0,$$

or entry by entry

$$P_{ij}^{(n)} = m_j \sum_{k=0}^{N} \lambda_k^n \phi_i^{(k)} \phi_j^{(k)}, \tag{2.11}$$

where λ_i are the diagonal entries of D and $\phi_i^{(k)} = M_{ik}/\sqrt{m_i}, i,k \in \mathcal{S}$ satisfy that

$$\sum_{k=0}^{N} \phi_i^{(k)} \phi_j^{(k)} = \frac{\delta_{ij}}{m_i}, \quad i,j \in \mathcal{S}.$$

A special case of discrete-time Markov chains are *birth–death chains* on the non-negative integers $\mathbb{N}_0 = \{0, 1, \ldots\}$. The transitions are only allowed between adjacent states, so the one-step transition probability matrix $P = (P_{ij})$ can be defined as

$$P_{ij} = \begin{cases} p_i, & \text{if} \quad j = i+1, \\ r_i, & \text{if} \quad j = i, \\ q_i, & \text{if} \quad j = i-1, \\ 0, & \text{if} \quad |i-j| > 1. \end{cases}$$

In other words, P is a tridiagonal matrix of the form

$$P = \begin{pmatrix} r_0 & p_0 & 0 & 0 & 0 & \cdots \\ q_1 & r_1 & p_1 & 0 & 0 & \cdots \\ 0 & q_2 & r_2 & p_2 & 0 & \cdots \\ 0 & 0 & q_3 & r_3 & p_3 & \cdots \\ \vdots & \vdots & \vdots & \ddots & \ddots & \ddots \end{pmatrix}. \tag{2.12}$$

A diagram of the transitions between states is

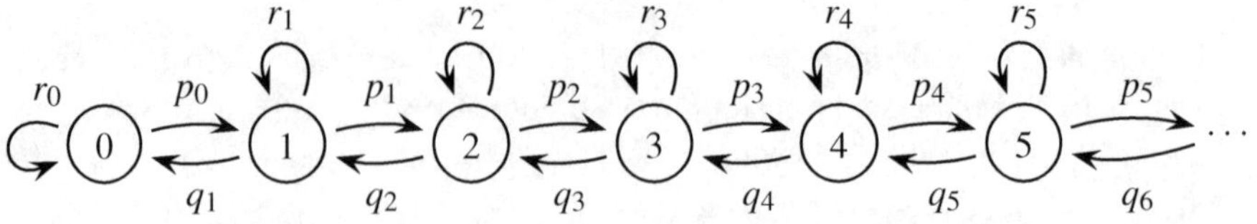

We will assume for the transition probabilities $p_i, q_{i+1} > 0, r_i \geq 0, i \geq 0$ and $q_i + r_i + p_i \leq 1, i \geq 0$. The possible inequality $q_i + r_i + p_i < 1$ is interpreted in terms of a permanent absorbing state i^* which can only be reached from state i

with probability $1-(q_i+r_i+p_i)$. We will focus on the cases where either P is stochastic (i.e. $r_0+p_0=1$ and $q_i+r_i+p_i=1, i\geq 1$), or P is strictly substochastic with only one absorbing state, denoted by -1, which can only be reached through state 0 with probability $q_0=1-p_0-r_0>0$.

The behavior of a birth–death chain is characterized by the behavior of the so-called *potential coefficients*, defined by

$$\pi_0=1, \quad \pi_n=\frac{p_0p_1\cdots p_{n-1}}{q_1q_2\cdots q_n}, \quad n\geq 1. \tag{2.13}$$

These potential coefficients are generated using the fact that P is always weakly symmetric or reversible (taking $m_i=\pi_i$), i.e. $\pi_iP_{ij}=\pi_jP_{ji}$. In the case of birth–death chains that means that we have the following symmetry equations:

$$\pi_np_n=\pi_{n+1}q_{n+1}, \quad n\geq 0. \tag{2.14}$$

But these equations are satisfied by $(\pi_n)_{n\in\mathbb{N}_0}$ in (2.13). Observe that if the transition probability matrix P in (2.12) is stochastic, a straightforward computation shows that π satisfies $\pi P=\pi$. Therefore it is an *invariant vector* of the birth–death chain, which will be a distribution if $\sum_{n=0}^{\infty}\pi_n<\infty$.

2.2 Karlin–McGregor Representation Formula

This integral representation was derived for the first time in [87] (inspired by the work of W. Feller and H. P. McKean). First we will derive it formally and then we will prove it in a more rigorous way. From the transition probability matrix P, let us define a sequence of polynomials $(Q_n)_n$ with $\deg Q_n=n$ in the following way:

$$\begin{aligned} Q_0(x)&=1, \quad Q_{-1}(x)=0,\\ xQ_0(x)&=p_0Q_1(x)+r_0Q_0(x),\\ xQ_n(x)&=p_nQ_{n+1}(x)+r_nQ_n(x)+q_nQ_{n-1}(x), \quad n\geq 1. \end{aligned} \tag{2.15}$$

Observe that in the case where $q_0=1-p_0-r_0>0$ the first polynomial is defined by $p_0Q_1(x)=x-r_0=x-1+p_0+q_0$ and all polynomials will depend on this parameter q_0.

If the state space is finite, i.e. $\mathcal{S}=\{0,1,\dots,N\}$, then the three-term recurrence relation is valid from $n=0,1,\dots,N-1$. The polynomial $R_{N+1}(x)=xQ_N(x)-r_nQ_N(x)-q_NQ_{N-1}(x)$ is a polynomial of degree $N+1$ whose zeros coincide with the eigenvalues of the transition probability matrix P (of size $(N+1)\times(N+1)$). This is a consequence of writing the three-term recurrence relation (2.15) in vector form as $xQ=PQ$ and looking at the last component. In this case, if we are able to find a collection of eigenvalues and eigenvectors of P we get the spectral representation (2.11).

In general, coming back to the state space $\mathcal{S} = \mathbb{N}_0$, we can always write the three-term recurrence relation (2.15) in vector form as $xQ(x) = PQ(x)$, where $Q(x) = (Q_0(x), Q_1(x), \ldots)^T$. Iterating this formula n times we get that $x^n Q(x) = P^n Q(x)$. Therefore, componentwise, we have

$$x^n Q_i(x) = \sum_{j=0}^{\infty} P_{ij}^n Q_j(x).$$

The spectral theorem (see Section 1.3) guarantees that there exists a unique probability measure ψ supported on the interval $[-1, 1]$ (see Theorem 2.5 below) such that the polynomials generated by (2.15) are orthogonal with respect to ψ.

Multiplying $x^n Q(x) = P^n Q(x)$ on the right by $Q^T(x)$ and integrating entry by entry with respect to the measure ψ we have, using the orthogonality, that all entries on the right-hand side outside the main diagonal are 0. Therefore we obtain the following expression:

$$P_{ij}^n \int_{-1}^{1} Q_j^2(x) d\psi(x) = \int_{-1}^{1} x^n Q_i(x) Q_j(x)\, d\psi(x).$$

Hence, since P_{ij}^n are equal to the probabilities $P_{ij}^{(n)}$, we get that

$$P_{ij}^{(n)} = \frac{\displaystyle\int_{-1}^{1} x^n Q_i(x) Q_j(x)\, d\psi(x)}{\displaystyle\int_{-1}^{1} Q_j^2(x)\, d\psi(x)}.$$

From (2.15) it is possible to derive the following formula:

$$\int_{-1}^{1} Q_j^2(x) d\psi(x) = \frac{q_1 q_2 \cdots q_j}{p_0 p_1 \cdots p_{j-1}} = \frac{1}{\pi_j}, \tag{2.16}$$

where $(\pi_n)_n$ are the potential coefficients defined by (2.13). Indeed, multiplying (2.15) by $Q_{j+1}(x)$ and integrating with respect to the measure ψ, we get

$$\int_{-1}^{1} x Q_j(x) Q_{j+1}(x) d\psi(x) = p_j \int_{-1}^{1} Q_{j+1}^2(x) d\psi(x).$$

Back to formula (2.15) for $n = j + 1$ and using again the orthogonality, we obtain

$$q_{j+1} \int_{-1}^{1} Q_j^2(x) d\psi(x) = p_j \int_{-1}^{1} Q_{j+1}^2(x) d\psi(x),$$

and as a consequence we get (2.16). Observe from (2.16) that we have

$$\pi_j = \left(\int_{-1}^{1} Q_j^2(x) d\psi(x) \right)^{-1}.$$

If the birth–death chain is positive recurrent, then we have

$$\tau_{jj} = \int_{-1}^{1} Q_j^2(x)\, d\psi(x),$$

where τ_{jj} is the expected return time defined by (2.9) (see also Theorem 2.3). Therefore we get (formally) the Karlin–McGregor integral representation formula for birth–death chains:

$$P_{ij}^{(n)} = \pi_j \int_{-1}^{1} x^n Q_i(x) Q_j(x)\, d\psi(x).$$

The key to proving this integral formula in a more rigorous way is to consider the spectral representation of the operator P acting on a certain Hilbert space. Let $\ell_\pi^2(\mathbb{N}_0)$ be the Hilbert space of all the sequences $f = (f_n)_n$ of complex numbers such that

$$\|f\|_\pi^2 = \sum_{n=0}^{\infty} |f_n|^2 \pi_n < \infty,$$

where $\pi = (\pi_n)_n$ are the potential coefficients (2.13). Here we are assuming that if $f \in \ell_\pi^2(\mathbb{N}_0)$ then the sequence f vanishes outside of $\mathbb{N}_0$. Let us consider P as an operator acting on $\ell_\pi^2(\mathbb{N}_0)$ in the following way:

$$(Pf)_n = \sum_{m=0}^{\infty} P_{nm} f_m = p_n f_{n+1} + r_n f_n + q_n f_{n-1}, \quad f \in \ell_\pi^2(\mathbb{N}_0).$$

Lemma 2.4 ([87, Lemma 1]) *The transformation $f \to Pf$ induces in $\ell_\pi^2(\mathbb{N}_0)$ a bounded self-adjoint linear operator of norm ≤ 1.*

Proof Linearity is trivial. If $f \in \ell_\pi^2(\mathbb{N}_0)$, then

$$\begin{aligned}\|Pf\|_\pi^2 &= \sum_{n=0}^{\infty} |q_n f_{n-1} + r_n f_n + p_n f_{n+1}|^2 \pi_n \\ &\leq \sum_{n=0}^{\infty} |f_{n-1}|^2 \pi_n + \sum_{n=0}^{\infty} |f_n|^2 \pi_n + \sum_{n=0}^{\infty} |f_{n+1}|^2 \pi_n < \infty.\end{aligned}$$

Therefore P is bounded. The norm is ≤ 1 as a consequence of the Perron–Frobenius theorem (see [72, Chapter 8] or more generally [130, Chapter 1]), since all eigenvalues λ of P satisfy $|\lambda| \leq 1$. Finally, the self-adjointness is a consequence of the symmetry equation $\pi_i p_i = \pi_{i+1} q_{i+1}$, since for all $f, g \in \ell_\pi^2(\mathbb{N}_0)$, we have

$$\begin{aligned}(Pf, g)_\pi &= \sum_{n=0}^{\infty} (q_n f_{n-1} + r_n f_n + p_n f_{n+1}) g_n \pi_n \\ &= \sum_{n=0}^{\infty} q_n f_{n-1} g_n \pi_n + \sum_{n=0}^{\infty} r_n f_n g_n \pi_n + \sum_{n=0}^{\infty} p_n f_{n+1} g_n \pi_n\end{aligned}$$

$$
\begin{aligned}
&= \sum_{n=0}^{\infty} p_{n-1} f_{n-1} g_n \pi_{n-1} + \sum_{n=0}^{\infty} r_n f_n g_n \pi_n + \sum_{n=0}^{\infty} q_{n+1} f_{n+1} g_n \pi_{n+1} \\
&= \sum_{n=0}^{\infty} p_n f_n g_{n+1} \pi_n + \sum_{n=0}^{\infty} r_n f_n g_n \pi_n + \sum_{n=0}^{\infty} q_n f_n g_{n-1} \pi_n \\
&= \sum_{n=0}^{\infty} f_n (q_n g_{n-1} + r_n g_n + p_n g_{n+1}) \pi_n \\
&= (f, Pg)_\pi,
\end{aligned}
$$

recalling that f and g vanish outside of $\mathbb{N}_0$. Therefore all eigenvalues are real and are contained in the interval $[-1,1]$. □

Therefore the matrix P extends to a bounded self-adjoint linear operator of norm ≤ 1 in the Hilbert space $\ell^2_\pi(\mathbb{N}_0)$, which we call again P abusing the notation.

We will follow the same steps as in the second part of Section 1.3 but applied to the Hilbert space $\ell^2_\pi(\mathbb{N}_0)$. Let $e^{(i)}, i \geq 0$, be the sequences in $\ell^2_\pi(\mathbb{N}_0)$ given by $e^{(i)}_j = \delta_{ij}/\pi_i$, or, in vectorial form, by

$$
e^{(i)} = (0, \dots, 0, 1/\pi_i, 0, \dots)^T,
$$

with all entries equal to 0 except the ith component. Using the symmetry equation $\pi_i p_i = \pi_{i+1} q_{i+1}$ we have that

$$
\begin{aligned}
Pe^{(i)} &= (0, \dots, p_{i-1}/\pi_i, r_i/\pi_i, q_{i+1}/\pi_i, \dots)^T = (0, \dots, q_i/\pi_{i-1}, r_i/\pi_i, p_i/\pi_{i+1}, \dots)^T \\
&= q_i e^{(i-1)} + r_i e^{(i)} + p_i e^{(i+1)}.
\end{aligned}
$$

Now, using the recurrence formula (2.15) and induction, we can see that $Q_n(P)e^{(0)} = e^{(n)}$. Indeed, for $n = 0, 1$, it is immediate. Assuming that it is true for all values $\leq n$, for $n+1$ we have, using the previous formula, that

$$
\begin{aligned}
Q_{n+1}(P)e^{(0)} &= \frac{1}{p_n}\left((P - r_n)Q_n(P)e^{(0)} - q_n Q_{n-1}(P)e^{(0)}\right) \\
&= \frac{1}{p_n}\left((P - r_n)e^{(n)} - q_n e^{(n-1)}\right) \\
&= \frac{1}{p_n}\left(Pe^{(n)} - r_n e^{(n)} - q_n e^{(n-1)}\right) = e^{(n+1)}.
\end{aligned}
$$

This result is also a consequence of Lemma 1.16 of Chapter 1. Now, notice that, since $P^n e^{(j)} = \frac{1}{\pi_j}(P^n_{0,j}, P^n_{1,j}, \dots)$, we have

$$
(P^n e^{(j)}, e^{(i)})_\pi = \frac{1}{\pi_j} \sum_{k=0}^{\infty} P^n_{kj} \frac{\delta_{ik}}{\pi_i} \pi_k = \frac{1}{\pi_j} P^{(n)}_{ij}.
$$

Therefore, using the above and the fact that P is self-adjoint, we get

$$P_{ij}^{(n)} = \pi_j(P^n Q_j(P)e^{(0)}, Q_i(P)e^{(0)})_\pi = \pi_j(P^n Q_j(P)Q_i(P)e^{(0)}, e^{(0)})_\pi.$$

Using the spectral theorem we know that there exists a probability measure ψ supported on the interval $[-1,1]$ (since the norm of P is ≤ 1) such that

$$\int_{-1}^{1} f(x)d\psi(x) = (f(P)e^{(0)}, e^{(0)})_\pi,$$

where f is any real and bounded function on $[-1,1]$. In particular,

$$P_{ij}^{(n)} = \pi_j \int_{-1}^{1} x^n Q_i(x)Q_j(x)d\psi(x) = \frac{(x^n Q_i, Q_j)_\psi}{(Q_j, Q_j)_\psi}.$$

Summarizing we obtain the following theorem:

Theorem 2.5 ([87, Theorem 1]) *Let* $\{X_n, n=0,1,\ldots\}$ *be a one-dimensional discrete-time birth–death chain with one-step transition probability matrix* P *given by* (2.12). *Then there exists a unique positive regular distribution* ψ *supported on the interval* $[-1,1]$ *such that for all* i,j,n, *we get the integral representation*

$$P_{ij}^{(n)} = \pi_j \int_{-1}^{1} x^n Q_i(x)Q_j(x)\, d\psi(x). \tag{2.17}$$

Proof We have already proved the existence. The uniqueness is a consequence of the fact that (2.17) with $i = j = 0$ determines all the moments of ψ and hence determines ψ uniquely (see the proof of Theorem 1.17). □

Formula (2.17) is usually known as the *Karlin–McGregor representation formula for birth–death chains*. Observe that formula (2.17) for $n=0$ implies that the polynomials built from (2.15) are orthogonal with respect to the measure ψ. This measure ψ will be called the *spectral or birth–death measure* while the polynomials $(Q_n)_n$ will be also called *birth–death polynomials*.

Up to now we have proved that for every discrete-time birth–death chain on $\mathbb{N}_0$ there exists a unique spectral measure supported on the interval $[-1,1]$ such that all entries of the n-step transition probability matrix can be represented as an integral by the Karlin–McGregor formula (2.17). Now we will study the reverse question, that is, given a measure ψ supported on the interval $[-1,1]$, we would like to characterize all birth–death chains on $\mathbb{N}_0$ with the same spectral measure ψ and under what conditions it is unique.

A partial result in this direction was given by [87] for *symmetric discrete-time birth–death chains* (i.e. $r_n = 0$, $n \geq 0$). In this case, the spectral measure ψ is symmetric around $x=0$, since $P_{00}^{(n)}$ is 0 or positive according to the parity of n (the Markov chain has period 2). Therefore the odd moments of ψ will be 0, which makes ψ symmetric. The inverse result is also true, as we will see in the following:

Proposition 2.6 ([87, Lemma 2]) *Every positive symmetric distribution function supported on the interval* $[-1,1]$ *not supported by a finite set of points, and with total mass equal to 1, is the spectral measure of a symmetric birth–death chain.*

Proof Let $Q_n(x)$ be the orthogonal polynomials with respect to ψ normalized in such a way that $Q_n(1) = 1$. Since the measure is symmetric, $Q_n(x)$ is even or odd according to whether n is even or odd. Therefore $r_n = \int_{-1}^{1} xQ_n^2(x)\,d\psi(x) = 0$ and this implies that the recurrence relation must be of the form

$$xQ_n(x) = q_nQ_{n-1} + p_nQ_{n+1}.$$

Since $Q_n(x) \geq 1$ for $x \geq 1$ (all zeros must be inside the interval $[-1,1]$) and $Q_n(1) = 1$, we have that the leading coefficient must be positive. Equating the leading coefficient of the recurrence relation gives that $p_n > 0$. In fact, for $x = 1$ we have that $p_n + q_n = 1$. It remains to prove that $q_n > 0$. But this is a consequence of multiplying the recurrence relation by x^{n-1} and using the orthogonality to see that $I_n = q_nI_{n-1}$ where $I_n = \int_{-1}^{1} x^nQ_n(x)\,d\psi(x)$. Since the leading coefficient of Q_n is positive, given by $(p_0 \cdots p_{n-1})^{-1}$, then $I_n > 0$ and therefore $q_n > 0$. □

In general, for any measure ψ with $\text{supp}(\psi) \subseteq [-1,1]$, the corresponding orthogonal polynomials will not give a sequence of birth–death polynomials. That is, not all measures supported on $[-1,1]$ will generate a discrete-time birth–death chain. This can be easily seen from the definition of the middle coefficient r_n of the three-term recurrence relation, given by

$$r_n = \int_{-1}^{1} xQ_n^2(x)\,d\psi(x), \quad n \geq 0,$$

which in general is not always nonnegative. In some sense the mass of ψ on the negative axis should not be greater than the mass on the positive axis, so that we may have $r_n \geq 0$ for all n. The conditions under which we have a birth–death measure are given in the following result:

Theorem 2.7 ([46, Theorem 1.2]) *Let* ψ *be a measure supported on the interval* $[-1,1]$ *and consider the monic family of orthogonal polynomials* $(\widehat{P}_n)_n$ *satisfying the three-term recurrence relation*

$$\widehat{P}_{n+1}(x) = (x-\alpha_n)\widehat{P}_n(x) - \beta_n\widehat{P}_{n-1}(x), \quad n \geq 1,$$
$$\widehat{P}_0(x) = 1, \quad \widehat{P}_1(x) = x - \alpha_0,$$

where $\alpha_n \geq 0$ *and* $\beta_{n+1} > 0, n \geq 0$. *Then* $(\widehat{P}_n)_n$ *is a sequence of birth–death polynomials if and only if the sequence* $(Q_n)_n$ *with* $Q_n(x) = \widehat{P}_n(x)/\widehat{P}_n(1)$ *satisfies a three-term recurrence relation as in* (2.15) *(*$q_0 = 0$*) where*

$$p_n, q_{n+1} > 0 \quad \text{and} \quad r_n \geq 0, \quad n \geq 0. \tag{2.18}$$

Proof According to Proposition 1.8 all zeros of $\widehat{P}_n(x)$ are located inside the interval $[-1,1]$. Therefore, since they are monic, we have that $\widehat{P}_n(1) > 0$ and $Q_n(x) = \widehat{P}_n(x)/\widehat{P}_n(1)$ is well defined. This new family of polynomials satisfies the recurrence relation

$$\begin{aligned} xQ_n(x) &= \frac{\widehat{P}_{n+1}(1)}{\widehat{P}_n(1)} Q_{n+1}(x) + \alpha_n Q_n(x) + \beta_n \frac{\widehat{P}_{n-1}(1)}{\widehat{P}_n(1)} Q_{n-1}(x), \\ Q_0(x) &= 1, \quad \widehat{P}_1(1) Q_1(x) = x - \alpha_0, \end{aligned}$$

which is of type (2.15) and satisfies (2.18), since $\alpha_n \geq 0$. In fact, by equating the leading coefficient in the three-term recurrence formula, we get $Q_n(x) = L_n x^n + \cdots$, where $L_n = (p_0 \cdots p_{n-1})^{-1} > 0$.

Conversely, let $(Q_n)_n$ be the sequence of polynomials satisfying (2.15) and (2.18). The corresponding monic polynomials have, as three-term recurrence relation coefficients, the following:

$$\alpha_n = r_n \geq 0 \quad \text{and} \quad \beta_{n+1} = \frac{L_n}{L_{n+1}} q_{n+1} = p_n q_{n+1} > 0, \quad n \geq 0.$$

Finally, the support of ψ is inside of $[-1,1]$ as a consequence of the following inequalities (see [16, Theorem II.4.5] and [44, Theorem 2]):

$$\inf \operatorname{supp}(\psi) \geq \inf_j \{2r_j - 1\} \geq -1, \quad \sup \operatorname{supp}(\psi) \leq \sup_j \{p_j + q_j + r_j\} \leq 1. \quad \square$$

For instance, if $\operatorname{supp}(\psi) \subseteq [0,1]$ then, according to the previous result, it will be a birth–death measure, and it will always be possible to compute the corresponding transition probabilities of the birth–death chain.

The previous theorem gives the necessary and sufficient conditions for a measure ψ to be the spectral measure of a birth–death chain. But this birth–death chain *does not need to be unique*. There are some situations in which a different normalization of the orthogonal polynomials can give a birth–death chain different from the original one (in fact, there is an infinite number of them), as we will see in the following result:

Theorem 2.8 ([22, Theorem 4.2]) *Let $\{X_n, n=0,1,\ldots\}$ be a birth–death chain with transition probability matrix P where $q_0 = 0$ and consider the sequence of polynomials $(Q_n)_n$ generated by the three-term recurrence relation* (2.15). *For any parameter $0 < \mu < 1$ there exists a sequence $(a_n)_n$ with $a_n > 0, n \geq 0$, such that $R_n(x) = a_n Q_n(x)$ satisfies the recurrence relation*

$$\begin{aligned} R_0(x) &= 1, \quad xR_0(x) = (1 - p_0' - \mu) R_0(x) + p_0' R_1(x), \\ xR_n(x) &= p_n' R_{n+1}(x) + (1 - p_n' - q_n') R_n(x) + q_n' R_{n-1}(x), \end{aligned} \tag{2.19}$$

with $0 < p_n', q_n' < 1$ *if and only if the series* $\sum_{n=0}^{\infty}(p_n\pi_n)^{-1}$ *is convergent and* μ *satisfies*

$$\mu \sum_{n=0}^{\infty} \frac{1}{p_n\pi_n} < 1.$$

Proof Let $R_n(x) = a_n Q_n(x)$. This new sequence of polynomials satisfies the recurrence relation

$$R_0(x) = a_0, \quad xR_0(x) = (1 - p_0)R_0(x) + p_0\frac{a_0}{a_1}R_1(x),$$
$$xR_n(x) = p_n\frac{a_n}{a_{n+1}}R_{n+1}(x) + (1 - p_n - q_n)R_n(x) + q_n\frac{a_n}{a_{n-1}}R_{n-1}(x).$$

This sequence is of the required type (2.19) if and only if

$$\begin{aligned} a_0 &= 1, \\ \mu + p_0\frac{a_0}{a_1} &= p_0, \\ p_n\frac{a_n}{a_{n+1}} + q_n\frac{a_n}{a_{n-1}} &= p_n + q_n. \end{aligned}$$

Consider now the sequences

$$\rho_0 = \frac{\mu}{p_0}, \quad \rho_n = \frac{q_n}{p_n}, \quad \sigma_n = \frac{a_n}{a_{n-1}},$$
$$\tau_n = \rho_0 + \rho_0\rho_1 + \cdots + \rho_0\rho_1\cdots\rho_n = \mu\sum_{k=0}^{n}\frac{1}{p_k\pi_k}.$$

Then the above relations may be written as

$$\begin{aligned} \rho_0 + \frac{1}{\sigma_1} &= 1, \\ \rho_n\sigma_n + \frac{1}{\sigma_{n+1}} &= 1 + \rho_n. \end{aligned}$$

From the above and using induction it is easy to prove that $\sigma_n > 1, n \geq 1$. Substituting σ_1 in σ_2 and so on, we have that

$$\sigma_{n+1} = 1 + \frac{\rho_0\rho_2\cdots\rho_n}{1 - \tau_n}.$$

Therefore the system has solution with $\sigma_n > 1, n \geq 1$, if and only if $\tau_n < 1, n \geq 0$, as requested. The values of σ_{n+1} are given by

$$\sigma_{n+1} = \frac{1 - \mu\displaystyle\sum_{k=0}^{n-1}\frac{1}{p_k\pi_k}}{1 - \mu\displaystyle\sum_{k=0}^{n}\frac{1}{p_k\pi_k}},$$

and from here we can compute $a_n = \sigma_1\sigma_2\cdots\sigma_n$. □

In Corollary 2.27 we will see that the condition for which the phenomenon described in the previous theorem occurs is that the birth–death chain is transient.

Remark 2.9 Theorem 2.8 was derived for the first time in [83] for continuous-time birth–death processes. Another proof using the notion of *canonical moments* can be found in [34, Chapter 8] (see also a brief comment about canonical moments at the end of Section 2.5). ◊

2.3 Properties of the Birth–Death Polynomials and Other Related Families

In this section we will derive some useful formulas for the sequence of polynomials $(Q_n)_n$ defined by a three-term recurrence relation as in (2.15). For that we will introduce *dual polynomials* and study some of their properties, including orthogonality properties. We will also introduce the *kth associated polynomials* and some of their properties, and in particular we will find important relations involving the Stieltjes transform. Finally, we will introduce the so-called *Derman–Vere–Jones transformation* consisting of moving the supremum of the support of the measure to the point $x = 1$. These notions will be important for the development of later sections.

2.3.1 Dual Polynomials

Let $(Q_n)_n$ be the polynomials generated by the three-term recurrence relation (2.15). We will study first the case of possible absorption at the state -1, i.e. $q_0 > 0$ and later the case where $q_0 = 0$.

Case $q_0 > 0$: Using that $q_n + r_n + p_n = 1, n \geq 1$, and the symmetry equations (2.14), the three-term recurrence relation can be written in the following way:

$$\begin{aligned} Q_0(x) &= 1, \\ (x-1)Q_0(x)\pi_0 &= p_0\pi_0[Q_1(x) - Q_0(x)] - q_0, \\ (x-1)Q_n(x)\pi_n &= p_n\pi_n[Q_{n+1}(x) - Q_n(x)] - p_{n-1}\pi_{n-1}[Q_n(x) - Q_{n-1}(x)], \quad n \geq 1. \end{aligned} \tag{2.20}$$

Let us define the *dual polynomials* as

$$Q_0^d(x) = 1, \quad Q_{n+1}^d(x) = \frac{p_n\pi_n}{q_0}[Q_{n+1}(x) - Q_n(x)], \quad n \geq 0. \tag{2.21}$$

Then from the previous relations we have that

$$\begin{aligned} Q_{n+1}^d(x) &= Q_n^d(x) + \frac{x-1}{q_0}Q_n(x)\pi_n = \cdots = Q_0^d(x) + \frac{x-1}{q_0}\sum_{j=0}^{n} Q_j(x)\pi_j \\ &= 1 + \frac{x-1}{q_0}\sum_{j=0}^{n} Q_j(x)\pi_j. \end{aligned}$$

Therefore $Q_n^d(1) = 1, n \geq 0$. Using the definition of $Q_n^d(x)$ in (2.21) and solving again the recurrence equation we get

$$Q_{n+1}(x) = Q_n(x) + \frac{q_0}{p_n\pi_n} - \frac{1-x}{p_n\pi_n}\sum_{j=0}^{n} Q_j(x)\pi_j = \cdots$$
$$= 1 + q_0\sum_{m=0}^{n}\frac{1}{p_m\pi_m} - (1-x)\sum_{m=0}^{n}\frac{1}{p_m\pi_m}\sum_{j=0}^{m} Q_j(x)\pi_j.$$

Therefore we get

$$Q_n(x) = 1 + q_0\sum_{m=0}^{n-1}\frac{1}{p_m\pi_m} - (1-x)\sum_{m=0}^{n-1}\frac{1}{p_m\pi_m}\sum_{j=0}^{m} Q_j(x)\pi_j. \tag{2.22}$$

From the above we have, evaluating at $x = 1$, that

$$Q_n(1) = 1 + q_0\sum_{k=0}^{n-1}\frac{1}{p_k\pi_k}. \tag{2.23}$$

We can also write Q_n in terms of Q_n^d. Indeed, we have that $Q_{n+1}^d(x) - Q_n^d(x) = \frac{x-1}{q_0}Q_n(x)\pi_n$. Therefore, from (2.21), we get

$$Q_n(x) = 1 + \sum_{k=0}^{n-1}\frac{q_0}{p_k\pi_k}Q_{k+1}^d(x).$$

From the modified three-term recurrence relation we can follow the same lines to derive the Christoffel–Darboux formula

$$(x-y)\sum_{k=0}^{n}\pi_k Q_k Q_k(y) = p_n\pi_n[Q_{n+1}(y)Q_n(x) - Q_{n+1}(x)Q_n(y)]$$
$$= q_0[Q_{n+1}^d(y)Q_n(x) - Q_{n+1}^d(x)Q_n(y)].$$

From here it follows that the zeros of $(Q_n^d)_n$ interlace with the zeros of $(Q_n)_n$.

An easy computation from the definition of $Q_n^d(x)$ in (2.21) gives that they satisfy the three-term recurrence relation

$$Q_0^d(x) = 1, \quad xQ_0^d(x) = q_0Q_1^d(x) + (1-q_0)Q_0^d(x),$$
$$xQ_{n+1}^d(x) = q_{n+1}Q_{n+2}^d(x) + (1-p_n-q_{n+1})Q_{n+1}^d(x) + p_nQ_n^d(x), \quad n \geq 0.$$

Therefore the associated Jacobi matrix is given by

$$P^d = \begin{pmatrix} 1-q_0 & q_0 & & & \\ p_0 & 1-p_0-q_1 & q_1 & & \\ & p_1 & 1-p_1-q_2 & q_2 & \\ & & \ddots & \ddots & \ddots \end{pmatrix}. \tag{2.24}$$

It should be noted that the polynomials $(Q_n^d)_n$ are not in general birth–death polynomials. Although the sum of every row of P^d is 1 and the off-diagonal entries are positive and less than 1, in general the diagonal entries need not be nonnegative. $(Q_n^d)_n$ constitutes a sequence of birth–death polynomials if and only if $p_n + q_{n+1} \leq 1$ for all $n \geq 0$. If that is the case, we call it the *dual birth–death chain* and denote it by $\{X_n^d, n = 0, 1, \ldots\}$, and observe that the matrix P^d is now stochastic ($q_0^d = 0$). We then have $p_n^d = q_n, q_{n+1}^d = p_n$, while the potential coefficients are given by

$$\pi_0^d = 1, \quad \pi_n^d = \frac{q_0 \cdots q_{n-1}}{p_0 \cdots p_{n-1}} = \frac{q_0}{p_{n-1}\pi_{n-1}} = \frac{q_0}{q_n \pi_n}, \quad n \geq 1.$$

We also have $(p_n^d \pi_n^d)^{-1} = \pi_n / q_0$. Therefore the character of the series $\sum_{n=0}^{\infty} (p_n \pi_n)^{-1}$ is equivalent to the character of the series $\sum_{n=0}^{\infty} \pi_n^d$, while the character of the series $\sum_{n=0}^{\infty} (p_n^d \pi_n^d)^{-1}$ is equivalent to the character of the series $\sum_{n=0}^{\infty} \pi_n$. From here we get the name dual polynomials.

Finally, dual polynomials (with or without being birth–death polynomials) form an orthogonal system, as we will see in the following result:

Theorem 2.10 *Let $q_0 > 0$. Then there exists a unique probability measure ψ^d supported on $[-1, 1]$ such that*

$$\int_{-1}^{1} Q_n^d(x) Q_m^d(x)\, d\psi^d(x) = \frac{\delta_{nm}}{\pi_n^d}. \tag{2.25}$$

If ψ^d is such measure then $d\psi(x) = (1-x)\, d\psi^d(x)/q_0$ is the spectral measure associated with the original moment problem for which

$$q_0 \int_{-1}^{1} \frac{d\psi(x)}{1-x} \leq 1. \tag{2.26}$$

Conversely, if ψ is the spectral measure of the original moment problem such that (2.26) *is valid, then the probability measure ψ^d is given by*

$$d\psi^d(x) = q_0 \frac{d\psi(x)}{1-x} \mathbf{1}_{[-1,1)} + \left(1 - q_0 \int_{-1}^{1} \frac{d\psi(x)}{1-x}\right) \delta_1(x), \tag{2.27}$$

where $\mathbf{1}_A$ is the indicator function, and it satisfies (2.25).

Proof Let ψ^d be the (unique) spectral measure associated with the transition probability matrix P^d in (2.24), and which makes the dual polynomials $(Q_n^d)_n$ orthogonal. Since $Q_n^d(x) - Q_{n+1}^d(x) = \frac{1-x}{q_0} Q_n(x)\pi_n$, we have, for $k = 0, 1, \ldots, n-1$, that

$$\pi_n \int_{-1}^{1} (1-x)^k Q_n(x) \frac{1-x}{q_0}\, d\psi^d(x) = \int_{-1}^{1} (1-x)^k [Q_n^d(x) - Q_{n+1}^d(x)]\, d\psi^d(x) = 0,$$

and for $k = n$ the integral is not zero by the orthogonality of $(Q_n^d)_n$. Therefore $d\psi(x) = (1-x)\, d\psi^d(x)/q_0$ and we have (2.26).

Alternatively, let ψ be the spectral measure associated with the original moment problem such that (2.26) is valid and let ψ^d be the measure given by (2.27). Then ψ^d is a positive probability measure and, since $Q_n^d(1) = 1$ for $n \geq 0$, we have

$$\int_{-1}^{1} Q_n^d(x) d\psi^d(x) = 1 - q_0 \int_{-1}^{1} \frac{d\psi(x)}{1-x} + \int_{-1}^{1} q_0 \left(1 + \frac{x-1}{q_0} \sum_{j=0}^{n-1} Q_j(x)\pi_j\right) \frac{d\psi(x)}{1-x}$$
$$= 1 - \sum_{j=0}^{n-1} \pi_j \int_{-1}^{1} Q_j(x) d\psi(x) = 0, \quad n \geq 1.$$

Moreover, if $1 \leq k < n$, we have, using (2.21), that

$$\int_{1}^{1} (1-x)^k Q_n^d(x)\, d\psi^d(x) = \int_{-1}^{1} q_0 (1-x)^{k-1} Q_n^d(x)\, d\psi(x)$$
$$= p_{n-1}\pi_{n-1} \int_{-1}^{1} (1-x)^{k-1} [Q_n(x) - Q_{n-1}(x)]\, d\psi(x) = 0.$$

□

Remark 2.11 The probability measure ψ^d in (2.27) is typically called a *Geronimus transformation* of the measure ψ, while the measure ψ is a *Christoffel transformation* of the measure ψ^d. This suggests that the Jacobi operators P and P^d are related by a *discrete Darboux transformation* centered at the point $x = 1$. Indeed, let U and L be the following upper and lower triangular matrices:

$$U = \begin{pmatrix} -1 & 1 & & & \\ 0 & -1/\pi_1 & 1/\pi_1 & & \\ & 0 & -1/\pi_2 & 1/\pi_2 & \\ & & \ddots & \ddots & \ddots \end{pmatrix},$$

$$L = \begin{pmatrix} q_0 & 0 & & & \\ -p_0\pi_0 & p_0\pi_0 & 0 & & \\ & -p_1\pi_1 & p_1\pi_1 & 0 & \\ & & \ddots & \ddots & \ddots \end{pmatrix}.$$

Then, a direct computation shows that $P = I + UL$ and $P^d = I + LU$. Therefore P^d is a discrete Darboux transformation of P. In the case where P^d is again a stochastic matrix, we will have produced, from a birth–death chain with positive probability of absorption, a new non-absorbing birth–death chain, which has a transition probability matrix given by P^d in (2.24). For more information about discrete Darboux transformations in the context of orthogonal polynomials see [67, 150]. For a probabilistic use of the discrete Darboux transformation, see [69, 70]. ◊

Case $q_0 = 0$: In this case we observe from the three-term recurrence relation (2.20) that the polynomial $Q_{n+1}(x) - Q_n(x)$ is always a multiple of the polynomial $x-1$. Therefore we define the *dual polynomials* as

$$Q_n^d(x) = \frac{p_n \pi_n}{x-1}[Q_{n+1}(x) - Q_n(x)], \quad n \geq 0. \tag{2.28}$$

We have now the relations

$$Q_n^d(x) = \sum_{k=0}^{n} \pi_k Q_k(x), \quad Q_n(x) = 1 + (x-1)\sum_{k=0}^{n-1} \frac{1}{p_k \pi_k} Q_k^d(x), \quad n \geq 0.$$

Therefore $Q_n(1) = 1, n \geq 0$ and $Q_n^d(1) = \sum_{k=0}^{n} \pi_k, n \geq 0$. We also have formula (2.22) setting $q_0 = 0$, as well as other similar relations to previous case. The dual polynomials $Q_n^d(x)$ in (2.28) satisfy the three-term recurrence relation

$$\begin{aligned} &Q_0^d(x) = 1, \quad xQ_0^d(x) = q_1 Q_1^d(x) + (1 - p_0 - q_1)Q_0^d(x), \\ &xQ_n^d(x) = q_{n+1}Q_{n+1}^d(x) + (1 - p_n - q_{n+1})Q_n^d(x) + p_n Q_{n-1}^d(x), \quad n \geq 1. \end{aligned}$$

Therefore the associated Jacobi matrix is given by

$$P^d = \begin{pmatrix} 1 - p_0 - q_1 & q_1 & & & \\ p_1 & 1 - p_1 - q_2 & q_2 & & \\ & p_2 & 1 - p_2 - q_3 & q_3 & \\ & & \ddots & \ddots & \ddots \end{pmatrix}.$$

Again, the dual polynomials $(Q_n^d)_n$ are not in general birth–death polynomials. They will be if and only if $p_n + q_{n+1} \leq 1$ for all $n \geq 0$. Observe now that the matrix P^d is not stochastic ($q_0^d > 0$). We then have $p_n^d = q_{n+1}, q_n^d = p_n$, while the potential coefficients are given by

$$\pi_0^d = 1, \quad \pi_n^d = \frac{q_1 \cdots q_n}{p_1 \cdots p_n} = \frac{p_0}{p_n \pi_n}, \quad n \geq 1.$$

We also have $(p_n^d \pi_n^d)^{-1} = \pi_{n+1}/p_0$ and the same interpretation as before in terms of the series $\sum_{n=0}^{\infty}(p_n\pi_n)^{-1}$ and $\sum_{n=0}^{\infty} \pi_n$. Finally, we also get the following result:

Theorem 2.12 *Let $q_0 = 0$. Then there exists a unique probability measure ψ^d supported on $[-1,1]$ such that*

$$\int_{-1}^{1} Q_n^d(x) Q_m^d(x)\, d\psi^d(x) = \frac{\delta_{nm}}{\pi_n^d}, \quad \text{and} \quad p_0 \int_{-1}^{1} \frac{d\psi(x)}{1-x} \leq 1.$$

If ψ^d is such a measure then

$$d\psi(x) = p_0 \frac{d\psi^d(x)}{1-x} \mathbf{1}_{[-1,1)} + \left(1 - p_0 \int_{-1}^{1} \frac{d\psi^d(x)}{1-x}\right) \delta_1(x)$$

is the spectral measure associated with the original moment problem. Conversely, if ψ is the spectral measure of the original moment problem, then $d\psi^d(x) = (1-x)\,d\psi(x)/p_0$ defines a probability measure which satisfies the above conditions.

Proof The proof follows the same steps as the proof of Theorem 2.10. □

Remark 2.13 As before, if we call U and L the upper and lower triangular matrices

$$U = \begin{pmatrix} -p_0\pi_0 & p_0\pi_0 & & & \\ 0 & -p_1\pi_1 & p_1\pi_1 & & \\ & 0 & -p_2\pi_2 & p_2\pi_2 & \\ & & \ddots & \ddots & \ddots \end{pmatrix},$$

$$L = \begin{pmatrix} 1 & 0 & & & \\ -1/\pi_1 & 1/\pi_1 & 0 & & \\ & -1/\pi_2 & 1/\pi_2 & 0 & \\ & & \ddots & \ddots & \ddots \end{pmatrix},$$

then, a direct computation shows that $P = I + LU$ and $P^d = I + UL$. Therefore P^d is a Darboux transformation of P but taking the LU decomposition instead of the UL decomposition as for the case where $q_0 > 0$. In this case, the original birth–death chain P is stochastic and the new birth–death chain P^d will have a positive probability of absorption at -1 (if it is a birth–death chain). ◊

2.3.2 *k*th Associated Polynomials

For any $k \geq 0$ consider the *kth discrete-time birth–death chain* $\{X_n^{(k)}, n = 0, 1, \ldots\}$ which has a transition probability matrix defined from the original transition probability matrix P by eliminating the first $k+1$ rows and columns, i.e.

$$P = \left(\begin{array}{cccc|cccc} r_0 & p_0 & & & & & & \\ q_1 & r_1 & p_1 & & & & & \\ & \ddots & \ddots & \ddots & & & & \\ & & q_k & r_k & p_k & & & \\ \hline & & & q_{k+1} & r_{k+1} & p_{k+1} & & \\ & & & & q_{k+2} & r_{k+2} & p_{k+2} & \\ & & & & & \ddots & \ddots & \ddots \end{array}\right).$$

This new birth–death chain has an absorption probability at the first state of probability q_{k+1}. In particular for the 0th birth–death chain, the probability of absorption is given by q_1. Define the sequence of polynomials $(Q_n^{(k)})_n$ using the following recurrence relation:

$$xQ_n^{(k)}(x) = \delta_{n,k} + p_n Q_{n+1}^{(k)}(x) + r_n Q_n^{(k)}(x) + q_n Q_{n-1}^{(k)}(x). \tag{2.29}$$

The polynomials $(Q_n^{(k)})_n$ have degrees $n-k-1$ if $n > k$ and satisfy $Q_n^{(k)}(x) = 0$ for $0 \leq n \leq k$. Moreover

$$Q_{k+1}^{(k)}(x) = -\frac{1}{p_k}. \tag{2.30}$$

These polynomials are usually called *kth associated polynomials*. It is easy to check that they can be written in the following way:

$$Q_n^{(k)}(x) = -\pi_k \int_{-1}^{1} Q_k(y) \frac{Q_n(y) - Q_n(x)}{y - x} d\psi(y). \tag{2.31}$$

Indeed, for $n = 0, 1, \ldots, k$, we have that $\frac{Q_n(y)-Q_n(x)}{y-x}$ is a polynomial in y of degree $n-1$. Therefore, by orthogonality, it is zero. For $n \geq k+1$, we just plug (2.31) into (2.29) and use the three-term recurrence relation for the sequence of polynomials $(Q_n)_n$.

An alternative way of defining these polynomials (as we can see for example in [31]) is to set the degrees back in such a way that the initial polynomial is constant. If we call

$$R_n^{(k)}(x) = -Q_{n+k+1}^{(k)}(x), \tag{2.32}$$

we have that the polynomials $R_n^{(k)}(x)$ satisfy the three-term recurrence relation

$$R_0^{(k)}(x) = 1/p_k, \quad R_{-1}^{(k)}(x) = 0,$$
$$xR_n^{(k)}(x) = p_{n+k+1} R_{n+1}^{(k)}(x) + r_{n+k+1} R_n^{(k)}(x) + q_{n+k+1} R_{n-1}^{(k)}(x), \quad n \geq 1.$$

For every $k \geq 0$ we have a new birth–death chain and by Theorem 2.5 there exists a unique probability measure $\psi^{(k)}$ associated with the kth birth–death chain $\{X_n^{(k)}, n = 0, 1, \ldots\}$. The goal now is to relate the Stieltjes transforms of these measures $B(z; \psi^{(k)})$ to the original one $B(z; \psi)$. Let us begin first with the relation between ψ and $\psi^{(0)}$. We will give a probabilistic proof of Theorem 1.5, or more generally formula (1.22) applied to the one-step transition probability matrix P in (2.12). Formula (2.33) below can be found in [87, p. 72].

Theorem 2.14 *Let ψ and $\psi^{(0)}$ be the spectral measures associated with the original and the 0th birth–death chains, respectively. Then we have*

$$\int_{-1}^{1} \frac{d\psi(x)}{1 - xs} = \frac{1}{1 - r_0 s - p_0 q_1 s^2 \displaystyle\int_{-1}^{1} \frac{d\psi^{(0)}(x)}{1 - xs}}, \quad |s| < 1. \tag{2.33}$$

Proof Let $P_{ij}(s)$ and $F_{ij}(s)$ be the generating functions defined by (2.6) associated with the transition probabilities $P_{ij}^{(n)}$ and the first passage time probabilities $f_{ij}^{(n)}$, respectively. Then, from the Karlin–McGregor integral representation (2.17), we obtain

$$\begin{aligned}P_{ij}(s) &= \sum_{n=0}^{\infty} P_{ij}^{(n)} s^n = \sum_{n=0}^{\infty} \pi_j \int_{-1}^{1} (xs)^n Q_i(x) Q_j(x)\, d\psi(x) \\ &= \pi_j \int_{-1}^{1} \sum_{n=0}^{\infty} (xs)^n Q_i(x) Q_j(x)\, d\psi(x) = \pi_j \int_{-1}^{1} \frac{Q_i(x)Q_j(x)}{1-xs}\, d\psi(x). \qquad (2.34)\end{aligned}$$

The change of integral and sum is possible since $|xs| < 1$. Therefore, from (2.8), we get

$$F_{ii}(s) = 1 - \frac{1}{\pi_i \displaystyle\int_{-1}^{1} \frac{Q_i^2(x)}{1-xs}\, d\psi(x)}, \qquad F_{ij}(s) = \frac{\displaystyle\int_{-1}^{1} \frac{Q_i(x)Q_j(x)}{1-xs}\, d\psi(x)}{\displaystyle\int_{-1}^{1} \frac{Q_j^2(x)}{1-xs}\, d\psi(x)}, \quad i \neq j. \qquad (2.35)$$

For a birth–death chain with an absorbing state -1 with probability $q_0 > 0$, we have, using (2.5), the following representation:

$$\begin{aligned}F_{i,-1}(s) &= \sum_{n=1}^{\infty} f_{i,-1}^{(n)} s^n = q_0 \sum_{n=1}^{\infty} P_{i,0}^{(n-1)} s^n = q_0 \sum_{n=1}^{\infty} \pi_0 \int_{-1}^{1} x^{n-1} Q_i(x)\, d\psi(x) s^n \\ &= q_0 s \int_{-1}^{1} \sum_{n=1}^{\infty} (xs)^{n-1} Q_i(x)\, d\psi(x) = q_0 s \int_{-1}^{1} \frac{Q_i(x)}{1-xs}\, d\psi(x). \qquad (2.36)\end{aligned}$$

Applied to the 0th birth–death chain (with absorption probability $q_1 > 0$), we have, using (2.36), that

$$F_{10}(s) = q_1 s \int_{-1}^{1} \frac{d\psi^{(0)}(x)}{1-xs}.$$

Using now (2.7) for $i = 1, j = 0$, (2.34) and the definition of $Q_1(x) = (x - r_0)/p_0$, we have that

$$F_{10}(s) = \frac{P_{10}(s)}{P_{00}(s)} = \frac{\displaystyle\int_{-1}^{1} \frac{Q_1(x)\, d\psi(x)}{1-xs}}{\displaystyle\int_{-1}^{1} \frac{d\psi(x)}{1-xs}} = \frac{1}{p_0} \left[\frac{\displaystyle\int_{-1}^{1} \frac{x\, d\psi(x)}{1-xs}}{\displaystyle\int_{-1}^{1} \frac{d\psi(x)}{1-xs}} - r_0 \right]$$

$$= \frac{1}{p_0}\left[\frac{\dfrac{1}{s}\displaystyle\int_{-1}^{1}\left[\frac{1}{1-xs}-1\right]d\psi(x)}{\displaystyle\int_{-1}^{1}\frac{d\psi(x)}{1-xs}} - r_0\right]$$

$$= \frac{1}{p_0}\left[\frac{1}{s}\left(1-\frac{1}{\displaystyle\int_{-1}^{1}\frac{d\psi(x)}{1-xs}}\right) - r_0\right].$$

Thus

$$\frac{1}{p_0}\left[\frac{1}{s}\left(1-\frac{1}{\displaystyle\int_{-1}^{1}\frac{d\psi(x)}{1-xs}}\right) - r_0\right] = q_1 s\int_{-1}^{1}\frac{d\psi^{(0)}(x)}{1-xs},$$

and reordering we get (2.33). □

Using (1.7) we can write formula (2.33) in terms of the Stieltjes transform $B(z;\psi)=\int_{-1}^{1}\frac{d\psi(x)}{x-z}$ for $z=1/s$ as

$$B(z;\psi) = -\frac{1}{z-r_0+p_0q_1B(z;\psi^{(0)})}. \tag{2.37}$$

Once we know one of these Stieltjes transformations, the Perron–Stieltjes inversion formula, given by (1.8), can be used to compute the other measure. Observe that we can iterate this formula as many times as we want. Let us now see how to obtain a relation between the Stieltjes transforms of ψ and $\psi^{(k)}, k\geq 0$, in terms of the original polynomials $(Q_n)_n$ and the 0th associated polynomials. The same approach for measures supported on $[0,\infty)$ can be found in [86] (see also Section 3.3.2).

Observe that for $k\geq 0$ the formula (2.37) can be extended to ($\psi=\psi^{(-1)}$)

$$B(z;\psi^{(k-1)}) = -\frac{1}{z-r_k+p_kq_{k+1}B(z;\psi^{(k)})} = -\frac{\dfrac{1}{p_k}}{\dfrac{z-r_k}{p_k}+q_{k+1}B(z;\psi^{(k)})}. \tag{2.38}$$

Iterating this formula we get that $B(z;\psi)$ can be written as

$$B(z;\psi) = \frac{\alpha_k B(z;\psi^{(k-1)})+\beta_k}{\gamma_k B(z;\psi^{(k-1)})+\delta_k}, \quad k\geq 1, \tag{2.39}$$

where $\alpha_k,\beta_k,\gamma_k,\delta_k$ are functions in z that do not necessarily have to be unique. Let us now try to write $B(z;\psi)$ in terms of $B(z;\psi^{(k-1)})$ using the polynomials $Q_k(x)$ and the 0th associated polynomials $Q_k^{(0)}(x)$. For $k=0$, we have that $\alpha_0=\delta_0=1$,

$\beta_0 = \gamma_0 = 0$, while for $k = 1$, from (2.37) and (2.38), a solution is $\alpha_1 = 0, \beta_1 = -1/p_0 = Q_1^{(0)}(z), \gamma_1 = q_1, \delta_1 = (z - r_0)/p_0 = Q_1(z)$. Thus

$$B(z;\psi) = \frac{Q_1^{(0)}}{q_1 B(z;\psi^{(0)}) + Q_1(z)}.$$

Substituting (2.38) in (2.39) and comparing the coefficients of the formula (2.39) for $k+1$, we get the following relations:

$$\alpha_{k+1} = q_{k+1}\beta_k, \quad p_k\beta_{k+1} = -\alpha_k + (z - r_k)\beta_k,$$
$$\gamma_{k+1} = q_{k+1}\delta_k, \quad p_k\delta_{k+1} = -\gamma_k + (z - r_k)\delta_k.$$

Substituting the first in the second we have that β_k satisfies the three-term recurrence relation

$$p_k\beta_{k+1} + r_k\beta_k + q_k\beta_{k-1} = z\beta_k, \quad \beta_0 = 0, \quad \beta_1 = -1/p_0 = Q_1^{(0)}(z).$$

The solution for β_k is given by the 0th associated polynomials $Q_k^{(0)}(x)$ defined by (2.31). Therefore $\beta_k = Q_k^{(0)}(z)$. Hence $\alpha_k = q_k Q_{k-1}^{(0)}(z)$. In the same way, the third and fourth equations can be solved taking into account the initial conditions, resulting in $\gamma_k = q_k Q_{k-1}(z)$ and $\delta_k = Q_k(z)$. Consequently (2.39) can be rewritten as

$$B(z;\psi) = \frac{q_k Q_{k-1}^{(0)}(z) B(z;\psi^{(k-1)}) + Q_k^{(0)}(z)}{q_k Q_{k-1}(z) B(z;\psi^{(k-1)}) + Q_k(z)}, \quad k \geq 1. \tag{2.40}$$

In this way we can iterate this formula until we find the Stieltjes transform of the original spectral measure.

Another use of the kth associated polynomials is that we can write the generating functions $P_{ij}(s)$ in terms only of these polynomials and the generating function $P_{00}(s)$. Indeed, let $P(s)$ be the matrix of generating functions given by (2.6). It is easy to see from the definition that

$$I + (sP)P(s) = P(s).$$

Entry by entry the relation above can be written as

$$\delta_{ij} + s[p_i P_{i+1,j}(s) + r_i P_{ij}(s) + q_i P_{i-1,j}(s)] = P_{ij}(s).$$

A general solution of the previous equation is given by the sum of a particular solution of $I + (sP)P(s) = P(s)$ and the general solution of $(sP)P(s) = P(s)$. On the one hand, a particular solution of $I + (sP)P(s) = P(s)$ is given by

$$P_{ij}(s) = s^{-1} Q_i^{(j)}(s^{-1}),$$

as can be easily seen directly from the recurrence relation (2.29). On the other hand, the general solution of $(sP)P(s) = P(s)$ is given by

$$P_{ij}(s) = g_j(s) Q_i(s^{-1}),$$

where $g_j(s)$ is any function (to be determined below). Therefore the general solution of $I+(sP)P(s)=P(s)$ is given by

$$P_{ij}(s)=s^{-1}Q_i^{(j)}(s^{-1})+g_j(s)Q_i(s^{-1}). \tag{2.41}$$

Now, since $Q_0^{(j)}=0$ and $Q_0=1$, the previous formula at $i=0$ gives an expression for $g_j(s)$, which is equal to $P_{0,j}(s)$. Using the symmetry $\pi_i P_{ij}(s)=\pi_j P_{ji}(s)$ (as a consequence of $\pi P^n=\pi$ or the Karlin–McGregor representation) in formula (2.41) for $i=0$ we arrive at

$$g_j(s)=P_{0,j}(s)=\pi_j P_{j,0}(s)=\pi_j\left(s^{-1}Q_j^{(0)}(s^{-1})+Q_j(s^{-1})P_{00}(s)\right).$$

Therefore

$$P_{ij}(s)=s^{-1}Q_i^{(j)}(s^{-1})+\pi_j Q_i(s^{-1})\left(s^{-1}Q_j^{(0)}(s^{-1})+Q_j(s^{-1})P_{00}(s)\right). \tag{2.42}$$

In particular, for $i\le j$, since $Q_i^{(j)}(y)=0$, we have that

$$P_{ij}(s)=\pi_j Q_i(s^{-1})\left(s^{-1}Q_j^{(0)}(s^{-1})+Q_j(s^{-1})P_{00}(s)\right),\quad i\le j. \tag{2.43}$$

Formula (2.42) gives other interesting relations. Using again the symmetry $\pi_i P_{ij}(s)=\pi_j P_{ji}(s)$ we get

$$\pi_i Q_i^{(j)}(s^{-1})+\pi_i\pi_j Q_i(s^{-1})Q_j^{(0)}(s^{-1})=\pi_j Q_j^{(i)}(s^{-1})+\pi_i\pi_j Q_j(s^{-1})Q_i^{(0)}(s^{-1}).$$

For $j>i$, for instance $j=n+i$, we have that $Q_i^{(n+i)}=0$, and therefore

$$\pi_i Q_i(s^{-1})Q_{n+i}^{(0)}(s^{-1})=Q_{n+i}^{(i)}(s^{-1})+\pi_i Q_{n+i}(s^{-1})Q_i^{(0)}(s^{-1}).$$

In particular, for $n=1$ and using (2.30), we have that

$$Q_{i+1}(s^{-1})Q_i^{(0)}(s^{-1})-Q_i(s^{-1})Q_{i+1}^{(0)}(s^{-1})=-\frac{Q_{i+1}^{(i)}(s^{-1})}{\pi_i}=\frac{1}{p_i\pi_i}.$$

This expression can be rewritten as

$$\frac{Q_{i+1}^{(0)}(s^{-1})}{Q_{i+1}(s^{-1})}=\frac{Q_i^{(0)}(s^{-1})}{Q_i(s^{-1})}-\frac{1}{p_i\pi_i Q_{i+1}(s^{-1})Q_i(s^{-1})},$$

so

$$\frac{Q_{i+1}^{(0)}(s^{-1})}{Q_{i+1}(s^{-1})}=-\sum_{k=0}^{i}\frac{1}{p_k\pi_k Q_{k+1}(s^{-1})Q_k(s^{-1})}.$$

Using Markov's theorem 1.11, we have

$$\lim_{n\to\infty}\frac{s^{-1}Q_{n+1}^{(0)}(s^{-1})}{Q_{n+1}(s^{-1})}=-\frac{1}{s}\int_{-1}^{1}\frac{d\psi(x)}{s^{-1}-x}=-\int_{-1}^{1}\frac{d\psi(x)}{1-xs},$$

and thus

$$\int_{-1}^{1}\frac{d\psi(x)}{1-xs}=\sum_{k=0}^{\infty}\frac{1}{p_k\pi_kQ_{k+1}(s^{-1})Q_k(s^{-1})}.$$

This last formula can be written, using the definition of $Q_n^d(x)$ in (2.21), as ($q_0>0$)

$$\int_{-1}^{1}\frac{d\psi(x)}{1-xs}=\frac{1}{q_0}\sum_{k=0}^{\infty}\frac{1}{Q_k^d(s^{-1})}\left[\frac{1}{Q_k(s^{-1})}-\frac{1}{Q_{k+1}(s^{-1})}\right].$$

Now, as $s\uparrow 1$ and using that $Q_k^d(1)=1$, we have the formula

$$\int_{-1}^{1}\frac{d\psi(x)}{1-x}=\frac{1}{q_0}\left(1-\lim_{n\to\infty}\frac{1}{Q_n(1)}\right).$$

Using the definition of $Q_n(1)$ in (2.23) we have that

$$\int_{-1}^{1}\frac{d\psi(x)}{1-x}=\frac{1}{q_0}\left(1-\frac{1}{1+q_0\sum_{n=0}^{\infty}\frac{1}{p_n\pi_n}}\right)=\frac{\sum_{n=0}^{\infty}\frac{1}{p_n\pi_n}}{1+q_0\sum_{n=0}^{\infty}\frac{1}{p_n\pi_n}},\tag{2.44}$$

which is valid even if $q_0=0$.

Remark 2.15 An application of the previous calculations is the computation of the first passage time probabilities to a certain state j starting from i ($i<j$). In that case we have to analyze the functions $F_{ij}(s)$. From (2.43) and (2.7), we obtain

$$F_{ij}(s)=\frac{Q_i(s^{-1})}{Q_j(s^{-1})},\quad i<j.$$

For the special case $j=i+1$ and using the three-term recurrence relation for the polynomials $Q_j(x)$ in (2.15), we have that

$$F_{i,i+1}(s)=\frac{p_is}{1-r_is-q_isF_{i-1,i}(s)},\quad i\geq 1,\quad F_{0,1}(s)=\frac{p_0s}{1-r_0s}.$$

Therefore we have a recurrent way of computing $F_{i,i+1}(s)$ for all i. An interesting result in this direction (see [141]) is that we can also find some results about the behavior of the random variable T_{ij}, given by the number of transitions of the birth–death chain to move from i to j. We have that, for $i>0$, the distribution of $T_{i-1,i}$ is given by a mixture of i different geometric distributions, while the distribution of $T_{0,i}$ is given by a convolution of i different geometric distributions (see [141] for more details). For more information about other formulas related to $P_{ij}(s)$ and $F_{ij}(s)$ see [30]. ◊

Remark 2.16 The polynomials $R_n^{(k)}(y)$ in (2.32) and their corresponding associated measures $\psi^{(k)}$ can be used to compute the first passage time probabilities $f_{ij}^{(n)}$ in (2.3). Indeed, we have (see [31, Theorem 1.1 and Corollary 1.2]) that

$$f_{i,0}^{(n)} = p_0 q_1 \int_{-1}^{1} y^{n-1} R_{i-1}^{(0)}(y) d\psi^{(0)}(y), \quad i,n \geq 1,$$

$$f_{0,0}^{(n)} = p_0 q_1 \int_{-1}^{1} y^{n-2} d\psi^{(0)}(y), \quad n \geq 2,$$

and

$$f_{i,j}^{(n)} = p_j q_{j+1} \int_{-1}^{1} y^{n-1} R_{i-j-1}^{(j)}(y) d\psi^{(j)}(y), \quad n \geq 1, \quad j \geq 0, \quad i > j.$$

◊

2.3.3 The Derman–Vere–Jones Transformation

Consider a birth–death chain $\{X_n, n=0,1,\ldots\}$ with transition probabilities p_j, r_j, q_j and associated spectral measure ψ. Let us denote

$$\xi = \inf \operatorname{supp}(\psi), \quad \eta = \sup \operatorname{supp}(\psi).$$

We have, following the proof of Theorem 2.7, that $-1 \leq \xi < \eta \leq 1$ and $\xi + \eta \geq 0$. The last assertion is a consequence of the inequalities (see [47, Lemma 2.3])

$$\inf_j \{r_j + r_{j+1}\} \leq \xi + \eta \leq \sup_j \{r_j + r_{j+1}\}. \tag{2.45}$$

The Derman–Vere–Jones transformation consists of mapping the original birth–death chain into a new one $\{\tilde{X}_n, n = 0,1,\ldots\}$ with transition probabilities $\tilde{p}_j, \tilde{r}_j, \tilde{q}_j$ and spectral measure $\tilde{\psi}$ such that $\tilde{q}_0 = 0$ and $\tilde{\eta} = 1$. It was introduced by Derman in [28] and generalized by Vere-Jones in [143]. The new transition probabilities are given by

$$\tilde{p}_j = \frac{Q_{j+1}(\eta)}{\eta Q_j(\eta)} p_j,$$

$$\tilde{r}_j = \frac{r_j}{\eta},$$

$$\tilde{q}_j = \frac{Q_{j-1}(\eta)}{\eta Q_j(\eta)} q_j.$$

Since $Q_j(\eta) > 0$ for all $j \geq 0$ (the zeros of $Q_j(x)$ must be inside the interval (ξ, η)) then $\tilde{p}_j, \tilde{q}_{j+1} > 0, j \geq 0$. Using the three-term recurrence relation (2.15) for the polynomials $Q_j(x)$, we have

$$\tilde{q}_0 = 1 - \tilde{p}_0 - \tilde{r}_0 = 1 - \frac{Q_1(\eta)}{\eta Q_0(\eta)} p_0 - \frac{r_0}{\eta} = 1 - \frac{1}{\eta Q_0(\eta)} (p_0 Q_1(\eta) + r_0 Q_0(\eta)) = 0,$$

and

$$\tilde{p}_j + \tilde{r}_j + \tilde{q}_j = \frac{1}{\eta Q_j(\eta)} \left(p_j Q_{j+1}(\eta) + r_j Q_j(\eta) + q_j Q_{j-1}(\eta) \right) = 1.$$

The corresponding sequence of birth–death polynomials and potential coefficients is given by

$$\tilde{Q}_j(x) = \frac{Q_j(x\eta)}{Q_j(\eta)}, \quad \tilde{\pi}_j = \pi_j Q_j^2(\eta),$$

while the spectral measure $\tilde{\psi}$ is defined by

$$\tilde{\psi}([-1,x]) = \psi([-\eta, x\eta]), \quad -1 \leq x \leq 1. \tag{2.46}$$

It is evident that $\tilde{\eta} = \sup \operatorname{supp}(\tilde{\psi}) = 1$, while $\tilde{\xi} = \inf \operatorname{supp}(\tilde{\psi}) = \xi/\eta \geq -1$. We also have the following relation of the n-step transition probabilities $\tilde{P}_{ij}^{(n)}$ in terms of $P_{ij}^{(n)}$:

$$\begin{aligned} \tilde{P}_{ij}^{(n)} &= \tilde{\pi}_j \int_{-1}^{1} x^n \tilde{Q}_i(x) \tilde{Q}_j(x) d\tilde{\psi}(x) \\ &= \frac{\pi_j Q_j(\eta)}{\eta^n Q_i(\eta)} \int_{-1}^{1} (x\eta)^n Q_i(x\eta) Q_j(x\eta)\, d\psi(x\eta) = \frac{Q_j(\eta)}{\eta^n Q_i(\eta)} P_{ij}^{(n)}. \end{aligned}$$

Therefore

$$P_{ij}^{(n)} = \frac{\eta^n Q_i(\eta)}{Q_j(\eta)} \tilde{P}_{ij}^{(n)}, \quad i,j \geq 0.$$

2.4 Examples

In this section we will study a collection of instructive and useful examples. We will begin with some examples where the state space is finite, i.e. $\mathcal{S} = \{0, 1, \ldots, N\}$. In that case we can use a standard eigenvalue decomposition as in (2.11). We will focus on examples that are related somehow with classical orthogonal polynomials. Second, we will study some examples of birth–death chains on $\mathcal{S} = \mathbb{N}_0$ where we do not know the spectral measure, but it can be computed using formula (2.37), which relates the Stieltjes transforms of the spectral measures of the original birth–death chain and the 0th associated birth–death chain or more generally formula (2.40). Finally, we will study an example related to the Jacobi polynomials, where the spectral measure is known and we can generate a birth–death chain for it on the state space $\mathcal{S} = \mathbb{N}_0$.

2.4.1 Examples with Finite State Space

Example 2.17 *Gambler's ruin with two absorbing barriers.* Consider the birth–death chain $\{X_n, n=0,1,\ldots\}$ given by gambler's ruin with state space $\{0,1,\ldots,N\}$ and transition probability matrix

$$P = \begin{pmatrix} 0 & p & & & & \\ q & 0 & p & & & \\ & \ddots & \ddots & \ddots & & \\ & & & q & 0 & p \\ & & & & q & 0 \end{pmatrix}.$$

A diagram of the one-step transitions is

$$0 \underset{q}{\overset{p}{\rightleftarrows}} 1 \underset{q}{\overset{p}{\rightleftarrows}} 2 \underset{q}{\overset{p}{\rightleftarrows}} \cdots \underset{q}{\overset{p}{\rightleftarrows}} N$$

The potential coefficients are given by

$$\pi_j = \left(\frac{p}{q}\right)^j, \quad j = 0,1,\ldots,N.$$

Then the matrix P is weakly symmetric. Therefore all the $N+1$ eigenvalues of P are located inside the interval $[-1,1]$. The three-term recurrence relation is given by

$$Q_{-1} = 0, \quad Q_0 = 1, \quad Q_1(x) = x/p,$$
$$xQ_n(x) = pQ_{n+1}(x) + qQ_{n-1}(x), \quad n = 1,\ldots,N-1.$$

If we make the change of variables $y = \dfrac{x}{2\sqrt{pq}}$ and call $Q_n(x) = (q/p)^{n/2}U_n(y)$, these new polynomials satisfy

$$U_0 = 1, \quad U_1(y) = 2y, \quad yU_n(y) = \frac{1}{2}U_{n+1}(y) + \frac{1}{2}U_{n-1}(y), \quad n = 1,\ldots,N-1,$$

which can be identified with the Chebychev polynomials of the second kind $(U_n)_n$ (see (1.87)). Therefore

$$Q_n(x) = \left(\frac{q}{p}\right)^{n/2} U_n\left(\frac{x}{2\sqrt{pq}}\right), \quad n = 0,1,\ldots,N.$$

As mentioned at the beginning of Section 2.2, $R_{N+1}(x) = xQ_N(x) - qQ_{N-1}(x)$ is a polynomial of degree $N+1$ whose zeros coincide with the eigenvalues of the matrix P. It is possible to see, using the three-term recurrence relation for the Chebychev polynomials of the second kind, that

$$R_{N+1}(x) = p\left(\frac{q}{p}\right)^{\frac{N+1}{2}} U_{N+1}\left(\frac{x}{2\sqrt{pq}}\right).$$

Therefore these eigenvalues are given by these zeros, which can be computed using (1.89). Indeed,

$$x_{N+1,k} = 2\sqrt{pq}\cos\left(\pi\frac{k+1}{N+2}\right), \quad k = 0,1,\dots,N.$$

Observe that the eigenvalues are located inside the interval $[-1,1]$. The values of the jumps at these nodes are related to the Christoffel numbers in (1.31) (see also (1.36)). In general, these numbers are difficult to compute, but in the case of the Chebychev polynomials of the second kind they admit an explicit formula (see [142, pp. 352–353]). Indeed,

$$\begin{aligned}\psi\left(\{x_{N+1,k}\}\right) &= \frac{2}{N+2}\sin^2\left(\pi\frac{k+1}{N+2}\right)\\ &= \frac{1}{2pq(N+2)}(4pq - x_{N+1,k}^2), \quad k = 0,1,\dots,N.\end{aligned}$$

Therefore the spectral measure is given by

$$\psi(x) = \sum_{k=0}^{N}\psi\left(\{x_{N+1,k}\}\right)\delta_{x_{N+1,k}}(x),$$

and the Karlin–McGregor representation is given by (see (2.11) and (2.17))

$$\begin{aligned}P_{ij}^{(n)} &= \frac{2\pi_j}{N+2}\sum_{k=0}^{N}x_{N+1,k}^n Q_i(x_{N+1,k})Q_j(x_{N+1,k})\sin^2\left(\pi\frac{k+1}{N+2}\right)\\ &= \frac{2\left(\sqrt{\frac{p}{q}}\right)^{j-i}}{N+2}\sum_{k=0}^{N}(2\sqrt{pq})^n\cos^n\left(\pi\frac{k+1}{N+2}\right)U_i\left(\cos\left(\pi\frac{k+1}{N+2}\right)\right)\\ &\quad\times U_j\left(\cos\left(\pi\frac{k+1}{N+2}\right)\right)\sin^2\left(\pi\frac{k+1}{N+2}\right)\\ &= \frac{2(2\sqrt{pq})^n}{(N+2)\left(\sqrt{\frac{p}{q}}\right)^{i-j}}\sum_{k=0}^{N}\cos^n\left(\pi\frac{k+1}{N+2}\right)\sin\left(\pi\frac{(i+1)(k+1)}{N+2}\right)\\ &\quad\times\sin\left(\pi\frac{(j+1)(k+1)}{N+2}\right).\end{aligned}$$

In the last part we used the relation of the Chebychev polynomials of the second kind with the cosine function given by (1.88). ◊

Example 2.18 *Gambler's ruin with two reflecting barriers.* Consider the birth–death chain $\{X_n, n=0,1,\dots\}$ given by gambler's ruin with state space $\mathcal{S} = \{0,1,\dots,N\}$ and transition probability matrix

$$P = \begin{pmatrix} 0 & 1 & & & & \\ q & 0 & p & & & \\ & \ddots & \ddots & \ddots & & \\ & & q & 0 & p \\ & & & 1 & 0 \end{pmatrix}.$$

A diagram of the one-step transitions is

$$\text{(0)} \underset{q}{\overset{1}{\rightleftarrows}} \text{(1)} \underset{q}{\overset{p}{\rightleftarrows}} \text{(2)} \underset{q}{\overset{p}{\rightleftarrows}} \cdots \underset{1}{\overset{p}{\rightleftarrows}} \text{(N)}$$

Observe now that the states 0 and N are reflecting, so the process will never stop. The potential coefficients are given by

$$\pi_0 = 1, \quad \pi_j = \frac{p^{j-1}}{q^j}, j = 1, \ldots, N-1, \quad \pi_N = \left(\frac{p}{q}\right)^{N-1}.$$

Then the matrix P is weakly symmetric. Therefore all the $N+1$ eigenvalues of P are located inside the interval $[-1, 1]$. The three-term recurrence relation is given by

$$Q_{-1} = 0, \quad Q_0 = 1, \quad Q_1(x) = x,$$
$$xQ_n(x) = pQ_{n+1}(x) + qQ_{n-1}(x), \quad n = 1, \ldots, N-1.$$

It is possible to see (see [101, p. 14]) that the polynomials are given by

$$Q_n(x) = \left(\frac{q}{p}\right)^{n/2} \left[(2-2p)T_n\left(\frac{x}{2\sqrt{pq}}\right) + (2p-1)U_n\left(\frac{x}{2\sqrt{pq}}\right)\right], \; n = 0, 1, \ldots, N,$$

where $(T_n)_n$ and $(U_n)_n$ are the Chebychev polynomials of the first and second kind, respectively (see (1.86) and (1.88)). The polynomial

$$R_{N+1}(x) = xQ_N(x) - Q_{N-1}(x),$$

is a polynomial of degree $N+1$ whose zeros coincide with the eigenvalues of the matrix P. However, in this case, it is not easy to get an explicit expression of these zeros (and the corresponding jumps at the nodes). Nevertheless, there is a special case where everything simplifies considerably. That case is given by $p = q = 1/2$, where the term of U_n in Q_n vanishes, and therefore $Q_n(x) = T_n(x)$. The polynomials $R_{N+1}(x)$ are then given by

$$R_{N+1}(x) = xT_N(x) - T_{N-1}(x).$$

Using the first relation in (1.90) and the three-term recurrence relation for the Chebychev polynomials of the first kind (see (1.85)), we have that

$$R_{N+1}(x) = xT_N(x) - 2xT_N(x) + T_{N+1}(x)$$
$$= -xT_N(x) + T_{N+1}(x) = -(1-x^2)U_{N-1}(x).$$

Therefore the zeros of $R_{N+1}(x)$ are $x = \pm 1$ and the rest can be computed using (1.89). Indeed,

$$x_{N+1,0} = 1, \quad x_{N+1,k} = \cos\left(\pi \frac{k}{N}\right), \quad k = 1, \ldots, N-1, \quad x_{N+1,N} = -1.$$

Observe that the eigenvalues are located inside the interval $[-1,1]$. The values of the jumps at these nodes are related to the Christoffel numbers in (1.31) (see also (1.36)). In general, these numbers are difficult to compute, but in the case of Chebychev polynomials of the first kind they admit an explicit formula (see [142, pp. 352–353]). Indeed,

$$\psi(\{1\}) = \frac{1}{2N}, \quad \psi\left(\{x_{N+1,k}\}\right) = \frac{1}{N}, \quad k = 1, \ldots, N-1, \quad \psi(\{-1\}) = \frac{1}{2N}.$$

Therefore the spectral measure is given by

$$\psi(x) = \frac{1}{N}\sum_{k=1}^{N-1} \delta_{x_{N+1,k}}(x) + \frac{1}{2N}(\delta_1(x) + \delta_{-1}(x)),$$

and the Karlin–McGregor representation is given by (see (2.11) and (2.17))

$$P_{ij}^{(n)} = \frac{2}{N}\sum_{k=1}^{N-1} \cos^n\left(\frac{k\pi}{N}\right) T_i\left(\cos\left(\frac{k\pi}{N}\right)\right) T_j\left(\cos\left(\frac{k\pi}{N}\right)\right)$$
$$+ \frac{1}{2N}\left(T_i(1)T_j(1) + T_i(-1)T_j(-1)\right)$$
$$= \frac{2}{N}\sum_{k=1}^{N-1} \cos^n\left(\frac{k\pi}{N}\right) \cos\left(\frac{ik\pi}{N}\right) \cos\left(\frac{jk\pi}{N}\right) + \frac{1}{2N}\left(1 + (-1)^{i+j}\right).$$

In the last part we used the relation of the Chebychev polynomials of the first kind with the cosine function given by (1.86) and that $T_n(1) = 1$ and $T_n(-1) = (-1)^n$. ◊

Example 2.19 *Ehrenfest urn model of heat exchange between two isolated bodies.* This birth–death chain lives on a finite state space $\{0, 1, 2, \ldots, 2N\}$. Here $2N$ is interpreted as the number of total balls, numbered and divided into two urns A and B, where initially there are j balls in A and $2N - j$ balls in B. The process evolves according to the following strategy: we choose a random number between 1 and $2N$. If that ball is in urn A, then we pass that ball into the urn B. Conversely, if the ball is in urn B, we pass it to urn A. The birth–death chain $\{X_n, n = 0, 1, \ldots\}$ counts the number of balls in A at time n. The transition probabilities of the birth–death chain are given by

$$p_j = \frac{2N-j}{2N}, \; j = 0, 1, \dots, 2N-1,$$

$$q_j = \frac{j}{2N}, \; j = 1, 2, \dots, 2N, \quad r_j = 0, \quad j = 0, 1, \dots, 2N,$$

while the transition probability matrix is given by

$$P = \begin{pmatrix} 0 & 1 & & & & & \\ \frac{1}{2N} & 0 & \frac{2N-1}{2N} & & & & \\ & \frac{2}{2N} & 0 & \frac{2N-2}{2N} & & & \\ & & \ddots & \ddots & \ddots & & \\ & & & \frac{2N-1}{2N} & 0 & \frac{1}{2N} \\ & & & & 1 & 0 \end{pmatrix}.$$

A diagram of the one-step transitions is

$$\textcircled{0} \underset{q_1}{\overset{1}{\rightleftarrows}} \textcircled{1} \underset{q_2}{\overset{p_1}{\rightleftarrows}} \textcircled{2} \underset{q_3}{\overset{p_2}{\rightleftarrows}} \cdots \underset{1}{\overset{p_{2N-1}}{\rightleftarrows}} \textcircled{2N}$$

The spectral representation in this case will consist of $2N+1$ eigenvalues, one of which is 1, since the matrix P is stochastic. In this case we proceed in a different way since we do not have any information of the eigenvalues a priori. The transition probability matrix is connected with the Jacobi matrix associated with the *Krawtchouk polynomials*. Indeed, from the three-term recurrence relation (1.115), for $p = q = 1/2$ and replacing N by $2N$, we have

$$\frac{2N-j}{2} K_{j+1}(x) - NK_j(x) + \frac{j}{2} K_{j-1}(x) = -xK_j(x),$$

where $K_j(x) = K_j(x; 1/2, 2N)$. Dividing by N we can write the previous expression as

$$\frac{2N-j}{2N} K_{j+1}(x) + \frac{j}{2N} K_{j-1}(x) = \left(1 - \frac{x}{N}\right) K_j(x).$$

This last relation can be written in matrix form as $PK(x) = (1 - x/N)K(x)$ where P is the transition probability matrix of the Ehrenfest model and $K(x) = (K_0(x), K_1(x), \dots, K_{2N}(x))^T$ is the vector of the Krawtchouk polynomials evaluated at x. This implies that the eigenvalues of P are given by $1 - x/N$, while the eigenvectors are given by $K(x)$ for $x = 0, 1, \dots, 2N$. The eigenvalues represent $2N+1$ points which are uniformly distributed in the interval $[-1, 1]$, i.e.

$$-1, -\frac{N-1}{N}, \dots, -\frac{1}{N}, 0, \frac{1}{N}, \dots, \frac{N-1}{N}, 1.$$

The (normalized) spectral measure of the Krawtchouk polynomials is given by the *binomial distribution*

$$\psi(x) = \frac{1}{2^{2N}} \sum_{x=0}^{2N} \binom{2N}{x} \delta_x,$$

while the potential coefficients (also the inverse of the norms of the Krawtchouk polynomials) are given by

$$\pi_j = \binom{2N}{j}, \quad j = 0, 1, \ldots, 2N.$$

Therefore the Karlin–McGregor integral representation is given by

$$P_{ij}^{(n)} = \frac{\binom{2N}{j}}{2^{2N}} \sum_{x=0}^{2N} \left(1 - \frac{x}{N}\right)^n K_i(x) K_j(x) \binom{2N}{x}.$$

In particular, for $i = j = 0$, it is easy to show that for $|s| < 1$, we get

$$P_{00}(s) = \sum_{n=0}^{\infty} s^n \sum_{x=0}^{2N} \left(1 - \frac{x}{N}\right)^n \frac{\binom{2N}{x}}{2^{2N}} = \sum_{x=0}^{2N} \frac{N}{N(1-s) + xs} \frac{\binom{2N}{x}}{2^{2N}}.$$

Using the Karlin–McGregor representation we can compute the expected return time to state 0 by using formula (2.10). Indeed, we have

$$\tau_{00} = \mathbb{E}(T_0 | X_0 = 0) = \lim_{s \uparrow 1} \frac{1}{(1-s) P_{00}(s)} = 2^{2N}.$$

That means that, if N is large enough, the expected return time to state 0 is quite large. For any state i we have that the expected return time is given by

$$\tau_{ii} = \mathbb{E}(T_i | X_0 = i) = \lim_{s \uparrow 1} \frac{1}{(1-s) P_{ii}(s)} = \frac{2^{2N}}{\binom{2N}{i}},$$

using that $K_i(0) = 1$ (see (1.115)). For instance, for the middle state $i = N$, we have

$$\tau_{N,N} = \frac{\sqrt{\pi}\,\Gamma(N+1)}{\Gamma(N+1/2)} \approx \sqrt{\pi N}, \quad \text{as } N \to \infty.$$

This model will be related to the *Orstein–Uhlenbeck process*, which is a diffusion process with a central force, as we will see in Chapter 4. ◊

Example 2.20 *The Bernoulli–Laplace urn model.* A model similar to the Ehrenfest model was proposed by Bernoulli as a probabilistic analogue for the flow of two incompressible liquids between two containers. We have two urns and a total of $\alpha + \beta$ balls, where α is the number of white balls and β the number of black balls. We will assume for simplicity that $\beta \geq \alpha$ (the other case can be treated in a similar way by interchanging the urns). The number of balls in urn A and B remains always equal to α and β, respectively. At each trial one ball is chosen from each urn and these two balls are interchanged. The state of the birth–death chain is described by the number of white balls in urn A. This model was considered in [57, p. 378] for the special case $\alpha = \beta$ (see also [62] for the general case).

Assume that initially we have j white balls and $\alpha - j$ black balls in urn A and $\alpha - j$ white balls and $\beta - \alpha + j$ black balls in urn B. Clearly the state space is discrete $\{0, 1, 2, \dots, \alpha\}$ and the transition probabilities are given by

$$p_j = \frac{\alpha - j}{\alpha}\frac{\alpha - j}{\beta}, \quad j = 0, 1, \dots, \alpha - 1,$$

$$q_j = \frac{j}{\alpha}\frac{\beta - \alpha + j}{\beta}, \quad j = 1, 2, \dots, \alpha,$$

$$r_j = \frac{j}{\alpha}\frac{\alpha - j}{\beta} + \frac{\alpha - j}{\alpha}\frac{\beta - \alpha + j}{\beta}, \quad j = 0, 1, \dots, \alpha.$$

The probability p_j occurs when we get a black ball from urn A and a white ball from urn B (so that we will have $j + 1$ white balls in urn A in the next trial), while the probability q_j occurs when we get a white ball from urn A and a black ball from urn B (so that we will have $j - 1$ white balls in urn A in the next trial). The probability r_j is the sum of the remaining two situations. Observe that α is a reflecting state, but not necessarily the state 0 (unless $\alpha = \beta$).

Again, the transition probability matrix P can be connected with the Jacobi matrix associated with the dual Hahn polynomials. Indeed, from the three-term recurrence relation (1.123), for $\gamma = -\alpha - 1, \delta = -\beta - 1$ and $N = \alpha$, we have

$$(\alpha - j)^2 R_{j+1}(\lambda(x)) - [(\alpha - j)^2 + j(\beta - \alpha + j)]R_{j+1}(\lambda(x)) \\ + j(\beta - \alpha + j)R_{j-1}(\lambda(x)) = \lambda(x)R_j(x),$$

where $R_j(x) = R_j(x; -\alpha - 1, -\beta - 1, \alpha)$ and $\lambda(x) = x(x - \alpha - \beta - 1)$. Therefore, dividing by $\alpha\beta$ we conclude that

$$p_j R_{j+1}(\lambda(x)) + r_j R_j(\lambda(x)) + q_j R_{j-1}(\lambda(x)) = \left(1 + \frac{\lambda(x)}{\alpha\beta}\right) R_j(\lambda(x)).$$

This last relation can be written in matrix form as $PR(\lambda(x)) = (1 + \lambda(x)/\alpha\beta)R(\lambda(x))$, where P is the transition probability matrix of the Bernoulli–Laplace urn model, and

$$R(\lambda(x)) = (R_0(\lambda(x)), R_1(\lambda(x)), \dots, R_\alpha(\lambda(x)))^T, \quad x = 0, 1, \dots, \alpha,$$

is the vector of the dual Hahn polynomials evaluated at $\lambda(x)$. This implies that the eigenvalues of P are given by

$$1 + \frac{x(x - \alpha - \beta - 1)}{\alpha\beta}, \quad x = 0, 1, \dots, \alpha,$$

while the eigenvectors are given by $R(\lambda(x))$. The eigenvalues are in this case quadratically distributed. Observe that, from the definition of the dual Hahn polynomials, in order to have a positive measure we need to impose either $\gamma,\delta > -1$ or $\gamma,\delta < -N$. The first condition is not possible since α and β are nonnegative integers. The second condition implies that $\beta \geq \alpha$, which was already assumed. Therefore the (normalized) spectral measure of the dual Hahn polynomials is given by (see (1.122))

$$\psi(x) = \sum_{x=0}^{\alpha} \psi_x \delta_x, \quad \psi_x = \frac{(2x-\alpha-\beta-1)(-\alpha)_x^2(\beta-\alpha+1)_\alpha}{(-1)^{x+\alpha}(x-\alpha-\beta-1)_{\alpha+1}(-\beta)_x x!}.$$

With these conditions the eigenvalues are always inside the interval $[-1,1]$. The potential coefficients (also the inverse of the norms of the dual Hahn polynomials) are given by

$$\pi_j = \frac{\binom{j-\alpha-1}{j}\binom{\alpha-\beta-j-1}{\alpha-j}}{\binom{\alpha-\beta-1}{\alpha}}, \quad j = 0,\dots,\alpha.$$

Therefore the Karlin–McGregor integral representation is given by

$$P_{ij}^{(n)} = \pi_j \sum_{x=0}^{\alpha} \left(1 + \frac{x(x-\alpha-\beta-1)}{\alpha\beta}\right)^n R_i(\lambda(x))R_j(\lambda(x))\psi_x.$$

In particular, for $i = j = 0$, it is easy to show that for $|s| < 1$, we get

$$P_{00}(s) = \sum_{n=0}^{\infty} s^n \sum_{x=0}^{\alpha} \left(1 + \frac{x(x-\alpha-\beta-1)}{\alpha\beta}\right)^n \psi_x = \sum_{x=0}^{\alpha} \frac{1}{1-s(1+\lambda(x))}\psi_x.$$

Using the Karlin–McGregor representation we can compute the expected return time to state 0 using formula (2.10). Indeed, we have

$$\tau_{00} = \mathbb{E}(T_0|X_0=0) = \lim_{s\uparrow 1} \frac{1}{(1-s)P_{00}(s)} = \frac{1}{\psi_0} = \frac{(\alpha+\beta)!\,(\beta-\alpha)!}{(\beta!)^2}.$$

For instance, for $\alpha = \beta$, we have

$$\tau_{00} = \frac{(2\alpha)!}{(\alpha!)^2} = \frac{2^{2\alpha}\Gamma(\alpha+1/2)}{\sqrt{\pi}\,\Gamma(\alpha+1)} \approx \frac{2^{2\alpha}}{\sqrt{\pi\alpha}}, \quad \text{as } \alpha \to \infty.$$

◊

2.4.2 Examples Using the Stieltjes Transform

Example 2.21 *Gambler's ruin on $\mathbb{N}_0$ with one absorbing barrier.* Consider the symmetric birth–death chain $\{X_n, n=0,1,\dots\}$ with constant transition probabilities

$p_n = p, n \geq 0$, $q_n = q, n \geq 1$ with $p + q = 1$ and $r_n = 0, n \geq 0$. The transition probability matrix is then

$$P = \begin{pmatrix} 0 & p & & & \\ q & 0 & p & & \\ & q & 0 & p & \\ & & \ddots & \ddots & \ddots \end{pmatrix},$$

and a diagram of the one-step transitions is

$$\textcircled{0} \underset{q}{\overset{p}{\rightleftarrows}} \textcircled{1} \underset{q}{\overset{p}{\rightleftarrows}} \textcircled{2} \underset{q}{\overset{p}{\rightleftarrows}} \textcircled{3} \underset{q}{\overset{p}{\rightleftarrows}} \textcircled{4} \underset{q}{\overset{p}{\rightleftarrows}} \textcircled{5} \underset{q}{\overset{p}{\rightleftarrows}} \cdots$$

This chain is known as gambler's ruin, where one player plays against an adversary which is infinitely rich. It is the extension of the finite state space studied in Example 2.17 and was studied for the first time in the context of orthogonal polynomials in [87, Example (i)]. We cannot study the spectrum as in Example 2.17, so we will try to compute the Stieltjes transform of the spectral measure and then use the Perron–Stieltjes inversion formula (1.10). Since the 0th birth–death chain is the same as the original one, formula (2.37) gives an expression for the Stieltjes transform $B(z;\psi) = B(z)$ of the corresponding spectral measure ψ. Indeed,

$$pqB^2(z) + zB(z) + 1 = 0,$$

and then

$$B(z) = \frac{-z + \sqrt{z^2 - 4pq}}{2pq},$$

where the branch of the square root is determined by analytical continuation of the positive values of real $z > 1$. We also discard the negative solution of $B(z)$ since as $z \to \infty$ the Stieltjes transform should vanish. A special case of this example for $p = q = 1/2$ was studied in Example 1.3.

Since $B(z)$ does not have any poles, the associated density will be absolutely continuous with respect to the Lebesgue measure, i.e. $d\psi(x) = \psi(x)\,dx$. From the Perron–Stieltjes inversion formula (1.10) we get

$$\begin{aligned} \psi(x) &= \frac{1}{\pi} \lim_{\varepsilon \downarrow 0} \mathrm{Im} B(x + i\varepsilon; \psi) \\ &= \frac{1}{2\pi pq} \lim_{\varepsilon \downarrow 0} \left(-\varepsilon + \mathrm{Im}\sqrt{x^2 - \varepsilon^2 + 2ix\varepsilon - 4pq} \right) = \frac{1}{2\pi pq} \mathrm{Im}\sqrt{x^2 - 4pq}. \end{aligned}$$

The last part has only imaginary part when $|x| < 2\sqrt{pq}$. Therefore

$$\psi(x) = \frac{1}{2\pi pq}\sqrt{4pq - x^2}, \quad |x| < 2\sqrt{pq}.$$

This measure is defined on a compact interval. If we make the change of variables $x^2 = 4pqy^2$ we get

$$\psi(x)\,dx = \frac{\sqrt{4pq - x^2}}{2\pi pq}\,dx = \frac{\sqrt{4pq - 4pqy^2}}{2\pi pq}2\sqrt{pq}\,dy = \frac{2}{\pi}\sqrt{1 - y^2}dy, \quad |y| < 1,$$

which is the *Wigner semicircle distribution* or also the measure for the Chebychev polynomials of the second kind $U_n(y)$, which are the Jacobi polynomials for $\alpha = \beta = 1/2$ (see (1.87)).

Another way to identify the spectral measure without using the Stieltjes transform is to try to identify the polynomials that are generated from the recursion formula $xQ(x) = PQ(x)$, which in this case is

$$Q_{-1} = 0, \quad Q_0 = 1, \quad Q_1(x) = x/p, \quad xQ_n(x) = pQ_{n+1}(x) + qQ_{n-1}(x), \quad n \geq 1.$$

It is possible to see, as in Example 2.17, that the polynomials can be written in terms of the Chebychev polynomials of the second kind $(U_n)_n$. Indeed,

$$Q_n(x) = \left(\frac{q}{p}\right)^{n/2} U_n\left(\frac{x}{2\sqrt{pq}}\right), \quad n \geq 0. \tag{2.47}$$

With all this information we can analyze some probabilistic aspects of the birth–death chain. Observe that, since $p + q = 1$, there exists a state -1 where the birth–death chain can be absorbed with probability q. The potential coefficients are given by

$$\pi_j = \left(\frac{p}{q}\right)^j, \quad j \geq 0.$$

Therefore the Karlin–McGregor formula (2.17) can be written as

$$\begin{aligned}
P_{ij}^{(n)} &= \left(\sqrt{\frac{p}{q}}\right)^{j-i} \int_{-2\sqrt{pq}}^{2\sqrt{pq}} x^n U_i\left(\frac{x}{2\sqrt{pq}}\right) \\
&\quad \times U_j\left(\frac{x}{2\sqrt{pq}}\right) \frac{\sqrt{4pq - x^2}}{2\pi pq}\,dx \qquad (x = 2\sqrt{pq}\cos\theta) \\
&= \frac{2\left(\sqrt{\frac{p}{q}}\right)^{j-i}}{\pi} \int_0^\pi (2\sqrt{pq}\cos\theta)^n U_i(\cos\theta) U_j(\cos\theta)\sqrt{1 - \cos^2\theta}\,\sin\theta\,d\theta \\
&= \frac{2(2\sqrt{pq})^n}{\pi\left(\sqrt{\frac{p}{q}}\right)^{i-j}} \int_0^\pi \cos^n\theta\,\sin((i+1)\theta)\,\sin((j+1)\theta)\,d\theta,
\end{aligned}$$

where, as in Example 2.17, we have used the relation of the Chebychev polynomials of the second kind with the cosine function given by (1.88). ◊

Example 2.22 Let us now consider the birth–death chain $\{X_n, n=0,1,\dots\}$ with transition probabilities

$$p_n = p, \quad q_n = q, \quad r_n = 0, \quad n \geq 1, \quad p+q=1,$$

with arbitrary p_0 and r_0. This example was studied for the first time in the context of orthogonal polynomials in [87, Example (ii)]. The transition probability matrix is then

$$P = \begin{pmatrix} r_0 & p_0 & 0 & 0 & \cdots \\ q & 0 & p & 0 & \cdots \\ 0 & q & 0 & p & \cdots \\ \vdots & \vdots & \ddots & \ddots & \ddots \end{pmatrix},$$

and a diagram of the one-step transitions is

The potential coefficients are given in this case by

$$\pi_0 = 1, \quad \pi_j = \frac{p_0 p^{j-1}}{q^j}, \quad j \geq 1.$$

Since the 0th birth–death chain can be identified with the previous example, formula (2.37) gives an expression of the Stieltjes transform $B(z;\psi)$ in terms of the previous spectral measure. Indeed, substituting and rationalizing, we obtain

$$B(z;\psi) = \frac{r_0 - (1-p_0/2p)z + (p_0/2p)\sqrt{z^2-4pq}}{(1-p_0/p)z^2 - 2r_0(1-p_0/2p)z + r_0^2 + p_0^2 q/p}. \tag{2.48}$$

From here we can try to deduce the spectral measure depending on the arbitrary parameters. We can identify two interesting cases:

1. Case $r_0 = 0$ and p_0 a free parameter. In this case the birth–death chain is symmetric and we have

$$\begin{aligned} B(z;\psi) &= \frac{-(2p-p_0)z + p_0\sqrt{z^2-4pq}}{2(p-p_0)z^2 + 2p_0^2 q} \\ &= \frac{-(2p-p_0)z + p_0\sqrt{z^2-4pq}}{2(p_0-p)(\gamma - z)(\gamma + z)}, \quad \gamma = \sqrt{\frac{p_0^2 q}{p_0 - p}}. \end{aligned}$$

Observe that there are points where the denominator vanishes. Depending on the parameters, this will create jumps located at certain discrete points. Using the Perron–Stieltjes inversion formula (1.10) and an argument similar to the previous example, we can compute the absolutely continuous part of the spectral measure, given by

$$\frac{1}{2\pi}\frac{p_0\sqrt{4pq-x^2}}{x^2(p-p_0)+p_0^2q}, \qquad |x|<2\sqrt{pq}.$$

Let us now compute the sizes of these two jumps. From formula (1.12) we have

$$\begin{aligned}
\psi(\{\gamma\}) &= \lim_{\varepsilon\downarrow 0}\varepsilon \mathrm{Im} B(\gamma+i\varepsilon;\psi)\\
&= \lim_{\varepsilon\downarrow 0}\mathrm{Im}\left[\frac{i}{2(p_0-p)}\frac{-(2p-p_0)(\gamma+i\varepsilon)+p_0\sqrt{(\gamma+i\varepsilon)^2-4pq}}{2\gamma+i\varepsilon}\right]\\
&= \mathrm{Im}\left[i\frac{-(2p-p_0)\gamma+p_0\sqrt{\gamma^2-4pq}}{4(p_0-p)\gamma}\right]\\
&= \mathrm{Im}\left[-i\frac{2p-p_0}{4(p_0-p)}+i\frac{p_0}{4(p_0-p)}\sqrt{1-\frac{4pq}{\gamma^2}}\right]\\
&= \mathrm{Im}\left[-i\frac{2p-p_0}{4(p_0-p)}+i\frac{p_0}{4(p_0-p)}\sqrt{\left(\frac{p_0-2p}{p_0}\right)^2}\right]\\
&= -\frac{2p-p_0}{4(p_0-p)}+\frac{p_0}{4(p_0-p)}\left|\frac{p_0-2p}{p_0}\right|\\
&= \frac{p_0-2p}{2(p_0-p)}\mathbf{1}_{\{p_0\geq 2p\}},
\end{aligned}$$

where $\mathbf{1}_A$ is the indicator function. Since the measure is symmetric, the size of the jump at $-\gamma$ is exactly the same. Observe that when $p_0 = 2p$ the sizes of these jumps, which are located at $\pm 2\sqrt{pq}$, vanish.

Observe that we could have previously analyzed the location and behavior of the jumps $\pm\gamma$ by its own definition. Since the support of the measure must be inside the interval $[-1,1]$, we have that $\gamma^2 \leq 1$, or in other words, $p_0 \geq \frac{p}{1-p}$. Since $p_0 \leq 1$, then we must have that $p \leq 1/2$. Therefore $p_0 \geq 2p$ (i.e. the range where the size of the jumps makes sense). We also observe that this jump is located outside the interval $(-2\sqrt{pq}, 2\sqrt{pq})$, but inside the interval $[-1,1]$, since $\gamma^2 \geq 4pq$, as we can check from the relation $(p_0-2p)^2 \geq 0$.

In summary the spectral measure is given by

$$\psi(x) = \frac{1}{2\pi}\frac{p_0\sqrt{4pq-x^2}}{x^2(p-p_0)+p_0^2q}\mathbf{1}_{\{|x|<2\sqrt{pq}\}} + \frac{p_0-2p}{2(p_0-p)}(\delta_\gamma(x)+\delta_{-\gamma}(x))\mathbf{1}_{\{p_0\geq 2p\}}.$$

There is a special case of the previous example when $p_0 = 1$ (*reflecting barrier*). In this case the spectral measure simplifies considerably:

$$\psi(x) = \frac{1}{2\pi q}\frac{\sqrt{4pq-x^2}}{1-x^2}\mathbf{1}_{\{|x|<2\sqrt{pq}\}} + \frac{1-2p}{2q}(\delta_1(x)+\delta_{-1}(x))\mathbf{1}_{\{p\leq 1/2\}}.$$

It is possible to see that the polynomials are given by (see [101, p. 14])

$$Q_n(x) = \left(\frac{q}{p}\right)^{n/2}\left[(2-2p)T_n\left(\frac{x}{2\sqrt{pq}}\right) + (2p-1)U_n\left(\frac{x}{2\sqrt{pq}}\right)\right], \tag{2.49}$$

where, as before, $U_n(x)$ and $T_n(x)$ are the Chebychev polynomials of the second and first kind, respectively. Therefore we can use the Karlin–McGregor formula (2.17) to compute $P_{ij}^{(n)}$. For the special case $p = q = 1/2$ we have that $Q_n(x) = T_n(x)$ and the spectral measure is $\psi(x) = \frac{1}{\pi}\frac{1}{\sqrt{1-x^2}}\mathbf{1}_{\{|x|<1\}}$. Then, since the potential coefficients are given by $\pi_0 = 1$ and $\pi_n = 2, n \geq 1$, we have, for $j \neq 0$, the Karlin–McGregor representation

$$\begin{aligned} P_{ij}^{(n)} &= \frac{2}{\pi}\int_{-1}^{1} x^n \frac{T_i(x)T_j(x)}{\sqrt{1-x^2}}\,dx = \frac{2}{\pi}\int_0^\pi \cos^n(x)T_i(\cos\theta)T_j(\cos\theta)\,d\theta \\ &= \frac{2}{\pi}\int_0^\pi \cos^n(x)\cos(i\theta)\cos(j\theta)\,d\theta, \quad j\neq 0, \end{aligned}$$

and similarly, for $j = 0$,

$$P_{i,0}^{(n)} = \frac{1}{\pi}\int_0^\pi \cos^n(x)\cos(i\theta)\,d\theta.$$

As a final comment about this example (for $p_0 = 1$ and any p), we can apply Theorem 2.8 to see if the birth–death chain associated with this spectral measure ψ is unique. This only happens when the series $\sum_{n=0}^{\infty}(p_n\pi_n)^{-1}$ is divergent. An easy computation shows

$$\sum_{n=0}^{\infty}\frac{1}{p_n\pi_n} = \sum_{n=0}^{\infty}\left(\frac{q}{p}\right)^n = \frac{p}{p-q},$$

as long as $q < p$, i.e. $p > 1/2$. So if $p \leq 1/2$, the birth–death chain associated with the spectral measure is unique.

2. Case $p_0 = p$ and $r_0 = q = 1-p$. In this case the birth–death chain is not symmetric. The Stieltjes transform is given by

$$B(z;\psi) = \frac{2(1-p) - z + \sqrt{z^2-4pq}}{2(1-p)(1-z)}.$$

As before, there is a point (at $z = 1$) where the denominator vanishes. Therefore there will be a jump at that point. Using again the Perron–Stieltjes inversion formula (1.10), we can compute the absolutely continuous part of the spectral measure, given by

$$\frac{1}{2\pi(1-p)}\frac{\sqrt{4pq-x^2}}{1-x}, \quad |x| < 2\sqrt{pq}.$$

The size of the jump, using formula (1.12) as before, is given by

$$\begin{aligned}
\psi(\{1\}) &= \lim_{\varepsilon\downarrow 0} \varepsilon \mathrm{Im} B(1+i\varepsilon;\psi) \\
&= \lim_{\varepsilon\downarrow 0} \mathrm{Im}\left[i\frac{2(1-p)-(1+i\varepsilon)+\sqrt{(1+i\varepsilon)^2-4pq}}{2(1-p)}\right] \\
&= \mathrm{Im}\left[i\frac{2(1-p)-1+\sqrt{1-4pq}}{2(1-p)}\right] = \mathrm{Im}\left[i\frac{1-2p+\sqrt{(1-2p)^2}}{2(1-p)}\right] \\
&= \frac{1-2p+|1-2p|}{2(1-p)} = \frac{1-2p}{1-p}\mathbf{1}_{\{p<1/2\}}.
\end{aligned}$$

Therefore the spectral measure is given by

$$\psi(x) = \frac{1}{2\pi(1-p)}\frac{\sqrt{4pq-x^2}}{1-x}\mathbf{1}_{\{|x|<2\sqrt{pq}\}} + \frac{1-2p}{1-p}\delta_1(x)\mathbf{1}_{\{p<1/2\}}. \tag{2.50}$$

It is possible to see that the polynomials are now given by

$$Q_n(x) = \left(\frac{q}{p}\right)^{n/2}\left[U_n\left(\frac{x}{2\sqrt{pq}}\right) - \sqrt{\frac{q}{p}}U_{n-1}\left(\frac{x}{2\sqrt{pq}}\right)\right],$$

where $(U_n)_n$ are the Chebychev polynomials of the second kind. The shape of the spectral measure (dividing by $1-x$ and adding a Dirac delta at $x=1$) as well as the orthogonal polynomials $Q_n(x)$ suggest that we might be in the situation of dual polynomials as in Theorem 2.10. Indeed, consider Example 2.21 with polynomials defined by (2.47), but changing p by q. Then, by definition of the dual polynomials in (2.21), we have that $Q_0^d(x) = 1$ and for $n \geq 0$,

$$\begin{aligned}
Q_{n+1}^d(x) &= \left(\frac{q}{p}\right)^{n+1}\left[\left(\frac{p}{q}\right)^{(n+1)/2} U_{n+1}\left(\frac{x}{2\sqrt{pq}}\right) - \left(\frac{p}{q}\right)^{n/2} U_n\left(\frac{x}{2\sqrt{pq}}\right)\right] \\
&= \left(\frac{q}{p}\right)^{n+1}\left(\frac{p}{q}\right)^{(n+1)/2}\left[U_{n+1}\left(\frac{x}{2\sqrt{pq}}\right) - \left(\frac{p}{q}\right)^{-1/2} U_n\left(\frac{x}{2\sqrt{pq}}\right)\right] \\
&= \left(\frac{q}{p}\right)^{(n+1)/2}\left[U_{n+1}\left(\frac{x}{2\sqrt{pq}}\right) - \sqrt{\frac{q}{p}}U_n\left(\frac{x}{2\sqrt{pq}}\right)\right].
\end{aligned}$$

Therefore the dual polynomials of Example 2.21 (changing p by q) are the polynomials of this example. Also Theorem 2.10 gives an expression of the dual spectral measure (see (2.27)) and it is exactly the spectral measure (2.50).

In [13] a concise study was made for the general case (2.48). The continuous part of the spectral measure is given by

$$\frac{p_0}{2\pi}\frac{\sqrt{4pq-x^2}}{(p-p_0)x^2-r_0(2p-p_0)x+pr_0^2+p_0^2q}dx, \quad |x|<2\sqrt{pq}.$$

To this continuous part one needs to add one or two possible mass points. In [13] all possible scenarios have been studied. Also the eigenfunctions in the general case are given by (see [16, p. 204])

$$Q_n(x)=\left(\frac{q}{p}\right)^{n/2}\left[\frac{2(p_0-p)}{p_0}T_n\left(\frac{x}{2\sqrt{pq}}\right)\right.$$
$$\left.+\frac{2p-p_0}{p_0}U_n\left(\frac{x}{2\sqrt{pq}}\right)-\frac{r_0}{p_0}\sqrt{\frac{p}{q}}U_{n-1}\left(\frac{x}{2\sqrt{pq}}\right)\right],$$

where $(U_n)_n$ and $(T_n)_n$ are the Chebychev polynomials of the second and first kind, respectively. A similar approach has been recently studied in [69] for a birth–death chain $\{X_n, n=0,1,\dots\}$ with transition probabilities $p_n = p, q_n = q, r_n = r, n \geq 1$, $p+q+r=1$, with arbitrary p_0 and r_0 (see also Exercise 2.10). ◊

Example 2.23 Consider $r_n = 0, n \geq 0$, and period 2 transition probabilities p_n and q_n, that is,

$$p_{2n}=p,\quad q_{2n}=q=1-p,\ n\geq 1,\quad p_{2n+1}=p_1,\ q_{2n+1}=q_1=1-p_1,\quad n\geq 0.$$

This example was studied for the first time in the context of orthogonal polynomials in [87, Example (iii)]. The transition probability matrix is then

$$P=\begin{pmatrix} 0 & p & 0 & 0 & 0 & \cdots \\ q_1 & 0 & p_1 & 0 & 0 & \cdots \\ 0 & q & 0 & p & 0 & \cdots \\ 0 & 0 & q_1 & 0 & p_1 & \cdots \\ \vdots & \vdots & \vdots & \ddots & \ddots & \ddots \end{pmatrix}.$$

We observe that the first birth–death chain is the same as the original one, so we can use formula (2.40) for $k=2$ to compute $B(z;\psi)=B(z;\psi^{(1)})=B(z)$. Indeed, in this case we have

$$Q_1^{(0)}(z)=-\frac{1}{p},\quad Q_2^{(0)}(z)=-\frac{z}{pp_1},\quad Q_1(z)=\frac{z}{p},\quad Q_2(z)=\frac{1}{p_1}\left(\frac{z^2}{p}-q_1\right).$$

Therefore, using that $q_2=q$, from formula (2.40) we get

$$B(z)=-\frac{p_1qB(z)+z}{p_1qzB(z)+z^2-pq_1},\quad\Rightarrow\quad zp_1qB^2(z)+(z^2-pq_1+p_1q)B(z)+z=0.$$

Solving we get

$$B(z) = \frac{-(z^2 - pq_1 + p_1q) + \sqrt{(z^2 - pq_1 + p_1q)^2 - 4p_1qz^2}}{2p_1qz}.$$

If $pq_1 \neq p_1q$, then $(x^2 - pq_1 + p_1q)^2 - 4p_1qx^2$ is negative in the intervals

$$J_1: \quad |\sqrt{pq_1} - \sqrt{p_1q}| < x < \sqrt{pq_1} + \sqrt{p_1q},$$
$$J_2: \quad -\sqrt{pq_1} - \sqrt{p_1q} < x < -|\sqrt{pq_1} - \sqrt{p_1q}|.$$

At $J_1 \cup J_2$ the measure ψ has an absolutely continuous part given by

$$\frac{\sqrt{4p_1qx^2 - (x^2 - pq_1 + p_1q)^2}}{2\pi p_1qx}, \qquad x \in J_1 \cup J_2.$$

There is a jump at $x = 0$. The size of the jump, using formula (1.12), is given by

$$\begin{aligned}\psi(\{0\}) &= \lim_{\varepsilon \downarrow 0} \varepsilon \mathrm{Im} B(i\varepsilon; \psi) \\ &= \lim_{\varepsilon \downarrow 0} \mathrm{Im}\left[-i\frac{\varepsilon^2 + pq_1 - p_1q + \sqrt{(-\varepsilon^2 + pq_1 - p_1q)^2 + 4p_1q\varepsilon^2}}{2p_1q}\right] \\ &= \mathrm{Im}\left[-i\frac{pq_1 - p_1q + \sqrt{(pq_1 - p_1q)^2}}{2p_1q}\right] = -\frac{pq_1 - p_1q + |pq_1 - p_1q|}{2p_1q} \\ &= \left(1 - \frac{pq_1}{p_1q}\right)\mathbf{1}_{\{pq_1 < p_1q\}}.\end{aligned}$$

Therefore the spectral measure is given by

$$\psi(x) = \frac{\sqrt{4p_1qx^2 - (x^2 - pq_1 + p_1q)^2}}{2\pi p_1qx}\mathbf{1}_{\{x \in J_1 \cup J_2\}} + \left(1 - \frac{pq_1}{p_1q}\right)\delta_0(x)\mathbf{1}_{\{pq_1 < p_1q\}},$$

which is obviously symmetric. The potential coefficients are given by

$$\pi_{2n} = \left(\frac{pp_1}{qq_1}\right)^n, \quad n \geq 0 \quad \text{and} \quad \pi_{2n+1} = \left(\frac{pp_1}{qq_1}\right)^n \frac{p}{q_1}, \quad n \geq 0.$$

The special case $pq_1 = p_1q$ is equivalent to $p = p_1$ and this is just Example 2.21, where there is no jump. The explicit expression of the associated orthogonal polynomials in terms of classical orthogonal polynomials is not known. ◊

2.4.3 Birth–Death Chain Generated by the Jacobi Polynomials

All the examples we have considered so far have started with the one-step transition probability matrix P, and from there we have tried to compute the spectral measure ψ. Now we will focus on the only family of classical polynomials that are orthogonal with respect to a measure supported on $[-1,1]$ (not counting the Chebychev polynomials, which have already appeared), namely the Jacobi polynomials. Traditionally, the Jacobi polynomials are considered in the interval $[-1,1]$ but in this section we will consider the Jacobi polynomials on the interval $[0,1]$.

If we consider the Jacobi polynomials on $[-1,1]$, then they are orthogonal with respect to the (normalized) measure

$$\psi(x) = \frac{\Gamma(\alpha+\beta+2)}{2^{\alpha+\beta+1}\Gamma(\alpha+1)\Gamma(\beta+1)}(1-x)^{\alpha}(1+x)^{\beta}, \quad x \in [-1,1], \quad \alpha,\beta > -1.$$

According to Theorem 2.7, these Jacobi polynomials will be a family of birth–death polynomials if the middle coefficient of the three-term recurrence relation is always nonnegative. Normalizing the polynomials by $P_n^{(\alpha,\beta)}(x)/P_n^{(\alpha,\beta)}(1)$ (see (1.77), (1.81) and (1.82)) then the coefficients of the three-term recurrence relation are given by

$$\begin{aligned} p_n &= \frac{2(n+\alpha+1)(n+\alpha+\beta+1)}{(2n+\alpha+\beta+1)(2n+\alpha+\beta+2)}, \\ r_n &= \frac{\beta^2-\alpha^2}{(2n+\alpha+\beta)(2n+\alpha+\beta+2)}, \\ q_n &= \frac{2n(n+\beta)}{(2n+\alpha+\beta)(2n+\alpha+\beta+1)}. \end{aligned} \tag{2.51}$$

According to Theorem 2.7, we clearly see that we need to have $\alpha=\beta$ (symmetric) or $\beta \geq |\alpha|$ in order to have a family of birth–death polynomials.

For any values α and β in these conditions, we have that $q_0 = 0$. Therefore we can apply Theorem 2.8 to see under what conditions the sequence of Jacobi polynomials for the birth–death chain is unique. For that, we need to have that $\sum_{n=0}^{\infty}(p_n\pi_n)^{-1} = \infty$. An easy computation, using the values of p_j and q_j, shows

$$\pi_n = \frac{(2n+\alpha+\beta+1)(\alpha+1)_n(\alpha+\beta+1)_n}{n!\,(\alpha+\beta+1)(\beta+1)_n}, \quad n \geq 0. \tag{2.52}$$

Therefore

$$\begin{aligned} \sum_{n=0}^{\infty}\frac{1}{p_n\pi_n} &= \sum_{n=0}^{\infty}\frac{(2n+\alpha+\beta+2)n!\,(\beta+1)_n}{2(\alpha+1)_{n+1}(\alpha+\beta+2)_n} \\ &= \frac{\Gamma(\alpha+\beta+2)\Gamma(\alpha+1)}{2\Gamma(\beta+1)}\sum_{n=0}^{\infty}\frac{(2n+\alpha+\beta+2)\Gamma(n+\beta+1)\Gamma(n+1)}{\Gamma(n+\alpha+\beta+2)\Gamma(n+\alpha+2)}. \end{aligned} \tag{2.53}$$

The previous series is not easy to compute so it is better to use formula (2.44) that gives (for $q_0=0$) $\sum_{n=0}^{\infty}(p_n\pi_n)^{-1} = \int_{-1}^{1}(1-x)^{-1}d\psi(x)$. Since we have an explicit expression of the spectral measure, we get

$$\int_{-1}^{1} \frac{d\psi(x)}{1-x} = \frac{\Gamma(\alpha+\beta+2)}{2^{\alpha+\beta+1}\Gamma(\alpha+1)\Gamma(\beta+1)},$$

$$\int_{-1}^{1} (1-x)^{\alpha-1}(1+x)^{\beta} dx = \begin{cases} \dfrac{\alpha+\beta+1}{2\alpha}, & \text{if} \quad \alpha > 0, \\ \infty, & \text{if} \quad \alpha \leq 0. \end{cases}$$

Therefore the discrete-time birth–death chain $\{X_n, n=0,1,\ldots\}$ associated with the Jacobi polynomials is unique if and only if $-1 < \alpha \leq 0$ and $\alpha = \beta$ or $\beta \geq -\alpha$. Otherwise we can always take a free parameter μ defined on the range

$$\mu \in \left[0, \frac{2\alpha}{\alpha+\beta+1}\right],$$

such that we can construct an infinite family of sequences of birth–death polynomials for ψ (see Theorem 2.8).

Let us now give one interpretation in terms of *urn models*. For that we will focus on the Jacobi polynomials defined on $[0,1]$ so that, according to Theorem 2.7, we will always have a birth–death measure. This (normalized) measure is now given by

$$\psi(x) = \frac{\Gamma(\alpha+\beta+2)}{\Gamma(\alpha+1)\Gamma(\beta+1)} x^{\alpha}(1-x)^{\beta}, \quad x \in [0,1].$$

Calling $Q_n^{(\alpha,\beta)}(x)$ the Jacobi polynomials normalized in such a way that $Q_n^{(\alpha,\beta)}(1) = 1$, we have that the coefficients of the three-term recurrence relation (and therefore the transition probabilities of the birth–death chain) are given by (see (1.84))

$$\begin{aligned} p_n &= \frac{(n+\beta+1)(n+1+\alpha+\beta)}{(2n+\alpha+\beta+1)(2n+2+\alpha+\beta)}, \\ r_n &= \frac{(n+\beta+1)(n+1)}{(2n+\alpha+\beta+1)(2n+2+\alpha+\beta)} + \frac{(n+\alpha)(n+\alpha+\beta)}{(2n+\alpha+\beta+1)(2n+\alpha+\beta)}, \\ q_n &= \frac{n(n+\alpha)}{(2n+\alpha+\beta+1)(2n+\alpha+\beta)}. \end{aligned} \tag{2.54}$$

In [66] one finds what is probably the first urn model going along with the Jacobi polynomials. This is a rather contrived model when compared to other familiar ones such as those of Ehrenfest or Bernoulli–Laplace, so we will follow the development made in [69].

Assume that α and β are nonnegative integers. Consider the birth–death chain $\{X_t : t = 0,1,\ldots\}$ on $\mathbb{N}_0$ whose one-step transition probability matrix P coincides with the probabilities given by (2.54). We will consider two experiments, Experiment 1 and Experiment 2. At times $t = 0,1,2,\ldots$ an urn contains n blue balls and this determines the state of the birth–death chain on $\mathbb{N}_0$ at that time. Each urn for both experiments sits in an environment, or bath for short, consisting of an infinite number of blue and red balls.

Experiment 1 consists of a discrete-time pure-birth chain on the nonnegative integers $\mathbb{N}_0$. Let us call this chain $\{X_t^{(1)} : t = 0, 1, \ldots\}$. If the state of the system is n blue balls ($n \geq 0$), take $\beta + 1$ blue balls and $n + \alpha$ red balls from the bath and add them to the urn. Draw one ball from the urn at random with uniform distribution. We have two possibilities:

- If we get a red ball then we remove all red balls in the urn and $\beta + 1$ blue balls from the urn (i.e. all balls we introduced in this first step) and start over. Therefore

$$\mathbb{P}\left(X_1^{(1)} = n | X_0^{(1)} = n\right) = \frac{n + \alpha}{2n + \alpha + \beta + 1}.$$

- If we get a blue ball then we remove all red balls and β blue balls from the urn (so that there are $n + 1$ blue balls in the urn) and start over. Therefore

$$\mathbb{P}\left(X_1^{(1)} = n + 1 | X_0^{(1)} = n\right) = \frac{n + \beta + 1}{2n + \alpha + \beta + 1}.$$

Experiment 2 consists of a discrete-time pure-death chain on $\mathbb{N}_0$. Let us call this chain $\{X_t^{(2)} : t = 0, 1, \ldots\}$. If the state of the system is n blue balls ($n \geq 0$), take $n + \alpha + \beta$ red balls from the bath and add them to the urn. As before, draw one ball from the urn at random. Again, we have two possibilities:

- If we get a red ball then we remove all red balls in the urn and start over. Therefore

$$\mathbb{P}\left(X_1^{(2)} = n | X_0^{(2)} = n\right) = \frac{n + \alpha + \beta}{2n + \alpha + \beta}.$$

- If we get a blue ball then we remove that blue ball and all red balls from the urn (so that there are $n - 1$ blue balls in the urn) and start over. Therefore

$$\mathbb{P}\left(X_1^{(2)} = n - 1 | X_0^{(2)} = n\right) = \frac{n}{2n + \alpha + \beta}.$$

If the urn is empty, we stop the experiment. Observe that if we have 0 blue balls in the urn the experiment will not change from that moment on.

The urn model for P will be the composition of Experiment 1 and then Experiment 2. If we perform Experiment 1 first, we will end up with an urn with either n (if we draw a red ball) or $n + 1$ (if we draw a blue ball) blue balls. Now we perform Experiment 2 with n or $n + 1$ blue balls, in which case we may have either $n - 1$ (if we draw a blue ball) or n (if we draw a red ball) blue balls, while for the $n + 1$ case we may have either n (if we draw a blue ball) or $n + 1$ (if we draw a red ball) blue balls. The combination of probabilities of these four cases gives

$$
\begin{aligned}
p_n &= \mathbb{P}\left(X_1^{(1)} = n+1 | X_0^{(1)} = n\right) \mathbb{P}\left(X_1^{(2)} = n+1 | X_0^{(2)} = n+1\right), \\
r_n &= \mathbb{P}\left(X_1^{(1)} = n+1 | X_0^{(1)} = n\right) \mathbb{P}\left(X_1^{(2)} = n | X_0^{(2)} = n+1\right) \\
&\quad + \mathbb{P}\left(X_1^{(1)} = n | X_0^{(1)} = n\right) \mathbb{P}\left(X_1^{(2)} = n | X_0^{(2)} = n\right), \\
q_n &= \mathbb{P}\left(X_1^{(1)} = n | X_0^{(1)} = n\right) \mathbb{P}\left(X_1^{(2)} = n-1 | X_0^{(2)} = n\right),
\end{aligned}
$$

which coincide with the coefficients given by (2.54). The following diagram shows a schematic of Experiments 1 and 2. The circled regions represent the state or urn, with $B_\bullet$ and $R_\bullet$ indicating the number of blue or red balls, respectively, contained within the urn. When a ball is drawn from the urn, this is shown as $B_\bullet^{\mathrm{d}}$ or $R_\bullet^{\mathrm{d}}$. The initial state is B_n.

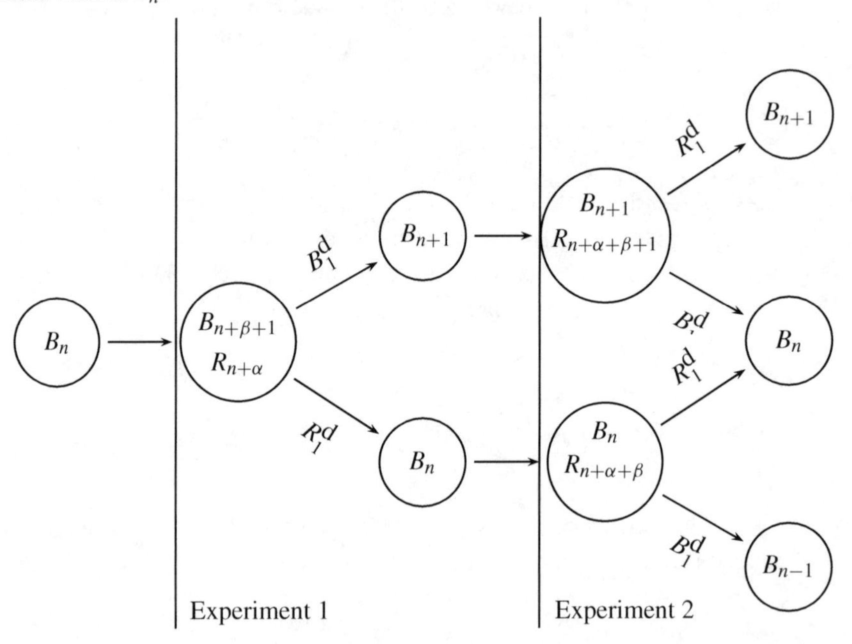

2.5 Applications to the Probabilistic Aspects of Discrete-Time Birth–Death Chains

In this section we will use the Karlin–McGregor integral representation formula (2.17) to study the probabilistic aspects associated with any discrete-time birth–death chain. In the first part we will start with the concept of recurrence (ergodic, null or transient) for $q_0 = 0$ (see Definition 2.1) and how to study it in terms of the spectral measure ψ and other formulas. If $q_0 > 0$ then the process will always be transient, since there is a positive probability that the birth–death chain is absorbed at the state -1. We will study which kind of absorption (certain, ergodic or transient) has the birth–death chain in terms also of the spectral measure ψ and other formulas. In the second part we will study limit theorems, meaning that we will be concerned with the behavior of $P_{ij}^{(n)}$ as $n \to \infty$. If the birth–death chain is ergodic then we already have the Convergence Theorem (Theorem 2.3). But if it is not ergodic we do not have so much information about the behavior as $n \to \infty$. We will see some results concerning this question, especially to the ratio $P_{ij}^{(n)}/P_{kl}^{(n)}$ as $n \to \infty$, even in cases where $q_0 > 0$.

2.5.1 Recurrence and Absorption

Consider an irreducible discrete-time birth–death chain $\{X_n, n=0,1,\ldots\}$ with transition probabilities p_n, r_n, q_n where $p_n + r_n + q_n = 1, n \geq 1$ and $p_0 + r_0 \leq 1$ (possibly $q_0 > 0$) and potential coefficients π_n given by (2.13). Since the birth–death chain is irreducible (i.e. $p_n, q_{n+1} > 0, n \geq 0$) all non-absorbing states in the birth–death chain form the same class of communication. Therefore in order to study recurrence we only need to study one single state, namely the state 0. The birth–death chain is *aperiodic* if $r_n > 0$ for some $n \geq 0$, and periodic of period 2 if $r_n = 0$ for all $n \geq 0$. We know from Theorem 2.5 that there exists a unique measure ψ supported on $[-1,1]$ such that we have the Karlin–McGregor representation formula (2.17). For birth–death chains (either if we have $q_0 = 0$ or $q_0 > 0$) we always have that $\sum_{n=0}^{\infty}\left((p_n\pi_n)^{-1} + \pi_n\right) = \infty$. This can be seen by definition of π_n or also as a consequence of the uniqueness of the *Hausdorff moment problem* (see [2, 137] for more details or Theorem 3.26 for the case of the *Stieltjes moment problem*).

Theorem 2.24 ([87, Section 2]) *Let $q_0 = 0$ (i.e. 0 is a reflecting state). Then the following are equivalent:*

1. *The birth–death chain is recurrent.*
2. $\displaystyle\int_{-1}^{1} \frac{d\psi(x)}{1-x} = \infty.$
3. $\displaystyle\sum_{n=0}^{\infty} \frac{1}{p_n\pi_n} = \infty.$

In addition, the following are equivalent:

1. *The birth–death chain is positive recurrent or ergodic.*
2. *The measure ψ has a finite jump at $x = 1$ of size* $\psi(\{1\}) = \left(\sum_{n=0}^{\infty}\pi_n\right)^{-1}.$
3. $\displaystyle\sum_{n=0}^{\infty}\pi_n < \infty.$

Proof By irreducibility it is enough to study recurrence for the state 0. The state 0 is recurrent if and only if $\lim_{s\uparrow 1} F_{00}(s) = 1$. By (2.35) (for $i = j = 0$) this will be possible if and only if $\lim_{s\uparrow 1}\int_{-1}^{1}(1-xs)^{-1}d\psi(x) = \infty$. Therefore this is equivalent to $\int_{-1}^{1}(1-x)^{-1}d\psi(x) = \infty$. The third relation is a consequence of formula (2.44) for $q_0 = 0$.

As for the positive recurrence, by definition, we need to see that the expected return time is finite. By the Convergence Theorem 2.3 we need to check the existence of the limit $\lim_{n\to\infty} P_{00}^{(n)} > 0$. If the birth–death chain is symmetric, then $\lim_{n\to\infty} P_{00}^{(n)}$ never exists since $P_{00}^{(2n)} > 0$ and $P_{00}^{(2n+1)} = 0$, for all $n \geq 0$. Then we assume that the birth–death chain is aperiodic, i.e. there exists some $n \geq 0$ such that $r_n > 0$). In particular, for the even indexes, we should have $\lim_{n\to\infty} P_{00}^{(2n)} > 0$. Since $x^{2n} \to 0$ monotonically in $x \in (-1,1)$, from (2.17) we see that the limit is positive if the spectral measure has positive jumps at $x = 1$ or at $x = -1$. However, there cannot be a jump at $x = -1$, since otherwise, the size of the jump would be

$$-\lim_{n\to\infty}\int_{-1}^{1} x^{2n+1}d\psi(x) = -\lim_{n\to\infty} P_{00}^{(2n+1)} = -\lim_{n\to\infty} P_{00}^{(2n)} \leq 0.$$

But this quantity must be positive, so there is no jump at $x = -1$. Therefore the birth–death chain is positive recurrent if and only if there is a jump at $x = 1$. The size of this jump $\psi(\{1\})$ can be computed using that the vector π is an invariant vector. Therefore $\pi_j = \sum_{i=0}^{\infty}\pi_i P_{ij}^{(n)}, n \geq 0$. Using the integral representation (2.17) for $j = i$ we get that

$$1 = \sum_{i=0}^{\infty}\pi_i \int_{-1}^{1} x^n Q_i^2(x)d\psi(x) \to \sum_{i=0}^{\infty}\pi_i Q_i^2(1)\psi(\{1\}), \quad n \to \infty.$$

Therefore $1/\psi(\{1\}) = \sum_{n=0}^{\infty} Q_n^2(1)\pi_n$. In the case where $q_0 = 0$ we have that $Q_n(1) = 1$ (see (2.23)). Then, in order to be positive recurrent the series $\sum_{n=0}^{\infty}\pi_n$ has to be convergent. □

Remark 2.25 Observe from the previous theorem that if $q_0 = 0$ and the birth–death chain is recurrent then $\eta = \sup\operatorname{supp}(\psi) = 1$, since $\int_{-1}^{1}(1-x)^{-1}d\psi(x) = \infty$. This is not necessarily true in reverse. ◇

Remark 2.26 Many of the formulas that appear in this and the following sections are extensions of formulas that were previously derived for continuous-time birth–death processes (which we will see in Chapter 3). Following [87], if $r_0 + p_0 = 1$ and $q_n + r_n + p_n = 1, n \geq 1$, the polynomials defined by $R_n(x) = Q_n(1-x)$ satisfy the recurrence relation

$$-xR_n(x) = q_n R_{n-1} - (p_n + q_n)R_n(x) + p_n R_{n+1}(x),$$

which are orthogonal in the interval $[0,2]$ with respect to a distribution θ that is obtained from ψ using the same change of variables. This three-term recurrence relation gives rise to the infinitesimal operator of a continuous-time birth–death process. The spectral properties of these processes were analyzed chronologically before those corresponding to discrete-time birth–death chains. ◇

A consequence of the previous theorem in combination with Theorem 2.8 is the following:

Corollary 2.27 *(i) The sequence $(Q_n)_n$ satisfying (2.15) with $q_0 = 0$ is the unique sequence of birth–death polynomials for a measure ψ satisfying (2.17) if and only if the integral $I = \int_{-1}^{1}(1-x)^{-1}d\psi(x) = \infty$, i.e. the birth–death chain is recurrent.*

(ii) If $I < \infty$ then there exists an infinite family of sequences of birth–death polynomials for ψ satisfying (2.15) parametrized by $\mu \in [0, I^{-1}]$.

Example 2.28 Consider Example 2.22 with $r_0 = 1-p$ and $p_0 = p$. Then we are in the conditions of Theorem 2.24. Since we know the spectral measure (given by (2.50)), we have in this case

$$\int_{-1}^{1}\frac{d\psi(x)}{1-x} = \frac{1}{2\pi(1-p)}\int_{-2\sqrt{pq}}^{2\sqrt{pq}}\frac{\sqrt{4pq-x^2}}{(1-x)^2}dx + \frac{1-2p}{1-p}\lim_{x\uparrow 1}\frac{1}{1-x}\mathbf{1}_{\{p<1/2\}}.$$

If $p < 1/2$ then $2\sqrt{pq} < 1$ so the integral is bounded. However, the discrete part is ∞. If $p = 1/2$ then the discrete part is 0 while the integral now is ∞. If $p \geq 1/2$ then the discrete part is 0 and the integral is bounded. Therefore if $p \leq 1/2$ the birth–death chain is recurrent. Otherwise it is transient. Another way to study recurrence is by looking to the convergence of the series

$$\sum_{n=0}^{\infty}\frac{1}{p_n\pi_n} = \sum_{n=0}^{\infty}\frac{q^n}{p^{n+1}} = \frac{1}{p}\sum_{n=0}^{\infty}\left(\frac{q}{p}\right)^n = \infty \quad \Leftrightarrow \quad \frac{q}{p} \geq 1 \quad \Leftrightarrow \quad p \leq 1/2.$$

As for the positive recurrence, according to Theorem 2.24, the measure has a finite jump at $x = 1$ if and only if $p < 1/2$. This jump was already given in (2.50) but another way to compute it is by using the second part of Theorem 2.24, i.e.

$$\begin{aligned}\psi(\{1\}) &= \left(\sum_{n=0}^{\infty}\pi_n\right)^{-1} = \left(\sum_{n=0}^{\infty}\left(\frac{p}{q}\right)^n\right)^{-1} = \left(\frac{q}{q-p}\right)^{-1} \\ &= \frac{1-2p}{1-p} > 0 \quad \Leftrightarrow \quad p < 1/2.\end{aligned}$$

The only case where we have that the birth–death chain is null recurrent is for $p = 1/2$. ◊

Example 2.29 Consider now the example in Section 2.4.3 of the Jacobi polynomials. According to Theorem 2.24 we need to study the behavior of $\sum_{n=0}^{\infty}(p_n\pi_n)^{-1}$. But observe that in this case this series is quite difficult to treat (see (2.53)). Nevertheless, the condition on the spectral measure is much easier to analyze. Indeed, as we did in Section 2.4.3, we have that

$$\int_{-1}^{1} \frac{d\psi(x)}{1-x} = \frac{\Gamma(\alpha+\beta+2)}{2^{\alpha+\beta+1}\Gamma(\alpha+1)\Gamma(\beta+1)} \int_{-1}^{1} (1-x)^{\alpha-1}(1+x)^{\beta} dx$$

$$= \begin{cases} \dfrac{\alpha+\beta+1}{2\alpha}, & \text{if} \quad \alpha > 0, \\ \infty, & \text{if} \quad \alpha \leq 0. \end{cases}$$

Therefore the birth–death chain is recurrent if and only if $-1 < \alpha \leq 0$ and $\alpha = \beta$ or $\beta \geq -\alpha$. Otherwise it is transient. Since the measure does not have a jump at $x = 1$, then the birth–death chain is never positive recurrent.

Observe that the above integral gives a condition under there exists an infinite family of sequences of birth–death polynomials for ψ, according to Corollary 2.27. If we choose the absorbing probability μ (see Theorem 2.8) in the range

$$\mu \in \left[0, \frac{2\alpha}{\alpha+\beta+1}\right],$$

then we can construct an infinite family of sequences of birth–death polynomials for ψ. This happens only when $\beta \geq \alpha > 0$ (i.e. it is transient). For example, for the Gegenbauer polynomials ($\beta = \alpha > 0$) we have that the birth–death chain generated by the transition probabilities (2.51) is not the only one generating a family of birth–death polynomials, but also by the birth–death chain with transition probabilities

$$p_n = \frac{n+1}{2n+2\alpha+1}, \quad q_n = \frac{n+2\alpha}{2n+2\alpha+1}, \quad r_n = 0, \quad n \geq 0.$$

Observe that $q_0 = 2\alpha/(2\alpha+1)$. ◊

Example 2.30 The examples with a finite number of states, such as gambler's ruin with reflecting barriers, the Ehrenfest urn model and the Bernoulli–Laplace urn model (see Examples 2.18, 2.19 and 2.20, respectively) are always positive recurrent. This is due to the fact that we can always compute an invariant distribution of the birth–death chain. Additionally we observe that the spectral measures associated with these three examples have always a finite jump at $x = 1$. ◊

Let us now analyze the case when $q_0 > 0$, i.e. there is a positive probability that the birth–death chain is absorbed at the state -1. Observe that the birth–death chain is automatically transient.

Definition 2.31 Let $\{X_n, n = 0, 1, \ldots\}$ be a discrete-time birth–death chain with $q_0 > 0$.

- We say that absorption at -1 is *certain* if the probability of eventual absorption equals 1, i.e.

$$\mathbb{P}(T_{-1} < \infty | X_0 = i) = 1, \quad \text{for all} \quad i \in \mathbb{N}_0,$$

where T_{-1} is defined by (2.4). If, additionally, the expected time to absorption is finite, i.e. $\tau_{i,-1} = \mathbb{E}(T_{-1}|X_0 = i) < \infty$ for all $i \in \mathbb{N}_0$ (see (2.9)), then absorption at -1 is *ergodic*.

- We say that absorption at -1 is *transient* if it is not certain, i.e.

$$\mathbb{P}(T_{-1} < \infty | X_0 = i) < 1, \quad \text{for some} \quad i \in \mathbb{N}_0. \qquad \diamondsuit$$

Theorem 2.32 ([47, Theorem 2.3]) *Let $\{X_n, n=0,1,\dots\}$ be a discrete-time birth–death chain with $q_0 > 0$. Then the following are equivalent:*

1. *Absorption at -1 is certain.*
2. $q_0 \displaystyle\int_{-1}^{1} \frac{d\psi(x)}{1-x} = 1.$
3. $\displaystyle\sum_{n=0}^{\infty} \frac{1}{p_n \pi_n} = \infty.$

In addition, the following are equivalent:

1. *Absorption at -1 is ergodic.*
2. $q_0 \displaystyle\int_{-1}^{1} \frac{d\psi(x)}{1-x} = 1$ *and* $\displaystyle\int_{-1}^{1} \frac{d\psi(x)}{(1-x)^2} < \infty.$
3. $\displaystyle\sum_{n=0}^{\infty} \pi_n < \infty.$

Proof By irreducibility we can focus only on the behavior at the state 0. Let us call $a_0 = \mathbb{P}(T_{-1} < \infty | X_0 = 0)$. According to (2.36) absorption at -1 depends on the value of $F_{0,-1}(s)$ as $s \uparrow 1$. Hence absorption at -1 is certain if and only if

$$a_0 = \lim_{s \uparrow 1} F_{i,-1}(s) = q_0 \int_{-1}^{1} \frac{d\psi(x)}{1-x} = 1. \tag{2.55}$$

From formula (2.22) for $x = 1$, (2.55) and (2.44) we get an expression of a_0 given by

$$a_0 = \frac{q_0 \displaystyle\sum_{n=0}^{\infty} \frac{1}{p_n \pi_n}}{1 + q_0 \displaystyle\sum_{n=0}^{\infty} \frac{1}{p_n \pi_n}}. \tag{2.56}$$

Therefore $a_0 = 1$ if and only if $\sum_{n=0}^{\infty} (p_n \pi_n)^{-1} = \infty$.

For the second part, we first notice that (see (2.9))

$$\tau_{0,-1} = \mathbb{E}(T_{-1}|X_0 = 0) = \lim_{s \uparrow 1} F'_{0,-1}(s).$$

Since

$$\frac{d}{ds}\left(\frac{s}{1-xs}\right) = \frac{1}{(1-xs)^2},$$

we have, using (2.36), that

$$\tau_{0,-1} = \mathbb{E}(T_{-1}|X_0 = 0) = q_0 \int_{-1}^{1} \frac{d\psi(x)}{(1-x)^2}.$$

Therefore this integral should be finite. Since we need absorption at -1 to be certain (otherwise $\tau_{0,-1} = \infty$) we have that $q_0 \int_{-1}^{1} (1-x)^{-1} d\psi(x) = 1$ and $\sum_{n=0}^{\infty} (p_n \pi_n)^{-1} = \infty$. Now, from Theorem 2.10 (since we are in the situation of $q_0 > 0$), the dual polynomials defined by (2.21) are orthogonal with respect to ψ^d defined by (2.27). Since absorption is certain, we have that $q_0 \int_{-1}^{1} (1-x)^{-1} d\psi(x) = 1$, so ψ^d does not have a jump at $x = 1$. Therefore

$$\int_{-1}^{1} \frac{d\psi^d(x)}{1-x} = q_0 \int_{-1}^{1} \frac{d\psi(x)}{(1-x)^2} = \tau_{0,-1}.$$

Alternatively, we can apply Theorem 2.24 to the dual measure ψ^d. Therefore we have

$$\int_{-1}^{1} \frac{d\psi^d(x)}{1-x} = \sum_{n=0}^{\infty} \frac{1}{p_n^d \pi_n^d} = \frac{1}{q_0} \sum_{n=0}^{\infty} \frac{p_{n-1} q_{n-1}}{q_n} = \frac{1}{q_0} \sum_{n=0}^{\infty} \pi_n.$$

Hence we get $q_0^2 \int_{-1}^{1} (1-x)^{-2} \, d\psi(x) = \sum_{n=0}^{\infty} \pi_n$ and as a consequence

$$\tau_{0,-1} = \mathbb{E}(T_{-1}|X_0 = 0) = \frac{1}{q_0} \sum_{n=0}^{\infty} \pi_n < \infty. \tag{2.57}$$

□

Remark 2.33 In general, if we call $a_i = \mathbb{P}(T_{-1} < \infty | X_0 = i)$, we have

$$a_i = \lim_{s \uparrow 1} F_{i,-1}(s) = q_0 \int_{-1}^{1} \frac{Q_i(x)}{1-x} \, d\psi(x). \tag{2.58}$$

Dividing (2.22) by $1 - x$, and integrating with respect to the spectral measure ψ, we get

$$a_i = a_0 - q_0(1 - a_0) \sum_{n=0}^{i-1} \frac{1}{p_n \pi_n}. \tag{2.59}$$

Substituting (2.56) in the previous formula, we get

$$a_i = \mathbb{P}(T_{-1} < \infty | X_0 = i) = \frac{q_0 \displaystyle\sum_{n=i}^{\infty} \frac{1}{p_n \pi_n}}{1 + q_0 \displaystyle\sum_{n=0}^{\infty} \frac{1}{p_n \pi_n}}, \quad i \geq 0. \tag{2.60}$$

Similarly, the excepted time to absorption from state i (see (2.9)) is given by

$$\tau_{i,-1} = \mathbb{E}(T_{-1}|X_0 = i) = q_0 \int_{-1}^{1} \frac{Q_i(x)}{(1-x)^2}\, d\psi(x).$$

The previous formula can also be written (using analog results from the ones given in [84]) as

$$\tau_{i,-1} = \mathbb{E}(T_{-1}|X_0 = i) = \sum_{j=0}^{i} \frac{1}{q_j \pi_j} \sum_{n=j}^{\infty} \pi_n. \tag{2.61}$$

◊

Remark 2.34 Recurrence and absorption are dual properties if we consider the relation between a birth–death chain and its dual. Indeed, if $q_0 > 0$, then the absorption of the birth–death chain at -1 is certain (not ergodic), certain ergodic or transient if and only if the dual birth–death chain (where $q_0^d = 0$) is null recurrent, ergodic or transient, respectively. This is due to to the relations

$$\pi_n^d = \frac{q_0}{q_n \pi_n}, \quad \frac{1}{p_n^d \pi_n^d} = \frac{\pi_n}{q_0}.$$

For instance, if the original birth–death chain has a certain (not ergodic) absorption at -1, then $\sum_{n=0}^{\infty}(p_n\pi_n)^{-1} = \infty$ and $\sum_{n=0}^{\infty}\pi_n = \infty$ (not ergodic). Therefore $\sum_{n=0}^{\infty}(p_n^d\pi_n^d)^{-1} = \infty$ and $\sum_{n=0}^{\infty}\pi_n^d = \infty$, i.e. the dual birth–death chain is null recurrent. The same can be applied to the remaining cases. If $q_0 = 0$ we can apply the same argument but now in the reverse order. Since

$$\pi_n^d = \frac{p_0}{p_n \pi_n}, \quad \frac{1}{p_n^d \pi_n^d} = \frac{\pi_{n+1}}{p_0},$$

we have that the birth–death chain is null recurrent, ergodic or transient if and only if the absorption of the dual birth–death chain at -1 is certain (not ergodic), certain ergodic or transient. ◊

Remark 2.35 Observe from the previous theorem that if $q_0 > 0$ and absorption of the birth–death chain is certain, but not ergodic, then $\eta = \sup \operatorname{supp}(\psi) = 1$, since $\int_{-1}^{1}(1-x)^{-2}d\psi(x) = \infty$. This is not necessarily true the other way around. ◊

Example 2.36 Consider the birth–death chain of Example 2.21 with $q_0 = q = 1-p$. According to Theorem 2.32 we can study absorption at -1 using the spectral measure or studying the divergence of certain series. The series can be computed directly since

$$\sum_{n=0}^{\infty} \frac{1}{p_n \pi_n} = \sum_{n=0}^{\infty} \frac{q^n}{p^{n+1}} = \frac{1}{p} \sum_{n=0}^{\infty} \left(\frac{q}{p}\right)^n = \infty \quad \Leftrightarrow \quad \frac{q}{p} \geq 1 \quad \Leftrightarrow \quad p \leq 1/2.$$

Therefore absorption at -1 is certain if and only if $p \leq 1/2$. Alternatively, using the spectral measure we have that

$$q_0 \int_{-1}^{1} \frac{d\psi(x)}{1-x} = \frac{1}{2\pi p} \int_{-2\sqrt{pq}}^{2\sqrt{pq}} \frac{\sqrt{4pq - x^2}}{1-x} dx.$$

This integral is given by

$$\frac{1}{2\pi p} \int_{-2\sqrt{pq}}^{2\sqrt{pq}} \frac{\sqrt{4pq - x^2}}{1-x} dx = \frac{1}{2p} (1 - |1 - 2p|) = \begin{cases} 1, & \text{if} \quad p \leq 1/2, \\ q/p, & \text{if} \quad p > 1/2. \end{cases}$$

Therefore, again, absorption at -1 is certain if and only if $p \leq 1/2$. The absorption at -1 is ergodic if and only if $p < 1/2$. For these values of p, we have that

$$\sum_{n=0}^{\infty} \pi_n = \sum_{n=0}^{\infty} \left(\frac{p}{q}\right)^n < \infty \quad \Leftrightarrow \quad \frac{p}{q} < 1 \quad \Leftrightarrow \quad p < 1/2.$$

Additionally we could have studied the convergence of the integral

$$\int_{-1}^{1} \frac{d\psi(x)}{(1-x)^2} = \frac{1}{2\pi pq} \int_{-2\sqrt{pq}}^{2\sqrt{pq}} \frac{\sqrt{4pq - x^2}}{(1-x)^2} dx.$$

If $p \neq 1/2$ this integral is always bounded since $2\sqrt{pq} < 1$. For $p = 1/2$, we have that

$$\frac{1}{2\pi pq} \int_{-2\sqrt{pq}}^{2\sqrt{pq}} \frac{\sqrt{4pq - x^2}}{(1-x)^2} dx = \frac{2}{\pi} \int_{-1}^{1} \sqrt{\frac{1+x}{(1-x)^3}} dx = \infty.$$

Therefore, again, absorption at -1 is ergodic if and only if $p < 1/2$. The expected time of absorption is given by (2.57), i.e.

$$\tau_{0,-1} = \mathbb{E}(T_{-1} | X_0 = 0) = \frac{1}{2\pi p} \int_{-2\sqrt{pq}}^{2\sqrt{pq}} \frac{\sqrt{4pq - x^2}}{(1-x)^2} dx = \frac{1}{q} \sum_{j=0}^{\infty} \left(\frac{p}{q}\right)^n = \frac{1}{1-2p}.$$

Observe that the closer p is to $1/2$ the longer the expected time of absorption is.

We can also compute the absorption probabilities starting from a state i, already given by (2.60). Indeed, since

$$\sum_{k=i}^{\infty} \frac{1}{p_k \pi_k} = \frac{1}{p} \left(\frac{q}{p}\right)^i \sum_{k=0}^{\infty} \left(\frac{q}{p}\right)^k = \begin{cases} \dfrac{1}{p-q} \left(\dfrac{q}{p}\right)^i, & \text{if} \quad p > 1/2, \\ \infty, & \text{if} \quad p \leq 1/2, \end{cases}$$

we have that

$$a_i = \mathbb{P}(T_{-1} < \infty | X_0 = i) = \begin{cases} \dfrac{\dfrac{q}{p-q}\left(\dfrac{q}{p}\right)^i}{1+\dfrac{q}{p-q}} = \left(\dfrac{q}{p}\right)^{i+1}, & \text{if } p > 1/2, \\ 1, & \text{if } p \leq 1/2. \end{cases}$$

Moreover, the expected absorption time (2.61) is given by

$$\tau_{i,-1} = \mathbb{E}(T_{-1}|X_0 = i) = \frac{1}{q}\sum_{j=0}^{i}\left(\frac{q}{p}\right)^j\left(\frac{p}{q}\right)^j\sum_{k=0}^{\infty}\left(\frac{p}{q}\right)^k = \begin{cases} \dfrac{i+1}{1-2p}, & \text{if } p < 1/2, \\ \infty, & \text{if } p \geq 1/2. \end{cases}$$

◊

2.5.2 Strong Ratio Limit Property

Let us now study the asymptotic behavior of $P_{ij}^{(n)}$ as $n \to \infty$. The only result we have about this is the well-known Convergence Theorem (Theorem 2.3), which states that

$$\lim_{n\to\infty} P_{ij}^{(n)} = \frac{1}{\tau_{jj}} = \pi_j,$$

where $\tau_{ij} = \mathbb{E}(T_j|X_0 = i)$ and $(\pi_n)_n$ are the potential coefficients. Therefore, for an aperiodic birth–death chain, we have that $\lim_{n\to\infty} P_{ij}^{(n)} > 0$ if and only if the birth–death chain is ergodic or positive recurrent. Otherwise (null recurrent or transient) we have that $P_{ij}^{(n)} \to 0$ as $n \to \infty$. If the birth–death chain is symmetric, then this limit never exists since $P_{00}^{(2n)} > 0$ but $P_{00}^{(2n+1)} = 0$. However, it is possible to see that $P_{ij}^{(2n)}$ is always convergent, as a consequence of the symmetry of the spectral measure and the corresponding orthogonal polynomials in the Karlin–McGregor representation (2.17). In fact, from the Karlin–McGregor representation

$$P_{ij}^{(n)} = \pi_j \int_{-1}^{1} x^n Q_i(x) Q_j(x) d\psi(x), \tag{2.62}$$

we clearly see that $x^n \to 0$ as $n \to \infty$ uniformly inside the interval $(-1, 1)$. Therefore, as we saw in Theorem 2.24, the only possibility that $\lim_{n\to\infty} P_{ij}^{(n)}$ exists and is positive is if the measure ψ has a finite positive jump at $x = 1$. So in this section we will assume that the birth–death chain is aperiodic, unless otherwise specified.

In cases where $P_{ij}^{(n)} \to 0$ as $n \to \infty$ it is better to study the asymptotic behavior of growth rates of the form

$$\lim_{n\to\infty} \frac{P_{ij}^{(n)}}{P_{kl}^{(n)}}, \quad i,j,k,l \in \mathbb{N}_0. \tag{2.63}$$

If these limits exist simultaneously then the birth–death chain is said to have the *strong ratio limit property*. These limits are in general difficult to study and in fact a satisfactory and comprehensive solution to the problem *is still lacking*. However, it is possible to find necessary and/or sufficient conditions to study the strong ratio limit property using the Karlin–McGregor formula. The contents of this section are based on the references [44, 47, 48, 87].

The main problem with the asymptotic behavior of the Karlin–McGregor integral representation (2.62) is to analyze the integral on the negative part of the interval $[-1,1]$. We know that, in some sense, the spectral measure ψ is more concentrated on the interval $[0,1]$ than on $[-1,0]$, but the behavior on $[-1,0]$ may be chaotic. For that we will split the integral in two and study both situations asymptotically. Let us denote

$$\xi = \inf \operatorname{supp}(\psi), \quad \eta = \sup \operatorname{supp}(\psi). \tag{2.64}$$

We know that in this situation we have that $\xi + \eta \geq 0$ (see [48, Lemma 2.3]). We will focus first on sequences of the form

$$L_n(f,\psi) = \frac{\displaystyle\int_{-1}^{1} x^n f(x)\, d\psi(x)}{\displaystyle\int_{-1}^{1} x^n d\psi(x)}, \tag{2.65}$$

where f is a continuous function on $[-\eta,\eta)$ for which $\lim_{x\uparrow\eta} f(x)$ exists. We also allow the possibility of $|f(\eta)| = \infty$, in which case we assume that the integral $\int_{-1}^{1} x^n f(x)\, d\psi(x)$ is well defined. The sequence $L_n(f,\psi)$ in (2.65) can be written as

$$\begin{aligned} L_n(f,\psi) &= \frac{\left(\displaystyle\int_{-1}^{0} x^n f(x)\, d\psi(x) + \int_{0}^{1} x^n f(x)\, d\psi(x)\right) \Big/ \displaystyle\int_{0}^{1} x^n d\psi(x)}{\left(\displaystyle\int_{0}^{1} x^n d\psi(x) + (-1)^n \int_{-1}^{0} (-x)^n d\psi(x)\right) \Big/ \displaystyle\int_{0}^{1} x^n d\psi(x)} \\ &= \frac{A_n(f,\psi) + B_n(f,\psi)}{1 + (-1)^n C_n(\psi)}, \end{aligned}$$

where

$$A_n(f,\psi) = \frac{\displaystyle\int_{-1}^{0} x^n f(x)\,d\psi(x)}{\displaystyle\int_0^1 x^n d\psi(x)}, \quad B_n(f,\psi) = \frac{\displaystyle\int_0^1 x^n f(x)\,d\psi(x)}{\displaystyle\int_0^1 x^n d\psi(x)},$$

$$C_n(\psi) = \frac{\displaystyle\int_{-1}^{0} (-x)^n d\psi(x)}{\displaystyle\int_0^1 x^n d\psi(x)}. \tag{2.66}$$

So the asymptotic behavior of $L_n(f,\psi)$ will depend on the asymptotic behavior of these three new sequences. First, we will need the following lemma:

Lemma 2.37 ([48, Lemmas 3.1–3.3]) *Let $A_n(f,\psi), B_n(f,\psi)$ and $C_n(\psi)$ be the sequences defined by* (2.66). *Then we have:*

1. $\lim_{n\to\infty} B_n(f,\psi) = f(\eta)$.
2. *If $(n_k)_k$ is a subsequence of $(n)_n$ such that $\lim_{k\to\infty} (-1)^{n_k} C_{n_k}(\psi) = c$, then $\lim_{k\to\infty} A_{n_k}(f,\psi) = cf(-\eta)$.*
3. *For all $n \geq 0$ we have $0 \leq C_{2n+1}(\psi) \leq 1$, and there exists $\delta > 0$ such that $C_{2n+1} \leq 1 - \delta$ for n sufficiently large. Moreover, we have $0 \leq C_{2n}(\psi) \leq \left(\int_0^1 x\,d\psi(x)\right)^{-1} < \infty$.*

Proof For the first part, if $\eta < 1$, then $B_n(f,\psi) = B_n(g,\tilde{\psi})$ where $g(x) = f(x\eta)$ and $\tilde{\psi}$ is the spectral measure of the Derman–Vere–Jones transformation defined by (2.46). Recalling this transformation we have that $\tilde{\eta} = 1$ and $\tilde{q}_0 = 0$, so it is enough to consider the case $\eta = 1$ and $q_0 = 0$. For that, we first observe, for all $0 < \varepsilon < 1$, that

$$\frac{\displaystyle\int_0^{1-\varepsilon} x^n d\psi(x)}{\displaystyle\int_{1-\varepsilon}^1 x^n d\psi(x)} \leq \frac{(1-\varepsilon)^n}{(1-\varepsilon/2)^n \int_{1-\varepsilon/2}^1 d\psi(x)} \to 0, \quad \text{as} \quad n \to \infty.$$

Therefore, as $n \to \infty$, we get

$$|B_n(f,\psi) - f(1)| = \left| \frac{\displaystyle\int_0^1 x^n (f(x) - f(1))\,d\psi(x)}{\displaystyle\int_0^1 x^n d\psi(x)} \right|$$

$$= \left| \frac{\displaystyle\int_{1-\varepsilon}^{1} x^n (f(x) - f(1))\, d\psi(x)}{\displaystyle\int_{1-\varepsilon}^{1} x^n d\psi(x)} (1 + \mathcal{O}(1)) \right|$$

$$\leq \frac{\displaystyle\int_{1-\varepsilon}^{1} |x^n| d\psi(x) \int_{1-\varepsilon}^{1} |f(x) - f(1)| d\psi(x)}{\left| \displaystyle\int_{1-\varepsilon}^{1} x^n d\psi(x) \right|} (1 + \mathcal{O}(1))$$

$$\leq |f(1-\varepsilon) - f(1)|(1 + \mathcal{O}(1)).$$

Taking $\varepsilon \uparrow 1$ and using that f is continuous, we get $\lim_{n\to\infty} B_n(f,\psi) = f(1)$.

For the second part, as before, we can restrict ourselves to the case $\eta = 1$ and $q_0 = 0$. First, we observe, for the subsequence $(n_k)_k$ and using the definition of $C_n(\psi)$ in (2.66), that

$$\left| \frac{\displaystyle\int_{-1}^{-1+\varepsilon} x^{n_k} d\psi(x)}{\displaystyle\int_0^1 x^{n_k} d\psi(x)} - (-1)^{n_k} C_{n_k}(\psi) \right| = \left| \frac{\displaystyle\int_{-1+\varepsilon}^{0} x^{n_k} d\psi(x)}{\displaystyle\int_0^1 x^{n_k} d\psi(x)} \right|$$

$$\leq \frac{(1-\varepsilon)^{n_k}}{(1-\varepsilon/2)^{n_k} \int_{1-\varepsilon/2}^{1} d\psi(x)} \to 0,$$

as $k \to \infty$. Therefore, as $k \to \infty$, we get

$$|A_{n_k}(f,\psi) - (-1)^{n_k} C_{n_k}(\psi) f(-1)| = \left| \frac{\displaystyle\int_{-1}^{0} x^{n_k} (f(x) - f(-1))\, d\psi(x)}{\displaystyle\int_0^1 x^{n_k} d\psi(x)} \right|$$

$$= \left| \frac{\displaystyle\int_{-1}^{-1+\varepsilon} x^{n_k} (f(x) - f(-1))\, d\psi(x)}{\displaystyle\int_0^1 x^{n_k} d\psi(x)} (1 + \mathcal{O}(1)) \right|$$

$$\leq \frac{\displaystyle\int_{-1}^{-1+\varepsilon} |x^{n_k}| d\psi(x) \int_{-1}^{-1+\varepsilon} |f(x) - f(-1)| d\psi(x)}{\left| \displaystyle\int_0^1 x^n d\psi(x) \right|}$$

$$\times (1 + \mathcal{O}(1))$$

$$\leq |f(-1+\varepsilon) - f(-1)|(1 + \mathcal{O}(1)),$$

and, again, taking $\varepsilon \downarrow 0$ and using that f is continuous, we get $\lim_{n\to\infty} A_{n_k}(f,\psi) = cf(-1)$.

For the third part, since $C_n(\psi) \geq 0$ and $1+(-1)^n C_n(\psi) = P_{00}^{(n)} / \int_0^1 x^n d\psi(x) \geq 0$ we obtain that $1+(-1)^{2n+1}C_{2n+1}(\psi) \geq 0$. Therefore $C_{2n+1}(\psi) \leq 1$. For the bound we have that there exists an odd k such that $P_{00}^{(k)} > 0$ (by irreducibility). Then, for n sufficiently large, we have

$$1+(-1)^{2n+1}C_{2n+1}(\psi) = \frac{P_{00}^{(2n+1)}}{\int_0^1 x^{2n+1} d\psi(x)} \geq \frac{P_{00}^{(k)} P_{00}^{(2n+1-k)}}{\int_0^1 x^{2n+1} d\psi(x)}$$

$$\geq P_{00}^{(k)} B_{2n+1}(x^{-k},\psi) \to \frac{P_{00}^{(k)}}{\eta^k} = \delta,$$

and we get the bound. The last part is a consequence of the Hölder inequality (see [46] for more details). □

Now we will focus on the asymptotic behavior of $L_n(f,\psi)$ as $n\to\infty$.

Proposition 2.38 ([48, Proposition 3.1]) *Let $L_n(f,\psi)$ be defined by* (2.65). *Then we have:*

(i) If $f(\eta) = f(-\eta)$ or $|f(\eta)| = \infty$, then $\lim_{n\to\infty} L_n(f,\psi) = f(\eta)$.
(ii) If $f(\eta) \neq f(-\eta)$ and $|f(\eta)| < \infty$ then $\lim_{n\to\infty} L_n(f,\psi) = f(\eta)$ if and only if $\lim_{n\to\infty} C_n(\psi) = 0$.

Proof Let c be a point of accumulation of the sequence $((-1)^n C_n(\psi))_n$ and $(n_k)_k$ a subsequence of $(n)_n$ such that $(-1)^{n_k} C_{n_k}(\psi) \to c$, as $k\to\infty$. From the third part of Lemma 2.37 we have that $-1 < c < \infty$. Therefore, by the first and second parts of Lemma 2.37, we get

$$\lim_{k\to\infty} L_{n_k}(f,\psi) = \lim_{k\to\infty} \frac{A_{n_k}(f,\psi) + B_{n_k}(f,\psi)}{1+(-1)^{n_k} C_{n_k}(\psi)} = \frac{cf(-\eta)+f(\eta)}{1+c}.$$

For the first part, if $f(\eta) = f(-\eta)$ then it is obvious. If $|f(\eta)| = \infty$, then this limit is also ∞ by the continuity of f at $x = -\eta$. These limits are independent of c, so $f(\eta)$ is the only point of accumulation of the sequence $(L_n(f,\psi))_n$.

For the second part, assuming that $f(\eta) \neq f(-\eta)$ and $|f(\eta)| < \infty$, we have, on one hand, that if $C_n(\psi) \to 0$ then $c = 0$, so $L_n(f,\psi) \to f(\eta)$ as $n \to \infty$. On the other hand, assume that $C_n(\psi)$ does not converge to 0 and we have two points of accumulation c_1 and c_2 with their respectives subsequences and bound properties. Then, since $\lim_{n\to\infty} L_n(f,\psi)$ exists and is unique, we must have

$$\frac{c_1 f(-\eta)+f(\eta)}{1+c_1} = \frac{c_2 f(-\eta)+f(\eta)}{1+c_2} \quad \Leftrightarrow \quad \frac{c_1-c_2}{(1+c_1)(1+c_2)}(f(-\eta)-f(\eta)) = 0.$$

Therefore $f(-\eta) = f(\eta)$ and we get a contradiction. □

Therefore, in order to get convergence of the limit of $L_n(f,\psi)$ (and therefore the strong ratio limit property (2.63), as we will see below) it is necessary and sufficient that $C_n(\psi) \to 0$ as $n \to \infty$. This condition is usually difficult to check, so it would be good to have sufficient criteria for that to hold. One criteria was already given in [87, Lemma 3]:

Proposition 2.39 *If the birth–death chain is aperiodic and recurrent then* $C_n(\psi) \to 0$ *as* $n \to \infty$.

Proof Assume that there exists a subsequence $(n_k)_k$ of $(n)_n$ such that $(-1)^{n_k} C_{n_k}(\psi) \to c \neq 0$, as $k \to \infty$. Let Q be any polynomial and write it as $Q = R + S$ where R is even and S is odd. Since the birth–death chain is recurrent, then $\eta = 1$. Using the first and second parts of Lemma 2.37, we obtain

$$\begin{aligned}
&\lim_{k\to\infty} \frac{\displaystyle\int_{-1}^{1} x^{n_k} Q(x)\, d\psi(x)}{\displaystyle\int_{-1}^{1} x^{n_k}\, d\psi(x)} \\
&= \lim_{k\to\infty} \frac{\displaystyle\int_{-1}^{0} x^{n_k}(R(x)+S(x))\, d\psi(x) + \int_{0}^{1} x^{n_k}(R(x)+S(x))\, d\psi(x)}{\displaystyle\int_{-1}^{1} x^{n_k}\, d\psi(x)} \\
&= \lim_{k\to\infty} \left(A_{n_k}(R+S,\psi) + B_{n_k}(R+S,\psi)\right) = c(R(1)-S(1)) + R(1) + S(1) \\
&= Q(1) + cQ(-1).
\end{aligned}$$

Similarly, using the symmetry of the polynomials R and S, we obtain

$$\lim_{k\to\infty} \frac{\displaystyle\int_{-1}^{1} x^{n_k+1} Q(x)\, d\psi(x)}{\displaystyle\int_{-1}^{1} x^{n_k}\, d\psi(x)} = Q(1) - cQ(-1).$$

Taking $Q = Q_i$ (the ith polynomial generated by the three-term recurrence relation (2.15)) and using that $\int_{-1}^{1} x^n Q_i(x) d\psi(x) = P_{i,0}^{(n)} \geq 0$, we get that $Q_i(1) - cQ_i(-1) \geq 0$ and $Q_i(1) + cQ_i(-1) \geq 0$. Therefore

$$|c| \leq \frac{Q_i(1)}{|Q_i(-1)|} = \frac{1}{|Q_i(-1)|},$$

since the birth–death chain is recurrent and $Q_i(1) = 1$. We will prove now that, if i is sufficiently large, then $|Q_i(-1)|$ diverges. Indeed, call $\alpha_i = (-1)^i Q_i(-1)$. Then, a straightforward computation using (2.20) (for $q_0 = 0$ and $x = -1$) gives

$$\alpha_0 = 1, \quad \alpha_{n+1} = 1 + 2\sum_{i=0}^{n} \frac{1}{p_i\pi_i} \sum_{k=0}^{i} r_k\pi_k\alpha_k, \quad n \geq 0.$$

This shows that $1 = \alpha_0 \leq \alpha_1 \leq \cdots$. Since the birth–death chain is irreducible (so there exists at least one i such that $r_i > 0$) and recurrent (so $\sum_{n=0}^{\infty}(p_n\pi_n)^{-1} = \infty$), then $\alpha_n \to \infty$ as $n \to \infty$. Therefore $|Q_i(-1)| = |(-1)^i Q_i(-1)| = |\alpha_i| \to \infty$ as $i \to \infty$, so $c = 0$, meaning that $C_n(\psi) \to 0$ as $n \to \infty$. □

The following corollary can be proved in a similar way to the previous proposition:

Corollary 2.40 *Let* $0 < \varepsilon < \eta$. *Then*

$$\lim_{n\to\infty} C_n(\psi) = 0 \quad \Leftrightarrow \quad \lim_{n\to\infty} \frac{\displaystyle\int_{-\eta}^{-\eta+\varepsilon} (-x)^n d\psi(x)}{\displaystyle\int_{\eta-\varepsilon}^{\eta} x^n d\psi(x)} = 0.$$

If the birth–death chain is not recurrent then we cannot apply the previous proposition so we need other criteria to determine if $C_n(\psi) \to 0$ as $n \to \infty$. In [48, Corollary 3.1, 3.2 and Theorem 3.2] and [44] there are four different criteria which we will summarize in the following proposition:

Proposition 2.41 *Assume that one of the following four conditions hold:*

- $\eta > -\xi$, *where* η *and* ξ *are defined by* (2.64).
- $r_n \geq \delta > 0$ *for* n *sufficiently large.*
- $|Q_n(-\eta)/Q_n(\eta)| \to \infty$ *as* $n \to \infty$.
- $\sum_{j=0}^{\infty}(p_j\pi_j)^{-1}\sum_{k=0}^{j} r_k\pi_k = \infty$.

Then $C_n(\psi) \to 0$ *as* $n \to \infty$.

Proof For the first condition, we observe that $\xi + \eta > \varepsilon > 0$. Then, using Corollary 2.40, we have that the integral in the numerator vanishes. Therefore $C_n(\psi) \to 0$. For the second condition, we use (2.45). Since $r_n \geq \delta > 0$ for n sufficiently large that means that $\xi + \eta > 0$ and we are back to the first condition. The third part is a consequence of the Derman–Vere–Jones transformation (see Section 2.3.3) and the proof of Proposition 2.39. The transformed family of polynomials is given by $\tilde{Q}_j(x) = Q_j(x\eta)/Q_j(\eta)$. In particular $\tilde{Q}_j(-1) = Q_j(-\eta)/Q_j(\eta)$. So, as in the proof of Proposition 2.39, if $\alpha_n = (-1)^n\tilde{Q}_n(-1)$, then $\alpha_n \to \infty$, and this implies that $C_n(\psi) \to 0$. The fourth condition has been recently proved in [44] and this implies the third one. □

Remark 2.42 In [48] it is conjectured that

$$\lim_{n\to\infty} C_n(\psi) = \lim_{n\to\infty} \frac{Q_n^2(\eta)}{Q_n^2(-\eta)}.$$ ◊

We are now ready to prove the strong ratio limit property.

Theorem 2.43 ([48, Theorem 3.1]) *Let ψ be the spectral measure of an aperiodic discrete-time birth–death chain. Then we have that*

$$\lim_{n\to\infty}\frac{P_{ij}^{(n)}}{P_{kl}^{(n)}}=\frac{\pi_j Q_i(\eta)Q_j(\eta)}{\pi_l Q_k(\eta)Q_l(\eta)} \tag{2.67}$$

if and only if $C_n(\psi)\to 0$ as $n\to\infty$.

Proof On the one hand, if $C_n(\psi)\to 0$ as $n\to\infty$, then

$$\frac{P_{ij}^{(n)}}{P_{kl}^{(n)}}=\frac{P_{ij}^{(n)}}{P_{00}^{(n)}}\frac{P_{00}^{(n)}}{P_{kl}^{(n)}}=\frac{L_n(\pi_j Q_i Q_j,\psi)}{L_n(\pi_l Q_k Q_l,\psi)},$$

where $L_n(f,\psi)$ is defined by (2.65). Since $Q_i(\eta)Q_j(\eta)\neq Q_i(-\eta)Q_j(-\eta)$ (for instance, for $i=1$ and $j=0$) and $|Q_i(\eta)Q_j(\eta)|<\infty$, then, by the second part of Proposition 2.38, we get

$$\lim_{n\to\infty}\frac{P_{ij}^{(n)}}{P_{kl}^{(n)}}=\lim_{n\to\infty}\frac{L_n(\pi_j Q_i Q_j,\psi)}{L_n(\pi_l Q_k Q_l,\psi)}=\frac{\pi_j Q_i(\eta)Q_j(\eta)}{\pi_l Q_k(\eta)Q_l(\eta)}.$$

On the other hand, if we assume that the limits (2.67) exist simultaneously and that $C_n(\psi)$ does not converge to 0 then, for instance, $Q_1(\eta)\neq Q_1(-\eta)$ with $|Q_1(\eta)|<\infty$, so we are in the conditions of the second part of Proposition 2.38. Since we are assuming that $C_n(\psi)$ does not converge to 0, then $\lim_{n\to\infty}L_n(Q_1,\psi)=P_{10}^{(n)}/P_{00}^{(n)}$ does not exist. But this is a contradiction with the hypothesis. □

Remark 2.44 Observe that in the particular case of a recurrent birth–death chain, we have that $\eta=1$, $Q_i(1)=1$ and, using Proposition 2.39, $C_n(\psi)\to 0$ as $n\to\infty$. Therefore the previous theorem implies that

$$\lim_{n\to\infty}\frac{P_{ij}^{(n)}}{P_{kl}^{(n)}}=\frac{\pi_j}{\pi_l},$$

which is the result proved in [87, Theorem 2]. ◊

Remark 2.45 If the process is not recurrent, i.e. transient, the limit (2.67) may still exist but does not necessarily have the value (2.67). This can be shown by looking at Example 2.22 for $r_0=1-p, p_0=p$ and $p>1/2$. Then we have that $P_{01}^{(2n)}/P_{00}^{(2n)}$ and $P_{01}^{(2n+1)}/P_{00}^{(2n+1)}$ converge to different limits. This can be seen by looking at the first powers of P and noticing that $P_{01}^{(2n)}$ and $P_{01}^{(2n+1)}$ converge to different limits, while $P_{00}^{(2n)}$ and $P_{00}^{(2n+1)}$ converge to the same limit. ◊

Remark 2.46 If the birth–death chain is symmetric (i.e. $r_n=0$ for all $n\geq 0$) then the limit fails to exist because $P_{ij}^{(n)}$ is zero if $i-j$ and n have different parity. Nevertheless we have (see [87, Theorem 3]) that

$$\lim_{n\to\infty}\frac{P_{ij}^{(2n)}}{P_{kl}^{(2n)}}=\frac{\pi_j Q_i(\eta)Q_j(\eta)}{\pi_l Q_k(\eta)Q_l(\eta)},\quad \text{if}\quad i-j\quad\text{and}\quad k-l\quad\text{are even},$$

$$\lim_{n\to\infty}\frac{P_{ij}^{(2n+1)}}{P_{kl}^{(2n+1)}}=\frac{\pi_j Q_i(\eta)Q_j(\eta)}{\pi_l Q_k(\eta)Q_l(\eta)},\quad \text{if}\quad i-j\quad\text{and}\quad k-l\quad\text{are odd.}\qquad\Diamond$$

Example 2.47 Consider the example of the birth–death chain associated with the Jacobi polynomials in Section 2.4.3. According to Theorem 2.7, we have a proper birth–death chain if $\alpha=\beta$ (symmetric) or $\beta\geq|\alpha|$. We want to study the strong ratio limit property. We will use the third condition in Proposition 2.41. Since we know the values of $P_n^{(\alpha,\beta)}(1)$ and $P_n^{(\alpha,\beta)}(-1)$ (see (1.82)), the birth–death polynomials $Q_n^{(\alpha,\beta)}(x)=P_n^{(\alpha,\beta)}(x)/P_n^{(\alpha,\beta)}(1)$ satisfy that

$$Q_n^{(\alpha,\beta)}(1)=1,\quad Q_n^{(\alpha,\beta)}(-1)=(-1)^n\frac{(n+\beta)_n}{(n+\alpha)_n}.$$

Therefore

$$\lim_{n\to\infty}\left|\frac{Q_n^{(\alpha,\beta)}(-1)}{Q_n^{(\alpha,\beta)}(1)}\right|=\begin{cases}\infty, & \text{if}\quad \beta>\alpha,\\ 1, & \text{if}\quad \beta=\alpha,\\ 0, & \text{if}\quad \beta<\alpha.\end{cases}$$

Therefore if $\beta>\alpha>-1$ then $C_n(\psi)\to 0$ as $n\to\infty$ and we can apply Theorem 2.43. Therefore

$$\lim_{n\to\infty}\frac{P_{ij}^{(n)}}{P_{kl}^{(n)}}=\frac{\pi_j}{\pi_l}.$$

The potential coefficients for this example were already computed in (2.52). Therefore

$$\begin{aligned}\lim_{n\to\infty}\frac{P_{ij}^{(n)}}{P_{kl}^{(n)}}&=\frac{l!\,(2j+\alpha+\beta+1)(\alpha+1)_j(\beta+1)_l(\alpha+\beta+1)_j}{j!\,(2l+\alpha+\beta+1)(\alpha+1)_l(\beta+1)_j(\alpha+\beta+1)_l}\\&=\frac{(2j+\alpha+\beta+1)\Gamma(l+1)\Gamma(\alpha+1+j)\Gamma(\alpha+\beta+1+j)\Gamma(\beta+1+l)}{(2l+\alpha+\beta+1)\Gamma(j+1)\Gamma(\alpha+1+l)\Gamma(\alpha+\beta+1+l)\Gamma(\beta+1+j)}.\end{aligned}$$

For instance, for $\beta=1$ and $\alpha=0$ we have that $\lim_{n\to\infty}P_{ij}^{(n)}/P_{kl}^{(n)}=(j+1)/(l+1)$. $\Diamond$

Example 2.48 Consider the example of Exercise 2.7 where $p_n=p, q_{n+1}=q$, $n\geq 0$ with $0<r_0\leq q<1$. The spectral measure is absolutely continuous when $|x|<2\sqrt{pq}$ except for $r_0>\sqrt{pq}$, where we get a jump at the point r_0+pq/r_0 of size $1-pq/r_0^2$. Using the notation in (2.64), we get that

$$\xi = -2\sqrt{pq}, \quad \eta = \begin{cases} 2\sqrt{pq}, & \text{if} \quad r_0 \leq \sqrt{pq}, \\ r_0 + pq/r_0, & \text{if} \quad r_0 > \sqrt{pq}. \end{cases}$$

Observe that $r_0 + pq/r_0 \geq 2\sqrt{pq}$ as a consequence of $(r_0 - \sqrt{pq})^2 \geq 0$. If $r_0 = 1 - p$ then $r_0 + pq/r_0 = 1$ and we go back to Example 2.22 (2). We can see, using the criterion of $\eta > -\xi$ in Proposition 2.41 and some analysis on the measure (see [48]), that

$$\lim_{n\to\infty} C_n(\psi) = \begin{cases} 0, & \text{if} \quad r_0 \geq \sqrt{pq}, \\ \left(\dfrac{\sqrt{pq} - r_0}{\sqrt{pq} + r_0}\right)^2, & \text{if} \quad r_0 < \sqrt{pq}. \end{cases}$$

The orthogonal polynomials associated with the birth–death chain can be written as

$$Q_j(x) = (q/p)^{j/2} P_j(x/2\sqrt{pq}),$$

where P_j satisfies the recurrence relation

$$P_0(x) = 1, \quad P_1(x) = 2x - r_0/\sqrt{pq},$$
$$2xP_j(x) = P_{j-1}(x) + P_{j+1}(x), \quad j \geq 1.$$

Therefore we can compute in a recurrent way the expression $P_j(1)$ and get

$$Q_j(2\sqrt{pq}) = (q/p)^{j/2}(1 + j(1 - r_0/\sqrt{pq})),$$
$$Q_j(-2\sqrt{pq}) = (-1)^j (q/p)^{j/2}(1 + j(1 + r_0/\sqrt{pq})). \tag{2.68}$$

Using that the potential coefficients are given by $\pi_n = (p/q)^n, n \geq 0$, we can compute an explicit expression of the limit (2.67). Indeed,

$$\lim_{n\to\infty} \frac{P_{ij}^{(n)}}{P_{kl}^{(n)}} = \left(\sqrt{\frac{p}{q}}\right)^{j-i+k-l} \frac{(1 + i(1 - r_0/\sqrt{pq}))(1 + j(1 - r_0/\sqrt{pq}))}{(1 + k(1 - r_0/\sqrt{pq}))(1 + l(1 - r_0/\sqrt{pq}))}.$$

For instance, if $p = q = 1/2$, we have that

$$\lim_{n\to\infty} \frac{P_{ij}^{(n)}}{P_{kl}^{(n)}} = \frac{(1 + i(1 - 2r_0))(1 + j(1 - 2r_0))}{(1 + k(1 - 2r_0))(1 + l(1 - 2r_0))}.$$

Observe that in this case the ratio limits depend on the initial states i and k. If, for instance $r_0 = 1/2$ (so the birth–death chain is recurrent), then we will have $\lim_{n\to\infty} P_{ij}^{(n)}/P_{kl}^{(n)} = \pi_j/\pi_l = 1$. ◊

2.5.3 Limiting Conditional Distribution

Let $\{X_n, n = 0, 1, \ldots\}$ be a discrete-time birth–death chain with transition probabilities $p_j, r_j, q_j, j \geq 0$. We are now interested in finding an initial distribution $(\nu_j)_j$ such that the probability of the birth–death chain being in state j at time n is independent of n for all j. If $q_0 = 0$ this initial distribution is given by the invariant measure $(\pi_j)_j$, which is defined by the potential coefficients (2.13) and satisfies that $\pi P = \pi$. Therefore $\pi P^n = \pi$ for all $n \geq 0$ and the vector π with the potential coefficients as components is the (not normalized) initial vector. But $(\pi_j)_j$ is not a proper distribution unless $\sum_{k=0}^{\infty} \pi_k < \infty$. In that case the initial distribution is given by

$$(\nu_j)_j, \quad \nu_j = \frac{\pi_j}{\sum_{k=0}^{\infty} \pi_k}, \quad j \geq 0.$$

That is the case of a positive recurrent or ergodic birth–death chain. If the birth–death chain is either null recurrent or transient, then $\sum_{k=0}^{\infty} \pi_k = \infty$ and $\nu_j = 0$ for all j, so it does not converge to a proper distribution.

We consider now the more general case of a birth–death chain where $q_0 = 1 - r_0 - p_0 > 0$, i.e. -1 is an absorbing state. We obviously have that the birth–death chain is automatically transient, so such an initial distribution will never exist. However, we wonder if there is a *quasi-stationary distribution*, i.e. an initial distribution $(\nu_j)_j$ such that the conditional probability of the birth–death chain being in state j at time n, *given that the absorption has not occurred by that time*, is independent of n for all j. For that we will focus first on finding conditions on the transition probabilities $p_j, r_j, q_j, j \geq 0$, such that the probabilities of arriving to state j starting from i at time n given that absorption has not occurred by that time, as $n \to \infty$, converge to a proper distribution. This distribution is usually called the *limiting conditional distribution* and is defined as

$$\lim_{n\to\infty} q_{ij}^{(n)} = \lim_{n\to\infty} \mathbb{P}(X_n = j \mid X_0 = i, n < T_{-1} < \infty), \tag{2.69}$$

where T_{-1} denotes the absorption time at state -1. Limiting conditional distributions of discrete-time Markov chains have been studied in [26, 104, 131]. The approach we present here for birth–death chains uses the Karlin–McGregor representation formula and was studied in [47, 48].

Let us call $a_i = \mathbb{P}(T_{-1} < \infty | X_0 = i)$ the probability of eventual absorption to the state -1 starting from state i. These probabilities were given by (2.58) and (2.60). From (2.69), a simple computation, using conditional probability and the Markov property, shows that

$$\begin{aligned} q_{ij}^{(n)} &= \frac{\mathbb{P}(X_n = j, X_0 = i, n < T_{-1} < \infty)}{\mathbb{P}(X_0 = i, n < T_{-1} < \infty)} = \frac{\mathbb{P}(X_n = j, n < T_{-1} < \infty | X_0 = i)}{\mathbb{P}(n < T_{-1} < \infty | X_0 = i)} \\ &= \frac{\mathbb{P}(n < T_{-1} < \infty | X_n = j, X_0 = i)\mathbb{P}(X_n = j | X_0 = i)}{\mathbb{P}(T_{-1} < \infty | X_0 = i) - \mathbb{P}(T_{-1} \leq n | X_0 = i)} \\ &= \frac{\mathbb{P}(T_{-1} < \infty | X_0 = j)\mathbb{P}(X_n = j | X_0 = i)}{\mathbb{P}(T_{-1} < \infty | X_0 = i) - \mathbb{P}(T_{-1} \leq n | X_0 = i)} = \frac{a_j P_{ij}^{(n)}}{a_i - P_{i,-1}^{(n)}}. \end{aligned} \tag{2.70}$$

By definition, we have $P_{i,-1}^{(n)} = q_0 \sum_{k=0}^{n-1} P_{i0}^{(k)}$. Using the integral representation (2.17), we get

$$P_{i,-1}^{(n)} = q_0 \sum_{k=0}^{n-1} \int_{-1}^{1} x^k Q_i(x)\, d\psi(x) = q_0 \int_{-1}^{1} \frac{1-x^n}{1-x} Q_i(x)\, d\psi(x)$$
$$= q_0 \int_{-1}^{1} \frac{Q_i(x)\, d\psi(x)}{1-x} - q_0 \int_{-1}^{1} \frac{x^n Q_i(x)\, d\psi(x)}{1-x} = a_i - q_0 \int_{-1}^{1} \frac{x^n Q_i(x)\, d\psi(x)}{1-x}.$$

Since $\lim_{n\to\infty} P_{i,-1}^{(n)} = a_i$ we get that $\lim_{n\to\infty} \int_{-1}^{1} (1-x)^{-1} x^n Q_i(x)\, d\psi(x) = 0$. Therefore $q_{ij}^{(n)}$ can be written, using the Karlin–McGregor representation (2.17), as

$$q_{ij}^{(n)} = \frac{a_j P_{ij}^{(n)}}{a_i - P_{i,-1}^{(n)}} = \frac{a_j \pi_j}{q_0} \frac{\displaystyle\int_{-1}^{1} x^n Q_i(x) Q_j(x)\, d\psi(x)}{\displaystyle\int_{-1}^{1} \frac{x^n Q_i(x)\, d\psi(x)}{1-x}} = \frac{a_j \pi_j}{q_0} \frac{L_n(Q_i Q_j, \psi)}{L_n\left(\dfrac{Q_i}{1-x}, \psi\right)}, \tag{2.71}$$

using the notation already introduced in (2.65). Recall that $\eta = \sup \operatorname{supp}(\psi)$ and the sequence $C_n(\psi)$ is defined by (2.66). Then we have the following:

Lemma 2.49 ([48, Lemma 4.1]) *Let i be an initial state of an irreducible discrete-time birth–death chain (periodic or aperiodic) with $q_0 > 0$.*

(i) If $\eta = 1$ then $\lim_{n\to\infty} q_{ij}^{(n)} = 0$.

(ii) If $\eta < 1$ then $\lim_{n\to\infty} q_{ij}^{(n)}$ exist for all i,j if and only if $\lim_{n\to\infty} C_n(\psi) = 0$, in which case we have

$$\lim_{n\to\infty} q_{ij}^{(n)} = \frac{1-\eta}{q_0} a_j \pi_j Q_j(\eta). \tag{2.72}$$

Proof Assume first that the birth–death chain is aperiodic. Then we can use Proposition 2.38. If $\eta = 1$, then $L_n(Q_i Q_j, \psi)$ is bounded, since in general $Q_i(1)Q_j(1) \neq Q_i(-1)Q_j(-1)$ and $|Q_i(1)Q_j(1)| < \infty$. Alternatively, $\lim_{x\uparrow 1}(1-x)^{-1} Q_i(x) = \infty$, so $L_n((1-x)^{-1} Q_i, \psi) \to \infty$ as $n \to \infty$. Therefore (2.71) gives that $q_{ij}^{(n)} \to 0$ as $n \to \infty$. If $\eta < 1$ we use again part 2 of Proposition 2.38 and from (2.71), we have

$$\lim_{n\to\infty} q_{ij}^{(n)} = \lim_{n\to\infty} \frac{a_j \pi_j}{q_0} \frac{L_n(Q_i Q_j, \psi)}{L_n\left(\dfrac{Q_i}{1-x}, \psi\right)} = \frac{a_j \pi_j Q_i(\eta) Q_j(\eta)}{q_0 Q_i(\eta)/(1-\eta)} = \frac{1-\eta}{q_0} a_j \pi_j Q_j(\eta).$$

Now assume that the birth–death chain is periodic. Then the spectral measure ψ is symmetric, so $Q_i(x) = (-1)^n Q_i(-x)$. We also have that $C_n(\psi) = 1$ for all $n \geq 0$, so the second part of this lemma should not hold. Using the symmetry of ψ, let us write $q_{ij}^{(n)}$ as

$$q_{ij}^{(n)} = \frac{a_j\pi_j}{q_0} \frac{\displaystyle\int_0^1 x^n(1+(-1)^{n+i+j})Q_i(x)Q_j(x)\,d\psi(x)}{\displaystyle\int_0^1 x^n\left(\frac{1}{1-x}+\frac{(-1)^{n+i}}{1-x}\right)Q_i(x)\,d\psi(x)}.$$

These integrals can be written in terms of $B_n(f,\psi)$ (see (2.66)). Indeed, for n even or odd, we have

$$q_{ij}^{(2n)} = \frac{a_j\pi_j}{q_0}\frac{B_{2n}(Q_iQ_j,\psi)+(-1)^{i+j}B_{2n}(Q_iQ_j,\psi)}{B_{2n}((1-x)^{-1}Q_i,\psi)+(-1)^iB_{2n}((1+x)^{-1}Q_i,\psi)},$$

$$q_{ij}^{(2n+1)} = \frac{a_j\pi_j}{q_0}\frac{B_{2n+1}(Q_iQ_j,\psi)-(-1)^{i+j}B_{2n+1}(Q_iQ_j,\psi)}{B_{2n+1}((1-x)^{-1}Q_i,\psi)-(-1)^iB_{2n+1}((1+x)^{-1}Q_i,\psi)}.$$

Therefore, using the first part of Lemma 2.37, we obtain

$$\begin{aligned}\lim_{n\to\infty} q_{ij}^{(2n)} &= \frac{a_j\pi_j}{q_0}\frac{Q_i(\eta)Q_j(\eta)(1+(-1)^{i+j})}{Q_i(\eta)((1-\eta)^{-1}+(-1)^i(1+\eta)^{-1})}\\ &= \frac{a_j\pi_j(1-\eta)^2Q_j(\eta)(1+(-1)^{i+j})}{q_0(1+\eta+(-1)^i(1-\eta))},\\ \lim_{n\to\infty} q_{ij}^{(2n+1)} &= \frac{a_j\pi_j}{q_0}\frac{Q_i(\eta)Q_j(\eta)(1-(-1)^{i+j})}{Q_i(\eta)((1-\eta)^{-1}-(-1)^i(1+\eta)^{-1})}\\ &= \frac{a_j\pi_j(1-\eta)^2Q_j(\eta)(1-(-1)^{i+j})}{q_0(1+\eta-(-1)^i(1-\eta))}.\end{aligned}$$

If $\eta < 1$ we clearly see that both limits are different. For instance, if $i = j = 0$, then $q_{00}^{(2n)} \to \frac{a_0}{q_0}(1-\eta)^2$ but $q_{00}^{(2n+1)} \to 0$. So the limit never exists, as we wanted to prove. If $\eta = 1$ we clearly see that both limits tend to 0, so we get convergence to 0. □

We are now ready to prove the limiting conditional distribution theorem:

Theorem 2.50 ([48, Theorem 4.1]) *Let i be an initial state of an irreducible discrete-time birth–death chain with $q_0 > 0$.*

(i) If $\eta = 1$ then $(q_{ij}^{(n)})_j$ does not converge to a proper distribution as $n \to \infty$.
(ii) If $\eta < 1$ then $(q_{ij}^{(n)})_j$ converge to a proper distribution as $n \to \infty$ if and only if $C_n(\psi) \to 0$ as $n \to \infty$, in which case we have

$$\lim_{n\to\infty} q_{ij}^{(n)} = \frac{a_j\pi_jQ_j(\eta)}{\displaystyle\sum_{k=0}^{\infty} a_k\pi_kQ_k(\eta)}. \tag{2.73}$$

Proof The first part is a consequence of the first part of Lemma 2.49. The second part, on one hand, if $C_n(\psi) \to 0$ as $n \to \infty$, then we are in the conditions of the second part of Lemma 2.49 and we get (2.72). Using now the formula (see remark after the proof of this theorem)

$$\frac{1-\eta}{q_0}\sum_{k=0}^{\infty} a_k\pi_k Q_k(\eta) = 1, \tag{2.74}$$

we get the result. On the other hand, if the birth–death chain is periodic, then we saw in the proof of Lemma 2.49 that $C_n(\psi) = 1$ and $\lim_{n\to\infty} q_{ij}^{(n)}$ does not exist. If the birth–death chain is aperiodic, using a similar argument to the proof of the second part of Proposition 2.38 (assuming that we have two points of accumulation c_1 and c_2 of the sequence $(-1)^n C_n(\psi)$ with their respective subsequences and bound properties), we should have that $Q_i(-\eta)(1+\eta) = Q_i(\eta)(1-\eta)$. But this is not true (for instance for $i=0$). Therefore we get a contradiction and $C_n(\psi) \to 0$ as $n \to \infty$. □

Remark 2.51 Observe that the previous theorem still holds for the case where absorption is certain, i.e. $a_i = 1$ for all i. ◊

Remark 2.52 Let us now give a justification of formula (2.74). The expression of $Q_n(x)$ given in formula (2.22) can be rewritten, changing the order of summation in the second part of the formula, as

$$Q_n(x) = 1 + q_0\sum_{k=0}^{n-1}\frac{1}{p_k\pi_k} + \frac{x-1}{q_0}\sum_{k=0}^{n-1}\pi_k Q_k(x)\sum_{j=k}^{n-1}\frac{q_0}{p_j\pi_j}.$$

Taking $n\to\infty$ in the previous formula, using (2.23) for $Q_n(1)$ and (2.60) for a_i, we obtain

$$Q_\infty(x) = Q_\infty(1) + \frac{x-1}{q_0}\sum_{k=0}^{\infty}\pi_k Q_k(x)\sum_{j=k}^{\infty}\frac{q_0}{p_j\pi_j},$$

where $Q_\infty(x) = \lim_{n\to\infty} Q_n(x)$ (whenever this makes sense, see below). Therefore

$$\frac{1-x}{q_0}\sum_{k=0}^{\infty} a_k\pi_k Q_k(x) = 1 - \frac{Q_\infty(x)}{Q_\infty(1)}.$$

Similarly, using the dual polynomials Q_n^d defined by (2.21), we have

$$\frac{1-x}{q_0}\sum_{k=0}^{\infty}\pi_k Q_k(x) = 1 - Q_\infty^d(x).$$

Then formula (2.74) will be proved if $Q_\infty(\eta) = 0$, i.e. η is a zero of the analytic function $Q_\infty(x)$, or also if $Q_\infty^d(\eta) = 0$ in the case where $a_i = 1$ for all $i \geq 0$. This is true as a consequence of [83, Lemma 4]. Indeed, let us call

$$A = \sum_{n=0}^{\infty} \frac{1}{p_n \pi_n}, \quad B = \sum_{n=0}^{\infty} \pi_n, \quad C = \sum_{n=0}^{\infty} \frac{1}{p_n \pi_n} \sum_{j=0}^{n} \pi_j, \quad D = \sum_{n=0}^{\infty} \pi_n \sum_{j=0}^{n} \frac{1}{p_j \pi_j}.$$

Observe that $AB = C + D$. In the case of birth–death chains we always have that $A + B = \infty$ (see the comment at the beginning of Section 2.5.1). Then $C + D = \infty$. We have then two possible situations (see [83, Lemma 4] for more details):

- If $a_i < 1$, then, according to Theorem 2.32, we have that $A < \infty$. Therefore $B = \infty$, $C < \infty$ (otherwise A would not be finite) and $D = \infty$. $C < \infty$ is a necessary and sufficient condition for the convergence of $(Q_n(x))_n$ uniformly on bounded sets of the complex plane to an entire function $Q_\infty(x)$ whose zeros are simple and are precisely η_i (see (1.28)). In particular $Q_\infty(\eta) = 0$ and we obtain the formula.
- If $a_i = 1$, then, according to Theorem 2.32, we have that $A = \infty$ and $B < \infty$ (since we are assuming that absorption eventually occurs in finite time). Therefore $C = \infty$ and $D < \infty$ (otherwise B would not be finite). $D < \infty$ is a necessary and sufficient condition for the convergence of the dual polynomials $(Q_n^d(x))_n$ uniformly on bounded sets of the complex plane to an entire function $Q_\infty^d(x)$ whose zeros are simple and are precisely η_i (see (1.28)). In particular $Q_\infty^d(\eta) = 0$ and we obtain again the formula (for $a_i = 1$).

In Chapter 3 we will give a probabilistic interpretation of the numbers A, B, C, D. ◊

The limiting conditional distribution is related to the concept of the *quasi-stationary distribution*. Indeed, let $(\nu_j)_j$ be an honest initial distribution and define

$$p_j^{(n)} = \mathbb{P}_\nu(X_n = j) \doteq \sum_{i=0}^{\infty} \nu_i P_{ij}^{(n)}, \quad a = \mathbb{P}_\nu(T_{-1} < \infty) \doteq \sum_{i=0}^{\infty} \nu_i a_i.$$

Then, a simple computation, using the definition of $q_{ij}^{(n)}$ in (2.70), gives that

$$q_j^{(n)} \doteq \frac{a_j p_j^{(n)}}{a - p_{-1}^{(n)}} = \sum_{i=0}^{\infty} q_{ij}^{(n)} \left(\frac{\nu_i (a_i - P_{i,-1}^{(n)})}{\sum_{k=0}^{\infty} \nu_k (a_k - P_{k,-1}^{(n)})} \right)$$

is a quasi-stationary distribution for the birth–death chain, as it was defined at the beginning of this section.

An extension of the limiting conditional distribution is the concept of the *doubly limiting conditional distribution*. In this case, we assume that the birth–death chain does not leave the state space $\{0, 1, 2, \ldots\}$ in the distant future, but the absorption eventually occurs with some probability. The definition is given by

$$\lim_{n\to\infty} \lim_{m\to\infty} q_{ij}^{(n,m)} = \lim_{n\to\infty} \lim_{m\to\infty} \mathbb{P}(X_n = j \mid X_0 = i, n + m < T_{-1} < \infty).$$

Following the same lines as for $q_{ij}^{(n)}$ in (2.70), we have

$$
\begin{aligned}
q_{ij}^{(n,m)} &= \frac{\mathbb{P}(X_n = j, X_0 = i, n+m < T_{-1} < \infty)}{\mathbb{P}(X_0 = i, n+m < T_{-1} < \infty)} \\
&= \frac{\mathbb{P}(n+m < T_{-1} < \infty | X_n = j, X_0 = i)\mathbb{P}(X_n = j | X_0 = i)}{\mathbb{P}(T_{-1} < \infty | \, X_0 = i) - \mathbb{P}(T_{-1} \leq n+m | \, X_0 = i)} \\
&= \frac{\mathbb{P}(m < T_{-1} < \infty | X_0 = j)\mathbb{P}(X_n = j | X_0 = i)}{\mathbb{P}(T_{-1} < \infty | \, X_0 = i) - \mathbb{P}(T_{-1} \leq n+m | \, X_0 = i)} \\
&= \frac{(\mathbb{P}(T_{-1} < \infty | X_0 = j) - \mathbb{P}(T_{-1} \leq m | X_0 = j))P_{ij}^{(n)}}{a_i - P_{i,-1}^{(n+m)}} = \frac{P_{ij}^{(n)}(a_j - P_{j,-1}^{(m)})}{a_i - P_{i,-1}^{(n+m)}}.
\end{aligned}
$$

Theorem 2.53 *Let i be an initial state of an aperiodic discrete-time birth–death chain with $q_0 > 0$. If $\eta < 1$ and $C_n(\psi) \to 0$ as $n \to \infty$, then $(q_{ij}^{(n,m)})_j$ converges to a proper distribution as $n, m \to \infty$, in which case we have*

$$
\lim_{n \to \infty} \lim_{m \to \infty} q_{ij}^{(n,m)} = \frac{\pi_j Q_j^2(\eta)}{\sum_{k=0}^{\infty} \pi_k Q_k^2(\eta)}.
$$

Proof As in the proof of Lemma 2.49, we have

$$
\begin{aligned}
\lim_{m \to \infty} q_{ij}^{(n,m)} &= P_{ij}^{(n)} \lim_{m \to \infty} \frac{q_0 \displaystyle\int_{-1}^{1} \frac{x^m Q_j(x) d\psi(x)}{1-x}}{q_0 \displaystyle\int_{-1}^{1} \frac{x^{n+m} Q_j(x) d\psi(x)}{1-x}} \\
&= P_{ij}^{(n)} \lim_{m \to \infty} \frac{L_m\left(\dfrac{Q_j}{1-x}, \psi\right)}{L_m\left(\dfrac{x^n Q_j}{1-x}, \psi\right)} = P_{ij}^{(n)} \frac{Q_j(\eta)}{\eta^n Q_i(\eta)}.
\end{aligned}
$$

Therefore

$$
\begin{aligned}
\lim_{n \to \infty} \lim_{m \to \infty} q_{ij}^{(n,m)} &= \lim_{n \to \infty} P_{ij}^{(n)} \frac{Q_j(\eta)}{\eta^n Q_i(\eta)} = \frac{Q_j(\eta)}{Q_i(\eta)} \pi_j \lim_{n \to \infty} \int_{-1}^{1} \left(\frac{x}{\eta}\right)^n Q_i(x) Q_j(x) d\psi(x) \\
&= Q_j^2(\eta) \pi_j \psi(\{\eta\}) = \frac{Q_j^2(\eta)\pi_j}{\sum_{k=0}^{\infty} Q_k^2(\eta)\pi_k}.
\end{aligned}
$$

The last step is a consequence of having $\eta < 1$ and $\psi(\{-\eta\}) = 0$ (see [46, Lemma 2.3]). □

Observe that this limit does not depend on a_i. If $\eta = 1$ then $(q_{ij}^{(n,m)})_j$ does not converge to a proper distribution because there is no jump at $x = 1$, since the birth–death chain is always transient (since $q_0 > 0$). Therefore $P_{ij}^{(n)} \to 0$ as $n \to \infty$.

Example 2.54 Consider the example of Exercise 2.7 where $p_n = p, q_{n+1} = q, n \geq 0$ with $0 < r_0 \leq q < 1$ (see also Example 2.48). Theorem 2.50 gives information about how to compute the limiting conditional distribution $q_{ij}^{(n)}$ (see (2.69)). We have different situations according to the value of η and the probabilities p and r_0:

1. $\eta = 1$ if and only if $p = 1/2$. In this case, according to Lemma 2.49 we have that $q_{ij}^{(n)} \to 0$ as $n \to \infty$.
2. $\eta < 1$ if and only if $p \neq 1/2$. In this case we have three situations:
   (a) $p > 1/2$. Then $q < 1/2$ and $q < \sqrt{pq}$ since $q^2 < (1-q)q$ if and only if $q < 1/2$. Since we are taking $0 < r_0 \leq q < \sqrt{pq}$, we have, according to Example 2.48, that $C_n(\psi) \to c \neq 0$ as $n \to \infty$. Then $\lim_{n\to\infty} q_{ij}^{(n)}$ does not exist.
   (b) $p < 1/2$ and $0 < r_0 < \sqrt{pq}$. As before, $C_n(\psi) \to c \neq 0$ as $n \to \infty$ and again $\lim_{n\to\infty} q_{ij}^{(n)}$ does not exist.
   (c) $p < 1/2$ and $q > r_0 \geq \sqrt{pq}$. In this case $C_n(\psi) \to 0$ as $n \to \infty$. Therefore we are in the conditions of the second part of Lemma 2.49 where $\eta = r_0 + pq/r_0$ (the point of the discrete part of the spectral measure). Since $a_i = 1$ in this case (see Exercise 2.7) then, following (2.72), we have

$$\lim_{n\to\infty} q_{ij}^{(n)} = \frac{1 - r_0 - pq/r_0}{q - r_0} \left(\frac{p}{q}\right)^j Q_j(r_0 + pq/r_0), \quad j \geq 0.$$

The distribution (2.73) is not easy to compute but in the particular case of $r_0 = \sqrt{pq}$ (and therefore $\eta = 2\sqrt{pq}$) we can use (2.68) to see that these are geometric distributions. Indeed, since

$$\begin{aligned}\frac{1 - r_0 - pq/r_0}{q - r_0} &= \frac{1 - 2\sqrt{pq}}{\sqrt{q}\left(\sqrt{q} - \sqrt{p}\right)} = \frac{(1 - 2\sqrt{pq})(\sqrt{q} + \sqrt{p})}{\sqrt{q}(q - p)} \\ &= \frac{1 - 2p + (1 - 2q)\sqrt{p/q}}{q - p} = 1 - \sqrt{\frac{p}{q}},\end{aligned}$$

we have that

$$\lim_{n\to\infty} q_{ij}^{(n)} = \left(1 - \sqrt{\frac{p}{q}}\right)\left(\sqrt{\frac{p}{q}}\right)^j, \quad j \geq 0.$$

◇

2.5.4 A Summary of Other Relevant Results

There are a number of interesting papers in the literature that study other aspects of discrete-time birth–death chains using the associated spectral measure and the Karlin–McGregor integral representation. Let us give some comments about some of them.

1. In [145], T. Whitehurst proves the necessary and sufficient conditions for the support of the spectral measure ψ to be strictly contained in the interval $[0,1]$. For that, the birth–death chain has to behave (from a probabilistic point of view) as a chain Y_{2n} where Y_n is a symmetric birth–death chain. He also gives the necessary and sufficient conditions in terms of the entries of the transition probability matrix P using *continued fractions*. This happens if $0 < h_n < 1, n \geq 1$, where

$$h_n = \cfrac{a_n}{1 - \cfrac{a_{n-1}}{1 - \cfrac{a_{n-2}}{\cdots - \cfrac{a_1}{1 - a_0}}}}.$$

Here $a_{2n} = p_n$ and $a_{2n-1} = q_n$. He also gives conditions for the spectral measure ψ to be purely atomic:

- ψ is purely atomic if $r_n \to 1$ as $n \to \infty$.
- If $\mathrm{supp}(\psi) \subset [0,1]$ and $r_n \to 0$ as $n \to \infty$ then ψ is purely atomic.
- If $\mathrm{supp}(\psi) \subset [a,b]$ and $r_n \to a$ or $r_n \to b$ as $n \to \infty$ then ψ is purely atomic.

2. In [47] E. van Doorn and P. Schrijner give results related to the limiting conditional distribution as seen in Theorem 2.50 but for $a_i = 1$ for all i. They are related to the concept of *geometric ergodicity* and quasi-stationary distributions (see the beginning of Section 2.5.3). The *decay parameter* γ of a birth–death chain is defined as a number in the interval $(0,1]$ that characterizes the speed in which the probabilities $P_{ij}^{(n)} \to p_j$ as $n \to \infty$. In other words, if

$$\gamma_{ij} = \inf\{\rho \geq 0 | P_{ij}^{(n)} - p_j = \mathcal{O}(\rho^n) \text{ as } n \to \infty\},$$

then $\gamma = \sup\{\gamma_{ij}\}$. If $\gamma < 1$ then this convergence is fast and we say that the birth–death chain is *geometrically ergodic*. The main results about geometric ergodicity may be summarized as follows

- If the birth–death chain is null recurrent, then $\gamma_{ij} = 1$ for all i,j, so that $\gamma = 1$ and the birth–death chain is not geometrically ergodic.
- If the birth–death chain is positive recurrent, then $\gamma < 1$ if $\gamma_{ii} < 1$ for some i. It may happen that $\gamma_{ij} < \gamma$ for some pairs (i,j).
- If the birth–death chain is transient, then

$$\lim_{n\to\infty} \left(P_{ij}^{(n)}\right)^{1/n} = \gamma, \quad i,j \geq 0,$$

so that $\gamma_{ij} = \gamma \leq 1$ for all i,j. Moreover $R = \gamma^{-1}$ is the radius of convergence of the generating functions $P_{ij}(s)$ of $P_{ij}^{(n)}$.

The transient case gives rise to a new classification of the process depending on the asymptotic behavior of $R^n P_{ij}^{(n)}$ as $n \to \infty$. If $P_{ij}(R)$ converges, then we say that the birth–death chain is R-recurrent. Moreover, if it is R-recurrent and $\lim_{n\to\infty} R^n P_{ij}^{(n)} = 0$, then it will be R-null recurrent, and if $\lim_{n\to\infty} R^n P_{ij}^{(n)} > 0$, then it will be R-positive recurrent. If the birth–death chain is aperiodic and transient, then its decay parameter γ is given by $\gamma = \eta = \sup \operatorname{supp}\psi$. A similar result about R-recurrence (like Theorem 2.24) is given. Quasi-stationarity is also studied. A distribution $(\theta_j)_{j\in\mathbb{N}_0}$ is quasi-stationary if and only if it satisfies the relations

$$
\begin{aligned}
(1 - q_0\theta_0)\theta_j &= p_{j-1}\theta_{j-1} + r_j\theta_j + q_{j+1}\theta_{j+1}, \quad j \geq 1, \\
(1 - q_0\theta_0)\theta_0 &= r_0\theta_0 + q_1\theta_1.
\end{aligned}
$$

They proved that all families of quasi-stationary distributions have the form

$$\theta_j = \frac{\pi_j}{q_0}(1 - x)Q_j(x), \quad j \geq 0, \quad \eta \leq x < 1,$$

where $\eta = \sup \operatorname{supp}\psi$.

3. In [34, Chapter 8], H. Dette and W. Studden study properties of birth–death measures in terms of the theory of *canonical moments*. Given a probability measure ψ on the interval $[a,b]$ the ordinary moments are given by $c_n = c_n(\psi) = \int_a^b x^n d\psi(x)$, $n \geq 1$. The canonical moments are defined in the following geometrical manner. Let

$$c_n^- = c_n^-(\psi) = \min_\eta c_n(\eta), \quad c_n^+ = c_n^+(\psi) = \max_\eta c_n(\eta),$$

where the minimum and maximum are taken over those probability measures η for which the moments up to degree $n-1$ coincide with the moments of ψ. The canonical moments are then given by

$$d_n = \frac{c_n - c_n^-}{c_n^+ - c_n^-}, \quad n \geq 1.$$

The canonical moments have a number of interesting properties. If we define the sequence of monic polynomials by the three-term recurrence relation (1.17) with coefficients $\alpha_n \in \mathbb{R}$ and $\beta_n > 0$ we have that the corresponding orthogonality measure ψ is supported on the nonnegative real axis if there exists a sequence $(\zeta_n)_{n\geq 1}$ of positive numbers such that

$$\beta_n = \zeta_{2n-1}\zeta_{2n}, \quad n \geq 1, \quad \alpha_n = \zeta_{2n} + \zeta_{2n+1}, \quad n \geq 0.$$

Additionally the measure ψ is concentrated on the interval $[0,1]$ is the sequence $(\zeta_n)_{n\geq 1}$ is a *chain sequence*, i.e. it can be decomposed as

$$\zeta_n = (1 - d_{n-1})d_n,$$

where $0 < d_n < 1$. In this case the sequence $(d_n)_n$ turns out to represent the canonical moments of ψ. We also have the following representation of the Stieltjes transform of the spectral measure ψ on the interval $[0,1]$:

$$\int_0^1 \frac{d\psi(x)}{x-z} = -\cfrac{1}{z - \cfrac{d_1}{1 - \cfrac{(1-d_1)d_2}{z - \cfrac{(1-d_2)d_3}{1 - \cdots}}}}.$$

Similar results hold for measures supported on $[-1,1]$. In [34, Chapter 8] it is proved, among other results, that a probability measure ψ supported on the interval $[-1,1]$ with canonical moments $(d_n)_n$ is the spectral measure of a birth–death chain if and only if

$$2d_{2n-1}d_{2n} + 2(1-d_{2n})(1-d_{2n+1}) \leq 1, \quad n \geq 1.$$

Many other results, already seen in this chapter, and other different, can be proved using the theory of canonical moments.

2.6 Discrete-Time Birth–Death Chains on the Integers

Let $\{X_n, n=0,1,\ldots\}$ be a discrete-time birth–death chain where the state space is given by the set of integer numbers, i.e. $\mathcal{S} = \mathbb{Z}$. Therefore the transition probability matrix P is tridiagonal and doubly infinite:

$$P = \left(\begin{array}{ccc|ccc} \ddots & \ddots & \ddots & & & \\ & q_{-1} & r_{-1} & p_{-1} & & \\ \hline & & q_0 & r_0 & p_0 & \\ & & & q_1 & r_1 & p_1 \\ & & & & \ddots & \ddots & \ddots \end{array}\right),$$

where we assume that $p_i, q_i > 0$ for all $i \in \mathbb{Z}$. A diagram of the transitions between states is

$$\cdots \underset{q_{-2}}{\overset{p_{-3}}{\rightleftarrows}} \overset{r_{-2}}{\circlearrowleft}(-2) \underset{q_{-1}}{\overset{p_{-2}}{\rightleftarrows}} \overset{r_{-1}}{\circlearrowleft}(-1) \underset{q_0}{\overset{p_{-1}}{\rightleftarrows}} \overset{r_0}{\circlearrowleft}(0) \underset{q_1}{\overset{p_0}{\rightleftarrows}} \overset{r_1}{\circlearrowleft}(1) \underset{q_2}{\overset{p_1}{\rightleftarrows}} \overset{r_2}{\circlearrowleft}(2) \underset{q_3}{\overset{p_2}{\rightleftarrows}} \cdots$$

We will follow the last section of [87] with some extended comments and examples. The eigenvalue equation $x\phi(x) = P\phi(x)$ has for each x real or complex two polynomial families of linearly independent solutions $Q_n^\alpha(x), \alpha = 1,2, n \in \mathbb{Z}$, depending on the initial values at $n = 0$ and $n = -1$. These polynomials are given by

$$\begin{aligned} &Q_0^1(x) = 1, \quad Q_0^2(x) = 0, \\ &Q_{-1}^1(x) = 0, \quad Q_{-1}^2(x) = 1, \\ &xQ_n^\alpha(x) = p_n Q_{n+1}^\alpha(x) + r_n Q_n^\alpha(x) + q_n Q_{n-1}^\alpha(x). \end{aligned} \tag{2.75}$$

Observe that

$$\begin{aligned} \deg(Q_n^1) &= n, \quad n \geq 0, \qquad \deg(Q_n^2) = n-1, \quad n \geq 1, \\ \deg(Q_{-n-1}^1) &= n-1, \quad n \geq 1, \quad \deg(Q_{-n-1}^2) = n, \quad n \geq 0. \end{aligned}$$

The polynomials can be computed recurrently using the three-term recurrence relation (2.75):

$$\begin{aligned} Q_{n+1}^\alpha(x) &= \frac{1}{p_n}((x-r_n)Q_n^\alpha(x) - q_n Q_{n-1}^\alpha(x)), \quad \text{if} \quad n \geq 0, \\ Q_{-n-1}^\alpha(x) &= \frac{1}{q_{-n}}((x-r_{-n})Q_{-n}^\alpha(x) - p_{-n} Q_{-n+1}^\alpha(x)), \quad \text{if} \quad n \geq 1. \end{aligned}$$

The doubly infinite vectors of polynomials $\phi_\alpha(x), \alpha = 1,2,$ are given by

$$\begin{aligned} \phi_1(x) &= \left(\cdots, -\frac{p_{-1}}{q_{-1}q_{-2}}(x-r_{-2}), -\frac{p_{-1}}{q_{-1}}, 0, 1, \frac{x-r_0}{p_0}, \cdots\right), \\ \phi_2(x) &= \left(\cdots, \frac{x-r_{-1}}{q_{-1}}, 1, 0, -\frac{q_0}{p_0}, -\frac{q_0}{p_0p_1}(x-r_1), \cdots\right). \end{aligned}$$

The solution of the symmetry equations $\pi_i P_{ij} = \pi_j P_{ji}$ is unique up to a constant, which we normalize by $\pi_0 = 1$. Therefore the *potential coefficients* are given by

$$\pi_0 = 1, \quad \pi_n = \frac{p_0 p_1 \cdots p_{n-1}}{q_1 q_2 \cdots q_n}, \quad \pi_{-n} = \frac{q_0 q_{-1} \cdots q_{-n+1}}{p_{-1} p_{-2} \cdots p_{-n}}, \quad n \geq 1. \tag{2.76}$$

In particular we have that $\pi P = \pi$, i.e. π is an invariant vector. In the Hilbert space $\ell_\pi^2(\mathbb{Z})$ the matrix P gives rise to a self-adjoint operator of norm ≤ 1, which we will denote by P, abusing the notation. Defining as in the case of $\ell_\pi^2(\mathbb{N}_0)$ the vectors $e^{(i)} = \delta_{ij}/\pi_j$, we have that $Pe^{(i)} = q_i e^{(i-1)} + r_i e^{(i)} + p_i e^{(i+1)}$ and using the recurrence for the polynomials Q_n^α, we can prove by induction that

$$Q_i^1(P)e^{(0)} + Q_i^2(P)e^{(-1)} = e^{(i)}, \quad i \in \mathbb{Z}.$$

Hence

$$P_{ij}^n = \pi_j (P^n e^{(j)}, e^{(i)})_\pi = \pi_j (P^n [Q_j^1(P)e^{(0)} + Q_j^2(P)e^{(-1)}], Q_i^1(P)e^{(0)} + Q_i^2(P)e^{(-1)})_\pi.$$

We need to apply now the spectral theorem *three times*, counting all combinations of the previous formula. In particular there exist three measures $\psi_{11}(x), \psi_{22}(x)$ and

$\psi_{12}(x)$ (since $\psi_{12}(x) = \psi_{21}(x)$ because P is self-adjoint and the symmetry of the inner product) supported on the interval $[-1,1]$ such that

$$\int_{-1}^{1} f(x)d\psi_{11}(x) = (f(P)e^{(0)}, e^{(0)})_\pi,$$

$$\int_{-1}^{1} f(x)d\psi_{12}(x) = (f(P)e^{(0)}, e^{(-1)})_\pi,$$

$$\int_{-1}^{1} f(x)d\psi_{22}(x) = (f(P)e^{(-1)}, e^{(-1)})_\pi,$$

where f is any real bounded function on $[-1,1]$. We have then the Karlin–McGregor integral representation

$$P_{ij}^{(n)} = \mathbb{P}(X_n = j | X_0 = i) = \pi_j \int_{-1}^{1} x^n \sum_{\alpha,\beta=1}^{2} Q_i^\alpha(x) Q_j^\beta(x) d\psi_{\alpha\beta}(x), \quad i,j \in \mathbb{Z}. \quad (2.77)$$

We can study closely the measures $\psi_{\alpha\beta}$ by taking $n = 0$ in the previous formula and some values of i,j:

- If $i = j = 0$ then, by definition of the polynomials, we have that $1 = \int_{-1}^{1} d\psi_{11}(x)$. Therefore ψ_{11} is a proper probability distribution which vanishes for $x < -1$ and it is 1 for $x > 1$.
- If $i = j = -1$ then again we have that $1 = \pi_{-1} \int_{-1}^{1} d\psi_{22}(x)$. Therefore ψ_{22} is a positive distribution which vanishes for $x < -1$ and it is $1/\pi_{-1}$ for $x > 1$.
- If $i = -1$ and $j = 0$ then we get $0 = \int_{-1}^{1} d\psi_{12}(x)$. That means that ψ_{12} is a signed distribution vanishing for $x < -1$ and $x > 1$.

These three measures can be written in matrix form as the 2×2 matrix

$$\Psi(x) = \begin{pmatrix} \psi_{11}(x) & \psi_{12}(x) \\ \psi_{12}(x) & \psi_{22}(x) \end{pmatrix}. \quad (2.78)$$

This matrix is always positive definite since for any $c_1, c_2 \in \mathbb{C}$, we have that the measure

$$\varphi(x) = (c_1, c_2)\Psi(x) \begin{pmatrix} c_1^* \\ c_2^* \end{pmatrix} = \sum_{\alpha,\beta=1}^{2} c_\alpha c_\beta^* \psi_{\alpha\beta}(x)$$

is the one corresponding to $c_1 e^{(0)} + c_2 e^{(-1)}$. The matrix $\Psi(x)$ is called the *spectral matrix* of the discrete-time birth–death chain on $\mathbb{Z}$.

Let us now see that the spectral analysis of the birth–death chain on $\mathbb{Z}$ is reduced to study two birth–death chains corresponding to the two directions to infinity. Let ψ^+ be the measure associated with the birth–death chain with state space $\{0,1,2,\dots\}$ whose probability transition matrix P^+ is given by $P_{ij}^+ = P_{ij}, i,j \geq 0$, i.e.

$$P^+ = \begin{pmatrix} r_0 & p_0 & & & \\ q_1 & r_1 & p_1 & & \\ & q_2 & r_2 & p_2 & \\ & & \ddots & \ddots & \ddots \end{pmatrix}.$$

The probability of absorption at state -1 is given by $q_0 > 0$. Similarly, ψ^- is the measure associated with the birth–death chain with state space $\{-1, -2, \dots\}$ (in that order) whose probability transition matrix P^- is given by $P^-_{ij} = P_{ij}, i,j \leq -1$, i.e.

$$P^- = \begin{pmatrix} r_{-1} & q_{-1} & & & \\ p_{-2} & r_{-2} & q_{-2} & & \\ & p_{-3} & r_{-3} & q_{-3} & \\ & & \ddots & \ddots & \ddots \end{pmatrix},$$

where now the probability of absorption at state 0 is given by $p_{-1} > 0$. Let $P^{\pm}_{ij}(s)$ and $F^{\pm}_{ij}(s)$ be the corresponding generating functions of both birth–death chains (see (2.6)). From the identities

$$F_{00}(s) = F^+_{00}(s) + q_0 s F_{-1,0}(s), \quad F_{-1,0}(s) = p_{-1} s P^-_{-1,-1}(s),$$
$$P_{00}(s) = 1 + F_{00}(s) P_{00}(s), \quad P^+_{00}(s) = 1 + F^+_{00}(s) P^+_{00}(s),$$

we obtain that

$$P_{00}(s) = \frac{P^+_{00}(s)}{1 - p_{-1} q_0 s^2 P^+_{00}(s) P^-_{-1,-1}(s)}.$$

Similarly, using

$$F_{-1,-1}(s) = F^-_{-1,-1}(s) + p_{-1} s F_{0,-1}(s), \quad F_{0,-1}(s) = q_0 s P^+_{00}(s),$$

we get

$$P_{-1,-1}(s) = \frac{P^-_{-1,-1}(s)}{1 - p_{-1} q_0 s^2 P^+_{00}(s) P^-_{-1,-1}(s)}.$$

Finally, from $P_{-1,0}(s) = F_{-1,0}(s) P_{00}(s)$, we have that

$$P_{-1,0}(s) = \frac{p_{-1} s P^+_{00}(s) P^-_{-1,-1}(s)}{1 - p_{-1} q_0 s^2 P^+_{00}(s) P^-_{-1,-1}(s)}.$$

Now, using the extension of the formula (2.34) to this situation, we get that

$$P_{ij}(s) = \pi_j \int_{-1}^{1} \frac{1}{1 - xs} \sum_{\alpha,\beta=1}^{2} Q_i^{\alpha}(x) Q_j^{\beta}(x) \, d\psi_{\alpha\beta}(x).$$

Using the definition of the Stieltjes transform $B(z;\psi)$ in (1.6), from the previous considerations we can deduce that the Stieltjes transforms of $\psi_{\alpha\beta}, \alpha, \beta = 1, 2$, are given by

$$\begin{aligned} B(z;\psi_{11}) &= \frac{B(z;\psi^+)}{1 - p_{-1}q_0 B(z;\psi^+)B(z;\psi^-)}, \\ \frac{q_0}{p_{-1}} B(z;\psi_{22}) &= \frac{B(z;\psi^-)}{1 - p_{-1}q_0 B(z;\psi^+)B(z;\psi^-)}, \\ B(z;\psi_{12}) &= \frac{-p_{-1}B(z;\psi^+)B(z;\psi^-)}{1 - p_{-1}q_0 B(z;\psi^+)B(z;\psi^-)}. \end{aligned} \tag{2.79}$$

The spectral analysis of birth–death chains on $\mathbb{Z}$ can also be studied using the 2×2 spectral matrix (2.78). The Karlin–McGregor formula (2.77) can be written in matrix form as

$$P_{ij}^{(n)} = \pi_j \int_{-1}^{1} x^n \left(Q_i^1(x), Q_i^2(x)\right) d\Psi(x) \begin{pmatrix} Q_j^1(x) \\ Q_j^2(x) \end{pmatrix}, \quad i,j \in \mathbb{Z}.$$

This formula is always valid for all $i,j \in \mathbb{Z}$. However, it is possible to relabel the states in such a way that all the information of P can be collected in a semi-infinite block tridiagonal matrix $\boldsymbol{P}$ with blocks of size 2×2. Indeed, after the new labeling

$$\{0,1,2,\dots\} \to \{0,2,4,\dots\}, \quad \text{and} \quad \{-1,-2,-3,\dots\} \to \{1,3,5,\dots\},$$

we have that P (doubly infinite tridiagonal) is equivalent to a matrix $\boldsymbol{P}$ (semi-infinite block tridiagonal)

$$\boldsymbol{P} = \begin{pmatrix} B_0 & A_0 & & & \\ C_1 & B_1 & A_1 & & \\ & C_2 & B_2 & A_2 & \\ & & \ddots & \ddots & \ddots \end{pmatrix},$$

where

$$B_0 = \begin{pmatrix} r_0 & q_0 \\ p_{-1} & r_{-1} \end{pmatrix}, \quad B_n = \begin{pmatrix} r_n & 0 \\ 0 & r_{-n-1} \end{pmatrix}, \quad n \geq 1,$$

$$A_n = \begin{pmatrix} p_n & 0 \\ 0 & q_{-n-1} \end{pmatrix}, \quad n \geq 0, \quad C_n = \begin{pmatrix} q_n & 0 \\ 0 & p_{-n-1} \end{pmatrix}, \quad n \geq 1.$$

The birth–death chain generated by $\boldsymbol{P}$ can be interpreted as a walk that takes values in the two-dimensional state space $\{0,1,2,\dots\} \times \{1,2\}$. An isomorphic diagram associated with the probability transitions (two-dimensional) is given by

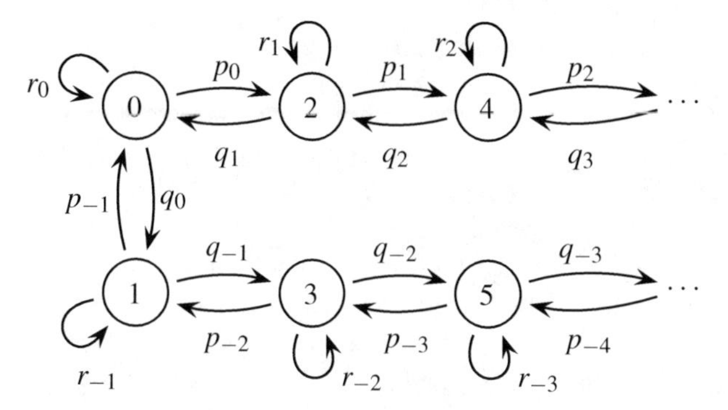

These types of processes are called discrete-time *quasi-birth–death processes*. The first component ($\mathbb{N}_0$) is usually called the *level*, while the second component ($\{1,2\}$ in this case but in general it can be any set $\{1,\dots,N\}$) is usually called the *phase*. In general, these processes allow transitions between all adjacent levels (see [110, 119] for a general reference).

We can now apply the spectral theorem of the matrix $\boldsymbol{P}$ but now applied to a 2×2 matrix of measures. This spectral matrix is precisely $d\Psi(x)$. If we define the matrix-valued polynomials $\boldsymbol{Q}_n(x)$ by

$$\boldsymbol{Q}_n(x) = \begin{pmatrix} Q_n^1(x) & Q_n^2(x) \\ Q_{-n-1}^1(x) & Q_{-n-1}^2(x) \end{pmatrix}, \quad n \geq 0, \tag{2.80}$$

then we have that they satisfy the three-term recurrence relation

$$\begin{aligned} x\boldsymbol{Q}_0(x) &= A_0\boldsymbol{Q}_1(x) + B_0\boldsymbol{Q}_0(x), \quad \boldsymbol{Q}_0(x) = I_{2\times 2}, \\ x\boldsymbol{Q}_n(x) &= A_n\boldsymbol{Q}_{n+1}(x) + B_n\boldsymbol{Q}_n(x) + C_n\boldsymbol{Q}_{n-1}(x), \quad n \geq 1, \end{aligned}$$

where $I_{2\times 2}$ denotes the 2×2 identity matrix. Observe that the leading coefficient of $\boldsymbol{Q}_n$ is a nonsingular matrix. The matrix orthogonality is defined in terms of the (matrix-valued) inner product

$$\int_{-1}^{1} \boldsymbol{Q}_n(x)\, d\Psi(x) \boldsymbol{Q}_m^*(x) = \begin{pmatrix} 1/\pi_n & 0 \\ 0 & 1/\pi_{-n-1} \end{pmatrix} \delta_{nm},$$

where A^* is the Hermitian transpose of a matrix A and $(\pi_n)_{n\in\mathbb{Z}}$ are the potential coefficients defined by (2.76). Therefore, the Karlin–McGregor formula applied to $\boldsymbol{P}$ is given by (see [32, 62]):

$$\boldsymbol{P}_{ij}^{(n)} = \left(\int_{-1}^{1} x^n \boldsymbol{Q}_i(x)\, d\Psi(x) \boldsymbol{Q}_j^*(x) \right) \begin{pmatrix} \pi_j & 0 \\ 0 & \pi_{-j-1} \end{pmatrix}, \quad i,j \in \mathbb{N}_0.$$

Observe now that the block entry (i,j) is a 2×2 matrix. The relation between this representation $\boldsymbol{P}_{ij}^{(n)}, i,j \in \mathbb{N}_0$, and the original one $P_{ij}^{(n)}, i,j \in \mathbb{Z}$, is given by

$$\boldsymbol{P}_{ij}^{(n)} = \begin{pmatrix} P_{ij}^{(n)} & P_{i,-j-1}^{(n)} \\ P_{-i-1,j}^{(n)} & P_{-i-1,-j-1}^{(n)} \end{pmatrix}, \quad i,j \in \mathbb{N}_0.$$

It is possible to see that the concept of recurrence can be extended in a natural way to discrete-time quasi-birth–death processes (see [32, Section 4.1]). For birth–death chains on $\mathbb{Z}$ we can extend the result about recurrence in Theorem 2.24. Indeed, the birth–death chain is recurrent if and only if

$$\int_{-1}^{1} \frac{d\psi_{11}(x)}{1-x} = \infty, \quad \text{or} \quad \int_{-1}^{1} \frac{d\psi_{22}(x)}{1-x} = \infty. \tag{2.81}$$

In a similar way, the birth–death chain is positive recurrent if and only if one the measures ψ_{11}, ψ_{22} has a jump at the point 1.

Example 2.55 Consider $\{X_n, n=0,1,\dots\}$ the birth–death chain on the integers $\mathbb{Z}$ with transition probabilities

$$r_n = 0, \quad p_n = p, \quad q_n = q, \quad n \in \mathbb{Z}, \quad p+q=1.$$

The associated transition probability matrix is then

$$P = \left(\begin{array}{ccc|ccc} \ddots & \ddots & \ddots & & & \\ & q & 0 & p & & \\ \hline & & q & 0 & p & \\ & & & q & 0 & p \\ & & & \ddots & \ddots & \ddots \end{array}\right).$$

A diagram of the transitions between states is

$$\cdots \underset{q}{\overset{p}{\rightleftarrows}} (-2) \underset{q}{\overset{p}{\rightleftarrows}} (-1) \underset{q}{\overset{p}{\rightleftarrows}} (0) \underset{q}{\overset{p}{\rightleftarrows}} (1) \underset{q}{\overset{p}{\rightleftarrows}} (2) \underset{q}{\overset{p}{\rightleftarrows}} \cdots$$

This birth–death chain may be regarded as gambler's ruin, but now both players are infinitely rich. The matrices P^+ and P^- are given by

$$P^+ = \begin{pmatrix} 0 & p & & & \\ q & 0 & p & & \\ & q & 0 & p & \\ & & \ddots & \ddots & \ddots \end{pmatrix}, \quad P^- = \begin{pmatrix} 0 & q & & & \\ p & 0 & q & & \\ & p & 0 & q & \\ & & \ddots & \ddots & \ddots \end{pmatrix}.$$

We observe that in this case the measures $\psi^{\pm}$ are equal to the ones in Example 2.21 (no matter the diagonal entries). Therefore $B(z;\psi_{11}) = \frac{q}{p}B(z;\psi_{22})$ and $B(z;\psi^{+}) = B(z;\psi^{-}) = B(z)$, where $pqB^2(z) + zB(z) + 1 = 0$ and its solution is given by

$$B(z) = \frac{-z + \sqrt{z^2 - 4pq}}{2pq}.$$

Therefore, following (2.79) for $B(z;\psi_{11})$ and rationalizing, we get that

$$\begin{aligned} B(z;\psi_{11}) &= \frac{B(z)}{1 - pqB^2(z)} = \frac{B(z)}{2 + zB(z)} = \frac{-z + \sqrt{z^2 - 4pq}}{4pq + z(-z + \sqrt{z^2 - 4pq})} \\ &= \frac{z^2 - (z^2 - 4pq)}{-4pq(z + \sqrt{z^2 - 4pq}) + z(z^2 - (z^2 - 4pq))} = -\frac{1}{\sqrt{z^2 - 4pq}}. \end{aligned}$$

Hence the absolutely continuous part, using the Perron–Stieltjes inversion formula (1.8), is given by

$$d\psi_{11}(x) = \frac{1}{\pi\sqrt{4pq - x^2}}\,dx, \quad |x| < 2\sqrt{pq}.$$

Since $\psi^{+} = \psi^{-}$, then $q\psi_{22} = p\psi_{11}$. Finally, substituting these values in the third equation in (2.79) and rationalizing, we obtain

$$\begin{aligned} B(z;\psi_{12}) &= \frac{-pB^2(z)}{1 - pqB^2(z)} = \frac{1}{q}\left(\frac{1 + zB(z)}{2 + zB(z)}\right) = \frac{1}{q}\left(\frac{2pq + z(-z + \sqrt{z^2 - 4pq})}{4pq + z(-z + \sqrt{z^2 - 4pq})}\right) \\ &= \frac{1}{q}\left(\frac{-2pq(z + \sqrt{z^2 - 4pq}) + 4pqz}{-4pq(z + \sqrt{z^2 - 4pq}) + 4pqz}\right) \\ &= \frac{1}{2q}\left(\frac{-z + \sqrt{z^2 - 4pq}}{\sqrt{z^2 - 4pq}}\right) = \frac{1}{2q}\left(1 - \frac{z}{\sqrt{z^2 - 4pq}}\right). \end{aligned}$$

Hence, the absolutely continuous part, using again the Perron–Stieltjes inversion formula (1.8), is given by

$$d\psi_{12}(x) = \frac{x}{2\pi q\sqrt{4pq - x^2}}\,dx, \quad |x| < 2\sqrt{pq}.$$

Observe that the jumps, if any, should be located at $x = \pm 2\sqrt{pq}$. However, using (1.12), we can see that the size of these jumps must be 0, so there are no jumps. The spectral matrix is then

$$d\Psi(x) = \frac{1}{\pi\sqrt{4pq - x^2}}\begin{pmatrix} 1 & x/2q \\ x/2q & p/q \end{pmatrix} dx, \quad |x| < 2\sqrt{pq}.$$

Using the recurrence relation (2.75), we can derive that (see [63])

$$Q_n^1(x) = \left(\frac{q}{p}\right)^{n/2} U_n\left(\frac{x}{2\sqrt{pq}}\right), \quad Q_{-n-1}^1(x) = -\left(\frac{p}{q}\right)^{(n+1)/2} U_{n-1}\left(\frac{x}{2\sqrt{pq}}\right), \quad n \geq 0,$$

$$Q_n^2(x) = -\left(\frac{q}{p}\right)^{(n+1)/2} U_{n-1}\left(\frac{x}{2\sqrt{pq}}\right), \quad Q_{-n-1}^2(x) = \left(\frac{p}{q}\right)^{n/2} U_n\left(\frac{x}{2\sqrt{pq}}\right), \quad n \geq 0,$$

where $(U_n)_n$ are the Chebychev polynomials of the second kind (see (1.87)). Therefore the matrix-valued orthogonal polynomials (2.80) are given by

$$\boldsymbol{Q}_n(x) = \begin{pmatrix} \left(\frac{q}{p}\right)^{n/2} U_n\left(\frac{x}{2\sqrt{pq}}\right) & -\left(\frac{q}{p}\right)^{(n+1)/2} U_{n-1}\left(\frac{x}{2\sqrt{pq}}\right) \\ -\left(\frac{p}{q}\right)^{(n+1)/2} U_{n-1}\left(\frac{x}{2\sqrt{pq}}\right) & \left(\frac{p}{q}\right)^{n/2} U_n\left(\frac{x}{2\sqrt{pq}}\right) \end{pmatrix}, \quad n \geq 0.$$

Since $4pq < 1$ unless $p = q = 1/2$, we have that both integrals in (2.81) are bounded. Therefore, if $p \neq q$, the birth–death chain is always transient. If $p = q = 1/2$, then the integrals in (2.81) are divergent. Therefore this is the only case where the birth–death chain is recurrent. Since the measures ψ_{11}, ψ_{22} do not have a jump at 1, the birth–death chain can never be positive recurrent. ◊

Example 2.56 Consider now a slight modification of the previous birth–death chain (and considered in [63, Section 6]). The transition probabilities are now given by

$$r_n = 0, \quad n \in \mathbb{Z}, \quad p_n = p, \quad q_n = q, \quad n \geq 0, \quad p_{-n} = q, \quad q_{-n} = q \quad , n \geq 1,$$

with $p + q = 1$. The transition probability matrix is then

$$P = \left(\begin{array}{cccc|ccc} \ddots & \ddots & \ddots & & & & \\ & p & 0 & q & & & \\ & & p & 0 & q & & \\ \hline & & & q & 0 & p & \\ & & & & q & 0 & p \\ & & & & \ddots & \ddots & \ddots \end{array}\right).$$

A diagram of the transitions between states is

$$\cdots \underset{p}{\overset{q}{\rightleftarrows}} (-2) \underset{p}{\overset{q}{\rightleftarrows}} (-1) \underset{p}{\overset{q}{\rightleftarrows}} (0) \underset{q}{\overset{p}{\rightleftarrows}} (1) \underset{q}{\overset{p}{\rightleftarrows}} (2) \underset{q}{\overset{p}{\rightleftarrows}} \cdots$$

Observe that the probabilities p and q are interchanged for nonnegative and negative states of the birth–death chain. Therefore, if $p < q$, we have a birth–death chain where the origin is an absorbing state. Conversely, if $p > q$, then the origin is a reflecting state.

The matrices P^+ and P^- are exactly the same, and are also equal to the one in Example 2.21. Since $p_{-1} = q_0 = q$, we have that $B(z;\psi_{11}) = B(z;\psi_{22}) = B(z)$, where $B(z)$ is the one of the previous example. Following the same lines as before and rationalizing, using (2.79), we get

$$B(z;\psi_{11}) = \frac{pB(z)}{1+qzB(z)} = \frac{z(1-2p)+\sqrt{z^2-4pq}}{2q(1-z^2)},$$
$$B(z;\psi_{12}) = \frac{1+zB(z)}{1+qzB(z)} = \frac{2q-z^2+z\sqrt{z^2-4pq}}{2q(1-z^2)}.$$

Using the Perron–Stieltjes inversion formula (1.8), we have that

$$d\psi_{11}(x) = \frac{\sqrt{4pq-x^2}}{2\pi q(1-x^2)}dx, \quad d\psi_{12}(x) = \frac{x\sqrt{4pq-x^2}}{2\pi q(1-x^2)}dx, \quad |x| < 2\sqrt{pq}.$$

The absolutely continuous part of the spectral matrix is given by

$$\frac{\sqrt{4pq-x^2}}{2\pi q(1-x^2)}\begin{pmatrix} 1 & x \\ x & 1 \end{pmatrix}, \quad |x| < 2\sqrt{pq}.$$

We observe that the Stieltjes transform has two single poles at $z = \pm 1$. Therefore there will be two jumps at $x = \pm 1$ for certain values of p. Indeed, using (1.12), we have that the sizes of the jumps at $x = 1$ are given by

$$\begin{aligned}
\psi_{11}(\{1\}) &= \psi_{22}(\{1\}) \\
&= \lim_{\varepsilon\downarrow 0} \varepsilon \mathrm{Im} B(1+i\varepsilon;\psi_{11}) = \frac{1-2p+|1-2p|}{4q} = \frac{1-2p}{2q}\mathbf{1}_{\{p<1/2\}}, \\
\psi_{12}(\{1\}) &= \lim_{\varepsilon\downarrow 0} \varepsilon \mathrm{Im} B(1+i\varepsilon;\psi_{12}) = \frac{2q-1+|1-2p|}{4q} = \frac{1-2p}{2q}\mathbf{1}_{\{p<1/2\}},
\end{aligned}$$

while the sizes of the jumps at $x = -1$ are given by

$$\begin{aligned}
\psi_{11}(\{-1\}) &= \psi_{22}(\{-1\}) = \lim_{\varepsilon\downarrow 0} \varepsilon \mathrm{Im} B(-1+i\varepsilon;\psi_{11}) \\
&= -\frac{-(1-2p)+|1-2p|}{4q} = \frac{1-2p}{2q}\mathbf{1}_{\{p<1/2\}}, \\
\psi_{12}(\{-1\}) &= \lim_{\varepsilon\downarrow 0} \varepsilon \mathrm{Im} B(-1+i\varepsilon;\psi_{12}) = -\frac{2q-1+|1-2p|}{4q} = -\frac{1-2p}{2q}\mathbf{1}_{\{p<1/2\}}.
\end{aligned}$$

Therefore the spectral matrix has a discrete part given by

$$\frac{1-2p}{2q}\left[\begin{pmatrix} 1 & -1 \\ -1 & 1 \end{pmatrix}\delta_{-1} + \begin{pmatrix} 1 & 1 \\ 1 & 1 \end{pmatrix}\delta_1\right]\mathbf{1}_{\{p<1/2\}}.$$

In summary, the spectral matrix is given by

$$\Psi(x) = \frac{\sqrt{4pq - x^2}}{2\pi q(1-x^2)} \begin{pmatrix} 1 & x \\ x & 1 \end{pmatrix} \mathbf{1}_{\{|x|<2\sqrt{pq}\}} + \frac{1-2p}{2q} \left[\begin{pmatrix} 1 & -1 \\ -1 & 1 \end{pmatrix} \delta_{-1} + \begin{pmatrix} 1 & 1 \\ 1 & 1 \end{pmatrix} \delta_1 \right] \mathbf{1}_{\{p<1/2\}}.$$

As in the previous example, we can study recurrence by looking at the spectral matrix. There will be three cases depending on the value of p:

- If $p > 1/2$, then both integrals in (2.81) are bounded since $4pq < 1$ and there are no jumps. Therefore the birth–death chain is transient.
- If $p = 1/2$, then $\psi_{11}(x) = \psi_{22}(x) = \frac{1}{\pi\sqrt{1-x^2}}$. Therefore both integrals in (2.81) are divergent. Hence the birth–death chain is null recurrent (since there is no jump at the point 1).
- If $p < 1/2$, then both spectral measures $\psi_{11} = \psi_{22}$ have a jump at the point 1. Therefore the birth–death chain is positive recurrent.

In the case of a positive recurrent birth–death chain ($p < 1/2$) we can compute an explicit expression of the invariant or stationary distribution. The potential coefficients in (2.76) are given by

$$\pi_0 = 1, \quad \pi_n = \left(\frac{p}{q}\right)^n, \quad \pi_{-n} = \left(\frac{p}{q}\right)^{n-1}, \quad n \geq 1.$$

For $p < 1/2$, we have

$$\sum_{n\in\mathbb{Z}} \pi_n = 2\sum_{n=0}^{\infty} \left(\frac{p}{q}\right)^n = \frac{2q}{q-p} = \frac{2(1-p)}{1-2p},$$

which is $\psi_{11}(\{1\})^{-1}$. Therefore the invariant distribution is given by

$$\pi = \frac{1-2p}{2(1-p)} \left(\dots, \left(\frac{p}{1-p}\right)^2, \frac{p}{1-p}, 1, 1, \frac{p}{1-p}, \left(\frac{p}{1-p}\right)^2, \dots \right).$$

◊

Remark 2.57 The method of folding the state space in such a way that we have a discrete-time quasi-birth–death process on $\mathbb{N}_0 \times \{1, 2, \dots, N\}$ can be applied to other graphs different from the one we consider here for birth–death chains on $\mathbb{Z}$. For instance, in [64, 65] the spectral analysis of a discrete version of *Walsh's spider* is considered, which is an extension of the birth–death chains on $\mathbb{Z}$ but with N legs that extend to infinity, instead of two legs. Other more complicated examples can be found in [18, 32, 62, 68]. ◊

2.7 Exercises

2.1 Find a relation between the n-step transition probabilities $P_{ij}^{(n)}$ and the n-step transition probabilities $(P^d)_{ij}^{(n)}$ of the dual polynomials and vice versa for both cases, $q_0 > 0$ and $q_0 = 0$.

2.2 Consider the birth–death chain $\{X_n, n=0,1,\dots\}$ given by gambler's ruin with state space $\{0,1,\dots,N\}$ and transition probability matrix

$$P = \begin{pmatrix} 1/2 & 1/2 & & & \\ 1/2 & 0 & 1/2 & & \\ & \ddots & \ddots & \ddots & \\ & & 1/2 & 0 & 1/2 \\ & & & 1/2 & 1/2 \end{pmatrix}.$$

Compute the eigenvalues and eigenvectors of P and find an expression for the probabilities $P_{ij}^{(n)}$. (For the general case, see [57, Chapter XVI, Section 3].)

2.3 Consider the birth–death chain $\{X_n, n=0,1,\dots\}$ on the circle with state space $\{0,1,\dots,N\}$ and transition probability matrix

$$P = \begin{pmatrix} 0 & p & 0 & 0 & \cdots & q \\ q & 0 & p & 0 & \cdots & 0 \\ 0 & q & 0 & p & \cdots & 0 \\ \vdots & \vdots & \ddots & \ddots & \ddots & \vdots \\ 0 & 0 & \cdots & q & 0 & p \\ p & 0 & \cdots & 0 & q & 0 \end{pmatrix}.$$

Find a formula for the n-step transition probabilities $P_{ij}^{(n)}$ using spectral methods.

2.4 Consider the Bernoulli–Laplace urn model studied in Example 2.20. Compute the expected return time from state i to the same state (i.e. τ_{ii}). For the case $\alpha = \beta$ and α even, analyze $\tau_{\alpha/2,\alpha/2}$ as $\alpha \to \infty$.

2.5 Consider the Example 2.22, case 1. The potential coefficients are given by $\pi_0 = 1$ and $\pi_n = p^{n-1}/q^n, n \geq 1$. Compute the absorption probabilities a_i in (2.60) and the expected absorption times $\tau_{i,-1}$ in (2.61) in terms of p_0.

2.6 Consider the Example 2.22, case 2. Compute the polynomials $Q_n(x)$ in terms of known classical families of orthogonal polynomials.

2.7 Consider the Example 2.22, case 2. Study the case where $p_0 = p$ with a free parameter r_0 satisfying $0 < r_0 \leq q < 1$. Prove that the absolutely continuous part of the spectral measure is given by

$$\frac{1}{2\pi} \frac{\sqrt{4pq - x^2}}{pq + r_0^2 - r_0 x}, \quad |x| < 2\sqrt{pq}.$$

For the values $r_0 > \sqrt{pq}$, we have to add to the spectral measure a Dirac delta located at $r_0 + pq/r_0$ of size $1 - pq/r_0^2$. Prove also that the absorption probabilities a_i in (2.60) are given by

$$a_i = \begin{cases} 1, & \text{if } \; p \leq 1/2, \\ \dfrac{q - r_0}{p - r_0}\left(\dfrac{q}{p}\right)^i, & \text{if } \; p > 1/2. \end{cases}$$

2.8 Consider Example 2.23. The potential coefficients of this birth–death chain are given by

$$\pi_{2n} = \left(\frac{pp_1}{qq_1}\right)^n, \quad \pi_{2n+1} = \frac{p}{q_1}\left(\frac{pp_1}{qq_1}\right)^n, \quad n \geq 0.$$

Compute the absorption probabilities a_i in (2.60) and the expected absorption times $\tau_{i,-1}$ in (2.61) in terms of p, p_1, q and q_1.

2.9 Prove formula (2.61), i.e.

$$\tau_{i,-1} = \sum_{j=0}^{i} \frac{1}{q_j \pi_j} \sum_{n=j}^{\infty} \pi_n,$$

by induction, using the three-term recurrence relation for the sequence of orthogonal polynomials $(Q_n)_n$.

2.10 Consider the birth–death chain $\{X_n, n=0,1,\ldots\}$ with one-step transition probability matrix given by

$$P = \begin{pmatrix} r_0 & p_0 & 0 & 0 & \\ q & r & p & 0 & \\ 0 & q & r & p & \\ & & \ddots & \ddots & \ddots \end{pmatrix},$$

where $r_0 + p_0 = 1$ and $p + r + q = 1$ (see [69, Section 4]). Prove that the spectral measure ψ associated with P is given by

$$\psi(x) = \psi_c(x) + \psi(\{1\})\,\delta_1(x) + \psi(\{\gamma\})\,\delta_\gamma(x),$$
$$\gamma = \frac{p_0 - p + pp_0 - p_0 q - p_0^2}{p_0 - p}.$$

The absolutely continuous part of the spectral measure is given by

$$\psi_c(x) = \frac{p_0\sqrt{(x - \sigma_-)(\sigma_+ - x)}}{2\pi(p_0 - p)(1 - x)(x - \gamma)}, \quad x \in [\sigma_-, \sigma_+] \subseteq [-1, 1],$$

where $\sigma_\pm = 1 - (\sqrt{p} \mp \sqrt{q})^2$, while the discrete masses are given by

$$\psi(\{1\}) = \frac{q - p}{p_0 + q - p}\mathbf{1}_{\{q>p\}}, \quad \psi(\{\gamma\}) = \frac{(p_0 - p)^2 - pq}{(p_0 - p)^2 - pq + p_0 q}\mathbf{1}_{\{(p_0-p)^2>pq\}}.$$

2.11 Find an expression of the matrix-valued orthogonal polynomials $\boldsymbol{Q}_n(x)$ in (2.80) for the Example 2.56 using the polynomials given by (2.49) (observe that in this case we have that $\psi = \psi_{11}$).

2.12 Consider the birth–death chain on $\mathbb{Z}$ with transition probabilities $r_n = 0$, $n \in \mathbb{Z}, p_n = p, q_n = q, n \geq 1$, $p_0 = x_1, q_0 = x_2$ and $p_{-n} = q, q_{-n} = q, n \geq 1$, with $p + q = 1$ and $x_1 + x_2 = 1$ (see [63, Section 7]). Prove that the spectral matrix is given by

$$\Psi(x) = \frac{\sqrt{4pq - x^2}}{1 - x^2} \begin{pmatrix} p(1-x_1) & p(1-x_1)x \\ p(1-x_1)x & (1-p)x_1 + (p-x_1)x^2 \end{pmatrix} \mathbf{1}_{\{|x|<2\sqrt{pq}\}}$$
$$+ (1-x_1)(1-2p)\pi \left[\begin{pmatrix} 1 & -1 \\ -1 & 1 \end{pmatrix} \delta_{-1} + \begin{pmatrix} 1 & 1 \\ 1 & 1 \end{pmatrix} \delta_1 \right] \mathbf{1}_{\{p<1/2\}},$$

and study the recurrence of the birth–death chain.

2.13 For $p = q = 1/2$, write the Karlin–McGregor formulas of Examples 2.55 and 2.56 in terms of an angle $\theta \in [0, \pi]$, in the same way as we did in Examples 2.21 and 2.22 (2).

2.14 Consider the birth–death chain on $\mathbb{Z}$ with transition probabilities $p_n = p$, $q_n = q, r_n = r, n \in \mathbb{Z}$, where $p + r + q = 1$. Prove that the spectral matrix is given by

$$\Psi(x) = \frac{1}{\pi\sqrt{(x-\sigma_-)(\sigma_+ - x)}} \begin{pmatrix} 1 & \dfrac{x-r}{2q} \\ \dfrac{x-r}{2q} & p/q \end{pmatrix}, \quad x \in [\sigma_-, \sigma_+],$$

where $\sigma_\pm = 1 - \left(\sqrt{p} \mp \sqrt{q}\right)^2$.

2.15 Consider the birth–death chain on $\mathbb{Z}$ with transition probabilities $p_n = p$, $q_n = q$, $n \geq 0$, $p_{-n} = q$, $q_{-n} = p$, $n \geq 1$, and $r_n = r, n \in \mathbb{Z}$, where $p + r + q = 1$. Prove that the spectral matrix is given by

$$\Psi(x) = \frac{(p+q)\sqrt{(x-\sigma_-)(\sigma_+ - x)}}{2\pi q(1-x)(x-2r+1)} \begin{pmatrix} 1 & \dfrac{x-r}{p+q} \\ \dfrac{x-r}{p+q} & 1 \end{pmatrix} \mathbf{1}_{\{x \in [\sigma_-, \sigma_+]\}}$$
$$+ \frac{q-p}{2q} \left[\begin{pmatrix} 1 & -1 \\ -1 & 1 \end{pmatrix} \delta_{2r-1}(x) + \begin{pmatrix} 1 & 1 \\ 1 & 1 \end{pmatrix} \delta_1(x) \right] \mathbf{1}_{\{q>p\}},$$

where $\sigma_\pm = 1 - \left(\sqrt{p} \mp \sqrt{q}\right)^2$, and study the recurrence of the birth–death chain.

3

Spectral Representation of Birth–Death Processes

In this chapter we describe the probabilistic and asymptotic aspects of any birth–death process on $\mathbb{N}_0$ in terms of the spectral representation of the corresponding infinitesimal operator of the process, which in this case is a tridiagonal matrix $\mathcal{A}$ with nonpositive diagonal entries, positive off-diagonal entries (called the birth–death rates) and the sum of each row is less than or equal to 0. Although these processes are a special type of continuous-time Markov *chain*, we will use the (most common) term birth–death process to describe them. After the first section on basic definitions and fundamental properties of continuous-time Markov chains, we will obtain in Section 3.2 the *Karlin–McGregor integral representation formula* of the transition probability functions $P_{ij}(t)$ in terms of orthogonal polynomials with respect to a probability measure ψ with support inside the interval $[0,\infty)$. Unlike the case of discrete-time birth–death chains, this measure ψ is not unique in general since we are dealing now with the *Stieltjes moment problem*. In Section 3.3 we will study some families of polynomials related to the original orthogonal polynomials, as we did in Chapter 2. Section 3.4 is devoted to proving that the Karlin–McGregor representation formula is in fact a transition probability function of a birth–death process, something that was not necessary for the case of discrete-time birth–death chains. In Section 3.5 we will study birth–death processes with killing and some of their properties. In Section 3.6 we will study a collection of examples related to orthogonal polynomials, where we can apply the spectral analysis introduced in the previous sections. We will focus on examples with finite state space and examples on $\mathbb{N}_0$ related to the Stieltjes transform or with classical orthogonal polynomials. Among these examples we will study the $M/M/k$ queue with $1 \leq k \leq \infty$ servers, the continuous-time Ehrenfest and Bernoulli–Laplace urn models, a genetic model of Moran and linear birth–death processes. In Section 3.7 we will apply the Karlin–McGregor integral representation formula to the study of the probabilistic and asymptotic aspects of any birth–death process. We will see how to study *recurrence and absorption*, including the so-called *strong ratio limit property* and the concept of *limiting conditional distribution*, as we did for discrete-time birth–death chains. The main difference now is that we have a satisfactory and comprehensive solution of the strong ratio limit property, unlike the case of discrete-time birth–death chains. We will also study the concepts of the *decay parameter* and *quasi-stationary distributions*, and give a short summary of other interesting results. Finally, in Section 3.8, we will study the spectral representation of birth–death processes on $\mathbb{Z}$ or *bilateral birth–death processes*, where there are not so many results as in the case of birth–death processes on $\mathbb{N}_0$. Chronologically speaking, the spectral analysis of birth–death processes was studied before discrete-time birth–death chains. Although many of the results are similar or equivalent to those of discrete-time birth–death chains, the methods and techniques are quite different. We will emphasize the main differences when necessary and other results will be left as exercises.

3.1 Continuous-Time Markov Chains

This section is mostly based on the book by W. J. Anderson [1]. Let $(\Omega, \mathcal{F}, \mathbb{P})$ be a probability space and consider a continuous-time Markov chain $\{X_t, t \geq 0\}$ with state space $\mathcal{S} \subseteq \mathbb{Z}$, which can be finite, infinite or doubly infinite. We will mainly use $\mathcal{S} = \{0, 1, 2, \ldots\} = \mathbb{N}_0$, unless otherwise specified. The Markov property in this case can be written as

$$\mathbb{P}(X_{s+t} = j | X_s = i, X_\tau, 0 \leq \tau < s) = \mathbb{P}(X_{s+t} = j | X_s = i) = P_{ij}(s, t), \quad 0 < s < t.$$

As in the case of discrete-time birth–death chains, we will assume that the Markov chain is *homogeneous* in time and therefore the previous probabilities depend only on the time difference $t - s$, i.e. $P_{ij}(s,t) = P_{ij}(0, t - s)$. In this case we will always write $P_{ij}(t)$ and call them *transition probability functions*. These probabilities can be written in the so-called *transition probability matrix*

$$P(t) = \begin{pmatrix} P_{00}(t) & P_{01}(t) & P_{02}(t) & \cdots \\ P_{10}(t) & P_{11}(t) & P_{12}(t) & \cdots \\ P_{20}(t) & P_{21}(t) & P_{22}(t) & \cdots \\ \vdots & \vdots & \vdots & \end{pmatrix}.$$

Since time is continuous, the analysis of $P(t)$ will be possible using differential relations with respect to t. In particular we will focus on how the process evolves infinitesimally in time. The transition matrix $P(t)$ has the following properties:

(1) $P_{ij}(t) \geq 0$ for all $i, j \in \mathcal{S}$ and $P_{ij}(0) = \delta_{ij}$, the Kronecker delta.
(2) $\sum_{j \in \mathcal{S}} P_{ij}(t) \leq 1$ for all $t \geq 0, i \in \mathcal{S}$. The process $P_{ij}(t)$ is said to be *honest* if $\sum_{j \in \mathcal{S}} P_{ij}(t) = 1$ for all $t \geq 0, i \in \mathcal{S}$, and *dishonest* otherwise.

(3) For $i, j \in \mathcal{S}$ and for all $s, t \geq 0$, we have the *Chapman–Kolmogorov equations* (or semigroup property):

$$P_{ij}(s+t) = \sum_{k \in \mathcal{S}} P_{ik}(s) P_{kj}(t). \tag{3.1}$$

The converse is also true, that is, if we have a matrix $P(t)$ with the three properties above, it is always possible to construct a probability space $(\Omega, \mathcal{F}, \mathbb{P})$ and a continuous-time Markov chain $\{X_t, t \geq 0\}$ such that the corresponding probability transitions coincide with the entries $P_{ij}(t)$ of the matrix $P(t)$. Thus, the process is totally determined by a transition matrix $P(t)$ that satisfies properties **(1)–(3)**. The processes we will focus on will always be *stable*, meaning that the trajectories of the process will consist of right-continuous step functions. This is equivalent to the transition functions $P_{ii}(t)$ having finite derivatives of any order at $t = 0$ for any $i \in \mathcal{S}$.

Apart from the three previous properties **(1)–(3)**, we will always assume that the transition functions are *standard*, that is, they have the property:

(4) $\lim_{t \downarrow 0} P_{ii}(t) = 1$, for all $i \in \mathcal{S}$ (and therefore, since $0 \leq \sum_{j \neq i} P_{ij}(t) \leq 1 - P_{ii}(t)$, we have that $P_{ij}(t) \to \delta_{ij}$ as $t \downarrow 0$, for all $i, j \in \mathcal{S}$).

The following proposition shows that $P_{ij}(t)$ with the properties **(1)–(4)** is always differentiable at $t = 0$ (and therefore for all $t \geq 0$ by the semigroup property).

Proposition 3.1 ([1, Proposition 1.2.2]) *Let $P_{ij}(t)$ be a transition function satisfying properties (1)–(4). Then for $i \in \mathcal{S}$, we have that:*

1. $-P'_{ii}(0) = \lim_{t \downarrow 0} \dfrac{1 - P_{ii}(t)}{t} = q_i$ *(but it can be $+\infty$).*
2. *If $q_i < \infty$ then* $P'_{ij}(0) = \lim_{t \downarrow 0} \dfrac{P_{ij}(t)}{t} = q_{ij} < \infty$*, for all $j \neq i$.*

Observe that by definition we have $q_i, q_{ij} \geq 0$. A state $i \in \mathcal{S}$ is *stable* if $q_i < \infty$ and *instantaneous* if $q_i = \infty$. The process is stable if all states are stable. All the processes we study in these notes are stable. A state i is an *absorbing state* if $q_i = 0$, in which case $P_{ii}(t) = 1$ for all $t \geq 0$. Observe that $q_{ij} \geq 0$ while $q_{ii} = -q_i \leq 0$. We will call the matrix $\mathcal{A} = (a_{ij})$ the *infinitesimal operator matrix* of the process where

$$a_{ij} = \begin{cases} q_{ij}, & i \neq j, \\ -q_i, & i = j. \end{cases} \tag{3.2}$$

In other words,

$$\mathcal{A} = \begin{pmatrix} -q_0 & q_{01} & q_{02} & \cdots \\ q_{10} & -q_1 & q_{12} & \cdots \\ q_{20} & q_{21} & -q_2 & \cdots \\ \vdots & \vdots & \vdots & \end{pmatrix} = P'(0). \tag{3.3}$$

The diagonal entries are all nonpositive or possibly infinite, while the off-diagonal entries are finite and nonnegative. The sum of each row, according to the property **(2)** (taking derivatives with respect to t) is nonpositive, i.e. $\sum_{j\in\mathcal{S}} a_{ij} \leq 0$ for all $i \in \mathcal{S}$. If all the diagonal entries are finite, then $\mathcal{A}$ is *stable*. If in addition all the rows add 0, i.e. $\sum_{j\in\mathcal{S}} a_{ij} = 0$ for all $i \in \mathcal{S}$, then $\mathcal{A}$ is *conservative*.

In the case of discrete-time Markov chains, the process is completely determined by the one-step transition matrix P and the initial distribution. The semigroup formed by the powers of P, i.e. $\{P^n, n = 0, 1, 2, \ldots\}$ has a "minimum" element P, apart from the identity I. However, for continuous-time Markov chains, this minimum element for the semigroup $\{P(t), t \geq 0\}$ is represented by the infinitesimal operator matrix $\mathcal{A}$. In fact, another way to write Proposition 3.1 is using that

$$\begin{aligned}
\mathbb{P}(X_{t+h} = i|X_t = i) &= P_{ii}(h) = 1 - q_i h + o(h), \quad h \downarrow 0, \\
\mathbb{P}(X_{t+h} = j|X_t = i) &= P_{ij}(h) = q_{ij} h + o(h), \quad h \downarrow 0, \quad i \neq j,
\end{aligned}$$

where $o(h)$ means that $o(h)/h \downarrow 0$ as $h \downarrow 0$. Therefore the "minimum" element of the semigroup $\{P(t), t \geq 0\}$ is given by the infinitesimal operator $\mathcal{A}$ in (3.3).

If the chain is stable and conservative the coefficients (3.2) have a meaning in terms of how to built the trajectories of the process. Define, for $i \in \mathcal{S}$ and $X_0 = i$, the holding time in state i as the random variable

$$H_i = \begin{cases} \inf\{t \geq 0 | X_t \neq i\}, & \text{if the set is not empty,} \\ +\infty, & \text{otherwise.} \end{cases} \tag{3.4}$$

In the case that $q_i = 0$, then from Proposition 3.1 we get that $P_{ii}(t) = 1, t \geq 0$, i.e. i is an absorbing state and $H_i = \infty$. If $q_i > 0$ we get the following properties (see [1, Proposition 1.2.8]):

$$\mathbb{P}(H_i > t|X_0 = i) = e^{-q_i t}, \quad t \geq 0, \quad \text{where} \quad q_i = -q_{ii}, \tag{3.5}$$

$$\mathbb{P}(X_{H_i} = j|X_0 = i) = \frac{q_{ij}}{q_i}, \quad j \neq i. \tag{3.6}$$

Property (3.5) follows using that $\mathbb{P}(H_i > t + s|X_0 = i) = \mathbb{P}(H_i > s|X_0 = i)\mathbb{P}(H_i > t|X_0 = i)$, so it must be exponentially distributed. Property (3.6) holds using that the process is homogeneous. Indeed,

$$\begin{aligned}
\mathbb{P}(X_{H_i} = j|X_0 = i) &= \lim_{h\downarrow 0} \mathbb{P}(X_{t+h} = j|X_t = i, X_{t+h} \neq i) \\
&= \lim_{h\downarrow 0} \frac{\mathbb{P}(X_h = j|X_0 = i)}{\mathbb{P}(X_h \neq i|X_0 = i)} = \lim_{h\downarrow 0} \left(\frac{P_{ij}(h)}{h}\right) \Big/ \left(\frac{1 - P_{ii}(h)}{h}\right) = \frac{q_{ij}}{q_i}.
\end{aligned}$$

The interpretation of (3.5) and (3.6) is as follows. The chain starts at a certain state i. If $q_i = 0$, then the chain will always remain in the same absorbing state. If $0 < q_i < \infty$, then, according to property (3.5), the process will remain at the same state an amount of time H_i, which is an exponentially distributed random variable with mean q_i^{-1}. When we arrive at H_i it makes a transition to another state j. From (3.6), the probability that it jumps to the state j, if $j \neq i$, is q_{ij}/q_i (which is nonnegative since $q_i \geq \sum_{k \neq i} q_{ik} \geq q_{ij}$). Due to the right-continuity of the trajectories, the chain will be in the state j at time H_i, i.e. $X_{H_i} = j$. From H_i, due to the homogeneity and the Markov property of the chain, the future behavior of the process is independent of what happened before H_i, so the process behaves as if it started at time 0 at state j. Next, the process will remain in the same state j an amount of time H_j until it jumps to another state k with probability q_{jk}/q_j, and so on.

Assume now that $\mathcal{A}$ is conservative and define

$$J_n = \begin{cases} 0, & \text{if} \quad n = 0, \\ \inf\{t > J_{n-1} | X_t \neq X_{t^-}\}, & \text{if} \quad n \geq 1, \end{cases} \tag{3.7}$$

and $X_n = X_{J_n}$. J_n is the time of the nth transition, i.e. $J_n = H_{i_1} + \cdots + H_{i_n}$. The process X_n is a discrete-time Markov chain with transition probabilities (since $\mathcal{A}$ is conservative)

$$P_{ij} = \begin{cases} \delta_{ij}, & \text{if} \quad q_i = 0, \\ 0, & \text{if} \quad q_i > 0, \quad j = i, \\ \dfrac{q_{ij}}{q_i}, & \text{if} \quad q_i > 0, \quad j \neq i. \end{cases} \tag{3.8}$$

This discrete-time Markov chain is called the *jump chain* or embedded Markov chain. It is clear that the chain X_t is completely determined by the processes J_n and X_n and vice versa.

The previous construction suggests that $J_\infty = \lim_{n\to\infty} J_n = \infty$. However, there are situations in which J_∞ is finite, called the *time of the first infinity or explosion*. In these cases, in a finite time interval there can be an infinite number of jumps. This situation is not possible for discrete-time Markov chains. The following result reflects the situation in which this may happen:

Lemma 3.2 ([1, Lemma 1.2.9]) *Let S_n be a sequence of independent and identically distributed exponential random variables with parameters $\lambda_n, n \geq 1$. Let $S = \sum_{n=1}^{\infty} S_n$. Then*

$$\mathbb{P}(S = \infty) = 1 \quad \textit{if and only if} \quad \sum_{n=1}^{\infty} \frac{1}{\lambda_n} = \infty.$$

In particular, $\mathbb{P}(S < \infty) = 0$ *or* $\mathbb{P}(S < \infty) = 1$.

For example, for a pure-birth process with

$$q_{ij} = \begin{cases} \lambda_i, & \text{if} \quad j = i + 1, \\ -\lambda_i, & \text{if} \quad j = i, \\ 0, & \text{otherwise}, \end{cases}$$

we have that

$$\mathbb{P}(J_\infty = \infty | X_0 = i) = 1 \Leftrightarrow \sum_{n=i}^{\infty} \frac{1}{\lambda_n} = \infty \Leftrightarrow \sum_{n=1}^{\infty} \frac{1}{\lambda_n} = \infty.$$

Therefore for the *Poisson process* (where $\lambda_n = \lambda > 0$, $n \geq 0$), or the *Yule–Furry process* or pure-birth linear process (where $\lambda_n = \lambda n, n \geq 0$), we have that $\mathbb{P}(J_\infty = \infty | X_0 = i) = 1$. However, for a quadratic pure-birth process (where $\lambda_n = \lambda n^2$, $n \geq 0$), we will have $\mathbb{P}(J_\infty < \infty | X_0 = i) = 1$. In this case it can be checked that the distribution of J_∞ is given by (see [1, pp. 20–21])

$$\mathbb{P}(J_\infty \leq t | X_0 = i) = 1 - 2\sum_{k=i}^{\infty} (-1)^{k+1} e^{-\lambda k^2 t}, \quad t \geq 0, \quad i \geq 1.$$

Although it may seem paradoxical, after this first explosion time J_∞ it is possible that the chain is still active. However, this information cannot be derived from the information contained in the infinitesimal operator matrix $\mathcal{A}$ (see more information in [1, Chapter 2]).

From now on we will assume that $\mathcal{A}$ is always stable, i.e. $q_i < \infty$, unless otherwise specified.

Proposition 3.3 ([1, Proposition 1.2.7]) *Let $\mathcal{A}$ be stable and q_{ij} the coefficients introduced in Proposition 3.1. Then*

$$P'_{ij}(t) = \sum_{k \in \mathcal{S}} q_{ik} P_{kj}(t), \quad t \geq 0, \quad i,j \in \mathcal{S}. \tag{3.9}$$

Moreover, if $\sum_{j \in \mathcal{S}} P_{ij}(t) = 1$ (honest), then (3.9) *holds if and only if $\mathcal{A}$ is conservative.*

Proof Using the Chapman–Kolmogorov equations (3.1) and Proposition 3.1 we have that

$$\begin{aligned} P'_{ij}(t) &= \lim_{s \downarrow 0} \frac{\partial}{\partial s} P_{ij}(s+t) = \lim_{s \downarrow 0} \frac{\partial}{\partial s} \sum_{k \in \mathcal{S}} P_{ik}(s) P_{kj}(t) \\ &= \lim_{s \downarrow 0} \sum_{k \in \mathcal{S}} P'_{ik}(s) P_{kj}(t) = \sum_{k \in \mathcal{S}} q_{ik} P_{kj}(t). \end{aligned}$$

Alternatively, from the honesty property, we have that $\sum_{j \in \mathcal{S}} P'_{ij}(t) = 0$. If the differential equations (3.9) hold, then

$$0 = \sum_{j \in \mathcal{S}} P'_{ij}(t) = \sum_{j \in \mathcal{S}} \sum_{k \in \mathcal{S}} q_{ik} P_{kj}(t) = \sum_{k \in \mathcal{S}} q_{ik} \sum_{j \in \mathcal{S}} P_{kj}(t) = \sum_{k \in \mathcal{S}} q_{ik}.$$

Therefore $\mathcal{A}$ is conservative. If $\mathcal{A}$ is conservative, then the first part of this proposition implies that the differential equations (3.9) hold. □

The differential equations (3.9) are usually called the *Kolmogorov backward equations*, and in all the situations we consider in this text they will always hold. Meanwhile, the differential equations

$$P'_{ij}(t) = \sum_{k \in \mathcal{S}} P_{ik}(t) q_{kj}, \quad t \geq 0, \quad i,j \in \mathcal{S}, \tag{3.10}$$

are usually called the *Kolmogorov forward equations* (also known as evolution or Fokker–Planck equation). Unlike the Kolmogorov backward equations, the Kolmogorov forward equations may fail since it is possible that before reaching time t the process has taken infinite jumps. This can happen even in cases where the process is conservative. In Proposition 3.5 below we will review some aspects concerning the existence and uniqueness of the Kolmogorov equations given an infinitesimal operator matrix $\mathcal{A}$. The Kolmogorov equations can be written more conveniently in matrix form as

$$P'(t) = \mathcal{A}P(t), \quad P'(t) = P(t)\mathcal{A}, \quad P(0) = I. \tag{3.11}$$

If we are given a transition function $P_{ij}(t)$ then it is always possible to derive the Kolmogorov equations. But in most practical situations we are given the infinitesimal operator $\mathcal{A}$ and wonder if there are (unique) solutions $f_{ij}(t)$ of the Kolmogorov equations that give a proper transition function. This will depend on the shape of $\mathcal{A}$. The matrix $\mathcal{A}$ will describe the process for a finite number of jumps, but it does not have to specify the process in a unique way. So the Kolmogorov equations may have more than one solution, if there are solutions. This situation does not happen in the case of discrete-time Markov chains, where solutions exist and are always unique.

Definition 3.4 Let $\mathcal{A}$ be stable. A transition function $P_{ij}(t)$ is called an *$\mathcal{A}$-function* if $\mathcal{A}$ is the infinitesimal operator matrix of a transition probability function $P(t)$, i.e. $P'(0) = \mathcal{A}$. ◊

Proposition 3.5 ([1, Theorem 2.2.2]) *Let $\mathcal{A}$ be stable but not necessarily conservative.*

1. *There exists a (possibly dishonest) transition function $f_{ij}(t)$ satisfying both Kolmogorov equations such that $f_{ij}(t)$ is minimal, in the sense that $f_{ij}(t) \leq P_{ij}(t)$ for all $i,j \in \mathcal{S}$ and $t \geq 0$, where $P_{ij}(t)$ is any other nonnegative solution (not necessarily a transition function) of the Kolmogorov equations. Also $f_{ij}(t)$ is the minimal $\mathcal{A}$-function in the sense that $f_{ij}(t) \leq P_{ij}(t)$ for all $i,j \in \mathcal{S}$ and $t \geq 0$, where $P_{ij}(t)$ is any other nonnegative $\mathcal{A}$-function (not necessarily a solution of the Kolmogorov equations).*
2. *Let $f_{ij}(t)$ be the transition function of the previous paragraph. Then if $f_{ij}(t)$ is honest then it is the unique solution of both Kolmogorov equations and in fact is the unique $\mathcal{A}$-function. If $\mathcal{A}$ is conservative, the minimal solution $f_{ij}(t)$ is unique if and only if it is honest. If $\mathcal{A}$ is not conservative then $f_{ij}(t)$ can never be honest.*

In general, it is not easy to compute a transition function $f_{ij}(t)$ given the infinitesimal operator $\mathcal{A}$. We will give a collection of necessary and sufficient conditions for the transition function $f_{ij}(t)$ to be unique in terms of the infinitesimal operator $\mathcal{A}$ (see [1, Chapter 2] for more details):

1. If $\mathcal{A}$ is conservative, the minimal solution $f_{ij}(t)$ of the Kolmogorov backward equation is unique if and only if the equation $\mathcal{A}x = \lambda x, 0 \leq x \leq 1$, that is,

$$\sum_{j \in \mathcal{S}} q_{ij} x_j = \lambda x_i, \quad 0 \leq x_i \leq 1, \quad i \in \mathcal{S}$$

   does not have a nontrivial solution for some (and therefore all) $\lambda > 0$. In this case we say that the matrix $\mathcal{A}$ is *regular*.
2. Let $\mathcal{A}$ be not necessarily conservative but *uniformly bounded* (i.e. $\sup_{i \in \mathcal{S}} q_i < \infty$). Then $f_{ij}(t)$ is the unique $\mathcal{A}$-function.
3. Let $\mathcal{A}$ be conservative and let X_n be the discrete-time Markov chain whose transition probability matrix P is given by (3.8). If all states in $\mathcal{S}$ are recurrent for X_n, then the minimal solution $f_{ij}(t)$ is honest and therefore unique.

If $\mathcal{A}$ is not conservative then there may be $\mathcal{A}$-functions that satisfy neither the backward nor the Kolmogorov forward equations (see [1, Chapter 4] for more details).

All the definitions we saw for discrete-time Markov chains in Chapter 1 (such as communication, communication classes, irreducibility, recurrence, transience, invariant vector, etc.) can be extended naturally to continuous-time Markov chains. Since the time is continuous, we cannot define the generating function associated with the transition functions $P_{ij}(t)$ (unless we consider them with respect to the state space). Therefore we replace the concept of the generating function with the *Laplace transform* of the transition functions $P_{ij}(t)$. Indeed,

$$\widehat{P}_{ij}(\lambda) = \int_0^\infty e^{-\lambda t} P_{ij}(t)\, dt, \quad \lambda > 0, \quad i,j \in \mathcal{S}. \tag{3.12}$$

This function, in terms of λ, is also called the *resolvent function*, and satisfies the following properties (see [1, Section 1.1.3]):

1. $\widehat{P}_{ij}(\lambda) \geq 0$ for all $i,j \in \mathcal{S}$ and $\lambda > 0$.
2. $\lambda \sum_{j \in \mathcal{S}} \widehat{P}_{ij}(\lambda) \leq 1$, for all $i \in \mathcal{S}$ and $\lambda > 0$.
3. $\widehat{P}_{ij}(\lambda) - \widehat{P}_{ij}(\mu) + (\lambda - \mu) \sum_{k \in \mathcal{S}} \widehat{P}_{ik}(\lambda)\widehat{P}_{kj}(\mu) = 0$ for all $i,j \in \mathcal{S}$ and $\lambda, \mu > 0$.
4. $\lim_{\lambda \to \infty} \lambda \widehat{P}_{ii}(\lambda) = 1$, for all $i \in \mathcal{S}$ (and therefore $\lim_{\lambda \to \infty} \lambda \widehat{P}_{ij}(\lambda) = \delta_{ij}$).

Many of the results about existence and uniqueness of the minimal transition functions satisfying the Kolmogorov equations (or other properties of $P_{ij}(t)$) can be given in terms of this resolvent function. As in the case of discrete-time Markov chains we can talk about *first passage time distributions*, defined by

$$F_{ij}(t) = \mathbb{P}(X_\tau = j \text{ for some } \tau, 0 < \tau \leq t | X_0 = i), \quad i \neq j, \tag{3.13}$$

$$F_{ii}(t) = \mathbb{P}(X_{\tau_1} \neq i, X_{\tau_2} = i \text{ for some } \tau_1, \tau_2, 0 < \tau_1 < \tau_2 \leq t | X_0 = i). \tag{3.14}$$

Applying (3.5) (enumeration of paths), the quantities $P_{ij}(t)$ and $F_{ij}(t)$ are related by the formulas

$$P_{ii}(t) = e^{-q_i t} + \int_0^t P_{ii}(t-s) dF_{ii}(s), \tag{3.15}$$

$$P_{ij}(t) = \int_0^t P_{jj}(t-s) dF_{ij}(s), \quad i \neq j. \tag{3.16}$$

Similarly, if we call $\widehat{F}_{ij}(\lambda)$ the Laplace transformation of the distributions $F_{ij}(t)$ defined by (3.13) and (3.14), i.e.

$$\widehat{F}_{ij}(\lambda) = \int_0^\infty e^{-\lambda t} dF_{ij}(t),$$

we have, using (3.15), (3.16) and the convolution property for Laplace transforms, the following relations:

$$\widehat{P}_{ii}(\lambda) = \frac{1}{\lambda + q_i} + \widehat{P}_{ii}(\lambda)\widehat{F}_{ii}(\lambda), \tag{3.17}$$

$$\widehat{P}_{ij}(\lambda) = \widehat{P}_{jj}(\lambda)\widehat{F}_{ij}(\lambda), \quad i \neq j. \tag{3.18}$$

Definition 3.6 With the previous notation we have the following:

- A state $i \in \mathcal{S}$ is *recurrent* if $\int_0^\infty P_{ii}(t)dt = \infty$ and *transient* otherwise. In terms of the distributions $F_{ii}(t)$, we must have that $\int_0^\infty dF_{ii}(t) = 1$ or equivalently $\lim_{\lambda \downarrow 0} \widehat{F}_{ii}(\lambda) = 1$ (the same applies for a transient state).
- A recurrent state $i \in \mathcal{S}$ is *positive recurrent* (or ergodic) if $\lim_{t\to\infty} P_{ii}(t) > 0$ and *null recurrent* otherwise. In terms of the distributions $F_{ii}(t)$, we must have that $\int_0^\infty t dF_{ii}(t) < \infty$ or equivalently $-\lim_{\lambda \downarrow 0} \frac{\partial}{\partial \lambda} \widehat{F}_{ii}(\lambda) < \infty$ (the same applies for a null recurrent state). ◊

A vector $\pi = (\pi_i)_{i \in \mathcal{S}}$ is said to be *invariant* (or stationary) for the continuous-time Markov chain if

$$\pi_j \geq 0, \quad \text{and} \quad \sum_{i \in \mathcal{S}} \pi_i P_{ij}(t) = \pi_j, \quad t \geq 0, \quad j \in \mathcal{S}, \tag{3.19}$$

or, in vector form, $\pi P(t) = \pi$, for all $t \geq 0$. In the case where this vector can be normalized to a probability vector (i.e. all components are nonnegative and $\sum_{i \in \mathcal{S}} \pi_i = 1$), then we say that π is an *invariant distribution* (or stationary) of the continuous-time Markov chain. The existence of this invariant distribution is equivalent to irreducibility and positive recurrence.

Due to the difficulty of computing $P(t)$ explicitly, the invariant vector is usually calculated using the infinitesimal operator matrix $\mathcal{A}$ in (3.3) and following the

Kolmogorov equations (3.10). This is applicable when $\mathcal{A}$ is stable, conservative and there is a minimal transition function. Taking derivatives with respect to t in (3.19), we have

$$0 = \pi P'(t) = \pi P(t)\mathcal{A} = \pi \mathcal{A}P(t) \quad \Rightarrow \quad \pi\mathcal{A} = 0.$$

Therefore if we look for a vector π such that $\pi\mathcal{A} = 0$ and with nonnegative components, we will have a candidate for the invariant vector (or distribution).

If the state space is finite, i.e. $\mathcal{S} = \{0, 1, \dots, N\}$, then we can derive a spectral representation of the transition functions $P_{ij}(t)$ in terms of the eigenvalues of $\mathcal{A}$. There will always be a unique $\mathcal{A}$-function for any $\mathcal{A}$ (not necessarily conservative) such that we can always solve the Kolmogorov equations. If $\mathcal{A}$ is similar to a simpler matrix J such that $\mathcal{A} = MJM^{-1}$ then

$$P(t) = Me^{tJ}M^{-1},$$

satisfies both the Kolmogorov equations with initial conditions. The simplest situation is when $\mathcal{A}$ has $N+1$ independent eigenvectors and J is diagonal with the corresponding eigenvalues. All eigenvalues λ of $\mathcal{A}$ satisfy that $\operatorname{Re}\lambda \le 0$. Indeed, since $\mathcal{A}$ is uniformly bounded (finite dimensional), call $\tau = \sup_{i\in\mathcal{S}} q_i \ge 0$. The matrix $I - \tau^{-1}\mathcal{A}$ is a nonnegative matrix (substochastic). Therefore all the eigenvalues $\tilde{\lambda}$ satisfy $|\tilde{\lambda}| \le 1$ (by the Perron–Frobenius theorem) and consequently the eigenvalues λ of $\mathcal{A}$ can be written as $\lambda = \tau(\tilde{\lambda} - 1)$. Then $\operatorname{Re}\lambda \le 0$.

There is a special case when $\mathcal{A}$ is assumed to be *weakly symmetric*, i.e. there exists a vector $m = (m_i)_{i=0}^N$ such that $m_i q_{ij} = m_j q_{ji}$ for all $i,j \in \mathcal{S}$ (this is also equivalent to $m_i P_{ij}(t) = m_j P_{ji}(t)$ for all $t \ge 0$ where $P_{ij}(t)$ is the minimal $\mathcal{A}$-function, see [1, Proposition 7.1.2]). In that case, if we define the diagonal matrix $\Pi = \operatorname{diag}\{\sqrt{m_0}, \sqrt{m_1}, \dots \sqrt{m_N}\}$, we have that $S = \Pi\mathcal{A}\Pi^{-1}$ is a symmetric matrix. Therefore all eigenvalues are real and located inside the interval $(-\infty, 0]$. Then we can write $S = MDM^T$, where M is an orthogonal real matrix and D a diagonal matrix with the eigenvalues d_k of $\mathcal{A}$ located in the main diagonal. Therefore we obtain

$$P(t) = \Pi^{-1}Me^{tD}M^T\Pi, \quad t \ge 0,$$

or entry by entry

$$P_{ij}(t) = m_j \sum_{k=0}^{N} e^{td_k}\phi_i^{(k)}\phi_j^{(k)}, \tag{3.20}$$

where d_i are the diagonal entries of D and $\phi_i^{(k)} = M_{ik}/\sqrt{m_i}, i,k \in \mathcal{S}$ satisfy that

$$\sum_{k=0}^{N} \phi_i^{(k)}\phi_j^{(k)} = \frac{\delta_{ij}}{m_i}, \quad i,j \in \mathcal{S}.$$

The simplest examples of continuous-time Markov chains are *birth–death processes*. A set of birth–death rates is a sequence $\{(\lambda_n, \mu_n), n \geq 0\}$ of pairs of numbers such that $\lambda_n > 0, n \geq 0$, $\mu_n > 0, n \geq 1$ and $\mu_0 \geq 0$. The transition function $P(t)$ satisfies the following conditions as $t \downarrow 0$:

$$P_{ij}(t) = \begin{cases} \lambda_i t + o(t), & \text{if} \quad j = i+1, \\ \mu_i t + o(t), & \text{if} \quad j = i-1, \\ 1 - (\lambda_i + \mu_i)t + o(t), & \text{if} \quad j = i. \end{cases}$$

We will focus on birth–death processes with state space $\mathcal{S} = \mathbb{N}_0$, but we will also find finite or doubly infinite state spaces (in which case they are called *bilateral* birth–death processes, as we will see in Section 3.8). The matrix corresponding to the infinitesimal operator $\mathcal{A}$ in (3.3) is

$$\mathcal{A} = \begin{pmatrix} -(\lambda_0 + \mu_0) & \lambda_0 & 0 & 0 & 0 & \cdots \\ \mu_1 & -(\lambda_1 + \mu_1) & \lambda_1 & 0 & 0 & \cdots \\ 0 & \mu_2 & -(\lambda_2 + \mu_2) & \lambda_2 & 0 & \cdots \\ 0 & 0 & \mu_3 & -(\lambda_3 + \mu_3) & \lambda_3 & \cdots \\ \vdots & \vdots & \vdots & \ddots & \ddots & \ddots \end{pmatrix}. \tag{3.21}$$

A diagram of the transitions between states is

$$\textcircled{0} \underset{\mu_1}{\overset{\lambda_0}{\rightleftarrows}} \textcircled{1} \underset{\mu_2}{\overset{\lambda_1}{\rightleftarrows}} \textcircled{2} \underset{\mu_3}{\overset{\lambda_2}{\rightleftarrows}} \textcircled{3} \underset{\mu_4}{\overset{\lambda_3}{\rightleftarrows}} \textcircled{4} \underset{\mu_5}{\overset{\lambda_4}{\rightleftarrows}} \textcircled{5} \underset{\mu_6}{\overset{\lambda_5}{\rightleftarrows}} \cdots$$

$\mathcal{A}$ is conservative if and only if $\mu_0 = 0$. In the case where $\mu_0 > 0$ we are allowing the state 0 to jump to an absorbing state with probability $\mu_0/(\lambda_0 + \mu_0)$, which we usually denote by -1.

The Kolmogorov backward equations (3.9) are given by

$$\begin{aligned} P'_{0,j}(t) &= -(\lambda_0 + \mu_0)P_{0,j}(t) + \lambda_0 P_{1,j}(t), \quad P_{ij}(0) = \delta_{ij}, \\ P'_{i,j}(t) &= \mu_i P_{i-1,j}(t) - (\lambda_i + \mu_i)P_{i,j}(t) + \lambda_i P_{i+1,j}(t), \quad i \geq 1, \end{aligned} \tag{3.22}$$

while the Kolmogorov forward equations (3.10) are given by

$$\begin{aligned} P'_{i,0}(t) &= -(\lambda_0 + \mu_0)P_{i,0}(t) + \mu_1 P_{i,1}(t), \quad P_{ij}(0) = \delta_{ij}, \\ P'_{i,j}(t) &= \lambda_{j-1} P_{i,j-1}(t) - (\lambda_j + \mu_j)P_{i,j}(t) + \mu_{j+1} P_{i,j+1}(t), \quad j \geq 1. \end{aligned} \tag{3.23}$$

As in the case of discrete-time birth–death chains, we can define the *potential coefficients* as follows:

$$\pi_0 = 1, \quad \pi_n = \frac{\lambda_0 \lambda_1 \cdots \lambda_{n-1}}{\mu_1 \mu_2 \cdots \mu_n}, \quad n \geq 1. \tag{3.24}$$

These potential coefficients satisfy the symmetry equations

$$\pi_n \lambda_n = \pi_{n+1} \mu_{n+1}, \quad n \geq 0, \tag{3.25}$$

as a result of looking for a vector π that makes the chain *reversible*, i.e. $\pi_i P_{ij}(t) = \pi_j P_{ji}(t)$ for all $t \geq 0$, and using the Kolmogorov equations (3.11). Observe that $\mathcal{A}$ is *weakly symmetric* taking $m_i = \pi_i$. If $\mathcal{A}$ is conservative then $\pi \mathcal{A} = 0$, and therefore π is an *invariant vector* of the birth–death process, which will become a distribution if $\sum_{n=0}^{\infty} \pi_n < \infty$.

The transition probability matrix P of the jump chain (3.8) is given by

$$P = \begin{pmatrix} 0 & \frac{\lambda_0}{\lambda_0+\mu_0} & 0 & 0 & 0 & \cdots \\ \frac{\mu_1}{\lambda_1+\mu_1} & 0 & \frac{\lambda_1}{\lambda_1+\mu_1} & 0 & 0 & \cdots \\ 0 & \frac{\mu_2}{\lambda_2+\mu_2} & 0 & \frac{\lambda_2}{\lambda_2+\mu_2} & 0 & \cdots \\ 0 & 0 & \frac{\mu_3}{\lambda_3+\mu_3} & 0 & \frac{\lambda_3}{\lambda_3+\mu_3} & \cdots \\ \vdots & \vdots & \vdots & \ddots & \ddots & \ddots \end{pmatrix}, \tag{3.26}$$

and therefore we get a symmetric discrete-time birth–death chain with transition probabilities

$$p_n = \frac{\lambda_n}{\lambda_n + \mu_n}, \quad r_n = 0, \quad q_n = \frac{\mu_n}{\lambda_n + \mu_n}, \quad n \geq 0.$$

The potential coefficients π_n^P of this jump chain are related to the potential coefficients (3.24) in the following form:

$$\pi_0^P = 1, \quad \pi_n^P = \pi_n \frac{\lambda_n + \mu_n}{\lambda_0 + \mu_0}, \quad n \geq 1. \tag{3.27}$$

The existence of a minimal transition function $f_{ij}(t)$ satisfying both Kolmogorov equations is guaranteed by Proposition 3.5. The uniqueness will depend strongly on the birth–death rates $\{(\lambda_n, \mu_n), n \geq 0\}$ of the infinitesimal operator matrix $\mathcal{A}$. To study uniqueness let us define the following series:

$$A = \sum_{n=0}^{\infty} \frac{1}{\lambda_n \pi_n}, \quad B = \sum_{n=0}^{\infty} \pi_n, \quad C = \sum_{n=0}^{\infty} \frac{1}{\lambda_n \pi_n} \sum_{m=0}^{n} \pi_m, \quad D = \sum_{n=1}^{\infty} \pi_n \sum_{m=0}^{n-1} \frac{1}{\lambda_m \pi_m}. \tag{3.28}$$

Observe that A can also be written as $A = \sum_{n=1}^{\infty} (\mu_n \pi_n)^{-1}$, while $C = \sum_{m=0}^{\infty} \pi_m \sum_{n=m}^{\infty} (\lambda_n \pi_n)^{-1}$ and $D = \sum_{m=0}^{\infty} (\lambda_m \pi_m)^{-1} \sum_{n=m+1}^{\infty} \pi_n$. We have that $C + D = AB$ and the following implications:

- $C < \infty \Rightarrow A < \infty$ and $A = \infty \Rightarrow C = \infty$.
- $D < \infty \Rightarrow B < \infty$ and $B = \infty \Rightarrow D = \infty$.

Then we have the following:

Proposition 3.7 ([1, Theorems 3.2.2 and 3.2.3]) *1. Let $\mathcal{A}$ defined by* (3.21) *and assume that $\lambda_n > 0$ for all $n \geq 1$. Then the minimal $\mathcal{A}$-function $f_{ij}(t)$ of the Kolmogorov backward equation is unique (if $\mu_0 = 0$ the unique $\mathcal{A}$-function) if and only if $C = \infty$ where C is given by* (3.28).

2. *Let $\mathcal{A}$ defined by* (3.21) *and assume that $\mu_n > 0$ for all $n \geq 1$. If the minimal solution $f_{ij}(t)$ associated with $\mathcal{A}$ of the Kolmogorov backward equation is honest, then it is unique. If it is dishonest, this minimal solution is unique if and only if $D = \infty$ where D is given by* (3.28).

Let us give an interpretation of the quantities C and D and after that we will give a classification of the boundary point ∞. Denote by

$$T_j = \inf\{t \geq 0\colon X_t = j\}, \quad j \in \mathbb{N}_0, \tag{3.29}$$

the *first passage time* of the birth–death process to the state j. Let us also denote by H_n the holding time at state n (see (3.4)), which is an exponential random variable of mean $(\lambda_n + \mu_n)^{-1}$ and $J_i, i \geq 0$, the time where the ith jump occurs (see (3.7)). Then, applying the law of total expectation, we have

$$\begin{aligned}
\mathbb{E}(T_{n+1} |\, X_0 = n) &= \mathbb{P}(X_{J_1} = n+1 |\, X_{J_0} = n)\mathbb{E}(H_n) \\
&\quad + \mathbb{P}(X_{J_1} = n-1 |\, X_{J_0} = n)(\mathbb{E}(H_n) + \mathbb{E}(T_n | X_0 = n-1) \\
&\quad + \mathbb{E}(T_{n+1} |\, X_0 = n)) \\
&= \frac{\lambda_n}{(\lambda_n + \mu_n)^2} + \frac{\mu_n}{(\lambda_n + \mu_n)} \\
&\quad \times \left(\frac{1}{(\lambda_n + \mu_n)} + \mathbb{E}(T_n | X_0 = n-1) + \mathbb{E}(T_{n+1} |\, X_0 = n) \right).
\end{aligned}$$

Assuming that $\mu_0 = 0$, we get $\mathbb{E}(T_1 | X_0 = 0) = 1/\lambda_0$ and for $n \geq 1$,

$$\mathbb{E}(T_{n+1} | X_0 = n) = \frac{1}{\lambda_n} + \frac{\mu_n}{\lambda_n}\mathbb{E}(T_n | X_0 = n-1).$$

Solving the recurrence, we obtain

$$\mathbb{E}(T_{n+1} | X_0 = n) = \frac{1}{\lambda_n \pi_n} \sum_{m=0}^{n} \pi_m, \quad n \geq 0.$$

Therefore, summing up in the index n, we have

$$\sum_{n=0}^{\infty} \mathbb{E}(T_{n+1} | X_0 = n) = \sum_{n=0}^{\infty} \frac{1}{\lambda_n \pi_n} \sum_{m=0}^{n} \pi_m = C.$$

So C is interpreted as the expected time to reach the state ∞ (as a boundary value of $\mathbb{N}_0$) from the state 0.

Similarly, if we consider the birth–death process on the state space $\mathcal{S} = \{0, 1, \dots, L\}$ with $\lambda_L = 0$, we obtain that $\mathbb{E}(T_{L-1}|X_0 = L) = 1/\mu_L$ and for $n = 1, \dots, L-1$,

$$\mathbb{E}(T_{n-1}|X_0 = n) = \frac{1}{\mu_n} + \frac{\lambda_n}{\mu_n}\mathbb{E}(T_n|X_0 = n+1).$$

Solving the recurrence, we obtain

$$\mathbb{E}(T_{n-1}|X_0 = n) = \frac{1}{\mu_n \pi_n}\sum_{m=n}^{L} \pi_m, \quad n = 1, \dots, L.$$

Therefore we have

$$\sum_{n=1}^{L} \mathbb{E}(T_{n-1}|X_0 = n) = \sum_{n=1}^{L} \frac{1}{\mu_n \pi_n}\sum_{m=n}^{L} \pi_m,$$

and, using the symmetry equations (3.25) and $L \to \infty$, we get the value of D. So D is interpreted as the expected time to reach the state 0 from the state ∞ (as a boundary value of $\mathcal{S}$).

Depending on the values of C and D we get the following classification of the boundary ∞ due to W. Feller [56]:

- The boundary ∞ is said to be *regular* if $C, D < \infty$ (this is equivalent to $A, B < \infty$).
- The boundary ∞ is said to be an *exit* if $C < \infty$ and $D = \infty$ (this is equivalent to $A < \infty, B = \infty, C < \infty$).
- The boundary ∞ is said to be an *entrance* if $C = \infty$ and $D < \infty$ (this is equivalent to $A = \infty, B < \infty, D < \infty$).
- The boundary ∞ is said to be *natural* if $C, D = \infty$ (this is equivalent to any other situation).

According to Proposition 3.7, if ∞ is a regular boundary (i.e. in finite time we can reach ∞ from 0 and vice versa) then both Kolmogorov equations have infinitely many solutions. If ∞ is an exit boundary (i.e. in finite time we can reach ∞ from 0 but not the other way around) then the Kolmogorov backward equation has infinitely many solutions but the forward equation only the minimal. If ∞ is an entrance boundary (i.e. in finite time we can reach 0 from ∞ but not the other way around) then the Kolmogorov backward equation is unique but the forward equation has only one solution if $\mu_0 = 0$ or infinitely many solutions otherwise. Finally, if ∞ is a natural boundary (i.e. the time to reach ∞ from 0 is infinity and vice versa) then both Kolmogorov equations have a unique $\mathcal{A}$-function as a solution and it is honest if and only if $\mu_0 = 0$.

3.2 Karlin–McGregor Representation Formula

This integral representation was derived for the first time in [82, 83] and later exploited in [84], although these representations were also discovered by Ledermann and Reuter (see [111]). These authors used a method of passage to the limit from a system with a finite number of states, and obtained the integral representation of the minimal solution. By a similar limiting process an integral representation of the transition matrix of a discrete-time birth–death chain was found in a number of interesting cases by Kac (see [80]). Similarly to the case of discrete-time birth–death chains we first will derive it formally and later prove it in a more rigorous way, different from the one we used in Chapter 1.

Let us define a sequence of polynomials $(Q_n)_n$ with $\deg Q_n = n$ recurrently using the infinitesimal operator $\mathcal{A}$ in (3.21) as follows:

$$\begin{aligned} Q_0(x) &= 1, \quad Q_{-1}(x) = 0, \\ -xQ_0(x) &= \lambda_0 Q_1(x) - (\lambda_0 + \mu_0)Q_0(x), \\ -xQ_n(x) &= \lambda_n Q_{n+1}(x) - (\lambda_n + \mu_n)Q_n(x) + \mu_n Q_{n-1}(x), \quad n \geq 1. \end{aligned} \tag{3.30}$$

In vector form all the relations can be written as $-xQ(x) = \mathcal{A}Q(x)$, where $Q(x) = (Q_0(x), Q_1(x), \ldots)^T$. Consider the vector

$$f(x,t) = P(t)Q(x). \tag{3.31}$$

Using the Kolmogorov forward equation from (3.11), this vector satisfies

$$\frac{\partial f(x,t)}{\partial t} = P'(t)Q(x) = P(t)\mathcal{A}Q(x) = -xf(x,t), \quad f(x,0) = Q(x).$$

Therefore the solution of the previous differential equation is given by

$$f(x,t) = e^{-xt}Q(x). \tag{3.32}$$

Component-wise, equating (3.31) and (3.32), we have that

$$e^{-xt}Q_i(x) = \sum_{j=0}^{\infty} P_{ij}(t)Q_j(x).$$

The spectral theorem (or Favard's theorem) guarantees that there exists at least one probability measure ψ supported on the interval $[0,\infty)$ such that the polynomials defined by (3.30) are orthogonal with respect to ψ.

Multiplying $e^{-xt}Q(x) = P(t)Q(x)$ on the right by $Q^T(x)$ and integrating entry by entry with respect to the measure ψ we observe by orthogonality that all the entries outside the main diagonal are 0, and in the main diagonal we obtain the following representation:

$$P_{ij}(t)\int_0^\infty Q_j^2(x)\,d\psi(x) = \int_0^\infty e^{-xt}Q_i(x)Q_j(x)\,d\psi(x).$$

Therefore, we have

$$P_{ij}(t) = \frac{\displaystyle\int_0^\infty e^{-xt} Q_i(x) Q_j(x)\, d\psi(x)}{\displaystyle\int_0^\infty Q_j^2(x)\, d\psi(x)}.$$

From (3.30) it is possible to deduce (in the same way as for discrete-time birth–death chains, see (2.16)) that

$$\int_0^\infty Q_j^2(x)\, d\psi(x) = \frac{\mu_1 \mu_2 \cdots \mu_j}{\lambda_0 \lambda_1 \cdots \lambda_{j-1}}.$$

Therefore

$$\pi_j = \left(\int_0^\infty Q_j^2(x)\, d\psi(x) \right)^{-1},$$

and we have the Karlin–McGregor integral representation formula

$$P_{ij}(t) = \mathbb{P}(X_t = j | X_0 = i) = \pi_j \int_0^\infty e^{-xt} Q_i(x) Q_j(x)\, d\psi(x).$$

In order to prove the previous formula in a more rigorous way, we will use the following result, also known as the *Kendall representation theorem*:

Theorem 3.8 ([1, Theorem 7.1.8]) *Let $f_{ij}(t)$ be the minimal $\mathcal{A}$-function. Then for each $i,j \in \mathbb{N}_0$, there exists a finite signed measure γ_{ij} (which is a probability measure if $i = j$) supported on $[0, \infty)$ such that*

$$f_{ij}(t) = \sqrt{\frac{\pi_j}{\pi_i}} \int_0^\infty e^{-tx} d\gamma_{ij}(x). \tag{3.33}$$

Proof The proof is based on the spectral analysis already seen in Section 1.3 and the spectral decomposition of truncated matrices of $\mathcal{A}$. Let $\mathcal{S} = \{0, 1, \ldots, N\}$ and consider the truncated matrix $\mathcal{A}_N$ taking from $\mathcal{A}$ the first $N+1$ rows and columns. $\mathcal{A}_N$ is weakly symmetric with the same potential coefficients π_i. From (3.20) we can get the spectral representation of the associated minimal solution of $\mathcal{A}_N$ (which is also weakly symmetric), given by

$$f_{ij}^N(t) = \sqrt{\frac{\pi_j}{\pi_i}} \sum_{k=0}^N e^{d_k t} \phi_i^{(k)} \phi_j^{(k)}, \tag{3.34}$$

where d_k, $\phi_i^{(k)}$ and $\phi_j^{(k)}$ depends on N and the vectors ϕ_i satisfy the orthogonality conditions

$$\sum_{k=0}^N \phi_i^{(k)} \phi_j^{(k)} = \delta_{ij}, \quad i,j = 0, 1, \ldots, N.$$

Now (3.34) can be rewritten as

$$f_{ij}^N(t) = \sqrt{\frac{\pi_j}{\pi_i}} \int_0^\infty e^{-tx} d\gamma_{ij}^N(x),$$

where γ_{ij}^N is a signed measure with Dirac deltas located at $-d_k, k = 0, 1, \dots, N$, and sizes of the jumps $\phi_i^{(k)}\phi_j^{(k)}$, that is,

$$d\gamma_{ij}^N(x) = \sum_{k=0}^N \phi_i^{(k)}\phi_j^{(k)}\delta_{-d_k}(x).$$

This measure is finite as a consequence of the Cauchy–Schwarz inequality. This signed measure (which does not have to be positive) can be decomposed in the following way:

$$\gamma_{ij}^N = \gamma_{ij}^{N,+} - \gamma_{ij}^{N,-},$$

where $\gamma_{ij}^{N,\pm}$ are positive measures. Each of these measures is uniformly bounded, so we can apply the Helly selection principle (as we did in Theorem 1.13) and there exists a subsequence for each $\gamma_{ij}^{N,\pm}$, which we call $\gamma_{ij}^{N_k,\pm}$, such that it is convergent to a distribution $\gamma_{ij}^{\pm}$ as $k \to \infty$. Calling $\gamma_{ij} = \gamma_{ij}^+ - \gamma_{ij}^-$, we have that

$$\begin{aligned} f_{ij}(t) &= \lim_{k\to\infty} f_{ij}^{N_k}(t) = \lim_{k\to\infty} \sqrt{\frac{\pi_j}{\pi_i}} \int_0^\infty e^{-xt} d\gamma_{ij}^{N_k}(x) \\ &= \sqrt{\frac{\pi_j}{\pi_i}} \int_0^\infty e^{-xt} d\gamma_{ij}^+(x) - \sqrt{\frac{\pi_j}{\pi_i}} \int_0^\infty e^{-xt} d\gamma_{ij}^-(x) = \sqrt{\frac{\pi_j}{\pi_i}} \int_0^\infty e^{-xt} d\gamma_{ij}(x), \end{aligned}$$

and we get the result. If $i = j$ we observe from (3.34) that $f_{ii}^N(t) \geq 0$. Therefore, taking limits, it will be a probability measure. □

As a consequence we have the following:

Theorem 3.9 ([1, Theorem 8.2.1]) *Let $\mathcal{A}$ be the matrix of the infinitesimal operator associated to a birth–death process and $P_{ij}(t)$ a weakly symmetric $\mathcal{A}$-function (for example, the minimal) that is a solution of both Kolmogorov equations* (3.22) *and* (3.23). *Then there exists a probability measure ψ supported on the interval* $[0, \infty)$ *such that we have the integral representation*

$$P_{ij}(t) = \pi_j \int_0^\infty e^{-xt} Q_i(x) Q_j(x)\, d\psi(x), \quad i,j \in \mathbb{N}_0, \quad t \geq 0. \tag{3.35}$$

This formula is also known as the Karlin–McGregor representation formula for birth–death processes and appeared for the first time in [83] (see also [82]). We denote this $\mathcal{A}$-function as $P_{ij}(t; \psi)$ and sometimes refer to this measure as $\psi_{\min}$.

Proof From Theorem 3.8 we get the representation

$$P_{00}(t) = \int_0^\infty e^{-tx} d\psi(x), \quad t \geq 0, \tag{3.36}$$

where ψ is a probability measure supported on $[0, \infty)$. To prove (3.35) we will use induction twice. First over the index i (when $j=0$) and second in general. For $i = j = 0$ we get the result from (3.36). Assume formula (3.35) is true for all $k \leq i$ with i fixed. Then

$$P'_{i,0}(t) = -\int_0^\infty xe^{-tx} Q_i(x) Q_0(x)\, d\psi(x).$$

Using the Kolmogorov backward equation from (3.22) for $j = 0$ and the recurrence relation (3.30), we get

$$\begin{aligned}
\lambda_i P_{i+1,0}(t) &= P'_{i,0}(t) - \mu_i P_{i-1,0}(t) + (\lambda_i + \mu_i) P_{i,0}(t) \\
&= \int_0^\infty e^{-tx} \left[-xQ_i(x) - \mu_i Q_{i-1}(x) + (\lambda_i + \mu_i) Q_i(x) \right] Q_0(x)\, d\psi(x) \\
&= \int_0^\infty e^{-tx} \lambda_i Q_{i+1}(x) Q_0(x)\, d\psi(x).
\end{aligned}$$

Therefore, cancelling λ_i, we have that

$$P_{i+1,0}(t) = \int_0^\infty e^{-tx} Q_{i+1}(x) Q_0(x)\, d\psi(x).$$

Let us see now the general case by induction on j. For $j = 0$ it is true. Assume formula (3.35) is true for all $k \leq j$ with j fixed. Then

$$P'_{i,j}(t) = -\pi_j \int_0^\infty xe^{-tx} Q_i(x) Q_j(x)\, d\psi(x).$$

Using the Kolmogorov forward equation from (3.23), the symmetry equation (3.25) and the recurrence relation (3.30), we have that

$$\begin{aligned}
&\mu_{j+1} P_{i,j+1}(t) \\
&= P'_{i,j}(t) - \lambda_{j-1} P_{i,j-1}(t) + (\lambda_j + \mu_j) P_{i,j}(t) \\
&= \int_0^\infty e^{-tx} Q_i(x) \left[-xQ_j(x)\pi_j - \lambda_{j-1}\pi_{j-1} Q_{j-1}(x) + (\lambda_j + \mu_j)\pi_j Q_j(x) \right] d\psi(x) \\
&= \int_0^\infty e^{-tx} Q_i(x) \left[-xQ_j(x)\pi_j - \mu_j \pi_j Q_{j-1}(x) + (\lambda_j + \mu_j)\pi_j Q_j(x) \right] d\psi(x) \\
&= \int_0^\infty e^{-tx} Q_i(x) \lambda_j \pi_j Q_{j+1}(x)\, d\psi(x) = \int_0^\infty e^{-tx} Q_i(x) \mu_{j+1} \pi_{j+1} Q_{j+1}(x)\, d\psi(x).
\end{aligned}$$

Therefore, cancelling μ_{j+1}, we get

$$P_{i,j+1}(t) = \pi_{j+1} \int_0^\infty e^{-tx} Q_i(x) Q_{j+1}(x)\, d\psi(x),$$

as requested. □

The spectral representation (3.35) clearly separates the variables i, j and t. Also, as $t \to \infty$, the integral tends to 0, except at the point $x = 0$, where there may be a jump. The existence of this jump is conditioned on the process being positive recurrent and the size of the jump is given as in the proof of Theorem 2.24, by

$$\psi(\{0\}) = \left(\sum_{i=0}^{\infty} \pi_i Q_i^2(0) \right)^{-1}. \tag{3.37}$$

If $\mu_0 > 0$, then the spectral measure ψ has no atom at $x = 0$. Observe also from (3.35) that for $t = 0$, we obtain the orthogonality of the polynomials $Q_n(x)$ defined by (3.30), i.e.

$$\int_0^\infty Q_i(x) Q_j(x)\, d\psi(x) = \frac{\delta_{ij}}{\pi_j}, \quad i, j \in \mathbb{N}_0.$$

A measure satisfying the previous conditions is called a *solution of the Stieltjes moment problem*. If the solution is not unique then there exists a *maximal* measure (usually denoted by $\psi_{\max}$) characterized by the fact that the mass at the point $x = 0$ is maximal, given by (3.37). The transition functions $P_{ij}(t; \psi_{\max})$ are not maximal in the sense that $P_{ij}(t; \psi_{\min})$ are minimal (see [83, Section 2] or [137, pp. 23–76] for more details).

Unlike the case of discrete-time birth–death chains, Theorem 3.9 only ensures the existence of at least one solution (for example, the minimal). There is a converse result. Assume that we have a sequence of real polynomials $(Q_n)_n$ orthogonal with respect to a probability measure ψ supported on $[0, \infty)$. Since the zeros of the polynomials are inside the interval $(0, \infty)$ (see Proposition 1.8) then we can always normalize the polynomials in such a way that $Q_n(0) = 1$ for every $n \geq 0$. The polynomials satisfy a three-term recurrence relation of the form

$$\begin{aligned} Q_0(x) = 1, \quad -xQ_0(x) &= A_0 Q_1(x) + B_0 Q_0(x), \\ -xQ_n(x) &= A_n Q_{n+1}(x) + B_n Q_n(x) + C_n Q_{n-1}(x), \quad n \geq 1, \end{aligned}$$

where A_n, B_n, C_n are real. Because of the normalization we have $A_0 + B_0 = 0$ and $A_n + B_n + C_n = 0, n \geq 1$. The leading coefficient of $Q_n(x)$ is given by $(-1)^n (A_0 A_1 \cdots A_n)^{-1}$ and since $Q_n(0) = 1$ this is equal to $(-1)^n x_{n,1} \cdots x_{n,n}$, where $x_{n,j}$ are the positive zeros of $Q_n(x)$. Therefore $A_n > 0$ for $n \geq 0$. We also have

$$\int_0^\infty (-x)^n Q_n(x) d\psi(x) = C_n/A_{n-1} \int_0^\infty (-x)^{n-1} Q_{n-1}(x) d\psi(x),$$

so $C_n > 0, n \geq 1$. Hence the three-term recurrence formula of the polynomials $Q_n(x)$ determines the birth–death rates of the infinitesimal operator $\mathcal{A}$ of a birth–death process with $\mu_0 = 0$. These polynomials are usually called *birth–death polynomials*. Unlike the case of discrete-time birth–death chains (see Theorem 2.7) there is a one-to-one correspondence between birth–death processes and solvable Stieltjes moment problems. As we saw in Theorem 2.8 for discrete-time birth–death chains, the conditions under which it is possible to renormalize the polynomials so that the recurrence formula determines a birth–death process with $\mu_0 > 0$, are given by the following result:

Theorem 3.10 ([83, Lemma 1]) *Consider a birth–death process with infinitesimal matrix $\mathcal{A}$ where $\mu_0 = 0$ and the sequence of polynomials $(Q_n)_n$ generated by the three-term recurrence relation (3.30). For any parameter $\mu > 0$, there exists a sequence $(a_n)_n$ with $a_n > 0, n \geq 0$, such that $R_n(x) = a_n Q_n(x)$ satisfies the recurrence relation*

$$R_0(x) = 1, \quad -xR_0(x) = -(\lambda_0' + \mu_0')R_0(x) + \lambda_0' R_1(x),$$
$$-xR_n(x) = \lambda_n' R_{n+1}(x) - (\lambda_n' + \mu_n')R_n(x) + \mu_n' R_{n-1}(x),$$

with $\mu_0' = \mu$ and $\lambda_n', \mu_n' > 0$ if and only if the series $\sum_{n=0}^\infty (\lambda_n \pi_n)^{-1}$ is convergent and μ satisfies

$$\mu \sum_{n=0}^\infty \frac{1}{\lambda_n \pi_n} < 1. \tag{3.38}$$

Proof The proof is identical to the one of Theorem 2.8 for discrete-time birth–death chains, with the obvious modifications. □

For discrete-time birth–death chains, the Karlin–McGregor representation formula is a unique way to represent the n-step transition probabilities $P_{ij}^{(n)}$. Now for birth–death processes, it is not clear if the Karlin–McGregor representation formula (3.35) is going to represent a transition probability function of the corresponding birth–death process $\{X_t, t \geq 0\}$ or even if it is going to be unique.

From the proof of Theorem 3.9 it is clear that $P_{ij}(t;\psi)$ satisfies the Kolmogorov equations with the initial conditions. But it is not clear, for instance, if $P_{ij}(t;\psi) \geq 0$ or if $\sum_{j\in\mathbb{N}_0} P_{ij}(t) \leq 1$ (even if we have convergence of the series) or if the Chapman–Kolmogorov equations hold or uniqueness questions. We will consider these problems below in Section 3.4, but before that we need to study some properties of the birth–death polynomials $Q_n(x)$ and introduce some useful related families.

3.3 Properties of the Birth–Death Polynomials and Other Related Families

In this section we will derive some useful formulas for the sequence of polynomials $(Q_n)_n$ defined by a three-term recurrence relation as in (3.30). For that, as in Chapter 2, we will study the *dual polynomials*, the *kth associated polynomials* and some of their properties. Since the formulas are based on the three-term recurrence relation for the polynomials $(Q_n)_n$, most of the formulas are very similar to the ones we saw in Section 2.3, so many details will be omitted.

3.3.1 Dual Polynomials

Let $(Q_n)_n$ be the polynomials generated by the three-term recurrence relation (3.30). We will study first the case of possible absorption at the state -1, i.e. $\mu_0 > 0$ and later the case where $\mu_0 = 0$.

Case $\mu_0 > 0$: As in the case of discrete-time birth–death chains, the three-term recurrence relation (3.30) can be rewritten, using the symmetry equations (3.25), as

$$\begin{aligned} Q_0(x) &= 1, \\ -xQ_0(x)\pi_0 &= \lambda_0\pi_0[Q_1(x) - Q_0(x)] - \mu_0, \\ -xQ_n(x)\pi_n &= \lambda_n\pi_n[Q_{n+1}(x) - Q_n(x)] - \lambda_{n-1}\pi_{n-1}[Q_n(x) - Q_{n-1}(x)], \quad n \geq 1. \end{aligned} \tag{3.39}$$

Let us define the *dual polynomials* as

$$Q_0^d(x) = 1, \quad Q_{n+1}^d(x) = \frac{\lambda_n\pi_n}{\mu_0}[Q_{n+1}(x) - Q_n(x)], \quad n \geq 0. \tag{3.40}$$

Then from the previous relations we have that

$$Q_{n+1}^d(x) = 1 - \frac{x}{\mu_0}\sum_{j=0}^{n} Q_j(x)\pi_j.$$

Therefore $Q_n^d(1) = 1, n \geq 0$. Using the definition of $Q_n^d(x)$ in (3.40) and solving again the recurrence equation, we get

$$Q_n(x) = 1 + \mu_0 \sum_{m=0}^{n-1} \frac{1}{\lambda_m\pi_m} - x\sum_{m=0}^{n-1} \frac{1}{\lambda_m\pi_m} \sum_{j=0}^{m} Q_j(x)\pi_j. \tag{3.41}$$

From the above we have, evaluating at $x = 0$, that

$$Q_n(0) = 1 + \mu_0 \sum_{k=0}^{n-1} \frac{1}{\lambda_k\pi_k}. \tag{3.42}$$

We can also write Q_n in terms of Q_n^d. Indeed, we have that

$$Q_{n+1}^d(x) - Q_n^d(x) = -\frac{x}{\mu_0} Q_n(x)\pi_n. \tag{3.43}$$

Therefore, from (3.40), we get

$$Q_n(x) = 1 + \sum_{k=0}^{n-1} \frac{\mu_0}{\lambda_k \pi_k} Q_{k+1}^d(x).$$

As the original polynomial $Q_n(x)$, $Q_n^d(x)$ has n positive, simple zeros denoted by $x_{n,i}^d, i = 1, \dots, n$, these satisfy the interlacing property (see Proposition 1.8). Theorem I.7.2 in [16] may be used to show that

$$0 < x_{n,i}^d < x_{n,i} < x_{n,i+1}^d < x_{n,i+1}, \quad i = 1, \dots, n-1.$$

We also have (see (1.28))

$$0 \le \xi_i^d \le \xi_i \le \xi_{i+1}^d < \infty, \quad i \ge 1, \tag{3.44}$$

where ξ_i^d is defined in the same way as ξ_i, i.e. $\xi_i^d = \lim_{n\to\infty} x_{n,i}^d$. A simple induction shows that

$$Q_{n+1}(x) > Q_n(x) \ge 1, \quad n \ge 0, \quad x < 0. \tag{3.45}$$

In fact, these inequalities hold for $x \le \inf \operatorname{supp}(\psi)$. As a consequence of the Christoffel–Darboux formula (1.23), we get

$$Q_n(x) > Q_n(y) \ge 1, \quad x < y \le 0, \quad n \ge 1.$$

An easy computation from the definition of $Q_n^d(x)$ in (3.40) gives that they satisfy the three-term recurrence relation

$$\begin{aligned} Q_0^d(x) = 1, \quad & -xQ_0^d(x) = \mu_0 Q_1^d(x) - \mu_0 Q_0^d(x), \\ -xQ_{n+1}^d(x) = \mu_{n+1}Q_{n+2}^d(x) & - (\mu_{n+1} + \lambda_n)Q_{n+1}^d(x) + \lambda_n Q_n^d(x), \quad n \ge 0. \end{aligned}$$

Therefore the associated Jacobi matrix is given by

$$\mathcal{A}^d = \begin{pmatrix} -\mu_0 & \mu_0 & & & \\ \lambda_0 & -\lambda_0 - \mu_1 & \mu_1 & & \\ & \lambda_1 & -\lambda_1 - \mu_2 & \mu_2 & \\ & & \ddots & \ddots & \ddots \end{pmatrix}.$$

Unlike the case of discrete-time birth–death chains, the dual polynomials $(Q_n^d)_n$ are always birth–death polynomials, so we will call this process the *dual birth–death process* and denote it by $\{X_t^d, t \geq 0\}$. Observe that $\mathcal{A}^d$ is conservative ($\mu_0^d = 0$) and the dual birth–death rates are $\{(\lambda_n^d, \mu_n^d), n \geq 0\}$ where $\lambda_n^d = \mu_n, \mu_{n+1}^d = \lambda_n$, while the potential coefficients are given by

$$\pi_0^d = 1, \quad \pi_n^d = \frac{\mu_0 \cdots \mu_{n-1}}{\lambda_0 \cdots \lambda_{n-1}} = \frac{\mu_0}{\lambda_{n-1}\pi_{n-1}} = \frac{\mu_0}{\mu_n \pi_n}, \quad n \geq 1.$$

We also have $(\lambda_n^d \pi_n^d)^{-1} = \pi_n/\mu_0$. If we call A^d, B^d, C^d, D^d the same analog series as in (3.28) but for the dual process, we obtain

$$A^d = \frac{B}{\mu_0}, \quad B^d = 1 + \mu_0 A, \quad C^d = \frac{B}{\mu_0} + D, \quad D^d = C.$$

Therefore the series A^d, B^d, C^d and D^d have the same character (convergent or divergent) as the series B, A, D and C, respectively. The dual polynomials form an orthogonal system, as we will see in the following:

Theorem 3.11 ([83, Lemma 2]) *Let $\mu_0 > 0$. Then there exists at least one probability measure ψ^d supported on $[0, \infty)$ such that*

$$\int_0^\infty Q_n^d(x) Q_m^d(x) d\psi^d(x) = \frac{\delta_{nm}}{\pi_n^d}. \tag{3.46}$$

If ψ^d is such measure then $d\psi(x) = x d\psi^d(x)/\mu_0$ is a solution of the original Stieltjes moment problem for which

$$\mu_0 \int_0^\infty \frac{d\psi(x)}{x} \leq 1. \tag{3.47}$$

Conversely, if ψ is a solution of the original Stieltjes moment problem such that (3.47) is valid, then the probability measure ψ^d is given by

$$d\psi^d(x) = \mu_0 \frac{d\psi(x)}{x} \mathbf{1}_{(0,\infty)} + \left(1 - \mu_0 \int_0^\infty \frac{d\psi(x)}{x}\right) \delta_0(x), \tag{3.48}$$

where $\mathbf{1}_A$ is the indicator function, and it satisfies (3.46).

Proof The proof follows exactly the same lines as the proof of Theorem 2.10 with the appropriate modifications. □

Remark 3.12 The probability measure ψ^d in (3.48) is typically called a *Geronimus transformation* of the measure ψ, while the measure ψ is a *Christoffel transformation* of the measure ψ^d. This suggests that the Jacobi operators $\mathcal{A}$ and $\mathcal{A}^d$ are related by a *discrete Darboux transformation* centered at the point $x = 0$. Indeed, let U and L be the following upper and lower triangular matrices:

$$U = \begin{pmatrix} 1 & -1 & & & \\ 0 & 1/\pi_1 & -1/\pi_1 & & \\ & 0 & 1/\pi_2 & -1/\pi_2 & \\ & & \ddots & \ddots & \ddots \end{pmatrix},$$

$$L = \begin{pmatrix} -\mu_0 & 0 & & & \\ \lambda_0\pi_0 & -\lambda_0\pi_0 & 0 & & \\ & \lambda_1\pi_1 & -\lambda_1\pi_1 & 0 & \\ & & \ddots & \ddots & \ddots \end{pmatrix}.$$

Then, a direct computation shows that $\mathcal{A} = UL$ and $\mathcal{A}^d = LU$. Therefore $\mathcal{A}^d$ is a discrete Darboux transformation of $\mathcal{A}$. ◊

Remark 3.13 Let us give a probabilistic interpretation of the quantities ξ_i, ξ_i^d in (3.44) in relation with the numbers A, B, C, D defined by (3.28). If we assume that $P_{ij}(t)$ satisfies both the backward and forward equations (3.22) and (3.23), then the birth–death process is uniquely determined by its rates, and is the minimal process (see Propositions 3.5 and 3.7) if and only if $A + B = \infty$, i.e. at least one of C and D diverges. That means, as we will see later in Theorem 3.26, that the spectral measure ψ is uniquely determined and $\xi_1 = \inf \operatorname{supp}(\psi)$. If $C = \infty$ (and hence $A + B = \infty$), then the birth–death process is non-explosive and absorption at -1 occurs with probability 1 if and only if $A = \infty$ (see Theorem 3.51 below). If $C < \infty$ and $D = \infty$, then again $A + B = \infty$, and the birth–death process is explosive and is absorbed either at ∞ or at -1. If $A + B < \infty$, i.e. both C and D converge, then the birth–death process is not determined uniquely by its rates (indeterminate). But the rates determine a one-parameter family of spectral measures and a one-parameter family of transition functions of the form (3.35), indexed by $\inf \operatorname{supp}(\psi)$ in the range $[\xi_1^d, \xi_1]$. If $\inf \operatorname{supp}(\psi) = \xi_1$, then we obtain the minimal process, where the boundary at ∞ is completely absorbing. If $\inf \operatorname{supp}(\psi) = \xi_1^d$, then the corresponding process is called the maximal process and the boundary at ∞ is completely reflecting. It is then the unique honest process. When $\inf \operatorname{supp}(\psi) \in (\xi_1^d, \xi_1)$, the boundary is said to be mixed. ◊

Case $\mu_0 = 0$: In this case we observe from the three-term recurrence relation (3.39) that the polynomial $Q_{n+1}(x) - Q_n(x)$ is always a multiple of the polynomial $-x$. Therefore we define the *dual polynomials* as

$$Q_n^d(x) = \frac{\lambda_n\pi_n}{-x}[Q_{n+1}(x) - Q_n(x)], \quad n \geq 0. \tag{3.49}$$

We have now the relations

$$Q_n^d(x) = \sum_{k=0}^{n} \pi_n Q_n(x), \quad Q_n(x) = 1 - x\sum_{k=0}^{n-1} \frac{1}{\lambda_k \pi_k} Q_k^d(x), \quad n \geq 0.$$

Therefore $Q_n(0) = 1, n \geq 0$, and $Q_n^d(0) = \sum_{k=0}^{n} \pi_k, n \geq 0$. We also have formula (3.41) setting $\mu_0 = 0$, as well as other similar relations. As before, $Q_n^d(x)$ has n positive, simple zeros denoted by $x_{n,i}^d, i = 1,\dots,n$, which satisfy the interlacing property (see Proposition 1.8), as well as

$$0 < x_{n,i} < x_{n,i}^d < x_{n,i+1} < x_{n,i+1}^d, \quad i = 1,\dots,n-1.$$

We also have (see (1.28))

$$0 \leq \xi_i \leq \xi_i^d \leq \xi_{i+1} < \infty, \quad i \geq 1,$$

where ξ_i^d is defined in the same way as ξ_i, i.e. $\xi_i^d = \lim_{n\to\infty} x_{n,i}^d$. The dual polynomials $Q_n^d(x)$ in (3.49) satisfy the three-term recurrence relation

$$\begin{aligned} Q_0^d(x) = 1, \quad -xQ_0^d(x) &= \mu_1 Q_1^d(x) - (\lambda_0 + \mu_1)Q_0^d(x), \\ -xQ_n^d(x) = \mu_{n+1}Q_{n+1}^d(x) &- (\lambda_n + \mu_{n+1})Q_n^d(x) + \lambda_n Q_{n-1}^d(x), \quad n \geq 1. \end{aligned}$$

Therefore the associated Jacobi matrix is given by

$$\mathcal{A}^d = \begin{pmatrix} -\lambda_0 - \mu_1 & \mu_1 & & & \\ \lambda_1 & -\lambda_1 - \mu_2 & \mu_2 & & \\ & \lambda_2 & -\lambda_2 - \mu_3 & \mu_3 & \\ & & \ddots & \ddots & \ddots \end{pmatrix}.$$

Observe now that the matrix $\mathcal{A}^d$ is not conservative ($\mu_0^d > 0$). The dual birth–death rates are given by $\lambda_n^d = \mu_{n+1}, \mu_n^d = \lambda_n$, while the potential coefficients are given by

$$\pi_0^d = 1, \quad \pi_n^d = \frac{\mu_1 \cdots \mu_n}{\lambda_1 \cdots \lambda_n} = \frac{\lambda_0}{\lambda_n \pi_n}, \quad n \geq 1.$$

We also have $(\lambda_n^d \pi_n^d)^{-1} = \pi_{n+1}/\lambda_0$. If we call A^d, B^d, C^d, D^d the same analog series as in (3.28) but for the dual process, we obtain

$$A^d = B - \frac{1}{\lambda_0}, \quad B^d = \lambda_0 A, \quad C^d = D, \quad D^d = C - A.$$

Therefore, the series A^d, B^d, C^d and D^d have the same character (convergent or divergent) as the series B, A, D and C, respectively. Finally, we also get the following:

Theorem 3.14 ([83, Lemma 3]) *Let* $\mu_0 = 0$. *Then there exists at least one probability measure* ψ^d *supported on* $[0,\infty)$ *such that*

$$\int_0^\infty Q_n^d(x)Q_m^d(x)d\psi^d(x) = \frac{\delta_{nm}}{\pi_n^d}, \quad \text{and} \quad \lambda_0 \int_0^\infty \frac{d\psi(x)}{x} \leq 1.$$

If ψ^d is such a measure then

$$d\psi(x) = \lambda_0 \frac{d\psi^d(x)}{x} \mathbf{1}_{(0,\infty)} + \left(1 - \lambda_0 \int_0^\infty \frac{d\psi^d(x)}{x}\right) \delta_0(x)$$

is a solution of the original Stieltjes moment problem. Conversely, if ψ is the spectral measure of the original moment problem, then $d\psi^d(x) = xd\psi(x)/\lambda_0$ defines a probability measure that satisfies the above conditions.

Remark 3.15 As before, if we call U and L the upper and lower triangular matrices

$$U = \begin{pmatrix} -\lambda_0\pi_0 & \lambda_0\pi_0 & & & \\ 0 & -\lambda_1\pi_1 & \lambda_1\pi_1 & & \\ & 0 & -\lambda_2\pi_2 & \lambda_2\pi_2 & \\ & & \ddots & \ddots & \ddots \end{pmatrix},$$

$$L = \begin{pmatrix} 1 & 0 & & & \\ -1/\pi_1 & 1/\pi_1 & 0 & & \\ & -1/\pi_2 & 1/\pi_2 & 0 & \\ & & \ddots & \ddots & \ddots \end{pmatrix},$$

then a direct computation shows that $\mathcal{A} = LU$ and $\mathcal{A}^d = UL$. Therefore $\mathcal{A}^d$ is a discrete Darboux transformation of $\mathcal{A}$ but taking the LU decomposition instead of the UL as before. ◊

Remark 3.16 We have the same considerations as in Remark 3.13 with the difference that there is no absorption at -1. Now, if $A + B < \infty$, then we have a one-parameter family of spectral measures and a one-parameter family of transition functions of the form (3.35), indexed by $\inf \operatorname{supp}(\psi)$ in the range $[0, \xi_1]$. If $\inf \operatorname{supp}(\psi) = \xi_1$, then we obtain the minimal process, where the boundary at ∞ is completely absorbing. If $\inf \operatorname{supp}(\psi) = 0$, then the corresponding process is called the maximal process and the boundary at ∞ is completely reflecting. ◊

In future sections we will be concerned about properties of the analytic functions (if they converge)

$$Q_\infty(x) \doteq \lim_{n\to\infty} Q_n(x) = \begin{cases} 1 + \displaystyle\sum_{j=0}^\infty \frac{\mu_0}{\lambda_j\pi_j} Q_{j+1}^d(x), & \text{if} \quad \mu_0 > 0, \\ 1 - x\displaystyle\sum_{j=0}^\infty \frac{1}{\lambda_j\pi_j} Q_j^d(x) & \text{if} \quad \mu_0 = 0, \end{cases} \tag{3.50}$$

$$Q_\infty^d(x) \doteq \lim_{n\to\infty} Q_n^d(x) = \begin{cases} 1 - \dfrac{x}{\mu_0}\displaystyle\sum_{j=0}^{\infty} \pi_j Q_j(x), & \text{if} \quad \mu_0 > 0, \\ \displaystyle\sum_{j=0}^{\infty} \pi_j Q_j(x) & \text{if} \quad \mu_0 = 0. \end{cases} \tag{3.51}$$

In [83, Lemma 4] the following equivalent statements are proven:

(a) $Q_n(x)$ converges as $n \to \infty$ for every complex x uniformly on $|x| \le R$.
(b) $Q_n(x)$ is bounded as $n \to \infty$ for at least one $x < 0$.
(c) The series C is convergent (see (3.28)).

The same applies for the dual polynomials but changing C by D. Since the only zeros of the polynomials are inside the interval $[0, \infty)$, it follows from Hurwitz's theorem that the only zeros of the limiting functions are inside the interval $[0, \infty)$. In fact these zeros are given by $\xi_i = \lim_{n\to\infty} x_{n,i}$ where $0 < x_{n,1} < x_{n,2} < \cdots < x_{n,n}$ are the zeros of the polynomial $Q_n(x)$ (the same applies for the dual function $Q_\infty^d(x)$ and the zeros $x_{n,i}^d$ of the dual polynomial $Q_n^d(x)$). These zeros are simple and $Q_\infty(x)$ and $Q_\infty^d(x)$ cannot have common zeros (by the interlacing property).

As a consequence of the previous statements, we have that $C = \infty$ if and only if $Q_n(x) \to \infty$ as $n \to \infty$ for every $x < 0$. Also we have $D = \infty$ if and only if $\sum_{n=0}^{\infty} \pi_n Q_n(x) \to \infty$ as $n \to \infty$ for every $x < 0$.

3.3.2 *k*th Associated Polynomials

For any $k \ge 0$ consider the *kth birth–death process* $\{X_t^{(k)}, t \ge 0\}$, whose infinitesimal operator matrix is defined from the original infinitesimal operator matrix $\mathcal{A}$ eliminating the first $k+1$ rows and columns of $\mathcal{A}$, i.e.

$$\mathcal{A} = \left(\begin{array}{cccc|ccc} -\mu_0 - \lambda_0 & \lambda_0 & & & & & \\ \mu_1 & -\mu_1 - \lambda_1 & \lambda_1 & & & & \\ & \ddots & \ddots & \ddots & & & \\ & & \mu_k & -\mu_k - \lambda_k & \lambda_k & & \\ \hline & & & \mu_{k+1} & -\mu_{k+1} - \lambda_{k+1} & \lambda_{k+1} & \\ & & & & \mu_{k+2} & -\mu_{k+2} - \lambda_{k+2} & \lambda_{k+2} \\ & & & & & \ddots & \ddots \quad \ddots \end{array}\right).$$

This new birth–death process has birth–death rates $\{(\lambda_{n+k+1}, \mu_{n+k+1}), n \ge 0\}$ and it is not conservative since $\mu_{k+1} > 0$. We will denote by $\psi^{(k)}$ the spectral measure associated with the kth associated birth–death process. Define the sequence of polynomials $(Q_n^{(k)})_n$ using the recurrence relation

$$-xQ_n^{(k)}(x) = \delta_{n,k} + \lambda_n Q_{n+1}^{(k)}(x) - (\lambda_n + \mu_n) Q_n^{(k)}(x) + \mu_n Q_{n-1}^{(k)}(x). \tag{3.52}$$

The polynomials $(Q_n^{(k)})_n$ have degrees $n-k-1$ if $n>k$ and satisfy $Q_n^{(k)}(x)=0$ for $0\le n\le k$. Moreover

$$Q_{k+1}^{(k)}(x)=-\frac{1}{\lambda_k}. \tag{3.53}$$

It is easy to see that these polynomials can be written

$$Q_n^{(k)}(x)=-\pi_k\int_0^\infty Q_k(y)\frac{Q_n(y)-Q_n(x)}{y-x}\,d\psi(y). \tag{3.54}$$

For every $k\ge 0$ we have a new birth–death process with spectral measure $\psi^{(k)}$. Let us see now how to relate the Stieltjes transforms $B(z;\psi^{(k)})$ to the original one $B(z;\psi)$. Let $F_{ij}(t)$ be the first passage time distributions defined by (3.13) and (3.14) and consider the Laplace transforms $\widehat{P}_{ij}(\lambda)$ of $P_{ij}(t)$ (see (3.12)) and $\widehat{F}_{ij}(\lambda)$ of $F_{ij}(t)$. Therefore, from (3.17) and (3.18), we get

$$\begin{aligned}\widehat{P}_{ii}(\lambda)&=\frac{1}{\lambda+\lambda_i+\mu_i}+\widehat{P}_{ii}(\lambda)\widehat{F}_{ii}(\lambda),\\ \widehat{P}_{ij}(\lambda)&=\widehat{P}_{jj}(\lambda)\widehat{F}_{ij}(\lambda),\quad i\neq j.\end{aligned} \tag{3.55}$$

In particular, using the Karlin–McGregor formula (3.35), we have that

$$\begin{aligned}\widehat{P}_{ij}(\lambda)&=\int_0^\infty e^{-\lambda t}P_{ij}(t)dt=\int_0^\infty e^{-\lambda t}\left(\pi_j\int_0^\infty e^{-xt}Q_i(x)Q_j(x)d\psi(x)\right)dt\\ &=\pi_j\int_0^\infty\left(\int_0^\infty e^{-(\lambda+x)t}dt\right)Q_i(x)Q_j(x)\,d\psi(x)\\ &=\pi_j\int_0^\infty\frac{Q_i(x)Q_j(x)}{\lambda+x}\,d\psi(x),\quad \lambda>0.\end{aligned} \tag{3.56}$$

Therefore, from (3.55), we get

$$\begin{aligned}\widehat{F}_{ii}(\lambda)&=1-\frac{1}{(\lambda+\lambda_i+\mu_i)\pi_i\displaystyle\int_0^\infty\frac{Q_i^2(x)}{\lambda+x}\,d\psi(x)},\\ \widehat{F}_{ij}(\lambda)&=\frac{\displaystyle\int_0^\infty\frac{Q_i(x)Q_j(x)}{\lambda+x}\,d\psi(x)}{\displaystyle\int_0^\infty\frac{Q_j^2(x)}{\lambda+x}\,d\psi(x)},\quad i\neq j.\end{aligned} \tag{3.57}$$

Another useful formula, using again the Karlin–McGregor representation (3.35), is the following:

$$F_{i,-1}(t)=\mu_0\int_0^t P_{i,0}(s)\,ds=\mu_0\int_0^\infty\frac{1-e^{-xt}}{x}Q_i(x)\,d\psi(x).$$

Using the same argument we also have

$$F_{10}(t) = \mu_1 \int_0^t P_{00}^{(0)}(s)\,ds = \mu_1 \int_0^\infty \frac{1-e^{-xt}}{x}\,d\psi^{(0)}(x), \tag{3.58}$$

where $P_{ij}^{(0)}(t) = P_{ij}^{(0)}(t;\psi^{(0)})$ is the transition function of the 0th associated birth–death process $\{X_t^{(0)}, t \geq 0\}$. The Laplace transformation of the previous formula gives

$$\begin{aligned}\widehat{F}_{10}(\lambda) &= \mu_1 \int_0^\infty e^{-\lambda t}\left(\int_0^\infty \frac{1-e^{-xt}}{x}\,d\psi^{(0)}(x)\right)\\ &= \mu_1 \int_0^\infty \frac{1}{x}\left(\int_0^\infty (e^{-\lambda t} - e^{-(\lambda+x)t})\,dt\right) d\psi^{(0)}(x) = \mu_1 \int_0^\infty \frac{d\psi^{(0)}(x)}{\lambda+x}.\end{aligned}$$

Alternatively, from (3.57) for $i = 1$ and $j = 0$ and using that $Q_1(x) = -(x-\lambda_0-\mu_0)/\lambda_0$, we obtain

$$\begin{aligned}\widehat{F}_{10}(\lambda) &= \frac{\displaystyle\int_0^\infty \frac{Q_1(x)Q_0(x)}{\lambda+x}\,d\psi(x)}{\displaystyle\int_0^\infty \frac{Q_0^2(x)}{\lambda+x}\,d\psi(x)} = -\frac{\displaystyle\int_0^\infty \frac{(x-\lambda_0-\mu_0)\,d\psi(x)}{\lambda_0(\lambda+x)}}{\displaystyle\int_0^\infty \frac{d\psi(x)}{\lambda+x}}\\ &= \frac{\lambda_0+\mu_0}{\lambda_0} + \frac{1}{\lambda_0}\frac{\displaystyle\int_0^\infty \left[\frac{\lambda}{\lambda+x} - 1\right] d\psi(x)}{\displaystyle\int_0^\infty \frac{d\psi(x)}{\lambda+x}}\\ &= \frac{1}{\lambda_0}\left[\lambda_0+\mu_0+\lambda - \left(\int_0^\infty \frac{d\psi(x)}{\lambda+x}\right)^{-1}\right].\end{aligned}$$

Combining these two results we get

$$\lambda_0\mu_1 \int_0^\infty \frac{d\psi^{(0)}(x)}{\lambda+x} = \left[\lambda_0+\mu_0+\lambda - \left(\int_0^\infty \frac{d\psi(x)}{\lambda+x}\right)^{-1}\right], \tag{3.59}$$

or, in other words,

$$\int_0^\infty \frac{d\psi(x)}{\lambda+x} = \left(\lambda_0+\mu_0+\lambda - \lambda_0\mu_1 \int_0^\infty \frac{d\psi^{(0)}(x)}{\lambda+x}\right)^{-1}.$$

Changing λ by $-z$, we obtain

$$B(z;\psi) = \frac{1}{\lambda_0+\mu_0-z-\lambda_0\mu_1 B(z;\psi^{(0)})}, \tag{3.60}$$

where $B(z;\psi)$ is the Stieltjes transform, defined by (1.6). We will use this formula later as a way of calculating some measures associated with certain birth–death processes. Once we know one of these Stieltjes transformations, the Perron–Stieltjes inversion formula, given by (1.8), can be used to compute the other measure.

Another interesting case is when ψ is a discrete distribution with masses $(\rho_n)_n$ located at the real nonnegative points $(x_n)_n$ with $0 \leq x_0 < x_1 < x_2 < \cdots < x_n \to \infty$. The Stieltjes transform of ψ is given by the series

$$B(z;\psi) = \int_0^\infty \frac{d\psi(x)}{x-z} = \sum_{n=0}^\infty \frac{\rho_n}{x_n - z}, \quad z \in \mathbb{C} \setminus \{x_0, x_1, \ldots\},$$

where $B(z;\psi)$ is a meromorphic function with simple poles at the points $(x_n)_n$. If we choose $x_n < z < x_{n+1}$ then $B(z;\psi)$ is an increasing function from $-\infty$ to ∞. Therefore there exists one single zero y_n in that interval such that $B(y_n;\psi) = 0$. From (3.60) (see also (3.59)), we can write

$$B(z;\psi^{(0)}) = \frac{1}{\lambda_0\mu_1}\left(\lambda_0 + \mu_0 - z - \frac{1}{B(z;\psi)}\right).$$

Therefore, the function $B(z;\psi^{(0)})$ is a meromorphic function whose simple poles are located at $(y_n)_n$ (see more details in [86]). Then $\psi^{(0)}$ is given by

$$\psi^{(0)}(x) = \sum_{n=0}^\infty \gamma_n \delta_{y_n}(x), \tag{3.61}$$

where $0 \leq x_0 < y_0 < x_1 < y_1 < x_2 < \cdots$ and the masses $(\gamma_n)_n$ are given by

$$\gamma_n = \psi^{(0)}(\{y_n\}) = -\frac{1}{\lambda_0\mu_1 B'(y_n;\psi)}, \quad n \geq 0. \tag{3.62}$$

We can iterate formula (3.60) as many times as we want. Let us now see how to obtain a relation between the Stieltjes transforms of ψ and $\psi^{(k)}, k \geq 0$, in terms of the original polynomials $(Q_n)_n$ and the 0th associated polynomials. First, observe that for $k \geq 0$ the formula (3.60) can be extended to ($\psi = \psi^{(-1)}$)

$$B(z;\psi^{(k-1)}) = \frac{1}{\lambda_k + \mu_k - z - \lambda_k\mu_{k+1}B(z;\psi^{(k)})} = \frac{\dfrac{1}{\lambda_k}}{\dfrac{\lambda_k + \mu_k - z}{\lambda_k} - \mu_{k+1}B(z;\psi^{(k)})}. \tag{3.63}$$

Iterating this formula we get that $B(z;\psi)$ can be written as

$$B(z;\psi) = \frac{\alpha_k B(z;\psi^{(k-1)}) + \beta_k}{\gamma_k B(z;\psi^{(k-1)}) + \delta_k}, \quad k \geq 0, \tag{3.64}$$

where $\alpha_k, \beta_k, \gamma_k, \delta_k$ are functions in z that do not necessarily have to be unique. Let us now try to write $B(z;\psi)$ in terms of $B(z;\psi^{(k-1)})$ using the polynomials $Q_k(x)$ and the 0th associated polynomials $Q_k^{(0)}(x)$. For $k = 0$, we have that $\alpha_0 = \delta_0 = 1$, $\beta_0 = \gamma_0 = 0$, while for $k = 1$, from (3.60) and (3.63), a solution is $\alpha_1 = 0, \beta_1 = -1/\lambda_0 = Q_1^{(0)}(z), \gamma_1 = \mu_1, \delta_1 = -(\lambda_0 + \mu_0 - z)/\lambda_0 = -Q_1(z)$. Thus

$$B(z;\psi) = \frac{Q_1^{(0)}}{\mu_1 B(z;\psi^{(0)}) - Q_1(z)}.$$

Substituting (3.63) in (3.64) and comparing the coefficients of the formula (3.64) for $k+1$, we get the following relations:

$$\alpha_{k+1} = -\mu_{k+1}\beta_k, \quad \lambda_k\beta_{k+1} = \alpha_k + (\lambda_k + \mu_k - z)\beta_k,$$
$$\gamma_{k+1} = -\mu_{k+1}\delta_k, \quad \lambda_k\delta_{k+1} = \gamma_k + (\lambda_k + \mu_k - z)\delta_k.$$

Substituting the first in the second we have that β_k satisfies the three-term recurrence relation

$$\lambda_k\beta_{k+1} = -\mu_k\beta_{k-1} + (\lambda_k + \mu_k - z)\beta_k, \quad \beta_0 = 0, \quad \beta_1 = -1/\lambda_0 = Q_1^{(0)}(z).$$

The solution for β_k is given by the associated polynomials $Q_k^{(0)}(x)$ defined by (3.54). Therefore $\beta_k = Q_k^{(0)}(z)$. Hence $\alpha_k = -\mu_k Q_{k-1}^{(0)}(z)$. In the same way, the third and fourth equations can be solved taking into account the initial conditions, resulting in $\gamma_k = \mu_k Q_{k-1}(z)$ and $\delta_k = -Q_k(z)$. Consequently (3.64) can be rewritten as

$$B(z;\psi) = \frac{-\mu_k Q_{k-1}^{(0)}(z) B(z;\psi^{(k-1)}) + Q_k^{(0)}(z)}{\mu_k Q_{k-1}(z) B(z;\psi^{(k-1)}) - Q_k(z)}, \quad k \geq 1. \tag{3.65}$$

In this way we can iterate this formula until we find the Stieltjes transform of the original spectral measure.

Another use for the kth associated polynomials is that we can write the Laplace transformation $\widehat{P}(\lambda)$ of the transition probability function $P(t)$ in terms only of these polynomials and $\widehat{P}_{00}(\lambda)$. Indeed, let $\widehat{P}(\lambda)$ be the Laplace transformation of the transition function $P(t)$. Then we have

$$-I + \lambda\widehat{P}(\lambda) = \mathcal{A}\widehat{P}(\lambda) = \widehat{P}(\lambda)\mathcal{A}.$$

A general solution of the previous equation is given by the sum of a particular solution of $-I + \lambda\widehat{P}(\lambda) = \mathcal{A}\widehat{P}(\lambda)$ and the general solution of $\lambda\widehat{P}(\lambda) = \mathcal{A}\widehat{P}(\lambda)$. On the one hand, a particular solution of $-I + \lambda\widehat{P}(\lambda) = \mathcal{A}\widehat{P}(\lambda)$ is given by

$$\widehat{P}_{ij}(\lambda) = Q_i^{(j)}(-\lambda),$$

as we can directly check from the recurrence relation (3.52). On the other hand, a general solution of $\lambda\widehat{P}(\lambda) = \mathcal{A}\widehat{P}(\lambda)$ is given by

$$\widehat{P}_{ij}(\lambda) = g_j(\lambda)Q_i(-\lambda),$$

where $g_j(\lambda)$ is any function (to be determined below). Therefore the general solution of $-I + \lambda\widehat{P}(\lambda) = \mathcal{A}\widehat{P}(\lambda)$ is given by

$$\widehat{P}_{ij}(\lambda) = Q_i^{(j)}(-\lambda) + g_j(\lambda)Q_i(-\lambda). \tag{3.66}$$

Now, since $Q_0^{(j)} = 0$ and $Q_0 = 1$, taking $i = 0$ in the previous formula, we get an expression for $g_j(\lambda)$, which is exaclty $\widehat{P}_{0,j}(\lambda)$. Using the symmetry $\pi_i \widehat{P}_{ij}(\lambda) = \pi_j \widehat{P}_{ji}(\lambda)$ in formula (3.66) for $i = 0$ we arrive at

$$g_j(s) = \widehat{P}_{0,j}(\lambda) = \pi_j \left(Q_j^{(0)}(-\lambda) + Q_j(-\lambda)\widehat{P}_{00}(\lambda) \right).$$

Therefore

$$\widehat{P}_{ij}(\lambda) = Q_i^{(j)}(-\lambda) + \pi_j Q_i(-\lambda) \left(Q_j^{(0)}(-\lambda) + Q_j(-\lambda)\widehat{P}_{00}(\lambda) \right). \tag{3.67}$$

In particular, for $i \leq j$, since $Q_k^{(l)}(y) = 0$, we have

$$\widehat{P}_{ij}(\lambda) = \pi_j Q_i(-\lambda) \left(Q_j^{(0)}(-\lambda) + Q_j(-\lambda)\widehat{P}_{00}(\lambda) \right), \quad i \leq j.$$

As a consequence we have, for $i < j$ and using (3.55), that

$$\widehat{F}_{ij}(\lambda) = \frac{\widehat{P}_{ij}(\lambda)}{\widehat{P}_{jj}(\lambda)} = \frac{Q_i(-\lambda)}{Q_j(-\lambda)}.$$

Formula (3.67) gives other interesting formulas. Using again the symmetry $\pi_i \widehat{P}_{ij}(\lambda) = \pi_j \widehat{P}_{ji}(\lambda)$, we get the following relation:

$$\pi_i Q_i^{(j)}(-\lambda) + \pi_i \pi_j Q_i(-\lambda) Q_j^{(0)}(-\lambda) = \pi_j Q_j^{(i)}(-\lambda) + \pi_i \pi_j Q_j(-\lambda) Q_i^{(0)}(-\lambda).$$

For $j > i$, for example $j = n + i$, we have that $Q_i^{(n+i)} = 0$, and therefore

$$\pi_i Q_i(-\lambda) Q_{n+i}^{(0)}(-\lambda) = Q_{n+i}^{(i)}(-\lambda) + \pi_i Q_{n+i}(-\lambda) Q_i^{(0)}(-\lambda).$$

In particular, for $n = 1$ and using (3.53), we have that

$$Q_{i+1}(-\lambda) Q_i^{(0)}(-\lambda) - Q_i(-\lambda) Q_{i+1}^{(0)}(-\lambda) = -\frac{Q_{i+1}^{(i)}(-\lambda)}{\pi_i} = \frac{1}{\lambda_i \pi_i}.$$

This expression can be rewritten as

$$\frac{Q_{i+1}^{(0)}(-\lambda)}{Q_{i+1}(-\lambda)} = \frac{Q_i^{(0)}(-\lambda)}{Q_i(-\lambda)} - \frac{1}{\lambda_i \pi_i Q_{i+1}(-\lambda) Q_i(-\lambda)},$$

so

$$\frac{Q_{i+1}^{(0)}(-\lambda)}{Q_{i+1}(-\lambda)} = -\sum_{k=0}^{i} \frac{1}{\lambda_k \pi_k Q_{k+1}(-\lambda) Q_k(-\lambda)}.$$

Applying Markov's theorem 1.11, we get

$$\int_0^\infty \frac{d\psi(x)}{x+\lambda} = \sum_{k=0}^\infty \frac{1}{\lambda_k \pi_k Q_{k+1}(-\lambda) Q_k(-\lambda)}. \tag{3.68}$$

This last formula can be rewritten as

$$\int_0^\infty \frac{d\psi(x)}{x+\lambda} = \frac{1}{\mu_0} \sum_{k=0}^\infty \frac{1}{Q_k^d(-\lambda)} \left[\frac{1}{Q_k(-\lambda)} - \frac{1}{Q_{k+1}(-\lambda)} \right].$$

Taking $\lambda \downarrow 0$ and using that $Q_k^d(0) = 1$, we get the formula

$$\int_0^\infty \frac{d\psi(x)}{x} = \frac{1}{\mu_0} \left(1 - \lim_{n\to\infty} \frac{1}{Q_n(0)} \right). \tag{3.69}$$

Using the definition of $Q_n(0)$ in (3.42) we have that

$$\int_0^\infty \frac{d\psi(x)}{x} = \frac{1}{\mu_0} \left(1 - \frac{1}{1 + \mu_0 \displaystyle\sum_{n=0}^\infty \frac{1}{\lambda_n \pi_n}} \right) = \frac{\displaystyle\sum_{n=0}^\infty \frac{1}{\lambda_n \pi_n}}{1 + \mu_0 \displaystyle\sum_{n=0}^\infty \frac{1}{\lambda_n \pi_n}}, \tag{3.70}$$

which is valid even if $\mu_0 = 0$.

Remark 3.17 The Laplace transform of the transition probability function of a birth–death process (also called the resolvent function, see (3.12)) is related to continued fractions (see [79, 117]). In particular, for $i = j = 0$, we have

$$\widehat{P}_{00}(\lambda) = \cfrac{1}{\lambda + \lambda_0 - \cfrac{\lambda_0 \mu_1}{\lambda + \lambda_1 + \mu_1 - \cfrac{\lambda_0 \mu_1}{\lambda + \lambda_2 + \mu_2 - \cfrac{\lambda_2 \mu_3}{\ddots}}}}.$$

For a more combinatorial approach see [58]. ◊

Finally, as in the case of discrete-time birth–death chains, we can make use of the so-called *Derman–Vere–Jones transformation*, consisting of moving the infimum of the support of the measure to the point $x = 0$. Let $\{X_t, t \geq 0\}$ be a birth–death process and denote by $\{X_t^{(\alpha)}, t \geq 0\}$ the α*-birth–death process*, which is defined by the following:

Proposition 3.18 ([1, Proposition 8.2.6]) *Let $\{(\lambda_n, \mu_n), n \geq 0\}$ be a set of birth–death rates and $Q_n(x)$ be the corresponding associated polynomials defined by* (3.30). *For any parameter $\alpha \leq \xi$, where $\xi = \inf \operatorname{supp}(\psi)$, we define a new set of birth–death rates $\{(\tilde{\lambda}_n, \tilde{\mu}_n), n \geq 0\}$ by*

$$\tilde{\lambda}_n = \lambda_n \frac{Q_{n+1}(\alpha)}{Q_n(\alpha)}, \quad n \geq 0, \quad \tilde{\mu}_n = \begin{cases} \mu_n \dfrac{Q_{n-1}(\alpha)}{Q_n(\alpha)}, & \text{if } n \geq 1, \\ 0, & \text{if } n = 0, \end{cases} \tag{3.71}$$

and the polynomials $\tilde{Q}_n(x)$ *by*

$$\tilde{Q}_n(x) = \frac{Q_n(x+\alpha)}{Q_n(\alpha)}, \quad n \geq 0.$$

This new process is called the α*-birth–death process. Then:*

1. $\tilde{Q}_n(x)$ *are the polynomials associated with* $\{(\tilde{\lambda}_n, \tilde{\mu}_n), n \geq 0\}$.
2. *If for a measure* ψ *supported on* $[0,\infty)$ *we define another measure* $\tilde{\psi}$ *supported on* $[0,\infty)$ *by*

$$\tilde{\psi}\left([0,x]\right) = \psi\left([\alpha, x+\alpha]\right), \quad x \geq 0,$$

then we have that ψ *is a solution of the Stieltjes moment problem for the set* $\{(\lambda_n, \mu_n), n \geq 0\}$ *if and only if* $\tilde{\psi}$ *is a solution of the Stieltjes moment problem for the set* $\{(\tilde{\lambda}_n, \tilde{\mu}_n), n \geq 0\}$.

3. *Let*

$$\tilde{P}_{ij}(t) = \tilde{\pi}_j \int_0^\infty e^{-xt} \tilde{Q}_i(x) \tilde{Q}_j(x) \, d\tilde{\psi}(x),$$

where $\tilde{\pi}_j = \pi_j Q_j^2(\alpha)$. *Then* $\tilde{P}_{ij}(t)$ *is a transition function satisfying*

$$P_{ij}(t) Q_j(\alpha) = e^{-\alpha t} Q_i(\alpha) \tilde{P}_{ij}(t). \tag{3.72}$$

Moreover, $\tilde{P}_{ij}(t)$ *is honest if and only if the vector* $(Q_0(\alpha), Q_1(\alpha), \ldots)$ *is an* α*-invariant vector of* $P_{ij}(t)$*, i.e.*

$$\sum_{j \in \mathbb{N}_0} P_{ij}(t) Q_j(\alpha) = e^{-\alpha t} Q_i(\alpha). \tag{3.73}$$

Proof Observe first that $\tilde{\lambda}_n, \tilde{\mu}_n$ are positive, since $Q_n(\alpha) \geq 1$. Substituting the definitions of $\tilde{\lambda}_n, \tilde{\mu}_n, \tilde{Q}_n(x)$ and using the recurrence relation for $Q_n(x)$ (at $x = x+\alpha$ and $x = \alpha$), we have

$$\begin{aligned}
&\tilde{\lambda}_n \tilde{Q}_{n+1}(x) + (x - \tilde{\lambda}_n - \tilde{\mu}_n)\tilde{Q}_n(x) + \tilde{\mu}_n \tilde{Q}_{n-1}(x) \\
&= \frac{1}{Q_n(\alpha)} \Bigg[\lambda_n Q_{n+1}(x+\alpha) + \left(x - \frac{\lambda_n Q_{n+1}(\alpha) + \mu_n Q_{n-1}(\alpha)}{Q_n(\alpha)} \right) Q_n(x+\alpha) \\
&\qquad + \mu_n Q_{n-1}(x+\alpha) \Bigg] \\
&= \frac{1}{Q_n(\alpha)} \big[\lambda_n Q_{n+1}(x+\alpha) + (x + \alpha - \lambda_n - \mu_n) Q_n(x+\alpha) \\
&\qquad + \mu_n Q_{n-1}(x+\alpha) \big] = 0.
\end{aligned}$$

As for the orthogonality, we have

$$\int_0^\infty \tilde{Q}_i(x)\tilde{Q}_j(x)\,d\tilde{\psi}(x) = \frac{1}{Q_i(\alpha)Q_j(\alpha)}\int_0^\infty Q_i(x+\alpha)Q_j(x+\alpha)\,d\tilde{\psi}(x)$$
$$= \frac{1}{Q_i(\alpha)Q_j(\alpha)}\int_0^\infty Q_i(x)Q_j(x)\,d\psi(x) = \frac{\delta_{ij}}{\pi_j Q_j^2(\alpha)},$$

and we obtain the equivalence between solutions of the Stieltjes moment problems. Similarly, we can check the relation

$$e^{-\alpha t}Q_i(\alpha)\tilde{P}_{ij}(t) = e^{-\alpha t}Q_i(\alpha)\tilde{\pi}_j\int_0^\infty e^{-xt}\tilde{Q}_i(x)\tilde{Q}_j(x)\,d\tilde{\psi}(x)$$
$$= Q_j(\alpha)\pi_j\int_0^\infty e^{-(x+\alpha)t}Q_i(x+\alpha)Q_j(x+\alpha)\,d\tilde{\psi}(x)$$
$$= Q_j(\alpha)\pi_j\int_0^\infty e^{-xt}Q_i(x)Q_j(x)\,d\psi(x) = Q_j(\alpha)P_{ij}(t).$$

If $\tilde{P}_{ij}(t)$ is honest, then $\sum_{j\in\mathbb{N}_0}\tilde{P}_{ij}(t) = 1$. Summing the relation above, we get that the vector $(Q_0(\alpha), Q_1(\alpha), \ldots)$ is an α-invariant vector of $P_{ij}(t)$. □

3.4 The Karlin–McGregor Formula as a Transition Probability Function

As we mentioned at the end of Section 3.2, Theorem 3.9 gives an integral representation of one solution $P_{ij}(t;\psi)$ (for example, the minimal) of both Kolmogorov equations with initial conditions. In this section we will see that the Karlin–McGregor formula (3.35) represents a proper transition probability function, i.e. it satisfies conditions **(1)–(3)** at the beginning of Section 3.1, apart from other properties, as we will see.

We will start the proof of the positivity of the Karlin–McGregor formula (3.35) with the following:

Proposition 3.19 ([83, Theorem 3]) *Let ψ be a measure supported on $[0,\infty)$ with finite moments and with infinite points in the support. Let $P_n(x)$ be a family of orthogonal polynomials with respect to ψ. Then*

$$\int_0^\infty e^{-xt}P_n(x)\,d\psi(x) > 0, \quad t > 0, \quad n \geq 0. \tag{3.74}$$

Moreover, if $n \geq 2$ and $p(x)$ is a polynomial of degree r with $0 \leq r < n$ whose roots are all real and separated for some subset of $r+1$ roots of $P_n(x)$, and $p(0) > 0$, then

$$\int_0^\infty e^{-xt}p(x)P_n(x)\,d\psi(x) > 0, \quad t > 0, \quad n \geq 0. \tag{3.75}$$

In particular, for $p(x) = P_m(x)$ *we have that*

$$\int_0^\infty e^{-xt} P_m(x)P_n(x)\, d\psi(x) > 0, \quad t > 0, \quad n,m \geq 0. \tag{3.76}$$

Proof Let $1 \leq r < n$. We know from Proposition 1.8 that all zeros of $P_n(x)$ are located inside the interval of orthogonality $[0,\infty)$. Let $x_{n,j}, j = 1,\dots,n$ be those zeros ordered in the following way: $0 < x_{n,1} < \cdots < x_{n,n}$. The polynomial $P_n(x)$ can be written as

$$P_n(x) = k_n(x_{n,1} - x)(x_{n,2} - x)\cdots(x_{n,n} - x),$$

where $k_n > 0$ (since $P_n(0) > 0$). For indices $1 \leq i_1 < i_2 < \cdots < i_r$, consider the r-factor of $P_n(x)$

$$A_{i_1,\dots,i_r}(x) = \prod_{k=1}^{r} (x_{n,i_k} - x).$$

We will prove first that

$$f_{i_1,\dots,i_r}(t) = \int_0^\infty e^{-xt} A_{i_1,\dots,i_r}(x) P_n(x)\, d\psi(x) > 0. \tag{3.77}$$

Denote $1 \leq j_1 < j_2 < \cdots < j_{n-r}$ the complementary set of the indices from 1 to n removing the indices $i_1,\dots,i_k$. Define, for any continuous differentiable function $g(t)$, the operator

$$D_j g(t) = e^{-x_{n,j}t} \frac{d}{dt}[e^{x_{n,j}t} g(t)] = g'(t) + x_{n,j} g(t).$$

This operator has the following property:

$$D_j g(t) > 0, \text{ for all } t > 0 \text{ and } g(0) = 0 \Rightarrow g(t) > 0 \text{ for all } t > 0. \tag{3.78}$$

Indeed, since $g(0) = 0$ and $D_j g(t) > 0$ if and only if $g'(t) + x_{n,j} g(t) > 0$, then we have $g'(0) > 0$, i.e. g is strictly increasing at 0. If there were a $t_0 > 0$ such that $g(t_0) \leq 0$, then $g'(t_0) > -g(t_0)x_{n,j} \geq 0$. Therefore g would be strictly increasing at t_0. This implies that there exists a zero z_0 of g in the interval $(0,t_0)$, i.e. $g(z_0) = 0$ and also $g'(z_0) < 0$. But that is a contradiction since $g'(z_0) + x_{n,j} g(z_0) > 0$.

Let j be one of the indices $j_1,\dots,j_{n-k}$. Then

$$D_j f_{i_1,\dots,i_r}(t) = \int_0^\infty e^{-xt}(-x + x_{n,j}) A_{i_1,\dots,i_r}(x) P_n(x)\, d\psi(x).$$

Observe that the zero $x_{n,j}$ is not one of the zeros of $A_{i_1,\dots,i_r}$. Iterating for the remaining indices $j_1,\dots,j_{n-k}$, we get

$$D_{j_1} D_{j_2} \cdots D_{j_{n-r}} f_{i_1,\dots,i_r}(t) = \frac{1}{k_n} \int_0^\infty e^{-xt} [P_n(x)]^2 d\psi(x) > 0.$$

Moreover, if $k < n - r$, by the orthogonality property of $P_n(x)$, we have that

$$D_{j_1} D_{j_2} \cdots D_{j_k} f_{i_1,\dots,i_r}(0) = 0.$$

Applying (3.78) $n - r$ times we get (3.77). Let us now see (3.74). For that, define

$$F_n(t) = \int_0^\infty e^{-xt} P_n(x)\, d\psi(x), \quad n \geq 0.$$

It is obvious that $F_0(t) > 0$. Since $P_1(x) = -(x + B_0)/C_0$ (called the three-term recurrence relation) where $C_0 > 0$, we get

$$\begin{aligned} \frac{d}{dt}[e^{-B_0 t} F_1(t)] &= e^{-B_0 t} \int_0^\infty e^{-xt} P_1(x)(-x - B_0)\, d\psi(x) \\ &= \frac{C_0}{P_0(x)} e^{-B_0 t} \int_0^\infty e^{-xt} [P_1(x)]^2 d\psi(x) > 0 \end{aligned}$$

and $F_1(0) = 0$. Then we can apply (3.78) to prove that $F_1(t) > 0$. For $n \geq 2$ we have that $F_n(0) = 0$ (orthogonality) and, using (3.77) for one index j, we obtain

$$D_j F_n(t) = e^{-x_{n,j} t} \frac{d}{dt}[e^{x_{n,j} t} F_n(t)] = f_j(t) > 0.$$

So for the same reason we get $F_n(t) > 0$ for all $t > 0$.

Let us now prove (3.75). Let $p(x)$ be as in the statement of the proposition and choose $i_1, \dots, i_{r+1}$, such that $x_{n,i_1}, \dots, x_{n,i_{r+1}}$ separate the r roots of $p(x)$. Then, if we denote by $A_k(x) = A_{i_1,\dots,i_{r+1}}(x)/(x_{n,i_k} - x)$, we have that, using partial fractions,

$$\frac{p(x)}{A_{i_1,\dots,i_{r+1}}(x)} = \sum_{k=1}^{r+1} \frac{\alpha_k}{x_{n,i_k} - x}, \quad \alpha_k = \frac{p(x_{n,i_k})}{A_k(x_{n,i_k})} \quad \Rightarrow p(x) = \sum_{k=1}^{r+1} \alpha_k A_k(x),$$

where the coefficients α_k are nonnegative and not all are zero, since the signs of $p(x_{n,i_k})$ and $A_k(x_{n,i_k})$ are always the same for all k. Therefore, for all $t > 0$ and using (3.77), we have that, using the first part,

$$\int_0^\infty e^{-xt} p(x) P_n(x)\, d\psi(x) = \sum_{k=1}^{r+1} \alpha_k \int_0^\infty e^{-xt} A_k(x) P_n(x)\, d\psi(x) > 0.$$

Formula (3.76) is a consequence of (3.75). For $m = n$ it is trivial. If $m \neq n$, for example $m < n$, by the separation property of the zeros of orthogonal polynomials, a subset of zeros of $P_n(x)$ separates the roots of $P_m(x)$ and also $P_m(0) > 0$, so we can apply (3.76) for $p(x) = P_m(x)$. □

Remark 3.20 Observe that formula (3.76) (for $P_n(x) = Q_n(x)$) is exactly $P_{mn}(t;\psi)/\pi_n$ where $P_{mn}(t;\psi)$ is the Karlin–McGregor formula (3.35). Therefore we have proved that $P_{ij}(t;\psi) > 0$ for all $i,j \in \mathbb{N}_0$ and $t > 0$. ◊

Let us now study convergence and the boundness of the series $\sum_{j=0}^\infty P_{ij}(t;\psi)$.

Proposition 3.21 ([83, Theorem 5]) *Let ψ be a solution of the Stieltjes moment problem with $\mu_0 = 0$ and consider the Karlin–McGregor formula* (3.35) *for $P_{ij}(t;\psi)$. Then we have*

$$0 < \sum_{j=0}^{n} P_{i+1,j}(t;\psi) < \sum_{j=0}^{n} P_{ij}(t;\psi) < 1. \tag{3.79}$$

Each of the series $\sum_{j=0}^{\infty} P_{ij}(t;\psi)$ converges uniformly on every $t \in [0, t_0]$ and satisfies

$$0 < \sum_{j=0}^{\infty} P_{i+1,j}(t;\psi) \leq \sum_{j=0}^{\infty} P_{ij}(t;\psi) \leq 1. \tag{3.80}$$

Proof Using the Karlin–McGregor formula (3.35) and the definition of the dual polynomials in (3.49) for $\mu_0 = 0$ we get

$$\sum_{j=0}^{n} P_{ij}(t;\psi) = \int_0^{\infty} e^{-xt} Q_i(x) \sum_{j=0}^{n} \pi_j Q_j(x)\, d\psi(x) = \int_0^{\infty} e^{-xt} Q_i(x) Q_n^d(x)\, d\psi(x).$$

Taking $i = 0$ and the derivative with respect to t in the previous expression, we obtain

$$\frac{d}{dt} \sum_{j=0}^{n} P_{0,j}(t;\psi) = -\int_0^{\infty} x e^{-xt} Q_n^d(x)\, d\psi(x) = -\lambda_0 \int_0^{\infty} e^{-xt} Q_n^d(x)\, d\psi^d(x) < 0,$$

where we have used the relation between the measures ψ and ψ^d given by Theorem 3.14. Therefore $\sum_{j=0}^{n} P_{0,j}(t;\psi)$ is strictly decreasing for $t \in [0,\infty)$. Since $\sum_{j=0}^{n} P_{0,j}(0;\psi) = 1$ we get

$$0 < \sum_{j=0}^{n} P_{0,j}(t;\psi) < 1, \quad n \geq 0, \quad t > 0.$$

We also have

$$\sum_{j=0}^{n} [P_{ij}(t;\psi) - P_{i+1,j}(t;\psi)] = \frac{\lambda_0}{\lambda_i \pi_i} \int_0^{\infty} e^{-xt} Q_i^d(x) Q_n^d(x)\, d\psi^d(x) > 0.$$

Therefore we get (3.79) and both series are convergent with sums satisfying (3.80). □

Remark 3.22 It is possible to prove (see [83, p. 509]) that the series $\sum_{j=0}^{\infty} P_{ij}(t;\psi)$ can be differentiated termwise any number of times and the derivatives also converge uniformly on every $[0, t_0]$. The case where $\mu_0 > 0$ is a little bit more difficult can be found in [83, pp. 512–514]. In this situation there may be some solutions of the Stieltjes moment problem for which $\sum_{j=0}^{\infty} P_{ij}(t;\psi) \leq 1$ does not hold. ◊

Let us now study the honesty of the series $\sum_{j=0}^{\infty} P_{ij}(t;\psi)$.

Proposition 3.23 ([83, Theorem 16]) *Let ψ be the minimal solution of the Stieltjes moment problem and consider the Karlin–McGregor formula* (3.35) *for $P_{ij}(t;\psi)$. Then we have*

$$\sum_{j=0}^{\infty} P_{ij}(t;\psi) = 1, \quad \textit{for all} \quad i \in \mathbb{N}_0, \quad t \geq 0 \quad \Leftrightarrow \quad \mu_0 = 0 \quad \textit{and} \quad C = \infty,$$

where C is defined by (3.28).

Proof We will prove first that if honesty holds for some state i then it will hold for all states $j \in \mathbb{N}_0$. Let us call

$$F_i(s) = \int_0^{\infty} e^{-st}\left(1 - \sum_{j=0}^{\infty} P_{ij}(t;\psi)\right) dt \tag{3.81}$$

and $F = (F_0, F_1, \dots)$. Then, using the Kolmogorov backward equation, integration by parts and that $\int_0^{\infty} se^{-st}dt = 1$, we obtain that $\mathcal{A}F = sF$. Therefore $F_i(s) = cQ_i(-s)$ for some c. But if we have honesty for at least one state i, then $c = 0$, and the honesty will hold for all states $j \in \mathbb{N}_0$. Alternatively, for $i = 0$, taking derivatives with respect to t and using the Kolmogorov backward equation, then

$$0 = \frac{d}{dt}\sum_{j=0}^{\infty} P_{0,j}(t;\psi) = -(\lambda_0 + \mu_0)\sum_{j=0}^{\infty} P_{0,j}(t;\psi) + \lambda_0 \sum_{j=0}^{\infty} P_{1,j}(t;\psi) = \mu_0.$$

So $\mu_0 = 0$ is necessary. Assume now that $C < \infty$. Then, according to **(a)**–**(c)** after formula (3.50) we obtain that $Q_{\infty}(x)$ converges and $Q_{\infty}(0) = 1$, so the first zero of $Q_{\infty}(x)$ is $\xi = 2\alpha > 0$ and ψ does not have a mass in $0 \leq x < 2\alpha$. Using Proposition 3.18 we have from (3.72) that

$$\sum_{j=0}^{\infty} P_{ij}(t;\psi) = e^{-\alpha t}\sum_{j=0}^{\infty} \frac{Q_i(\alpha)}{Q_j(\alpha)}\tilde{P}_{ij}(t;\tilde{\psi}) \leq \frac{e^{-\alpha t}}{Q_{\infty}(\alpha)} < 1,$$

for t sufficiently large. The last inequality comes from $0 < Q_{\infty}(\alpha) < Q_n(\alpha) < 1$ and that $\sum_{j=0}^{\infty} \tilde{P}_{ij}(t;\tilde{\psi}) \leq 1$. Alternatively, assume that $C = \infty$. Then, using again **(a)**–**(c)** after formula (3.50) we get that $Q_i(-\lambda) \to \infty$ as $i \to \infty$ for some $\lambda > 0$. From (3.81) we see that $F_i(\lambda) = cQ_i(-\lambda)$ must be bounded as $i \to \infty$. Therefore $F_i(\lambda) = 0$ for all i and we get honesty. □

Let us see now under what conditions $P_{ij}(t;\psi)$ satisfies the Chapman–Kolmogorov equations (3.1).

Proposition 3.24 ([83, Theorem 9]) *Let ψ be a solution of the Stieltjes moment problem and $P_{ij}(t;\psi)$ be the Karlin–McGregor formula* (3.35). *Then the series*

$$\sum_{k=0}^{\infty} P_{ik}(t;\psi)P_{kj}(s;\psi)$$

converges for all $i,j \in \mathbb{N}_0$ *and* $s,t \geq 0$ *and it is equal to* $P_{ij}(s+t;\psi)$ *if and only if* ψ *is extremal (see* (1.35)*), i.e. if Parseval's identity holds.*

Proof Consider $F_i(x,t) = e^{-xt}Q_i(x)$. We have that $F_i(x,t) \in L^2_\psi(0,\infty)$ for all $t \geq 0$. Consider the orthonormal system of polynomials $\left(\pi_n^{1/2}Q_n\right)_n$. Then the Fourier coefficients of $F_i(x,t)$ and $F_j(x,s)$ are given by

$$a_k(i,t) = \int_0^\infty e^{-xt}Q_i(x)(\pi_k)^{1/2}Q_k(x)\,d\psi(x) = \pi_k^{-1/2}P_{ik}(t;\psi),$$

$$a_k(j,s) = \int_0^\infty e^{-xs}Q_j(x)(\pi_k)^{1/2}Q_k(x)\,d\psi(x) = \frac{\pi_k^{1/2}}{\pi_j}P_{kj}(s;\psi).$$

Therefore

$$\sum_{k=0}^{\infty} a_k(i,t)a_k(j,s) = \frac{1}{\pi_j}\sum_{k=0}^{\infty} P_{ik}(t;\psi)P_{kj}(s;\psi)$$

is convergent by the usual convergence of Fourier coefficients and we get the first part of this proposition. Now, if we assume that ψ is extremal, then Parseval's identity (1.35) holds. That is,

$$\sum_{k=0}^{\infty} a_k(i,t)a_k(j,s) = \int_0^\infty F_i(x,t)F_j(x,s)\,d\psi(x)$$

$$= \int_0^\infty e^{-x(t+s)}Q_i(x)Q_j(x)\,d\psi(x) = \frac{1}{\pi_j}P_{ij}(t+s;\psi),$$

so we get the Chapman–Kolmogorov equations. Alternatively, if $P_{ij}(t;\psi)$ satisfies the Chapman–Kolmogorov equation and $F(x,t) = e^{-xt}R(x)$, where R is a polynomial that can be written as $R = \sum \alpha_i Q_i$, then

$$\int_0^\infty |F(x,t)|^2 d\psi(x) = \sum_{i,j} \alpha_i\bar{\alpha}_j \int_0^\infty e^{-2xt}Q_i(x)Q_j(x)\,d\psi(x) = \sum_{i,j} \alpha_i\bar{\alpha}_j \frac{P_{ij}(2t;\psi)}{\pi_j}$$

$$= \sum_{i,j} \alpha_i\bar{\alpha}_j \sum_{k=0}^{\infty} \frac{P_{ik}(t;\psi)P_{kj}(t;\psi)}{\pi_k} = \sum_{k=0}^{\infty}\left|\sum_i \alpha_i \frac{P_{ik}(t;\psi)}{\pi_k^{1/2}}\right|^2$$

$$= \sum_{k=0}^{\infty}\left|\int_0^\infty F(x,t)Q_k(x)\pi_k^{1/2}d\psi(x)\right|^2.$$

Therefore Parseval's identity holds for the functions $F(x,t)$. Since these functions are dense in $L^2_\psi(0,\infty)$ then it holds for every $f \in L^2_\psi(0,\infty)$ and we get the result. □

Remark 3.25 The previous result also holds for measures ψ which are solutions of the *Hamburger moment problem* (not necessarily supported on $[0,\infty)$) but with left bounded support. ◊

For uniqueness, we have the following result:

Theorem 3.26 ([83, Theorems 14 and 15]) *Let $\{(\lambda_n,\mu_n), n\geq 0\}$ be a set of birth–death rates with potential coefficients $(\pi_n)_n$ as in (3.24). Then:*

1. If $\mu_0 = 0$, the solution of the Stieltjes moment problem is unique if and only if

$$A+B=\sum_{n=0}^{\infty}\left(\frac{1}{\lambda_n\pi_n}+\pi_n\right)=\infty. \tag{3.82}$$

2. If $\mu_0 > 0$, the solution of the Stieltjes moment problem is unique if and only if

$$\sum_{n=0}^{\infty}\pi_n Q_n^2(0)=\sum_{n=0}^{\infty}\pi_n\left(1+\mu_0\sum_{k=0}^{n-1}\frac{1}{\lambda_k\pi_k}\right)^2=\infty. \tag{3.83}$$

Proof Assume first that $\mu_0=0$. By Proposition 3.21, for any spectral measure generated by the birth–death rates $\{(\lambda_n,\mu_n), n\geq 0\}$, there exists a corresponding transition function $P_{ij}(t;\psi)$. If $A+B=\infty$ then we have that either C or D (defined by (3.28)) are ∞. By Proposition 3.7 the backward or the forward (or both) Kolmogorov equation has a unique solution (the minimal $\mathcal{A}$-function). If ψ_1 and ψ_2 are two solutions of the Stieltjes moment problem and $P_{ij}^{(1)}(t;\psi_1)$ and $P_{ij}^{(2)}(t;\psi_2)$ are the corresponding transition functions, then both solutions must be the same. In particular, $P_{00}^{(1)}(t;\psi_1)=P_{00}^{(2)}(t;\psi_2)$, and by the uniqueness theorem for Laplace transforms we must have $\psi_1=\psi_2$. Alternatively, assume that (3.82) is not true. Then $A+B<\infty$, i.e. both A and B are finite. Then $\psi_{\max}$ has an atom located at $x=0$ of mass B^{-1} (see discussion after (3.37)). Since $AB=C+D$ (see (3.28)) then C and D are finite. Following the three conditions after (3.51) we have, using formula (3.68), that $\int_0^\infty x^{-1}d\psi_{\min}(x)$ is convergent. Since $\int_0^\infty x^{-1}d\psi_{\max}(x)$ is not defined at $x=0$ that means that $\psi_{\max}\neq\psi_{\min}$ and the solution of the Stieltjes moment problem is not unique, so we get a contradiction.

Let us assume now that $\mu_0>0$. Then, by Proposition 3.18, the uniqueness of the solution of the Stieltjes moment problem for the set $\{(\lambda_n,\mu_n), n\geq 0\}$ is equivalent to the uniqueness of the solution of the Stieltjes moment problem for the set $\{(\tilde{\lambda}_n,\tilde{\mu}_n), n\geq 0\}$ (here we are taking $\alpha=0$ in that proposition). Hence, by the first part of this theorem, since $\tilde{\mu}_0=0$, we must have

$$\sum_{n=0}^{\infty}\left(\tilde{\pi}_n+\frac{1}{\tilde{\lambda}_n\tilde{\pi}_n}\right)=\infty.$$

Since $\tilde{\pi}_n = \pi_n Q_n^2(0)$ and recalling the definition of $\tilde{\lambda}_n$ and $\tilde{\mu}_n$ in (3.71), the previous is equivalent to

$$\sum_{n=0}^{\infty}\left(\pi_n Q_n^2(0) + \frac{1}{\lambda_n \pi_n Q_n(0) Q_{n+1}(0)}\right) = \infty.$$

From (3.68) and (3.69) we see, taking $\lambda = 0$, that the second series is bounded. Therefore we must have (3.83). □

Remark 3.27 In [83, Theorem 15] it is also shown that, for the case $\mu_0 > 0$, the uniqueness of a transition function $P_{ij}(t)$ satisfying both Kolmogorov equations, $P_{ij}(t) \geq 0$ and $\sum_{j=0}^{\infty} P_{ij}(t) \leq 1$, is equivalent to either that the solution of the Stieltjes moment problem is unique or $\mu_0 \int_0^{\infty} x^{-1} d\psi_{\min}(x) = 1$; or equivalently that the single condition $A + B = \infty$ holds. ◊

Remark 3.28 In [83, Chapter V] (see also [81]) Karlin and McGregor treat the concept of *total positivity*. Let $B = (b_{ij})$ be a matrix (possibly of infinite dimension). Let $i_1, i_2, \dots, i_p$ be integers satisfying $1 \leq i_1 < i_2 < \cdots < i_p$ and let $j_1, j_2, \dots, j_p$, be integers satisfying $1 \leq j_1 < j_2 < \cdots < j_p$. Call

$$B\begin{pmatrix} i_1, i_2, \dots, i_p \\ j_1, j_2, \dots, j_p \end{pmatrix} = \begin{vmatrix} b_{i_1 j_1} & b_{i_1 j_2} & \cdots & b_{i_1 j_p} \\ b_{i_2 j_1} & b_{i_2 j_2} & \cdots & b_{i_2 j_p} \\ \vdots & \vdots & \ddots & \vdots \\ b_{i_p j_1} & b_{i_p j_2} & \cdots & b_{i_p j_p} \end{vmatrix}$$

the determinant of the $p \times p$ matrix obtained from rows $i_1, i_2, \dots, i_p$, and columns $j_1, j_2, \dots, j_p$, of B. If $i_1 = j_1, i_2 = j_2, \dots, i_p = j_p$ then it is called the *principal subdeterminant of order p*. The matrix B is a *totally positive matrix* of order r if each subdeterminant of order less than or equal to r is nonnegative. In [83, Theorem 20] we find that if ψ is an extremal solution of the Stieltjes moment problem then the transition probability function matrix $P(t;\psi)$ is strictly totally positive for each $t > 0$. ◊

3.5 Birth–Death Processes with Killing

Let $\{X_t, t \geq 0\}$ be a birth–death process on $\mathbb{N}_0$ with birth–death rates given by $\{(\lambda_n, \mu_n), n \geq 0\}$ and assume that for each state n there exists a *rate of absorption* $\gamma_n \geq 0$, or *killing*, into a fictitious state which we will call ∂. We will also assume that $\mu_0 = 0$, since we already have a rate of absorption γ_0 at state 0. A transition to the absorbing state ∂ is sometimes called a *total catastrophe*. The entries of the infinitesimal operator $\mathcal{A} = (a_{ij})$ in (3.3) are given by

$$\begin{aligned} a_{n,n+1} &= \lambda_n, \quad a_{n+1,n} = \mu_{n+1}, \\ a_{nn} &= -(\lambda_n + \mu_n + \gamma_n), \quad n \geq 0, \quad a_{ij} = 0, \quad |i-j| > 1. \end{aligned} \tag{3.84}$$

These processes were first considered by S. Karlin and S. Tavaré in [98], where they studied the case of linear rates, but a long time passed until E. van Doorn and A. Zeifman covered the problem in a couple of papers [49, 50]. In this section we will follow these last two works.

We will assume that the transition probability function $P_{ij}(t)$ is the minimal $\mathcal{A}$-function (see Definition 3.4). Let us consider the polynomials $(Q_n)_n$ using the infinitesimal operator $\mathcal{A}$ with coefficients as in (3.84), as follows:

$$\begin{aligned} Q_0(x) = 1, \quad & -xQ_0(x) = \lambda_0 Q_1(x) - (\lambda_0 + \gamma_0)Q_0(x), \\ -xQ_n(x) &= \lambda_n Q_{n+1}(x) - (\lambda_n + \mu_n + \gamma_n)Q_n(x) + \mu_n Q_{n-1}(x), \quad n \geq 1. \end{aligned} \tag{3.85}$$

If $\gamma_n = 0, n \geq 1$, except possibly for $\gamma_0 \geq 0$, we go back to the regular birth–death process that we studied in the previous sections, so we get the usual Karlin–McGregor formula (3.35). We will see now that the same formula holds if we assume the general case of $\gamma_n \geq 0, n \geq 0$. For that we will consider a *similar* birth–death process $\{\widetilde{X}_t, t \geq 0\}$ with birth–death rates $\{(\tilde{\lambda}_n, \tilde{\mu}_n), n \geq 0\}$ given by

$$\begin{aligned} \tilde{\mu}_0 &= 0, \quad \tilde{\lambda}_0 = \lambda_0 + \gamma_0, \\ \tilde{\mu}_n &= (\lambda_{n-1}/\tilde{\lambda}_{n-1})\mu_n, \quad \tilde{\lambda}_n = \lambda_n + \gamma_n + \mu_n - \tilde{\mu}_n, \quad n \geq 1. \end{aligned} \tag{3.86}$$

It is easy to see using induction that $\tilde{\lambda}_n \geq \lambda_n + \gamma_n \geq \lambda_n > 0$ and $\tilde{\mu}_{n+1} > 0$ for $n \geq 0$. For this new (conservative) birth–death process consider the corresponding infinitesimal operator $\tilde{\mathcal{A}}$ and potential coefficients $(\tilde{\pi}_n)_n$. Then, from the definition, we have

$$\tilde{\Pi}\tilde{\mathcal{A}}\tilde{\Pi}^{-1} = \Pi\mathcal{A}\Pi^{-1}, \tag{3.87}$$

where Π and $\tilde{\Pi}$ are diagonal matrices with entries $\sqrt{\pi_n}$ and $\sqrt{\tilde{\pi}_n}$, respectively. We will assume that $\{\widetilde{X}_t, t \geq 0\}$ is the minimal $\tilde{\mathcal{A}}$-process represented by a minimal $\tilde{\mathcal{A}}$-function $\tilde{P}_{ij}(t)$ which satisfies the Karlin–McGregor formula (3.35) for a certain spectral measure $\tilde{\psi}$ on $[0, \infty)$ and corresponding orthogonal polynomials $(\tilde{Q}_n)_n$ built from the birth–death rates $\{(\tilde{\lambda}_n, \tilde{\mu}_n), n \geq 0\}$ given by (3.86). In fact, by induction and using the recurrence relations for the polynomials $(Q_n)_n$ and $(\tilde{Q}_n)_n$, it is easy to see that

$$\tilde{Q}_n(x) = \frac{\lambda_0\lambda_1 \cdots \lambda_{n-1}}{\tilde{\lambda}_0\tilde{\lambda}_1 \cdots \tilde{\lambda}_{n-1}} Q_n(x) = \sqrt{\frac{\pi_n}{\tilde{\pi}_n}} Q_n(x), \quad n \geq 0.$$

Therefore they are equal up to a multiplicative constant, so the polynomials $(Q_n)_n$ defined by (3.85) are orthogonal with respect to the measure $\tilde{\psi}$. Then we have the following:

Theorem 3.29 ([49, Theorem 1]) *The minimal $\mathcal{A}$-function $P(t)$ and the minimal $\tilde{\mathcal{A}}$-function $\tilde{P}(t)$ satisfy*

$$P(t) = \Pi^{-1}\tilde{\Pi}\tilde{P}(t)\tilde{\Pi}^{-1}\Pi, \quad t \geq 0. \tag{3.88}$$

Therefore, we have

$$\begin{aligned} P_{ij}(t) = \pi_j \int_0^{\infty} e^{-xt} Q_i(x) Q_j(x)\, d\psi(x) &= \sqrt{\frac{\tilde{\pi}_i \pi_j}{\pi_i \tilde{\pi}_j}} \tilde{P}_{ij}(t) \\ &= \sqrt{\frac{\tilde{\pi}_i}{\pi_i} \pi_j \tilde{\pi}_j} \int_0^{\infty} e^{-xt} \tilde{Q}_i(x) \tilde{Q}_j(x)\, d\tilde{\psi}(x). \end{aligned} \tag{3.89}$$

Proof From (3.87) the right-hand side of (3.88), denoted by $F(t)$, satisfies the system of backward and Kolmogorov forward equations for the process $\{X_t, t \geq 0\}$. $F(t)$ is also minimal since $\tilde{P}(t)$ is minimal by assumption (see Proposition 3.5). Then, using again Proposition 3.5, we have that $F(t)$ is the minimal $\mathcal{A}$-function and therefore $F(t) = P(t), t \geq 0$. Formula (3.89) is a consequence of the definition. □

Theorem 4 of [49] gives the necessary and sufficient conditions in order that there is only one $\mathcal{A}$-function $P(t)$ satisfying the system of Kolmogorov backward and forward equations. For that, at least one of the two conditions

$$\lim_{n\to\infty} Q_n(0) = \infty,$$

or

$$\sum_{n=0}^{\infty} \left(\pi_n + \frac{1}{\lambda_n \pi_n} \right) = \infty, \tag{3.90}$$

needs to be satisfied. Observe that the second condition is the condition for the uniqueness of the Stieltjes moment problem (see Theorem 3.26).

Let us now denote T_∂ the *killing time*, i.e. the (possible defective) random variable representing the time at which absorption in the absorbing state ∂ occurs. We are interested in the functions

$$\tau_i(t) = \mathbb{P}(T_\partial \leq t | X_0 = i),$$

as well as their limits

$$\tau_i = \lim_{t\to\infty} \tau_i(t).$$

We will call $\tau_i(t)$ and τ_i the *extinction probability* at time t and the *eventual extinction probability*, respectively, when the initial state is i. By using the forward equations, the Karlin–McGregor formula (3.89) and interchanging the integrals, we obtain

$$\tau_i(t) = \sum_{j=0}^{\infty} \gamma_j \int_0^t P_{i,j}(u)\,du = \sum_{j=0}^{\infty} \gamma_j \pi_j \int_0^t \int_0^{\infty} e^{-xu} Q_i(x) Q_j(x)\,d\psi(x)\,du$$

$$= \sum_{j=0}^{\infty} \gamma_j \pi_j \int_0^{\infty} \frac{1-e^{-xt}}{x} Q_i(x) Q_j(x)\,d\psi(x), \quad i \in \mathbb{N}_0, \quad t \geq 0.$$

Taking limits as $t \to \infty$, we get

$$\tau_i = \sum_{j=0}^{\infty} \gamma_j \pi_j \int_0^{\infty} Q_i(x) Q_j(x) \frac{d\psi(x)}{x}, \quad i \in \mathbb{N}_0, \quad t \geq 0. \tag{3.91}$$

Therefore the extinction probabilities can be written as

$$\tau_i(t) = \tau_i - \sum_{j=0}^{\infty} \gamma_j \pi_j \int_0^{\infty} e^{-xt} Q_i(x) Q_j(x) \frac{d\psi(x)}{x}, \quad i \in \mathbb{N}_0, \quad t \geq 0.$$

Let us now give a criterion to decide if the eventual extinction probabilities τ_i are certain in terms of the evaluation of the polynomials $(Q_n)_n$ at the point $x = 0$. First observe that, following a similar argument to (3.41), we obtain

$$Q_n(x) = 1 + \sum_{k=0}^{n-1} \frac{1}{\lambda_k \pi_k} \sum_{j=0}^{k} (\gamma_j - x) \pi_j Q_j(x), \quad n \geq 1.$$

If we call $q_n = Q_n(0)$, then we have

$$q_0 = 1, \quad q_n = 1 + \sum_{k=0}^{n-1} \frac{1}{\lambda_k \pi_k} \sum_{j=0}^{k} \gamma_j \pi_j q_j, \quad n \geq 1, \tag{3.92}$$

and define

$$q_\infty \doteq \lim_{n\to\infty} q_n = 1 + \sum_{k=0}^{\infty} \frac{1}{\lambda_k \pi_k} \sum_{j=0}^{k} \gamma_j \pi_j q_j. \tag{3.93}$$

Lemma 3.30 ([50, Lemma 1]) *We have $q_\infty = \infty$ if and only if*

$$\sum_{k=0}^{\infty} \frac{1}{\lambda_k \pi_k} \sum_{j=0}^{k} \gamma_j \pi_j = \infty. \tag{3.94}$$

Proof If $q_\infty = \infty$ then (3.94) is ∞ because $q_n \geq 1$. Alternatively, call $\beta_k = \frac{1}{\lambda_k \pi_k} \sum_{j=0}^{k} \gamma_j \pi_j$ and assume that $\sum_{k=0}^{\infty} \beta_k < \infty$. Then we have, by definition of q_n in (3.92) and that $q_n \geq 1$, the following:

$$q_{n+1} = q_n + \frac{1}{\lambda_n \pi_n} \sum_{j=0}^{n} \gamma_j \pi_j q_j \leq q_n(1+\beta_n) \leq \prod_{k=0}^{n} (1+\beta_k).$$

Since $\sum_{k=0}^{\infty} \beta_k < \infty$ then $\prod_{k=0}^{\infty}(1+\beta_k) < \infty$, so $q_\infty < \infty$, as requested. □

The sequence $(q_n)_n$ plays an important role in the behavior of the eventual extinction probabilities $(\tau_n)_n$ in (3.91). First, observe from the representation (3.91), the three-term recurrence relation (3.85) and orthogonality properties, that $(\tau_n)_n$ satisfies the following recurrence relations:

$$(\lambda_i + \mu_i + \gamma_i)\tau_i = \lambda_i \tau_{i+1} + \mu_i \tau_{i-1} + \gamma_i, \quad i \geq 1,$$
$$(\lambda_0 + \gamma_0)\tau_0 = \lambda_0 \tau_1 + \gamma_0.$$

Alternatively, using the same three-term recurrence relation (3.85), the sequence $(q_n)_n$ satisfies the following recurrence relations:

$$(\lambda_i + \mu_i + \gamma_i)q_i = \lambda_i q_{i+1} + \mu_i q_{i-1}, \quad i \geq 1,$$
$$\lambda_0 + \gamma_0 = \lambda_0 q_1, \quad q_0 = 1.$$

Using the previous two relations and an argument of induction we get the following formula:

$$1 - \tau_i = (1 - \tau_0) q_i, \quad i \in \mathbb{N}_0.$$

Therefore τ_i may be expressed in terms of τ_0 and q_i. Using Lemma 3.1 of [11], who studied eventual extinction probabilities in a more general setting (see also [1, Section 9.2]), gives $\tau_i = 1 - q_i/q_\infty$, with the interpretation that $\tau_i = 1$ if $q_\infty = \infty$. Therefore we have the following result:

Theorem 3.31 ([50, Theorem 1]) *If* (3.94) *is satisfied, then* $\tau_i = 1$ *for all* $i \in \mathbb{N}_0$. *Otherwise, the eventual probabilities satisfy*

$$\tau_i = 1 - \frac{q_i}{q_\infty},$$

where q_i *and* q_∞ *are given by* (3.92) *and* (3.93), *respectively.*

For more information about birth–death processes with killing see [49, 50]. For an example, see Example 3.41 below.

3.6 Examples

In this section we will study a collection of instructive and useful examples. From now on we will assume that if $\mu_0 = 0$, then $A + B = \infty$, meaning that the solution ψ of the Stieltjes moment problem is unique and therefore we will have a unique transition function $P_{ij}(t;\psi)$ satisfying both Kolmogorov equations, $P_{ij}(t) \geq 0$ and $\sum_{j=0}^{\infty} P_{ij}(t) \leq 1$. If $\mu_0 > 0$, a sufficient condition for (3.83) to hold is that $\sum_{n=1}^{\infty} 1/\mu_n = \infty$. All examples we will see in this section (with infinite state space) will automatically satisfy these conditions.

We will begin with some examples where the state space is finite, i.e. $\mathcal{S} = \{0, 1, \ldots, N\}$. In this case we can use a standard eigenvalue decomposition as in (3.20). As in the case of discrete-time birth–death chains, we will focus on examples that are related somehow to classical orthogonal polynomials. Second, we will study some examples of birth–death processes on $\mathcal{S} = \mathbb{N}_0$ where we do not know the spectral measure, but it can be identified using one of the classical families of orthogonal polynomials or computed using formula (3.60), which relates the spectral measures of the original birth–death process and the 0th associated birth–death process. We will also use the more general formula (3.65) or different special functions methods, as in Example 3.39.

In this section we will follow *Kendall's notation* used in *queuing theory* (see [102]), based on three factors written as $A/S/c$. A denotes the time between arrivals to the queue, S the service time distribution and c the number of servers at the node. We will mostly use $A = M$, which means Markovian or memoryless, i.e. exponential inter-arrival times. Also $S = M$ means Markovian or memoryless, and we have an exponential service time. This notation has been extended to $A/S/c/K/N/D$, where K is the capacity of the queue, N is the size of the population of jobs to be served and D is the queueing discipline.

3.6.1 Examples with Finite State Space

Example 3.32 *The $M/M/1/N$ queue.* In this case both the arrival and the service times in the queue are always exponential with constant parameter $(\lambda + \mu)^{-1} > 0$, there is only one server and the capacity of the queue or the maximum number of customers allowed in the system including those in service is a finite number N. The birth–death rates are constant numbers $\lambda, \mu > 0$, the state space of the birth–death process is finite $\mathcal{S} = \{0, 1, \dots, N\}$ and the infinitesimal operator is

$$\mathcal{A} = \begin{pmatrix} -\lambda-\mu & \lambda & & & \\ \mu & -\lambda-\mu & \lambda & & \\ & \ddots & \ddots & \ddots & \\ & & \mu & -\lambda-\mu & \lambda \\ & & & \mu & -\mu-\lambda \end{pmatrix}.$$

A diagram of the one-step transitions is

$$0 \underset{\mu}{\overset{\lambda}{\rightleftarrows}} 1 \underset{\mu}{\overset{\lambda}{\rightleftarrows}} 2 \underset{\mu}{\overset{\lambda}{\rightleftarrows}} \cdots \underset{\mu}{\overset{\lambda}{\rightleftarrows}} N$$

Observe that $\mathcal{A}$ is not conservative. The potential coefficients are given by

$$\pi_j = \left(\frac{\lambda}{\mu}\right)^j, \quad j = 0, 1, \dots, N.$$

The three-term recurrence relation satisfied by the polynomials is given by $Q_{-1}=0$, $Q_0 = 1$ and

$$\begin{aligned} Q_1(x) &= -(x-\lambda-\mu)/\lambda, \quad -xQ_n(x) \\ &= \lambda Q_{n+1}(x) - (\lambda+\mu)Q_n(x) + \mu Q_{n-1}(x), \quad n = 1,\dots,N-1. \end{aligned}$$

If we take the change of variables $y = \frac{\lambda+\mu-x}{2\sqrt{\lambda\mu}}$ and call $Q_n(x) = (\mu/\lambda)^{n/2}U_n(y)$, these new polynomials are the Chebychev polynomials of the second kind $(U_n)_n$ (see (1.87)). Therefore

$$Q_n(x) = \left(\frac{\mu}{\lambda}\right)^{n/2} U_n\left(\frac{\lambda+\mu-x}{2\sqrt{\lambda\mu}}\right), \quad n = 0,1,\dots,N.$$

A similar argument to the one used in Example 2.17, gives that the polynomial

$$R_{N+1}(x) = -(x+\lambda-\mu)Q_N(x) - \mu Q_{N-1}(x) = \lambda\left(\frac{\mu}{\lambda}\right)^{\frac{N+1}{2}} U_{N+1}\left(\frac{\lambda+\mu-x}{2\sqrt{\lambda\mu}}\right)$$

is a polynomial of degree $N+1$ whose zeros coincide with the eigenvalues of the matrix $\mathcal{A}$. In this case we have, using (1.89), that

$$x_{N+1,k} = \lambda+\mu-2\sqrt{\lambda\mu}\cos\left(\pi\frac{k+1}{N+2}\right), \quad k = 0,1,\dots,N.$$

Observe that the eigenvalues are located inside the interval $[0,\infty)$. As in the case of discrete-time birth–death chains, the values of the jumps are given by

$$\psi\left(\{x_{N+1,k}\}\right) = \frac{2}{N+2}\sin^2\left(\pi\frac{k+1}{N+2}\right), \quad k = 0,1,\dots,N.$$

Therefore the spectral measure is given by

$$\psi(x) = \sum_{k=0}^{N} \psi\left(\{x_{N+1,k}\}\right)\delta_{x_{N+1,k}}(x),$$

and the Karlin–McGregor representation is given by

$$\begin{aligned} P_{ij}(t) &= \frac{2\pi_j}{N+2}\sum_{k=0}^{N} e^{-x_{N+1,k}t}Q_i(x_{N+1,k})Q_j(x_{N+1,k})\sin^2\left(\pi\frac{k+1}{N+2}\right) \\ &= \frac{2e^{-(\lambda+\mu)t}}{(N+2)\left(\sqrt{\frac{p}{q}}\right)^{j-i}}\sum_{k=0}^{N} e^{2t\sqrt{\lambda\mu}\cos\left(\pi\frac{k+1}{N+2}\right)}U_i\left(\cos\left(\pi\frac{k+1}{N+2}\right)\right) \\ &\quad \times U_j\left(\cos\left(\pi\frac{k+1}{N+2}\right)\right)\sin^2\left(\pi\frac{k+1}{N+2}\right). \end{aligned}$$

$$= \frac{2e^{-(\lambda+\mu)t}}{(N+2)\left(\sqrt{\frac{p}{q}}\right)^{i-j}} \sum_{k=0}^{N} e^{2t\sqrt{\lambda\mu}\cos\left(\pi\frac{k+1}{N+2}\right)} \sin\left(\pi\frac{(i+1)(k+1)}{N+2}\right)$$
$$\times \sin\left(\pi\frac{(j+1)(k+1)}{N+2}\right).$$

In the last part we used the relation of the Chebychev polynomials of the second kind with the cosine function given by (1.88). The case of a conservative $M/M/1/N$ queue can be found in Exercise 3.3 (see also [4] or [134]). ◊

Example 3.33 *Ehrenfest and Bernoulli–Laplace urn models.* As in the case of discrete-time birth–death chains (see Examples 2.19 and 2.20) there are two well-known models related to classical orthogonal polynomials, namely, the Ehrenfest urn model (Krawtchouk polynomials) and the Bernoulli–Laplace urn model (dual Hahn polynomials). Both have finite state space $\mathcal{S} = \{0, 1, \ldots, N\}$. The main difference in the continuous-time case is that we do not have to rescale the eigenvalues of the corresponding infinitesimal operator, since the polynomials are defined automatically in terms of these operators.

For the Ehrenfest urn model the birth–death rates are given by

$$\lambda_n = (N-n)p, \quad \mu_n = nq, \quad 0 \le n \le N, \quad 0 < p < 1, \quad q = 1-p,$$

and the infinitesimal operator matrix is given by

$$\mathcal{A} = \begin{pmatrix} -Np & Np & & & \\ q & -q-(N-1)p & (N-1)p & & \\ & \ddots & \ddots & \ddots & \\ & & (N-1)q & -(N-1)q-p & p \\ & & & Nq & -Nq \end{pmatrix}.$$

The recurrence relation is identified directly with the Krawtchouk polynomials in (1.115), so

$$Q_n(x) = K_n(x; p, N), \quad n \ge 0.$$

The potential coefficients are given by

$$\pi_j = \binom{N}{j}\left(\frac{p}{q}\right)^j, \quad j = 0, 1, \ldots, N.$$

The Karlin–McGregor formula is then

$$P_{ij}(t) = \pi_j \sum_{x=0}^{N} e^{-xt} K_i(x; p, N) K_j(x; p, N) \binom{N}{x} p^x q^{N-x}.$$

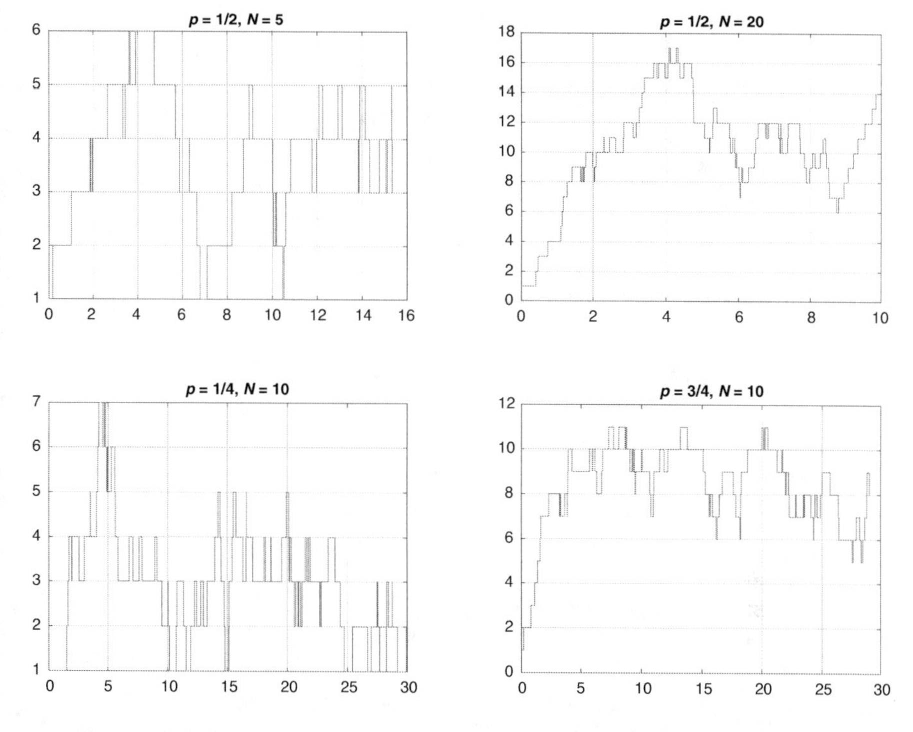

Figure 3.1 Trajectories of continuous-time Ehrenfest urn model for different values of p and N starting at $X_0 = 1$ with state space $\mathcal{S} = \{1, 2, \ldots, N+1\}$.

Using the generating function for the Krawtchouk polynomials in (1.114) twice we see that

$$\sum_{j=0}^{N} P_{ij}(t)s^j = \left[s + q(1-s)(1-e^{-t})\right]^i \left[1 - p(1-s)(1-e^{-t})\right]^{N-i}.$$

In Figure 3.1 we can see some trajectories of the process for different values of p and N. For a more general treatment of the Ehrenfest urn models, see [92].

For the Bernoulli–Laplace urn model, the birth–death rates are now quadratic in n:

$$\lambda_n = (N-n)(a-n), \quad \mu_n = n(b-N+n), \quad 0 \leq n \leq N, \quad a, b > N-1,$$

and the infinitesimal operator matrix is now

$$\mathcal{A} = \begin{pmatrix} -Na & Na & & & \\ b-N+1 & -\lambda_1 - \mu_1 & (N-1)(a-1) & & \\ & \ddots & \ddots & \ddots & \\ & & (N-1)(b+1) & -\lambda_{N-1} - \mu_{N-1} & a-N+1 \\ & & & Nb & -Nb \end{pmatrix}.$$

The recurrence relation is identified directly with the dual Hahn polynomials in (1.123) by calling $a = -\gamma - 1$ and $b = -\delta - 1$, so

$$Q_n(x) = R_n(\lambda(x); a, b, N), \quad n \geq 0, \quad \lambda(x) = x(x - a - b - 1).$$

The potential coefficients are given by

$$\pi_j = \binom{a}{j}\binom{b}{N-j}, \quad j = 0, 1, \ldots, N.$$

The Karlin–McGregor formula is then

$$P_{ij}(t) = \pi_j \sum_{x=0}^{N} e^{\lambda(x)t} R_i(x; a, b, N) R_j(x; a, b, N) \psi_x,$$

where ψ_x is the normalization of the measure for the dual Hahn polynomials given by (1.122). It is not difficult to see that this process has as its stationary distribution the hypergeometric distribution. Indeed,

$$\lim_{t\to\infty} \mathbb{P}(X_t = j | X_0 = i) = \lim_{t\to\infty} P_{ij}(t) = \frac{\binom{a}{j}\binom{b}{N-j}}{\binom{a+b}{N}}.$$

Another example with quadratic birth–death rates and related to Racah polynomials (see Section 1.6) can be found in [4]. ◊

Example 3.34 *A genetic model of Moran.* This model was introduced by S. Karlin and J. McGregor in [91] as an extension of other genetic models studied by S. Wright [149] or W. Feller [53]. The formulation of the model is as follows. Assume that there are N gametes which are either of type a or A. Denote by X_t the number of a-gametes at time t and assume that the total population size N remains constant for all $t \geq 0$. Therefore the state space of the process is $\mathcal{S} = \{0, 1, \ldots, N\}$. A change of state occurs (for an exponentially distributed time of mean $1/\lambda$) when a single individual reproduces and is replaced by a new individual. Furthermore, it is assumed that the probability of two or more matings occurring in a small time interval is negligible. The mechanism of mating has the following structure: an individual is chosen at random to be fertilized. Another individual (fertilizer), also chosen at random, does the fertilizing (self-fertilization is allowed) and this individual dominates the outcome. For instante, if an A-gamete fertilizes an a-gamete or an A-gamete, then the result is always an A-gamete (the same applies for an a-gamete as fertilizer). Immediately after fertilization the progeny may mutate into the other type of gamete. Let us denote by γ_1 the probability that an a-gamete mutates to an A-gamete (similarly for γ_2 but the other way around). If at time t we have $X_t = n$, then the conditional probability of having $n+1$ a-gametes in an infinitesimal time is given by

$$\left(1-\frac{n}{N}\right)\frac{n}{N}(1-\gamma_1)+\left(1-\frac{n}{N}\right)\left(1-\frac{n}{N}\right)\gamma_2.$$

The first term of the previous sum follows from a fertilized A-gamete which transforms into an a-gamete. The chance of selecting an A-gamete when $X_t = n$ is $1 - n/N$. Now if an a-gamete fertilizes the selected A-gamete, then the outcome is a new a-gamete provided no mutation occurs. The second term follows in a similar manner when an A-gamete fertilizes an A-gamete but the progeny mutates into an a-gamete. In a similar way, the conditional probability of having $n-1$ a-gametes in an infinitesimal time is given by

$$\frac{n}{N}\left(1-\frac{n}{N}\right)(1-\gamma_2)+\frac{n}{N}\frac{n}{N}\gamma_1.$$

Therefore, the stochastic process described above can be identified as an example of a birth–death process with rates

$$\begin{aligned}\lambda_n &= \lambda\left(1-\frac{n}{N}\right)\left[\frac{n}{N}(1-\gamma_1)+\left(1-\frac{n}{N}\right)\gamma_2\right],\\ \mu_n &= \lambda\frac{n}{N}\left[\frac{n}{N}\gamma_1+\left(1-\frac{n}{N}\right)(1-\gamma_2)\right],\end{aligned} \tag{3.95}$$

where $n = 0, 1, \ldots, N$. Observe that $\mu_0 = 0$ and $\lambda_N = 0$, so the corresponding infinitesimal operator $\mathcal{A}$ is conservative. Call $R_n(x), n = 0, 1, \ldots N$, the polynomials generated by the three-term recurrence relation

$$\begin{aligned}R_0(x) &= 1, \quad R_{-1}(x) = 0,\\ -xR_n(x) &= \lambda_n R_{n+1}(x) - (\lambda_n+\mu_n)R_n(x) + \mu_n R_{n-1}(x), \quad n = 0, 1, \ldots N,\end{aligned} \tag{3.96}$$

and denote by π_n the corresponding potential coefficients defined by (3.24). Depending on the values of γ_1, γ_2, we will have three different cases connected with classical discrete orthogonal polynomials.

Case 1. $\gamma_1, \gamma_2 > 0, 1-\gamma_1-\gamma_2 > 0$. Here the two linear factors contained in λ_n and μ_n in (3.95) oppose each other, one exhibiting attraction, the other repulsion towards the same end state. Let us call

$$\alpha = \frac{N\gamma_2}{1-\gamma_1-\gamma_2} - 1, \quad \beta = \frac{N\gamma_1}{1-\gamma_1-\gamma_2} - 1,$$

$$\lambda_n = n(n+\alpha+\beta+1), \quad c = \frac{\lambda(1-\gamma_1-\gamma_2)}{N^2}.$$

Using (1.118), if we call

$$R_n(c\lambda_k) = Q_k(n;\alpha,\beta,N), \quad n,k = 0, 1, \ldots, N,$$

where $Q_k(n;\alpha,\beta,N)$ are the Hahn polynomials defined by (1.117), then the polynomials $R_n(x)$ satisfy the three-term recurrence relation (3.96) with birth–death rates

λ_n, μ_n defined by (3.95). Therefore the Karlin–McGregor representation formula can be written as

$$P_{ij}(t) = \pi_j \binom{\beta+N}{N} \sum_{k=0}^{N} e^{-c\lambda_k t} R_i(c\lambda_k) R_j(c\lambda_k) \psi_k(\alpha, \beta, N),$$

where ψ_x is defined by (1.122).

Case 2. $\gamma_1, \gamma_2 > 0, \gamma_1 + \gamma_2 - 1 > 0$. Now the two linear factors in λ_n and μ_n in (3.95) extend their force in the same direction. Again, let us call

$$\alpha = -(N + \beta' + 2), \quad \beta = -(N + \alpha' + 2),$$
$$\alpha' = \frac{N(1-\gamma_2)}{\gamma_1 + \gamma_2 - 1} - 1, \quad \beta' = \frac{N(1-\gamma_1)}{\gamma_1 + \gamma_2 - 1} - 1,$$
$$\lambda_n' = n(2N + \alpha' + \beta' + 3 - n), \quad c = \frac{\lambda(\gamma_1 + \gamma_2 - 1)}{N^2}.$$

Using (1.118), if we call

$$R_n(c\lambda_k') = Q_k(n; \alpha, \beta, N), \quad n, k = 0, 1, \ldots, N,$$

where $Q_k(n; \alpha, \beta, N)$ are again the Hahn polynomials defined by (1.117), then the polynomials $R_n(x)$ satisfies the three-term recurrence relation (3.96) with birth–death rates λ_n, μ_n defined by (3.95). Therefore the Karlin–McGregor representation formula can be written as

$$P_{ij}(t) = \pi_j \binom{\beta+N}{N} \sum_{k=0}^{N} e^{-c\lambda_k' t} R_i(c\lambda_k') R_j(c\lambda_k') \psi_k(\alpha, \beta, N),$$

where ψ_x is defined by (1.122).

Case 3. $\gamma_1, \gamma_2 > 0, \gamma_1 + \gamma_2 = 1$. In this case the birth–death rates (3.95) become linear in n rather than quadratic, i.e.

$$\lambda_n = \frac{\lambda \gamma_2}{N^2}(N - n), \quad \mu_n = \frac{\lambda}{N^2}(1 - \gamma_2)n.$$

The polynomials $R_n(x)$ satisfying the three-term recurrence relation (3.96) can be identified, using (1.112), with the Krawtchouk polynomials, defined by (1.111). Specifically,

$$R_n\left(\frac{\lambda x}{N^2}\right) = K_n(x; \gamma_2, N).$$

Therefore the Karlin–McGregor representation formula can be written as

$$P_{ij}(t) = \pi_j \sum_{n=0}^{N} e^{-\lambda n t/N^2} K_n(i; \gamma_2, N) K_n(j; \gamma_2, N) \psi_{n,N},$$

where now

$$\psi_{n,N} = \binom{N}{n} \gamma_2^{N-n}(1-\gamma_2)^n, \quad \pi_j = \binom{N}{j} \left(\frac{\gamma_2}{1-\gamma_2}\right)^n.$$

There are also some limiting cases, such as $\gamma_1 = \gamma_2 = 0$, $\gamma_1 = 0, 0 < \gamma_2 < 1$ or $\gamma_2 = 0, 0 < \gamma_1 < 1$, which can be studied using dual polynomials (see [91] for details). ◊

3.6.2 Examples with State Space $\mathbb{N}_0$

Example 3.35 *The $M/M/1$ queue.* Consider the birth–death process $\{X_t, t \geq 0\}$ with constant birth–death rates $\lambda, \mu > 0$, and infinitesimal operator given by

$$\mathcal{A} = \begin{pmatrix} -\lambda-\mu & \lambda & & & \\ \mu & -\lambda-\mu & \lambda & & \\ & \mu & -\lambda-\mu & \lambda & \\ & & \ddots & \ddots & \ddots \end{pmatrix}. \tag{3.97}$$

A diagram of the transitions is

$$\textcircled{0} \underset{\mu}{\overset{\lambda}{\rightleftarrows}} \textcircled{1} \underset{\mu}{\overset{\lambda}{\rightleftarrows}} \textcircled{2} \underset{\mu}{\overset{\lambda}{\rightleftarrows}} \textcircled{3} \underset{\mu}{\overset{\lambda}{\rightleftarrows}} \textcircled{4} \underset{\mu}{\overset{\lambda}{\rightleftarrows}} \textcircled{5} \underset{\mu}{\overset{\lambda}{\rightleftarrows}} \cdots$$

This birth–death process models the $M/M/1$ queue with only one single server but the queue can be infinitely long. We observe that the Stieltjes transform of the spectral measure associated with $\mathcal{A}$ is the same as the Stieltjes transform of the spectral measure of the 0th birth–death process. Therefore, using (3.60), we get

$$B(z;\psi) = \frac{\lambda+\mu-z-\sqrt{(\lambda+\mu-z)^2-4\lambda\mu}}{2\lambda\mu}, \tag{3.98}$$

where the square root is taken positive for $z < 0$. The spectral measure has only an absolutely continuous part given by

$$\psi(x) = \frac{\sqrt{4\lambda\mu-(\lambda+\mu-x)^2}}{2\pi\lambda\mu}, \quad |\lambda+\mu-x| \leq 2\sqrt{\lambda\mu}.$$

If we define $\sigma_\pm = (\sqrt{\lambda} \pm \sqrt{\mu})^2$, the support of orthogonality of ψ is the interval $[\sigma_-, \sigma_+]$, which is bounded and contained in the interval $[0,\infty)$. The spectral measure can then be written as

$$\psi(x) = \frac{\sqrt{(x-\sigma_-)(\sigma_+-x)}}{2\pi\lambda\mu}, \quad x \in [\sigma_-, \sigma_+].$$

It is easy to see that the polynomials generated by the three-term recurrence relation are given by

$$Q_n(x) = \left(\frac{\mu}{\lambda}\right)^{n/2} U_n\left(\frac{\lambda+\mu-x}{2\sqrt{\lambda\mu}}\right),$$

where $(U_n)_n$ are the Chebychev polynomials of the second kind. The Karlin–McGregor formula (3.35) can be written as

$$\begin{aligned}
P_{ij}(t) &= \left(\sqrt{\frac{\lambda}{\mu}}\right)^{j-i} \int_{(\sqrt{\lambda}-\sqrt{\mu})^2}^{(\sqrt{\lambda}+\sqrt{\mu})^2} e^{-xt} U_i\left(\frac{\lambda+\mu-x}{2\sqrt{\lambda\mu}}\right) U_j\left(\frac{\lambda+\mu-x}{2\sqrt{\lambda\mu}}\right) \\
&\quad \times \frac{\sqrt{4\lambda\mu-(\lambda+\mu-x)^2}}{2\pi\lambda\mu}\, dx \\
&= \frac{2\left(\sqrt{\frac{\lambda}{\mu}}\right)^{j-i}}{\pi} \int_0^{\pi} e^{-(\lambda+\mu-2\sqrt{\lambda\mu}\cos\theta)t} U_i(\cos\theta) U_j(\cos\theta) \sqrt{1-\cos^2\theta}\, \sin\theta\, d\theta \\
&= \frac{2e^{-(\lambda+\mu)t}}{\pi\left(\sqrt{\frac{\lambda}{\mu}}\right)^{i-j}} \int_0^{\pi} e^{2t\sqrt{\lambda\mu}\cos\theta} \sin((i+1)\theta)\sin((j+1)\theta)\, d\theta.
\end{aligned}$$

We have made the change of variables $x = \lambda+\mu-2\sqrt{\lambda\mu}\cos\theta$ and used the relation of the Chebychev polynomials of the second kind with the cosine function given by (1.88).

If we study the conservative case (i.e. in the entry $(0,0)$ of (3.97) only appears $-\lambda$ instead of $-\lambda-\mu$) we recover the analog of Example 2.22 (case 2). In that case the Stieltjes transform, substituting (3.98) in (3.60), is

$$B(z;\psi) = -\frac{-\lambda+\mu+z+\sqrt{(\lambda+\mu-z)^2-4\lambda\mu}}{2\mu z}.$$

The absolutely continuous part of the spectral measure is given by

$$\frac{\sqrt{4\lambda\mu-(\lambda+\mu-x)^2}}{2\pi\mu x}, \quad |\lambda+\mu-x| \le 2\sqrt{\lambda\mu}.$$

Moreover, it has a jump at $x=0$ of size

$$\psi(\{0\}) = \lim_{\varepsilon\downarrow 0} \varepsilon \operatorname{Im} B(i\varepsilon;\psi) = \frac{1}{2\mu}[(\mu-\lambda)+|\mu-\lambda|] = \begin{cases} 0, & \text{if } \mu \le \lambda, \\ \dfrac{\mu-\lambda}{\mu}, & \text{if } \mu > \lambda. \end{cases}$$

Therefore the spectral measure is given by

$$\psi(x) = \frac{\sqrt{4\lambda\mu - (\lambda + \mu - x)^2}}{2\pi \mu x} \mathbf{1}_{\{|\lambda+\mu-x|\leq 2\sqrt{\lambda\mu}\}} + \frac{\mu - \lambda}{\mu} \delta_0(x) \mathbf{1}_{\{\mu>\lambda\}}. \quad (3.99)$$

It is possible to see that the polynomials are now given by

$$Q_n(x) = \left(\frac{\mu}{\lambda}\right)^{n/2} \left[U_n\left(\frac{\lambda + \mu - x}{2\sqrt{\lambda\mu}}\right) - \sqrt{\frac{\mu}{\lambda}} U_{n-1}\left(\frac{\lambda + \mu - x}{2\sqrt{\lambda\mu}}\right)\right].$$

As in the case of discrete-time birth–death chains, the shape of the spectral measure (dividing by x and adding a Dirac delta at $x = 0$) as well as the orthogonal polynomials $Q_n(x)$ suggest that we might be in the situation of dual polynomials as in Theorem 3.11. This is true if we start with the previous case (where it is not conservative) but replace λ with μ. The details are left as an exercise. ◊

Example 3.36 *The $M/M/\infty$ queue.* In this case both the arrival and the service times in the queue are Markovian (exponentially distributed) but it is assumed that there are infinite servers. This example was studied for the first time in the context of orthogonal polynomials in [86, Section 3]. The birth–death rates are then given by

$$\lambda_n = \lambda, \quad \mu_n = \mu n, \quad n \geq 0, \quad \lambda, \mu > 0.$$

The infinitesimal operator is

$$\mathcal{A} = \begin{pmatrix} -\lambda & \lambda & & & \\ \mu & -\lambda - \mu & \lambda & & \\ & 2\mu & -\lambda - 2\mu & \lambda & \\ & & \ddots & \ddots & \ddots \end{pmatrix}.$$

A diagram of the transitions is

$$\text{(0)} \underset{\mu}{\overset{\lambda}{\rightleftarrows}} \text{(1)} \underset{2\mu}{\overset{\lambda}{\rightleftarrows}} \text{(2)} \underset{3\mu}{\overset{\lambda}{\rightleftarrows}} \text{(3)} \underset{4\mu}{\overset{\lambda}{\rightleftarrows}} \text{(4)} \underset{5\mu}{\overset{\lambda}{\rightleftarrows}} \text{(5)} \underset{6\mu}{\overset{\lambda}{\rightleftarrows}} \cdots$$

The potential coefficients are given by

$$\pi_j = \frac{(\lambda/\mu)^j}{j!}, \quad j \geq 0.$$

The three-term recurrence relation for the polynomials $Q_n(x)$ is

$$\lambda Q_{n+1}(x) - (\lambda + \mu n) Q_n(x) + \mu n Q_{n-1}(x) = -x Q_n(x), \quad Q_0(x) = 1, \quad Q_{-1}(x) = 0.$$

Immediately we realize that this recurrence relation is very similar to that of the Charlier polynomials (see (1.103)). Dividing by μ and calling $a = \lambda/\mu$, we get that

$$a Q_{n+1}(x) - (a + n) Q_n(x) + n Q_{n-1}(x) = -\frac{x}{\mu} Q_n(x).$$

Therefore

$$Q_n(x) = C_n\left(\frac{x}{\mu};\frac{\lambda}{\mu}\right), \quad n \geq 0,$$

where $C_n(x;a)$ are the Charlier polynomials defined by (1.99). The associated spectral measure is the (discrete) Poisson distribution, which is given by $\psi_x = e^{-a}a^x/x!, x = 0, 1, \ldots$. Therefore the spectral measure is given by

$$\psi(x) = \sum_{n=0}^{\infty} \frac{e^{-a}a^n}{n!}\delta_{x_n}, \quad x_n = \mu n. \tag{3.100}$$

The Karlin–McGregor integral representation is then

$$P_{ij}(t) = \pi_j \sum_{n=0}^{\infty} e^{-t\mu n} Q_i(\mu n) Q_j(\mu n) \frac{e^{-a}a^n}{n!}.$$

In particular, using the generating function for the Charlier polynomials in (1.102) and the dual property $Q_n(k\mu) = Q_k(n\mu)$, we have that

$$P_{i,0}(t) = e^{-a} \sum_{n=0}^{\infty} Q_n(i\mu) \frac{(ae^{-\mu t})^n}{n!} = e^{-a(1-e^{-\mu t})}(1-e^{-\mu t})^i.$$

We also have

$$\begin{aligned}\sum_{j=0}^{\infty} P_{ij}(t) z^j &= e^{-a} \sum_{n=0}^{\infty} Q_n(i\mu) \frac{(ae^{-\mu t})^n}{n!} \sum_{j=0}^{\infty} Q_j(n\mu) \frac{(az)^j}{j!} \\ &= e^{-a(1-z)(1-e^{-\mu t})}[1-(1-z)e^{-\mu t}]^i.\end{aligned}$$

These formulas were previously known using generating function techniques (see [57, p. 396]). Figure 3.2 shows some trajectories of this process for different values of λ and μ.

If we consider the 0th associated process (that is, removing the first row and column), now the process is a queue with infinite servers, but there is a probability $\mu/(\lambda+\mu)$ to end the process when there are no clients in the queue. The Stieltjes transform of the spectral measure $\psi^{(0)}$ of this new process is given by (3.61), where $(y_n)_n$ are the zeros of the meromorphic function

$$B(z;\psi) = \sum_{n=0}^{\infty} \frac{a^n e^{-a}}{n!\,(n\mu - z)},$$

and the masses $(\gamma_n)_n$ (see (3.62)) are given by

$$\psi^{(0)}(\{y_n\}) = \gamma_n = \frac{1}{\lambda\mu B'(y_n;\psi)}.$$

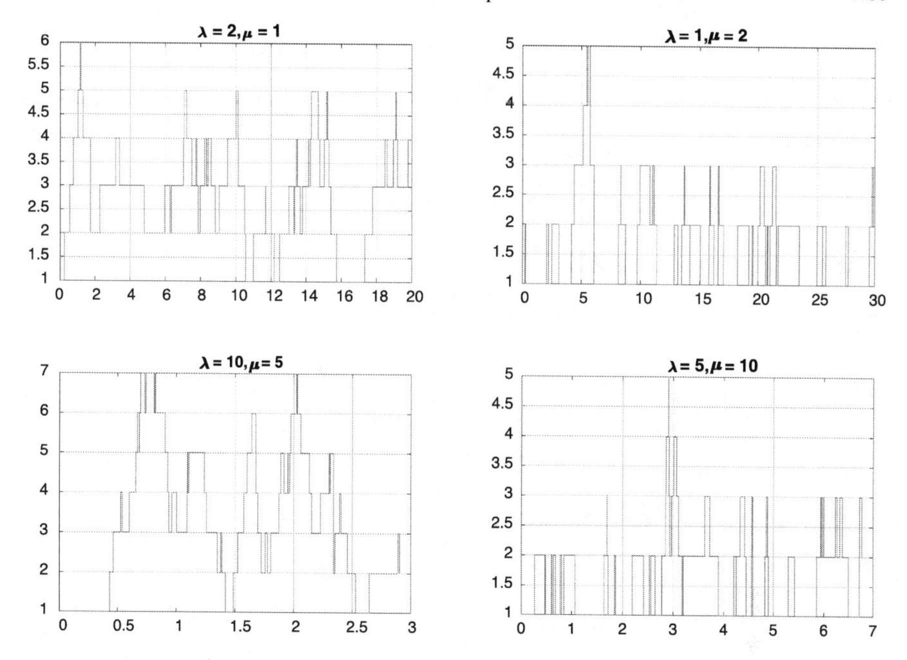

Figure 3.2 Trajectories of the $M/M/\infty$ queue for different values of λ and μ starting at $X_0 = 1$ with state space $\mathcal{S} = \mathbb{N}$.

On a practical level it is enough to know the first values of y_n and γ_n. For example, in this case, using (3.58) and $\int_0^\infty x^{-1} d\psi^{(0)}(x) = 1/\mu$, we have that the distribution of the length of a busy period is given by

$$F_{10}(t) = \mathbb{P}(X_\tau = 0 \text{ for some } \tau, 0 < \tau \leq t | X_0 = 1) = \mu \int_0^\infty \frac{1 - e^{-xt}}{x} d\psi^{(0)}(x)$$
$$= 1 - \mu \sum_{n=0}^\infty \frac{\gamma_n}{y_n} e^{-y_n t},$$

and we observe that, for t large, the most significant terms in the sum are the first ones. In [86] one can find more details about how to compute $(y_n)_n$ and $(\gamma_n)_n$ numerically. ◊

Example 3.37 *The $M/M/k$ queue*. In this case we have a queue with a finite number of k servers. This example was studied for the first time in the context of orthogonal polynomials in [86]. The birth–death rates are the same as for the $M/M/\infty$ queue, but limiting the number of servers, i.e.

$$\lambda_n = \lambda, \quad \mu_n = \begin{cases} n\mu, & \text{if } \ n \leq k \\ k\mu, & \text{if } \ n > k \end{cases}, \quad \lambda, \mu > 0.$$

The first k polynomials are the same as for the $M/M/\infty$ queue, that is, the Charlier polynomials. Then

$$Q_n(x) = C_n\left(\frac{x}{\mu};\frac{\lambda}{\mu}\right), \quad n \le k.$$

We will try to obtain an explicit expression of the spectral measure ψ. For that we use (3.65) for $n = k$. Then

$$B(z;\psi) = \frac{-k\mu Q_{k-1}^{(0)}(z)B(z;\psi^{(k-1)}) + Q_k^{(0)}(z)}{k\mu Q_{k-1}(z)B(z;\psi^{(k-1)}) - Q_k(z)}. \tag{3.101}$$

If we remove the first k rows and columns, the infinitesimal operator matrix $\mathcal{A}$ is the same from that moment on, that is, $B(z;\psi^{(k-1)}) = B(z;\psi^{(k)})$. Therefore, solving in (3.63), we get that

$$B(z;\psi^{(k-1)}) = \frac{\lambda + k\mu - z - \sqrt{(\lambda + k\mu - z)^2 - 4k\lambda\mu}}{2k\lambda\mu},$$

where the square root is chosen such that it is positive for $z < 0$. Substituting $B(z;\psi^{(k-1)})$ in (3.101) and rationalizing, using $\lambda_{k-1}\pi_{k-1}[Q_k Q_{k-1}^{(0)} - Q_k^{(0)} Q_{k-1}] = 1$, where $\lambda_k\pi_k = \lambda(\lambda/\mu)^k/k!$, and the three-term recurrence formula, we arrive at

$$B(z;\psi) = -\frac{L_k(z)}{H_k(z)}, \tag{3.102}$$

where

$$\begin{aligned} L_k(z) &= 4\lambda^2 Q_k(z)Q_k^{(0)}(z) + 4k\lambda\mu Q_{k-1}(z)Q_{k-1}^{(0)}(z) \\ &\quad - 2\lambda(\lambda + k\mu - z)[Q_k(z)Q_{k-1}^{(0)}(z) + Q_{k-1}(z)Q_k^{(0)}(z)] \\ &\quad + 2(k-1)!\left(\frac{\mu}{\lambda}\right)^{k-1}\sqrt{(\lambda + k\mu - z)^2 - 4k\lambda\mu} \\ &= 4\lambda^2\left[Q_k(z)Q_k^{(0)}(z) - \frac{1}{2}\left(Q_{k+1}(z)Q_{k-1}^{(0)}(z) + Q_{k-1}(z)Q_{k+1}^{(0)}(z)\right)\right] \\ &\quad + 2(k-1)!\left(\frac{\mu}{\lambda}\right)^{k-1}\sqrt{(\lambda + k\mu - z)^2 - 4k\lambda\mu}, \end{aligned}$$

and

$$\begin{aligned} H_k(z) &= 4\lambda^2[Q_k^2(z) - Q_{k-1}(z)Q_{k+1}(z)] \\ &= 4k\lambda\mu[Q_{k-1}^2(z) - Q_{k-2}(z)Q_k(z)] - 4\lambda\mu Q_k(z)[Q_{k-1}(z) - Q_{k-2}(z)]. \end{aligned}$$

Observe that $H_k(z)$ is a polynomial in z of degree $2k - 1$ and the polynomial part of $L_k(z)$ has degree $2k - 2$. The Perron–Stieltjes inversion formula shows that $\operatorname{Im} B(x + i\varepsilon;\psi)$ converges uniformly to 0 as $\varepsilon \downarrow 0$, provided that x is in an interval that does not contain any root of $H_k(x)$ and disjoint to the interval $|\lambda + k\mu - x| \leq 2\sqrt{k\lambda\mu}$. Therefore the absolutely continuous part of the spectral measure is given by

$$\frac{(k-1)!}{2\pi\lambda^2}\left(\frac{\mu}{\lambda}\right)^{k-1}\frac{\sqrt{4k\lambda\mu-(\lambda+k\mu-x)^2}}{Q_k^2(x)-Q_{k-1}(x)Q_{k+1}(x)}. \tag{3.103}$$

Additionally the spectral measure ψ can have discrete masses at some or all zeros of $H_k(x)$.

We will compute now the polynomials $Q_n(x)$ for $n \geq k$. Call $R_n = Q_{n+k}(x)$, $n \geq -1$. Then we have that

$$\begin{aligned} R_0(x) &= Q_k(x), \quad R_{-1}(x) = Q_{k-1}(x), \\ -xR_n(x) &= k\mu R_{n-1} - (\lambda + k\mu)R_n(x) + \lambda R_{n+1}(x), \quad n \geq 0, \end{aligned} \tag{3.104}$$

where we observe that the coefficients of the recurrence relation are independent of n. This indicates that the polynomials $R_n(x)$ can be written as a linear combination of Chebychev polynomials $U_n(x)$, $T_n(x)$ that satisfy

$$\begin{aligned} &T_0(x) = 1, \quad T_1(x) = x, \quad U_0(x) = 1, \quad U_1(x) = 2x, \\ &xP_n(x) = \frac{1}{2}P_{n-1} + \frac{1}{2}P_{n+1}(x), \quad n \geq 1, \quad P_n(x) = \{T_n(x), U_n(x)\}. \end{aligned}$$

Indeed, if we call

$$V_n(x) = \left(\frac{k\mu}{\lambda}\right)^{n/2} T_n\left(\frac{\lambda + k\mu - x}{2\sqrt{k\lambda\mu}}\right), \quad W_n(x) = \left(\frac{k\mu}{\lambda}\right)^{n/2} U_n\left(\frac{\lambda + k\mu - x}{2\sqrt{k\lambda\mu}}\right),$$

we observe that $V_n(x)$ and $W_n(x)$ are solutions of (3.104) where

$$V_0(x) = 1, \quad V_{-1}(x) = \frac{\lambda + k\mu - x}{2\sqrt{k\lambda\mu}}, \quad W_0(x) = 1, \quad W_{-1}(x) = 0.$$

Therefore $R_n(x) = v(x)V_n(x) + w(x)W_n(x)$ and solving for $n = 0, -1$, we get that

$$\begin{aligned} R_n(x) &= \frac{2k\mu}{\lambda + k\mu - x}Q_{k-1}(x)V_n(x) + \left[Q_k(x) - \frac{2k\mu}{\lambda + k\mu - x}Q_{k-1}(x)\right]W_n(x) \\ &= Q_k(x)W_n(x) - \frac{k\mu}{\lambda}Q_{k-1}(x)W_{n-1}(x), \quad n \geq 0. \end{aligned}$$

The last equality is a consequence of the second formula in (1.91) applied to $V_n(x)$ and $W_n(x)$. Therefore,

$$\begin{aligned} Q_{n+k}(x) = \left(\frac{k\mu}{\lambda}\right)^{n/2} &\left[Q_k(x)U_n\left(\frac{\lambda + k\mu - x}{2\sqrt{k\lambda\mu}}\right)\right. \\ &\left. - \sqrt{\frac{k\mu}{\lambda}}Q_{k-1}(x)U_{n-1}\left(\frac{\lambda + k\mu - x}{2\sqrt{k\lambda\mu}}\right)\right], \quad n \geq 0. \end{aligned}$$

In a similar way, we can compute the 0th associated polynomials

$$Q_{n+k}^{(0)}(x) = \left(\frac{k\mu}{\lambda}\right)^{n/2} \left[Q_k^{(0)}(x)U_n\left(\frac{\lambda+k\mu-x}{2\sqrt{k\lambda\mu}}\right) - \sqrt{\frac{k\mu}{\lambda}}Q_{k-1}^{(0)}(x)U_{n-1}\left(\frac{\lambda+k\mu-x}{2\sqrt{k\lambda\mu}}\right)\right], \quad n \geq 0.$$

The case of one single server ($k = 1$) was studied in Example 3.35 (conservative case, see (3.99)). The case of two servers ($k = 2$) is a little bit more complicated. The explicit expressions of $Q_n, Q_n^{(0)}$ for $n = 0, 1, 2,$ are given by

$$Q_0(x) = 1, \quad Q_1(x) = \frac{\lambda - x}{\lambda}, \quad Q_2(x) = \frac{(\lambda+\mu-x)(\lambda-x)}{\lambda^2} - \frac{\mu}{\lambda},$$

$$Q_0^{(0)}(x) = 0, \quad Q_1^{(0)}(x) = -\frac{1}{\lambda}, \quad Q_2^{(0)}(x) = -\frac{\lambda+\mu-x}{\lambda^2}.$$

With this we can compute directly the functions $L_2(z)$ and $H_2(z)$ in (3.102), given by

$$L_2(z) = \frac{2\mu}{\lambda^2}\left[(\lambda - z)(2\mu - \lambda - 2z) + \lambda\sqrt{(\lambda+2\mu-z)^2 - 8\lambda\mu}\right],$$

$$H_2(z) = \frac{4\mu z}{\lambda^2}[z^2 - (2\lambda+\mu)z + \lambda(\lambda+2\mu)].$$

Therefore,

$$B(z;\psi) = -\frac{(\lambda - z)(2\mu - \lambda - 2z) + \lambda\sqrt{(\lambda+2\mu-z)^2 - 8\lambda\mu}}{2z[z^2 - (2\lambda+\mu)z + \lambda(\lambda+2\mu)]}.$$

Hence, following (3.103), the absolutely continuous part of the spectral measure ψ is given by

$$\frac{\lambda\sqrt{8\lambda\mu - (\lambda+2\mu-x)^2}}{2\pi x[x^2 - (2\lambda+\mu)x + \lambda(\lambda+2\mu)]}, \quad |\lambda + 2\mu - x| \leq \sqrt{8\lambda\mu}.$$

Observe that the support of this continuous part is given by the interval $[(\sqrt{\lambda} - \sqrt{2\mu})^2, (\sqrt{\lambda} + \sqrt{2\mu})^2]$. The spectral measure can also have localized jumps at the zeros of the denominator of the previous expression. A simple calculation shows that these zeros are given by

$$s_0 = 0, \quad s_1 = \frac{2\lambda+\mu}{2} - \frac{1}{2}\sqrt{\mu(\mu - 4\lambda)}, \quad s_2 = \frac{2\lambda+\mu}{2} + \frac{1}{2}\sqrt{\mu(\mu-4\lambda)}.$$

The size of the jump at $x = s_0 = 0$ is given by

$$\psi(\{0\}) = \lim_{\varepsilon \downarrow 0} \varepsilon \mathrm{Im} B(i\varepsilon;\psi) = \frac{(2\mu - \lambda) + \sqrt{(2\mu-\lambda)^2}}{2(2\mu+\lambda)} = \begin{cases} 0, & \text{if } 2\mu \leq \lambda, \\ \dfrac{2\mu - \lambda}{2\mu+\lambda}, & \text{if } 2\mu > \lambda. \end{cases}$$

The other two zeros s_1, s_2 are real as long as $\mu \geq 4\lambda$. Since $B(z;\psi)$ cannot have poles that are not real (by definition of the transform), we can assume that $\mu \geq 4\lambda$. When $\mu = 4\lambda$ there is a double zero at $z = 3\lambda$. However, in this case it is easy to see that the numerator of $B(z;\psi)$ also has a double zero at $z = 3\lambda$, so there would be no jump.

If $\mu > 4\lambda$, the size of the jump at $x = s_2$ is given by

$$\begin{aligned}
\psi(\{s_2\}) &= \lim_{\varepsilon \downarrow 0} \varepsilon \operatorname{Im} B(s_2 + i\varepsilon;\psi) \\
&= -\frac{(\lambda - s_2)(2\mu - \lambda - 2s_2) + \lambda\sqrt{(\lambda + 2\mu - s_2)^2 - 8\lambda\mu}}{2s_2(s_1 - s_2)} \\
&= \frac{\frac{\lambda}{2}\left[\mu - 3\sqrt{\mu(\mu - 4\lambda)} - |\mu - 3\sqrt{\mu(\mu - 4\lambda)}|\right]}{\sqrt{\mu(\mu - 4\lambda)}\left[2\lambda + \mu + \sqrt{\mu(\mu - 4\lambda)}\right]} \\
&= \begin{cases} 0, & \text{if} \quad \mu \geq 3\sqrt{\mu(\mu - 4\lambda)}, \\ \dfrac{\lambda\left[3\sqrt{\mu(\mu - 4\lambda)} - \mu\right]}{\sqrt{\mu(\mu - 4\lambda)}\left[2\lambda + \mu + \sqrt{\mu(\mu - 4\lambda)}\right]}, & \text{if} \quad \mu < 3\sqrt{\mu(\mu - 4\lambda)}. \end{cases}
\end{aligned}$$

The condition $\mu < 3\sqrt{\mu(\mu - 4\lambda)}$ is equivalent to $\mu/\lambda > 9/2$. A similar calculation shows that there is never a jump at the point $x = s_1$ because it does not meet the condition that $\mu > 4\lambda$. If $\mu > 4\lambda$ we have $s_2 < (\sqrt{\lambda} - \sqrt{2\mu})^2$ (that is, outside the support of the absolutely continuous part of the spectral measure), and the equality is obtained precisely for $\mu = 4\lambda$, in which case there is no jump, as we have already seen. Therefore the spectral measure is given by

$$\begin{aligned}
\psi(x) = &\frac{\lambda\sqrt{8\lambda\mu - (\lambda + 2\mu - x)^2}}{2\pi x[x^2 - (2\lambda + \mu)x + \lambda(\lambda + 2\mu)]}\mathbf{1}_{\{|\lambda + 2\mu - x| \leq \sqrt{8\lambda\mu}\}} + \frac{2\mu - \lambda}{2\mu + \lambda}\delta_0(x)\mathbf{1}_{\{2\mu > \lambda\}} \\
&+ \frac{\lambda\left[3\sqrt{\mu(\mu - 4\lambda)} - \mu\right]}{\sqrt{\mu(\mu - 4\lambda)}\left[2\lambda + \mu + \sqrt{\mu(\mu - 4\lambda)}\right]}\delta_{s_2}(x)\mathbf{1}_{\{2\mu > 9\lambda\}}.
\end{aligned} \tag{3.105}$$

In [86, Sections 6, 7 and 8] we can find the spectral analysis or some probability distributions of various random quantities associated with these processes, such as the distribution of waiting time of a customer arriving at time t, the distribution of the length of a busy period and other related probabilistic problems (see also the Exercises at the end of this chapter). ◊

Example 3.38 *Linear birth–death process.* This example was studied for the first time in the context of orthogonal polynomials in [85] (see also [1, Section 8.8.4]). In this case it is assumed that the birth–death rates are specific linear functions (not all cases), as follows:

$$\lambda_n = (n + \beta)\lambda, \quad \mu_n = \mu n, \quad n \geq 0, \quad \lambda, \mu, \beta > 0.$$

The infinitesimal operator is then conservative and is given by

$$\mathcal{A} = \begin{pmatrix} -\lambda\beta & \beta\lambda & & & \\ \mu & -(\beta+1)\lambda-\mu & (\beta+1)\lambda & & \\ & 2\mu & -(\beta+2)\lambda-2\mu & (\beta+2)\lambda & \\ & & \ddots & \ddots & \ddots \end{pmatrix}.$$

The factor λn (μn) represents the growth (reduction) of the population due to the current size of the population, while the constant factor $\lambda\beta$ may be interpreted as the increase of the state of the system due to an external source. The potential coefficients are given by

$$\pi_j = \frac{(\beta)_j(\lambda/\mu)^j}{j!}, \quad j \geq 0.$$

The three-term recurrence relation satisfied by the polynomials $Q_n(x)$ is

$$\begin{aligned}(n+\beta)\lambda Q_{n+1}(x) &- ((n+\beta)\lambda+\mu n)Q_n(x)\\ &+ \mu n Q_{n-1}(x) = -xQ_n(x), \quad Q_0(x)=1, \quad Q_{-1}(x)=0. \end{aligned} \tag{3.106}$$

There are several cases, depending on the values of λ, μ.

Case 1 ($\lambda < \mu$). Dividing the recurrence relation by λ and calling $c = \lambda/\mu < 1$, we have that

$$(n+\beta)Q_{n+1}(x) - \left[n + \frac{n}{c} + \beta\right]Q_n(x) + \frac{n}{c}Q_{n-1}(x) = -\frac{x}{\mu-\lambda}\left(\frac{1-c}{c}\right)Q_n(x).$$

This recurrence relation is identified by the same recurrence relation associated with the Meixner polynomials in (1.109) if we take

$$Q_n(x) = M_n\left(\frac{x}{\mu-\lambda};\beta,\frac{\lambda}{\mu}\right), \quad n \geq 0.$$

Since the Meixner polynomials are orthogonal with respect to the normalized discrete measure $\psi_x = (1-c)^\beta \frac{(\beta)_x}{x!}c^x, x = 0,1,\ldots,$ then the polynomials $Q_n(x)$ are orthogonal with respect the same measure but evaluated at the points $(\mu-\lambda)n, n \geq 0$, that is,

$$\psi(x) = \sum_{n=0}^{\infty}(1-c)^\beta \frac{(\beta)_n}{n!}c^n\delta_{x_n}, \quad x_n = (\mu-\lambda)n. \tag{3.107}$$

The Karlin–McGregor integral representation is then, using the duality of the Meixner polynomials and the generating function (1.108) ($i \leq j$),

$$
\begin{aligned}
P_{ij}(t) &= \pi_j \sum_{n=0}^{\infty} e^{-t(\mu-\lambda)n} Q_i((\mu-\lambda)n) Q_j((\mu-\lambda)n)(1-c)^{\beta} \frac{(\beta)_n}{n!} c^n \\
&= \pi_j \sum_{n=0}^{\infty} e^{-t(\mu-\lambda)n} M_i\left(n;\beta,\frac{\lambda}{\mu}\right) M_j\left(n;\beta,\frac{\lambda}{\mu}\right)\left(1-\frac{\lambda}{\mu}\right)^{\beta} \frac{(\beta)_n}{n!}\left(\frac{\lambda}{\mu}\right)^n \\
&= \pi_j \left(1-\frac{\lambda}{\mu}\right)^{\beta} \sum_{n=0}^{\infty} M_n\left(i;\beta,\frac{\lambda}{\mu}\right) M_n\left(j;\beta,\frac{\lambda}{\mu}\right) \frac{(\beta)_n}{n!}\left(\frac{\lambda}{\mu} e^{-(\mu-\lambda)t}\right)^n \\
&= \left(\frac{\lambda}{\mu}\right)^j \left(1-\frac{\lambda}{\mu}\right)^{\beta} \frac{(1-e^{-(\mu-\lambda)t})^{i+j}}{(1-\frac{\lambda}{\mu}e^{-(\mu-\lambda)t})^{i+j+\beta}} \\
&\quad \times \sum_{k=0}^{i} \binom{i}{k} (-1)^k \left(\frac{1-\frac{\mu}{\lambda}e^{-(\mu-\lambda)t}}{1-e^{-(\mu-\lambda)t}}\right)^k \left(\frac{1-\frac{\lambda}{\mu}e^{-(\mu-\lambda)t}}{1-e^{-(\mu-\lambda)t}}\right)^k \frac{(i+\beta)_{j-k}}{(j-k)!}.
\end{aligned}
$$

For $i>j$ we use that $\pi_i P_{ij}(t)=\pi_j P_{ji}(t)$. The dual process is also related to the Meixner polynomials. Indeed, the birth–death rates are given by

$$
\lambda_n^d = (n+1)\mu, \quad \mu_n^d = (n+\beta-1)\lambda, \quad n \geq 0, \quad \mu > \lambda > 0, \beta > 0.
$$

The points where the dual spectral measure is supported are given by $(\mu-\lambda)(n+1)$, $n \geq 0$ (see Theorem 3.14), while the corresponding dual polynomials are given by (see (3.49))

$$
Q_n^d(x) = \frac{(\beta)_n}{n!}\left(\frac{\lambda}{\mu}\right)^n M_n\left(\frac{x}{\mu-\lambda}-1;\beta,\frac{\lambda}{\mu}\right).
$$

The transition function of the dual process $P_{ij}^d(t)$ can be computed using Exercise 3.2.

Case 2 ($\lambda > \mu$). In this case we multiply (3.106) by λ^n/μ^{n+1} and call $c=\mu/\lambda<1$. Then

$$
\begin{aligned}
(n+\beta)\left(\frac{\lambda}{\mu}\right)^{n+1} Q_{n+1}(x) - \left(\frac{\lambda}{\mu}\right)^n \left[n+\frac{n}{c}+\beta\right] Q_n(x) + \frac{n}{c}\left(\frac{\lambda}{\mu}\right)^{n-1} Q_{n-1}(x) \\
= -\left(\frac{\lambda}{\mu}\right)^n \left(\frac{x}{\lambda-\mu}-\beta\right)\left(\frac{1-c}{c}\right) Q_n(x).
\end{aligned}
$$

Again, this recurrence relation is identified with the Meixner polynomials in (1.109) if we take

$$
Q_n(x) = \left(\frac{\mu}{\lambda}\right)^n M_n\left(\frac{x}{\lambda-\mu}-\beta;\beta,\frac{\mu}{\lambda}\right).
$$

The polynomials $Q_n(x)$ are orthogonal with respect to the Meixner measure but evaluated at the points $(\lambda-\mu)(n+\beta), n \geq 0$, that is,

$$\psi(x) = \sum_{n=0}^{\infty}(1-c)^{\beta}\frac{(\beta)_n}{n!}c^n\delta_{x_n}, \quad x_n = (\lambda-\mu)(n+\beta). \tag{3.108}$$

As before, the Karlin–McGregor integral representation is given by

$$\begin{aligned}P_{ij}(t) &= \pi_j \sum_{n=0}^{\infty} e^{-t(\lambda-\mu)(n+\beta)} Q_i((\lambda-\mu)(n+\beta))Q_j((\lambda-\mu)(n+\beta))(1-c)^{\beta}\frac{(\beta)_n}{n!}c^n \\ &= \pi_j\left(\frac{\mu}{\lambda}\right)^{i+j}\left[\left(1-\frac{\mu}{\lambda}\right)e^{-(\lambda-\mu)t}\right]^{\beta} \\ &\quad\times \sum_{n=0}^{\infty} M_n\left(i;\beta,\frac{\mu}{\lambda}\right)M_n\left(j;\beta,\frac{\mu}{\lambda}\right)\frac{(\beta)_n}{n!}\left(\frac{\mu}{\lambda}e^{-(\lambda-\mu)t}\right)^n,\end{aligned}$$

and again we can use the generating function (1.108) to compute an explicit expression. The dual process is also related to the Meixner polynomials. Indeed, the birth–death rates are given by

$$\lambda_n^d = (n+1)\mu, \quad \mu_n^d = (n+\beta-1)\lambda, \quad n \geq 0, \quad \lambda > \mu > 0, \beta > 0. \tag{3.109}$$

The points where the dual spectral measure is supported are given by $(\lambda-\mu)(n+\beta-1), n \geq 0$ (see Theorem 3.14), while the corresponding dual polynomials are given by (see (3.49))

$$Q_n^d(x) = \frac{(\beta)_n}{n!}M_n\left(\frac{x}{\lambda-\mu}-\beta+1;\beta,\frac{\mu}{\lambda}\right).$$

The transition function of the dual process $P_{ij}^d(t)$ can be computed using Exercise 3.2.

Case 3 ($\lambda = \mu$). The recurrence relation is in this case

$$(n+\beta)\lambda Q_{n+1}(x) - (2n+\beta)\lambda Q_n(x) + \lambda n Q_{n-1}(x) = -xQ_n(x).$$

A simple computation, using the recurrence relation for the Laguerre polynomials in (1.74), gives that

$$Q_n(x) = \frac{n!}{(\beta)_n}L_n^{(\beta-1)}\left(\frac{x}{\lambda}\right).$$

The associated spectral measure is a normalization of the Laguerre measure, that is,

$$\psi(x) = \frac{1}{\Gamma(\beta)\lambda^{\beta}}x^{\beta-1}e^{-x/\lambda}. \tag{3.110}$$

Therefore the Karlin–McGregor formula is given by

$$\begin{aligned}P_{ij}(t) &= \pi_j\int_0^{\infty} e^{-xt}Q_i(x)Q_j(x)\frac{1}{\Gamma(\beta)\lambda^{\beta}}x^{\beta-1}e^{-x/\lambda}dx \\ &= \frac{i!}{(\beta)_i\Gamma(\beta)\lambda^{\beta}}\int_0^{\infty} e^{-x(t+1/\lambda)}L_i^{(\beta-1)}\left(\frac{x}{\lambda}\right)L_j^{(\beta-1)}\left(\frac{x}{\lambda}\right)x^{\beta-1}dx.\end{aligned}$$

The last integral expression can be explicitly computed using (1.76), giving

$$P_{ij}(t) = \frac{(\beta+i)_j}{\lambda^\beta j!} \frac{t^{i+j}}{(t+1/\lambda)^{i+j+\beta}} \times {}_2F_1\left(\begin{matrix} -i, -j \\ -i-j-\beta+1 \end{matrix} ; 1 - \frac{1}{(t\lambda)^2} \right),$$

where ${}_2F_1$ is the Gauss hypergeometric function (1.4). The dual process is also related to the Laguerre polynomials. Indeed, the birth–death rates are given by

$$\lambda_n^d = (n+1)\lambda, \quad \mu_n^d = (n+\beta-1)\lambda, \quad n \geq 0, \quad \lambda, \beta > 0.$$

The spectral measure is the Laguerre measure but taking β instead of $\beta - 1$ (see Theorem 3.14), i.e.

$$\psi^d(x) = \frac{1}{\Gamma(\beta+1)\lambda^{\beta+1}} x^\beta e^{-x/\lambda}, \quad \beta > 0.$$

The corresponding dual polynomials are given by (see (3.49))

$$Q_n^d(x) = L_n^{(\beta)}\left(\frac{x}{\lambda}\right).$$

Figure 3.3 shows some trajectories of the linear birth–death process for different values of λ, μ and β. ◊

Example 3.39 *Linear birth–death process II*. The linear birth–death process from Example 3.38 does not cover all possible cases. An extension of this case is the process with rates

$$\lambda_n = \lambda(n+\alpha+c+1), \quad \mu_{n+1} = n+c+1, \quad n \geq 0, \quad \mu_0 = 0 \text{ or } c, \quad \lambda, c, \alpha > 0.$$

If $c = 0$ we have the Laguerre polynomials when $\lambda = 1$ and the Meixner polynomials when $\lambda \neq 1$. If $\lambda = 1, c > 0$, the polynomials are the so-called *associated Laguerre polynomials* (not to be confused with the kth associated polynomials). This and the case $\lambda \neq 1, c > 0$, were treated in [76].

We will give an alternative way of computing the spectral measure ψ of this birth–death process when $\lambda = 1, c > 0$. Consider the generating function of the transition probabilities $P_{mn}(t)$, given by

$$P_m(t,s) = \sum_{n=0}^{\infty} P_{mn}(t) s^n.$$

Using the notation $F_n(x) = \pi_n Q_n(x)$ and the Karlin–McGregor integral representation we obtain

$$\pi_m P_m(t,s) = \int_0^\infty e^{-xt} F_m(x) F(x,s)\, d\psi(x), \tag{3.111}$$

where $F(x,s)$ is given by

$$F(x,s) = \sum_{n=0}^{\infty} F_n(x) s^n.$$

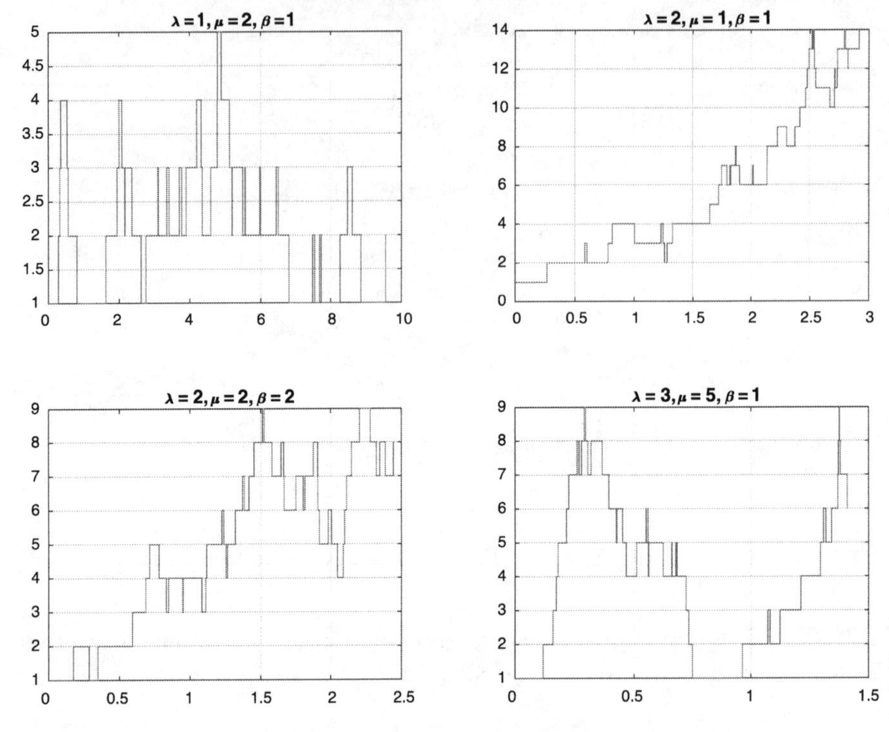

Figure 3.3 Trajectories of the linear birth–death process for different values of λ, μ and β starting at $X_0 = 1$ with state space $\mathcal{S} = \mathbb{N}$.

Using the Kolmogorov forward equation for $P_{mn}(t)$, multiplying by s^n and summing over n, it turns out that $P_m(t,s)$ satisfies the following partial differential equation (valid for any birth–death process):

$$\frac{\partial}{\partial t} P_m(t,s) = (1-s)\left[\frac{1}{s}\mu_\delta - \lambda_\delta\right] P_m(t,s)$$
$$+ \left[\widetilde{\mu}_0 \frac{s-1}{s} - \mu_0\right] P_{m,0}(t), \quad \delta = s\frac{\partial}{\partial s}, \quad \widetilde{\mu}_0 = \lim_{n \downarrow 0} \mu_n.$$

Using the Karlin–McGregor formula and (3.111) we get for $F(x,s)$ that

$$\left[(1-s)\left(\frac{1}{s}\mu_\delta - \lambda_\delta\right) + x\right] F(x,s) = \mu_0 + \widetilde{\mu}_0 \frac{1-s}{s}.$$

Using the definition of λ_n and μ_n for this example, we have

$$s(1-s)^2 \frac{\partial F}{\partial s} + [(1-s)(c - s(c+\alpha+1)) + xs]F = c(1-s)^\eta,$$

where

$$\eta = \begin{cases} 0, & \text{if} \quad \mu_0 = c, \\ 1, & \text{if} \quad \mu_0 = 0. \end{cases}$$

In addition, we have the boundary condition $\int_0^\infty F(x,s)d\psi(x) = 1$. This first-order partial differential equation can be solved using the method of constant variations, whose general bounded solution at $s = 0$ is given by

$$F(x,s) = \frac{1}{s^c(1-s)^{\alpha+1}} e^{-\frac{x}{1-s}} \left[C + c \int_a^s u^{c-1}(1-u)^{\eta+\alpha-1} e^{\frac{x}{1-u}} \, du \right].$$

The initial conditions $F(x,0) = 1$ give $C = 0$ and $a = 0$. Therefore the solution can be written as

$$F(x,s) = \frac{c}{s^c(1-s)^{\alpha+1}} \int_0^s u^{c-1}(1-u)^{\eta+\alpha-1} \exp\left[\frac{xu}{1-u} - \frac{xs}{1-s}\right] du.$$

By the change of variables

$$z = \frac{s}{1-s}, \quad v = \frac{u}{1-u},$$

the integral representation is then

$$F(x, z/(1+z)) = \frac{c(1+z)^{c+\alpha+1}}{z^c} \int_0^z \frac{v^{c-1}}{(1+v)^{\eta+\alpha+c}} e^{-x(z-v)} dv.$$

Integrating both sides with respect to the spectral measure ψ and using that $\int_0^\infty F(x,s)d\psi(x) = 1$, we have that

$$\frac{z^c}{(1+z)^{c+\alpha+1}} = c \int_0^\infty \left[\int_0^z \frac{v^{c-1}}{(1+v)^{\eta+\alpha+c}} e^{-x(z-v)} dv \right] d\psi(x). \tag{3.112}$$

We observe that the internal integral of the right-hand part of the previous formula is the convolution of two functions. Indeed,

$$\int_0^z \frac{v^{c-1}}{(1+v)^{\eta+\alpha+c}} e^{-x(z-v)} dv = \int_0^z f(v)g(z-v)\, dv = (f * g)(z),$$

where

$$f(v) = v^{c-1}(1+v)^{-\eta-\alpha-c}, \quad g(v) = e^{-xv}.$$

Therefore, applying the Laplace transform to (3.112) and using the convolution theorem, we get that

$$\int_0^\infty \frac{d\psi(x;\eta)}{x+p} = \frac{\Psi(c+1, 1-\alpha; p)}{\Psi(c, 1-\alpha-\eta; p)}, \quad \eta = 0, 1,$$

where

$$\Psi(a,b;x) = \frac{1}{\Gamma(a)}\int_0^\infty e^{-xt}t^{a-1}(1+t)^{b-a-1}dt, \quad \mathrm{Re}(a) > 0, \quad -\frac{\pi}{2} < \arg x < \frac{\pi}{2}. \tag{3.113}$$

This function Ψ is known in the theory of special functions as the *Tricomi function* (see [52, vol. 1, Section 6.5]), which is one of the solutions of the *confluent differential equation* at the origin

$$x\frac{d^2y}{dx^2} + (b-x)\frac{dy}{dx} - ay = 0.$$

Using some special functions techniques, and the Perron–Stieltjes inversion formula, it is possible to prove that the spectral measure ψ is absolutely continuous when $c \geq 0$ and $\alpha > -1$, in which case it is given by

$$\psi(x;\eta) = \frac{x^\alpha e^{-x}}{\Gamma(c+1)\Gamma(c+\alpha++1)}|\Psi(c,1-\alpha-\eta;xe^{-i\pi})|^{-2}. \qquad \Diamond$$

Remark 3.40 A similar analysis to the last example for the case of quadratic birth–death rates was studied in [77], which is related to the continuous dual Hahn polynomials (see Section 1.6). In [73] the author studied one example of a queueing model related to q-polynomials (see [74] for more information about these families of orthogonal polynomials). Also, in [123] a birth–death process was related to the Rogers–Ramanujan continued fraction. Finally, many examples related to orthogonal polynomials, especially for q-polynomials, can be found in [128], using an approach coming from quantum mechanics. $\Diamond$

Example 3.41 *Linear birth–death process with killing.* If the killing rates in (3.84) are constant, i.e. $\gamma_n = \gamma$ for all $n \geq 0$, then by conditioning, the transition probability functions $P_{ij}(t)$ of the birth–death process with killing can simply be expressed in terms of the transition probability functions $\hat{P}_{ij}(t)$ of the process with the same birth–death rates, but zero killing rates. Namely

$$P_{ij}(t) = e^{-\gamma t}\hat{P}_{ij}(t), \quad i,j \in \mathbb{N}_0, \quad t \geq 0.$$

Therefore, interesting cases arise when the killing rates are state dependent. Consider the birth–death process with linear birth, death and killing rates of the form

$$\lambda_n = \lambda n + \theta, \quad \mu_n = n\mu, \quad \gamma_n = n\gamma, \quad n \in \mathbb{N}_0,$$

with $\lambda,\theta,\mu,\gamma > 0$. These rates satisfy (3.90), so the birth–death process is uniquely determined by its rates. The spectral representation of this example was studied for the first time by S. Karlin and S. Tavaré in [98] and later by E. van Doorn and A. Zeifman in [49].

The recurrence relation (3.85) becomes

$$Q_0(x) = 1, \quad -x = \theta Q_1(x) - \theta,$$
$$-xQ_n(x) = (\lambda n + \theta)Q_{n+1}(x) - (\theta + (\lambda + \mu + \gamma)n)Q_n(x) + \mu n Q_{n-1}(x), \quad n \geq 1.$$

Write

$$\beta = \frac{\theta}{\lambda}, \quad \rho = \sqrt{(\lambda + \mu + \gamma)^2 - 4\lambda\mu}, \quad \kappa = \theta\left(1 - \frac{2\mu}{\lambda + \mu + \gamma + \rho}\right),$$

and

$$P_n(x) = \left(\frac{2\lambda}{\lambda + \mu + \gamma - \rho}\right)^n Q_n(\rho x + \kappa), \quad n \geq 0.$$

After some computations we get that $(P_n)_n$ satisfies the recurrence relation

$$c(n + \beta)P_{n+1}(x) = ((c-1)x + (c+1)n + c\beta)P_n(x) - nP_{n-1}(x),$$

where

$$c = \frac{\lambda + \mu + \gamma - \rho}{\lambda + \mu + \gamma + \rho}.$$

We clearly see, from (1.109), that $P_n(x) = M_n(x; \beta, c)$, i.e. the Meixner polynomials, which are orthogonal with respect to the discrete Meixner measure $\psi_x = (1-c)^b \dfrac{(b)_x}{x!} c^x, x = 0, 1, \ldots$. Therefore, after some computations, the Karlin–McGregor representation formula (3.89) is given by

$$P_{ij}(t) = \left(\frac{\lambda}{\mu}\right)^j \frac{(\beta)_j}{j!}(1-c)^\beta \sum_{x=0}^{\infty} e^{-t(\rho x + \kappa)} Q_i(\rho x + \kappa)Q_j(\rho x + \kappa)\frac{c^x(\beta)_x}{x!}. \qquad \Diamond$$

3.7 Applications to the Probabilistic Aspects of Birth–Death Processes

In this section we will use the Karlin–McGregor integral representation formula (3.35) to study the probabilistic aspects associated with any birth–death process. As in the previous section, we will assume that if $\mu_0 = 0$, then $A + B = \infty$, meaning that the solution ψ of the Stieltjes moment problem is unique and therefore we will have a unique transition probability function $P_{ij}(t; \psi)$ satisfying both Kolmogorov equations, $P_{ij}(t) \geq 0$ and $\sum_{j=0}^{\infty} P_{ij}(t) \leq 1$. If $\mu_0 > 0$ a sufficient condition for uniqueness is that $\sum_{n=1}^{\infty} 1/\mu_n = \infty$ (see (3.83)). We will follow the same lines as in Chapter 2 but some of the techniques will be different. We will start with the concept of recurrence (ergodic, null or transient) for $\mu_0 = 0$ and how to study it in terms of the spectral measure ψ and other formulas. If $\mu_0 > 0$ then the process will always be transient. We will study which kind of absorption (certain, ergodic or transient) has the birth–death process in terms also of the spectral measure ψ and other formulas. In the second part we will study limit theorems, meaning that we will be concerned with the behavior of $P(t)$ as $t \to \infty$. The treatment of these properties will be different as in the case of discrete-time birth–death chains, due to the fact that the integrals involved will have more regular behavior. Finally, in the third part, we will study the concept of the decay parameter and its relationship with quasi-stationary distributions for birth–death processes.

3.7.1 Recurrence and Absorption

Let $\{X_t, t \geq 0\}$ be an irreducible birth–death process with a set of birth–death rates $\{(\lambda_n, \mu_n), n \geq 0\}$. Therefore all states belong to the same class of communication. In order to study recurrence we only need to study one single state, namely the state 0. The following result is a combination of [84, Theorems 1 and 2] and [41, Theorem 5.1]:

Theorem 3.42 *Let $\mu_0 = 0$ (i.e. $\mathcal{A}$ is conservative). Then the following are equivalent:*

1. *The birth–death process is recurrent.*
2. $\displaystyle\int_0^\infty \frac{d\psi(x)}{x} = \infty.$
3. $\displaystyle\sum_{n=0}^\infty \frac{1}{\lambda_n \pi_n} = \infty.$
4. $\displaystyle\int_0^\infty \frac{d\psi^{(0)}(x)}{x} = 1/\mu_1$, *where $\psi^{(0)}$ is the spectral measure of the* 0*th associated birth–death process.*

In addition, the following are equivalent:

1. *The birth–death process is positive recurrent.*
2. *The spectral measure ψ has a finite jump at $x = 0$ of size $\psi(\{0\}) = \left(\sum_{n=0}^\infty \pi_n\right)^{-1}$.*
3. $\displaystyle\sum_{n=0}^\infty \pi_n < \infty.$
4. $\displaystyle\int_0^\infty \frac{d\psi^{(0)}(x)}{x} = 1/\mu_1$ *and* $\displaystyle\int_0^\infty \frac{d\psi^{(0)}(x)}{x^2} < \infty.$

Proof Since the process is irreducible it is enough to study recurrence at a single state, for example 0. By Definition 3.6 state 0 is recurrent if and only if $\lim_{\lambda\downarrow 0} \widehat{F}_{00}(\lambda) = \int_0^\infty dF_{00}(t) = 1$. By (3.57) for $i = j = 0$, this is possible if and only if $\lim_{\lambda\downarrow 0} \int_0^\infty (\lambda+x)^{-1} d\psi(x) = \infty$. Therefore we must have $\int_0^\infty x^{-1} d\psi(x) = \infty$. For the third relation we use (3.69) and (3.42) for $\mu_0 = 0$. The last part is a consequence of (3.59) taking $\lambda \downarrow 0$ and $\mu_0 = 0$.

As for the positive recurrence, by definition, we need to see that the expected return time is finite. This is equivalent to $\int_0^\infty t dF_{00}(t) < \infty$. If $\sum_{n=0}^\infty \pi_n < \infty$ then ψ

has an atom at $x=0$ of size $\psi(\{0\}) = \left(\sum_{n=0}^{\infty}\pi_n\right)^{-1}$ (see (3.37)) and $\lim_{t\to\infty}P_{ij}(t) = \psi(\{0\})\pi_j$. The first moment of $F_{00}(t)$ is given by $-\lim_{\lambda\downarrow 0}\widehat{F}'_{00}(\lambda)$. Therefore, using (3.55), we get that

$$-\widehat{F}'_{00}(\lambda) = -\frac{\widehat{P}'_{00}(\lambda)}{(\lambda+\lambda_0)\widehat{P}^2_{00}(\lambda)} - \frac{1}{(\lambda+\lambda_0)^2\widehat{P}_{00}(\lambda)}.$$

The second term tends to 0 since the process is recurrent, and then $\widehat{P}_{00}(\lambda)\to\infty$ as $\lambda\downarrow 0$. On the one hand,

$$-\frac{\widehat{P}'_{00}(\lambda)}{\widehat{P}^2_{00}(\lambda)} = \frac{\displaystyle\int_0^\infty \frac{d\psi(x)}{(\lambda+x)^2}}{\left(\displaystyle\int_0^\infty \frac{d\psi(x)}{\lambda+x}\right)^2} = \frac{\displaystyle\frac{\psi(\{0\})}{\lambda^2} + \int_0^\infty \frac{d\psi^+(x)}{(\lambda+x)^2}}{\left(\displaystyle\frac{\psi(\{0\})}{\lambda} + \int_0^\infty \frac{d\psi^+(x)}{\lambda+x}\right)^2},$$

where ψ^+ denotes the measure obtained ψ removing the atom at $x=0$. Therefore ψ^+ is continuous at $x=0$. So multiplying and dividing by λ^2, we get that

$$\int_0^\infty t dF_{00}(t) = -\lim_{\lambda\downarrow 0}\widehat{F}'_{00}(\lambda) = \frac{1}{\lambda_0\psi(\{0\})},$$

which is finite. On the other hand, if the process is positive recurrent, then the Markov chain $P_{ii}(nh)$ with fixed h is positive recurrent. Then $\lim_{n\to\infty}P_{ii}(nh) = \psi(\{0\})\pi_j > 0$. Since $\psi(\{0\}) = \left(\sum_{n=0}^{\infty}\pi_n\right)^{-1}$ then $\sum_{n=0}^{\infty}\pi_n < \infty$. For the last part, we observe from (3.60) and (1.12) that

$$\begin{aligned}\psi(\{0\}) &= -\lim_{z\downarrow 0} zB(z;\psi) = \lim_{\lambda\downarrow 0}\frac{\lambda}{\lambda-\lambda_0+\lambda_0\mu_1\displaystyle\int_0^\infty\frac{d\psi^{(0)}(x)}{x+\lambda}}\\ &= \frac{1}{1+\lambda_0\mu_1\displaystyle\int_0^\infty\frac{d\psi^{(0)}(x)}{x^2}}.\end{aligned}$$

The last part is a consequence of L'Hôpital's rule. Therefore $\psi(\{0\}) > 0$ if and only if $\int_0^\infty x^{-2}d\psi^{(0)}(x) < \infty$. The last formula also gives an alternative way of computing $\psi(\{0\})$ in the case that the process is positive recurrent. □

Let us now give some comments about transient birth–death process ($\mu_0=0$). According to Theorem 3.42 this is equivalent to $\sum_{n=0}^{\infty}(\lambda_n\pi_n)^{-1} < \infty$ and $\sum_{n=0}^{\infty}\pi_n = \infty$. We will see that this kind of process tends to ∞ with probability 1, so the state ∞ is a permanent absorbing state. There are two types of transient birth–death process. The case when for some $t>0$ and some $i\in\mathbb{N}_0$, we have

$$\sum_{j=0}^{\infty}P_{ij}(t) < 1, \tag{3.114}$$

in which case it is called of type 1, and of type 2 if (3.114) is not satisfied. From the semigroup property and the Chapman–Kolmogorov equations, condition (3.114) is satisfied for all $i \in \mathbb{N}_0$. From Proposition 3.23 we obtain that a transient birth–death process is of type 2 if and only if $C = \infty$, where C is defined by (3.28). Recall that C is interpreted as the mean time to reach the state ∞ from any interior and finite state (see the end of Section 3.1).

Proposition 3.43 ([84, Theorem 8]) *For any transient birth–death process, we have*

$$\mathbb{P}\left(\lim_{t\to\infty} X_t = \infty | X_0 = i\right) = 1.$$

Proof Fix an initial state $i \in \mathbb{N}_0$ and let E_j denote the set of all trajectories for which the particle spends an infinite amount of time in state $j \in \mathbb{N}_0$. Since the time spent at j is exponentially distributed with parameter $\lambda_j + \mu_j$, the set E_j differs only by a set of measure zero from the set of all trajectories which visit j infinitely often. If $\mathbb{P}(E_j) > 0$ then $\int_0^\infty P_{ij}(t)dt = \infty$ because the expected occupation time of state j is infinite. By the Karlin–McGregor formula we then have that $\int_0^\infty Q_iQ_j d\psi/x = \infty$ and therefore $\sum_{n=0}^\infty (\lambda_n\pi_n)^{-1} = \infty$, i.e. the process is recurrent, contrary to the hypothesis. Hence $\mathbb{P}(E_j) = 0$ and $\mathbb{P}(\cup_j E_j) = 0$. If a trajectory of the process is not in $\cup_j E_j$ then $X_t \to \infty$ as $t \to \infty$ and we get the result. □

Let J_∞ be the time of the first explosion (see (3.7)) and call

$$H_i(t) = \mathbb{P}(J_\infty \le t | X_0 = i) = 1 - \sum_{j=0}^\infty P_{ij}(t).$$

For transient birth–death processes of type 2, we have that $\mathbb{P}(J_\infty = \infty | X_0 = i) = 1$ for all $i \in \mathbb{N}_0$. For processes of type 1, we have the following result:

Proposition 3.44 ([84, Theorem 9]) *If* $\mathbb{P}(J_\infty < \infty | X_0 = i) > 0$, *then* $\mathbb{P}(J_\infty < \infty | X_0 = i) = 1$ *and* $1 - H_i(t)$ *tends exponentially to zero as* $t \to \infty$.

Proof If $\mathbb{P}(J_\infty < \infty | X_0 = i) > 0$ then $1 - \sum_{j=0}^\infty P_{ij}(t) > 0$ and $C < \infty$. Then $-Q_n'(0)$ is bounded (see the statements **(a)** – **(c)** after (3.51)). Following the same steps as in the proof of Proposition 3.23, we obtain that

$$\sum_{j=0}^\infty P_{ij}(t) \le \frac{e^{-bt}}{Q_\infty(b)},$$

where $0 < b < \xi$, and $\xi = \inf \operatorname{supp}(\psi)$. Therefore $1 - H_i(t)$ tends exponentially to zero as $t \to \infty$ and we get the result. □

The following result gives the necessary and sufficient conditions for the existence of an invariant measure (see (3.19)):

Proposition 3.45 *Let $(v_j)_j$ be a nonnegative solution of the equation*

$$v_j = \sum_{i=0}^{\infty} v_i P_{ij}(t).$$

Then there exists a unique positive solution of the previous equation if and only if $C = \infty$, where C is defined by (3.28). In that case the solution is a constant multiple of $(\pi_j)_j$, i.e. the potential coefficients defined by (3.24).

Proof $C = \infty$ is equivalent to $\sum_{j=0}^{\infty} P_{ij}(t) = 1$, by Proposition 3.23. Then $v_j = \pi_j$ is a solution since $\pi_i P_{ij}(t) = \pi_j P_{ji}(t)$. Assume we have another solution $(v_j)_j$ different from $(\pi_j)_j$. Since $P_{ij}(t) > 0$ for all $t > 0$ (see Proposition 3.19) it follows that $v_j > 0$ and we can take $v_0 = 1$. Define

$$Q_{ij}(t) = \frac{v_j}{v_i} P_{ji}(t).$$

$Q_{ij}(t)$ defines the transition probabilities of a birth–death process with rates $\lambda_i' = v_{i+1}\mu_{i+1}/v_i$ and $\mu_i' = v_{i-1}\lambda_{i-1}/v_i, i \geq 1$. These new rates satisfy

$$\lambda_i' + \mu_i' = \frac{v_{i+1}}{v_i}\mu_{i+1} + \frac{v_{i-1}}{v_i}\lambda_{i-1}.$$

It can be verified by induction that the only solution of the previous equation is given by $v_j = \pi_j$. Conversely, if we have a solution $(v_j)_j$, then we obviously have $\sum_{j=0}^{\infty} Q_{ij}(t) = 1$ and again because of the previous equation we find that $v_j = \pi_j$. Then $P_{ij}(t) = Q_{ij}(t)$, so we get the result. □

Example 3.46 Consider Example 3.35 of the conservative $M/M/1$ queue ($\mu_0 = 0$). Then we are in the conditions of Theorem 3.42. Since we know the spectral measure (given by (3.99)), we have

$$\int_0^{\infty} \frac{d\psi(x)}{x} = \frac{1}{2\pi\lambda\mu} \int_{(\sqrt{\lambda}-\sqrt{\mu})^2}^{(\sqrt{\lambda}+\sqrt{\mu})^2} \frac{\sqrt{4\lambda\mu - (\lambda+\mu-x)^2}}{x^2} + \frac{\mu-\lambda}{\mu} \lim_{x\downarrow 0} \frac{1}{x} \mathbf{1}_{\{\mu>\lambda\}}.$$

If $\lambda \neq \mu$ then $(\sqrt{\lambda} - \sqrt{\mu})^2 > 0$ so the part with the integral is bounded. On the one hand, if $\lambda < \mu$, the discrete part is ∞. On the other hand, if $\lambda > \mu$ then the discrete part is 0. If $\lambda = \mu$ then the discrete part is 0 while the integral now is ∞. Therefore if $\lambda \leq \mu$ the birth–death process is recurrent. Otherwise it is transient. Another way to study recurrence is by looking to the convergence of the series

$$\sum_{n=0}^{\infty} \frac{1}{\lambda_n \pi_n} = \sum_{n=0}^{\infty} \frac{\mu^n}{\lambda^{n+1}} = \frac{1}{\lambda}\sum_{n=0}^{\infty}\left(\frac{\mu}{\lambda}\right)^n = \infty \quad \Leftrightarrow \quad \frac{\mu}{\lambda} \geq 1 \quad \Leftrightarrow \quad \mu \geq \lambda.$$

As for positive recurrence, according to Theorem 3.42, the spectral measure has a finite jump at $x = 0$ if and only if $\mu > \lambda$. This jump was already given in (3.99) but another way to compute it is

$$\psi(\{0\}) = \left(\sum_{n=0}^{\infty} \pi_n\right)^{-1} = \left(\sum_{n=0}^{\infty} \left(\frac{\lambda}{\mu}\right)^n\right)^{-1} = \left(\frac{\mu}{\mu - \lambda}\right)^{-1}$$

$$= \frac{\mu - \lambda}{\mu} > 0 \quad \Leftrightarrow \quad \mu > \lambda.$$

The only case when the conservative $M/M/1$ queue is null recurrent is for $\mu = \lambda$.

The same analysis can be made for the $M/M/2$ queue of Example 3.37 (see (3.105)). Now we have that if $\lambda \leq 2\mu$ then the $M/M/2$ queue is recurrent. Otherwise it is transient. Also, if $\lambda < 2\mu$ then the $M/M/2$ queue is positive recurrent. Otherwise, if $\lambda = 2\mu$, it is null recurrent. The same can be applied for the $M/M/k$ queue, i.e. the process will be recurrent if $\lambda \leq k\mu$. ◊

Example 3.47 Consider Example 3.36 of the $M/M/\infty$ queue. This process is always positive recurrent. On the one hand, from the spectral measure (3.100) (the Poisson distribution), we have

$$\int_0^{\infty} \frac{d\psi(x)}{x} = \sum_{n=0}^{\infty} \frac{e^{-\lambda/\mu}(\lambda/\mu)^n}{n!\,\mu n} = \infty.$$

On the other hand, the series

$$\sum_{n=0}^{\infty} \frac{1}{\lambda_n \pi_n} = \sum_{n=0}^{\infty} \frac{\mu^n n!}{\lambda^{n+1}} = \infty.$$

Therefore, Theorem 3.42 gives that the $M/M/\infty$ queue is recurrent. The same theorem also says that it is positive recurrent, since

$$\psi(\{0\}) = \left(\sum_{n=0}^{\infty} \pi_n\right)^{-1} = \left(\sum_{n=0}^{\infty} \frac{(\lambda/\mu)^n}{n!}\right)^{-1} = e^{-\lambda/\mu} > 0.$$

The previous value can also be computed by looking at the jump of the spectral measure (3.100) at $x = 0$. Therefore, the stationary distribution is given by

$$\pi = e^{-\lambda/\mu}\left(1, \frac{\lambda}{\mu}, \frac{\lambda^2}{2\mu^2}, \frac{\lambda^3}{3!\,\mu^3}, \cdots\right),$$

i.e. a Poisson distribution of parameter λ/μ. ◊

Example 3.48 Consider the linear birth–death process of Example 3.38, which is always conservative. There are three situations:

- If $\lambda < \mu$, then the process is always positive recurrent. From the spectral measure (3.107), we have

$$\int_0^{\infty} \frac{d\psi(x)}{x} = \sum_{n=0}^{\infty} (1-c)^{\beta} \frac{(\beta)_n}{n!\,(\mu - \lambda)n} c^n = \infty, \quad c = \lambda/\mu.$$

Also the series $\sum_{n=0}^{\infty}(\lambda_n \pi_n)^{-1}$ is divergent. Therefore, Theorem 3.42 gives that the process is recurrent. The same theorem also says that it is positive recurrent, since

$$\psi(\{0\}) = \left(\sum_{n=0}^{\infty} \pi_n\right)^{-1} = \left(1 - \frac{\lambda}{\mu}\right)^{\beta} > 0.$$

- If $\lambda > \mu$, then the process is always transient. It is clear from the expression of the spectral measure (3.108) that

$$\int_0^{\infty} \frac{d\psi(x)}{x} = \sum_{n=0}^{\infty} (1-c)^{\beta} \frac{(\beta)_n}{n!\,(\lambda-\mu)(n+\beta)} c^n \leq \frac{1}{\beta(\lambda-\mu)} < \infty, \quad c = \mu/\lambda.$$

- If $\lambda = \mu$, then the recurrence of the process depends on the parameter β. From the spectral measure (3.110), we have

$$\int_0^{\infty} \frac{d\psi(x)}{x} = \frac{1}{\Gamma(\beta)\lambda^{\beta}} \int_0^{\infty} x^{\beta-2} e^{-x/\lambda} = \begin{cases} \infty, & \text{if} \quad 0 < \beta \leq 1, \\ \dfrac{1}{\lambda(\beta-1)}, & \text{if} \quad \beta > 1. \end{cases}$$

Therefore Theorem 3.42 gives that the process is recurrent if $0 < \beta \leq 1$. Otherwise, if $\beta > 1$, it is transient. The process is always null recurrent, since the spectral measure does not have a finite jump at $x = 0$. Another way to see this is checking that $\sum_{n=0}^{\infty} \pi_n = \infty$, so the invariant measure can never be a proper distribution. ◊

Example 3.49 The examples with a finite number of states, such as the $M/M/1/N$ queue with reflecting barriers (see Exercise 3.3), the Ehrenfest and Bernoulli–Laplace urn models (see Example 3.33) and the genetic model of Moran (see Example 3.34), are always positive recurrent. This is due to the fact that we can always compute an invariant distribution of the birth–death process. Additionally we observe that the spectral measures associated with these examples always have a finite jump at $x = 0$ (since 0 is always an eigenvalue of $\mathcal{A}$). The rate of convergence to the stationary distributions is of the order of magnitude $\exp(-\lambda_1 t)$, where λ_1 is the smallest positive point in the spectrum of the spectral measure. ◊

Let us now study the case where -1 is an absorbing state, that is, $\mu_0 > 0$, i.e. there is a positive probability that the birth–death process is absorbed at the state -1. In this case the process is always transient.

Definition 3.50 Let $\{X_t, t \geq 0\}$ be a birth–death process with $\mu_0 > 0$.

- We say that absorption at -1 is *certain* if the probability of eventual absorption equals 1, i.e.

$$\mathbb{P}(T_{-1} < \infty | X_0 = i) = 1, \quad \text{for all} \quad i \in \mathbb{N}_0,$$

where T_{-1} is defined by (3.29). If, additionally, the expected time to absorption is finite, i.e. $\mathbb{E}(T_{-1}|X_0 = i) < \infty$ for all $i \in \mathbb{N}_0$, then absorption at -1 is *ergodic*.

- We say that absorption at -1 is *transient* if it is not certain, i.e.

$$\mathbb{P}(T_{-1} < \infty | X_0 = i) < 1, \quad \text{for some} \quad i \in \mathbb{N}_0. \qquad \diamond$$

Theorem 3.51 ([84, Section 5]) *Let $\mu_0 > 0$. Then the following are equivalent:*

1. *Absorption at -1 is certain.*
2. $\mu_0 \int_0^\infty \frac{d\psi(x)}{x} = 1.$
3. $\sum_{n=0}^{\infty} \frac{1}{\lambda_n \pi_n} = \infty.$
4. *$Q_n(0) \to \infty$ as $n \to \infty$.*

In addition, the following are equivalent:

1. *Absorption at -1 is ergodic.*
2. $\mu_0 \int_0^\infty \frac{d\psi(x)}{x} = 1$ *and* $\int_0^\infty \frac{d\psi(x)}{x^2} < \infty.$
3. $\sum_{n=0}^{\infty} \pi_n < \infty.$

Proof By irreducibility we can focus only on the behavior at state 0. For any state $i \in \mathbb{N}_0$, let $G_i(t) = \mathbb{P}(T_{-1} \leq t | X_0 = i)$, i.e. the probability that, starting at i, absorption occurs before time t. An argument similar to the one we saw in (3.58) implies that

$$\begin{aligned} G_i(t) &= \mu_0 \int_0^t P_{i0}(s)\, ds = \mu_0 \int_0^t ds \int_0^\infty e^{-xs} Q_i(x)\, d\psi(x) \\ &= \mu_0 \int_0^\infty \frac{1 - e^{-xt}}{x} Q_i(x)\, d\psi(x). \end{aligned} \tag{3.115}$$

Therefore the density of $G_i(t)$ can be written as $dG_i(t) = \mu_0 P_{i0}(t)\, dt$. To compute the eventual absorption probability, we apply the Laplace transform to this density. Indeed,

$$\begin{aligned} \widehat{G}_i(\lambda) &= \int_0^\infty e^{-\lambda t} dG_i(t) = \mu_0 \int_0^\infty dt \int_0^\infty e^{-t(\lambda + x)} Q_i(x)\, d\psi(x) \\ &= \mu_0 \int_0^\infty \frac{Q_i(x)}{\lambda + x}\, d\psi(x). \end{aligned}$$

The eventual absorption probability is given then by (the formula for $\int_0^\infty Q_i(x)x^{-1}d\psi(x)$ can be derived in a similar way to (2.59))

$$a_i = \mathbb{P}(T_{-1} < \infty | X_0 = i) = \int_0^\infty dG_i(t) = \lim_{\lambda \downarrow 0} \widehat{G}_i(\lambda) = \mu_0 \int_0^\infty \frac{Q_i(x)}{x}\, d\psi(x)$$

$$= \frac{\mu_0 \sum_{n=i}^\infty \frac{1}{\lambda_n \pi_n}}{1 + \mu_0 \sum_{n=0}^\infty \frac{1}{\lambda_n \pi_n}} = 1 - \frac{Q_i(0)}{\lim_{n\to\infty} Q_n(0)}. \tag{3.116}$$

Absorption at -1 is certain is equivalent to $a_0 = 1$, which is equivalent to $\mu_0 \int_0^\infty x^{-1} d\psi(x) = 1$, which is equivalent to $\sum_{n=0}^\infty (\lambda_n \pi_n)^{-1} = \infty$ and finally to $Q_n(0) \to \infty$ as $n \to \infty$.

For the second part, the expected return time starting from state i is given by

$$\mathbb{E}(T_{-1} | X_0 = i) = \int_0^\infty t\, dG_i(t) = -\lim_{\lambda \downarrow 0} \frac{d}{d\lambda} \widehat{G}_i(\lambda) = \mu_0 \lim_{\lambda \downarrow 0} \int_0^\infty \frac{Q_i(x)}{(\lambda + x)^2}\, d\psi(x)$$

$$= \mu_0 \int_0^\infty \frac{Q_i(x)}{x^2}\, d\psi(x).$$

In a similar way we can compute all the moments of $dG_i(t)$. Therefore this integral is finite if and only if the integral $\int_0^\infty x^{-2} Q_i(x)\, d\psi(x)$ is finite and in particular for $i = 0$. Now, from Theorem 3.11 (since we are in the situation of $\mu_0 > 0$), the dual polynomials defined by (3.40) are orthogonal with respect to ψ^d defined by (3.48). Since absorption is certain, we have that $\mu_0 \int_0^\infty x^{-1} d\psi(x) = 1$, so ψ^d does not have a jump at $x = 0$. Therefore,

$$\int_0^\infty \frac{d\psi^d(x)}{x} = \mu_0 \int_0^\infty \frac{d\psi(x)}{x^2} = \mathbb{E}(T_{-1} | X_0 = 0).$$

Alternatively, we can apply Theorem 3.42 to the dual measure ψ^d. Therefore we have

$$\int_0^\infty \frac{d\psi^d(x)}{x} = \sum_{n=0}^\infty \frac{1}{\lambda_n^d \pi_n^d} = \frac{1}{\mu_0} \sum_{n=0}^\infty \frac{\lambda_{n-1}\mu_{n-1}}{\mu_n} = \frac{1}{\mu_0} \sum_{n=0}^\infty \pi_n.$$

Hence $\sum_{n=0}^\infty \pi_n < \infty$. In general, we have that (see [84])

$$\mathbb{E}(T_{-1} | X_0 = i) = \sum_{j=0}^{i} \frac{1}{\mu_j \pi_j} \sum_{n=j}^\infty \pi_n. \tag{3.117}$$

□

Remark 3.52 As in the case of discrete-time birth–death chains, recurrence and absorption are dual properties if we consider the relation between a birth–death process and its dual. For instance, if the original birth–death process has a certain (not ergodic) absorption at -1, then the dual birth–death process is null recurrent. The same can be applied to the remaining cases either if $\mu_0 > 0$ or $\mu_0 = 0$. ◊

An interesting random variable in the context of birth–death processes (assuming that absorption is certain) is the number of transitions that occur before absorption. Let N be that random variable and denote

$$R_i^n = \mathbb{P}(N = n|\, X_0 = i), \quad i \geq 0, \quad n \geq 1. \tag{3.118}$$

Recalling the jump chain associated with the birth–death process, defined by (3.26), with probabilities $p_n = \lambda_n/(\lambda_n + \mu_n)$ and $q_n = \mu_n/(\lambda_n + \mu_n)$, we have that

$$R_0^n = p_0 R_1^{n-1}, \quad R_i^n = q_i R_{i-1}^{n-1} + p_i R_{i+1}^{n-1}, \quad i \geq 1,$$

for $n \geq 2$ with initial condition

$$R_i^1 = \begin{cases} q_0, & \text{if} \quad i = 0, \\ 0, & \text{if} \quad i \geq 1. \end{cases}$$

The relation of R_i^n with the transition probability matrix (3.26) is given by

$$R_i^n = q_0 P_{i,0}^{(n-1)}. \tag{3.119}$$

Call $S_i = \sum_{n=1}^{\infty} R_i^n$. From the previous relation, we have

$$S_0 = p_0 S_1 + q_0, \quad S_i = q_i S_{i-1} + p_i S_{i+1}, \quad i \geq 1.$$

Therefore,

$$S_{i+1} - S_i = \frac{q_i}{p_i}(S_i - S_{i-1}) = \frac{q_0 \cdots q_i}{p_0 \cdots p_i}(S_0 - 1) = \mu_0 \frac{1}{\lambda_n \pi_n}(S_0 - 1)$$

and solving the recurrence, we get

$$S_{n+1} = 1 + \left(1 + \mu_0 \sum_{i=0}^{n} \frac{1}{\lambda_i \pi_i}\right)(S_0 - 1).$$

If $\sum_{n=0}^{\infty} (\lambda_n \pi_n)^{-1} = \infty$, then $S_0 = 1$ (since the absorption is certain) and $S_n = 1$ for all n. However, if $\sum_{n=0}^{\infty} (\lambda_n \pi_n)^{-1} < \infty$, there is a positive probability that the process is never absorbed.

Computing the probabilities S_n is reduced to computing S_0. Let us see how. Starting from (3.119) and using the tools in Chapter 2, we can define the polynomials $T_n(x)$ by the three-term recurrence relation

$$T_0(x) = 1, \quad x T_0(x) = p_0 T_1(x),$$
$$x T_n(x) = q_n T_{n-1}(x) + p_n T_{n+1}(x),$$

which is orthogonal with respect to some measure α supported on $[-1,1]$. The Karlin–McGregor formula is given by

$$P_{ij}^{(n)} = \pi_j^P \int_{-1}^{1} x^n T_i(x) T_j(x) d\alpha(x),$$

where π_j^P is defined by (3.27). Therefore, as we saw in (2.61), we have that

$$\sum_{n=1}^{\infty} R_i^n = q_0 \int_{-1}^{1} \frac{T_i(x)}{1-x} d\alpha(x), \quad \sum_{n=1}^{\infty} n R_i^n = q_0 \int_{-1}^{1} \frac{T_i(x)}{(1-x)^2} d\alpha(x).$$

In particular, for $i = 0$, and using (2.60), we have that

$$S_0 = \sum_{n=1}^{\infty} R_0^n = \frac{q_0 \sum_{n=0}^{\infty} \frac{1}{p_n \pi_n^P}}{1 + q_0 \sum_{n=0}^{\infty} \frac{1}{p_n \pi_n^P}} = \frac{\mu_0 \sum_{n=0}^{\infty} \frac{1}{\lambda_n \pi_n}}{1 + \mu_0 \sum_{n=0}^{\infty} \frac{1}{\lambda_n \pi_n}} = \int_0^{\infty} dG_0(t) = a_0.$$

Alternatively, using (2.57), we have that the expected value of the number of transitions before absorption starting at the state 0 is given by

$$\begin{aligned} \sum_{n=1}^{\infty} n R_0^n &= q_0 \int_{-1}^{1} \frac{d\alpha(x)}{(1-x)^2} = \frac{1}{q_0} \sum_{n=0}^{\infty} \pi_n^P = \frac{\lambda_0 + \mu_0}{\mu_0} \sum_{n=0}^{\infty} \pi_n \frac{\lambda_n + \mu_n}{\lambda_0 + \mu_0} \\ &= \frac{1}{\mu_0} \sum_{n=0}^{\infty} \pi_n \lambda_n + \frac{1}{\mu_0} \sum_{n=0}^{\infty} \pi_n \mu_n \\ &= \frac{1}{\mu_0} \sum_{n=0}^{\infty} \pi_n \lambda_n + \frac{1}{\mu_0} \left(\mu_0 + \sum_{n=0}^{\infty} \pi_{n+1} \mu_{n+1} \right) = 1 + \frac{2}{\mu_0} \sum_{n=0}^{\infty} \lambda_n \pi_n. \end{aligned} \tag{3.120}$$

Therefore, this number is finite if and only if $\sum_{n=0}^{\infty} \lambda_n \pi_n < \infty$.

Example 3.53 Consider the nonconservative $M/M/1$ queue of Example 3.35. According to Theorem 3.51 we can study absorption at -1 using the spectral measure or studying the divergence of certain series. The series can be computed directly since

$$\sum_{n=0}^{\infty} \frac{1}{\lambda_n \pi_n} = \sum_{n=0}^{\infty} \frac{\mu^n}{\lambda^{n+1}} = \frac{1}{\lambda} \sum_{n=0}^{\infty} \left(\frac{\mu}{\lambda}\right)^n = \infty \quad \Leftrightarrow \quad \frac{\mu}{\lambda} \geq 1 \quad \Leftrightarrow \quad \mu \geq \lambda.$$

Therefore absorption at -1 is certain if and only if $\mu \geq \lambda$. Alternatively, using the spectral measure, we have that

$$\mu_0 \int_0^{\infty} \frac{d\psi(x)}{x} = \frac{1}{2\pi\lambda} \int_{(\sqrt{\lambda}-\sqrt{\mu})^2}^{(\sqrt{\lambda}+\sqrt{\mu})^2} \frac{\sqrt{4\lambda\mu - (\lambda+\mu-x)^2}}{x} dx.$$

This integral is given by

$$\frac{1}{2\pi\lambda}\int_{(\sqrt{\lambda}-\sqrt{\mu})^2}^{(\sqrt{\lambda}+\sqrt{\mu})^2}\frac{\sqrt{4\lambda\mu-(\lambda+\mu-x)^2}}{x}\,dx = \frac{1}{2\lambda}(\lambda+\mu-|\lambda-\mu|)$$
$$= \begin{cases} 1, & \text{if } \lambda \le \mu, \\ \mu/\lambda, & \text{if } \lambda > \mu. \end{cases}$$

Therefore, again, absorption at -1 is certain if and only if $\lambda \le \mu$. The absorption at -1 is ergodic if and only if $\lambda < \mu$. For these values, we have that

$$\sum_{n=0}^{\infty}\pi_n = \sum_{n=0}^{\infty}\left(\frac{\lambda}{\mu}\right)^n < \infty \quad \Leftrightarrow \quad \frac{\lambda}{\mu} < 1 \quad \Leftrightarrow \quad \lambda < \mu.$$

Additionally, we could have studied the convergence of the integral

$$\int_0^{\infty}\frac{d\psi(x)}{x^2} = \frac{1}{2\pi\lambda\mu}\int_{(\sqrt{\lambda}-\sqrt{\mu})^2}^{(\sqrt{\lambda}+\sqrt{\mu})^2}\frac{\sqrt{4\lambda\mu-(\lambda+\mu-x)^2}}{x^2}\,dx.$$

If $\lambda \neq \mu$ this integral is always bounded since $(\sqrt{\lambda}-\sqrt{\mu})^2 > 0$ so there is no problem at $x = 0$. For $\lambda = \mu$ we have that

$$\int_0^{\infty}\frac{d\psi(x)}{x^2} = \frac{1}{2\pi\lambda^2}\int_0^{4\lambda}\sqrt{\frac{4\lambda - x}{x^3}}\,dx = \infty.$$

Therefore, again, absorption at -1 is ergodic if and only if $\lambda < \mu$. The expected time of absorption is given by (3.117) for $i = 0$, i.e.

$$\mathbb{E}(T_{-1}|X_0 = 0) = \frac{1}{2\pi\lambda}\int_{(\sqrt{\lambda}-\sqrt{\mu})^2}^{(\sqrt{\lambda}+\sqrt{\mu})^2}\frac{\sqrt{4\lambda\mu-(\lambda+\mu-x)^2}}{x^2}\,dx$$
$$= \frac{1}{\mu}\sum_{j=0}^{\infty}\left(\frac{\lambda}{\mu}\right)^n = \frac{1}{\mu-\lambda}.$$

We can also compute the absorption probabilities starting from a state i, already given by (3.116). Indeed, since

$$\sum_{k=i}^{\infty}\frac{1}{\lambda_k\pi_k} = \frac{1}{\lambda}\left(\frac{\mu}{\lambda}\right)^i\sum_{k=0}^{\infty}\left(\frac{\mu}{\lambda}\right)^k = \begin{cases} \dfrac{1}{\lambda-\mu}\left(\dfrac{\mu}{\lambda}\right)^i, & \text{if } \lambda > \mu, \\ \infty, & \text{if } \lambda \le \mu, \end{cases}$$

we have that

$$a_i = \mathbb{P}(T_{-1} < \infty|X_0 = i) = \begin{cases} \dfrac{\dfrac{\mu}{\lambda-\mu}\left(\dfrac{\mu}{\lambda}\right)^i}{1+\dfrac{\mu}{\lambda-\mu}} = \left(\dfrac{\mu}{\lambda}\right)^{i+1}, & \text{if } \lambda > \mu, \\ 1, & \text{if } \lambda \le \mu. \end{cases}$$

Moreover, the expected absorption time (3.117) is given by

$$\mathbb{E}(T_{-1}|X_0 = i) = \frac{1}{\mu}\sum_{j=0}^{i}\left(\frac{\mu}{\lambda}\right)^j\left(\frac{\lambda}{\mu}\right)^j\sum_{k=0}^{\infty}\left(\frac{\lambda}{\mu}\right)^k = \begin{cases} \dfrac{i+1}{\mu-\lambda}, & \text{if } \lambda < \mu, \\ \infty, & \text{if } \lambda \geq \mu. \end{cases}$$

From (3.120) we can also compute the expected value of the number of transitions before absorption starting at the state 0, given by

$$1 + \frac{2\lambda}{\mu}\sum_{n=0}^{\infty}\left(\frac{\lambda}{\mu}\right)^n = \begin{cases} \dfrac{\mu+\lambda}{\mu-\lambda}, & \text{if } \lambda < \mu, \\ \infty, & \text{if } \lambda \geq \mu. \end{cases}$$

◊

3.7.2 Strong Ratio Limit Property and Limiting Conditional Distribution

Similar to the case of discrete-time birth–death chains we will now study the behavior of $P_{ij}(t)$ as $t \to \infty$. In this section we will be concerned first with the limits

$$\lim_{t\to\infty}\frac{P_{ij}(t)}{P_{kl}(t)}, \quad i,j,k,l \in \mathbb{N}_0.$$

Clearly, from the Karlin–McGregor formula, we have that if $k = j$ and $l = i$ then $P_{ij}(t)/P_{ji}(t) = \pi_j/\pi_i$ is independent of t. If these limits exist simultaneously then the birth–death process is said to have the *strong ratio limit property*. Unlike the case of discrete-time birth–death chains this question is completely answered by the following result:

Theorem 3.54 ([84, Theorem 11]) *Let* $\xi = \inf \operatorname{supp}(\psi) \geq 0$. *Then* $\lim_{t\to\infty} P_{ij}(t)/P_{kl}(t)$ *exists, is finite and positive, and is given by*

$$\lim_{t\to\infty}\frac{P_{ij}(t)}{P_{kl}(t)} = \frac{\pi_j Q_i(\xi)Q_j(\xi)}{\pi_l Q_k(\xi)Q_l(\xi)}.$$

In particular, for $\mu_0 = 0$ *and* $\xi = 0$, *we have that*

$$\lim_{t\to\infty}\frac{P_{ij}(t)}{P_{kl}(t)} = \frac{\pi_j}{\pi_l}.$$

Proof Since $Q_n(\xi) > 0$ for all $n \geq 0$, we can write

$$P_{ij}(t) = e^{-\xi t}\pi_j Q_i(\xi)Q_j(\xi)\int_{\xi}^{\infty} e^{-(x-\xi)t}f(x)d\psi(x),$$

where

$$f(x) = \frac{Q_i(x)Q_j(x)}{Q_i(\xi)Q_j(\xi)}.$$

Let us study the ratio

$$\frac{\displaystyle\int_\xi^\infty e^{-(x-\xi)t} f(x)\, d\psi(x)}{\displaystyle\int_\xi^\infty e^{-(x-\xi)t} d\psi(x)} = 1 + \frac{\displaystyle\int_\xi^\infty e^{-(x-\xi)t}[f(x)-1]\, d\psi(x)}{\displaystyle\int_\xi^\infty e^{-(x-\xi)t} d\psi(x)} = 1 + R.$$

Since f is continuous and $f(\xi) = 1$ we have that for any ε there exists some $\delta > 0$ such that $|f(x) - 1| < \varepsilon$ when $\xi \leq x \leq \xi + \delta$. Therefore

$$|R| \leq \frac{\left|\displaystyle\int_\xi^{\xi+\delta} e^{-(x-\xi)t}[f(x)-1]\, d\psi(x)\right| + \left|\displaystyle\int_{\xi+\delta}^\infty e^{-(x-\xi)t}[f(x)-1]\, d\psi(x)\right|}{\displaystyle\int_\xi^\infty e^{-(x-\xi)t} d\psi(x)}$$

$$\leq \varepsilon + \frac{\displaystyle\int_{\xi+\delta}^\infty e^{-(x-\xi)t}|f(x)-1|\, d\psi(x)}{\displaystyle\int_\xi^{\xi+\delta/2} e^{-(x-\xi)t} d\psi(x)} \leq \varepsilon + ce^{\delta t/2}\int_{\xi+\delta}^\infty e^{-(x-\xi)t}|f(x)-1|\, d\psi(x),$$

where $c = \left[\int_\xi^{\xi+\delta/2} d\psi(x)\right]^{-1}$, which is positive and finite. Therefore $\limsup_{t\to\infty} |R| \leq \varepsilon$ and $R \downarrow 0$ as $t \to \infty$. Using this ratio for $P_{ij}(t)$ and $P_{kl}(t)$, we obtain the result. □

Remark 3.55 Observe from the previous proof that if f is any continuous function and $\xi = \inf \operatorname{supp}(\psi) \geq 0$, then we have

$$\lim_{t\to\infty} \frac{\displaystyle\int_0^\infty e^{-xt} f(x)\, d\psi(x)}{\displaystyle\int_0^\infty e^{-xt} d\psi(x)} = f(\xi). \tag{3.121}$$

◊

As a consequence of the previous theorem, we can study a classical result of Doeblin, which states that

$$0 < \lim_{t\to\infty} \frac{\displaystyle\int_0^t P_{ij}(s)\, ds}{\displaystyle\int_0^t P_{kl}(s)\, ds} < \infty, \quad i,j,k,l \in \mathbb{N}_0.$$

This Doeblin ratio theorem is an abelian form of the above formula in the recurrent case. In the ergodic case ($\mu_0 = 0$ and $\lim_{t\to\infty} P_{ij}(t) = \pi_j \psi(\{0\}) > 0$) we have, using L'Hôpital's rule, that $t^{-1}\int_0^t P_{ij}(s)\, ds \to \pi_j \psi(\{0\})$ as $t \to \infty$. Therefore,

$$\lim_{t\to\infty}\frac{\displaystyle\int_0^t P_{ij}(s)\,ds}{\displaystyle\int_0^t P_{kl}(s)\,ds}=\frac{\pi_j}{\pi_l}=\lim_{t\to\infty}\frac{P_{ij}(t)}{P_{kl}(t)}.$$

Something similar happens for the null recurrent case ($\mu_0=0$ and then $\xi=0$) and we get the same result. In the transient case, the ratios of the limits may be different. If we assume that $\mu_0=0$, from formula (3.56) (see Exercise 3.17), we can deduce that

$$\int_0^\infty P_{ij}(t)\,dt=\pi_j\int_0^\infty\frac{Q_i(x)Q_j(x)}{x}\,d\psi(x)=\pi_j\sum_{n=j}^{\infty}\frac{1}{\lambda_n\pi_n},\qquad i\le j.$$

Therefore, if $i\le j$ and $k\le l$, we have that

$$\lim_{t\to\infty}\frac{\displaystyle\int_0^t P_{ij}(s)\,ds}{\displaystyle\int_0^t P_{kl}(s)\,ds}=\frac{\displaystyle\pi_j\sum_{n=j}^{\infty}\frac{1}{\lambda_n\pi_n}}{\displaystyle\pi_l\sum_{n=l}^{\infty}\frac{1}{\lambda_n\pi_n}}.\tag{3.122}$$

As in the case of discrete-time birth–death chains, let us now study the concept of *limiting conditional distribution*. We will distinguish two cases. First the *evanescent or absorbing* case ($\mu_0>0$) and second the *irreducible or reflecting* case ($\mu_0=0$). Both cases were studied originally in [105] although some particular cases were considered before in [40]. For simplicity we will assume that there is no explosion at finite time, i.e. $C=\infty$ where C is defined by (3.28).

The evanescent or absorbing case ($\mu_0>0$)

Let $\{X_t,t\ge 0\}$ be a birth–death process where $\mu_0>0$. Let T_{-1} be the random variable that denotes the absorption time at state -1. We consider the conditional probabilities

$$q_{ij}(t)=\mathbb{P}(X_t=j\mid X_0=i,t<T_{-1}<\infty).\tag{3.123}$$

We are interested in studying if $\lim_{t\to\infty}q_{ij}(t)$ converges to some distribution. This will be the limiting probability of being in the state j at time t given that the process has not yet been absorbed.

Similarly, we can study the *doubly limiting conditional distribution* $\lim_{t\to\infty}\lim_{s\to\infty}q_{ij}(t,s)$, where

$$q_{ij}(t,s)=\mathbb{P}(X_t=j\mid X_0=i,t+s<T_{-1}<\infty).\tag{3.124}$$

This will be the limiting probability of being in state j given that the process will not leave $\{0, 1, 2, \dots\}$ in a distant future. We have the following result:

Theorem 3.56 ([105, Theorem 3.4]) *Let i be an initial state of a birth–death process with $\mu_0 > 0$ and call $\xi = \inf \operatorname{supp}(\psi)$. Then, for the limiting conditional probabilities* (3.123) *and* (3.124), *we have that*

$$\lim_{t\to\infty} q_{ij}(t) = \frac{a_j \pi_j Q_j(\xi)}{\sum_{k=0}^{\infty} a_k \pi_k Q_k(\xi)}, \qquad \lim_{t\to\infty} \lim_{s\to\infty} q_{ij}(t,s) = \frac{\pi_j Q_j^2(\xi)}{\sum_{k=0}^{\infty} \pi_k Q_k^2(\xi)},$$

where $a_i = \mathbb{P}(T_{-1} < \infty | X_0 = i)$.

Proof First we observe, using an argument similar to the one given in (2.70), that

$$q_{ij}(t) = \frac{a_j P_{ij}(t)}{a_i - P_{i,-1}(t)}.$$

The probabilities $P_{i,-1}(t)$ are the probabilities $G_i(t)$ in (3.115), given by

$$P_{i,-1}(t) = \mu_0 \int_0^t P_{i0}(s)\, ds = \mu_0 \int_0^\infty \frac{Q_i(x)}{x}\, d\psi(x) - \mu_0 \int_0^\infty e^{-xt} \frac{Q_i(x)}{x}\, d\psi(x).$$

Since $a_i = \mu_0 \int_0^\infty x^{-1} Q_i(x)\, d\psi(x)$ (see (3.116)) then we have

$$q_{ij}(t) = \frac{a_j \pi_j}{\mu_0} \frac{\int_0^\infty e^{-xt} Q_i(x) Q_j(x)\, d\psi(x)}{\int_0^\infty e^{-xt} \frac{Q_i(x)}{x}\, d\psi(x)}.$$

Multiplying and dividing by $\int_0^\infty e^{-xt}\, d\psi(x)$, and using (3.121), we have that

$$\lim_{n\to\infty} q_{ij}(t) = \frac{a_j \pi_j}{\mu_0} \frac{Q_i(\xi) Q_j(\xi)}{\frac{Q_i(\xi)}{\xi}} = \frac{\pi_j}{\mu_0} \xi Q_j(\xi).$$

If $\xi = 0$ then it converges to 0 and it is not a proper distribution. For $\xi > 0$, we use the same arguments as in Remark 2.52 (see also [105, Section 3.2] or [59, Lemma]) to conclude that

$$\frac{\xi}{\mu_0} \sum_{k=0}^{\infty} a_k Q_k(\xi) \pi_k = 1, \tag{3.125}$$

either for the case of certain absorption ($a_i = 1$) or not ($a_i < 1$).

For the doubly limiting case, using an argument similar to the one we used in Theorem 2.53, we have that

$$q_{ij}(t,s) = \frac{P_{ij}(t)(a_j - P_{j,-1}(s))}{a_i - P_{i,-1}(t+s)}.$$

Using (3.121), we get that

$$\lim_{s\to\infty} q_{ij}(t,s) = P_{ij}(t)\frac{Q_j(\xi)}{e^{-\xi t}Q_i(\xi)}.$$

Therefore

$$\begin{aligned}\lim_{t\to\infty}\lim_{s\to\infty} q_{ij}(t,s) &= \frac{Q_j(\xi)}{Q_i(\xi)}\pi_j \lim_{t\to\infty}\int_0^\infty e^{-t(x-\xi)}Q_i(x)Q_j(x)d\psi(x)\\ &= \frac{Q_j(\xi)}{Q_i(\xi)}\pi_j\left[Q_i(\xi)Q_j(\xi)\psi(\{\xi\}) + \int_\xi^\infty e^{-t(x-\xi)}Q_i(x)Q_j(x)d\psi(x)\right]\\ &= Q_j^2(\xi)\pi_j\psi(\{\xi\}) = \frac{Q_j^2(\xi)\pi_j}{\sum_{k=0}^\infty Q_k^2(\xi)\pi_k}.\end{aligned}$$

□

Example 3.57 Consider the linear birth–death process of Example 3.38, case 2, changing λ by μ and writing $\beta = 2$. The birth–death rates are given by (see (3.109))

$$\lambda_n = \lambda(n+1), \quad \mu_n = \mu(n+1), \quad n \geq 0.$$

Initially, if $\lambda < \mu$, then absorption is certain. The polynomials $Q_n(x)$ can be written in terms of the Meixner polynomials and are orthogonal with respect to a measure supported at the points $(\mu - \lambda)(n+1), n \geq 0$. Therefore in this case we have $\xi = \mu - \lambda$. We can easily see that

$$\pi_n = \left(\frac{\lambda}{\mu}\right)^n \frac{1}{n+1}, \quad Q_n(\xi) = n+1.$$

Therefore,

$$\lim_{t\to\infty} q_{ij}(t) = \frac{\left(\frac{\lambda}{\mu}\right)^j}{\sum_{k=0}^\infty \left(\frac{\lambda}{\mu}\right)^k} = \left(\frac{\lambda}{\mu}\right)^j\left(1-\frac{\lambda}{\mu}\right),$$

and

$$\lim_{t\to\infty}\lim_{s\to\infty} q_{ij}(t,s) = \frac{\left(\frac{\lambda}{\mu}\right)^j (j+1)}{\sum_{k=0}^\infty \left(\frac{\lambda}{\mu}\right)^k (k+1)} = (j+1)\left(\frac{\lambda}{\mu}\right)^j\left(1-\frac{\lambda}{\mu}\right)^2.$$

The first is a geometric distribution and the second is the Pascal distribution. If $\lambda > \mu$ a straightforward computation shows that the absorption probabilities are given by

$$a_i = \frac{1}{\lambda}\left(\frac{\mu}{\lambda}\right)^i.$$

But in this case we have that $a_i\pi_i Q_i(\xi) = 1/\lambda$. Therefore the corresponding series diverges and $q_{ij}(t)$ does not converge to a proper distribution. ◊

The irreducible or reflecting case ($\mu_0 = 0$)

Now it does not make sense to consider absorption probabilities, so we have to define something else to substitute the analysis. Let S denote the random variable that gives the last exit time from the state 0 before going to infinity, i.e. the time at which the birth–death process does not return to 0 again. We will work with transient birth–death processes. Otherwise $S = \infty$. We will study the limit $\lim_{t\to\infty} \bar{q}_{ij}(t)$ of the conditioned probabilities

$$\bar{q}_{ij}(t) = \mathbb{P}(X_t = j \mid X_0 = i, t < S). \tag{3.126}$$

This will be the limiting probability of being in state j given that the process has not yet started to drift to infinity. We will also study the doubly limiting conditional probabilities

$$\bar{q}_{ij}(t,s) = \mathbb{P}(X_t = j \mid X_0 = i, t+s < S), \tag{3.127}$$

which are the limiting probabilities of being in state j given that the process will not drift to infinity in a distant future. Before the main result we will need the following lemma:

Lemma 3.58 ([105, Lemma 4.1]) *Assume that the birth–death process $\{X_t, t \geq 0\}$ is transient and define*

$$K_i(t) = \mathbb{P}(t < S|X_0 = i) = \mathbb{P}(X_{t+s} = 0 \textit{ for some } s > 0|X_0 = i).$$

Then we have

$$K_i(t) = \frac{\displaystyle\int_t^\infty P_{i0}(s)\, ds}{\displaystyle\int_0^\infty P_{00}(s)\, ds} = \frac{\displaystyle\int_0^\infty e^{-xt} x^{-1} Q_i(x)\, d\psi(x)}{\displaystyle\int_0^\infty x^{-1} d\psi(x)}. \tag{3.128}$$

Proof Let $V_0(t)$ denote the total time spent at state 0 during the time interval (t,∞). Then, since the process is transient, we have

$$\mathbb{E}(V_0(t)|X_0 = i) = \mathbb{E}\left(\int_t^\infty \mathbf{1}_{\{X_u=0\}}\, du|X_0 = i\right) = \int_t^\infty P_{i,0}(s)\, ds < \infty.$$

By the Markov property, we have

$$\int_t^\infty P_{i,0}(s)\, ds = \mathbb{E}(V_0(t)|X_0 = i) = K_i(t)\mathbb{E}(V_0(0)|X_0 = 0) = K_i(t)\int_0^\infty P_{00}(s)\, ds.$$

Therefore, using the Karlin–McGregor representation, we obtain (3.128). □

Using (3.141) (see Exercise 3.17) for $j=0, k=1$ and (3.70), we have that

$$c_i = K_i(0) = \mathbb{P}(0 < S|X_0 = i) = \frac{\displaystyle\sum_{n=i}^{\infty} \frac{1}{\lambda_n \pi_n}}{\displaystyle\sum_{n=0}^{\infty} \frac{1}{\lambda_n \pi_n}}. \tag{3.129}$$

Theorem 3.59 ([105, Theorem 4.4]) *Let i be an initial state of a transient birth–death process with $\mu_0 = 0$ and call $\xi = \inf \operatorname{supp}(\psi)$. Then, for the limiting conditional probabilities* (3.126) *and* (3.127), *we have that*

$$\lim_{t\to\infty} \bar{q}_{ij}(t) = \frac{c_j \pi_j Q_j(\xi)}{\displaystyle\sum_{k=0}^{\infty} c_k \pi_k Q_k(\xi)}, \quad \lim_{t\to\infty}\lim_{s\to\infty} \bar{q}_{ij}(t,s) = \frac{\pi_j Q_j^2(\xi)}{\displaystyle\sum_{k=0}^{\infty} \pi_k Q_k^2(\xi)},$$

where c_i are the probabilities defined by (3.129).

Proof The probabilities $\bar{q}_{ij}(t)$ can be written as

$$\begin{aligned}\bar{q}_{ij}(t) &= \mathbb{P}(X_t = j|X_0 = i, t < S) = \frac{\mathbb{P}(X_t = j, X_0 = i, t < S)}{\mathbb{P}(X_0 = i, t < S)} \\ &= \frac{\mathbb{P}(X_t = j, t < S|X_0 = i)}{\mathbb{P}(t < S|X_0 = i)} = \frac{c_j \mathbb{P}(X_t = j|X_0 = i)}{\mathbb{P}(t < S|X_0 = i)} = \frac{c_j P_{ij}(t)}{K_i(t)},\end{aligned}$$

where $K_i(t)$ and c_i are defined by (3.128) and (3.129), respectively. Multiplying and dividing by $\int_0^\infty e^{-xt} d\psi(x)$, and using (3.121), we get that

$$\lim_{t\to\infty} \bar{q}_{ij}(t) = c_j \pi_j \frac{Q_i(\xi) Q_j(\xi)}{\frac{Q_i(\xi)}{\xi}} \left(\int_0^\infty \frac{d\psi(x)}{x} \right) = c_j \pi_j Q_j(\xi) \xi \left(\int_0^\infty \frac{d\psi(x)}{x} \right).$$

The proof is completed using the following formula (using arguments similar to the proof of the previous theorem, see also [105, Section 4.2]):

$$\xi \left(\int_0^\infty \frac{d\psi(x)}{x} \right) = \left(\sum_{k=0}^{\infty} c_k \pi_k Q_k(\xi) \right)^{-1}.$$

For the doubly limiting case we follow the same argument, recalling that

$$\begin{aligned}\bar{q}_{ij}(t,s) &= \mathbb{P}(X_t = j|X_0 = i, t+s < S) = P_{ij}(t) \frac{\mathbb{P}(s < S|X_0 = j)}{\mathbb{P}(t+s < S|X_0 = i)} \\ &= P_{ij}(t) \frac{K_j(s)}{K_i(t+s)} = P_{ij}(t) \frac{\displaystyle\int_0^\infty e^{-sx} Q_j(x) x^{-1} d\psi(x)}{\displaystyle\int_0^\infty e^{-(s+t)x} Q_i(x) x^{-1} d\psi(x)},\end{aligned}$$

and from here we see that the limits are the same as in the case of $q_{ij}(t,s)$. □

Remark 3.60 If the birth–death process is recurrent, then $c_i = 1$ for all $i \in \mathbb{N}_0$. Therefore $\bar{q}_{ij}(t)$ behaves as $q_{ij}(t)$ as $t \to \infty$. In the previous theorem, since $C = \infty$, the minimal process is unique. In general, these results can be extended to non-minimal processes using similar arguments. ◊

Example 3.61 Consider Example 3.38 (linear rates) with $\beta = 1$ and $\lambda > \mu$ (transient). The birth–death rates are given by

$$\lambda_n = \lambda(n+1), \quad \mu_n = \mu n, \quad n \geq 0.$$

The measure is located at the points $(n+1)(\lambda - \mu)$, so $\xi = \lambda - \mu$, and again

$$\pi_n = \left(\frac{\lambda}{\mu}\right)^n, \quad Q_n(\xi) = \left(\frac{\mu}{\lambda}\right)^n.$$

As before, we have that

$$\lim_{t\to\infty} \lim_{s\to\infty} \bar{q}_{ij}(t,s) = \left(\frac{\mu}{\lambda}\right)^j \left(1 - \frac{\mu}{\lambda}\right).$$

Let us try to compute now $\lim_{t\to\infty} \bar{q}_{ij}(t)$. For this we need to calculate the probabilities c_i defined by (3.129). By definition, we have that

$$c_j = \frac{\displaystyle\sum_{n=j}^{\infty} \frac{1}{\lambda_n \pi_n}}{\displaystyle\sum_{n=0}^{\infty} \frac{1}{\lambda_n \pi_n}} = \frac{\displaystyle\sum_{n=j}^{\infty} \frac{(\mu/\lambda)^n}{n+1}}{\displaystyle\sum_{n=0}^{\infty} \frac{(\mu/\lambda)^n}{n+1}} = \left(\frac{\mu}{\lambda}\right)^j \frac{\displaystyle\sum_{n=0}^{\infty} \frac{(\mu/\lambda)^n}{n+j+1}}{\displaystyle\sum_{n=0}^{\infty} \frac{(\mu/\lambda)^n}{n+1}}$$

$$= \left(\frac{\mu}{\lambda}\right)^j \frac{L\left(\frac{\mu}{\lambda}, 1, j+1\right)}{\log\left(\dfrac{\lambda}{\lambda - \mu}\right)}, \quad j \geq 1, \quad c_0 = 1,$$

where

$$L(z, a, v) = \sum_{n=0}^{\infty} \frac{z^n}{(v+n)^a}$$

is the *Lerch zeta function* or *Lerch transcendent*. This function does not have an explicit expression in general. However, for some particular cases it can be computed. The limit is then given by

$$\lim_{t\to\infty} \bar{q}_{ij}(t) = \frac{c_j}{\displaystyle\sum_{k=0}^{\infty} c_k} = \frac{\left(\frac{\mu}{\lambda}\right)^j L\left(\frac{\mu}{\lambda}, 1, j+1\right)}{\displaystyle\sum_{k=0}^{\infty} \left(\frac{\mu}{\lambda}\right)^k L\left(\frac{\mu}{\lambda}, 1, k+1\right)}.$$

Using the definition of the Lerch zeta function and changing the order of summation it is easy to see that the denominator converges to $\lambda/(\lambda-\mu)$. Therefore,

$$\lim_{t\to\infty} \bar{q}_{ij}(t) = \frac{\lambda-\mu}{\lambda}\left(\frac{\mu}{\lambda}\right)^j L\left(\frac{\mu}{\lambda},1,j+1\right).$$

Also the Lerch zeta function can be generated recurrently using the following formulas:

$$L\left(\frac{\mu}{\lambda},1,1\right) = -\frac{\lambda}{\mu}\log\left(\frac{\lambda-\mu}{\lambda}\right),$$

$$L\left(\frac{\mu}{\lambda},1,j+1\right) = \frac{\lambda}{\mu}\left(L\left(\frac{\mu}{\lambda},1,j\right)-\frac{1}{j}\right), \quad j\geq 1.$$

An easy computation shows that

$$\lim_{t\to\infty} \bar{q}_{i0}(t) = \frac{\lambda-\mu}{\mu}\log\left(\frac{\lambda}{\lambda-\mu}\right), \quad \lim_{t\to\infty} \bar{q}_{i1}(t) = \frac{\lambda-\mu}{\mu}\log\left(\frac{\lambda}{\lambda-\mu}\right)-\frac{\lambda-\mu}{\lambda},$$

and in general

$$\lim_{t\to\infty} \bar{q}_{ij}(t) = \frac{\lambda-\mu}{\mu}\left[\log\left(\frac{\lambda}{\lambda-\mu}\right)-\sum_{k=1}^{j}\frac{1}{k}\left(\frac{\mu}{\lambda}\right)^k\right], \quad j\geq 0.$$

If $\lambda \leq \mu$ then the birth–death process is recurrent, so $\bar{q}_{ij}(t)$ does not converge to a proper distribution. ◊

3.7.3 Decay Parameter and Quasi-Stationary Distributions

Let $\{X_t, t\geq 0\}$ be an irreducible birth–death process on $\mathbb{N}_0$ with a set of birth–death rates $\{(\lambda_n,\mu_n), n\geq 0\}$ and assume that $\mu_0>0$ and $A=\infty$ where A is defined by (3.28). That means that the birth–death rates uniquely determine the process $\{X_t, t\geq 0\}$ (see Theorem 3.26). Let T_{-1} be the random variable that denotes the absorption time to the state -1 (see (3.29)). Since $A=\infty$ that means that $\mathbb{P}(T_{-1}<\infty|X_0=i)=1$ for all $i\in\mathbb{N}_0$, i.e. absorption at -1 is certain (see Theorem 3.51). A proper distribution $\nu=(\nu_j)_{j\in\mathbb{N}_0}$ is called a *quasi-stationary distribution* for the birth–death process $\{X_t, t\geq 0\}$ if

$$\mathbb{P}_\nu(X_t=j|T_{-1}>t)=\nu_j, \quad \text{for all} \quad j\in\mathbb{N}_0,$$

where, for any event A, we have

$$\mathbb{P}_\nu(A) \doteq \sum_{i=0}^{\infty}\mathbb{P}(A|X_0=i)\nu_i.$$

Since $\mathbb{P}_\nu(X_t = j, T_{-1} > t) = \mathbb{P}_\nu(T_{-1} > t)$, another way to write the previous definition is

$$\mathbb{P}_\nu(X_t = j) = \nu_j \mathbb{P}_\nu(T_{-1} > t), \quad \text{for all} \quad j \in \mathbb{N}_0.$$

If ν is a quasi-stationary distribution then there exists a nonnegative number α_ν such that

$$\mathbb{P}_\nu(T_{-1} > t) = e^{-\alpha_\nu t}, \quad t \geq 0.$$

This is a consequence of the Markov property and the fact that $\mathbb{P}_\nu(T_{-1} > t + s) = \mathbb{P}_\nu(T_{-1} > t)\mathbb{P}_\nu(T_{-1} > s)$ (see [21, Theorem 2.2] for more details). The coefficient α_ν satisfies

$$\alpha_\nu = -\frac{1}{t} \log \mathbb{P}_\nu(T_{-1} > t), \quad t > 0.$$

So, when the process starts from ν, α_ν is the *exponential rate of survival*, which is stationary in time, so the limit as $t \to \infty$ (if it exists) is independent of the initial distribution. Let us call

$$\alpha^* \doteq -\liminf_{t \to \infty} \frac{1}{t} \log \mathbb{P}_i(T_{-1} > t). \tag{3.130}$$

The value of α_ν is in fact given by

$$\alpha_\nu = \mu_0 \nu_0.$$

Indeed, from the forward equation (3.23) evaluated at $j = -1$ we obtain $P'_{i,-1}(t) = \mu_0 P_{i0}(t)$. Solving this equation, multiplying by ν_i and summing over i, we obtain

$$\sum_{i=0}^{\infty} \nu_i P_{i,-1}(t) = \mu_0 \sum_{i=0}^{\infty} \int_0^t \nu_i P_{i0}(u)\, du.$$

The left-hand side is the definition of $\mathbb{P}_\nu(T_{-1} \leq t) = 1 - e^{-\alpha_\nu t}$, while the right-hand side is, after integration, $\frac{\mu_0 \nu_0}{\alpha_\nu}(1 - e^{-\alpha_\nu t})$. Therefore we get $\alpha_\nu = \mu_0 \nu_0 > 0$.

From the definition and the previous observation, we have that a quasi-stationary distribution ν will always satisfy the following relation:

$$\sum_{i=0}^{\infty} \nu_i P_{ij}(t) = e^{-\mu_0 \nu_0 t} \mu_j, \quad j \in \mathbb{N}_0, \quad t \geq 0.$$

This is equivalent (see [118]) to the condition that, for some $\mu = \mu_0 \nu_0$, ν is a μ-*invariant measure* on $\mathbb{N}_0$ for $P(t)$ (see [1, Section 5.2] or Proposition 3.18). Since $P'_{ij}(0^+) = a_{ij}$, where $\mathcal{A} = (a_{ij})$ is the infinitesimal operator, a formal argument suggests that a quasi-stationary distribution ν should also satisfy

$$\sum_{i=0}^{\infty} \nu_i a_{ij} = -\mu_0 \nu_0 \nu_j, \quad j \in \mathbb{N}_0, \quad \nu_{-1} = 0.$$

In other words, in the case of birth–death processes, we have (see [40, Theorem 3.1])

$$\lambda_{i-1}\nu_{i-1} - (\lambda_i + \mu_i)\nu_i + \mu_{i+1}\nu_{i+1} = -\mu_0\nu_0\nu_i, \quad i \in \mathbb{N}_0, \quad \nu_{-1} = 0. \tag{3.131}$$

We can also consider the *exponential rate of the transition probabilities* $P_{ij}(t)$ given by

$$\alpha = -\lim_{t\to\infty} \frac{1}{t} \log P_{ij}(t), \tag{3.132}$$

also known as *Kingman's parameter*. This parameter is independent of $i,j \in \mathbb{N}_0$ (see [12, 107, 108]). This value α is also called the *decay parameter* of the process $\{X_t, t \geq 0\}$ and it measures how fast $P_{ij}(t)$ tends to its limit (0 in this case) as $t \to \infty$. If $\alpha > 0$ then we say that the birth–death process is *exponentially ergodic*. From [21, Proposition 4.12] (see also [78, Theorem 3.3.2]), we have that

$$\alpha^* = -\lim_{t\to\infty} \frac{1}{t} \log \mathbb{P}_i(T_{-1} > t) = \alpha.$$

Consider now the vector of polynomials $Q(x) = (Q_0(x), Q_1(x), \ldots)^T$ generated by the eigenvalue equation $-xQ(x) = \mathcal{A}Q(x)$ (see (3.30)) and the corresponding spectral measure ψ. We know from Proposition 1.8 that all zeros of Q_n are simple and strictly positive. Let us denote them by $x_{n,j}, j = 1, \ldots, n$, and we arrange them in the following form: $0 < x_{n,1} < x_{n,2} < \cdots < x_{n,n}$. As in (1.28), we will call

$$\xi_i = \lim_{n\to\infty} x_{n,i}. \tag{3.133}$$

The interlacing property of the zeros (see Proposition 1.8) also gives

$$\xi_1 = \sup\{x \geq 0\colon Q_n(x) > 0, \text{for all } n \geq 0\}.$$

From [21, Proposition 4.12], we know that α^* defined by (3.130) can also be given by

$$\alpha^* = \inf \operatorname{supp}(\psi).$$

Therefore the decay parameter α in (3.132) is also represented by $\inf \operatorname{supp}(\psi)$. We will prove now that, in fact, α coincides with ξ_1.

Proposition 3.62 ([21, Proposition 5.1]) *The decay parameter α defined by* (3.132) *of a birth–death process can be expressed as*

$$\alpha = \xi_1.$$

Proof We already knew that $\alpha = \alpha^*$. By definition we always have $\xi_1 \leq \alpha^*$. Alternatively, for a fixed N and denoting $T_{\geq N} = \inf\{t \geq 0\colon X_t \geq N\}$, we always have that there exists some constant $c_N > 0$ such that

$$\mathbb{P}(X_t = 0, t < T_{-1} \wedge T_{\geq N} | X_0 = 0) = \left(e^{\mathcal{A}_N t}\right)_{00} \geq c_N e^{-x_{N+1,1}t},$$

where $\mathcal{A}_N$ is the truncated matrix of the infinitesimal operator $\mathcal{A}$. Since $\alpha^* = \inf \operatorname{supp}(\psi)$, we deduce that $\alpha^* \leq x_{N+1,1}$, so, taking $N \to \infty$, we get $\xi_1 \geq \alpha^*$. □

Remark 3.63 The previous proposition was actually proved in a more general setting in [38, Lemma 3.2]), defining the decay parameter α as

$$\alpha = \sup\{x \geq 0 | P_{ij}(t) - p_j = \mathcal{O}(e^{-xt}) \text{ as } t \to \infty\}, \quad i,j \in \mathbb{N}_0,$$

where

$$p_j \doteq \lim_{t\to\infty} P_{ij}(t) = \begin{cases} \pi_j \left(\sum_{n=0}^{\infty} \pi_n\right)^{-1}, & \text{if } \mu_0 = 0 \text{ and } \sum_{n=0}^{\infty} \pi_n < \infty, \\ 0, & \text{otherwise.} \end{cases}$$

In this case, the decay parameter α is given by

$$\alpha = \begin{cases} \xi_2, & \text{if } \xi_2 > \xi_1 = 0, \\ \xi_1, & \text{otherwise.} \end{cases}$$

In this context, the decay parameter is related to the concept of a *spectral gap* of a birth–death process, which is identified as the second largest eigenvalue of the infinitesimal operator $\mathcal{A}$. If $\mu_0 = 0$ then the vector of all components equal to 1 is a right-eigenvector of $\mathcal{A}$ with eigenvalue 0. Then the spectral gap (or decay parameter) will be ξ_2 (see [15]). For lower and upper bounds of this spectral gap, see [38, Section 4]. In this section we are assuming that $\mu_0 > 0$, so $\xi_1 > 0$ will be the decay parameter but not the spectral gap. The spectral gap in this situation will be given by the *decay parameter of the conditioned process*, which will be the quantity $\xi_2 - \xi_1$ (see [40, Theorem 5]). New representations of the decay parameter using the Courant–Fischer theorem can be found in [43]. ◊

Let us now study the structure of the set of *quasi-stationary distributions*. Up to now we know that each vector $Q(x)$ is a right-eigenvector of $\mathcal{A}$ with eigenvalue $-x$ and it is unique up to a multiplicative constant. From the equality $\alpha^* = \xi_1$ and the property (3.45), the right-eigenvector $Q(x)$ is strictly positive (all components are positive) when $x \leq \alpha^*$. Since the birth–death process is reversible with respect to the potential coefficients π_n in (3.24) (see also (3.25)), we have that for every $x \leq \alpha^*$ the measure $\nu(x) = (\nu_j(x))_{j\in\mathbb{N}_0}$, where $\nu_i(x) = C(x)Q_i(x)\pi_i$, is a strictly positive left-eigenvector of $\mathcal{A}$ with eigenvalue $-x$ for some constant $C(x) > 0$. Evaluating at $i = 0$, we actually get that $C(x) = \nu_0(x)$. Therefore

$$\nu_i(x) = \nu_0(x)Q_i(x)\pi_i, \tag{3.134}$$

and $\nu^T(x)\mathcal{A} = -x\nu^T(x)$. As we saw in (3.131), in order for $\nu(x)$ to be a quasi-stationary distribution we must have $x = \nu_0(x)\mu_0$. Then, for $x \leq 0$, $\nu(x)$ is a positive measure but it cannot be a quasi-stationary distribution (since $\nu_0(0), \mu_0 > 0$). Observe that, in the case $x = 0$, the left-eigenvector is given by

$$\nu_i(0) = \nu_0(0)\pi_i \left(1 + \sum_{j=0}^{i-1} \frac{\mu_0 \cdots \mu_j}{\lambda_0 \cdots \lambda_j}\right), \quad i \geq 1,$$

but this is not a quasi-stationary distribution. So the candidates to be quasi-stationary distributions are the measures $\nu(x)$ such that $x \in (0, \alpha^*]$. We will characterize all quasi-stationary distributions in the next result, which can be found in [40, Theorem 3.2] (see also [21, Theorem 5.4]). For other cases, see [105, Theorems 3.5 and 4.5].

Theorem 3.64 *Let $A = \infty$ and assume that $\alpha^* > 0$.*

1. *If $D < \infty$ (∞ is an entrance boundary), then $\nu(\alpha^*)$ is the unique quasi-invariant distribution.*
2. *If $D = \infty$ (∞ is an natural boundary), then the set of quasi-stationary distributions is given by the continuum*

$$(\nu(x) \colon x \in (0, \alpha^*]).$$

Proof We showed above this theorem that $\nu(x)$ is a strictly positive vector and it can only be a quasi-stationary distribution if $x \in (0, \alpha^*]$. From (3.134) and (3.43), and using $x = \nu_0(x)\mu_0$, we have

$$\nu_i(x) = Q_i^d(x) - Q_{i+1}^d(x), \quad i \geq 1,$$

where $Q_n^d(x)$ are the dual polynomials defined by (3.40). Therefore, since $Q_0^d(x) = 1$, we get

$$\sum_{i=0}^{N} \nu_i(x) = 1 - Q_{N+1}^d(x).$$

Hence $\nu(x)$ is a probability measure if and only if $Q_\infty^d(x) = \lim_{N\to\infty} Q_N^d(x) = 0$ for all $x \in (0, \alpha^*]$. In (3.51) we showed that

$$Q_\infty^d(x) = 1 - \frac{x}{\mu_0} \sum_{j=0}^{\infty} \pi_j Q_j(x),$$

and this series converges uniformly for every complex x on $|x| \leq R$ if and only if $D < \infty$. So, if $D < \infty$, then $Q_\infty^d(x)$ is an entire function. But in [105, Section 3.2 (Case 2)] it is proved that

$$0 < \xi_i^d = \xi_i < \xi_{i+1}^d, \quad i \geq 1,$$

where ξ_i^d are defined in the same way as in (3.133) but for the zeros of the dual polynomials $Q_\infty^d(x)$. Since the zeros of the entire function $Q_\infty^d(x)$ are given by ξ_i^d, we conclude that $Q^d_\infty(x) = 0$ if and only if $x = \xi_i = \xi_i^d$. Since quasi-stationary distributions can only exist when $x \in (0,\alpha^*]$, the only value for which $\nu(x)$ is a quasi-stationary distribution is for $x = \xi_1 = \alpha^*$. Alternatively, if $D = \infty$, then we can use (3.125) (which is valid for $x > 0$, and in particular for $x \in (0,\alpha^*]$) to conclude that $Q^d_\infty(x) = 0$. Therefore we have a continuum of quasi-stationary distributions. □

Remark 3.65 The concept of quasi-stationary distribution is related to the concept of limiting conditional distribution that we saw in the previous section. From the proof of Theorem 3.56 we learnt that the limiting conditional probability $q_{ij}(t)$ defined by (3.123) is given by

$$\lim_{n\to\infty} q_{ij}(t) = \frac{\pi_j}{\mu_0}\xi Q_j(\xi),$$

where $\xi = \xi_1 = \alpha^*$. But these are precisely the components of the quasi-invariant distribution introduced in (3.134) taking $x = \xi$ (and since $\xi = \nu_0(\xi)\mu_0$). Therefore the limiting conditional distribution is a quasi-stationary distribution for the birth–death process (also called the *extremal* quasi-stationary distribution). ◊

Let us now give some comments about the conditional limiting behavior when the spectrum of the infinitesimal operator $\mathcal{A}$ has a gap after $\alpha^* > 0$ and $B = \sum_{n=0}^\infty \pi_n < \infty$. In this case $D < \infty$, so according to the previous theorem $\nu(\alpha^*)$ will be the only (extremal) quasi-invariant distribution. We say that $\mathcal{A}$ has a *spectral gap after* α^* if

$$\operatorname{supp}(\psi) \cap [\alpha^*, \alpha^* + \varepsilon) = \{\alpha^*\}, \quad \text{for some } \varepsilon > 0.$$

In [21, Proposition 5.20] it is proved that, under these conditions, the conditional evolution of every probability measure ρ on $\mathbb{N}_0$ that satisfies the condition

$$\sum_{j=0}^\infty \frac{\rho_j^2}{\pi_j} < \infty,$$

converges to the quasi-stationary distribution $\nu(\alpha^*)$. In other words,

$$\mathbb{P}_\rho(X_t = j | T > t) = \nu_j(\alpha^*), \quad j \geq 0.$$

One condition to ensure that this happens is if the spectrum of the infinitesimal operator $\mathcal{A}$ is purely discrete. As shown by E. van Doorn in [38], $\operatorname{supp}(\psi)$ is discrete, i.e. constituted by isolated points, if and only if $\sigma = \infty$, where

$$\sigma = \lim_{k\to\infty} \xi_k,$$

and ξ_i are defined by (3.133). For the lower and upper bounds of σ, see [38, Section 5].

The right-eigenvector $(Q_0(\xi_1), Q_1(\xi_1), \dots)^T$ is strictly positive and all other right-eigenvectors $(Q_0(\xi_k), Q_1(\xi_k), \dots)^T, k \geq 2)$ do not have a defined sign. The following conditions imply that the infinitesimal operator $\mathcal{A}$ has a discrete spectrum (for a proof, see [21, Proposition 5.22]):

- $\lim_{n\to\infty} \lambda_n = \infty$, or $\lim_{n\to\infty} \mu_n = \infty$,
- $\lim_{n\to\infty} \dfrac{\mu_n}{\lambda_n} = q \neq 1$,
- $\dfrac{\lambda_n}{\lambda_{n+1}} = 1 + \mathcal{O}(1/n)$,
- $\dfrac{\mu_n}{\mu_{n+1}} = 1 + \mathcal{O}(1/n)$.

For more information about quasi-stationary distributions (also for diffusion processes), see [21] or the survey [45].

Quasi-stationary distributions have also been studied for birth–death processes with killing (see Section 3.5) in a couple of papers [23, 42]. Let $\gamma_n \geq 0$ be the killing rates. If absorption is certain and

$$0 < \alpha < \liminf_{n\to\infty} \gamma_n,$$

where α is given by (3.132), then there exists a quasi-invariant distribution for the birth–death processes with killing. Conversely, if absorption is certain and

$$\alpha > \limsup_{n\to\infty} \gamma_n,$$

then a quasi-stationary distribution for the birth–death processes with killing exists if and only if the *unkilled process* (the birth–death process one obtains setting $\gamma_n = 0$ for all $n \geq 0$) is recurrent. For a proof of these two facts see [42, Theorems 1 and 2].

3.7.4 A Summary of Other Relevant Results

As in the case of discrete-time birth–death chains, there are several interesting papers in the literature that study other aspects of birth–death processes using the associated spectral measure and the Karlin–McGregor integral representation. In fact, the literature is more extensive for birth–death processes than for discrete-time birth–death chains. Let us give some comments about some of them.

1. In [113] D. P. Maki explores birth–death processes with rational growth birth–death rates of the form

$$\lambda_n = \frac{A_p(n)}{B_q(n)}, \quad \mu_n = \frac{C_r(n)}{D_s(n)},$$

where A_p, B_q, C_r, D_s are polynomials of degree p, q, r, s respectively. The author is interested in finding conditions on these degrees in such a way that there exists a unique transition function $P_{ij}(t, \psi)$ satisfying both the Kolmogorov equations,

$P_{ij}(t,\psi) \geq 0$, $\sum_{j=0}^{\infty} P_{ij}(t) \leq 1$ and the Chapman–Kolmogorov equations. The distribution function is also given in terms of a continued fraction. In each of the following cases there is a unique solution:

- $p > q$ and $r \leq s$.
- $p \leq q$ and $r > s$.
- $p > q, r > s$ and $p - q \neq r - s$.
- $p > q, r > s, p-q = r-s$, but $ad \neq bc$, where a,b,c,d are the leading coefficients of the polynomials A_p, B_q, C_r, D_s.

2. In [35] (see also the monograph [36]), the concept of *stochastic monotonicity* using spectral methods is treated for the case of natural birth–death process (i.e. $C = D = \infty$, see (3.28)). For any birth–death process $\{X_t, t \geq 0\}$ with initial distribution $u_i = \mathbb{P}(X_0 = i)$, we say that it is *stochastically increasing* on the interval (t_1,t_2) if and only if for all τ_1, τ_2 such that $t_1 \leq \tau_1 < \tau_2 < t_2$, we have that

$$\sum_{i=0}^{\infty} u_i \sum_{j=0}^{k} P_{ij}(\tau_2) < \sum_{i=0}^{\infty} u_i \sum_{j=0}^{k} P_{ij}(\tau_1) \quad \text{for all} \quad k \geq 0.$$

The same definition holds for stochastically decreasing birth–death process with the obvious changes. For a more general definition of stochastic monotonicity, see [1, Section 7.3]. The concept of stochastic monotonicity is related to the dual polynomials. In [35] it is proved that for $\mu_0 = 0$ and $u_n = \delta_{in}, i \geq 1$ (i.e. the process that starts at state i), a birth–death process is stochastically increasing on the interval (t_1,t_2) with $t_1 > 0$ if and only if

$$\int_0^{\infty} e^{-xt_1} Q_i(x) x \, d\psi(x) \geq 0.$$

It is also proved that if the birth–death process is transient or null-recurrent then it is stochastically increasing on the interval (τ_i, ∞) and nowhere else, where τ_i is the unique solution of the equation

$$\int_0^{\infty} e^{-xt} Q_i(x) x \, d\psi(x) = 0, \quad t \geq 0.$$

3. The concept of decay parameter (see Section 3.7.3) is related to the concept of α-recurrence (see [41]). We say that a birth–death process is *α-recurrent* if for some state $i \in \mathbb{N}_0$, we have that

$$\int_0^{\infty} e^{\alpha t} P_{ii}(t) dt = \infty,$$

or *α-transient* otherwise. Similarly, if the birth–death process is α-recurrent, then we say that it is *α-positive recurrent* if for some state $i \in \mathbb{N}_0$, we have that

$$\lim_{t\to\infty} e^{\alpha t} P_{ii}(t) > 0,$$

or *α-null recurrent* otherwise. When $\alpha = 0$ we recover the usual definitions of recurrence.

In order to study α-recurrence we use the α-birth–death process, already defined in Proposition 3.18. It is easy to see that the birth–death process $\{X_t, t \geq 0\}$ is α-recurrent (α-positive recurrent) if and only if the α-birth–death process $\{X_t^{(\alpha)}, t \geq 0\}$ is recurrent (positive recurrent). A combination of Proposition 3.18 and Theorem 3.42 gives that, for $\mu_0 = 0$, a birth–death process is α-recurrent if and only if

$$\int_\alpha^\infty \frac{d\psi(x)}{x-\alpha} = \infty \quad \Leftrightarrow \quad \sum_{n=0}^{\infty} \frac{1}{\lambda_n \pi_n Q_n(\alpha) Q_{n+1}(\alpha)} = \infty,$$

and it is α-positive recurrent if and only if

$$\psi(\{\alpha\}) = \left(\sum_{n=0}^{\infty} \pi_n Q_n^2(\alpha)\right)^{-1} > 0 \quad \Leftrightarrow \quad \sum_{n=0}^{\infty} \pi_n Q_n^2(\alpha) < \infty.$$

3.8 Bilateral Birth–Death Processes

Let $\{X_t, t \geq 0\}$ be a birth–death process where the state space is given by the set integer numbers, i.e. $\mathcal{S} = \mathbb{Z}$. The infinitesimal operator matrix $\mathcal{A}$ is a doubly infinite tridiagonal matrix

$$\mathcal{A} = \left(\begin{array}{cccc|ccc} \ddots & \ddots & & \ddots & & & \\ & \mu_{-1} & & -\mu_{-1}-\lambda_{-1} & \lambda_{-1} & & \\ \hline & & & \mu_0 & -\mu_0-\lambda_0 & \lambda_0 & \\ & & & & \mu_1 & -\mu_1-\lambda_1 & \lambda_1 \\ & & & & & \ddots & \ddots \quad \ddots \end{array}\right),$$

where $\lambda_i, \mu_i > 0$ for all $i \in \mathbb{Z}$. These processes are also called *bilateral birth–death processes* (see [124]). A diagram of the transitions between states is

$$\cdots \underset{\mu_{-2}}{\overset{\lambda_{-3}}{\rightleftarrows}} (-2) \underset{\mu_{-1}}{\overset{\lambda_{-2}}{\rightleftarrows}} (-1) \underset{\mu_{0}}{\overset{\lambda_{-1}}{\rightleftarrows}} (0) \underset{\mu_{1}}{\overset{\lambda_{0}}{\rightleftarrows}} (1) \underset{\mu_{2}}{\overset{\lambda_{1}}{\rightleftarrows}} (2) \underset{\mu_{3}}{\overset{\lambda_{2}}{\rightleftarrows}} \cdots$$

Define the potential coefficients as usual by

$$\pi_0 = 1, \quad \pi_n = \frac{\lambda_0 \lambda_1 \cdots \lambda_{n-1}}{\mu_1 \mu_2 \cdots \mu_n}, \quad \pi_{-n} = \frac{\mu_0 \mu_{-1} \cdots \mu_{-n+1}}{\lambda_{-1} \lambda_{-2} \cdots \lambda_{-n}}, \quad n \geq 1. \tag{3.135}$$

We will work in the Hilbert space $\ell^2_\pi(\mathbb{Z})$. In this space it is easy to see that the domain $D(\mathcal{A})$ of the operator $\mathcal{A}$ acting on $\ell^2_\pi(\mathbb{Z})$ is symmetric and negative, i.e. $D(\mathcal{A})$ is dense in $\ell^2_\pi(\mathbb{Z})$, $(u,\mathcal{A}v) = (\mathcal{A}u, v)$ for all $u, v \in D(\mathcal{A})$ and $(u,\mathcal{A}u) < 0$ if $u \neq 0$. As all symmetric operators, it has a closed minimal symmetric extension, so we can assume that $\mathcal{A}$ is a closed, symmetric and negative operator. We also assume that $\mathcal{A}$ is self-adjoint, so the Hamburger moment problem is unique. A necessary and sufficient condition for this, following Theorem 3.26, is that

$$\sum_{n=0}^{\infty}\pi_n\left(1+\mu_0\sum_{k=0}^{n-1}\frac{1}{\lambda_k\pi_k}\right)^2=\infty,\quad \sum_{n=0}^{\infty}\pi_{-n}\left(1+\lambda_0\sum_{k=0}^{n-1}\frac{1}{\mu_k\pi_k}\right)^2=\infty.$$

Define now the operator $P(t)=e^{t\mathcal{A}}$, where the exponential is defined in a functional way for self-adjoint operators (see more details in [139]). If we define the vectors $e^{(i)}=\delta_{ij}/\pi_j$ in $\ell^2_\pi(\mathbb{Z})$, we have that $\mathcal{A}e^{(i)}=\mu_i e^{(i-1)}-(\lambda_i+\mu_i)e^{(i)}+\lambda_i e^{(i+1)}$. The eigenvalue equation $-x\phi(x)=\mathcal{A}\phi(x)$ has, for each real or complex x, two linearly independent polynomials families of solutions $Q_n^\alpha(x), \alpha=1,2, n\in\mathbb{Z}$, depending on the initial values at $n=0$ and $n=-1$. These polynomials are given by

$$\begin{aligned} Q_0^1(x)&=1,\quad Q_0^2(x)=0,\\ Q_{-1}^1(x)&=0,\quad Q_{-1}^2(x)=1,\\ -xQ_n^\alpha(x)&=\lambda_n Q_{n+1}^\alpha(x)-(\lambda_n+\mu_n)Q_n^\alpha(x)+\mu_n Q_{n-1}^\alpha(x). \end{aligned}\tag{3.136}$$

As in the case of discrete-time birth–death chains, we have

$$\begin{aligned} &\deg(Q_n^1)=n,\quad n\geq 0, &&\deg(Q_n^2)=n-1,\quad n\geq 1,\\ &\deg(Q_{-n-1}^1)=n-1,\quad n\geq 1, &&\deg(Q_{-n-1}^2)=n,\quad n\geq 0, \end{aligned}$$

and the two families of polynomial solutions can be computed using the three-term recurrence relation (3.136):

$$Q_{n+1}^\alpha(x)=\frac{1}{\lambda_n}((-x+\lambda_n+\mu_n)Q_n^\alpha(x)-\mu_n Q_{n-1}^\alpha(x)),\quad\text{if}\quad n\geq 0,$$

$$Q_{-n-1}^\alpha(x)=\frac{1}{\mu_{-n}}((-x+\lambda_{-n}+\mu_{-n})Q_{-n}^\alpha(x)-\lambda_{-n}Q_{-n+1}^\alpha(x)),\quad\text{if}\quad n\geq 1.$$

Using the recurrence for the polynomials Q_n^α, we can prove using induction that

$$Q_i^1(-\mathcal{A})e^{(0)}+Q_i^2(-\mathcal{A})e^{(-1)}=e^{(i)},\quad i\in\mathbb{Z}.$$

Therefore,

$$\begin{aligned} P_{ij}(t)&=\pi_j(e^{t\mathcal{A}}e^{(j)},e^{(i)})_\pi\\ &=\pi_j(e^{t\mathcal{A}}[Q_j^1(-\mathcal{A})e^{(0)}+Q_j^2(-\mathcal{A})e^{(-1)}],Q_i^1(-\mathcal{A})e^{(0)}+Q_i^2(-\mathcal{A})e^{(-1)})_\pi. \end{aligned}$$

We need to apply now the spectral theorem *three times*. In particular, there exist three measures $\psi_{11}(x)$, $\psi_{22}(x)$ and $\psi_{12}(x)$ (since $\psi_{12}(x) = \psi_{21}(x)$ because $\mathcal{A}$ is self-adjoint and the symmetry of the inner product) supported on the interval $[0,\infty)$ such that

$$\int_0^\infty f(x)\,d\psi_{11}(x) = (f(-\mathcal{A})e^{(0)}, e^{(0)})_\pi,$$

$$\int_0^\infty f(x)\,d\psi_{12}(x) = (f(-\mathcal{A})e^{(0)}, e^{(-1)})_\pi,$$

$$\int_0^\infty f(x)\,d\psi_{22}(x) = (f(-\mathcal{A})e^{(-1)}, e^{(-1)})_\pi,$$

where f is any bounded real function defined on $[0,\infty)$. We then have the Karlin–McGregor representation formula

$$P_{ij}(t) = \mathbb{P}(X_t = j \mid X_0 = i) = \pi_j \int_0^\infty e^{-xt} \sum_{\alpha,\beta=1}^{2} Q_i^\alpha(x) Q_j^\beta(x)\,d\psi_{\alpha\beta}(x), \quad i,j \in \mathbb{Z}. \tag{3.137}$$

The method used in [124] to compute $P_{ij}(t)$ is by the Laplace transform of $P(t)$ (resolvent function, see (3.12)) and using functional analysis methods. By taking $n = 0$ in the previous formula and some values of i,j we can give more information about the measures $\psi_{11}(x)$, $\psi_{22}(x)$ and $\psi_{12}(x)$:

- If $i = j = 0$ then, by definition of the polynomials, we have that $1 = \int_0^\infty d\psi_{11}(x)$. Therefore ψ_{11} is a proper probability distribution which vanishes for $x < 0$.
- If $i = j = -1$ then again we have that $1 = \pi_{-1} \int_0^\infty d\psi_{22}(x)$. Therefore ψ_{22} is a positive distribution which vanishes for $x < 0$ and it is $1/\pi_{-1}$ evaluated at the whole interval $[0,\infty)$.
- If $i = -1$ and $j = 0$ then we get $0 = \int_0^\infty d\psi_{12}(x)$. That means that ψ_{12} is a signed distribution vanishing for $x < 0$.

To compute the measures $\psi_{\alpha\beta}, \alpha = \beta = 1,2$, we can use a similar argument to the one we used for discrete-time birth–death chains, consisting in splitting $\mathcal{A}$ into two regular birth–death processes with state space $\{0,1,\ldots\}$, where the infinitesimal operator matrices are given by $\mathcal{A}_{ij}^+ = \mathcal{A}_{ij}, i,j \geq 0$ and $\mathcal{A}_{ij}^- = \mathcal{A}_{ij}, i,j \leq -1$. If we denote by $\psi^\pm$ the spectral measures associated with the matrices $\mathcal{A}^\pm$, and using (3.55) and (3.56), we get that the Stieltjes transforms of $\psi_{\alpha\beta}, \alpha,\beta = 1,2$, are given by

$$\begin{aligned} B(z;\psi_{11}) &= \frac{B(z;\psi^+)}{1 - \lambda_{-1}\mu_0 B(z;\psi^+)B(z;\psi^-)}, \\ \frac{\mu_0}{\lambda_{-1}} B(z;\psi_{22}) &= \frac{B(z;\psi^-)}{1 - \lambda_{-1}\mu_0 B(z;\psi^+)B(z;\psi^-)}, \\ B(z;\psi_{12}) &= \frac{\lambda_{-1} B(z;\psi^+)B(z;\psi^-)}{1 - \lambda_{-1}\mu_0 B(z;\psi^+)B(z;\psi^-)}. \end{aligned} \tag{3.138}$$

As in the case of discrete-time birth–death chains, these three measures can be grouped in a positive definite 2×2 matrix, called the *spectral matrix* of the bilateral birth–death process:

$$\Psi(x) = \begin{pmatrix} \psi_{11}(x) & \psi_{12}(x) \\ \psi_{12}(x) & \psi_{22}(x) \end{pmatrix}.$$

The spectral analysis on $\mathbb{Z}$ can be studied using the 2×2 spectral matrix. The Karlin–McGregor formula (3.137) can be written in matrix form as

$$P_{ij}(t) = \pi_j \int_0^\infty e^{-xt} \left(Q_i^1(x), Q_i^2(x) \right) d\Psi(x) \begin{pmatrix} Q_j^1(x) \\ Q_j^2(x) \end{pmatrix}, \quad i,j \in \mathbb{Z}.$$

Relabeling the states as

$$\{0,1,2,\dots\} \to \{0,2,4,\dots\}, \quad \text{and} \quad \{-1,-2,-3,\dots\} \to \{1,3,5,\dots\},$$

we have that the matrix $\mathcal{A}$ (doubly infinite) is equivalent to a semi-infinite block matrix $\boldsymbol{\mathcal{A}}$, given by

$$\boldsymbol{\mathcal{A}} = \begin{pmatrix} B_0 & A_0 & & & \\ C_1 & B_1 & A_1 & & \\ & C_2 & B_2 & A_2 & \\ & & \ddots & \ddots & \ddots \end{pmatrix},$$

where

$$B_0 = \begin{pmatrix} -\lambda_0 - \mu_0 & \mu_0 \\ \lambda_{-1} & -\lambda_{-1} - \mu_{-1} \end{pmatrix},$$

$$B_n = \begin{pmatrix} -\lambda_n - \mu_n & 0 \\ 0 & -\lambda_{-n-1} - \mu_{-n-1} \end{pmatrix}, \quad n \geq 1,$$

$$A_n = \begin{pmatrix} \lambda_n & 0 \\ 0 & \mu_{-n-1} \end{pmatrix}, \quad n \geq 0, \quad C_n = \begin{pmatrix} \mu_n & 0 \\ 0 & \lambda_{-n-1} \end{pmatrix}, \quad n \geq 1.$$

The birth–death process generated by $\boldsymbol{\mathcal{A}}$ can be interpreted as a process that takes values in a two-dimensional state space $\{0,1,2,\dots\} \times \{1,2\}$. An isomorphic diagram associated with the probability transitions (two-dimensional) is given by

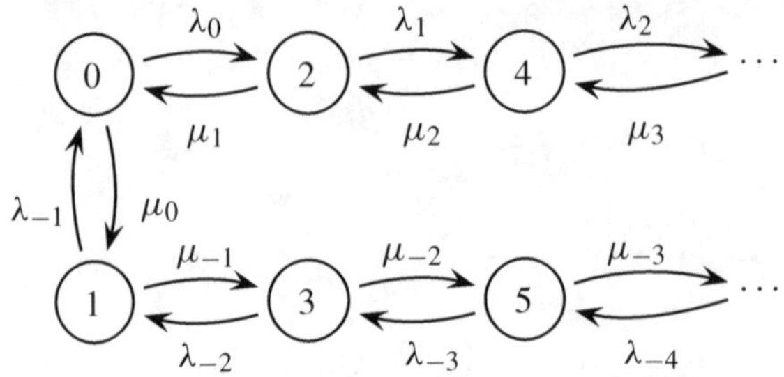

These types of processes are called continuous-time *quasi-birth–death processes*. The first component ($\mathbb{N}_0$) is usually called the *level*, while the second component ($\{1,2\}$ in this case but in general it can be any set $\{1,\dots,N\}$) is usually called the *phase*. In general, these processes allow transitions between all adjacent levels (see [110, 119] for a general reference).

We can apply the spectral theorem to the matrix $\boldsymbol{\mathcal{A}}$, but now applied to a 2×2 matrix of measures, given precisely by $d\Psi(x)$. If we define the matrix-valued polynomials $\boldsymbol{Q}_n(x)$ by

$$\boldsymbol{Q}_n(x) = \begin{pmatrix} Q_n^1(x) & Q_n^2(x) \\ Q_{-n-1}^1(x) & Q_{-n-1}^2(x) \end{pmatrix}, \quad n \geq 0, \tag{3.139}$$

then we have that they satisfy the three-term recurrence relation

$$\begin{aligned} -x\boldsymbol{Q}_0(x) &= A_0\boldsymbol{Q}_1(x) + B_0\boldsymbol{Q}_0(x), \quad \boldsymbol{Q}_0(x) = I_{2\times 2}, \\ -x\boldsymbol{Q}_n(x) &= A_n\boldsymbol{Q}_{n+1}(x) + B_n\boldsymbol{Q}_n(x) + C_n\boldsymbol{Q}_{n-1}(x), \quad n \geq 1, \end{aligned}$$

where $I_{2\times 2}$ denotes the 2×2 identity matrix. Observe that the leading coefficient of $\boldsymbol{Q}_n$ is a nonsingular matrix. The matrix orthogonality is defined in terms of the (matrix-valued) inner product

$$\int_0^\infty \boldsymbol{Q}_n(x)d\Psi(x)\boldsymbol{Q}_m^*(x) = \begin{pmatrix} 1/\pi_n & 0 \\ 0 & 1/\pi_{-n-1} \end{pmatrix} \delta_{nm},$$

where A^* is the Hermitian transpose of a matrix A and $(\pi_n)_{n\in\mathbb{Z}}$ are the potential coefficients defined by (3.135). The Karlin–McGregor formula applied $\boldsymbol{\mathcal{A}}$ is then given by (see [33])

$$\boldsymbol{P}_{ij}(t) = \left(\int_0^\infty e^{-xt}\boldsymbol{Q}_i(x)d\Psi(x)\boldsymbol{Q}_j^*(x)\right)\begin{pmatrix} \pi_j & 0 \\ 0 & \pi_{-j-1} \end{pmatrix}, \quad i,j \in \mathbb{N}_0.$$

Observe now that the block entry (i,j) is a 2×2 matrix. The relation between this representation $\boldsymbol{P}_{ij}(t), i,j \in \mathbb{N}$, and the original one $P_{ij}(t), i,j \in \mathbb{Z}$, is given by

$$\boldsymbol{P}_{ij}(t) = \begin{pmatrix} P_{ij}(t) & P_{i,-j-1}(t) \\ P_{-i-1,j}(t) & P_{-i-1,-j-1}(t) \end{pmatrix}, \quad i,j \in \mathbb{N}_0.$$

As in the case of discrete-time birth–death chains on $\mathbb{Z}$, it is possible to see that the concept of recurrence can be extended in a natural way to continuous-time quasi-birth–death processes (see [33]). We can extend the result about recurrence in Theorem 3.42. Indeed, the birth–death process is recurrent if and only if

$$\int_0^\infty \frac{d\psi_{11}(x)}{x} = \infty, \quad \text{or} \quad \int_0^\infty \frac{d\psi_{22}(x)}{x} = \infty. \tag{3.140}$$

In a similar way, the birth–death process is positive recurrent if and only if one the measures ψ_{11}, ψ_{22} has a jump at the point 0.

Example 3.66 Consider $\{X_t, t \geq 0\}$ the birth–death process on the integers $\mathbb{Z}$ with constant birth–death rates, i.e.

$$\lambda_n = \lambda, \quad \mu_n = \mu, \quad n \in \mathbb{Z}, \quad \lambda, \mu > 0.$$

This example was considered for the first time in the last section of [75]. The matrix $\mathcal{A}^+$ is the same as the one of the $M/M/1$ queue in Example 3.35 (see (3.97)), while $\mathcal{A}^-$ is the symmetric matrix of $\mathcal{A}^+$. Therefore both processes generate the same Stieltjes transform given by (3.98). Following (3.138), for $B(z; \psi_{11})$ and $B(z; \psi_{12})$ and rationalizing, we obtain that

$$B(z; \psi_{11}) = \frac{1}{\sqrt{(\lambda + \mu - z)^2 - 4\lambda\mu}},$$

$$B(z; \psi_{12}) = \frac{1}{2\mu}\left(-1 + \frac{\lambda + \mu - z}{\sqrt{(\lambda + \mu - z)^2 - 4\lambda\mu}}\right).$$

Since $\psi^+ = \psi^-$, then $\mu\psi_{22} = \lambda\psi_{11}$. Observe that the jumps, if any, should be located at $x = (\sqrt{\lambda} \pm \sqrt{\mu})^2$. However, using (1.12), we can see that the size of these jumps must be 0, so there are no jumps. The spectral matrix is then

$$d\Psi(x) = \frac{1}{\pi\sqrt{4\lambda\mu - (\lambda + \mu - x)^2}} \begin{pmatrix} 1 & (\lambda + \mu - x)/2\mu \\ (\lambda + \mu - x)/2\mu & \lambda/\mu \end{pmatrix} dx,$$

defined on the interval $[(\sqrt{\lambda} - \sqrt{\mu})^2, (\sqrt{\lambda} + \sqrt{\mu})^2]$. The polynomials are normalizations of the Chebychev polynomials of the second kind, as we had in the case of discrete-time birth–death chains.

Since $(\sqrt{\lambda} - \sqrt{\mu})^2 > 0$ unless $\lambda = \mu$, we have that both integrals in (3.140) are bounded. Therefore, if $\lambda \neq \mu$, the birth–death process is always transient. If $\lambda = \mu$, then the integrals in (3.140) are divergent. Therefore this is the only case where the birth–death process is recurrent. Since the measures ψ_{11}, ψ_{22} do not have a jump at 0, the birth–death process can never be positive recurrent. ◊

Remark 3.67 As in the case of discrete-time birth–death chains on $\mathbb{Z}$, many other examples can be studied using the same techniques. However, there are not so many examples studied in the literature, as in the case of discrete-time birth–death chains on $\mathbb{Z}$. In the Exercises we include some of them, but they have not been studied before. ◊

3.9 Exercises

3.1 Consider a quadratic pure birth process $\{X_t, t \geq 0\}$ where $\lambda_n = \lambda n^2, n \geq 0$. Prove that $\mathbb{P}(J_\infty < \infty | X_0 = i) = 1$, where J_∞ is the time of the first explosion. Prove also that the distribution function of J_∞ is given by

$$\mathbb{P}(J_\infty \leq t | X_0 = i) = 1 - 2\sum_{k=i}^{\infty}(-1)^{k+1}e^{-\lambda k^2 t}, \quad t \geq 0, \quad i \geq 1.$$

3.2 Assume that $C = D = \infty$ (see (3.28)). Let $f_{ij}(t)$ and $f_{ij}^d(t)$ be the unique minimal $\mathcal{A}$-functions of a birth–death process $\{X_t, t \geq 0\}$ with $\mu_0 = 0$ and its dual $\{X_t^d, t \geq 0\}$, respectively. Prove that

$$f_{ij}^d(t) = \sum_{k=0}^{i}\left[f_{jk}(t) - f_{j+1,k}(t)\right], \quad f_{ij}(t) = \sum_{k=i}^{\infty}\left[f_{jk}^d(t) - f_{j-1,k}^d(t)\right], \quad i,j \geq 0,$$

with the convention that $f_{-1,k}^d(t) = 0$ for all $k \geq 0$.

3.3 Consider the birth–death process $\{X_t, t \geq 0\}$ with infinitesimal operator matrix given by

$$\mathcal{A} = \begin{pmatrix} -\lambda & \lambda & & & \\ \mu & -\lambda-\mu & \lambda & & \\ & \ddots & \ddots & \ddots & \\ & & \mu & -\lambda-\mu & \lambda \\ & & & \mu & -\mu \end{pmatrix}.$$

Compute the eigenvalues and eigenvectors of $\mathcal{A}$ and find an expression for the transition probability functions $P_{ij}(t)$.

3.4 Consider the modified $M/M/1$ queue with infinitesimal operator matrix given by

$$\mathcal{A} = \begin{pmatrix} -\lambda_0-\mu_0 & \lambda_0 & & & \\ \mu & -\lambda-\mu & \lambda & & \\ & \mu & -\lambda-\mu & \lambda & \\ & & \ddots & \ddots & \ddots \end{pmatrix}.$$

Find an expression of the spectral measure and the corresponding orthogonal polynomials.

3.5 Consider the conservative $M/M/1$ queue from Example 3.35 for $\lambda = \mu$. Prove that

$$P_{00}(t) \to \frac{1}{\sqrt{\pi \mu t}}, \quad \text{as} \quad t \to \infty,$$

and that the distribution of the length of a busy period (or the distribution of the first passage time) is given by

$$F_{10}(t) = \frac{1}{2\pi\lambda}\int_0^{4\lambda} \frac{1-e^{-xt}}{x}\sqrt{x(4\lambda - x)}\,dx.$$

3.6 Consider the $M/M/\infty$ queue from Example 3.36. For the case $\mu = 1$ compute numerically the first α_n, s_n (see [86, p. 94]).

3.7 Consider the $M/M/k$ queue from Example 3.37 with $k = 2$ servers. Let T be the time at which the process first reaches the state 1 starting from state 2 (busy period time if both servers are busy) and T^* the time at which the process first reaches the state 0 starting from state 1 (busy period time if at least one server is busy). Prove that

$$\mathbb{P}(T < t) = \frac{1}{2\lambda}\int_{\lambda+2\mu-\sqrt{8\lambda\mu}}^{\lambda+2\mu+\sqrt{8\lambda\mu}} \frac{1-e^{-xt}}{x}\sqrt{8\lambda\mu - (\lambda+2\mu-x)^2}\,dx,$$

$$\mathbb{P}(T^* < t) = \frac{1-e^{-t(\mu-\lambda)}}{(\mu-2\lambda)^{-1}(\mu-\lambda)}\mathbf{1}_{\{\mu>2\lambda\}} + \frac{1}{2\pi}\int_{\lambda+2\mu-\sqrt{8\lambda\mu}}^{\lambda+2\mu+\sqrt{8\lambda\mu}} \frac{1-e^{-xt}}{x}\,\frac{\sqrt{8\lambda\mu-(\lambda+2\mu-x)^2}}{x-\mu+\lambda}\,dx.$$

3.8 Consider the $M/M/k$ queue from Example 3.37 with $k = 2$ servers and T^* the busy period time if at least one server is busy (see the previous exercise). Compute the probability distribution of the number N^* of customers arriving during the busy period T^* in terms of the spectral measure associated with the discrete-time birth–death chain on $\mathbb{N}$ of the jump chain with probabilities

$$p_1 = \frac{\lambda}{\lambda+\mu}, \quad p_n = p = \frac{\lambda}{\lambda+2\mu}, \quad n \geq 2, \quad q_n = q = \frac{\mu}{\lambda+2\mu}, \quad n \geq 1.$$

Prove also that this spectral measure is given by

$$\psi(x) = \frac{\gamma}{\pi}\frac{\sqrt{4pq - x^2}}{4pq - (1-(\gamma-1)^2)x^2}, \quad |x| < 2\sqrt{pq}, \quad \gamma = 2p/p_1.$$

3.9 Consider the $M/M/k$ queue from Example 3.37. Study the case of $k = 3$ servers. Prove, using some computational software, that $L_3(z)$ and $H_3(z)$ are given by

$$L_3(z) = \frac{4\mu}{\lambda^4}\Big[z^4 - (4\mu+3\lambda)z^3 + (5\mu^2+3\lambda^2+9\mu\lambda)z^2$$
$$- (2\mu^3+\lambda^3+4\mu\lambda^2+6\mu^2\lambda)z - \lambda^2\mu(\lambda-3\mu)\Big]$$
$$+ \frac{4\mu^2}{\lambda^2}\sqrt{(\lambda+3\mu-z)^2 - 12\lambda\mu},$$
$$H_3(z) = -\frac{4\mu z}{\lambda^4}\Big[z^4 - 4(\lambda+\mu)z^3 + (6\mu\lambda+6\lambda^2+5\mu^2)z^2$$
$$-(2\mu^2\lambda+4\lambda^3+2\mu^3)z + \lambda^2(\lambda^2-2\mu\lambda-6\mu^2)\Big].$$

Compute the spectral measure ψ as well as all possible atoms and sizes for the special case $\lambda = \mu$.

3.10 By using the generating function for the Laguerre polynomials (1.70) prove that the transition function $P_{ij}(t)$ of the linear birth–death process of Example 3.38 (for $\lambda = \mu$) satisfies the following relations (see [85, Section 4]):

$$\sum_{j=0}^{\infty} P_{ij}(t)s^j = \frac{(\lambda t)^i}{(1+\lambda t)^{\beta+i}} \frac{\left(1 + \dfrac{1-\lambda t}{\lambda t}s\right)^i}{\left(1 - \dfrac{\lambda t}{1+\lambda t}s\right)^{\beta+i}},$$

and for $i \leq j$,

$$P_{ij}(t) = \frac{(\lambda t)^i}{(1+\lambda t)^{\beta+i}} \sum_{k=0}^{i} \binom{i}{k} \left(\frac{1-\lambda t}{\lambda t}\right)^k \left(\frac{\lambda t}{1+\lambda t}\right)^{j-k} \frac{(\beta+i)_{j-k}}{(j-k)!}.$$

3.11 Consider the jump chain (3.26) associated with the linear birth–death process of Example 3.38 for $\lambda = \mu$ (absorbing case). Find the spectral measure associated with the corresponding one-step transition probability matrix and compute the probabilities R_i^n given by (3.118) in terms of this spectral measure.

3.12 Consider the birth–death process on $\mathbb{N}_0$ with birth–death rates given by

$$\lambda_n = \frac{\lambda}{n+1}, \quad n \geq 0, \quad \mu_0 = 0, \quad \mu_n = \mu, \quad n \geq 1, \quad \lambda, \mu > 0.$$

This model serves as a single-server queueing model where potential customers are discouraged by queue length (see [17, 37]). Prove that the spectral measure associated with this example is a discrete measure with jumps located at the points $0, \mu, \mu a_k, \mu b_k, k \geq 1$, where

$$a_k = 1 - \frac{\sqrt{\nu^2+4k\nu} - \nu}{2k}, \quad b_k = \frac{1}{a_k} = 1 + \frac{\sqrt{\nu^2+4k\nu}+\nu}{2k}, \quad \nu = \lambda/\mu.$$

Follow [37] to compute the corresponding jumps and orthogonal polynomials, as well as a Karlin–McGregor representation formula.

3.13 Prove the following formulas for the first two moments of $F_{ii}(t)$:

$$\int_0^\infty t dF_{ii}(t) = \frac{1}{\psi(\{0\})\pi_i(\lambda_i+\mu_i)},$$

$$\int_0^\infty t^2 dF_{ii}(t) = \frac{2}{\psi(\{0\})\pi_i(\lambda_i+\mu_i)^2} + \frac{2}{\pi_i(\lambda_i+\mu_i)} \times \left[\sum_{n=i}^\infty \frac{1}{\lambda_n\pi_n}\left(\sum_{j=n}^\infty \pi_{j+1}\right)^2 + \sum_{n=0}^{i-1}\frac{1}{\lambda_n\pi_n}\left(\sum_{j=n}^\infty \pi_j\right)^2\right].$$

3.14 Use the same analysis as in the previous exercise to compute the moments of $F_{ij}(t), i \neq j$, using now the formula

$$\int_0^\infty e^{-\lambda t} dF_{ij}(t) = 1 - \frac{\lambda}{\psi(\{0\})}\frac{\displaystyle\int_0^\infty \frac{1}{x+\lambda}[Q_j^2(x) - Q_i(x)Q_j(x)]\,d\psi^+(x)}{1+\displaystyle\frac{1}{\psi(\{0\})}\int_0^\infty \frac{\lambda}{x+\lambda}Q_j^2(x)\,d\psi^+(x)}$$

(see more details in [84, Theorem 5]).

3.15 Show that the first moment of $G_i(t)$ in (3.115) (if finite) can be written as (see more details in [84, (9.20)])

$$\int_0^\infty t dG_i(t) = \frac{1}{\mu_0}\sum_{k=0}^\infty \pi_k + \sum_{j=0}^{i-1}\frac{1}{\lambda_j\pi_j}\sum_{r=j+1}^\infty \pi_r.$$

3.16 Let $\sum_{n=0}^\infty (\lambda_n\pi_n)^{-1} < \infty$ and the solution of the Stieltjes moment problem be unique. Prove that if $\int_0^\infty x^{-k}d\psi(x) < \infty$ for some $k \geq 0$ then

$$\lim_{n\to\infty}\int_0^\infty \frac{Q_n(x)}{x^k}\,d\psi(x) = 0.$$

3.17 Let $\mu_0 = 0$ and $\sum_{n=0}^\infty \pi_n = \infty$. Use the previous exercise to prove that

$$\int_0^\infty \frac{Q_i(x)Q_j(x)}{x^k}d\psi(x) = \int_0^\infty \frac{Q_j(x)}{x^k} - \sum_{n=0}^{i-1}\frac{1}{\lambda_n\pi_n}\sum_{h=0}^n \pi_h \int_0^\infty \frac{Q_h(x)Q_j(x)}{x^{k-1}}\,d\psi(x). \tag{3.141}$$

In particular, prove that if $i \leq j$, then

$$\int_0^\infty \frac{Q_i(x)Q_j(x)}{x}d\psi(x) = \pi_j\sum_{n=j}^\infty \frac{1}{\lambda_n\pi_n}.$$

3.18 Study recurrence and absorption for the linear birth–death process of Example 3.39 in terms of the parameters $\lambda, c, \alpha > 0$ using Theorems 3.42 and 3.51.

3.19 Compute the Doeblin ratio (3.122) for the Examples 3.35, 3.36 and 3.38 in the cases where the birth–death process is transient.

3.20 For $\lambda = \mu$, find an expression of the matrix-valued orthogonal polynomials of the Example 3.66 and write the Karlin–McGregor formula in terms of an angle $\theta \in [0, \pi]$, in the same way as we saw in Example 3.35.

3.21 For each of the following bilateral birth–death processes with birth–death rates $\{(\lambda_n, \mu_n), n \in \mathbb{Z}\}$ find the spectral matrix, the corresponding matrix-valued orthogonal polynomials and study the recurrence ($\lambda, \widetilde{\lambda}, \mu, \widetilde{\mu}, \alpha, \beta > 0$):

(a) $\lambda_n = \lambda, \mu_n = \mu, n \geq 0, \quad \lambda_{-n} = \mu, \mu_{-n} = \lambda, n \geq 1.$

(b) $\lambda_n = \lambda, \mu_n = \mu, n \neq 0, \quad \lambda_0 \neq \lambda, \quad \mu_0 \neq \mu.$

(c) $\lambda_n = \lambda, \mu_n = \mu, n \geq 1, \quad \lambda_{-n} = \mu, \mu_{-n} = \lambda, n \geq 1, \quad \lambda_0 \neq \lambda, \mu_0 \neq \mu.$

(d) $\lambda_{2n} = \mu_{2n} = \lambda, \quad \lambda_{2n+1} = \mu_{2n+1} = \mu, \quad n \in \mathbb{Z}.$

(e) $\lambda_n = \lambda, \mu_{n+1} = \mu, n \geq 0, \quad \lambda_{-n-1} = \widetilde{\lambda}, \mu_{-n} = \widetilde{\mu}, n \geq 1, \quad \lambda_{-1} = \alpha,$ $\mu_0 = \beta.$

4

Spectral Representation of Diffusion Processes

The goal of this chapter is to describe the probabilistic properties of any one-dimensional diffusion process in terms of the corresponding infinitesimal operator of the process, which in this case is a second-order differential operator. After a first section of basic definitions and fundamental properties about one-dimensional diffusion processes, we will obtain in Section 4.2 a spectral representation of the transition probability density $p(t; x, y)$ in terms of the orthogonal eigenfunctions of the corresponding infinitesimal operator of the diffusion process. The spectrum of the infinitesimal operator can now be a set of countable eigenvalues or an absolutely continuous spectrum (or a mix of them). In Section 4.3 we will analyze the behavior at the boundary points of the state space. This will give boundary conditions for the eigenvalue problem of the second-order differential operator corresponding to the infinitesimal operator, also called a *Sturm–Liouville problem*. In Section 4.4 we will analyze the spectral decomposition of a diffusion process with a killing coefficient in terms of the spectral decomposition of a regular diffusion process. Next, in Section 4.5 we will give a collection of useful examples, divided into four categories: first we will study Brownian motions with different drifts and scalings defined in different state spaces and with different boundary behaviors; then we will study examples related to classical orthogonal polynomials, such as the Orstein–Uhlenbeck process, a population growth model, the Wright–Fisher model, the Jacobi diffusion model and other related models; then we will see examples of radial N-dimensional diffusion processes such as the Bessel process and the radial N-dimensional Orstein–Uhlenbeck process; and finally we will include a couple of examples with both absolutely continuous and discrete spectrum. In Section 4.6 we will study the concept of quasi-stationary distributions for one-dimensional diffusion processes, for which the spectral representation will play an important role. This chapter is intended to give a general overview of the main tools to develop spectral representations of diffusion processes rather than a concise and rigorous description of the subject.

4.1 Diffusion Processes

This section is mostly based on references [5, 101]. Let $(\Omega, \mathcal{F}, \mathbb{P})$ be a probability space and consider a diffusion process $\{X_t, t \geq 0\}$ (one-dimensional and without restrictions), which is a continuous-time Markov process with state space $\mathcal{S} = (a,b)$, $-\infty \leq a < b \leq \infty$ and continuous trajectories. The simplest example of diffusion process is *Brownian motion* (also known as the *Wiener process*), which we denote by $\{B_t, t \geq 0\}$. For small intervals of time the expected value and the variance of the increment $B_{t+s} - B_s$ are given by

$$\mathbb{E}(B_{s+t} - B_s | B_s = x) = o(t), \quad t \downarrow 0,$$
$$\mathbb{E}((B_{s+t} - B_s)^2 | B_s = x) = t + o(t), \quad t \downarrow 0,$$

where $o(t)$ is an infinitesimal time satisfying $\lim_{t \downarrow 0} o(t)/t = 0$. The state space is given by $\mathcal{S} = \mathbb{R}$ and the transition probability distribution $p(t;x,dy)$ of B_{t+s} given that $B_s = x$ follows a Gaussian distribution, i.e.

$$p(t;x,y) = \frac{1}{\sqrt{2\pi t}} e^{-\frac{1}{2t}(y-x)^2}.$$

The Brownian motion has independent increments and is homogeneous in time (stationary), so the previous density does not depend on s.

Now consider a Markov process $\{X_t, t \geq 0\}$ which is homogeneous in time, with continuous trajectories, but does not necessarily have independent increments. Assume that for $X_s = x$, for small time values t, the displacement $X_{t+s} - X_s = X_{t+s} - x$ has a mean and variance approximately $t\mu(x)$ and $t\sigma^2(x)$, respectively. The function $\mu(x)$ is called the *drift coefficient* while $\sigma^2(x)$ is called the *diffusion coefficient*. These coefficients may depend on the state x, unlike what happened in the case of Brownian motion (where $\mu(x) = 0$ and $\sigma^2(x) = 1$). More precisely, we assume that as $t \downarrow 0$ and for each $x \in \mathcal{S}$, we have that

$$\begin{aligned}\mathbb{E}(X_{s+t} - X_s | X_s = x) &= t\mu(x) + o(t), \\ \mathbb{E}((X_{s+t} - X_s)^2 | X_s = x) &= t\sigma^2(x) + o(t), \\ \mathbb{E}(|X_{s+t} - X_s|^3 | X_s = x) &= o(t).\end{aligned} \tag{4.1}$$

There are other more general ways to define a diffusion process, such as the following one. For all $\varepsilon > 0$ and as $t \downarrow 0$, we assume that

$$\begin{aligned}\mathbb{E}((X_{s+t} - X_s)\mathbf{1}_{\{|X_{s+t} - X_s| \leq \varepsilon\}} | X_s = x) &= t\mu(x) + o(t), \\ \mathbb{E}((X_{s+t} - X_s)^2 \mathbf{1}_{\{|X_{s+t} - X_s| \leq \varepsilon\}} | X_s = x) &= t\sigma^2(x) + o(t), \\ \mathbb{P}(|X_{s+t} - X_s| > \varepsilon | X_s = x) &= o(t),\end{aligned} \tag{4.2}$$

where $\mathbf{1}_B(x)$ is the indicator function on any Borel set B. Relations (4.1) imply (4.2), but the converse is not true in general. Therefore a Markov process $\{X_t, t \geq 0\}$

with state space $\mathcal{S}$ is a diffusion process with drift coefficient $\mu(x)$ and diffusion coefficient $\sigma^2(x) > 0$, if it has continuous trajectories and relations (4.2) hold for all $x \in \mathcal{S}$. If both drift and diffusion coefficients are constant then we will call $\{X_t, t \geq 0\}$ a *Brownian motion with drift and scaling*.

Let $p(t;x,dy)$ be the transition probability distribution of the diffusion process defined by

$$p(t;x,dy) = \mathbb{P}_x(X_t \in dy) \doteq \mathbb{P}(X_t \in dy | X_0 = x).$$

If $p(t;x,dy)$ has a density $p(t;x,y)$, then for any Borel set $B \subset \mathcal{S}$ we have that

$$p(t;x,B) = \int_B p(t;x,y)\,dy.$$

This density exists if $\mu(x), \sigma(x) \in C^1(\mathcal{S})$, i.e. continuously differentiable functions, with bounded derivatives on $\mathcal{S}$, and $\sigma''(x)$ exists, is continuous and $\sigma^2(x) > 0$ for all $x \in \mathcal{S}$.

Consider the set $\mathcal{B}(\mathcal{S})$ of all measurable Borel real and bounded functions over $\mathcal{S}$. In this space we define the following *transition operator*:

$$(T_t f)(x) = \mathbb{E}_x(f(X_t)) \doteq \mathbb{E}(f(X_t)|X_0 = x) = \int_{\mathcal{S}} f(y)p(t;x,dy), \quad t > 0. \tag{4.3}$$

This operator replaces the concept of transition probability matrix for discrete-time birth–death chains or transition probability functions for birth–death processes. For instance, if $f = \mathbf{1}_B(x)$, then

$$(T_t \mathbf{1}_B)(x) = \mathbb{P}(X_t \in B | X_0 = x).$$

The transition operator $T_t \colon \mathcal{B}(\mathcal{S}) \to \mathcal{B}(\mathcal{S})$ is a linear bounded operator in $\mathcal{B}(\mathcal{S})$ with the supremum norm $\|f\| = \sup\{|f(y)| \colon y \in \mathcal{S}\}$ and $(T_0 f)(x) = x$. Indeed, T_t is a contraction, i.e. $\|T_t f\| \leq \|f\|$ for all $f \in \mathcal{B}(\mathcal{S})$, since

$$|(T_t f)(x)| \leq \int_{\mathcal{S}} |f(y)| p(t;x,dy) \leq \|f\|.$$

The family of transition operators $\{T_t \colon t > 0\}$ has the *semigroup property* $T_{s+t} = T_s T_t$. Indeed, using the well-known property $\mathbb{E}(X) = \mathbb{E}(X|Y)$, we have that

$$\begin{aligned}(T_{s+t} f)(x) &= \mathbb{E}(f(X_{s+t})|X_0 = x) = \mathbb{E}[\mathbb{E}(f(X_{s+t})|X_s)|X_0 = x] = \mathbb{E}[(T_t f)(X_s)|X_0 = x] \\ &= T_s(T_t f)(x).\end{aligned} \tag{4.4}$$

As a consequence we have that $T_s T_t = T_t T_s$. This semigroup property implies that the behavior of T_t is completely determined by the behavior of T_t as $t \downarrow 0$ (similar to the case of continuous-time Markov chains).

Definition 4.1 Let $\{T_t, t > 0\}$ be the transition operator (4.3) of a diffusion process $\{X_t, t \geq 0\}$. We define the *infinitesimal operator* $\mathcal{A}$ of $\{T_t, t > 0\}$ as a linear operator of the form

$$(\mathcal{A}f)(x) = \lim_{t \downarrow 0} \frac{(T_t f)(x) - f(x)}{t},$$

for all $f \in \mathcal{B}(\mathcal{S})$ such that the right-hand part converges uniformly in x to some function. The class of all these functions f forms the domain $D_{\mathcal{A}}$ of $\mathcal{A}$. ◊

Proposition 4.2 ([5, Proposition V.2.1]) *Let $f \in C^2(\mathcal{S})$ and bounded on $\mathcal{S}$. For $x \in \mathcal{S}$ we have that*

$$(\mathcal{A}f)(x) = \lim_{t \downarrow 0} \frac{(T_t f)(x) - f(x)}{t} = \mu(x)f'(x) + \frac{1}{2}\sigma^2(x)f''(x), \tag{4.5}$$

where $\mu(x)$ and $\sigma^2(x)$ are the drift and diffusion coefficients, respectively, defined by (4.2).

Proof For $x \in \mathcal{S}$ and $\delta > 0$, let $\varepsilon > 0$ be such that $|f''(x) - f''(y)| \leq \delta$ for all y satisfying that $|y - x| \leq \varepsilon$. Write

$$(T_t f)(x) = \mathbb{E}(f(X_t)\mathbf{1}_{\{|X_t - x| \leq \varepsilon\}} | X_0 = x) + \mathbb{E}(f(X_t)\mathbf{1}_{\{|X_t - x| > \varepsilon\}} | X_0 = x).$$

By Taylor's expansion of $f(X_t)$ around x and using the Lagrange form of the remainder, we have that

$$\begin{aligned}\mathbb{E}(f(X_t)\mathbf{1}_{\{|X_t - x| \leq \varepsilon\}} | X_0 = x) = \mathbb{E}\Bigg[\Bigg(f(x) &+ (X_t - x)f'(x) + \frac{1}{2}(X_t - x)^2 f''(x) \\ &+ \frac{1}{2}(X_t - x)^2(f''(\xi_t) - f''(x))\Bigg)\mathbf{1}_{\{|X_t - x| \leq \varepsilon\}} | X_0 = x\Bigg],\end{aligned}$$

where ξ_t is located between x and X_t. Using the first two relations of (4.2) we have that the first three summands of the previous formula converge to

$$f(x) + t\mu(x)f'(x) + \frac{1}{2}t\sigma^2(x)f''(x) + o(t), \quad t \downarrow 0, \tag{4.6}$$

while the remainder is less than or equal to

$$\frac{1}{2}\delta\mathbb{E}\left((X_t - x)^2\mathbf{1}_{\{|X_t - x| \leq \varepsilon\}} | X_0 = x\right) = \frac{1}{2}\delta\sigma^2(x)t + o(t), \quad t \downarrow 0.$$

Therefore we get

$$\begin{aligned}\limsup_{t \downarrow 0}&\left|\frac{(T_t f)(x) - f(x)}{t} - \left(\mu(x)f'(x) + \frac{1}{2}\sigma^2(x)f''(x)\right)\right| \\ &\leq \frac{1}{2}\delta\sigma^2(x) + \limsup_{t \downarrow 0}\frac{\|f\|}{t}\mathbb{E}\left(\mathbf{1}_{\{|X_t - x| > \varepsilon\}} | X_0 = x\right).\end{aligned}$$

The limit on the right is 0 is a consequence of the third relation in (4.2). Since $\delta > 0$ is arbitrary, we get (4.5). □

The previous proof does not prove that any $C^2(\mathcal{S})$ and bounded function f belongs to the domain $D_{\mathcal{A}}$, since we did not prove that (4.5) is uniform in x. We may have three sources of non-uniformity. The first one is in terms of $o(t)$ in (4.2). The second one is the term $o(t)$ in (4.6), which may not be uniform in x, even if $o(t)$ of (4.2) is. Finally, the third one comes from the fact that, given $\delta > 0$, there may exist some ε independent of x such that $|f(y) - f(x)| < \delta$ for all x, y satisfying $|x - y| < \varepsilon$. To avoid these problems it is enough to assume that $f \in C^2(\mathcal{S})$ and vanishes outside of any closed and bounded subinterval of $\mathcal{S}$. In this situation we have that $f \in D_{\mathcal{A}}$ and for that f, the infinitesimal operator $\mathcal{A}$ is the second-order differential operator

$$(\mathcal{A}f)(x) = \mu(x)f'(x) + \frac{1}{2}\sigma^2(x)f''(x). \tag{4.7}$$

Proposition 4.3 ([5, Proposition V.2.2]) *Let $\{T_t, t > 0\}$ be a family of transition operators of a Markov process with state space $\mathcal{S}$. If $f \in D_{\mathcal{A}}$, then the following partial differential equation is satisfied:*

$$\frac{\partial}{\partial t}(T_t f)(x) = \mathcal{A}(T_t f)(x), \quad t > 0, \quad x \in \mathcal{S}. \tag{4.8}$$

Proof Let us prove that $T_t f \in D_{\mathcal{A}}$. If $f \in D_{\mathcal{A}}$, using that T_t and T_s commute and that T_t is a contraction, we have, as $s \downarrow 0$, that

$$\left\| \frac{T_s(T_t f) - T_t f}{s} - T_t(\mathcal{A}f) \right\| = \left\| T_t \left(\frac{T_s f - f}{s} - \mathcal{A}f \right) \right\| \le \left\| \frac{T_s f - f}{s} - \mathcal{A}f \right\| \to 0.$$

Then $T_t f \in D_{\mathcal{A}}$, since it converges to $T_t(\mathcal{A}f)$, and therefore, by definition of $\mathcal{A}$, we get (4.8). □

Remark 4.4 Observe from the proof of the previous proposition that we have $\mathcal{A}(T_t f) = T_t(\mathcal{A}f)$, i.e. T_t and $\mathcal{A}$ commute in $D_{\mathcal{A}}$. ◊

Let $p(t; x, y)$ be the transition probability density of a diffusion process $\{X_t, t \ge 0\}$ and consider $f = \mathbf{1}_B$ (indicator function) in (4.4). We have then

$$\begin{aligned}(T_{s+t} f)(x) &= \int_B p(s+t; x, y)\, dy = \int_{\mathcal{S}} \left(\int_B p(t; z, y)\, dy \right) p(s; x, z)\, dz \\ &= \int_B \left(\int_{\mathcal{S}} p(t; z, y) p(s; x, z)\, dz \right) dy\end{aligned}$$

for any Borel set B in $\mathcal{S}$. This implies the *Chapman–Kolmogorov equation*

$$p(s+t; x, y) = \int_{\mathcal{S}} p(t; z, y) p(s; x, z)\, dz = (T_s f)(x),$$

where $f(z) = p(t; z, y)$. Applying (4.5) to this f we have the *Kolmogorov backward equation* for the transition probability density $p(t; x, y)$

$$\frac{\partial p(t;x,y)}{\partial t} = \mu(x)\frac{\partial p(t;x,y)}{\partial x} + \frac{1}{2}\sigma^2(x)\frac{\partial^2 p(t;x,y)}{\partial x^2}.$$

The solution of this backward equation will depend on some initial conditions, especially how the process behaves at the boundary points of the state space $\mathcal{S}$.

The *Kolmogorov forward equation* (or the *evolution or Fokker–Planck equation*) gives information about the future behavior of the diffusion process X_t when X_0 has an arbitrary initial probability density. Assume that we call g this initial density. We will try to derive the Kolmogorov forward equation in terms of the adjoint of the operator T_t. The density of X_t is then given by

$$(T_t^* g)(y) = \int_{\mathcal{S}} g(x)p(t;x,y)\,dx.$$

The operator T_t^* transforms a density g into another probability density. This operator is the *adjoint* of T_t with respect to the inner product $(u,v) = \int u(x)v(x)\,dx$. Indeed,

$$(T_t^* g,f) = \int_{\mathcal{S}} (T_t^* g)(y)f(y)\,dy = \int_{\mathcal{S}} g(x)(T_t f)(x)\,dx = (g,T_t f).$$

If $f \in C^2(\mathcal{S})$ and vanishes outside of a compact subset of $\mathcal{S}$, we can take derivatives with respect to t in the previous relation and, using (4.8) and that $\mathcal{A}$ commutes with T_t, we get that

$$\left(\frac{\partial}{\partial t}T_t^* g,f\right) = \left(g,\frac{\partial}{\partial t}T_t f\right) = (g,T_t\mathcal{A}f) = (T_t^* g,\mathcal{A}f) = \int_{\mathcal{S}} (T_t^* g)(y)(\mathcal{A}f)(y)\,dy. \tag{4.9}$$

Now, if $f,h \in C^2(\mathcal{S})$ and vanishing outside of a finite interval, we have, integrating by parts, that

$$\begin{aligned}(h,\mathcal{A}f) &= \int_{\mathcal{S}} h(y)\left[\mu(y)f'(y) + \frac{1}{2}\sigma^2(y)f''(y)\right]dy\\ &= \int_{\mathcal{S}}\left[-\frac{d}{dy}(\mu(y)h(y)) + \frac{1}{2}\frac{d^2}{dy^2}(\sigma^2(y)h(y))\right]f(y)\,dy = (\mathcal{A}^* h,f),\end{aligned}$$

where $\mathcal{A}^*$ is the *formal adjoint operator* of the differential operator $\mathcal{A}$. Therefore, applied to (4.9), we get that

$$\left(\frac{\partial}{\partial t}T_t^* g,f\right) = (\mathcal{A}^* T_t^* g,f).$$

Since this relation holds for a large class of functions vanishing outside of a closed interval, we have the partial differential equation

$$\frac{\partial}{\partial t}(T_t^* g)(y) = \mathcal{A}^*(T_t^* g)(y). \tag{4.10}$$

In other words,

$$\int_S \frac{\partial p(t;x,y)}{\partial t} g(x)\, dx = \int_S \left(-\frac{\partial}{\partial y} [\mu(y)p(t;x,y)] + \frac{1}{2}\frac{\partial^2}{\partial y^2} \left[\sigma^2(y)p(t;x,y)\right]\right) g(x)\, dx.$$

Therefore, the Kolmogorov forward equation for the transition probability density $p(t;x,y)$ is given by

$$\frac{\partial p(t;x,y)}{\partial t} = -\frac{\partial}{\partial y} [\mu(y)p(t;x,y)] + \frac{1}{2}\frac{\partial^2}{\partial y^2} \left[\sigma^2(y)p(t;x,y)\right], \quad t > 0. \tag{4.11}$$

Again, the solution of this equation will depend on the behavior of the diffusion process at the boundary points of the state space $\mathcal{S}$.

Let $\{X_t, t \geq 0\}$ be a diffusion process on a state space $\mathcal{S} = [a,b]$ with $-\infty \leq a < x < b \leq \infty$. We denote by τ_y the *first hitting or passage time* to a state y defined by

$$\tau_y = \inf\{t \geq 0\colon X_t = y\}. \tag{4.12}$$

For $[c,d] \subset \mathcal{S}$, we also call

$$\tau = \tau_{c,d} = \inf\{t \geq 0\colon X_t \notin (c,d)\} = \tau_c \wedge \tau_d. \tag{4.13}$$

There are some interesting functionals that we can analyze using the infinitesimal operator $\mathcal{A}$. These functionals will play a very important role in order to study recurrence and classification of boundaries. Consider first a diffusion process $\{X_t, t \geq 0\}$ with $X_0 = x$ and a closed interval $[c,d] \subset \mathcal{S}$, $c < d$. Call

$$\upsilon(x) = \upsilon_{c,d}(x) = \mathbb{P}(\tau_c < \tau_d |\, X_0 = x), \quad c \leq x \leq d, \tag{4.14}$$

the probability that the process reaches c before d, where τ_y is the first passage time to y defined by (4.12). For a small time interval h, using the law of total expectation and Taylor's theorem, and assuming that $\upsilon \in C^2(\mathcal{S})$, we have that

$$\begin{aligned}
\upsilon(x) &= \mathbb{E}(\upsilon(X_h)|\, X_0 = x) + o(h) = \mathbb{E}(\upsilon(x + X_h - x)|X_0 = x) \\
&= \mathbb{E}\left(\upsilon(x) + (X_h - x)\upsilon'(x) + \frac{1}{2}(X_h - x)^2\upsilon''(x)|X_0 = x\right) + o(h) \\
&= \upsilon(x) + \mu(x)h\upsilon'(x) + \frac{1}{2}\sigma^2(x)h\upsilon''(x) + o(h).
\end{aligned}$$

We have then the following:

Proposition 4.5 *The probabilities* $\upsilon(x)$ *in* (4.14) *are solutions of the following second-order differential equation with boundary conditions:*

$$\frac{1}{2}\sigma^2(x)\upsilon''(x) + \mu(x)\upsilon'(x) = 0, \quad c < x < d, \tag{4.15}$$

$$\upsilon(c) = 1, \quad \upsilon(d) = 0.$$

Proof See [5, p. 415] or [101, p. 193]. □

We want to compute now the *mean exit time* $E(x)$ to reach either c or d of a diffusion process starting at $X_0 = x$, with $x \in (c, d)$, i.e.

$$E(x) = E_{c,d}(x) = \mathbb{E}(\tau_{c,d}|X_0 = x),$$

where $\tau_{c,d}$ is defined by (4.13). As before, for a small interval of time h and assuming that $E \in C^2(\mathcal{S})$, we have that

$$\begin{aligned} E(x) &= h + \mathbb{E}(E(X_h)|\, X_0 = x) + o(h) \\ &= h + E(x) + \mu(x)hE'(x) + \frac{1}{2}\sigma^2(x)hE''(x) + o(h). \end{aligned}$$

We have then the following:

Proposition 4.6 *The mean exit time $E(x)$ is a solution of the following second-order differential equation with boundary conditions:*

$$\frac{1}{2}\sigma^2(x)E''(x) + \mu(x)E'(x) = -1, \quad c < x < d, \tag{4.16}$$

$$E(c) = E(d) = 0.$$

Proof See [5, p. 421] or [101, p. 194]. □

Call

$$\rho_{xy} = \mathbb{P}(\tau_y < \infty|X_0 = x). \tag{4.17}$$

Definition 4.7 With the previous notation we have that:

- A state y is *recurrent* if $\rho_{xy} = 1$ for all $x \in \mathcal{S}$ such that $\rho_{yx} > 0$, and *transient* otherwise.
- A recurrent diffusion process is *positive recurrent* if $\mathbb{E}(\tau_y|X_0 = x) < \infty$ for all $x, y \in \mathcal{S}$, and *null recurrent* otherwise. ◊

If all states of $\mathcal{S}$ are recurrent, then the diffusion process is recurrent. In cases where the diffusion process is positive recurrent, there will be an *invariant or stationary distribution* $\pi(y)$ satisfying

$$\pi(y)\,dy = (T_t^*\pi)(y)\,dy = \int_{\mathcal{S}} \pi(x)p(t; x, dy)\,dx \doteq \mathbb{P}_\pi(X_t \in dy), \quad t > 0. \tag{4.18}$$

As for Markov chains, this invariant distribution satisfies

$$\lim_{t\to\infty} p(t; x, y) = \pi(y).$$

Therefore, using this on the Kolmogorov forward equation (4.11) or (4.10), we obtain that the invariant distribution satisfies $\mathcal{A}^*\pi(y) = 0$, i.e.

$$\frac{1}{2}\frac{d^2}{dy^2}\left[\sigma^2(y)\pi(y)\right] - \frac{d}{dy}\left[\mu(y)\pi(y)\right] = 0, \quad y \in \mathcal{S},$$

as long as all boundary points cannot be reached from any point inside the state space. In other words,

$$\frac{d}{dy}\left[\sigma^2(y)\pi(y)\right] = 2\mu(y)\pi(y), \quad y \in \mathcal{S}.$$

Observe the similarity with the Pearson equation (1.49) in Chapter 1. This invariant distribution will be given precisely by the spectral measure associated with the self-adjoint infinitesimal operator $\mathcal{A}$.

Finally, for any diffusion process $\{X_t, t \geq 0\}$ it is always possible to transform it into another diffusion process with another state space by a transformation function.

Proposition 4.8 *Let $\{X_t, t \geq 0\}$ be a diffusion process with state space $\mathcal{S} = (c,d)$ and drift and diffusion coefficients $\mu(x)$ and $\sigma^2(x)$, respectively. If ϕ is a three-times continuously differentiable function from (c,d) to (a,b) such that ϕ' is strictly positive or negative, then the process $\{Z_t = \phi(X_t), t \geq 0\}$ is a diffusion process with state space (a,b) and drift and diffusion coefficients given by*

$$\widetilde{\mu}(z) = \phi'(\phi^{-1}(z))\mu(\phi^{-1}(z)) + \frac{1}{2}\phi''(\phi^{-1}(z))\sigma^2(\phi^{-1}(z)),$$

$$\widetilde{\sigma}^2(z) = (\phi'(\phi^{-1}(z)))^2\sigma^2(\phi^{-1}(z)), \quad z \in (a,b).$$

Proof A proof can be found in [5, Proposition V.3.1] or [101]. □

4.2 Spectral Representation of the Transition Probability Density

Let $\{X_t, t \geq 0\}$ be a diffusion process with state space $\mathcal{S}$ and infinitesimal operator $\mathcal{A}$ given by (4.7), with drift coefficient $\mu(x)$ and diffusion coefficient $\sigma^2(x)$. We will give a spectral representation of the transition probability density in terms of the eigenfunctions and eigenvalues of $\mathcal{A}$. This kind of representation for a special class of one-dimensional diffusion processes was found by Hille [71], and later studied also by Barrett and Lampard [6] and McKean [116].

Following [5, Section V.5], for an arbitrary x_0, define the function

$$I(x_0, x) = \int_{x_0}^{x} \frac{2\mu(z)}{\sigma^2(z)}\, dz. \tag{4.19}$$

Observe that

$$\frac{d}{dx} I(x_0, x) = \frac{2\mu(x)}{\sigma^2(x)}.$$

For this function, define, up to a multiplicative constant, the measure

$$\pi(x) = \frac{2}{\sigma^2(x)} e^{I(x_0,x)}. \tag{4.20}$$

The choice of x_0 is independent of the behavior of the diffusion process. We have then that the infinitesimal operator can be written in Sturm–Liouville form as

$$\mathcal{A} = \frac{1}{\pi(x)} \frac{d}{dx} \left(\frac{\sigma^2(x)}{2} \pi(x) \frac{d}{dx} \right). \tag{4.21}$$

Indeed, let $f \in C^2(\mathcal{S})$. Then

$$\begin{aligned}
\frac{1}{\pi(x)} \frac{d}{dx} \left(\frac{\sigma^2(x)}{2} \pi(x) \frac{d}{dx} \right) f &= \frac{1}{\pi(x)} \frac{d}{dx} \left(e^{I(x_0,x)} \frac{d}{dx} \right) f \\
&= \frac{1}{\pi(x)} \left(\left[\frac{d}{dx} e^{I(x_0,x)} \right] \frac{d}{dx} + e^{I(x_0,x)} \frac{d^2}{dx^2} \right) f \\
&= \frac{\sigma^2(x)}{2} e^{-I(x_0,x)} \left(e^{I(x_0,x)} \left[\frac{2\mu(x)}{\sigma^2(x)} \frac{d}{dx} + \frac{d^2}{dx^2} \right] \right) f \\
&= \frac{1}{2} \sigma^2(x) f''(x) + \mu(x) f'(x) = (\mathcal{A}f)(x).
\end{aligned}$$

Consider the Hilbert space $L^2_\pi(\mathcal{S})$ with the inner product

$$(f,g)_\pi = \int_{\mathcal{S}} f(x) g(x) \pi(x)\, dx. \tag{4.22}$$

By Proposition 1.18 and using that $\mathcal{A}$ can be rewritten in the form (4.21), we have that $\mathcal{A}$ is self-adjoint with respect to the inner product defined by π, i.e. $(\mathcal{A}f, g)_\pi = (f, \mathcal{A}g)_\pi$ for all $f, g \in L^2_\pi(\mathcal{S})$. In particular, the measure $\pi(x)$ satisfies the *Pearson equation*

$$\frac{1}{2} (\sigma^2(x)\pi(x))' = \mu(x)\pi(x).$$

Therefore, in case of existence of an invariant distribution, this should be the (normalized) measure $\pi(x)$.

Let $f \in C^0(\mathcal{S})$, bounded on $\mathcal{S}$ and consider the function

$$u(t,x) = (T_t f)(x) = \mathbb{E}(f(X_t) | X_0 = x), \tag{4.23}$$

where T_t is the transition operator (4.3). We know that this function satisfies the Kolmogorov backward equation (4.8), i.e.

$$\frac{\partial}{\partial t} u(t,x) = \mathcal{A} u(t,x), \quad u(0,x) = f(x). \tag{4.24}$$

The idea is to try to compute this function $u(t,x)$ by using the *method of separation of variables*. Assume that

$$u(t,x) = c(t)\phi(x).$$

Substituting in the Kolmogorov backward equation, we obtain

$$\frac{c'(t)}{c(t)} = \frac{\mathcal{A}\phi(x)}{\phi(x)} = -\lambda,$$

where λ is constant. The differential equation for $c(t)$ is easy to solve from the initial conditions. Then $c(t) = e^{-\lambda t}$. The second relation is a second-order differential equation of the form $\mathcal{A}\phi(x) = -\lambda\phi(x)$. If $-\lambda$ is such that $\mathcal{A}\phi(x) = -\lambda\phi(x)$ has a nontrivial solution, then $-\lambda$ is called an eigenvalue of $\mathcal{A}$ and $\phi(x) = \phi_\lambda(x)$ is called the corresponding (nontrivial) eigenfunction. If we can find such a solution, then a solution of $u(t,x)$ will be given by

$$u(t,x) = e^{-\lambda t}\phi(x).$$

Assume now that we have a countable set of eigenvalues of $\mathcal{A}$ (counting multiplicities) and call them

$$-\lambda_0, -\lambda_1, -\lambda_2, \ldots$$

with eigenfunctions

$$\phi_0(x), \phi_1(x), \phi_2(x), \ldots$$

such that they are normalized by the inner product (4.22), i.e. $\|\phi_n\|_\pi = 1, n \geq 0$. As we saw in Theorem 1.19 if $\lambda_i \neq \lambda_j$, then $\phi_i(x)$ and $\phi_j(x)$ are orthogonal with respect to the inner product (4.22), i.e. $(\phi_i, \phi_j)_\pi = 0, i \neq j$. Assume that this sequence of orthonormal eigenfunctions $(\phi_n)_n$ is *complete* in $L^2_\pi(\mathcal{S})$ (this condition will hold for all examples we will see in this chapter). Therefore, for any $f \in L^2_\pi(\mathcal{S})$, using the Fourier expansion with respect to the basis $(\phi_n)_n$, we have that

$$f(x) = \sum_{n=0}^{\infty} (f, \phi_n)_\pi \phi_n(x).$$

By the principle of superposition, each of these solutions is a solution of (4.24). Therefore,

$$u(t,x) = \sum_{n=0}^{\infty} e^{-\lambda_n t}(f, \phi_n)_\pi \phi_n(x), \quad u(0,x) = f(x).$$

By the definition of $(T_t f)(x)$ in (4.3) and since it satisfies the Kolmogorov backward equation for a large class of functions f, we have that $(T_t f)(x) = u(t,x)$. Therefore,

$$\begin{aligned} u(t,x) &= \int_S f(y)p(t;x,dy) = \sum_{n=0}^{\infty} e^{-\lambda_n t}(f,\phi_n)_\pi \phi_n(x) \\ &= \sum_{n=0}^{\infty} e^{-\lambda_n t}\left(\int_S f(y)\phi_n(y)\pi(y)dy\right)\phi_n(x) \\ &= \int_S f(y)\left(\sum_{n=0}^{\infty} e^{-\lambda_n t}\phi_n(x)\phi_n(y)\pi(y)\right)dy. \end{aligned}$$

We can interchange integral and sum in the last formula since $\lambda_n \geq 0$, as we will see below. As a consequence, the transition probability distribution $p(t;x,dy)$ will have a density $p(t;x,y)$ given by

$$p(t;x,y) = \pi(y)\sum_{n=0}^{\infty} e^{-\lambda_n t}\phi_n(x)\phi_n(y). \tag{4.25}$$

In most situations this formula will be valid, but it will depend on the initial conditions of the Kolmogorov backward equation. In most cases we will have to solve a second-order differential equation for $u(t,x)$ (4.23) of the Sturm–Liouville type on $\mathcal{S} = [a,b]$ with boundary conditions of the form (see [89])

$$\begin{aligned} \mathcal{A}\phi &= \frac{1}{\pi(x)}\frac{d}{dx}\left(\frac{\sigma^2(x)}{2}\pi(x)\phi'(x)\right) = -\lambda\phi(x), \\ c_{1,1}\phi(a) &+ c_{1,2}\left(\frac{\sigma^2(x)}{2}\pi(x)\phi'(x)\right)(a) = 0, \\ c_{2,1}\phi(b) &+ c_{2,2}\left(\frac{\sigma^2(x)}{2}\pi(x)\phi'(x)\right)(b) = 0, \end{aligned} \tag{4.26}$$

for certain constants $c_{ij}, i,j = 1,2$. From the general theory of Sturm–Liouville problems we know that there is an infinite but countable number of eigenvalues of $\mathcal{A}$ if a and b are finite (see Theorem 1.19). Following (1.45) we have that

$$\lambda_n \geq 0, \quad n \geq 0,$$

and these can be ordered in the following way:

$$0 \leq \lambda_0 < \lambda_1 < \cdots < \lambda_n \to \infty.$$

In particular, if $c_{1,1} = c_{2,1} = 0$ then $\phi_0 = 1$ is an eigenfunction with eigenvalue $-\lambda_0 = 0$.

If one or both boundary points are infinite, we may have that the spectrum of $\mathcal{A}$ is discrete, absolutely continuous or a combination of both. Call $\Xi \subset \mathbb{R}$ the absolutely continuous spectrum of $\mathcal{A}$. The orthonormality of $\phi(x,\lambda)$ is understood in this case as

$$\int_{\mathcal{S}} \phi(x,\lambda)\phi(x,\mu)\pi(x)\,dx = \delta(\lambda-\mu). \tag{4.27}$$

If $f \in L^2_\pi(\mathcal{S})$, then there exists a function $\widehat{f}$ such that

$$f(x) = \int_\Xi \widehat{f}(\lambda)\phi(x,\lambda)\,d\lambda.$$

To compute $\widehat{f}(\lambda)$, we multiply the previous expression by $\phi(x,\mu)\pi(x)$, integrate over $\mathcal{S}$ and use the orthonormality (4.27). Indeed,

$$\begin{aligned}\int_{\mathcal{S}} f(x)\phi(x,\mu)\pi(x)\,dx &= \int_{\mathcal{S}}\left[\int_\Xi \widehat{f}(\lambda)\phi(x,\lambda)\,d\lambda\right]\phi(x,\mu)\pi(x)\,dx \\ &= \int_\Xi \widehat{f}(\lambda)\left[\int_{\mathcal{S}} \phi(x,\lambda)\phi(x,\mu)\pi(x)\,dx\right]d\lambda \\ &= \int_\Xi \widehat{f}(\lambda)\delta(\lambda-\mu)\,d\lambda = \widehat{f}(\mu).\end{aligned}$$

The solutions $u(t,x)$ in (4.23) can be written now as

$$u(t,x) = \int_\Xi e^{-\lambda t}\widehat{f}(\lambda)\phi(x,\lambda)\,d\lambda.$$

By definition of $(T_t f)(x)$ in (4.3) and since it satisfies the Kolmogorov backward equation for a large class of functions f, we will have that $(T_t f)(x) = u(t,x)$. Therefore

$$\begin{aligned}u(t,x) &= \int_{\mathcal{S}} f(y)p(t;x,dy) = \int_\Xi e^{-\lambda t}\widehat{f}(\lambda)\phi(x,\lambda)\,d\lambda \\ &= \int_\Xi e^{-\lambda t}\left[\int_{\mathcal{S}} f(y)\phi(y,\lambda)\pi(y)\,dy\right]\phi(x,\lambda)\,d\lambda \\ &= \int_{\mathcal{S}} f(y)\pi(y)\left[\int_\Xi e^{-\lambda t}\phi(x,\lambda)\phi(y,\lambda)\,d\lambda\right]dy.\end{aligned}$$

As a consequence, the transition probability distribution $p(t;x,dy)$ will have a density $p(t;x,y)$ given by (properly normalized)

$$p(t;x,y) = \pi(y)\int_\Xi e^{-\lambda t}\phi(x,\lambda)\phi(y,\lambda)\,d\lambda.$$

If we have an absolutely continuous range of eigenvalues we can take cases with finite boundaries (so we have a discrete spectrum) and consider the limiting situation when one or both boundaries tend to infinity to conclude also that $\lambda \geq 0$.

Finally, if the spectrum of the infinitesimal operator has both absolutely continuous Ξ and discrete Θ parts, then the transition probability density $p(t;x,y)$ will be given by

$$p(t;x,y) = \pi(y)\left[\sum_{n\in\Theta} e^{-\lambda_n t}\phi_n(x)\phi_n(y) + \int_{\Xi} e^{-\lambda t}\phi(x,\lambda)\phi(y,\lambda)\,d\lambda\right]. \quad (4.28)$$

In Section 4.5 we will see a collection of examples where we can give explicit expressions of the eigenvalues and eigenfunctions.

Remark 4.9 One interesting question is if we have a transition function as in (4.28) then when is this function going to be the transition probability density of some diffusion process? According to [89] (see also [55]) if the two boundary conditions are of *dissipative type* (i.e. $0 \leq c_{1,1}/c_{1,2} \leq \infty$ and $-\infty \leq c_{2,1}/c_{2,2} \leq 0$) then the function

$$P(t;x,B) = \int_B p(t;x,y)\,dy,$$

for any Borel set B, where $p(t;x,y)$ is given by (4.28), is the transition probability function of a stationary Markov process $\{X_t, t \geq 0\}$ in the interior of $\mathcal{S}$ (where the Chapman–Kolmogorov equation is valid). If $c_{1,2} = c_{2,2} = 0$, then both boundaries are absorbing (see section below) and it is known that the trajectories of the process $\{X_t, t \geq 0\}$ are continuous. For other boundary conditions it is known that the trajectories are continuous in the interior of $\mathcal{S}$, but the behavior at the boundaries may be uncertain. In [89, 90] the authors also treat the concept of *total positivity* (see also Remark 3.28 for the case of birth–death processes). The analytic property of total positivity is related to the stochastic property that the path functions of the process are continuous everywhere, including the boundary points (see [88]). ◊

Remark 4.10 In [60] the authors analyze different uses of the spectral representation in mathematical finance. In particular, they observe that the knowledge of the spectral representation of the transition probability density is a main step for the analysis of nonlinear dynamics of diffusion processes. For instance, if $\{X_t, t \geq 0\}$ is a regular diffusion process, the conditional probability density of X_t given X_{t-1} can be written as

$$\pi(X_t|\, X_{t-1}) = \pi(X_t)\sum_{n=0}^{\infty} e^{-\lambda_n t}\phi_n(X_t)\phi_n(X_{t-1}).$$

This series can be used to approximate the ratio of the conditional probability density function and the marginal one by well-chosen eigenfunctions evaluated at X_t and X_{t-1}. In particular, if the marginal probability density function has a simple expression but the conditional probability density function has no tractable form, the truncated expansion can be used to approximate maximum likelihood methods. The spectral decomposition is also useful for computing nonlinear predictions at different horizons. For any function g, we have

$$\mathbb{E}(g(X_{t+h})|\, X_t) = \sum_{n=0}^{\infty} e^{-h\lambda_n} g_n \phi_n(X_t), \quad g_n = \int_{\mathcal{S}} g(y)\phi_n(y)\pi(y)\, dy,$$

which provides the dependence of the prediction with respect to the horizon. It is especially useful for studying term structures of interest rates or term structures of volatilities, or for analyzing the multiplier effects in a nonlinear framework (see [6, 60] and references therein for more details). ◊

4.3 Classification of Boundary Points

We have seen in the previous section that in order to find a spectral representation of the transition probability density of a diffusion process we need to solve a certain regular Sturm–Liouville problem with separated boundary conditions (see (4.26)). In this section we will see how to choose the boundary conditions in such a way that the diffusion process has certain behavior at one or both boundary points. If $p(t;x,y)$ is the transition probability density of a diffusion process, the Sturm–Liouville problem (4.26) comes from the Kolmogorov backward equation with initial conditions

$$\mathcal{A}p = \frac{1}{\pi(x)}\frac{\partial}{\partial x}\left(\frac{\sigma^2(x)}{2}\pi(x)\frac{\partial p(t;x,y)}{\partial x}\right) = \frac{\partial p(t;x,y)}{\partial t},$$

$$c_{1,1}p(t;a,y) + c_{1,2}\left(\frac{\sigma^2(x)}{2}\pi(x)\frac{\partial p(t;x,y)}{\partial x}\right)(a) = 0,$$

$$c_{2,1}p(t;b,y) + c_{2,2}\left(\frac{\sigma^2(x)}{2}\pi(x)\frac{\partial p(t;x,y)}{\partial x}\right)(b) = 0.$$

4.3.1 Reflecting and Absorbing Boundaries

Let $\mathcal{S} = [a,\infty)$ be the state space of a diffusion process $\{X_t, t \geq 0\}$. We say that the boundary point a is *inaccessible* if it cannot be reached from any point inside the state space. This means that if some trajectory of the process is close to an inaccessible state, it immediately sends the process inside the state space. In order to continue the process after it reaches a boundary point, we must specify the boundary condition consistent with the requirement that the process is Markovian.

A diffusion process on $\mathcal{S} = [a,\infty)$ whose transition probability density $p(t;x,y)$ satisfies

$$\mathcal{A}p = \frac{1}{2}\sigma^2(x)\frac{\partial^2 p(t;x,y)}{\partial x^2} + \mu(x)\frac{\partial p(t;x,y)}{\partial x} = \frac{\partial p(t;x,y)}{\partial t}, \quad t > 0, x > a, y \geq a, \tag{4.29}$$

with boundary conditions

$$\left.\frac{\partial p(t;x,y)}{\partial x}\right|_{x=a} = 0, \quad t > 0, y \geq a, \tag{4.30}$$

is called a *reflecting diffusion* on $[a,\infty)$ having drift $\mu(x) = (\sigma^2(x)\pi(x))'/2\pi(x)$ and diffusion coefficient $\sigma^2(x)$. The point a is then called a *reflecting boundary point*.

This particular boundary condition (4.30) is also known as the *Neumann boundary condition*. Let us give a justification of this behavior. Consider $\{X_t, t \geq 0\}$ a diffusion process on $\mathbb{R}$ with drift coefficient satisfying $\mu(-x) = \mu(x), x \geq 0$ and diffusion coefficient satisfying $\sigma^2(-x) = \sigma^2(x), x \geq 0$. Let $q(t;x,y)$ be the corresponding transition probability density. Then we have (see [5, Theorem 6.1]) that $\{|X_t|, t \geq 0\}$ is a Markov process on the state space $[0,\infty)$, whose transition probability density $p(t;x,y)$ is given by

$$p(t;x,y) = q(t;x,y) + q(t;-x,y), \quad x,y \in [0,\infty).$$

From here it is easy to see that $p(t;x,y)$ satisfies the Kolmogorov backward equation (4.29) and the boundary condition (4.30), with 0 replaced by a.

By similar arguments we can derive the Kolmogorov forward equation (4.11) for a reflecting diffusion on $[a,\infty)$, where now the boundary condition at $x = a$ is derived by taking derivatives with respect to t in the following equation:

$$1 = \int_a^\infty p(t;x,y)\,dy.$$

Then we obtain, using (4.11),

$$0 = \int_a^\infty \frac{\partial p(t;x,y)}{\partial t}\,dy = \int_a^\infty \frac{\partial}{\partial y}\left(-\mu(y)p(t;x,y) + \frac{1}{2}\frac{\partial[\sigma^2(y)p(t;x,y)]}{\partial y}\right)dy.$$

Hence, using the fact that $p(t;x,y)$ is a probability density, the *forward boundary condition* is given by

$$\left.\frac{1}{2}\frac{\partial[\sigma^2(y)p(t;x,y)]}{\partial y}\right|_{y=a} - \mu(a)p(t;x,a) = 0.$$

The same analysis for the boundary conditions holds if we consider a bounded state space $\mathcal{S} = [a,b]$ with both reflecting boundaries.

Let us now study the behavior of an absorbing boundary. Let $\mathcal{S} = [a,\infty)$ be the state space of a diffusion process $\{X_t, t \geq 0\}$. We say that the boundary point a is *absorbing* if upon arrival at the boundary point a the process remains in that state for all times thereafter. If we consider a diffusion $\{\widetilde{X}_t, t \geq 0\}$ on $\mathbb{R}$, the diffusion process $\{X_t, t \geq 0\}$ with an absorbing boundary at a starting at $x \in [a, \infty)$ can be defined as

$$X_t = \begin{cases} \widetilde{X}_t, & \text{if} \quad t < \tau_a, \\ a = \widetilde{X}_{\tau_a}, & \text{if} \quad t \geq \tau_a, \end{cases}$$

where τ_a is the first passage time to a defined by (4.12). In [5, Theorem 7.1] it is proven that this process is a time-homogeneous Markov process. The transition probability distribution $p(t;x,dy)$ is divided in two. One distribution $p^0(t;x,dy)$ defined on the interval (a,∞) and an absorption probability of being absorbed at the boundary a. Therefore, we have

$$p(t;x,B) = \begin{cases} p^0(t;x,B) = \mathbb{P}(\widetilde{X}_t \in B \text{ and } \tau_a > t | X_0 = x) & \\ \qquad = \displaystyle\int_B p^0(t;x,y)\,dy, & \text{if} \quad B \subset (a,\infty), \\ p(t;x,\{a\}) = \mathbb{P}(\tau_a \leq t | X_0 = x), & \text{if} \quad B = \{a\}. \end{cases}$$

It is possible to see (see [5, Section V.15]) that $p^0(t;x,y)$ satisfies the Kolmogorov backward equation (4.29) and the *Dirichlet boundary condition*

$$\lim_{x \downarrow a} p^0(t;x,y) = 0, \quad t > 0, y > a. \tag{4.31}$$

Indeed, we have

$$\int_a^\infty p^0(t;x,y)\,dy = \mathbb{P}(\tau_a > t | X_0 = x) = 1 - \mathbb{P}(\tau_a \leq t | X_0 = x), \quad t > 0, x > a,$$

so as $x \downarrow a$ the probability on the right-hand side goes to 0 (assuming that $p^0(t;x,y)$ has a limit as $x \downarrow a$). The same can be applied to the Kolmogorov forward equation (4.11), where the boundary condition at $y = a$ will be given by

$$\lim_{y \downarrow a} p^0(t;x,y) = 0, \quad t > 0, x > a.$$

If we have a diffusion process $\{X_t, t \geq 0\}$ on the state space $\mathcal{S} = [a,b]$ with two absorbing boundaries, then the process, starting at some $x \in (a,b)$, can be defined now as

$$X_t = \begin{cases} \widetilde{X}_t, & \text{if} \quad t < \tau, \\ \widetilde{X}_\tau, & \text{if} \quad t \geq \tau, \end{cases}$$

where $\tau = \tau_a \wedge \tau_b$ is the first passage time to the boundary defined by (4.13). Again the transition probability density $p(t;x,y)$ is given by a density $p^0(t;x,y)$ restricted to the interior (a,b) and the probability $\mathbb{P}(\tau \leq t | X_0 = x)$. The density

$p^0(t;x,y)$ satisfies again the Kolmogorov backward equation (4.29) and the Dirichlet boundary conditions

$$\lim_{x\downarrow a} p^0(t;x,y) = 0, \quad \lim_{x\uparrow b} p^0(t;x,y) = 0, \quad t > 0, y \in (a,b).$$

Unlike the one-point boundary case, $p^0(t;x,y)$ does not completely determine $p(t;x,y)$ and we have to determine $\mathbb{P}(\tau \le t|X_0 = x) = p(t;x,\{a\}) + p(t;x,\{b\})$. It is possible to see (see [5, Proposition 7.2]) that the absorption probability $p(t;x,\{a\})$ can be written in terms of the probabilities $v(x) = \mathbb{P}(\tau_a < \tau_b|\, X_0 = x)$ and $p^0(t;x,y)$. Indeed,

$$p(t;x,\{a\}) = v(x) - \int_a^b v(y)p^0(t;x,y)\,dy, \quad t > 0, \quad a < x < b. \tag{4.32}$$

Once we have $p^0(t;x,y)$ and $p(t;x,\{a\})$, we can compute directly $p(t;x,\{b\})$. The same analysis applies for the boundary conditions of the Kolmogorov forward equation.

Remark 4.11 It is possible to have a mixed two-point boundary situation where one of the boundaries is absorbing and the other reflecting. In that case we have to consider both boundary conditions (4.30) and (4.31). ◊

4.3.2 Feller's Classification of Boundaries

Let $\{X_t, t \ge 0\}$ be a regular diffusion process on $\mathcal{S} = [a,b]$ with drift $\mu(x)$ and diffusion coefficient $\sigma^2(x)$. For $x_0 \in \mathcal{S}$ we define the *scale function* as

$$S(x) = S(x_0,x) = \int_{x_0}^x e^{-I(x_0,z)}dz, \quad x \in \mathcal{S}. \tag{4.33}$$

We also define the *speed function* as

$$M(x) = M(x_0,x) = \int_{x_0}^x \frac{2}{\sigma^2(z)} e^{I(x_0,z)}dz = \int_{x_0}^x \pi(z)dz, \quad x \in \mathcal{S}. \tag{4.34}$$

Observe that $M'(x) = \pi(x)$. With these new definitions, the infinitesimal operator $\mathcal{A}$ can be rewritten as

$$\mathcal{A} = \frac{d}{dM}\left(\frac{d}{dS}\right). \tag{4.35}$$

Indeed, for any $f \in C^2(\mathcal{S})$, we have, by the chain rule, that

$$\frac{df(x)}{dS(x)} = \frac{df(x)}{dx}\frac{dx}{dS(x)} = \frac{1}{S'(x)}f'(x) = e^{I(x_0,x)}f'(x) = \frac{1}{2}\sigma^2(x)\pi(x)f'(x),$$
$$\frac{df(x)}{dM(x)} = \frac{df(x)}{dx}\frac{dx}{dM(x)} = \frac{1}{M'(x)}f'(x) = \frac{1}{\pi(x)}f'(x).$$

Therefore

$$\begin{aligned}\frac{d}{dM(x)}\left(\frac{df(x)}{dS(x)}\right) &= \frac{d}{dM(x)}\left(\frac{1}{S'(x)}f'(x)\right) = \frac{d}{dM(x)}\left(e^{I(x_0,x)}f'(x)\right)\\ &= \frac{1}{M'(x)}\left[\frac{2\mu(x)}{\sigma^2(x)}e^{I(x_0,x)}f'(x) + e^{I(x_0,x)}f''(x)\right]\\ &= \frac{\sigma^2(x)}{2}e^{-I(x_0,x)}\left[\frac{2\mu(x)}{\sigma^2(x)}e^{I(x_0,x)}f'(x) + e^{I(x_0,x)}f''(x)\right]\\ &= \frac{1}{2}\sigma^2(x)f''(x) + \mu(x)f'(x) = (\mathcal{A}f)(x).\end{aligned}$$

With this notation we can easily solve the differential equation (4.15) for $v(x)$ defined by (4.14). Using (4.35), the equation (4.15) can be written as

$$\frac{d}{dM(x)}\left(\frac{dv(x)}{dS(x)}\right) = 0.$$

Therefore, solving, we get that

$$\frac{dv(x)}{dS(x)} = c_1, \quad v(x) = c_1 S(x) + c_2.$$

The initial conditions $v(c) = 1$ and $v(d) = 0$ give the constants c_1, c_2:

$$c_1 = -\frac{1}{S(d) - S(c)}, \quad c_2 = -c_1 S(d) = \frac{S(d)}{S(d) - S(c)}.$$

Therefore,

$$v(x) = \frac{S(d) - S(x)}{S(d) - S(c)}. \tag{4.36}$$

Choosing $x_0 = c$, we have that

$$v(x) = \frac{\int_x^d e^{-I(c,z)}dz}{\int_c^d e^{-I(c,z)}dz}.$$

We can see now the meaning of the name of scale function for $S(x)$. If the distance is measured in this scale, then $S(y) - S(x)$ is the distance between x and y ($x < y$). Then the probability of reaching c before d starting at x is proportional to the distance $S(d) - S(x)$ between x and d. In particular, if we start from the middle point of the interval (c,d) *in this scale* the probability of reaching one of the extremes is $1/2$. Thus, in general, the probability of reaching d before c is given by $1 - v(x)$.

When $S(x) - S(a) = \infty$ that means that any interior point x is "far" from a (in this scale) and the probability that the process reaches a before any other point d with

$d > x$ is almost 0 (or close to 1 if $S(x) - S(a) < \infty$). Then we have an alternative way of defining absorbing and reflecting boundaries in terms of the scale function. Indeed:

1. The boundary point a is said to be *absorbing* if $S(x) - S(a) < \infty$ independently of x.
2. The boundary point a is said to be *reflecting* if $S(x) - S(a) = \infty$ independently of x.

The internal point x is not important for the behavior at the boundary a since the scale function is additive. Also the scale function gives one way of determining if a diffusion process on $\mathcal{S} = (a, b)$ is recurrent (see Definition 4.7). Then a diffusion process is recurrent on $\mathcal{S} = (a, b)$ if and only $S(a) = -\infty$ and $S(b) = \infty$ (here the value of $S(a)$ or $S(b)$ is computed taking the limit $S(a) = \lim_{c \downarrow a} S(c)$).

Let us now solve the differential equation with boundary values given in (4.16) for the mean exit time $E(x)$. Again, using (4.35), we have to solve

$$\frac{d}{dM(x)}\left(\frac{dE(x)}{dS(x)}\right) = -1. \tag{4.37}$$

Therefore

$$\frac{dE(x)}{dS(x)} = -M(x) + c_1, \quad E(x) = c_1 S(x) - \int_c^x M(y)\, dS(y) + c_2.$$

The initial conditions $E(c) = 0$ and $E(d) = 0$ give the constants c_1, c_2 :

$$c_1 = \frac{\int_c^d M(y)\, dS(y)}{S(d) - S(c)}, \quad c_2 = -c_1 S(c).$$

Therefore, choosing $x_0 = c$, using the definition for $v(x)$ in (4.36) and changing the order of integration, we get

$$\begin{aligned}
E(x) &= \frac{S(c;x)}{S(c;d)} \int_c^d M(c,y) S'(c,y)\, dy - \int_c^x M(c,y) S'(c,y)\, dy \\
&= (1 - v(x)) \int_c^d M(y)\, dS(y) - \int_c^x M(y) v\, dS(y) \\
&= (1 - v(x)) \int_c^d \left[\int_c^y \pi(z)\, dz\right] dS(y) - \int_c^x \left[\int_c^y \pi(z)\, dz\right] dS(y) \\
&= (1 - v(x)) \int_c^d \left[\int_z^d dS(y)\right] \pi(z)\, dz - \int_c^x \left[\int_z^x dS(y)\right] \pi(z)\, dz \\
&= (1 - v(x)) \int_c^d [S(d) - S(z)]\, dM(z) + \int_c^x [S(z) - S(x)]\, dM(z) \\
&= (1 - v(x)) \int_x^d [S(d) - S(z)]\, dM(z) + v(x) \int_c^x [S(z) - S(c)]\, dM(z).
\end{aligned}$$

Therefore

$$E(x) = (1 - v(x)) \int_x^d [S(d) - S(z)]\, dM(z) + v(x) \int_c^x [S(z) - S(c)]\, dM(z). \quad (4.38)$$

We can see now the meaning of the name of speed function for $M(x)$. Assume that the diffusion process is rescaled so that $S(x) = x$. Then the differential equation (4.37) is equivalent to $E''(x) = -M'(x)$. From this relationship we see that the larger $M'(x)$, the slower the exit speed. Now it is possible to see (see [5, Proposition V.10.2]) that a recurrent diffusion process is positive recurrent on $\mathcal{S} = (a,b)$ if and only $M(a) > -\infty$ and $M(b) < \infty$. Otherwise it is null recurrent.

In order to refine the behavior at the boundary point a, we need to have more information about the expected first passage time to that boundary (finite or infinite). Let us call

$$E_{a,d}(x) = \lim_{c \downarrow a} E_{c,d}(x) = \lim_{c \downarrow a} \mathbb{E}(\tau_c \wedge \tau_d | X_0 = x), \quad x < d < b.$$

Assume that a is attracting (and therefore $S(x) - S(a) < \infty$). Then,

$$v_{a,d}(x) = \lim_{c \downarrow a} v_{c,d}(x) = \lim_{c \downarrow a} \frac{S(d) - S(x)}{S(d) - S(c)} < \infty,$$

and also it is positive. Similar for $1 - v_{a,d}(x)$. Therefore, from $E(x) = E_{a,d}(x)$ in (4.38) we have that this quantity if finite if the right integral is finite (the left integral does not play any role at the boundary a). Define then

$$\Sigma(a) = \lim_{c \downarrow a} \int_c^x [S(z) - S(c)]\, dM(z). \quad (4.39)$$

Abusing notation, $\Sigma(a)$ can be written, changing the order of integration, as

$$\Sigma(a) = \int_a^x [S(z) - S(a)]\, dM(z) = \int_c^x \left[\int_a^z dS(y) \right] dM(z)$$
$$= \int_a^x \left[\int_y^x dM(z) \right] dS(y) = \int_a^x [M(x) - M(y)]\, dS(y).$$

The value of $\Sigma(a)$ measures the time it takes to reach the boundary a from an interior point x. Then we have the following extra description of the boundary point a:

1. The boundary point a is said to be *attainable* if $\Sigma(a) < \infty$.
2. The boundary point a is said to be *unattainable* if $\Sigma(a) = \infty$.

Observe that if a is attainable then $S(x) - S(a) < \infty$, so it is attracting. However, if a is unattainable, the boundary a can be attracting or reflecting. Consider now

$$N(a) = \int_a^x [S(x) - S(z)]\, dM(z) = \int_a^x [M(y) - M(a)]\, dS(y). \quad (4.40)$$

$M(y) - M(a)$ measures the speed with which the diffusion process reaches the boundary a and $N(a)$ is the time it takes to reach an interior point $x \in (a,b)$, *if the process starts at a*. If $N(a) < \infty$ that means that we can leave off the boundary a if we start there. We have the following implications:

Lemma 4.12 ([101, Lemma 15.6.3]) *(i)* $S(x) - S(a) = \infty \quad \Rightarrow \quad \Sigma(a) = \infty$,
(ii) $\Sigma(a) < \infty \quad \Rightarrow \quad S(x) - S(a) < \infty$,
(iii) $M(x) - M(a) = \infty \quad \Rightarrow \quad N(a) = \infty$,
(iv) $N(a) < \infty \quad \Rightarrow \quad N(x) - N(a) < \infty$,
(v) $\Sigma(a) + N(a) = (S(x) - S(a))(M(x) - M(a))$.

Depending on whether $\Sigma(a)$ and/or $N(a)$ is finite or infinite, we have *Feller's boundary classification for regular diffusion processes* (compare with the same classification for birth–death processes at the end of Section 3.1):

- *Regular boundary*: $\Sigma(a) < \infty, N(a) < \infty$. A diffusion process can both enter or leave from a regular boundary point. The behavior at the boundary depends on the speed $M(a)$. If $M(a) = 0$ then it is reflecting. If $M(a) = \infty$ then it will be attracting. The values between absorption and reflection is what is called the *sticky barrier phenomenon*, where a strictly positive time is spent at the boundary. Another possibility is to restart the process in the interior of the state space. But in this case the trajectories will not be continuous. However, the transition probability density will satisfy the Kolmogorov backward equation, so we can use the tools given in this chapter. As a consequence of Lemma 4.12, in order to establish that a boundary is regular it is enough to check that $S(x) - S(a) < \infty$ and $M(x) - M(a) < \infty$.
- *Exit boundary*: $\Sigma(a) < \infty, N(a) = \infty$. Starting at a it is impossible to reach any interior point in (a,b) no matter how close is that point from a. It is possible that after a time at a the process starts over from an interior point, but again the trajectories would not be continuous. In order to establish if a boundary is exit it is enough to check that $\Sigma(a) < \infty$ and $M(x) - M(a) = \infty$.
- *Entrance boundary*: $\Sigma(a) = \infty, N(a) < \infty$. The boundary a cannot be reached from the interior of the state space, but it is possible to consider the process starting there. The process moves quickly to the interior and never returns to boundary a. In order to establish if a boundary is entrance it is enough to check that $S(x) - S(a) = \infty$ and $N(a) < \infty$.
- *Natural boundary*: $\Sigma(a) = \infty, N(a) = \infty$. The diffusion process can neither reach in finite mean time nor start from that boundary point a. Therefore the boundary point does not belong to the state space. In order to establish if a boundary is natural it is enough to check that $S(x) - S(a) = \infty$ and $N(a) = \infty$.

If we want to study the boundary behavior at the point b then we have to evaluate $S(b) - S(x)$, $M(b) - M(x)$ and

$$\Sigma(b) = \int_x^b [S(b) - S(z)]\, dM(z) = \int_x^b [M(y) - M(x)]\, dS(y), \tag{4.41}$$

$$N(b) = \int_x^b [S(z) - S(x)]\, dM(z) = \int_x^b [M(b) - M(y)]\, dS(y). \tag{4.42}$$

This classification was given by Feller in [54] but in [101] a table was proposed that presents the relation between Feller's and Russian's boundaries classifications. For further classifications see [101, Sections 15.7 and 15.8].

4.4 Diffusion Processes with Killing

We consider a special type of diffusion processes whose trajectories are the same as those of a regular diffusion process until some random time ξ (possibly infinite) where the process is killed and stays in that state from that time on. These processes are usually called *diffusion processes with killing*. In this case the process can be written as $\{X_t, 0 \leq t < \xi\}$. If $\xi = \infty$ starting at $X_0 = x$ we say that the process is *conservative*. Sometimes an additional state Δ (called the "cemetery") is added to the original state space $\mathcal{S}$. For a diffusion, for each point $x \in \mathcal{S}$ there exists a probability $k(x)\, dt + o(t)$ such that the process is killed in an infinitesimal interval of time $(t, t + dt)$ and a probability $1 - k(x)\, dt + o(t)$ of not being killed. This killing coefficient $k(x)$, which is always positive, may also depend on t, but we only consider here killing coefficients independent of t. Then

$$\lim_{h \downarrow 0} \frac{1}{h} \mathbb{P}(t < \xi < t + h | X_t = x) = k(x).$$

The infinitesimal operator associated with diffusion processes with killing is closely related to the infinitesimal operator of the same diffusion process but eliminating the killing coefficient $k(x)$. Also, its semigroup structure is related in the same way. Therefore the infinitesimal operator will be given by

$$\mathcal{A} = \frac{1}{2}\sigma^2(x)\frac{d^2}{dx^2} + \mu(x)\frac{d}{dx} - k(x). \tag{4.43}$$

It is possible to write the Kolmogorov equations for this kind of diffusion using the so-called *Kac functional*. For a nonnegative function $k(x)$ defined on $\mathcal{S} = (a, b)$ and a bounded and continuous function $g(x)$ on $\mathcal{S}$, the Kac functional is defined by

$$w(x,t) = \mathbb{E}\left[\exp\left(-\int_0^t k(X_s)\, ds\right) g(X_t) | X_0 = x\right].$$

It is possible to see (see [101, p. 223]) that $w(x,t)$ satisfies the backward differential equation

$$\frac{\partial w}{\partial t}(x,t) = \frac{1}{2}\sigma^2(x)\frac{\partial^2 w}{\partial x^2}(x,t) + \mu(x)\frac{\partial w}{\partial x}(x,t) - k(x)w(x,t), \quad w(x,0) = g(x).$$

The function $w(x,t)$ is usually called the *Feynman–Kac formula.*

It is possible to write a spectral representation if we have information about the eigenvalues and eigenfunctions of $\mathcal{A}$ but with the coefficient $k(x)$. In general, this is difficult, but if we have a spectral representation of a regular diffusion process, then it is possible, after transformations, to get the infinitesimal operator of a diffusion process with killing, under certain restrictions. We will transform the pair $\{\pi, \mathcal{A}\}$ of a regular diffusion process into another pair $\{\widetilde{\pi}, \widetilde{\mathcal{A}}\}$, where $\widetilde{\mathcal{A}}$ is the infinitesimal operator of a diffusion process with killing.

Let $\{X_t, t \geq 0\}$ be a regular diffusion process with state space $\mathcal{S}$ and infinitesimal operator $\mathcal{A}$ given by (4.7). Assume that there exist eigenvalues $-\lambda_n \leq 0$ and (orthonormal) eigenfunctions ϕ_n associated with $\mathcal{A}$ such that the transition probability density admits a spectral representation as in (4.25), i.e.

$$p(t;x,y) = \pi(y)\sum_{n=0}^{\infty} e^{-\lambda_n t}\phi_n(x)\phi_n(y).$$

Let $s \in C^2(\mathcal{S})$, invertible and not vanishing on $\mathcal{S}$. Consider the following transformation of the spectral measure:

$$\widetilde{\pi}(x) = s^2(x)\pi(x).$$

In the space $L^2_{\widetilde{\pi}}(\mathcal{S})$ (which is isomorphic to $L^2_{\pi}(\mathcal{S})$) the orthonormal eigenfunctions are given by

$$\widetilde{\phi}_n(x) = \frac{\phi_n(x)}{s(x)},$$

since $\|\widetilde{\phi}_n\|^2_{\widetilde{\pi}} = \|\phi_n\|^2_{\pi} = 1$. The differential equation for $\widetilde{\phi}_n$ can be derived from the differential equation for the family ϕ_n, since it is a *gauge transformation* of the solutions (see Exercise 1.6). Indeed,

$$\widetilde{\mathcal{A}}\widetilde{\phi}_n = \frac{1}{2}\sigma^2(x)\widetilde{\phi}''_n(x) + \widetilde{\mu}(x)\widetilde{\phi}'_n(x) - k(x)\widetilde{\phi}_n(x) = -\lambda_n\widetilde{\phi}_n(x), \tag{4.44}$$

where

$$\widetilde{\mu}(x) = \mu(x) + \sigma^2(x)\frac{s'(x)}{s(x)}, \tag{4.45}$$

$$k(x) = -\left[\frac{1}{2}\sigma^2(x)\frac{s''(x)}{s(x)} + \mu(x)\frac{s'(x)}{s(x)}\right]. \tag{4.46}$$

That is, $\widetilde{\phi}_n$ are the eigenfunctions of the second-order differential operator (4.44) with an independent coefficient given by $k(x)$. This second-order differential operator is also symmetric with respect to the measure $\widetilde{\pi}(x)$. In general, we do not have any information about the function $k(x)$, since it depends strongly on the function $s(x)$ and the drift and diffusion coefficients, so we cannot claim that $\widetilde{\mathcal{A}}$ is the infinitesimal operator of a diffusion process with killing. Nevertheless it will be possible if the function $k(x)$ is *bounded from below*.

Let us call

$$c = \inf\{k(x), x \in \mathcal{S}\},$$

and assume that $c > -\infty$. We may have two cases:

1. $c \geq 0$. In this case $k(x) \geq 0$, so we will immediately have a diffusion process with killing where the killing factor is given by $k(x)$ and we have all the eigenvalues and eigenfunctions of the corresponding infinitesimal operator, so we can get a spectral representation as in (4.25) of the transition probability density given that the process has not been killed yet, i.e. $\xi > t$. If we call $\widetilde{p}^0(t;x,y)$ this transition probability density, then we have

$$\widetilde{p}^0(t;x,y) = \widetilde{\pi}(y) \sum_{n=0}^{\infty} e^{-\lambda_n t} \widetilde{\phi}_n(x) \widetilde{\phi}_n(y)$$

$$= s^2(y)\pi(y) \sum_{n=0}^{\infty} e^{-\lambda_n t} \frac{\phi_n(x)}{s(x)} \frac{\phi_n(y)}{s(y)} = \frac{s(y)}{s(x)} p(t;x,y).$$

Therefore,

$$\mathbb{P}(\xi > t | X_0 = x) = \int_{\mathcal{S}} \widetilde{p}^0(t;x,y)\, dy = \frac{1}{s(x)} \int_{\mathcal{S}} s(y) p(t;x,y)\, dy.$$

From here, we can approximate this distribution and compute the *mean killing time*, given by

$$\mathbb{E}(\xi | X_0 = x) = \int_0^{\infty} \mathbb{P}(\xi > t | X_0 = x)\, dt = \frac{1}{s(x)} \int_0^{\infty} \left(\int_{\mathcal{S}} s(y) p(t;x,y)\, dy \right) dt.$$

2. $c < 0$. In this case we can write $k(x)$ in the following form:

$$k(x) = \widetilde{k}(x) + c, \quad c < 0,$$

where $\widetilde{k}(x) \geq 0$. The constant c can be moved to the eigenvalue part, so the differential equation (4.44) can be rewritten as

$$\frac{1}{2}\sigma^2(x)\widetilde{\phi}_n''(x) + \widetilde{\mu}(x)\widetilde{\phi}_n'(x) - \widetilde{k}(x)\widetilde{\phi}_n(x) = -(\lambda_n - c)\widetilde{\phi}_n(x), \quad c < 0.$$

Therefore we have a diffusion process with killing where the killing factor is given by $\widetilde{k}(x)$ and the eigenvalues are given by $-(\lambda_n - c) < 0$, so we can get a spectral representation of the transition probability density given that the process has not been killed yet, i.e. $\xi > t$. If we call this transition probability density $\widetilde{p}^0(t;x,y)$ then we have

$$\widetilde{p}^0(t;x,y) = \widetilde{\pi}(y)\sum_{n=0}^{\infty} e^{-(\lambda_n - c)t}\widetilde{\phi}_n(x)\widetilde{\phi}_n(y) = s^2(y)\pi(y)\sum_{n=0}^{\infty} e^{-(\lambda_n - c)t}\frac{\phi_n(x)}{s(x)}\frac{\phi_n(y)}{s(y)}$$
$$= \frac{s(y)}{s(x)}e^{ct}p(t;x,y).$$

Therefore

$$\mathbb{P}(\xi > t|X_0 = x) = \int_{\mathcal{S}} \widetilde{p}^0(t;x,y)\,dy = \frac{e^{ct}}{s(x)}\int_{\mathcal{S}} s(y)p(t;x,y)\,dy.$$

From here we can approximate this distribution and compute the *mean killing time*, given by

$$\mathbb{E}(\xi|X_0 = x) = \int_0^{\infty} \mathbb{P}(\xi > t|X_0 = x)\,dt = \frac{1}{s(x)}\int_0^{\infty} e^{ct}\left[\int_{\mathcal{S}} s(y)p(t;x,y)\,dy\right]dt.$$

In conclusion, in order to use the equivalence between diffusion processes, it is necessary that $k(x)$ in (4.46) is bounded from below. The coefficient $\widetilde{\mu}(x)$ in (4.45) does not add any additional problem. We could also raise the inverse problem, that is, given a killing diffusion process if it is possible to relate it to an example for which we know its spectral representation. This way is much more difficult and will not guarantee success, since the transformation function $s(x)$ that relates both measures has to satisfy certain differential equations generated by (4.45) and (4.46) and that all coefficients match. We will see in Example 4.24 below a case in which this can be done, based on an example introduced in [99]. Another example can be found in the last section of [148].

4.5 Examples

In this section we will study several examples. Many of them can be found in the book of S. Karlin and H. M. Taylor [101] (see also [89]) but we will also study other examples included in [5, 99, 148]. We will start with diffusion processes with constant coefficients (Brownian motions with drift and scaling) and with different behavior at the boundary points. After that we will study some examples related to classical orthogonal polynomials, such as the Orstein–Uhlenbeck process, population growth models and the Wright–Fisher model. We will also see a couple of examples related to radial processes of N-dimensional Brownian motions and finally we will see a couple of examples with absolutely continuous and discrete spectrum.

4.5.1 Examples with Constant Drift and Diffusion Coefficients

Example 4.13 *Brownian motion with drift 0 and scaling σ on $[0,d]$ with two reflecting boundaries.* In this case $\mathcal{S} = [0,d]$ and the Kolmogorov backward equation for the transition probability density $p(t;x,y)$ is given by

$$\frac{\partial p}{\partial t} = \frac{1}{2}\sigma^2\frac{\partial^2 p}{\partial x^2}, \quad t > 0, \quad 0 < x < d, \quad y \in \mathcal{S},$$

with the Neumann boundary conditions (see (4.30))

$$\left.\frac{\partial p}{\partial x}\right|_{x=0,\,d} = 0, \quad t > 0, \quad y \in \mathcal{S}.$$

We try to find eigenvalues and eigenfunctions of the infinitesimal operator $\mathcal{A}$ given in this case by

$$\mathcal{A} = \frac{1}{2}\sigma^2\frac{d^2}{dx^2}.$$

In other words, from (4.26), we need to solve the Sturm–Liouville problem

$$\frac{1}{2}\sigma^2\phi''(x) = -\lambda\phi(x),$$
$$\phi'(0) = \phi'(d) = 0.$$

The general solution of this differential equation is given in terms of exponentials. Indeed,

$$\phi(x) = C_1 e^{\frac{\sqrt{-2\lambda}x}{\sigma}} + C_2 e^{-\frac{\sqrt{-2\lambda}x}{\sigma}}.$$

The first boundary condition $\phi'(0) = 0$ gives that $C_1 = C_2$, while the second boundary condition $\phi'(d) = 0$ gives

$$e^{\frac{\sqrt{-2\lambda}d}{\sigma}} - e^{-\frac{\sqrt{-2\lambda}d}{\sigma}} = 0.$$

If $\lambda \leq 0$ then the only solution is $\lambda = 0$. Alternatively, if $\lambda > 0$ the previous formula can be written in terms of the sine function, i.e.

$$\sin\left(\frac{\sqrt{2\lambda}d}{\sigma}\right) = 0.$$

Therefore the zeros are integer multiples of π, i.e.

$$\frac{\sqrt{2\lambda_n}d}{\sigma} = n\pi, \quad n \in \mathbb{Z}.$$

Hence we have

$$\lambda_n = \frac{n^2\pi^2\sigma^2}{2d^2}, \quad n \geq 0.$$

Substituting in $\phi(x)$, for each n, the eigenfunctions are given by

$$\phi_n(x) = b_n \cos\left(\frac{n\pi x}{d}\right), \quad n \geq 0.$$

In this case, and since $I(x_0, x) = 0$, the normalized measure $\pi(x)$ in (4.20) can be written as $\pi(x) = 1/d$. We now normalize the eigenfunctions ϕ_n. Since

$$\int_0^d \frac{1}{d}\, dx = 1, \quad \int_0^d \frac{1}{d} \cos^2\left(\frac{n\pi x}{d}\right) = \frac{1}{2},$$

then the normalizing constants b_n are given by

$$b_n = \sqrt{2}, \quad n \geq 1, \quad b_0 = 1.$$

It is well known (by the theory of Fourier series) that the system of eigenfunctions $(\phi_n)_n$ is complete in $L^2_\pi([0,d])$. Therefore, for $t > 0$ and $0 < x, y < d$, we have that

$$p(t; x, y) = \frac{1}{d} + \frac{2}{d} \sum_{n=1}^{\infty} \exp\left(-\frac{\sigma^2 \pi^2 n^2}{2d^2} t\right) \cos\left(\frac{n\pi x}{d}\right) \cos\left(\frac{n\pi y}{d}\right). \tag{4.47}$$

In this case the functions $S(x)$ and $M(x)$ in (4.33) and (4.34) respectively are polynomials of degree 1 in x. Since we are working in a bounded interval, the values of $\Sigma(\alpha)$ and $N(\alpha)$ defined in (4.39) and (4.40), respectively, with $\alpha = 0, d,$ are finite. Therefore the boundaries are regular, that is, the process can enter and leave the boundaries as many times as wanted.

Observe that in this case

$$\lim_{t \to \infty} p(t; x, y) = \frac{1}{d}.$$

This is an exponential and uniform convergence with respect to x and y. Therefore the invariant distribution converges to the uniform distribution on $[0, d]$, as expected. ◊

Example 4.14 *Brownian motion with drift 0 and scaling σ on $[0,\infty)$ with one reflecting boundary.* In this case we have $S = [0, \infty)$ and the Kolmogorov backward equation for the transition probability density $p(t; x, y)$ is given by

$$\frac{\partial p}{\partial t} = \frac{1}{2}\sigma^2 \frac{\partial^2 p}{\partial x^2}, \quad t > 0, \quad x > 0, \quad y \geq 0,$$

with the Neumann boundary conditions (see (4.30))

$$\left.\frac{\partial p}{\partial x}\right|_{x=0} = 0, \quad t > 0, \quad y \leq 0.$$

This diffusion process is the limiting form of the previous example as $d \to \infty$. Indeed, using (4.47), we have

$$\begin{aligned}
p(t;x,y) &= \lim_{d\to\infty} \frac{1}{d} \sum_{n=-\infty}^{\infty} \exp\left(-\frac{\sigma^2\pi^2 n^2}{2d^2}t\right) \cos\left(\frac{n\pi x}{d}\right) \cos\left(\frac{n\pi y}{d}\right) \\
&= \int_{-\infty}^{\infty} \exp\left(-\frac{t\sigma^2\pi^2 u^2}{2}\right) \cos(\pi x u)\cos(\pi y u)\, du \qquad \text{(Riemann integral)} \\
&= \frac{1}{2}\int_{-\infty}^{\infty} \exp\left(-\frac{t\sigma^2\pi^2 u^2}{2}\right) \\
&\quad \times [\cos(\pi(x+y)u) + \cos(\pi(x-y)u)]\, du \qquad \text{(trigonometric form)} \\
&= \frac{1}{2\pi}\int_{-\infty}^{\infty} \exp\left(-\frac{t\sigma^2\xi^2}{2}\right) \\
&\quad \times \left[e^{-i\xi(x+y)} + e^{-i\xi(x-y)}\right] d\xi \qquad (\xi = \pi u \text{ and imparity}) \\
&= \frac{1}{\sqrt{2\pi t\sigma^2}}\left(\exp\left(-\frac{(x+y)^2}{2t\sigma^2}\right) + \exp\left(-\frac{(x-y)^2}{2t\sigma^2}\right)\right),
\end{aligned}$$

where the last equality follows from the Fourier inversion formula and the fact that $\exp(-t\sigma^2\xi^2/2)$ is the characteristic function of the Gaussian distribution with mean 0 and variance $t\sigma^2$. As before, the boundary 0 is regular, while the boundary ∞ is natural, that is, it cannot be reached in finite time. A different derivation of this example can be found in [148, Example A]. ◊

Example 4.15 *Brownian motion with drift and scaling on* $[0,d]$ *with two absorbing boundaries.* In this case $\mathcal{S} = [0,d]$ and the Kolmogorov backward equation for the transition probability density $p^0(t;x,y)$ given that the process has not been absorbed is given by

$$\frac{\partial p^0}{\partial t} = \frac{1}{2}\sigma^2\frac{\partial^2 p^0}{\partial x^2} + \mu\frac{\partial p^0}{\partial x}, \quad t > 0, \quad 0 < x < d, \quad y \in \mathcal{S},$$

with the Dirichlet boundary conditions (see (4.31))

$$\lim_{x\downarrow 0} p^0(t;x,y) = 0, \quad \lim_{x\uparrow d} p^0(t;x,y) = 0, \quad t > 0, \quad y \in \mathcal{S}.$$

We seek eigenvalues and eigenfunctions of the differential operator $\mathcal{A}$, which in this case is given by

$$\mathcal{A} = \frac{1}{2}\sigma^2\frac{d^2}{dx^2} + \mu\frac{d}{dx}.$$

In other words,

$$\frac{1}{2}\sigma^2\phi''(x) + \mu\phi'(x) = -\lambda\phi(x),$$
$$\phi(0) = \phi(d) = 0.$$

The general solution of this differential equation can be written in terms of exponential functions. Indeed,

$$\phi(x) = e^{-\frac{\mu x}{\sigma^2}}\left(C_1 e^{\frac{\sqrt{\mu^2-2\lambda\sigma^2}}{\sigma^2}x} + C_2 e^{-\frac{\sqrt{\mu^2-2\lambda\sigma^2}}{\sigma^2}x}\right).$$

The first boundary condition $\phi(0) = 0$ gives that $C_1 = -C_2$, while the second boundary condition $\phi(d) = 0$ gives that

$$e^{\frac{\sqrt{\mu^2-2\lambda\sigma^2}}{\sigma^2}d} - e^{-\frac{\sqrt{\mu^2-2\lambda\sigma^2}}{\sigma^2}d} = 0.$$

If $\lambda \le \lambda_0 = \mu^2/2\sigma^2$ then the only solution is $\lambda = \lambda_0$. Alternatively, if $\lambda > \lambda_0$ the previous formula can be written in terms of the sine function, i.e.

$$\sin\left(\frac{\sqrt{-\mu^2 + 2\lambda\sigma^2}}{\sigma^2}d\right) = 0.$$

Therefore the zeros are integer multiples of π, i.e.

$$\frac{\sqrt{-\mu^2 + 2\lambda\sigma^2}}{\sigma^2}d = n\pi, \quad n \in \mathbb{Z}.$$

Therefore we get

$$\lambda_n = \frac{n^2\pi^2\sigma^2}{2d^2} + \frac{\mu^2}{2\sigma^2}, \quad n \ge 0. \tag{4.48}$$

Substituting in $\phi(x)$, for each n, the eigenfunctions are given by

$$\phi_n(x) = b_n \exp\left(-\frac{\mu x}{\sigma^2}\right)\sin\left(\frac{n\pi x}{d}\right), \quad n \ge 1. \tag{4.49}$$

In this case, the measure $\pi(x)$ in (4.20) can be written (up to a normalizing constant) as

$$\pi(x) = \exp\left(\frac{2\mu x}{\sigma^2}\right), \quad 0 \le x \le d.$$

The normalizing constants b_n are given by

$$b_n = \sqrt{\frac{2}{d}}, \quad n \ge 1.$$

For $t > 0$ and $0 < x, y < d$, the spectral representation of $p^0(t; x, y)$ is given by

$$p^0(t;x,y) = \frac{2}{d}\exp\left(\frac{\mu(y-x)}{\sigma^2}\right)\exp\left(-\frac{\mu^2 t}{2\sigma^2}\right) \times \sum_{n=1}^{\infty}\exp\left(-\frac{\sigma^2\pi^2 n^2 t}{2d^2}\right)\sin\left(\frac{n\pi x}{d}\right)\sin\left(\frac{n\pi y}{d}\right). \tag{4.50}$$

Since the boundaries are absorbing it remains to compute the absorption probabilities

$$p(t;x,\{0\}) = \mathbb{P}(X_t = 0|X_0 = x), \quad p(t;x,\{d\}) = \mathbb{P}(X_t = d|X_0 = x), \quad 0 < x < d.$$

Since

$$p(t;x,\{0\}) + p(t;x,\{d\}) = 1 - \int_0^d p^0(t;x,y)\,dy,$$

we have that

$$p(t;x,\{0\}) + p(t;x,\{d\}) = 1 - \frac{2}{d}\exp\left(-\frac{\mu x}{\sigma^2}\right)\sum_{n=1}^{\infty} a_n \exp\left(-\frac{\sigma^2\pi^2 n^2 t}{2d^2}\right)\sin\left(\frac{n\pi x}{d}\right),$$

where

$$a_n = \int_0^d \exp\left(\frac{\mu y}{\sigma^2}\right)\sin\left(\frac{n\pi y}{d}\right)dy = \frac{n\,d\pi\sigma^4}{\mu^2 d^2 + n^2\pi^2\sigma^4}\left(1 - (-1)^n \exp\left(\frac{d\mu}{\sigma^2}\right)\right).$$

Therefore, it is enough to determine $p(t;x,\{0\})$ and from the previous formula we can compute $p(t;x,\{d\})$. Following (4.32) we have that

$$p(t;x,\{0\}) = v(x) - \int_0^d v(y)p^0(t;x,y)\,dy, \tag{4.51}$$

where $v(x) = \mathbb{P}(\tau_0 < \tau_d |\, X_0 = x)$ is defined by (4.14). This function $v(x)$ was already computed in (4.36) and is given by

$$v(x) = \frac{S(d) - S(x)}{S(d) - S(0)},$$

where $S(x)$ is the scale function defined by (4.33). In this case, choosing $x_0 = 0$, we have that

$$S(x) = \frac{\sigma^2}{2\mu}\left(1 - \exp\left(-\frac{2\mu x}{\sigma^2}\right)\right).$$

After some straightforward computation, we obtain

$$v(x) = \frac{1 - \exp\left(\frac{2\mu}{\sigma^2}(d-x)\right)}{1 - \exp\left(\frac{2d\mu}{\sigma^2}\right)}.$$

In the particular case of $\mu = 0$, we have that $v(x) = 1 - x/d$. Therefore we can approximate $p(t; x, \{0\})$ using (4.51) and after that $p(t; x, \{d\})$. Again, the diffusion process is defined on a bounded interval, and the functions $S(x)$ and $M(x)$ have no problems at the boundary points. Therefore the integrals $\Sigma(\alpha)$ and $M(\alpha)$ with $\alpha = 0, d$, are bounded, and both boundaries are regular. ◊

Example 4.16 *Brownian motion with drift and scaling on* $[0, \infty)$ *with one absorbing boundary.* As before, it is enough to take limits in (4.50) as $d \to \infty$, in which case we have (following the same lines as in Example 4.14)

$$
\begin{aligned}
p^0(t; x, y) &= 2\exp\left(\frac{\mu(y-x)}{\sigma^2} - \frac{\mu^2 t}{2\sigma^2}\right) \int_0^\infty \exp\left(-\frac{\pi^2\sigma^2 t u^2}{2}\right) \sin(\pi x u) \sin(\pi y u)\, du \\
&= \frac{2}{\pi}\exp\left(\frac{\mu(y-x)}{\sigma^2} - \frac{\mu^2 t}{2\sigma^2}\right) \int_0^\infty \exp\left(-\frac{\sigma^2 t\xi^2}{2}\right) \sin(\xi x) \sin(\xi y)\, d\xi \\
&= \frac{1}{\pi}\exp\left(\frac{\mu(y-x)}{\sigma^2} - \frac{\mu^2 t}{2\sigma^2}\right) \\
&\quad \times \int_0^\infty \exp\left(-\frac{\sigma^2 t\xi^2}{2}\right) [\cos(\xi(x-y)) - \cos(\xi(x+y))]\, d\xi \\
&= \frac{1}{2\pi}\exp\left(\frac{\mu(y-x)}{\sigma^2} - \frac{\mu^2 t}{2\sigma^2}\right) \\
&\quad \times \int_{-\infty}^\infty \exp\left(-\frac{\sigma^2 t\xi^2}{2}\right) \left[e^{-i\xi(x-y)} - e^{-i\xi(x+y)}\right] d\xi \\
&= \exp\left(\frac{\mu(y-x)}{\sigma^2} - \frac{\mu^2 t}{2\sigma^2}\right) \\
&\quad \times \frac{1}{\sqrt{2\pi\sigma^2 t}}\left[\exp\left(-\frac{(x-y)^2}{2\sigma^2 t}\right) - \exp\left(-\frac{(x+y)^2}{2\sigma^2 t}\right)\right],
\end{aligned}
$$

where, as before, the last equality follows from the Fourier inversion formula and the fact that $\exp(-t\sigma^2\xi^2/2)$ is the characteristic function of the Gaussian distribution with mean 0 and variance $t\sigma^2$. Integration of $p^0(t; x, y)$ with respect to y over $(0, \infty)$ gives $p(t; x, \{0\}^c)$, which is the probability that, starting at x, the first passage time to 0 is greater than t. ◊

Example 4.17 *Brownian motion.* This example can be found in [101, p. 337]. The spectrum of the infinitesimal operator now is absolutely continuous. Indeed, we seek eigenfunctions satisfying

$$\frac{1}{2}\phi''(x) = -\widetilde{\lambda}\phi(x),$$

with the condition that $\phi(x)$ is bounded for all $x \in \mathbb{R}$. This condition forces $\widetilde{\lambda}$ to be positive. Call $\widetilde{\lambda} = \lambda^2/2$. Then we seek eigenfunctions $\phi(x)$ satisfying

$$\phi''(x) = -\lambda^2\phi(x),$$

with the same initial conditions. The general solution of this differential equation is given by

$$\phi(x,\lambda) = c_1 \sin(\lambda x) + c_2 \cos(\lambda x), \quad \lambda, x \in \mathbb{R},$$

which is always bounded. Since $I(x_0,x)=0$, we can choose the measure as $\pi(x) = 1$. In $L^2(\mathbb{R})$ (now with the inner product $(f,g) = \int_{\mathbb{R}} f(x)\overline{g(x)}\,dx$) we have that for each $\lambda \in \mathbb{R}$, the normalized eigenfunctions can be written as

$$\phi(x,\lambda) = \frac{1}{\sqrt{2\pi}}e^{ix\lambda}.$$

The orthogonality implies that

$$\int_{\mathbb{R}} \phi(x,\lambda)\overline{\phi(x,\mu)}\,dx = \int_{\mathbb{R}} e^{ix(\lambda-\mu)}dx = \delta(\lambda-\mu).$$

Therefore, a spectral representation of the transition probability density is

$$p(t;x,y) = \frac{1}{2\pi}\int_{-\infty}^{\infty} e^{-\lambda^2 t/2}e^{ix\lambda}e^{-iy\lambda}d\lambda,$$

which is the inverse Fourier transform of the Gaussian kernel, and then

$$p(t;x,y) = \frac{1}{\sqrt{2\pi}}e^{-(x-y)^2/2t}, \quad x,y \in \mathbb{R}, \quad t \geq 0.$$

In this case, we can choose $x_0 = 0$, $S(x) = x$ and $M(x) = x$. Therefore both boundary points $\pm\infty$ are natural and do not belong to the state space. We also observe that the Brownian motion is null recurrent, since $S(\pm\infty) = \pm\infty$, but the expected return time is $M(\pm\infty) = \pm\infty$. ◊

4.5.2 Examples Related to Classical Orthogonal Polynomials

All examples included in this subsection can be found in [101, Section 15.13] (see also [89]) and the last example in [99]. In most of them we have included some extra details, especially related to the behavior at the boundary points.

Example 4.18 *The Orstein–Uhlenbeck diffusion process.* The drift and diffusion coefficients of this process are given by

$$\mu(x) = -x, \quad \sigma^2(x) = 1, \quad x \in \mathbb{R}.$$

The state space is interpreted as the speed of a particle following a Brownian motion but with a restoring force directed towards the origin and of magnitude proportional to the distance. Taking $x_0 = 0$, we have that $I(0,x) = -x^2$ (see (4.19)). Therefore the measure $\pi(x)$ in (4.20) can be chosen (up to a normalization constant) as

$$\pi(x) = e^{-x^2}. \tag{4.52}$$

The Kolmogorov backward equation for the transition probability density $p(t;x,y)$ is given by

$$\frac{\partial p}{\partial t} = \frac{1}{2}\frac{\partial^2 p}{\partial x^2} - x\frac{\partial p}{\partial x}, \quad t > 0.$$

We seek then eigenfunctions and discrete eigenvalues of the infinitesimal operator

$$\mathcal{A} = \frac{1}{2}\frac{d^2}{dx^2} - x\frac{d}{dx}.$$

In other words,

$$\frac{1}{2}\phi''(x) - x\phi'(x) = -\lambda\phi(x), \quad x \in \mathbb{R},$$

such that $\int_{-\infty}^{\infty} e^{-x^2}[\phi(x)]^2 dx < \infty$. Since the measure $\pi(x)$ vanishes at the endpoints of the state space $\mathcal{S} = \mathbb{R}$, the boundary conditions of the Sturm–Liouville problem (4.26) hold (the boundary points are reflecting). This differential equation is identified as the one corresponding with the Hermite polynomials $(H_n)_n$ (see (1.61)) with

$$\lambda_n = n, \quad n \geq 0.$$

The norms of $H_n(x)$ are given by (1.62). Therefore, the spectral representation of the transition probability density $p(t;x,y)$ can be written as

$$p(t;x,y) = e^{-y^2}\sum_{n=0}^{\infty} e^{-nt}\frac{H_n(x)H_n(y)}{2^n n!\sqrt{\pi}} = \frac{e^{-y^2}}{\sqrt{\pi}}\sum_{n=0}^{\infty}\frac{H_n(x)H_n(y)}{n!}\left(\frac{e^{-t}}{2}\right)^n$$

using (1.64)

$$= \frac{e^{-y^2}}{\sqrt{\pi}\sqrt{1-e^{-2t}}}\exp\left(\frac{2e^{-t}xy}{1+e^{-t}} - \frac{e^{-2t}(x-y)^2}{1-e^{-2t}}\right)$$

$$= \frac{e^{-y^2}}{\sqrt{\pi}\sqrt{1-e^{-2t}}}\exp\left(-\frac{e^{-2t}(x^2+y^2)}{1-e^{-2t}}\right)\exp\left(\frac{2xye^{-t}}{1-e^{-2t}}\right).$$

The probability density of the Orstein–Uhlenbeck diffusion process is well known and can be computed using other methods (see [5, 101]). Figure 4.1 shows a couple of trajectories of the Orstein–Uhlenbeck process starting at different values of X_0. To study the behavior at the boundary points $\pm\infty$, we observe that

$$dS(y) = e^{y^2}dy, \quad dM(z) = e^{-z^2}dz.$$

For example, for ∞ (similar for $-\infty$), using (4.42), we have that

$$N(\infty) = \int_x^{\infty}\left[\int_y^{\infty} dM(z)\right]dS(y) = \int_x^{\infty}\left[\int_y^{\infty} e^{-z^2}dz\right]e^{y^2}dy.$$

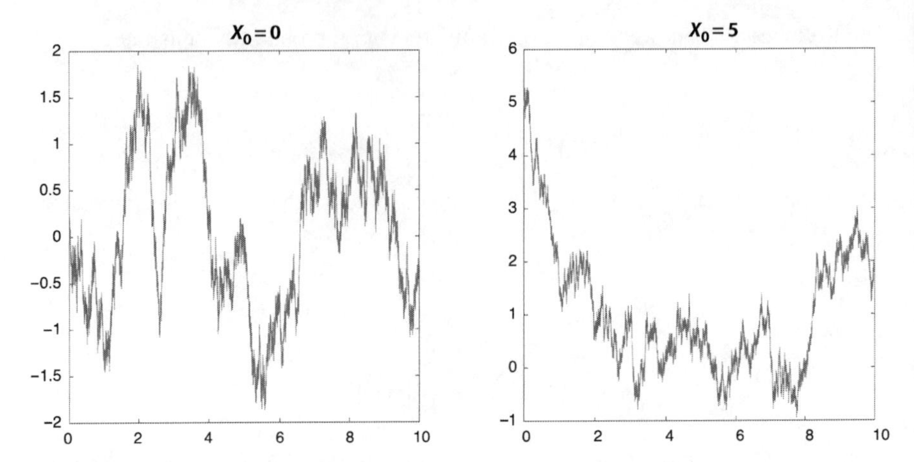

Figure 4.1 Trajectories of the Orstein–Uhlenbeck process starting at different values of X_0.

Using integration by parts, the previous integral is

$$\int_y^\infty e^{-z^2} dz = \frac{1}{2}\int_y^\infty \frac{1}{z} 2ze^{-z^2} dz = \frac{1}{2y}e^{-y^2} - \frac{1}{2}\int_y^\infty \frac{e^{-z^2}}{z^2} dz$$
$$= \frac{1}{2y}e^{-y^2} - \frac{1}{4y^2}e^{-y^2} + \frac{3}{4}\int_y^\infty \frac{e^{-z^2}}{z^4} dz \sim \frac{1}{2y}e^{-y^2}, \quad y \to \infty.$$

We see that the integral is divergent, therefore $N(\infty) = \infty$. Similarly, we can prove, using (4.41), that $\Sigma(\infty) = \infty$. Therefore both boundary points are natural. The main difference with Brownian motion is that the Orstein–Uhlenbeck process is positive recurrent, since $S(\infty) = \infty$, but $M(\infty) < \infty$ (similarly for $-\infty$), while the Brownian motion is null recurrent. The stationary distribution for the Orstein–Uhlenbeck process is given by the normalization of the measure π in (4.52), i.e.

$$\pi(y) = \frac{1}{\sqrt{\pi}}e^{-y^2}, \quad y \in \mathbb{R}.$$

◊

Example 4.19 *Population growth model.* The drift and diffusion coefficients of this process are given by

$$\mu(x) = bx + c, \quad \sigma^2(x) = 2ax, \quad 0 < x < \infty,$$

where $a, b, c,$ are constants with $a > 0$. For $x_0 = 1$, we have that

$$I(1, x) = \int_1^x \frac{bz + c}{az} dz = \frac{b}{a}(x - 1) + \frac{c}{a}\log x.$$

Therefore the measure $\pi(x)$ in (4.20) can be chosen as

$$\pi(x) = x^{\alpha} e^{bx/a}, \quad \alpha = \frac{c}{a} - 1, \quad x > 0.$$

The Kolmogorov backward equation for the transition probability density $p(t;x,y)$ is given by

$$\frac{\partial p}{\partial t} = ax\frac{\partial^2 p}{\partial x^2} + (bx+c)\frac{\partial p}{\partial x}, \quad t > 0.$$

We seek then eigenfunctions and discrete eigenvalues of the infinitesimal operator

$$\mathcal{A} = ax\frac{d^2}{dx^2} + (bx+c)\frac{d}{dx}.$$

In other words,

$$ax\phi''(x) + (bx+c)\phi'(x) = -\lambda\phi(x), \quad x > 0,$$

such that $\int_0^{\infty} x^{\alpha} e^{bx/a}[\phi(x)]^2 dx < \infty$. In order to have reflecting boundaries we need to impose the following boundary conditions (see (4.26)):

$$ax^{\alpha+1}e^{bx/a}\phi'(x)\Big|_{x=0} = 0, \quad ax^{\alpha+1}e^{bx/a}\phi'(x)\Big|_{x=\infty} = 0.$$

Therefore we need $\alpha > -1, b < 0$ and $c > 0$, since $a > 0$. The differential equation is identified as the one corresponding with the Laguerre polynomials $(L_n^{(\alpha)})_n$ (see (1.67)). In particular, the eigenfunctions are given by

$$\phi_n(x) = L_n^{(\alpha)}\left(-\frac{bx}{a}\right), \quad n \geq 0.$$

The corresponding values of λ_n, substituting the polynomial into the differential equation (1.67), are given by

$$\lambda_n = -bn, \quad n \geq 0.$$

Using the explicit expression of the norms of $L_n^{(\alpha)}(x)$ in (1.68) we get the spectral representation of the transition probability density $p(t;x,y)$:

$$\begin{aligned} p(t;x,y) &= \frac{(|b|/a)^{\alpha+1} y^{\alpha} e^{by/a}}{\Gamma(\alpha+1)} \sum_{n=0}^{\infty} e^{bnt} \frac{n!}{\Gamma(n+\alpha+1)} L_n^{(\alpha)}\left(-\frac{bx}{a}\right) L_n^{(\alpha)}\left(-\frac{by}{a}\right) \\ &= \frac{(|b|/a)^{\alpha+1} y^{\alpha} e^{by/a}}{\Gamma(\alpha+1)} \frac{\exp\left(\frac{be^{bt}(x+y)}{a(1-e^{bt})}\right) I_{\alpha}\left(\frac{2|b|\sqrt{xye^{bt}}}{a(1-e^{bt})}\right)}{\left(\frac{b^2}{a^2}xye^{bt}\right)^{\alpha/2}(1-e^{bt})} \\ &= \frac{|b|e^{by/a}e^{-bt\alpha/2}}{a\Gamma(\alpha+1)(1-e^{bt})}\left(\frac{y}{x}\right)^{\alpha/2} \exp\left(\frac{be^{bt}(x+y)}{a(1-e^{bt})}\right) I_{\alpha}\left(\frac{2|b|\sqrt{xye^{bt}}}{a(1-e^{bt})}\right). \end{aligned}$$

We have used formula (1.71) in the last series, where we can find the definition of the Bessel function $I_\alpha(z)$. As for the boundary points, a simple calculation gives

$$dS(y) = y^{-\alpha-1}e^{-by/a}dy, \quad dM(z) = z^\alpha e^{bz/a}dz.$$

At 0 we have, choosing $x = 1$,

$$\Sigma(0) = \int_0^1 \left[\int_y^1 dM(z)\right] dS(y) = \int_0^1 \left[\int_y^1 z^\alpha e^{bz/a}dz\right] y^{-\alpha-1}e^{-by/a}dy$$
$$\sim \int_0^1 (c_1 y^{\alpha+1} + c_2)y^{-\alpha-1}dy, \qquad y \downarrow 0.$$

Therefore, this integral is convergent if and only if $-\alpha - 1 > -1$, i.e. $\alpha < 0$. Otherwise it is divergent. A similar argument works for $N(0)$, if we choose $x = 1$. Indeed,

$$N(0) = \int_0^1 \left[\int_z^1 dS(y)\right] dM(z) = \int_0^1 \left[\int_z^1 y^{-\alpha-1}e^{-by/a}dy\right] z^\alpha e^{bz/a}dz$$
$$\sim \int_0^1 (c_1 z^{-\alpha} + c_2)z^\alpha dz, \qquad z \downarrow 0.$$

Therefore, this integral is finite if and only if $\alpha > -1$. If $\alpha = -1$ then it is divergent. We have then three cases:

- 0 is an exit boundary if $\alpha = -1$, in which case we have $\Sigma(0) < \infty$ and $N(0) = \infty$.
- 0 is a regular boundary if $-1 < \alpha < 0$, in which case we have $\Sigma(0) < \infty$ and $N(0) < \infty$.
- 0 is an entrance boundary if $\alpha \geq 0$, in which case we have $\Sigma(0) = \infty$ and $N(0) < \infty$.

As for the boundary ∞, we have, choosing $x = 1$ and using (4.41), that

$$\Sigma(\infty) = \int_1^\infty \left[\int_1^y dM(z)\right] dS(y) = \int_1^\infty \left[\int_1^y z^\alpha e^{bz/a}dz\right] y^{-\alpha-1}e^{-by/a}dy$$
$$\sim \int_1^\infty (c_1 y^{\alpha+1}e^{by/a} + c_2)y^{-\alpha-1}e^{-by/a}dy \to \infty.$$

Alternatively, choosing $x = 1$ and using (4.42), we have that

$$N(\infty) = \int_1^\infty \left[\int_1^z dS(y)\right] dM(z) = \int_1^\infty \left[\int_1^z y^{-\alpha-1}e^{-by/a}dy\right] z^\alpha e^{bz/a}dz$$
$$\sim \int_1^\infty (c_1 z^{-\alpha}e^{-bz/a} + c_2)z^\alpha e^{bz/a}dz \to \infty.$$

Therefore, the boundary ∞ is natural. Figure 4.2 shows some trajectories of this process for different values of a, b and c starting at $X_0 = 5$.

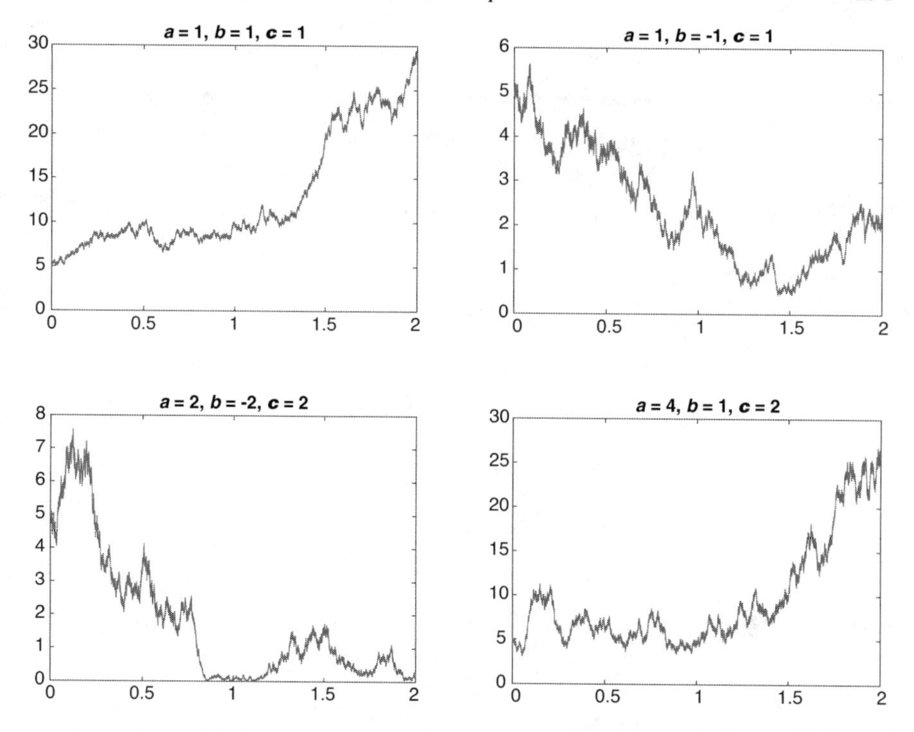

Figure 4.2 Trajectories of the population growth model for different values of a, b and c starting at $X_0 = 5$.

Remark 4.20 This model is also used in mathematical finance to describe the evolution of interest rates. It is also known as the *Cox–Ingersoll–Ross model* [24, 25]. The drift and diffusion coefficients are given by (see the similarity with the population growth model)

$$\mu(x) = a(b - x), \quad \sigma^2(x) = \sigma^2 x, \quad 0 < x < \infty,$$

where a corresponds to the speed of adjustment, b to the mean and σ to volatility. This process can also be defined as a sum of squared Ornstein–Uhlenbeck processes. For a more general approach applied to the spectral representation for branching processes on $[0, \infty)$ see [122] (see also [93, 94, 95]). ◊

Example 4.21 *The Wright–Fisher model.* This model was derived as a limiting case of a stochastic model on genetics to study the fluctuations of gene frequency under the influence of mutation or selection (see [53, 149] or Example 3.34). In [91] the authors consider other situations, not necessarily related to orthogonal polynomials (see also [101]). The Wright–Fisher model has drift and diffusion coefficients given by

$$\mu(x) = \gamma_2(1-x) - \gamma_1 x, \quad \sigma^2(x) = x(1-x), \quad 0 < x < 1, \quad \gamma_1, \gamma_2 > 0.$$

The state space is $\mathcal{S} = [0,1]$. The coefficients γ_1, γ_2 are interpreted as the intensity of mutation from the population A to another population B and vice versa, which are both competing.

This example can be identified with the Jacobi polynomials defined on the interval $[0,1]$ (see (1.83)). The intensities of mutation γ_1, γ_2, are identified with the parameters $\alpha, \beta > -1$ by

$$\gamma_1 = \frac{\beta+1}{2}, \quad \gamma_2 = \frac{\alpha+1}{2}. \tag{4.53}$$

The function $I(x_0, x)$ can be explicitly computed:

$$\begin{aligned} I(x_0,x) &= \int_{x_0}^{x} \frac{2\gamma_2(1-z) - 2\gamma_1 z}{z(1-z)}\, dz \\ &= \int_{x_0}^{x} \left(\frac{2\gamma_2}{z} - \frac{2\gamma_1}{1-z} \right) dz = \log\left[x^{2\gamma_2}(1-x)^{2\gamma_1} \right] + C. \end{aligned}$$

Therefore the measure $\pi(x)$ in (4.20) can be written as

$$\pi(x) = x^{2\gamma_2 - 1}(1-x)^{2\gamma_1 - 1}, \quad 0 < x < 1. \tag{4.54}$$

The Kolmogorov backward equation for the transition probability density $p(t;x,y)$ is given by

$$\frac{\partial p}{\partial t} = \frac{1}{2} x(1-x) \frac{\partial^2 p}{\partial x^2} + [\gamma_2(1-x) - \gamma_1 x] \frac{\partial p}{\partial x}, \quad t > 0.$$

We seek a discrete set of eigenfunctions and eigenvalues of the infinitesimal operator

$$\mathcal{A} = \frac{1}{2} x(1-x) \frac{d^2}{dx^2} + [\gamma_2(1-x) - \gamma_1 x] \frac{d}{dx}.$$

In other words,

$$\frac{1}{2} x(1-x)\phi''(x) + [\gamma_2(1-x) - \gamma_1 x]\, \phi'(x) = -\lambda \phi(x), \quad 0 < x < 1,$$

such that the boundary points are reflecting (see (4.26)), i.e.

$$\frac{1}{2} x^{2\gamma_2}(1-x)^{2\gamma_1} \phi'(x) \Big|_{x=0} = 0, \quad \frac{1}{2} x^{2\gamma_2}(1-x)^{2\gamma_1} \phi'(x) \Big|_{x=1} = 0.$$

The solution of the differential equation can be written in terms of the Jacobi polynomials $(P_n^{(\alpha,\beta)}(y))_n$ on $[-1,1]$ (see (1.77)) after a change of variables:

$$x = \frac{1+y}{2}.$$

The corresponding values of λ_n (see (1.78)) are given by

$$\lambda_n = \frac{1}{2}n(n+\alpha+\beta+1), \quad n \geq 0.$$

We have then the spectral representation of the transition probability density $p(t;x,y)$, given by

$$\begin{aligned}p(t;x,y) = &\frac{y^{2\gamma_2-1}(1-y)^{2\gamma_1-1}}{\mathrm{B}(2\gamma_2,2\gamma_1)}\\ &\times \sum_{n=0}^{\infty} e^{-n(n+2\gamma_1+2\gamma_2-1)t/2} P_n^{(2\gamma_1-1,2\gamma_2-1)}(2x-1) P_n^{(2\gamma_1-1,2\gamma_2-1)}(2y-1)\\ &\times \frac{\mathrm{B}(2\gamma_2,2\gamma_1)(2n+2\gamma_1+2\gamma_2-1)n!\,\Gamma(n+2\gamma_1+2\gamma_2-1)}{\Gamma(n+2\gamma_2)\Gamma(n+2\gamma_1)},\end{aligned}$$

where $\mathrm{B}(x,y)$ is the Beta function defined in (1.3). Unlike the Hermite and Laguerre examples, it is not known an explicit expression of this series in terms of elementary functions. As for the behavior of the boundary point, a simple computation gives

$$dS(y) = \frac{1}{y^{2\gamma_2}(1-y)^{2\gamma_1}}dy, \quad dM(z) = z^{2\gamma_2-1}(1-z)^{2\gamma_1-1}dz.$$

Let us focus on the boundary 0 (similar for 1). In this case we have, if we choose $x = 1/2$, that

$$\begin{aligned}\Sigma(0) &= \int_0^{1/2}\left[\int_y^{1/2} dM(z)\right]dS(y) = \int_0^{1/2}\left[\int_y^{1/2} z^{2\gamma_2-1}(1-z)^{2\gamma_1-1}dz\right]dS(y)\\ &\sim \int_0^{1/2}(c_1y^{2\gamma_2}+c_2)y^{-2\gamma_2}dy, \qquad y \downarrow 0.\end{aligned}$$

Therefore this integral is convergent if and only if $-2\gamma_2 > -1$, i.e. $2\gamma_2 < 1$. In all other cases it is divergent. A similar argument works for $N(0)$, if we choose $x = 1/2$. Indeed,

$$\begin{aligned}N(0) &= \int_0^{1/2}\left[\int_z^{1/2} dS(y)\right]dM(z) = \int_0^{1/2}\left[\int_z^{1/2} \frac{1}{y^{2\gamma_2}(1-y)^{2\gamma_1}}dy\right]dM(z)\\ &\sim \int_0^{1/2}(c_1z^{-2\gamma_2+1}+c_2)z^{2\gamma_2-1}dz, \qquad z \downarrow 0.\end{aligned}$$

Therefore this integral is finite if and only if $2\gamma_2 - 1 > -1$, i.e. $\gamma_2 > 0$. If $\gamma_2 = 0$ then it is divergent. We have then three cases:

- 0 is an exit boundary if $\gamma_2 = 0$, in which case we have $\Sigma(0) < \infty$ and $N(0) = \infty$.
- 0 is a regular boundary if $0 < \gamma_2 < 1/2$, in which case we have $\Sigma(0) < \infty$ and $N(0) < \infty$.

- 0 is an entrance boundary if $\gamma_2 \geq 1/2$, in which case we have $\Sigma(0) = \infty$ and $N(0) < \infty$.

The same results hold for the boundary 1 but changing γ_2 by γ_1.

If $\gamma_1, \gamma_2 \geq 1/2$ (both entrance boundaries) the process converges to a stationary distribution given by the normalization of the measure π in (4.54), i.e.

$$\pi(y) = \frac{\Gamma(2\gamma_1)\Gamma(2\gamma_2)}{\Gamma(2\gamma_1 + 2\gamma_2)} y^{2\gamma_2 - 1}(1-y)^{2\gamma_1 - 1}.$$

If $0 < \gamma_1, \gamma_2 < 1/2$ (both regular boundaries), the stationary distribution is still valid, but we have to decide what happens after the process reaches one the boundary points. From the application point of view, the most common assumption is that we have reflecting boundaries. But sometimes it may make sense to have an exponentially distributed holding time at the boundary and after that starting again the process from an interior point of $(0, 1)$. The right-stationary distribution should have atoms with positive jumps at the boundaries 0 and 1, apart from the absolutely continuous part $\pi(y)$. Figure 4.3 shows some trajectories of the Wright–Fisher model for different values of γ_1 and γ_2 starting at $X_0 = 1/2$.

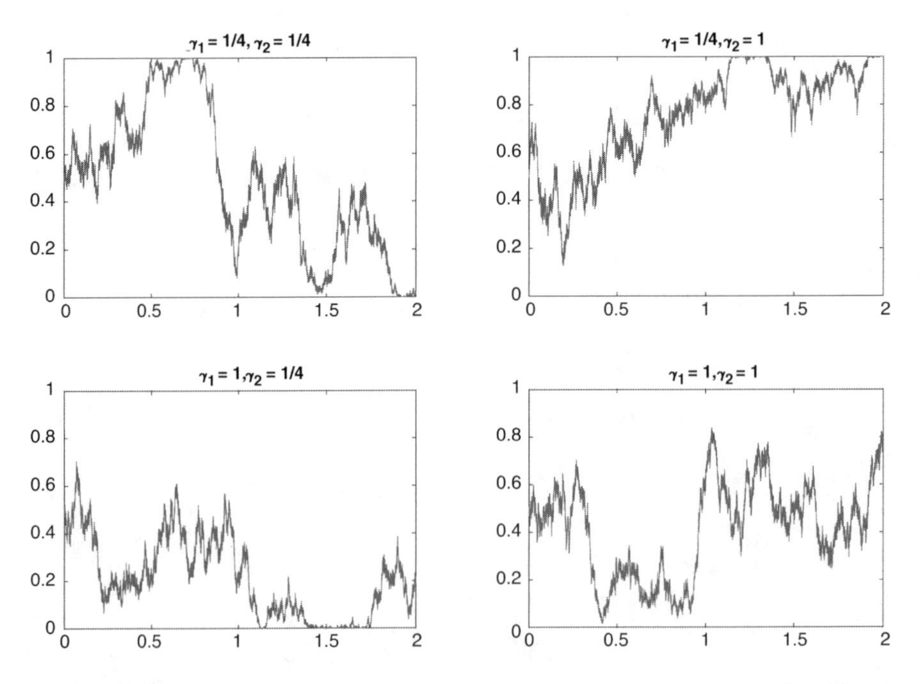

Figure 4.3 Trajectories of the Wright–Fisher model for different values of γ_1 and γ_2 starting at $X_0 = 1/2$.

A limiting case of the previous example is the situation in which there are no mutation effects, i.e. $\gamma_1 \downarrow 0$ and $\gamma_2 \downarrow 0$ (both are absorbing boundaries). In this case the drift and diffusion coefficients are

$$\mu(x) = 0, \quad \sigma^2(x) = 2\gamma x(1-x), \quad 0 \le x \le 1, \quad \gamma > 0.$$

The Kolmogorov backward equation is given by

$$\frac{\partial p}{\partial t} = \gamma x(1-x)\frac{\partial^2 p}{\partial x^2}, \quad t > 0,$$

with initial conditions $p(t;0,y) = p(t;1,y) = 0$. Therefore we seek a discrete set of eigenfunctions of $\mathcal{A}$ such that

$$\gamma x(1-x)\phi''(x) = -\lambda\phi(x), \quad 0 \le x \le 1,$$

conditioned on $\phi(0) = \phi(1) = 0$ (absorbing boundaries). A candidate for a solution is

$$\phi_n(x) = x(1-x)P_{n-1}^{(1,1)}(1-2x), \quad n \ge 1,$$

where $P_n^{(\alpha,\beta)}$ are the Jacobi polynomials on $[-1,1]$ normalized such that $P_n^{(\alpha,\beta)}(1) = 1$, and

$$\lambda_n = \gamma n(n+1).$$

Observe that the polynomials ϕ_n have degree $n+1$ and are orthogonal with respect to the measure $\pi(x) = x^{-1}(1-x)^{-1}$ on the interval $[0,1]$. Therefore the spectral representation of the transition probability density is given by

$$\begin{aligned} p(t;x,y) &= \frac{1}{y(1-y)}\sum_{n=1}^{\infty} e^{-\gamma n(n+1)t} n(n+1)(2n+1)\phi_n(x)\phi_n(y) \\ &= x(1-x)\sum_{n=1}^{\infty} e^{-\gamma n(n+1)t} n(n+1)(2n+1)P_{n-1}^{(1,1)}(1-2x)P_{n-1}^{(1,1)}(1-2y). \end{aligned}$$

Observe that $p(t;x,y)$ tends to 0 exponentially $e^{-2\gamma t}$ as $t \to \infty$, since we have two absorbing boundaries. This density converges to the uniform distribution on $(0,1)$ conditioned that absorption has not occurred yet, i.e.

$$\lim_{t\to\infty} \mathbb{P}(y < X_t < y + dy | X_0 = x, X_t \ne 0, 1) = dy.$$

◊

Example 4.22 *The Jacobi diffusion process.* The drift and diffusion coefficients are given by

$$\mu(x) = \frac{1}{2}\left[(\beta+1)(1-x) - (\alpha+1)(1+x)\right], \quad \sigma^2(x) = 1 - x^2, \quad -1 < x < 1,$$

where α, β are constants. For $x_0 = 0$, we have that

$$
\begin{aligned}
I(0,x) &= \int_0^x \frac{(\beta+1)(1-z)-(\alpha+1)(1+z)}{(1-z)(1+z)}\, dz \\
&= \int_0^x \left(\frac{\beta+1}{1+z} - \frac{\alpha+1}{1-z} \right) dz = \log\left[(1+x)^{\beta+1}(1-x)^{\alpha+1} \right].
\end{aligned}
$$

Therefore the measure $\pi(x)$ in (4.20) can be chosen as

$$
\pi(x) = (1-x)^{\alpha}(1+x)^{\beta}, \qquad -1 < x < 1.
$$

The integrability conditions on $-1 < x < 1$ force that $\alpha, \beta > -1$. The Kolmogorov backward equation for the transition probability density $p(t;x,y)$ is given by

$$
\frac{\partial p}{\partial t} = (1-x^2)\frac{\partial^2 p}{\partial x^2} + \frac{1}{2}\left[(\beta+1)(1-x)-(\alpha+1)(1+x)\right]\frac{\partial p}{\partial x}, \qquad t > 0.
$$

We seek then a discrete set of eigenfunctions and eigenvalues of the infinitesimal operator

$$
\mathcal{A} = (1-x^2)\frac{d^2}{dx^2} + \frac{1}{2}\left[(\beta+1)(1-x)-(\alpha+1)(1+x)\right]\frac{d}{dx}.
$$

In other words,

$$
\begin{aligned}
&(1-x^2)\phi''(x) + \frac{1}{2}\left[(\beta+1)(1-x)-(\alpha+1)(1+x)\right]\phi'(x) \\
&\quad = -\lambda\phi(x), \qquad -1 < x < 1,
\end{aligned}
$$

such that the boundary points are reflecting (see (4.26)), i.e.

$$
\frac{1}{2}(1-x)^{\alpha+1}(1+x)^{\beta+1}\phi'(x)\bigg|_{x=1} = 0, \qquad \frac{1}{2}(1-x)^{\alpha+1}(1+x)^{\beta+1}\phi'(x)\bigg|_{x=1} = 0.
$$

The solution of this differential equation is given in terms of the Jacobi polynomials $(P_n^{(\alpha,\beta)})_n$ defined by (1.77). The corresponding values of λ_n (see (1.78)), are given by

$$
\lambda_n = \frac{1}{2}n(n+\alpha+\beta+1), \qquad n \geq 0.
$$

Normalizing the polynomials in such a way that $P_n^{(\alpha,\beta)}(1) = 1$ (see (1.79) and (1.82)) we have the following spectral representation of the transition probability density $p(t;x,y)$:

$$
\begin{aligned}
p(t;x,y) &= \frac{(1-y)^{\alpha}(1+y)^{\beta}}{2^{\alpha+\beta+1}\Gamma(\alpha+1)^2}\sum_{n=0}^{\infty} e^{-n(n+\alpha+\beta+1)t/2} P_n^{(\alpha,\beta)}(x) P_n^{(\alpha,\beta)}(y) \\
&\quad \times \frac{(2n+\alpha+\beta+1)\Gamma(n+\alpha+1)\Gamma(n+\alpha+\beta+1)}{n!\,\Gamma(n+\beta+1)}.
\end{aligned} \tag{4.55}
$$

Again it is not known if this formula can be written explicitly in terms of the elementary functions. The behavior at the boundaries ± 1 is exactly the same as the behavior of the boundaries in the Wright–Fisher model in Example 4.21 considering the change of variables (4.53). $\diamondsuit$

Example 4.23 *Brownian motion on the sphere and Gegenbauer polynomials.* The spectral representation of this example was studied for the first time in [89] (see also [101]). Let Ω represent a sphere of radius 1 inside an Euclidean space of dimension $N+2$. We will consider Brownian motion on the sphere and relate it to the transition probability (4.55) for the special case of Gegenbauer polynomials (for $\lambda = N/2$, see Section 1.4.3), i.e.

$$\alpha = \beta = \frac{N-1}{2}, \quad N = 0, 1, 2, \ldots$$

Let $\xi_1, \ldots, \xi_{N+2}$, be the cartesian coordinates on $\mathbb{R}^{N+2}$. The unit radius sphere ($\xi_1^2 + \cdots + \xi_{N+2}^2 = 1$) can be parametrized using spherical coordinates $\theta_1, \ldots, \theta_N, \phi$, in the following form:

$$\begin{aligned}
\xi_1 &= \cos\theta_1, \\
\xi_2 &= \sin\theta_1 \cos\theta_2, \\
&\vdots \\
\xi_N &= \sin\theta_1 \cdots \sin\theta_{N-1} \cos\theta_N, \\
\xi_{N+1} &= \sin\theta_1 \cdots \sin\theta_N \cos\phi, \\
\xi_{N+2} &= \sin\theta_1 \cdots \sin\theta_N \sin\phi,
\end{aligned}$$

where $0 \leq \theta_i \leq \pi$ and $0 \leq \phi \leq 2\pi$. The *Laplace–Beltrami operator* Δ (the analog of the Laplace operator in N-dimensional spaces) applied on the sphere can be written in the following form:

$$\begin{aligned}
\Delta u = {} & (\sin\theta_1)^{-N} \frac{\partial}{\partial\theta_1}\left[(\sin\theta_1)^N \frac{\partial u}{\partial\theta_1}\right] + (\sin\theta_1)^{-2}(\sin\theta_2)^{1-N} \frac{\partial}{\partial\theta_2}\left[(\sin\theta_2)^{N-1} \frac{\partial u}{\partial\theta_2}\right] \\
& + \cdots + (\sin\theta_1 \cdots \sin\theta_{N-1})^{-2}(\sin\theta_N)^{-1} \frac{\partial}{\partial\theta_N}\left[\sin\theta_N \frac{\partial u}{\partial\theta_N}\right] \\
& + (\sin\theta_1 \cdots \sin\theta_N)^{-2} \frac{\partial^2 u}{\partial\phi^2}.
\end{aligned}$$

The diffusion equation $\partial u/\partial t = \Delta u$ has a unique fundamental solution on Ω, which we denote by $p(t; \boldsymbol{\xi}, \boldsymbol{\eta})$. It represents the density of a stationary Markov process on Ω (the density refers in this case to an element of the surface $d\omega$ on Ω). We will assume that $\boldsymbol{\xi}(t)$, called Brownian motion on Ω, has continuous trajectories. The spectral representation is then given by

$$p(t;\boldsymbol{\xi},\boldsymbol{\eta}) = \sum_{n=0}^{\infty} e^{-\lambda_n t} \sum_{l=0}^{h(n)} S_n^{(l)}(\boldsymbol{\xi}) S_n^{(l)}(\boldsymbol{\eta}),$$

where

$$\lambda_n = n(n+N), \quad h(n) = \frac{2n+N}{N}\binom{n+N-1}{n},$$

and $S_n^{(l)}(\boldsymbol{\xi}), l = 1,2,\ldots,h(n)$ is a complete orthonormal set of *spherical harmonics* of degree n (see [52, vol. 2]). These spherical harmonics are the eigenfunctions of the Laplace–Beltrami operator on the sphere Ω. By the addition theorem for spherical harmonics (see [52, p. 243, vol. 2]), the spectral representation above can be rewritten as

$$p(t;\boldsymbol{\xi},\boldsymbol{\eta}) = \sum_{n=0}^{\infty} e^{-\lambda_n t} \frac{h(n)}{\omega(N)} C_n^{(N/2)}\left((\boldsymbol{\xi},\boldsymbol{\eta})\right), \tag{4.56}$$

where $(\boldsymbol{\xi},\boldsymbol{\eta}) = \xi_1\eta_1 + \cdots + \xi_{N+2}\eta_{N+2}$ is the cosine of the angle between two unitary vectors $\boldsymbol{\xi}$ and $\boldsymbol{\eta}$,

$$\omega(N) = \frac{2\pi^{(N+3)/2}}{\Gamma(\frac{N+3}{2})}$$

is the area of the unit radius sphere on $\mathbb{R}^{N+2}$ and $C_n^{(\lambda)}$ are the Gegenbauer polynomials.

Let $\boldsymbol{\xi}(t)$ be the Brownian motion over Ω with an initial distribution symmetric with respect to the ξ_1 axis and with no masses at the points $\xi_1 = \pm 1$. The distribution of $\boldsymbol{\xi}(t), t > 0$, will be symmetric with respect to the ξ_1 axis. The random variable $X_t = \xi_1(t)$ (projection of $\boldsymbol{\xi}(t)$ on the ξ_1 axis) is a Markov process with continuous trajectories and state space $\mathcal{S} = [-1,1]$. Let us compute the transition probability density of X_t. For that, consider the spherical coordinates of $\boldsymbol{\xi}$ and $\boldsymbol{\eta}$

$$\boldsymbol{\xi} = (\theta_1,\ldots,\theta_N,\phi), \quad \boldsymbol{\eta} = (\theta_1',\ldots,\theta_N',\phi').$$

The transition probabilities are given then by

$$p(t;x,[-1,y]) = \mathbb{P}(X_t \le y | X_0 = x) = \int_{-1\le\eta_1\le y} p(t;\boldsymbol{\xi},\boldsymbol{\eta})d\omega_{\eta},$$

where $\boldsymbol{\xi}$ is a fixed unit vector with $\xi_1 = x$. By symmetry we can take, without loss of generality,

$$\theta_1 = \cos^{-1} x, \quad \theta_2 = \theta_3 = \cdots = \theta_N = 0,$$

so

$$(\boldsymbol{\xi},\boldsymbol{\eta}) = \cos\theta_1 \cos\theta_1' + \sin\theta_1 \sin\theta_1' \cos\theta_2'.$$

Using the area element

$$d\omega_\eta = (\sin\theta_1')^N (\sin\theta_2')^{N-1} \cdots (\sin\theta_N')\, d\theta_1'\, d\theta_2' \cdots d\theta_N'\, d\phi,$$

and the representation (4.56) we arrive at

$$p(t;x,[-1,y]) = \int_{\cos\theta_1' \le y} (\sin\theta_1')^N d\theta_1' \sum_{n=0}^{\infty} e^{-\lambda_n t} \frac{\omega(N-2)h(n)}{\omega(N)} f_n(\theta_1,\theta_1'),$$

where (there is a different result for $N=0$)

$$f_n(\theta_1,\theta_1') = \int_0^\pi C_n^{(N/2)}(\cos\theta_1 \cos\theta_1' + \sin\theta_1 \sin\theta_1' \cos\theta_2')(\sin\theta_2')^{N-1} d\theta_2.$$

Using a formula due to Gegenbauer (which can be found in [144, p. 369]), we have that

$$f_n(\theta_1,\theta_1') = C_n^{(N/2)}(\cos\theta_1) C_n^{(N/2)}(\cos\theta_1') \int_0^\pi (\sin\theta_2)^{N-1} d\theta_2.$$

Therefore, rearranging the variables, we have that

$$p(t;x,[-1,y]) = \frac{1}{c_{N-1}} \int_{-1}^{y} \tilde{p}(t;x,z)(1-z^2)^{(N-1)/2} dz,$$

where $\tilde{p}(t;x,z)$ is the spectral representation of the transition probability density of the Jacobi process (4.55) for $\alpha = \beta = (N-1)/2$ and $c_N = \int_{-1}^{1}(1-y^2)^{N/2} dy$. ◊

Example 4.24 *A population genetics diffusion process with killing.* A genetic model related to the Laguerre polynomials appeared in [99]. Certain gene population is evolving in each generation but at some point the process stops. In [99] the authors start with the discrete model and then they take conveniently the limit so that the process converges to a diffusion process with killing (see (4.43)). In particular, when the killing of the process is certain, we have the drift, diffusion and killing coefficients given by

$$\sigma^2(x) = x, \quad \mu(x) = 0, \quad k(x) = \frac{\theta^2 x}{2}, \quad x > 0, \quad \theta \ge 0.$$

The state space is $\mathcal{S} = [0,\infty)$. Since $\mu(x) = 0$, we have that $I(x_0,x) = 0$. Therefore, the measure $\pi(x)$ in (4.20) is given by

$$\pi(x) = \frac{1}{x}, \quad x > 0.$$

The boundary 0 is an exit boundary, since in this case

$$dS(y) = dy, \quad dM(z) = \frac{dz}{z},$$

and therefore, if we choose $x = 1$, we have

$$\Sigma(0) = \int_0^1 \left[\int_y^1 dM(z) \right] dS(y) = \int_0^1 \left[\int_y^1 \frac{dz}{z} \right] dy = -\int_0^1 \log y \, dy < \infty.$$

For $N(0)$, if we choose $x = 1$, we have that

$$N(0) = \int_0^1 \left[\int_z^1 dS(y) \right] dM(z) = \int_0^1 \left[\int_z^1 dy \right] \frac{dz}{z} = \int_0^1 \frac{1-z}{z} dz = \infty.$$

It is easy to see that the boundary ∞ is natural.

Therefore, in this diffusion process, there are two different events that may stop the trajectory, the boundary point 0 (which is absorbing or exit) and the time where the process stops (depending on the killing coefficient). Let us call $\xi = \tau_0 \wedge \tau_\kappa$, where τ_0 is the first passage time to the absorbing state 0 and τ_κ is the killing time. In order to get the spectral representation of the transition probability density of $\{X_t, 0 \le t < \xi\}$ we have to seek the eigenfunctions and eigenvalues of the infinitesimal operator

$$\mathcal{A} = \frac{x}{2} \frac{d^2}{dx^2} - \frac{\theta^2 x}{2}. \tag{4.57}$$

In other words,

$$\frac{x}{2} \phi''(x) - \frac{\theta^2 x}{2} \phi(x) = -\lambda \phi(x), \quad x > 0.$$

This differential equation can be related to the one for the Laguerre polynomials for $\alpha = 1$. Indeed, defining the functions

$$\phi_n(x) = x e^{-\theta x} L_n^{(1)}(2\theta x), \quad x \ge 0, \quad n \ge 0,$$

we have, using the differential equation for the Laguerre polynomials for $\alpha = 1$ (see (1.67)), that

$$\begin{aligned}
&\frac{x}{2} \phi_n''(x) - \frac{\theta^2 x}{2} \phi_n(x) \\
&\quad = x\theta e^{-\theta x} \left[2\theta x (L_n^{(1)})''(2\theta x) + (2 - 2\theta x)(L_n^{(1)})'(2\theta x) - L_n^{(1)}(2\theta x) \right] \\
&\quad = -\theta(n+1) x e^{-\theta x} L_n^{(1)}(2\theta x) = -\theta(n+1)\phi_n(x).
\end{aligned}$$

The system of orthogonal functions $(\phi_n)_n$ is complete in the space $L^2_\pi(\mathcal{S})$, since the Laguerre polynomials for $\alpha = 1$ are complete in $L^2_{xe^{-x}}(\mathcal{S})$ (both spaces are isomorphic). The norms can be computed from the norms of the Laguerre polynomials. Indeed,

$$\|\phi_n\|_\pi^2 = \int_0^\infty \phi_n^2(x) \frac{1}{x} dx = \int_0^\infty x e^{-2\theta x} L_n^{(1)}(2\theta x)^2 dx = \frac{n+1}{4\theta^2}.$$

Therefore the spectral representation of the transition probability density $p^0(t;x,y)$ given that $\xi > t$ is given by

$$\begin{aligned}
p^0(t;x,y) &= \pi(y)\sum_{n=0}^{\infty} e^{-\lambda_n t}\phi_n(x)\phi_n(y)\pi_n \\
&= \frac{1}{y}\sum_{n=0}^{\infty} e^{-\theta(n+1)t} xe^{-\theta x} L_n^{(1)}(2\theta x) ye^{-\theta y} L_n^{(1)}(2\theta y)\frac{4\theta^2}{n+1} \\
&= 4\theta^2 xe^{-\theta(x+y)}\sum_{n=0}^{\infty} e^{-\theta(n+1)t}\frac{L_n^{(1)}(2\theta x)L_n^{(1)}(2\theta y)}{n+1}.
\end{aligned}$$

We could have used what we did in Section 4.4 to analyze this example. In this case the transformation functions $s(x)$ will be

$$s(x) = \frac{1}{x}e^{\theta x}.$$

It is easy to see that the coefficients $\widetilde{\mu}(x)$ and $k(x)$ in (4.45) and (4.46) respectively, are given by

$$\widetilde{\mu}(x) = 0, \quad k(x) = \frac{x\theta^2}{2} - \theta.$$

The killing coefficient is bounded from below by $c = -\theta$. Therefore this constant goes to the eigenvalue part and the coefficients are the same as this example.

One of the main questions for this kind of diffusion with killing is to determine the probabilities that the process ends and the killing mean time, either because it is absorbed at 0 or because the process is killed by the effect of $k(x)$. This probability is given by

$$\mathbb{P}(\xi > t | X_0 = x) = \int_0^{\infty} p^0(t;x,y)\,dy.$$

Since we have an explicit expression for $p^0(t;x,y)$, we can compute these probabilities. For that, we use the following formulas, which can be found in [52, vol 2, pp. 191 and 215]:

$$\int_0^{\infty} e^{-py}L_n^{(1)}(y)dy = 1 - \left(1 - \frac{1}{p}\right)^{n+1}, \quad p > 0, \qquad \sum_{n=0}^{\infty}\frac{z^{n+1}L_n^{(1)}(x)}{n+1} = \frac{1 - e^{xz/(z-1)}}{x}.$$

If we change y by $2\theta y$ and choose $p = 1/2$ in the first formula, and change x by $2\theta x$ and choose $z = e^{-\theta t}$ in the second formula, we obtain

$$\begin{aligned}
\mathbb{P}(\xi > t | X_0 = x) &= \int_0^{\infty} p^0(t;x,y)\,dy \\
&= 4\theta^2 xe^{-\theta x}\sum_{n=0}^{\infty} e^{-\theta(n+1)t}\frac{L_n^{(1)}(2\theta x)}{n+1}\int_0^{\infty} e^{-\theta y}L_n^{(1)}(2\theta y)\,dy
\end{aligned}$$

$$= 2\theta x e^{-\theta x} \sum_{n=0}^{\infty} e^{-\theta(n+1)t} \frac{L_n^{(1)}(2\theta x)}{n+1} \left(1 - (-1)^{n+1}\right)$$

$$= 2\theta x e^{-\theta x} \left[\frac{1 - \exp\left(-\frac{2\theta x e^{-\theta t}}{1 - e^{-\theta t}}\right)}{2\theta x} - \frac{1 - \exp\left(\frac{2\theta x e^{-\theta t}}{1 + e^{-\theta t}}\right)}{2\theta x} \right]$$

$$= \exp\left(-x\theta \frac{1 - e^{-\theta t}}{1 + e^{-\theta t}}\right) - \exp\left(-x\theta \frac{1 + e^{-\theta t}}{1 - e^{-\theta t}}\right)$$

$$= e^{-x\theta b(t)} - e^{-x\theta / b(t)},$$

where $b(t) = (1 - e^{-\theta t})/(1 + e^{-\theta t})$. Observe that as $t \to \infty$ this probability tends to zero, so the process must be killed in finite time. A straightforward computation also gives

$$\int_0^{\infty} p^0(t; x, y)\, dt = \begin{cases} \dfrac{e^{-\theta x}}{\theta y}(e^{\theta y} - e^{-\theta y}), & \text{if } \ y < x, \\ \dfrac{e^{-\theta y}}{\theta y}(e^{\theta x} - e^{-\theta x}), & \text{if } \ y \geq x. \end{cases}$$

We can use the previous two formulas to compute the killing mean time or mean time to termination. Indeed,

$$\begin{aligned} \mathbb{E}(\xi | X_0 = x) &= \int_0^{\infty} \mathbb{P}(\xi > t | X_0 = x)\, dt \\ &= \int_0^{\infty} \left[\exp\left(-x\theta \frac{1 - e^{-\theta t}}{1 + e^{-\theta t}}\right) - \exp\left(-x\theta \frac{1 + e^{-\theta t}}{1 - e^{-\theta t}}\right) \right] dt, \qquad (4.58) \end{aligned}$$

or another way to compute it is

$$\begin{aligned} \mathbb{E}(\xi | X_0 = x) &= \int_0^{\infty} \int_0^{\infty} p^0(t; x, y)\, dt\, dy \\ &= \int_0^{x} \frac{e^{-\theta x}}{\theta y}(e^{\theta y} - e^{-\theta y})\, dy + \int_x^{\infty} \frac{e^{-\theta y}}{\theta y}(e^{\theta x} - e^{-\theta x})\, dy. \qquad (4.59) \end{aligned}$$

The above integral can be written in terms of *exponential integrals* (see Exercise 4.11). The following table gives some numerical analysis about the explicit value of $\mathbb{E}(\xi | X_0 = x)$ depending on x and θ:

	$x = 1/2$	$x = 1$	$x = 3/2$	$x = 2$
$\theta = 1/2$	1.8367	2.3968	2.5816	2.5870
$\theta = 1$	1.1984	1.2935	1.1848	1.0310
$\theta = 3/2$	0.8605	0.7899	0.6372	0.5044
$\theta = 2$	0.6476	0.5159	0.3783	0.2829

We can also compute the probabilities $v(x) = \mathbb{P}(\tau_0 < \tau_\kappa | X_0 = x)$ in terms of the infinitesimal operator $\mathcal{A}$ in (4.57) in the same way as for regular diffusion processes. In this case the solution cannot be written in terms of the scale function $S(y)$, since the differential operator is different. However, $v(x)$ is a solution of the following second-order differential equation with boundary values:

$$v''(x) - \theta^2 v(x) = 0, \quad v(0) = 1, \quad v(x) \quad \text{positive decreasing.}$$

The solution is given by

$$v(x) = e^{-\theta x}, \quad x \geq 0.$$

Therefore the probability that the process is killed before it is absorbed is given by

$$1 - v(x) = 1 - e^{-\theta x}, \quad x \geq 0.$$

In [96, 97, 99] other diffusion models related to population genetics with killing are treated, where the killing coefficient is now a quadratic polynomial. The eigenfunctions are then related to *Airy functions.* ◊

Remark 4.25 It can be proved (see [147] or [7]) that the only diffusion processes having a representation formula as in (4.25) such that the corresponding eigenfunctions $(\phi_n)_n$ constitute a complete set of orthonormal polynomials in the space $L^2_\pi(\mathcal{S})$ are affine transformations of the Orstein–Uhlenbeck process (Hermite polynomials), the population growth model (Laguerre polynomials) and the Jacobi process (Jacobi polynomials). In [8] a step forward has been made to characterize orthonormal polynomial eigenfunctions of two-dimensional diffusion processes. ◊

4.5.3 Radial Diffusion Processes

Example 4.26 *Radial Brownian motion in N dimensions for a particle starting inside the unit sphere and absorbing at the surface of the unit sphere.* Let $\{B_i(t), t \geq 0\}, i = 1, \ldots, N$ be standard and independent Brownian motions. We consider the radial process

$$Z_t = B_1^2(t) + \cdots + B_N^2(t).$$

Let us compute heuristically the drift and diffusion coefficients of Z_t conditioned on $B_i(t) = y_i, i = 1, \ldots, N$. Since $B_i(t + \Delta t) = y_i + \Delta B_i$ then $Z_{t+\Delta t} = x + \Delta Z$, where $x = y_1^2 + \cdots + y_N^2$. Then

$$\begin{aligned}\Delta Z &= B_1(t + \Delta t)^2 - y_1^2 + \cdots + B_N(t + \Delta t)^2 - y_N^2 \\ &= 2(y_1 \Delta B_1 + \cdots + y_N \Delta B_N) + (\Delta B_1)^2 + \cdots + (\Delta B_N)^2.\end{aligned}$$

Then, using that $\Delta B_i \sim \mathcal{N}(0, \Delta t)$ (Gaussian distribution), we have

$$\mathbb{E}(\Delta Z \mid Z_t = x) = N\Delta t,$$
$$\mathbb{E}((\Delta Z)^2 \mid Z_t = x) = 4x\Delta t + o(\Delta t).$$

Therefore, Z_t is a diffusion process with drift and diffusion coefficients given by

$$\mu_Z(x) = N, \quad \sigma_Z^2(x) = 4x, \quad x > 0.$$

The *Bessel process* is defined as the Euclidean distance of a N-dimensional Brownian motion, i.e.

$$Y_t = \sqrt{Z_t} = \sqrt{B_1^2(t) + \cdots + B_N^2(t)}.$$

We can use Proposition 4.8 to compute the drift and diffusion coefficients for the Bessel process using the function $\phi(x) = \sqrt{x}$. Then the new coefficients are given by

$$\mu(x) = \frac{N-1}{2x}, \quad \sigma^2(x) = 1, \quad x > 0.$$

For $N = 1$ we have the reflected Brownian motion $|B_t|$. Taking $x_0 = 1$, we have that

$$I(1,x) = \int_1^x \frac{N-1}{z}\, dz = (N-1)\log x.$$

Therefore, the measure $\pi(x)$ in (4.20) can be chosen as

$$\pi(x) = x^{N-1}, \quad x > 0.$$

Consider now the Bessel process on the state space $(0, 1]$ where there is an absorbing state at $x = 1$ (in the N-dimensional space, the state space would be the interior of the unit ball including the surface). The Kolmogorov backward equation is then

$$\frac{\partial p}{\partial t} = \frac{1}{2}\frac{\partial^2 p}{\partial x^2} + \frac{N-1}{2x}\frac{\partial p}{\partial x} = \frac{1}{2}\frac{1}{x^{N-1}}\frac{\partial}{\partial x}\left(x^{N-1}\frac{\partial p}{\partial x}\right), \quad t > 0,$$

with initial conditions

$$p(t; 1, y) = 0, \quad x^{N-1}\left.\frac{\partial p}{\partial x}\right|_{x=0} = 0,$$

i.e. the state 0 is reflecting while the state 1 is absorbing. Therefore, we seek eigenvalues and eigenfunctions of the infinitesimal operator $\mathcal{A}$ such that

$$\frac{1}{2}\phi''(x) + \frac{N-1}{2x}\phi'(x) = -\lambda\phi(x), \quad 0 \le x \le 1,$$

conditioned on

$$\phi(1) = 0, \quad x^{N-1}\,\phi'(x)\big|_{x=0} = 0.$$

The eigenfunctions of this differential equation are identified with modifications of the Bessel functions $J_\alpha(x)$ defined by (1.72). Indeed,

$$\phi_n(x) = x^{-(N-2)/2} J_{(N-2)/2}\left(x\sqrt{\lambda_n}\right), \quad n \geq 0,$$

where $\sqrt{\lambda_n}$ is the sequence of positive zeros of $J_{(N-2)/2}(x)$. In fact the eigenvalues are given by $-2\lambda_n$. The identification with the differential equation follows using (1.73). Therefore the spectral representation of the transition probability density $p(t;x,y)$ can be written as

$$\begin{aligned} p(t;x,y) &= y^{N-1} \sum_{n=0}^{\infty} e^{-2\lambda_n t} \phi_n(x)\phi_n(y)\pi_n \\ &= y\left(\frac{y}{x}\right)^{\frac{N-2}{2}} \sum_{n=0}^{\infty} e^{-2\lambda_n t} J_{\frac{N-2}{2}}\left(x\sqrt{\lambda_n}\right) J_{\frac{N-2}{2}}\left(y\sqrt{\lambda_n}\right)\pi_n, \end{aligned}$$

where $\pi_n^{-1} = \int_0^1 \phi_n^2(x)x^{N-1}dx = \int_0^1 xJ^2_{(N-2)/2}\left(x\sqrt{\lambda_n}\right)dx$.

A similar analysis to the one we did in Example 4.14, taking the boundary to infinity, gives the spectral representation of the density for radial Brownian motion on $\mathbb{R}^N$. In this case the spectrum is absolutely continuous and the spectral representation of the transition probability density $p(t;x,y)$ is given by

$$p(t;x,y) = \left[\frac{\Gamma((N+1)/2)}{\Gamma(N/2)}\right]^2 y^{N-1} \int_0^{\infty} e^{-\lambda^2 t/2} J_{\frac{N-2}{2}}(\lambda x) J_{\frac{N-2}{2}}(\lambda y)\lambda^{N-1} d\lambda.$$

As for the boundary points, a simple computation shows

$$dS(y) = y^{1-N}dy, \quad dM(z) = z^{N-1}dz.$$

Let us see the behavior at 0 (similar for 1). In this case we have, choosing $x = 1/2$,

$$\begin{aligned} \Sigma(0) &= \int_0^{1/2}\left[\int_y^{1/2} dM(z)\right] dS(y) = \int_0^{1/2}\left[\int_y^{1/2} z^{N-1}dz\right] y^{1-N}dy \\ &= \frac{1}{N}\int_0^{1/2}\left(\frac{1}{2^N} - y^N\right) y^{1-N}dy = \frac{1}{N}\int_0^{1/2} \frac{y^{1-N}}{2^N} - y\,dy. \end{aligned}$$

This integral is convergent if and only if $N < 2$ and divergent for $N \geq 2$. A similar argument works for $N(0)$ if we choose $x = 1/2$. Indeed,

$$\begin{aligned} N(0) &= \int_0^{1/2}\left[\int_z^{1/2} dS(y)\right] dM(z) = \int_0^{1/2}\left[\int_z^{1/2} y^{1-N}dy\right] z^{N-1}dz \\ &= \begin{cases} -\displaystyle\int_0^{1/2} \log(2z)z^{N-1}dz, & \text{if } N = 2, \\ \dfrac{1}{2-N}\displaystyle\int_0^{1/2}\left(\frac{z^{N-1}}{2^{2-N}} - z\right) dz, & \text{if } N \neq 2. \end{cases} \end{aligned}$$

Therefore, this integral is finite (in both cases) if and only if $N > 0$ and divergent for $N = 0$. We have then three cases:

- 0 is an exit boundary if $N = 0$, in which case we have $\Sigma(0) < \infty$ and $N(0) = \infty$.
- 0 is a regular boundary if $0 < N < 2$, in which case we have $\Sigma(0) < \infty$ and $N(0) < \infty$.
- 0 is an entrance boundary if $N \geq 2$, in which case we have $\Sigma(0) = \infty$ and $N(0) < \infty$.

For the boundary 1 we have, from (4.41) and (4.42), that $\Sigma(1) < \infty$ and $N(1) < \infty$. Therefore it is a regular boundary. Same if the right boundary is ∞, where $\Sigma(\infty) = \infty$ and $N(\infty) = \infty$, therefore we have a natural boundary. From the previous analysis we can conclude a well-known result for Brownian motions on $\mathbb{R}^N$. For $N = 1$ the boundary 0 is regular, so the process can visit 0 as many times as wanted, i.e. it is recurrent. However, for $N \geq 2$ the boundary 0 behaves like an entrance boundary, so the process cannot be reached from the interior, i.e. the process is transient. Figure 4.4 shows some trajectories of the Bessel process for

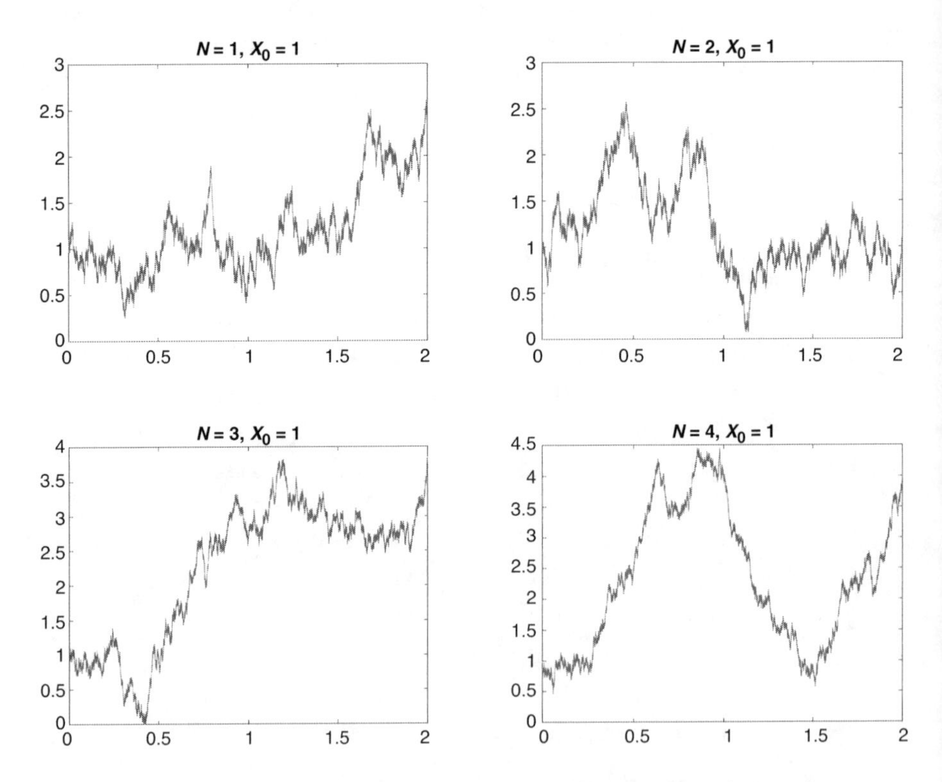

Figure 4.4 Trajectories of the Bessel process for different values of N starting at $X_0 = 1$.

different values of N starting at $X_0 = 1$. For a recent application of the spectral representation of the Bessel process to compute the density functions of the first exit times, see [132]. ◊

Example 4.27 *Radial Orstein–Uhlenbeck process in N dimensions.* We take, as before, the Bessel process but in this case we include an attractive force proportional to the distance to the origin. The drift and diffusion coefficients are given by

$$\mu(x) = \frac{N-1}{2x} - x, \quad \sigma^2(x) = 1, \quad x > 0.$$

For $x_0 = 1$, we have that

$$I(1,x) = \int_1^x \left(\frac{N-1}{z} - z \right) dz = (N-1)\log x - \frac{x^2}{2} + \frac{1}{2}.$$

Therefore the measure $\pi(x)$ in (4.20) can be chosen as

$$\pi(x) = x^{N-1} e^{-x^2}, \quad x > 0.$$

The Kolmogorov backward equation is given by

$$\frac{\partial p}{\partial t} = \frac{1}{2}\frac{\partial^2 p}{\partial x^2} + \left(\frac{N-1}{2x} - x \right) \frac{\partial p}{\partial x}, \quad t > 0,$$

with the boundary conditions

$$x^{N-1} \frac{\partial p}{\partial x}\bigg|_{x=0} = 0,$$

i.e. the state 0 is reflecting. Therefore we seek eigenvalues and eigenfunctions of the infinitesimal operator $\mathcal{A}$ such that

$$\frac{1}{2}\phi''(x) + \left(\frac{N-1}{2x} - x \right) \phi'(x) = -\lambda\phi(x), \quad x > 0,$$

conditioned on

$$x^{N-1}\, \phi'(x)\big|_{x=0} = 0.$$

The eigenfunctions can be written in terms of the Laguerre polynomials evaluated at $y = x^2$ and with parameter

$$\alpha = \frac{N-2}{2}.$$

Therefore the spectral representation of the transition probability density $p(t;x,y)$ can be written as

$$p(t;x,y) = 2y^{N-1}e^{-y^2}\sum_{n=0}^{\infty} e^{-2nt}\frac{n!}{\Gamma(n+\alpha+1)}L_n^{(\alpha)}(x^2)L_n^{(\alpha)}(y^2)$$
$$= \frac{2y^{N-1}e^{-y^2}}{(xye^{-t})^{\alpha}(1-e^{-2t})}\exp\left(-\frac{e^{-2t}(x^2+y^2)}{1-e^{-2t}}\right)I_{\alpha}\left(\frac{2xye^{-t}}{1-e^{-2t}}\right),$$

where we are using the formula for the Poisson kernel (1.71) for the Laguerre polynomials and $I_{\alpha}(z) = i^{-\alpha}J_{\alpha}(iz)$, where J_{α} is the standard Bessel function defined by (1.72). The case of $N = 1$ gives $\alpha = -1/2$. Using the relationship between the Laguerre and Hermite polynomials in (1.75), we go back to the regular Orstein–Uhlenbeck process of Example 4.18.

The behavior at the boundaries 0 and ∞ is exactly the same as in the previous case since the multiplication by e^{x^2} in $dS(x)$ and e^{-x^2} in $dM(x)$ does not affect the value of the limits at the boundary points. ◊

4.5.4 Examples with Absolutely Continuous and Discrete Spectrum

In Section 1.4 of Chapter 1 we studied the classification of the classical orthogonal polynomials. Apart from the classical families of Hermite, Laguerre and Jacobi (which are the only orthogonal families on the real line such that they are eigenfunctions of a second-order differential operator of the Sturm–Liouville type) we studied two other families, the Bessel and Romanovski polynomials. These polynomials are only defined for a finite number of degrees, so there is a finite number of finite moments with respect to the measure. However, these two examples will give two instances of diffusion processes where the spectrum of the infinitesimal operator will be given by a discrete (finite) and an absolutely continuous part (see [148]).

Example 4.28 Consider the Case E in [148] of the infinitesimal operator (1.53) for the values $b = 0$ and $a = \alpha + 1/2$. Therefore $\alpha > 0$. In this case the (normalized) measure is given by

$$\pi(x) = \frac{\Gamma(\alpha+1/2)}{\sqrt{\pi}\,\Gamma(\alpha)}(1+x^2)^{-(\alpha+1/2)}, \quad x \in \mathbb{R}, \quad \alpha > 0. \tag{4.60}$$

The drift and diffusion coefficients are given by

$$\mu(x) = -(2\alpha-1)x, \quad \sigma^2(x) = 2(1+x^2), \quad x \in \mathbb{R}.$$

Therefore, the Kolmogorov backward equation is

$$\frac{\partial p}{\partial t} = (1+x^2)\frac{\partial^2 p}{\partial x^2} - (2\alpha-1)x\frac{\partial p}{\partial x}, \quad t > 0.$$

The Sturm–Liouville equation has $N+1$ discrete eigenvalues, where $\alpha-1 \leq N < \alpha$, and an absolutely continuous range of eigenvalues. More precisely,

$$\lambda_n = -n(n-2\alpha), \quad n = 0, 1, \ldots, N, \quad \lambda = (\alpha^2 + \mu^2), \quad \mu \geq 0.$$

Therefore, the transition probability density $p(t; x, y)$ can be written as

$$p(t; x, y) = (1+y^2)^{-(\alpha+1/2)} \left[\frac{1}{\pi} \sum_{n=0}^{N} \frac{\alpha - n}{n!\,\Gamma(2\alpha+1-n)} e^{n(n-2\alpha)t} \theta_n(x) \theta_n(y) \right.$$
$$\left. + \frac{1}{2\pi} \int_0^\infty e^{-(\alpha^2+\mu^2)t} \left(\psi(\mu, x)\psi(-\mu, y) + \psi(-\mu, x)\psi(\mu, y) \right) d\mu \right],$$

where $\theta_n(x)$ are the Romanovski polynomials of degree n defined by Rodrigues' formula

$$\theta_n(x) = 2^{\alpha-n}\Gamma(\alpha-n+1/2)(-1)^n(1+x^2)^{\alpha+1/2} \frac{d^n}{dx^n} \left[(1+x^2)^{n-\alpha-1/2} \right],$$

and $\psi(\mu, x)$ is the function defined by

$$\psi(\mu, x) = (x + \sqrt{1+x^2})^{i\mu} \sqrt{1+x^2}\, {}_2F_1 \left(-\alpha, \alpha+1; 1+i\mu; \frac{1}{2} + \frac{1}{2} \frac{x}{\sqrt{1+x^2}} \right),$$

where ${}_2F_1$ is the Gauss hypergeometric function defined by (1.4).

For $\alpha = K$ a positive integer, the integral part of $p(t; x, y)$ can be computed explicitly, in which case we have

$$p(t; x, y) = (1+y^2)^{-(\alpha+1/2)} \left[(1+x^2)^{K/2}(1+y^2)^{K/2} \frac{1}{2\sqrt{\pi t}} e^{-K^2 t} e^{-u^2} \right.$$
$$\left. + \frac{1}{\pi} \sum_{n=0}^{K-1} \frac{K-n}{n!\,\Gamma(2K+1-n)} e^{n(n-2K)t} \theta_n(x)\theta_n(y) f_n(t; x, y) \right],$$

where

$$u = u(t; x, y) = \frac{\operatorname{arc\,sinh} y - \operatorname{arc\,sinh} x}{2\sqrt{t}}, \quad f_n(t; x, y) = \frac{1}{\sqrt{\pi}} \int_{u-(K-n)\sqrt{t}}^{u+(K-n)\sqrt{t}} e^{-z^2} dz.$$

Observe that for $\alpha = 1/2$ the distribution (4.60) is given by $\pi(x) = \frac{1}{\pi}(1+x^2)^{-1}$, i.e. the *Cauchy distribution*. As for the behavior of the boundary points $\pm\infty$, we observe that

$$dS(y) = (1+y^2)^{\alpha-1/2} dy, \quad dM(z) = (1+z^2)^{-(\alpha+1/2)} dz.$$

For ∞ (and similarly for $-\infty$), using (4.41), we get that

$$\Sigma(\infty) = \int_x^\infty \left[\int_x^y dM(z)\right] dS(y) = \int_x^\infty \left[\int_x^y (1+z^2)^{-(\alpha+1/2)} dz\right] (1+y^2)^{\alpha-1/2} dy$$
$$\sim \int_x^\infty \left[c_1 y \times {}_2F_1\left(1/2, \alpha+1/2; 3/2; -y^2\right) + c_2\right] (1+y^2)^{\alpha-1/2} dy.$$

The integral part, without the hypergeometric function, is convergent if and only if $-2\alpha+1 > 1$, i.e. $\alpha < 0$. Since $\alpha > 0$, we have that $\Sigma(\infty) = \infty$. Alternatively, using (4.42), we have that

$$N(\infty) = \int_x^\infty \left[\int_x^z dS(y)\right] dM(z) = \int_x^\infty \left[\int_x^y (1+y^2)^{\alpha-1/2} dy\right] (1+z^2)^{-(\alpha+1/2)} dz$$
$$\sim \int_x^\infty \left[c_1 z \times {}_2F_1\left(1/2, -\alpha+1/2; 3/2; -z^2\right) + c_2\right] (1+z^2)^{-\alpha-1/2} dz.$$

In this case the second part of the integral is always convergent. However, a numerical analysis of the first part gives that this integral is always divergent for $\alpha > 0$. Therefore $N(\infty) = \infty$ and both boundary points are natural. $\Diamond$

Example 4.29 Consider now the Case F in [148] of the infinitesimal operator (1.54) for the values $b = 1$ and $a = 2\alpha + 1$. Therefore $\alpha > 0$. In this case the (normalized) measure is given by

$$\pi(x) = \frac{1}{\Gamma(2\alpha)} x^{-(2\alpha+1)} e^{-1/x}, \quad x > 0, \quad \alpha > 0. \tag{4.61}$$

The drift and diffusion coefficients are given by

$$\mu(x) = 1 - (2\alpha - 1)x, \quad \sigma^2(x) = 2x^2, \quad x > 0.$$

Therefore the Kolmogorov backward equation is

$$\frac{\partial p}{\partial t} = x^2 \frac{\partial^2 p}{\partial x^2} + [1 - (2\alpha - 1)x] \frac{\partial p}{\partial x}, \quad t > 0.$$

Again, the Sturm–Liouville equation has $N+1$ discrete eigenvalues, where $\alpha - 1 \le N < \alpha$, and an absolutely continuous range of eigenvalues. More precisely,

$$\lambda_n = -n(n - 2\alpha), \quad n = 0, 1, \ldots, N, \quad \lambda = (\alpha^2 + \mu^2), \quad \mu \ge 0.$$

Therefore the transition probability density $p(t; x, y)$ can be written as

$$p(t; x, y) = y^{-(2\alpha+1)} e^{-1/y} \left[\frac{1}{\pi} \sum_{n=0}^{N} \frac{2(\alpha - n)}{n!\,\Gamma(2\alpha + 1 - n)} e^{n(n-2\alpha)t} B_n(x) B_n(y) \right.$$
$$\left. + \frac{1}{2\pi} \int_0^\infty e^{-(\alpha^2+\mu^2)t} A(\mu) \phi(\mu, x) \phi(\mu, y) d\mu \right],$$

where $\beta_n(x)$ are the Bessel polynomials of degree n defined by Rodrigues' formula

$$\beta_n(x) = (-1)^n x^{2\alpha+1} e^{-1/x} \frac{d^n}{dx^n} \left[x^{2n-2\alpha-1} e^{-1/x} \right],$$

$A(\mu)$ is a normalization factor given by

$$A(\mu) = \frac{\Gamma(-\alpha + i\mu)\Gamma(-\alpha - i\mu)}{\Gamma(i\mu)\Gamma(-i\mu)},$$

and $\phi(\mu, x)$ is the function given by

$$\phi(\mu, x) = {}_2F_0\left(-\alpha - i\mu, -\alpha + i\mu; -; -x\right) = \sum_{n=0}^{\infty} \frac{(-1)^n}{n!} \prod_{k=0}^{n-1} [(\alpha - k)^2 + \mu^2] x^n,$$

where ${}_2F_0$ is the hypergeometric function (1.4). This function can also be written in terms of the *Tricomi function* (or confluent hypergeometric function) defined by (3.113). Indeed,

$$\phi(\mu, x) = x^{\alpha + i\mu} \Psi\left(-\frac{\alpha}{2} - i\mu, 1 - 2i\mu; \frac{1}{x} \right).$$

Observe that for some values of α, the distribution (4.61) is also well known. In particular, for $\alpha = 1/2$ we get $\pi(x) = \frac{1}{\pi} x^{-2} e^{-1/x}$, which is the *Fréchet distribution*. As for the behavior of the boundary points 0 and ∞, we observe that

$$dS(y) = y^{2\alpha-1} e^{1/y} dy, \quad dM(z) = z^{-2\alpha-1} e^{-1/z} dz.$$

For 0, using (4.39), we have that

$$\begin{aligned} \Sigma(0) &= \int_0^x \left[\int_y^x dM(z) \right] dS(y) = \int_0^x \left[\int_y^x z^{-2\alpha-1} e^{-1/z}\, dz \right] y^{2\alpha-1} e^{1/y} dy \\ &\sim \int_0^x [c_1 F(y) + c_2]\, y^{2\alpha-1} e^{1/y} dy, \end{aligned}$$

where $F(y)$ is a certain combination of *Whittaker functions*. The second part of the integral is divergent by the action of $e^{1/y}$. Therefore $\Sigma(0) = \infty$. Alternatively, using (4.40), we have (assuming that $\alpha \neq k/2$ for $k \in \mathbb{N}$, although if not, it also has a similar behavior) that

$$\begin{aligned} N(0) &= \int_0^x \left[\int_z^x dS(y) \right] dM(z) = \int_0^x \left[\int_z^x y^{2\alpha-1} e^{1/y} dy \right] z^{-2\alpha-1} e^{-1/z} dz \\ &\sim \int_x^{\infty} \left[c_1 z^{-2\alpha} \times {}_1F_1\left(-2\alpha; 1 - 2\alpha; -z\right) + c_2 \right] z^{-2\alpha-1} e^{-1/z} dz. \end{aligned}$$

The second part of the integral is always convergent. The first part of the integral, after a numerical analysis, is also convergent. Therefore $N(0) < \infty$ and the point 0 is an entrance boundary. It is possible to see that the boundary point ∞ is a natural boundary, i.e. $\Sigma(\infty) = \infty$ y $N(\infty) = \infty$. $\Diamond$

4.6 Quasi-Stationary Distributions

Let $\{X_t, t \geq 0\}$ be a regular diffusion process with state space $\mathcal{S}$. Consider $\partial\mathcal{S}$ a set of forbidden states and $\mathcal{S}^0 = \mathcal{S} \setminus \partial\mathcal{S}$ the set of allowed states. Let $\tau = \tau_{\partial\mathcal{S}}$ be the first passage time to the set $\partial\mathcal{S}$ defined in a similar way to (4.12). We will assume that $\mathbb{P}_x(\tau < \infty) \doteq \mathbb{P}(\tau < \infty | X_0 = x) = 1$ for all $x \in \mathcal{S}^0$. We are interested in studying the process $\{X_t, 0 \leq t < \tau\}$ and $X_t = X_\tau$ for all $t \geq \tau$. In particular, in this section, we will study the concept of *quasi-stationary distributions*, which are distributions that are invariant under time evolution when the process is conditioned on survive. More precisely, a probability measure ν on $\mathcal{S}^0$ is called a quasi-stationary distribution for the process killed at $\partial\mathcal{S}$ if for any Borel set $B \subset \mathcal{S}^0$, we have

$$\mathbb{P}_\nu(X_t \in B | \tau > t) = \nu(B), \quad t \geq 0,$$

where, for any event A, we have

$$\mathbb{P}_\nu(A) \doteq \int_{\mathcal{S}^0} \mathbb{P}(A | X_0 = x)\, d\nu(x).$$

Therefore, ν is a quasi-stationary distribution if

$$\mathbb{P}_\nu(X_t \in B, \tau > t) = \nu(B)\mathbb{P}_\nu(\tau > t).$$

Since $\mathbb{P}_\nu(X_t \in B, \tau > t) = \mathbb{P}_\nu(X_t \in B)$ for any Borel set B in $\mathcal{S}^0$, the condition of quasi-stationary distribution takes the form

$$\mathbb{P}_\nu(X_t \in B) = \nu(B)\mathbb{P}_\nu(\tau > t).$$

Since $\mathbb{P}_x(\tau < \infty) > 0$ for all $x \in \mathcal{S}$ then a quasi-stationary distribution can never be a stationary distribution, because in that case we should have $\mathbb{P}_\nu(X_t \in B) = \nu(B)$ (see (4.18)) and the previous formula would not hold.

The concept of quasi-stationary distributions for diffusion processes was introduced by P. Mandl in [114]. In this section we will see that we can study these distributions using the spectral representation of the transition probability density associated with this diffusion process killed at $\partial\mathcal{S}$. We will follow Chapter 6 of the book by P. Collett *et al.* [21]. For simplicity, we will show the simplest case of a regular diffusion living inside a bounded interval $(0, d)$, where both ends 0 and d are regular points for the diffusion process X_t. We will make at the end a few comments about the case of regular diffusions on $[0, \infty)$.

Let then $\mathcal{S} = [0, d]$ be the state space of the diffusion $\{X_t, 0 \leq t < \tau\}$ where $X_t = X_\tau$ for all $t \geq \tau$ and $\partial\mathcal{S} = \{0, d\}$, the forbidden states, so that $\mathcal{S}^0 = (0, d)$. Consider the transition operator T_t^0 (see (4.3)) given by

$$(T_t^0 f)(x) = \mathbb{E}_x(f(X_t), t < \tau), \tag{4.62}$$

acting on all bounded and measurable functions f on $\mathcal{S}^0$. As for the case of regular diffusions this transition operator has the semigroup property and we will denote $p^0(t;x,y)$ the corresponding transition density. Associated with T_t^0 we will assume that the infinitesimal operator (see (4.5)) is given by

$$\mathcal{A}^0 = \frac{1}{2}\frac{d^2}{dx^2} - \alpha(x)\frac{d}{dx}, \tag{4.63}$$

when applied to C^1 functions with compact support contained in $(0,d)$. Along this section we will assume that $\alpha \in C^1(\mathcal{S})$. A regular diffusion $\{Y_t, t \geq 0\}$ on a bounded state space with infinitesimal operator as in (4.5) with drift $\mu(x)$ and diffusion coefficient $\sigma^2(x)$ can always be reduced to a diffusion with infinitesimal operator of the form (4.63) by assuming that $\sigma \in C^1(\mathcal{S})$ and positive on $\mathcal{S}$. This *Liouville transformation* (see Exercise 1.7) is given by $X_t = F(Y_t)$ where $F(y) = \int_0^y 1/\sigma(u)\,du$ (see [21, p. 117] for more details). Then $F \in C^2(\mathcal{S})$ and F^{-1} exists. Therefore

$$\alpha(x) = \frac{\sigma'(F^{-1}(x))}{2} - \frac{\mu(F^{-1}(x))}{\sigma(F^{-1}(x))}. \tag{4.64}$$

The first passage time τ is the same for both processes so it is enough to study the infinitesimal operator (4.63) for our purposes.

We will show now that the semigroup (4.62) has a discrete spectrum. Consider the eigenfunction ϕ, the solution of $\mathcal{A}^0\phi = -\lambda\phi$ with the Dirichlet boundary conditions $\phi(0) = 0$ and $\phi(d) = 0$. Since α is regular at 0 and d, the $C^2(\mathcal{S})$ functions with compact support contained in $[0,d]$, vanishing at 0 and d, belong to the domain of $\mathcal{A}^0$. In a similar way we can consider the eigenfunction φ of the adjoint operator $(\mathcal{A}^0)^*$ with the same initial conditions. These eigenfunctions are related by the formula $\varphi(x) = e^{I(x)}\phi(x)$, where $I(x) = I(0,x) = -2\int_0^x \alpha(z)\,dz$ is defined by (4.19). Observe that $\mathcal{A}^0$ is formally self-adjoint in $L^2_\pi(\mathcal{S})$, where $\pi(x)$ is defined by (4.20) (see Section 4.2) while $(\mathcal{A}^0)^*$ is formally self-adjoint in $L^2_{S'}(\mathcal{S})$, where S is the scale function defined by (4.33).

We will use the theory developed in [19, Chapter 7], which applies to differential operators of the form

$$\mathcal{L} = \frac{1}{2}\frac{d^2}{dx^2} + \frac{1}{2}(\alpha'(x) - \alpha^2(x)).$$

Notice that $\mathcal{L}$ is an unbounded self-adjoint operator on $L^2(\mathcal{S})$. If $u(x)$ is an eigenfunction of $\mathcal{L}$, i.e. $\mathcal{L}u = -\lambda u$ with initial conditions $u(0) = u(d) = 0$, then the relation between this eigenfunction and the one of $\mathcal{A}^0$ is given by

$$\phi(x) = u(x)e^{-\frac{1}{2}I(x)}. \tag{4.65}$$

In [19, Chapter 7] it is proved that there exists a complete orthonormal basis of eigenfunctions for the operator $\mathcal{L}$, which we call $u_n(x)$ with eigenvalues given by $-\lambda_n$. Each λ_n is simple and the only possible accumulation point of the set $\{\lambda_n, n \geq 0\}$ is ∞ (see [21, Lemma 6.1]). Therefore we can order the eigenvalues in the following way:

$$0 < \lambda_0 < \lambda_1 < \cdots < \lambda_n < \cdots .$$

Using the relation (4.65), it is possible to derive a spectral representation of the transition operator T_t^0 in (4.62) in the same way as we did in Section 4.2. Therefore

$$(T_t^0 f)(x) = \mathbb{E}_x(f(X_t), t < \tau) = \sum_{n=0}^{\infty} e^{-\lambda_n t} \left(\int_0^d f(y)\phi_n(y)\phi_n(x)\pi(y)\, dy \right), \tag{4.66}$$

where $\phi_n(x)$ are the normalized eigenfunctions of $\mathcal{A}^0$ and the density $p^0(t; x, y)$ is given by (see [21, Proposition 6.2])

$$p^0(t; x, y) = \sum_{n=0}^{\infty} e^{-\lambda_n t} \phi_n(x)\phi_n(y)\pi(y).$$

This representation gives

$$\lim_{t\to\infty} e^{\lambda_0 t} p^0(t; x, y) = \phi_0(x)\phi_0(y)\pi(y),$$

uniformly in x and y. In particular, ϕ_0 is positive in $(0, d)$. Also, for all $f \in L^2_\pi(S)$, we have

$$\lim_{t\to\infty} e^{\lambda_0 t} \mathbb{E}_x(f(X_t), t < \tau) = \phi_0(x) \int_0^d f(y)\phi_0(y)\pi(y)\, dy,$$

uniformly in x. In particular, if we take $f = 1$, we have

$$\lim_{t\to\infty} e^{\lambda_0 t} \mathbb{P}_x(t < \tau) = \phi_0(x) \int_0^d \phi_0(y)\pi(y)\, dy. \tag{4.67}$$

Also we can obtain an expression of the quasi-stationary distribution, which is given by

$$d\nu(x) = \frac{\phi_0(x)\pi(x)\, dx}{\displaystyle\int_0^d \phi_0(y)\pi(y)\, dy}. \tag{4.68}$$

Indeed, taking $f = \mathbf{1}_B(x)$ in (4.66), where B is any Borel set of $(0,d)$, we get

$$\mathbb{P}_x(X_t \in B | t < \tau)$$

$$= \frac{\mathbb{P}_x(X_t \in B, t < \tau)}{\mathbb{P}_x(t < \tau)} = \frac{\mathbb{P}_x(X_t \in B, t < \tau)}{\mathbb{P}_x(X_t \in B, t < \tau)} = \frac{\sum_{n=0}^{\infty} e^{-\lambda_n t}\phi_n(x)\left(\int_B \phi_n(y)\pi(y)\,dy\right)}{\sum_{n=0}^{\infty} e^{-\lambda_n t}\phi_n(x)\left(\int_0^d \phi_n(y)\pi(y)\,dy\right)}$$

$$= \frac{e^{-\lambda_0 t}\phi_0(x)\int_B \phi_0(y)\pi(y)\,dy}{\sum_{n=0}^{\infty} e^{-\lambda_n t}\phi_n(x)\left(\int_0^d \phi_n(y)\pi(y)\,dy\right)} + \frac{e^{-\lambda_1 t}\phi_1(x)\int_B \phi_1(y)\pi(y)\,dy}{\sum_{n=0}^{\infty} e^{-\lambda_n t}\phi_n(x)\left(\int_0^d \phi_n(y)\pi(y)\,dy\right)} + \cdots$$

$$= \frac{\phi_0(x)\int_B \phi_0(y)\pi(y)\,dy}{\phi_0(x)\int_0^d \phi_0(y)\pi(y)\,dy + \sum_{n=1}^{\infty} e^{-(\lambda_n-\lambda_0)t}\phi_n(x)\left(\int_0^d \phi_n(y)\pi(y)\,dy\right)}$$

$$+ \frac{\phi_1(x)\int_B \phi_0(y)\pi(y)\,dy}{e^{-(\lambda_0-\lambda_1)t}\phi_0(x)\int_0^d \phi_0\pi\,dy + \phi_1(x)\int_0^d \phi_1\pi\,dy + \sum_{n=2}^{\infty} e^{-(\lambda_n-\lambda_0)t}\phi_n(x)\left(\int_0^d \phi_n\pi\,dy\right)}$$

$$+ \cdots \xrightarrow{t\to\infty} \frac{\int_B \phi_0(y)\pi(y)\,dy}{\int_0^d \phi_0(y)\pi(y)\,dy} = \nu(B).$$

Therefore, extending this to any bounded and measurable function f on $\mathcal{S}^0$ we get the so-called *Yaglom limit*

$$\lim_{t\to\infty} \mathbb{E}_x(f(X_t)|t < \tau) = \int_0^d f(y)\,d\nu(y),$$

where ν is the quasi-stationary distribution (4.68). This limit is also valid for any initial probability distribution (see more technical details in [21, Theorem 6.4]). Finally, we also obtain the *ratio limits*

$$\lim_{t\to\infty} \frac{p^0(t+s;x,y)}{p^0(t;z,w)} = e^{-\lambda_0 s}\frac{\phi_0(x)\phi_0(y)\pi(y)}{\phi_0(z)\phi_0(w)\pi(w)},$$

uniformly on compacts, and from (4.67) we get

$$\lim_{t\to\infty} \frac{\mathbb{P}_x(t < \tau)}{\mathbb{P}_y(t < \tau)} = \frac{\phi_0(x)}{\phi_0(y)}.$$

Example 4.30 Consider Example 4.15 of Brownian motion with drift and scaling on $[0,d]$ with two absorbing boundaries. Assume that $\sigma = 1$. The eigenvalue equation with Dirichlet initial conditions to be solved is the following:

$$\frac{1}{2}\phi''(x) + \mu\phi'(x) = -\lambda\phi(x),$$
$$\phi(0) = \phi(d) = 0.$$

The eigenvalues are given by (4.48) while the eigenfunctions are given by (4.49). According to (4.68), the quasi-stationary distribution ν is given by the normalization of the first (nontrivial) eigenfunction, which in this case is $\phi_1(x)$ in (4.49). Therefore

$$d\nu(x) = \frac{\pi^2 + d^2\mu^2}{d(1+e^{\mu d})} e^{\mu x} \sin\left(\frac{\pi x}{d}\right) dx.$$

In Figure 4.5 we can see a couple of plots of this quasi-stationary distribution for the values of $\mu = 3$ and $\mu = -3$ (with $d = 1$). ◊

For diffusion processes on $\mathcal{S} = \mathbb{R}^+ = [0,\infty)$ (see [21, Section 6.2] for details) most of the results that we have shown for the case $\mathcal{S} = [0,d]$ hold if we take $d \to \infty$. In this situation it is usually assumed that 0 is a regular boundary and $+\infty$ is a natural boundary, i.e. $\Sigma(\infty) = N(\infty) = \infty$, where Σ and N are defined in (4.39) and (4.40), respectively. That means that the process cannot explode to ∞ in a finite time, i.e. $\mathbb{P}_x(\tau_\infty < \infty) = 0$. The semigroup (4.62) is defined in the same way for $\tau = \tau_0$ as well as the infinitesimal operator $\mathcal{A}^0$ in (4.63), where $\alpha \in C^1(\mathbb{R}^+)$. Now the eigenfunctions ϕ_λ and φ_λ of $\mathcal{A}^0$ and $(\mathcal{A}^0)^*$ are subject to the initial conditions

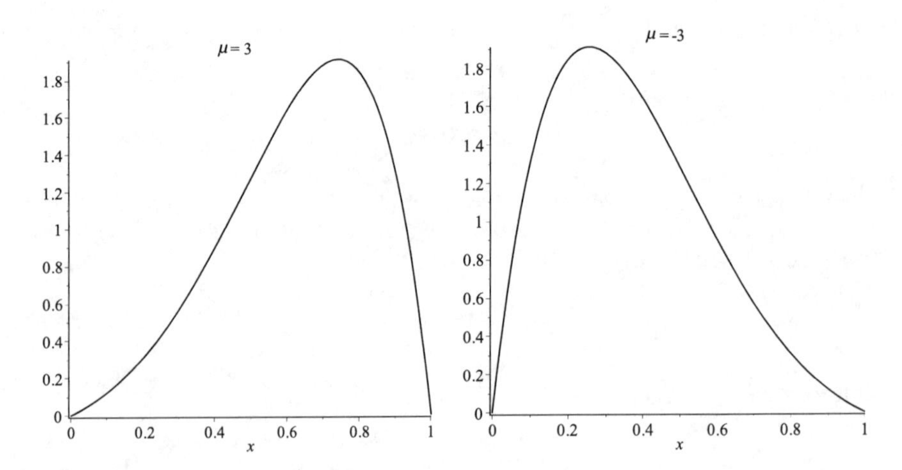

Figure 4.5 Quasi-stationary distribution for Brownian motion with drift and scaling on [0, 1] with two absorbing boundaries with different values of the drift μ.

$\phi_\lambda(0) = 0, \phi'_\lambda(0) = 1$ and $\varphi_\lambda(0) = 0, \varphi'_\lambda(0) = 1$. The spectral representation for the semigroup (4.62) still holds

$$(T_t^0 f)(x) = \mathbb{E}_x(f(X_t), t < \tau) = \int_0^\infty e^{-\lambda t}\phi_\lambda(x)\left(\int_0^\infty f(y)\phi_\lambda(y)\pi(y)\,dy\right)d\Gamma(\lambda),$$

where $\Gamma(\lambda)$ is the distribution function of the eigenvalues over $(0,\infty)$, where now the spectrum can be discrete and/or absolutely continuous (see (4.28)). If we denote by $\underline{\lambda}$ the infimum of the closed support for Γ and

$$\lambda^+ = \sup\{\lambda : \phi_\lambda \geq 0\},$$

then we have (see [21, Lemma 6.7] and [115])

$$\underline{\lambda} = \lambda^+ = \zeta,$$

where ζ is the exponential decay for the absorption probability, i.e.

$$\zeta = -\lim_{t\to\infty}\frac{1}{t}\log\mathbb{P}_x(\tau > t),$$

which exists and is independent of $x > 0$ (see [20, Theorem A]). Notice that $\phi_{\underline{\lambda}} > 0$ for $x > 0$. Under all these hypotheses, we also have (see [21, Proposition 6.13])

$$\mathbb{P}_x(t < \tau) = \int_{\underline{\lambda}}^\infty \frac{e^{-\lambda t}}{2\lambda}\phi_\lambda(x)\,d\Gamma(\lambda). \tag{4.69}$$

Finally, in this situation, a characterization of the quasi-stationary distribution family is obtained (see [21, Theorem 6.34]). The family of quasi-stationary distributions is empty when $\underline{\lambda} = 0$. If $\underline{\lambda} > 0$ then for every $\lambda \in (0, \underline{\lambda}]$, we have that

$$d\nu_\lambda(x) = \frac{\phi_\lambda(x)\pi(x)\,dx}{\displaystyle\int_0^\infty \phi_\lambda(y)\pi(y)\,dy}$$

is a quasi-stationary distribution for the process. Using (4.69) the family of quasi-stationary distributions can also be written as

$$d\nu_\lambda(x) = 2\lambda\phi_\lambda(x)\pi(x)\,dx.$$

Example 4.31 Consider Example 4.16 of Brownian motion with drift and scaling on $[0,\infty)$ with one absorbing boundary. Assume that $\sigma = 1$ and $\mu < 0$ (otherwise $\mathbb{P}_x(\tau < \infty) < 1$). In this case there is a continuum of eigenvalues. Following the same lines as in Example 4.15, the solution of the eigenvalue problem

$$\frac{1}{2}\phi''(x) + \mu\phi'(x) = -\lambda\phi(x),$$
$$\phi(0) = 0, \phi'(0) = 1,$$

is given by

$$\phi(x) = \frac{1}{2\sqrt{\mu^2 - 2\lambda}} e^{-\mu x} \left(e^{x\sqrt{\mu^2-2\lambda}} - e^{-x\sqrt{\mu^2-2\lambda}} \right).$$

We have three situations:

1. If $\mu^2 - 2\lambda > 0$, then, for every $\lambda \in (0, \mu^2/2]$ and since $\pi(x) = e^{2\mu x}$, we have that

$$d\nu_\lambda(x) = \frac{2\lambda}{2\sqrt{\mu^2 - 2\lambda}} e^{\mu x} \sinh\left(x\sqrt{\mu^2 - 2\lambda} \right) dx,$$

   so we have a family of quasi-stationary distributions.
2. If $\mu^2 - 2\lambda = 0$, then $\lambda = \mu^2/2$ and the quasi-stationary distribution is given by

$$d\nu_\lambda(x) = \mu^2 x e^{\mu x} dx.$$

3. If $\mu^2 - 2\lambda < 0$, then the eigenfunctions are given by

$$\phi(x) = \frac{1}{\sqrt{2\lambda - \mu^2}} e^{-\mu x} \sin\left(x\sqrt{2\lambda - \mu^2} \right).$$

   But we clearly see that these eigenfunctions change signs in $[0, \infty)$. Therefore, there are no quasi-stationary distributions.

In Figure 4.6 we can see a couple of plots of this quasi-stationary distribution for the values of $\mu = -1, \lambda = 1/2$ and $\mu = -1, \lambda = 1/4$. ◊

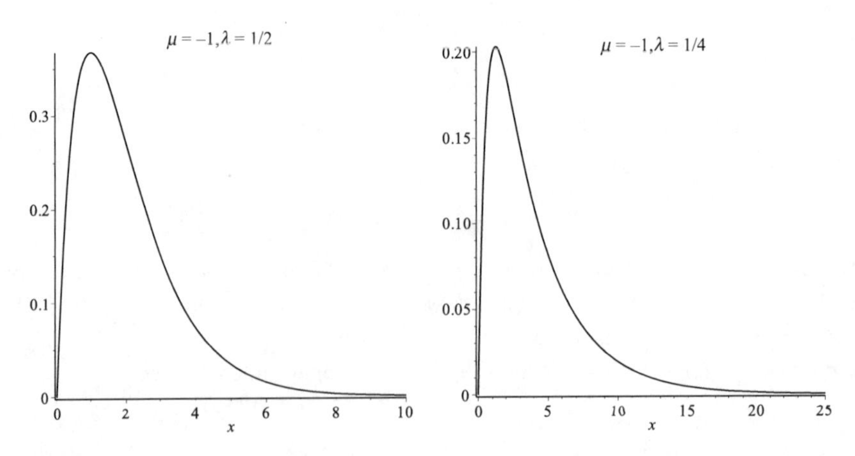

Figure 4.6 Quasi-stationary distribution for Brownian motion with drift and scaling on $[0, \infty)$ with one absorbing boundary.

Example 4.32 Consider the Orstein–Uhlenbeck diffusion studied in Example 4.18, but the state space is given by $[0, \infty)$ with an absorbing boundary at $x = 0$. In this case the solution of the eigenvalue problem

$$\frac{1}{2}\phi''(x) - x\phi'(x) = -\lambda\phi(x),$$
$$\phi(0) = 0, \phi'(0) = 1$$

can be written in terms of *parabolic cylindric functions*. From [112, Lemma 3.6] we know that for any $\lambda \in (0, 1]$ we have a quasi-stationary distribution given by

$$d\nu_\lambda(x) = 2\phi_\lambda(x)e^{-x^2}dx,$$

where $\phi_\lambda(x)$ is a solution of the eigenvalue problem. For instance, if we take $\lambda = 1$, then the quasi-stationary distribution is given by

$$d\nu(x) = 2xe^{-x^2}dx.$$

There is another simple case if we choose $\lambda = 1/2$. Then the quasi-stationary distribution given by

$$d\nu(x) = \frac{1+i}{2}\Gamma(3/4)x^{3/2}e^{-x^2/2}\left(I_{-1/4}(-x^2/2) + I_{3/4}(-x^2/2)\right)dx,$$

where $I_\alpha(z)$ is the *modified Bessel function* defined by $I_\alpha(z) = i^{-\alpha}J_\alpha(iz)$ where J_α is the Bessel function defined by (1.72). The shape of these distributions is similar to the ones displayed in Figure 4.6. ◇

In [21, Chapter 7] (see also [14]) the case where $+\infty$ is an entrance boundary so that $N(\infty) < \infty$ is studied. This holds, for instance, for functions α in (4.63) of the form

$$\alpha(x) = \frac{1}{2x} - \frac{rx}{2} + \frac{c\sigma x^3}{8},$$

obtained from models in ecology and economics. Under the assumption that

$$-\inf_{y\in(0,\infty)} \alpha^2(y) - \alpha'(y) < \infty, \quad \text{and} \quad \lim_{y\to\infty} \alpha^2(y) - \alpha'(y) = \infty,$$

it is possible to prove that the infinitesimal operator $\mathcal{A}^0$ in (4.63) has a purely discrete spectrum and we get a spectral representation as in (4.66).

Finally, the concept of quasi-stationary distributions for diffusion processes with killing has been studied in [138].

4.7 Exercises

4.1 Use Proposition 4.8 to answer the following questions:

(a) Let $\{X_t, t \geq 0\}$ be a regular diffusion process on $(0,1)$ with diffusion coefficient $\sigma^2(x) = x^2(1-x)^2$. Show that the diffusion $Y_t = \log(X_t/(1-X_t))$ on $\mathbb{R}$ has a constant diffusion coefficient.

(b) Let $\{X_t, t \geq 0\}$ be a regular diffusion process on $(0,\infty)$ with drift $\mu(x) = bx + c, c > 0$, and diffusion coefficient $\sigma^2(x) = 4x$. Compute the drift and diffusion coefficients for the diffusion $Y_t = \sqrt{X_t}$.

(c) Let $\{X_t, t \geq 0\}$ be a regular diffusion process on $(0,\infty)$ with drift $\mu(x) = cx, c > 0$, and diffusion coefficient $\sigma^2(x) = x^\alpha, \alpha \neq 2$. What choice of β gives a constant diffusion coefficient for the diffusion $Y_t = (X_t)^\beta$?

4.2 Let $\{X_t, t \geq 0\}$ be a regular diffusion process on $(0,1)$ with drift $\mu(x) = x^2(1-x)(\alpha - \beta x^2)$ and diffusion coefficient $\sigma^2(x) = \beta x^4(1-x)x^2$. If $0 < \alpha < \beta/2$, show that there exists a unique stationary distribution and compute it.

4.3 Let $\{X_t, t \geq 0\}$ be a regular diffusion process on $(0,\infty)$ with drift $\mu(x) = \gamma x^\alpha$, $\alpha > 0, \gamma \in \mathbb{R}$, and diffusion coefficient $\sigma^2(x) = \eta x^\beta, \eta, \beta > 0$. Classify the boundaries 0 and ∞ in terms of α, β, γ and η using Feller's classification of boundaries.

4.4 Consider Brownian motion with drift μ and diffusion coefficient σ^2 on $S = [0,d]$ (as in Example 4.13) where both boundaries are reflecting. Find the eigenvalues and eigenfunctions of the corresponding infinitesimal operator, as well as the spectral representation of the transition probability density.

4.5 Consider Brownian motion with drift μ and diffusion coefficient σ^2 on $S = [0,d]$ with 0 an absorbing boundary and d a reflecting boundary. Solve the corresponding Sturm–Liouville problem and find the eigenvalues and eigenfunctions of the corresponding infinitesimal operator, as well as the spectral representation of the transition probability density.

4.6 Consider the standard Brownian motion on $S = [0,d]$ and the corresponding Sturm–Liouville with boundary conditions

$$\phi(0) = 0, \quad h\phi(d) + \phi'(d) = 0, \quad h > 0.$$

Solve the Sturm–Liouville problem, finding the eigenvalues and eigenfunctions, and give an expression of the transition probability density.

4.7 Let $\{X_t, t \geq 0\}$ be the diffusion process with drift $\mu(x) = x$ and diffusion coefficient $\sigma^2(x) = 2x^2$ on $S = [1,d]$. Solve the corresponding Sturm–Liouville problem with boundary conditions

$$x^2\phi''(x) + x\phi'(x) = -\lambda\phi(x), \quad 1 < x < d,$$
$$\phi(1) = \phi(d) = 0.$$

4.8 Consider Example 4.15 of Brownian motion with drift and scaling on $[0,d]$ with two absorbing boundaries with $\mu = 0$. Prove that the probabilities $p(t;x,\{0\})$ and $p(t;x,\{d\})$ are given by

$$p(t;x,\{0\}) = 1 - \frac{x}{d} - \frac{2}{\pi}\sum_{n=1}^{\infty}\frac{1}{n}\exp\left(-\frac{\sigma^2\pi^2n^2t}{2d^2}\right)\sin\left(\frac{n\pi x}{d}\right),$$

$$p(t;x,\{d\}) = \frac{x}{d} - \frac{2}{\pi}\sum_{n=1}^{\infty}\frac{(-1)^{n+1}}{n}\exp\left(-\frac{\sigma^2\pi^2n^2t}{2d^2}\right)\sin\left(\frac{n\pi x}{d}\right).$$

4.9 Consider Example 4.16 of Brownian motion with drift and scaling on $[0,\infty)$ with one absorbing boundary. Find an approximation of the distribution of the first passage time τ_0 to the absorbing state 0.

4.10 Consider the Orstein–Uhlenbeck diffusion process with drift and diffusion coefficients given by

$$\mu(x) = -\alpha x, \quad \alpha > 0, \quad \sigma^2(x) = \sigma^2, \quad x \in \mathbb{R}.$$

Find the eigenvalues and eigenfunctions of the corresponding infinitesimal operator in terms of the Hermite polynomials, as well as an explicit expression of the transition probability density.

4.11 Consider Example 4.24 of the population genetics diffusion process with killing. Find an explicit expression of $\mathbb{E}(\xi|X_0 = x)$ in (4.58) or (4.59) in terms of the *exponential integral* for positive values of the real part of the argument, defined by

$$E_1(z) = \int_1^{\infty}\frac{e^{-xz}}{x}\,dx, \qquad \mathrm{Re}(z) \geq 0.$$

4.12 Consider the spectral representation for $p(t;x,y)$ for the Examples 4.28 and 4.29. Using differential properties and other formulas for the special functions that appear in those representations (see Chapter 1) prove that they satisfy the corresponding Kolmogorov backward equations.

4.13 Considering the Liouville transformation at the beginning of Section 4.6 and the transformation of the drift given by (4.64), find quasi-stationary distributions for the population growth model given in Example 4.19 and the Wright–Fisher model given in Example 4.21.

References

[1] Anderson, W. J., *Continuous-time Markov Chains: An Applications-oriented Approach*, Springer series in Statistics, Springer, New York, 1991. (Cited on pp. 147, 148, 149, 150, 151, 152, 153, 155, 158, 161, 162, 178, 191, 207, 236 and 242.)

[2] Akhiezer, N. I., *The Classical Moment Problem and Some Related Questions in Analysis*, University Mathematical Monographs, Oliver & Boyd, Edimburgh and London, 1961. (Cited on pp. 18 and 105.)

[3] Andrews, G. E., Askey, R. and Roy, R., *Special Functions*, Encyclopedia of Mathematics, No. 71, Cambridge University Press, Cambridge, 1999. (Cited on pp. 1, 3, 16 and 17.)

[4] van Assche, W., Parthasarathy, P. R. and Lenin, R. B. Spectral representation of four finite birth and death processes, *Math. Sci.* **24** (1999), 105–112. (Cited on pp. 194 and 196.)

[5] Bhattacharya, R. N. and Waymire, E. C., *Stochastic Processes with Applications*, Wiley Series in Probability and Mathematical Statistics: Applied Probability and Statistics, John Wiley & Sons, Inc., New York, 1990. (Cited on pp. 58, 255, 257, 258, 260, 261, 262, 269, 270, 271, 274, 279 and 287.)

[6] Barrett, J. F. and Lampard, D. G., An expansion for some second-order probability distributions and its application to noise problems, *IRE Trans. Inf. Th.* **1** (1955), 10–15. (Cited on pp. ix, 262 and 268.)

[7] Bakry, D. and Mazet, O., *Characterization of Markov Semigroups on* $\mathbb{R}$ *Associated to Some Families of Orthogonal Polynomials*, Séminaire de Probabilités XXXVII, Lecture Notes in Math., 1832, Springer, Berlin, 2003, pp. 60–80. (Cited on p. 303.)

[8] Bakry, D., Orevkov, S. and Zani, M., *Orthogonal Polynomials and Diffusion Operators*, arXiv:1309.5632v2. (Cited on p. 303.)

[9] Beals, R. and Wong, R., *Special Functions: A Graduate Text*, Cambridge University Press, New York, 2010. (Cited on pp. 1, 31, 45 and 55.)

[10] Bochner, S., Über Sturm-Liouvillesche Polynomsysteme, *Math Z.* **29** (1929), 730–736. (Cited on p. 27.)

[11] Brockwell, P. J., The extinction time of a general birth and death process with catastrophes, *J. Appl. Prob.* **23** (1986), 851–858. (Cited on p. 191.)

[12] Callaert, H., On the rate of convergence in birth-and-death processes, *Bull. Soc. Math. Belg.* **26**, (1974), 173–184. (Cited on p. 237.)

[13] Castro, M. M. and Grünbaum, F. A., On a seminal paper by Karlin and McGregor, *SIGMA* **9** (2013), 020. (Cited on pp. 57 and 99.)

[14] Cattiaux, P., Collet, P., Lambert, A., Martínez, S., Méléard, S. and San Martín, J., Quasi-stationary distributions and diffusions models in population dynamics, *Ann. Prob.* **37** (2009), 1926–1969. (Cited on p. 319.)

[15] Chen, M. F., Exponential L2-convergence and L2- spectral gap for Markov processes, *Acta Math. Sinica* **7** (1991), 19–37. (Cited on p. 238.)

[16] Chihara, T. S., *An Introduction to Orthogonal Polynomials*, Gordon and Breach, London, 1968. (Cited on pp. 1, 14, 16, 18, 69, 99 and 167.)

[17] Chihara, T. S. and Ismail, M. E. H., Orthogonal polynomials suggested by a queueing model, *Adv. Appl. Math.* **3** (1982), 441–462. (Cited on p. 251.)

[18] Clayton, A., Quasi-birth-and-death processes and matrix-valued orthogonal polynomials, *SIAM J. Matrix Anal. Appl.* **31** (2010), 2239–2260. (Cited on p. 142.)

[19] Coddington, E. A. and Levinson, N., *Theory of Ordinary Differential Equations*, McGraw-Hill, New York, 1955. (Cited on pp. 313 and 314.)

[20] Collet, P., Martínez, S. and San Martín, J., Asymptotic laws for one-dimensional diffusions conditioned to nonabsorption, *Ann. Prob.* **23** (1995), 1300–1314. (Cited on p. 317.)

[21] Collet, P., Martínez, S. and San Martín, J., *Quasi-stationary Distributions*, Probability and its Applications, Springer, New York, 2013. (Cited on pp. 236, 237, 239, 240, 241, 312, 313, 314, 315, 316, 317 and 319.)

[22] Coolen-Schrijner, P. and van Doorn, E. A., Analysis of random walks using orthogonal polynomials, *J. Comp. Appl. Math.* **99** (1998), 387–399. (Cited on pp. 57 and 69.)

[23] Coolen-Schrijner, P. and van Doorn, E. A., Quasi-stationary distributions for birth-death processes with killing, *J. Appl. Math. Stoch. Anal.* 2006, Art. ID 84640. (Cited on p. 241.)

[24] Cox J., Ingersoll J. and Ross, S., An intertemporal general equilibrium model of asset prices, *Econometrica* **53** (1985), 363–384. (Cited on p. 291.)

[25] Cox J., Ingersoll J. and Ross, S., A theory of the term structure of interest rates, *Econometrica* **53** (1985), 385–407. (Cited on p. 291.)

[26] Darroch, J. N. and Seneta, E., On quasi-stationary distributions in absorbing discrete-time finite Markov chains, *J. Appl. Prob.* **2** (1965), 88–100. (Cited on p. 123.)

[27] Deift, P. A., *Orthogonal Polynomials and Random Matrices: A Riemann-Hilbert Approach*, Courant Lecture Notes in Mathematics, 3, American Mathematical Society, Providence, RI, 1999. (Cited on pp. 18, 21 and 24.)

[28] Derman, C., A solution to a set of fundamental equations in Markov chains, *Proc. Amer. Math. Soc.* **5** (1954), 332–334. (Cited on p. 83.)

[29] Dette, H., On a generalization of the Ehrenfest urn model, *J. Appl. Prob.* **31** (1994), 930–939.

[30] Dette, H., On the generating functions of a random walk on the non-negative integers, *J. Appl. Prob.* **33** (1996), 1033–1052. (Cited on pp. 57 and 82.)

[31] Dette, H., First return probabilities of birth and death chains and associated orthogonal polynomials, *Proc. Amer. Math. Soc.* **129** (2000), 1805–1815. (Cited on pp. 16, 77 and 83.)

[32] Dette, H., Reuther, B., Studden, W. and Zygmunt, M., Matrix measures and random walks with a block tridiagonal transition matrix, *SIAM J. Matrix Anal. Applic.* **29** (2006), 117–142. (Cited on pp. 137, 138 and 142.)

[33] Dette, H. and Reuther, B., Some comments on quasi-birth-and-death processes and matrix measures, *J. Prob. Stat.* **2010** (2010), https://doi.org/10.1155/2010/730543. (Cited on p. 247.)

[34] Dette, H. and Studden, W. J., *The Theory of Canonical Moments with Applications in Statistics, Probability and Analysis*, John Wiley & Sons, Inc., New York, 1997. (Cited on pp. 16, 71, 131 and 132.)

[35] van Doorn, E. A., Stochastic monotonicity of birth-death processes, *Adv. Appl. Prob.* **12** (1980), 59–80. (Cited on p. 242.)

[36] van Doorn, E. A., *Stochastic Monotonicity and Queueing Applications of Birth-Death Processes*, Lecture Notes in Statistics, 4, Springer, New York, 1981. (Cited on p. 242.)

[37] van Doorn, E. A., The transient state probabilities for a queueing model where potential customers are discouraged by queue length, *Adv. Appl. Prob.* **17** (1985) 514–530. (Cited on pp. 251 and 252.)

[38] van Doorn, E. A., Conditions for exponential ergodicity and bounds for the decay parameter of a birth-death process, *Adv. Appl. Prob.* **17** (1985) 514–530. (Cited on pp. 238 and 240.)

[39] van Doorn, E. A., Representations and bounds for zeros of orthogonal polynomials and eigenvalues of sign-symmetric tridiagonal matrices, *J. Approx. Theory* **51** (1987) 254–266.

[40] van Doorn, E. A., Quasi-stationary distributions and convergence for quasi-stationarity of birth-death processes, *Adv. Appl. Prob.* **23** (1991), 683–700. (Cited on pp. 229, 237, 238 and 239.)

[41] van Doorn, E. A., Birth-death processes and associated polynomials, *J. Comp. Appl. Math.* **23** (2003), 497–506. (Cited on pp. 216 and 242.)

[42] van Doorn, E. A., Conditions for the existence of quasi-stationary distributions for birth-death processes with killing, *Stochastic Process. Appl.* **122** (2012), 2400–2410. (Cited on p. 241.)

[43] van Doorn, E. A., Representations for the decay parameter of a birth-death process based on the Courant-Fischer theorem, *J. Appl. Prob.* **52** (2015), 278–289. (Cited on p. 238.)

[44] van Doorn, E. A., On the strong ratio limit property for discrete-time birth-death processes, *SIGMA* **14** (2018), 047. (Cited on pp. 69, 114 and 119.)

[45] van Doorn, E. A. and Pollett, P. K., Quasi-stationary distributions for discrete-state models, *European J. Oper. Res.* **230** (2013), 1–14. (Cited on p. 241.)

[46] van Doorn, E. A. and Schrijner, P., Random walk polynomials and random walk measures, *J. Comp. Appl. Math* **49** (1993), 289–296. (Cited on pp. 57, 68, 117 and 128.)

[47] van Doorn, E. A. and Schrijner, P., Geometric ergodicity and quasi-stationarity in discrete-time birth-death processes, *J. Austral. Math. Soc. Ser. B* **37** (1995), 121–144. (Cited on pp. 83, 109, 114, 123 and 130.)

[48] van Doorn, E. A. and Schrijner, P., Ratio limits and limiting conditional distributions for discrete-time birth-death processes, *J. Math. Anal. Appl.* **190** (1995), 263–284. (Cited on pp. 114, 115, 117, 119, 120, 122, 123, 124 and 125.)

[49] van Doorn, E. A. and Zeifman, A. I., Birth-death processes with killing, *Statist. Prob. Lett.* **72** (2005), 33–42. (Cited on pp. 188, 189, 191 and 214.)

[50] van Doorn, E. A. and Zeifman, A. I., Extinction probability in a birth-death process with killing, *J. Appl. Prob.* **42** (2005), 185–198. (Cited on pp. 188, 190 and 191.)

[51] Dunford, N. and Schwartz, J. T., *Linear Operators, Part II: Spectral Theory*, Wiley-Interscience, New York, 1988. (Cited on pp. 3 and 21.)

[52] Erdelyi, A., Magnus, W., Oberhettinger, F. and Tricomi, F. G., *Higher Transcendental Functions* (3 vols.), McGraw-Hill, New York, 1953. (Cited on pp. 214, 298 and 301.)

[53] Feller, W., Diffusion processes in genetics, *Proc. Second Berkeley Symposium on Mathematical Statistics and Probability* (Berkeley, 1951), pp. 227–246. (Cited on pp. ix, 196 and 291.)

[54] Feller, W., The parabolic differential equations and the associated semi groups of transformations, *Ann. Math.* **55** (1952), 468–519. (Cited on p. 276.)

[55] Feller, W., Diffusion processes in one dimension, *Trans. Amer. Math. Soc.* **77** (1954), 1–31. (Cited on p. 267.)

[56] Feller, W., The birth and death processes as diffusion processes, *J. Math. Pure Appl.* **38** (1959), 301–345. (Cited on pp. ix and 159.)

[57] Feller, W., *An Introduction to Probability Theory and its Applications*, vol. 1, John Wiley & Sons Inc, New York, 1968. (Cited on pp. 58, 60, 91, 143 and 202.)

[58] Flajolet, P. and Guillemin, F., The formal theory of birth-and-death processes, lattice path combinatorics and continued fractions, *Adv. Appl. Prob.* **32** (2000), 750–778. (Cited on p. 178.)

[59] Good, P., The limiting behavior of transient birth and death processes conditioned on survival, *J. Austral. Math. Soc.* **8** (1968), 716–722. (Cited on p. 230.)

[60] Gouriéroux, C., Renault, E. and Valéry, P., Diffusion processes with polynomial eigenfunctions, *Annales d'Économie et de Statistique* **85** (2007), 115–130. (Cited on pp. 267 and 268.)

[61] Grosjean, C. C., The weight functions, generating functions and miscellaneous properties of the sequences of orthogonal polynomials of the second kind associated with the Jacobi and the Gegenbauer polynomials, *J. Comp. Appl. Math.* **16** (1986), 259–307. (Cited on p. 16.)

[62] Grünbaum, F. A., Random walks and orthogonal polynomials: some challenges, in M. Pinsky and B. Birnir (eds.) *Probability, Geometry and Integrable Systems*, MSRI Publication, vol. **55**, Cambridge University Press, Cambridge, 2008, pp. 241–260. See also arXiv: math.PR/0703375v1. (Cited on pp. 57, 91, 137 and 142.)

[63] Grünbaum, F. A., QBD processes and matrix orthogonal polynomials: some new explicit examples, in D. Bini, B. Meini, V. Ramaswami, M. A. Remiche and P. Taylor (eds.), *Numerical Methods for Structured Markov Chains*, Dagstuhl Seminar Proceedings, 2008. (Cited on pp. 140 and 145.)

[64] Grünbaum, F. A., The Karlin–McGregor formula for a variant of a discrete version of Walsh's spider, *J. Phys. A* **42** (2009), no. 45, 454010. (Cited on p. 142.)

[65] Grünbaum, F. A., Gohberg, I. and Ball, J. A., A spectral weight matrix for a discrete version of Walsh's spider, in Operator Theory Advances and Applications *Topics in Operator Theory, vol. 1: Operators, Matrices and Analytic Functions*, Operator Theory: Advances and Applications, vol. 202, Birkhäuser Verlag, Basel, 2010 pp. 253–264. (Cited on p. 142.)

[66] Grünbaum, F. A., Un urn model associated with Jacobi polynomials, *Comm. App. Math. Comp. Sci.* **5** (2010), 55–63. (Cited on p. 102.)

[67] Grünbaum, F. A. and Haine, L., Orthogonal polynomials satisfying differential equations: the role of the Darboux transformation, in D. Levi, L. Vinet and P.

Winternitz (eds.), *Symmetries and Integrability of Differential Equations*, CRM Proc. Lecture Notes, vol. 9, American Mathematical Society, Providence, RI, 1996, pp. 143–154. (Cited on p. 74.)

[68] Grünbaum, F. A. and de la Iglesia, M. D., Matrix valued orthogonal polynomials arising from group representation theory and a family of quasi-birth-and-death processes, *SIAM J. Matrix Anal. Appl.* **30** (2008), 741–761. (Cited on p. 142.)

[69] Grünbaum, F. A. and de la Iglesia, M. D., Stochastic LU factorizations, Darboux transformations and urn models, *J. Appl. Prob.* **55** (2018), 862–886. (Cited on pp. 57, 74, 99, 102 and 144.)

[70] Heyman, D. P., A decomposition theorem for infinite stochastic matrices, *J. Appl. Prob.* **32** (1995), 893–903. (Cited on p. 74.)

[71] Hille, E., The abstract Cauchy problem and Cauchy's problem for parabolic differential equations, *J. Anal. Math.* **3** (1953), 83–196. (Cited on pp. ix and 262.)

[72] Horn, R. A. and Johnson, C. R., *Matrix Analysis*, Cambridge University Press, Cambridge 1990. (Cited on p. 65.)

[73] Ismail, M. E. H., A queueing model and a set of orthogonal polynomials, *J. Math. Anal. Appl.* **108** (1985), 575–594. (Cited on p. 214.)

[74] Ismail, M. E. H., *Classical and Quantum Orthogonal Polynomials in one Variable*, Encyclopedia of Mathematics and its Applications, vol. 98, Cambridge University Press, Cambridge, 2005. (Cited on pp. 1, 15, 18, 19, 20, 30 and 214.)

[75] Ismail, M. E. H., Letessier, J., Masson, D. and Valent, G., Birth and death processes and orthogonal polynomials, in P. Nevai (ed.), *Orthogonal Polynomials*, Kluwer Academic Publishers, Boston, MA, 1990, pp. 229–255. (Cited on p. 248.)

[76] Ismail, M. E. H., Letessier, J. and Valent, G., Linear birth and death models and associated Laguerre and Meixner polynomials, *J. Approx. Theory* **55** (1988), 337–348. (Cited on p. 211.)

[77] Ismail, M. E. H., Letessier, J. and Valent, G., Quadratic birth and death processes and associated continuous dual Hahn polynomials, *SIAM J. Math. Anal.* **20** (1989), 727–737. (Cited on p. 214.)

[78] Jacka, S. D. and Roberts, G. O., Weak convergence of conditioned process on a countable state space, *J. Appl. Prob.* **32** (1995), 902–916. (Cited on p. 237.)

[79] Jones, W. B. and Magnus, A., Application of Stieltjes fractions to birth-death processes, in E. B. Saff and R. S. Varga (eds.), *Padé and Rational Approximation*, Academic Press, New York, 1977, pp. 173–179. (Cited on p. 178.)

[80] Kac, M., Random walk and the theory of Brownian motion, *American Math. Monthly* **54** (1947), 369–391. (Cited on pp. ix and 160.)

[81] Karlin, S., *Total Positivity, vol. I*. Stanford University Press, Stanford, CA, 1968. (Cited on p. 187.)

[82] Karlin, S. and McGregor, J., Representation of a class of stochastic processes, *Proc. Nat. Acad. Sci.* **41** (1955), 387–391. (Cited on pp. ix, 160 and 162.)

[83] Karlin, S. and McGregor, J., The differential equations of birth and death processes, and the Stieltjes moment problem, *Trans. Amer. Math. Soc.* **85** (1957), 489–546. (Cited on pp. 71, 126, 127, 160, 162, 164, 165, 168, 170, 172, 180, 183, 184, 186 and 187.)

[84] Karlin, S. and McGregor, J., The classification of birth-and-death processes, *Trans. Amer. Math. Soc.* **86** (1957), 366–400. (Cited on pp. 111, 160, 216, 218, 222, 223, 227 and 252.)

[85] Karlin, S. and McGregor, J., Linear growth, birth and death processes, *J. Math. Mech.* **7** (1958), 643–662. (Cited on pp. 207 and 251.)

[86] Karlin, S. and McGregor, J., Many server queueing processes with Poisson input and exponential service times, *Pacific J. Math.* **8** (1958), 87–118. (Cited on pp. 79, 175, 201, 203, 207 and 250.)

[87] Karlin, S. and McGregor, J., Random walks, *Illinois J. Math.* **3** (1959), 66–81. (Cited on pp. ix, 57, 63, 65, 67, 68, 77, 93, 95, 99, 105, 106, 114, 118, 120 and 132.)

[88] Karlin, S. and McGregor, J., Coincidence probabilities, *Pacific J. Math.* **9** (1959), 1141–1164. (Cited on p. 267.)

[89] Karlin, S. and McGregor, J., Classical diffusion processes and total positivity, *J. Math. Anal. Appl.* **1** (1960), 163–183. (Cited on pp. ix, 265, 267, 279, 286 and 297.)

[90] Karlin, S. and McGregor, J., Total positivity of fundamental solutions of parabolic equations, *Proc. Am. Math. Soc.* **13** (1962), 136–139. (Cited on p. 267.)

[91] Karlin, S. and McGregor, J., On a genetics model of Moran, *Math. Proc. Camb. Phil. Soc.* **58** (1962), 299–311. (Cited on pp. 196, 199 and 291.)

[92] Karlin, S. and McGregor, J., Ehrenfest urn models, *J. Appl. Prob.* **2** (1965), 352–376. (Cited on p. 195.)

[93] Karlin, S. and McGregor, J., Spectral theory of branching processes. I. The case of discrete spectrum, *Z. Wahrscheinlichkeitstheorie und Verw. Gebiete* **5** (1966), 6–33. (Cited on p. 291.)

[94] Karlin, S. and McGregor, J., Spectral theory of branching processes. II. Case of continuous spectrum, *Z. Wahrscheinlichkeitstheorie und Verw. Gebiete* **5** (1966), 34–54. (Cited on p. 291.)

[95] Karlin, S. and McGregor, J., On the spectral representation of branching processes with mean one, *J. Math. Anal. Appl.* **21** (1968), 485–495. (Cited on p. 291.)

[96] Karlin, S. and Tavaré, S., The detection of a recessive visible gene in finite populations, I, *Genetics Res.*, **37** (1981), 33–46. (Cited on p. 303.)

[97] Karlin, S. and Tavaré, S., The detection of particular genotypes in finite populations, II, *Theoret. Pop. Biol.* **19** (1981), 215–229. (Cited on p. 303.)

[98] Karlin, S. and Tavaré, S., Linear birth and death processes with killing, *J. Appl. Prob* **19** (1982), 477–487. (Cited on pp. 188 and 214.)

[99] Karlin, S. and Tavaré, S., A class of diffusion processes with killing arising in population genetics, *SIAM J. Appl. Math.* **43** (1983), 31–41. (Cited on pp. 279, 286, 299 and 303.)

[100] Karlin, S. and Taylor, H., *A First Course in Stochastic Processes*, Academic Press, New York, 1975. (Cited on p. 58.)

[101] Karlin, S. and Taylor, H., *A Second Course in Stochastic Processes*, Academic Press, New York, 1981. (Cited on pp. 87, 97, 255, 260, 261, 262, 275, 276, 277, 279, 285, 286, 287, 291 and 297.)

[102] Kendall, D. G., Stochastic processes occurring in the theory of queues and their analysis by the method of the imbedded Markov chain, *Ann. Math. Statist.* **24** (1953), 338–354. (Cited on p. 192.)

[103] Kendall, D. G., Unitary dilations of one-parameter semigroups of Markov transition operators, and the corresponding integral representation for Markov processes with a countable infinity of states, *Proc. London Math. Soc.* **3** (1959), 417–432. (Cited on p. ix.)

[104] Kijima, M., On the existence of quasi-stationary distributions in denumerable *R*-transient Markov chains, *J. Appl. Prob.* **29** (1992), 21–36; correction **30** (1993), 496. (Cited on p. 123.)

[105] Kijima, M., Nair, M. G., Pollett, P. K. and van Doorn, E. A., Limiting conditional distributions for birth-death processes, *Adv. Appl. Prob.* **29** (1997), 185–204. (Cited on pp. 229, 230, 232, 233 and 239.)

[106] Koekoek, R., Lesky, P. A. and Swarttouw, R. F., *Hypergeometric Orthogonal Polynomials and their q-Analogues*, Springer Monographs in Mathematics, Springer, Berlin, 2010. (Cited on pp. 52 and 53.)

[107] Kingman, J. F. C., The exponential decay of Markov transition probabilities, *Proc. London Math. Soc.* **13** (1963), 337–358. (Cited on p. 237.)

[108] Kingman, J. F. C., Ergodic properties of continuous-time Markov processes and their discrete skeletons, *Proc. London Math. Soc.* **13** (1963), 593–604. (Cited on p. 237.)

[109] Koelink, E., Spectral theory and special functions, Summer School, Laredo, Spain, July 24–28, 2000, http://arxiv.org/abs/math/0107036v1. (Cited on pp. 18, 21 and 24.)

[110] Latouche, G. and Ramaswami, V., *Introduction to Matrix Analytic Methods in Stochastic Modeling*, ASA-SIAM Series on Statistics and Applied Probability, Society for Industrial and Applied Mathematics, Philadelphia, PA, 1999. (Cited on pp. 137 and 247.)

[111] Ledermann, W. and Reuter, G. E., Spectral theory for the differential equations of simple birth and death processes, *Phil. Trans. Roy. Soc. London, Ser. A.* **246** (1954), 321–369. (Cited on pp. ix and 160.)

[112] Lladser, M. and San Martín, J., Domain of attraction of the quasi-stationary distributions for the Ornstein-Uhlenbeck process, *J. Appl. Prob.* **37** (2000), 511–520. (Cited on p. 319.)

[113] Maki, D. P., On birth-death processes with rational growth rates, *SIAM J. Math. Anal.* **7** (1976), 29–36. (Cited on p. 241.)

[114] Mandl, P., Spectral theory of semi-groups connected with diffusion processes and its application, *Czechoslovak Math. J.* **11** (1961), 558–569. (Cited on p. 312.)

[115] Martínez, S. and San Martín, J., Rates of decay and h-processes for one dimensional diffusions conditional on non-absorption, *J. Theoret. Prob.* **14** (2001), 199–212. (Cited on p. 317.)

[116] McKean, H. P. Jr., Elementary solutions for certain parabolic partial differential equations, *Trans. Amer. Math. Soc.* **82** (1956), 519–548. (Cited on pp. ix and 262.)

[117] Murphy, J. A. and O'Donohoe, J. L., Some properties of continued fractions with applications in Markov processes, *J. Inst. Math. Appl.* **16** (1975), 57–71. (Cited on p. 178.)

[118] Nair, M. G. and Pollett, P. K., On the relationship between μ-invariant measures and quasi-stationary distributions for continuous-time Markov chains, *Adv. Appl. Prob.* **25** (1993), 82–102. (Cited on p. 236.)

[119] Neuts, M. F., *Structured Stochastic Matrices of $M/G/1$ Type and Their Applications*, Marcel Dekker, New York, 1989. (Cited on pp. 137 and 247.)

[120] Norris, J. R., *Markov Chains*, Cambridge University Press, Cambridge, 1997. (Cited on pp. 58 and 61.)

[121] *NIST Digital Library of Mathematical Functions*. http://dlmf.nist.gov/, Release 1.0.20 of 2018-09-15. F. W. J. Olver, A. B. Olde Daalhuis, D. W. Lozier, B. I. Schneider, R. F. Boisvert, C. W. Clark, B. R. Miller, and B. V. Saunders (eds.). (Cited on pp. 52 and 53.)

[122] Ogura, Y., Spectral representation for branching processes on the real half line, *Publ. Res. Inst. Math. Sci.* **5** (1969/1970), 423–441. (Cited on p. 291.)

[123] Parthasarathy, P. R., Lenin, R. B., Schoutens, W. and van Assche, W., A birth and death process related to the Rogers-Ramanujan continued fraction, *J. Math. Anal. Appl.* **224** (1998), 297–315. (Cited on p. 214.)

[124] Pruitt, W., Bilateral birth and death processes, *Trans. Amer. Math. Soc.* **107** (1962), 508–525. (Cited on pp. 243 and 245.)

[125] Reed, M. and Simon, B., *Methods of Modern Mathematical Physics. I. Functional Analysis*, Academic Press, New York, 1972. (Cited on pp. 20, 21 and 24.)

[126] Routh, E., On some properties of certain solutions of a differential equation of the second order, *Proc. London Math. Soc.* **16** (1884), 245–261. (Cited on pp. 27 and 29.)

[127] Rudin, W., *Functional Analysis*, McGraw-Hill, New York, 1973. (Cited on pp. 20 and 21.)

[128] Sasaki, R., Exactly solvable birth and death processes, *J. Math. Phys.* **50** (2009), 103509. (Cited on p. 214.)

[129] Schoutens, W., *Stochastic Processes and Orthogonal Polynomials*, Lecture Notes in Statistics 146, Springer, New York, 2000. (Cited on p. ix.)

[130] Seneta, E. *Non-negative Matrices and Markov Chains*, 2nd edition, Springer Series in Statistics XVI, Springer, New York, 1981. (Cited on p. 65.)

[131] Seneta, E. and Vere-Jones, D., On quasi-stationary distributions in discrete-time Markov chains with a denumerable infinity of states, *J. Appl. Prob.* **3** (1966), 403–434. (Cited on p. 123.)

[132] Serafin, G., Exit times densities of the Bessel process, *Proc. Amer. Math. Soc.* **145** (2017), 3165–3178. (Cited on p. 307.)

[133] Simon, B., The classical moment problem as a self-adjoint finite difference operator, *Adv. Math.* **137** (1998), 82–203. (Cited on p. 18.)

[134] Sharma, O. P. and Gupta, U. C., Transient behaviour of an $M/M/1/N$ queue, *Stoch. Processes Appl.* **13** (1982), 327–331. (Cited on p. 194.)

[135] Shen, J., Tang, T. and Wang, L., *Spectral Methods: Algorithms, Analysis and Applications*, Springer Series in Computational Mathematics 41, Springer, Berlin, 2011. (Cited on p. 1.)

[136] Sherman, Y. J., On the numerators of the convergents of the Stieltjes continued fractions, *Trans. Amer. Math. Soc.* **35** (1933), 64–87. (Cited on p. 16.)

[137] Shohat, J. and Tamarkin, J., *The Problem of Moments*, AMS Mathematical Surveys, 1, American Mathematical Society, Providence, RI, 1943. (Cited on pp. 1, 18, 19, 105 and 164.)

[138] Steinsaltz, D. and Evans, S.N, Quasi stationary distributions for one-dimensional diffusions with killing, *Trans. Amer. Math. Soc.* **359** (12007), 1285–1324. (Cited on p. 319.)

[139] Stone, M. H., *Linear Transformations in Hilbert Spaces and their Applications to Analysis*, Colloquium Publications, vol. 15, American Mathematical Society, Providence, RI, 1932. (Cited on p. 244.)

[140] Stroock, D. W., *An Introduction to Markov Processes*, Graduate Texts in Mathematics, 230, Springer, Berlin, 2005. (Cited on p. 58.)

[141] Sumita, U. and Masuda, Y., On first passage time structure of random walks, *Stoch. Processes Appl.* **20** (1985), 133–147. (Cited on p. 82.)

[142] Szegö, G., *Orthogonal Polynomials*, AMS Colloquium Publications, vol. 23, American Mathematical Society, Providence, RI, 1939. (Cited on pp. 1, 86 and 88.)

[143] Vere-Jones, D., Geometric ergodicity in denumerable Markov chains, *Quart. J. Math. Oxford* **13** (1962), 7–28. (Cited on p. 83.)

[144] Watson, G. N., *A Treatise on the Theory of Bessel Functions*, Cambridge University Press, Cambridge, 1922. (Cited on p. 299.)

[145] Whitehurst, T., An application of orthogonal polynomials to random walks, *Pacific J. Math.* **99** (1982), 205–213. (Cited on p. 130.)

[146] Widder, D., *The Laplace Transform*, Princeton University Press, Princeton, NJ, 1941. (Cited on p. 6.)

[147] Wong, E. and Thomas, J. B., On polynomial expansions as second-order distributions, *J. Soc. Indust. Appl. Math.* **10** (1962), 507–516. (Cited on p. 303.)

[148] Wong, E., The construction of a class of stationary Markoff processes, in R. Bellman (ed.), *Stochastic Processes in Mathematical Physics and Engineering*, American Mathematical Society, Providence, RI, pp. 264–276, 1964. (Cited on pp. 279, 282, 308 and 310.)

[149] Wright, S., The genetical structure of populations, *Ann. Eugenics* **15** (1951), 323–354. (Cited on pp. 196 and 291.)

[150] Yoon, G. J., Darboux transforms and orthogonal polynomials, *Bull. Korean Math. Soc.* **39** (2002), 359–376. (Cited on p. 74.)

[151] Zettl, A., *Sturm–Liouville Theory*, AMS Mathematical Surveys and Monographs vol. 121, American Mathematical Society, Providence, RI, 2005. (Cited on p. 26.)

Index

Printed in the United States
by Baker & Taylor Publisher Services

III-V Integrated Circuit Fabrication Technology

Shiban Tiku
Dhrubes Biswas

Pan Stanford Publishing

Published by

Pan Stanford Publishing Pte. Ltd.
Penthouse Level, Suntec Tower 3
8 Temasek Boulevard
Singapore 038988

Email: editorial@panstanford.com
Web: www.panstanford.com

British Library Cataloguing-in-Publication Data
A catalogue record for this book is available from the British Library.

III–V Integrated Circuit Fabrication Technology

ISBN 978-981-4669-30-6 (Hardcover)
ISBN 978-981-4669-31-3 (eBook)

Printed in the USA

Contents

Preface

In this Internet age, practicing engineers still need a book that they can keep on their desk. This book is aimed for them and also graduate students and engineers new to the field of III–V semiconductor integrated circuit (IC) processing. This book specifically addresses the needs of students who know semiconductor theory but lack detailed processing knowledge. The content is chosen on the basis of the needs of students as seen by a teacher and the needs of practicing engineers dealing with processing issues as seen by an experienced process engineer. GaAs processing has reached a mature stage, a long way from a few decades ago, when it was more of an art than a science. New semiconductor compounds are emerging that will dominate future materials and device research; however, the processing techniques used for GaAs will still remain relevant. This book covers all aspects of the current state of the art of III–V processing, with emphasis on heterojunction bipolar transistors (HBTs), the volume leader technology, having grown due to the explosive growth of wireless technology. The book's primary purpose is to discuss processing; only necessary equations are derived and device behavior is discussed for the purpose of understanding device figures of merit and electrical parameters that engineers need to understand and control. All aspects of processing of active and passive devices, from crystal growth to backside processing, including lithography, etching, and film deposition, are covered. New material systems based on GaN are playing a larger role on the development side; although the etching chemistries, deposition materials, and temperature regimes are different, similar principles apply. The most promising structures of these material systems and devices are covered in the book.

The book covers semiconductor material basics, physics of devices used in semiconductor IC processing, and all the processing technologies used in III–V semiconductor fabrication. In the discussion, differences with silicon IC processes are emphasized. Crystal growth and particularly epitaxy are discussed in depth because of the special role played by them and device structures made possible by them. Photolithography, ion implantation, wet and plasma etching, and deposition of films are covered in detail. Thermal processes and diffusion are discussed to the level needed for III–V processing. Schottky and ohmic contact physics and processing are discussed from a practical point of view for controlling these in high-volume production. All the device technologies currently in use in the III–V semiconductor marketplace are discussed in depth, including recently introduced bipolar field-effect transistor (BiFET) and bipolar high-electron-mobility transistor (BiHEMT) technologies. Device types that are emerging and expected to be important in the near future, like metal–oxide–semiconductor field-effect transistors (MOSFETs), are also introduced. Passive devices and interconnects are covered, being integral to monolithic microwave integrated circuit (MMIC) fabrication. Also, backside processing, which is absolutely necessary for high efficiency, is described in detail and wafer-scale bumping is introduced, being critical to future higher-frequency needs. Characterization of films and semiconductor layers, as well as device parameter measurement, is covered in detail. Reliability issues relevant to III–V semiconductors are discussed. Finally, emerging GaN devices and microelectromechanical systems (MEMS) are briefly described.

Most published books on the market emphasize III–V device physics. No new processing book has been published in a decade. Published books are old and cover mostly FET processing. Ralph Williams's book, *Modern GaAs Processing*, is over 20 years old and does not cover processing technologies in detail. S. K. Ghandhi's book, *VLSI Fabrication Principles: Silicon and Gallium Arsenide*, covers processing techniques in detail but IC processing very briefly. This book is also old, published in the 1980s. Fazal Ali's book, *HEMTs and HBTs: Devices, Fabrication and Circuits*, covers fabrication very broadly and was also published in 1991, over 20 years ago. Baca and Ashby's book, *Fabrication of GaAs Devices*, has a narrow focus, specializing in cleaning and passivation; basic IC processing techniques are not covered. It was published in 2005, a decade old now.

This (present) book covers all aspects of processing, from crystal growth to backside processing. It covers the current volume production device types, HBTs, HEMTs, etc. The book is not restricted to GaAs; other emerging III–V materials are covered, too.

Epigrowth, device structure, and processing discussions are connected together through different chapters. Processing techniques relevant to III–V IC fabrication are described as they are used in III–V processing facilities in high-volume production. Process flows are illustrated by step-by-step block diagrams. Scanning electron microscopy (SEM) pictures of actual devices are included, where needed. This is one book to find any topic relevant to III–V processing. Practical process problems and ways to handle these are described.

The current understanding of III–V processing has come a long way from the era when GaAs processing was based on practical knowledge and company trade secrets. This book attempts to connect practice on the fabrication floor to current scientific understanding.

Shiban Tiku

Dhrubes Biswas

Acknowledgments

The earlier books on GaAs and III–V semiconductor materials processing were written in an era when GaAs was considered the "technology of the future." This present work, although inspired by those, is aimed at fulfilling the needs of this era in which the technology is well established and perhaps becoming the "technology of the past," while paving the way for future technologies. Many minds and hands have contributed to this work. I am indebted to all my teachers over the years, who left an indelible mark on my life. I also thank the people of India for the almost free education I received.

This book would not have been possible without support from Skyworks Solutions' management, Ravi Ramanathan, Nercy Ebrahimi, and Andy Hunt, and IP council Donald Bollella, who weighed the benefits of contributing to the III–V industry worldwide over the risk of disclosing trade secrets. A lot of data came from my fellow engineers at Skyworks and the CS MANTECH community in general. Early feedback from Martin Brophy (Avago) and Peter Asbeck (UCSD) encouraged me. In particular, help from the following Skyworks colleagues is acknowledged: Heather Knoedler, Jens Riege, Dave Crawford, Ravi Ramanathan, Mike Sun, Jiro Yota, Pete Zampardi, Lance Rushing, Cristian Cismaru, Sam Mony, Lam Luu, and Manjeet Singh. Constant support from my wife, Sushma, and son, Vikram, helped me during difficult times. I am also thankful to Archana Ziradkar of Pan Stanford Publishing for systematic editorial help and to Barron Miller for drafting of many of the figures.

Shiban Tiku

Chapter 1

Semiconductor Basics

1.1 Introduction

Gallium arsenide as a semiconductor material was originally investigated because of its superior electronic transport properties and other related advantageous material properties. Electron mobility in GaAs and other III–V compounds is higher than in silicon and these materials are useful in optical devices because of the nature of their band structure. GaAs can be made in semi-insulating form and this makes it possible to make monolithic circuits with ease on GaAs substrates. Radiation hardness was another driving force behind the original funding of research by the government and the defense industry. However, silicon devices have made tremendous progress in speed and complexity and this makes it difficult for GaAs circuits to compete in digital circuits. However, because of the simplicity and shortness of the GaAs integrated circuit (IC) fabrication process and the fact that high power and speed can be achieved simultaneously, GaAs circuits have established a niche in analog ICs. Future possibilities of combining optical and electronic functions into single chips and the possibility of combining superfast III–V compounds into silicon substrates keep the interest in III–V electronics alive.

Silicon and germanium are elemental semiconductors. Silicon happens to have properties that are well suited to large-scale

III–V Integrated Circuit Fabrication Technology
Shiban Tiku and Dhrubes Biswas

ISBN 978-981-4669-30-6 (Hardcover), 978-981-4669-31-3 (eBook)
www.panstanford.com

integration. GaAs is a compound and its properties make it difficult to process it. Loss of arsenic due to dissociation and lack of a good native oxide make it less attractive from the device processing point of view. Good insulating behavior more or less makes up for the problems of lower thermal conductivity. Low hole mobility in GaAs leads to slower p-type channel devices. However, deft processing innovations have turned some of these disadvantages into advantages. The lack of oxide has been addressed by use of Schottky gates, simplifying the process and helping radiation hardness. Enhancement and depletion mode field-effect transistor (FET) circuits have been designed and built. III–V devices are commonly made using epitaxial wafers and the advances in epitaxial growth have led to devices not otherwise possible with diffusion-dominated silicon-like processing.

Defense and space applications were the reason for funding of GaAs process development by the government and the defense industry. As the industry matured, emphasis shifted to commercial products like front-end receivers. Emphasis shifted from digital to mixed-signal and analog circuits with the advent of the wireless era. Use of high frequencies to avoid spectrum crowding, new modulation, and channel division techniques needing linear amplifiers finally established a niche for III–V semiconductors that cannot be filled by silicon-based systems.

GaAs is a direct bandgap material and is used for fabricating optical devices like light-emitting diodes (LEDs) and lasers. Some of the processing methods, from crystal growth to packaging, are common to IC fabrication and thus drive the technology. Despite the recent downturn of the telecommunications industry (2005), the market for high-frequency devices operating near 40 GHz is bound to make a comeback.

1.1.1 GaAs Device Applications

Here are a few applications of GaAs devices:

- GaAs metal semiconductor FETs (MESFETs) and epi-FETs: Front-end receiver (FER) gate arrays, low-noise amplifier (LNA), X and Ku band applications
- GaAs pseudomorphic high-electron-mobility transistors (PHEMTs): Power amplifier (PA) switches, low-noise amplifiers

- GaAs heterojunction bipolar transistors (HBTs): PA, prescalers, multiplexer (mux), demultiplexer (demux), A to D converters
- InP HBTs: 40 GHz optical applications
- InP HEMTs: 300 GHz
- GaN diodes, FETs: Power conversion
- GaN HEMTs: High-power amplifiers

1.2 GaAs Crystal Structure [1, 2]

GaAs is formed by combining group III gallium with group V arsenic to form a single-crystalline semiconductor compound. Solid materials can be classified into three broad categories: amorphous, polycrystalline, and crystalline. Most semiconductor materials are single crystalline, although some practical applications require amorphous or polycrystalline material because of form requirements like flexibility and cost. In amorphous solids, there is no geometrical regularity or periodicity. The atoms are randomly distributed without any long- or short-range order and the bonding to neighbors is not uniform, although the solid is tightly bound. In a crystalline material, the structure has perfect order and the periodicity extends to the edges of the solid, with only a few imperfections or impurities in the whole solid. Polycrystalline solids are in between these two in structure. Smaller single-crystal grains are spread through the solid in random order, with grain boundaries in between (Fig. 1.1).

The basic building block of a crystalline solid is called a unit cell. Figure 1.2 shows the unit cell for a cubic solid. The unit cell for most semiconductors is of the face-centered cubic (fcc) type, where the unit cell has one atom on the corners of the cube and one atom in the center of each face. III–V compounds like GaAs have the zinc-blende structure, which can be regarded as two interpenetrating fcc lattices (Fig. 1.3), one of Ga and the other of As. For silicon, which has a diamond structure, the two sublattices are identical. The lattice constant is defined as the distance between the corners of the unit cell. It can be seen in Fig. 1.3 that each Ga or As has four neighbor atoms forming a tetrahedron. The unit cell size of GaAs is 5.65 Å. Crystalline and few other properties of GaAs are compared to silicon, listed in Table 1.1.

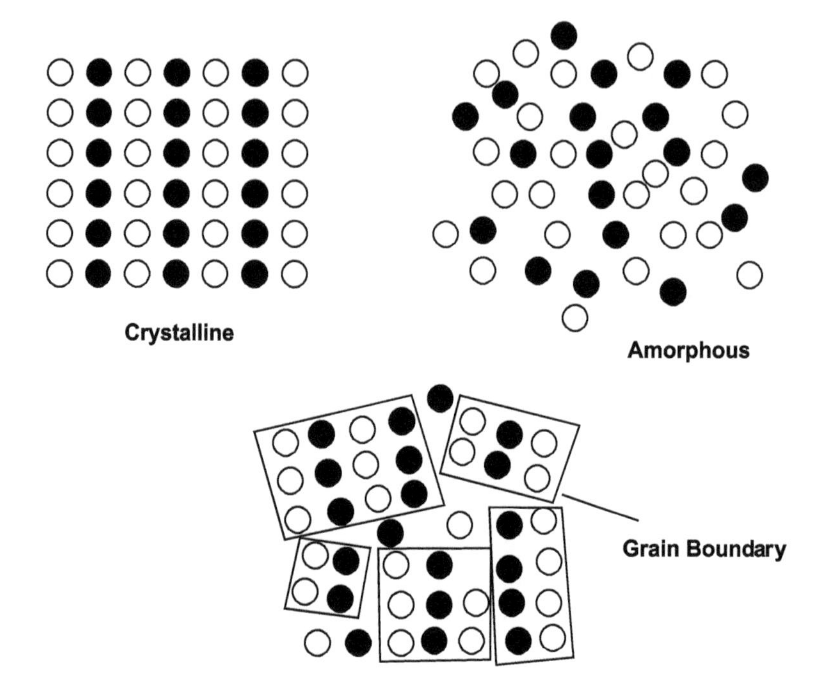

Figure 1.1 Simplified 2D representations of crystalline, amorphous, and polycrystalline solids.

Table 1.1 Comparison of silicon and GaAs

	GaAs	Si
Crystal structure	Zinc blende	Diamond
Lattice constant	5.646	5.431 Å
Density	5.32	2.328 g/cm^3
Melting point	1238	1412°C
Thermal expansion coefficient	6.86	2.6 (10^{-6} @ 300K)
Thermal conductivity	0.46	1.5 W/cm-°C
Energy gap	1.435	1.1 eV
Dielectric constant	13.1	11.9
Intrinsic carrier concentration	8×10^6	1.45×10^{10}/cm^3

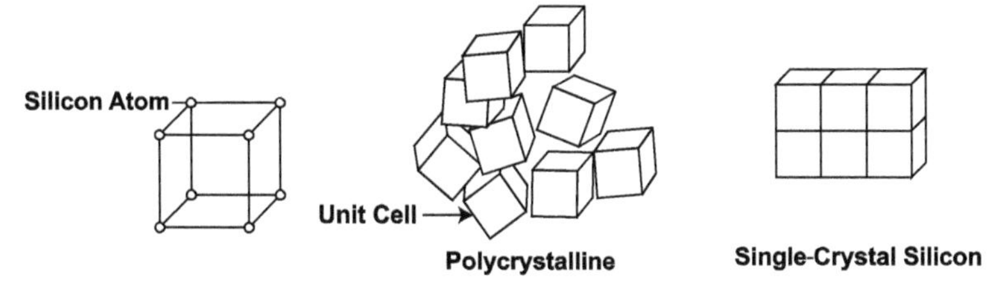

Figure 1.2 Arrangement of unit cells in single-crystal and polycrystalline materials.

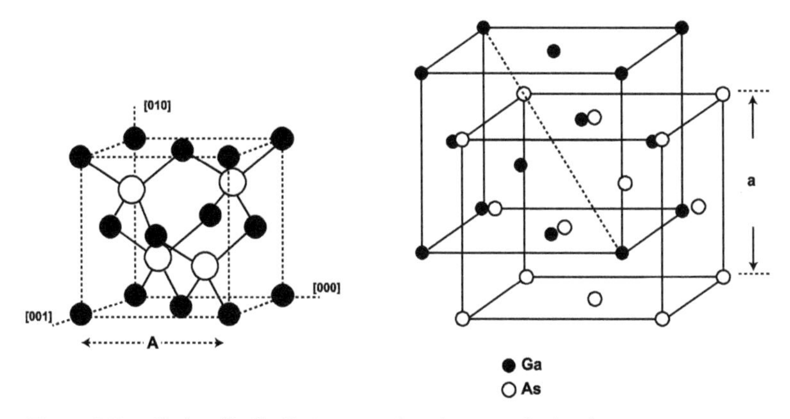

Figure 1.3 Unit cell of a GaAs crystal and as two fcc lattices.

Crystal growth and etching behavior of solids can be better explained if the structure in different directions and along different planes is well understood. Figures 1.4 and 1.5 show the zinc-blende unit cell truncated along the face diagonal and the body diagonal. The terminology of Miller indices is used to describe directions at planes within the crystal. A set of three integers enclosed in square brackets is used to specify direction in the lattice. [abc] defines a direction whose vector is **ax^** + **by^** + **cz^**, where **x^**, **y^**, and **z^** are unit Cartesian vectors along *x*, *y*, and *z*. Surfaces perpendicular to [abc] are designated as (abc). Some common crystalline directions and planes are shown in Fig. 1.4.

Also, <abc> indicates a family of [abc] directions that are equivalent. {abc} indicates a family of planes equivalent to (abc).

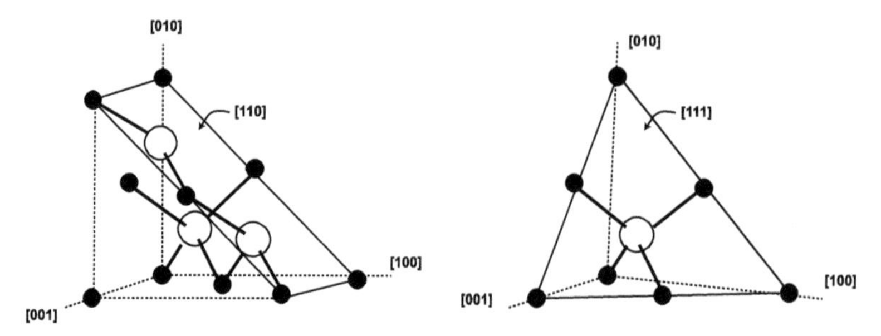

Figure 1.4 Truncation of a GaAs unit cube by the (110) plane and the (111) plane.

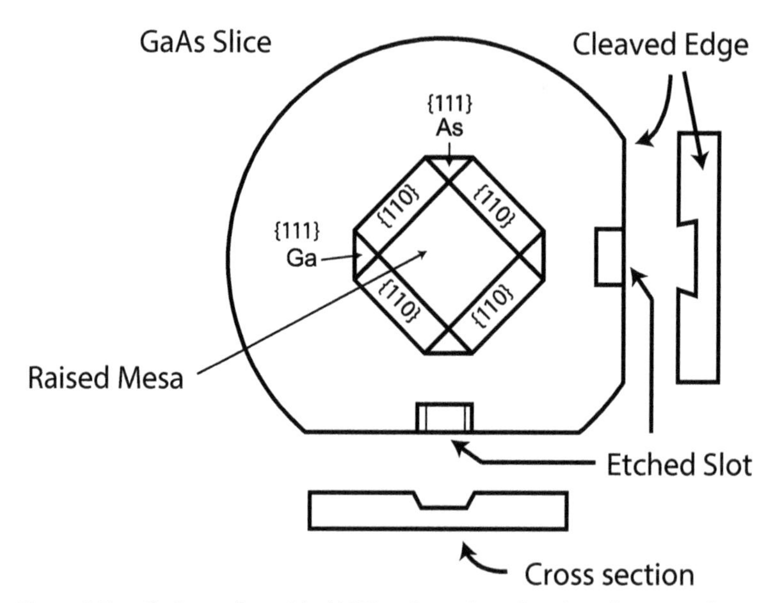

Figure 1.5 GaAs wafer with (100) orientation showing cleavage planes and anisotropy in etch cross sections.

Single-crystal boules of semiconductors are grown and then sliced into wafers for circuit fabrication along certain orientations. The most commonly used wafer orientation for GaAs is (100). The {111} family of planes contains only one type of atoms, either Ga or As. The letter A or B is attached to the plane family designation to denote Ga or As planes. {111}A contains only Ga atoms and {111}B contains only As atoms. This distinction is important for understanding etching and other directional properties of III–V semiconductors. GaAs wafers can be easily cleaved or broken into a die by scribing along the crystal orientations. Figure 1.5 shows the top view and cross sections of a wafer oriented along the (100) plane [7]. The two cross sections show different behaviors in the two perpendicular directions due to atomic density differences, as shown in Fig. 1.6, which will be discussed further in Chapter 6 on wet etching. The density of atoms in different directions is also important for epigrowth and ion implantation.

Other properties of GaAs that are useful for circuit fabrication are listed in Table 1.2. It should be noted that while the thermal conductivity of GaAs is lower than that of silicon, the thermal expansion coefficient is higher for the former.

Table 1.2 Room temperature properties of GaAs

Property	Parameter
Crystal structure	Zinc blende
Lattice constant	5.65 Å
Density	5.32 g/cm^3
Atomic density	4.5×10^{22} atoms/cm^3
Molecular weight	144.64
Bulk modulus	7.55×10^{11} dyn/cm^2
Sheer modulus	3.26×10^{11} dyn/cm^2
Coefficient of thermal expansion	5.8×10^{-6} K^{-1}
Specific heat	0.327 J/g-K
Lattice thermal conductivity	0.55 W/cm-°C
Dielectric constant	12.85
Bandgap	1.42 eV
Threshold field	3.3 kV/cm
Peak drift velocity	2.1×10^7 cm/sec
Electron mobility (undoped)	8500 cm^2/V-sec
Hole mobility (undoped)	400 cm^2/V-sec
Melting point	1238°C

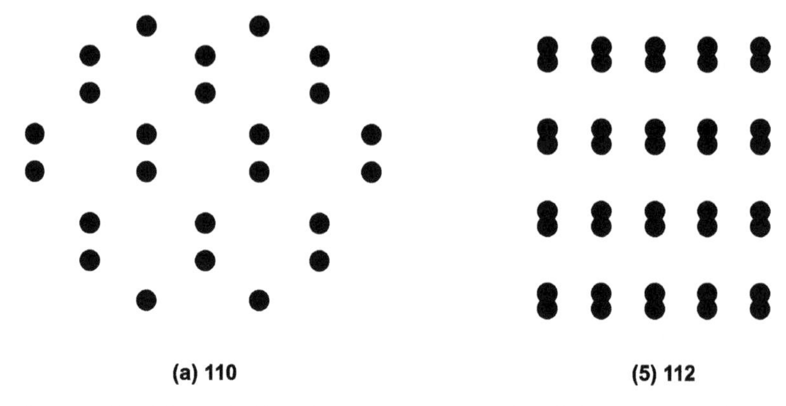

Figure 1.6 View of a GaAs lattice from two different directions (a) along the [110] axis and (b) along the [112] axis.

1.3 Bonding in III–V Semiconductors

Elemental semiconductors are held together by covalent bonds, in which valence electrons are shared by neighboring atoms. Insulating materials are generally ionic crystals, where the bonding is due to attraction between ions; in NaCl the ions are created by the transfer of an electron from Na to Cl. In III–V compounds, the bonding is mostly covalent but does have considerable ionic character. The ionic nature goes up for II–VI semiconductors. Figure 1.7 shows a schematic bonding diagram [9]. All Ga and As atoms have eight shared electrons surrounding them (five from As and three from Ga). At higher temperatures thermal energy excites some electrons into higher energy states where they are free to move around as carriers. This free-electron concentration, which goes up with temperature, is called intrinsic carrier concentration. This concentration is very small, of the order of $10^{10}/cm^3$ at room temperature (compared to $10^{23}/cm^3$ atomic concentration).

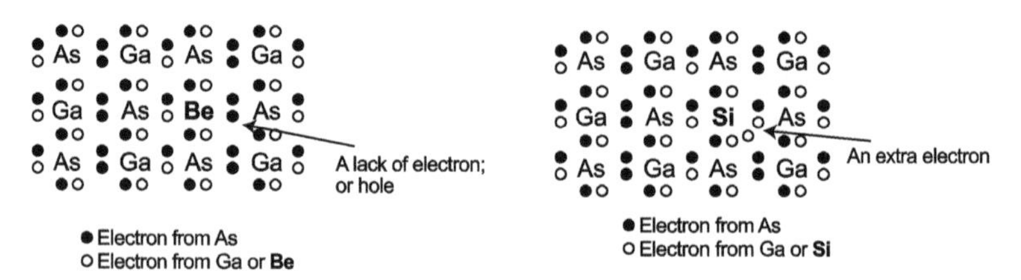

Figure 1.7 Atomic bonding in (a) p-type Be-doped GaAs and (b) n-type Si-doped GaAs.

1.3.1 Bonding in a Doped Crystal

Large concentrations of carriers are introduced in a controlled manner in semiconductors by using doping techniques. (Doping methods will be discussed in detail elsewhere, Chapters 3 and 4, on crystal growth and epitaxy, respectively). If a silicon atom is placed in place of a gallium atom on a GaAs lattice, the extra electron from the outer shell of Si is relatively free to move around at ordinary temperatures. This gives rise to an n-type semiconductor.

On the other hand, if a Ga atom is replaced by a Be atom, which has only two electrons in the outer shell, it results in a missing electron or a hole. Electrons from a neighboring atom can jump into this hole and thus the hole can move around. This creates a p-type semiconductor, where the conduction is said to take place by movement of holes. The dynamics and mobility of charge carriers vary with material and crystal structure and can be better discussed after the introduction of band structure.

Only necessary details will be given here for understanding of the material that will be presented in this book. For details of semiconductor physics, refer to solid-state and semiconductor physics textbooks [1, 2].

1.4 Energy Band Structure [4]

Electrons in free space can have a continuous range of energies. In an isolated atom, electrons can have only discrete energy values, which can be determined by quantum mechanics. As atoms are brought close together to form molecules and crystals, the energy levels get split into bands of energies (see Fig. 1.8). The Pauli exclusion principle is still followed—no two electrons can occupy the same quantum state. The bands of interest in semiconductors are the ones formed by the outer shells of electrons. These are called the valence and the conduction band and are separated by the energy bandgap. The size of the gap determines if a material is an insulator, a semiconductor, or a conductor (Fig. 1.9). In a semiconductor at 0 K temperature, the electrons are confined to the valence band and the material behaves as an insulator. At higher temperature, the same electrons have sufficient thermal energy to make a transition to the conduction band, where they are free to move and carry a current.

The probability of an electron having enough energy to make the transition is given by the Fermi distribution function. Figure 1.10 shows the Fermi level within the bandgap of a semiconductor. The Fermi level, E_f, is defined as the energy at which the probability function is equal to one-half. For intrinsic semiconductors, E_f is at the center of the gap. For doped crystals, n- or p-type extrinsic semiconductors, the Fermi level is near the conduction band or the valence band. The energy band diagram is referenced to a potential called the vacuum potential. The electron affinity, χ, is the energy required to excite the electron from the conduction band to the vacuum level (Fig. 1.10).

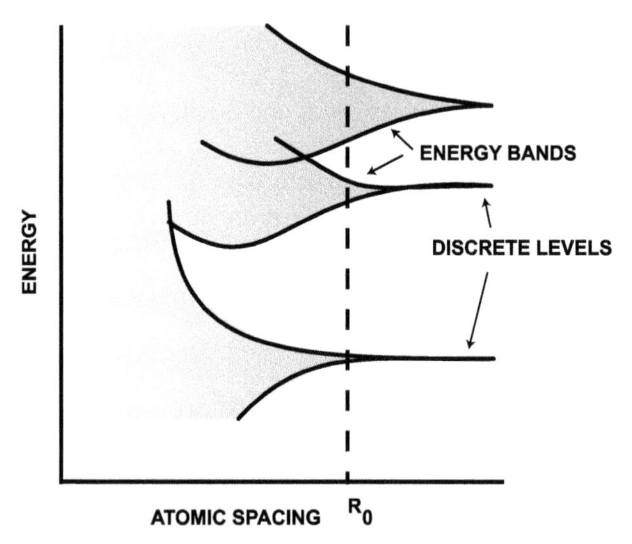

Figure 1.8 Energy band diagram showing creation of bands as discrete atoms come together to form a solid.

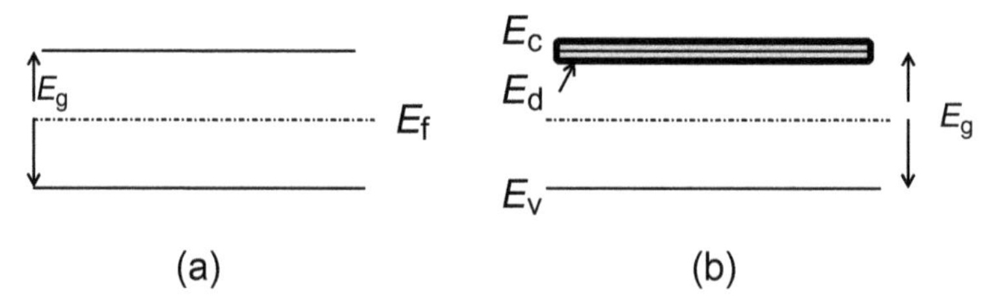

Figure 1.9 Energy band diagram of undoped (a) and heavily n-type doped semiconductor (b).

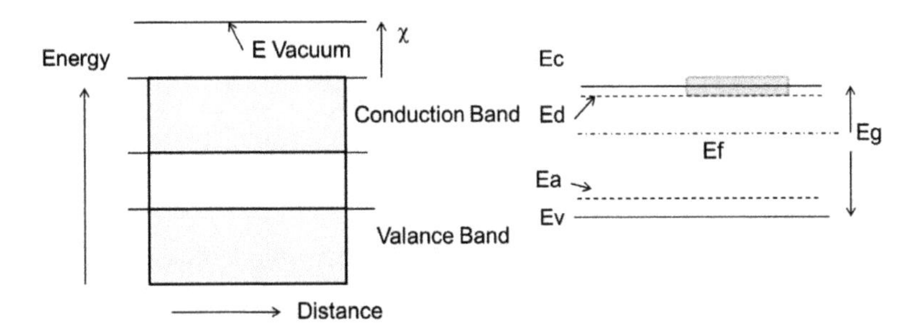

Figure 1.10 Energy band diagrams of a semiconductor showing electron affinity and Fermi level.

The energy band diagrams of three types of materials are shown in Fig. 1.11. In insulators, the magnitude of the gap is of the order of 5 eV and above. In metals, the conduction and valence bands overlap and or are partially filled, so electrons can move freely into other states. Since there is no gap, the number of electrons is large and conductivity is high. In semiconductors, the situation is between these—the gap is small. As mentioned earlier, just as the effective mass of electrons varies depending upon the crystal direction, the bandgap also varies. Band diagrams for GaAs and Si are shown in Fig. 1.12 along [111] and [100] crystal directions. In GaAs, a direct-bandgap material, the lowest gap is seen to be lowest at $k = 0$. In Si the lowest gap is along [100], and the gap is indirect. The details of the band structure are very important for the understanding of optical and electronic devices.

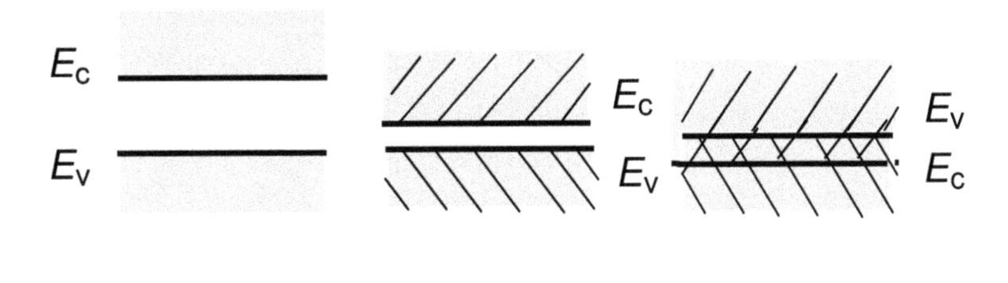

Figure 1.11 Energy band diagrams of three types of materials.

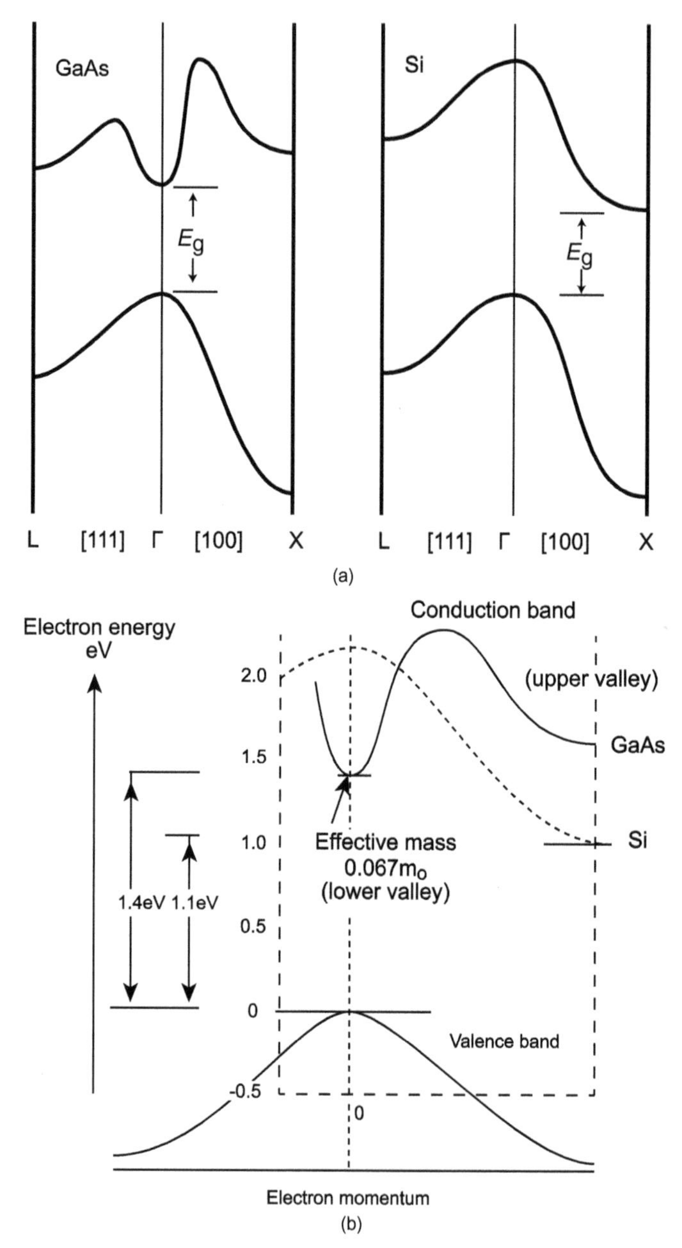

Figure 1.12 (a) Energy band diagrams along two crystal directions for GaAs (left) and Si (right). (b) Comparison of energy bandgaps of Si and GaAs along [100]; the effective electron mass in the lower valley of the GaAs conduction band is also shown.

1.4.1 Band Structure and Mobility

When an electron moves in free space its speed or momentum is determined by the applied field. In a crystal, an electron also encounters the periodic potential of the atoms, which varies along different directions in the crystal. An easy approach to deal with this complex problem is to assume the electron to have an effective mass m_e, which differs from the mass in free space. The kinetic energy of the electron, E_k, is given by

$$E_k = \frac{P^2}{2m_e} \tag{1.1}$$

where P is the electron's momentum.

The effective masses of electrons (and holes) can be different in different semiconducting materials. In GaAs, the electron wave is accelerated with respect to the lattice due to the applied field, and the effective mass is 0.067 m_e, whereas the holes are decelerated, or the holes are heavy. Electron velocities in devices based on n-type GaAs are higher and result in a better high-frequency response. The energy band structures of GaAs and Si are shown in Fig. 1.12b. Drift velocity of electrons in GaAs is shown in Fig. 1.13 [13]. Peak mobility

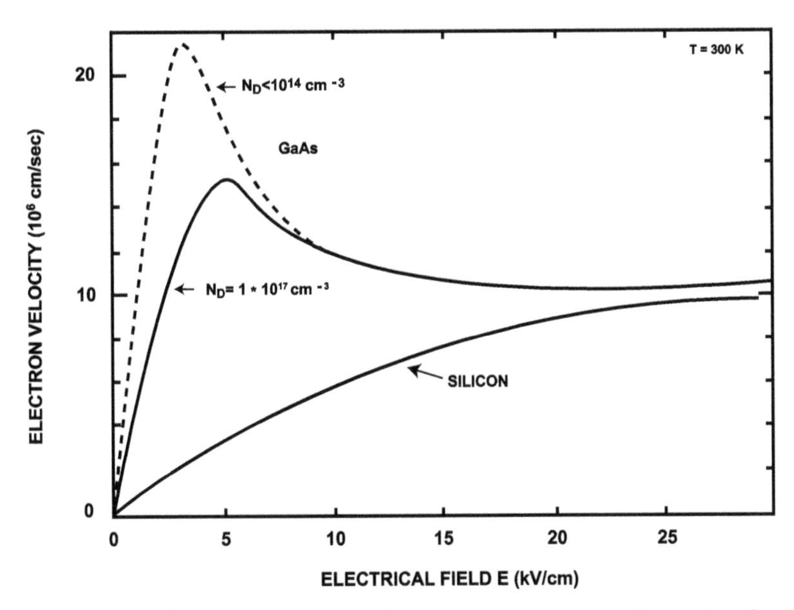

Figure 1.13 Electron velocity as a function of electric field, illustrating the mobility differences between silicon and GaAs (at two different doping levels).

of GaAs in the linear region can be about six times greater than silicon. At typical fields may be a factor of 2 higher. Electron mobility is also influenced by impurity scattering in doped semiconductors and will be discussed further under HEMTs.

1.4.2 Free Carrier Concentration and Fermi Level [4, 11]

The concentration of electrons and holes in a semiconductor is determined by the distribution of electrons in the valence and conduction bands and the concentration of donors and acceptors and the location of their levels in the energy bandgap. The Pauli exclusion principle leads to the Fermi–Dirac distribution (Fig. 1.14). Figure

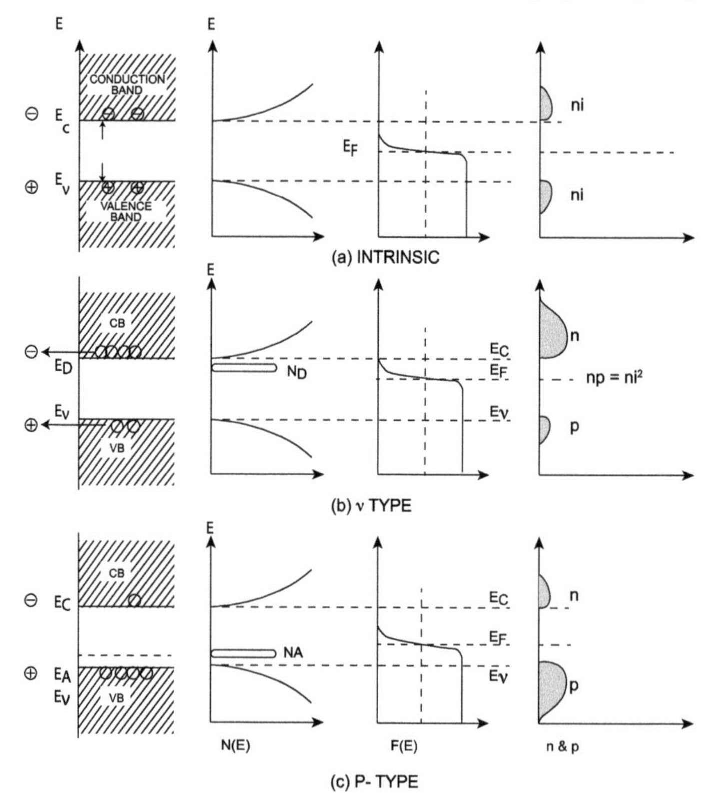

Figure 1.14 Schematic diagram for an intrinsic (a), n-type (b), and p-type (c) semiconductor, showing a band diagram, density of states, Fermi–Dirac distribution, and carrier concentration. From Ref. [4]. Copyright © 1985. Reproduced with permission of John Wiley & Sons, Inc.

1.15a shows the *E–k* diagram for electrons in a semiconductor in which the band is almost empty, with its bottom near levels that are full, like a donor. Figure 1.15b shows the corresponding diagram for a band that is full, the top of which is near energy levels of acceptors, as in a p-type semiconductor. The shape of the energy curve varies with the direction in a crystal lattice. In the diagram, the *x* direction is shown. In the following discussion, electron concentrations are supposed to be low (nondegenerate), and equilibrium is assumed.

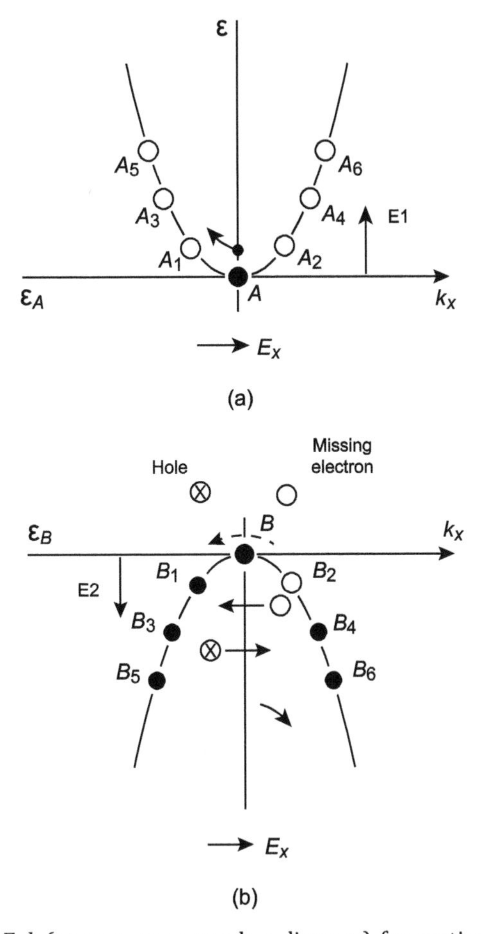

Figure 1.15 *E–k* (energy–wave number diagram) for motion of electrons in the conduction band (a), which is almost empty, and in the valance band (b), which is almost full, except for some holes. From Wang, *Fundamentals of Semiconductor Theory & Device Physics*, 1st Ed. ©1989 Printed and reproduced by permission of Pearson Education, Inc., New York.

The number of electrons is calculated by integrating the product of $N(E)$, the density of states, and the electron distribution over the whole energy range. In the case of electrons in the conduction band, that is, from E_c to infinity (∞)

$$n = \int_{E_c}^{\infty} N(E)F(E)dE \tag{1.2}$$

The electron concentration is given by the Fermi–Dirac distribution function.

The electron distribution function

$$Fe(E) \approx \exp\left\{-\frac{E_c - E_f + E}{kT}\right\} \tag{1.3}$$

In the case of the conduction band, the electrons are mostly near the bottom of the conduction band (see Fig. 1.15 for E_1 and E_2 definitions).

$$E_1 = E - E_c = \frac{h^2k^2}{2m_e} \tag{1.4}$$

where m_e is the effective mass of electrons.

For holes, similarly (see Fig. 1.15b)

$$E = E_v - E_2 = E_v - \frac{h^2k^2}{2m_h} \tag{1.5}$$

For holes the distribution represents the probability that a valence band state with energy E is vacant.

$$1 - F_p(E) = \exp\left\{-\frac{E_f - E}{kT}\right\}$$

$$1/\{1 + \exp(E_f - E)/kT\} \cong \exp\{-(E_f - E_v + E_2)/kT\} \tag{1.6}$$

$$\exp\left\{-\frac{E_f - E_v + E_2}{kT}\right\}$$

where $E = E_v - E_2$.

On the basis of the above equations and assumptions, the electron concentration can be shown to be (for full derivation see Ref. [4] or [11])

$$n = N_c \left\{-\frac{E_c - E_f}{kT}\right\} \tag{1.7}$$

where $N_c = 2\{2\pi m_e^* kT/h^2\}^{3/2}$

And similarly for holes

$$p = N_v \exp\left\{-\frac{E_f - E_v}{kT}\right\} \tag{1.8}$$

where $N_v = 2\{2\pi m_h kT/h^2\}^{3/2}$.

The product pn is given by

$$pn = N_c N_v \exp\left\{-\frac{E_c - E_v}{kT}\right\} \tag{1.9}$$

In a pure semiconductor, where NA = ND = 0, electrons and holes are created by thermal excitation, so $n = p$. Using Eqs. 1.7 and 1.8

$$n = p = n_i \tag{1.10}$$

$$n_i = \sqrt{(N_c N_v).\exp\left\{\frac{-Eg}{2kT}\right\}} \tag{1.11}$$

where $E_c - E_v = E_g$ is the bandgap.

Also from Eqs. 1.7, 1.8, and 1.10

$$E_f = E_i = \frac{E_c + E_v}{2} + \frac{3kT}{4} \ln \frac{m_e}{m_h} \tag{1.12}$$

The Fermi level is close to the center of the energy bandgap, the deviation being a function of the ratio of the effective mass of holes and electrons.

1.4.3 Energy Levels in Doped Semiconductors

As described earlier, donor or acceptor levels are introduced into a pure semiconductor to make these n or p type. These levels are close to the conduction band or valence band. A donor atom in GaAs, like Si (see Fig. 1.7b) is neutral when filled with its electron but has a positive charge when empty. Similarly a Be atom is neutral when empty but has a negative charge when it picks up an electron. The donor atom can be considered as a hydrogen atom, so using the hydrogen atom model

$$E_H = \frac{m_0 e^4}{32\pi^2 \varepsilon_0^2 h^2} = 13.6 \text{ eV} \tag{1.13}$$

where ε_0 is the permittivity of free space and m_0 is the electron mass. The donor level in a semiconductor can be derived to be

$$\frac{m_{\text{ce}}}{m_0}Ed = \left(\frac{\varepsilon_0}{\varepsilon_s}\right)^2 E_{\text{H}} \tag{1.14}$$

where m_{ce} is the effective mass of donor electrons. For GaAs this works out to E_{d} = 0.007 eV. This number is close to the measured values of shallow donors. In practice, Shubnikov–de-Hass and cyclotron resonance measurements are used to calculate values of effective mass.

When a dopant is introduced, the Fermi level must adjust to maintain charge neutrality:

$$n = N_{\text{d}}^{+} + p$$

Therefore, from Eqs. 1.7 and 1.8

$$N_{\text{c}} \exp\left\{-\frac{E_{\text{c}} - E_{\text{f}}}{kT}\right\} = \frac{N_{\text{D}}}{1 + 2\exp\frac{E_{\text{f}} - E_{\text{c}}}{kT}} + N\text{vexp}\left\{-\frac{E_{\text{f}} - E_{\text{v}}}{kT}\right\} \tag{1.15}$$

Here N_{D}^{+}, the number of ionized donors, is given by [4]

$$N_{\text{D}}^{+} = \frac{N_{\text{D}}}{1 + 2\exp\frac{E_{\text{f}} - E_{\text{D}}}{kT}} \tag{1.16}$$

The Fermi level can be determined from the above equation (numerically) using values of N_{c}, N_{D}, N_{v}, E_{c}, E_{d}, E_{v}, and temperature.

The electron concentration can be derived as

$$n \cong \frac{N_{\text{D}} - N_{\text{A}}}{2N_{\text{A}}} N_{\text{C}} \exp\left\{-\frac{E_4}{kT}\right\} \tag{1.17}$$

where $E_{\text{d}} = E_{\text{c}} - E_{\text{D}}$.

For $N_{\text{D}} >> N_{\text{A}}$

$$n \cong \sqrt{\frac{N_{\text{C}} N_{\text{D}}}{2}} \exp\left\{-\frac{E_{\text{D}}}{2kT}\right\} \tag{1.18}$$

For $N_{\text{D}} >> ½ N_{\text{c}} \exp(-E_{\text{d}}/kT) >> N_{\text{A}}$.

This function for n is plotted for Si in Fig. 1.16a [11]. Only within a certain temperature range, the electron density is equal to the donor concentration, $n \cong N_D$. At high temperatures, the intrinsic electrons dominate, and at low temperatures, carrier freeze-out takes place.

The concentration of electrons in an n-type material in equilibrium is

$$n_{n0} \cong N_D$$

From Eq. 1.11 for n_i above

$$np = n_i^2 = \exp\left\{-\frac{E_g}{kT}\right\} \tag{1.19}$$

and from charge neutrality $n + N_A = p + N_D$.

Hole concentration in an n-type material at equilibrium is

$$p_{n0} \cong n_i^2/N_D \tag{1.20}$$

Useful equations for doped semiconductors for the position of the Fermi level and carrier concentrations are listed below:

$$E_c - E_f = kT \ln\left(\frac{N_C}{N_D}\right) \tag{1.21}$$

$$E_f = (E_c + E_v)/2 \tag{1.22}$$

And using Eq. 1.10 again

$$E_f - E_i = \frac{kT}{e} \ln\left(\frac{n_{n0}}{n_i}\right) \tag{1.23}$$

Similarly for p-type materials

$$p_{p0} \cong N_A \tag{1.24}$$

$$n_{p0} = n_i^2/p_{p0} = n_i^2/N_A \tag{1.25}$$

$$E_f - E_v = kT \ln \frac{N_v}{N_A} \tag{1.26}$$

$$E_i - E_f = \frac{kT}{e} \ln\left(\frac{p_{p0}}{n_i}\right) \tag{1.27}$$

These equations are useful for discussion of semiconductor junction behavior, etc., in the chapters that follow. Figure 1.16b shows the position of the Fermi level for GaAs as a function of temperature for various n- and p-type doping levels, depending on the above equations. Note that at higher temperatures and low doping levels the Fermi level approaches the energy band center.

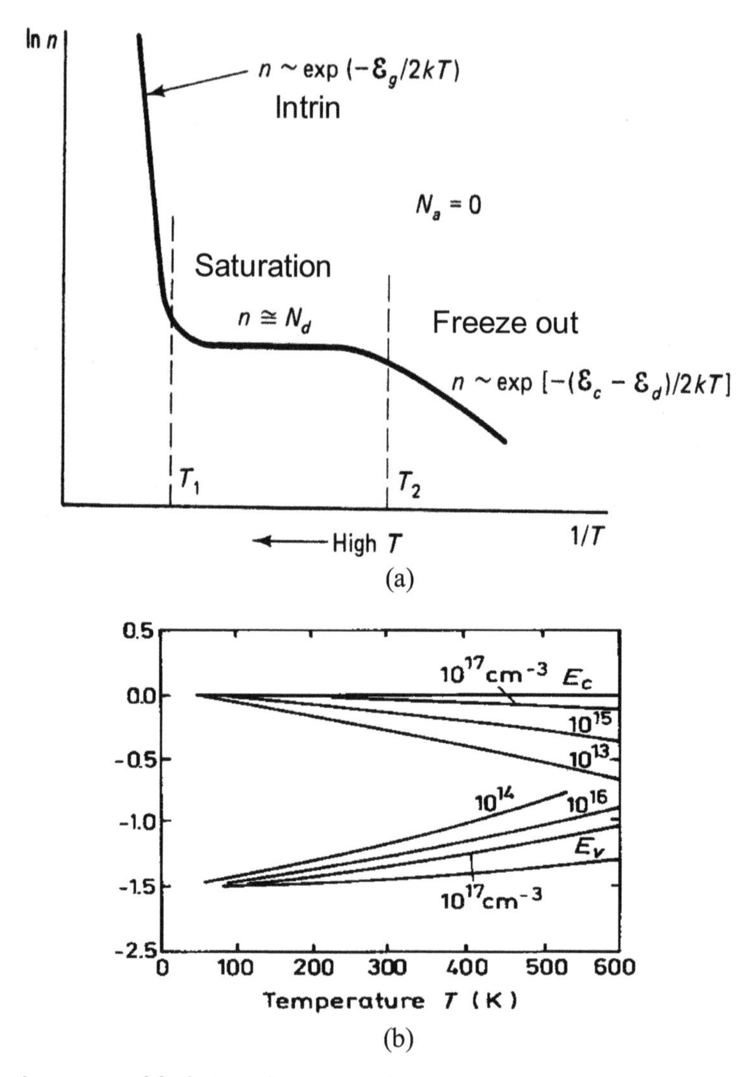

Figure 1.16 (a) Electron density as a function of temperature for an n-type semiconductor. (b) Fermi level position for GaAs as a function of temperature (http://www.ioffe.ru/SVA/NSM/Semicond/GaAs/bandstr.html).

1.4.4 Impurities in GaAs

As mentioned earlier, n-type extrinsic semiconductors are produced by incorporating donors into the lattice. These donors introduce excess electrons that are freed up with small energy excitation, so the energy states introduced by these (E_d) are very close to the conduction band, only a few kT below. See Fig. 1.17a, where shallow dopants are defined as those within 3 kT. Conversely acceptors introduce states (E_a) near the valence band. It is easy for a valence band electron to jump into one of these states and create a hole. Thus E_a states are close to the valence band. If the doping concentrations are high, the states turn into a narrow band of states and the energy gap becomes very narrow. Carrier freeze-out at low temperatures disappears under such conditions. Positions of common impurities of interest in GaAs are shown in Fig. 1.17b.

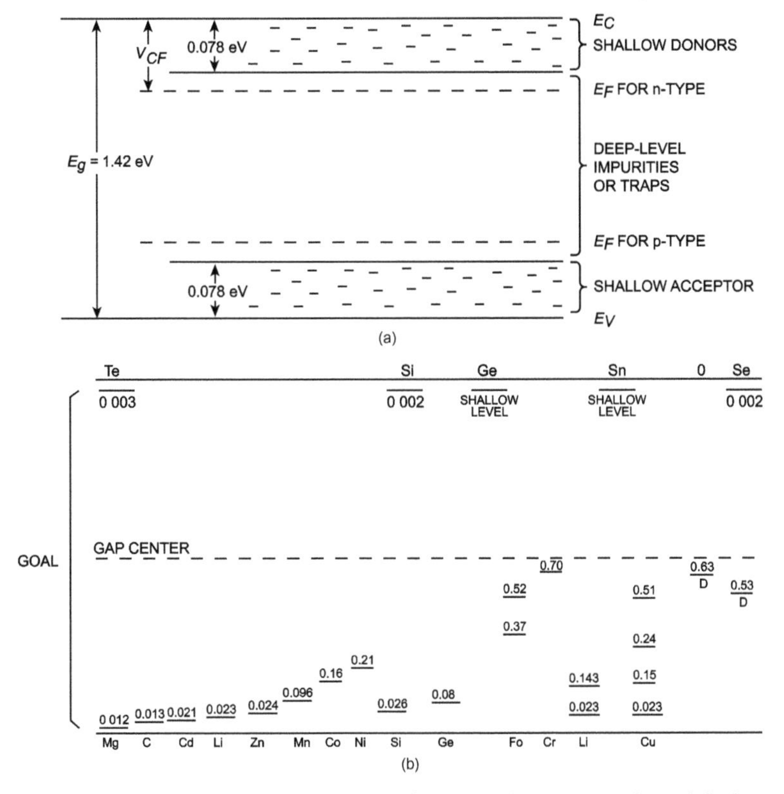

Figure 1.17 (a) Energy band diagram of GaAs with impurities (simplified to show only the crystal lattice center). (b) Measured ionization energies of common impurities of interest in GaAs. From Ref. [4]. Copyright © 1985. Reproduced with permission of John Wiley & Sons, Inc.

Group VI elements, like S, Se, and Te, are incorporated substitutionally on the As sublattice. Having one more electron than As, these act as n-type donors and contribute one electron to the conduction band. These impurities are shallow in the energy gap, as shown in Fig. 1.17b and listed in Table 1.3. Group IV impurities Ge, Si, Sn, etc., are incorporated substitutionally, mostly on the Ga sublattice, but are amphoteric like carbon and do go on the As sublattice. The net free n-type carrier concentration depends on the compensation on the As sublattice, which further depends upon the temperature of processing. An example of the effect of temperature on Si activation in GaAs is discussed under ion implant activation in Chapter 9 on ion implantation and activation.

Table 1.3 Ionization energy of shallow impurities in GaAs

		Ionization energy (eV)	
Impurity	**Type**	**From conduction band**	**From valence band**
S	n	0.0061	
Se	n	0.0059	
Te	n	0.0058	
Sn	n	0.0060	
C	n/p	0.0060	~0.026
Ge	n/p	0.0061	0.040
Si	n/p	0.0058	~0.035
Cd	p		0.035
Zn	p		0.031
Be	p		0.028
Mg	p		0.028
Li	p		0.023, 0.05

Source: From Ref. [10]. Copyright © 1983. Reproduced with permission of John Wiley & Sons, Inc.

1.4.4.1 Specific impurities

- Silicon: Si is always present in III–V starting materials and crystals due to the fact that quartz or silica is used for processing. Silicon is commonly used as an n-type dopant. As discussed above, this is an amphoteric dopant, so activation efficiency depends upon how the Si is incorporated.

- Se and Te: Se is a good dopant because it goes only on the As site, so the carrier concentration can be high after ion implantation. Te is heavier and is not a good candidate for implantation; however, it is a good dopant for heavily doped n-type epigrown ternary layers.
- Tin, Sn: This is always n type and was used for liquid-phase epitaxy (LPE) and for diffusion.
- Carbon: Carbon is a shallow acceptor and a deep donor (on the Ga site). High levels of carbon can be incorporated. This dopant will be discussed in more detail in other chapters.
- Beryllium: This has good solid solubility, but it is a fast diffuser. More discussion will follow in Chapter 22 on reliability.
- Copper: This is a deep triple acceptor and a fast diffuser. Traditionally, copper has been carefully avoided in the fabrication process, and good barriers are used to avoid device degradation if copper is used as an interconnect or a backside metal.
- Chromium: Historically Cr doping was used to make semi-insulating GaAs, because Cr provides a level in the center of the energy band diagram. However, the adoption of carbon has replaced Cr.
- Oxygen: Oxygen is hard to avoid as an impurity during the fabrication process of crystals and wafers. It also goes to both Ga and As sites. Carbon and oxygen may be present in GaAs in the 10^{15} to $10^{16}/cm^3$ range but still have electron concentration levels below $10^{14}/cm^3$. In spite of this, it is better to avoid these impurities, because the electron mobility goes down as the background compensated impurity concentration goes up.

1.5 Crystal Defects

So far we have discussed ideal semiconductors. Real semiconductor materials contain defects of the point, line, and surface type. These defects have a strong effect on device performance and reliability. Defects are produced during growth of the material or introduced during processing. Defects can also be intentional like those due to doping. A few types of common defects are described below. For more details readers should refer to textbooks, etc. [3, 4].

1.5.1 Point Defects

Point defects can be missing atoms, vacancies, extra atoms, interstitials, misplaced atoms, or impurities. These have important effects on electronic properties as well as diffusion behavior.

Simple point defects are listed below:

- Vacancies: These are missing atoms, for example, Ga or As for GaAs, also called Schottky defects.
- Interstitials: These are extra atoms, for example, an extra Si atom, as shown in Fig. 1.18.

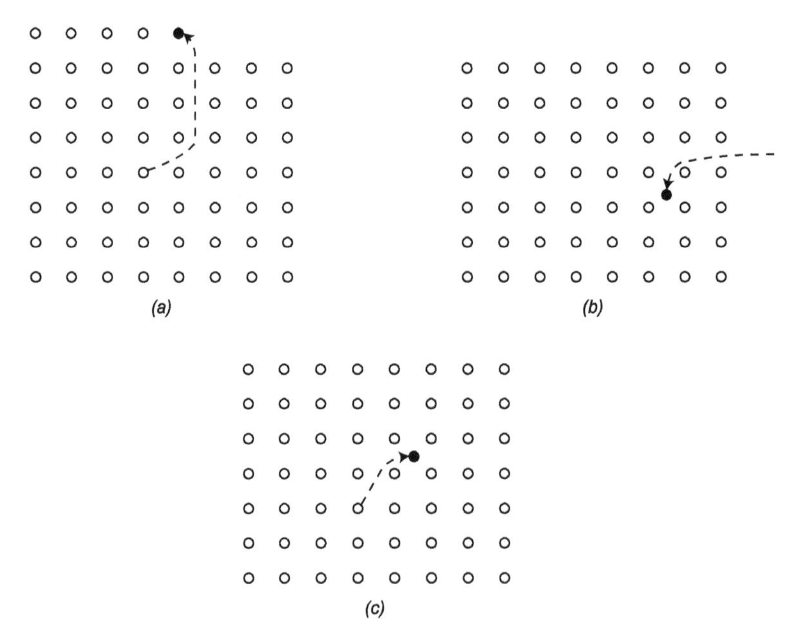

Figure 1.18 Simple point defects in crystals. Vacancy (a), interstitial (b), and Frenkel pair (c).

- Frenkel defects: This is a pair of vacancy and interstitial defects close to each other, for example, a Si vacancy and a Si interstitial atom in a Si crystal.

- Antistructure defect: It is possible for a Ga atom to be on an As site and vice versa.

Impurities present in the crystal are also point defects.

In III–V compounds the type of vacancies depends upon the constituents and the growth conditions. Thus P vacancies are expected in InP growth due to higher vapor pressure of phosphorus. In GaAs, Ga and As vacancies are present depending upon the processing conditions. Ga and As vacancies act as deep acceptors and deep donors, respectively. EL2 is an important defect in GaAs that is present in crystals grown from As-rich melt or in epilayers grown under As-rich conditions. This defect causes levels in the middle of the energy bandgap of GaAs, thus creating an electron trap [5]. A lightly doped p-type material becomes semi-insulating in the presence of these defects. Another deep-level defect complex, known as the DX center, was first seen in donor-doped AlGaAs and exhibits metastable behavior and persistent photoconductivity. All these defects interfere with device behavior and must be minimized.

1.5.2 Dislocations

A dislocation is an array of point defects forming a line in a perfect crystal. These are formed due to stress during growth or by enhanced point diffusion under thermal or mechanical stress. These defects cause electron trapping and affect device performance. The presence of these defects causes reduction of electron mobility in HEMT-type devices, recombination centers in HBT, and lower quantum efficiency in LEDs and laser devices. Also, the presence of these causes enhanced rate of diffusion and thus affects reliability. During development of crystal growth and epitaxial processes, considerable attention is given to defect reduction.

There are two main types of dislocations, edge and screw. An edge dislocation is an extra plane of atoms in an otherwise perfect crystal. An extra plane ABCD, as shown in Fig. 1.19, results in a line dislocation AD. Distortion in concentrated along this line. An edge dislocation is produced by applying shear force; the plane along which the force is applied in known as the slip plane.

Application of shear stress can also cause creation of a line defect, known as screw dislocation, illustrated in Fig. 1.20.

Dislocations can move along a slip plane under application of stress. They can also climb, which is move 90° to the slip plane, by displacement of atoms to interstitial sites.

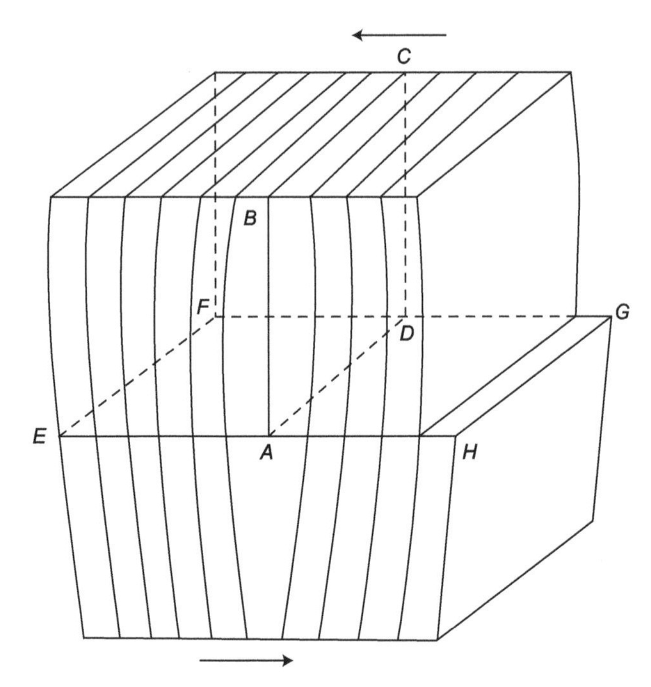

Figure 1.19 Schematic diagram of an edge dislocation along the line AD.

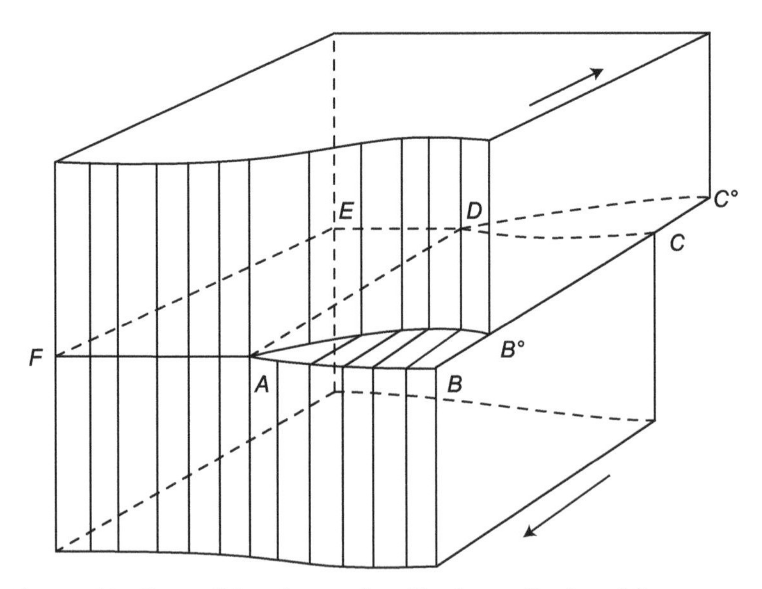

Figure 1.20 Screw dislocation produced by the application of shear stress.

1.5.3 Other Defects

A high degree of dislocations may lead to formation of large defects, a common one being twins. When one portion of a crystal is not oriented exactly with the rest, a twin is formed, as shown in Fig. 1.21. The atoms at the boundary are in intimate contact with others, but a clear discontinuity exists. Twinning occurs when a portion of the growing crystal is not free to move during growth, but is somehow restricted, for example, by the boat or vertical container. If a large number of broken bonds is present, and the orientation difference is over a certain limit, a grain boundary is formed. A low-angle grain boundary is shown in Fig. 1.22 [9]. A number of these may be present in a large crystal.

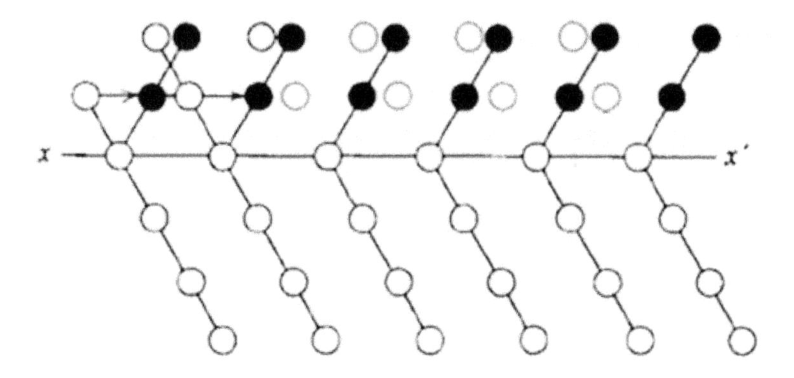

Figure 1.21 A twin produced by misorientation in a crystal.

1.6 Other Properties

Thermal characteristics: Thermal conductivity of GaAs is low, one-third that of silicon, 0.55W/cm-K. High thermal resistance limits the packing density of devices on GaAs. Too high a packing density would cause the temperature of the junction region to be too high for long-term, stable performance. Thus power-handling capability and reliability are related to the junction temperature during normal operation. Analog circuits that handle high power must be modeled and tested for thermal considerations.

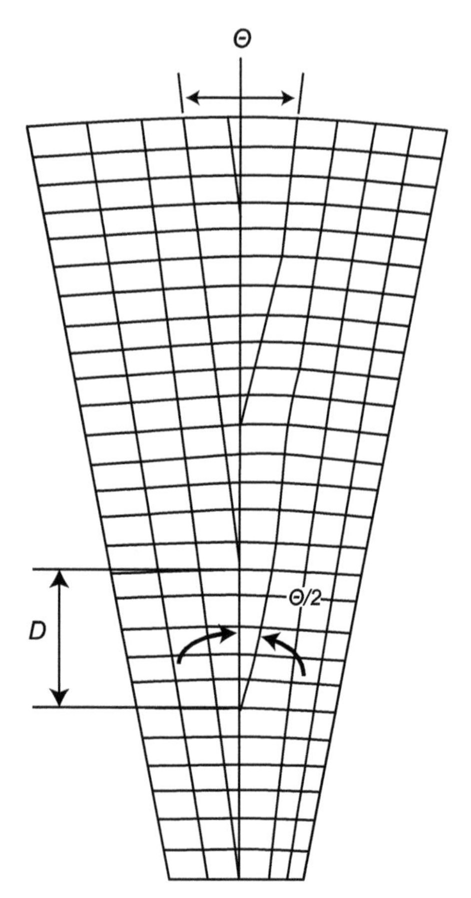

Figure 1.22 Low-angle grain boundary between two sections of a crystal.

The coefficient of thermal expansion for GaAs is also larger, 5.8 × $10^{\wedge -6}$ /K, so a mismatch to packaging materials is more likely.

For a list of properties of GaAs and a comparison to Si refer to Tables 1.1 and 1.2.

References

1. C. Kittel, *Solid State Physics*, John Wiley and Sons (1971).
2. D. Biswas and D. A. Neamen, *Semiconductor Physics and Devices*, 4th Ed., Mc Graw Hill, Special Indian Edition (2012).
3. F. A. Kroger, *The Chemistry of Imperfect Crystals*, Vol. 1, North-Holland, Amsterdam; American Elsevier, New York (1973).

4. S. M. Sze, *Semiconductor Device Physics and Technology*, John Wiley, New York (1985).
5. R. F. Pierret, *Semiconductor Fundamentals*, Addison-Wesley, New York (1989).
6. S. Markram-Ebied, Nature of EL2: the main native midgap electron trap in VPE and bulk GaAs, in *Semi-Insulating III-V Materials*, Ed., D. Look, Shiva, England (1984).
7. J. S. Blakemore, *J. Appl. Phys.*, **53**, R123 (1982).
8. R. Williams, *Modern GaAs Processing Methods*, Artech House (1990).
9. W. Liu, *Handbook of III-V Heterojunction Bipolar Transistors*, John Wiley and Sons, New York (1998).
10. S. K. Ghandhi, *VLSI Fabrication Principles, Silicon and Gallium Arsenide*, Wiley Interscience, New York (1983).
11. S. Wang, *Fundamentals of Semiconductor Theory and Device Physics*, Prentice Hall International (1989).
12. A. Dolittle, Georgia Tech University ECE 4813.
13. P. Asbeck et al., GaAs based heterojunction bipolar transistors for very high performance electronic circuits, *Proc. IEEE*, **81**, 1709 (1993).

Chapter 2

GaAs Devices

In this chapter the basic device configuration or structure and operation of GaAs devices is described. The discussion is phenomenological in nature and covers most commonly used III–V electronic device types. Relevant equations that are important for discussion of fabrication and process parameters will be introduced without complete derivation. Two-terminal and three-terminal devices of unipolar and bipolar type will be discussed, starting with the simple p–n and metal-semiconductor junctions. No attempt will be made to describe these in a historical context and original references (e.g., Shockley) will not be mentioned. For details of device physics, the reader is referred to numerous excellent books on the subject matter [1–4].

2.1 p–n and Metal–Semiconductor Junctions

2.1.1 p–n Junction Physics

The p–n junction is fundamental to the discussion of semiconductor devices, and its theory will be discussed here, with relevance to the processing of the respective family of devices.

The p–n junction devices are commonly used in III–V integrated circuits (ICs) for making level shift or electrostatic discharge (ESD) protection diodes. Knowledge of the p–n junction physics

III–V Integrated Circuit Fabrication Technology
Shiban Tiku and Dhrubes Biswas

ISBN 978-981-4669-30-6 (Hardcover), 978-981-4669-31-3 (eBook)
www.panstanford.com

is important also for understanding of three-terminal bipolar transistors and optical devices like light-emitting diodes (LEDs) and lasers.

Figure 2.1 shows a p-type and an n-type bar of semiconductor joined together. The ideal band diagram for a p–n junction is shown in Fig. 2.1b. The two Fermi levels line up as soon as equilibrium is reached in the process, creating a potential barrier, V_{bi}. Electrons and holes experience this potential barrier for going across the junction.

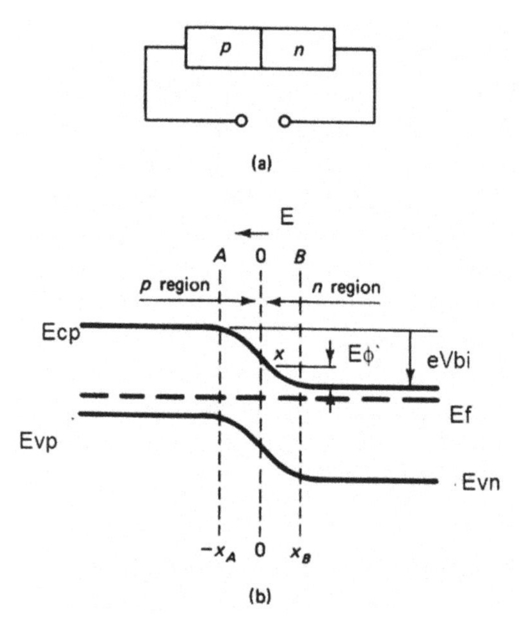

Figure 2.1 A semiconductor junction made by placing adjacent p-type and n-type semiconductors together (a) and the corresponding energy band diagram at equilibrium (b).

The electron and hole concentrations are given by (refer to Chapter 1)

$$\text{On the p side: } n_{p0} = N_c \exp\left\{-\frac{E_{cp} - E_f}{kT}\right\} \tag{2.1}$$

$$p_{p0} = N_v \exp-\left\{\frac{E_f - E_{vp}}{kT}\right\} \tag{2.2}$$

$$\text{On the n side: } n_{n0} = N_c \exp\left\{-\frac{E_{cn} - E_f}{kT}\right\} \tag{2.3}$$

$$p_{n0} = N_v \exp\left\{-\frac{E_f - E_{vn}}{kT}\right\} \quad (2.4)$$

Subscript 0 is for equilibrium; see Fig. 2.1 for definitions of various energy levels. N_c and N_v are the effective densities of states in the conduction and valence bands in the n-type and p-type regions.

Note that the potential barrier $e.V_{bi} = E_{cp} - E_{cn}$ (2.5)

Using the above Eqs. 2.1 to 2.5, under equilibrium conditions

$$\frac{E_{cp} - E_{cn}}{kT} = e\frac{V_{bi}}{kT} = \frac{\ln n_{n0}}{\ln n_{p0}} = \frac{\ln p_{p0}}{\ln p_{n0}} \quad (2.6)$$

Referring to Fig. 2.1b, the electron concentration at any point x in the junction region is given by

$$N = N_c \exp\left\{-\frac{E_{cn} + e\varphi' - E_f}{kT}\right\} \quad (2.7)$$

$$= n_{n0}\left\{-\frac{-e\varphi'}{kT}\right\} \quad (2.8)$$

The electric field in this region is given by

$$Ex = -\partial V/\partial x = \partial\phi'/\partial x \quad (2.9)$$

Consider the electron (n) flow first. Electrons flow from right to left and holes flow from left to right due to diffusion and due to high electron concentration on the n side and high hole concentration on the p side. Under thermal equilibrium, no current flows. So the current due to the built-in electric field must balance the diffusion flow. The net current must be zero.

Flow due to electric field = $e\mu_n n E_x$

Flow due to diffusion = $eD_n \frac{dn}{dx}$

$$\text{Net current } J_n = e\mu_n n E_x + eD_n \frac{dn}{dx} = 0 \quad (2.10)$$

Keep in mind the Einstein relationship between the two:

$$\mu_n = \frac{eDn}{kT} \quad (2.11)$$

2.1.1.1 *I–V* characteristics

The *I–V* behavior of the p–n junction is the result of the potential barrier.

The same discussion applies to holes (p) as well.

Figure 2.2 shows the dopant distribution, electric field distribution, and potential and energy band diagram for a p–n junction under equilibrium [1]. The built-in potential is equal to

$$eV_{bi} = E_g - \{eV_n + eV_p\}$$

$$= kT \ln \frac{N_c N_v}{n_i^2} \left\{ kT \ln \frac{N_c}{n_{n0}} \right\} kT \ln \frac{N_v}{p_{p0}} \quad (2.12)$$

$$= kT \ln \left\{ \frac{n_{n0} n_{p0}}{n_i^2} \right\} \approx kT \ln \frac{N_A N_D}{n_i^2}$$

Assuming $n_{n0} = N_D$ and $p_{p0} = N_A$, at equilibrium, on both p and n sides

$$n_{n0} \cdot p_{n0} = p_{p0} \cdot p_{p0} = n_i^2 \quad (2.13)$$

$$V_{bi} = \frac{kT}{e} \ln \frac{n_{n0}}{p_{p0}} = \frac{kT}{e} \ln \frac{n_{n0}}{p_{p0}} \quad (2.14)$$

From Eq. 2.14, the electron and hole densities on either side of the junction can be written as

$$p_{n0} = p_{p0} \exp\left\{ -\frac{eV_{bi}}{kT} \right\} \quad (2.15)$$

$$\text{And } n_{p0} = n_{n0} \exp\left\{ -\frac{eV_{bi}}{kT} \right\} \quad (2.16)$$

The current densities can be written as

$$I = A \exp\left\{ -\frac{eV_{bi}}{kT} \right\} B,$$

where A and B are constants.

Under no applied field, current is zero; therefore

$$B = A \exp\left\{ -\frac{eV_{bi}}{kT} \right\} \quad (2.17)$$

$$\text{Or} \quad A = B \exp\left\{ \frac{eV_{bi}}{kT} \right\} \quad (2.18)$$

Under applied voltage V_a, the current is equal to

$$I = A \exp\left\{ -\frac{e(V_{bi} - V_a)}{kT} \right\} - B \quad (2.19)$$

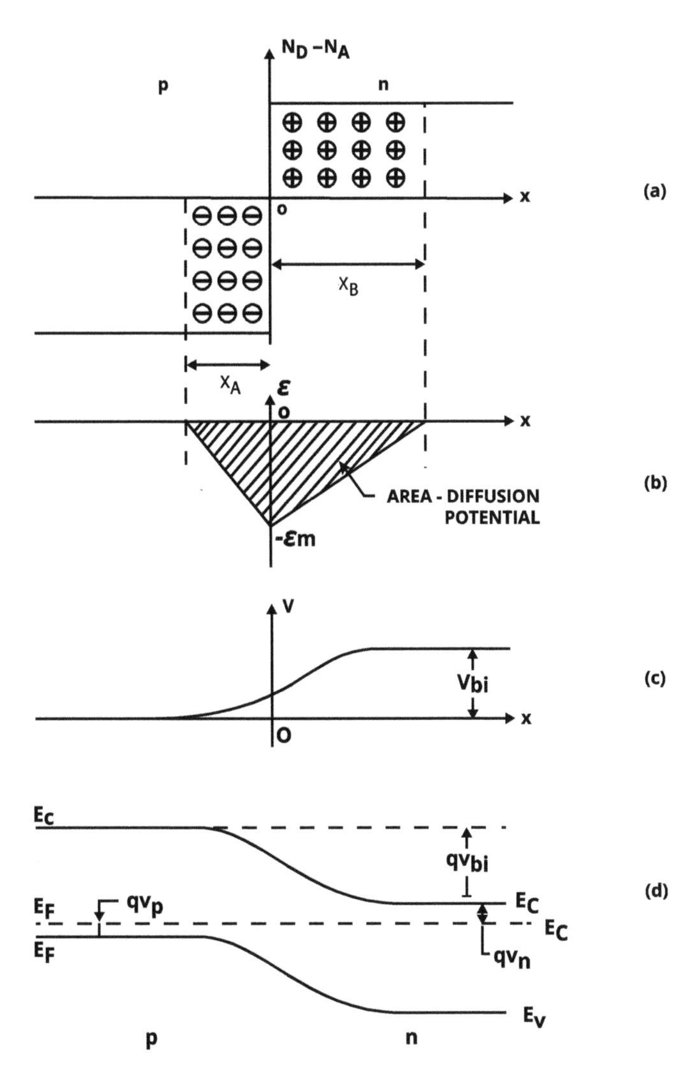

Figure 2.2 Schematic diagram for a p–n junction showing doping distribution (a) and electric field (b), potential (c), and energy band diagram (d), all under thermal equilibrium.

Using Eq. 2.18

$$I = B\left\{\exp\frac{eVa}{kT} - 1\right\} \tag{2.20}$$

I–V characteristics of a p–n junction follow Eq. 2.20 and are shown in Fig. 2.3.

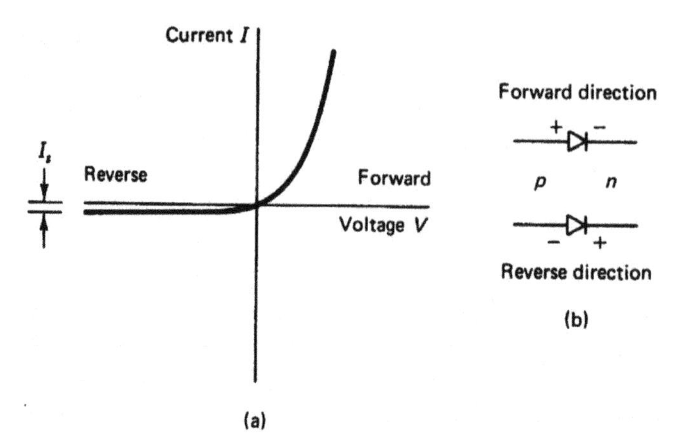

Figure 2.3 I–V characteristics of a p–n junction as a result of the built-in potential V_{bi}.

2.1.1.2 Space charge and junction capacitance

The equilibrium concentration of electrons is given by (Eq. 2.8 above):

$N = n_{n0} \exp\left\{-\frac{eVbi}{kT}\right\}$, and a similar equation for holes. The net charge at any point x in the junction (referring to Fig. 2.1) $\rho(x)$ is given by

$$\rho(x) = e(p + N_D^+ - n - N_A^-)$$

N_A and N_D are number of acceptors and donors at any point. N_1 in the p region is given by

$$N_1 = N_A^- - N_D^+ \tag{2.21}$$

Similarly, N_2 in the p region is given by

$$N_2 = N_D^+ - N_A^- \tag{2.22}$$

Using Eqs. 2.8 and 2.22

$$\rho(x) = eN_2\ 1 - \exp\left\{-\frac{eVbi}{kT}\right\} \tag{2.23}$$

Electrostatic potential is obtained by solving Poisson's equation for the one-dimensional case:

$$\frac{d^2V}{dx^2} = -\rho(x)\ \frac{\rho}{\varepsilon} \tag{2.24}$$

where ε is the dielectric constant for the semiconductor. For a uniformly doped n region, $\rho(x) = eN_2$ is a constant. Therefore integration of Poisson's equation (Eq. 2.24) yields

$$\frac{dV}{dx} = -\frac{eN_2}{\varepsilon}(x - x_B) \tag{2.25}$$

(See Fig. 2.1 for positions.)

Integration once more leads to V.

$$V_B = -\frac{eN_2}{2\varepsilon}\{x(x-2x_B)\} \text{ for } 0 < x < x_B \tag{2.26}$$

On the p side, similarly

$$\frac{dV}{dx} = \frac{eN_1}{\varepsilon} - (x + x_A) \tag{2.27}$$

and

$$V_A = -\frac{eN_1}{2\varepsilon}\{x(x+2x_A)\} \text{ for } -x_A < x < 0 \tag{2.28}$$

Figure 2.2 shows the results in graphical form. These charges form a capacitor, the total capacitance being

$$C_T = \left|\frac{dQ}{dVa}\right| = A\frac{\varepsilon}{w}$$

The total width of the capacitor is $W = x_A + x_B$?

The charges on the two sides are $eN_2x_B = eN_1x_A$.

And using $V_d = V_A + V_B$ and using Eqs. 2.26 and 2.28

$$V_d = \frac{e}{2\varepsilon}\{N_1x_A^2 + N_2x_B^2\} \tag{2.29}$$

$$x_A = \sqrt{\frac{2\varepsilon V_d N_2}{eN_1}\frac{1}{(N_1+N_2)}} \tag{2.30}$$

$$x_B = \sqrt{\frac{2\varepsilon V_d N_1}{eN_2}\frac{1}{(N_1+N_2)}} \tag{2.31}$$

The total depletion width under equilibrium conditions

$$W = x_{A+}x_B = \sqrt{\frac{2\varepsilon V_d}{e}\frac{(N_1+N_2)}{N_1N_2}} \tag{2.32}$$

Or

$$= x_{A+}x_B = \sqrt{\frac{2\varepsilon V_d}{e}\left(\frac{1}{N_1}+\frac{1}{N_2}\right)} \tag{2.32}$$

Under nonequilibrium conditions, when a voltage V_a is applied, the above equations are valid with V_d replaced by $(V_d + V_a)$ as long as $e(V_d + V_a) >> kT$.

The charge Q per unit area $A = eN_2x_B = eN_1x_A$.

$$Q/A = \sqrt{\frac{2e\varepsilon}{e}\frac{(N_1N_2)}{N_1+N_2}}\cdot\sqrt{V_d+V_a} \tag{2.33}$$

$$Q/A = \text{W}.\ \frac{e(N_1N_2)}{N_1+N_2} \quad \text{(using Eq. 2.32)} \tag{2.34}$$

The capacitance can be shown to be

$$C_T = \left|\frac{dQ}{dVa}\right|$$

$$C_T = \frac{A}{2}\sqrt{\frac{2e\varepsilon}{e}\frac{(N_1N_2)}{N_1+N_2}}\ .\ (V_d+V_a)^{-1/2} \tag{2.35}$$

These equations can be simplified in many practical applications. If one side has a very high doping concentration, like a p+/n− junction (base–collector junction of a heterojunction bipolar transistor [HBT]), the depletion width on the p+ side ($N_1 >> N_2$) will be very small. The total depletion width may be written as

$$\text{W} = \sqrt{\frac{2\varepsilon V_d}{eN_2}} \tag{2.36}$$

This is similar to the case of the Schottky diode to be discussed in the next section.

2.1.2 Metal–Semiconductor Junctions

Metal layers are deposited on semiconductors as wires to connect a semiconductor device to other circuit elements or to form junctions with the semiconductors as a diode or as a gate to control transistor current. The first solid-state device reported in 1875 was a metal–semiconductor device called the whisker contact rectifier. It consisted of a metal whisker touching a lead sulfide (II–VI compound) crystal. Metal–semiconductor rectifiers and junctions, involving III–V semiconductors, continue to be used today in high-frequency applications. An understanding of the junction is important for understanding two-terminal and three-terminal III–V devices.

2.1.2.1 Junction physics

Figure 2.4a shows the energy band diagram of a semiconductor. For GaAs, and other III–V semiconductors, the crystal structure at the surface is not perfect. Atoms at the surface or interface do not have ideal neighbors to complete all bonds. Surface atoms may thus become positively or negatively charged. Surface states cause the energy bands to bend. The position of the Fermi level at the surface should depend upon the electron affinity (χ) and the work function of the semiconductor ϕ_s. The Fermi level is actually pinned at the surface, as shown in Fig. 2.4b and the ϕ_b is not directly related to χ and ϕ_s.

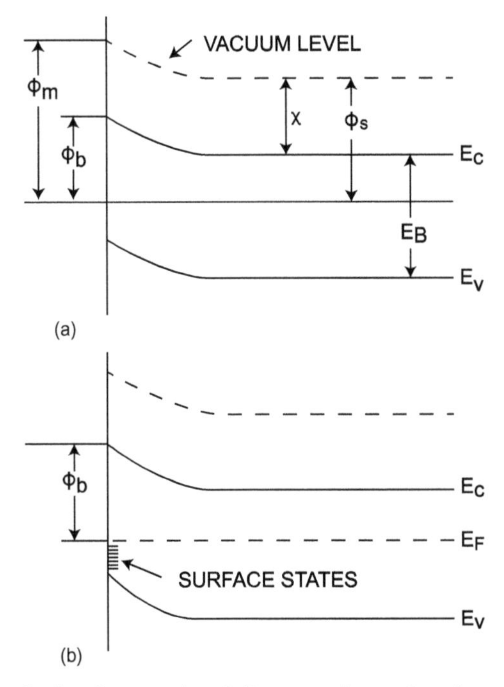

Figure 2.4 Idealized energy band diagram of a semiconductor surface (a) and realistic diagram for GaAs where surface states pin the Fermi level and consequent barrier height (b).

Figure 2.5a shows the energy band diagram of a metal and an n-type semiconductor. The Fermi level of the metal is shown $e\phi_m$ below the vacuum level, where ϕ_m is the metal work function, the energy required to remove an electron from the metal to vacuum potential. Most metals used in III–V circuits have work functions between 4 and 5.5 eV, larger than the electron affinity (χ) of GaAs. When the metal and semiconductor are brought into intimate contact ($\chi + V_{cf} < \phi_m$) electrons will diffuse from the semiconductor to the metal, where the Fermi level is V_{cf} below the conduction band. As electrons leave, positive charge builds up until equilibrium is reached, where the positive charge exerts a force on the electrons that opposes the diffusion current. The width of this region depleted of charge is called the depletion width and is a function of doping density, etc. The energy bands in the semiconductor bend at the interface, as shown in Fig. 2.5b, creating a built-in potential V_{bi}. For an electron to cross from the semiconductor to the metal it must overcome V_{bi}. An electron moving from metal to semiconductor must overcome ϕ_b. For GaAs, the crystal structure at the interface is not perfect. The barrier potential under these conditions is not dependent upon the metal work function but is somewhat independent of the metal. The actual experimental data is shown in Fig. 2.6 and is only weakly correlated to ϕ_m.

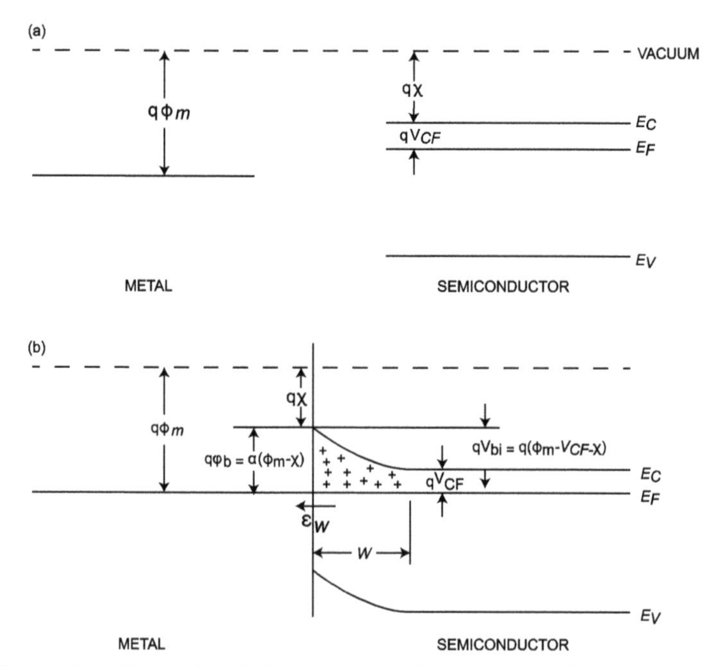

Figure 2.5 Energy band diagram of metal and semiconductor separately (a) and in contact (b).

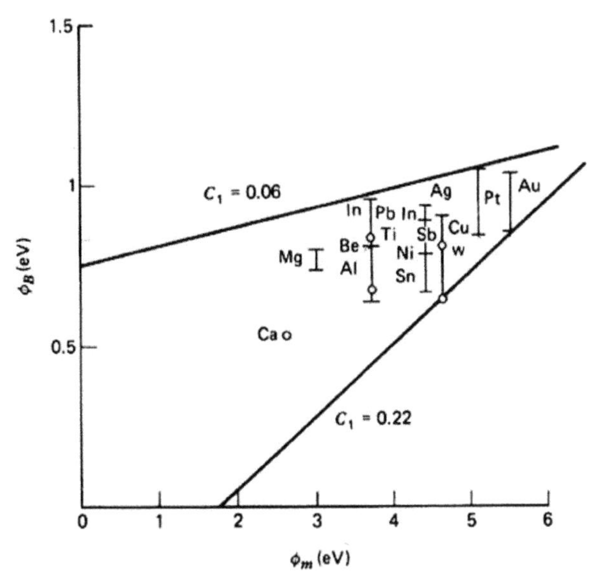

Figure 2.6 Actual measured potential barrier for various metals (φ_B), showing only weak dependence on φ_m. Reproduced from Ref. [15] with permission from *Solid State Technology*.

Table 2.1 lists barrier heights of some common metals and alloys on GaAs.

Table 2.1 Experimental barrier heights in eV at 300 K

Metal	Barrier height
Al	0.80
Ag	0.88
Au	0.86
Bi	0.89
Cr	0.77
Cu	0.82
In	0.82
Ni	0.83
Pt	0.86
Sb	0.86
Ti	0.82
W	0.64

Source: Reproduced from Ref. [15] with permission from *Solid State Technology*.

When an electric field is applied to a junction, the equilibrium is disturbed. If a positive potential V_{FB} is applied to the metal and the circuit completed using an ohmic contact to the semiconductor, which is described elsewhere, it will reduce the electron barrier by V_{FB} and a current will flow (see Fig. 2.7a). If a negative potential, reverse bias, is applied to the metal, the external potential will add to the field produced by band bending and prevent diffusion current from flowing, thus forming a rectifier or diode (Fig. 2.7b). Figure 2.8 shows a schematic diagram of a metal–semiconductor junction called a Schottky diode. See Fig. 2.9 for typical *I*–*V* characteristics.

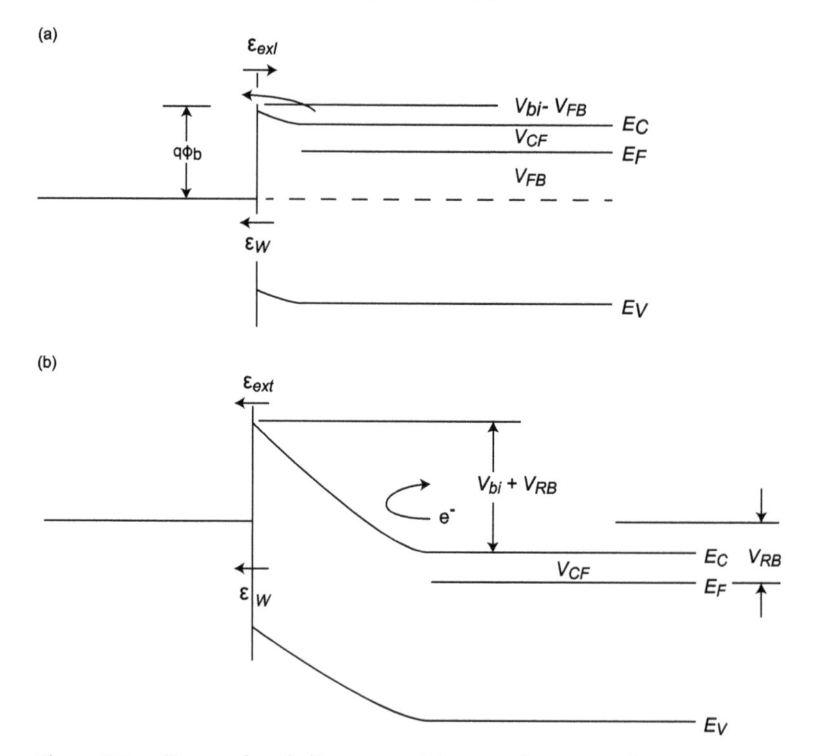

Figure 2.7 Energy band diagrams of the metal–semiconductor junction under forward bias (a) and reverse bias (b).

The electrical behavior of semiconductor devices is predicted by analyzing the devices with the use of basic equations from electromagnetic theory, like Poisson's equation, semiconductor equilibrium statistics, current conduction under an electric field, diffusion, and charge continuity.

For a detailed discussion the reader is referred to basic device physics textbooks [1–3].

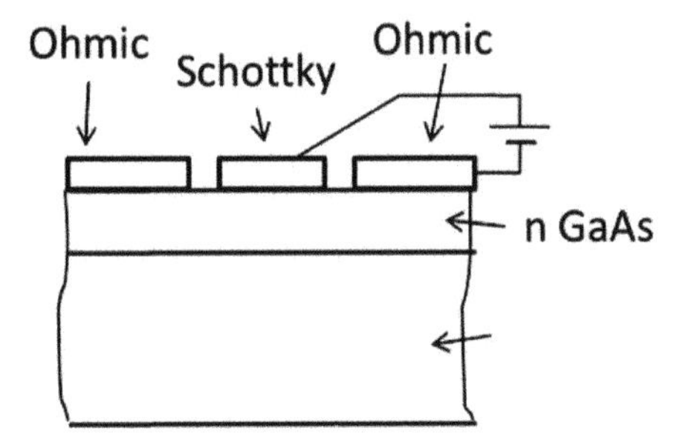

Figure 2.8 Schematic cross section of a Schottky diode.

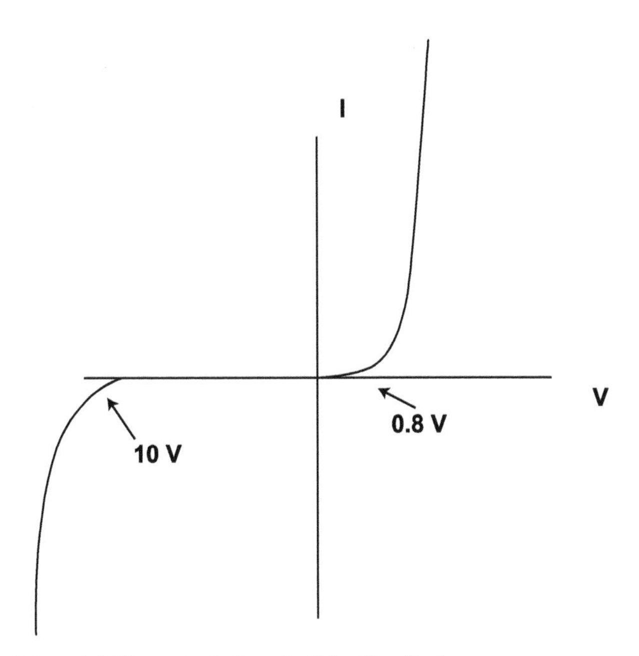

Figure 2.9 *I–V* Characteristics of a Schottky diode.

2.1.2.2 Junction characteristics

The depletion width is derived by solving Poisson's equation for the Schottky barrier formed at the metal–semiconductor junction, with the boundary conditions set by the lining up of the Fermi levels ([1], p. 370):

$$W = \sqrt{\frac{2\varepsilon_{\rm s}}{qN_{\rm d}}\left(V_{\rm bi} - V - \frac{kT}{e}\right)} \tag{2.37}$$

where N_d is the donor concentration, ε_s is the dielectric constant of the semiconductor ($\varepsilon_s = \varepsilon_0 . \varepsilon_r$), and V is the applied voltage. The depletion width is small if N_d is large, as in highly doped layers, and varies inversely with applied voltage, being low for positive-bias conditions and large for negative-bias conditions. (The depletion width estimate for typical metal–semiconductor field-effect transistor [MESFET] channel doping in the 10^{17} per cm^2 range is about 500 A.) The depletion region creates a capacitor, whose capacitance is given by $C = \varepsilon_s/w$. Since w depends upon the applied voltage, the capacitance varies inversely with the applied voltage. Large-area field-effect transistors (FETs) with high doping levels can be utilized as varactor diodes in voltage-controlled oscillators (VCOs).

2.1.2.2.1 *Current flow*

An understanding of the *I–V* curve for a Schottky diode, shown in Fig. 2.8, is important for device design and process monitoring. The current transport of a metal–semiconductor barrier is mainly due to majority carriers in contrast to p–n junctions, where minority carriers are key. The current flow through a junction can be described by the following equation, derived by incorporating thermionic emission and diffusion [1]:

$$J = J_0 \exp\left\{-\frac{q\phi_{\rm b}}{kT}\right\} \cdot \exp\left\{\frac{qV}{kT} - 1\right\} \tag{2.38}$$

where $J_0 = A^{**}.T^2 J_0$ and A^{**} is the effective Richardson constant for thermionic emission into a vacuum and is a function of temperature and doping concentration. The current J is an exponential function of voltage and barrier height. When current in a practical Schottky diode is measured, the current fits an exponential function of the type

$$J = \exp\left\{\frac{qV}{nkT}\right\} \tag{2.39}$$

where the parameter n is called the *ideality factor*. An ideal diode would have an n factor of 1, and nonideal diodes have n greater than 1. When the *I–V* curve is plotted on a log scale, a line can be fit to the data. The n factor can be derived from the slope and the barrier height can be calculated from the intercept on the I axis, J_S.

$$\phi_b = \frac{kT}{q} \ln\left(\frac{A^{**}T^2}{J_s}\right) \tag{2.40}$$

The series resistance of the diode is the sum of the contact resistance and the junction resistance.

The junction resistance is related to the slope of the I–V curve and is given by

$$R_j = \frac{nkT}{qJA_j} \tag{2.41}$$

where A_j is the junction area.

An important figure of merit for Schottky diodes for high-frequency applications is the forward bias cutoff frequency:

$$f_c = = \frac{1}{2\pi R_f C_f} \tag{2.42}$$

where R_f is the total series resistance and C_f is the junction capacitance. R_f and C_f must be minimized for achieving high cutoff frequency. GaAs Schottky diodes on an n-type material are capable of high f_c due to high electron mobility (low R_f. C_f product can be achieved).

2.2 MESFETs

2.2.1 Basic MESFETs

The MESFET was the dominant GaAs device till the mid-1990s, when HBTs were introduced for power amplifier applications. These devices were and are still used for power amplifiers, both high-power and low-noise type, front-end receivers, and digital circuits. Although GaAs junction field-effect transistor (JFET) devices were first proposed and built, the MESFET, proposed in 1966 [5], became the most popular device because of its processing simplicity. A schematic diagram of a FET is shown in Fig. 2.10. The FET consists of a channel made of an n-type semiconductor material with heavily doped n+ regions, called source and drain, at the two ends. Ohmic contacts are made to these regions to allow the device to be connected to the circuit. Between the source and the drain, a metal line is deposited that acts as a gate to control the current flow. This contact is made with metals that form a rectifying or Schottky contact, as discussed previously. When a voltage is applied between the source and the drain, a current flows, which increases linearly at first and then saturates due to the fact that electron velocity in the semiconductor reaches the saturation value. (See Fig 2.11 for an ungated FET.) The current is given by

$$I_d = w.a.q\, n\, v \tag{2.43}$$

where w is the width of the channel, a is the depth of the channel, n is the electron density, and v is the electron velocity. As the velocity v saturates, I_d will saturate to a maximum value. Once the gate metal is deposited, a depletion layer will be created, the depth of the depletion depending upon the voltage applied to the gate. At 0 V, the depletion depth is more than that for a passivated free surface and the channel depth decreases. The current that can flow is reduced. The voltage on the drain side is more positive, making the depletion depth even deeper on that side. The depletion depth and therefore the current flowing through the channel can be varied by applying a bias on the gate. When the applied voltage is negative and large enough the depletion depth will equal the channel depth and the current will go to zero. The gate voltage at which this happens is called the pinch-off voltage. Solving Poisson's equation for this case V_p can be derived:

$$V_p = \frac{qN_d}{2\varepsilon_s}.a^2 \tag{2.44}$$

where N_d is the doping concentration and ε_s is the dielectric constant of the semiconductor. The magnitude of the current flowing at zero gate bias depends upon the doping level N_d. If the doping level is low, the FET is normally off and can be turned on only if a positive voltage is applied to the gate. If the doping level is high, the MESFET is normally on and a large negative V_p is needed to turn it off. I–V curves for a MESFET are shown in Fig. 2.12. A small voltage signal on the gate can be used to control a large drain current, and thus the device act as a voltage-controlled current source and be used as an amplifier. The transconductance of the MESFET is defined as

$$g_m = \frac{\partial Id}{\partial Vg} \tag{2.45}$$

(change of drain current output with applied voltage signal)

$$g_m = 2w.\mu q N_d(y_2 - y_1) \tag{2.46}$$

where y_1 and y_2 are the depletion depths on the two ends of the channel (uniformly doped channels with a doping concentration of N_d). See Fig. 2.13.

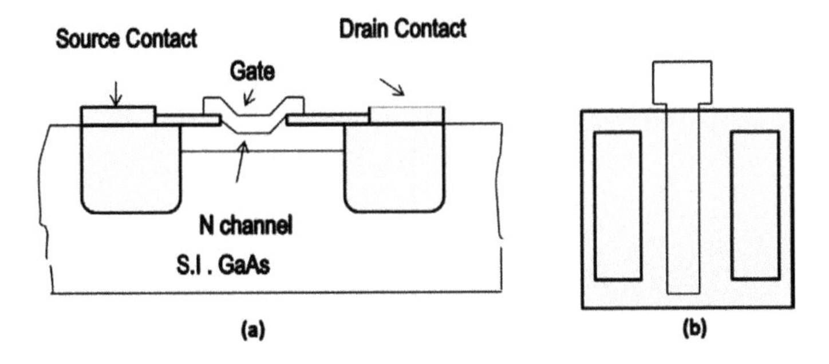

Figure 2.10 Schematic cross section of a MESFET (a) and top view (b) fabricated on a semi-insulating (S.I.) GaAs substrate.

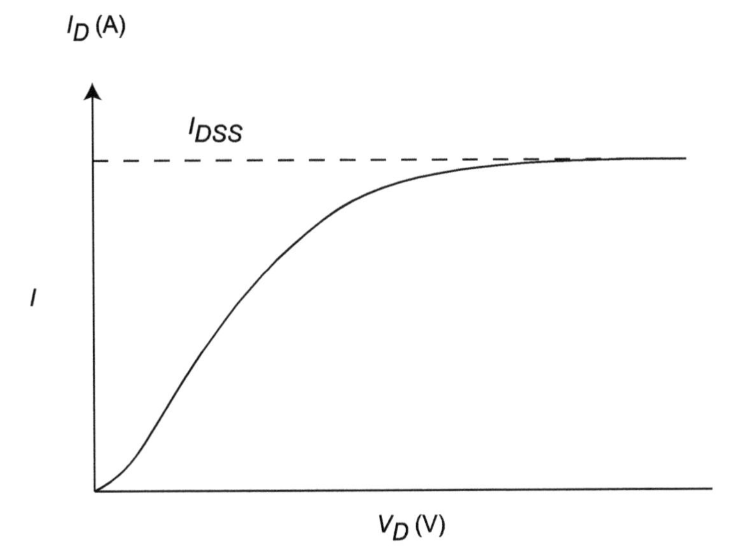

Figure 2.11 *I–V* characteristics of an ungated MESFET.

The maximum value of transconductance can be

$$g_m = \frac{qN_d a\mu w}{L} \tag{2.47}$$

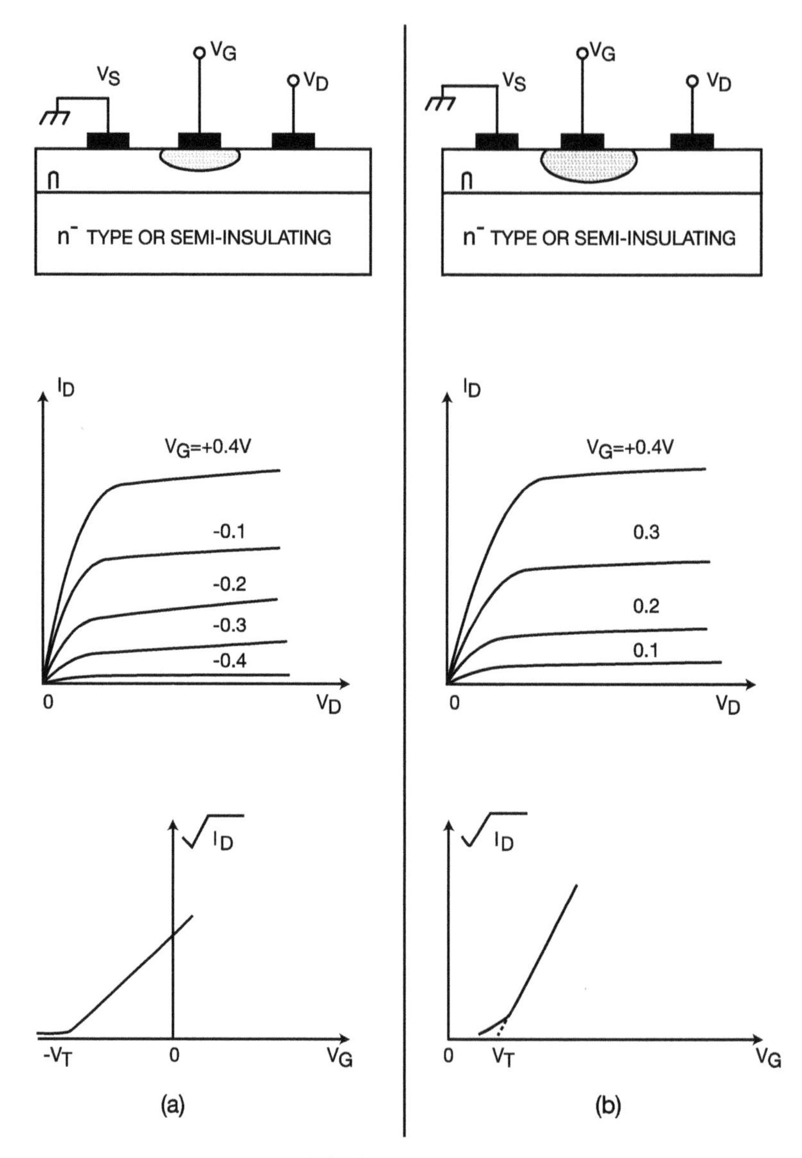

Figure 2.12 Comparison of depletion region and *I–V* characteristics of a D-mode (normally on) and E-mode (normally off) MESFET.

From this equation, it can be seen that g_m increases as μ increases and L decreases. Electron mobility in the channel must be high and the gate length must be small for high transconductance. The extrinsic transconductance can be reduced by parasitic resistances like source resistance (resistance between the gated channel and the source contact and the contact resistance). This can be derived as

$$g_{\mathrm{m}}(\mathrm{ext}) = \frac{g_{\mathrm{m}}}{\sqrt{(1+R.g_{\mathrm{m}})}} \qquad (2.48)$$

Effort is made in the MESFET layout and processing to minimize this parasitic resistance.

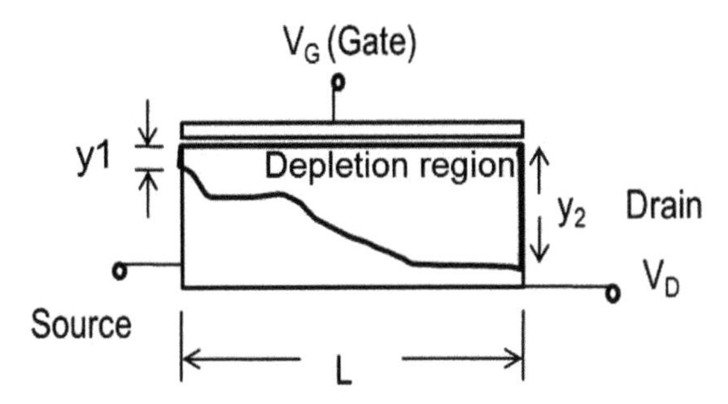

Figure 2.13 Cross-sectional view of a MESFET gate showing the depletion region.

For microwave transistors, the commonly used figures of merit are the gain-bandwidth product, the maximum frequency of oscillation, and the cutoff frequency. The cutoff frequency, f_{t}, is related to the electron transit time (τ) across the channel:

$$f_{\mathrm{t}} = \frac{1}{2\pi\tau} = \frac{V_{\mathrm{sat}}}{2\pi L} \qquad (2.49)$$

For a high cutoff frequency, semiconductor materials with high V_{sat} must be used for the channel and the gate length (L) must be small, below 1 μm for GaAs and other III–V materials. The maximum frequency of oscillation f_{max}, defined as the frequency where unilateral gain reaches a value of 1, is related to f_{t} by

$$f_{\mathrm{max}} = \frac{f_{\mathrm{t}}}{2\sqrt{(r_1 + f_{\mathrm{t}}.\tau_3)}} \qquad (2.50) \text{ From Ref. [1]}$$

where $r_1 = \dfrac{R_{\mathrm{g}} + R_{\mathrm{i}} + R_{\mathrm{s}}}{Rds}$ (2.51)

and $\tau_3 = 2\pi R_{\mathrm{g}} C_{\mathrm{gd}}$

where R_i is the input resistance, R_g and R_s are extrinsic resistances, and C_{gd} is the feedback capacitance.

The parasitics are shown in Fig. 2.14, which shows the equivalent circuit of a MESFET.

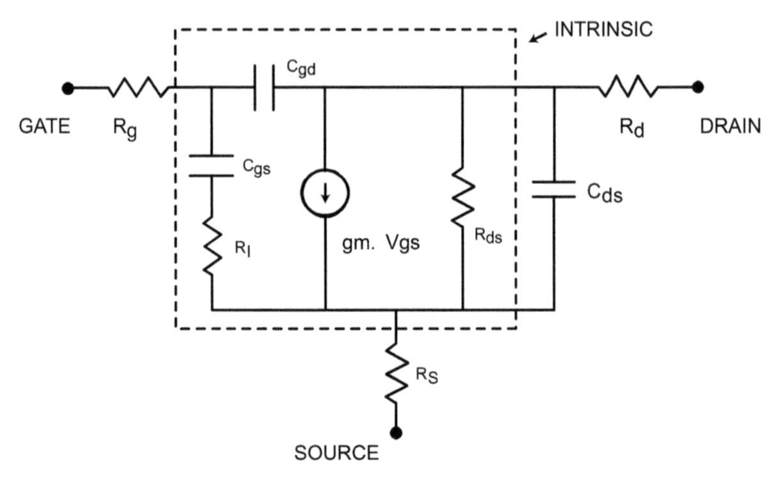

Figure 2.14 Equivalent circuit of a MESFET.

R_i is related to the n factor of the Schottky diode ideality factor.

To maximize f_{max}, f_t must be maximized and all the parasitics, R_g, R_s, and C_{gd}, must be minimized.

$$f_{max} \approx \frac{f_t}{2} \cdot \sqrt{\left(\frac{Rds}{R_g}\right)} \tag{2.52}$$

A practical limit for the gate length may be 0.1 μm from the consideration of electron beam lithography limits. For effective gate control, the channel depth must be less than 0.1 μm. To achieve high currents and high f_{max}, the channel carrier concentration must be high.

2.2.2 Low-Noise FETs

These FETs are optimized to achieve a low-noise figure, which is the ratio of signal to noise at the input to the noise ratio at the output. MESFETs, being majority carrier devices, are inherently low-noise devices.

$$F_o = 1 + f_L \sqrt{g_m \frac{R_s + R_g}{4}} \quad \text{(2.53) From Ref. [4]}$$

For low noise, the parasitic must be minimized, as expected. Low-noise FETs are biased to operate at nearly 15% of Idss, and high I_d contributes to noise because of electron scattering. To reduce R_s further, the gate may be placed closer to the source side.

2.2.3 FETs for Digital Logic Circuits

Digital circuits generally employ E/D logic, where both enhancement-mode (E) and depletion-mode (D) FETs are used. The E-FET is used as the switching transistor for achieving low power (normally off) and high speed. At $V_g = 0$, the channel is not conducting as it is totally depleted. When a positive voltage greater than the threshold voltage (V_t) is applied, channel current flows ($V_t \approx V_{bi} - V_p$, where V_p is the pinch-off voltage).

The drain current is given by the following square law equation:

$$I_d = \left(w\mu\varepsilon_s \frac{(V_g - V_t)^2}{2aL} \right) \quad \text{(2.54) From Ref. [1]}$$

I–V curves for E- and D-type FETs are shown in Fig. 2.12. The basic current–voltage characteristics of E and D devices are similar, generally represented by

$$I_d = K(V_g - V_t)^2 \tag{2.55}$$

$$K = \frac{w\mu\varepsilon_s}{2aL} \tag{2.56}$$

The K factor is used as a figure of merit for digital FETs. For high K, you need a large μ and a small L and a (channel length and depth).

For microwave circuits, transconductance ($g_m = 2K[V_g - V_t]$) and transit frequency are important device parameters, but for logic applications, the K factor is the most important figure of merit. Process development of FET technology for digital circuits is focused on achieving a high K factor.

The transit time in a FET is related to the saturation velocity v_s

$$\tau = L/v_s$$

and the frequency f_t can be written as

$$f_t = \frac{1}{2\pi\tau} = \frac{g_m}{2\pi C} \tag{2.57}$$

For GaAs logic circuits, the interconnect capacitance is generally much larger than the device capacitance, and the transconductance does not saturate within the voltage swing. Under those conditions, the gate delay is given by

$$\Delta t = \frac{2C_L}{K(V_{on} - V_t)} \tag{2.58}$$

where V_{on} is the turn-on voltage.

Logic delay is inversely proportional to the K factor. Again a high K is desired for short logic delay.

2.3 HEMTs and PHEMTs

In the discussion of the figure of merit, g_m, for MESFETs it is clear that it can be improved by increasing the doping in the channel and making the channel depth small. And for high f_{max} and for low noise, high mobilities and high carrier concentrations at the lowest doping concentrations are desired. The mobility of electrons in GaAs channels made with ion implantation or even by epitaxial growth decreases as the doping density is increased. High-electron-mobility transistors (HEMTs) and pseudomorphic high-electron-mobility transistors (PHEMTs) made possible by epigrowth techniques overcome this limitation. Drop in the cost of epiwafers has contributed to the rapid development and introduction of epigrown devices into production for all low-noise and high-gain applications in the millimeter-wave frequencies.

In an HEMT device (see Fig. 2.15), a larger bandgap layer (AlGaAs) is grown near the channel and the doping is place in this layer, the channel being left undoped. The electrons from the AlGaAs layer flow into the undoped GaAs layer and form the channel. Figure 2.16 depicts the energy band diagram for an AlGaAs /GaAs junction, showing the bandgap discontinuity and the position of the charge left behind on the donor that electrostatically attracts the electrons to the boundary. A potential barrier confines the carriers to a sheet of charge. These electrons are referred to as two-dimensional electron gas (2DEG).

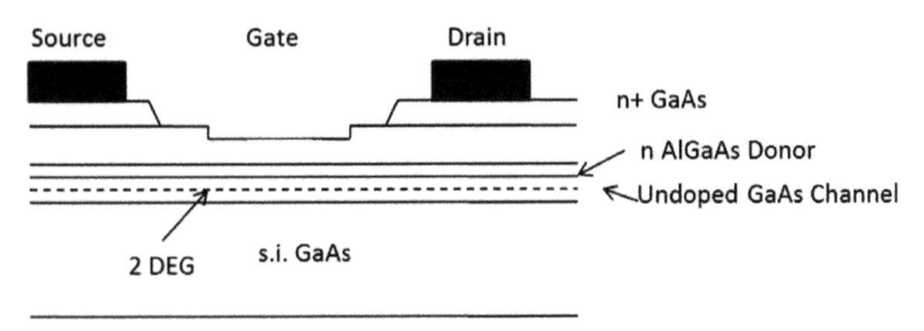

Figure 2.15 Schematic cross section of an HEMT device.

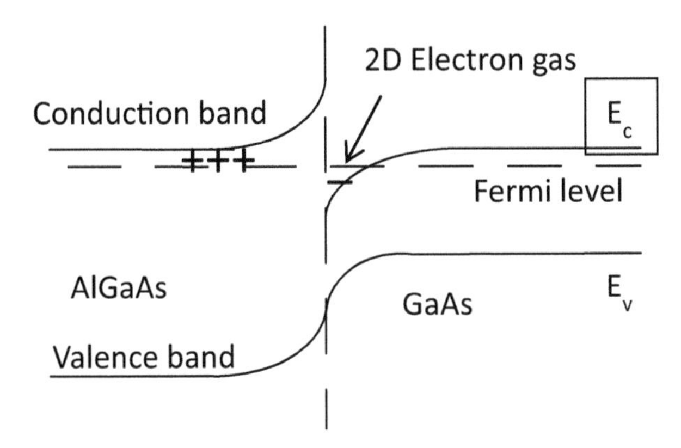

Figure 2.16 HEMT energy band diagram.

Figure 2.15 shows the schematic cross section for an HEMT device, including the epitaxial layer structure. For devices made using a GaAs substrate, the heterojunction is made with AlGaAs, with the Al mole fraction between 0.2 to 0.3. Heavily doped n+ cap layers are used to make low-resistance contacts possible. In the case of PHEMTs the channel is replaced by InGaAs, with an In mole fraction of less than 0.3, in which the channel mobility is even higher than that in GaAs. However, InGaAs has a lattice mismatch with GaAs and AlGaAs (see Fig. 4.11) and its thickness must be restricted to values at which the stress is still tolerable.These structures that use strained layers are called pseudomorphic.

2.3.1 Device Operation

Details of the epitaxial and device structure will be discussed under epitaxial growth. The operation of HEMT and PHEMT devices is similar to MESFETs. When a negative bias voltage is applied to the gate, the AlGaAs layer under the gate becomes depleted. With a further increase in bias, the 2DEG is depleted and pinched off. Intrinsic transconductance, g_{m0}, is given by

$$g_{m0} = \frac{Vsat.w.\varepsilon_s}{d} \tag{2.59}$$

where d is the distance from the gate to the 2DEG [10].

Compare this to $g_m = qN_d\, a\, \mu\, w/L$ (Eq. 2.47 for MESFETs.

HEMTs show higher saturation electron velocity compared to FETs. Parasitic source and drain resistances are lower because of higher mobility in the 2DEG layer. f_t and f_{max} are higher and the noise figure is lower for the same gate length.

The most important device parameter that determines f_{max} is total delay time and is given by

$$T_{tot} = t_{par} + t_{gd} + t_i + t_{RC} + t_{drain} \tag{2.60}$$

where t_i is the intrinsic transit time and t_{RC} is the channel charging time. The intrinsic transit time is a function of the gate length. The other transit times, t_{par} (parasitic) and t_{gd} (gate drain), can be reduced by self-alignment techniques.

Other equations of importance to process engineers are the following:

$$f_t = \frac{g_{m0}}{2\pi(C_{gs} + C_{gd})} \tag{2.61}$$

$$f_{max} = \frac{f_t}{2\sqrt{(R_i + R_s + R_g).g_{m0}} + 2\pi(f_t.R_g.C_{gd})} \tag{2.62}$$

As in the case of FET-type devices discussed above, the importance of reducing the different parasitics is obvious.

2.4 Bipolar Junction Transistors

The semiconductor era started with the invention of the transistor by a research team at Bell Labs in 1948. Since then there has been an explosive growth of device types and integrated circuits based on those. Processing of semiconductors has become a major industry worldwide. The bipolar device is a simple device physically, but conceptually it is not as simple as the FET. The physics of these devices is covered in numerous books [1–4, 10].

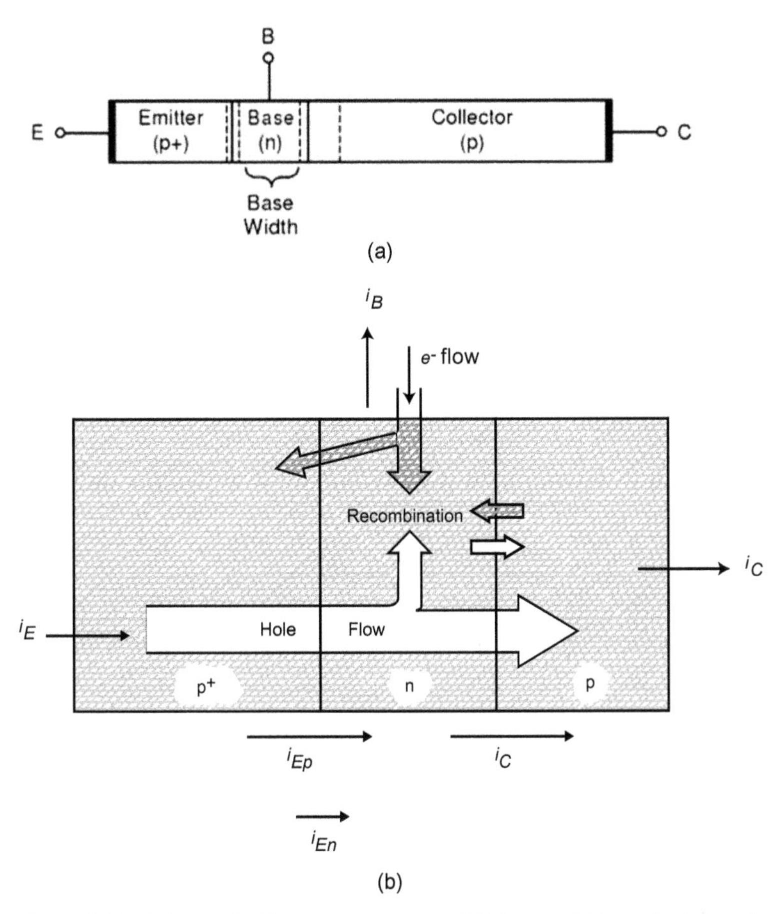

Figure 2.17 Schematic diagram of a p–n–p BJT device (silicon device with holes as majority carriers). Current and hole flow in a p–n–p bipolar device.

2.4.1 Phenomenological Description of the BJT

A simple qualitative view is presented here (from Ref. [11]). Since GaAs bipolar junction transistors (BJTs) are not used in practice, the discussion of the BJT is limited to that required for an explanation of the need for HBTs. A schematic diagram of a silicon p–n–p transistor is shown in Fig. 2.18. The emitter junction is forward-biased to inject holes across the junction and the reverse-biased collector junction collects the injected holes. See Fig. 2.19 for an energy band diagram of the BJT. The key to transistor operation is the very thin n-type base region, its width being much less than the diffusion length of holes $(D_p T_p)^{1/2}$. Since these carriers are minority carriers in the base, these would be lost to recombination in a long base region; however, in a thin base layer, most escape capture. Electrons that reach the base collector depletion region are swept across by the electric field there. Most of the holes injected are collected by the collector; however, a fraction is lost to recombination, and the difference is made up by the base current. Excess bound holes in the base region repel the free holes trying to cross the base. By applying a negative bias to the base lead, some of the bound carriers can be sucked out. The transistor works as an amplifier because the large current flow to the collector is controlled by a small base current.

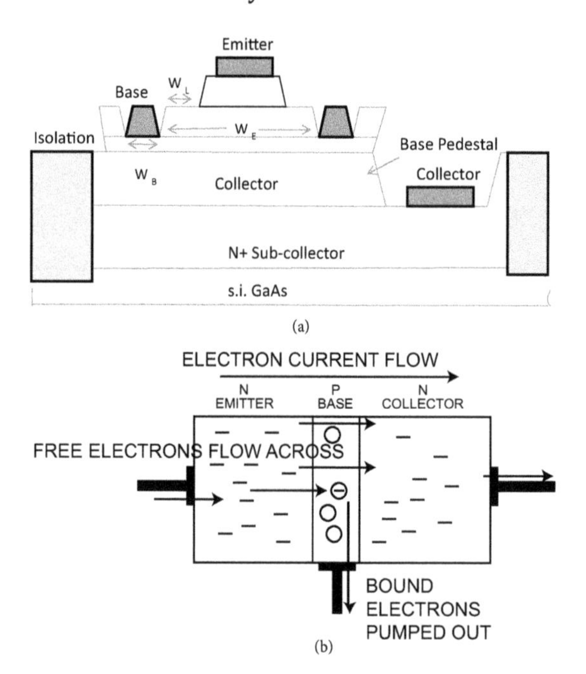

Figure 2.18 Cross-sectional diagram of a GaAs HBT (a) and components of current flow (b).

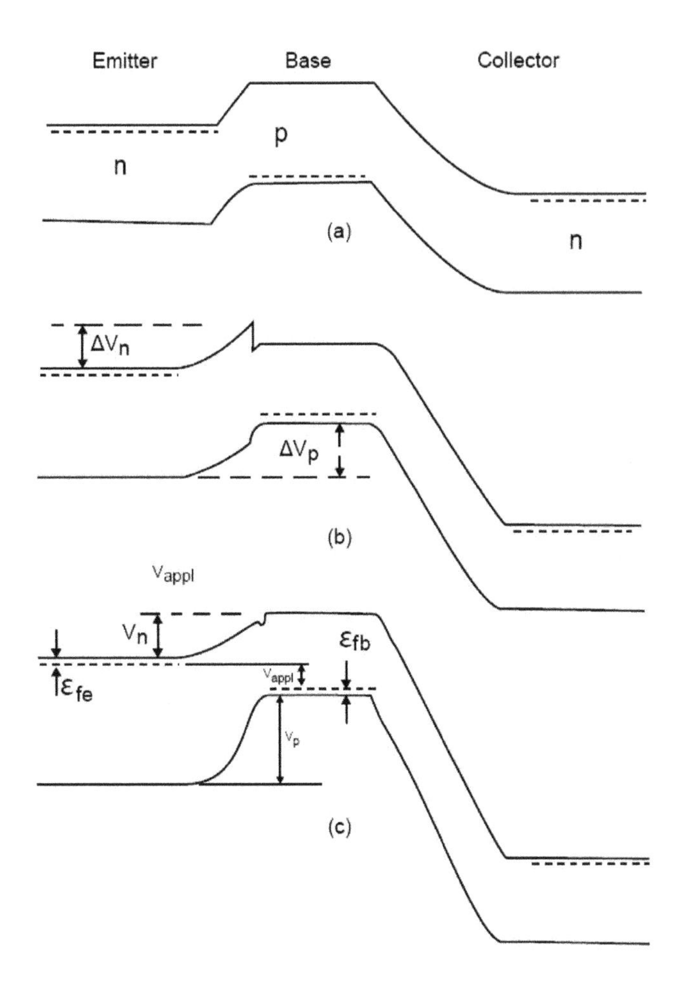

Figure 2.19 Energy band diagram of an n–p–n homojunction bipolar transistor (BJT) (a), for an n–p–n heterojunction bipolar transistor (HBT) (b), and a smooth transition emitter–base junction HBT under forward bias (c) [11].

For a GaAs HBT, the n–p–n configuration is commonly used because of the high electron mobility and poor hole mobility in GaAs. Figure 2.18 shows a cross section of such a device. Therefore the discussion is shifted to electrons as the majority carriers. For large gain, the base current I_b must be small. The major components of the base current are

- Recombination in the base
- Electrons injected from the base to emitter (minority carriers in this case)

Not dominant is a small current due to electrons swept into the base at the reverse-biased collector junction.

The current transfer ratio, α, defined as

$$\frac{I_c}{I_E} = \alpha \tag{2.63}$$

$$\frac{I_c}{I_b} = \beta = \frac{\alpha}{1-\alpha} \tag{2.64}$$

Figure 2.20 shows the equivalent circuit diagram of a BJT device connected in the common emitter configuration and *I–V* characteristics as a Gummel plot (Fig. 2.20a) and the more common I_c family of curves as a function of V_{ce}, as I_b is stepped, for a simple BJT and HBT (Fig. 2.21). Note the offset in the turn on voltage, the V_{ce} offset for the HBT. These will be discussed in a little more detail later.

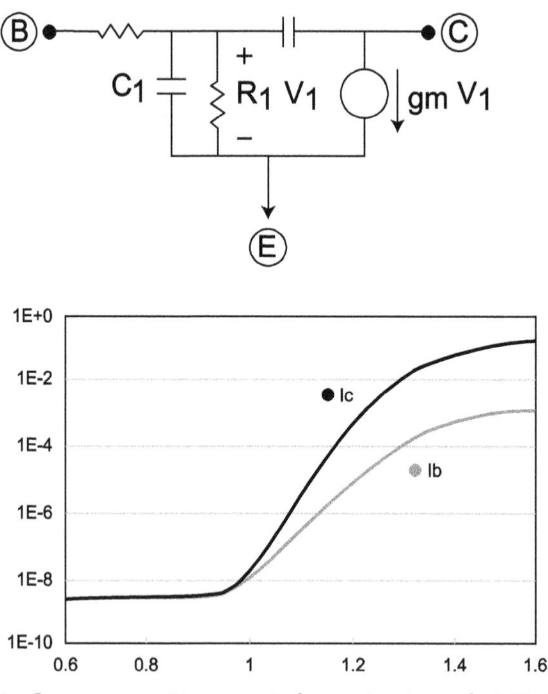

Figure 2.20 Common emitter equivalent circuit and *I–V* (Gummel) characteristics of an HBT.

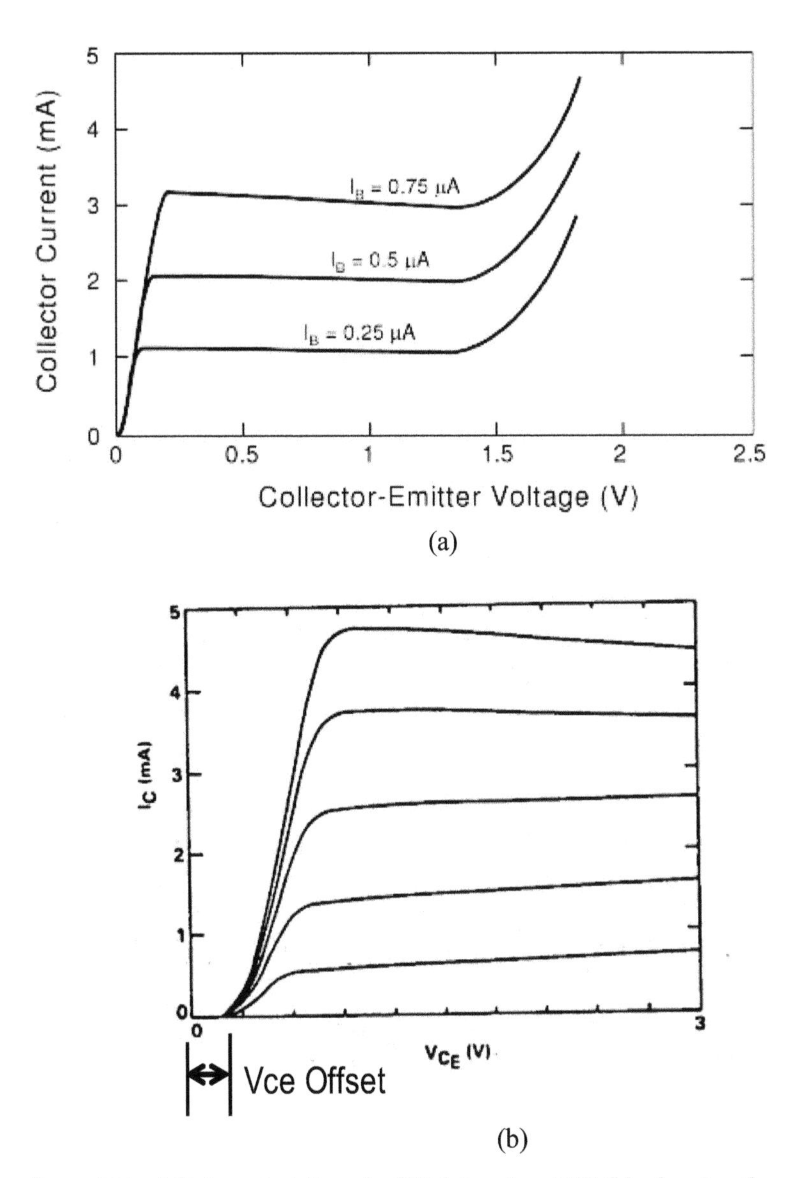

Figure 2.21 *I–V* characteristics of a BJT (a) and an HBT (b), showing the collector current as a function of V_{ce} as the base current is stepped. Note the offset in the turn-on voltage for the HBT (V_{ce} offset).

2.4.2 Current–Voltage Characteristics of a BJT

Static characteristics of the transistor can be derived from p–n junction theory, discussed in detail in various textbooks [1, 3].

For a p–n–p transistor, holes diffuse from the emitter to the collector through the base.

$$I_c = I_1 \exp\left(\frac{qV_{be}}{kT}\right) + I_2 \tag{2.65}$$

where I_2 is the saturation current [1].

For a graded base, where minority carriers can drift in addition to diffusion, the Gummel–Poon model is more accurate.

$$I_c \approx \exp\left(\frac{qV_{be}}{2kT}\right) \tag{2.66}$$

$$\text{The current gain} = I_c / I_b = \frac{\exp\left(\frac{qV_{be}}{2kT}\right)}{\exp\left(\frac{qV_{be}}{kT}\right)} \sim \exp\left(\frac{-qV_{be}}{2kT}\right) \quad \text{........2.4.5}$$

Current-voltage characteristics of a BJT in the form of the log of the collector and base current versus V_{eb} (similar to the one shown for an HBT in Fig. 2.20) is called a Gummel plot.

For a low V_{eb}, the Gummel–Poon model accounts for generation-recombination current by accounting for such effects by the diode ideality factor n:

$$I_b \approx \exp\left(\frac{qV_{be}}{nkT}\right) \tag{2.67}$$

$$\text{Current gain } \beta = I_c/I_b = \exp\left(\frac{qV_{be}}{\left(1-\frac{1}{n}\right)kT}\right) \tag{2.68}$$

$$\beta \approx \exp\{-(1 - 1/n)\} \tag{2.69}$$

The HBT device used extensively in III–V circuit applications will be discussed next.

2.5 HBT Principles of Operation

The cross section of an HBT is shown in Fig. 2.18. The details of the epistructure will be discussed in the chapter on epitaxial growth. The main features are a wide bandgap emitter, for example, n-AlGaAs, and a thin, heavily p+-doped GaAs base layer. Details of processing, for example, matching the lattice constants, etc., are discussed in the chapter on HBT processing.The band diagram of this structure is shown in Fig. 2.19. The emitter bandgap is wider than that of GaAs by at least 10 times kT (250 mV at room temperature). The band diagram is shown under transistor operating conditions, where the emitter–base junction is forward-biased and the base–collector junction is reverse-biased. Electrons are injected from emitter to base, travel across the base by diffusion or drift, and are swept across the base–collector junction by the high field. As in the case of the homojunction transistor, there is a flow of holes from the base to the emitter, adding to the undesired base current. To suppress this, the base doping must be limited, resulting in high base resistance. High base resistance decreases $f_{\max}$ (see Eq. 2.97 later). In an HBT, the wider-bandgap emitter causes a large barrier for the holes, the bandgap discontinuity being larger in the valance band than the conduction band. Thus the base current is suppressed, making it possible to increase the base doping drastically and reduce the base resistance. The emitter doping level can be reduced also, reducing emitter–base capacitance because of increased depletion layer width.

Other inherent advantages that attracted the development of HBTs are radiation hardness and compatibility with opto-electronic devices (direct-bandgap materials). As the cellular phone and other wireless markets appeared, the advantages of HBTs for power amplification and other mixed signal applications were explored and exploited. A brief summary of these follows:

- High maximum frequency of oscillation
- Better linearity (gain vs. power level)
- Excellent threshold control (V_{be}) due to improved epigrowth processes
- High process yields due to the fact that submicron features are not needed like in FET-type devices

2.5.1 Basic Transport Equations [11, 12, 16]

As mentioned earlier, the flow of holes back into the emitter from the base is suppressed by the introduction of a heterojunction between emitter and base. This improves the injection efficiency. A simplified derivation follows. The potential barriers for holes, ΔV_p, and for electrons, ΔV_n, are shown in Fig. 2.19b, which shows energy band diagram details, under low-forward-bias conditions.

For an n-type semiconductor, under equilibrium conditions, the electron concentration in the conduction band, according to Fermi–Dirac distribution, is given by [1]

$$n = N_c \exp\left(\frac{-(E_c - E_f)}{kT}\right) \tag{2.70}$$

where E_f is the Fermi level and E_c is the conduction band energy. Similarly, the hole concentration in the valance band is given by

$$p = N_v \exp\left(\frac{-(E_f - E_v)}{kT}\right) \tag{2.71}$$

The local concentration of electrons $n(x)$ in an n-type semiconductor of fixed doping and composition, under quasi-equilibrium, is given by

$$n(x) = N_c \exp\left(\frac{-\{E_c - qV_{(x)} - E_f(x)\}}{kT}\right) \tag{2.72}$$

where N_c is the effective density of states in the conduction band. In a heterostructure, the material composition varies; E_c and N_c are functions of position. Current density in a semiconductor with a small perturbation from equilibrium is given by

$$J_n = n\,\mu_n \frac{dE_x}{dx} \tag{2.73}$$

where μ_n is the electron mobility.

From Eq. 2.72 above, applying natural log on both sides

$$kT \ln\left\{\frac{n(x)}{Nc(x)}\right\} = -E_c + qV_{(x)} + E_f(x) \tag{2.74}$$

or

$$E_f(x) = E_c - qV(x) + kT \ln\left\{\frac{n(x)}{Nc(x)}\right\} \tag{2.75}$$

Differentiating Eq. 2.75

$$\frac{\mathrm{d}E_\mathrm{f}(x)}{\mathrm{d}x} = \frac{\mathrm{d}E_\mathrm{c}(x)}{\mathrm{d}x}\ q\ \frac{\mathrm{d}V(x)}{\mathrm{d}x}\ kT\left\{\frac{N_\mathrm{c}(x)}{n(x)}\right\} \tag{2.76}$$

Equation 2.73 can be written as

$$J_\mathrm{n} = n\mu_\mathrm{n}\left[-q\frac{\mathrm{d}V}{\mathrm{d}x}+\frac{dE_\mathrm{c}(x)}{dx}\right] + \mu_\mathrm{n}kT\left[\frac{\mathrm{d}n}{\mathrm{d}x}-\frac{n}{N_\mathrm{c}(x)}\left\{d\frac{N_\mathrm{c}(x)}{dx}\right\}\right] \tag{2.77}$$

In the expression above, different terms for different components of current flow can be seen, electron flow by drift by the term $\mathrm{d}V/\mathrm{d}x$, diffusion current by $\mathrm{d}n/\mathrm{d}x$ (keeping in mind the Einstein equation $Dn = \mu_\mathrm{n}kT/q$), and quasi-field due to composition variation by $\mathrm{d}E_\mathrm{c}(x)/\mathrm{d}x$.

Similar equations can be written for holes. These equations, combined with Poisson's equation along with electron and hole continuity equations, form the basic set of equations for an HBT.

2.5.2 Current Gain and Injection Efficiency [11, 12, 15]

In an HBT, holes in the base vastly outnumber the electrons injected from the emitter. For a device in which the bandgap varies linearly from a large emitter (say AlGaAs) to a base (GaAs), see Fig. 2.19c with no abrupt step in the conduction band, under forward-bias conditions

$$J_\mathrm{p} = \mu_\mathrm{p}p\left[-q\frac{\mathrm{d}V}{\mathrm{d}x}+\frac{dE_\mathrm{v}}{dv}\right] \tag{2.78}$$

The collector current density, J_c, is given by

$$J_\mathrm{c} = qn(b_0)v_\mathrm{n} \tag{2.79}$$

where $n(b_0)$ is the number of electrons at the emitter edge of the base and v_n is the effective velocity toward the collector (mostly due to diffusion but includes drift due to the derivative of the bandgap from emitter to base).

$$v_\mathrm{n} = \mu_\mathrm{n}\varepsilon_\mathrm{eff} - \frac{D}{Wb} \tag{2.80}$$

where ε_eff is the effective quasi-electric field from bandgap grading. Similarly, the hole current density J_b is given by

$$J_\mathrm{b} = qp(e_0)v_\mathrm{p} \tag{2.81}$$

where $p(e_0)$ is the number of holes at the edge of the emitter and v_p is the velocity of holes. v_p is governed by diffusion in the presence of recombination:

$$v_p = D_p/L_p \tag{2.82}$$

Current gain is given by

$$\beta = J_c/J_b = \left\{\frac{n(b_0)}{p(e_0)}\right\}\left\{\frac{V_n}{V_p}\right\} \tag{2.83}$$

or

$$\beta = \frac{V_n L_p D_n}{V_p L_n D_p}$$ (2.84) (using Eq. 2.82)

The carrier concentrations are given by equations of the form 2.70 and 2.71:

$$n(b_0) = N_{DE} \exp\left\{-\frac{V_n}{kT}\right\} \tag{2.86}$$

and

$$p(e_0) = N_{AB} \exp\left\{-\frac{V_p}{kT}\right\} \tag{2.87}$$

where N_{DE} and N_{AB} are the doping concentration in the emitter and base, respectively. From Fig. 2.19c

$$V_n = E_{gb} - V_{app} - \varepsilon_{fb} - \varepsilon_{fe} \tag{2.88}$$

$$V_p = E_{ge} - V_{app} - \varepsilon_{fb} - \varepsilon_{fe} \tag{2.89}$$

where E_{gb} and E_{ge} are the energy bandgaps of the base and emitter, respectively, and ε_{fb} and ε_{fe} are the Fermi level positions at the base and emitter band edges, respectively. Combining Eqs. 2.84 to 2.89 into Eq. 2.83, β can be derived as

$$\beta = \frac{V_n}{V_p} \cdot \frac{V}{kT} \cdot \frac{N_{DE}}{N_{AB}} \exp\left\{-\frac{E_{gb} - E_{ge}}{kT}\right\} \tag{2.90}$$

In practice, the linear grading assumption at the emitter–base junction may not hold, or grading may be changed on purpose; therefore there may be a discontinuity in the band diagram at the emitter–base junction, and the actual β may be different.

The bandgap difference between the emitter and the base, $E_{gb} - E_{ge}$, called ΔE_g, is given by

$$\Delta E_g = E_{gb} - E_{ge} = q(\Delta V_p - \Delta V_n) \quad (2.91)$$

Using Eq. 2.82, $V_p = D_p/L_p$ and similar equation for electrons, $V_n = D_n/L_e$, and Eq. 2.89 in Eq. 2.88 and replacing L_e by W_b for short base thickness

$$\beta \frac{L_p}{W_b} \cdot \frac{D_n}{D_p} \cdot \frac{N_{DE}}{N_{AB}} \cdot \exp\left\{\frac{\Delta Eg}{kT}\right\} \quad (2.91)$$

This ratio is called the injection efficiency and depends exponentially on ΔE_g. The term $\exp(\Delta E_g/kT)$ for AlGaAs with an Al mole fraction of 0.3 is 2×10^6. Since the N_{AB}/N_{DE} ratio is large from the injection efficiency optimization, ΔE_g has to be over 240 mV at room temperature to achieve high β (over 100).

Therefore the N_{DE}/N_{AB} ratio can be small, allowing use of high values for $N_{AB.}$ Therefore the doping in the base can be increased to lower down R_B to achieve high f_{max} (figure of merit).

2.5.3 Figures of Merit for HBTs [11, 12]

The most common figures of merit for microwave bipolar transistors, as in the case of FETs described in Section 2.2, are f_t, the cutoff frequency, and f_{max}, the maximum frequency of oscillation. The cutoff frequency corresponds to the frequency at which the incremental short-circuit current gain falls to unity.

$$f_t = \frac{1}{2\pi t} \quad (2.92)$$

where t is the transit time of carriers, across the device control regions, emitter, base, and collector. Over a wide range of frequencies, the absolute value of G_i is,

$$|G_i| = f_t/f \quad (2.93)$$

The maximum frequency of oscillation is defined as the frequency at which the power gain falls to unity. Over a wide range of frequencies, the absolute value of power gain G_p is given by

$$|G_p| \cong f_{max}{}^2/f^2 \quad (2.94)$$

Under optimal input and output matching conditions, the maximum achievable power gain

$$G_p \cong \frac{G_i^2}{4} \frac{Re(Z_{out})}{Re(Z_{in})} \quad (2.95)$$

where Z_{out} and Z_{in} are the output and input impedances of the device, respectively. The equivalent circuit of the HBT is shown in Fig. 2.21a. At high frequencies, the input impedance in $Z_{in} = R_b$. The output impedance depends upon the base–collector junction capacitance C_{bc}:

$$R_e\{Z_{out}\} = \frac{1}{\omega C_{bc}} = \frac{1}{2\pi f_t C_{bc}} \tag{2.96}$$

Using these impedances in Eq. 2.94, f_{max} can be derived as

$$f_{max} = \sqrt{\frac{f_t}{8\pi R_b C_{bc}}} \tag{2.97}$$

To maximize f_{max}, C_{bc} and R_b must be minimized. R_b can be reduced by reducing the sheet resistivity of the base by increasing the doping concentration (without adversely affecting mobility). C_{bc} can be reduced by reducing the area of the base–collector junction and keeping the collector doping low (depletion width high).

Other figures of merit that are also important for power amplifier performance are the V_{ce} offset, on-resistance, and linearity.

The V_{ce} offset, shown in Figs. 2.21b and 2.22, should be small because the knee voltage will be large if the turn-on voltage is large. To get large injection efficiency, a large ΔE_g is needed at the heterojunction. This leads to a high V_{ce} offset. A compromise must be made for overall optimization. Double-heterojunction bipolar transistors have a zero V_{ce} offset; however, epigrowth and device processing are more complicated.

The on-resistance, R_{ON}, should be low as in the case of any device to be used as an amplifier. This resistance depends upon the emitter resistance and the collector resistance. Techniques to reduce these will be discussed in processing chapters.

2.6 PIN Diodes

PIN diodes have been used for high-power-switching applications and photodetectors. GaAs p–n junctions are not a viable device because of low hole mobility, which limits the maximum frequency. In the PIN diode, as p+ and n+ regions are used to contact the semi-insulating or n- GaAs region, the p+ and n+ regions can be considered as metal electrodes. These devices exhibit breakdown voltages as high as 70 V and series resistance of a few ohms. The development of epistructures and fabrication processes for HBTs has led to advances in PIN diode performance.

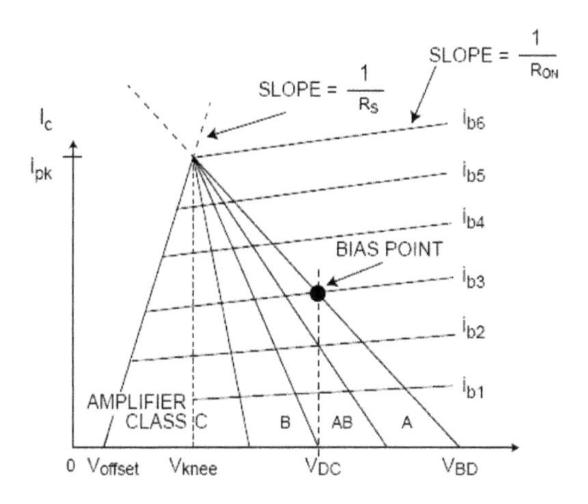

Figure 2.22 Idealized *I–V* family of curves for an HBT, showing the load line and Vce offset. The wasted power area is shaded dark. Courtesy of Skyworks Solutions.

Figure 2.23 shows a schematic and the depletion region, etc.

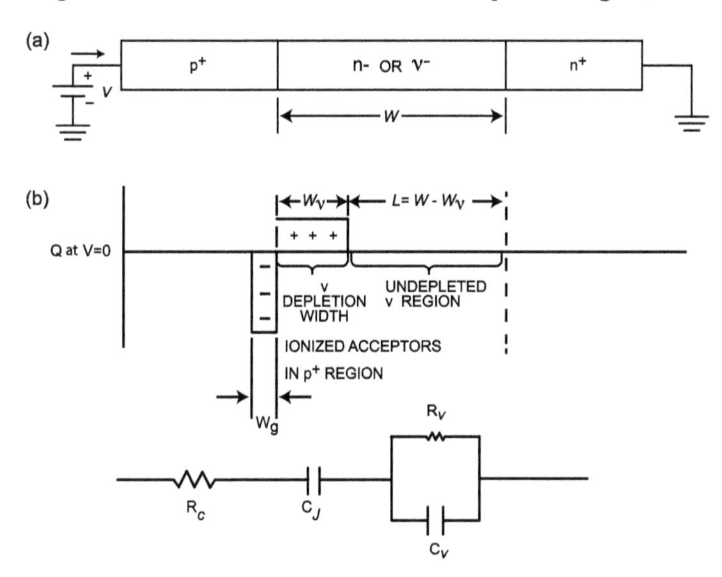

Figure 2.23 PIN diode schematic and depletion region.

2.7 IMPATT

Impact ionization avalanche transit time (IMPATT) diodes are used as solid-state sources of microwave power, the capacitive diode being placed in an inductive cavity to form the resonant system. These diodes employ impact ionization and transit time properties of semiconductors to produce negative resistance behavior. A p–n junction can be operated in IMPATT mode when it is reverse-biased into avalanche breakdown. When this device, which is capacitive in nature, is mounted in a cavity it forms a resonant system at microwave frequencies. IMPATT devices are made in many structural configurations. A basic type is shown in Fig. 2.24 [1–4, 10].

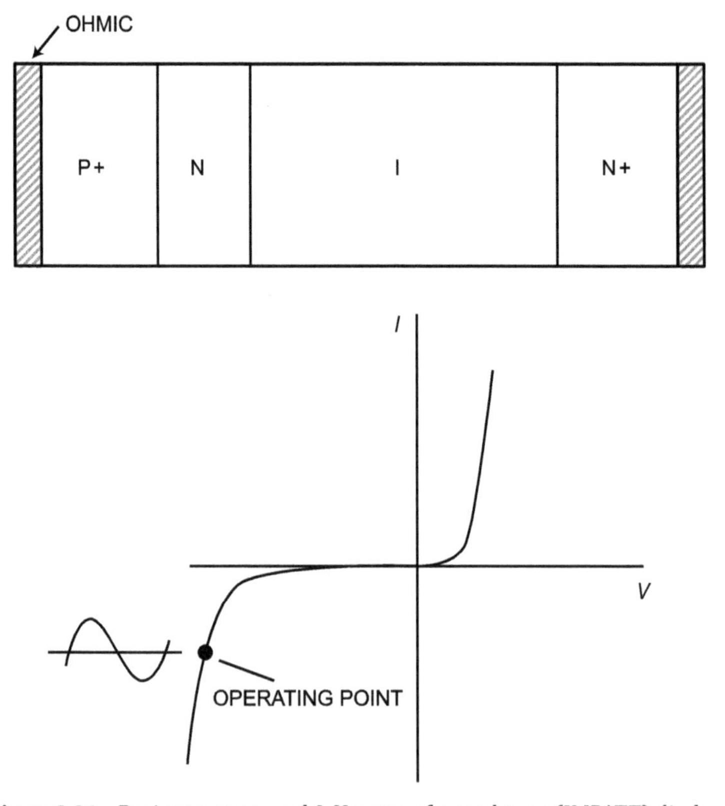

Figure 2.24 Basic structure and *I–V* curve of a read-type (IMPATT) diode.

2.7.1 Read-Type IMPATT

The diode shown in Fig. 2.24 is a reverse-biased p+/n junction that is biased into avalanche breakdown [10]. This also depletes the long intrinsic region. The electrons generated by the avalanche drift across the long intrinsic region and are collected at the n+ contact. The diode geometry is such that the avalanche drift times are such that the radio frequency (RF) current is half cycle out of phase with the RF voltage. As part of the resonant circuit, the diode supplies energy at radio frequencies. Advances in fabrication technology have made it possible to make diodes that work over 60 GHz.

2.8 Gunn Diodes

These devices are not diodes but two terminal devices that exhibit oscillations due to negative differential resistance caused by a transferred electron mechanism. In this mechanism, conduction electrons in the semiconductor are accelerated to such high energy that they shift to another satellite valley of the conduction band that has low mobility. Figure 2.25 shows GaAs that has such a band structure. This type of device was first reported by Gunn (1963) and hence the name. When an electric field is applied to a bar of GaAs, electrons continue to gain energy until they transfer to the L valley, where the effective mass is an order of magnitude larger. The electron velocity drops, as shown in Fig. 2.26, and charge begins to accumulate. Thus a space charge dipole is formed. The electric field is mostly across this domain and no other dipole can form until this one grows and drifts out. In this mode of operation, the frequency is determined by the diode parameters. In practice, other modes of operation are used, where the frequency is controlled by an external circuit. Gunn and other devices utilizing the transferred electron effect can be made with materials that have a band structure similar to GaAs, like InP. The devices are simple to make either in bulk materials or using epilayers where the thickness can be controlled better and high mobility can be achieved by limiting defects. Typical devices use Au Ge/Ni contacts on n+–n–n++ sandwiches.

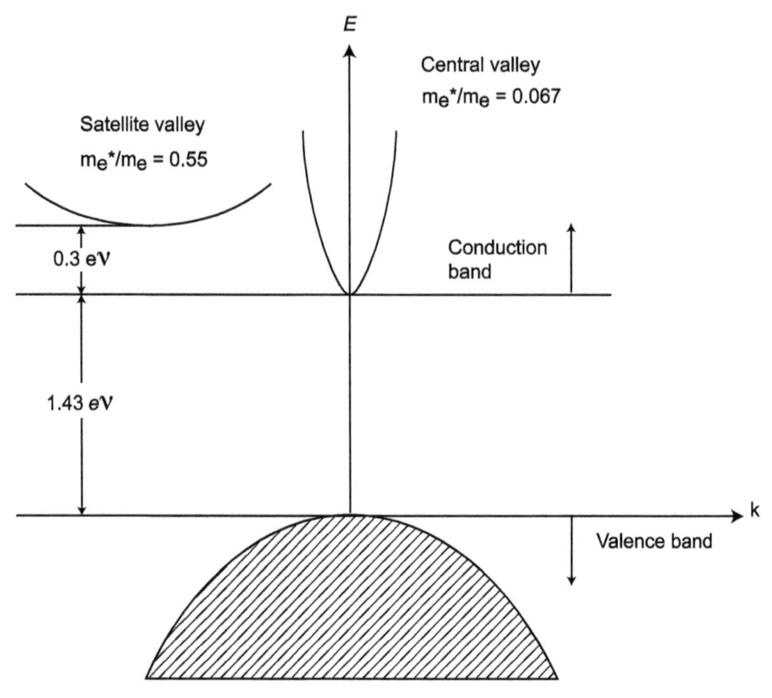

Figure 2.25 Details of the band structure of GaAs showing a satellite valley.

2.9 MOSFET

Complementary metal–oxide semiconductor (CMOS) technology is the most common technology used for silicon ICs. A metal–oxide–semiconductor field-effect transistor (MOSFET) is the basic device type used in CMOS. The basic reason for success of this device type is the fact that native oxide grown on silicon is stable, is an excellent insulator, and has a very low surface state density due to perfect oxide growth. CMOS uses both p-channel and n-channel devices. For III–V materials, native oxide of good quality is difficult to grow (discussed in the chapter on III–V MOSFETs). Therefore, an insulator is deposited or grown by epitaxial methods. MOS-type devices provide numerous advantages over simple MESFET-type devices, the most important ones being high drive currents and low gate leakage. A phenomenological discussion of an MIS device follows. For details, the reader is referred to device physics textbooks [1–4, 13].

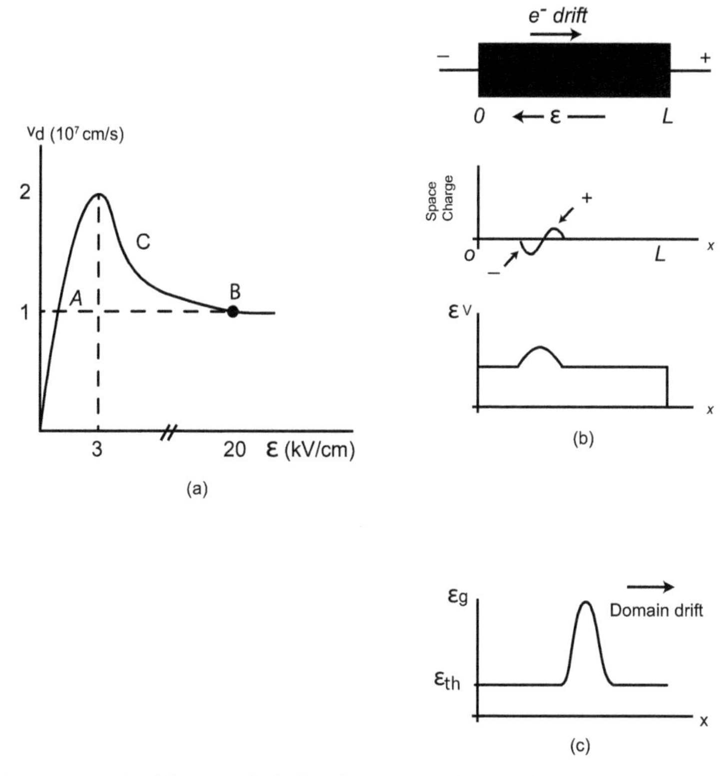

Figure 2.26 Build-up and drift of a space charge domain in GaAs. (a) Velocity-field characteristic for an n-type GaAs, (b) formation of a dipole, and (c) growth and drift of a domain.

2.9.1 Metal–Insulator–Semiconductor Devices [13]

Device structures in which an insulator separates the gate metal from the semiconductor channel are called metal–insulator–semiconductor (MIS) devices. A common example would be a MOS device where the insulator is an oxide. In an ideal MIS device, when a negative voltage is applied to the metal (see Fig. 2.27), a positive charge is induced in the surface space charge region. For an n-type semiconductor, the positive charge will cause inversion and the surface will become p type. If a positive voltage is applied to the metal, a negative charge will be induced. For an n-type semiconductor this will result in accumulation. This space charge region will form a capacitor (C_T). Thus by application of voltage on the metal, the semiconductor space charge region can be varied continuously from inversion to accumulation. Under accumulation, the capacitance of the semiconductor drops to zero. The semiconductor surface works like a capacitor, the whole device being the insulator capacitor in series with a variable semiconductor capacitor. For an equivalent circuit of this structure, see Fig. 2.28a,b.

The capacitance of the semiconductor is given by

$$C_T = C_T \cdot \frac{1+\omega^2 C_T^2 R^2}{\omega^2 C_T^2 R^2}. \tag{2.98}$$

At low frequencies, C_T' approaches infinity; therefore the total capacitance is C_I, the insulator capacitance. At high frequencies, C_T' approaches C_T, so the total capacitance is C_I and C_T in parallel:

$$C = \frac{C_T . C_I}{C_T + C_I} \tag{2.99}$$

The capacitance change with applied voltage is shown in Fig. 2.28c.

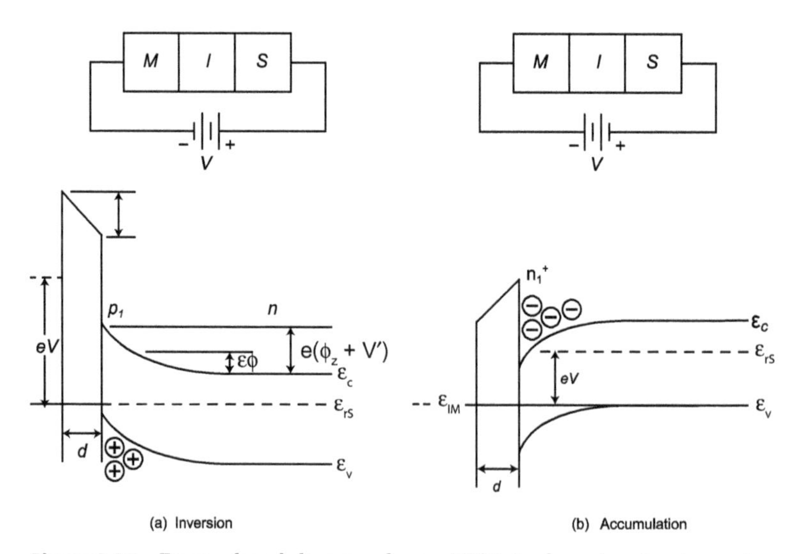

Figure 2.27 Energy band diagram for an MIS interface showing inversion and accumulation for an n-type semiconductor under negative and positive voltage on the gate. From Wang, *Fundamentals of Semiconductor Theory & Device Physics*, 1st Ed. ©1989 Printed and reproduced by permission of Pearson Education, Inc., New York.

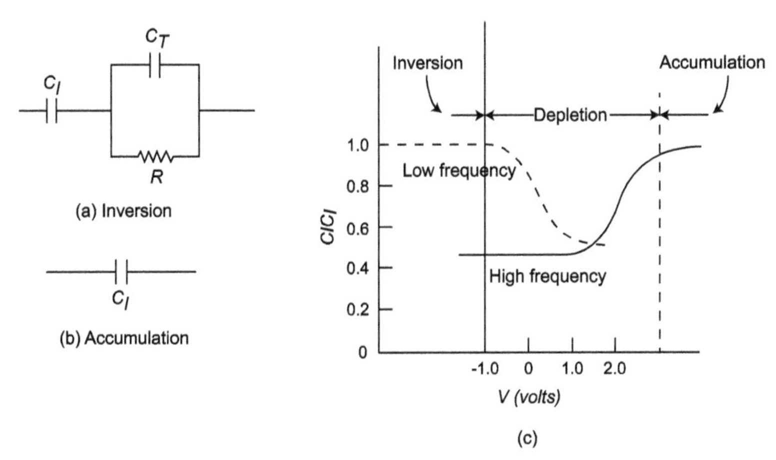

Figure 2.28 Capacitance associated with an MIS interface under inversion (a) and accumulation (b) and capacitance as a function of gate voltage at low and high frequency. From Wang, *Fundamentals of Semiconductor Theory & Device Physics*, 1st Ed. ©1989 Printed and reproduced by permission of Pearson Education, Inc., New York.

2.9.2 *I–V* Characteristics [13]

Figure 2.29 shows a MOS device made on n-type silicon with p+ source and drain contacts [13]. If a sufficient gate voltage V_g is applied, in this case a negative voltage, an inversion layer is formed, in this case a p-type layer (in the n-substrate channel) under the oxide. This layer forms a conductive channel between the two p+ regions. If $V(x)$ is the voltage at a point x in the channel, the voltage drop across the oxide is $V_G - V(x)$. $V(x)$ varies from zero at the source to $-V_D$ at the drain.

From Gauss's theorem, the field E across the insulator (oxide) is $E = \rho_s/\varepsilon_I$, where ρ_s is the total induced charge, which consists of two parts, Q1 stored in the p–n junction and Q2 stored in the p+ channel, which leads to a current.

$$\rho_{s1} = eN_d W \tag{2.100}$$

where W is the width of the inversion layer.

$$\rho_C = \rho_s - \rho_{s1} = \varepsilon_I \frac{-V_G + V(x)}{d} - eN_d W \tag{2.101}$$

For a definition of dimensions, see Fig. 2.29. L is the length of the channel, d is the thickness of the dielectric (or oxide), and Z is the width of the gate (in the depth direction in the diagram).

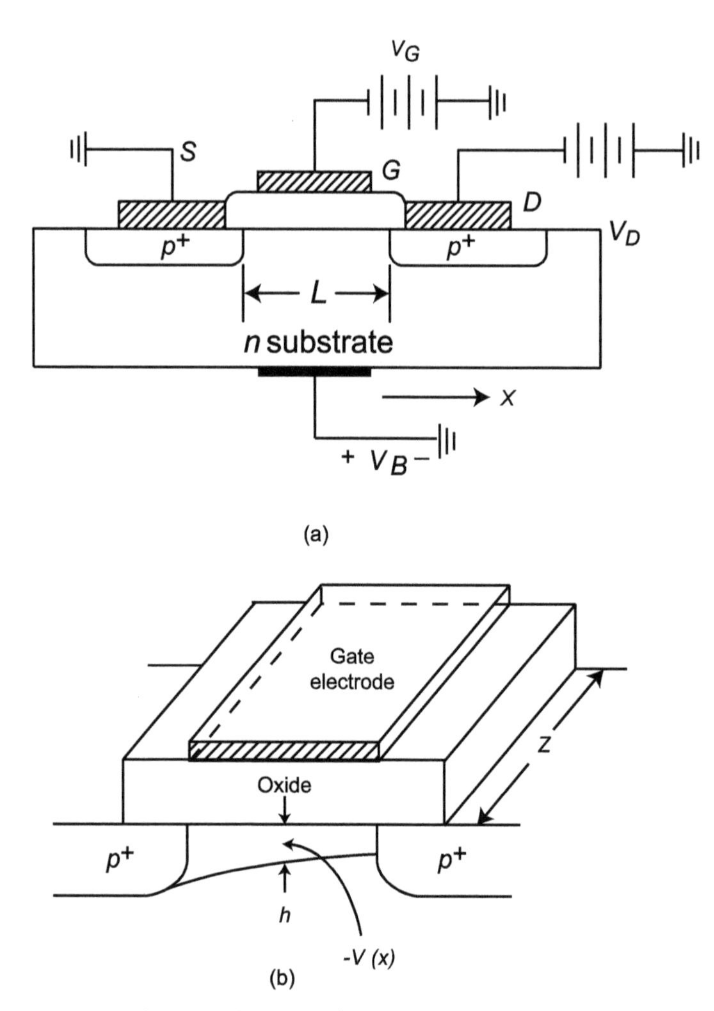

Figure 2.29 Schematic diagram of a MOSFET on an n-silicon substrate (a) and channel formation by inversion (b).

$$I_D = \int_0^h JZdy = \int_0^h ep\mu_p ExZdy \tag{2.102}$$

Hole concentration per cubic centimeter at the surface is

$$\rho_C = \int_0^h e.p.dy \tag{2.103}$$

Using Eqs. 2.101 and 2.103 in Eq. 2.102

$$I_{\mathrm{D}} = Z\mu_{\mathrm{p}}E_{\mathrm{x}}\left\{\varepsilon_1 \frac{-V_{\mathrm{G}}+V(x)}{d} - eN_dW\right\} \tag{2.104}$$

Since $E_{\mathrm{x}} = \dfrac{\mathrm{d}V(x)}{\mathrm{d}x}$ (2.105)

$$I_{\mathrm{D}}\,.\,\mathrm{d}x = \frac{\varepsilon_{\mathrm{I}}Z\mu_{\mathrm{p}}}{d}\{V_{\mathrm{g}} - V(x) + V_0\}\, d\,V(x) \tag{2.106}$$

where $V_0 = \dfrac{eN_{\mathrm{d}}Wd}{\varepsilon_{\mathrm{I}}}$ is defined as the voltage threshold. Integrating over x from source to drain

$$I_{\mathrm{D}} = \frac{\varepsilon_{\mathrm{I}}Z\mu_{\mathrm{p}}}{2Ld}\{2V_{\mathrm{G}} + 2V_{\mathrm{o}} - V_{\mathrm{D}}\}V_{\mathrm{D}} \tag{2.107}$$

Like the FET, the insulated gate FET parameters of importance are the channel conductance, g_{D}, and the transconductance, g_{m}.

$$g_{\mathrm{D}} = \frac{\partial I_{\mathrm{D}}}{\partial V_{\mathrm{D}}} = \frac{\varepsilon_{\mathrm{I}}Z\mu_{\mathrm{p}}}{2Ld}\{V_{\mathrm{G}} + V_{\mathrm{o}} - V_{\mathrm{D}}\} \tag{2.108}$$

and the transconductance $g_{\mathrm{m}} = \dfrac{\partial I_{\mathrm{D}}}{\partial V_{\mathrm{G}}} = \dfrac{\varepsilon_{\mathrm{I}}Z\mu_{\mathrm{p}}}{Ld}V_{\mathrm{D}}$ (2.108)

The I–V characteristics of the MOSFET are similar to those of a GaAs MESFET, except for the fact that much higher voltages can be applied to the MOSFET gate before gate leakage starts. At saturation $g_{\mathrm{D}} = 0$ and $V_{\mathrm{D}} = V_{\mathrm{DS}}$.

$$I_{\mathrm{DS}} = \frac{\varepsilon_{\mathrm{I}}Z\mu_{\mathrm{p}}}{2Ld}V_{\mathrm{DS}}^2 \tag{2.109}$$

In the case of a III–V MOSFET (Fig. 2.30), the insulator is expected to be a deposited layer, for example, atomic layer deposition (ALD)-deposited Al_2O_3. The equations describing an n-type III–V MOSFET are similar to those above.

$$I_{\mathrm{D}} = \frac{\varepsilon_{\mathrm{I}}Z\mu_{\mathrm{n}}}{2Ld}\{2V_{\mathrm{G}} + 2V_{\mathrm{o}} - V_{\mathrm{D}}\}\,V_{\mathrm{D}} \tag{2.110}$$

$$I_{\mathrm{DS}} = \frac{\varepsilon_{\mathrm{I}}Z\mu_{\mathrm{p}}}{2Ld}V_{\mathrm{DS}}^2 \tag{2.111}$$

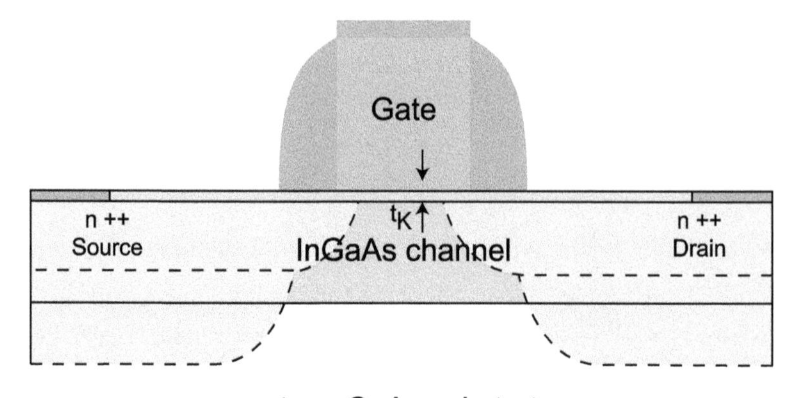

Figure 2.30 Schematic cross section of a III–V insulated gate MOSFET.

From these equations, it can be seen that high tranconductance can be achieved by high μ_n, high ε_I, and small L and d, which are all possible with epitaxially grown InGaAs layers.

2.10 Remarks on Applications

There are numerous applications for all the device types discussed in this chapter. MESFETs, pHEMTs, and HBTs have found their niches. MOSFET-type devices have potential for future new applications.

For low noise, the pHEMT is the best choice. FET-type devices benefit from the capacitive coupling that provides feedback from drain to gate, and HEMT-type devices, have low thermal-diffusion-type noise due to no doping in the channel. The best noise performance is obtained by minimizing the source resistance and maximizing f_t. InP-based devices with InGaAs channels have the best potential in this regard.

For cellular power amplifier applications, InGaP/GaAs HBTs are the best choice currently because of excellent linearity and good power efficiency. For higher frequencies needed in the near future, InP-based HBTs will be needed. Reduction of V_{be} and knee voltage will be needed for reduction of efficiency for good battery life.

For very-high-power applications, up to 150 W, needed for base stations, GaN-based devices are a good candidate.

Appendix

Some Extra Useful HBT Equations from the Processing Point of View

$$V_{\text{be}} = \frac{E_{\text{gb}}}{q} - \frac{kT}{q}\ln\frac{QD_{\text{n}}N_{\text{c}}N_{\text{v}}}{Jcp_{\text{b}}w_{\text{b}}}$$

$$V_{\text{be}} = \frac{kT}{q}\ln\left(\frac{p_{\text{b}}w_{\text{b}}I_{\text{c}}}{qD_{\text{n}}ni^{2}A}\right) + \frac{R_{\text{b}}I_{\text{c}}}{\beta} + \frac{\text{R}_{\text{e}}I_{\text{c}}(\beta+1)}{\beta} \quad [17]$$

Collector current J_{c}

$$J_{\text{c}} = \frac{qD_{\text{n}}nib^{2}}{p_{\text{b}}w_{\text{b}}}\exp\frac{(qV_{\text{be}})}{kT}$$

$$I_{\text{c}} = A_{\text{e}}J_{\text{c}}$$

where A_{e} = emitter area.

$$\beta = \frac{v(n)}{v(p)}\frac{Nde}{qAab} - \exp\left(\frac{-(E_{\text{gb}} - E_{\text{gc}})}{kT}\right)$$

β is proportional to $\exp\left(\frac{\Delta E_{\text{g}}}{qkT}\right)$

where $\Delta E_{\text{g}} = E_{\text{gb}} - E_{\text{gc}}$.

References

1. S. M. Sze, *Physics of Semiconductor Devices*, John Wiley and Sons (1970).
2. E. S. Yang, *Fundamentals of Semiconductor Devices*, McGraw Hill, New York (1978).
3. B. G. Streetman, *Solid State Electronic Devices*, Prentice Hall, New Jersey (1980).
4. D. Biswas and D. A. Neamen, *Semiconductor Physics and Devices*, 4th Ed., McGraw Hill, Special Indian Edition (2012).
5. R. Williams, *Modern GaAs Processing Methods*, Artech House, Boston (1990).
6. C. A. Mead, Schottky barrier gate field effect transistor, *Proc. IEEE*, **54**, 307 (1966).
7. H. Fukui, Optimalnoise figure in microwave GaAs MESFET, *IEEE Trans. Electron. Devices*, **ED-26**, 1032 (1979).
8. J. Munn, Ed., *GaAs Integrated Circuits*, Macmillan, New York (1988).

9. F. Ali and A. Gupta, Eds., *HEMTs and HBTs, Devices, Fabrication and Circuits*, Artech House, Boston (1991).
10. S. Kayali, G. Ponchak and R. Shaw, GaAs MMIC reliability assurance guideline for space applications, JPL Publication 96-25 (1996).
11. M. F. Chang and P. M. Asbeck, III-V heterojunction bipolar technology, *Int. J. High Speed Electron.*, **1**, 245 (1990).
12. M. Frank Chang et al. *Current Trends in Heterojunction Bipolar Transistors*, p. 54, Google Books (1996).
13. S. Wang, *Fundamentals of Semiconductor Theory and Device Physics*, Prentice Hall International (1989).
14. P. Zampardi, Personal communication (2009).
15. B. L. Sharma and S. C Gupta, Metal semiconductor Schottky barrier junctions, *Solid State Technol.*, p.97 Part I (May 1980) and p.90 part II (June 1980).
16. P. M. Asbeck, M. F. Chang, K.C Wang and D. L. Miller, Chapter 4, in *Introduction to Semiconductor Technology, GaAs and Related Compounds*, Ed., Cheng T. Wang, Wiley, New Jersey (1990).
17. W. Liu, *III-V Heterojunction Bipolar Transistors*, Wiley Interscience, New York (1998).

Chapter 3

Phase Diagrams and Crystal Growth of Compound Semiconductors

3.1 Phase Diagrams

3.1.1 Introduction [1, 2]

Different semiconductor and metallic materials are used in semiconductor manufacturing. Binary and ternary semiconductor compounds are designed into the devices. Metallization schemes, involving multiple layers of different metals, are used for contacts and interconnects. These layers should not intermix to form undesired combinations or phases. Doping is done on purpose, so an understanding of solid solubility is needed. Therefore, knowledge of phase diagrams is useful for semiconductor engineers.

A *phase* is a state in which a material may exist and exhibit a set of uniform properties. A material usually changes phase with temperature or pressure. Water (H_2O) is a good example of this. Unitary-phase diagrams can be drawn for this material with vapor pressure and temperature as the axes, with areas showing solid, liquid, and gaseous phases. A phase diagram usually is a map of the equilibrium phases with temperature and composition as axes. Phase data is collected ideally under equilibrium conditions, but for solid materials, quasi-equilibrium is more realistic.

III–V Integrated Circuit Fabrication Technology
Shiban Tiku and Dhrubes Biswas

ISBN 978-981-4669-30-6 (Hardcover), 978-981-4669-31-3 (eBook)
www.panstanford.com

Binary-phase diagrams are most commonly used in the study of metal and semiconductor material systems. Figure 3.1 shows a hypothetical diagram for explaining the terminology. The two components A and B are the ends of the x axis. The y axis is the temperature. Pressure is set to 1 atm. At the two ends, close to pure A or B, as the temperature goes up, the solid phase (α or β) changes to liquid as it crosses the liquidus line. At any composition (% A), as the temperature goes up, the solid phase changes to a mixture of either phase, including the liquid phase, and then to just the liquid phase. The lowest melting point is seen at point E, the eutectic point.

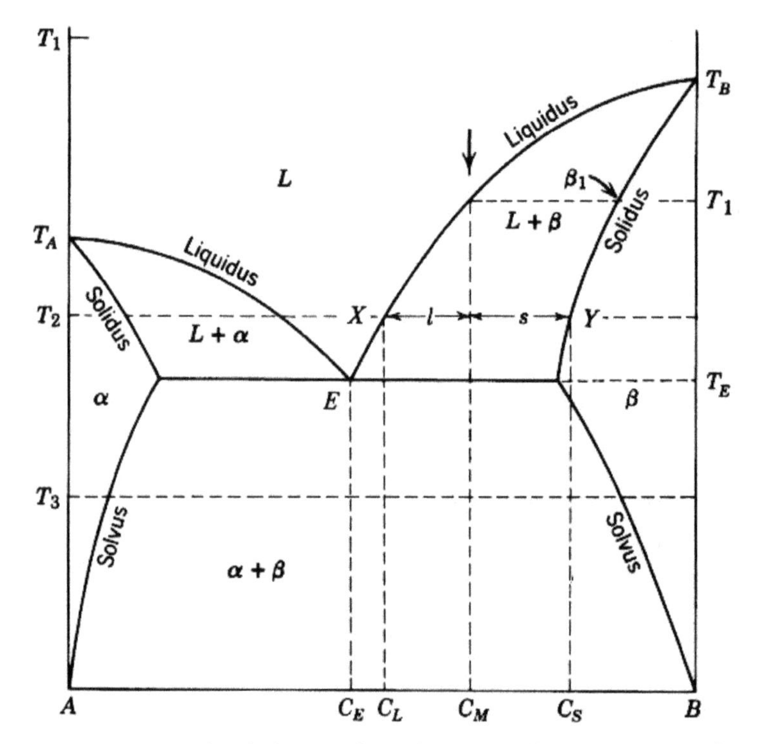

Figure 3.1 Example of a binary-phase diagram, with composition as the x axis and temperature as the y axis. α and β are the solid phases of A and B; L is the liquid.

3.1.2 Phase Diagram Types

A few simple phase diagrams are discussed below that are more relevant to semiconductor processing.

3.1.2.1 Isomorphous phase diagram

This type of phase diagram is shown in Fig. 3.2, which shows the phase diagram for the GaAs–GaP system [3, 1]. This type of diagram is characteristic of materials that are completely soluble in each other. Ga being common, As and P occupy the group V sites randomly, the atomic radii being within 15% of each other. Si and Ge form a similar phase diagram. For such material combinations, other properties may vary monotonically also, such as energy bandgap, lattice constant, etc.

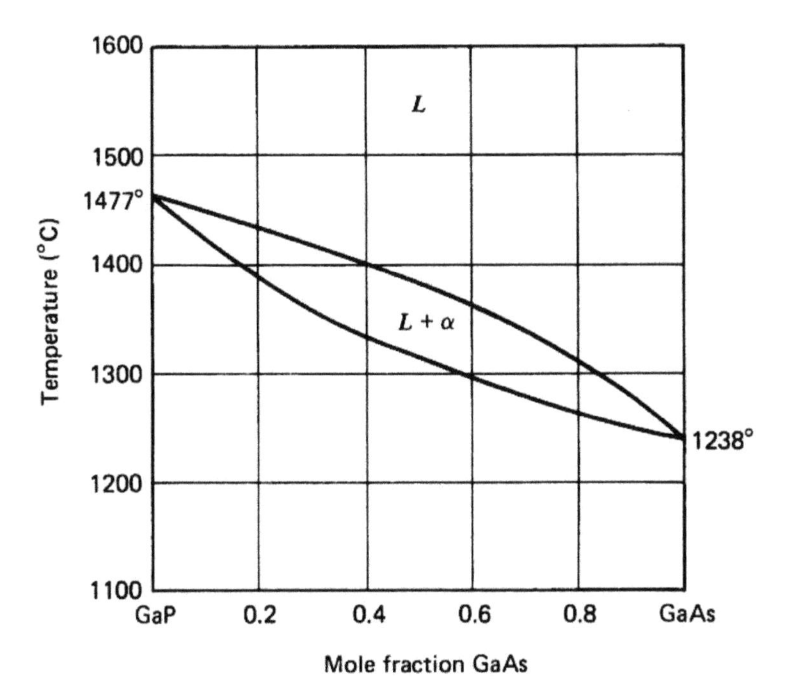

Figure 3.2 The GaAs–GaP system. From Ref. [1]. Copyright © 1983. Reproduced with permission of John Wiley & Sons, Inc.

3.1.2.2 Eutectic diagrams

These phase diagrams result when addition of a component drops the melting point of the other one, as shown in the generic binary phase diagram of Fig. 3.1. The minimum value is called the eutectic point. The most common examples known to semiconductor engineers are the Au–Si and Au–Ge systems, the latter shown in Fig. 3.3. The components are completely miscible in each other in the liquid phase but, unlike the isomorphous case, are not fully soluble in the solid state. The drop in melting point in the Au–Si and Au–Ge systems is huge, so these alloys are useful as solders and for making contacts at reasonably low temperature.

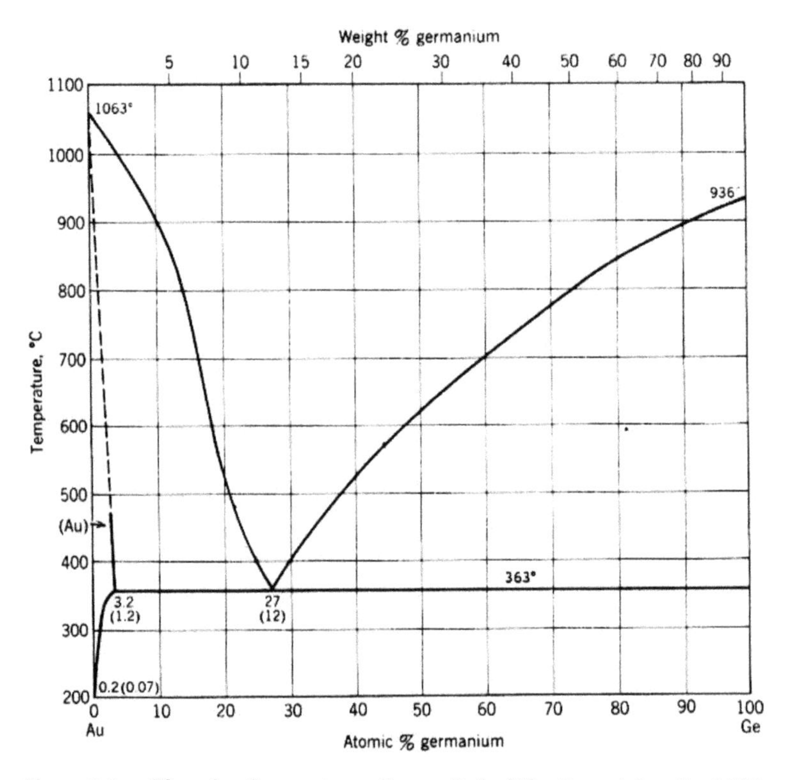

Figure 3.3 The Au–Ge system. From Ref. [1]. Copyright © 1983. Reproduced with permission of John Wiley & Sons, Inc.

3.1.2.3 Peritectic diagrams

When multiple compounds are formed in a binary system, a peritectic phase diagram can result, as shown in Fig. 3.4 for a Au–Ga system. In this case, a eutectic low point occurs between Au and AuGa and another one between AuGa and $AuGa_2$.

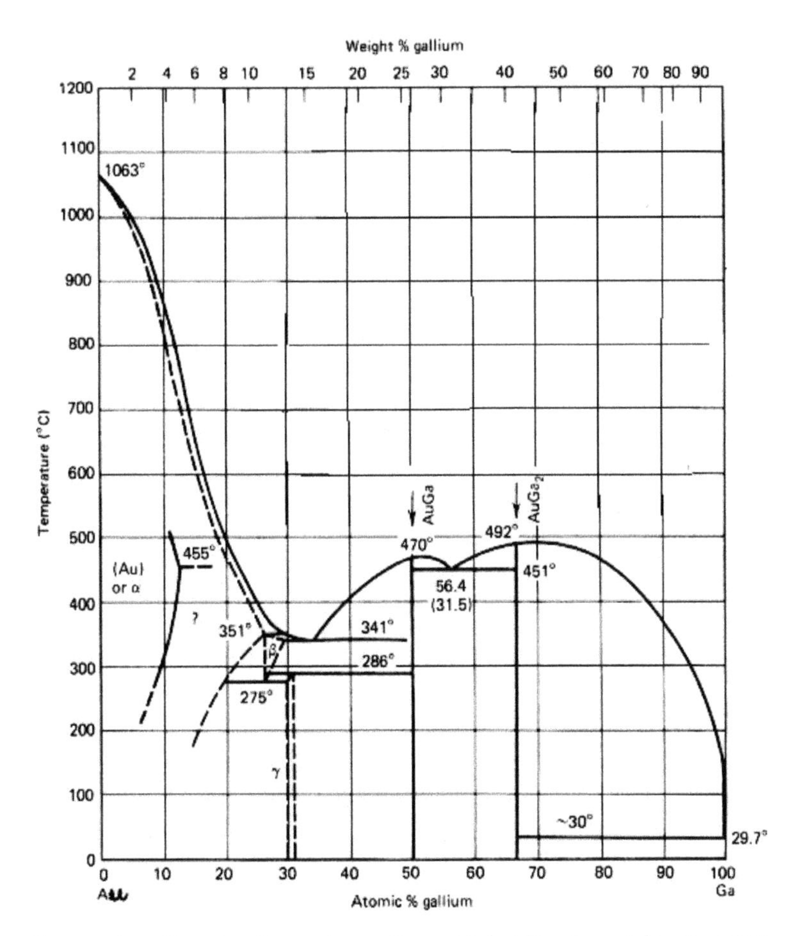

Figure 3.4 The Au–Ga system. From Ref. [1]. Copyright © 1983. Reproduced with permission of John Wiley & Sons, Inc.

3.1.3 Congruent Transformation

III–V compound semiconductors are not mixtures of group III and group V elements but actual compounds that exist in an extremely low composition range. The phase diagram of the Ga–As system is shown in Fig. 3.5. A discreet line represents GaAs. Here one phase changes to another without any change in composition. The vertical GaAs line effectively splits the phase diagram into two diagrams, which could be studied separately. This type of phase transformation is called congruent transformation.

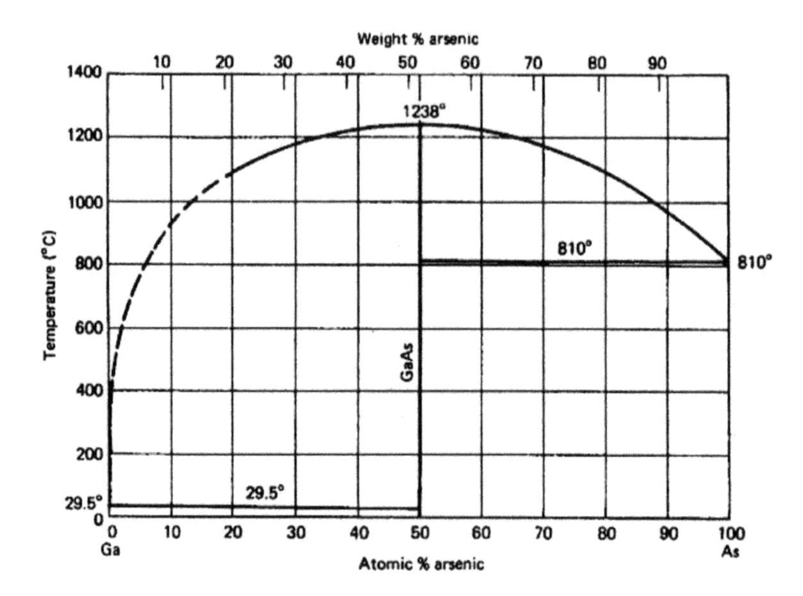

Figure 3.5 Gallium–arsenic phase diagram. From Ref. [1]. Copyright © 1983. Reproduced with permission of John Wiley & Sons, Inc.

In reality, in III–V compounds, the congruent melting line actually has a finite width. So, on an expanded scale, shown in Fig. 3.6 for GaAs, the solid phase has a finite width, with small excess of either Ga or As. This excess may be only a few hundred parts per million (ppm) but is too high for an ideal semiconductor. The excess of one component, for example, Ga, means there are vacancies of the other one, As. These act as electronic donors or acceptors, as discussed in the section on defects in Chapter 1.

3.2 Crystal Growth

Semiconducting materials used in integrated circuit (IC) fabrication are single-crystal wafers that are sliced from boules or ingots. The boules are grown in sizes that can be up to 150 mm (6 inches) in diameter for III–V materials and up to 300 mm for silicon. There are numerous methods of growing these large crystals that basically fall into a few categories. These will be discussed in this chapter, along with the necessary concepts needed for understanding the preparation and properties of pure and doped materials.

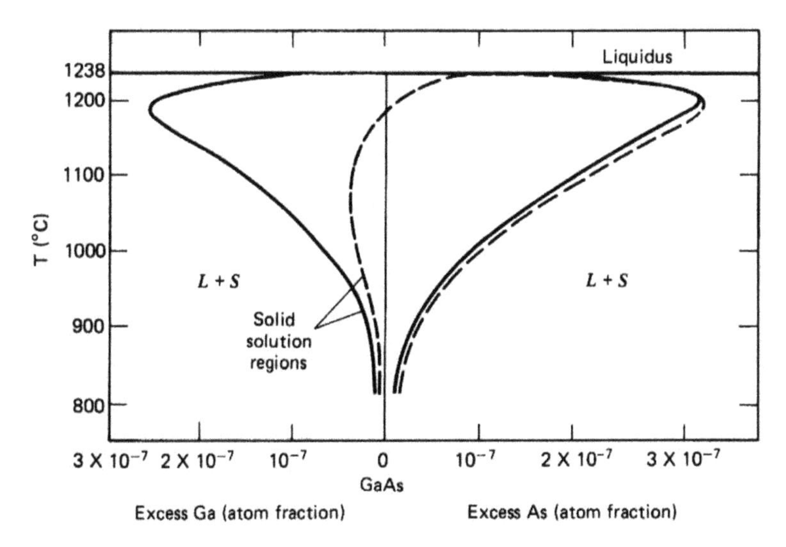

Figure 3.6 Details of the GaAs phase diagram with composition scale expanded near the congruent melting line. From Ref. [1]. Copyright © 1983. Reproduced with permission of John Wiley & Sons, Inc.

3.2.1 Starting Materials and Compounding Method [1]

The starting materials for making GaAs crystals are chemically pure gallium and arsenic. The phase diagram of the gallium–arsenic system is shown in Fig. 3.5 [2]. As can be seen in the phase diagram, the melting point of GaAs is 1238°C. At this temperature the formation reaction from Ga and As is violently exothermic. The vapor pressure of arsenic is much higher than that of gallium, so synthesis of GaAs must be carried out under overpressure of arsenic. Figure 3.7 shows the pressure versus temperature diagram for different species over GaAs. It is seen that at least 1 atm. pressure is needed at the melting point [3]. The reaction can be carried out in a sealed tube (as it was historically) or by providing arsenic overpressure using a separate arsenic source at lower temperature. The temperature of the arsenic source needed can be figured out from the *P*–*T* diagram to be about 620°C; see Fig. 3.8 for the relevant portion of the *P*–*T* diagram that shows pressure of As vapor over As melt [4]. Gallium arsenide polycrystalline ingots weighing several kilograms can be made in this manner.

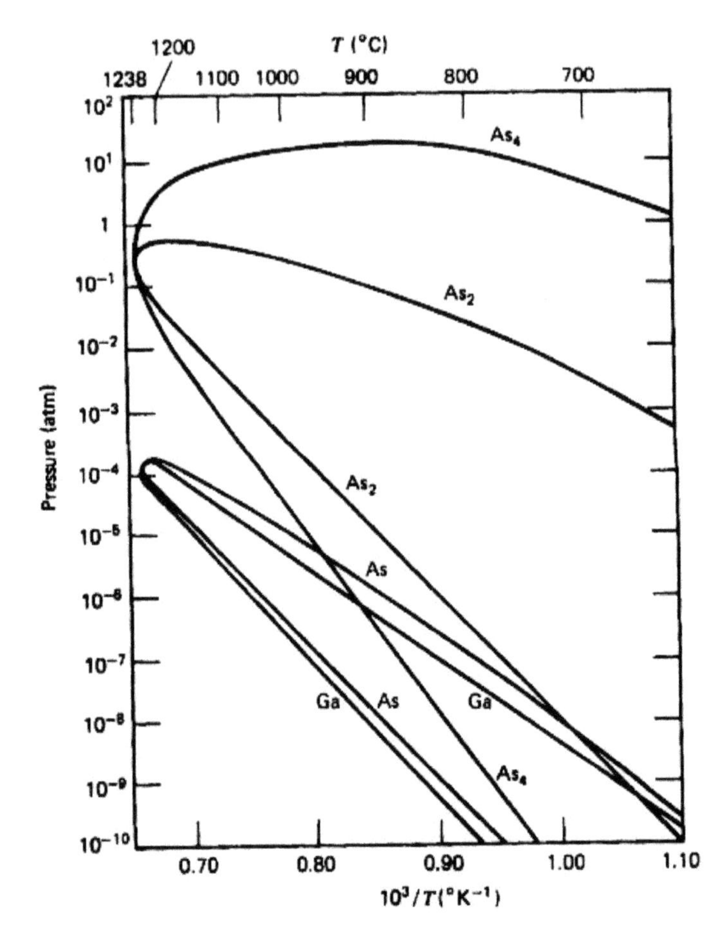

Figure 3.7 *P–T* diagram shows partial pressure of gallium and arsenic over gallium arsenide as a function of temperature. Reprinted from Ref. [4], Copyright (1967), with permission from Elsevier.

3.2.2 Single-Crystal Growth

Single crystals of GaAs and other III–V materials are grown from a melt by using a single-crystal seed to start the growth. Historically, two basic growth techniques were employed, the Bridgman and the Czochralski method [1, 6, 7]. Modern growth techniques are improvements over these basic methods, improvements being necessary for defect and resistivity control of large-size boules and scaling for cost reduction.

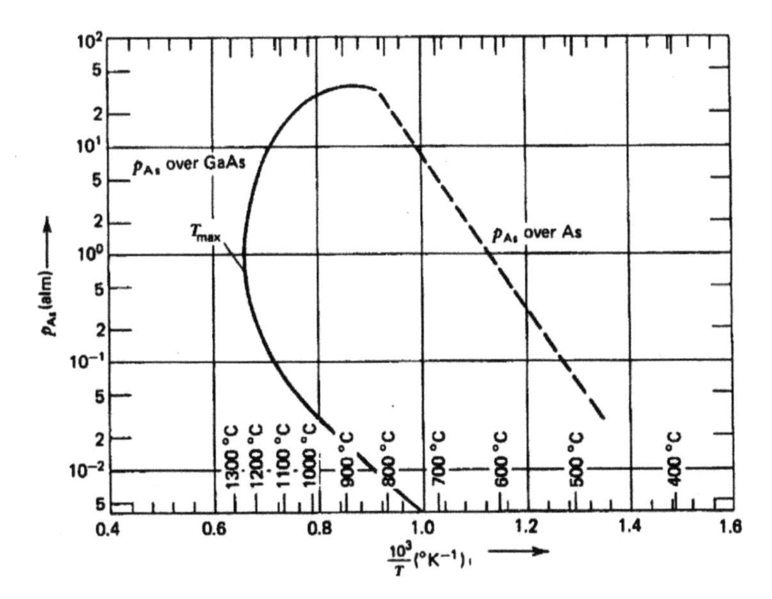

Figure 3.8 Details from the *P–T* diagram for gallium arsenide. From Ref. [1]. Copyright © 1983. Reproduced with permission of John Wiley & Sons, Inc.

3.2.2.1 Bridgman/gradient freeze technique

A polycrystalline ingot is melted in a long vitreous carbon boat placed inside a sealed quartz ampoule. The melt is cooled from one end, which is necked down to minimize nucleation of more than one single crystal. A single-crystal seed is placed on the cool end and the ampoule cooled by moving it through the temperature gradient to the cold zone of the furnace. Figure 3.9 shows a schematic of the apparatus and the temperature profile.

The polycrystalline ingot can be replaced by elemental starting materials for saving processing steps. Temperature gradients of the order of 10°C/cm and travel speeds of a few microns/sec are used typically. Historically, horizontal Bridgman crystals with a noncircular shape used to be grown along the <111> orientation and then sawed at an angle to result in (100) surface wafers. Even wafers with the shape of the letter D were used in research and development of GaAs ICs. Other disadvantages of this technique, in addition to the noncircular shape, are the difficulty of growing semi-insulating material and larger-size wafers. Today 100 and 150 mm diameter wafers are grown by the vertical gradient freeze

(VGF) technique. Lowest etch pit densities are reported by the VGF technique compared to the liquid-encapsulated Czochralski (LEC) techniques. Pioneered by AT&T Bell Laboratories, and later improvised independently to commercial perfection by AXT, this has become the workhorse substrate of the epitaxial high-volume manufacturing industry. This is mainly due to scalability, low stress, mechanical strength, and of course the lowest etch pit density (EPD) of the order of under 1000/cm^2 (around 40,000/cm^2 for light-emitting diode [LED] substrates). This has tremendous impact on the high reliability of heterojunction bipolar transistor ICs [8]. Currently, the feed material may be presynthesized in a high-pressure furnace [6]. A crucible of pyrolytic boron nitride (PBN) is used and dopant concentration is controlled by the partial pressure of oxygen and the activity of carbon (to be discussed later in this chapter). State-of-the-art material with low defects (EPD) and excellent electrical properties control have been produced.

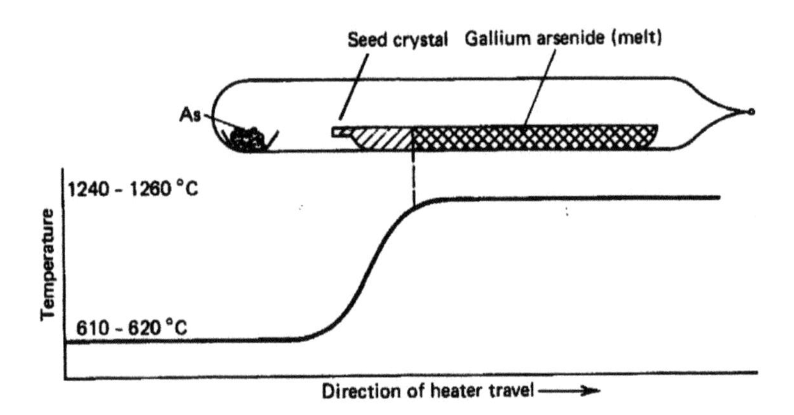

Figure 3.9 Schematic diagram of the gradient freeze system for crystal growth.

3.2.2.2 Liquid-encapsulated Czochralski method [9]

In boat-grown methods the choice of the boat material is critical. As the crystal is cooled after growth, it expands compared to the boat, causing defects to form near the periphery. In the Czochralski method, shown in Fig. 3.10a, the growing crystal is pulled from the melt in the vertical direction, so no crucible or boat is required.

However, for GaAs, arsenic overpressure is still required. This can be done by keeping the growth apparatus at over 600°C; however, the choice of materials to be used for the apparatus is limited because of the reactive nature of arsenic. The LEC method has been developed to overcome these limitations [7]. In this method, a liquid of an inert, light material is used to cover the melt and an inert gas used to provide pressure. See Fig. 3.10b for details of the melt and the encapsulating material. B_2O_3 has been found to be perfect for this application. LEC and its variations have been used for decades as GaAs and other III–V materials have progressed in size from 2-inch to 6-inch wafers.

3.2.2.3 Vertical boat and vertical gradient freeze methods

As the crystal diameter increases, the generation of defects and slip dislocations increases. The crystals tend to twin. One method to reduce this is to reduce the temperature gradient at the solid–liquid interface, and this has been done by enclosing the growing crystal in a boat, hence the name vertical boat (VB). Availability of PBN crucibles has made this practical. A schematic diagram of the VB method is shown in Fig. 3.11; the differences between the VB method and conventional LEC can be seen in this diagram. GaAs presynthesized crystals are charged into a PBN crucible, which is mounted on a crucible support and heated in a vertical furnace. Notice the position of the seed at the bottom in comparison to Fig. 3.10 for the LEC method. The ingot is melted and moved downward relative to the furnace, resulting in the growth of <100> single crystals. The reduced temperature gradient results in less strain and better defect-free crystals. The containment of the crystal in the crucible also eliminates the need for diameter control. The resistivity, background impurities, and defects can be controlled as in other methods.

In the more modern version of this method, commonly known as the VGF method, a silica vessel is used. See VGF in Fig. 3.12 for details; notice the position of the seed and the addition of B_2O_3 encapsulation. Numerical modeling is used to optimize the temperature gradient. N_2 pressure is used, if needed (e.g., for InP).

No rotation is used (unlike LEC), and there is no mechanical movement of the crystal (only the furnace moves at an extremely low rate).

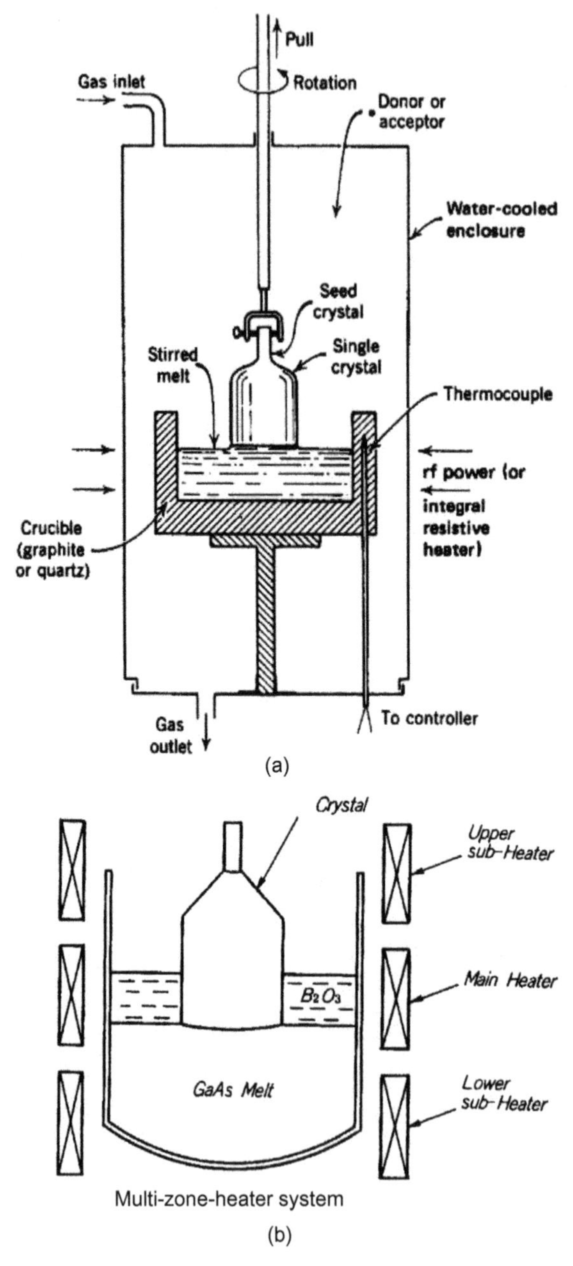

Figure 3.10 (a) Czochralski crystal growth system. (b) Czochralski crystal growth system showing details of the melt and liquid encapsulation.

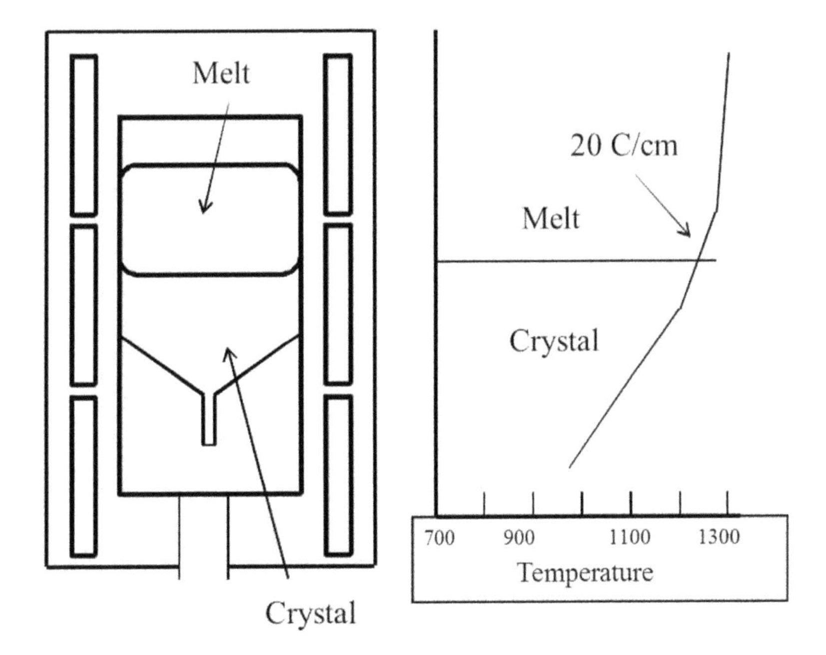

Figure 3.11 Schematic diagram of the vertical boat method.

3.2.2.4 Vapor pressure–controlled Czochralski method

In the vapor pressure-controlled Czochralski (VCZ) method, the vapor pressure is controlled to suppress nonstoichiometery due to high vapor pressure of one constituent. Otherwise the method is similar to LEC. In the case of InP growth, a heated source of phosphorus is provided and overpressure provided by N_2.

A comparison of the three main methods of crystal growth is shown in Fig. 3.12.

3.2.3 InP Crystal Growth [10, 11]

Ternary and quaternary III–V compounds are being studied for numerous electronic and opto-electronic applications tor very high speeds over 30 GHz. InGaAs and InGaAsP are good examples of these materials. These can be grown lattice-matched to InP (to be discussed in more detail in Chapter 4). InP growth is made difficult by the high vapor pressure of phosphorus, so dissociation needs to be suppressed. VCZ-type techniques are needed. Figure 3.13 illustrates the use of a pressure vessel, a phosphorous vapor source, and B_2O_3 encapsulation. One-hundred-millimeter-diameter <100> crystals have been grown. A low EPD with a temperature gradient of 10°C/cm and a growth rate of 0.4 mm/hour has been demonstrated [11]. Large-diameter crystals are being grown by the VGF technique in production volumes at 100 mm in growth systems, as shown in Fig. 3.12.

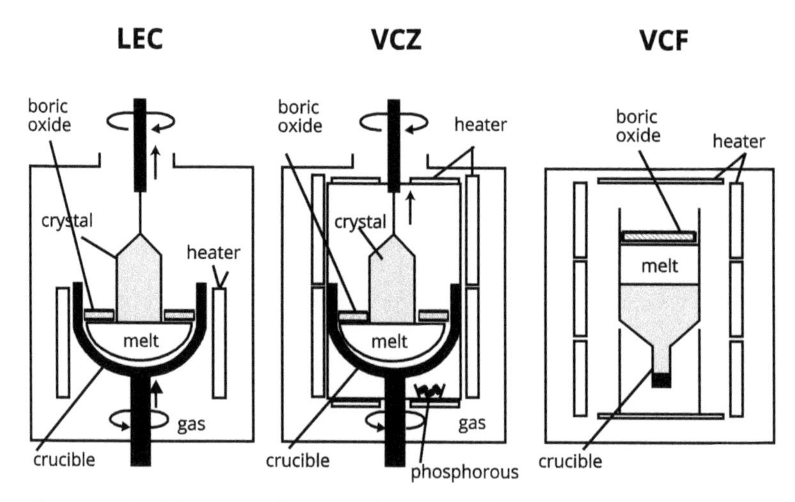

Figure 3.12 Schematic diagram showing differences between LEC, VCZ, and VGF crystal growth techniques. © 2002 IEEE. Reprinted, with permission, from Ref. [10].

3.3 Doping and Resistivity Control

As mentioned previously, bulk GaAs cannot be made in pure form because of the vapor pressure differences of As and Ga over molten GaAs. It is also difficult to avoid unintentional doping by Si, O, and C from the materials used for the growth apparatus, for example, quartz. Historically Cr was intentionally added to compensate for an n-type Si dopant and achieve a semi-insulating material for LEC-grown crystals. Cr is a deep-level impurity in GaAs. With the use PBN, EL2 defects (which are formed as a vacancy complex) serve to compensate for carbon; this material is called undoped, but carbon control becomes critical. This will be discussed further under field-effect transistor (FET) processing. Resistivity control is an important issue in producing GaAs substrates. In general, wafers must be semi-insulating with a resistivity greater than 10^7 ohm-cm. For ion-implanted FET processes, exact resistivity control is more important for threshold voltage control. Resistivity of the currently grown material depends upon carbon and EL2 defect concentration. By controlling carbon concentration, for example, by controlling partial pressure of CO over the melt, desired resistivity can be achieved [12]. Figure 3.14 shows the resistivity versus carbon concentration data for some production boules [13].

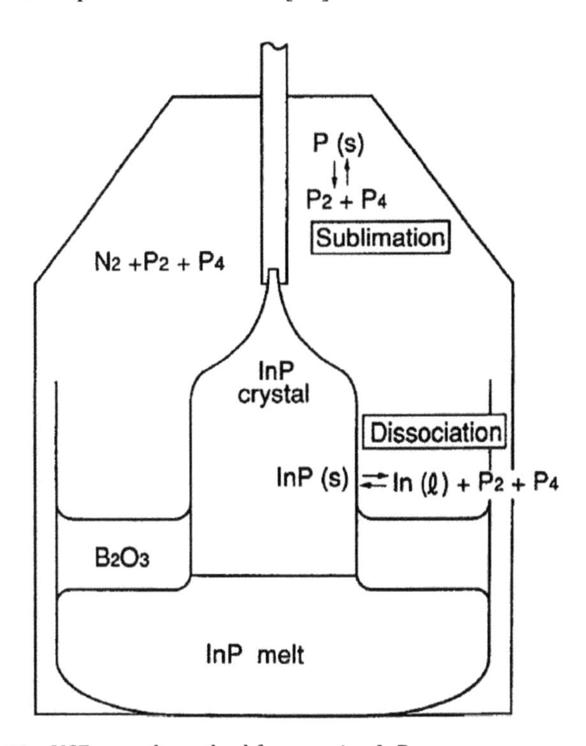

Figure 3.13 VCZ growth method for growing InP.

Grown crystals are annealed in tube furnaces for better uniform quality. The crystal is annealed at high enough temperature, at which defects and dopants are homogenously distributed. Under ideal conditions of no thermal gradients or shock, annealing can take too long. For economic reasons, temperature ramp rates are carefully chosen by modeling.

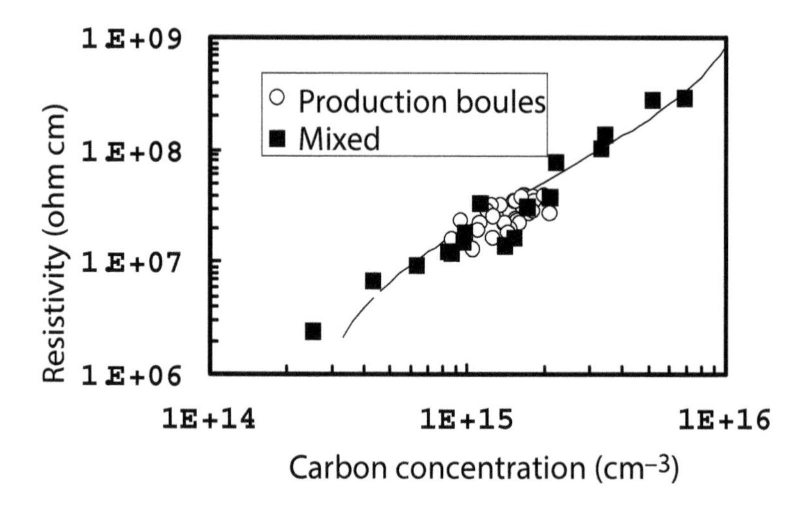

Figure 3.14 Electrical resistivity as a function of carbon concentration for GaAs. Reprinted from Ref. [13] with permission. Copyright © 1999 GaAs MANTECH.

3.3.1 n- and p-Type Crystals

Dopant quantities needed, for even highly doped n+ and p+ crystals, are still minute compared to the melt weight. Therefore the doping is done in steps. Dopants are added in the form of a highly doped polycrystal or powder to undoped material. The doped polycrystal must have a controlled known concentration. In the case of GaAs, the dopant may be dissolved in gallium. Zn and Cd are commonly used dopants for p type, and Si, Te, and Se are used for n type. The concentration of the dopants in the solid is quite different from that in the liquid melt and depends on the distribution coefficient, $k = C_s/C_l$, where C_s and C_l are concentrations in solid and liquid, respectively. Values of k are much less than unity. Excess solute is rejected at the growth interface and the melt gets richer as growth proceeds, resulting in a crystal of varying doping. Zone-refining and zone-leveling techniques used in silicon are not used for GaAs because of As overpressure requirement. Postgrowth annealing is an option.

Growth of larger-size GaAs wafers has been demonstrated; however, to date, there is no commercial interest in 8-inch or larger wafers.

In the case of InP, sulfur doping is used for n-type crystals. Semi-insulating InP is grown by compensating the unintentional Si doping from the silica container, by the addition of Fe, which acts as a deep acceptor. Large 6-inch (150 mm) diameter InP crystals are needed for high-volume production of higher-frequency devices.

References

1. S. K. Ghandhi, *VLSI Fabrication Principles, Silicon and Gallium Arsenide*, Wiley Interscience, New York (1983).
2. M. Hansen and A. Anderko, *Constitution of Binary Alloys*, McGraw Hill Book (1958).
3. G. A. Antypas, The Ga-GaP-GaAs ternary phase diagram, *J. Electrochem. Soc.*, **117**, 700 (1970).
4. J. R. Arthur, Vapor pressures and phase equilibria in the Ga-As system, *J. Phys. Chem. Solids*, **28**, 2257 (1967).
5. J. van den Boomgard and K. Scholl, The P-T-x phase diagrams of the systems In-As, Ga-As and In-P, *Philips Res. Rep.*, **12**, 127 (1957).
6. C. H. L. Goodman, Ed., *Crystal Growth: Theory and Techniques*, Vol. 1, Plenum Press, New York (1974).
7. N. B. Hannay, Ed., *Semiconductors*, Reinhold, New York (1959).
8. T. S. Low et al. *The Role of Substrates Dislocations in Causing Infant Failures in High Complexity InGap/GaAs HBT ICs*, CS MANTECH (2007).
9. J. B. Mullin, B. W. Straughan and W. S. Brickell, Pulling of gallium phosphide crystals by liquid encapsulation, *J. Phys. Chem. Solids*, **26**, 782 (1965).
10. U. Sahr and G. Muller, Growth of InP substrate crystals by VGF technique, *IEEE XII Int. Conf. Semicond. Insulat. Mater.*, 13 (2002).
11. T. Asahi et al., Growth of 100-mm-diameter <100> InP single crystals by the vertical gradient freezing method, *Jpn. J. Appl. Phys.*, **38**, 977 (1999).
12. T. Bunger et al., *Active Carbon Control during VGF Growth of Semi-Insulating GaAs*, CS MANTECH (2003).
13. Tomohiro Kawase et al., *Properties of 6-inch Semi-insulating GaAs Substrates Manufactured by Vertical Boat Method*, GaAs MANTECH (1999).

Chapter 4

Epitaxy

There are three ways of forming undoped or n- and p-type semiconducting layers for devices or IC fabrication on III–V semiconductor wafers. Diffusion is the oldest process, but as explained elsewhere (Chapter 10), it is not commonly used in III–V processing. Ion implantation is a low-cost process and will be discussed in detail in Chapter 9. The third method, epitaxial growth, provides the means to fabricate devices not possible with diffusion or ion implantation and improve device quality and performance at the same time.

Epitaxy is the growth of single-crystalline layers on top of single-crystal substrates, preserving the single-crystal nature of the substrate through the grown layer. Epitaxy can be divided into two broad categories: heteroepitaxy, where the grown layer is a different material from the substrate, for example, GaAs on sapphire; and epitaxy or autoepitaxy, where the material is of the same or similar crystal structure but may have a different chemical composition, for example, AlGaAs on GaAs. The mechanism and process for heteroepitaxy are not straightforward and ways of making the transition from the substrate to the epilayer must be developed. Also, the substrate must be inert to the growth environment and closely matched in thermal expansion, otherwise excess stress can build upon cooling after growth. Further discussion will be limited to epitaxy and heteroepitaxy of similar materials.

III–V Integrated Circuit Fabrication Technology
Shiban Tiku and Dhrubes Biswas

ISBN 978-981-4669-30-6 (Hardcover), 978-981-4669-31-3 (eBook)
www.panstanford.com

There are three main techniques for growing epitaxial layers. These are molecular beam epitaxy (MBE), vapor-phase epitaxy (VPE), and liquid-phase epitaxy (LPE). These methods differ in the delivery of the reactant materials to the substrate. MBE has the most direct transport of materials in vacuum and can produce layers of immense variety and complexity. VPE has many variants, MOCVD (also called metal–organic vapor-phase epitaxy [MOVPE]) being the most common for III–V materials. LPE is the oldest technique and will be discussed first.

4.1 Liquid-Phase Epitaxy

Liquid-phase epitaxy (LPE) was used in the early work on GaAs and is not used anymore for producing integrated circuit (IC)-grade material or devices. It is an inexpensive method and as such is useful for growing thick layers over tens of microns thick. This growth method has found applications in light-emitting diodes (LEDs), photodetectors, solar cells, and now superconducting materials.

This method involves growth by direct precipitation of crystalline layers from the liquid phase [1]. The key to the success of this process is to find a solvent in which to dissolve the material to be deposited. For GaAs, gallium metal and tin are used. Both have high solubility at temperatures below the dissociation temperature of GaAs. Deep-level impurity incorporation in this method can be very low because of the low distribution coefficient of most impurities (discussed in Chapter 3). The main disadvantages of this technique are poor thickness control, inability to grow thin layers in the micron range, and low throughput.

Basic LPE processes consist of dipping a substrate in a solution and pulling it out at a steady rate and wiping off the excess solution. This can be done in vertical, inclined, or horizontal systems and may involve in situ cleaning or etch back, doping by impurity addition, and sequential deposition of more than one film. The most popular systems are horizontal sliding boat type. For GaAs an additional source that saturates the solution with arsenic before the substrate comes in contact for growth can be added. Multiple layers can be grown in a single operation by adding more sources. Thickness is controlled by controlling the linear speed over the different sources. Thickness control is the main drawback of this growth technique. This, combined with poor morphology and defects, make this technique unsuitable for manufacture of epilayers for ICs.

4.2 Vapor-Phase Epitaxy [2]

In this growth process, reactants needed for growth are transported to the growing surface by vapor phase, either as constituent elements or their compounds. The substrate is held at high temperature, and the reactant species move over the surface and attach to the growing crystal. High-purity reactants are used and growth systems are designed to avoid reactions and nucleation in the gas phase.

4.2.1 System Configuration

Three basic types of reactors have been used for growth of silicon and III–V semiconductors [3], as shown in Fig. 4.1. The simplest system consists of a horizontal quartz tube. Substrates are held on a susceptor heated by radio frequency (RF) induction. In this type of reactor gases flow parallel to the substrate; as reactants get depleted, the flux drops, resulting in growth layer variation along the length of the tube. Uniformity can be improved somewhat by using a rectangular tube and tilting the wafer holder (See Fig. 4.1a).

Vertical reactors of the type shown in Fig. 4.1b solve this problem of flux nonuniformity. The substrate holder is rotated to improve uniformity. The number of wafers is limited in this geometry. Commercial versions of these reactors have been developed for metal–organic vapor-phase epitaxy (MOVPE) to grow large batch sizes. The barrel reactor, Fig. 4.1c, is designed to handle more wafers for volume production. Here the wafers are held on the tilted faces of a vertical barrel and the gases are fed from a shower above and flow parallel to the wafers. Growth temperature is optimized for good crystal quality. Systems are operated under low-pressure chemical vapor deposition (LPCVD) to improve laminar flow and deposition uniformity.

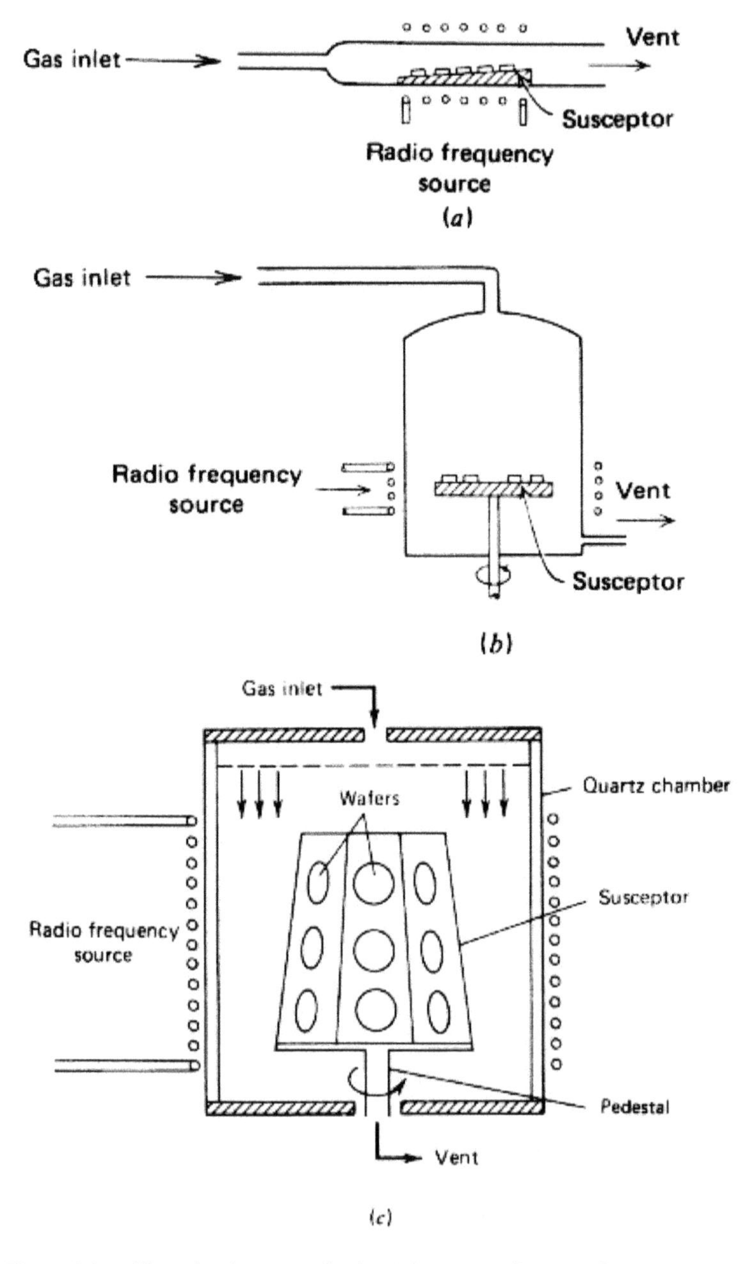

Figure 4.1 Three basic types of epitaxial reactors showing the progression of reactor geometries: (a) horizontal reactor, (b) vertical reactor, and (c) barrel reactor. From Ref. [3]. Copyright © 1983. Reproduced with permission of John Wiley & Sons, Inc.

4.2.2 VPE Chemistries for GaAs

As discussed in Chapter 3, GaAs decomposes at high temperatures and its direct transport is not possible. Although arsenic has a high enough vapor pressure at reasonable temperatures, processes using chlorides and hydrides are more practical, so systems have been developed to deal with the poisonous nature of $AsCl_3$ and AsH_3. Gallium vapor pressure is extremely low at temperatures at which the source can be held; therefore transport by volatile compounds, $GaCl_x$, is chosen for growth. Hydrogen is the most commonly used carrier gas. It is purified from commercial grade to six 9s purity in a purifier close to the system inlet. In both horizontal and vertical geometry reactors, the substrate is heated by RF or regular furnaces. RF systems have the advantage of cold walls, thus avoiding premature reactions on the wall surfaces. These systems also have low thermal inertia and are suitable for multiple layer processes requiring change of temperature.

4.2.2.1 Substrate orientation

GaAs ICs are fabricated on (100) substrates. Epigrowth on this orientation does not have any issues compared to the (111) orientation, which has more bonds on the wafer growth plane and would have an inherent barrier to nucleation. See Fig. 4.2a. Wafers for epigrowth are cut 2 to 3 degrees off-axis to make nucleation even easier. Figure 4.2b shows an exaggerated view of the creation of nucleation sites when a crystal is cut at a slight angle. This also reduces stacking faults.

4.2.2.2 Halide process Ga–$AsCl_3$–H_2 [3, 4]

Ga and $AsCl_3$ are available in high-purity form (six 9s) so that high-purity epilayers can be produced by this process. Figure 4.3 shows the schematic for a two-zone, hot-wall reactor, where the Ga source is held at a higher temperature of about 800°C–850°C and the substrate temperature can be independently varied. Gallium chloride is deliquescent and cannot be handled in the atmosphere, so it must be produced in situ. Growth temperature is set around 750°C. Hydrogen gas is bubbled through the $AsCl_3$ source, which is maintained at a controlled low temperature.

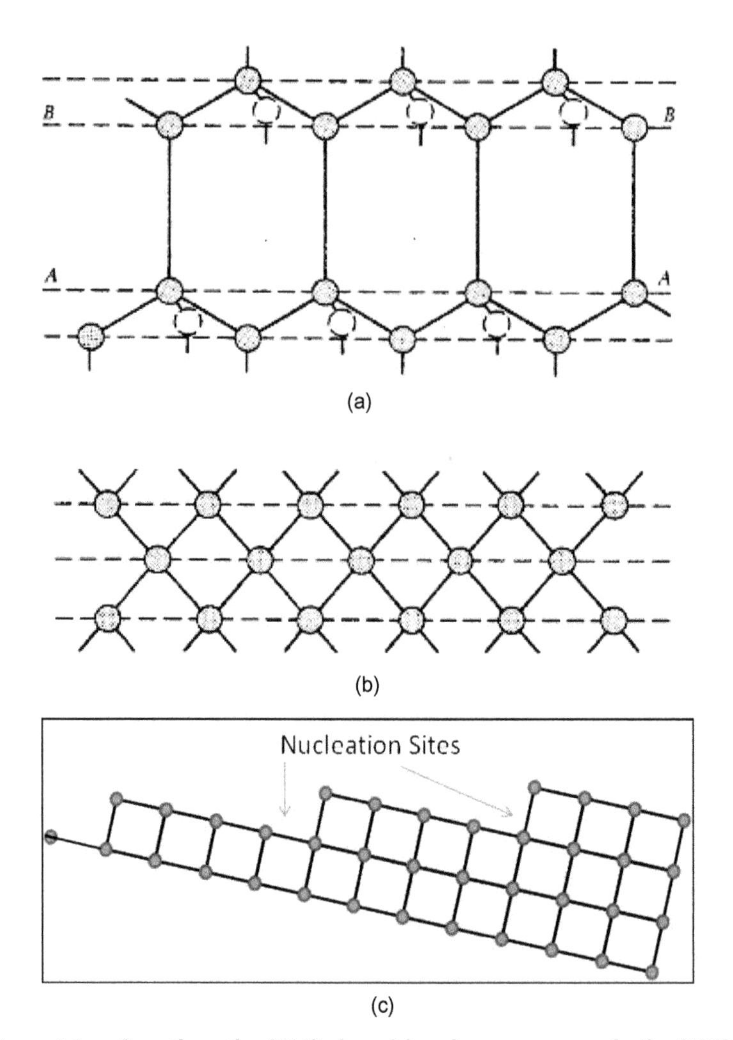

Figure 4.2 Growth on the (111) plane (a) and, more commonly, the (100) plane. Cutting the crystal surface at an angle creates nucleation sites (c)—angle exaggerated here.

$AsCl_3 + 3/2H_2 \Leftrightarrow \frac{1}{4}As_4 + 3HCl$

Arsenic (As_4) formed in this reaction reacts with Ga to form GaAs and HCl reacts with GaAs to convert it to transportable GaCl.

$GaAs + HCl \Leftrightarrow GaCl + 1/2\ H_2 + \frac{1}{4}As_4$

This reaction proceeds in the forward direction at a high source temperature and backward at the substrate temperature. The Ga source must be fully saturated with As and the substrate must be etched back in situ for good crystalline layer quality. The growth rate is controlled by carrier gas flow through the $AsCl_3$ bubbler.

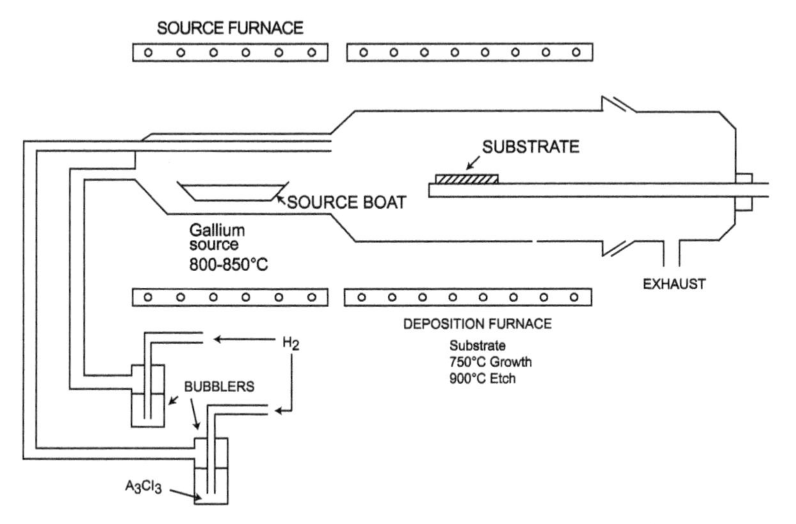

Figure 4.3 Schematic diagram of a VPE system using the halide process.

4.2.2.3 Hydride process Ga–AsH_3–HCl–H_2 [5]

This process chemistry allows better control of As and Ga species as these are formed independently, thus allowing better process control. GaCl is generated the same way as in the halide process. As_4 is formed by the decomposition of AsH_3 over 400°C as it enters the hot source zone. Growth reaction and kinetics are identical to the halide process. Growth rates around 0.5 μm/min can be achieved.

In VPE systems, for multilayer growth, more sources must be provided. Layer and doping control have limitations in these processes.

4.2.3 MOCVD [6, 7]

VPE methods described so far have many shortcomings. The precision of layer thickness control is limited and abrupt junctions cannot be made. Many of these shortcomings can be overcome by replacing the group III source by an organometallic source. This method called MOCVD or MOVPE has become one of the most popular methods for growth of epilayers in III–V circuit fabrication processes.

MOCVD began with the work of Manasevit [8–11] and progressed slowly at first compared to molecular beam epitaxy (MBE) because of its complexity. The deposition studies of Stringfellow [6] and Dupuis and Dapkus [12] accelerated the development, as potential in light-emitting structures was demonstrated. It became a dominant growth system since the 1980s with the production of lasers, detectors, solar cells, and high-speed ICs. Among the advantages over other methods are unequalled flexibility, truly large-scale production with automatic wafer handling, and growth of materials not possible with bulk techniques.

A suite of organometallic compounds is available for growing different III–V materials and as sources for n- and p-type dopants. A major driving force for the popularity of MOCVD for heterojunction bipolar transistors (HBTs) has been the availability of carbon doping for the base p+ layer. Carbon doping has resolved all the issues associated with previously used Be doping, namely stability, reliability, etc. (discussed in more detail in Chapter 22). The reaction kinetics and operation of MOCVD is more complex. Advantages of a lower cost per unit area and low defect density make up for a lack of precision compared to MBE. Interface sharpness and growth of precise thin layers are progressing due to system design improvements. All the concerns regarding safety of gases like AsH_3 and other volume production issues have been resolved.

The basic chemical reaction in MOCVD is between the metal–organic compound and the hydride. For example, for growth of a GaAs layer, trimethyl gallium and arsine react at high temperature to form GaAs, as in the reaction below:

$$(CH_3)_3Ga + AsH_3 \rightarrow GaAs + 3CH_4$$

Rate control is easy and accurate with mass flow controllers (MFCs) compared to other growth methods where, for example, source temperature may be used. Organocompounds for most species needed are available in high-purity liquid form. These are held in stainless steel bubblers placed in constant-temperature baths. Flow of hydrogen, used as a carrier gas, is controlled by MFCs.

A typical system diagram is shown in Fig. 4.4. The MOCVD chamber can be one of any of the types discussed under Section 4.2 (Fig. 4.1). Most dopants are available in gas form as hydrides or as organometallics so that bubbler sources can be used.

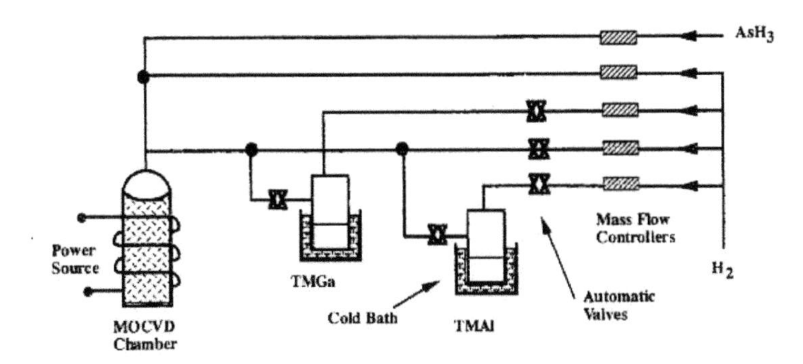

Figure 4.4 Schematic diagram of a MOCVD system showing group III sources for GaAs and AlGaAs growth and arsine as a source for As.

4.2.3.1 Process control and mechanisms [6]

The reaction between AsH_3 and the metal–organic compound may proceed in multiple steps of pyrolytic reactions of the following type:

$$AsH_3 \rightarrow AsH_2 + H$$

$$(CH_3)_3Ga \rightarrow (CH_3)_2Ga + CH_3$$

$$(CH_3)_2Ga \rightarrow CH_3Ga + CH_3$$

$$CH_3Ga + As\,H_2 \rightarrow GaAs + CH_4 + H$$

The by-products of the pyrolytic reactions are unstable, but a lot of reactive hydrogen is available during the growth process. Thus, hydrogen can get incorporated in the growing crystal. Hydrogen can be trapped in interstitial sites or form complexes with dopants like Si and carbon, thus reducing the active dopant concentration (discussed also in Chapter 9). Growth process recipes must be optimized for low hydrogen incorporation. The other fear of excessive carbon incorporation from organometallics has been addressed, carbon being mostly removed as CH_4, and levels of the order of $10^{16}/cm^3$ or less are normal. For carbon incorporation as a dopant, a special growth strategy must be adopted (to be discussed under p-type doping in Section 4.3.2). Excess arsenic conditions are maintained, like in MBE, so that the growth rate is primarily determined by the concentration of Ga species, which, in turn, is a function of the flow rate and the vapor pressure of Ga in the source. These can be controlled by MFCs and the bubbler temperature. The level of dopants is controlled in a similar manner.

The fundamental growth process can be divided into thermodynamic and kinetic components, which determine the rate. Fluid dynamics and mass transport control the rate of transport of material to the growing surface. A study of the growth rate on temperature and flow rates gives an insight into the growth mechanism. MOCVD process reactions are exothermic, increasing the temperature results in a decrease in the thermodynamically limited growth rate. On the basis of the growth rate versus temperature data, processes can be categorized as limited by mass transport, surface kinetics, or thermodynamics [11, 13]. If the reaction rate is kinetically limited, the growth rate increases with temperature. Figure 4.5 shows the growth rate dependence on temperature (plotted as $1/T$) for a general CVD reaction.

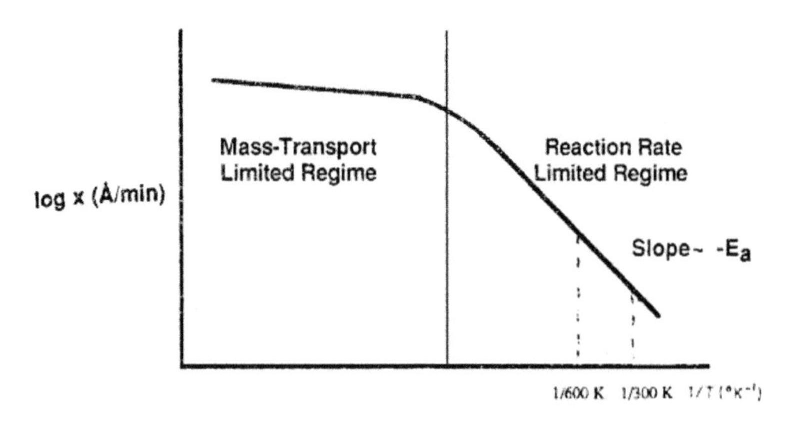

Figure 4.5 Temperature dependence of growth rate for CVD.

The surface reaction rate has the usual dependence on temperature:

$$R = k_0 \exp\left\{-\frac{\Delta E_a}{RT}\right\} \tag{4.1}$$

The activation energy for deposition, ΔE_a, is between 25 and 100 kcal/mole (where 1 kcal = 0.0434 eV).

Diffusion-limited processes are nearly independent of substrate temperature. Other factors such as total flow rate, substrate orientation, and other geometrical factors can also have an effect and have been reported (see, for example, Ref. [6]).

For GaAs growth from TMGa and AsH_3, the growth efficiency first increases from 550°C and then decreases above 750°C, as shown by the data in Fig. 4.6 [6, 46, 47, 48]. Growth rate dependence on the TMGa (alkyl) flow rate is shown in Fig. 4.7 [6, 49, 50]. These and other studies have shown that growth rate is mass transport limited in the temperature range of interest. The reactants are delivered to the substrate surface by transport in the gas stream. The velocity of this stream at the surface is zero, because of friction, and velocity is very low in a boundary layer near the surface. This layer is really stagnant and reactants must diffuse across it for growth to proceed. Figure 4.8 shows examples of stagnant layers for a horizontal and a vertical system [3]. Under conditions of laminar flow (below velocities for turbulence), the boundary layer thickness is given by

$$\delta = \sqrt{\frac{\mu x}{\rho v}} \tag{4.2}$$

where μ is the absolute viscosity, ρ is the density of the fluid, $\mathbf{v}$ is the velocity, and x is the distance (see Fig. 4.8c).

Reactants have to diffuse across this layer, the flux being given by

$$j = -D\frac{dn}{dx} = \frac{D}{\delta}(N_g - N_0) \tag{4.3}$$

where δ is the effective boundary layer thickness. N_g is the concentration of the diffusing species in the gas, and N_0 is at the surface. The above equation is also written in the form of Henry's law (where h is the gas-phase mass transfer coefficient):

$$j = h(N_g - N_0) \tag{4.4}$$

Typically, $D \approx T^m$, where m is between 1.75 and 2. Reactant flux is inversely proportional to the boundary layer thickness.

Control of the boundary layer is an important consideration in the design of reactors and optimization of the deposition process. A more detailed discussion of reaction rates based on diffusion, rate kinetics, etc., can be found in textbooks on VPE, etc. (e.g., Ref. [3]).

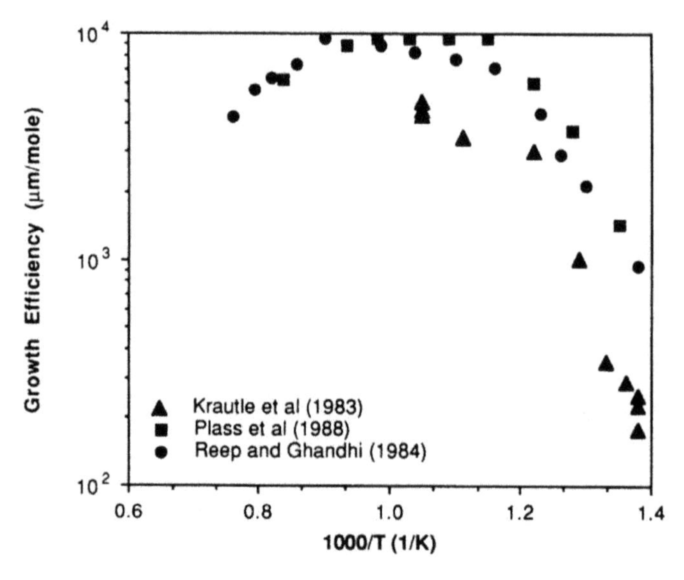

Figure 4.6 Growth efficiency (growth rate/TMGa flow) as a function of reciprocal temperature for GaAs growth using TMGa and arsine [6, 46, 47, 48]. Reprinted from Ref. [6], Copyright (1989), with permission from Elsevier.

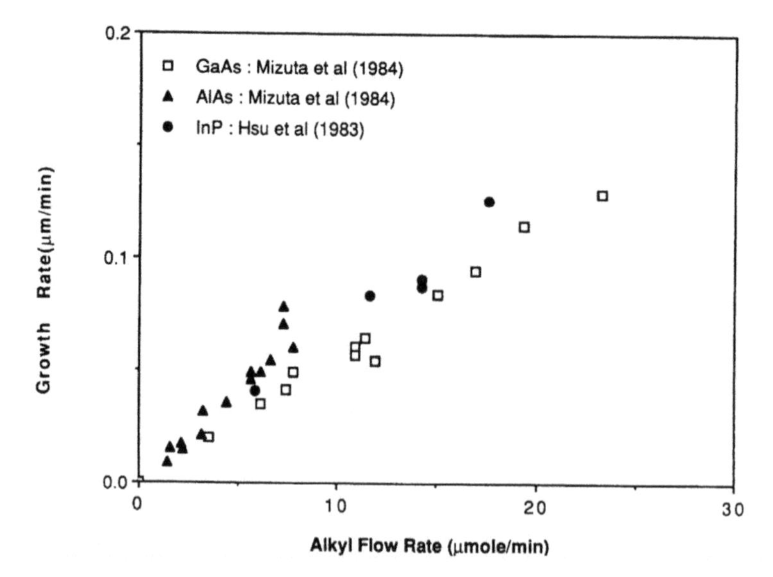

Figure 4.7 Growth rate versus group III alkyl flow rate for GaAs using TMGa and arsine ($AsH3$) [6, 49, 50]. Reprinted from Ref. [6], Copyright (1989), with permission from Elsevier.

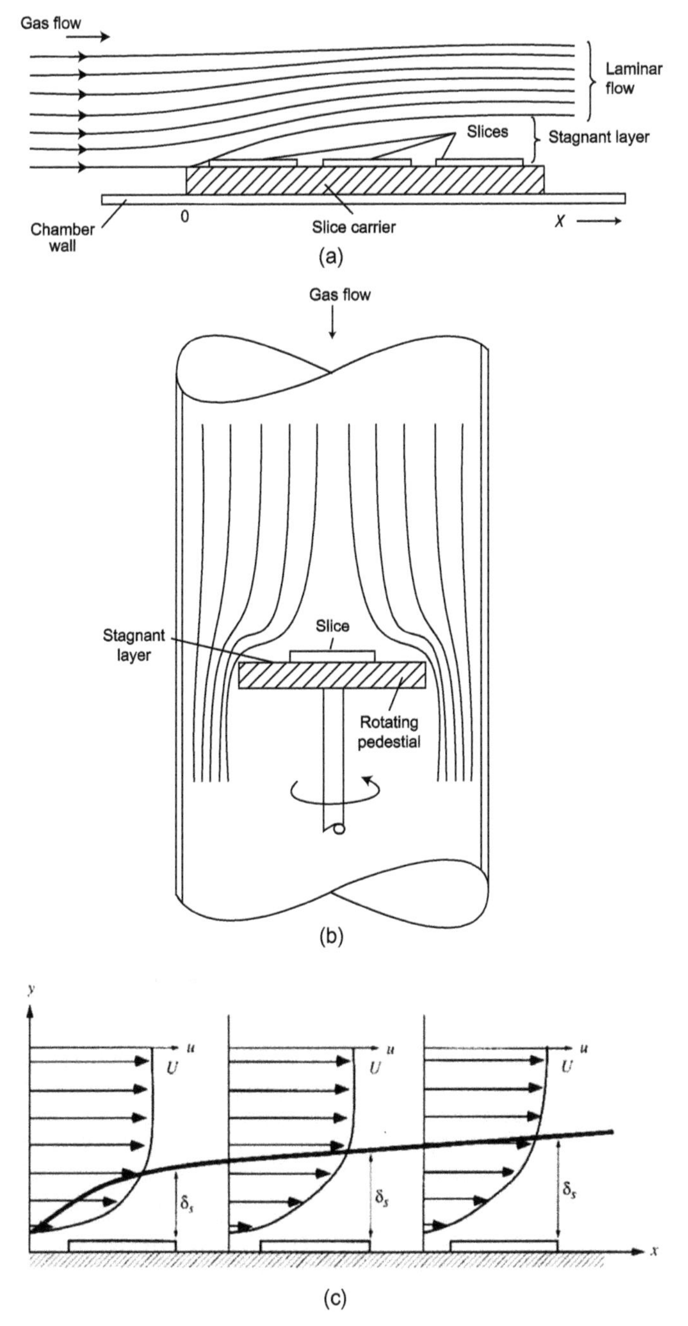

Figure 4.8 Examples of stagnant layer for horizontal (a) and vertical flow geometries (b) and details of boundary layer formation (c).

4.2.3.2 MOCVD sources

Chemical precursors must fulfill the following requirements. The vapor pressure should be convenient at a reasonable bubbler temperature, must be available in high purity, and must decompose at growth temperature without creating unwanted impurity incorporation. Table 4.1 lists some compounds commonly used in the IC epi-industry. Figure 4.9 shows the dependence of vapor pressure on temperature. Vapor pressures for many group II, III, V, and VI precursors can be found in Ref. [6].

Table 4.1 Boiling points and vapor pressures of some common metal-organic sources

Compound	Boiling point (°C)	Vapor pressure (Torr) $\log p = a - b/T$	
		a (K)	b (K)
Group II			
DMZn	46	7.8	1560
DEZn	118	8.28	2109
Group III			
TMAl	126	8.22	2135
TMGa	143	8.50	1824
TMIn	135.8	10.52	3014
Group V			
TMP	38	7.76	1518
TMAs	50	7.40	1456
Group VI			
TEAs	140		
DM Te	92	7.97	1865

Source: Adapted from Ref. [6].

4.2.3.3 Doping

Common n-type dopants for GaAs epilayers are the same as those used for bulk crystals, group IV elements Si, Ge, and Sn, or group V elements, like Se and Te. For Si doping, a disilane source is better than silane, the incorporation efficiency being independent of reactor pressure and temperature [14]. Group VI compound sources, H_2Se, and DETe adsorb on reactor walls, so abrupt doping profiles are hard to produce.

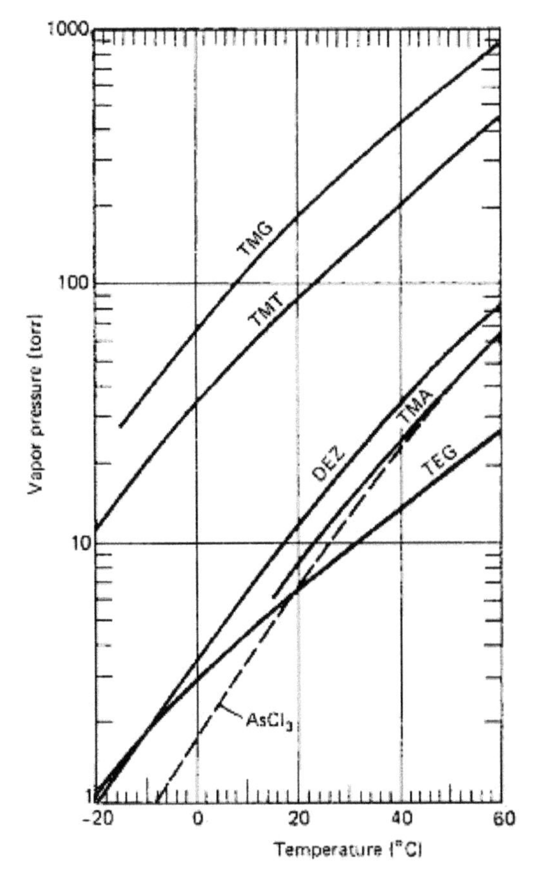

Figure 4.9 Vapor pressure as a function of temperature for reactants used in gallium arsenide growth. Reprinted from Ref. [6], Copyright (1989), with permission from Elsevier.

For p-type dopants, diethyl zinc (DEZ) is a good source of zinc, which was the most common dopant, but it has the disadvantage of diffusing at high doping levels. Carbon has the advantage of being much less mobile; in addition hole mobility as high as that with Be can be achieved, as seen in Fig. 4.10, which shows hole mobility as a function of hole concentration in GaAs. CCl_4 can be used for carbon doping; its by-product Cl_2 is not inert and etches GaAs. Trimethyl arsine is a better source of carbon [15]. However, the organometallic compounds being used for group III sources already contain a lot of carbon, so it is possible to eliminate the need for a separate source of carbon. Weyers et al. [16] demonstrated controlled carbon doping in the range from 10^{14} to $10^{10}/cm^3$ in an organometallic MBE system. Kuech et al. [17] demonstrated similar results in a conventional low-pressure MOCVD system under a proper V/III ratio.

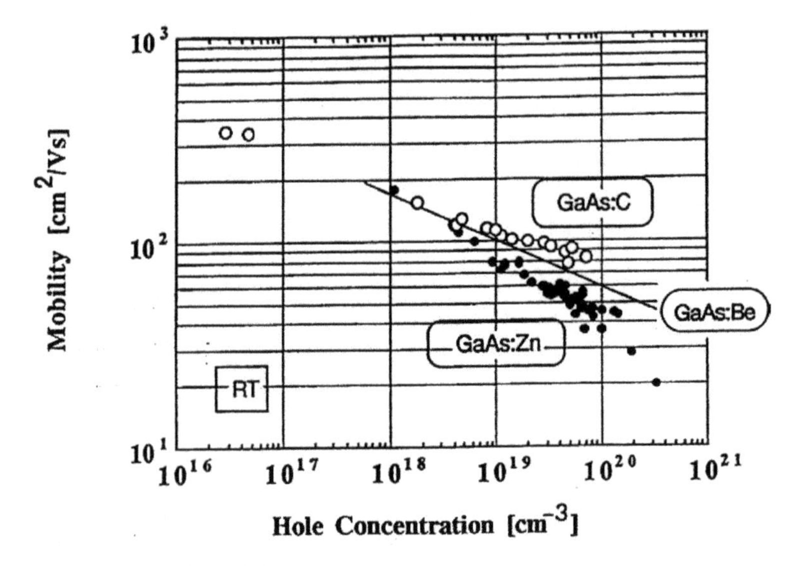

Figure 4.10 Hole mobility dependence on hole concentration for Zn, Be, and C dopants. Courtesy of the Furukawa Electric Co.

Carbon is a preferred dopant, as discussed before, because of a low diffusion constant, about 2 orders of magnitude smaller than Be and Zn. For growing a p+-doped layer the following growth conditions can be used:

- Temperature: 500°C–600°C
- Pressure: 10–100 Torr
- V/III ratio: 10–30
- Growth rate: 0.1–2 μm/hr

Under these V/III ratio conditions, TEAs can be used as a source for As, as well as provide carbon for doping. Carbon doping can be monitored by monitoring the lattice constant, which is measured by double-crystal X-ray diffraction (XRD) [18]. Very high carbon concentration levels can be achieved by this technique. At such high levels the percentage of active carbon drops to about 50%. High carbon levels can also be achieved by V/III ratio control in the TMG/AsH_3 system.

Growth of GaN and other nitride-containing compound semiconductors has been at the forefront for the last few years, with specific differences with other As/P epitaxy due to the differences of growth kinetics of nitrides. Primarily two types of technology—planar planetary MOVPE systems, such as typical Aixtron reactors, and close-coupled showerhead designs like those of Thomas Swan and other manufacturers—are currently used in GaN-based epitaxy.

In the former type, due to differences of decomposition temperatures of ammonia as the most common precursor and the group III sources as TMAl, TMGa, and TMIn, high V–III ratios are maintained to prevent dissociation of the growing crystalline layer, along with higher flow rates of ammonia, leading to implication to gas-phase transport and flow dynamics. Other aspects of the sensitivity of the growth process with respect to the type of carrier gas (N_2 or H_2) impact complex formation and gas-phase nucleation chemistry on the group III incorporation efficiencies. The majority of ammonia is decomposed on the growth surface catalytically (well above 800°C–900°C) during growth temperatures of 1000°C–1100°C, through dissociative adsorption in conjunction with adsorbed group III compounds. Device structures requiring thick buffer layers, such as in LEDs or high-electron-mobility transistors (HEMTs), about 70% of the growth time is typically consumed by these buffers. In scaled-up reactors, intentional separation of group V and group III gas flows at the central injector allows avoidance of undesired premature gas-phase reactions between group III metalorganics and ammonia. The gas flow pattern and point of mixing are carefully determined by appropriate design of the reactor geometry. The separation of the two gas flows also extends the point at which susceptor deposition starts to occur away from the inlet by diffusing through the layer of ammonia and the carrier gas. This allows the adjustment of the point of onset of the deposition in the radial direction. Measurement of Fabry–Perot interferences and black-body-based pyrometry are used as in situ monitoring techniques.

Historically, though, the latter type of closed-couple showerhead technology has been the frontrunner for GaN epitaxy [8–10]. Nakamura et al. [19] have made tremendous advances on a special variant of this technology, which has since been scaled up to address the high-volume manufacturing ability of the LED business. This technology is based on a boundary layer of gas, arising from a uniform flow of gas stalling against a perpendicular flat surface, through which the velocity increases from zero at the surface to eventually free-stream velocity. The growth rate and composition of the alloys depend on the diffusion of the precursors occurring through the boundary layer, while its invariance on the position of the susceptor ideally gives good uniformity, although in reality the perturbation of the boundary layers makes it somewhat less stable. This is addressed by introducing the reagents across the whole of the reaction zone from a multiplicity of several injector tubes, embodied in a water-cooled showerhead, located very close to the growing wafers. These are also known as *short showerhead design* employed for the growth of varying alloy compositions of AlGaN alloys. Optical pyrometry and laser interferometry are also used in such epitaxial processes. During GaN growth, coalescence of GaN grains with facets angled with respect to the (001) direction are more effective at bending over threading dislocations and preventing them from reaching the top surface than are grains with vertical sidewalls [20].

4.2.3.4 HBT growth

The growth of epistructures for HBTs has been one of the driving forces for the development of MOCVD. HBT structures were grown by MBE first. Growth by MOCVD is more challenging because of the need for controlling p–n junction sharpness, etc. Choice of materials has been discussed in Chapter 2. To grow defect-free device structures, lattice-matched material systems are used, for example, AlGaAs:GaAs, where the Al fraction is near 25%. For InGaP, the fraction of In is chosen to match the lattice constant of GaAs. See Fig. 4.11, which shows an energy band gap versus lattice constant diagram for elemental and compound semiconductors. Junction sharpness and avoidance of dopant movement are the most important factors in growth recipe development. The temperature of growth should be as low as possible for post-base-layer growth. Sometimes that is unavoidable, for example, AlGaAs emitter growth needs a higher temperature, so the time at this high temperature must be minimized.

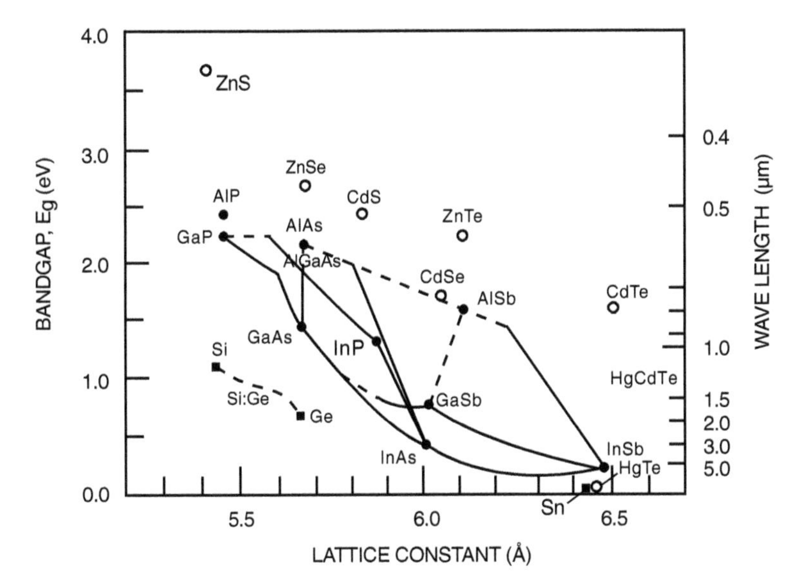

Figure 4.11 Room-temperature, E_g, energy bandgap as a function of lattice constant for Si, Ge, and some common III–V and II–VI compound semiconductors. Solid lines connecting two compounds represent alloy compositions with a direct bandgap; dashed line represents an indirect bandgap.

Forming an abrupt junction in MOCVD is more difficult than in MBE. The system pressure is high, reacting species have a large residence time, and the gas lines feeding the system have a finite length from the valve to the growth system inlet. The growth rates are high, so a junction sharper than 20 Å or so is hard to make. This results in natural grading, which is not detrimental to HBT operation. A sharpness of about 10 Å may be the limit for this growth method.

4.2.3.5 Volume production [21–24]

The most popular large-volume reactors have been the barrel reactor, similar to large-scale silicon epireactors, and the horizontal parallel-plate reactor. These reactors have been improved by the addition of careful hydrodynamic design, reactant diffusers, substrate rotation, and even planetary rotation, etc. See, for example, Fig. 4.12 [23]. The reactant streams are kept separate until just above wafer surface, minimizing pre-reactions. Systems are designed for low particulate generation for achieving a low count of large-area epidefects, uniformity over a large area up to 6 inches, and run-to-run reproducibility.

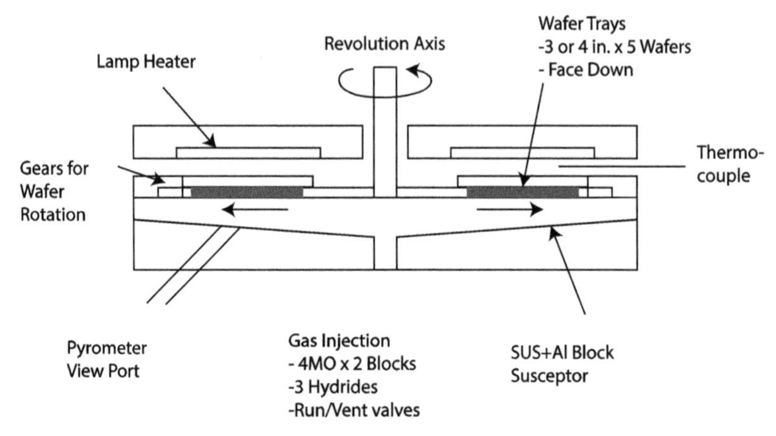

Figure 4.12 Details of chamber in which gases are fed from below and wafers have planetary rotation.

Aixtron has combined all known state-of-the-art techniques, as well as providing in situ monitoring [22]. A schematic cross section of a horizontal reactor is shown in Fig. 4.13 for the Aixtron reactor. A description of the components is given in the figure caption. The substrates are held on a graphite disk held between quartz disks. The graphite disk contains satellite wafer holders that can turn around concentric tubes. The system is designed for temperature uniformity and equal distribution of reactants in all radial directions. The combination of these with satellite rotation gives excellent thickness uniformity of +/−1%. Run-to-run thickness and doping uniformity can be +/−7% and +/−4%, respectively.

Other features of Aixtron reactors:

- The chamber volume is kept low.
- The hot zone extends to the exit of the growth chamber, avoiding unwanted deposits.
- The smallest thickness achievable has been steadily coming down to numbers below 20 Å, approaching MBE. Thus etch stop layers as thin as 10 Å are possible.

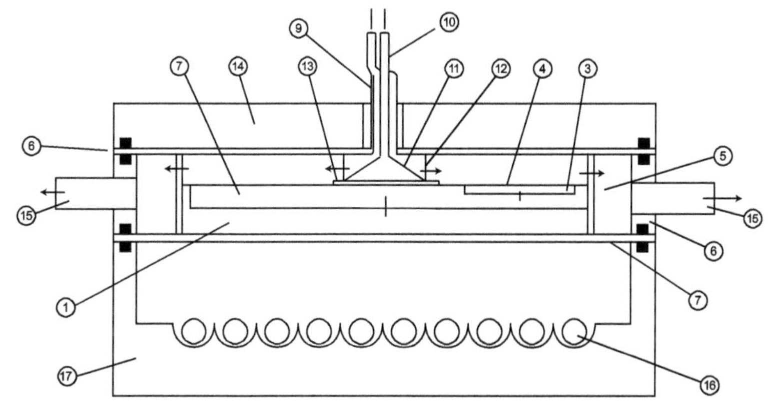

Figure 4.13 Cross section of Aixtron growth chamber. Description of parts: 1. Base 2. Main rotating platform 3. Satellite-rotating platens 4. Wafer 5. Perforated quartz ring 6. Water cooled stainless steel ring 7. Lower quartz disk 8. Upper quartz disk 9. Organometallic inlet tube (concentric) 10. Inner tube for hydrides $+H_2$ 11. Cone-shaped gas dispersal 12. Cylindrical entrance grating 13. Molybdenum reflector ring 14. Water-cooled Al top plate 15. Exhaust 16. IR tube lamps 17. IR reflector. Reprinted from Ref. [22] with permission from AIXTRON.

- Threshold voltage control of pseudomorphic high-electron-mobility transistor (PHEMT) wafers for 6-inch wafers, depending on possible thickness and doping control, has come down to 50 mV for a 6-inch wafer.
- Particle levels are also down to less than a few per cm^2 (for 0.8 μm^2 size).

4.2.3.6 Specific materials

A few examples of materials commonly used in mixed-signal IC products are briefly discussed below.

4.2.3.6.1 *AlGaAs*

This material is more difficult to grow than GaAs because of the reactivity of aluminum. The AlAs lattice constant is close to that of GaAs. So the fraction of Al can be varied to change the energy bandgap. The ratio of Al to Ga can be controlled simply by varying the TMAl-to-TMGa ratio. Aluminum has an affinity for oxygen as well as carbon; therefore growth conditions are optimized to make sure these contaminants are minimized. A higher V/III ratio and high substrate temperatures (~680°C) are needed. Layer quality is optimized by secondary ion mass spectrometry (SIMS) analysis and maintained by photoluminescence (PL) spectroscopy, which gives fast feedback.

4.2.3.6.2 *InGaAs*

A small bandgap, high electron mobility, and lattice match to InP (at Ga = 0.47) make this an attractive material. Referring to Fig. 4.11, which shows the energy bandgap versus lattice constant diagram for common III–V materials, it can be seen that at a lattice constant of about 5.57 Å and a Ga fraction of 0.47, the lattice constant matches. TMIn has been proven to be better than TEIn as a source for indium. High temperature is necessary for high quality, but too high a temperature cannot be used if this layer is used as the top emitter cap layer in the HBT structure in order to avoid junction smearing. High doping levels are needed for cap layers. At such high levels of doping with silicon, the morphology is rough; with tellurium it is better.

Heavily carbon- and zinc-doped InGaAs layers have been grown by MOCVD on InP substrates for use in HBTs. Hydrogen from the AsH_3 and carrier gas will get incorporated if the ratio of AsH_3 to total flow of metal–organic compounds is high (V/III ratio). For carbon doping, hydrogen passivation can be avoided by lowering the V/III ratio to a range of 5–10 (compared to 30–50 for Zn), as shown by Chelli et al. [25]. These authors used an Emcore reactor with TMIn and TMGa as sources and CBr_4 and DEZn as sources for doping.

Although carbon is better for reliability because of its stability, higher performance is achieved with Zn (quality not limited by the need for low growth temperature as in the case of carbon).

4.2.3.6.3 *InP [26]*

This material is gaining interest for microwave and optoelectronic applications for 40 GHz products in optical communication. As mentioned above, the highest-electron-mobility material, InGaAs, can be lattice-matched to InP. MOCVD is uniquely capable of growing phosphorus-containing materials (see also MBE). TMIn and PH_3 are used as precursors for InP growth. TMIn has higher vapor pressure than TEIn. With TEIn, growth must be carried out at low pressure (76 Torr) to avoid gas-phase interactions between the reactants [6]. Again a high V/III ratio is needed for the highest electron mobility. SiH_4 is used for n-type doping and DETe and DMZn for p-type doping. Special care must be taken in reactors using phosphorus sources, because of the potential for fire at the exhaust. In-line scrubbers are needed for exhaust lines.

4.2.3.6.4 InGaP

This compound can be lattice-matched to GaAs and is used as the emitter layer in HBTs because of improved reliability. On a GaAs substrate, In can be near 25% for lattice matching; the percentage of In can be higher on InP. Thin, strained layers can have a much higher In percentage, close to 50%. Growth can be carried out in a low-pressure reactor from TEGa, TEIn, and PH_3 sources. A growth temperature of 620°C–640°C and a high V/III ratio can be used (~45). Zn and Se can be used as p- and n-type dopants.

4.2.3.7 Selective epitaxy [13, 25]

MOCVD has the potential to fabricate structures in which material may be needed only in selective regions on the wafer, thus making possible integration of different device types on the same wafer or growth of special structures needing layer overgrowth after a wafer has been partially processed, for example, in growth of highly doped layers in source and drain regions for field-effect transistor (FET)-type devices. This process, called selective epitaxy, can reduce the overall system cost by eliminating the need for hybrid modules, for example, by combining light detectors and HBT ICs. In selective epitaxy a mask is used to define the areas where the epitaxial layer must be deposited on a wafer that may already have devices on it. The choice of source chemistry and deposition conditions is made to promote single-crystal growth only where required.

Other reasons for using selective epitaxy are to avoid ion implantation and to avoid annealing for deposition of source drain areas after a refractory gate has been formed.

4.2.3.8 In situ monitoring of epigrowth

In situ monitoring by techniques that work in ultrahigh vacuum (UHV) has been a major advantage of MBE over MOCVD. A few techniques can be used in MOCVD, for example, reflection anisotropy spectroscopy (RAS). This technique gives valuable information about the doping concentration, composition, and crystalline quality in real time so that epiprocess recipes can be optimized. The technique is already used in production, for example, at IQE [21].

4.3 Molecular Beam Epitaxy

MBE was conceived for growing high-purity epitaxial layers of compound semiconductors [27, 28]. Since the 1970s, it has been developed into one of the most common techniques for growing III–V compound semiconductors for optical and electronic applications. MBE can produce high-purity, high-crystalline-quality materials with abrupt interfaces and layers having an extremely well-controlled composition, doping levels, and thickness. Because of a high degree of control, reaching almost monoatomic layers, it is used for devices demanding excellent quality, necessary where minority carrier behavior is important, like the HBT or excellent crystal perfection and accuracy of dopant placement like in HEMT devices.

In MBE, single crystals are formed layer by layer by directing molecular beams in UHV to a substrate held at high temperature. MBE differs from ordinary evaporation by the use of UHV of the order of 10^{-9} Torr and the use of independent sources for the constituent elements of the compound to be formed. Growth rates are low, of the order of a few angstroms per second, and the source beams can be interrupted by shutters in less than a second, thus creating abrupt transitions or controlled grading, if needed.

4.3.1 System Description

A typical production MBE system consists of a UHV-capable growth chamber, a load lock, and maybe a buffer chamber. The load lock is needed to ease transfer of wafers, while maintaining good vacuum in the growth chamber. A diagram of the growth chamber is shown in Fig. 4.14. Wafers are loaded on a heated substrate holder capable of being rotated on two axes. A beam flux monitor is mounted on the backside of the substrate holder and can measure beam-equivalent pressure (BEP) when flipped into position. A liquid nitrogen cryo-shroud is located between the walls and the wafer holder assembly. This shroud along with the cryo-pumps used to pump the system can reduce the partial pressure of common residual gases like H_2O to less than 10^{-10} Torr. The mean free path of atoms in this low-pressure vacuum is very long, over hundreds of meters (see Fig. 4.15). The residual gases are monitored by residual gas analysis (RGA). All the parts that get hot are made of molybdenum, tantalum, or pyrolytic boron nitride (PBN), which do not outgas. The material sources, as many as eight, are fusion cell types and are mounted facing the substrate and have independent temperature control, typically to less than +/−1°C. Computer-controlled shutters are located in front of the sources.

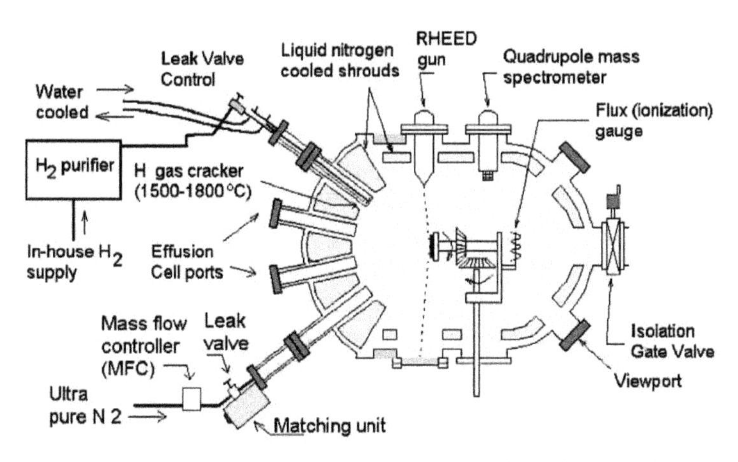

Figure 4.14 Typical MBE system growth chamber (MIT open courseware).

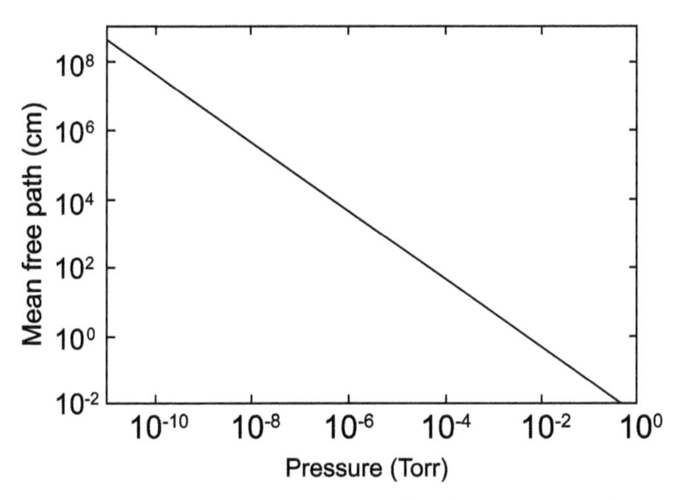

Figure 4.15 Mean free path for nitrogen molecules at 300°C as a function of pressure.

4.3.2 MBE Sources

The sources are designed to work as Knudsen cells; the vapor pressure and the flux coming into the growth chamber are a function of temperature. An ideal Knudsen cell is a large enclosure in which the source solid and the gas are in thermodynamic equilibrium, and the orifice is so small that the equilibrium is not disturbed. For arsenic, a valved cracker source is used [29]. Flux from a heated arsenic ingot is controlled by a valve and passes through a cracking tube, which converts the molecular species from As_4 to As_2. This incorporates more efficiently and results in better crystalline growth. The sources are also enclosed in LN2 shrouds to reduce source-to-source interference. One of the major advantages of MBE has been that reflection high-energy electron diffraction (RHEED) can be used under high vacuum to monitor growth. It can be used to make sure the surface is clean and free of oxide, calibrate the growth rate, provide instantaneous feedback on surface morphology, and help understand growth kinetics. The RHEED gun emits electrons at about 10 keV, which strike the substrate at a glancing angle (see the MBE chamber diagram, Fig. 4.14). The reflected electrons strike a phosphor screen, forming a specular reflection and diffraction pattern, which is recorded by a camera. A mass spectrometer is used to ensure that the system is clean before growth starts, free of H_2O, CO, CO_2, N_2, etc. [29, 30]. In gas-source MBE (GSMBE), gaseous sources like phosphine, arsine, and gaseous dopant sources are used. The rate is controlled by controlling flow rates by MFCs.

The substrate temperature is optimized for material type—for GaAs it is about 600°C. The evaporated source beams do not collide with each other but land directly on the growing surface (no gas-phase reactions). The moderately high substrate temperature promotes movement of the atoms on the surface until they find proper crystalline sites. The substrates are rotated during growth to improve uniformity of all constituents arriving at different angles.

MBE growth is particularly suitable for III–V compounds, while atomic layer epitaxy (ALE) (discussed later in this chapter) may be better suited for II–VI compounds. When atoms or molecules arrive at the surface, these can adsorb, migrate around, or desorb. The sticking coefficient of As species is proportional to Ga flux and approaches unity for As_2 and 0.5 for As_4 [30]. Group V atoms, like As , arriving at the substrate, immediately stick to the surface if a group III element is already there, until an atomic layer of group V is formed. Group III elements can aggregate on the surface to form a group III–rich surface, resulting in nonstoichiometry and defects. Therefore As overpressure is used to ensure defect-free growth. RHEED patterns are used to understand the growth patterns and determine optimum fluxes and growth temperatures of different materials [31]. Excess As flux does not affect growth rate (like NH_3 in silicon nitride by PECVD). Excess As cannot be incorporated for an As-to-Ga flux ratio of 10 to 20 as measured by a BEP gauge.

4.3.2.1 RHEED intensity oscillation

The intensity of the central spot of the RHEED pattern can be measured by a photodiode and recorded as an oscillation pattern. The oscillation frequency corresponds to the monolayer growth rate (one layer of Ga and As). The oscillation can be explained by a layer-by-layer growth mode. The oscillation results from a change in the roughness through each monolayer. RHEED oscillations from GaAs and AlAs growth runs on GaAs are shown in Fig. 4.16. Intensities vary as layers are grown. In spite of the uncertainties, the growth rate can be measured within a few percent accuracy. In production MBE systems, computer control is utilized for controlling substrate and source temperatures. Shutters, etc., are used to control highly complicated layer structures with monoatomic layer accuracy.

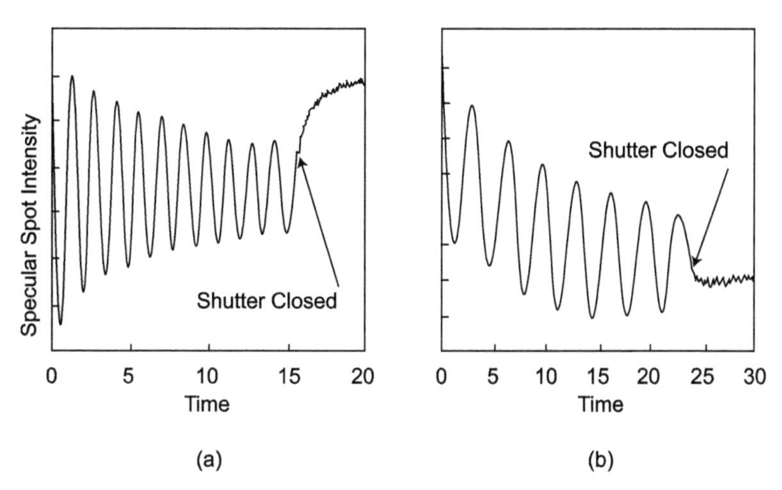

Figure 4.16 RHEED intensity oscillations at the start of growth of GaAs (a) and AlAs (b) on a GaAs substrate.

4.3.3 Specific Materials

Important growth details are given below for a few III–V compounds.

4.3.3.1 AlGaAs

For GaAs the sticking coefficient of Ga is unity to about 650°C. The sticking coefficient of Al on AlGaAs is near unity even above 700°C. The growth rates are determined by the beam fluxes. High substrate temperature and lower group V flux produce smooth layers with good stoichiometry and hence high electron mobility. With ternary compounds, the temperature window of good growth conditions is smaller, and the upper temperature may be limited by the layers that are already grown. Growing layers slowly may produce smoother films under lower-than-optimum temperature.

4.3.3.2 InGaAs

This layer is used as the contacting layer for the emitter because of its small bandgap and low resistivity and can be used as the base layer in InP HBTs because of its high mobility, as discussed under MOCVD. In MBE, the growth temperature is limited by the indium desorption temperature of 540°C. High-quality layers have been reported with a substrate temperature in the range of 450°C–520°C [6]. Incorporating In into GaAs (or GaP) decreases the bandgap and increases the lattice constant and thus reduces mismatch with the InP substrate (see Fig. 4.11).

4.3.3.3 InGaAlAs

This quartenary compound has the advantage of being capable of matching the InP substrate in the lattice constant and has a lower bandgap and high electron mobility (indium content of 53%). To achieve good growth, conditions discussed under InGaAs with additional constraints due to InAlAs must be utilized. At a substrate temperature of 520°C and arsenic overpressure of 1.5×10^{-5} Torr, optimum growth has been reported [32, 33].

4.3.3.4 GaN and related alloys

Both plasma-assisted MBE (PAMBE) and ammonia MBE are used for GaN epitaxy and related alloys. While the corrosive nature of ammonia affects the MBE system, particularly pumping systems, effusion cells, and other finer parts of the MBE infrastructure, it allows flow rates of up to 100 sccm, while still maintaining pressure at 10^{-5} Torr. Cold-lipped effusion cells have been used to address the creeping of aluminum in the presence of NH_3. Though PAMBE is free from such deleterious effects of ammonia, it has the inability to have higher flow rates at lower growth pressures due to the inability of the cryo-shroud to appreciably pump N_2. The reduced N_2 flux requires a lower growth temperature, which is actually advantageous to the indium-based compounds, which limits the desorption of indium from the substrate during the growth, which is a disadvantage for NH_3-based MBE. Ammonia-based MBE did not suffer from a low group V flux, resulting in much higher temperatures of around 900°C, while maintaining high growth rates of greater than 1 μm/hr but generally rough surfaces with a pronounced faceted grain structure. On the other hand, growth in PAMBE is carried out at temperatures close to 700°C and in Ga-rich conditions, with a growth rate controlled by group V flux. Unlike in conventional P/As MBE and NH_3 MBE, the growth rate is determined by group III flux. PAMBE can achieve atomically flat surfaces under Ga-rich conditions. The AlN nucleation layer promotes a Ga face, which is important to achieve a useful piezoelectric effect, whereas otherwise the N face is obtained. It may be noted that NH_3 MBE has little sensitivity to substrate and buffer layer conditions unlike that of PAMBE.

4.3.4 Doping

The most common type dopants in MBE growth are Si, Sn, and Te because of a small activation energy (with respect to conduction energy band) and a high sticking coefficient. Silicon doping is generally limited to the $10^{18}/cm^3$ range in ion implantation due to the amphoteric nature of Si, but in MBE, levels up to $10^{19}/cm^3$ can be achieved. In InGaAs the activation efficiency of Si is even worse. Sn and Te can be incorporated up to $10^{20}/cm^3$.

The most common p-type dopant in MBE has been beryllium (Be) because of its near-unity sticking coefficient [34]. In InP, redistribution occurs over $2 \times 10^{18}/cm^3$. Zinc shows similar redistribution behavior [35] because of a high diffusion coefficient at growth temperatures. In InGaAs also, the maximum Be concentration that can be incorporated is a strong function of temperature [32]. Beryllium interstitials form at high temperatures and enhanced diffusion takes place. For a Be level of 2E19 in InP, only 5E18 is electrically active. Elemental Be is toxic and gaseous sources (e.g., diethyl beryllium) have been tried, but mobilities are lower.

Carbon has become the most common dopant in GaAs (advantages already discussed under MOCVD), particularly because of improved reliability. In MBE a CCl_4 gaseous source can be used for GaAs and InGaAs. CCl_4 and CBr_4 have been shown to be better than TMG as a source for carbon [34]. Hole concentrations up to 1E21 have been achieved.

4.3.5 HBT Growth [36]

Improvements in HBT performance achieved in the last two decades can be attributed to improvements in epigrowth, material purity, and composition, as well as thickness control. HBTs, being a minority carrier device, are one of the most demanding device structures to grow. The emitter and base layers and the emitter–base junction are particularly important. Excellent control of doping, layer thickness, and junction sharpness is needed, without introduction of unwanted defects or impurities that might cause minority carrier lifetime problems. Low doping in the collector demands high purity. Growth temperature of each layer must be optimized for the quality of that layer without causing diffusion and junction smearing. Excellent control of dopant and matrix species is needed for graded junctions.

MBE was the first method to be used to produce HBTs due to these advantages [32].

4.3.5.1 AlGaAs HBT

Optimum conditions for the growth of each layer cannot be used for the growth of the total HBT structure because the layers and interfaces that have already been grown may be affected by the conditions of the growth. Growth of device structures is started after in situ cleaning by opening arsenic and gallium shutters with source cell temperatures stabilized to the desired growth rate. The GaAs collector and base are doped by selectively opening the silicon and Be (or carbon) source shutters. The AlGaAs emitter growth is started by closing the Be shutter and opening the Al and Si shutters. By varying the Al flux, a graded or abrupt heterojunction is formed. Two methods can be used for grading. In thermal grading, the Al flux is varied by changing the temperature; however, this places a limitation on grading sharpness due to the thermal inertia of the source furnace system. The second method, called chirp grading, uses the cycling of the shutter between on and off positions and varying the duty cycle to adjust the Al percentage. Chirp grading can result in sharp junctions of the order of 10 Å.

4.3.5.2 InGaP HBT [36–38]

To grow InGaP, a phosphorus source is needed. As discussed earlier, solid phosphorus cannot be used. A GaP solid source or a gaseous source like tertiary butyl phosphine must be used [36, 38]. A GaP source generates a P_2 (phosphorus dimer) beam, which has a high sticking coefficient at a temperature of ~450°C. Si and Be are used as n-type and p-type dopants, respectively. The emitter contact layer is heavily doped InGaAs, as in the case of the AlGaAs/GaAs HBT. The GaAs base layer can be replaced by a graded In(0.05)Ga(0.95)As layer for improved performance (current gain) [38].

4.3.5.3 InP HBT

A high-mobility base layer GaInAs lattice-matched to InP has been used on an InP substrate to make HBTs or double HBTs (see Fig. 4.11). InP is used as the emitter in a single HBT on an InP substrate or as the emitter and collector in a double HBT. Because of the difficulties of getting a solid phosphorus source, a gaseous source of phosphorous is used. The indium sticking coefficient is unity below a 540°C substrate temperature. The III–V ratio is kept low for smooth morphology. Growth rates are independent of the substrate.

Growth parameters for the subcollector and collector have to be such that impurities in the collector are minimized, because the doping level in the collector is very low (1E16). Base and base-emitter junction growth are critical to performance. If beryllium is used as the dopant, separation (spacer ~50 Å) is needed to keep the diffusing Be away from the emitter. A low base growth temperature is also needed, as low as 300°C in MBE. In metal organic molecular beam epitaxy (MOMBE), a higher temperature of 450°C may be necessary to break the precursors; this is practical because carbon diffusion is not an issue.

InP wafers are also available as epiready, so wet etching is not necessary. The incongruent sublimation temperature of InP is only 360°C, but oxide removal requires nearly 500°C. If phosphorus overpressure cannot be used, arsenic overpressure can be used to remove the oxide [28].

4.3.5.4 GaN HBT

Significant advancements have been made in epitaxy of GaN HBTs by Mishra et al. at the University of California, Santa Barbara, and by Jana et al. at the University of Notre Dame, with technology innovations at achieving p-type doped GaN, which is at the heart of constraints in n–p–n HBTs. Mg-doped GaN has been achieved with innovation in InGaN-graded base HBTs as the future direction for such devices. There is also active interest in high-GaN-collector-breakdown-voltage-based GaN/GaAs base HBTs as well.

4.3.6 PHEMTs

MBE has become a dominant manufacturing technology for the growth of PHEMT epistructures for microwave power amplifiers and switches. As discussed in Chapter 2, in PHEMT devices structures, a high-purity, high-mobility channel layer is grown between two large-bandgap doped layers or near a large-bandgap layer to create a two-dimensional electron gas (2DEG) layer. For double-heterojunction devices, a GaAs buffer layer is grown first, followed by AlGaAs, an undoped InGaAs channel, doped AlGaAs, and an n+ GaAs contact layer. The two sides of the InGaAs channel layer are doped with Si in the concentration of ~1E18, (called delta doping if it is an abrupt spike). AlAs etch stop may be grown between the contact layer and the AlGaAs Schottky layer to facilitate recess etching and hence threshold voltage control. When the recess etch is done, the etch chemistry is chosen to stop etching when it reaches AlAs. This layer is then removed just before gate metal deposition. (Details are discussed in Chapter 13). The nitride MBE technique (both PAMBE and NH_3 MBE) has been a useful tool for AlGaN/GaN HEMT devices, mostly on SiC, sapphire, and silicon substrates. The smallest lattice mismatch and high thermal conductivity make SiC the most promising candidate for GaN FETs but the high cost of such substrates is a limitation. Epitaxial GaN FETs on Si is a solution, but special growth techniques with AlN-based nucleation layers with careful nitridation of the silicon substrate are essential to achieve proper device performance without cracking of the epilayer.

4.4 Atomic Layer Epitaxy [39]

ALE has been proposed to improve uniformity and reproducibility of structures grown by MOCVD [40, 41]. In ALE, the growth species are sequentially delivered to the growth surface. For example, for GaAs, the surface would be first exposed to Ga species and then the reactor purged before the surface is exposed to arsenic. The temperature of the substrate is just hot enough for the process to be self-limiting, accepting only one monolayer at a time. Thus a thick layer of Ga will not grow at the growth temperature; only one monolayer of Ga bonded to As below it is stable. At this point the arsenic source is admitted to form one monolayer of As. The process is continued and the number of cycles is used to control the thickness. ALE is a low-temperature process compared to MOCVD and suffers from impurity incorporation. The process has advantages for FETs and other charge control devices, where threshold control is most important.

The advantage of relatively low-temperature growth can be exploited for growing a gate insulator for III–V compound semiconductors. Al_2O_3 has been grown on a GaN device on a SiC substrate for an HEMT-type device using atomic layer deposition (ALD) after ohmic contacts have been already made [42]

During the last two decades, tremendous effort has been made to improve oxide quality on GaAs and other ternary epilayers. However, surface trap density has not come down to a low enough level for metal–oxide–semiconductor field-effect transistors (MOSFETs). A number of materials have been deposited by various techniques, with similar results. ALD provides a better approach to achieving a MOSFET-type device on n-type GaAs or other III–V material layer. In a recent work, using ALE of a $La_{1.8}Y_{0.2}O_3$ layer followed by an Al_2O_3 cap layer, on a (111) GaAs substrate, very low interface state density has been achieved ($3 \times 10^{-12}/cm^2$-eV) [43]. Using this layer under a 0.5 um gate, a high on-state current of 3356 mA/mm has been demonstrated. Preoxide treatment with ammonium sulfide and postgrowth annealing by rapid thermal processing (RTP) at 860°C reduce the trap density.

4.4.1 GaAs on Silicon Substrates

GaAs grown on silicon substrates has a huge market potential due to the lower cost of silicon and availability of large-size wafers and equipment to handle those. This has been the goal of numerous research attempts in the last two decades. However, the technical challenges are considerable due to the 4% mismatch in the lattice constant and the large difference in the thermal expansion coefficient. The current research has been dominated by finding a pathway using various kinds of buffer layers that have an intermediate lattice constant, using Ge, GaAsP, etc. No clear choice has emerged yet [44].

Growth of GaAs on silicon, a dream of many epigrowers, has been hampered by such challenges. Past and present major research programs' thrusts on integration of compound semiconductor on Si across the globe are very active currently. The COSMOS program was driven by DARPA with compound semiconductor industry key players leading the show, which was later followed by the DAHI program. Microassembly, monolithic epitaxial growth, and epitaxy layer bonding have been the final goals of the integration schemes within these efforts. Elsewhere, such as in India, major efforts have been underway by the government at IIT Kharagpur (extensive laboratory setup by the coauthor, Dr. Dhrubes Biswas) through focus on heterointegration by cluster tool MBE and MOCVD for electronics and optical devices, respectively, based on a 6-inch manufacturing platform (Fig. 4.17) [45]. Bandgap engineering and metamorphic growth are the drivers here for creation of high-performance compound semiconductor heterostructures combining optimization of high power, high linearity, high frequency, high efficiency, low R_{on}, and low noise performances of III–V devices on silicon for "beyond Moore's epitaxial devices."

(a)

(b)

Figure 4.17 (a) Cluster tool MBE for integration of III–V on Si. (b) Integrated MOCVD for integration of III–V on Si.

4.5 Epilayer Characterization

Epilayer characterization is discussed in detail in Chapter 21. A short summary of different methods used follows here.

The semiconducting epilayer grown by any of the techniques must be qualified for use in the growth of the device. Before growing a total device structure, a recipe for each layer is therefore developed separately and then optimized for the total device structure. Qualification of a recipe on different epireactors or MBE systems is also necessary for meeting performance and reliability requirements of complicated devices being used currently.

Thickness, sheet resistance, and carrier concentration are basic measurements needed for all layers. Resistivity can be measured by four-point contact methods or noncontact sheet resistance methods. Mobility and doping levels are measured by the van der paw Hall method. Doping can be determined by both SIMS and the effective carrier concentration by the electrochemical capacitance–voltage (*C–V*) method. Photoreflectance and PL are used for measuring the composition of ternary compounds, as well as for band discontinuities. Carbon concentration can be measured by SIMS as well as monitored by XRD. Impurities like O, C, and H are detected by SIMS. Four-crystal high-resolution XRD with reciprocal space mapping and X-ray reflectivity is used to find alloy composition, alloy and binary thickness, and residual strain determination. For thickness calibration of thin layers, scanning electron microscopy (SEM) or transmission electron microscopy (TEM) may have to be used.

4.6 Concluding Remarks

Practically all well-known epitaxy methods are in current use in manufacturing of III–V devices. MOCVD and MBE are dominant techniques for monolithic microwave integrated circuits (MMICs) for GaAs, InP, and GaN. MBE has the advantage of excellent control of thin layers, superlattice structures, composition, and carrier concentration control. MOCVD is better suited to larger-volume production and larger-area circuits due to lower defect density and has dominated the growth for HBTs due to the success of the

InGaP HBT for wireless applications. There is no better technology overall; the choice depends upon the application. Table 4.2 shows a summary comparison of MBE versus MOCVD [45]. It is conceivable that future HBTs may benefit from the advantages of MBE.

Table 4.2 Comparison of MOVPE and MBE

Category	Characteristics of MOVPE	Characteristics of MBE
Technical	High growth rate for bulk layers	Fast switching for superior interfaces
	Growth near thermodynamic equilibrium, excellent quality/ crystallinity	Ability to grow thermodynamically forbidden materials
	Ability to explicitly control background doping	No hydrogen passivation, no burn-in inherent to MOVPE
		Uniformity easier to tune, largely set by reactor geometry
Commercial	Shorter maintenance periods	Longer individual campaigns, less setup variability
	More flexibility for source and reactor configuration changes	
	Higher safety risk, increasing scrutiny of legislative bodies worldwide	Lower material cost per wafer
	Economic to idle, overhead cost scaling with run rate	Overhead not scaling with run rate, contribution per wafer increasing with wafer volume

Source: Reprinted from Ref. [44] with permission. Copyright © 2013 CS MANTECH.

References

1. L. R. Dawson, Liquid phase epitaxy, in *Progress in Solid State Physics*, Vol. 7, Eds., H. Reiss and J.O. McCaldin, Pergamon, New York (1972).
2. J. R. Knight, D. Effer, and P.R. Evans, The preparation of high purity GaAs by vapor phase epitaxial growth, *Solid State Electron.*, **8**, 78 (1965).
3. S. K. Ghandhi, *VLSI Fabrication Principles, Silicon and Gallium Arsenide*, Wiley Interscience, New York (1983).

4. S. Sawada et al., Investigation of InP epitaxial films on GaAs substrates grown by chloride process, *IEEE Int. Conf. InP Relat. Mater.* (2001).
5. J. J. Tietjen and J. A. Amick, The preparation and properties of vapor-deposited epitaxial gaasp using arsine and phosphine, *J. Electrochem. Soc.*, **113**, 724 (1966).
6. G. B. Stringfellow, *Organometallic Vapor Phase Epitaxy, Theory and Practice*, Academic Press, San Diego (1989).
7. T. F. Kuech, Recent advances in metal-organic vapor phase epitaxy, *Proc. IEEE*, **80**, 1609 (1992).
8. H. M. Manasevit, *Appl. Phys. Lett.*, **116**, 1725 (1969).
9. H. M. Manasevit and W. I. Simpson, *J. Electrochem. Soc.*, **12**, 156 (1968).
10. H. M. Manasevit, *J. Cryst. Growth*, **13/14**, 306 (1972).
11. D. W. Shaw, *J. Cryst. Growth*, **31**, 130 (1972).
12. R. D. Dupuis and P. D. Dapkus, *IEEE J. Quantum Electron.*, **15**, 127 (1979).
13. T. Kuech, Recent advances in MO-VPE, *Proc. IEEE*, **80**, 1609 (1992).
14. E. Veuhoff, T. F. Kuech, and B. S. Myerson, *J. Electrochem. Soc.*, **132**, 1958 (1985).
15. Furukawa company brochure, Furukawa Electric Co.
16. M. Weyers et al., *J. Electron. Mater.*, **15**, 57 (1986).
17. T. F. Kuech et. al., *Appl. Phys. Lett.*, **53**, 1317 (1988).
18. T. J. De Lyon et al., *Appl. Phys. Lett.*, **56**, 1040 (1990).
19. S. Nakamura, M. Senoh, and T. Mukai, High power InGaN/GaN double heterostructure violet light emitting diodes, *Appl. Phys. Lett.*, **62**, 2390–2392 (1993).
20. P. Gibart, Metal organic vapor phase epitaxy of GaN and lateral overgrowth, *Rep. Prog. Phys.*, **67**(5), 667–715 (2004).
21. K. Christiansen et al., Advances in MOCVD technology for research, development and mass production of compound semi-conductor devices, *Opto-Electron. Rev.*, **10**, 237 (2002).
22. P. M. Frijlink, A new versatile large area size MOVPE reactor, *J. Cryst. Growth*, **93**, 207 (1988).
23. H. Tanaka et al., Multi-wafer growth of HEMT LSI quality AlGaAs/GaAs heterostructures by MOCVD, *Jpn. J. Appl. Phys.*, **26**, L1456, (1987).
24. Descriptions of production systems on company websites, for example, Emcore, Aixtron, etc.

25. S. Lourdudoss et al., Heteroepitaxy and selective epitaxy for discrete and integrated devices, *Conf. Optoelectron. Microelectron. Mater. Devices*, 309 (2006).
26. R. Jakomin et al., MOVPE of InP based III-V compound semiconductors, dspace-unipr.cilea.it.
27. Al Cho, *Molecular Beam Epitaxy*, Springer (1994).
28. T. Mattord et al., *J. Vac. Sci. Technol. B*, **11**, 1050 (1993).
29. C. T. Foxon, MBE growth of GaAs and III-V alloys, *J. Vac. Sci. Technol. B*, **1**, 293 (1983).
30. V. Swaminnathan and A. MacRander, *Materials Aspects of GaAs and InP Based Structures*, Prentice Hall, p. 191 (1991).
31. B. Jalali and S. J. Pearton (Eds.), *InP HBTs: Growth, Processing and Applications*, Artech House, Boston (1995)
32. A.Kohzenetal, *J. Cryst. Growth*, **107**, 932 (1991).
33. M. B. Panish et al., *J. Cryst. Growth*, **112**, 343 (1991).
34. S. J. Pearton and N. J. Shah, in *High Speed Semiconductor Devices*, Ed., S. M. Sze, Wiley (1990).
35. C. R. Abernathy et al., Composition of intrinsic....GaAs and AlGaAs grown by MBE and related materials, *J. Vac. Sci. Technol. A*, **12**, 1186 (1994).
36. H. Sai, H. Fujikura and H. Hasegawa, Growth of high quality InGaP on GaAs by gas source MBE, *Int. Conf. InP Relat. Mater.* (1997).
37. J. Hyon Joe and M. Missous, High-gain InGaP/GaAs HBTs with compositionally graded $In_xGa_{1-x}As$ bases grown by molecular beam epitaxy, *10th IEEE Int. Symp. Electron. Devices Microwave Optoelectron. Appl.*, 237 (2002).
38. S. Oktyabrisky and P. D. Ye, *Fundamentals of III-V semiconductor MOSFETs*, Springer-Verlag, New York (2010).
39. T. Suntola, Atomic layer epitaxy, *Mater. Sci. Rep.*, **4**, 261–312 (1998).
40. M. Ozeki, Atomic layer epitaxy of III-V compounds using metal organic and hydride sources, *Mater. Sci. Rep.*, **8**, 97–146 (1992).
41. H. Wang et al., *Al_2O_3 Deposited Using ALD*, CS MANTECH, p. 185 (2010).
42. L. Dong et al., *IEEE Electron Device Lett.*, **34**, 487 (2013).
43. Yu B. Bolkhovityanov and D.P. Pchelyakov, *Phys. Uspekhi*, **51**, 437 (2008).
44. R. Pelzel, *A Comparison of MOVPE and MBE Growth Technologies for III-V Epitaxial Structures*, CS MANTECH, p. 105 (2013).

45. D. Biswas et al., IIT Kharagpur, India (2014).
46. H. Kraulte et al., *J. Electron. Mater.*, **12**, 215 (1983).
47. C. Plass et al., *J. Cryst. Growth*, **88**, 455 (1988).
48. D. H. Reep and S. K. Ghandhi, *J. Electrochem. Soc.*, **131**, 2697 (1984).
49. M. Mizuta et al., *J. Cryst. Growth*, **68**, 142 (1984). Figure 7 data (Stringfellow [5]).
50. C. C. Hsu et al., *J. Cryst. Growth*, **63**, 8 (1983). Figure 7 data (Stringfellow [5]).

Chapter 5

Photolithography

5.1 Introduction

Photolithography processes are used to define active and passive devices, as well as interconnect wiring between these on semiconducting wafers. In general lithographic processes are used for numerous patterning steps, including the following: create semiconducting regions by ion implantation, isolate devices from each other, remove material, and add conducting layers. Photolithographic processes are used to transfer an image from a reticle or mask to the wafer. In III–V processing, 10 to 15 lithographic steps are needed to complete integrated circuit (IC) fabrication. This may involve a few layers for active devices, for example, contacts and gates, a few more for thin-film resistors and interconnect layers, and a few more for backside processing. The cost and cycle time of the whole fabrication process depend upon the number of lithography steps. Photolithography plays a central role in IC fabrication. The nature of other process steps used in the overall fabrication varies a lot from step to step, for example, from ion implantation to plasma etching and technology type (high-electron-mobility transistor [HEMT] vs. heterojunction bipolar transistor [HBT]), but the photolithography part can be common.

The photolithography process based on the use of masks or reticles will be discussed here in detail; however, creation of the pattern directly on the wafer will be discussed only briefly.

III–V Integrated Circuit Fabrication Technology
Shiban Tiku and Dhrubes Biswas

ISBN 978-981-4669-30-6 (Hardcover), 978-981-4669-31-3 (eBook)
www.panstanford.com

5.2 Mask Making

Circuit designers produce a circuit layout that is a drawing broken into geometrical components and saved, historically, on a computer tape and now on other media. The design is broken into multiple masks or layers, one for each layer, starting from the first contact all the way to the top bond pad layer. The designs are transferred by the mask shop onto glass plates by using special patterning process. These masks are then used as a master to make more masks for contact printing or used directly in projection printing. Extreme care must be taken in creating the mask without defects and ensuring it is free from particles or damage. The masks consist of chromium film on low-expansion, stable glass. The plates are coated with a photoresist and a pattern is generated by electron beam or ultraviolet (UV) light exposure, followed by etching. Electron beam techniques are necessary for submicron geometries. If resolution of near 0.1 μm is needed, that requires a resolution of 0.5 μm on the mask for a 5-to-1 projection. In this case optical deep ultraviolet (DUV) exposure may be acceptable. For typical critical dimension (CD) specifications of 0.25 μm or less on the mask, DUV lithography is not acceptable. The fundamental limit of optical lithography is determined by diffraction of light and therefore depends on the wavelength of light used. The electron beam does not suffer from a diffraction limit. The electron beam writing system is similar to a scanning electron microscope with the addition of beam blanking, computer-controlled deflection, laser-controlled stage, and fiducial mark detectors. The rectangular and other geometrical shapes are translated into lines. These lines are transferred to the photoresist by the scanning mechanism. There are two ways of scanning the electron beam. In the raster scan method, the electron beam is scanned over the photoresist area, a few millimeters square at a time. In the vector scan method, the electron beam is used to scan a feature and move directly to another feature and continued to finish the circuit. This tends to be faster because no time is wasted in feature-less blank areas.

Ordinary soda lime glass is not acceptable for mask making. Low-thermal-expansion borosilicate glass or quartz is used. To ensure transparency to the wavelength of light, quartz is used for masks to be used for DUV exposure. The plates are polished and cleaned before a layer of chromium (~1000 Å) is deposited by sputtering or evaporation. The chrome layer must have good adhesion and uniformity. An electron beam resist (polymethylmethacrylate [PMMA]) is applied and soft-baked before exposure. After exposure the resist is developed by a spray process. Chrome is etched by wet etching using a commercially available chrome etchant (see Chapter 6). Dry etching by chlorine-based plasmas using Cl_2 + O_2, CCl_4 + O_2, or CF_4 + O_2 has been investigated for better CD control at small dimensions. Inductively coupled plasma (ICP)-type etchers provide a high etch rate and are preferred.

The resolution of the mask depends on the electron beam size. For finer resolution a small beam size is needed. The writing time increases as the beam size is reduced.

Masks can be made in two types, dark-field and light-field image. See Fig. 5.1. Dark-field masks are less prone to particles, but the choice of the mask type also depends upon the type of resist and whether the etch or lift-off patterning process is used.

Masks for micron and submicron geometries are mounted into pellicles. A pellicle is a transparent membrane that seals the mask and protects it from particles and damage, mechanical and electrostatic discharge (ESD) type. The membrane is mounted on a metal frame and attached to the chrome side of the mask. A minimum stand-off distance is required to ensure that particles, etc., on the pellicle are not in the focal plane and are not imaged.

5.3 Basics of Printing/Imaging [1–2, 3]

The printing process consists of transferring the image of the circuit from the reticle to the wafer surface. In optical lithography, this is done by exposing a photosensitive resist that is coated on the wafer, to this image. This process is followed by etching to remove material or addition of thin-film materials to form contacts or interconnect lines. This process must be carried out on each wafer for all mask layers in successive steps. Since this process is repeated numerous times, the amount of time it takes becomes important in production. Optical lithography techniques are used down to half-micron-dimension geometries, needed for III–V circuit fabrication, and will be discussed in detail. More expensive and time-consuming electron beam lithography is used only when needed for gate sizes below 0.5 μm.

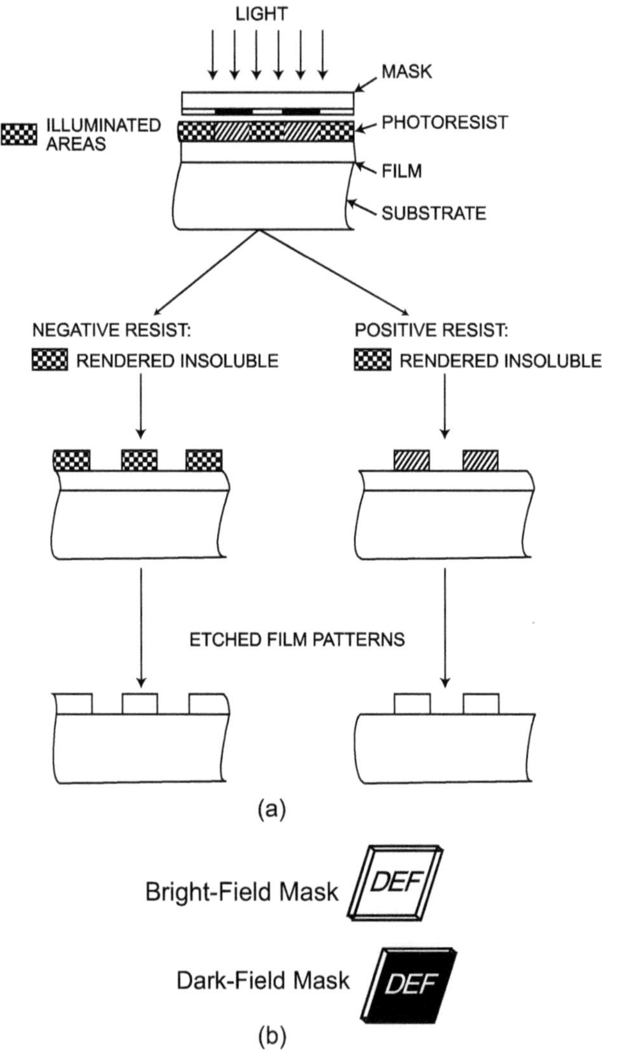

Figure 5.1 Positive and negative photoresist processes and bright-field and dark-field masks.

Figure 5.2 shows schematic diagrams for lithographic techniques. The oldest photolithographic technique is contact printing. In this method the mask is the same magnification, feature size, as the circuit. The mask is placed close to the wafer surface with the resist on it and aligned using alignment marks. The mask is then pressed against the wafer and exposed to UV light. The mask gets dirty and damaged in this process because of continuous contacting between mask and photoresist. In an alternate approach, called proximity printing, the mask is separated by a small gap in order to minimize damage on the mask, at the expense of resolution (Fig. 5.2b) [1, 4]. Projection lithography was developed in the early 1970s to avoid these problems. Perkin Elmer Co. first introduced the process in 1973, and it has become the standard process in the semiconductor industry since then. In this technique the mask image is projected onto the wafer by means of a lens system. A small area, 10 to 20 mm wide, is exposed and the wafer is stepped to the next position until the whole large-area wafer is exposed. Step and repeat systems (steppers) have been developed for 5-to-1 or 10-to-1 reduction optics for better resolution and alignment accuracy. (See Fig. 5.2c,d for more details to be discussed later.)

Resolution determines the size of the smallest feature that can be printed and is limited by the wavelength of the light. Well-designed exposure systems can achieve down to about the same size as the wavelength. Over the years, steppers have shifted to shorter and shorter wavelengths. Figure 5.3 shows the wavelength of different spectral lines from a mercury lamp. Most systems currently in production use G-line (436 nm) or i-line (365 nm). Half-micron (500 nm) features are possible with these. For sub-half-micron features, DUV systems with special lenses and correction systems must be utilized.

Photoresists have been used in the printed circuit board industry for a long time. The first modern photoresists were developed by the Eastman Kodak Company and were based on cyclized rubbers with cinnamic acid derivatives as photosensitive cross-linking agents [1]. Earlier resists were negative resists, the image on the plate being the opposite of the mask. A resist exposed to light remains on the wafer after the developing process, so the film being etched remains under the resist. See Fig. 5.1 for positive and negative resist pattern transfer. Historically, negative resists were not suitable for micron-size geometries because of limitations imposed by swelling. Positive resists, which behave in the opposite manner, have better resolution. Resist chemistries will be discussed in the next section.

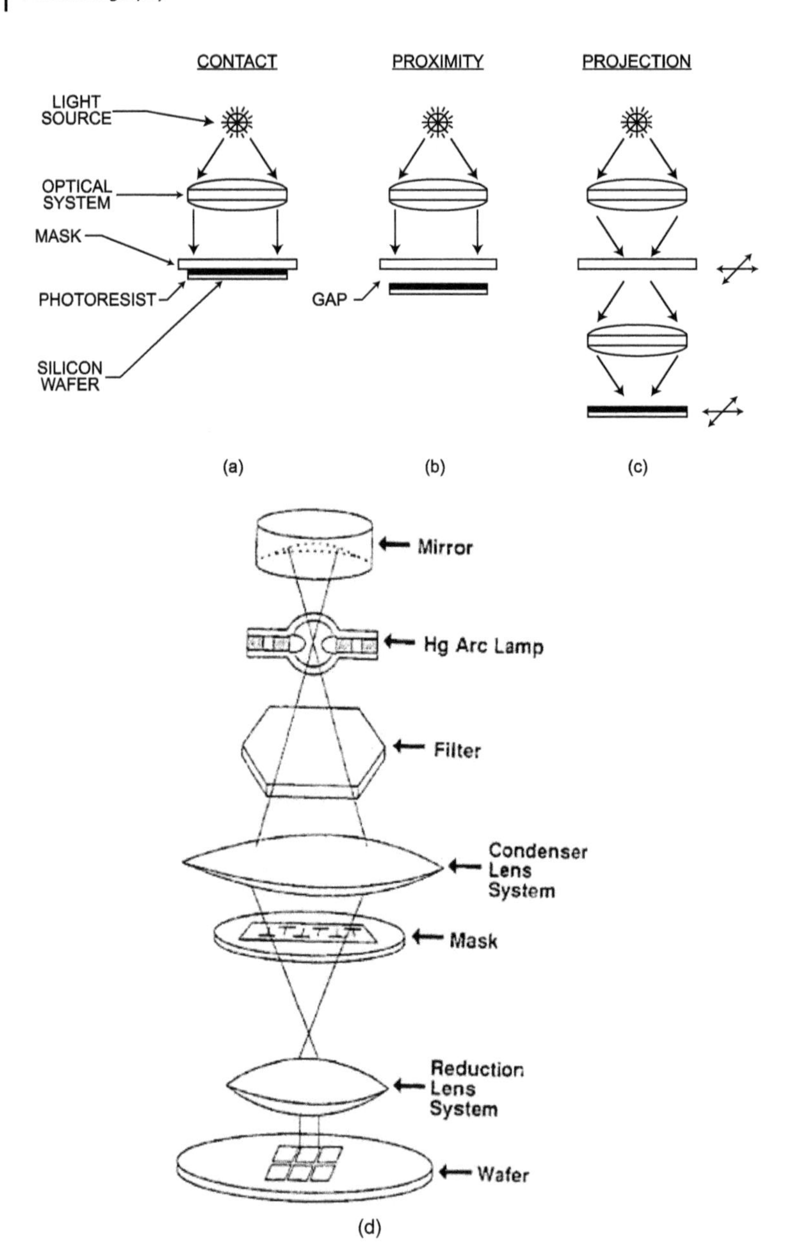

Figure 5.2 Three basic optical lithography techniques: (a) contact, (b) proximity, and (c) projection. (d) Schematic of a reduction step and repeat system with refractive optics (Reprinted with permission from Ref. [1], Copyright (1994) American Chemical Society).

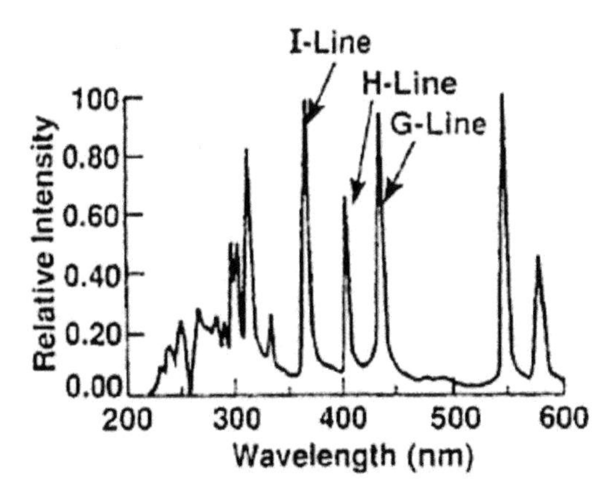

Figure 5.3 High-pressure mercury-arc lamp spectrum. Reprinted with permission from Ref. [1], Copyright (1994) American Chemical Society.

5.3.1 Typical Etch Photoresist Process

Process sequence for patterning a deposited film on part of a semiconductor by etching and lift-off is shown in Fig. 5.4. The corresponding process flow for the etch process is shown in Fig 5.5a. A cleaning process step may be used before the photoprocess is started in order to make sure that the wafers are clean and free of particles and contamination. The adhesion promoter and photoresist are coated on the wafer and followed by a bake operation (prebake). These may be done on a track system in sequence. The wafers are then "exposed" to UV light on a stepper to chemically change the exposed regions and then developed to dissolve away the exposed (positive) photoresist. A plasma de-scum step is used to remove minor organic contamination from the open areas, and the wafers are etched to remove the film to be patterned. After the etching is complete, the photoresist is stripped off using solvents or the plasma process.

5.3.2 Lift-Off Photoresist Process

The lift-off technique has been invaluable for production of GaAs and other III–V ICs because of the difficulty of patterning gold, platinum, and other thin-film materials like nichrome, used extensively for contacts, resistors, and interconnect wiring. The problems associated with etching these materials are discussed further in Chapters 6 and 7. A flow diagram for this process is shown in Fig. 5.4 and the process flow is shown in Fig. 5.5b. In this scheme using a positive photoresist, the image is reversed after the stepper exposure by using one of several image reversal processes. After developing, these processes provide a proper resist profile. The film to be patterned is deposited next and it contacts the substrate only in the open regions. The photoresist is finally removed using a lift-off process, where it is dissolved and removed along with the film on top of it. Lift-off is possible because of a gap between the required contact or line and the resist at the bottom. Creation of this gap involves development of sloped walls in the resist or the use of multiple layers of the resist. See Fig. 5.4 for a comparison of the basic etch and lift-off processes. Details of the image reverse process are discussed in a later section.

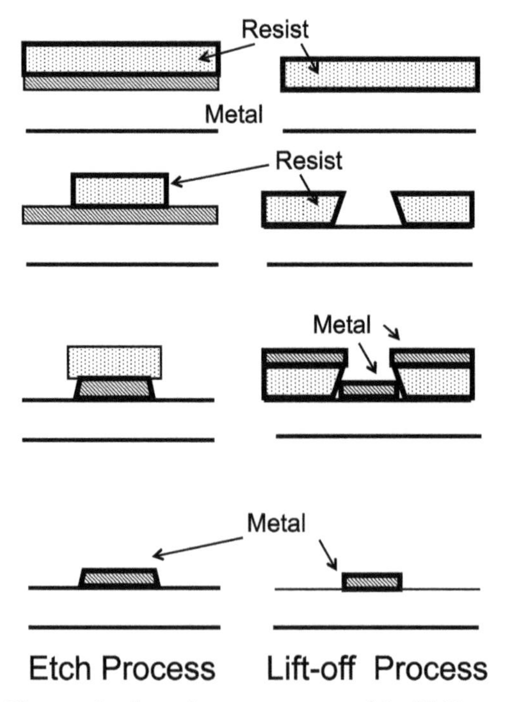

Figure 5.4 The two basic resist processes used in III–V processing: the etch process used heavily in silicon and the lift-off process used mostly in III–V processing.

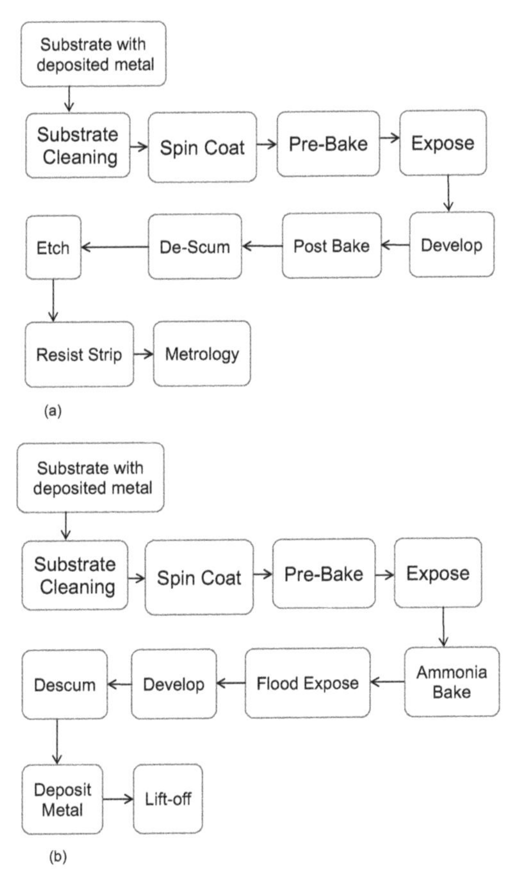

Figure 5.5 (a) Typical photoprocess (etch) and (b) typical photoprocess (lift-off).

The photoresist along with unwanted metal may also be removed in the lift-off process by using tape lift-off. In this process a large sheet of tape is used to remove the majority of the metal and resist, and the remaining resist pieces and residue are removed using solvent spray, like N-methylpyrrolidone (NMP). This process needs less pressure and may produce less damage to the metal features being patterned.

5.4 Photoresist

Photoresist materials respond to optical radiation and change in such a way that a pattern can be replicated in them. Numerous requirements must be met by photoresist materials. Photoresists must be highly sensitive to the radiation being used but resist attack by etchants and withstand the whole photoprocess. Table 5.1 lists the major characteristics of positive and negative photoresists. Among these, high resolution, good adhesion, chemical stability through wet and dry etching, and the ability to be stripped completely without leaving contamination behind are important for GaAs processing. To accomplish all these requirements, many components are added together to formulate the resist. The main three components are:

- Resin, which forms the inert matrix and fulfils the mechanical and chemical requirements
- Sensitizer, or the photoactive compound (PAC)
- Solvent, which keeps the components in the liquid state

Examples of PACs and the sequence of photochemical transformations in positive photoresists are shown in Fig. 5.6. Other components are added for light filtering, adhesion, etc., and numerous other components that are proprietary may be added to fulfill the large list of requirements shown in Table 5.1. A few of the main characteristics are described below.

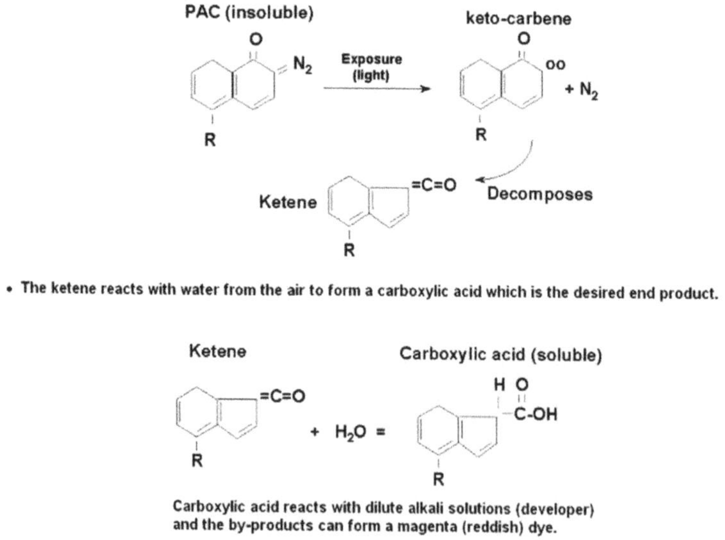

Figure 5.6 Example of the sequence of photochemical transformations in a positive photoresist.

5.4.1 Resolution and Contrast

Resolution defines the ability to print small dimensions, which is measured as line width. The resolution of the photolithographic process is determined by the exposure equipment and the resist process. To build submicron features consistently in production, the overall process must be able to print the minimum dimension over the whole range of processing conditions.

The contrast of a resist directly influences resolution and wall angle. In an ideal resist, the developed film thickness would drop in a step function as the exposure reaches a certain value. In a real resist, the transition region has a certain slope and the contrast is a function of this slope. Figure 5.7 shows a response curve for a positive photoresist, where the normalized film thickness after development has been plotted as a function of the exposure dose on a log scale. In positive resists the contrast is related to the rate of chain scission and change of solubility at a constant irradiation dose.

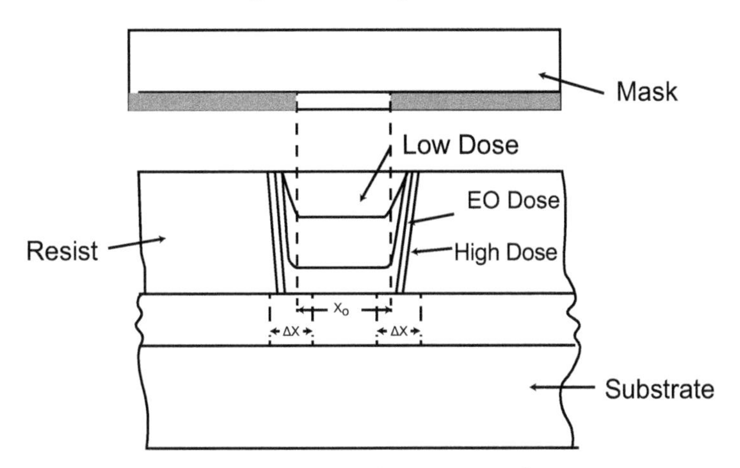

Figure 5.7 Determination of the E_0 characteristic of a resist.

Resolution is also a function of resist swelling during the processing, the thickness of the resist, and the proximity effect of features on the mask that are close to the feature being measured. Positive resists do not swell much. An isolated via is easier to resolve in a positive-resist system compared to densely packed vias or lines. The features that are most difficult should be monitored for process control. Resolution is improved as the resist is made thinner in order to keep the aspect ratio (w/t) high. Effort should be made to optimize resist thickness to meet the requirements of the etch or the lift-off process and keep the resist as thin as possible. Both contrast and sensitivity of resists are measured by exposing resist layers of known thickness to different radiation doses and then determining the remaining film thickness after development (see Fig. 5.7). The data is plotted in the form of a response curve (Fig. 5.8). Contrast as well as E_0, a characteristic of the resist, can be determined and monitored for resist batches.

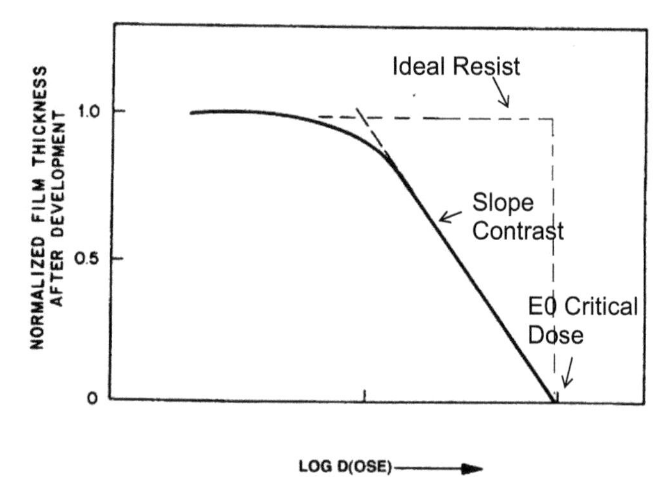

Figure 5.8 Response curve for a positive photoresist.

5.4.2 Sensitivity

Sensitivity can be defined as the ratio of chemical response to input energy. The resist must be matched to the wavelength of UV light being used. The information is provided in the form of a spectral response curve by the vendors. High sensitivity is favorable in order to keep the exposure time short for high throughput.

Adhesion is improved by the addition of dehydration bakes in the process flow and using a very thin layer of hexamethyldisilazane (HMDS) as the adhesion promoter. Resist viscosity depends upon the solid-content-to-solvent ratio. Along with spin speed it determines the resist thickness and must be controlled well.

Resist constituents do degrade with time. The shelf life of a resist can be from six weeks to months. A resist must be stored properly and its age monitored.

5.4.3 Optical Photoresist Reaction Mechanism

To understand how a photoresist works, an example of one type of positive resist is discussed below. As shown in Fig 5.5, the PAC in the resist, a diazoquinone (DQ), upon exposure to radiation, undergoes a molecular rearrangement and then in the presence of water forms carboxylic acid. So, it transforms from an insoluble to a dissolution enhancer. The matrix is a novalac resin that dissolves in an aqueous base. Carboxylic acid readily dissolves in a basic developer. Unexposed areas are not permeated by the developer.

5.4.4 Image Reversal of a Positive Photoresist

Image reversal is used extensively in III–V IC processing to pattern gold-based metallization layers by the lift-off process. Another advantage is that reversal allows use of dark-field masks, which reduce light-scattering effects and are less sensitive to particles and defects.

In this process, the exposed wafers are subjected to ammonia vapor. The vapor diffuses into the resist and reacts with the carboxylic acid formed there as a result of normal positive-resist exposure to radiation and water. This is followed by a high-dose flood exposure, which now reverses the soluble and insoluble regions. Figure 5.9 shows the sequence of these processes. This process provides good resolution possible with a positive photoresist, good CD control, and adequate depth of focus. The resist wall angle can be controlled, making it ideal for lift-off. Another approach to image reversal is the thermal reversal approach (developed before the image reverse process) [5]. In this process, the positive photoresist is doped with a material (monazoline) that makes the carboxylic acid degrade during heat treatment. In this process ammonia is not needed; the rest of the process is similar [6].

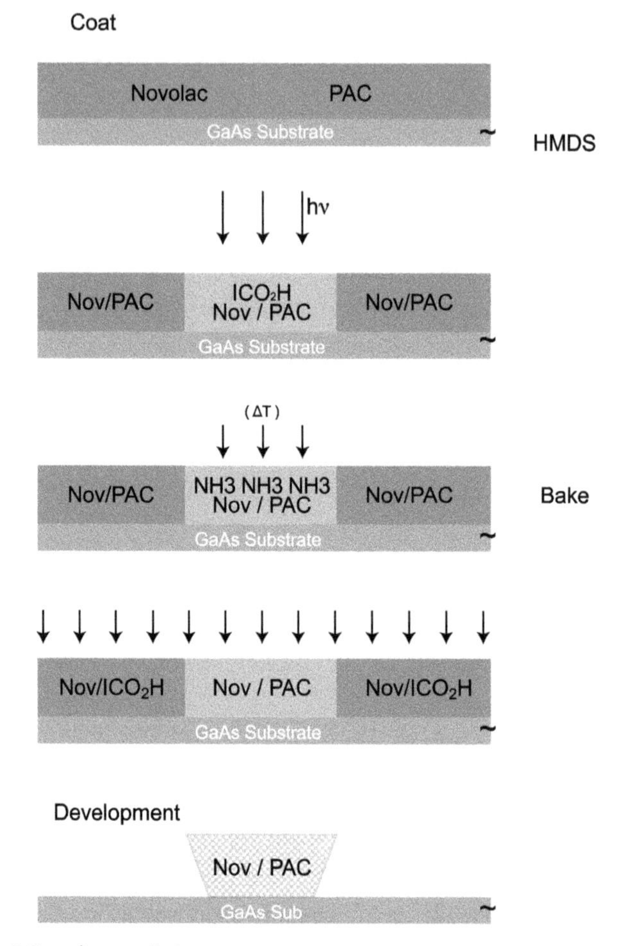

Figure 5.9 Ammonia image reverse process steps.

5.4.5 Negative Resists

Negative photoresists are based on cross-linking of main chains of synthetic rubbers upon exposure to radiation. A commonly used two-component resist is bis(aryl) azide rubber in a cyclized poly(cis-isoprene) matrix. Cross-linking makes negative resists insoluble in organic solvents like xylene. Major disadvantages of negative resists are listed in Table 5.1. These resists are sensitive to oxidation from the atmosphere, need organic developers (solvents), and have been historically limited to 2–3 μm resolution. Negative resists are used

Table 5.1 Comparison of negative and positive photoresists [3]

	Resist type	
Characteristic	**Positive resist**	**Negative resist**
Adhesion to Si	Fair	Excellent
Available compositions	Many	Vast
Contrast γ	Higher (e.g., 2.2)	Lower (e.g., 1.5)
Cost	More expensive	Less expensive
Developer	Aqueous based (ecologically sound)	Organic solvent
Developer process window	Small	Very wide, insensitive to overdeveloping
Image width to resist thickness	1:1	3:1
Influence of oxygen	No	Yes
Lift-off	Yes	No
Minimum feature	0.5 μm and below	±2 μm
Opaque dirt on clear portion of mask	Not very sensitive to it	Causes printing of pinholes
Photo speed	Slower	Faster
Pinhole count	Higher	Lower
Pinholes in mask	Prints mask pinholes	Not so sensitive to mask pinholes
Plasma etch resistance	Very good	Not very good
Proximity effect	Prints isolated holes or trenches better	Prints isolated lines better
Residue after development	Mostly at <1 μm and high aspect ratio	Often a problem
Sensitizer quantum yield Φ	0.2–0.3	0.5–1
Step coverage	Better	Lower
Strippers to resist over		
Oxide steps	Acid	Acid
Metal steps	Simple solvents	Chlorinated solvent compounds
Swelling in developer	No	Yes
Thermal stability	Good	Fair
Wet chemical resistance	Fair	Excellent

because of low cost and higher sensitivity (speed). New negative resists have been formulated to overcome the limitations mentioned above. nLOF from AZ is one of them. It uses tetramethyl aluminum hydroxide (TMAH) as the developer, which is water soluble, and the resist is compatible with an i-line resist. When this resist process is used for lift-off, an ammonia oven and a flood exposure system are not needed. This resist has a high (135°C) softening point and is not limited to 110°C like commonly used positive resists.

5.4.6 Resolution Improvement

Multilayer resist processes are sometimes used to improve resolution over topological steps or achieve features smaller than 0.5 μm. A common material used as an underlayer is PMMA, which is sensitive to DUV. A thin layer of a second resist that is sensitive to UV is deposited on the top and patterned. The wafer is flood- exposed to expose the PMMA. Upon development the image is transferred to PMMA with better resolution. Other approaches use even more layers as part of the imaging process.

5.5 Physics of Photolithograpy [1, 2]

5.5.1 Diffraction

Optical systems are designed with sources that have spatial coherency. These systems have limited angular distribution, and light is in phase at all points. However, due to the wave nature of light, the image of an edge on the mask is not sharp on the wafer but is diffused due to diffraction. The intensity distribution for a shadow according to geometrical optics and actual with diffraction are shown in Fig. 5.10a,b.

Similarly for a slit, the image is not a sharp rectangle, but the intensity distribution shows an airy pattern. The amplitude varies with the diffraction angle θ according to the equation

$$A(\theta) = \frac{A_0 \sin\beta}{\beta} \tag{5.1}$$

where $\beta = \pi b \sin(\theta)/\lambda$, λ is the wavelength, and b is the slit width.

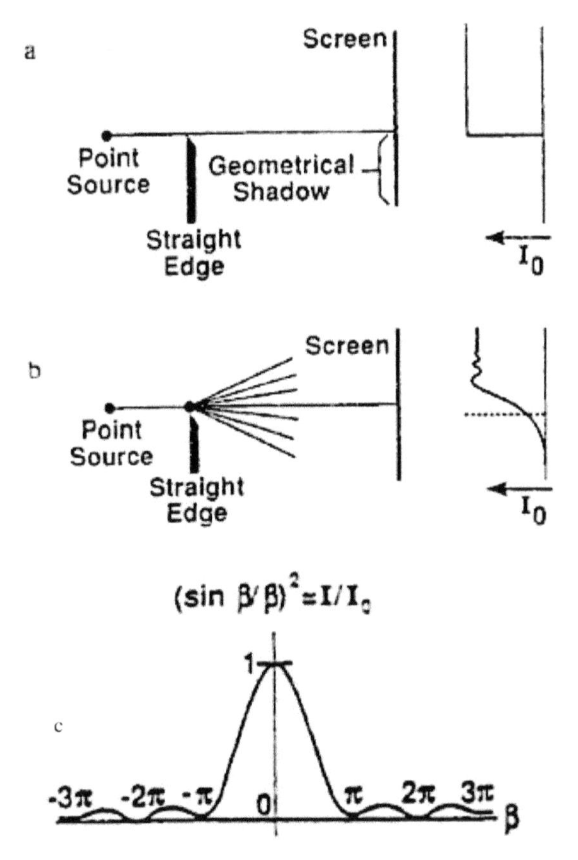

Figure 5.10 Intensity distribution produced by a spatially coherent ray of light as it passes by a straight edge (a) according to geometrical optics and (b) with diffraction. (c) Intensity distribution of Fraunhofer diffraction pattern of a single slit.

Intensity is the square of the amplitude

$$I(\theta) = \frac{I_0 \sin^2 \beta}{\beta^2} \tag{5.2}$$

The intensity pattern is shown in Fig. 5.10c. The intensity of the higher-order peaks drops as the order increases. Perfect image reconstruction requires that the light from all orders be collected by the projection lens and recombined at the image plane. To collect higher and higher orders, the acceptance angle of the lens must be large.

The major factor that limits the resolution capability is the design of the objective lens and its numerical aperture (NA). The NA is a measure of the capability of the lens to collect diffracted light from the photomask and project it onto the wafer. See Fig. 5.11 and Refs. [1] and [2] for details.

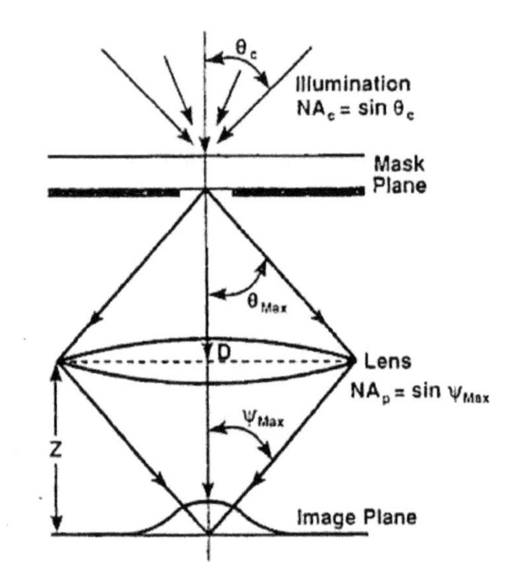

$$\text{Effective f-number (F)} = \frac{\text{Image Distance (Z)}}{\text{Clear Aperture (D)}} = \frac{1}{2NA_p}$$

Figure 5.11 Schematic diagram of illumination and projection optics, showing numerical aperture parameters.

$$NA = n \sin \theta \tag{5.3}$$

where n is the refractive index (1.0 for air). Typical NA values are 0.16 to 0.40. Resolution, w, depends on the wavelength and NA as given by the following relationship:

$$w = \frac{\lambda}{2(NA)} = \frac{k\lambda}{NA} \tag{5.4}$$

where factor k depends upon the resist technology process. For complex resist processes, k is about 0.6. As the NA increases, the resolution gets better, with a larger angle of collection. However, the depth of focus is inversely proportional to the square root of NA

$$d = \frac{\lambda}{(\mathrm{NA})^2} \tag{5.5}$$

If the depth of focus is too small, differences in height and flatness of wafer surface cannot be tolerated. A compromise must be made between high resolution and depth of focus.

The modulation transfer function (MTF) measures the capability of a projection system to reproduce a mask image on the wafer surface. If a mask with a line and space pattern is projected onto the wafer, the intensity of the light will show a wavy pattern, as shown in Fig. 5.12 [2].

$$\mathrm{MTF} = \frac{(I_{\max} - I_{\min})}{(I_{\max} + I_{\min})} \tag{5.6}$$

where $I_{\max}$ and $I_{\min}$ are the intensities of the brightest and darkest positions. M is unity for an ideal system. As spacing decreases, the MTF will approach a small number. The MTF depends on the NA, λ, mask feature size, and spatial coherency of the UV radiation.

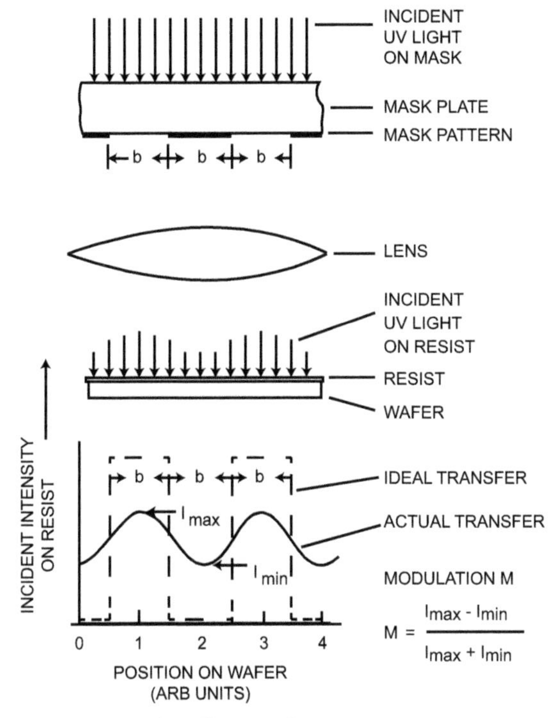

Figure 5.12 Image transfer efficiency for a 1:1 projection printer. Reprinted from Ref. [2] with permission from Lattice Press.

5.6 Step and Repeat Projection Aligner

The three main methods of pattern transfer were introduced earlier in Section 5.3 on the basics of printing. Projection printing used in GaAs large-scale integration (LSI) circuit fabrication will be discussed in a little more detail here. See Fig. 5.2d for details. This type of projection exposure system uses refractive optics to project the mask image onto the wafer (as opposed to reflective optics). These systems are designed to project an image onto a small portion of the wafer, called a field (square 20 mm side). Once a field is exposed, the process is repeated over until the whole wafer is covered. This way, wafer size is not limited by optics. Other advantages of this approach are higher image resolution, and corrections for wafer distortion can be made. However, precision control of the mechanical stage is needed. Masks that are 10x times the actual were used but 5x is becoming more common.

5.7 Pattern Registration

Next to CD control, the overlay accuracy is the most important process deliverable of lithography. In an IC process involving over 10 masks, the patterns must be aligned with very high accuracy. For critical layers, for example, the gate in an field-effect transistor (FET) process, the placement of the gate exactly between source and drain with an error of less than 0.1 μm may be required. Similarly in making contacts to underlying layers, a minimum overlap of vias may be required. These design rules are provided by the circuit and process engineers, and an alignment scheme must be designed to accomplish these in production. Box-in-a-box structures are used to measure the alignment in x and y directions. Factors that contribute to the overlay error are CD size or dimension difference, which is a function of the resist and etch process, alignment inaccuracy due to masks, registration error, and machine-to-machine variability, and process-related problems that may cause the edges to be different, for example, nonsymmetrical undercuts etc. Alignment targets are placed on masks and projected onto the wafer. Automatic alignment systems that control the stepper superimpose the image on a previously placed target on the wafer, which is viewed by a TV camera. Software algorithms are used to do the alignment from the captured images.

Alignment can be done globally and locally. Global alignment means rotational and translational alignment over the entire wafer. Local alignment is done within exposure fields. For supercritical layers, a die-by-die alignment can be used at the expense of throughput.

5.8 Resist Processing

The process sequence of typical etch and lift-off patterning was discussed briefly earlier and summarized for etch and lift-off process types in Fig. 5.5. Many of the process steps are discussed in detail in other chapters on cleaning and dry etching, etc. A few of the process steps are described below.

5.8.1 Prebake Dehydration

Wafers should ideally be processed through a photolithography process immediately after the completion of film deposition or other process steps. In practice, wafers are stored in the fab for days, even weeks. Therefore the wafers are cleaned routinely and then dried using a dehydration step for good resist adhesion. This step drives out moisture from the surface; typically surface water is released at 150°C–200°C. Much higher temperatures may be needed for driving out moisture completely but may not be practical.

5.8.2 Adhesion Promoter

Wafers are generally primed using an adhesion promoter before resist application. The most common adhesion promoter is HMDS. This compound works by providing a bridge to bond the organic resist to an inorganic surface that has a native oxide on it. This layer is deposited by spin coating on a track system or *vapor priming* in an oven.

5.8.3 Resist Coating

The resist is coated usually on the wafers by a spinning process. The wafers are placed on a vacuum chuck, a small amount of resist is placed on the wafer, and the wafer is spun at fast speed (>1000 rpm). Spin coating produces a uniform, defect-free film, the thickness being a function of the viscosity of the resist and the spin speed. The spin-coating process can be broken down into multiple steps, dispensing of the resist, acceleration, and uniform spin. All these can be optimized and controlled. The thickness drops as speed as goes up, as shown in Fig. 5.13, for a certain viscosity, which again can be a function of temperature. The thickness is generally higher at the edges; a thicker edge bead forms at the very edge. This edge bead can be removed by a solvent spray.

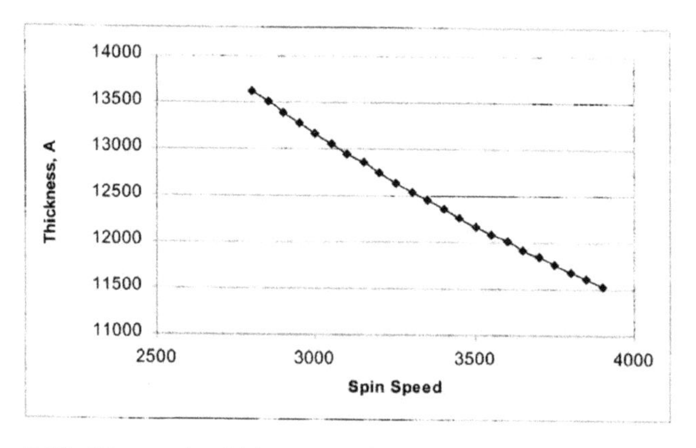

Figure 5.13 Photoresist thickness as a function of spinner speed.

Resist film thickness is measured with optical instruments, like ellipsometers and interferometers. These measurements will be discussed in more detail in the chapter on characterization. Resist processing is carried out in constant humidity to avoid thickness variations, condensation of water vapor on wafers, and excessive resist drying.

5.8.4 Soft Bake

Immediately after spinning, the wafer goes through a soft bake step. In this step the solvent is driven off. The amount of resist that remains in the resist is carefully controlled by determining proper temperature and time. This in effect controls the rate of attack of the resist by the developer and how the resist can stand etching, etc. An overbaked resist can lose sensitivity by loss of the PAC due to reaction. Conventional ovens as well as hot-plate systems are used for soft bake. Hot plates have advantages and can be part of an automated track system. The resist is heated from inside out, and the cycle time for each wafer can be about a minute. Hot plates are maintained at a fixed temperature. The temperature profile that the wafer sees is controlled by the rate of dropping of the wafer and hold time. Bubble formation, shrinkage, and developing behavior can be controlled by designing a proper temperature profile. Figure 5.14 shows desired temperature profiles for hot plates on phototracks.

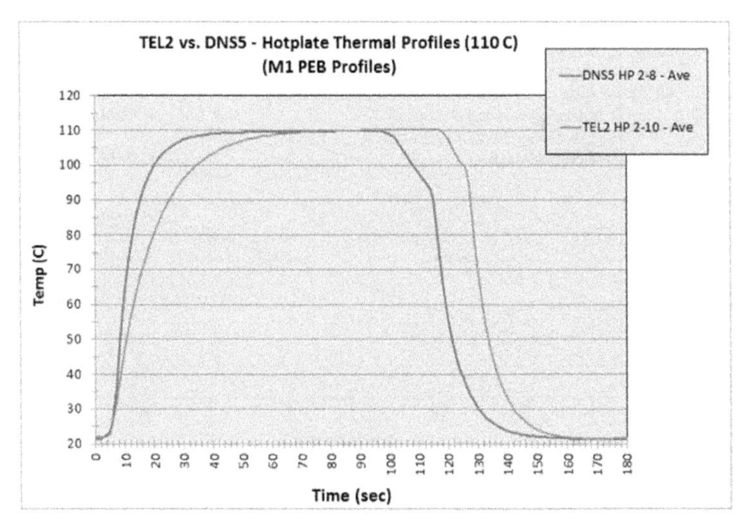

Figure 5.14 Examples of a soft-bake temperature profile of phototrack hot plates, showing a controlled temperature profile.

5.8.5 Exposure

This is the step where the resist is exposed to radiation on a stepper or other exposure system. The amount of energy or the dose is measured and controlled by an energy integrator, which controls the exposure process to achieve the photochemical effect in the shortest time in a reproducible manner. Since wafers are processed one at a time, field by field, at multiple steps, the stepper process must be optimized for high throughput. The whole photolithography process is designed around the exposure system. The dimensions of the features are a function of all the photolithography process steps; however, the line width can be controlled to some extent by the exposure dose. Figure 5.15 shows the change of line width with the exposure dose and resist thickness. For processes used in III–V semiconductors, topology is always an issue because mesa are used and features must be opened at multiple levels. Therefore, focus becomes an issue. The photoprocess is optimized for exposure and focus setting by using Bossung curves, as shown in Fig. 5.16. Focus setting and exposure dose are chosen to pick the flattest response. Resist thickness can be controlled to the order of 0.01 μm. Line widths can be controlled to +/− 0.2 μm on an i-line system.

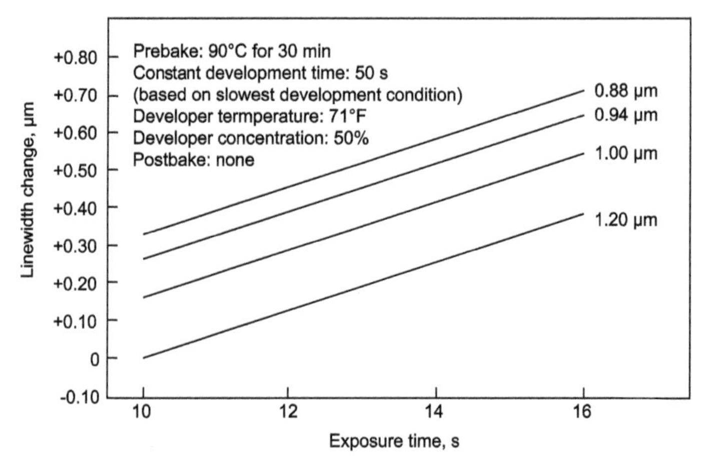

Figure 5.15 Line width change as a function of resist thickness and exposure time.

5.8.6 Standing Waves and Other Interference Effects

In addition to diffraction and the resist, there are other factors that determine line width, related to the resist topology and the layer underneath. The intensity distribution in the resist is a function of standing waves caused by the reflective properties of the underlying layer. Light incident on the resist film is absorbed as it passes through it, and what reaches the substrate is partially reflected back into the resist. The reflected waves constructively and destructively interfere with the incident waves and create standing waves. This results in a periodic variation of the light intensity and a wavy pattern in the resist (Fig. 5.17). This effect is strong over highly reflective underlayers and causes line width variations over steps. This and other interference effects are shown in Fig. 5.18 [3]. These effects also cause nonvertical resist profiles, reflective notching, scumming, and alignment issues. Antireflective coatings (ARCs) and addition of absorptive dies into the resist are used to overcome these effects. However, ARCs add processing steps and complexity.

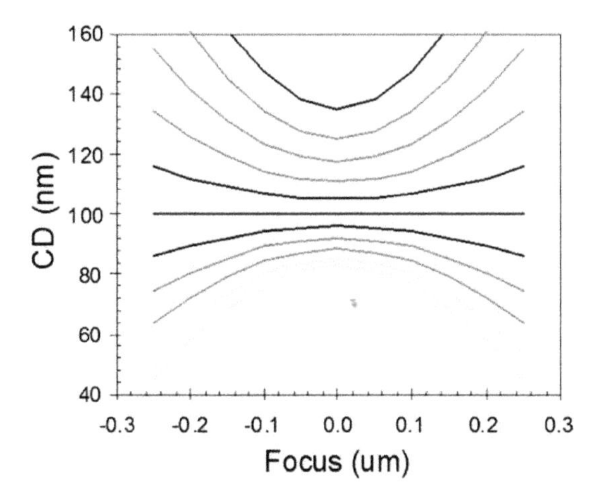

Figure 5.16 Example of Bossung curve used for optimizing focus and exposure dose.

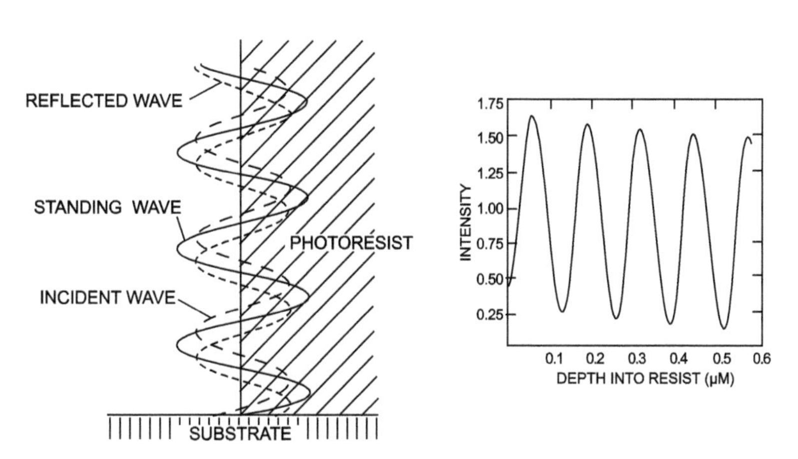

Figure 5.17 Standing waves as a function of resist thickness.

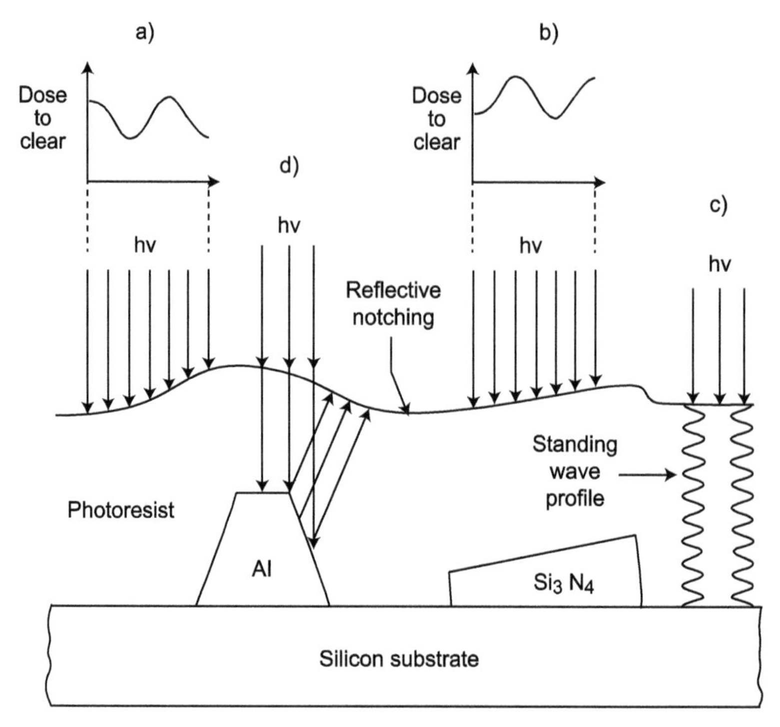

Figure 5.18 Thin-film interference effects showing light reflected from resist/Si and SiN/Si interfaces through different resist thicknesses and dose necessary to clear the resist. Standing waves and reflective notching are also shown.

5.8.7 Developing

The developing process is complex and needs to be optimized empirically for optimum CD control. The developing process should not reduce the resist film thickness, should not cause distortion or swelling, and should have high throughput. Liquid developers, preferably water soluble, are used in production. The process can be spray, puddle, or immersion. Spray developing provides good uniformity and reproducibility. Process variables, including developer concentration, time, temperature, and rinse times, must all be optimized.

5.8.8 De-scum

After developing, the edges of the open pattern may show some resist residue or the edges of the resist may be ragged. These cannot be completely eliminated by expending the developing process without compromising the developing process itself. For good adhesion, the surface of the layer underneath where contact has to be made must be clean before the next metal layer is deposited. An oxygen plasma step is used to remove this unwanted resist, or "scum." This plasma can change the dimension of the lifted-off metal lines by rounding off corners of the resist. Details of de-scum process will be discussed in Chapter 7.

5.8.9 Postbake

This step is used to remove the solvents and harden the resist film so that it can withstand the subsequent etch or deposit step. This step is carried out at a higher temperature than soft bake, because the image has already been defined. This temperature can be optimized for good adhesion and intentional polymer flow that will cause resist to change shape and size.

5.8.10 Stripping

After the etch or deposition process is complete, the resist must be removed by stripping or lift-off. This step appears to be noncritical because the critical parts of pattern definition are complete at this point. However, this step has yield and cost consequences. If the strip is not completed properly, it may be impossible to remove the cross-linked materials or complex compounds and crusts formed as a result of the dry etching or ion implant processes. In many cases a combination of wet and dry processes may have to be utilized. A brief description of these is given below.

Organic strippers that work by dissolving the resist are easy to use; however, their use is becoming difficult due to their disposal problems. A number of new commercial strippers are available that are easier to dispose of or reclaim, but the cost can still be prohibitive. Oxidizing-type liquid chemical strippers used in the silicon IC industry are solutions of sulfuric acid and hydrogen peroxide oxidizer. These are not useful in III–V processing for obvious incompatibility with the substrates. Dry stripping offers many advantages, such as reduced operating cost and no pollution or environmental issues. At the same time process problems like resist streaks or defects like metal stringers can be eliminated. Details of dry stripping using oxygen plasma are discussed in Chapter 7. The effect of energetic plasmas on devices, wherever sensitive device layers are exposed, must be carefully considered to avoid low-energy plasma damage. Damage at all prepassivation photolithography steps, for example, ledge in HBTs or pregate processing in pseudomorphic high-electron-mobility transistor (PHEMT), must be carefully considered and avoided.

5.9 Electron Beam Lithography [1]

Modern optical DUV steppers have made fabrication of submicron devices routinely possible. Most III–V device processes require dimensions near 1 mm; only some contacts for HBTs and gates for FET-type devices require half-micron and sub-half-micron feature sizes. For special high-frequency applications over a few gigahertz, the small gate size and its placement accuracy make nonoptical lithography a requirement.

Electron beam lithography is also used to form patterns on photomasks and for exposing resists directly on wafers, without using a mask, in direct wafer writing. The advantages of electron beam lithography, in addition to high resolution, are excellent registration, focus, and no need of masks. Disadvantages are the need to sustain a special process, high cost of equipment, and low throughput. In case of semi-insulating wafers like GaAs, wafer charging may slow down the process further.

A focused beam of electrons can be deflected accurately by electrical and magnetic fields. Thus a beam can be steered over a surface of a wafer and turned on and off to "write" a pattern. Electrons have wave- and particle-like properties, the wavelength being orders of magnitude shorter than UV light. Beams can be focused to a small size, the size of the written feature being limited by the scattering in the resist film. The electron energy is between 10 and 25 keV and the spot size is about 0.1 μm. The exposure is controlled by adjusting the spot size, beam current, and scan speed. The beam interaction with the resist causes of loss of energy, similar to ion implantation and resist damage. Bonds break, making the resist more soluble in the developer. Some amount of lateral scattering and reflection from the substrate can take place, diffusing the size of the exposed spot. The size of the developed resist opening at the bottom of the resist can be varied with exposure dose (more than the opening at the top). This gives control over the resist profile, which can be utilized in the lift-off patterning of the gate stripes. Details of gate formation are discussed in Chapters 12 and 15.

The definition of the smallest geometries requires a small spot size and dynamic adjustments for shape and size as the beam is being deflected. Good control can be maintained for small field sizes. Each field is aligned and exposed and then the stage is moved to the next field to repeat the process. The alignment marks should be well defined and be defined in the layer to which the first-order alignment is needed, for example, source and drain ohmic contacts for gate placement.

PMMA is a classical electron beam resist. It is a positive resist with high resolution and wide processing latitude. Its disadvantages are low sensitivity, which results in low production throughput. Other acrylates, styrenes, etc., have been tried. Properties of some electron beam resists are discussed in reference books, for example, Ref. [1] and other publications [7].

5.10 X-Ray Lithography [8, 9]

It may be noted that X-ray lithography is also becoming popular in fabrication of sub-0.1 μm gate PHEMTs. Low wavelengths of practically a few angstroms theoretically eliminate all diffraction limitation, with masks that can be placed in proximity to the wafer with no backscattering or refection from the substrate. This allows exposing attractively very fine lines in very thick resists with vertical walled patterns, that is, excellent resolution with a limitless depth of field (Fig. 5.19). The entire wafer can be exposed in one go, even giant 1-inch steps compared to optical steppers. Economical and sufficiently strong X-ray sources are hard to achieve. Extended sources limit the advantages of no-diffraction effects, and synchrotron sources are attractive but excessively expensive. Fragility of masks made out of gold on substrates such as boron nitride, silicon carbide, diamond, polyimide, or even etched-away silicon poses operational problems. Since X-rays cannot be produced at high enough intensity and without collimation, resist technologies are also challenging. The emission line of the X-ray source must match the absorption of a given resist. Negative resists are available with sensitivities of the order of 10–25 mJ/cm^2, but unfortunately they suffer from swelling and poor resolution. However, positive resists are not as sensitive. It is attractive to use trilevel resists as only minimum dosages are necessary to expose only thin resists. This also solves the problem of a uncollimated beam, which otherwise will expose lines at an angle that is proportionate to the distance from the center axis. Using the thin resist as a pattern will correct this problem, while dry-etching techniques can be used to produce a vertical pattern.

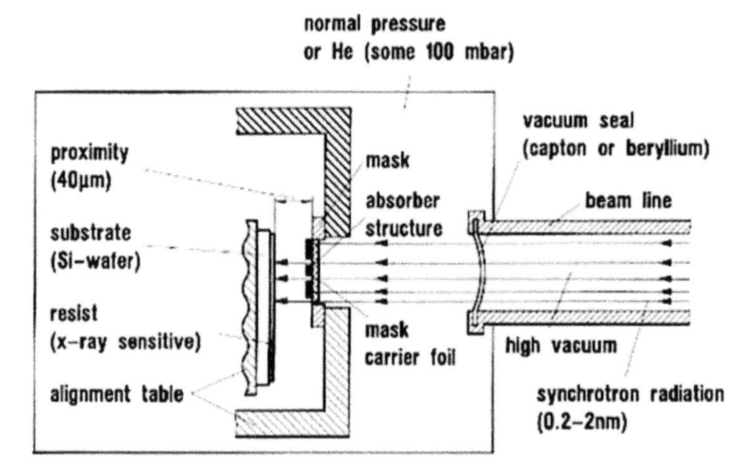

Figure 5.19 X-ray lithography. Courtesy of the Karlsruhe Institute of Technology, Germany.

5.11 Process Monitoring [10, 11, 12]

Photoresist processing can be reworked, so inspections are done frequently to catch mistakes and defects. The process is monitored by measuring the line width, called CD, and the alignment accuracy at the end of the development process. The CD must be controlled over the whole wafer and across multiple wafers in the lot. So, a good sample must be taken. If the line width or the alignment in the x or y direction does not meet the specification for the layer being processed, the photoprocess can be reworked. As design rules keep shrinking, maintaining the photoprocess can be a challenge. Automatic overlay measurement systems that improve yield through measurement of multiple sites on the wafer are becoming available.

Line width measurement systems use visible light optics or He–Ne laser light or electron beam techniques. For geometries over 1 μm, optical techniques are adequate, but for smaller geometries scanning electron microscopy (SEM) is needed.

5.11.1 Optical Systems

In the scanning slit technique a narrow slit is moved across the image of a magnified feature. The light intensity profile is acquired with a photomultiplier tube. Scanning along the feature length, autofocus algorithms can be used to obtain the best image. The profile is analyzed by edge-sensing algorithms to compute the line width of the feature. In video-based systems, the image is projected onto a video screen and the operator moves a cursor to the edges. The optical intensity profile is collected and analyzed in a manner similar to the scanning slit method. In laser-based systems, a 1 μm spot laser is scanned across the feature and the reflected light detected by photodetectors or a high-resolution interferometer. Resolution capability of 0.25 μm and repeatability of 0.1 μm can be achieved.

5.11.2 SEM

Better accuracy is needed for line width measurement of critical features, so in-line SEM is used for monitoring these critical processes. Low-voltage SEM that uses 20 keV provides line width data without the need to cleave and thus sacrifice a wafer.

Line width and alignment measurements must be made over properly designed features that represent the actual process. Measurements should also be made over steps and topology during the development of the process.

5.11.3 Advanced Photolithography Techniques

Numerous techniques have been developed to push the limits of photolithography to sub-0.5 μm dimensions. Phase-shift techniques, X-rays, and ion beam lithography have been developed [1, 2]. Such techniques are not yet utilized in III–V processing but may be needed when high-mobility semiconducting layers and small-dimension geometries are utilized for next-generation high-speed circuits.

References

1. L. F. Thompson, C. G. Wilson and M. J. Bowden, *Introduction to Microlithography*, ACS Professional Reference Book (1994).
2. S. Wolf and R. N. Tauber, *Silicon Processing for the VLSI Era*, Lattice Press, Sunset Beach, California (1986).
3. M. Madou, *Fundamentals of Microfabrication*, CRC Press, Washington, D.C. (1997).
4. D. A. McGillis, Lithography, in *VLSI Technology*, Ed., S. M. Sze, McGraw Hill, New York (1983).
5. E. Alling and C. Stauffer, Image reversal of positive photoresist, *Proc. SPIEI*, **539** (1985).
6. R. Williams, *Modern Gallium Arsenide Processing Methods*, Artech House (1990).
7. E. B. Hryhorenko, *A Positive Approach to Resist Process Characterization for Linewidth Control*, Eastman Kodak, Rochester, New York (1980).
8. J. P. Silverman, Challenges and progress in X-ray lithography, *J. Vac. Sci. Technol. B*, **16**, 3137–3140 (1998).
9. S. Ohki and S. Ishihara, An overview of X-ray lithography, *Microelectron. Eng.*, **30**, 171–178 (1996).
10. G. W. Martel and W. B. Thomson, *Semicond. Int.*, **Jan.–Feb.**, 69 (1979).
11. J. D. Cuthbert, Standing waves in photoresist, *Solid State Technol.*, **50**, 59 (1977).
12. M. W. Horn, *State Technol.*, 57 (1991).

Chapter 6

Wet Etching, Cleaning, and Passivation

6.1 Introduction

Wet etching continues to be an important part of wafer fabrication in spite of the numerous advantages offered by dry-etching processes. Wet etching, being a lower-cost process, is preferred wherever use of dry etching is not mandated by process control or reliability requirements, like those encountered in control of submicron features or alignment. Wet-etching processes are used for removing unwanted layers from parts of the wafer surface in order to create device structures, make contact paths to different layers underneath, and sometimes make larger mechanical elements like microelectromechanical systems (MEMS).

Wet-etching processes involve three subprocesses: transport of the etchant to the surface, the reaction process, and removal of the by-products of the reaction. Etching processes in the semiconductor industry need to be carried out at a controlled rate. The rate is determined by the slowest of the processes involved. If the rate is limited by the diffusion of reactants through a stagnant layer on the surface, the reaction is referred to as *diffusion limited*. On the other hand if the overall rate is determined by the reaction rate itself, the process is *reaction* or *kinetically limited*. Adsorption and desorption can also affect the rate of wet etching. Wet-etching processes are limited often by dissolution of the reaction products by the chemical

III–V Integrated Circuit Fabrication Technology
Shiban Tiku and Dhrubes Biswas

ISBN 978-981-4669-30-6 (Hardcover), 978-981-4669-31-3 (eBook)
www.panstanford.com

solution or removal of gaseous products in the form of bubbles. Agitation of the solution increases the overall rate in such cases.

Amorphous materials etch in an isotropic manner; however, most III–V processing involves etching of single crystals (amorphous silicon nitride is etched by dry etching). In this case, the etch processes tend to be anisotropic and delineate crystal planes, resist or mask edges, as well as surface defects. The chemical reaction rate is sensitive to temperature. Reaction-limited processes are therefore affected dramatically by temperature changes. Freshly mixed chemicals may be hot and the concentration of active species may change with time. Consideration of all these factors is necessary for understanding and controlling of wet-etching processes. In addition, a small amount of contaminants may affect etch behavior. Because of this reason and the general need of avoiding surface contamination by ions, high-purity chemicals (semiconductor grade) must be used. Also, H_2O, mentioned as part of an etch solution, is interpreted as de-ionized (DI) water.

6.1.1 Wet Etch Advantages

Wet-etching processes have the following advantages:

- Equipment cost is low and processes can be very high throughput.
- Photoresist masks can be used.
- Large-volume processes can be automated by robots, sprays tools, etc.

6.1.2 Wet Etch Disadvantages

Wet etches tend to be isotropic and will generally result in an undercut under the photoresist mask (see Fig. 6.1). The degree of anisotropy cannot be changed as in dry etching. Therefore small features cannot be etched by wet processing. The exact amount of undercut is determined by the lateral etch ratio, defined as the ratio of the etch rate in the horizontal direction to that in the vertical direction. For an isotropic etch this ratio is 1. Selectivity control is also tougher in wet etching. Figure 6.1b shows the etch profile for a perfectly selective etch, where layer 2 under the layer to be etched is not etched at all. In practice, the underlayer may be etched as shown in Fig. 6.1c.

Chemical safety and environmental issues are more difficult to handle for wet etch processes.

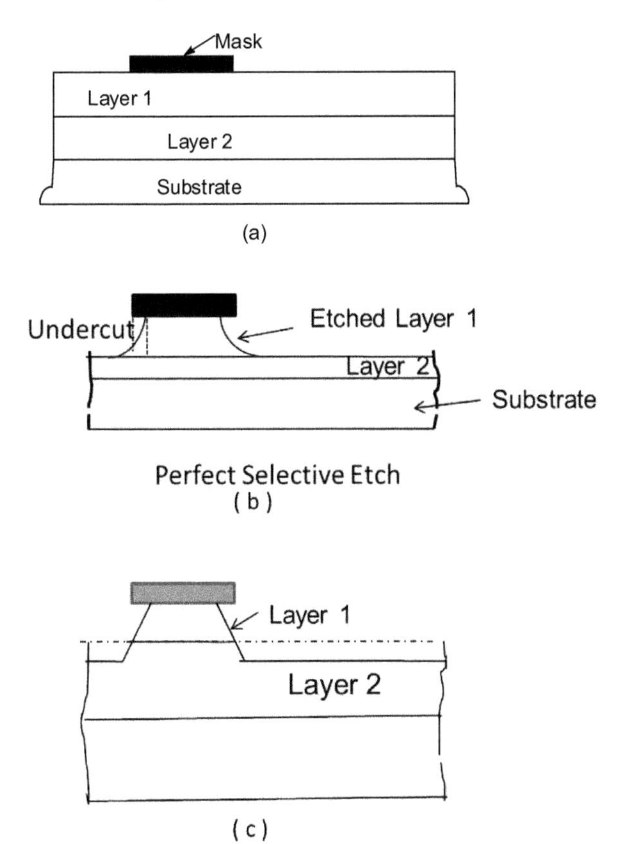

Figure 6.1 Wet etching of layer 1 over layer 2 using a photoresist mask (a) showing an isotropic etch with no etching of underlayer 2 (b) and etching of the underlayer (c).

6.2 GaAs Etching Basics [1–3]

Silicon and GaAs both have a similar crystal structure and their behavior during wet etching is similar except for the difference brought about by the differences in Ga and As sublattices in the GaAs case. The etching reaction consists of two steps, oxidation followed by dissolution. These processes proceed simultaneously in the etch mixture. The etch chemistry is chosen to achieve specific objectives for the process step, for example, anisotropy, fast rate, etc. The oxidation process is similar to anodic oxidation. Points on the semiconductor behave as anodes and cathodes, forming localized electrolytic cells. Over a certain time a certain area acts as a cathode and anode, the times being roughly equal for isotropic etching. Different factors can affect this process: impurities or additives in the etchant, temperature, defects, as well as presence of metal features on the wafer. The electrochemical nature of the etching also makes the reaction process sensitive to light by changing the availability of electrons and holes.

GaAs etch chemistries consist of an oxidizing agent (H_2O_2) mixed with a dissolution agent such as H_3PO_4 (phosphoric acid). GaAs does not etch in pure acid but dissolves rapidly in a H_3PO_4:H_2O_2:H_2O mixture. When GaAs is immersed in an etchant, electrons are transferred from the semiconductor to the electrolyte (or holes are added) until the Fermi levels in the energy band diagram line up. (See Fig. 11.14 in Chapter 11.)

$$GaAs + 6\,h^+ \rightarrow Ga^{3+} + As^{3+}$$

These ions react with $(OH)^-$ ions in the electrolyte to form oxides that are subsequently dissolved away [4].

A variety of wet etches have been investigated for III–V semiconductor etching. As expected, getting an isotropic etch is difficult. Unlike silicon, the III–V compounds have different atomic faces (e.g., Ga and As faces for GaAs), which have different surface activities, which leads to different etch profiles in different crystal orientations when a mask is used to define the etch openings. A rectangular mesa being etched on a (100) wafer surface shows a positive (V-shaped) slope when placed along the [$01\bar{1}$] orientation and a negative (dovetailed) slope in the [011] orientation. (See Fig. 6.2.) Circuit design rules generally dictate placement of device structures along major or minor flats to facilitate fabrication and ensure device performance. Also, placement of the die along crystal cleavage planes makes scribe and break processes possible (rather than sawing). In one direction, which gives the negative dovetail profile, metal lines cannot cross over without having seams and cracks. In the other direction, metal step coverage is not a problem. If the circuits are placed at a 45-degree angle to the major flat the etch profile is vertical.

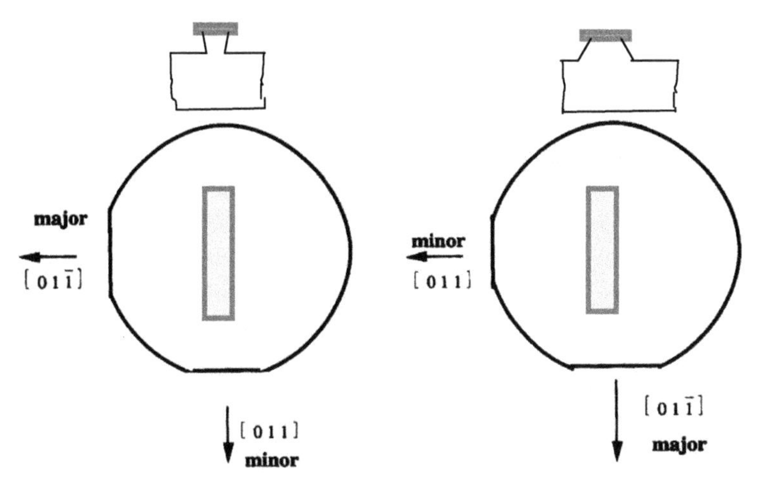

Figure 6.2 Etch profiles at different crystalline orientations (major and minor flats—USOrientation).

6.2.1 Mechanism

The lack of symmetry on the Ga and As sublattices gives rise to the directional nature of the etch behavior. Whenever small features are etched in GaAs, {111} planes are revealed. As discussed earlier, in the chapter on semiconductor materials, {111} planes in GaAs are all Ga atoms or all As atoms referred to as {111} A and {111} B planes. Along the [111]-type axes, the bonding is as shown in Fig. 6.3. The Ga and As planes are bonded by sets of three bonds or one bond. It is much easier to break one bond than three. (The density of atoms to be removed when removing this plane is a third of the next plane.) Assuming electrical charge neutrality, the surface As atoms have two free electrons and the Ga atoms do not have any. Thus the As face is expected to be more reactive. The sequence of planes in the order of etch rates, based on this type of argument, is (111)As, (100) (110), and (111)Ga. This type of etch dependence on crystal planes has been verified experimentally [5]. The exact shapes of edges produced by wet etching can be predicted for different wafer orientations by a procedure that models the relative etch rates of different planes. A detailed description of this procedure is described by Shaw [6]. The approach is based on constructing normals to radius vectors, which represent dissolution rates in a given direction. The resulting edge profile is represented by the envelope formed by the assemblage of normals. Such a construction is called a Wulff plot—an example is shown in Fig. 6.4—which shows a good match to the experimental edge profile obtained in the H_2SO_4:H_2O_2:H_2O system. Such an approach can be used to explain experimentally observed profiles and to guide in device layout. The mechanism given above explains etch behavior over large areas in an ideal crystal structure. In practice smoothness of the original surface, presence of defects, and surface contamination can affect the surface. Local lateral etching can cause microfaceting and result in a rough surface [7].

In reaction-limited etch processes, the etch rate is temperature dependent.

$$R \approx K \exp\left\{\frac{E_a}{kT}\right\}$$

where E_a is the activation energy and k is the Boltzman constant. Etch processes that do not rely on vigorous agitation (diffusion limited) are desirable for etching large wafers in a batch. Etch rates controlled by temperature, pH, and concentration are more desirable [8].

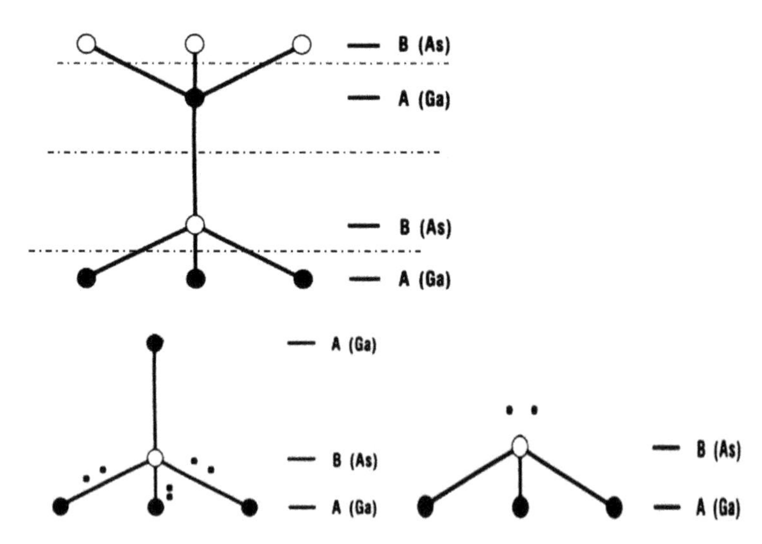

Figure 6.3 Planes of atoms showing alternating single and triple bonds.

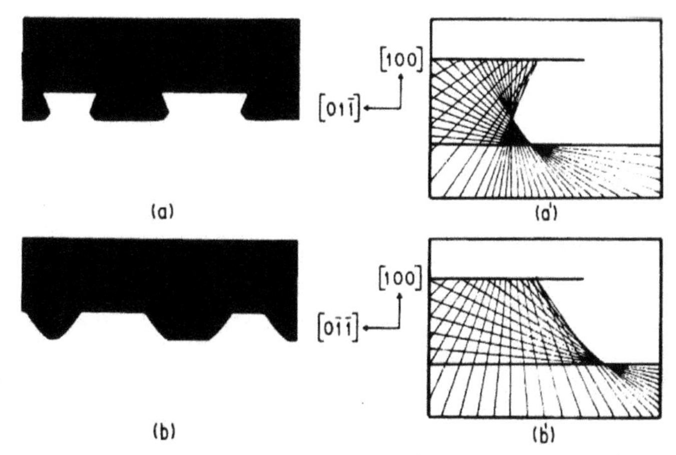

Figure 6.4 Comparison of computed and experimental etch profiles produced using a H_2SO_4:H_2O_2;H_2O etchant. Reprinted from Ref. [6], Copyright (1979), with permission from Elsevier.

6.3 GaAs Etch Chemical Systems

Numerous wet-etching chemistries have been reported for GaAs and III–V compound semiconductors in general. An extensive listing (for both wet and dry etching) has been compiled by Clawson [9]. A few of the more commonly used chemistries for crystallographic etches are listed in Table 6.1. A few etch systems used for etching product wafers in production are described below to explain their general behavior.

Table 6.1 Crystallographic etches for GaAs

Formulation	Remarks
1 mL Br_2	Distinguishes between (111) Ga and (111) As planes
100 mL CH_3OH	Polishing etch
1 mL HF:2 mL H2O:	Etch pits on (100) and (110) planes
8 mL $AgNO_3$:1 g CrO_3	
A: 40 mL HF	A–B dislocation etch, separate parts
40 mL H_2O: 0.3 g $AgNO_3$	stored indefinitely, used for delineation of epitaxial layers
B: 40 mL H_2O: 40 g CrO_3	Mixed in a 1:1 ratio before use
1 g $K_3Fe(CN)_6$ in 50 mL H_2O	Used for the delineation of epitaxial layers; mixed
12 mL NH_4OH in 36 mL H_2O	in a 1:1 ratio before use

Source: From Ref. [2]. Copyright © 1983. Reproduced with permission of John Wiley & Sons, Inc.

6.3.1 Hydrogen Peroxide–Based Etches

6.3.1.1 H_2SO_4:H_2O_2:H_2O system

This system can be used from large-area wafer polishing to small-area gate recess etching. Etch rates vary with formulation and with temperature. Isoetch curves are shown in Fig. 6.5 for a temperature of 0°C [10]. Etch rates are zero for the corners for pure constituents. Etchants with high concentrations of H_2SO_4 (region D in Fig. 6.5) are diffusion limited and are suitable for polishing. Region B results in a slow etch rate, which is good for surface defect delineation. In region A, where the etch rate is fast, cloudy etched surfaces are produced. Etch profiles for (100) wafers along two orientations are shown in Fig. 6.6.

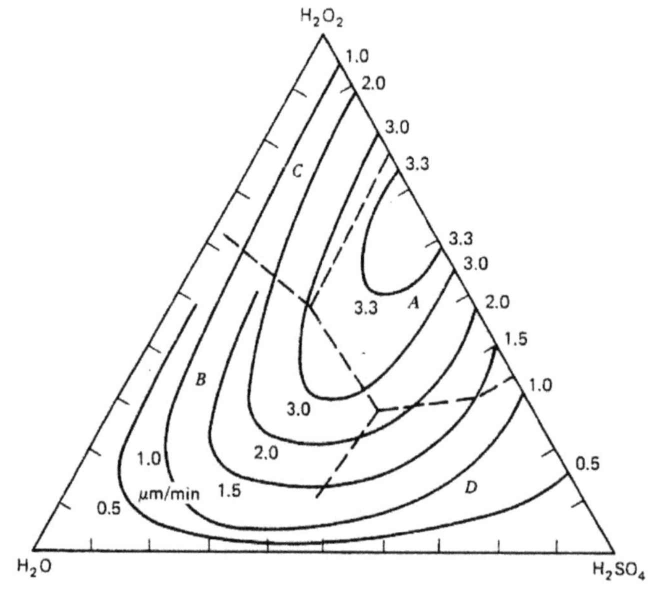

Figure 6.5 Isoetch curves for GaAs etching by the sulfuric acid:peroxide (H_2SO_4:H_2O_2) system. From Ref. [10]. Reproduced by permission of ECS—The Electrochemical Society.

6.3.1.2 H_3PO_4:H_2O_2:H_2O system

The isoetch curves for this system are similar to the sulfuric acid system, except for differences in the nature of the etched surface, which is different because of the more viscous nature of H_3PO_4.

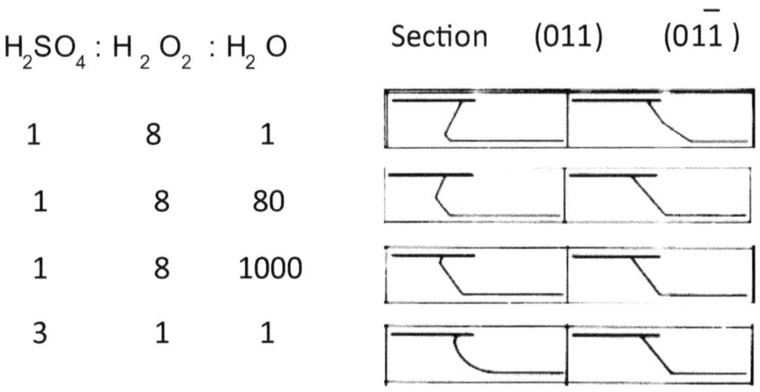

H_2SO_4	H_2O_2	H_2O	Section (011)	$(01\bar{1})$
1	8	1		
1	8	80		
1	8	1000		
3	1	1		

Figure 6.6 Etch profiles with varying concentrations of acid and peroxide in the H_2SO_4:H_2O_2system. The profiles are similar with HCl but at higher HCl concentrations.

Figure 6.7 shows the isoetch diagram. In region B, which has a high concentration of H_3PO_4, the etch rate is not linear with time but varies as its square root. In regions A, D, and C the etch rates for all principal planes are equal except for the (111) Ga plane, which etches at about half the rate. Anisotropic etching occurs in region C, which has the fastest etch rate. In large-volume production, the concentration of peroxide in the bath will change over time, so the etch bath life must be determined and the bath changed after that time. Agitation is needed to avoid the effect of bubble formation on small features. Bath temperature control of +/– 1°C may be required.

6.3.1.3 Citric acid system ($C_3H_4(OH)(COOH)_3$ H_2O:H_2O_2:H_2O)

This system has the advantage of good etch depth control for shallow etches, because the etch rate can be varied over a wide range by changing the composition. A mixture of 50% by weight of citric acid with H_2O_2 has been shown to give etch rates from 1 to 100 Å/sec [12]. At low etch rates, the rate is sensitive to agitation and spray etching may be preferred in practice.

6.3.1.4 Ammonia peroxide system (NH_4OH:H_2O_2:H_2O)

This system with a very low concentration of NH_4OH etches all faces at about the same rate (0.3 μm/min) and is closest to an isotropic etch. This bath also ages less in comparison to bromine–

methanol polishing etches. Other formulations of this system can be anisotropic.

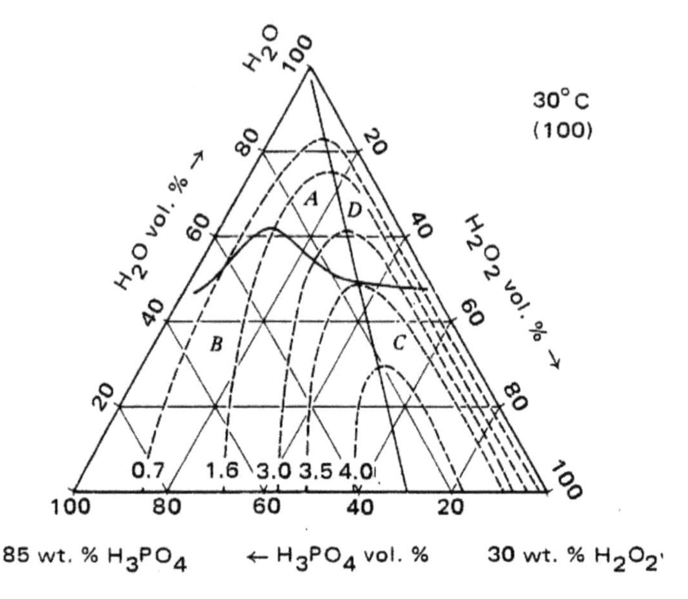

Figure 6.7 Isoetch curves for gallium arsenide using the phosphoric: peroxide (H_3PO_4:H_2O_2) system. From Ref. [11]. Reproduced by permission of ECS—The Electrochemical Society.

6.3.1.5 HCl-based systems

These etches have a tendency to form curved etch planes and are nearly isotropic at low HCl concentrations. High HCl concentrations are not isotropic but show a minimal mask undercut (see Fig. 6.6). This system can be used for polishing and stress relief of wafers after grinding.

6.3.2 Special Etches

6.3.2.1 Polishing etches

It is difficult to develop a polishing etch for GaAs because of its polar nature, as described above. Bromine–methanol chemistry is considered to be a polishing etch, although the etch rates of different crystal planes vary by 6:5:4.6:1` for {110}:{111} As:{100}:{111} Ga. This etch system is not compatible with photoresists. H_3PO_4:H_2O_2

and citric acid–based etches are used in practice, in spite of the difficulty posed by anisotropy, the processes being optimized by temperature, concentration, etc., to work in the diffusion-limited regime.

6.3.2.2 Crystallographic etches

Crystallographic etches are used for delineating regions so that these can be differentiated in visual or microscopic observations. <100> and <110> directions are useful for electro-optical devices. Dislocations are present in wafers, from starting grown boules, created during epigrowth or processing. These etch faster in localized regions if the overall etch rate is slow and reaction limited. Fast and polishing etches obscure these effects. For GaAs it is easy to differentiate the {111} Ga face from the fast-etching {111} As face. Etch pits on these faces are triangular, whereas these are rectangular on the {001} face. A widely used crystallographic etch in III–V compound semiconductors is known as the A–B etch [12]. Properties of crystallographic etches are listed in Table 6.1 [2]. Common etches for III–V compounds are listed in Table 6.2.

6.3.3 Wet Etches

6.3.3.1 InP

Common wet etches are based on HCl chemistry. HCl is combined with just DI H_2O or other acids like HNO_3, H_3PO_4, and H_2O_2 for faster etches. Diluted HCl is used for device fabrication where the rate needs to be moderately high and control of rate and profile need to be more accurate. For a very slow etch rate, the H_2SO_4:H_2O_2:H_2O system may be used.

6.3.3.2 InGaP

InGaP with ~0.5 In is lattice-matched to GaAs and is commonly used in high-electron-mobility transistor (HEMT) and heterojunction bipolar transistor (HBT) fabrication. H_3PO_4:HCl:H_2O chemistry is used for etching. The etch rate increases with HCl concentration and is fastest with addition of a little H_3PO_4. The rate is reaction limited, so it is sensitive to temperature.

6.3.3.3 InGaAs

This ternary is etched by phosphoric, sulfuric, or nitric acid–based systems, like GaAs. Aluminum-containing ternaries are being investigated for III–V metal–oxide–semiconductor field-effect transistor (MOSFET)-type devices where InGaAs is used as the channel. $HCl{:}H_2O$-based etches are used to etch InAlAs and AlInP. Table 6.2 [9, 23, 24] lists etch chemistries for other III–V compounds.

Table 6.2 Wet-chemical etches for other III–V materials

Material	Etch solution	Remarks
GaAs/AlGaAs	$H_2SO_4{:}H_2O_2{:}H_2O$ $HCl{:}H_2O_2{:}H_2O$ Citric acid:$H_2O_2{:}H_2O$	Nonselective etches $H_3PO_4{:}H_2O_2{:}H_2O$, $NH_4OH\text{-}H_2O_2\text{-}H_2O$
InGaAs	$H_2SO_4{:}H_2O_2{:}H_2O$ (1:1:20)	Selective over InP
InP	$HCL{:}H_2O$ $HCl{:}HNO_3$ $HCl{:}H_3PO_4{:}H_2O$ $HBr{:}Br_2{:}H_2O$	Selective over InGaAs Rate very temperature sensitive 900 Å/min for 1:4:20
InAlAs	$HCl{:}H_2O$ Citric acid	Selective over InGaAs Selective over InP
InGaP	$H_3PO_4{:}HCl{:}H_2O$ $H_3PO_4{:}HCl{:}H_2O$	Selective over GaAs
AlInP	$HCl{:}H_2O$	Selective over GaAs
GaSb	HCl (hot) $HNO_3{:}HF$	H_2O_2 or H_2O added
GaN	NaOH	Must be heated (80°C)
GaP	$HNO_3{:}HF$ or HCl $H_2SO_4{:}H_2O_2$	H_2O can be added
InSb	$HNO_3{:}HF$ or HCl $H_2SO_4{:}H_2O_2$	H_2O can be added
InAs	HCl $HNO_3{:}HF$ or HCl	H_2O can be added
AlAs	H_2O	

Source: Reprinted from Ref. [9], Copyright (2001), with permission from Elsevier.

6.3.4 Wet Etching of GaN/AlN

III–N compound semiconductors are known for their chemical stability. Wet etching has been investigated using a variety of chemicals [13] and nothing etches GaN except molten NaOH and KOH. These etchants are not compatible with device processing and are used only for defect studies, like dislocations. The etch rates are polarity dependent, *N* polar faces being more reactive.

6.3.5 Etching of Other Materials

Metal and dielectric films are used in contact and interconnect fabrication. Wet etches are used to pattern these in some of the process steps where small dimension control is not needed. Wet etches may be needed for removal of sacrificial or protective layers or patterning of streets where high precision is not needed. Removal of seed layers for plating is an example. For example, the TiW seed layer deposited over the whole wafer can be removed around the plated lines by wet etching by hydrogen peroxide. Wet etches are also used for diagnostic purposes, process issue solving, failure analysis, and process control, an example of which is monitoring of silicon nitride density by measuring the wet-etch rate. Silicon nitride and silicon dioxide are frequently used in III–V processing; these are etched using hydrofluoric acid (HF) or HF buffered by addition of NH_4F, called BHF. Etch rates of silicon nitride in BHF are low and conc. HF is not compatible with photoresists. Availability of dry-etching systems for silicon nitride has eliminated this problem. Out of all the wet-etching systems only gold etchants are used heavily for street etches or etching of large areas. Gold can be etched in aqua regia but is avoided in favor of etches based on KI (potassium iodide). Proprietary gold etchants are available for this purpose.

6.4 Wet Etching in Production

Etch rates can vary a lot with time as a bath ages. This is particularly relevant to chemistries that use H_2O_2. Therefore etch times may have to be adjusted to account for bath aging. Otherwise a fresh bath has to be poured every time a lot has to be run. In some cases where small openings must be etched, wetting may have to be improved by addition of surfactants to ensure etching proceeds uniformly in small and large openings. Loading effects due to the number of wafers and the area being etched must be accounted for in large-volume production. Wafer cassettes may have to be loaded in alternate slots to ensure wafer-to-wafer uniformity. III–V compound etch rates are sensitive to light, so bright light should be avoided and the first wafer in a cassette, which is exposed to more light, may have to be shielded by a dummy wafer.

Agitation can increase the etch rate and improve uniformity over large-area wafers but it is difficult to implement in high-volume production. Ultrasonic agitation is better but not used much in GaAs processing because of breakage. Spray etching is more compatible with III–V volume production. Premixed chemicals are sprayed on a rotating wafer, thus delivering a uniform concentration of etchants to all areas of the wafer. Wafers can be cleaned with DI water and dried in situ. Commercial systems are available for this purpose and have been used for etching of thin layers, for example, gate recess etching. Control of etch rates and uniformity must be achieved through numerous process parameters like spray nozzle configuration, pressure, rotation speed, temperature, concentration, etc. Etch depth control of tens of angstroms is possible; beyond this etch stop layers must be used. Chemical delivery systems can be utilized, giving the added benefit of using temperature control, mixing accuracy, and use of in-line filters.

6.4.1 Wet Etch Application Examples

6.4.1.1 Ion damage avoidance

In HEMT gate recess etching, plasma etching (with chlorine-based chemistries) cannot be used because it degrades the two-dimensional electron gas due to creation of traps in the AlGaAs by ion bombardment. Wet etching is more suitable if a more suitable chemistry/process can be developed that etches the GaAs cap layer without affecting the AlGaAs layer. Wet etches based on the popular H_3PO_4:H_2O_2 or NH_4OH:H_2O_2 chemistries do not provide required selectivity or attack the polymers (e.g., polymethylmethacrylate [PMMA]) used for T-gate formation. Citric acid:H_2O_2 mixtures have been shown to have good selectivity [3, 13]. For citric-acid-to-H_2O_2 ratios of over 2, the etching is chemical reaction rate limited at 20°C. Selectivity of over 100 has been achieved with a 4:1 mixture for an Al fraction of $x > 0.3$, thereby improving etch depth control to a few angstroms and the corresponding threshold voltage control to less than 25 mV (over a 100 mm wafer). Process control can be further improved by the insertion of an AlAs etch stop layer.

Table 6.3 Simple etchants for some common metals and dielectrics

Metals	
Aluminum	$HCl:H_2O$ (1:4) NaOH solutions $H_3PO_4:HNO_3:H_2O$ (74:7:19) (rapid)
Chromium	HCl HNO_3
Gold	$HCl:HNO_3$ (3:1) $KI:I_2:H_2O$ (4g:1g:40 mL)
Indium	
Molybdenum	$H_3PO_4:HNO_3:H_2O$ (5:3:2) $H_2SO_4:HNO_3:H_2O$ (1:1:1–5)
Nickel	$HNO_3:CH_3COOH:H_2SO_4$ (5:5:2) (plus water) $HNO_3:HCl:H_2O$ (1:1:3)
Platinum	$HNO_3:HCl:H_2O$ (1:7:8)
Silver	$HNO_3:H_2O$ (5–9:1–5) $Fe(NO_3)_3:H_2O$ (11 g:9 mL)
Tantalum	$H_2SO_4:HNO_3:HF$ (5:2:2) $HNO_3:HF:H_2O$ (1–2:1:1–2)
Tin	$HNO_3:C_2H_5OH$ (1:49) $FeCl_3$ solutions
Titanium	$HF:H_2O$ (1:9) $HNO_3:HF:H_2O$ (2:1:7)
Tungsten	$KH_2PO_4:KOH:K_3Fe(CN)_6$ (34 g:13.4 g:33 g) + water to make 1 L
Dielectrics	
SiO_2	HF, BHF (buffered HF)
Si_3N_4	HF, BHF, H_3PO_4
$Si_xO_yN_z$	HF, BHF
$Si_xN_yH_z$	HF, BHF, H_3PO_4
TiO_2 (amorphous)	HF, BHF, H_3PO_4, H_2SO_4
Ti_2O_5 (amorphous)	HF, BHF, $KOH:H_2O_2:H_2O$
Al_2O_3 (amorphous)	HF, BHF, H_3PO_4

Source: Reprinted from Ref. [3], Copyright (1978), with permission from Elsevier.

Wet etching is used in HBT processing for process steps where dry etching would be throughput limiting and does not provide any process control or reliability advantage. A good example is collector etching, which involves over 1 μm deep etches. Many processing steps involve a combination of dry and wet etches. Etching of a multilayer emitter falls into this category. Control of etch bath temperature is critical to wet processing for controlling etch profiles.

6.4.1.2 Wet etching of multilayer III–V compounds

Device processing of III–V compound circuits involves etching of different binary and ternary compounds. Selective etches are needed for stopping on different layers, for example, ledge or base layers in HBT. For the InGaP/GaAs HBT, the interface between InGaP and GaAs may be a complex GaInAsP of varying composition. In other device processes like the bipolar field-effect transistor (BiFET) process, there may be multiple interfaces between GaAs and InGaP. Special chemistry may be needed to remove these layers fully so that underetched islands are not left behind. In InP processing, an etch for InGaAs may be needed that stops on InP. A few excellent references are available for determining etch chemistries [9, 14]. Table 6.4 is a chart useful for selecting etch chemistry for selective etches.

Table 6.4 Material selectivity table for some common etch chemistries (http://terpconnect.umd.edu/~browns/wetetch.html)

	Material Selectivity											
Etchant	GaAs	InP	InGaAs	InGaAsP	GaInP	GaAsP	AlGaP	AlGaAs	AlInP	InAlAs	InGaAlAs	SiO_2
$HCL:H_3PO_4$	S	E	S	S	E				E			
$H_3PO_4:H_2O_2:H_2O$	E	S	E		S							
$H_2SO_4:H_2O_2:H_2O$	E	S	E	E								
$C_6H_8O_7:H_2O_2$	CD	S	CD					CD		CD		
$HCL:HNO_3:H_2O$	E	E					E		E			
$HNO_3:H_2SO_4:H_2O$	E					E						
$HCL:H_2O_2:H_2O$	E	E				E						
$HCL:H_2O$	S	E	S						E	CD	CD	
$BHF:H_2O$								CD				E

Legend			
Etches	Selective Stops	Composition Dependent	No Data
E	S	CD	

6.5 Cleaning

Wafer-cleaning processes are used throughout integrated circuit fabrication from wafer sawing all the way to assembly and packaging. Cleaning, like etch processes, can be wet or dry plasma. Cleaning is necessary to remove organic and inorganic contaminants. These may be organic residues from lithography, lift-off, or dry-etching processes that create metal–organic complexes. Air-borne particles, bacteria, films that collect on the surface during storage, as well as oxide layers that grow must be cleaned. Cleaning processes involve physical removal, organic solvent dissolution, and inorganic chemical cleaning. The quality of the surface of semiconductors is important for reproducible device performance. The presence of even monolayers of contaminants and surface oxide determines the quality of compound semiconductors.

GaAs wafers need to be cleaned after they have gone through the fabrication process of sawing and polishing to remove abrasives and polishing compounds or additives. Hot organic solvents like trichloroethylene (TCE) and xylene combined with mechanical scrubbing are used to remove heavy organics. Hot solvents are followed by acetone, methanol, and DI water to remove the solvents. Waxes or polymer-bonding compounds used for backside processing can also be cleaned by a similar solvent sequence. Solvents like acetone that are not miscible in water should not be used last, before DI cleaning.

Organic contaminants that end up on the surface of wafers during processing are routinely removed by de-scuming or ashing processes. Ashing processes will be discussed more under plasma cleaning. Ozone is a good source of reactive oxygen. Commercial ozone systems with proper ventilation for safety are available for cleaning and gentle oxidation.

Ionic contaminants, particularly heavy metals, must be removed next from the surface by use of inorganic chemicals. Strong oxidizing chemicals can be used for silicon wafers but are not used in III–V wafer processing as they react with these materials. Mobile ions like Na and K are a major issue in silicon MOS devices but are not a major issue in III–V processing. However, HCl that is used for removing heavy ions also removes these [2, 15, 16].

A freshly cleaved GaAs surface develops a surface oxide in air. A layer of 15 to 20 Å thickness is formed within an hour [2]. Oxide growth continues but is self-limiting beyond about 30 Å. This oxide is a nonstoichiometric mixture of As_2O_3 and Ga_2O_3 with an excess of As_2O_3. This oxide acts like Ga_2O_3 electronically, with As traps within the energy bandgap of Ga_2O_3. Schottky diodes formed on oxidized GaAs show poor *I–V* characteristics with an ideality factor much greater than unity ($n > 1$). HCl and NH_4OH are used to remove this oxide. After 1:1 HCl:H_2O treatment, the Ga-to-As ratio on the surface is seen to be 0.78–0.82 [2]. After 1:1 NH_4OH:H_2O treatment, the surface is a bit better with a Ga-to-As ratio of 0.84 to 0.94. Other acids are not suitable for producing good surfaces; H_2SO_4, for example, produces a very arsenic-rich surface. Surface state densities are high, in the 10^{14} /cm^2-eV range. Before deposition of metal layers on cleaned surfaces, the surfaces are rinsed with DI water and N_2-dried and loaded in vacuum systems within minutes. Close coupling of the different steps is closely followed in production. A tiny amount of gallium suboxide and elemental arsenic have been reported after ammonia treatment [17, 18]. Alkaline-treated surfaces, that are –OH terminated, are hydrophilic, while acid-treated surfaces that are –H terminated are hydrophobic. NH_4OH-treated surfaces are best for growing Al_2O_3 on GaAs (by the atomic layer deposition [ALD] process to achieve lowest surface state density [17, 19]. If HCl is used to remove metal ions or other contamination on GaAs, it is best to follow it with NH_4OH treatment for the lowest surface state density and making of good Schottky contacts. Surface state densities in the range of 10^{12}–10^{13} /cm^2-eV are possible for Ga-rich surfaces.

For InGaP, or other compounds that react with the Schottky metal, subsequent temperature steps may have to be restricted to retain good Schottky diode behavior. Figure 6.8 shows the effect of three chemical treatments to achieve low surface states [19].

Epiready wafers are supplied by substrate vendors by using NH_4OH and HCl chemistries. As an example, an NH_4OH mixture was used by Song et al. [20] to remove particles and metals as a first step of wet-chemical cleaning and an acidic peroxide mixture to remove residual metallic contaminants and control the components of the oxide layer as a second step of wet-chemical cleaning. The particle-free, metal-free, and thin As-oxide-rich surface is achieved by use of an NH_4OH mixture followed by a HCl mixture. Metal contamination was analyzed by total reflection X-ray fluorescence spectroscopy. Characterization of the oxide composition was carried out by X-ray photoelectron spectroscopy (XPS). Surface morphology and thickness of the oxide layer were observed by atomic force microscopy and an ellipsometer, which show a very flat surface roughness of 0.95 nm and a uniform oxide thickness of about 2 nm, respectively.

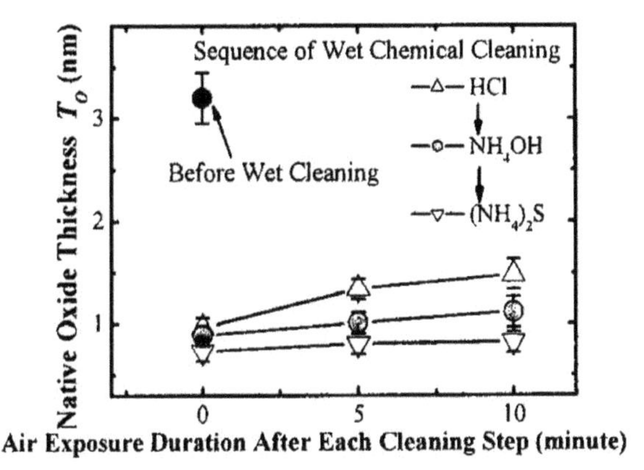

Figure 6.8 Native oxide thickness as a function of HCl, NH_4OH cleaning, and ammonium sulfide treatment. © [2008] IEEE. Reprinted, with permission, from Ref. [19].

6.5.1 Plasma Cleaning

Oxygen plasma cleaning, or *ashing*, is used to remove organic contamination from wafers during processing and to remove any leftover photoresist from exposed areas in a process called de-scumming. In this process the surface does get oxidized. Therefore oxides must be removed before proceeding to process steps that are sensitive to oxides on the surface. Details of plasma processes are discussed elsewhere.

6.6 Surface Passivation

6.6.1 Wet-Chemical Passivation

At the surface of III–V compound semiconductors, the continuity of the bulk is broken, with introduction of dangling bonds, surface reconstruction, defects, native surface oxides, contaminants, etc., which introduce energy levels in the energy bandgap. With any of the cleaning procedures described above, metal or dielectric deposition must be carried out immediately to preserve as much of the surface as possible. Surface passivation is carried out with the idea of modifying the surface so that least possible surface states or best possible electronic properties can be preserved and be stable over the life of the device.

6.6.2 Chalcogenide Passivation [21]

Sulfur and selenium have been known to modify the surface properties of GaAs and provide electronic and physical passivation [22]. A 20% basic solution of ammonium sulfide with a few percent sulfur added may be used. An acidic solution should not be used as it produces poisonous H_2S. Wafers must be rinsed properly before further processing. XPS studies confirm Ga–S and As–S bonds on the surface. The electronic passivation provided by sulfur using ammonium sulfide passivation to retard oxidation is shown in Fig. 6.8 above. Sulfidation can also be done using the gas phase, using UV photodissociation of sulfur or selenium. A modest current gain increase has been seen (beta increase from 90 to 95) for AlGaAs/GaAs HBTs [21]. Better MOSFET current densities have also been achieved with ammonium sulfide passivation before atomic layer epitaxy (ALE) (Ref. [46] in Chapter 4).

6.6.3 Dielectric Passivation

Surface state density must be reduced below 10^{12} /cm^2-eV levels for good device performance and life time. Such lower values can be achieved and retained through the life of the device by surface passivation using dielectric passivation. Dielectric layers are commonly deposited by plasma-enhanced chemical vapor deposition. Low-energy depositions activated by UV do not produce films of good enough quality. Therefore in practice, the surface of the semiconductor is exposed to plasma before the deposition starts. Energetic ions, chemical radicals, UV light, etc., all cause damage. Ion bombardment must be minimized. Displacement energy for creation of defects by ions is about 40 eV. Careful design of the plasma system is needed to reduce bias voltage, for example, by making the powered electrode area larger, optimizing radio frequency (RF) power, etc. (See Chapter 7 for the voltage-to-electrode-area relationship). Deposition temperature and deposition rate also need to be optimized to maintain a high-quality dielectric without introducing excessive damage. Electron cyclotron resonance (ECR) systems can be used to reduce damage. Details will be discussed in Chapter 8.

References

1. R. Williams, *Modern GaAs Processing Methods*, Artech House, Boston, London (1990).
2. S. K. Ghandhi, *VLSI Fabrication Principles, Silicon and Gallium Arsenide*, Wiley Interscience, New York (1983).
3. W. Kern and C. A. Deckert, in *Thin Film Processes*, Eds., J. L. Vossen and W. Kern, Academic Press, New York (1978).
4. D. S. Stirland and B. W. Straughan, A review of etching and defect characterization of GaAs substrate material, *Thin Solid Films*, **31**, 139 (1976).
5. G. Y. Tarui et al., Preferential etching and etched profile of GaAs, *J. Electrochem. Soc.*, **118**, 118 (1971).
6. D. W. Shaw, Morphology analysis in localized crystal growth and dissociation, *J. Cryst. Growth*, **47**, 509 (1979).
7. D. N. MacFadyen, *J. Electrochem. Soc.*, **130**, 1934 (1983).
8. S. P. Murarka and M. C. Peckerar, *Electronic Materials Science and Technology*, Academic Press, New York (1990).
9. A. R. Clawson, Guide to references on III-V semiconductor chemical etching, *Mater. Sci. Eng., R*, **31**, 1–438 (2001).
10. S. Lida and K. Ito, Selective etching of gallium arsenide crystals in $H_2SO_4/H_2O_2/H_2O$ system, *J. Electrochem. Soc.*, **118**, 768 (1971).
11. M. Otsubo et al., Preferential etching of GaAs through photoresist masks, *J. Electrochem. Soc.*, **123**, 676 (1975).
12. M. Tong et al., Selective dry etching of AlGaAs-GaAs surfaces, *J. Electron. Mater.*, **21**, 9 (1992).
13. D. Zhuang and J. H. Edgar, Wet etching of GaN, AlN and SiC: a review, *Mater. Sci. Eng.*, **R48**, 1–46 (2005).
14. B. Jalali and S. J. Pearton (Eds.), *InP HBTs: Growth, Processing and Applications*, Artech House, Boston (1995).

15. C. Chang, P. H. Citrin and B. Schwartz, Chemical preparation of GaAs surfaces and their characterization, *J. Vac. Sci. Technol.*, **14**, 943 (1977).

16. S. Osakabe and S. Adachi, Study of GaAs (100) surface treated in aqueous HCl solution, *Jpn. J. Appl. Phys.*, **36**, 7119 (1997).

17. Yi Xuan et al., Simplified surface preparation of GaAs, *IEEE Trans. Electron. Devices*, **54**, 1811 (2007).

18. M. V. Lebedev et al., Simplified surface preparation for GaAs passivation, *Appl. Surf. Sci.*, **229**, 226 (2004).

19. H. C. Chin et al., A new silane-ammonia surface passivation technology for realizing inversion type surface-channel GaAs MOSFET with 160 nm gate length and high quality metal-gate/high K dielectric stack, *IEDM Electron. Devices Meeting* (2008), also *IEEE Electron. Device Lett.*, **30** (2008).

20. J. S. Song et al., Wet cleaning process of GaAs substrates for ready-to-use, *J. Cryst. Growth*, **264**, 98 (2004).

21. G. Baca and C. I. H. Ashby, *Fabrication of GaAs Devices*, IEE, London (2005).

22. J. W. Seo et al., *Appl. Phys. Lett.*, **60**, 1114 (1992).

23. S. J. Pearton, *Wet and Dry Etching of Compound Semiconductors*, www.kepu.dicp.acx.cn.

24. Y. Mori and N. Watanabe, A new etching system, H_3PO_4-H_2O_2-H_2O for GaAs and its kinetics, *J. Electrochem. Soc.*, **125**, 1510 (1978).

Chapter 7

Plasma Processing and Dry Etching

Chemical reactions in plasmas were studied for decades before the first practical use of oxygen plasma for removal of photoresists in microelectronics fabrication. Fluorine-containing plasmas were used in the early seventies for etching in silicon device fabrication. Heavy use of plasmas for deposition and etching of dielectrics and metals followed well before the science of the process was understood. In modern IC fabrication about a third of the processing steps use plasmas. For a review of the historical aspects, the reader is referred to books on the subject [1–4].

7.1 Plasma Processing

7.1.1 Plasma Basics

A broad definition of plasma, as it applies to physical sciences, is a partially ionized gas composed of negatively charged electrons, positively charged ions, and neutrals whose collective properties and behavior are dominated by collective long-range Coulomb interactions [2]. Plasmas in the gas phase are generated by a variety of mechanisms such as ionization, radiation, electron beams, and more commonly application of an electric field. These plasma types

III–V Integrated Circuit Fabrication Technology
Shiban Tiku and Dhrubes Biswas

ISBN 978-981-4669-30-6 (Hardcover), 978-981-4669-31-3 (eBook)
www.panstanford.com

differ by electron concentration and average electron energy, kT_e, where k is the Boltzman constant and T_e is the electron temperature. The average energy of plasma species is expressed in terms of this electron temperature, T_e. Plasmas may be classified as thermal or nonthermal. In thermal, or hot, plasmas, all three constituents, electrons, ions, and neutrals, have the same energy or temperature. Plasma filling free space has a density of $n_e \approx 10/\text{cm}^3$ and a kT_e of about 25 eV. In nonthermal plasmas, the energy is mainly in the electrons, the plasma electrons being hotter than the other species. Nonthermal plasmas are used for technological applications in which the temperature may be close to room temperature, electron densities are of the order of 10^8 to $10^{13}/\text{cm}^3$, and electron energies are of the order of few to 10 eV. Glow discharge plasmas are nonequilibrium plasmas because the electron temperature is greater than the gas temperature. Plasma is weakly ionized and as a whole is almost neutral. Atoms, molecules, and ions are responsible for the sputtering or etching reactions. Only in high-density plasmas can the ionization reach 10%. Plasmas with densities of $10^{10}/\text{cm}^3$ and higher are referred to as high-density plasmas and are used for plasma deposition and etching in the semiconductor industry. Such high densities of electrons are needed to produce high densities of reactive radicals that are needed for these processes.

7.1.2 Glow Discharge Plasma [3, 4]

An understanding of the creation of a glow discharge is fundamental to plasma systems and can be understood by examining a simple direct current (DC) diode system, as shown in Fig. 7.1. A long glass tube with two electrodes is evacuated and then filled with an inert gas like Ar at low pressure. When a large DC voltage is applied across the two electrodes, a gas molecule can get ionized—for example, by external radiative excitation, the free electron generated gets accelerated by the electric field and given enough distance can attain enough energy to excite more atoms and thus lead to an avalanche breakdown. When sufficient electrons are available, the discharge is self-sustained. At equilibrium, the voltage across the tube drops to a much lower value than at start, due to the fact that the plasma has higher conductance. The current saturates to some value, depending upon the external series resistance. The glow discharge has a particular structure, as shown in Fig. 7.1a. Most notable is the dark space in front of the cathode, called the Crookes dark space. The density of the electrons and the ions is shown in Fig. 7.1b. Electrons near the cathode are accelerated away from it due to their light mass; the higher-mass, lower-mobility ions are accelerated toward the cathode but attain lower velocities. Thus a space charge builds in front of the cathode. The electric field increases in front of the cathode and screens the rest of the discharge from the cathode. The resulting field in the rest of the tube is low. The net voltage drop across the tube is divided up, as shown in Fig. 7.1c.

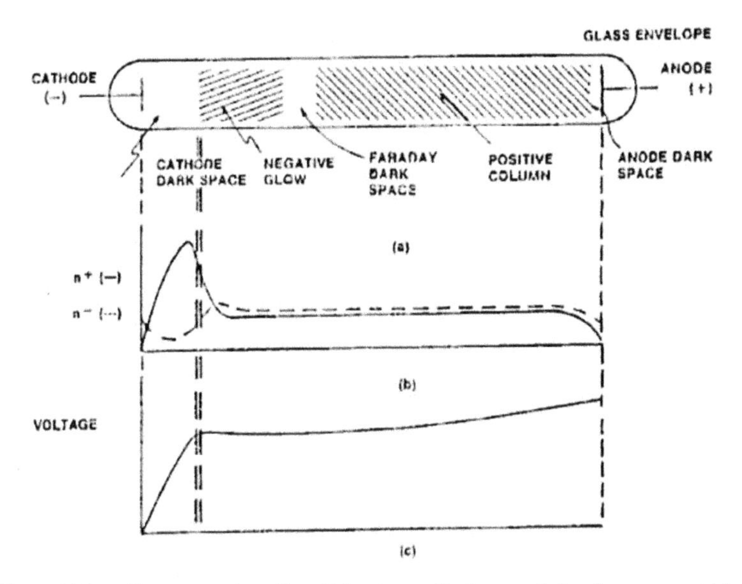

Figure 7.1 Components of a DC glow discharge (a), charge particle concentration (b), and voltage variation across the tube from cathode to anode. Reprinted from Ref. [3] with permission from Lattice Press.

The electron density in the dark-field region is low, but the average energy is high. Collision events in this region cause more ionization rather than excitation that could generate light. Electrons that reach the cathode cause emission of secondary electrons, which are accelerated away to sustain the plasma. If the pressure in the tube is reduced, the electrons can travel farther before colliding, and the dark space lengthens. At very low pressure, the dark space reaches the anode and the discharge gets extinguished.The breakdown voltage depends upon the gap distance, pressure, and the gas. Paschen's law describes the basic behavior.

$$V = f(P.d) \tag{7.1}$$

Figure 7.2 shows the breakdown voltage as a function of the gap. At low distance, the spacing is too small for breakdown to start. On the high ($P.d$) product side, the collisions are too frequent, so the electrons cannot gather enough momentum to overcome the ionization potential [5].

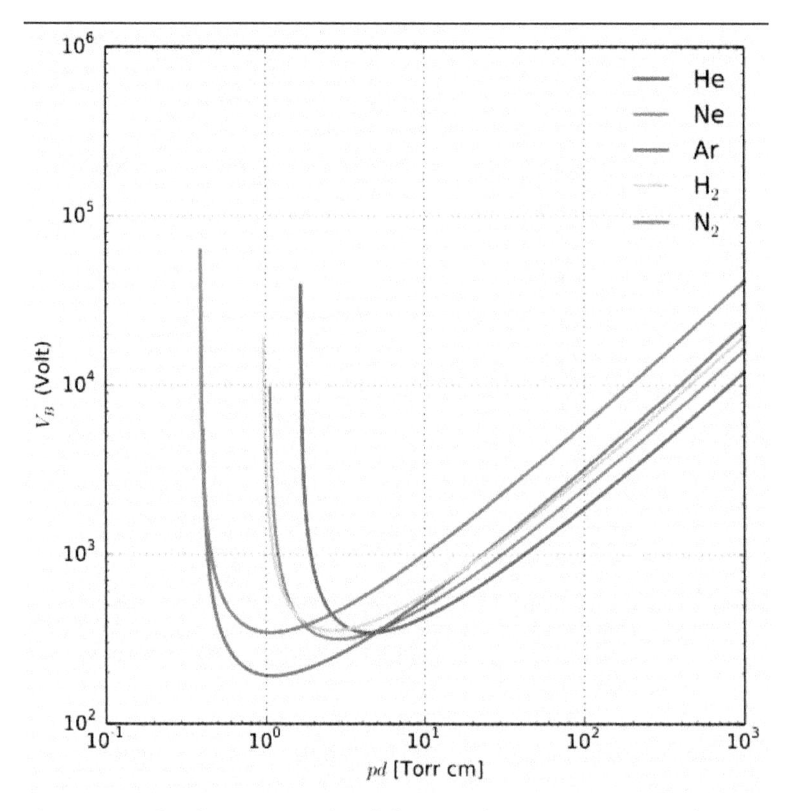

Figure 7.2 Paschen curve, breakdown voltage vs. pressure-distance product for some pure gases. Source: Wikipedia.

Understanding the basics of sustaining a discharge or avoiding glow where not needed is important is designing practical plasma systems. Electrons and ions are lost on the electrodes as well as on all the other surfaces within chambers, resulting in loss of energy as heat. Chamber geometry is important in determining the potentials that develop in the system and the energy with which the etch or sputtering process takes place.

7.1.3 Voltage Distribution

For experimental studies as well as practical applications, tube-type chambers were used first, followed by the parallel-plate reactor. Since then a variety of reactor configurations, discharge excitation mechanisms, and means of varying the chemical and ionic properties of the plasma have been introduced. The basic glow discharge process as it applies to etching and film deposition can be understood by considering the simple parallel-plate geometry. Figure 7.3a shows the geometry and the potential distribution is shown in Fig. 7.3b [6]. The plasma is the most positive and the cathode is the most negative, the largest field being near the cathode. Ions are accelerated across the cathode sheath and strike the surface. As soon as the plasma starts and becomes positively charged with respect to one electrode, the electron current rises and the the electrode/dark space/plasma system behaves as a diode. The dark space has limited conductance and it can be modeled as a capacitor.

$$C = A/D \tag{7.2}$$

where A is the area of the electrode and D is the sheath thickness. Figure 7.4 shows the equivalent circuit. If there is a blocking capacitor, the voltages are equal. See Fig. 7.5a. When an external blocking electrode is added, as shown in Fig. 7.5b, the voltage ratio depends on the area ratio. The voltage ratio has been estimated on the basis of the following assumptions [7]:

Current density of positive ions, with mass m_i, follows the space charge–limited behavior described by the Child–Langmuir equation:

$$J_i = \frac{KV^{3/2}}{\sqrt{m_i}\,D^2} \tag{7.3}$$

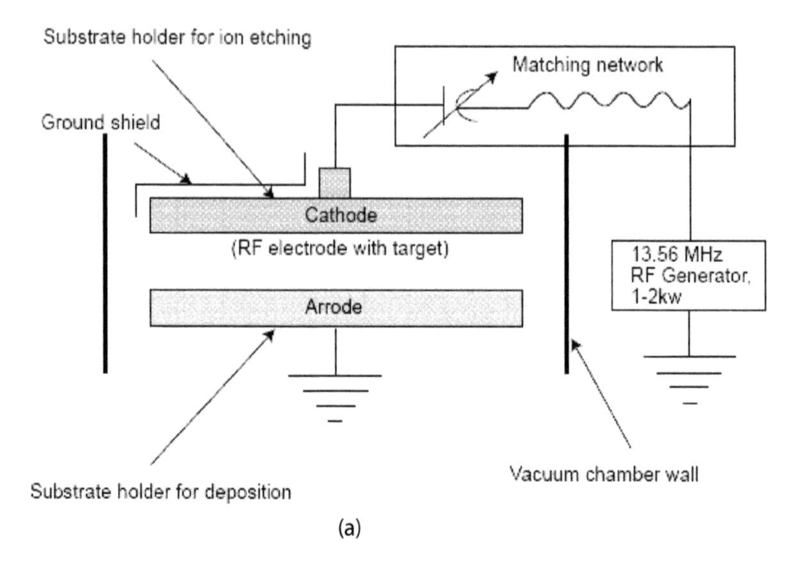

(a)

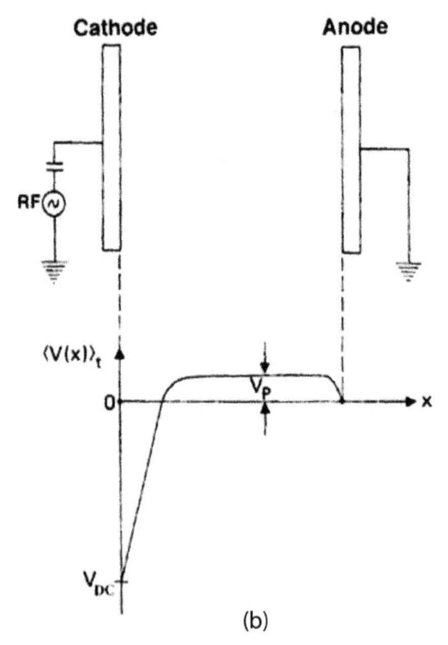

(b)

Figure 7.3 (a) A two-electrode parallel-plate system for plasma deposition or etching. For deposition, substrates are put on the anode, and for etching, substrates are put on the cathode. (b) Time-averaged potential distribution for a capacitively coupled planar RF glow discharge reactor.

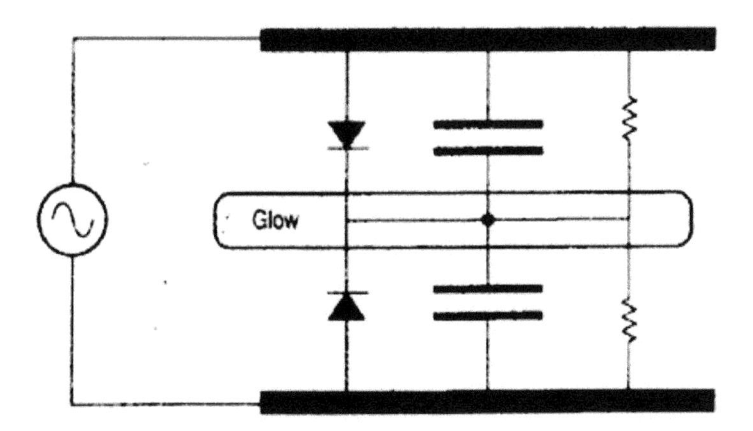

Figure 7.4 Equivalent circuit of a plasma.

If the current density of positive ions is uniform and equal at both electrodes

$$\frac{V_1^{3/2}}{D_1^2} = \frac{V_2^{3/2}}{D_2^2} \tag{7.4}$$

Using $C \approx A/D$ (Eq. 7.2) and $V_1/V_2 = C_2/C_1$ because the radio frequency (RF) voltage is capacitively divided

$$\frac{V_1}{V_2} = \frac{A_2 D_1}{A_1 D_2} \tag{7.5}$$

Combining Eqs. 7.4 and 7.5

$$\frac{V_1}{V_2} = \left\{\frac{A_2}{A_1}\right\}^4 \tag{7.6}$$

This exponent of 4 means that a very large voltage split takes place. If the assumption of equal current density is incorrect and in practice may be equal ion flux, rather than current density, the factor changes from 4 to 2. So, the voltage is inversely proportional to square of the area. Thus, large sheath voltages appear on the smaller electrode. Plasma system geometry is critical to voltages and needs careful attention. In etch systems, where ion damage may be more critical, reliance on the area may not be enough, so high-density, low-bias sources are utilized.

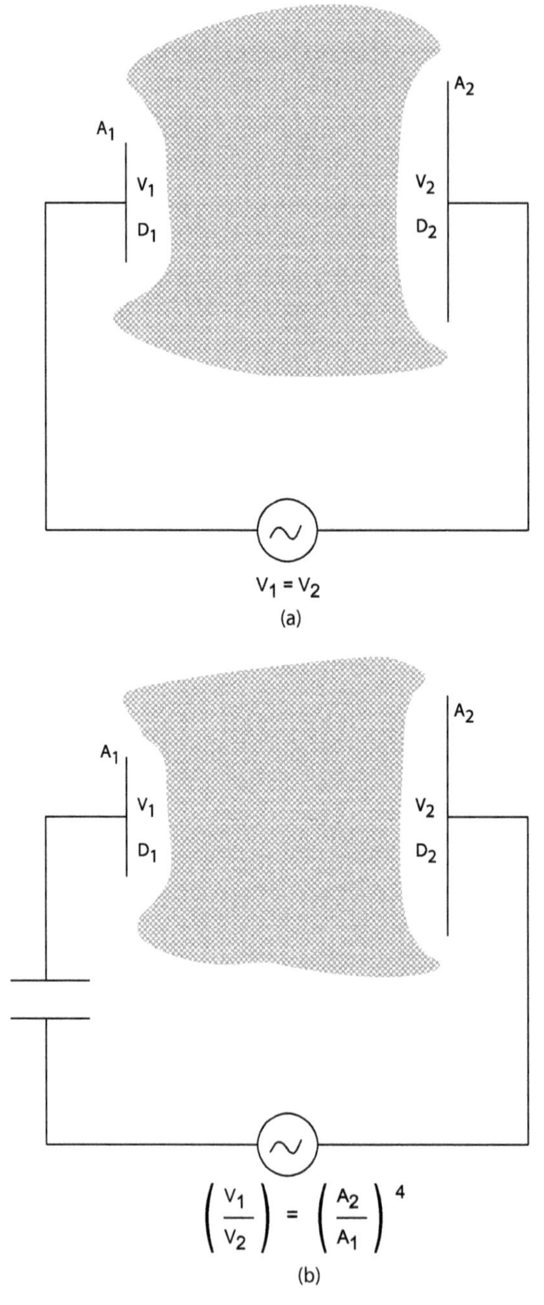

Figure 7.5 (a) Voltage distribution in a plasma system without a blocking capacitor. (b) Voltage distribution with a blocking capacitor.

7.1.4 Interaction of Ions with a Surface/Sputter Yield [3]

The cathode or target surface is exposed to a variety of particles and radiation, as shown in Fig. 7.6. When an ion hits a target surface, the following processes may occur:

- The ion may be reflected back as an ion or a neutral.
- The impact may cause secondary electron emission.
- The ion may cause damage to the surface, causing atomic rearrangement.
- The ion may eject target atoms in a process called sputtering.

For low-energy ions, with energy *Ex*, nuclear stopping power depends upon the masses of the impinging ion and the target ion and is given by

$$S(E) = \frac{m_i . m_t}{(m_i + m_t)^2} . Ex \tag{7.7}$$

From this sputter yield can be derived:

$$S \approx \frac{m_i . m_t}{(m_i + m_t)^2} \alpha . EU_0 \tag{7.8}$$

where α is a function of the m_t/m_i ratio and U_0 is the surface binding energy.

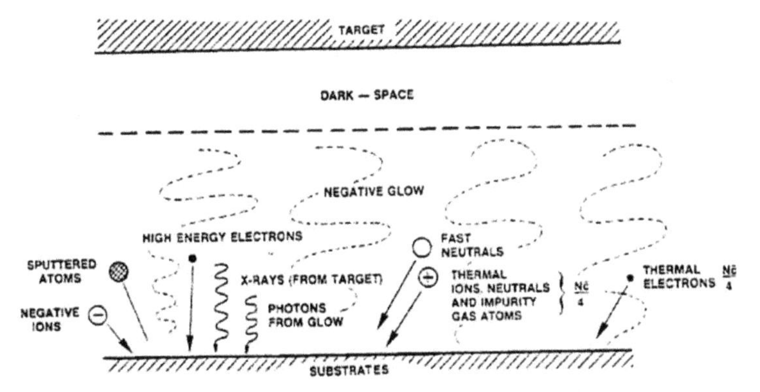

Figure 7.6 Species arriving at the cathode in a plasma system. Reprinted from Ref. [3] with permission from Lattice Press.

Sputter yields are high if the mass of the target atom is high and the binding energy is low. For energies over 1 keV, the process is more like ion implantation and sputter yield also depends upon the atomic number, in addition to the atomic mass (the equation for stopping power is described in Chapter 9). Figure 7.7 shows the sputter yield dependence on the mass of the sputter ions. Table 7.1 lists the sputtering yields for commonly used metals in argon plasma. Table 7.2 lists the ion milling rates for common metals and materials used in the semiconductor industry [8–10].

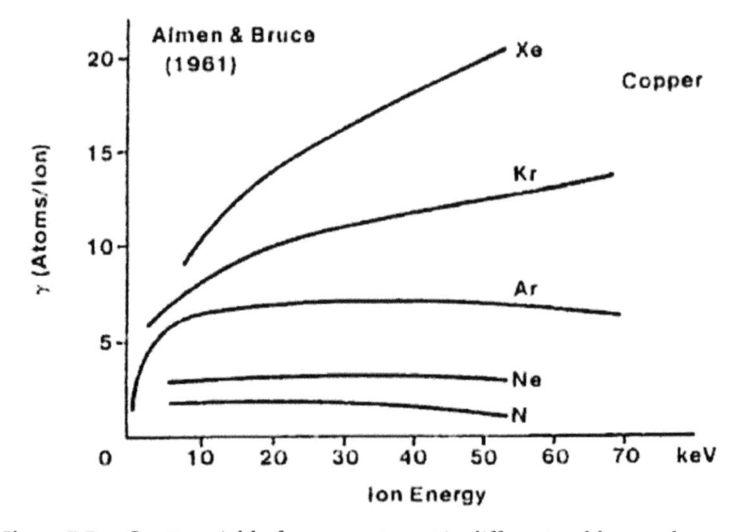

Figure 7.7 Sputter yield of a copper target in different noble gas plasmas. Reprinted from Ref. [11], Copyright (1961), with permission from Elsevier.

Table 7.1 Sputter yields of different metals in an argon plasma (in atoms/Ar)

Target	AA.Wt/Dens	100 eV	300 eV	600 eV	1000 eV	2000 eV
Al	10	0.11	0.65	1.2	1.9	2.0
Au	10.2	0.32	1.65	2.8	3.6	5.6
Cu	7.09	0.5	1.6	2.3	3.2	4.3
Ni	6.6	0.28	0.95	1.5	2.1	
Pt	9.12	0.2	0.75	1.6		
Si	12.05	0.07	0.31	0.5	0.6	0.9
Ta	10.9	0.1	0.4	0.6	0.9	
Ti	10.62	0.08	0.33	0.41	0.7	
W	14.06	0.12	0.41	0.75		

Table 7.2 Ion milling rates for various materials

METALS:	Rate (A/min)	ion energy
Aluminum	300–700	500
	450–750	1000
Gold	1050–1500	500
	1600–2150	1000
Tungsten	180	500
Tantalum	150–330	500
Titanium	200	500
	200	1000
Molybdenum	230	500
	400	1000
Copper	450	500
Chromium	200–400	1000
Zirconium	320	1000
Silver	2000	1000
Maganese	270	1000
Vanadium	220	1000
permalloy	330–450	500
Iron	320	1000
Niobium	300	1000
DIELECTRICS:		
Sio_2	280–420	500
	380–670	1000
Al_2O_3	83	500
	130	1000
$LiNbO_3$	640	1000
SEMICONDUCTORS:		
Silicon	215–420	500
	360–750	1000
GaAs	650	500
	2600	1000
RESIST MATERIALS:		
AZ1350	240–420	500
	600	1000
Riston 14 (Du Pont)	250	500
KFTR (Kodak)	390	1000
PMMA	840	1000

Source: Reprinted from Ref. [9], Copyright (1978), with permission from Elsevier.

Concepts described in this chapter are applicable to both deposition and etch processes used in the semiconductor industry, which will be described in detail in the next section and in Chapter 8.

7.2 Dry Etching

7.2.1 Problems with Wet Etching

Wet etching has been used in the GaAs integrated circuit (IC) fabrication for decades and is still used for deep semiconductor etches. As soon as dry or plasma etching was introduced in silicon processing, it was adopted for GaAs processing, first for dielectric etching and then, as etch chemistries were developed, for semiconductor etching. Transition of semiconductor ICs to submicron geometries has made use of wet etching impossible in critical process steps. In III–V processing, wet etches give different profiles in the two orthogonal directions on the wafer and also lead to undercuts; therefore emitter mesas and collector contacts near the 1 μm size cannot be fabricated. In addition the reliability of devices made with wet etching cannot be guaranteed because reliability depends upon critical dimension (CD) control, particularly that of the heterojunction bipolar transistor (HBT) ledge.

Strong corrosive, acidic, or alkaline etchants are needed for III–V compounds. The photoresist loses adhesion during etches by these compounds. Wet etches are not directional, do not etch only vertically, but undercut in the horizontal direction, so for submicron lines or spacing, CD control becomes an issue; therefore wet etching is not an option. For III–V compounds, the etch profile in one direction has a negative slope (dovetail shape), and metal step coverage over this step is difficult to achieve, so contact lines must be restricted to only one ({011} V-groove) direction (discussed in Chapter 6). In many cases wet etches are unavailable or problematic, for example, for gold etching on GaAs, where the wet etchants are not compatible with the substrate.

7.2.2 Advantages of Dry Etching

Plasma etching solves most of these problems mentioned above. The problem of selectivity between different materials needs to be addressed but can be tackled with selection of chemicals, referred to as gas chemistry, and recipe parameters, better than with wet etching. Plasma processes are capable of anisotropic etching; the etch rate in the vertical direction can be made much faster than in the horizontal direction. Figure 7.8 shows etch profiles for three cases, from the vertical one for the ion-dominated case to the isotropic chemical etch in a liquid or reactive gas. The vertical via has a very small process bias and produces an exact replica of the photoresist image.

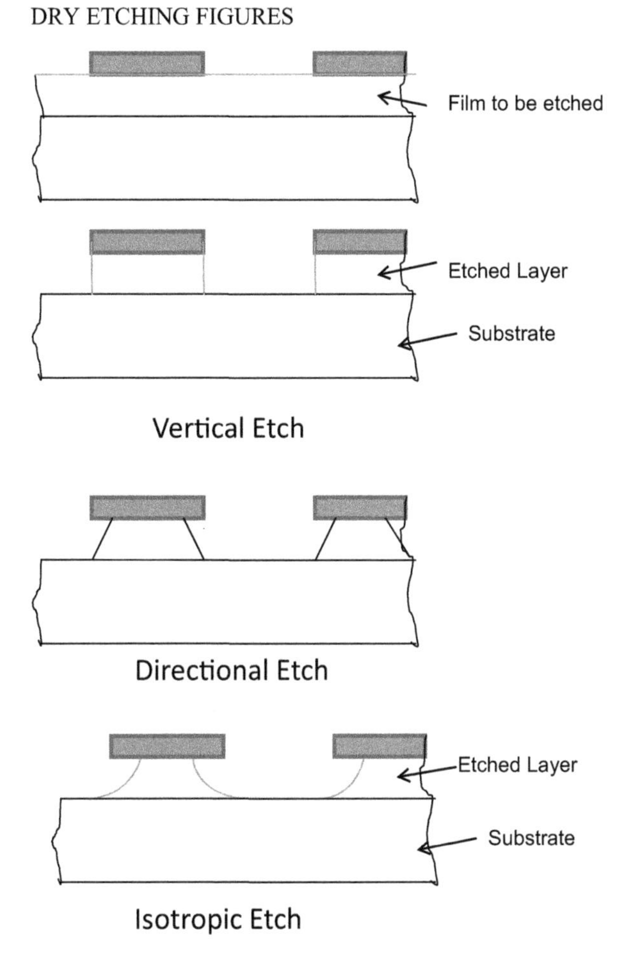

Figure 7.8 Etch geometries showing directionality from vertical to isotropic.

Expensive corrosive chemicals are not needed. Disposal of the used chemicals can be very expensive for batch processing, where whole baths must be dumped after each use. Dry etching can be cheaper, particularly for photoresist stripping with oxygen plasma. Single-wafer processing is possible with dry etching. Etch process modules can be added to cluster tools, making it possible to integrate etch, cleaning, and deposition process steps.

7.3 Plasma Etch Systems

7.3.1 Reaction Basics [1–4, 12, 13]

Tube-type chambers were used first because of historical reasons and simple geometry. As the technology evolved, parallel-plate reactors were introduced. Since then a variety of reactor configurations, discharge excitation mechanisms, and means of varying the chemical and ionic properties of plasmas have been introduced. The basic glow discharge process as it applies to etching and film deposition can be understood by considering the simple parallel-plate geometry. As explained in Section 7.1, because of the mobility difference between the electrons and ions, a depletion region or plasma sheath forms near the electrodes and this sheath is wider at the cathode. The energy of the ions and electrons is determined by the potentials inside the chamber. The potential distribution is shown in Fig. 7.9 [6]. See also Fig. 7.3b for voltage distribution in an RF discharge. The plasma is the most positive and the cathode is the most negative, the largest field being near the cathode. Ions are accelerated across the cathode sheath and strike the surface. Chemical reactions taking place in plasmas are very complex. In the gas phase, free radicals, metastable species, and ions are generated, as shown in Fig. 7.10 [13].

The gas-phase reactions generate free radicals, ions, and metastable species. Reactions of the following type occur:

Excitation $A_2 + e^- \rightarrow A_2^* + e^-$

Dissociative attachment $A_2 + e^- \rightarrow A^- + A^+ + e^-$

Dissociation $A_2 + e^- \rightarrow 2A + e^-$

Ionization $A_2 + e^- \rightarrow A_2^+ + 2e^-$

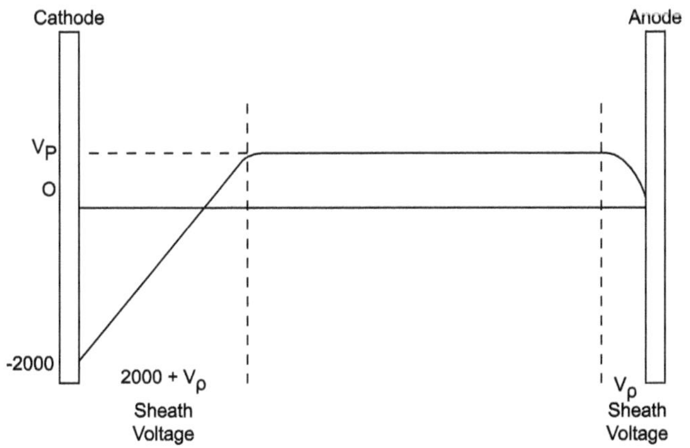

Figure 7.9 Voltage distribution in the DC glow discharge process. Reprinted from Ref. [6] with permission from Ferrotec.

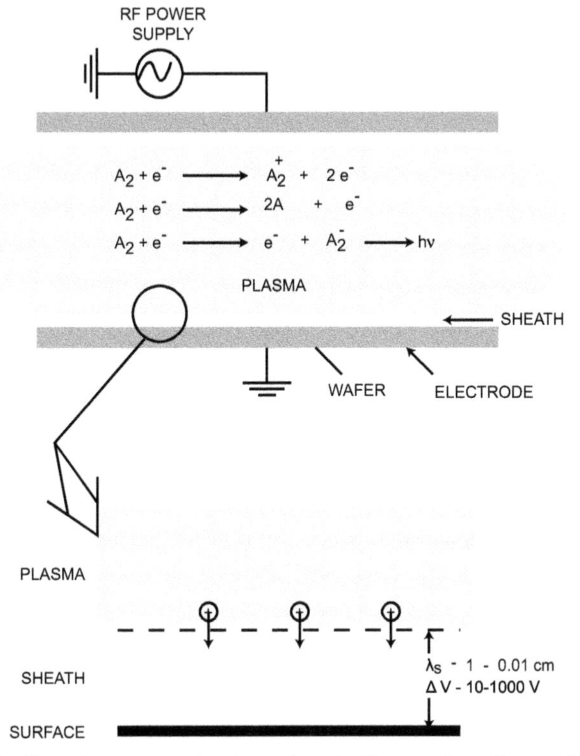

Figure 7.10 Reactions occurring in a chemically reactive plasma. Typical values for sheath thickness (λ_s) and sheath potential (ΔV) are shown. Reprinted with permission from Ref. [13]. Copyright (1983) American Chemical Society.

For reactive gases, atoms have electronegative character, and electron attachment can take place, for example, in the case of dissociative attachment for chlorine:

$$Cl_2 + e^- \rightarrow Cl^- + Cl$$

Electron impact interactions occur at a rate determined by mass action, the rate being proportional to concentrations, $R = kn_eN_A$. The proportionality constant is a function of the reaction cross section and the electron energy distribution function. These can be calculated in principle but are determined empirically [13].

7.3.2 Rate Equation [12]

The etch rate can be calculated as a function of the active species generation rate (G), the recombination rate (τ), the exposed area (A), and the volume (V) of the plasma. The dependence is of the form

$$R(n/G) = \frac{\tau k}{1+\frac{\tau kA}{V}} \qquad (7.9)\ [12]$$

If the factor $\tau kA/V$ is much smaller than 1, the etch rate is independent of the area; however, if this factor is larger than 1, the etch rate drops exponentially with area.

The temperature dependence is also exponential, being of the usual Arrhenius form:

$$K = k_0\exp\left\{-\frac{E_a}{kT}\right\} \qquad (7.10)$$

where E_a is the activation energy for the reaction.

On the substrate surface reactions take place that lead to etching. Plasma etching can be broken down into six steps (see Fig. 7.11) [13].

1. Electron–molecule reactions create reactive species, for example, F^- from CF_4.
2. Etchant species diffuse to the surface.
3. Etchant species get absorbed on the surface, depending upon the sticking coefficient.

4. Radicals react or migrate around till they react with the surface.
5. The product molecules, which are volatile, get desorbed.
6. The product vapor diffuses away from the surface, so it can be pumped away.

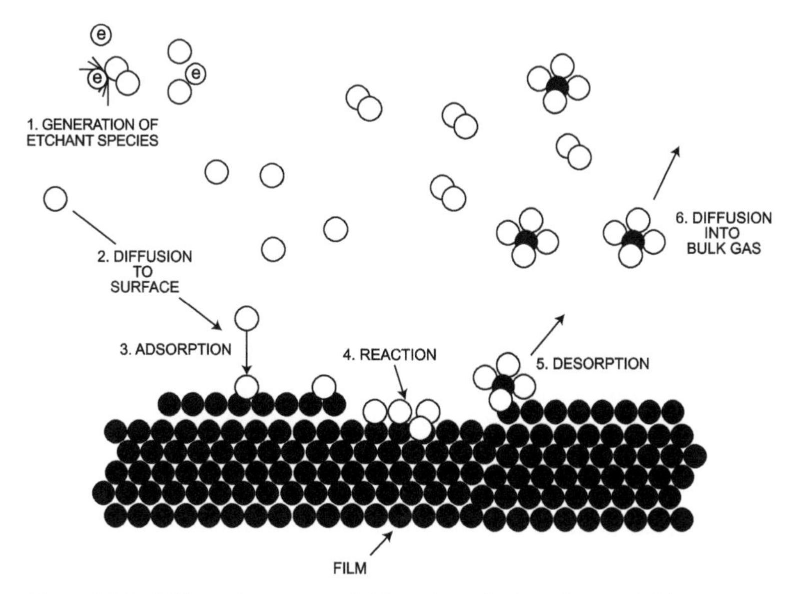

Figure 7.11 Different processes taking place during plasma etching.

7.3.3 Process Parameters [13]

A large number of parameters affects the plasma etching process. These can be divided into two types, external factors, or operating variables, and internal plasma properties (see Fig. 7.12 for a detailed list). The process engineer can vary the external process parameters to optimize the process but must watch out for changes in the internal plasma parameters. How these parameters change, for example, the ion bombardment energy, may depend upon the exact design of the plasma system. Some of these concerns and parameters may have to be considered when selecting the type of etch system and the geometry. The choice of materials used for walls, baffles, focus rings, etc., may depend upon the particular application.

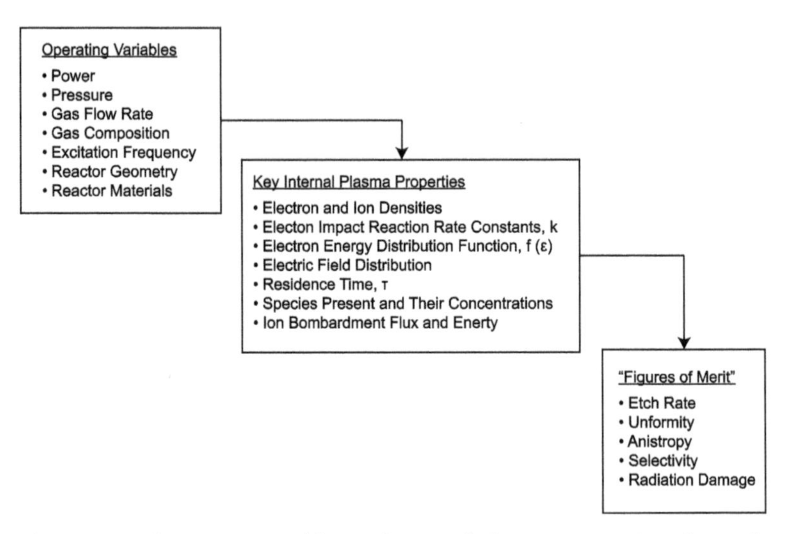

Figure 7.12 Operating variables and internal plasma properties of an etch system/process.

7.3.4 Plasma Etch System Types

Plasma etch systems evolved from the basic tube- or barrel-type geometry to parallel-plate or more complicated geometries. These can be categorized into few basic types: barrel, parallel plate, downstream, and high density inductively coupled. The basic components needed are a vacuum chamber and a pumping system, power supplies, gas handling and control, and in some cases endpoint monitoring. The basic system types are discussed in this section.

7.3.4.1 Barrel reactor

This uses a cylindrical quartz chamber in which the gases flow along the tube length and the RF field is applied by a coil or electrodes, as shown in Fig. 7.13. The wafers are placed in a quartz boat. For good uniformity over multiple wafers and to avoid radiation damage, a metallic, perforated cylinder is placed inside the barrel. Etch reactions are mostly of a chemical nature and isotropic. These systems are used for cleaning and ashing and noncritical etch steps where undercutting is not important. These are used in GaAs processing wherever ion damage needs to be avoided or an undercut is desired.

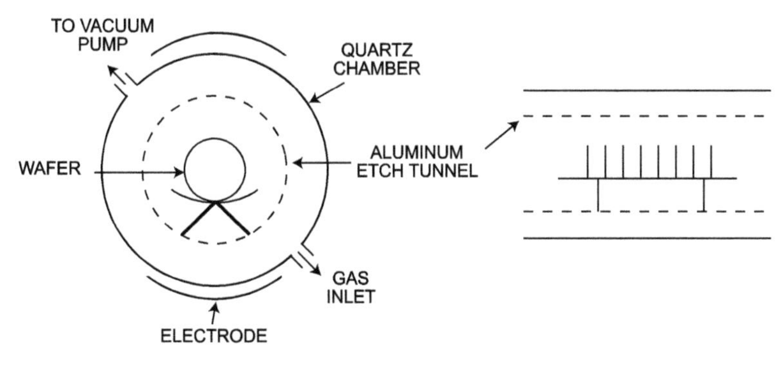

Figure 7.13 Schematic diagram of a barrel-type reactor.

7.3.4.2 Parallel-plate planar reactor

In this configuration, the reactor is similar to the ones used in sputter deposition, as shown in Fig. 7.14. Wafers are fully exposed to the plasma and the gas flow is parallel to the surface. This type of system is more suitable for submicron etches where etch control is more critical. These systems were originally designed as multiple-wafer batch systems but now have migrated to the single-wafer design. The reactors are designed to control the potentials of the substrate platform and the intensity of the plasma. The planar systems were originally built to work in the plasma etch mode, where the wafers were placed on a grounded electrode and the upper electrode was connected to the RF power. In newer systems, substrates are placed on the powered electrode so that the wafers are being "sputtered," as shown in Fig. 7.15. This mode is called reactive ion etching (RIE) mode, although the dominant mechanism is ion etching and not reaction. The chamber walls are grounded. Ion bombardment is more directional, and like sputtering, ions have more energy, being accelerated through the sheath before they strike the wafer. Ion energies of a few hundred volts are easily possible with a pressure below 100 mTorr.

Being a diode-type system, the ion energy and the bias cannot be controlled independently. In GaAs and other III–V semiconductors, ion damage minimization is important while increasing the rates by raising plasma density. So ways of increasing plasma power and ion flux while keeping the bias low have been devised. One example of this is magnetically enhanced RIE. A static magnetic field is applied parallel to the powered electrode, just like in sputter deposition systems. This field confines the electrons and increases plasma intensity without increasing the bias; however, the uniformity of etching is affected.

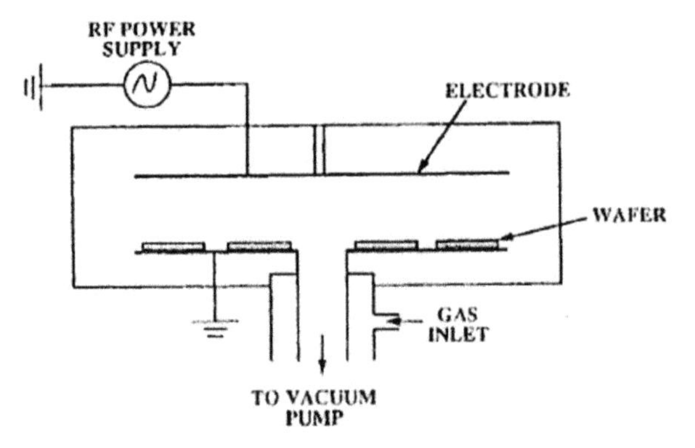

Figure 7.14 Schematic diagram of a multiwafer parallel-plate etching system.

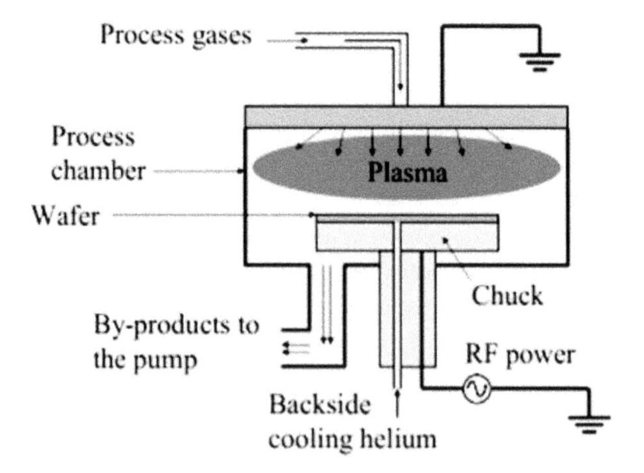

Figure 7.15 RIE reactor configuration. Substrates are placed on the powered electrode.

7.3.4.3 Downstream reactor [14]

Another approach for reduction of plasma ion and radiation damage is to generate the plasma in a separate chamber upstream and pass the activated gas stream over the wafers. See Fig. 7.16. The plasma is activated by an RF coil or electrode plates. Very few ions reach the wafer surface; only sufficiently long-lived radicals reach the surface of the wafer to react and etch it, especially if a baffle is placed in the path. Another advantage of this system is better temperature control due to no direct heating from the plasma. However, the lower etch rate and more isotropic etch can be a disadvantage.

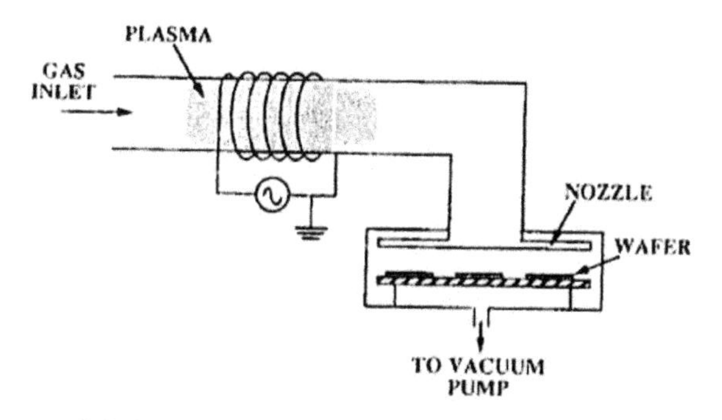

Figure 7.16 Downstream plasma etch system.

7.3.4.4 High-density plasma reactor [15]

Plasma and RIE reactors described so far provide low etch rates and limited anisotropy of the etch profile. These deficiencies make them unsuitable for submicron and high-throughput applications. New reactors have been designed to improve anisotropy and achieve high etch rates without introducing ion damage. These new reactors include electron cyclotron resonance (ECR) and inductively coupled plasma (ICP)-type systems. High-density plasma reactors provide high ion densities at low pressure (<10 mTorr) by decoupling plasma generation from substrate biasing. Power is applied to the plasma through a dielectric window by creating electromagnetic waves in the discharge. Thus power is coupled to electrons that create high-density ion flux at low ion energies. A few main reactor types of this family are discussed in a little more detail next.

7.3.4.5 ECR

In ECR reactors, RF power at microwave frequency is coupled into the plasma generation chamber in the presence of a static magnetic field. Electrons are accelerated to high speeds in circular paths in phase with the applied frequency.

$$\text{Frequency } \omega_{\text{ce}} = e\mathbf{B}/m_{\text{e}} \qquad (7.11) \text{ (see Chapter 9)}$$

where **B** is the magnetic field and m_e is the electron mass. A frequency of 2.45 GHz is used. Electrons gain energy from the microwave field and transfer that to the atoms and ions in the plasma. Figure 7.17 shows how the microwave power and magnets are positioned to create the high-density plasma. Gases are fed from above and flow through this plasma zone to the region where the wafer is held. The chamber is pumped by a high-speed pump to achieve low operating pressure. Different types of ECR reactors, based on the geometry and manner of excitation, are used in the semiconductor industry.

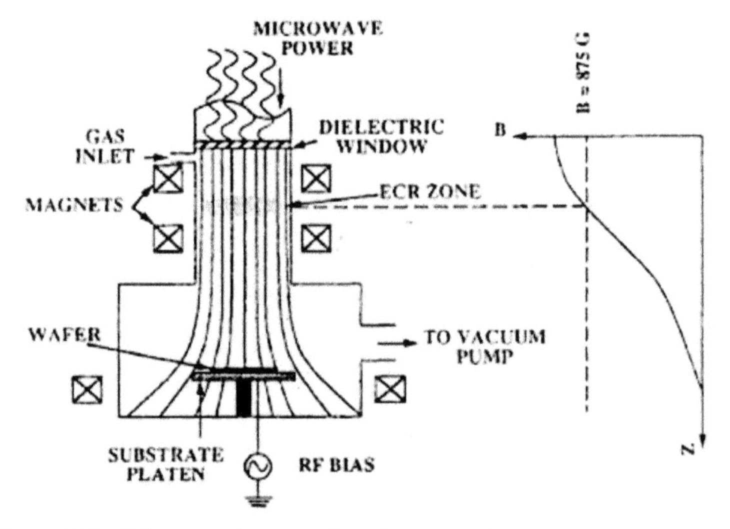

Figure 7.17 Schematic diagram of an electron cyclotron resonance system. Reprinted with permission from Ref. [13]. Copyright (1983) American Chemical Society.

7.3.4.6 ICP

An ICP system could be a simple tube reactor with RF coil excitation. However, the name is used for a special class of high-density reactors in which the coil is planar and is used to excite a high-density plasma (10^{12} /cm^3) in a low-pressure chamber [11, 13]. Figure 7.18 shows a schematic diagram for an ICP system. The shape of the coil can be round, square, or spiral, and it sits on the quartz window. An extra magnetic field may be used to confine the plasma. Plasma potentials are low, less than 40 V. Technically it is necessary to address the dual problem of transport impediments of neutral in/from small features as well as countering the effects of charging by ions and electrons with reduced spacing between sidewalls. Wafers are placed on an electrode that can be biased independently to control the ion energy. Thus ICP and RIE power can be controlled separately. Chamber volume can be small because the rf RF coil creates an intense plasma near the quartz plate. Therefore charge losses and the effect of the system walls on the process can be minimized. ICP etchers have made etching of deep mesas and vias practical in III–V processing. Details of deep through-wafer via (TWV) etching are discussed in Chapter 19.

7.3.4.7 Ion milling

Ion milling is conceptually the simplest way to use a beam of inert ions to sputter off the layer to be removed. It is a purely physical process, and no chemical reactions are involved. In this technique, positive ions are generated in a glow discharge and directed toward the wafer held at negative bias. Ions are accelerated and hit the wafer with sufficient energy to cause sputtering. Ion sources for milling are designed for high flux and long life. Advantages of this process are vertical etch profiles and no need for volatility of the reaction products. This process is well suited to metals like gold that have a high sputter rate. Disadvantages are that wafers need to be cooled, the by-products redeposit everywhere and are not removed from the chamber, the build-up causing particle generation, and redeposition on the photoresist walls can cause flags.

Ion milling can be done with a reactive gas to improve selectivity in a process called reactive ion beam etching.

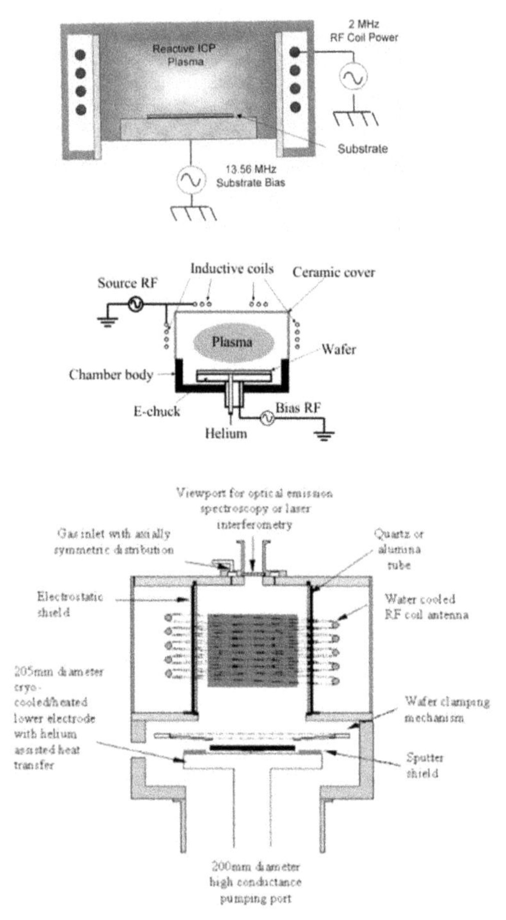

Figure 7.18 (a) ICP system schematic. (b) ICP system diagram from Oxford Instruments. ICP chamber diagram courtesy of Oxford Instruments Plasma Technology.

7.4 Etch Processes

Etch processes, especially for III–V compound processing, are extremely complex. Process development involves the choice of etch system or type, gas chemistry, and process conditions. The number of parameters is very large and process optimization may involve a number of designed experiments (see Fig. 7.12). The output parameters of these experiments are etch rate, selectivity, anisotropy, CD control, and other qualitative factors like surface texture or "grass" or pillar formation. Because of the high cost of capital equipment, throughput must be high without introduction of plasma damage.

7.4.1 Etch Rate and Selectivity

The reactor type is the most important factor that determines etch rate. After the reactor type, etch rates are determined by chemistry, power levels, and substrate temperature. Careful selection of etch gases is important for rate and selectivity. For example, fluorine atoms etch silicon at a high rate compared to SiO_2, and Cl atoms etch GaAs at a high rate compared to Si_3N_4. Etch chemistries will be discussed in a later section. Etch rates depend upon ion flux

$$R = Ak_R\alpha \tag{7.12}$$

where A is the rate at which molecules hit the surface, a strong function of pressure (P) and temperature (T_e), given by

$$A = \frac{KP}{\sqrt{\left(\frac{M}{T_e}\right)}} \tag{7.13}$$

P is the partial pressure, M is the molecular weight, and α is the sticking coefficient. The factor k_R is determined by the activation energy and follows the usual Arrhenius-type relationship:

$$k_R \approx \exp\left\{-\frac{E_a}{kT}\right\} \tag{7.14}$$

The activation energy E_a depends upon the material–reactant combination.

7.4.1.1 Loading

The etch rate also depends upon the area to be etched and loss to the walls and other surfaces. If the exposed surface area is increased beyond a certain value, where the concentration of the active species falls and generation cannot keep up, the etch rate will drop. Systems are designed to make sure the wafer area is below this critical area.

Locally on the wafer, microloading effects may be present and the etch rate may vary if there are large differences in the open area density. Microloading occurs when the reactant concentration depletes due to higher local consumption. If the pattern is dense and variable on the wafer, the etch rate may have to be increased and reactant flow ensured to reduce microloading.

7.4.1.2 Selectivity

Selectivity depends upon the relative ratio of etch rates. Selectivity with respect to the mask is needed to maintain the feature size (see Fig. 7.19). Selectivity with respect to the underlying substrate is needed to prevent loss of substrate thickness, while allowing an adequate overetch, which is needed to ensure the etch is complete in spite of nonuniformity (expected over large-area wafers). Selectivity depends strongly on temperature. Of the three components of etching, chemical, ion-assisted, and simple sputtering, only the first two can be used to affect selectivity; sputtering being purely physical gives very low selectivity (<3:1). Selection of proper chemistry is the most important. Ion energy is also a controlling factor and can be tuned to achieve high selectivity in high-density plasmas.

7.4.1.3 Uniformity

Uniform etch rates over large-area wafers and from wafer to wafer in batch systems are desired in order to achieve good control over CDs. If the etch has a close to vertical profile, an overetch could be used to compensate for nonuniformities, but in general excessive overetching must be avoided to reduce CD variation, undercuts, punch-through to underlying epilayers, and plasma damage. As wafer size increases, single-wafer etchers make more sense. In these systems, radial gradients must still be minimized.

Nonuniformities are caused by depletion of reactant ions, temperature differencesm and ion energy differences due to reactor geometry, magnetic fields, or electric fields. Uniformity can be improved by increasing the flow rate, optimizing pressure, improving ion transport or nonuniformity by adding focusing rings, and improving cooling so that wafer temperature is uniform. Attention must also be paid to substrate holder nonuniformities due to holes, channels, etc., in the chuck.

7.4.1.4 Microuniformity

Etch rates depend upon feature size, pattern density, and aspect ratio (depth-to-width ratio); they reduce as the feature size decreases or the aspect ratio increases [16, 17]. Etch rate differences on the microscopic scale can be due to a large number of factors, the most dominant ones being diffusion limitations through the gas and over the surface, shadowing or local charging, or bias. By reducing the surface reaction rate and increasing the transport rate of reactants, etch dependence on both size and aspect ratio can be minimized. As feature sizes become small, process optimization for pressure (generally pressure reduction) and temperature is needed.

7.4.2 CD and Etch Profile

Etch dimensions and profiles depend on the degree of anisotropy. For a perfectly anisotropic process, the exact size of the resist mask is achieved, as shown in Fig. 7.8, and the etch profile is vertical. For an isotropic etch, the profile is circular, and the size depends upon the etch time, more overetch resulting in a larger feature size and undercut. In addition if the selectivity to the underlying layer (substrate) is not very high, the etch depth will vary with the degree of overetch. If the resist erodes, or deforms during the overetch, it can cause additional CD control problems. However, resist erosion can be exploited to control the etch profile to achieve a more sloped via shape preferred for better step coverage of the next metal layer.

Anisotropic etching is an ion-dominated process. Starting from basic argon ion sputtering [9] it has been developed into ion etching by reactive species in modern RIE and high-density plasma systems. Simple sputtering has the disadvantage of lack of selectivity and redeposition. Ion bombardment by chemically reactive species causes faster and selective removal of material, while sidewalls receive very little direct bombardment and in practice may be covered by complex compounds as a result of redeposition. Thus high etch rates and anisotropy can be achieved. Deliberate sidewall or polymer formation may be promoted for via profile control as in deep TWV etching. System design and process optimization are still critical to achieving controlled anisotropy; the reactive ion distribution, flux, and direction are a function of geometry and plasma parameters.

7.5 Plasma Etching of Materials Used in III–V IC Processing

Fluorine-based etchants like CF_4 and SF_6 are used in silicon etching because product compounds, fluorides likes SiF_2 and SiF_4, are highly volatile. Different etch chemistries are used for silicon IC fabrication—a short list is given in Table 7.3 [5]. Oxygen is added to CF_4 to increase the concentration of available fluorine atoms in the plasma in order to attain high etch rates (see the section on silicon nitride etching below for the mechanism). Chlorine- and bromine-based chemistries are used for silicon etching whenever a high degree of anisotropy control is needed. III–V compound dry etching is not possible with fluorine-containing plasmas because group III fluorides are not volatile. Most III–V etching is done with Cl-generating plasmas, like BCl_3 and CCl_2F_2. Etch rates are determined by the rate of vaporization. Conditions that promote

Table 7.3 Frequently used materials and plasma etch chemistries used in the silicon IC industry

Film/Underlayer (F/U)	Etch Ratios- F/U
Si_3N_4 over AZ-2400 resist	CF,-10
SiO_2 over AZ-1350 resist	CF_4/H_2-lO
Poly Si over PBS resist	CF_4-15
PSG ovcrAZ-1350	SF_4-10
SiO_2 over Si underlayer	CF_3 Cl-30
Si_3N4 over SiO_2 underlayer	NF_3-50
Poly-Si over SiO_2 underlayer	$CC1_4$-10
Si over SiO_2 underlayer	SF_4-30
Etchant	**Material and Etch Rate (A/min)**
Ar	Si-124, Al-166, resist-185, quartz-159
CCl_2F_2	Si-2200, Al-1624, resist-410,quartz-533
CF_4	Si 900, PSG-200, thermal $SiO_2$50,CVD SiO_2-75
C_2F_6-Cl_2	Undoped Si-600, thermal SiO_2-100
CCl_3F	Si-1670

desorption enhance the etch rate, for example, lower pressure, high temperature, high ion bombardment, and addition of O_2 or CCl_2F. CH_4/H_2 plasmas have also been used to etch GaAs and InP [18], which work by generating volatile hydrides or organometallic compounds. Etching by Br plasma has also been reported [19]. The chemical composition of different planes in GaAs varies; therefore crystallographic etch patterns can be observed if chemical processes dominate. A list of GaAs etchants is included in Table 7.4, which also lists etchants for other semiconductors.

7.5.1 Selective Etches [20–22]

Selective etches are needed for device fabrication of HBTs as well as field-effect transistor (FET)-type devices. For details of these selective etches, the reader is referred to the discussion in Chapter 14. Problems due to lack of etch selectivity can be seen in Fig. 7.19, which shows loss of CD control and unwanted etching of the layer underneath. In the HBT process, highly conducting contact layers or wide-bandgap emitter layers must be removed without punching into the layers below. There is a need for etching GaAs over AlGaAs in an AlGaAs/GaAs HBT and also the opposite, etching AlGaAs over GaAs. Similar challenges are offered by other device structures utilizing InGaP, etc. A few of these scenarios are discussed below.

- GaAs on AlGaAs: Fluorine must be added to chlorine chemistry to provide a mechanism for stopping on AlGaAs. When F reacts with Al, it forms AlF_x, a nonvolatile group III fluoride; the reaction stops when AlGaAs is revealed. Typical gas chemistries include BCl_3/SF_6 and $SiCl_4/SiF_4$. Under RIE conditions, selectivity of over 80 for BCl_3/SF_6 (20%) has been achieved [18]. Etch rates can be increased without introduction of plasma damage by addition of ICP power; however, the selectivity gets worse. Lower selectivity is seen when SF_6 is replaced by NF_3. There is currently no selective etch for AlGaAs over GaAs, so endpoint detection must be used for process control.
- GaAs/InGaP: CH_4/H_2 can remove InGaP from GaAs. High selectivity (>80) is obtained for RIE and drops by an order of magnitude when 200 W of ICP power is added. If ICP is used to achieve high throughput, a two-step process may be necessary, where the ICP power is shut off toward the end.

Table 7.4 Plasma etch chemistries based on Cl and Br used in etching of semiconducting materials

Materials	ClKI	Antsoiropy	Selectivity
Si	Cl_2, CCl_4, CF_2Cl_2, CF_3C_l, Cl_2/C_2F 6 Cl_2/CCl_4,Br_2, CF,Br	Anisotropic a t high and low ion energies, isotropic for doped Si under high-pressure conditions with pure Cl_2	High over SiO_2
GaAs	Cl_2, $COCl_2$,CCl_4, CCl_2F_2. PCl_3 HCl, Cl_2/lBCl_3,CF_2Cl_2/O_2/Ar, CCl_4/O_2, Br_2	Anisotropic at high and low (for chlorocarbon and BCl_3 mixtures) ion energies, isotropic for Br_2, Cl_2, HCl under some conditions	High over SiO_2, Al_3O_3, Cr, MgO
InP, GalnAsP, GaAlAsl GalnAsP, GaAlAs	Cl_2, CCl_4, CF_2Cl_2, CCl_4/O_2, Cl_2/O_2, Br_2	Anisotropic at high ion energies	High over SiO_2, $A1_2O_3$, Cr, MgO
Al	Cl_2, CCl_4, BCl_3, Cl_2/CCl_4, Cl_2/BCl_3	Anisotropic at high and low (for chlorocarbon mixtures and BCl_3) ion energies, isotropic for Cl_2 without surface inhibitor	High over Al_2O_3) and some photoresists
Ti	Br_2	—	—
Cr	Cl_2/O_2/Ar, CCl_4/O_2/Ar	Anisotropic at low ion energies	1:1 vs. CrO_2 at $\geq$ 20% O_2
CrO_2	Cl2/Ar, CCl_4/Ar	—	High vs. Cr
Au	$C_2Cl_2F_4$, Cl_2	—	—

Source: Ref. [15] and other sources

- InN/GaN: Cl_2 or BCl_3 is used with a selectivity of less than 5. With BI_3 selectivity of over 40 has been achieved.
- InP: In indium-based materials, the surface is rough to start with and gets degraded further by BCl_3 plasma. $InCl_3$ is relatively nonvolatile and needs low pressure, over 135 C, or high ion flux for desorption to take place. Indium droplets or islands block the etch and cause a rough surface.
- Role of carbon: Carbon from the hydrocarbon molecules used for plasma etching reacts with surface atoms and forms Si–C bonds on silicon-based compounds or Ga–C-type bonds in III–V compounds. This carbon build-up must be removed before etching can continue. It can be etched by halogen atoms, ion bombardment, or addition of oxygen. If carbon is not removed it leads to polymer formation. Addition of hydrogen can also lead to polymerization.

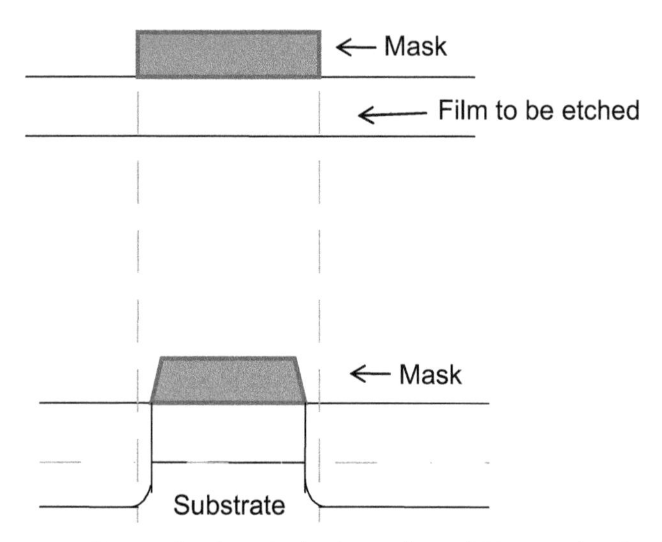

Figure 7.19 Poor etch selectivity leading to loss of CD control and etching of substrate.

7.5.2 Silicon Nitride and Oxide Etching

Silicon nitride is used extensively in GaAs processing for passivation, as an interlevel insulator, and as a dielectric in metal–insulator–metal (MIM) capacitors. The technical challenges for etching silicon nitride and oxide are different from those in silicon processing.

Etch selectivity for silicon is not a factor and selectivity for GaAs-type materials is high for commonly used fluorine chemistry (CF_4 + O_2). The fluorine free radical is the active species for etching silicon compounds and is produced in the plasma by reactions of the type

$$CF_4 + e^- \rightarrow CF_3 + F^* + 2e^-$$

F^* reacts with Si or Si_3N_4 to form SiF_4, which is volatile. Addition of a small quantity of oxygen increases the atomic fluorine rapidly by a reaction between the oxygen and CF_x-type radicals [12]. Beyond 15% oxygen, the rate of etching falls because too much oxygen leads to the formation of Si–O–Si (siloxane) linkages on the surface.

Other mixtures, like CF_4/H_2 and CHF_3, needed for achieving good selectivity between SiO_2, Si_3N_4, and silicon are not needed for these films over GaAs. CHF_3 may be added for control of sidewall polymerization, undercutting, and CD control. Silicon nitride etching behavior characteristics are in between silicon and silicon dioxide. The etch rate of silicon nitride, particularly plasma-enhanced chemical vapor deposition (PECVD)-deposited films used in GaAs processing, is very high. If selectivity for etching SiN over SiO_2 is desired, it is easily achievable with addition of CHF_3 or $CBrF_3$. However, etching SiO_2 selectively over SiN is difficult.

If plasma damage needs to be avoided, as in opening of nitride over FET channels for gate formation or Schottky diodes, helium gas may be added. Inert gas atoms can increase the concentration of radicals without increasing the plasma potential or bias on the substrate.

7.5.3 Metal Etching

Metal etching is mostly used for gate formation in III–V circuit fabrication because the gold-based interconnects are patterned by lift-off or plating. Gold can be etched by Cl-based plasmas, which are very aggressive and incompatible with III–V substrates, or by ion milling, which is a low-throughput, high-particulate process. Aluminum metallization if used for interconnects in digital circuits does use chlorine-based dry etching in commercial silicon-type systems. For interconnect patterning, etching has to stop on the oxide or nitride so that good selectivity is possible. Most other etch chemistries for other metals are also dominated by chlorine, except for refractory metals, like W, Ta, or their silicides or nitrides, where fluorine plasmas are used. Table 7.5 lists common etchants for metals.

Table 7.5 Plasma etchants used for metal etching

Material	Common Etch Gases	Dominant reactive species	Product	Comment	Vapor Pressure (Torr at 25°C)
Aluminum	Chlorine-based	Cl, Cl_2	$AlCl_3$	Toxic gas and corrosive gases	7×10^{-5}
Copper	Forms low pressure compounds	Cl, Cl_2	$CuCl_2$	Toxic gas and corrosive gases	5×10^{-2}
Molybednum	Fluorine-based	F	MoF_6	—	530
Polymers of carbon and photoresists (PMMA and polystyrene)	Oxygen	O	H_2O, CO, CO_1	Explosive hazard	H_2O =26, CO, CO_2 >1 atm
III-V and II-VI compounds	Alkanes	—	—	Flammable gas	—
Silicon	Fluorine-or chlorine-based	F, Cl, Cl_2	SiF_4,$SiCl_4$	Toxic gas	SiF_4 > 1 atm, $SiCl_4$ = 240
SiO_2	CF_4, CHF_4, C_2F_6, and C_3F_6	CF_x	SiF_4, CO, CO_2	—	SiF_4 > 1 atm, CO_2, > 1 atm
Tantalum	Fluorine-based	F	TaF_3	—	3
Titanium	Fluorine-or chlorine-based	F, Cl, Cl_2	TiF_4, TiF_3, $TiCl_4$	—	TiF4 = 2.10^{-4} $TiCl_4$ = 16
Tungsten	Fluorine-containing	F	WF_6	—	1000

Source: Ref. [5]

7.5.3.1 Refractory metals

Tungsten is a good example of refractory metals used in III–V compound semiconductor fabrication. It can be easily etched by fluorine-containing plasmas such as CF_4, SF_6, and NF_3 at very high etch rates. Oxygen is added for the usual reasons (increase F atom concentration and prevent polymer build-up). The etch is very anisotropic and is well suited for gate formation. CHF_3 may be added to achieve sloped sidewalls. Other metals, Ta, Mo, and their silicides and nitrides can also be etched by similar chemistry.

7.5.3.2 Aluminum

Aluminum-based interconnects have been used for digital circuits on GaAs substrates. Chlorine plasmas must be used to etch Al and the native surface oxide must be removed before etching can proceed. Enhanced ion etching or addition of compounds like BCl_3 is required. Since Al is used with SiO_2 as the dielectric, good selectivity to SiO_2 is needed. Tools and techniques developed for silicon ICs are available. Also, process modifications need to be adopted for getting clean etches with commercial Al targets used for interconnects that contain Si and Cu for better electromigration. These modifications may include increasing ion bombardment or substrate temperature.

The chlorine residue must be completely removed from the wafer before proceeding with further processing, otherwise corrosion and reliability problems may occur. In situ dry-cleans followed by rinses are required.

7.5.3.3 Gold/copper

Gold can be etched by CCl_2F_4 or $CClF_3$. Good etch selectivity is tough to achieve and a hard mask may have to be used. Also sidewall deposition can be an issue.

Copper is an alternative to aluminum in silicon IC fabrication. Copper is one of the most difficult materials to etch. Chlorine-based plasma etching is used in the silicon industry for copper etching. RIE in CCl_4/Ar, $SiCl_4/N_2$, and $SiCl_4$/Ar has been reported [13]. However, a temperature over 200°C is desired. Copper chlorides are not volatile at room temperature. For de-sorption to take place, temperatures over 150°C and Ar bombardment are needed. Copper front-side interconnects are being introduced in production for GaAs processing, but lift-off patterning is currently used. $SiCl_4/N_2$ may be used anisotropically to etch Cu with good selectivity to polyimide.

7.5.3.4 Organic films

Organic films like polyimide, polybenzoxazole (PBO), benzocyclobutene (BCB), etc., are used as interlevel dielectrics in III–V fabrication. These films can be easily etched in oxygen plasma. Oxygen plasma etch rates for photoresists are about the same as these materials, so resist erosion can lead to larger CDs and sloped sidewalls. Vias with sloped sidewalls are desirable for better step coverage; however, the increase in size can be of the order of a micron and must be allowed for in the layout.

In most etch processes, a low etch rate of the resist would be desirable so that the selectivity to the layer being etched is high. Addition of CHF_3, C_2H_4, and H_2 is used to achieve this. To reduce resist loss a multilayer resist can be used where the top layer has been specially formulated for a low oxygen etch rate. A spin-on layer of hexamethyldisilazane (HMDS) may be added to the top after a pattern is formed.

7.6 High-Aspect-Ratio Etching

Microelectromechanical systems (MEMS) technology is used to form small structures with dimensions in the micron range. Structures are formed using techniques used in IC fabrication, like lithography, etching, and deposition. Complex machines with moving parts are made in silicon; III–V semiconductors are used to form simpler structures that need dots, pillars, walls, and other 3D structures. All these require high-aspect-ratio etching. High-aspect-ratio III–V nanostructures are finding applications in light-emitting diodes (LEDs), solar cells, FETs, optoelectronics, nanophotonics, etc. These applications, like waveguides, generally require high-aspect-ratio geometries. Sidewalls must be vertical. These structures can be formed by additive growth or by etching.

7.6.1 Through-Wafer Via Etching

High-aspect-ratio etching is also needed for TWVs in III–V processing (discussed in Chapter 19).

7.6.1.1 Etch chemistry for profile control

Chlorine (Cl_2) chemistry is used for GaAs layer etching; for deep etching higher etch rates are desired, so in addition to high power density achieved by ICP, etch chemistries are also optimized for higher etch rates as well as profile control. BCl_3, Ar, and CF_4 gases are added for deep high-aspect-ratio etches. For deep vertical vias, the etch rate drops as the aspect ratio increases, because of reactant transport through narrower openings. Figure 7.20 shows the etch rate with an aspect ratio increase.

For TWVs where a sloped wall profile is desired, the etch chemistry and ICP power are chosen to encourage photoresist erosion. Figure 7.21 illustrates the resist erosion mechanism that promotes tapered vias.

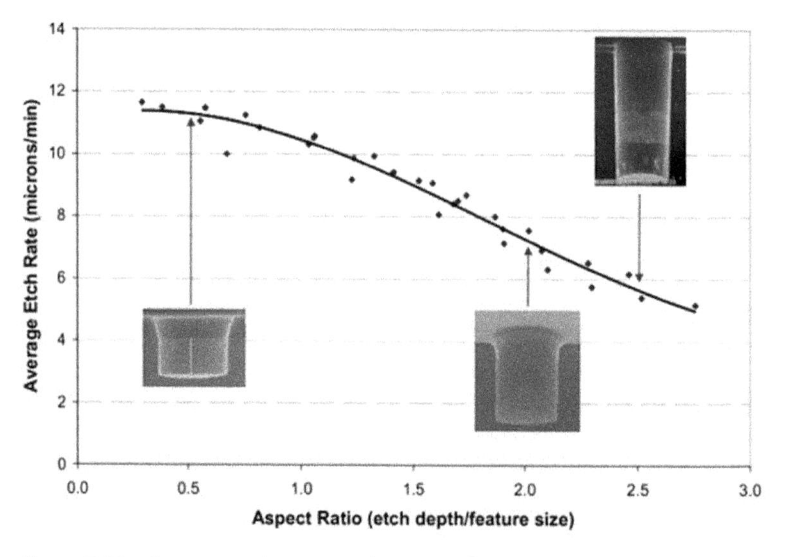

Figure 7.20 Average etch rate as a function of aspect ratio for TWV etching. Courtesy of Plasma-Therm LLC.

7.6.2 Wet Etching

Dry plasma or ion etching, like RIE, is common for top-down etching; however, for some applications plasma damage or rough surfaces are not acceptable. Metal-assisted chemical etching [23] is a simple directional wet etching method used for silicon. In this method a patterned metal catalyst is used to produce anisotropic etching in an oxidizing acid solution. In GaAs and other III–V materials, etching takes place in oxidizing solutions without the presence of a metal. However, in weaker oxidizing agents like $KMnO_4$, etching under the metal takes place at a reasonable rate, while the etch rate is low in the un-metallized region (Fig. 7.22) [23].

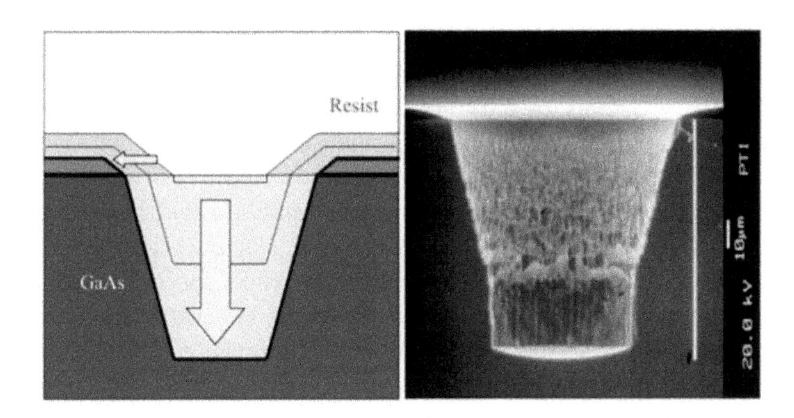

Figure 7.21 Deep via etching with tapered sidewall for better metal coverage of TWVs. Courtesy of Plasma-Therm LLC.

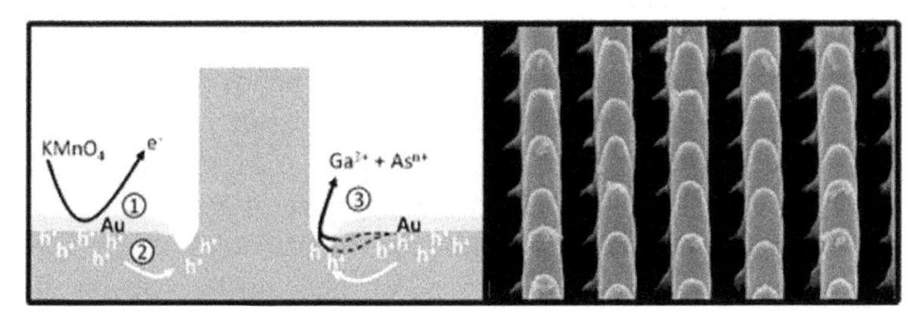

Figure 7.22 MacEtch using a thin layer of gold as a catalyst. Reprinted with permission from Ref. [23]. Copyright (2011) American Chemical Society.

Nanoscale gold mesh patterns are generated using lithography in the following manner: SiN is deposited all over and then patterned with SU8 and polymethylmethacrylate (PMMA) to form an array of holes or stripes in the silicon nitride. With SU8 still on, 200 nm gold is deposited and lifted off (dielectric-assisted lift process). Etching is done in sulfuric/$KMnO_4$ solution to create nanopillars and stripes.

If the structures are too large, etching will take place only laterally near the edges of the metal and not under the metal dots.

7.6.3 Aspect-Ratio-Dependent Etching [24]

In a plasma etch process, features of different sizes next to each other etch at different rates; also, etch rates depend upon pattern density. These effects are called aspect-ratio-dependent etching (ARDE) and microloading, respectively. Holes and trenches with small dimensions etch slower than large dimensions. In many materials/chemistry systems the etch rate scales with the aspect ratio (depth-to-width ratio). In this case, the mechanism responsible can be ion/neutral radical shadowing and Knudsen transport of neutrals into the feature. If the etch rate does not scale with the aspect ratio, other mechanisms like surface and bulk diffusion may be responsible for the etch rate difference. Reducing the pressure and temperature improves microuniformity in general. As the surface reaction rate is decreased, it becomes rate limiting; consequently, the effects of geometry and transport of reactants are reduced. For high-aspect-ratio etching, anisotropy must be increased, generally by increasing ion bombardment. Pure sputtering or ion milling provides ion bombardment but does not work well because of the nonselective nature of these, slow etch rates, and redeposition. Processes using ion bombardment in the presence of chemical etchants can be better optimized. Ion bombardment breaks crystal bonds on the surface and makes the material removal by etching faster, particularly on the bottom surface if the pressure is low and the ions are directed vertically. The sidewalls etch at normal chemical etch rates. Thus anisotropy can be increased. Deliberate sidewall inhibition is used to provide more anisotropy (e.g., see TWV etching). By changing the chemistry, deposition of polymers can be promoted, for example, by adding CHF_3 or CH_4 to CF_4.

In one of the early applications for waveguides, Vodjdani and Parrens [25] used polymer formation by addition of CH_4 in chlorine and demonstrated polymer control for chemistries based on gases like CCl_4, CF_2Cl_2, $SiCl_4$, BCl_3, and Cl_2. Hydrogen from CH_4 reacts with chlorine radicals and provides carbon for polymer formation. These polymers form a passivating film on the mask material (like a photoresist) and on the sidewalls. The etching takes place mainly by ions, and the rate drops. Contaminants are reduced by the hydrogen reaction with unsaturated compounds. They also added argon to increase ion bombardment. As discussed earlier, Ga and As have different volatilities; the addition of these gases balances the removal rate, thus minimizing formation of rough surfaces. Very high selectivity and vertical micron size geometries were demonstrated.

For submicron-size geometries, ICP etching has advantages. Photonic crystals and periodic structures like an array of deep submicron holes with an aspect ratio greater than 8 have been fabricated in GaInAsP by Cl_2/Xe ICP etching [26]. The authors used a Ni/Ti mask that was etched by CF_4 RIE and Ar ECR and high substrate temperature. As discussed earlier, ICP provides high density of plasma and a high degree of ionization. Vertical walls down to 0.3 μm were achieved.

Another approach for achieving high selectivity and smooth walls is the use of a very-high-voltage, low-current ion beam etching. A high-voltage-to-current-ratio ion beam, assisted by chemical etching, has been used to fabricate photonic crystals in InGaAsP and AlGaAs/GaAs [27].

In addition to H_2, N_2 also has been used for control of the sidewall angle in ICP etching [28]. Cl_2/BCl_3/Ar with a controlled N_2 fraction was used. Electron beam lithography was used for patterning and fabrication of nano-waveguides in a GaAs/AlGaAs system. GaAs nanowires have been fabricated by ICP-RIE using a Cl_2–N_2 gas system [29]. Figure 7.23 shows scanning electron microscopy (SEM) images of features achieved by use of a Ni mask with a Cl_2:N_2 ratio of 8:1 in the millimeter-Torr pressure regime, with a 60 W RF platen power and a 500 W ICP source power.

A more recent technique uses a time-multiplexed Inductively coupled plasma (ICP) process with $SiCl_4$ and O_2 steps, with gas feed chopping, thus separating the etch and passivastion steps in time [30]. Submicron features with an aspect ratio of 10 for holes, 17 for trenches, and 30 for stripes and a selectivity of 200 to 1 (SiO_2 particles used as a mask) were achieved. Oxygen passivation of sidewalls was verified by optical emission spectroscopy. Oxygen cycling has also been used for fabrication of photonic crystals in InP (using reactive ion beam etching in a CHF_3/N_2 gas mixture for the mask) by RIE with a CH_4/H_2 mixture and O_2 cycling [31].

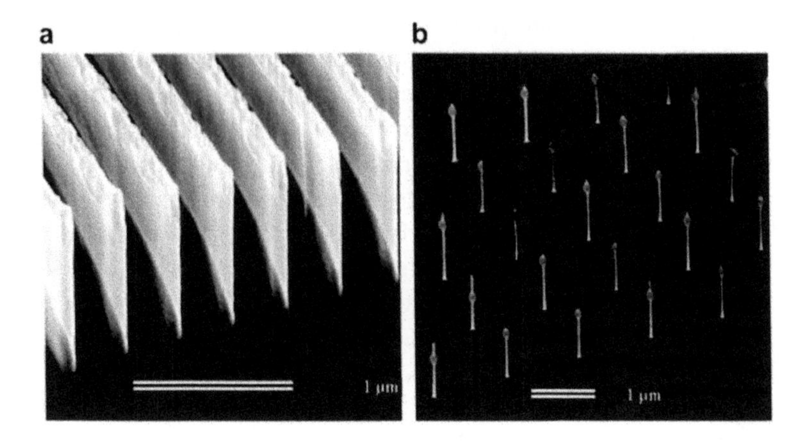

Figure 7.23 Features produced by ICP/RIE in Cl2/N2 chemistry. (a) Gratings of 930 mm and 130 nm width. (b) Nanocolumns 1 mm high and 30 nm in diameter. Reprinted from Ref. [29], Copyright (2008), with permission from Elsevier.

7.7 Plasma Damage [2]

Plasma processes are used extensively in semiconductor fabrication in the front end of the process, where devices can be degraded due to damage to the semiconductor. Therefore etch processes are always designed with ion damage in mind. Energetic particles cause damage similar to ion implantation, creation of vacancies, interstitials, other complex or larger defects, and embedding of atoms from the plasma. Defects created serve as trapping sites and degrade the semiconducting properties of the material layer exposed. Both electronic and optical properties are affected. The damage introduced can change carrier concentration, mobility, and luminescence properties. Ion damage has been studied extensively (see, for example, Refs. [32, 33]).The degree of damage is seen to be proportional to the ion mass and energy. Carrier concentration can drop up to 1000 Å deep from the surface, much deeper than expected from low-energy atoms, perhaps like channeling in ion implantation. Hydrogen passivation can also take place because it is almost always present in etch systems. Figure 7.24 shows the Schottky barrier height and ideality factor of a Ti/Pt/Au contact on n-type GaAs etched in a CCl_2F_2/O_2 plasma. As the power increases, the bias voltage and impinging energy increase, which affects both the ideality factor and the barrier height [32]. Some damage can be annealed out by subsequent processing, like rapid thermal annealing (RTA) alloying or other high-temperature processing steps used in the interconnect fabrication process. However, some damage cannot be annealed out because there is a limitation on the temperature that the devices can tolerate; therefore care must be taken to avoid high-bias processes. The choice of reactor geometry, process parameters like pressure, or the use of baffles may be investigated and optimized. The use of hot chucks to raise the temperature during etching could limit damage due to self-annealing, but this is not practical generally because the etch rates could be too high or the high temperature is not compatible with the photoresist mask.

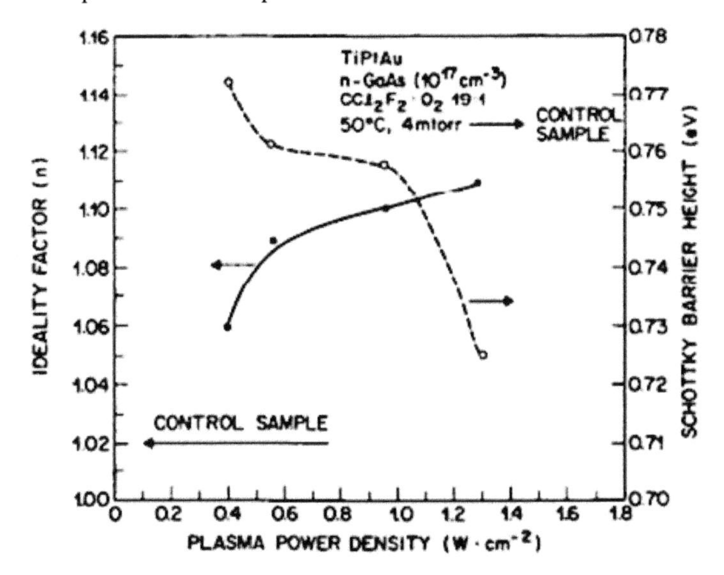

Figure 7.24 Schottky barrier height and ideality factor for TiPtAu Schottky diodes on n-type GaAs etched in a 19:1 CCl_2F_2:O_2 discharge as a function of plasma power density during RIE treatment. Reprinted from Ref. [39], Copyright (1996), with permission from Elsevier.

Even at low energies, as low as 25 V, damage can be deep and not repairable by either removal or by allowable annealing. Such damage can be seen in degradation of metal–semiconductor diode properties [34] and FET performance. Plasma damage can occur even in low-damage systems like ECR [35]. Dielectric materials exposed to plasma can also be damaged, resulting in a drop in the breakdown strength of MIM capacitors. In HBT fabrication, de-scum, ashing, and resist strips in oxygen plasma and dry etching of emitter mesa are all potential concerns [36]. The most sensitive steps from the point of view of ion damage are those in which the ledge of the HBT, that is, the area between the emitter and the base, is exposed and unprotected. There have been reports of damage by N_2, Ar, and Cl_2 plasmas, and even a few minutes of ashing in 300 W in an RIE-type reactor can cause damage, evident from beta measurement [37].

The damage threshold of FET- and HEMT-type devices has been determined and reported in the literature. Plasma processes with a bias voltage of over 150 V should be avoided. HBTs, being vertical current devices, may appear to be less sensitive, but adverse effects have been seen on the reliability of unprotected devices, the damage being cumulative rather than occurring over a certain threshold [36].

7.7.1 Particle and Veil Generation

Plasma systems get dirty with use—deposits build on walls, electrodes, and other surfaces. Immediately after periodic maintenance (PM), maintenance involving cleaning, the system may be clean and is conditioned before processing of production wafers is started. As the number of wafers processed increases, the system may behave differently, and particle generation may go up. Particles can get knocked off the walls, get suspended in the plasma, and eventually land on the wafer.

A veil generation mechanism like that of sidewall polymer deposition is due to carbon–metal complex formation and is affected by the degree of carbon deposits. In practice the process must be optimized by multivariable experiments.

7.8 Etch Process Monitoring [32]

Plasma etch processes are complex and involve numerous process and equipment parameters. Tight control of these is important for control of etch rates and undercuts, etc., from wafer to wafer, from run to run, and between maintenance events. If the process can be controlled well, the need for overetching can be reduced, thus improving overall control. Etch rates change even from run to run under well-controlled conditions, and there are always variations in the layer to be etched; therefore it is desirable to employ diagnostic tools, in particular for etch endpoints. Systems have been developed to monitor etch and product species and also the film being etched. The most common method to control the etch process is to monitor the etch rate by measuring the etch depths on the wafer at a few predetermined points on the wafer and entering these in a statistical process control (SPC) chart. The etch depth is measured by measuring the step height on a stylus surface profile measuring system. This process has its limitations: individual wafers in a lot may not have etched like the sample wafer and a lot of wafers may have to be scrapped. Therefore other in situ techniques are preferred for wafer-at-a-time systems for large wafers.

7.8.1 Film Monitoring

Optical reflection, interferometry, or ellipsometery developed for deposition systems can also be used for in situ monitoring of dielectric films. This technique is discussed in detail in Chapter 21. In an etch system, the window through which the laser beam passes must be kept clean.

7.8.2 Gas-Phase Monitoring

Mass spectrometry is used for residual gas analysis in vacuum systems for diagnostic purposes. The same system can be utilized for monitoring species in an affluent sample taken from a plasma system. Mass spectrometry can be used for understanding the reaction mechanism as well as predicting the endpoint.

7.8.3 Optical Emission

Optical emission from a glow discharge has been used to detect and quantify species in the plasma and to get a better understanding of the plasma etch mechanism. By using a monochromator and a photon detector, optical emission spectra can be collected as a function of wavelength in the visible-to-ultraviolet region. Data can be analyzed by assigning peaks to known species. Endpoint detection of some etch processes can be very straightforward, for example, SiO_2 etching over Si, where the SiF_3 peak intensity changes abruptly when the etching stops, due to excellent selectivity. However, in III–V compounds, where selectivity may not be very high, more sophisticated schemes for detecting changes and predicting the endpoint may be needed. For AlInAs/GaInAs nonselective etching, Ga emission at 417.2 nm can be used. For etching AlGaAs/GaAs, where selective etch does not exist, Al intensity cannot be used because the system walls are made of aluminum, the Ga intensity change can be monitored, and the differential of the intensity signal with time can be used to determine the endpoint. An example of a typical output trace is shown in Fig. 7.25 [38].

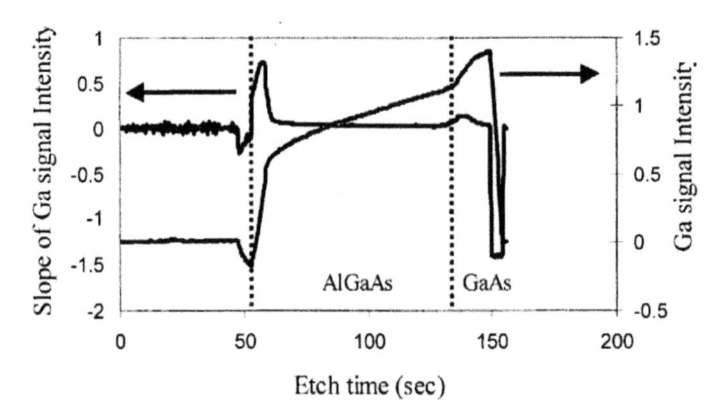

Figure 7.25 Example of signals from an endpoint detector.

References

1. J. L. Vossen and W. Kern, *Thin Film Processes*, Academic Press, New York (1978).
2. K. Becker and A. Belkind, Introduction to plasmas, *Vac. Technol. Coat.*, **8–10**, 32 (2007).
3. S. Wolf and R. N. Tauber, *Silicon Processing for the VLSI Era*, Lattice Press, Sunset Beach, California (1986).
4. B. Chapman, *Glow Discharge Processes, Sputtering and Plasma Etching*, Wiley Interscience, New York (1980).

5. (a) M. Madou, *Fundamentals of Microfabrication*, CRC Press (1997); (b) C. Torres et al., *J. Phys. Conf. Ser.*, **370**, 012067 (2012).
6. J. W. Coburn, *Plasma Etching and Reactive Ion Etching, AVS Monograph Ser.* (1982).
7. H. R. Koening and L. I. Maissel, *IBM J. Res. Dev.*, **14**, 168 (1970).
8. R. Williams, *Modern GaAs Processing Methods*, Artech House, Boston (1990).
9. C. M. Milliar-Smith and C. J. Mogab, in *Thin Film Processes*, Eds., J. L. Vossen and W. Kern, Academic Press, New York (1991).
10. J. W. Coburn and H. F. Winters, *J. Vac. Sci. Technol.*, **16**, 391 (1979).
11. O. Aleman and G. Bruce, Sputtering experiments in the high energy region, *Nucl. Instrum. Methods*, **11**, 279 (1961).
12. S. K. Ghandhi, *VLSI Principles Silicon and Gallium Arsenide,* Wiley Interscience New York (1983).
13. J. A. Mucha, D. W. Hess and E. S. Aydil, Plasma etching, in *Introduction to Microlithography,* Eds., L. Thomson, G. Wilson and M. J. Bowden, *Advances in Chemistry Ser*ies, **219**, ACS Washington, D.C. (1994).
14. J. E. Spencer et al., *J. Vac. Sci. Technol.*, **A7**, 676 (1989).
15. D. L. Flamm, High density plasma reactors, *Solid State Technol.*, **34(3)**, 47 (1991).
16. J. Hopwood et al., Self bias characteristics of inductively coupled plasmas, *J. Vac. Sci. Technol.*, **A11**, 152 (1993).
17. R. A. Gottscho, C. W. Jurgensen and D. J. Vitkavage, *J. Vac. Sci. Technol.*, **B10**, 2133 (1992).
18. R. Cheung et al., *Electron. Lett.*, **23**, 16 (1987).
19. D. E. Ibbotson et al., *J. Vac. Sci. Technol.*, **20**, 489 (1982).
20. D. C. Hayes et al., *Selective Dry Etching of the GaN/InN/AlN, GaAs/AlGaAs and GaAs/InGaP Systems*, http://www.mse.ufl.edu/~spear/.
21. W. T. Lim et al., Planar inductively coupled BCl3 plasma etching of III-V semiconductors, *J. Electrochem. Soc.*, **151**, G343 (2004).
22. P. Werbaneth et al., *Reactive Ion Etching of Gold on GaAs in High Density Plasma Etcher,* GaAs MANTECH (1999).
23. M. DeJarld et al., Formation of high aspect ratio GaAs nanowire structures with metal assisted chemical etching, *Nano Lett.*, **11**, 5259 (2011).

24. L. F. Thompson, C. G. Wilson and M. J. Bowden, *Introduction to Microlithgraphy*, ACS Professional Reference Book (2007).
25. N. Vodjdani and P. Parrens, *J. Vac. Sci. Technol. B*, **5**, 1591 (1987).
26. K. Inoshita, T. Izumi and T. Baba, Fabrication of air hole type GaInAsP photonic crystals by ICP etching, *Microprocessor Nano Technol. Conf. Digest Pap.*, 144 (2002).
27. M. V. Kotlyar et al., *J. Vac. Sci. Technol. B: Microelectron. Nanostruct.*, **22**, 1788 (2004).
28. M. Volatier et al., High aspect ratio GaAs nanowirs made by ICP-RIE etching using Cl2/N2 chemistry, *Nano Technol.*, **21**, 134014 (2010).
29. L. Jalabert et al., High aspect ratio GaAs nanowires made by ICP-RIE etching using Cl2/N2 chemistry, *Microelectron. Eng.*, **85**, 1173 (2008).
30. S. Golka et al., *J. Vac. Sci. Technol. B: Microelectron. Nanostruct.*, **27**, 2270 (2009).
31. L. J. Martinez et al., *J. Vac. Sci. Technol. B: Microelectron. Nanostruct.*, **27**, 1801 (2009).
32. R. J. Shul and S. J. Pearton, *Handbook of Advanced Plasma Processing Techniques*, books.google.com (2000).
33. S. J. Pearton, U. K. Chakrabarty and W. S. Hobson, *J. Appl. Phys.*, **66**, 2061 (1989).
34. S. Ashok, T. P. Chow and B. J. Baliga, *Appl. Phys. Lett.*, **42**, 687 (1983).
35. T. Hara et al., *J. Appl. Phys.*, **67**, 2836 (1990).
36. W. Liu, *Handbook of Heterojunction Bipolar Transistors*, John Wiley (1998).
37. S. O'Neil, S. Tiku et al., *Oxygen Plasma Damage Study of InGaP/GaAs HBTs*, GaAs MANTECH (2005).
38. J. W. Lee, *Development of End Point Techniques for Advanced Plasma Etching,* www.mse.ufl.edu/~spear/recent papers/40401.htm.
39. S. J. Pearton, High ion density dry etching of compound semiconductors, *Mater. Sci. Eng.*, B40, 101 (1996).

Chapter 8

Deposition Processes

8.1 Physical Vapor Deposition: Introduction

Out of the numerous thin-film deposition and coating processes a few are heavily used in the semiconductor industry. A broad classification is based on whether the process is based on vacuum or a wet-chemical system.

- **Vacuum environment**:
  - Physical vapor deposition (PVD):
    - Evaporation
    - Sputtering and different enhancements
  - Chemical vapor deposition (CVD):
    - Plasma-enhanced chemical vapor deposition (PECVD)
  - Thermal growth (oxidation)
  - Diffusion
- **Wet chemical**:
  - Electroplating
  - Spin or spray coating (photoresist, polyimide, etc.)
  - Anodization

Processes based on PVD in a vacuum will be discussed in this chapter.

III–V Integrated Circuit Fabrication Technology
Shiban Tiku and Dhrubes Biswas

ISBN 978-981-4669-30-6 (Hardcover), 978-981-4669-31-3 (eBook)
www.panstanford.com

PVD consists of vaporization of a material and its condensation on a substrate, where a film grows. This process may be divided into three steps: vaporization or ejection of the material from the source; transport through a vacuum or low-pressure region; and nucleation, reaction, and growth. The most common vaporization mechanisms used in the semiconductor industry are thermal evaporation and sputtering. These two processes will be discussed here in detail, but since these processes use vacuum systems, vacuum basics will be discussed first.

8.2 Vacuum Basics [1, 2]

Most semiconductor fabrication occurs in processes that involve use of reduced pressure and therefore vacuum systems. These range from extremely low pressure like in molecular beam epitaxy (MBE) growth or reduced pressure like CVD. An understanding of vacuum creation, properties of gases, and gas flow is essential for understanding of semiconductor processing. Properties of gases are described well by ideal gas behavior, in which the molecules in a gas can be treated as spheres that travel along straight lines and make perfectly elastic collisions. Using these assumptions, some of the properties of interest in vacuum processing are given below.

The average velocity of gas molecules is given by

$$V = \sqrt{\frac{8kT}{\pi m}} \tag{8.1}$$

where k is the Boltzman constant, T is the absolute temperature, and m is the mass of the molecule. The actual velocity is given by the Boltzman distribution. The pressure is given by

$$P = \frac{nmv^2}{3} \tag{8.2}$$

where n is the concentration of gas molecules. If the gas is a mixture, the total pressure in the system is given by the sum of the partial pressures of the gases in the mixture. The mean free path, which is the average of the distance travelled by molecules between collisions, is given by

$$\lambda = \frac{1}{\sqrt{2}\pi d_0^2 n} \tag{8.3}$$

where d_0 is the molecular diameter. For air at 300 K

$$\lambda = 0.05/P \tag{8.4}$$

where P is in Torr.

Mean free paths for different pressure regimes used in processing are given in Table 8.1. In evaporation, where pressure is below 10^{-6} Torr, the mean free path is over 3 km long [3]. Figure 8.1 shows the relationship between pressure, molecular density, mean free path, and surface impingement rate for nitrogen.

Table 8.1 Vacuum ranges and mean free path for deposition and etch processes

Process	Typical pressure range	Mean-free-path range
Deposition		
Evaporation	$<10^{-8}$Torr	>3 km
Sputter Deposition	1–10 mTorr	15–0.3 cm
Low-Pressure CVD	10 Torr	3 μm to 15 μm
Plasma CVD	0.2–2 Torr	0.6–μm
Patterning		
Sputter Etch	1–10 mTorr	15–0.3 cm
Ion Beam Etch	<1 mTorr	>3 cm
Plasma Etch	10–500 mTorr	30–0.6 mm

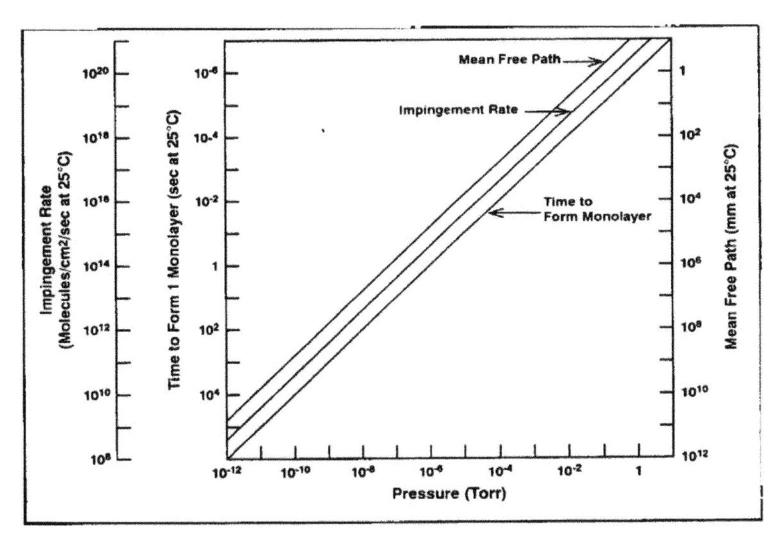

Figure 8.1 Relationship between pressure, mean free path, and surface impingement rate for nitrogen at 25°C. Reprinted from Ref. [3] with permission from Ferrotec.

8.2.1 Flow Regimes

In vacuum systems, gases are continuously being pumped out and in many cases being fed in for the process. So, there is a dynamic situation, the nature of which depends on the pressure, flow rates, dimensions of the vacuum chamber, etc. At high pressure the mean path is short, and the gas flow is in the viscous regime, which can be turbulent or laminar like a fluid. At low pressure, the mean free path is long, particle collisions are less frequent, and flow is in the molecular regime. In this regime, gas molecules or oil vapor can diffuse back upstream. If the characteristic dimension *d* of the system or pipe (pumping port) is much larger than the mean free path, viscous flow occurs. If the mean free path is much larger, molecular flow occurs. For intermediate cases, the flow is called Knudsen flow. When a vacuum system is being pumped on its own, it goes from viscous flow to molecular flow. As the pressure drops, surfaces start outgassing and may limit the pumping rate. At even lower pressure, the pumping rate goes into the diffusion regime. Figure 8.2 shows an example of a pump-down curve.

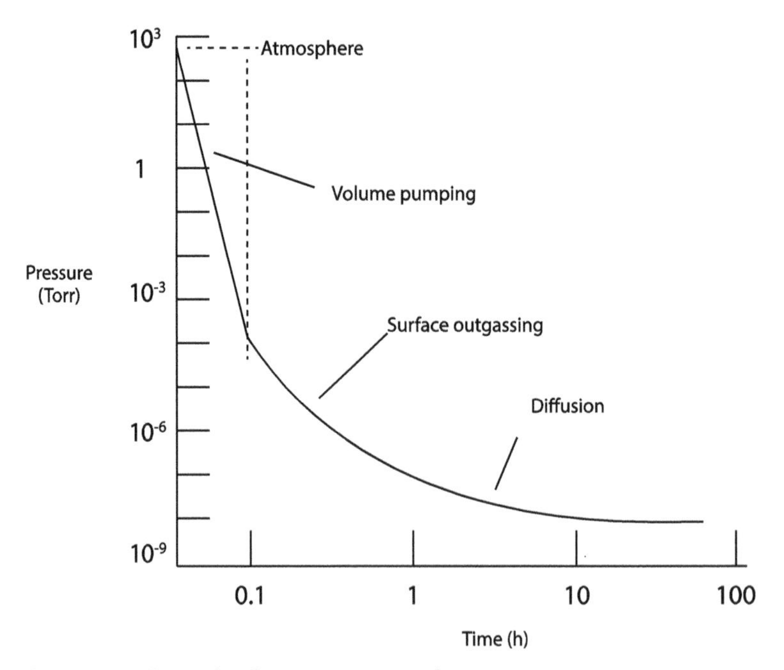

Figure 8.2 Example of a vacuum pump-down curve.

8.3 Pumping Systems for Semiconductor Processing

Vacuum processing plays a key role in semiconductor processing. Steady progress has been made over the decades since the beginning of electrical and electronics industries, but rapid progress has been made more recently due to the explosive growth of microprocessors, flat-panel displays, optodevices, solar cells, and numerous other applications. Vacuum systems used in semiconductor processing need high pumping speed and may be needed to handle a large gas load and aggressive chemicals. Systems for deposition, particularly evaporation, have a large throw distance and a large volume. Bell jar or box coater–type batch systems are used for evaporation. The gas load on the pumping system is high due to the large volume and surface area and also the heat load from substrate heating or the evaporation process itself. To avoid oil contamination, oil diffusion pumps are avoided, so cryogenic (cryo-) pumps are generally used. Oil pumps also seize up if used in plasma etch systems that use corrosive gases. Dry rotary pumps are used, which need tight running clearances and multiple stages to reach needed vacuum levels. Different pumping mechanisms are used, and these can be of the Roots, claw, or screw types. For large sputter deposition systems, Roots blowers may be required. These are high-capacity pumps that reduce oil contamination and cover the pressure range between mechanical pumps and high-vacuum pumps. Mechanical pumps are located far and isolated to some extent by foreline traps for oil in order to avoid back-diffusion of oil. Mechanical pumps that use polyfluoro polyethers (PFPEs), for example, Fomblin®, are preferred because these have very low vapor pressure. Liquid-nitrogen traps or other refrigerated coils may be used to aid in pumping water vapor and reduce pump-down time.

8.3.1 Cryogenic Pumps

These pumps are the preferred choice in integrated circuit (IC) fabrication systems wherever their use is possible. Cryogenic pumps are based on closed-circuit helium refrigeration pumping and work by capturing gases by condensation or cryo-sorption.

Backing or foreline pumps are not needed, so the system can be oil free. Mechanical pumps are used for rough pumping only and in the regeneration of the cryo-pump. The limitations of these pumps are that very-low-boiling-point gases cannot be pumped, such as H, He, and Ne. If these gases are used in the process, a supplemental turbo pump must be provided. These pumps are not recommended for high levels of oxygen or other corrosive gases, particularly in a plasma system where ozone could be generated and trapped, which could be a hazard during the regeneration process.

These systems use three pumping zones: 77 K, 15 K for condensation, and a 15 K cryo-sorption surface. The cryo-sorption surface must be shielded from the direct molecular path. The whole pump must be shielded from heat radiation coming from the vacuum chamber by baffles. Among many advantages, in addition to being oil free, are high pumping speed, ability to take high load momentarily, and recovery from power bumps.

8.3.2 Turbomolecular Pumps

These pumps use high-speed rotating turbine-like blades to transfer momentum to gas molecules and compress them into the foreline where they are pumped away by mechanical pumping. Conductance is governed by blade speed and molecular velocities. Speeds of greater than 24,000 rpm are used. Performance is limited by the maximum speed that the bearings and the rotor can handle. Multiple stages can be used to achieve the required compression ratio and pumping speed. These pumps have the advantage of no oil back-streaming, like cryo-pumps, but have the added advantages of being able to pump hydrogen and helium, although with less efficiency, and not needing regeneration like cryo-pumps. These pumps are more expensive to maintain because of the mechanical complexity. However, recent technological improvement in a bearing-free levitated turbine assembly has made these pumps very reliable. It has become the workhorse for high-vacuum processing in an all-out competition against cryogenic pumps.

Cryo- and turbo pumps achieve speeds and vacuum levels required for semiconductor processing. Only MBE growth systems require pumping systems that achieve lower pressure.

8.4 Pressure Measurement

A wide range of pressure measurements are made in semiconductor processing, from 10^{-12} Torr to 1 atmosphere (760 Torr). Figure 8.3 shows a list of different types used for pressure measurement. Different types of instruments are used for different pressure ranges, so a typical vacuum system may have two to three gauges to cover the pressure range needed for system and process control. Gauges used in process control fall into three ranges:

- Low-vacuum gauges: 10 Pa to 1 atmosphere; capacitance manometers
- Medium-vacuum gauges: 0.1 to 100 Pa; thermocouple and Pirani gauges
- High-vacuum gauges: 10^{-6} to 1 Pa; ionization gauges

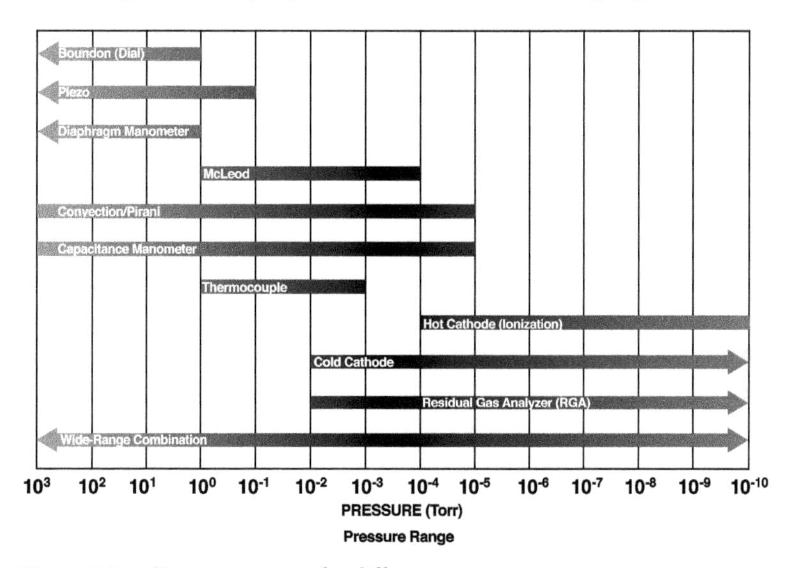

Figure 8.3 Pressure gauges for different pressure ranges.

Capacitance manometers: In these gauges the capacitance between a fixed plate and a diaphragm that moves with pressure is measured. These can have temperature error because the diaphragm can expand as a result of temperature. In the range of 10 Pa to 1 atmosphere the gauges have a linear response.

Ionization gauges: In these gauges a hot filament triode-type vacuum tube is used. Electrons from the grid ionize gas molecules, which are then collected by a negatively charged cathode, the resulting current being related to the pressure. Different types of pressure gauges are in current use for different high-vacuum gauges. These gauges are calibrated for nitrogen gas and do not work well for low-ionizing gases like hydrogen and helium.

8.5 Evaporation

In this process vaporization of the material to be deposited is achieved by direct resistance heating in a boat or heating in a crucible by electron beam or arc discharge. The pressure is generally below 10^{-5} Torr. Usually the pressure is in the 10^{-6} Torr range, where the mean free path is over meters long. The evaporant atoms travel to the substrate by line of sight without any collisions with gas molecules, arriving at the substrate at an angle determined by the system geometry and tooling. Figure 8.4 shows a schematic diagram for an evaporation system used in metallization of semiconductor wafers. In this type of system the thickness uniformity depends upon the flux distribution pattern of the evaporated material. For lift-off-compatible processing, a lift-off dome holds the wafers, evaporant atoms arrive at close to a 90° angle. Substrate planetary or blocking shapers are used to improve thickness uniformity. Uniformity of a few percent may be acceptable for most metal interconnect or contact layers, but better uniformity may be needed for critical layers, for example, base contact layers for heterojunction bipolar transistors (HBTs) or evaporated thin-film resistors. In these cases,the process parameters and tooling may have to be optimized. If true planetary substrate tooling is used, the material will arrive at the substrate at multiple angles and lift-off may not be possible unless a special photo or etch undercut process is utilized.

Evaporation has been the preferred process for metal film deposition in III–V processing due to the fact that gold-based metallization schemes are used in these circuits, and gold etching is not very suitable for patterning on III–V substrates. The electron beam process, in addition to the lift-off compatibility, offers the following advantages: high deposition rates, high material utilization, low or no film damage due to ions, alloy composition and purity control, and excellent uniformity control over wafer sizes used in III–V processing (upto 150 mm). The disadvantages are poor nodule (particle) or microstructure control, poor adhesion for certain materials, poor step coverage, and shadowing.

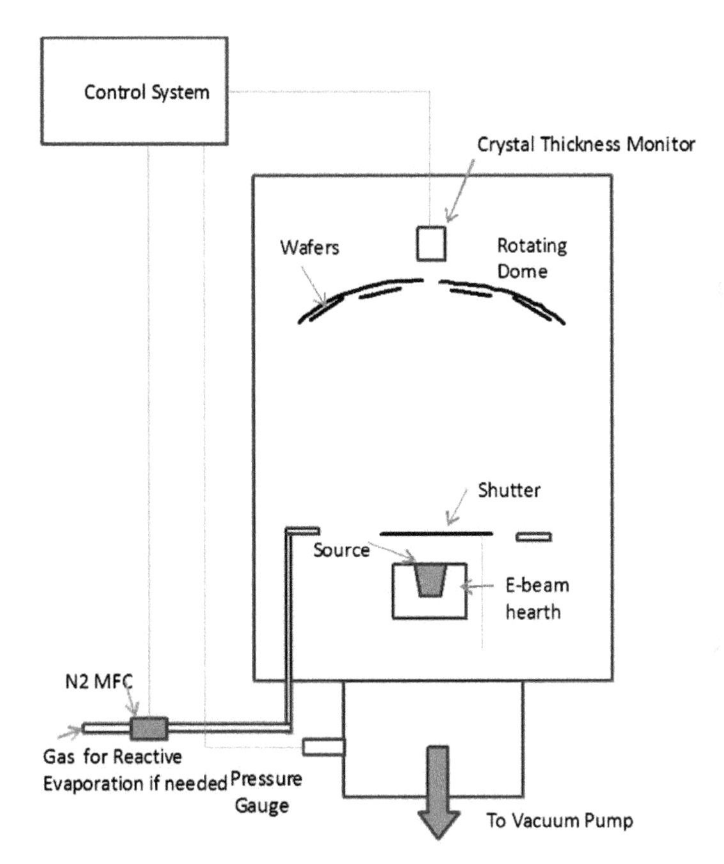

Figure 8.4 Evaporation system schematic diagram.

8.5.1 Evaporation Sources

A number of different source types are used for evaporation of materials. For large volume and throughput, electron beam, induction, or arc sources are used. For semiconductor applications, electron beam and thermal boats are common. The simplest thermal sources are resistance-heated wires made of tungsten or boats made of refractory metals. These can be used for small volumes of material, particularly for alloys where the whole charge needs to be evaporated to maintain correct composition of the film. If the material to be evaporated reacts with common refractory metals, a coating of ceramics like aluminum oxide may be needed. Material compatibility and evaporation data are tabulated in common handbooks and vendor brochures.

Many materials that do not melt, but sublime with sufficient vapor pressure, are used in semiconductor processing. Cr and SiO are good examples of these. In these cases special sources are designed to reduce contact area between the heater and the material. If gases are generated too fast, spitting and nodule generation can occur. The line of sight from the source to the substrate is avoided by the use of baffles or multiple box-type sources [1].

8.5.1.1 Electron beam sources

Thermal sources have a few drawbacks—control is very difficult, and some materials react with the boat material. In electron beam sources, beam power and scan pattern can be controlled over a wide range. The crucible is water-cooled and liners can be used, thus avoiding contamination. A thermionic electron gun is used to generate a beam of electrons, which is accelerated by high voltage applied to the filament with respect to the source, which is held at ground potential. A magnetic field is used to bend the beam, so the filament assembly does not receive any deposit directly. Figure 8.5 shows a simplified diagram for an electron beam gun.

Other mechanisms of heating are used for larger sources and high deposition rates. In induction heating, a radio frequency (RF) coil is used to induce eddy currents in the material itself or in a graphite susceptor (BN or TiB_2) if the material is insulating. In arc evaporation, an arc is struck between an anode and a target at cathode potential. A magnetic field can be use to confine the arc to the desired source area.

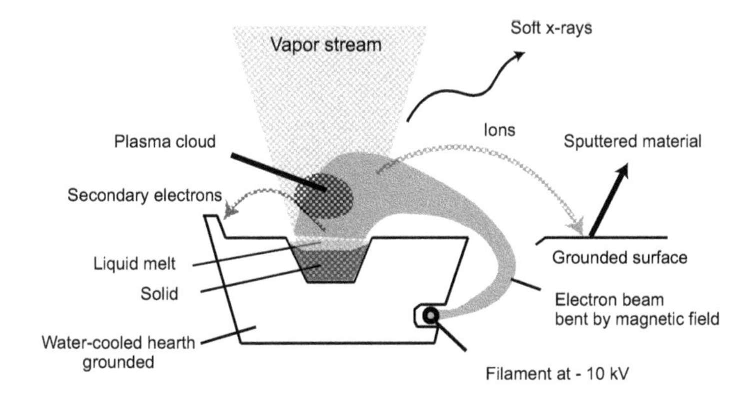

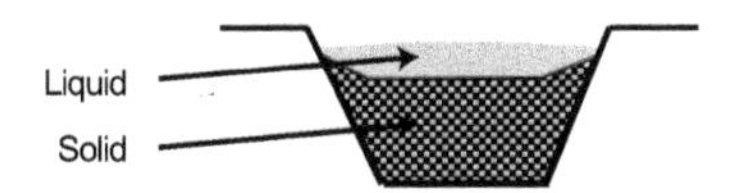

Figure 8.5 Schematic of an electron beam gun.

8.5.2 Deposition Rate [2]

8.5.2.1 Vapor pressure

The amount of heat required to vaporize a material (say a mole) is called the latent heat of evaporation, designated as ΔH_e. The vapor pressure, p, is related to ΔH_e by the Clausius–Clapeyron equation:

$$\frac{dp}{dt} = \frac{\Delta H_e}{T(V_g - V_c)} \tag{8.5}$$

where V_g and V_c are volumes of gaseous and condensed phases, respectively. The vapor pressure in an evaporator behaves like an ideal gas:

$$V_g = \frac{RT}{p} = \frac{N_{Av}kT}{p^*} \tag{8.6}$$

where R is the universal gas constant, k is the Boltzmann constant, and N_{Av} is Avagadro's number.

Integrating Eq. 8.5 and using Eq. 8.6,

$$\ln p^* = C - \frac{\Delta H_e}{RT} \tag{8.7}$$

where C is a constant of integration. This can be also be written as

$$p^* = P_0 \exp\left(-\frac{\Delta H_e}{kT}\right) \tag{8.8}$$

Vapor pressure curves follow this general behavior and are available (see, for example, Ref. [3]); some common ones of interest are shown in Fig. 8.6. Note that the vertical axis is on the log scale.

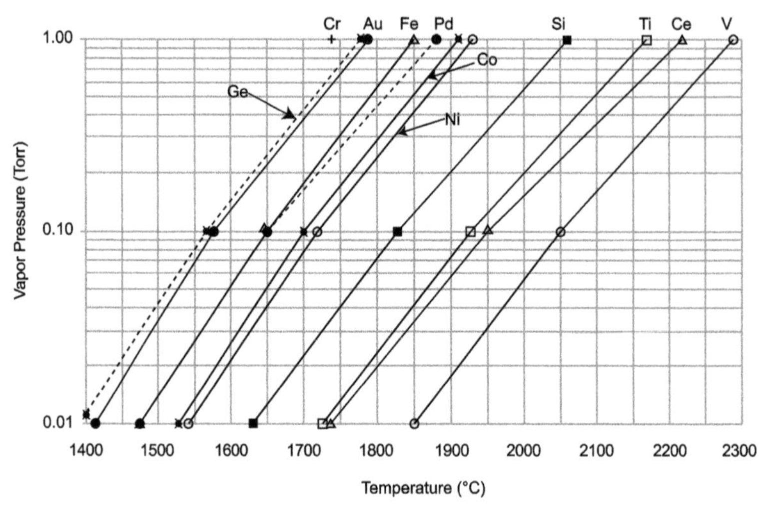

Figure 8.6 Evaporation temperatures for some common materials with moderate vapor pressure. Reprinted from Ref. [3] with permission from Ferrotec.

8.5.2.2 Evaporation rate

The molecular evaporation rate at a specified vapor pressure was observed, first by Hertz, not to exceed the impingement rate [4]. The impingement rate is the strike rate of molecules in a gas at rest. If the gas is ideal, and obeys Maxwell–Boltzmann statistics, the impingement rate can be shown to be

$$\left(\frac{1}{A}\right)\cdot\frac{\mathrm{d}Ni}{\mathrm{d}t} = \frac{p^*}{\sqrt{2\pi mkT}} \tag{8.9}$$

$$\text{The molecular evaporation rate} = \left|\left(\frac{1}{A}\right)\frac{\mathrm{d}Ne}{\mathrm{d}t}\right|_{\max} \tag{8.10}$$

where p^* is the vapor pressure, A is the surface area, and m is the mass of the molecule. The maximum evaporation is not achieved because some molecules return to the condensed phase and others are reflected from the evaporant flux to the gas phase. On the basis of this, the Hertz–Knudsen equation for the evaporation rate R_{ev}, which accounts for these effects, is given by

$$\text{Rev} = \left(\frac{1}{A}\right) = \alpha_v \frac{p^* - p}{\sqrt{2\pi mkT}} \tag{8.11}$$

Here α_v is the fraction that makes it from the liquid to the vapor phase.

The growth rate of the film on the substrate with mass m and density ρ is given by

$$G = \text{Rev}.\frac{m}{\rho} \tag{8.12}$$

Using Eqs. 8.11 and 8.12

$$G = m.\alpha_v \frac{p^* - p}{\rho.\sqrt{2\pi mkT}} \tag{8.13}$$

$$G = \sqrt{\frac{m}{2\pi\, k\, T}} \cdot \alpha_v \frac{(p^* - p)}{\rho} \tag{8.14}$$

The rate G is a strong function of temperature because p^* is also a function of temperature (Eq. 8.8 above).

8.5.2.3 Film thickness variation

Film thickness is a function of the rate equation above and the distance of the substrate and its orientation with respect to the source. Consider a point source that has spherical symmetry and a portion of the substrate as shown in Fig. 8.7. Assuming evaporant molecules have a Maxwell–Boltzmann velocity distribution, the evaporant mass will be equally distributed inside the surface of a sphere. The subtended area dA is given by

$$\mathrm{d}A = r.\,\mathrm{d}\Omega,$$

where Ω is the subtended solid angle. If the normal to the deposition area makes an angle ϕ with the line connecting the source to the deposition area

$$t = \sqrt{\frac{m}{2\pi\, k\, T}} \cdot \alpha_v \frac{(p^* - p)\cos\Phi}{4\pi r^2} \tag{8.15}$$

and $\cos\phi = \frac{D}{\sqrt{D^2 + l^2}}$ (8.16)

From this $\frac{t}{t_0} = \frac{1}{\left(1 + \frac{l}{(D)^2}\right)^{3/2}}$ (8.17)

where $t_0 = \sqrt{\frac{m}{2\pi\, k\, T}} \cdot \alpha_v \frac{(p^* - p)\cos\Phi}{4\pi\rho\, D^2}$ (8.18)

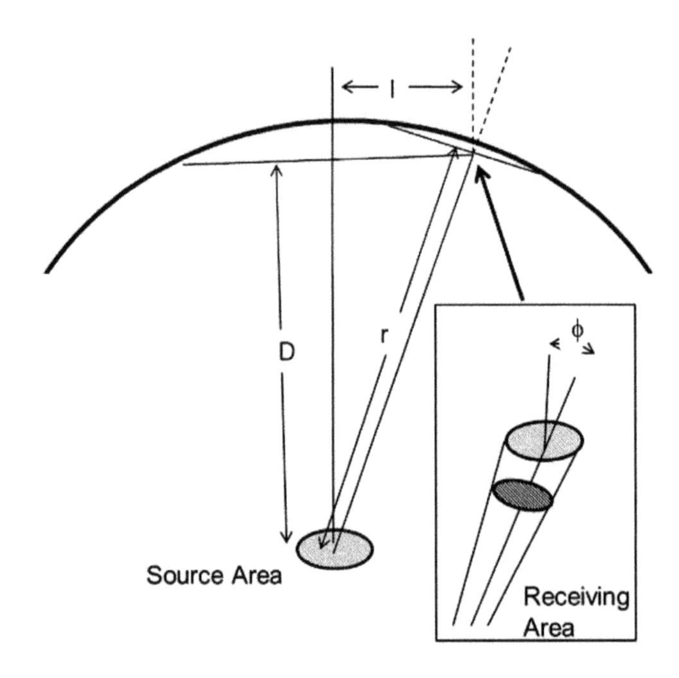

Figure 8.7 Geometry of evaporation from a source to a substrate.

This distribution is plotted in Fig. 8.8 for a point source. In practice, the source is a small surface, which can be considered to be flat (may be convex or concave in practice), but the deviation is not much, as seen in Fig. 8.8.

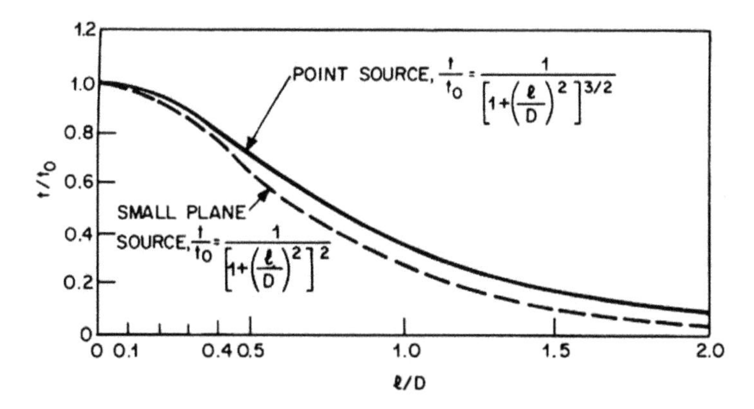

Figure 8.8 Deposited film thickness from an ideal point evaporation source and a small flat surface [4]. Glaser, *Integrated Circuit Engrg: Des Fabrictn Appl*, 1st, ©1977. Reproduced by permission of Pearson Education, Inc., New York, New York.

8.5.3 Deposition Rate Monitors

Vapor pressures of some common materials are given in Fig. 8.6. As the temperature goes up, the vapor pressure goes up exponentially. The deposition rate varies a lot with set evaporation parameters and must therefore be monitored and controlled. Quartz crystal monitors are generally used for thickness monitoring. These are placed in the line of sight in the center of the substrate tooling to collect the evaporated material. As the thickness of the deposit increases, the oscillation frequency of the crystal changes. The frequency change is related to the thickness. The output signal can be calibrated for different materials. Optical monitors are used for optical films, which are based on measuring the reflectance of the growing film or interference effects. These can be used for transparent or semitransparent films. Again, line-of-sight placement of the monitor wafer to the source is required. The detector is aimed at the monitor wafer. All thickness monitoring tools are programmed for automatically controlling the deposition rate and total thickness. For thin-film resistor layers, thickness can be controlled by additional resistance monitoring of the film deposited on the glass slide with contacts already deposited. This technique is useful for alloy films, like nichrome, whose composition may vary slightly and controlling thickness alone may not be effective for process control. The resistance of the growing film is monitored, and the deposition stopped, when the resistance reaches previously determined target value.

8.5.4 Alloy Deposition

There is a need for deposition of metal alloys or composite materials. For example, the AuGe eutectic alloy is used for ohmic contacts, NiCr alloys are used for resistor films, and CiSiO-type cermet materials are used for high-value resistors. Typically vapor pressures of the constituents are different; as a result, the composition of the vapor and the deposited films can be different from the starting alloy material and vary through the thickness as the film grows. Several approaches are used to solve this problem. A conceptually simple one is to use two sources situated as close as possible to the center; however, the composition uniformity across the whole dome, on all wafers, and the total thickness uniformity may be hard to optimize. Nonvertical arrival of the evaporant at the substrate may cause shadowing problems. Another approach is to use flash evaporation, in which the material is evaporated quickly to completion, or from a small boat. For thicker films, material pellets can be dropped continuously on a superheated tungsten strip. This type of procedure is also useful for refractory materials.

8.5.5 Film Growth Mechanism

PVD consists of vaporization of material and its condensation on a substrate where the film grows. The steps may be divided into:

(i) Vaporization of the material from the source
(ii) Transport through vacuum
(iii) Nucleation and growth on the substrate

In evaporation, the kinetic energy of the atoms is very low, only about 1 eV. In sputtering, on the other hand, the atoms are knocked out with higher kinetic energy and the ions get thermalized in the plasma, so the energy of the impinging ions and neutrals can be high. When the growth first starts, the deposited atoms form nuclei, the number and density depending upon wetting or bonding characteristics, the presence of nucleation sites, adatom mobility, etc. Generally it is advantageous to have high nucleation density for strong chemical interaction between the film and the substrate. The interface between the substrate and the film may be abrupt or diffused or a compound may form. An example of compound formation is titanium on an oxide surface, which forms a range of oxides, depending upon how much oxygen is available. Gold on the other hand does not react with most surfaces that have oxides or contamination. It does not bond to the surface and de-wets to form islands. Titanium is therefore used to form the "glue" or adhesion layer. When gold is deposited on top of the titanium, without exposure to atmosphere, Ti and Au alloy together, resulting in good adhesion.The interface or the nucleation region is also important for texture, film stress, and electrical properties. All these can be altered or improved by increasing the reaction probability of the adatoms. Some examples of these are in situ cleaning by plasmas, sputter deposition of the seed/barrier layer, substrate heat, ion bombardment, etc.

The morphology of a growing thin film depends upon substrate smoothness, nature of the nucleated layer, and surface mobility of the atoms being added. Thin films deposited by PVD processes are in general polycrystalline or amorphous. The texture is determined by nature of the substrate, rate of deposition, temperature of the substrate in comparison with the melting point of the material, pressure, angle of arrival, etc. The film texture depends on the nucleation process discussed above. Three-dimensional nuclei are formed on the surface first and these grow laterally as the film thickness increases. For some material/substrate combinations, a continuous film may not form until the thickness reaches over 100 Å. Once a continuous film is formed, the texture depends upon the factors mentioned above. As the film grows, the surface roughness on an atomic scale starts giving rise to high and low spots. The high spots receive more atoms and start to shadow the valleys. This mechanism continues and gives rise to a columnar structure. The size of these columns may be of the order of a micron. The column shapes are dominated by geometrical effects, but may depend upon the nature of the material, for example, crystalline or amorphous, and the growth conditions. As the substrate temperature is raised, or energy is added to the system (e.g., through ion bombardment), the surface or the lateral mobility of the adatoms increases and the valleys can get filled. This type of structural model was first proposed by Movchan and Demchishin [1, 4, 5a] for evaporated pure metals and subsequently modified by others for other materials and deposition types, like sputtering. An example of the Movchan–Demchishin (M–D) structure zone model is shown in Fig. 8.9 for evaporation. If deliberate ion bombardment is added, that would be equivalent to higher temperature. By controlling the texture, film stress, density, etch rates, and electrical properties can be controlled.

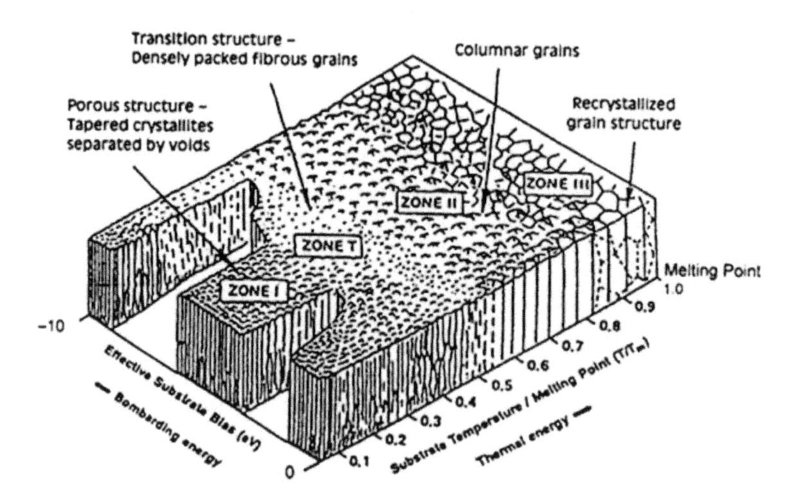

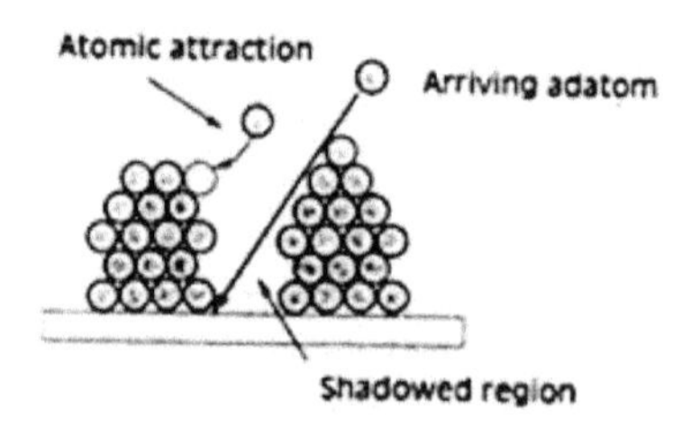

Figure 8.9 The M–D structure zone model for as a function of the ratio of substrate to melting temperature and the bombarding energy or substrate bias. Reprinted with permission from Ref. [5b]. Copyright © 1974, AIP Publishing LLC.

8.6 Sputter Deposition [6]

Sputter deposition refers to use of a sputtering process to knock out or eject particles from a target source and their deposition on a substrate. The process takes place in a plasma created in a vacuum system, discussed in more detail in Chapter 7. Most common depositions take place in an argon plasma, but for reactive deposition, N_2 or other gases may be used. A simple sputter deposition system consists of parallel-plate geometry, as shown in Fig. 8.10, where the target and the substrate holder are separated by a few centimeters.

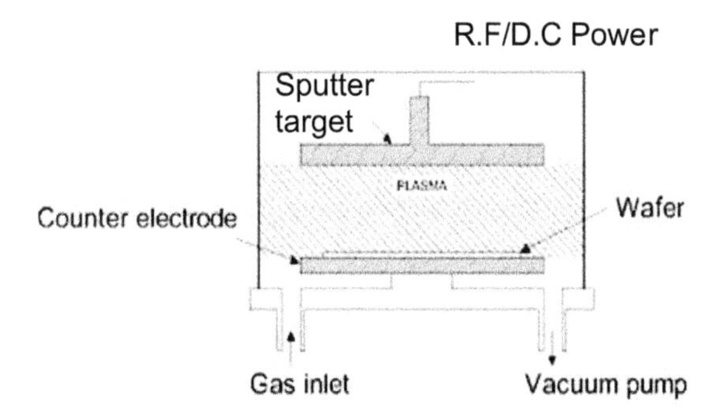

Figure 8.10 Schematic diagram of a planar diode sputter system.

8.6.1 Advantages of Sputter Deposition

Sputter deposition is a more universal process and does not have limitations of evaporation. It has numerous advantages over other deposition processes. It is a physical momentum transfer process and does not have chemical or thermal issues. Simple direct current (DC) voltage can be used for conducting materials, and RF is needed for insulating materials. Co-sputtering from multiple targets can be used for alloys. For oxides, nitrides, and other compounds, reactive sputtering can be used, the reactive gas being added to the argon plasma. For complex compounds, mosaic targets can be made for achieving any composition. Multilayer deposits, for example, TiW/Au, can be deposited by sequential deposition in the same chamber or in multiple chambers in wafer-at-a-time systems. Wafers can be cleaned in situ using an argon plasma and reversing the polarity to make the substrate holder the cathode. Deposition rates are very predictable, so thickness control is easy. Material utilization is also better than evaporation. Film density and stress can be controlled better by controlling ion bombardment or adding back-sputtering, again by adding a bias to the substrate holder. Film adhesion is excellent, as explained later.

The major dis-advantage of sputtering is that the substrates can get hot due to bombardment and the process is not lift-off-compatible due to the excellent step coverage. Substrates are immersed in a plasma and the operating voltages are high, which could be an issue for films/devices in which ion damage needs to be avoided.

8.6.2 Deposition System Types

Sputter deposition takes place in a pressure of a few milli-Torr so that the pumping system does not have to be capable of ultrahigh vacuum. Vacuum systems capable of achieving 10^{-6} Torr are used. The systems are throttled and a sputtering gas supplied through a mass flow controller (MFC) to control the deposition pressure in the milli-Torr range. Basic parallel-plate, cylindrical, and box-type systems are used in semiconductor manufacturing. Most modern systems have load locks, so targets are not exposed to the atmosphere. Automatic wafer handling is added, particularly for single-wafer systems, where manual handling would be impractical.

8.6.2.1 Planar diode

This is the simplest system, shown in Fig. 8.10. Cathodes can be round or rectangular, operated with DC or RF power. The target-to-anode spacing is such that a plasma can be sustained (discussed in detail in Chapter 7). The target is surrounded by a ground shield, so the plasma is contained, and erosion takes place only from the front surface of the target.

8.6.2.2 Triode

These systems utilize an additional electrode to sustain the plasma discharge, for example, a hot cathode to increase the ionization level and deposition rate. These systems are not easily scalable like magnetron systems discussed below.

8.6.2.3 Magnetron sputtering

There are many configurations of magnetrons, planar, S-Gun type or cylindrical [1]. The magnetic field is applied in such a way that it bends the electron trajectory into circular paths, thus increasing the ionization probability and raising the plasma intensity.

Figure 8.11 shows the system configuration, with the magnets behind the target and the field generated by these. An intense plasma is confined to the region near the target in the area that has the highest magnetic field, typically a closed track on the target. The impedance of the plasma drops, so the magnetron runs at a lower voltage. This leads to high power density, higher deposition rate, low substrate bombardment, and better uniformity control. One disadvantage of magnetrons is uneven erosion of the target, which leads to poor target material utilization. However, new schemes for improving target utilization have been devised. These include inset targets, with removable inserts in high-erosion areas to improve utilization and moving magnets to improve erosion. These systems operate at lower pressure, as low as 1 mTorr.

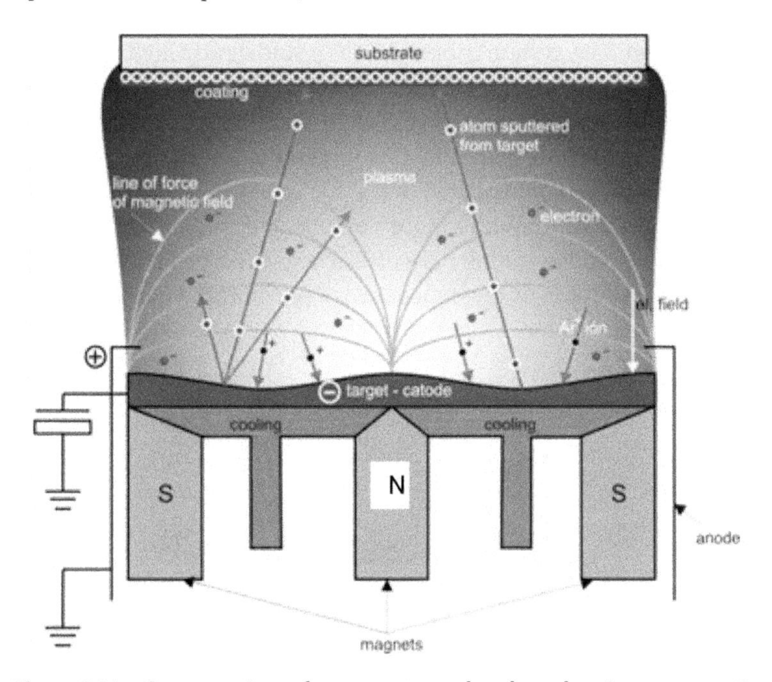

Figure 8.11 Cross section of a magnetron chamber showing a magnetic field.

Gun-type magnetrons are self-contained magnetrons and can be used in arrays to cover large-area substrates. Multiple guns can be aimed at a substrate to deposit alloys.

8.6.3 RF Sputtering

DC magnetron sputtering cannot be used to deposit insulating materials because a charge accumulates on the target surface and impedes current flow. If the DC power supply is replaced by an RF power supply, all kinds of insulating, semiconducting, and conducting materials can be sputtered. Deposition of silicon nitride, silicon dioxide, and numerous other materials is carried out in reactive RF sputter systems. In an RF sputter system the target is capacitively coupled to the glow discharge and a negative self-bias voltage develops on its surface with respect to the plasma. The reason for this is the difference in mobility between electrons and ions. Figure 8.12 shows a schematic diagram of a typical RF sputter system. The RF power supply is connected to the target through a tuning network, and the substrates are placed on a large-area electrode that is grounded. The system is operated at a sufficiently high frequency (13.56 MHz) so that ion charge accumulation does not occur. The substrate electrode is made large, so the sheath capacitance is large. Details of voltage distribution as a function of electrode geometry, etc., are discussed in Chapter 7.

RF sputtering requires more complicated equipment than DC sputtering because tuning networks are needed to match the load to the RF generator, as the impedance of the material plasma system changes.

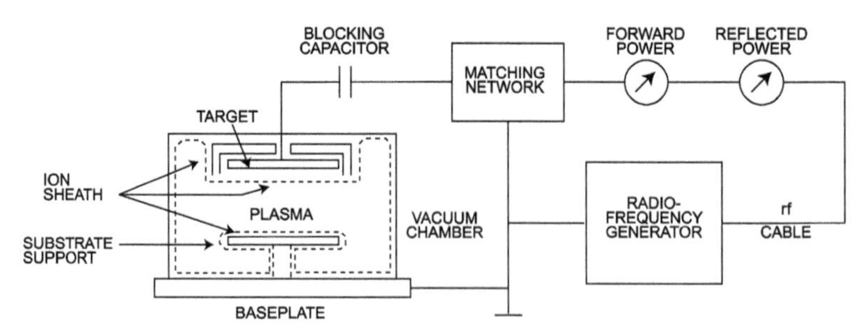

Figure 8.12 Simplified diagram of an RF sputter system.

8.6.4 Reactive Sputtering

Reactive-sputtering processes are used to deposit compound material films by sputtering an elemental material in a reactive gas plasma. Common examples are TaNx, indium tin oxide (ITO), AlN, and Si_3N_4. The target is a pure metal, an alloy, or a compound with one high-vapor-pressure component that can be supplanted by additional gas. Compound targets are difficult to make and their sputter yield can be low, so using a metal makes target preparation easy and deposition rates can be high. A common configuration for reactive sputtering is a single-target planar magnetron system in which the reactive gas can be evenly distributed by using a gas distribution ring or manifold. The control and distribution of gas can still be problematic, particularly in larger drum-type systems. If the target material can be "poisoned" by formation of an insulating surface layer, development of a process capable of a fast rate and proper electrical properties can be challenging. The partial pressure of reactive gas can be monitored by residual gas analysis (RGA).

In reactive sputtering, the deposited layer must react with the gaseous component rapidly for good stoichiometry control. The reaction rate depends upon availability of reactant, which depends upon pressure, flow rate, temperature, and ion bombardment that can increase the surface mobility of reacting ions. The concentration of the reacting gas ion, for example, N^+, can be increased by increasing the plasma intensity, as in magnetron sputtering, or by providing extra ion sources. For insulating material deposition, for example, SiN, anode surfaces can get coated with insulating layers and cause plasma instability, so some part of the anode must be shielded from deposition.

8.6.5 Bias Sputtering [6]

For better step coverage and denser films, loosely bound material and protrusions must be removed or back-sputtered. This can be done by applying a bias to the substrate, making it to act partially as a cathode, so that a fraction of the deposited material is etched back.

In a DC system, an insulating substrate will charge up to floating potential. So, only when depositing conducting films can bias be applied. By using RF, insulting materials can be deposited by this process. Using this process, bias-sputtered quartz (BSQ) has been deposited with a low dielectric constant and loss, a good indicator of dense film quality [7].

8.6.5.1 System selection

Sputter deposition systems are selected for production based on the application. The type of film, deposition rate that determines throughput, wafer size, and batch or single wafer are the main considerations. For metal layers a high deposition rate and low particle generation may be key factors. If the metal layer has to be patterned by lift-off, like gold-based metallization in III–V devices or TaN thin-film resistors, some amount of collimation or longer-than-normal throw distance may be needed. Otherwise, other process enhancements like dielectric-assisted lift-off may be necessary. If coverage inside deeper vias is needed, systems that are capable of ion plating may be necessary. In such systems additional bias is applied to the substrate to increase the ion flow into the vias. Good step coverage and coverage in holes and deeper vias is possible with ion plating. Figure 8.13 shows coverage at the bottom of trenches normally not possible without bias or ion assist.

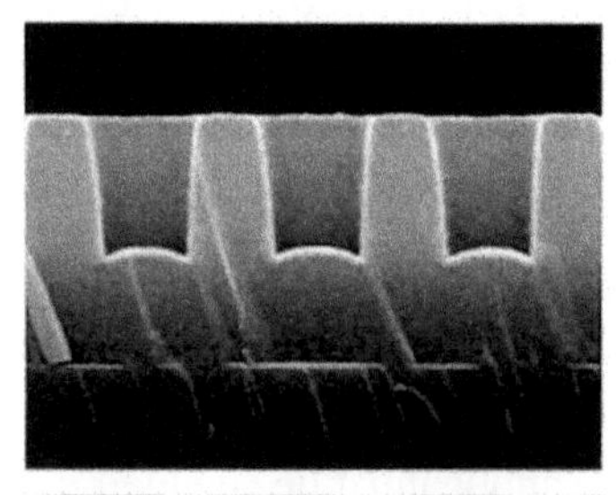

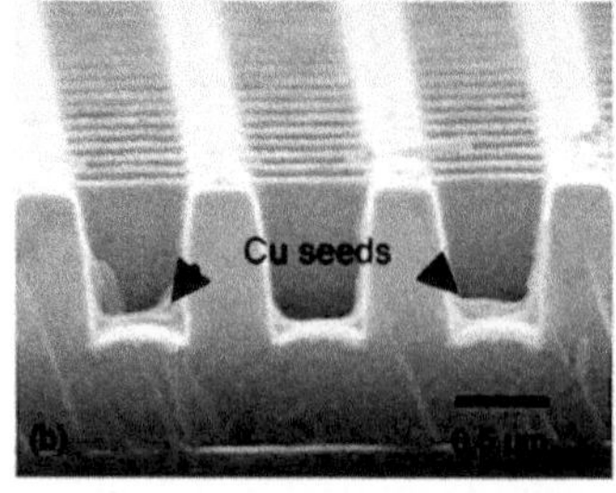

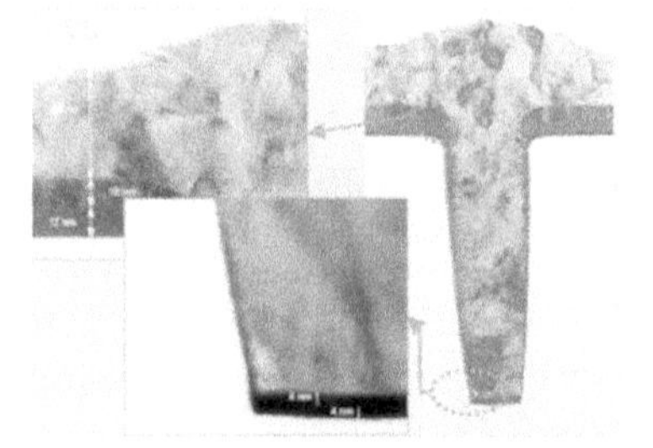

Figure 8.13 Trench bottom filling by ion beam–assisted deposition showing good coverage of corners at the bottom [8]. (Right) Reactive ion deposition showing excellent metal coverage in a 4:1 aspect ratio trench (using the Tango Ion Plating Deposition system from Axcela). Reprinted from Ref. [8], Copyright (2006), with permission from Elsevier.

If the metal is used for Schottky gate deposition, damage to the field-effect transistor (FET) channel layer must be avoided by using a low-bias deposition process.

Sputter processes are relatively easy to optimize as a number of process parameters are available for process optimization. Common parameters are pressure, power density, target-to-substrate spacing, target voltage, substrate temperature, and plasma bombardment conditions, which may need change of magnetic field strength for magnetrons. Excellent run-to-run and wafer uniformity can be achieved over 150 mm size in sputter systems.

8.6.6 Mechanism and Rates

In a plasma, energetic ions and neutrals collide with the target and eject atoms from it by momentum transfer. Most of the energy is lost as heat, but enough atoms are ejected to make sputter deposition practical for most materials. The sputter yield depends on the target species and the sputtering gas. The yield is high when the mass of the bombarding particle is close to that of the target atom. Argon is the most common inert sputter gas, and sputtering yields have been determined experimentally. Table 8.2 lists sputter yields for some common materials under argon ion bombardment. Variations of the yields as a function of the ion energy are shown in Fig. 8.14 [11]. Yields for most materials are near unity at ion energies of a few hundred electron volts, values typical of deposition systems. Nobel metals tend to have higher yields. Compared to evaporation yields are similar for materials that may have vapor pressures different by orders of magnitude. For alloy deposition,the flux of the different species varies only a little compared to evaporation. For an alloy target, the surface concentration will change until equilibrium is reached, and the composition of the film matches the target.

The atoms ejected from the surface are ejected in random directions, the distribution being nearly a cosine function (see Fig. 8.15). The mean free path of the atoms is also short at the pressures used. Thus the atoms arrive at the substrate at different angles. This results in good step coverage. The energy of the arriving species is of the order of 10 eV, more than an order of magnitude larger than that for evaporation. This results in much better adhesion.

Table 8.2 Sputtering yields of various materials under Ar ion bombardment

	Ion Energy (eV)						
Target	**200**	**600**	**1000**	**2000**	**5000**	**10,000**	**Heat of sublimation**
	Sputtering yields (atoms/ion)						**eV/atom**
Ag	1.6	3.4	–	–	–	8.8	2.94
Al	0.35	1.2	–	–	2.0	–	3.33
Au	1.1	2.8	3.6	5.6	7.9	–	3.92
C	0.05*	0.2*	–	–	–	–	7.39
Co	0.6	1.4	–	–	–	–	4.40
Cr	0.7	1.3	–	–	–	–	4.11
Cu	1.1	2.3	3.2	4.3	5.5	6.6	3.50
Fe	0.5	1.3	1.4	2.0**	2.5**	–	4.13
Ge	0.5	1.2	1.5	2.0	3.0	–	3.98
Mo	0.4	0.9	1.1	–	1.5	2.2	6.88
Nb	0.25	0.65	–	–	–	–	–
Ni	0.7	1.5	2.1	–	–	–	4.45
Os	0.4	0.95	–	–	–	–	8.19
Pd	1	2.4	–	–	–	–	3.90
Pt	0.6	1.6	–	–	–	–	5.95
Re	0.4	0.9	–	–	–	–	8.06
Rh	0.55	1.5	–	–	–	–	5.76
Si	0.2	0.5	0.6	0.9	1.4	–	4.68
Ta	0.3	0.6	–	–	1.05	–	8.10
Th	0.3	0.7	–	–	–	–	5.97
Ti	0.2	0.6	–	1.1	1.7	2.1	4.86
U	0.35	1.00	–	–	–	–	5.00
W	0.30	0.60	–	–	1.1	–	8.80
Zr	0.3	0.75	–	–	–	–	6.34
Compound	**Sputtering yields (molecules/ion)**						
CdS(1010)	0.5	1.2	–	–	–	–	
GaSa(110)	0.4	0.9	–	–	–	–	
GaP(111)	0.4	1.0	–	–	–	–	
GaSb(111)	0.4	0.9	1.2	–	–	–	
InSb(110)	0.25	0.55	–	–	–	–	
PbTe(110)	0.6	1.4	–	–	–	–	
SiC(0001)	–	0.45	–	–	–	–	
SiO_2	–	–	0.13	0.4	–	–	
Al_2O_3	–	–	0.04	0.11	–	–	

**Type 304 stainless steel

Source: Reprinted from Ref. [6] with permission from Lattice Press.

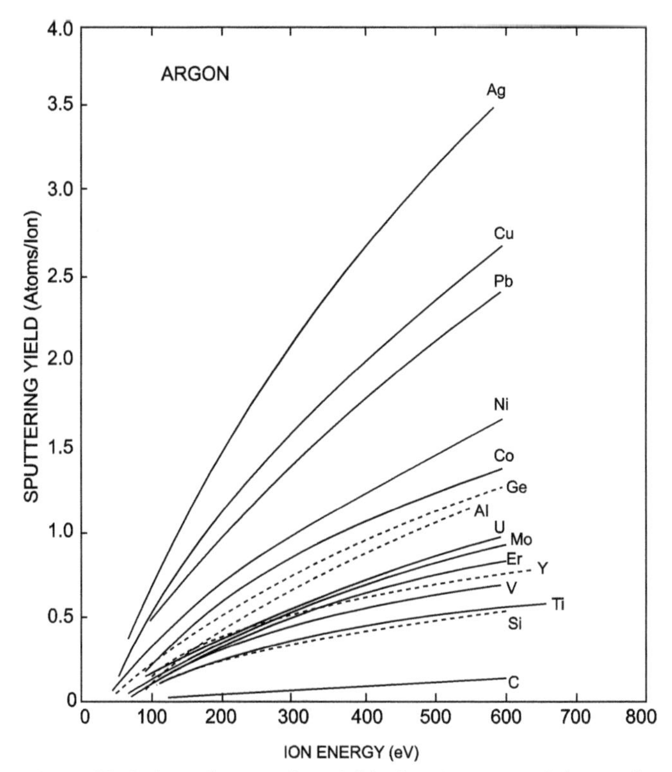

Figure 8.14 Variation of sputtering yield of various materials as a function of Ar^+ ion energy at a normal angle of incidence. Reprinted with permission from Ref. [11]. Copyright © 2004, AIP Publishing LLC.

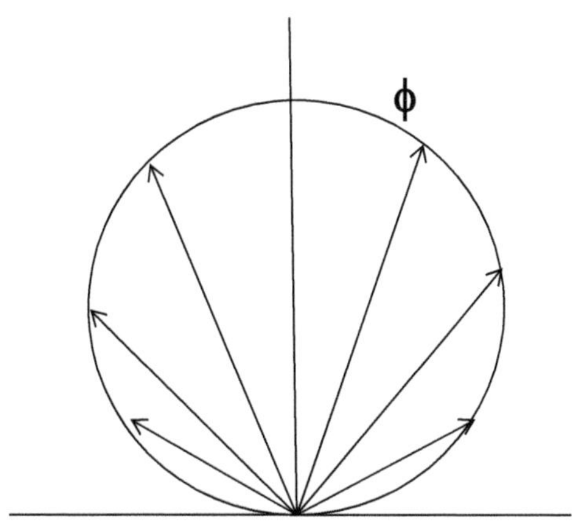

Figure 8.15 Ideal cosine distribution of atoms ejected from the surface of a sputter target.

8.7 Plasma-Enhanced Chemical Vapor Deposition [9]

8.7.1 Film Requirements

Insulating films are needed for passivating devices or protecting the semiconductor surface during processing and also for interlevel isolation of conducting layers. In silicon IC technology, these layers are used for masking during diffusion. In III–V processing, these films are needed for annealing caps. General requirements for these films are:

- No pinholes
- High density, impervious to diffusion, particularly of arsenic, alkali ions, etc.
- Low or controlled stress
- Good dielectric properties, low loss
- Etchable by plasma etching, so fine geometries or vias can be opened
- Stability for good reliability
- Moisture resistance

Thicknesses in the range of a few hundred angstroms to a few microns are used. Deposition rates of 100 to 1000 Å/min are needed to cover the range of dielectric films used in III–V processing. Silicon nitride meets all these requirements and is the material of choice for most III–V processing. Sputtering and PECVD processes are used for its deposition. CVD-type processes are not suitable for III–V processing because of the need for high temperature beyond the dissociation temperature.

Barrel-type plasma reactors have been used for deposition of dielectric films, oxides, and nitrides. A large batch of wafers loaded in a quartz boat are run in these systems. Wafer-to-wafer and within-wafer uniformity of such systems are poor. Parallel-plate planar reactors can also run a fairly large batch with better uniformity. These systems have become popular in III–V processing. System cleaning is needed for maintaining a reproducible deposition rate and cleanliness for low particles.

Both high- and low-dielectric-constant materials are needed in fabrication of ICs. Low-dielectric-constant materials for interlayer insulation and high-dielectric-constant materials (high *K*) for metal–

insulator–metal (MIM)- or gate metal–oxide–semiconductor (MOS)-type layers. For MIM capacitors various high-dielectric-constant materials have been investigated. Ta_2O_5, which is used for making thin-film capacitors, has been a good candidate with a *K* value three times that of silicon nitride; however, developing a deposition process that can actually deliver an MIM in practice with three times the capacitance has not been easy. Films with thickness below 1000 Å are too leaky. Dielectric losses and poor breakdown behavior have prevented introduction of Ta_2O_5. Other materials like TiO_2 and PZT (lead zirconate titanate), which have higher dielectric constants, have even higher losses and have higher challenges of process control. Silicon nitride has, therefore, become the dielectric of choice for GaAs and III–V processing. It provides a good compromise for use as interlevel isolation and MIM dielectric layer, and a single deposition system can be used for both these and also the overcoat for the moisture barrier. Layers thinner than 1 μm can make IC die highly accelerated stress test (HAST) compatible.

8.7.2 CVD Systems

Some basic concepts of CVD were discussed in Chapter 4. In this process, reactants are introduced into a reaction chamber and a solid film is deposited by the reaction of gaseous reactants on a substrate placed in the reaction chamber. The reaction takes place on the substrate and the by-products are removed from the chamber. CVD systems have the following components: gas delivery systems with manifolds and MFCs, a reaction chamber with substrate heating, a vacuum and pressure control system, RF generators, and matching networks for PECVD systems.

8.7.2.1 CVD reactor types

CVD reactors can be categorized on the basis of operating pressure, hot or cold wall and excitation type, plasma, laser, UV, etc. CVD reactor types are shown in Fig. 8.16. Hot-wall reactors of the tube type are used in the semiconductor industry, but these had major limitations for III–V compound semiconductor use, among these being thickness nonuniformity, discussed earlier, wafer handling, high temperature, and long cycle time. Since, III–V fabs are smaller and do not have uninterrupted power, scrap due to power outages is a big issue for such systems.

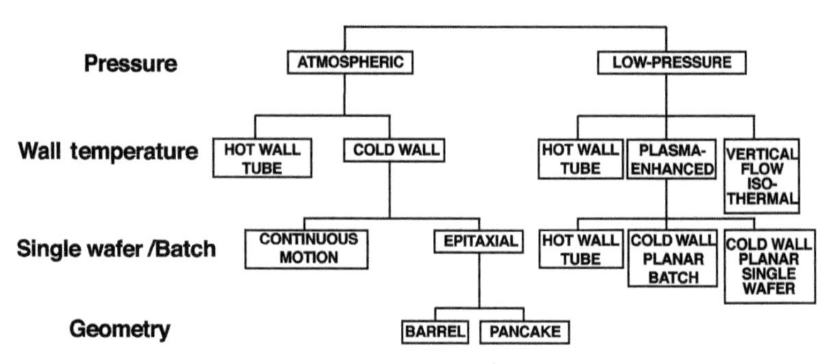

Figure 8.16 CVD reactor types.

Atmospheric pressure reactors operate in the mass transport–limited regime, so the reactor design must be such that all wafers see the same flux over the whole wafer. It is best to lay the wafers horizontally, but this reduces throughput. Low-pressure chemical vapor deposition (LPCVD) reactors overcome this limitation, so wafers can be stacked vertically in boats, but uniformity is still an issue. Typical pressure used in these systems is in the 0.2 to 2 Torr range, and typical temperatures are in the 500°C–600°C range. Heating methods are resistance heating in hot-wall reactors or substrate platen heating by RF induction. Characteristics, advantages, and disadvantages of CVD reactors are listed in Table 8.3.

Table 8.3 Characteristics and applications of CVD reactors

Process	Advantages	Disadvantages	Applications
APCVD (low temperature)	Simple reactor, fast depositions, low temperature	Poor step coverage, particle contamination	Low-temperature oxides, both doped and undoped
LPCVD	Excellent purity and uniformity, conformal step coverage, large wafer capacity	High temperature, low deposition rate	High-temperature oxides, both doped and undoped, silicon, nitride, poly-Si, W, WSi_2
PECVD	Low temperature, fast deposition, good step coverage	Chemical (e.g., H_2) and particulate contamination	Low-temperature insulators over metals, passivation (nitride)

Source: Reprinted from Ref. [6] with permission from Lattice Press.

8.7.3 Plasma-Enhanced CVD [9, 10]

The deposition temperature of CVD systems, even the low-pressure ones, is too high for III–V processing. To reduce the deposition temperature, glow discharge is used to provide extra energy for the creation of reactive species and promote the reaction. Because of this advantage of PECVD, deposition of silicon nitride and oxide on GaAs circuits, deposition systems, and processes has been developed for achieving all the required film properties, good adhesion, low pinhole density, good step coverage, excellent dielectric properties, and moisture resistance. The two basic configurations, tube and parallel plate, are used for PECVD. The parallel-plate configuration is most common for batch systems and is also used for large-area wafers in single-wafer-at-a-time systems. Details of plasma generation, etc., are discussed in Chapter 7. One difference between sputter deposition and PECVD systems is in the design of the tuning network for the RF power. In PECVD, the powered electrode should not get sputtered away and must be at about the same potential as the substrate holder. Therefore a high-frequency shunt is used in the tuning network. A schematic diagram of the radial parallel-plate reactor developed by Reinberg [13] is shown in Fig. 8.17 [10]. RF power at 13.56 MHz is applied to the upper electrode and the wafers are loaded on the bottom electrode, which may be rotated during deposition. For higher ion bombardment, a lower frequency of 450 kHz may be used. The spacing between the electrodes is 5 to 10 cm. Uniform deposition can be achieved by optimizing plasma power, gas flow, and pressure (typically 0.1 to a few Torr).

As the demand for higher volumes and better MIM uniformity grew, the disadvantages of the basic Reinberg reactor became a limitation. The system throughput is limited by thickness and uniformity control because the deposition rate varies with the buildup of insulating material on the walls and substrate platen.

8.7.4 Production Multistation System

These systems are designed to overcome most of the disadvantages of the basic Reinberg reactor. These are cold-wall reactors, with deposition taking place in multiple sublayers deposited from showerheads placed in the top electrode that are powered with high frequency (13.56 MHz). The substrates are placed on the bottom heated electrode platen that is powered by a low-frequency generator at 450 kHz. See the schematic diagram of a Novellus system in Fig. 8.18. Wafers are moved under the deposition showerheads in discrete steps. Thus the deposited film has six sublayers in a seven-station system. Electrode spacing is small, a few centimeters. The power to the two generators can be controlled independently to control the ion bombardment, which, in turn, controls film properties like density and stress. An example of film stress control is shown in Fig. 8.19. These films have inherently lower pinhole density due to the multiple layering, but the lateral wet-etch rate is higher because of the interface between the sublayers. The problem of insulator buildup is solved by having an in situ cleanup after every run. The hydrogen content of these films depends upon the substrate temperature. Since the deposition temperature in GaAs processing is close to 300°C, hydrogen incorporation is still an issue. Higher-density films with compressive stress (and N_2 incorporation) can also be achieved by He dilution of the N_2 gas in PECVD [14].

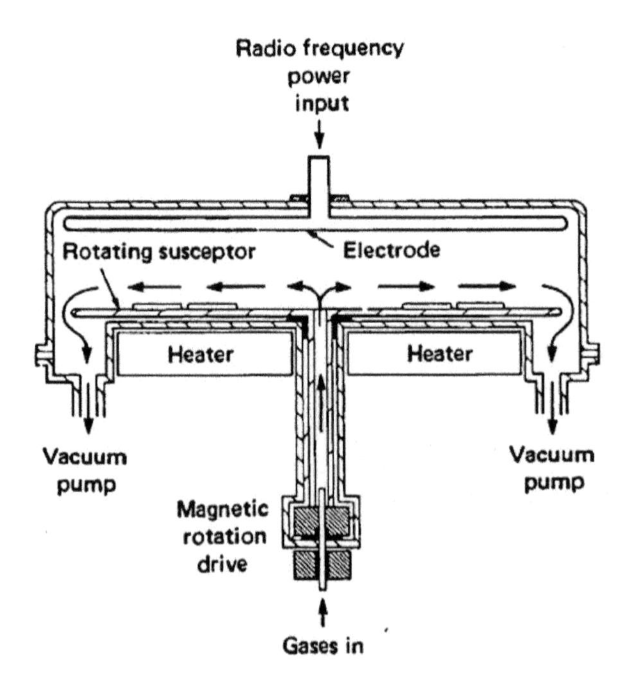

Figure 8.17 Plasma-enhanced CVD system (Reinberg geometry). Reprinted from Ref. [13], Copyright (1981), with permission from Elsevier.

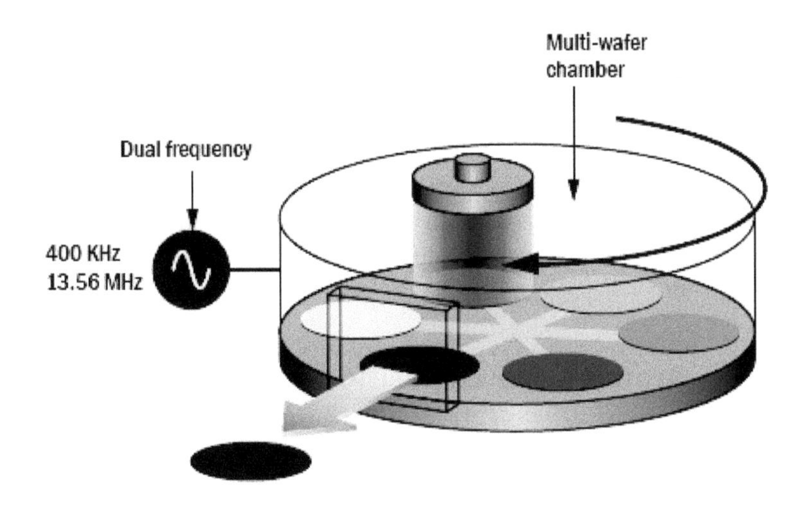

Figure 8.18 Simplified diagram of a Lam Research PECVD deposition system. Courtesy of Lam Research Corporation.

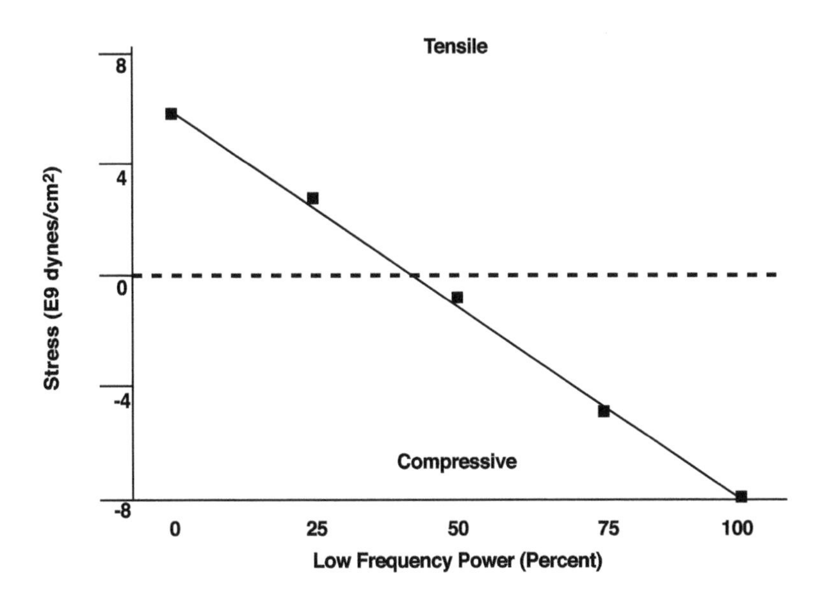

Figure 8.19 Film stress as a function of low-frequency power percentage in a Novellus system.

8.7.4.1 High-density plasma systems

Higher-density plasma systems, like inductively coupled plasma (ICP), have been developed to reduce hydrogen content and increase film density [15]. Better breakdown behavior has been verified. These processes must be used with caution for layers that are used for device-level passivation, for example, an HBT ledge or pseudomorphic high-electron-mobility transistor (PHEMT) postgate passivation.

8.8 Atomic Layer Deposition

8.8.1 ALD Principles

Atomic layer deposition (ALD) has attracted widespread attention because of high K dielectrics needed for the next-generation silicon complementary metal–oxide semiconductor (CMOS). (The dielectric constant for Al_2O_3 is 11 compared to 3.9 for SiO_2). For III–V compounds, ALD-deposited dielectrics could be used for high-capacitance-density MIM capacitors, as well as gate insulators for metal–oxide–semiconductor field-effect transistor (MOSFET) devices. The self-cleaning aspect of III–V semiconductor surfaces by ALD, before growth of the insulator begins, provides an additional impetus for intense research for an ALD-deposited insulator [16]. High-performance MOSFET devices using ALD have been demonstrated on InGAs channels with an interface trap density as low as 10^{11}–10^{12} /cm^2·eV and drain currents over 500 mA/mm [17].

In ALD, a binary compound AB is grown on a substrate by depositing one individual layer, say, A, followed by an individual layer of B. Figure 8.20 shows the growth on one monolayer. This sequence is repeated many times, depending upon the thickness required. The growth is linear with the number of AB cycles. The advantage is near-perfect growth and excellent thickness control due to the self-limiting nature of each layer as it grows due to the A-B reaction. The deposition temperature is high enough, that multiple layers of individual elements, for example, AA, are not stable; only the compound AB can form. Precursors used for the components do not stick to themselves but de-sorb to find an open spot, thus producing a very conformal and continuous film.

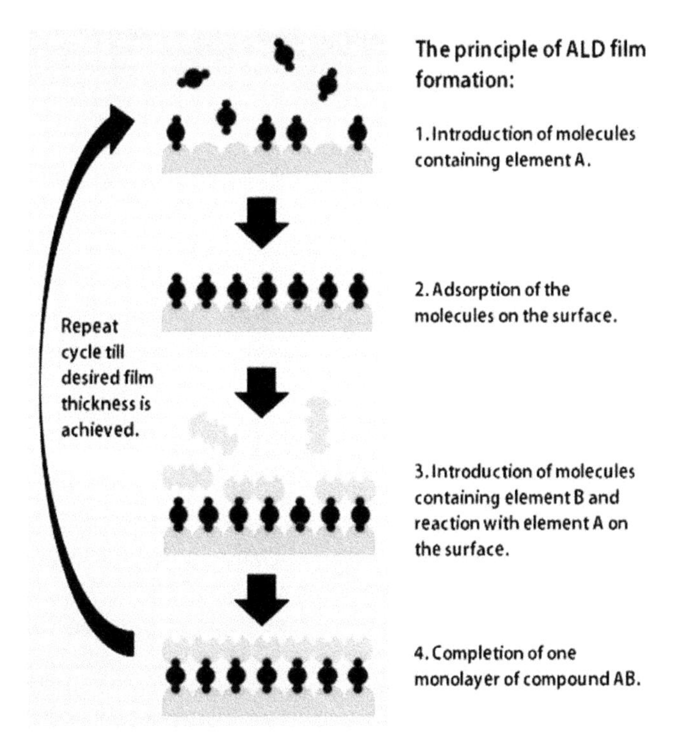

Figure 8.20 Schematic diagram showing formation of monolayers in ALD. Figure © Picosun Oy.

ALD started as atomic layer epitaxy (ALE) of ZnS for electroluminescent displays, with Al_2O_3 as the insulator also deposited by ALE [18]. For its history and a more detailed description, the reader is referred to a review by George [19]. The advantages of ALD or ALE are not easily applicable to III–V compound growth because Ga, Al, and In tend to accumulate on the surface. Also, the growth rates of ALE are low due to the fact that the reactant species have to be completely purged out between successive layers.

The most common dielectric of interest, Al_2O_3, is briefly discussed here [20]. Deposition is performed by using trimethyl aluminum (TMA) and H_2O as the precursors, with ozone added more recently. The substrate is held at 300°C in a vacuum chamber, and TMA is admitted to the chamber first. $AlCH_3$ from TMA sticks to the surface. H_2O is provided next, which reacts with the $AlCH_3$ to form AlOH.

$$AlCH_3^* + H_2O \rightarrow AlOH + CH_4$$

$$AlOH^* + Al(CH_3)_3 \rightarrow AlO\ Al(CH_3)_2^* + 3CH_4$$

The overall reaction is

$$2Al(CH_3)_3 + 3H_2O \rightarrow Al_2O_3 + 3CH_4$$

Complexes are formed from the precursors that eventually react on the surface to form Al_2O_3 and liberate CH_4. In the presence of ozone (O_3), the formation of Al_2O_3 is accelerated. The electrical properties are good. An electrically-low-leakage film, with no pinholes and a dielectric constant of 7.0 is produced. Other compounds like HfO_2, TiO_2, and ZnO can be deposited by ALD using precursors commonly used for CVD.

8.8.2 ALD Reactors

Historically Suntola's patents from 1977 and 1983 covered revolving substrate holders that rotated the substrate in and out of A and B reactant streams. More common now are CVD-type reactors, as shown in Fig. 8.21, in which one reactant is added and then removed from the chamber by purging with an inert gas. The other reactant is then added and after residence time is removed. Most systems operate with N_2 carrier/purge gas, the system pressure being in the viscous flow range (Torr). Production systems are made that are designed to deposit a batch of wafers at a time (http://www.picosun.com/en/home/).

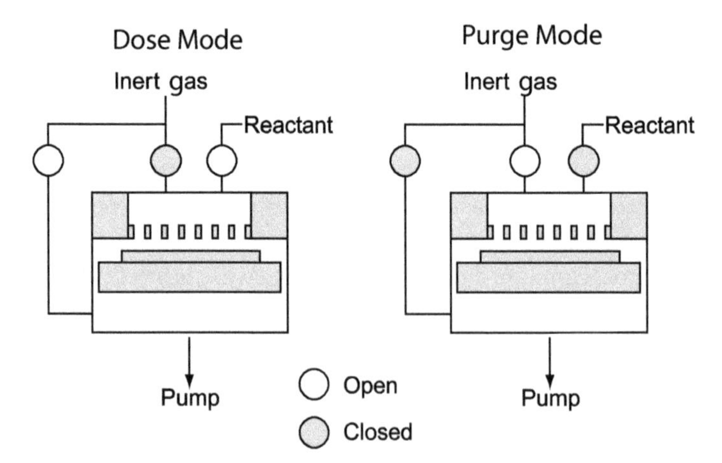

Figure 8.21 Schematic ALD system showing reactant flow on and off with inert gas purge on.

References

1. R. F. Bunshah, *Handbook of Deposition Technologies for Films and Coatings*, Noyes (1994).
2. (a) L. J. Maissel and R. Glang, *Handbook of Thin Film Technology*, McGraw Hill, New York (1970); (b) R. Glang, R. Holmwood and J. Kurtz, High vacuum technology, in *Handbook of Thin Film Technology*, McGraw Hill, New York (1970).
3. R. J. Hill, *Physical Vapor Deposition*, Temescal (1976).
4. A. B. Glaser and G. E. Subak-Sharpe, *Integrated Circuit Engineering*, Addison Wesley (1979).
5. (a) B. A. Movchan and A. V. Demchishin, Structure zone models for thin film growth, *Fiz. Met.*, **28**, 653 (1969) (translated from Russian); (b) John A. Thornton, Influence of apparatus geometry and deposition conditions on the structure and topography of thick sputtered coatings, *J. Vac. Sci. Technol.*, **11**(4), 666 (1974).
6. S. Wolf and R. N. Tauber, *Silicon Processing for the VLSI Era*, Lattice Press, Sunset Beach, California (1986).
7. Brian Chapman, *Glow Discharge Processes, Sputtering and Plasma Etching*, John Wiley and Sons (1980).
8. Shen Han et al., Trench gap-filling by ion beam sputter deposition, *Mater. Chem. Phys.*, **97**, 19 (2006).
9. J. R. Hollohan and R. S. Rosler, in *Thin Film Processes*, Eds., J. L. Vossen and W. Kern, Academic Press, New York (1978).
10. S. K. Ghandhi, *VLSI Fabrication Principles, Silicon and Gallium Arsenide*, Wiley Interscience, New York (1983).
11. R. V. Stuart and G. K. Wehner, Absolute sputtering yield measurement methods: a review, *J. Appl. Phys.*, **33**, 1842 (1962).
12. R. Behrisch, *Ergeb. Exakt. Naturw.*, **35**, 295 (1964) in Ref. [1].
13. B. Mattson, CVD films for interlayer dielectrics, *Solid State Technol.*, 60 (1980). See also [8].
14. K. D. Mackenzie et al., *High Throughput Stress Controlled Silicon Nitride Deposition for Compound Semiconductor Device Manufacturing*, CS MANTECH (2004).
15. Y. C. Chou et al., *On the Development of High Density Nitrides for MMICs*, CS MANTECH (2003).
16. S. Oktyabrsky and P. D. Ye, *Fundamentals of III-V Semiconductor MOSFETs*, Springer Science and Business Media LLC (2010).

17. Y. Q. Wu et al., First experimental demonstration of inversion mode InGaAs MOSFET, *Dig. Dec.*, 331 (2009), also *IEEE Electron. Device Lett.*, **30**, 700 (2009).
18. T. Suntola, *Thin Solid Films*, **84**, 216 (1992).
19. S. M. George, *Chem. Rev.*, **110**, 1 (2010).
20. B. L. Puurunen, *J. Appl. Phys.*, **97**, 121301 (2005).
21. D. M. Mattox, *Handbook of Physical Vapor Deposition (PVD) Processing*, William Andrews Publishing (1998).

Chapter 9

Ion Implantation and Device Isolation

9.1 Introduction [1, 2]

Ion implantation and epigrowth are the two main techniques for creating semiconducting layers in III–V processing. Ion implantation is most commonly used in III–V device processing because of the difficulties with diffusion (discussed in Chapter 10). It is an economical method to create doped semiconductors where epitaxial processes are not necessary. Ion implantation is used for the creation of channel and source–drain regions in field-effect transistor (FET)-type devices. Also in III–V processing, ion implantation is extensively used for isolation of active devices from the remainder of the processed device area. Junction isolation and oxide growth, used extensively in silicon processing for device isolation, are not used in GaAs processing; instead ion implant isolation combined with mesa etching is utilized. Isolation is needed for all FET- and heterojunction bipolar transistor (HBT)-type circuits.

In the ion implantation process, suitable dopant atoms are loaded into a source, where these are ionized and then accelerated to high velocity and directed at a target substrate, with specifications depending on the difference of dopant atoms and the necessary depth requirements for a particular device process. The atoms have high energy to penetrate the crystal lattice, collide with host atoms, and come to rest at different depths. The average penetration depth

III–V Integrated Circuit Fabrication Technology
Shiban Tiku and Dhrubes Biswas

ISBN 978-981-4669-30-6 (Hardcover), 978-981-4669-31-3 (eBook)
www.panstanford.com

or range depends upon the acceleration energy; the amount, called dose, is controlled by counting the ions by total charge measurement.

9.1.1 Advantages

A wide range of species can be implanted. Species that cannot be diffused or incorporated in large concentration in an equilibrium process can be ion-implanted. Mass separation techniques can be used to obtain a pure beam. A single machine can be used with multiple sources to implant different species. The possible concentration range is wide. Doses from 10^{11} to 10^{17} ions/cm^2 and ion concentrations from 10^{11} to 10^{19} ions/cm^3, with excellent run-to-run and within-wafer uniformity control, are possible. Control of very low doses makes it possible to control the threshold voltage of FET devices. Dopant profiles can be varied by changing the dose, energy, and number of implants. A variety of masks, including photoresists, can be used. Since ion implantation is generally a single-wafer process, large yield losses that occur in batch-type diffusion systems can be avoided.

9.1.2 Disadvantages

Ion implantation causes damage to the crystal, so it loses its semiconducting properties. High temperature is needed to heal the damage. For III–V materials, this annealing must be done under overpressure conditions (arsenic for GaAs) or a cap material must be deposited before anneal. Some species may diffuse during the anneal process and result in smearing of the profile toward lower peaks and larger tails. Sharp profiles like those achieved by epigrowth techniques cannot be achieved by ion implantation.

The process of ion implantation, dopant activation by thermal processes, and applications of ion implantation for active layers and device isolation are discussed in this chapter.

9.2 Ion Implantation: Theory

The basic theory needed for the process of ion implantation and associated calculation of the implant schedule or "recipes" have been presented.

9.2.1 Theory of Ion Stopping [2, 3]

There are two basic mechanisms for stopping of implanted ions in a material, electronic and ionic. When an energetic ion enters a semiconductor target, it interacts with the electrons and nuclei of the crystal. Implanted ions undergo a series of collisions until they finally stop. An example of trajectories of ions (calculated by Monte Carlo simulation) is shown in Fig. 9.1a [4]. The collision processes cause transfer of energy to the semiconductor, resulting in deflection of the projectile ions and dislodging of the target nuclei. Nuclear stopping also results in damage, and creation of point defects, line defects, and amorphous regions. A summarized version of the calculating range, etc., is given below. For details of the derivations, the reader is referred to textbooks on ion implantation (e.g., Refs. [1, 2]). If E is the energy of an ion at any point x along its path, the stopping power can be defined as $S_n = \left(\frac{dE}{dx}\right)$n energy loss per unit length for nuclear stopping and $S_e = \left(\frac{dE}{dx}\right)$ for electronic stopping.

Electronic interaction causes generation of electron–hole pairs as the projectile ion energy is transferred to the crystal. The average rate of energy loss with distance can be written as

$$\frac{dE}{dx} = N[S_n(E) + S_e(E)] \tag{9.1}$$

where N is the number of target atoms per unit volume of the semiconductor. Figure 9.2 shows these two terms as a function of ion velocity (or energy). At typical ion energies used in ion implantation, nuclear stopping dominates. As energies increase into the mega-electron volt range, electronic stopping takes over.

The nuclear stopping power can be treated as collisions between hard spheres. A more accurate model assumes the scattering to be Coulombic force at a distance interaction. The most successful model for predicting implantation profile is the Lindhard, Scharff, and Schiott (LSS) model [5], which utilizes a modified Thomas–Fermi screened potential for scattering. Calculations based on this show that nuclear stopping increases linearly at low energies, reaches a maximum, and then decreases at high energies, as shown in Fig. 9.2. Electronic stopping power is similar to viscous drag force and has been shown to be proportional to the ion velocity

$$S_e(E) = k\sqrt{E} \tag{9.2}$$

where the value of k depends upon the projectile and target material. For GaAs (nonion channeling case) $k \approx 0.52\ 10^{-15}\ (\text{eV})^{1/2}\ \text{cm}^2$.

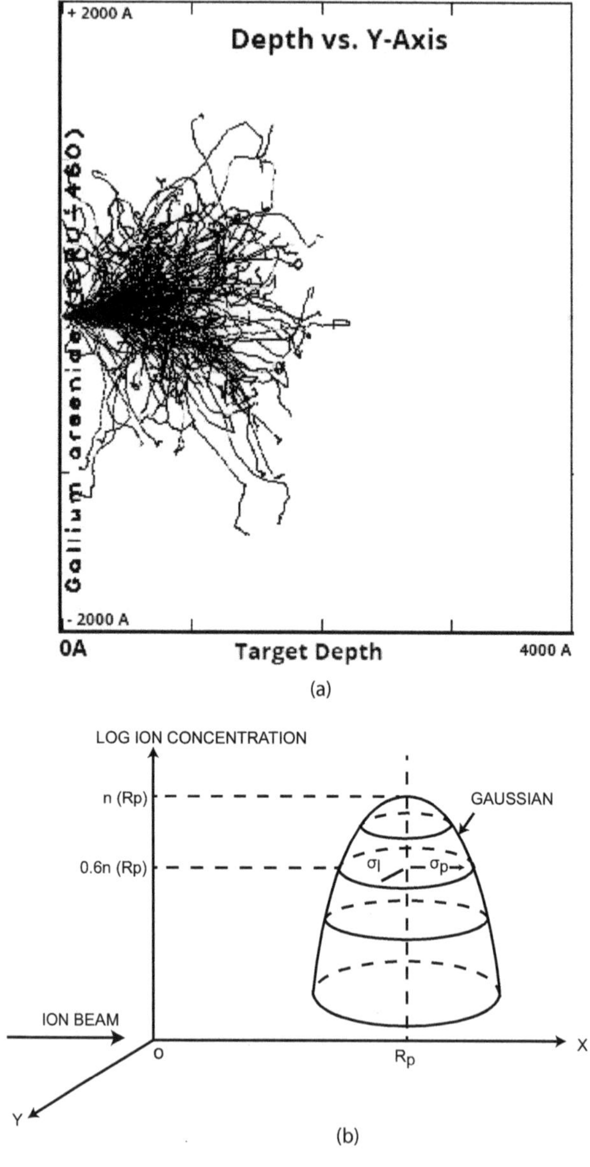

Figure 9.1 (a) Monte Carlo calculation using ion trajectories for 100 keV silicon implanted into GaAs using SRIM. (b) Schematic diagram of ion range and two-dimensional view of the Gaussian profile.

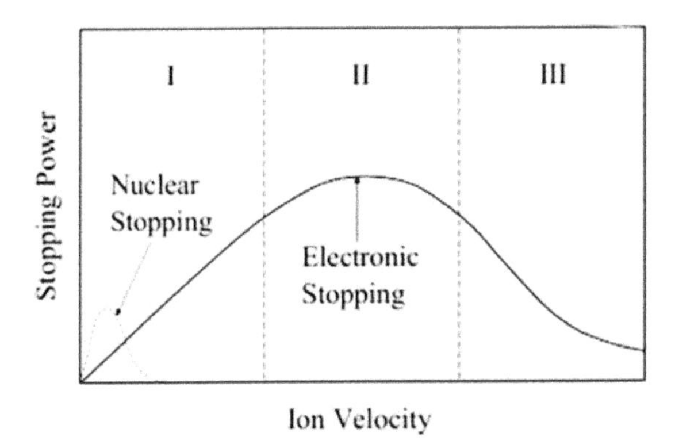

Figure 9.2 Simplified view of nuclear and electronic components of the ion-stopping power as a function of ion velocity, based on LSS theory.

Ions implanted into a target traverse different lengths. The average total path length is called the range, *R*. The average depth of the ions (component perpendicular to the surface) is called the projected range (R_p) (see Fig. 9.1b). Equations for computing the projected range have been derived in the literature [3, 6–8] and have been verified for silicon and GaAs. Simulation software is available from equipment and software vendors (e.g., SRIM [4]). The total distance travelled by the ion before coming to rest is given by (recall Eq. 9.1 from above)

$$\int_0^R dx = \frac{1}{N} \cdot \left[\int_0^{E_0} \frac{1}{Sn(E) + Se(E)} dE \right]. \tag{9.3}$$

where E_0 is the initial energy.

If the initial energy of the implanted ion is low (see Fig 9.2), then

$$R \approx \frac{E_0}{NSn} K_1 E_0 \tag{9.4}$$

where K_1 is a constant that depends upon the target and the projectile ion. If the energy is much higher, the range can be based on electronic stopping alone.

$$R \approx K_2 E_0^{1/2} \tag{9.5}$$

Equations for nuclear and electronic stopping have been derived in detail in terms of the mass and atomic numbers of the target and projectile species. Using energy loss due to nuclear stopping and the electronic stopping power approximated by Eq. 9.5 above, the range and straggle can be calculated (see, for example, Ref. [3]) to an approximation,

$$R_\mathrm{p} = \frac{R}{\left(1+\dfrac{M_2}{3M_1}\right)} \tag{9.6}$$

where M_1 and M_2 are the masses of the projectile and target species, respectively.

And the straggle, which is the spread along the normal direction, is

$$\Delta R_\mathrm{p} = \frac{2R_\mathrm{p}}{3} \cdot \frac{\sqrt{M_1\, M_2}}{M_1 + M_2} \tag{9.7}$$

For semiconductor processing where a large number of ions is implanted, a considerable fraction of the atom density of the crystal, the projected range, and its straggle can be calculated from the distribution that is shown in Fig. 9.1b. The ion beam and damage also spread laterally and determine the distribution of dopants and damage near resist edges. For a large-area implant, with a large number of ions, the statistical distribution can be described by a one-dimensional Gausssian distribution. This function has a maximum at R_p and falls off exponentially on either side. For crystalline materials, the distribution deviates from this ideal due to channeling. Channeling happens when the ions find channels along crystal axes, and atoms can travel farther compared to the case of an amorphous target. This results in a tail, as shown in Fig. 9.3.

The assumption of Gaussian distribution has been seen to be very close to experimental data for amorphous materials or crystalline materials under nonchanneling conditions. Experimental data for the range and straggle for some common dopants in GaAs is shown in Fig. 9.4a,b [2].

In the early days of ion implantation, range and straggle data were published in the form of tables for common dopants into silicon and GaAs and other semiconductor materials. These have been replaced by simulation software, which can be used to predict range as well as damage profiles. Monte Carlo methods are used to predict profiles.

For example, in the simulation program SUPREM, in which a Pearson IV distribution function is used to describe the profile, four moments describe the characteristics of the implant profile: the mean range, μ; square of the straggle, μ_2; skewness, ν; and kurtosis or tail aspect, β [2]. Skewness is caused by back-scattering, which is more likely with lighter atoms. This simulation has good agreement with all implanted species into amorphous targets and crystalline materials with an amorphous surface layer.

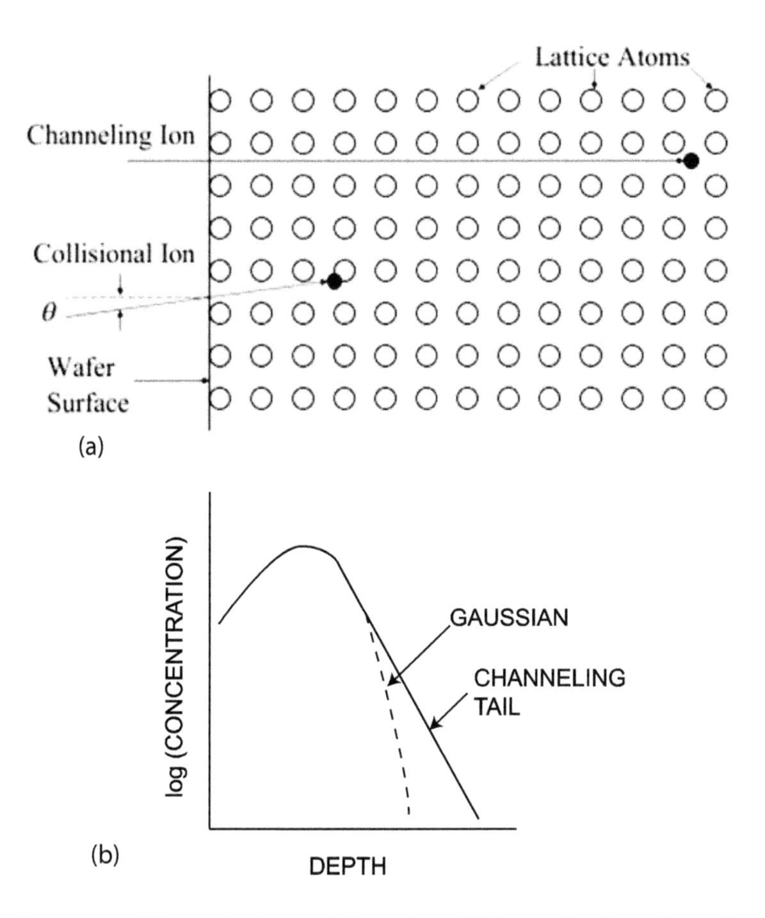

Figure 9.3 Schematic views of channeling. (a) Ion paths through a cubic lattice showing channeled and nonchanneled paths. (b) Ideal distribution and tail due to channeling.

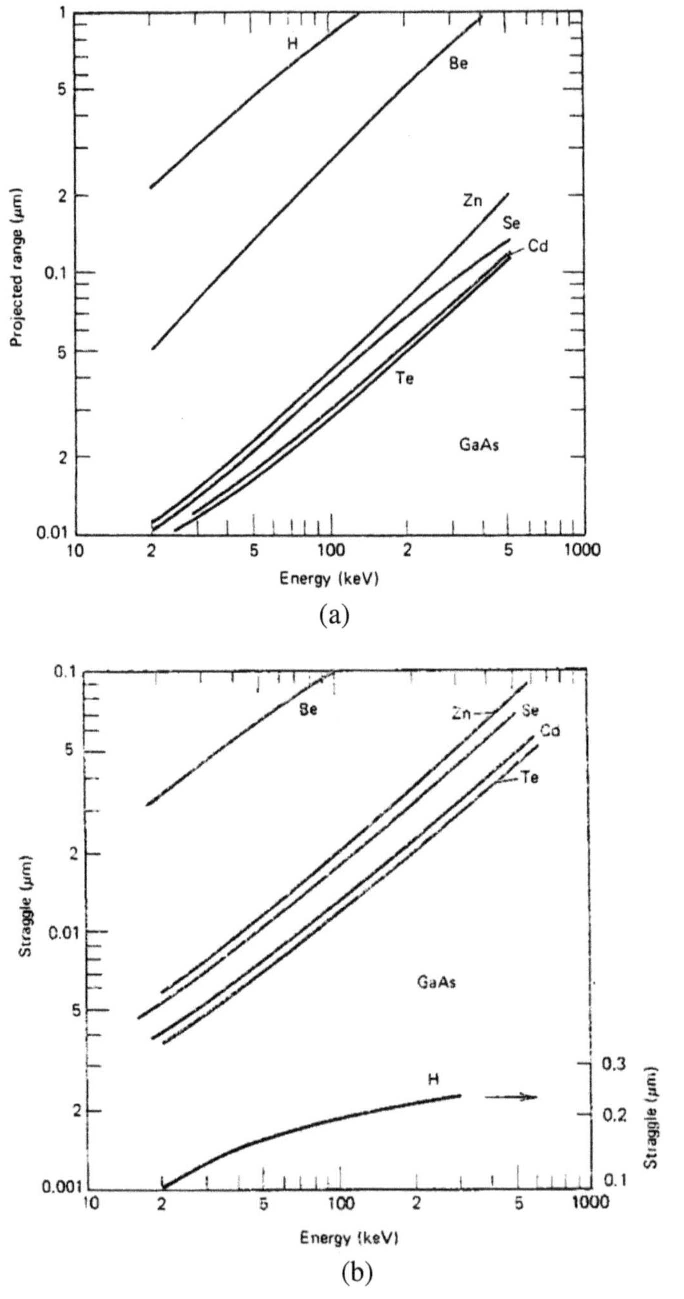

Figure 9.4 (a) Projected range of dopants in GaAs. (b) Straggle for dopants in GaAs. From Ref. [2]. Copyright © 1983. Reproduced with permission of John Wiley & Sons, Inc.

9.2.2 Channeling

In III–V semiconductor processing, ion implantation is done into single-crystal materials, not amorphous materials where LSS theory predicts near-Gaussian profiles. In single-crystal materials, there are directions, channels along which projectile ions will not encounter any target nuclei. The most likely channeling directions are the ones perpendicular to the surface used for circuit fabrication, like (100). Along these directions only electronic stopping is effective and the range can be several times larger than in an amorphous target. The most common approach to avoid channeling is to tilt the wafer surface relative to the beam so that the lattice projects a uniformly dense distribution of atoms. Even with tilt, wafers can still project symmetrical arrays or planes, leading to planar channeling.

This can be avoided by adding a twist to the wafers with respect to the axes of beam scanning. Another common method of avoiding channeling is to deposit an amorphous film on the surface, for example, a silicon nitride film for GaAs or SiO_2 on silicon. Also a low-energy implant can be used first to turn the surface layer into an amorphous structure.

9.2.3 Transverse Effects

Ion implantation is done in semiconductor processing into areas defined by photoresist masks. Transverse straggle defines the lateral spread of the ions in the target and is important at the edges of the implanted region or for understanding what happens in narrow slits defined by photoresist or other masks. See Fig. 9.5. The ion concentration can be shown to fall off to 50% at the edge of the mask and falls off following a complementary error function (erfc) function profile. (See Chapter 10 for an error function and complementary error function explanation). Figure 9.6 shows the straggle for a few common ions in GaAs as a function of implantation voltage [9]. It can be seen that the straggle can start to reach a micron range as energy reaches 100 keV. Therefore, it is important for making features smaller than a few microns in width. This behavior is important for understanding the processing of narrow implanted resistors and to provide guidelines for layout of semiconductor devices in general, for example, to ensure proper FET scaling. For implants used for isolation, enough space must be provided to avoid partial damage of regions of semiconductor devices, especially if junctions are nearby.

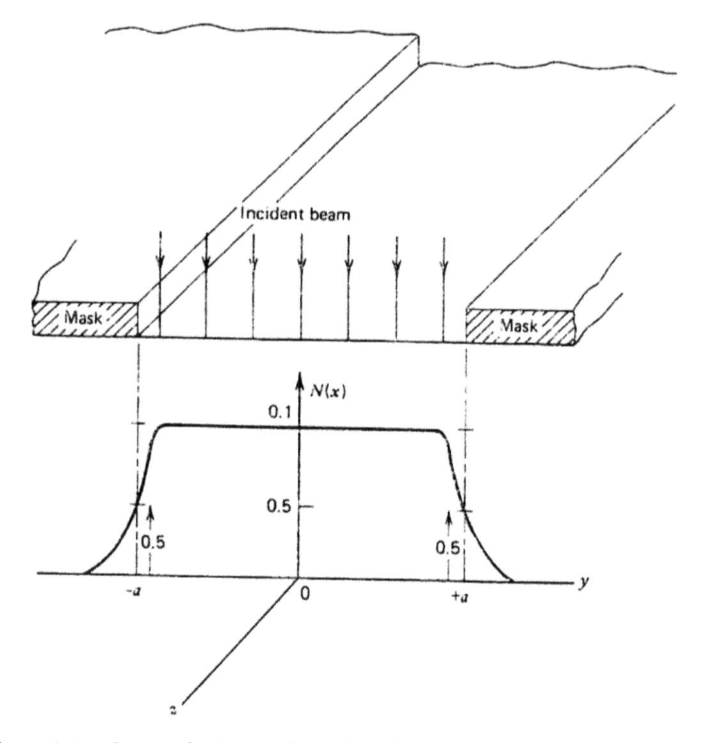

Figure 9.5 Ion implantation through a slit.

9.2.4 Implant Damage

Ion beam collisions with the nuclei of the target crystal displace the target atoms and produce defects. Defects can also be produced by recoil atoms. Electronic interaction does not produce damage. Frenkel defects (vacancy–interstitial pair), vacancies, and interstitials are produced by ion damage as well as more complex combinations. However, if the damage is heavy, it may lead to formation of amorphous regions. The number of defects created is larger than the number of implanted ions. The defects reduce electron mobility and introduce defect levels within the energy bandgap. The resistivity of the material goes up by several orders of magnitude for heavily damaging, high-dose isolation implants.

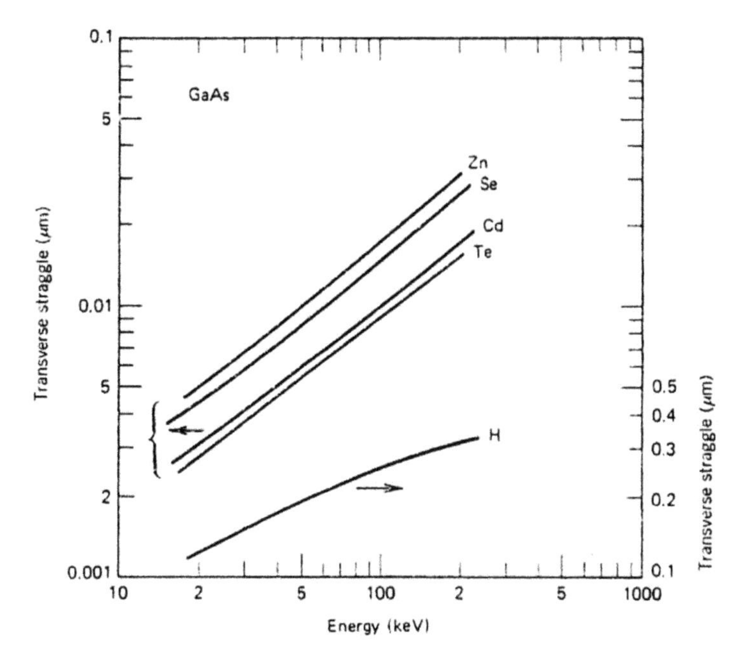

Figure 9.6 Transverse straggle for dopants in GaAs. From Ref. [2]. Copyright © 1983. Reproduced with permission of John Wiley & Sons, Inc.

If the energy required to dislodge an atom is E_d, the projectile ion must have an energy greater than $2E_d$ to cause displacement. The displacement energy is estimated to be about 15 eV. If the implant energy is 100 keV, thousands of atoms are displaced.

The damage profile from light-ion implantation is different from those of heavy ions. For light ions the initial energy loss is due to electronic interaction, so no atom displacement takes place; after the energy drops to lower levels, nuclear stopping dominates. The damage produced by light and heavy ions is therefore different from the doping profile. A light ion transfers only a small amount of energy to the target atoms and creates a long path of less concentrated damage. This kind of damage may self-anneal if the dose is high and the target gets hot. When heavy ions are implanted, the energy loss and stopping are due to nuclear collisions and substantial damage occurs close to the surface. Figure 9.7 shows this in a schematic fashion. Heavily damaged GaAs becomes polycrystalline with numerous defects (different from silicon, which becomes amorphous). Ga and As recoil differently [10], producing an arsenic-rich surface. This nonstoichiometry leads to excessive defects near the surface. The material has to be annealed to recover semiconducting properties. If nondoping atoms are used, III–V materials easily turn into the semi-insulating type, with a carrier concentration below $10^{11}/cm^3$ and very low mobility. This effect is used for creating insulating regions between devices.

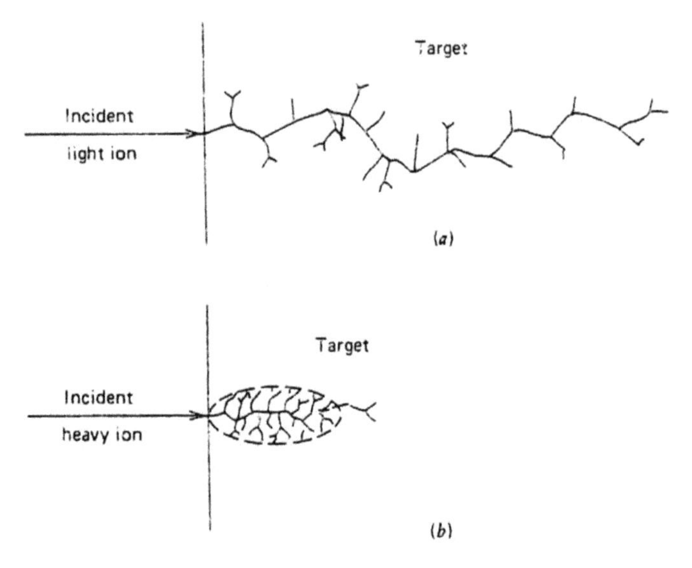

Figure 9.7 Trajectory and damage due to light ions (a) and heavy ions (b).

9.3 Ion Implantation Systems

Ion implantation systems have evolved at a fast pace for the silicon semiconductor industry and compound semiconductor fabrication has benefited as a result. Automation has been added to these systems to manage beam setup, recipe management, and wafer handling. Ion implants for III–V compounds cover a wide range of energies and doses. Implanters that can handle high energy or high current (implant rate) have been developed (generally not both at the same time). These systems are very complex and have a large number of subsystems. A wide range of implanting species, energy, and doses can be accomplished with ion implanters. Figure 9.8 shows a schematic diagram of a typical commercial ion implant system. Ions are created in a source, separated in an analyzer magnet, accelerated, and finally scanned before hitting the target in the end station. The various parts of the system are discussed with relevance to the latest available process technologies.

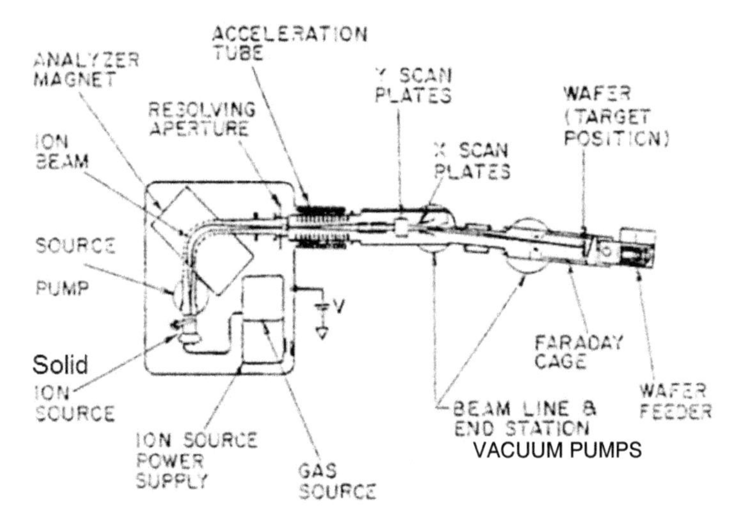

Figure 9.8 Schematic diagram of a typical commercial ion implantation system.

9.3.1 Implantation System Parts

9.3.1.1 Ion source

An ion source supplies the ions needed and may be designed to handle a gaseous source or a solid feed. Most ions used in GaAs fabrication can be generated from gas sources, like SiH_4, H_2Se, He, etc. Solid sources may require the use of a vaporizer. Many different kinds of sources have been developed, but most are based on the concept shown in Fig. 9.9 for a Nielson-type source [2]. Ionization of the source material is accomplished by passing the vapor over a hot cathode electronic discharge. The anode is held at about 100 V and a magnetic field is provided to force the electrons to move in a spiral trajectory and thus increase the mean free path. Modifications have been made to this simple design to increase the beam current capability and source life. When doubly ionized ions are used for achieving double energy, filament life can suffer due to the fact that the source must be driven harder to achieve high beam current for the doubly ionized species. The Bernas source, for example, uses a two-and-a-half-turn pigtail filament to increase the source life. Other modifications can be made to the source geometry to avoid common failure mechanisms of the filament or cathode shorting to the anode.

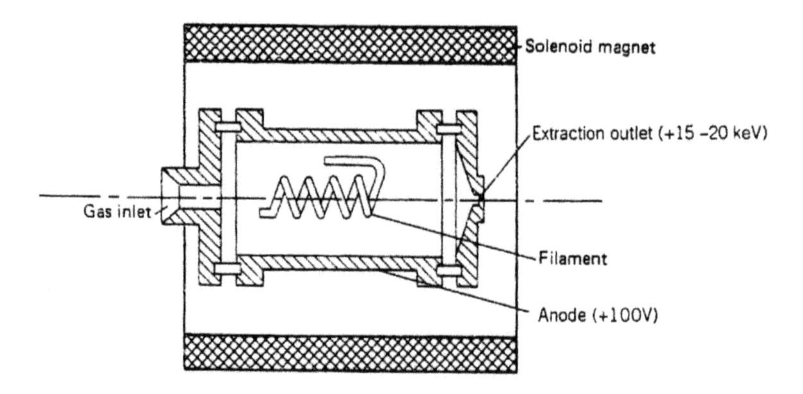

Figure 9.9 A Nielson-type ion implant gaseous source.

9.3.1.2 Ion extraction and analyzing device

Positive ions are extracted from the source discharge by an anode at 15–40 keV and then sent through a magnetic analyzer that works on the principle of separating mass/charge ratio through a slit (Fig. 9.10). Ions move in circular path in a uniform magnetic field, with the radius dependent upon the mass-to-charge ratio. The radius, derived by equating the centrifugal force to the force exerted by the magnetic field, is given by

$$r = \frac{143.95}{H} \cdot \sqrt{MV/n} \tag{9.8}$$

where **H** is the magnetic field in Gauss, M is the ionic mass in amu, V is the acceleration voltage, and n is the charge state of the ion. For a given acceleration voltage and magnetic flux density, the path radius is directly proportional to the square root of mass, M. Figure 9.10 shows trajectorythe for three different masses being accelerated by a magnetic field. Only ions of a particular mass (m_2 in the diagram) can be extracted. The analyzing magnet can separate 1 amu, so even isotopes, Si^{29} versus Si^{28}, can be separated. The slit aperture can be eroded by the beam and ions from it may end up on the wafer. Graphite that has a low sputter yield is used in the silicon industry, but aluminum may be better for GaAs if carbon (a p-type impurity) needs to be avoided.

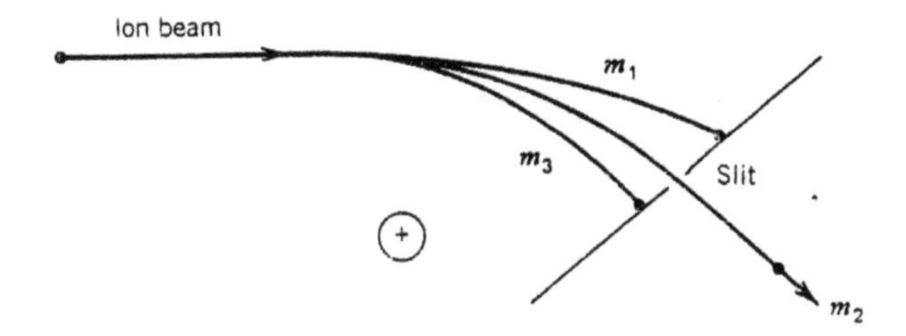

Figure 9.10 Mass separation through a slit. Magnetic field is at 90 degrees, depicted by the + sign.

9.3.1.3 Accelerator tube

In this section, ions are accelerated by applying a high electric field, which can range from 50 kV to mega-electron volts. The ion beam is also focused to the desired shape.

9.3.1.4 Beam scanning system

Beam scanning systems are used to distribute the ions uniformly over the whole wafer. In a single-wafer system, electrostatic scanning may be used in both x and y directions. The implantation time should be long enough to allow uniform scanning over the whole wafer at a reasonable rate for acceptable throughput. The beam current should not be too high, because it can cause wafer heating and nonuniform dose over the wafer.

In electromechanical scanning systems, mechanical motion of the wafer is provided in one direction and the beam is scanned electrostatically in the other direction. For a large batch system, a number of wafers may be loaded on a rotating disc to provide mechanical scanning along one axis.

9.3.1.5 System end station

After the beam is accelerated and scanned it impinges on the target wafer held on a substrate holder in the end station. The end station includes a robot for loading wafers, a chuck that may be electrostatic and also a means for dose monitoring. The dose monitoring is done by the use of an area-defining aperture, a Faraday cup or cups for collecting charge, and a current integrator. The whole target area is surrounded by a biased Faraday cage so that secondary electrons ejected from the target are repelled back and not collected to cause dose errors.

The different parts of the implantation system, source, acceleration beam line, and end station are pumped by vacuum pumps. The pressure is maintained below10^{-6} Torr in order to reduce formation of neutrals by loss of charge of the ions to the gas molecules. Turbo pumps are preferred for the source side to handle the gas load.

9.3.2 Ion Implanter Types

Ion implant applications in the III–V semiconductor fabrication vary from low dose with low energy needed for channel implants to high dose with high energy needed for isolation. Implants from 30 to 400 keV are common, although a few applications need mega-electron-volt implants. Medium current implanters, which process one wafer at a time, are a good compromise for GaAs processing.

Medium-current implanters have a maximum beam current capability of about 2 mA. The scanning is usually all electrostatic, and the maximum energy is limited to 200 keV (newer Eaton machines are available up to 250 keV). The maximum practical dose for these machines is in the range of 10^{14} ions/cm^2 for implantation times of the order of a minute per wafer. These implanters are ideal for active dopant implants and can be used for isolation.

High-current implanters have been developed for heavy implants with doses over 10^{14} per cm^2 used in silicon processing. Electromechanical scanning is used for batch processing of wafers.

High-energy implanters have maximum energies from 400 keV to a few mega-electron volts but relatively low beam current of a fraction of a milliampere. These are used for deep isolation and can be used for creation of buried layers under devices or isolation layers under collectors in HBT-type devices.

9.4 System/Process Issues

The maximum beam current is limited by the type of source and the ion species being generated. If the source gas is easily ionizable, a large current can be generated. For doubly ionized implants, stripping the second electron takes a lot of energy and thus the source current is limited. For high throughput large beam currents are desired. It may be noted that excessive heating of the substrate (limited by photoresist) may place an upper limit on the beam current.

Dose monitoring may be flawed because of beam neutralization, due to poor vacuum along the beam line or outgassing from the wafer. At extremely high currents, wafer charging may cause local dose errors.

The thickness and profile of the photoresist must be carefully chosen. For isolation implants, which are light ions, a sufficiently thick photoresist must be used so that active devices are not damaged. Simulation programs should be used to verify that the photoresist thickness is sufficient. Ion implants are done at large tilt angles to avoid channeling. This causes dose nonuniformity, the farther side or portion of the wafer getting fewer doses. Dose compensation is automatically applied in modern implanters but should be used if available as a feature.

9.4.1 Masking Considerations

A photoresist is used to define the pattern for ion implantation; however, other layers or masks may also be used for stopping ions from going into the substrate areas that need to be shielded. Silicon nitride and oxide are common mask layers for GaAs and are generally used in conjunction with a photoresist. In most FET processing approaches, Si_3N_4 or SiO_2 is deposited all over the wafer and then windows are opened in this layer prior to implantation or implantation is done through a cap layer using only the photoresist as the mask. This way, channeling is avoided and the semiconductor surface is not exposed to processing chemicals or harsh resist-stripping plasma processes.

9.4.2 Doubly Ionized Species

In HBT processing, isolation must be achieved for depths of over 2 μm. Energies over 400 keV are required for helium to achieve this depth. Ion implanters that have capability up to 250 keV are more common for medium-current implanters. By using doubly ionized species (He^{++}), energy is multiplied by a factor of 2. The atomic dose is, of course, half of the charge accumulated in the Faraday cage. As discussed earlier, the source life is affected in the process of generating doubly ionized species.

9.5 Common Ion Implant Species for GaAs

9.5.1 n-Type Dopants

Most GaAs FET-type devices are n-type devices because of the high electron mobility and poor hole mobility in this semiconductor material. Common dopants that have shallow levels in the energy bandgap (See Chapter 1) are Si, Ge, Se,Te, Sn, and S. Common dopant species are shown in Table 9.1. Developing ion implantation and anneal processes for many of these dopants has been challenging. The most popular one in production use is Si. Si as well as Sn are amphoteric in GaAs. Si has been preferred because of its lower mass over Se, so a single species can be used for shallow as well as deeper implants and source and drain regions. Si has a mass of 28, which coincides with N_2. Si^{29} (isotope with mass 29) is generally chosen for ion implantation to avoid N_2 (N_2 molecular mass 28) implantation.

Table 9.1 Characteristics of ion-implanted species in GaAs

	Donors
Ion	**Comments**
Si	Low mass means good activation for RT implant, versatile range
Se	Good activation for low-dose RT implant
Te	Poor activation for RT implant
Ge	Amphoteric species, poor activation, difficult to implant
S	Diffuses during annealing
Sn	Amphoteric, diffuses during annealing

Acceptors	
Ion	**Comments**
Be	Very light mass means good activation at low annealing temperature, easy to implant, versatile range, at high concentrations diffuses during annealing
Mg	Good activation, versatile range, difficult implant source
Zn	Reasonable activation, at high concentrations diffuses during annealing
Cd	Reasonable activation, diffuses during implantation at RT, difficult implant source

Activation of Si is limited by self-compensation. Silicon activation in GaAs has been studied extensively [11, 12]. Activation efficiency has been determined by the rate at which the implanted atoms react with vacancies on the Ga sublattice (V_{Ga}) to form Si_{Ga}. The activation is controlled by the anneal temperature, which determines the equilibrium between the donor Si^{+}_{Ga} and the acceptor species Si^{-}_{As}. At lower temperatures Si donors dominate, whereas at high temperature it is the Si acceptors that are thermodynamically favored. The activation behavior is initially kinetically controlled. For long-duration anneals, the thermodynamics of activation dominate. The activation efficiency can be over 90% for low-dose implants below 800°C anneal. For high-dose implants, the efficiency drops to below 80%, even with high-temperature anneals near 900°C. The drop of activation efficiency is shown in Fig. 9.11 for high-dose implants at 160 keV [12]. Efforts have been made to improve activation efficiency by co-implantation of species that go on the As sublattice, like P. However, the efficiency is still low [13].

Sulfur activation is low and departs from the predicted Gaussian profile, particularly for shallow implants due to outdiffusion. Se implants, on the other hand, are well behaved and profiles after anneal are close to LSS theory. Se implants should be preferred over Si for channel implants where threshold control makes repeatable activation important. Figure 9.12 shows the difference in the profiles between Se and S. [14]. Se goes on the As sublattice only, so there is no self-compensation. Minor As loss during the annealing process in not a significant factor. Tellurium behaves the same way but is heavier so needs an even higher energy ion implant voltage.

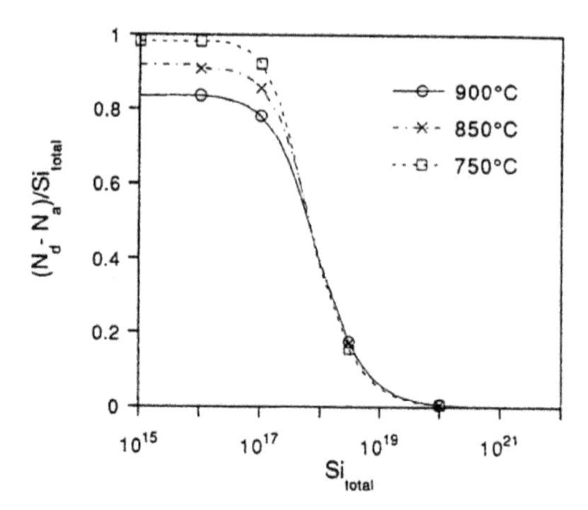

Figure 9.11 Active fraction vs. total Si, showing effect of increased compensation for high-dose implants in GaAs at different temperatures. From Ref. [12]. Reproduced by permission of ECS—The Electrochemical Society.

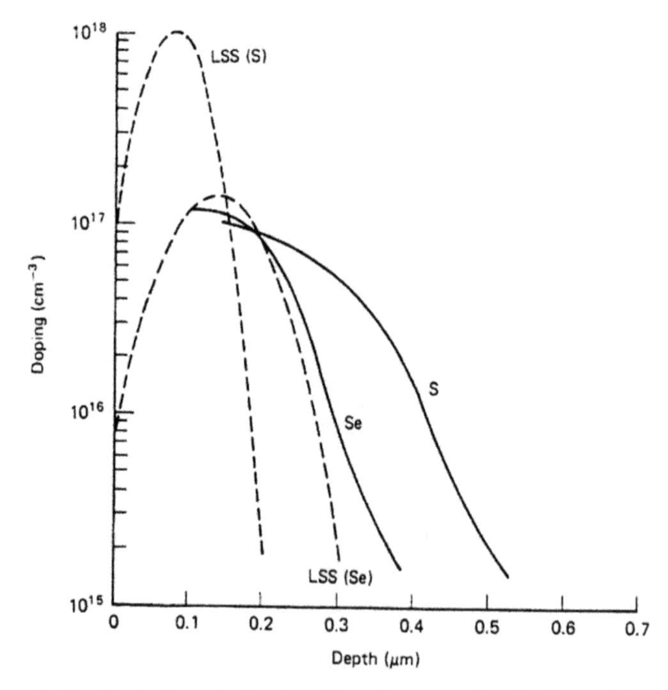

Figure 9.12 Deviation of actual profiles after annealing for Se and S implants in GaAs. From Ref. [2]. Copyright © 1983. Reproduced with permission of John Wiley & Sons, Inc..

9.5.2 p-Type Dopants

The p-type dopants are used for making heavily doped contact regions for optoelectronic devices, HBTs, and buried layers for n-channel devices. The p-type dopants that provide shallow levels in the bandgap Zn, Cd, Be, C, and Mg have been investigated. The hole mobility in GaAs is low, so there is not much interest in p-channel ion implantation. Buried p layers under n-channel devices are used to cut off or compensate the n implant tail to sharpen the channel. Beryllium, Be, has low mass and can be implanted without much damage. However, it diffuses fast and gives rise to profile smearing at relatively low anneal temperatures of 800°C [15]. The annealing behavior of Be has been found to be a strong function of dose and temperature. The mechanism is, however, different from that of Si in GaAs, being dominated by the fact that diffusion of Be increases rapidly with dose and temperature. Care should be taken in handling of sources because of toxicity of Be. Zinc and Cd show similar activation behavior. At high concentrations, Zn shows anomalous diffusion. Magnesium shows rapid initial diffusion for even low concentrations but becomes relatively stable. It also activates at low temperatures below 800°C and is a good candidate for buried p layers [16].

9.5.3 Implants for Isolation [10, 17]

A number of ions have been used for implant isolation. Common among these are H (proton), helium, oxygen, and boron. These ions are used to convert the semiconductor into an insulator by introducing heavy damage and deep levels in the energy bandgap (see, for example, Ref. [18]). The damage goes up with mass of the ions and dose. Very high doses are needed for H^+, and there are concerns about reliability. Oxygen introduces deep levels. Both boron and oxygen need higher-energy implanters. Isolation resistance is optimized by annealing. Details will be discussed in Section 9.9, "Device Isolation."

9.6 Ion Implant Characterization

Ion implant systems must be routinely monitored for dose accuracy and uniformity. Occasional monitoring of implant depth, damage profile, and annealing behavior is also desirable. In silicon fabrication facilities, dose monitoring is routinely done by Thermawave [19] type of systems. These systems measure the change of reflectivity due to implant damage and are nondestructive. In III–V fabrication, electrical characterization is more common because the resistivity measurements are more dependable.

9.6.1 Sheet Resistivity Monitoring

In silicon fabrication substrates are generally conducting, and resistivity measurements after ion implantation are not accurate. In GaAs processing, substrates are semi-insulating, so implants can be monitored quickly by implantation and rapid thermal annealing (RTA) annealing, followed by contactless sheet resistance measurement.

9.6.2 Optical Dosimetry

Optical dosimetry is based on measurement of darkening of polymer layer ion implantation. For GaAs, low doses that cannot be detected or quantified by contactless sheet resistivity measurement, can be monitored by this method. Thus doses in the range of 10^{11} ions/cm ^{2}, below the 10^{12} range used for enhancement-mode (E-mode) implants, can be monitored. Transparent substrates of glass, coated with a polymer or photoresist with excellent uniformity, can be used for this purpose. Wafers are scanned before and after implantation. The difference is plotted as a contour map. This is a quick monitoring method and works well for isolation implants that cannot be measured by sheet resistance methods. Ion implanter malfunction can usually be quickly diagnosed by optical techniques using the patterns produced on the contour maps.

9.6.3 *C–V* Method

Doping profiles can also be determined by the capacitance–voltage (*C–V*) technique, for low-dose active implants, discussed in detail in Chapter 21. This method is lower because the monitoring wafer needs to be annealed and data is harder to interpret, since the result is a profile rather than just a sheet resistance number that can be plotted on a statistical process control (SPC) chart.

9.7 Implant Activation

Thermal processes are used for annealing GaAs wafers after ion implantation of FET-type device wafers. Thermal processes are also used for diffusion and in the making of contacts, which will be discussed in Chapters 10 and 11. These processes will be discussed briefly in this chapter. RTA, which has become dominant for silicon as well as III–V processing, will be discussed in more detail.

9.7.1 Annealing [2]

The semiconductor materials damaged by ion implantation must be restored back to pre-implant conditions. Most of the implanted dopant ions may occupy interstitial positions and must be moved to substitutional sites, for example, Se needs to occupy an arsenic site. Thermal treatment is needed to restore semiconductor properties and activate the dopant. The temperature and time depend upon the degree of recovery needed. In III–V compounds temperatures beyond the dissociation temperature are required, so precautions must be taken to avoid group V loss. Annealing studies were historically done in furnace systems. Now rapid thermal systems are used because of higher throughput, sharper dopant profiles, and ease of cassette-to-cassette handling. Electron and laser beam annealing have also been investigated but have not been introduced into production. Laser annealing does not provide any advantage for GaAs annealing and is very hard to implement. Historically, quality of the substrate as well as background doping used to be factors in implant activation but are no longer an issue.

The emphasis in the development of annealing processes has been on carrier activation and achieving high electron mobility, particularly in n-channel layers. The most common dopant material in GaAs is silicon; this dopant does not cause extensive damage during ion implantation as other heavier ions like Se, Te, or Sn do, but still the material becomes completely semi-insulating. The surface is not stoichiometric and heavily damaged. The bulk of the crystal is not amorphous but polycrystalline. Reordering of the crystal cannot take place by regrowth but by diffusion of displaced atoms to proper sites, not by local rearrangement but by diffusion over larger lengths. In contrast to silicon, there is no well- defined activation energy associated with this process. This process takes high temperature as well as a long time. Annealing below 700°C has no effect. For high-dose implants, the activation efficiency is low even at 800°C. The typical furnace anneal temperature is 850°C for 20 to 30 minutes. Under these conditions low-dose Se and Si implant resistivity reaches a saturation value, showing close to complete activation (Fig. 9.13). At higher dose levels, the activation is lower and a higher annealing temperature may be necessary. If GaAs is implanted at elevated temperatures, self-annealing during implantation reduces the damage and reduces the time–temperature budget during the anneal. In contrast, lower temperatures are preferred in silicon to form amorphous material and reduce the thermal budget needed for recovery. In any case, anneal temperatures over 850°C are needed for GaAs annealing, particularly for RTA. At high temperatures, loss of stoichiometry due to As loss can be a problem, and a means of reducing As loss must be provided. Early annealing work was done under arsenic overpressure [2, 20]; however, this method is not production friendly. Use of capping layers solves this problem.

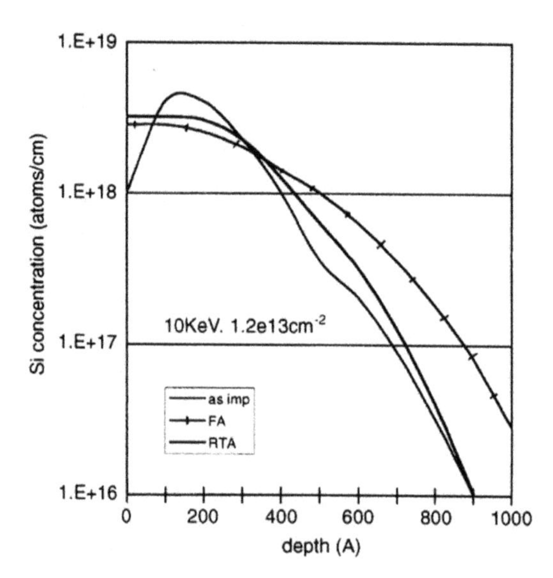

Figure 9.13 SIMS profiles of ion-implanted Si into GaAs and the diffusion effect of annealing treatments.

9.7.2 Encapsulation for Annealing

To suppress arsenic loss a cap layer is generally used. Cap layers also solve the problem of ion implant channeling if the cap layer is deposited before ion implantation. Numerous papers have been published on the variety of cap layers. Deposition processes for cap layers are discussed in detail in Chapter 8. The most common cap material is silicon nitride. Other materials, SiO_2, Al_2O_3, silicon oxynitrides, etc., have been tried. Oxide cap layers promote Ga outdiffusion and may increase silicon activation by providing more Ga vacancy sites. Activation efficiency also depends upon which site the dopant occupies, so this cap will not help Se activation. A double layer of nitride through the implant cap and oxide can be used to balance stress and achieve good activation. The oxide layer can be removed after anneal. Capped annealing is compatible high-volume production and it does not have the drawbacks of arsenic overpressure annealing.

9.7.3 Rapid Thermal Annealing

High-temperature process steps are less prevalent in III–V IC processing compared to silicon processing. However, high-temperature process steps are needed for ion implants, particularly FET-type devices, and for alloying of contacts of all types of devices. High-temperature process steps are avoided in HBT device fabrication. Thermal treatments cause diffusion and smearing of dopant profiles (see Fig. 9.13). Figure 9.14 shows a comparison of activation for RTA in comparison to furnace annealing. Thermal treatments are optimized to reduce diffusion and dopant smearing. Shallow-channel FETs with abrupt doping profiles are essential for E-mode FETs. Epigrown FET, high-electron-mobility transistor (HEMT), and HBT devices are even more sensitive to thermal steps and high temperatures are restricted even more. RTA has been developed to address many of these issues. RTA is used for contact sintering and alloying, dopant activation and damage repair, silicide gate formation, drive-in of dopants from surface-deposited layers, etc. RTA processes are fast and reliable enough to be used for monitoring of ion implant processes.

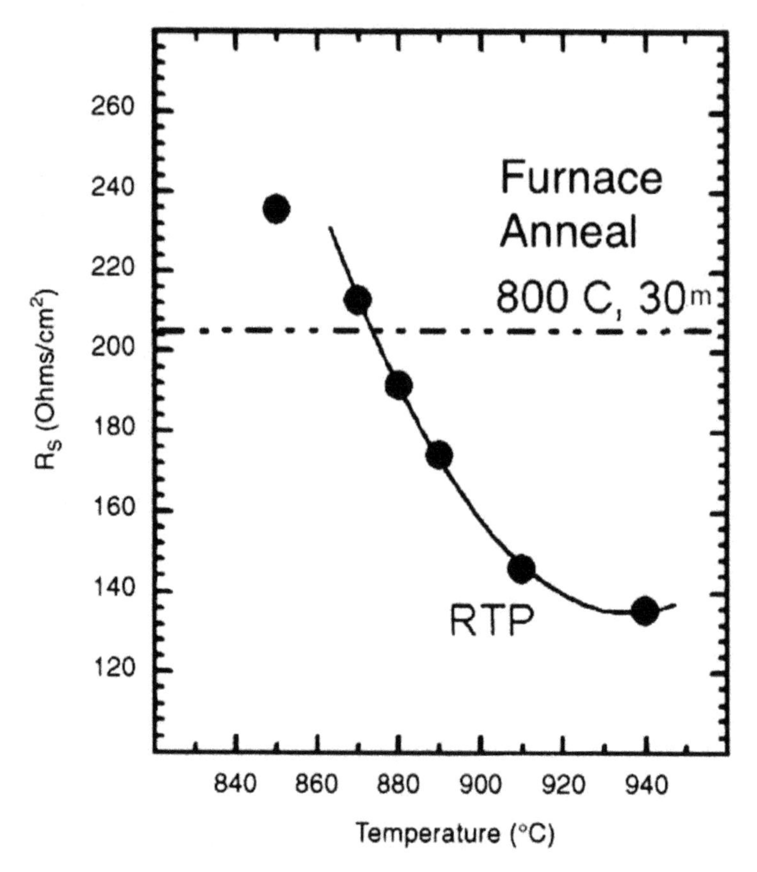

Figure 9.14 Higher activation of Si implant in GaAs shown by lower sheet resistivity achieved by RTP in comparison to 800°C furnace anneal.

In RTA, wafers are subjected to high temperature for several seconds to a few minutes, to a temperature higher than would be used for furnace processing, but for a short time, only long enough to achieve the desired process result. The total thermal input is such that unwanted diffusion is minimized. In conventional batch-type systems, the thermal mass of the wafers and the system is high, so rapid processing is not possible. RTA systems are designed for fast ramp-up and ramp-down of temperature. RTA systems take up much less space than furnace systems, where the systems become very large as the wafer diameter goes up.

9.7.3.1 History of development [3]

Early work on RTA was done using lasers. Lasers provided a means for sub-one-second heating, but resulted in huge temperature gradients. The other disadvantage of laser annealing or processing was long scan times. Although interesting from a physics or materials science point of view, laser processing was not commercially viable for III–V processing. Other rapid heating sources like arc lamps, graphite strips, and radio frequency (RF)-heated susceptors were tired. Temperature uniformity and control continued to be difficult to achieve until heating by radiant light from halogen lamps was introduced. These sources allow rapid heating and cooling in properly designed, low-thermal-mass photon boxes, where wafers are thermally isolated. Radiant heating and cooling are dominant. Heating is done in an inert atmosphere (N_2 or N_2/H_2 mixture) unless the reaction is desired as part of the process. With modern RTA systems, precise temperature control is possible. By controlling, temperature range rates and durations, thermal process can be precisely controlled.

RTA has been widely used for annealing, alloying diffusion, oxidation, silicidation, and other processes. A wide range of temperature, from a few hundred degrees to over 1000°C, is possible. An RTA cycle of 900°C for less than one minute can activate ion-implanted dopants in GaAs as effectively as a 30-minute furnace anneal. See Fig. 9.14b for the effectiveness of RTA in achieving activation of an n+ implant in GaAs in comparison to furnace annealing.

9.7.3.2 System description

Figure 9.15 shows a schematic diagram of an RTA system. The heart of the system is a cool-wall quartz chamber, which is surrounded by two banks of tungsten-halogen lamps. The total power input can be 5 to 10 kW. The system atmosphere can be controlled by the use of vacuum- and gas-handling systems. N_2 and forming gas capability is generally provided for annealing, reactive gases can be hooked up for oxidation, etc. The system is effectively a photon box [21], so all the energy is reflected into the wafers.

Temperature uniformity is controlled by controlling the power to the lamps, each lamp independently or in zones or sections [22].

Lateral gradients are minimized by use of an edge guard ring or a graphite susceptor. The susceptor should be larger in diameter than the wafer and may be coated with SiC to avoid particle generation. This is especially important as the wafer size goes up. Otherwise, thermal gradients can cause stress and warping. It may be noted that addition of the susceptor limits the rate at which the wafer can be heated or cooled. Unsupported wafers may also sag under high temperature. Slip lines may be generated due to temperature gradients, especially during cool-down, so the cooling rate should be controlled by ramping the power down in steps [23].

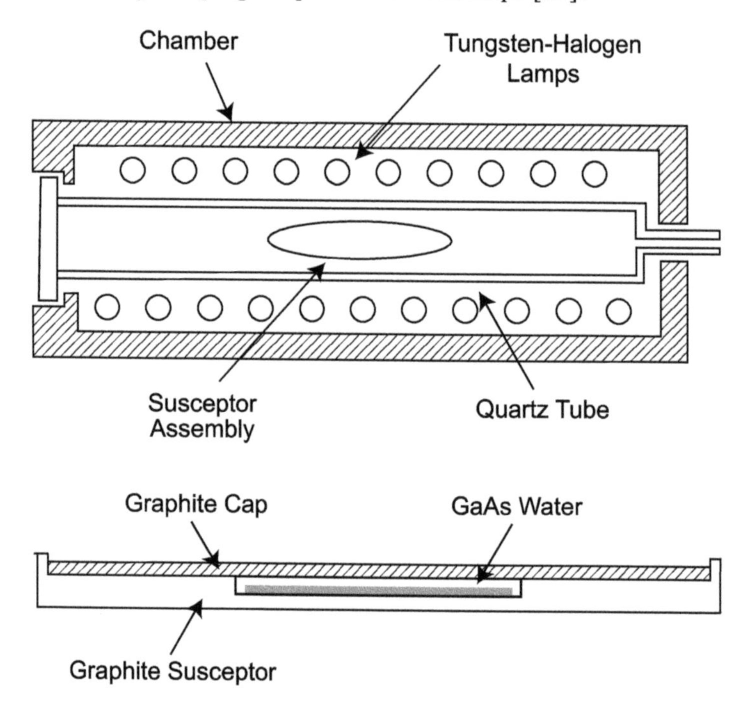

Figure 9.15 Schematic diagram of RTP system for III–V wafers.

9.7.3.3 Temperature control

Infrared (IR) sensors, pointed through IR transparent windows at the susceptor, are used to monitor the temperature. These work well at high temperature but can be a challenge at temperatures below 400°C, so a special extended-range, low-temperature IR pyrometer must be used.

9.7.4 Process Description

Wafers are capped before ion implant activation, as discussed earlier under Section 9.7.1. Silicon nitride caps are preferred in production. Capless annealing can be done in RTA, but the susceptor must be loaded with arsenic by the use of a sacrificial uncapped GaAs wafer. The system cavity must be purged of air before the process can begin, so a vacuum pump is used to evaluate the chamber and then forming gas or N_2 flow is started.

9.8 Activation of Dopants

9.8.1 n-Type Dopants

Electrical activation of ion-implanted dopants in GaAs has been studied by numerous authors [24]. Electrical characterization is performed by measuring sheet resistivity, Hall mobility, doping profiles by the C–V technique, and secondary ion mass spectrometry (SIMS) analysis. In silicon processing, the time required for dopant activation falls fast as temperature goes up, because activation is controlled by local rearrangement, whereas in GaAs processing, longer-range rearrangement is needed; therefore, the process time cannot be very short at high temperatures. For activation of Se, where the lattice damage is extensive, anneal times of over 2 minutes may be needed at 900°C [25, 26]. Se is not amphoteric and high-temperature annealing or high doses are not a problem. Activation is controlled by diffusion processes, which have a dependence of $\sqrt{D.t}$ where the diffusion coefficient has the usual $\exp(-E_a/kT)$ dependence on temperature and activation energy (see Chapter 10). A small increase in temperature can make a large change in the activation kinetics. For Se activation, the activation energy of the vacancies on the As sublattice are applicable. Tellurium being larger is harder to activate than Se.

Very high temperature is avoided because of failure of cap layers and creation of slip lines due to a gradient at very high temperatures. As has been discussed under Section 9.7.1, Si is amphoteric InGaAs, so self-compensation is a problem in achieving high activation efficiency. High-temperature, short-time annealing does not work well for silicon dopants because the Si prefers acceptor sites at high temperature (discussed earlier in this chapter). Therefore an optimum RTA cycle that uses moderate temperature (850°C) and enough time is used in practice to achieve high, reproducible activation. Figure 9.16 shows a typical RTA cycle for annealing Si-implanted GaAs [21]. Electrical profiles of dopants can be determined by *C–V* measurements and compared to those obtained by SIMS to determine activation percentage. Figure 9.17 shows the activation efficiency as a function of dose at 850°C [27, 28]. Electrical activation is seen to be higher for Si than SiF^+ for a given dose. SiF is easier to implant than Si[29] and gives the same activated implant levels as Si in furnace annealing. Activation efficiency improves for deeper, higher-energy implants because the fraction inside the surface-depleted layer goes down. As discussed previously, self-compensation dominates at high doses (used for n+ regions).

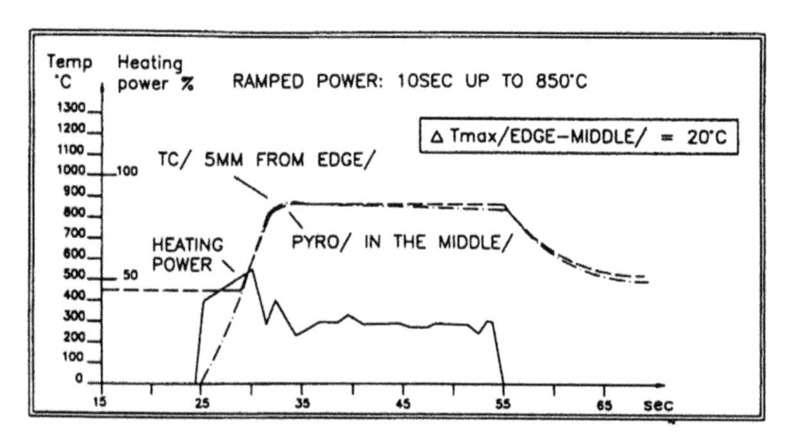

Figure 9.16 Typical thermal cycle used for annealing Si-implanted GaAs (also shows reduced photon-box effect by ramped heating power). From Ref. [21]. Reproduced by permission of ECS—The Electrochemical Society.

Sulfur activation is also difficult, compounded by diffusion during annealing. Ge and Sn are both amphoteric and exhibit poor activation.

For n+ regions, required for source and drain areas, high doping levels are needed. n+ doping levels up to 4 $\times 10^{18}/cm^3$ have been achieved by ion implantation. It may be noted that saturation of electrical activation occurs over 2–3 $\times$ $10^{18}/cm^3$. With co-implantation of Se and Si, sheet resistivity comparable to epilayers has been demonstrated [29]. Even at such high n+ doping levels, ohmic contact to n+ layers cannot be made by using simple metal contacts. Contact must be made by alloying techniques.

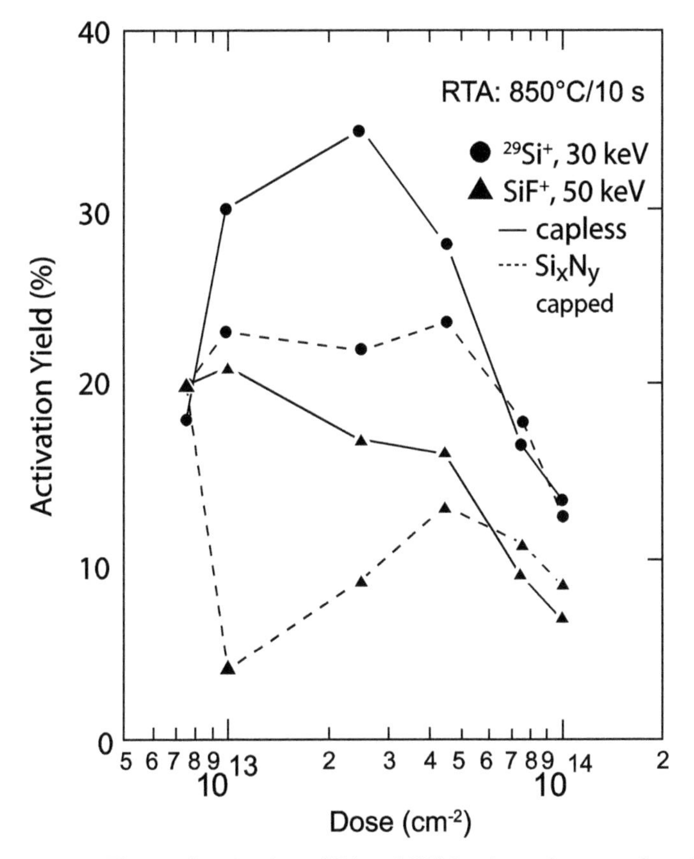

Figure 9.17 Electrical activation of Si^+ and SiF^+ implantations as a function of implant dose after RTA at 850°C. © [1992] IEEE. Reprinted, with permission, from Ref. [27].

9.8.2 p-Type Dopant Activation

Beryllium (Be), Mg, and Zn are common choices for p-type dopants. Being light, its activation efficiency is high at even low annealing temperatures. Be is a fast diffuser and as discussed previously has reliability and toxicity issues. Mg and Zn are better candidates [30] for buried layers and contact regions, respectively. Magnesium activates well up to a 10^{15} per cm^2 dose in the 815°C–850°C temperature range. It shows initial rapid diffusion followed by slow diffusion (interstitials in the tail of the distribution diffuse fast until they end up in substitutional sites, and then diffusion slows down). Anomalous diffusion takes place at concentrations above 10^{15} per cm^2.

Zinc is heavier and needs a higher activation temperature, closer to 900°C. As is well known, it shows concentration-dependent diffusion. (Details of Zn diffusion are discussed in Chapter 10.) Diffusion tails appear over 900°C anneal temperature. Cadmium shows reasonable activation. It also diffuses, and implant sources are difficult to manage.

RTA can be used to drive in zinc from deposited surface layers for making low-resistance contacts to p-type regions [31] at relatively low temperatures, because Zn is a very fast diffuser. Refractory metal layers with Zn can be deposited on the surface over p-contact areas and then subjected to RTA. Such processes may also be used to make contacts through a surface layer to the p+ layer underneath.

9.9 Device Isolation

9.9.1 Introduction

Isolation is used to confine electrical conduction to specific areas of a semiconductor-integrated circuit (IC) or active parts of the die area, so adjacent devices do not get connected, except by intentional wiring added on top. This can be accomplished either by etching away the surrounding material or by disrupting the crystal structure using ion implantation. Ideally, all areas outside of transistors, diodes, or semiconductor resistors must be converted from semiconducting- to insulating-type material. GaAs wafers have an inherent advantage over silicon in this regard because the starting wafers can be made semi-insulating. Thus for devices grown directly on GaAs, only shallow regions may need to be insulated. However, in the case of HBTs or other devices on epitaxially grown layers, for isolation to be effective, it must penetrate to larger depths through the epilayers.

In addition to separating active devices from each other, isolation reduces parasitic capacitance and leakage at high frequencies under bond pads, interconnects, and transmission lines, so large areas of wafers may be used for passive components. In analog or RF ICs, these components need to be on the insulating material in order to avoid losses due to induction.

9.9.2 Isolation by Etching

The portions of the wafer not needed for active devices can be etched away by wet or dry etching, leaving behind mesas. Practical mesa thicknesses are limited to about 0.5 μm because of step coverage issues, as interconnect metals must cross over the step. Wet etching results in V groove and dovetail profiles and metal lines must be routed only in one direction for reliable interconnects (see Chapter 6). Epigrown FET-type devices, including pseudomorphic HEMTs (PHEMTs), may be isolated by wet etching. For details of etching chemistries, the reader is referred to Chapter 6. Plasma etching can be used to make mesa that have the V groove slope in both directions and the slope of the mesa can be controlled. Therefore for shallow devices, dry etching may be utilized. For HBT devices, that use deep epilayers of the order of 2 to 3 microns, a combination of mesa etching (wet or dry) and ion implantation must be used for isolation. Alternately, mesas can be made in two or more steps, with a little flat area in between.

9.9.3 Ion Implant Isolation

Wet etches create large undercuts and result in step coverage problems. Removal of semiconductor material may also cause thermal problems due to reduced heat dissipation. Ion implant isolation overcomes these issues. Active areas are masked by photoresist patterning for protection during implantation. For example, Fig. 9.18 shows a schematic diagram for implant isolation of a planar FET-type device.

For GaAs and other III–V compounds, oxygen, boron, hydrogen, fluorine, and helium have been used as the isolating ions [32]. Heavy ions such as B^+, O^+, and F^+ are stopped relatively quickly within the crystal and can be used only for thin layers. Lighter ions such as H^+ and He^+ can penetrate deeper and hence are needed for thick epilayers to make sure that conducting interface layers below the epilayer are also converted. Crystal damage can be repaired by high-temperature annealing, so the implant isolation must be done after the annealing step used for activation of doped layers in FET processing but before any metal layers, which might block implantation, are deposited. Optimum high resistance may be achieved only after some intermediate temperature annealing, which can be done after implantation but before ohmic contact formation.

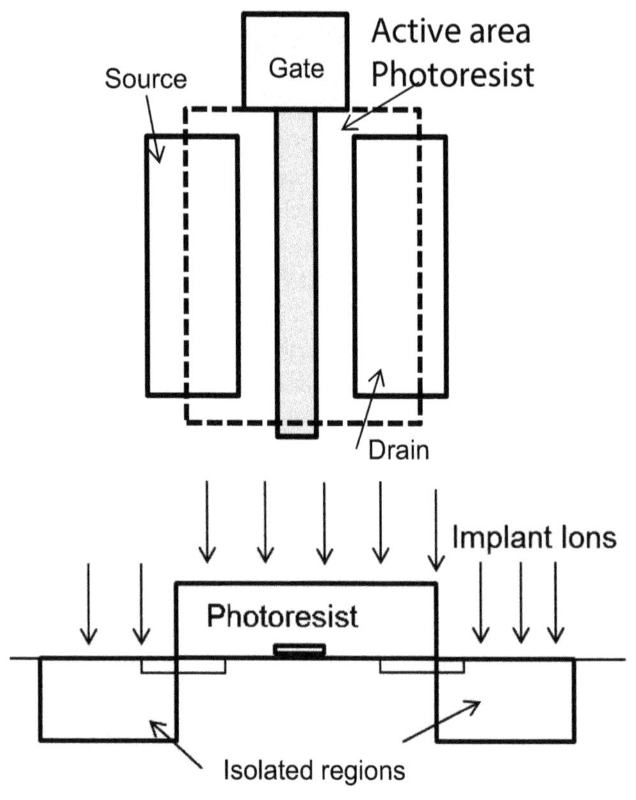

Figure 9.18 Schematic diagram showing isolation of active regions by ion implantation.

A combination of mesa and implant isolation can be used to isolate deeper epilayers. Figure 9.19 shows a schematic cross section for an HBT device where the emitter, base, and part of the collector layer have been removed by etching to facilitate implant isolation of deeper epilayers.

Energies of 50 to 150 keV are used for isolation of FETs by proton implantation. Usual doses are in the 10^{14} to 10^{15} per cm^2 range. A minimum threshold dose is needed for each species to reach the highest resistivity level possible by creating enough traps and reducing the electron mobility. If the dose accumulation is continued, the resistance reaches a plateau and eventually starts to decrease. The decrease of resistivity has been explained by enhanced hopping conduction [33]. In this mechanism, the defect sites become too numerous and close to each other, so electrons can hop from one site to another and give rise to conduction. Proton implants are stable up to 400°C, while other heavier species have stability to higher temperatures. Hot implants have been investigated for improving the stability of isolated material by dynamic annealing. As the implant temperature is increased from room temperature to 100°C and 200°C, the stability of the sheet resistance is improved to 400°C, 550°C, and 700°C, respectively [34]. At higher annealing temperatures, the implant damage can be recovered to turn the material back to semiconducting.

Doses needed for boron and oxygen are generally a tenth of those needed for protons. These ions are very stable because of low diffusion constants at ordinary temperatures but do need additional postimplant annealing to achieve optimum levels of isolation. For protons or He, normal high-temperature processing steps may be enough for optimum isolation resistance. For high throughput and optimum isolation, including low surface leakage, damage profiles must be matched to the doped layer profile that is being isolated. Simulation software like TRIM [35] is useful when determining the optimum implant recipe, especially if multiple implants are needed.

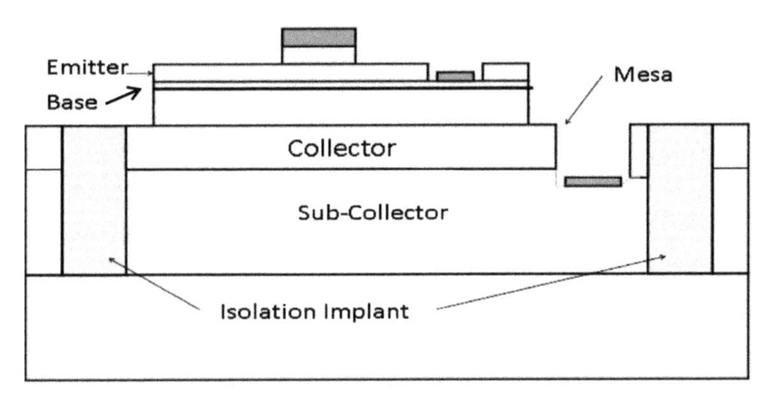

Figure 9.19 Ion implant isolation in an HBT-type device showing a combination of mesa and ion implantation.

9.9.4 Mechanism

In ion implantation, the bombardment of semiconductor material by ions causes damage, trapping implanted species and creating electron traps. Implantation damage, caused by inert atoms like He, for example, creates Frenkel pairs, interstitials, and vacancies in equal concentration. The Fermi level is pinned in the middle of the bandgap. Partial annealing of some defect complexes results in the reduction of electron hopping and the elimination of more mobile defects. In the case of H^+ implantation, protons may form complexes with donors like Si that pin the Fermi level. Pearton et al. have measured dissociation energies of donor–H and acceptor–H complexes in GaAs, by Arrhenius plots, to be about 1.25 eV for Si–H [36]. For carbon-doped material, the energies are much lower, about 0.6 eV [37].

9.9.5 Isolation-Related Reliability Issues for HBT

Proton implantation does have an advantage over He for HBT isolation. It eliminates the need for expensive high-voltage ion implantation systems. Presence of hydrogen has been related to reliability issues in GaAs ICs (see Chapter 22).

Proton isolation has been re-evaluated by Sealy et al. [38] for GaAs, InP, and InGaAs. They have found that implanting into substrates at high temperatures reduces the threshold dose to about 10^{15} per cm^2 and increases stability as well as resistance. The optimum annealing temperature was reduced to 350°C by using hot implants. In production practice, where implants have to be done using a photoresist that must be stripped easily after implantation, even 100°C is not practical.

Proton implants are also used for the reduction of base–collector capacitance (C_{bc}) in HBTs to increase f_{max} (recall $f_{max} = \sqrt{\frac{f_t}{8\pi R_b C_{bc}}}$ Eq. 2.97, Chapter 2), which increases device gain and bandwidth [39].

Helium ions may not reach deep enough below the base, and the implant may have to be complemented by protons, as suggested by Min-Chung Ho [39]. Use of the lowest possible H^+ dose is recommended to avoid hydrogen passivation of carbon in the base region.

A combination of deep proton and shallow He implants can be used with less reliability risk. The hydrogen dose should be minimized for achieving optimum isolation resistance for the thermal processing seen by HBTs in the process, for example, the polyimide cure cycle, 300°C for 30 minutes.

Diffusivity of H^+ in GaAs has been reported to be [40]

$$D_{H+} = 1.5\times10^{-5}\exp\left(-\frac{0.62}{kT}\right) \quad (9.9)$$

An activation energy of 0.66 eV has been reported by Johnson et al. [41]. At 525°C, used by Pearton et. al. [40] diffusion lengths of tens of microns are expected.

At 300°C, the diffusion length drops to the micron range. If the proton concentration is low, the number of protons diffusing into the base region from a distance of 5 μm (isolation region to active base) can be an order of magnitude below $1E19/cm^3$, the concentration of hydrogen being already present there.

During operation, at a junction temperature of 225°C, the diffusion length can be below 1 micron and the amount of available hydrogen below detectable levels. By optimizing hydrogen and He and possibly other shallow ions like boron, isolation or capacitance reduction can be achieved without affecting reliability.

References

1 J. W. Mayer, L. Erikson and J. A. Davies, *Ion Implantation in Semiconductors*, Academic Press (1970).

2 S. K. Ghandhi, *VLSI Fabrication Principles, Silicon and Gallium Arsenide*, Wiley Interscience, New York (1983).

3 S. Wolf and R. N. Tauber, *Silicon Processing for the VLSI Era*, Lattice Press (1986).

4 J. Lindhard, M. Scharff and H. Schiott, Range concepts and heavy ion ranges, *Mat-Fys. Med. Dan. Vid. Selsk.*, **33**, 1 (1963).

5 G. Carter and W. A. Grant, *Ion Implantation in Semiconductors*, Wiley, New York (1976).

6 J. F. Gibbons, W. S. Johnson and S. W. Mylrioe, *Projected Range Statistics, Semiconductor and Related Materials*, John Wiley and Sons (1975).

7 D. K. Brice, *Ion Implantation Range and Energy Distributions*, Plenum, New York (1975).

8 *SRIM Stopping and Range of Ions in Materials*, www.SRIM.org.
9 S. Furukawa, H. Matsumura and H. Ishiwara, *Jpn. J. Appl. Phys.*, **43**, 2973 (1972).
10 S. J. Pearton, W. S. Hobson and C. R. Abernathy, Ion implantation processing of GaAs and related compounds, in *Ion Beam Processing of Advanced Electronic Materials, Proc. Mater. Res. Soc.*, **147**, 261 (1989).
11 S. Tiku and W. Duncan, *J. Electrochem. Soc.*, **132**, 2237 (1985).
12 L. S. Vanasupa, M. D. Deal and J. D. Plummer, Modelling activation of implanted Si in GaAs, *J. Electrochem. Soc.*, **138**, 2134 (1991).
13 T. Taniguchi, in *Advanced Thermal Processing of Semiconductors, 9th Int. Conf. RTP*, 25 (2001).
14 F. H. Eisen, Ion implantation in III–V compounds, *Rad. Eff.*, **47**, 99 (1980).
15 W. V. McLevige et al., *J. Appl. Phys.*, **48,** 3342 (1977).
16 S. Tiwari, J. C. De Luca and V. R. Deline, *Inst. Phys. Conf. Ser.*, **74**, 83 (1984).
17 J. P. de Souza, I. Danilon and H. Boudinov, *Appl. Phys. Lett.*, **68**, 535 (1996).
18 S. M. Sze, *Semiconductor Device Physics and Technology*, Wiley, New York (1985).
19 E. Rimini, *Ion Implantation: Basics to Device fabrication*, Kluwer Academic (1995).
20 J. Kashara, M. Arai and T. Watanabe, *J. Electrochem. Soc.*, **126**, 1997 (1978).
21 Z. Nenyei, H. Walk and T. Knarr, Defect-guarded rapid thermal processing, *J. Electrochem. Soc.*, **140**, 1728 (1993).
22 A. Tillman, *Mater. Res. Soc. Symp. Proc.*, **342**, 376 (1994).
23 T. Sakurda et al., *Int. Conf. InP Relat. Mater.*, 425 (2001).
24 S. J. Pearton, Ion implantation in III-V semiconductor technology, *Int. J. Mod. Phys. B*, **7**, 4687 (1993).
25 S. Tiku, Rockwell International, unpublished data.
26 Z. Jianghong et al., *Microwave and Millimeter Wave Technology Proc. ICMMT'98*, 714 (1998).
27 J. P. DeSouza and D. K. Sadana, *IEEE Trans. Electron. Devices*, **39**, 166 (1992).
28 J. P. de Souza, D. K. Sadana and H. J. Hovel, *Mat. Res. Soc. Symp. Proc.*, **144,** 495 (1989).

29 M. Sun, S. Tiku et al., *Electron. Lett.*, **34,** 1155 (1998).

30 S. Tiwari, J. C. DeLuca and V. R. Deline, *Appl. Phys. Lett.*, **45**, 1204 (1984).

31 S. Tiku, *Electron. Lett.*, **21**, 1091 (1985).

32 S. J. Pearton et al., *J. Appl. Phys.*, **74,** 6580 (1993).

33 J. P. DeSouza, I. Danilov and H. Boudinov, *Appl. Phys. Lett.*, **68**, 535 (1996).

34 S. Ahmed, R. Gwillian and B. J. Sealy, *Semicond. Sci. Technol.*, **16**, L64 (2001).

35 *TRIM and SRIM, Stopping and Range of Ions in Matter*, www. Srim.org.

36 S. J. Pearton, C. R. Abernathy and J. Lopata, *Appl. Phys. Lett.*, **59**, 3571 (1991).

37 H. Fushimi and K. Wada, *J. Cryst. Growth*, **145**, 420 (1994).

38 B. J. Sealy et al., Proton implantation revisited, *III-V Rev.*, **16**(9) (2003/2004).

39 M. C. Ho, PhD thesis, University of California, San Diego (1995).

40 S. J. Pearton et al., Comparison of H+ and He implant Isolation, *J. Vac. Sci. Technol.*, **13**, 1 (1995).

41 N. M. Johnson, *Mat. Res. Soc. Symp. Proc.*, **262**, 369 (1992).

Chapter 10

Diffusion in III–V Compound Semiconductors

10.1 Introduction

Diffusion processes are fundamental to an understanding of semiconductor processing, particularly for silicon devices, where diffusion processes are used for doping. In III–V semiconductors diffusion is not so useful and ion implantation or epigrowth are used for making n- and p-type layers. However, p+-type layers are made in III–V devices by diffusion and are used in optoelectronic product processing. Additionally, avoidance of diffusion is critical to making good and reliable III–V electronic and optical devices. An understanding of diffusion phenomena is useful for an understanding of common issues of unwanted diffusion, alloying, interface smearing, dopant redistribution, migration of impurities, etc., many of which are relevant for improving device reliability.

Diffusion is responsible for the movement of impurity or dopant atoms in semiconductor crystals as well as the movement of host crystal atoms by self-diffusion. It takes place by different mechanisms, the dominant ones being interstitial and substitutional diffusion. Diffusion requires the presence of defects like vacancies and thermal excitation and always follows exponential behavior of the Arrhenius form $D = D_0 \exp(-E_A/kT)$ [1]. Radioactive tracer

III–V Integrated Circuit Fabrication Technology
Shiban Tiku and Dhrubes Biswas

ISBN 978-981-4669-30-6 (Hardcover), 978-981-4669-31-3 (eBook)
www.panstanford.com

analysis of the diffusion of Ga and As in GaAs gives the following values for the diffusion coefficients [2]:

$$D_{\text{Ga}} = 0.1 \exp\left(-\frac{3.2}{kT}\right)$$

$$D_{\text{As}} = 0.7 \exp\left(-\frac{5.6}{kT}\right)$$

10.1.1 Rate Equations

A large number of reactions or phenomena in semiconductor processing are temperature-dependent rate processes. These generally follow an Arrhenius type of equation, which has an exponential (E/kT) behavior. For a process to proceed there is generally an energy hump to overcome for going from one state to another. The same is true for a reaction to proceed from reactants to products or an atom to move by crossing a constriction in the lattice. See Fig. 10.1 for a simple schematic diagram. All these processes have temperature dependence. The probability of an atom to be at energy E^*, with enough energy to cross the hump, is given by the probability function of the type below:

$$\text{Probability} \propto \exp\left\{-\frac{(E^* - E)}{kT}\right\} \tag{10.1}$$

where k is the Boltzman constant and T is the absolute temperature.

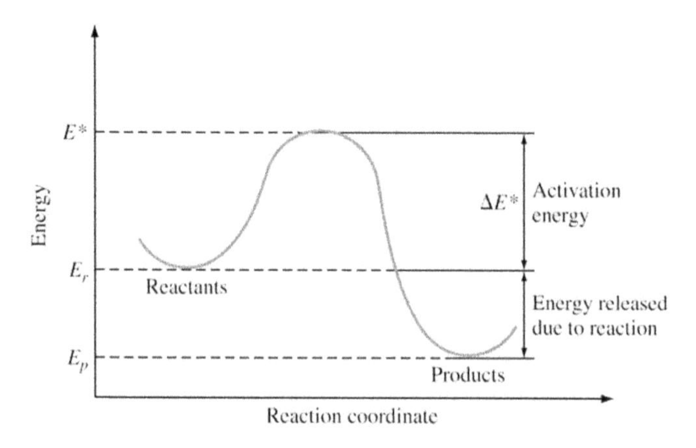

Figure 10.1 Activation energy concept: hump between two states, reactants to products.

The fraction of atoms that have this energy E^* is given by

$$\frac{n}{N} = \exp\left\{-\frac{E^*}{kT}\right\} \tag{10.2}$$

Therefore the rate of reaction takes the form

$$\sim C \exp\left(-\frac{E_{\mathrm{A}}}{kT}\right) \tag{10.3}$$

where E_{A} is commonly known as the activation energy.

10.2 Diffusion Basics

Fick's first law of diffusion states that the flux of a diffusing species is proportional to the concentration gradient of that species.

$$J = -D\frac{\mathrm{d}C}{\mathrm{d}x} \tag{10.4}$$

where J is the flux in atoms/cm^2/sec, D is a constant under fixed conditions, and $\mathrm{d}C/\mathrm{d}x$ is the concentration gradient. The diffusivity or the diffusion coefficient, D, depends upon the mechanism, temperature and crystal or material structure (including presence of defects, dislocations, etc.). The diffusion constant is particularly strongly dependent upon temperature and has the form described in Section 10.1 (Eq. 10.3):

$$D = D_0 \exp\left(-\frac{E}{kT}\right) \tag{10.5}$$

10.2.1 Basic Mechanisms

Diffusion is a result of atomic fluctuations and wandering of lattice atoms or impurities that cause a net movement of material atoms down a concentration gradient. There are two basic mechanisms and a few complicated mechanisms that are prevalent is semiconductor or metal material systems.

10.2.1.1 Interstitial mechanism

The first one is the interstitial mechanism, shown in Fig. 10.2a. Here an interstitial atom can move into a neighboring interstitial site and keep moving by repeating the process without needing any vacancies or substitutional sites. In the zinc-blende face-centered cubic (fcc) structure, the interstitial sites are in a tetrahedral arrangement (see Fig. 1.3) between Si atoms in a silicon lattice or between Ga and As atoms for GaAs. An interstitial atom has to overcome the energy barrier to jump to an adjacent interstitial. The atomic fluctuation magnitude has to be large enough to accomplish this. Therefore the jump frequency v_i is proportional to $\exp(-E_m/kT)$.

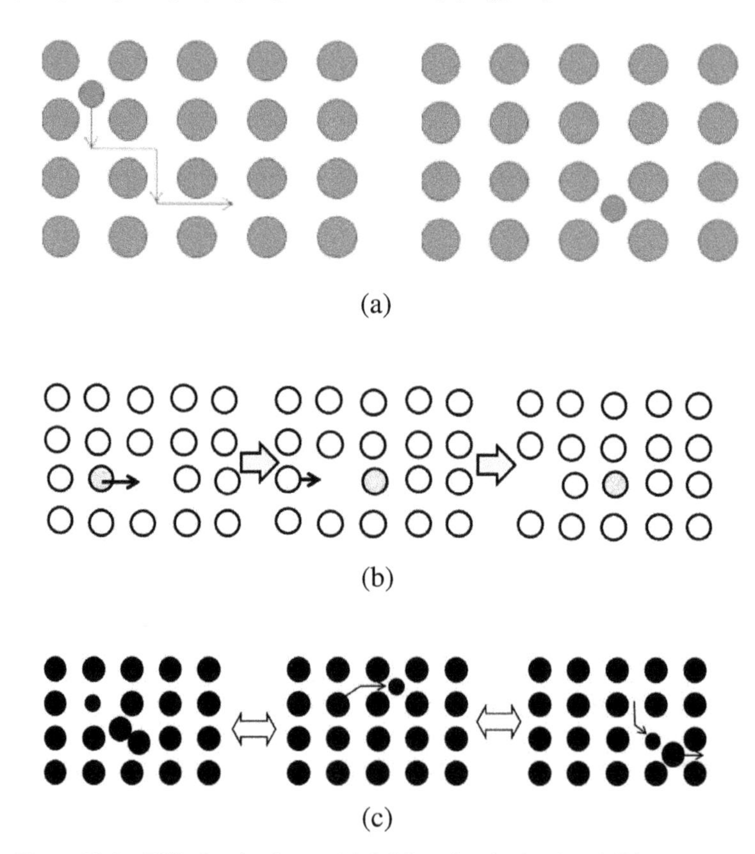

Figure 10.2 Diffusion by interstitial (a) and substitutional (b) movement and kick-out mechanism (c).

10.2.1.2 Substitutional mechanism

For diffusion to occur by this mechanism, a vacancy has to be created first and then the atom must overcome the energy barrier to jump (see Fig. 10.2b). Therefore the jump frequency is given by

$$V_s \approx \exp\{-(E_m + E_f)\} \tag{10.6}$$

The activation energy is the sum of the formation and migration energies.

10.2.1.3 Kick-out mechanism

A number of impurities like Zn, Be, and Mg diffuse in GaAs by the so-called kick-out mechanism. In these cases, the diffusing element kicks out a Ga atom from its substitutional site into an interstitial, as shown in Fig. 10.2c. This mechanism proceeds at high concentrations of the diffusing species. The charge balance equation is shown below:

$$\mathrm{Zn_{Ga}}^- + \mathrm{I_{Ga}}^{2+} \Leftrightarrow \mathrm{Zn_I}^+$$

10.2.1.4 Interstitial-substitutional mechanism

In III–V semiconductors many impurities can occupy both substitutional and interstitial sites. Zn is the most common example. The rate of diffusion along the interstitial mechanism is much faster than the other one. The effective jump frequency is a combination of the two mechanisms in parallel:

$$V_{\mathrm{eff}} = vs\frac{Ns}{(Ni+Ns)} + vi\frac{Ni}{(Ni+Ns)} \tag{10.7}$$

Dissociative diffusion: In semiconductor crystals, a substitutional atom can dissociate into an interstitial and a vacancy. Therefore the overall diffusion rate is a function of both interstitial and vacancy concentrations. The rate is slower than pure interstitial diffusion. The effective diffusion coefficients are correspondingly

$$D_{\mathrm{eff}} = D_{\mathrm{i}}\frac{Ni}{(Ni+Ns)} \tag{10.8}$$

for very fast interstitial diffusion, or

$$D_{\mathrm{eff}} = D_{\mathrm{v}}\frac{Nv}{(Ni+Ns)} \tag{10.9}$$

for fast substitutional diffusion.

Diffusion can also act over dislocations and grain boundaries. The rate can be very high because of the availability of open sites.

10.2.2 Impurity Diffusion Rates in GaAs

In silicon, the diffusion of vacancies is the same as Si self-diffusion.

$$D_v = 9000\ \exp\left(-\frac{5.13}{kT}\right)$$

In III–V compounds, vacancies on the two sublattices have different diffusion coefficients. These are determined by annealing studies to be (see page 122 in Ref. [2])

$$D(V_{Ga}) = 2.1 \times 10^{-3}\ \exp\left(-\frac{2.1}{kT}\right) \quad (10.10)$$

$$D(V_{As}) = 7.9 \times 10^{3}\ \exp\left(-\frac{4.0}{kT}\right) \quad (10.11)$$

Radioactive tracer analysis is used to determine diffusion coefficients of Ga and As. The diffusion coefficients are given by [2]

$$D_{Ga} = 0.1\ \exp\left(-\frac{3.2}{kT}\right) \quad (10.12)$$

Recent data $E_A = 4$ eV [3]

$$D_{As} = 0.7\ \exp\left(-\frac{5.6}{kT}\right)$$

Recent data $E_A = 4.4$ eV [3]

More recent data for Ga has the activation energy near 4.0 eV [3]. Impurities that substitute the Ga site, Al, Si, Au, Cu, etc., diffuse by substitutional mechanism assisted by vacancies (extent depending upon the concentration of vacancies). Therefore the activation energy is near 4 eV. At high concentrations the interstitial-substitutional mechanism may take over and the diffusion can be very rapid.

Impurities on the As sublattice move only on the As sublattice. Se, S, and Te are slow diffusers. More recent data for As self-diffusion between 900°C and 1000°C, measured by using phosphorus and antimony as tracers, shows an activation energy of 4.4 eV [4]. Diffusion constants and activation energy for some common elements of interest are shown in Table 10.1.

Table 10.1 Diffusion constants and activation energy of diffusion of common dopants and impurities in GaAs

Impurity	D_0 (cm^2/s)	E_0 (eV)	
Au	2.9×10^{1}	2.64	
Be	7.3×10^{-6}	1.2	
Cr	4.3×10^{3}	3.4	
Cu	3×10^{-2}	0.53	
Li	5.3×10^{-1}	1.0	
Mg	2.6×10^{-2}	2.7	
Mn	6.5×10^{-1}	2.49	
O	2×10^{-3}	1.1	
S	1.85×10^{-2}	2.6	
Se	3.0×10^{3}	4.16	
Sn	3.8×10^{-2}	2.7	
Zn	96.8	4.07	[5]
Zn (RTP)		2.6 eV at [2e^{19}/cm^3]	

10.3 Diffusion Equations for III–V Semiconductor Processing

Fick's first law and derivation of the second law are illustrated in Fig. 10.3 assuming a portion of the crystal with area A between two planes is separated by a small distance dx. The rate of accumulation in the region between these planes is equal to $A.(\partial N/\partial t)$.d$x$. The net flux entering the region is $-A$dJ.

Using the continuity equation

$$A\left(\frac{\partial N}{\partial t}\right).dx = -A\mathrm{d}J \tag{10.13}$$

Writing dJ as $(\partial J/\partial x)$.dx, Fick's first law (Eq. 10.3) can be written as

$$\frac{\partial N}{\partial t} = \frac{\partial}{\partial x}\left\{D\cdot\frac{\partial N}{\partial x}\right\} \tag{10.14}$$

This equation is the general form of Fick's second law.

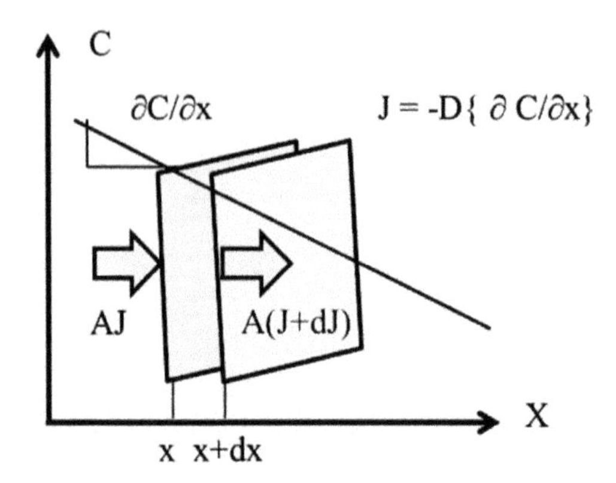

Figure 10.3 Diffusion down a gradient and diffusing flux.

10.3.1 Constant Diffusion Coefficient

The diffusion equation for the case of the constant diffusion coefficient is discussed first. This assumption is valid at low concentrations of dopants or impurities. In this case, Eq. 10.14 reduces to

$$\frac{\partial N}{\partial t} = D.\frac{\partial^2 N}{\partial x^2} \tag{10.15}$$

where N is the volume concentration (atoms/cm^3) and D is the diffusion coefficient in cm^2/sec. In the presence of an electric field another term dependent on the field strength and mobility can be added to this equation $\left(\mu E.\frac{\partial N}{\partial x}\right)$.

10.3.1.1 Thin-film solution

The integration of Fick's second law equation is not easy. Its solution is inferred indirectly as follows:

If a quantity α of an impurity is deposited as a thin film on a long rod, as shown in Fig. 10.4, and the rod is annealed for time t, the concentration of the impurity along the rod is given by

$$C(x, t) = \frac{\alpha}{2\sqrt{\pi D t}}.\exp\left(-\frac{x^2}{4Dt}\right) \tag{10.16}$$

That this is the correct solution of Eq. 10.15 is proven by differentiating this equation and applying boundary conditions [1].

For a pair of semi-infinite solids, with an infinite number of thin films (α), the concentration can be derived by integrating α from 0 to infinity.

$$C(x, t) = \frac{C0}{2} \sqrt{\pi Dt}.\int_0^\infty \exp\left\{-\frac{(x-\alpha)^2}{4Dt}\right\} dx \quad (10.17)$$

Let $\frac{x-\alpha}{2\sqrt{Dt}} = \eta$

The above Eq. 10.17 can be written as

$$C(x, t) = \frac{C0}{p}. \int_{-\infty}^{\pi/(2\sqrt{Dt}))} \exp\left\{-\frac{\eta^2}{4Dt}\right\} d\eta \quad (10.18)$$

(integration from $-\infty$ to $\pi/2\sqrt{(Dt)}$)

This integral appears often in problems of this kind and is not easy to solve, but its values are available in tabular form called the error function.

$$\text{erf}(z) = \frac{2}{\sqrt{\pi}}.\int_0^z \exp\{-\eta^2\} d\eta \quad (10.19)$$

$$\text{erf}(\infty) = 1$$

Equation 10.18 can be written as:

$$C(x, t) = \frac{C0}{2}. \left\{1+\text{erf}\left(\frac{x}{2\sqrt{Dt}}\right)\right\} \quad (10.20)$$

Note that all the diffusion equations show the dependence on $\sqrt{Dt}$. Therefore, the product of diffusion coefficient and time is an important process parameter.

10.3.1.2 Diffusion from a constant source

In a diffusion process for doping, wafer is exposed to a constant concentration from the surface. In this case the concentration is given by

$$N(x, t) = N_0 \,\text{erfc}\left(\frac{x}{2\sqrt{Dt}}\right) \quad (10.21)$$

where the complementary error function $\text{erfc}(x) = 1 - \text{erf}(x)$.

The form of the distribution profile is shown in Fig. 10.4b.

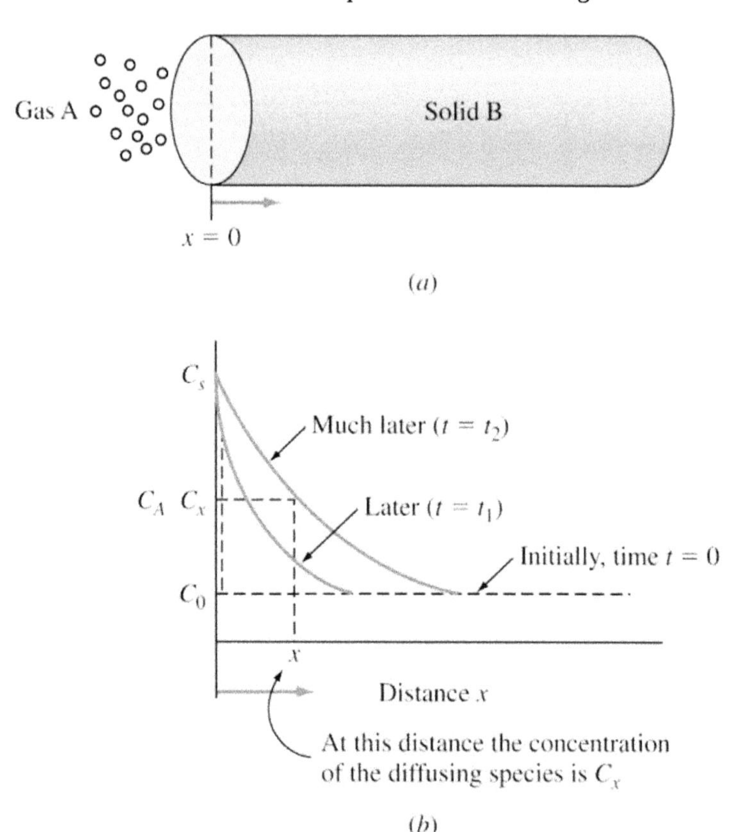

Figure 10.4 Diffusion profiles from a constant source, showing a complementary error function profile.

10.3.1.3 Diffusion from a limited source

In the case of limited source volume, the solution of the form given by Eq. 10.16, discussed above under the thin-film case [2]

$$N(x, t) = \frac{Q_0}{\sqrt{\pi D t}} \exp\left\{\left(-\frac{x}{2\sqrt{Dt}}\right)^2\right\} \tag{10.22}$$

This has the usual Gaussian form, shown in Fig. 10.5, for different times. Note that the peak concentration drops, but the total integral under the curve remains constant.

10.3.1.4 Concentration-dependent diffusion coefficient

The simple cases of the constant diffusion coefficient do not represent the majority of real diffusion processes. In general, the diffusion coefficient goes up with concentration of the impurity. As the diffusion coefficient dependence gets stronger, the diffusion front becomes sharper. The dependence is simplified as the following equation:

$$D = KD_0N^x \tag{10.23}$$

A few simple cases are shown in Fig. 10.6.

10.3.2 Interstitial-Substitutional Diffusion [7]

Here the diffusion rate is the sum of two mechanisms (refer to Section 10.2.1.4 and equations in the previous section):

$$\frac{\partial(Ns+Ni)}{\partial t} = \frac{\partial}{\partial x}\left(\mathrm{Ds}\frac{\partial Ns}{\partial x} + \mathrm{Di}\frac{\partial Ni}{\partial x}\right) \tag{10.24}$$

where "s" and "i" are for substitutional and interstitial atoms, respectively. In III–V semiconductors, for example, for Zn diffusion, it is $N_s >> N_i$. Therefore

$$\frac{\partial(Ns)}{\partial t} \approx \frac{\partial}{\partial x}\left(Ds + Di\frac{\partial Ni}{\partial Ns}\right)\frac{\partial Ns}{\partial x}$$

Effective diffusivity is

$$D \approx Ds + Di\frac{\partial Ni}{\partial Ns}$$

$$\approx Di\frac{\partial Ni}{\partial Ns} \tag{10.25}$$

because $D_i >> D_s$.

10.4 Measurement of Diffused Layers

Diffused layers can be analyzed by secondary ion mass spectrometry (SIMS) analysis for determining the doping profile. However, electrical measurements are faster for routine monitoring of diffusion processes. Contactless or four-point probe measurements are made to determine the overall conductivity and infer the carrier concentrations. Sheet resistivity, R_S, of the diffused layer is not constant but varies, the concentration dropping from the surface value. The sheet resistivity in ohms/square is a measure of the whole doped layer. The inverse of R_S, dj, is a measure of the conductivity.

10.5 Diffusion in GaAs

10.5.1 Diffusion by Periodic Table Groups

Ga and impurities that substitute for it migrate along the Ga sublattice, the same being true for As. Group III impurities move on the Ga sublattice and group V ones move along the As sublattice. Impurities for group II, Be, Cd, Zn, etc., are shallow p-type dopants and move along the Ga sublattice. Since each Ga has four As neighbors, it is generally accepted that these impurities move by the interstitial-substitutional or the kick-out mechanism, as described above, which show a concentration-dependent diffusion coefficient. Therefore, rapid movement is seen in high-concentration regions and slow movement in low-concentration regions.

Impurities from group VI, Se, S, and Te, are shallow n-type and move along the As sublattice. These impurities diffuse extremely slowly and are believed to follow a di-vacancy mechanism. More recent studies with S are pointing toward the kick-out mechanism [8]. Group IV impurities like C, Ge, Si, and Sn can be n- or p-type and move on both sublattices. These are also slow diffusers. Table 10.1 lists diffusion constants and activation energies for some common impurities of concern in GaAs. Arsenic loss takes place over 600°C in GaAs, and under this temperature, the diffusion coefficients are extremely small. Diffusion therefore is not used for incorporating dopants in GaAs except for fast diffusers like Zn. This fast diffusion is utilized, for example, in the making of contacts from the surface to the base layer of heterojunction bipolar transistors (HBTs).

Silicon is an important dopant in III–V compounds. Its diffusion depends upon electron concentration. Diffusion is very slow because it is believed to require movement of the following complex [3]:

$$Si_{Ga}^{+} + V_{Ga}^{3-} \Leftrightarrow (Si_{Ga}\,V_{Ga})^{2-}$$

From the perspective of impurity avoidance, copper, being a fast diffuser, must be avoided.

10.5.2 Zn Diffusion in GaAs

Zinc is used commonly as a p-type dopant where large p+-type contact layers are needed. Substitutional zinc acts as a singly ionized acceptor, whereas interstitial Zn is a singly ionized donor. For this situation, the dissociation reaction can be written as

$$Zn_s^- \Leftrightarrow V_{Ga}^0 + Zn_i^+ + 2e^- \quad (10.26)$$

Applying the mass action principle to this equation

$$K_1 = n^2 \cdot \frac{[Zn_i^+][V_{Ga}^0]}{[Zn_s^-]} \quad (10.27)$$

The hole concentration is due to substitutional Zn, $p = [Zn_s]$, and $np = n_i^2$.

$$[Zn_i] = K_2 \cdot \frac{[Zn_s^-]}{[V_{Ga}^0]} \frac{1}{n^2} \quad (10.28)$$

The electron concentration $n = [Zn_s]$.

$$[Zn_i] = K_2 \cdot \frac{[Zn_s^-]^3}{[V_{Ga}^0]} \quad (10.29)$$

Differentiating this equation and combining with Eq. 10.25

$$D \approx D_i \frac{\partial Ni}{\partial Ns} = K_3 \cdot D_i \cdot \frac{[Zn_s^-]^2}{[V_{Ga}^0]} \quad (10.30)$$

Therefore the effective diffusion coefficient is

$$D \propto D_i \cdot [Zn_s]^2 \quad (10.31)$$

This N^2 dependence is shown in Fig. 10.5 (as $D \approx N_A^2$) and is seen to have a sharp diffusion front. This sharp diffusion front is seen in practice. The N^2 dependence is a consequence of charge exchange of two, from Zn_s^- to Zn_i^+.

Diffusion by the kick-out mechanism, by the movement of neutral Zn_i and positively charge V_{Ga}^+, has been observed for diffusion from a dilute Ga–Zn source and not by doubly or triply charged Ga interstitials [5]. The charge balance equation in this case is the following:

$$Zn^- \Leftrightarrow V_{Ga}^+ + Zn_i^0 + 2e^- \quad (10.32)$$

Diffusion coefficients are in the 10^{-10} range at 700°C and 10^{-9} at 1000°C [9].

The mechanism is believed to be the same for Zn in InP [10].

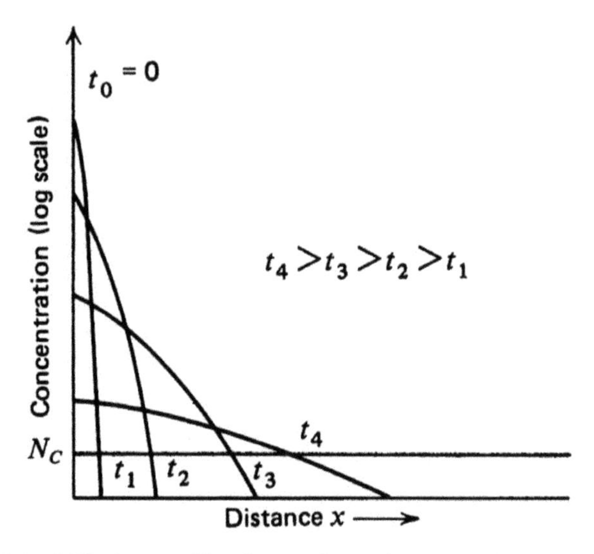

Figure 10.5 Diffusion profiles from a limited source, showing changing Gaussian profiles. From Ref. [2]. Copyright © 1983. Reproduced with permission of John Wiley & Sons, Inc.

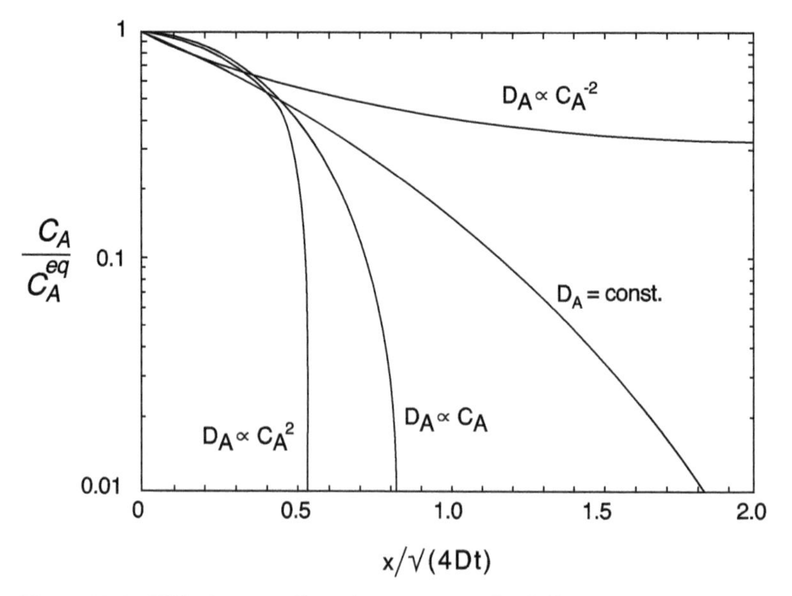

Figure 10.6 Diffusion profiles for cases with different concentration dependence compared to constant diffusion coefficient.

10.5.3 Sulfur Diffusion in GaAs

Of the n-type group VI elements, S, Se, and Te, sulfur has the highest diffusion constant at a reasonable process temperature. Si, Ge, and Sn are amphoteric; therefore the activation efficiency of Sn as an n-type dopant is low. A lot of work has been done on sulfur diffusion over the years, but practical applications have been limited. The most common source for sulfur is GaS. A temperature over 800°C is needed for diffusion; therefore arsenic overpressure is needed. At a temperature between 860°C and 900°C, a diffusion coefficient of 10^{-12} cm^2/sec has been measured from a sulfur dopant in the tube and 2×10^{-13} for a 10:1 GaS–As source [2].

Since Zn is used for p+-type diffusion for GaAs/GaAlAs diode lasers, sulfur from GaS has been used to create an n+ layer and suppress defects due to high Zn concentration [11]. GaS was used in a sealed quartz tube at 825°C as the source, with an arsenic overpressure at 1 atm. The diffusion time was 9 hours. The diffusion coefficient was determined to be 3.5×10^{-13} cm^2/sec in GaAs and 5.5×10^{-12} in $Ga_{0.7}AlAs$. Double-heterostructure lasers were fabricated on liquid-phase epitaxy (LPE)-grown structures.

Nitrogen overpressure is more practical and has been used instead of As overpressure. In 30 minutes at 850°C, n-type layer with a concentration of $8 \times 10^{12}/cm^3$ and mobility of 4200 cm^2/sec was achieved in liquid-encapsulated Czochralski (LEC)-grown material [12].

The diffusion mechanism has been verified to be the kick-out mechanism using an arsenic interstitial [8, 13].

10.6 Diffusion Systems

Production systems for diffusion in silicon consist of large quartz tubes that can hold a large number of wafers at high temperature under controlled atmospheric conditions. The temperature profile is controlled very accurately so that a large batch of wafers can be processed at the same time. Since diffusion is not used for III–V integrated circuit (IC) fabrication, production systems for large volume are not available.

Requirements for diffusion systems are controlled incorporation being over many orders of magnitude, similar to ion implantation;

material on the surface being removable; and preferably system having high throughput. Historically, diffusion in III–V compounds has been done in sealed quartz tubes that are inserted into a furnace for the required time. Diffusion time is long for steady-state conditions to be established. The system is then cooled down and the tube broken to unload the material.

Open-tube systems are preferred in production. High-purity quartz tubes are used that can hold a large number of wafers. One side of the tube is used for loading and the other side for the gas inlet, temperature monitors, etc. To ovoid thermal gradients, the movement of the boat in and out is very slow. Inert carrier gases are used to pressurize the tube. Solid diffusion sources are placed inside the tube. It may be noted that if a dopant is available in liquid or gaseous form (diethyl zinc [DEZ] for Zn), it can be admitted with a carrier gas. Figure 10.7 shows a schematic diagram for an open-tube system.

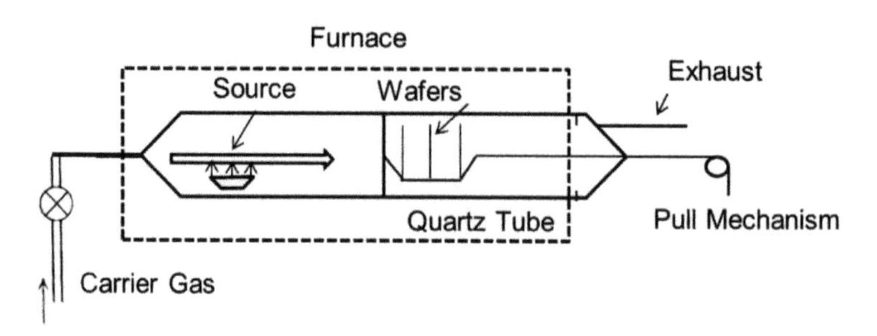

Figure 10.7 Open-tube diffusion system.

With III–V compounds, the loss of a group V element can be a concern in spite of overpressure. Therefore a cap layer may be used (SiO_2). Elemental dopants can cause surface damage, so dilute sources may be employed. For Zn, this can be $ZnAs_2$ and for sulfur it can be GaS.

Ternary-phase diagrams are used for determining temperatures at which the source and the wafers are held. Liquid phases are avoided for fear of corrosion.

A much safer process involves the use of doped glasses. Spin-on glasses (SOGs) for various dopants are available. These protect the surface and provide a steady-state source at the surface. Using a

$ZnO–SiO_2$ source on the surface, in an open-tube system, with a N_2/H_2 atmosphere, junction depth data has been collected. Diffusion has been shown to be extremely rapid and shows $\sqrt{Dt}$ dependence. See Fig. 10.8 [14].

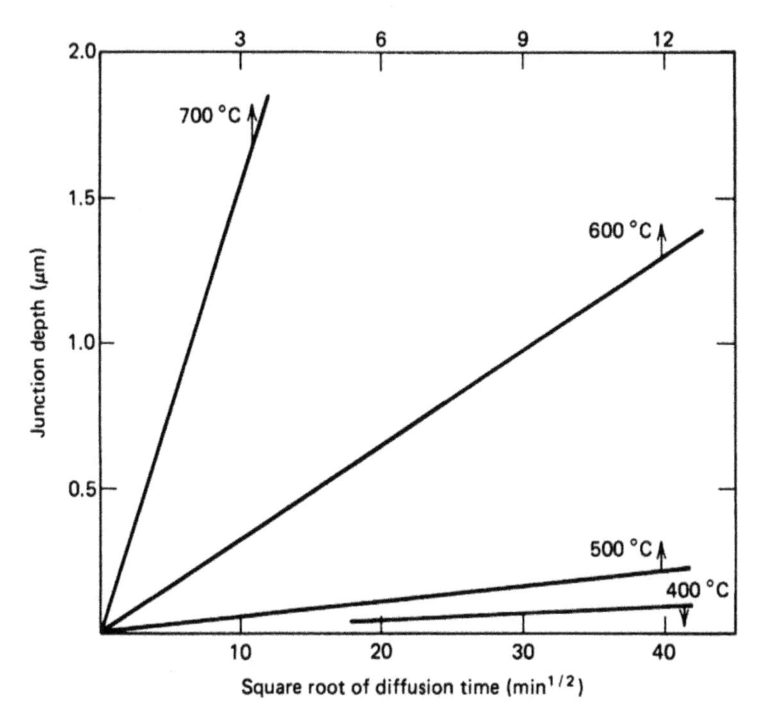

Figure 10.8 Junction depth as a function of square root of time for open-tube Zn diffusion in GaAs. From Ref. [2]. Copyright © 1983. Reproduced with permission of John Wiley & Sons, Inc.

Steep doping profiles are seen for Zn diffusion with a surface concentration of $10^{20}/cm^3$. Surface concentration versus sheet conductivity data is shown in Fig. 10.9 [9].

For n-type dopants like sulfur, the doping profile shows erfc diffusion. Sulfur has been diffused using sulfur in a sealed tube or Ga_2S from a gas source or solid GaS and As in a tube system at 860°C–910°C. For a GaS–As source, diffusion coefficients of 10^{-12} cm^2/sec for a 2:1 source and 2×10^{-13} for a 10:1 source have been determined. Dependence on $\sqrt{t}$ was verified, as shown in Fig. 10.10 [15].

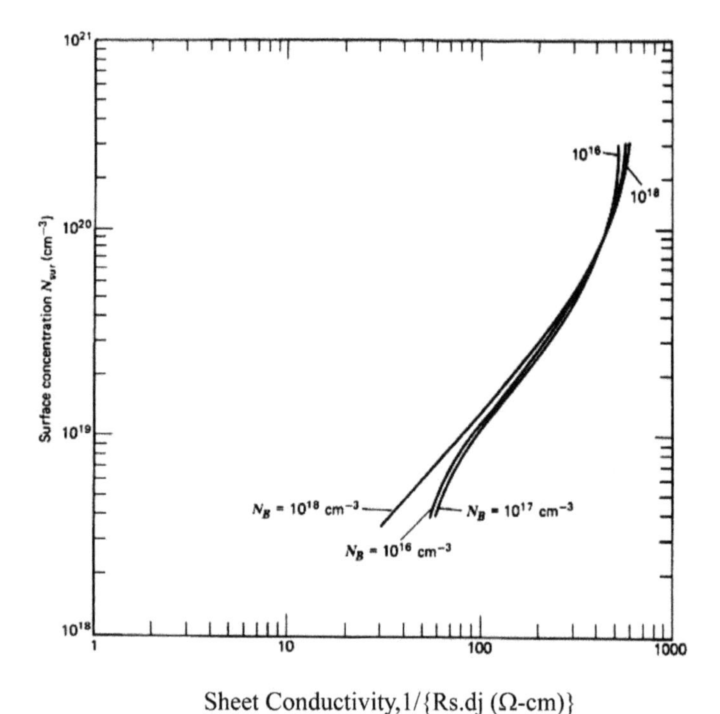

Figure 10.9 Surface concentration vs. sheet conductivity data for Zn. From Ref. [2]. Copyright © 1983. Reproduced with permission of John Wiley & Sons, Inc.

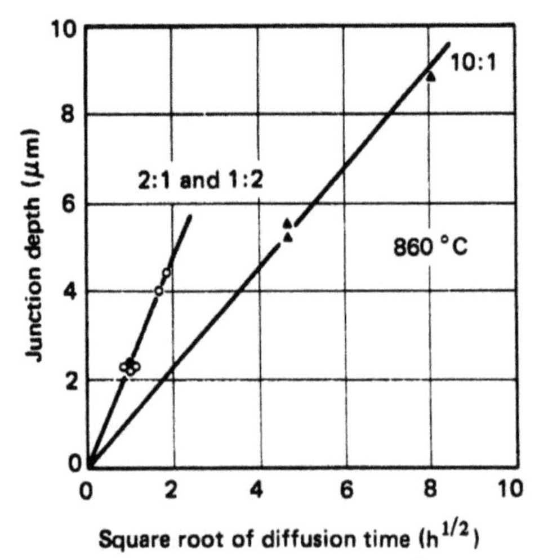

Figure 10.10 Sulfur diffusion in GaAs, showing $\sqrt{t}$ dependence. Reprinted from Ref. [15], Copyright (1974), with permission from Elsevier.

10.7 Rapid Thermal Diffusion

Diffusion using a rapid thermal processing (RTP) system is called rapid thermal diffusion. Details of the system are described in Chapter 9. In silicon processing, phosphorus and arsenic are diffused for doping using rapid thermal diffusion. In III–V optoelectronic devices, the use of rapid thermal diffusion is also prevalent. However, in III–V IC processing rapid thermal diffusion is used only for p+ contacts.

Zinc diffusion from a W–Zn source can be used to drive in Z to make contact to the p+ base layer of HBTs [16]. Zn diffusion into GaAs has been performed by using a source of $ZnSiO_3$ on the surface and using an RTP system for drive-in [17]. The diffusion activation energy was seen to shift from 1.1 eV at low temperatures to 2.6 eV above 790°C. The diffusion mechanism has been verified to be the kick-out mechanism involving Zn_i and $[V_{Ga}]$. The diffusion coefficient is a strong function of concentration and temperature. At moderately high temperature, it takes only a few seconds to achieve required junction depths. The diffusion coefficient is in the range of 10^{-10} cm^2/sec at a $2e^{19}/cm^3$ concentration at 700°C.

On the n-type side, sulfur is the only element that has a high enough diffusion coefficient to be a candidate for rapid thermal diffusion. An $(NH_4)_2S_x$ source (on the surface) has been used to produce a channel layer for metal–semiconductor field-effect transistors (MESFETs), with a transconductance of 190 mS/mm. The diffusion coefficient was measured to be $5e^{-12}$ at 900°C [18]. This value agrees well with activation energy of 2.6 eV.

References

1. P. G. Shewmon, *Diffusion in Solids*, McGraw Hill (1971).
2. S. K. Ghandhi, *VLSI Fabrication Principles, Silicon and Gallium Arsenide*, Wiley Interscience, New York (1983).
3. U. Gosele et al., Diffusion in GaAs and related compounds, *Defect Diffusion Forum*, **143–147**, 1079 (1997).
4. R. F. Scholz and U. Gosele, *J. Appl. Phys.*, **87**, 704 (2000).
5. H. Bracht and S. Brotzman, *Phys. Rev. B*, **71**, 1152116 (2005).
6. D. Shaw, Ed., *Atomic Diffusion in Semiconductors*, Plenum Press, New York (1973).

7. L. R. Weisberg and J. Blanc, Diffusion with interstitial-substitutional equilibrium, Zn in GaAs, *Phys. Rev.*, **131**, 1548 (1963).
8. B. Tuck and R. G. Powel, Sulfur diffusion by kick out mechanism, *Appl. Phys.*, **14**, 1317 (1981).
9. C. H Ting et al., *J. Appl. Phys.*, **42**, 2247 (1971).
10. D. L. Kendall, *Diffusion in Semiconductors and Semimetals*, Vol. 4, Eds., R. K. Willardson and A. C. Beer, Academic, New York (1968).
11. H. Nishi, Sulfur diffusion in GaAs/GaAlAs, *Appl. Phys. A*, **67**, 579 (1998).
12. F. C. Prince, M. Oren and M. Lam, *Appl. Phys. Lett.*, **48**, 546 (1986).
13. M. Uematsu et al., *Appl. Phys. Lett.*, **67**, 2863 (1995).
14. R. J. Field and S. K. Ghandhi, Open tube diffusion of Zn in GaAs, *J. Electrochem. Soc.*, **129**, 1567 (1982).
15. H. Matino, Reproducible sulfur diffusion into GaAs, *Solid State Electron.*, **17**, 35 (1974).
16. S. Tiku, Zn diffusion for self-aligned contact to HBT, *Electron. Lett.*, **21**, 1091 (1985).
17. G. Rajeshwaran, K. B. Kahen and D. J. Lawrence, *J. Appl. Phys.*, **69**, 1359 (1991).
18. J. L. Lee, *J. Appl. Phys.*, **85**, 807 (1999).

Chapter 11

Ohmic Contacts

11.1 Introduction

Ohmic contacts are needed in semiconductor circuit fabrication for allowing the current to flow in and out of semiconductor devices with insertion of negligible resistance. The electric contact must be *ohmic*, that is, its *I–V* curve must follow Ohm's law in forward and reverse directions. Ideal ohmic contacts are impossible to achieve in practice; however, contact schemes that do not introduce a resistance high enough to directly impact device performance have been developed for Si and III–V compound semiconductors. Ohmic contact metallurgical systems must adhere well to the semiconductor, introduce minimal stress, be compatible with the interconnect metallization, have low specific contact resistivity so as to be compatible with micron-size devices, and be stable through subsequent processing. The morphology should be smooth and the metal must not spike in through thin epilayers like the heterojunction bipolar transistor (HBT) base. Achieving these requirements may be a simple task in silicon, but it has been a major stumbling block in III–V semiconductors. When aluminum is deposited on silicon, and heat-treated, it forms a highly doped p+ region at the interface and an excellent ohmic contact. Most metals when deposited on III–V

III–V Integrated Circuit Fabrication Technology
Shiban Tiku and Dhrubes Biswas

ISBN 978-981-4669-30-6 (Hardcover), 978-981-4669-31-3 (eBook)
www.panstanford.com

semiconductors form a rectifying contact. Metal systems and contact formation procedures for ohmic contacts on III–V semiconductors are therefore more complicated and are the subject of this chapter.

11.2 History

In the 1960s, when Gunn devices were first introduced, low contact resistance was not critical, simple low-temperature alloying metal layers were used. The researchers' focused attention on Au–Ge systems as these provide a low alloying temperature. As newer devices like metal–semiconductor field-effect transistors (MESFETs) were being developed in the 1970s, the need for lower contact resistance became apparent. With the introduction of high-electron-mobility transistors (HEMTs) and HBTs, requirements became even more demanding. The most common metal systems show electrical degradation with further processing and time. Ohmic contacts introduce parasitic resistance in field-effect transistors (FETs) and HBTs, which is not desired for high-frequency performance and low-noise figures. Improvements have been sought over the last few decades, achieving perfect low resistance, and temperature stability has been an elusive goal. However, contact metallurgies that fulfil the needs of III–V and III–N compound semiconductor circuits have been developed and are in production.

The most extensively used technique to make ohmic contacts to GaAs is to deposit metal alloys containing dopants on the surface and then diffuse the dopants in by a process of melting called *alloying*. If the metal does not melt, then the contact is sintered to achieve intermixing. To improve high-temperature stability, metal systems that use high-temperature alloying or sintering, as well as refractory and barrier metals, have been used to suppress continued diffusion (e.g., that of Au) during operation. Table 11.1 lists some common metallization systems that have been tried for making contacts to GaAs. For extensive reviews of ohmic contact techniques and mechanisms, readers are referred to reviews by Rideout [1], Yoder [2], Braslau [3], and Murakami [4].

Table 11.1 Ohmic contact metallization systems

Material/s	r_c (ohm.cm^2)	Reference
n-Type GaAs:		
AuGe (12 wt% Ge)	10^{-5}	[5]
Au:Ge/Ni	5×10^{-7}	
Au:Sn	5×10^{-5}	
AuSn		
Ag:Sn	3×10^{-7}	
InNi		
Pd/Ge	5×10^{-5}	
Ni In W	10^{-6}	[6]
p-Type GaAs		
Ag:Zn		
Au:Be		
Au:Zn	3×10^{-6}	[7]
Ti/Pt/Au	10^{-6}	
Pt/Ti/Pt/Au	3×10^{-7}	[8]
Ni/Ti/Pt	low 10^{-7}	[9]

For references to other systems, refer to review articles [1–4] and a historical summary in Baca et al. [10].

11.3 Theory of Metal–Semiconductor Ohmic Contacts

A brief review of the theory of metal–semiconductor contacts is given here. For details, readers should refer to semiconductor texts (e.g., Ref. [11]).

When a metal contacts an n-type semiconductor surface, the valance and conduction energy bands of the semiconductor bend to level the Fermi levels of the two. See Fig. 11.1. The depletion width (discussed in the section on Schottky diodes in Chapter 12) varies with donor concentration. The carrier transport across this depends upon the depletion width and the temperature. For a lightly doped case (see Fig. 11.1a), N_d < 1E17/cm^3, the depletion width is wide, and electrons can cross only by thermionic emission over the energy barrier. When the semiconductor is highly doped (over 1E18/cm^3), the depletion width is narrow, and the electrons can tunnel through the barrier by a process called *field emission*. For intermediate cases of doping, both mechanisms may be operative. Contact resistance has been derived for these cases based on *I–V* relationships and verified by experiments [7, 12]. The current density for field emission has the form [11]

$$J \approx \exp\left\{-\frac{q\phi_b}{E_{00}}\right\} \tag{11.1}$$

where ϕ_b is the barrier height (shown in Fig. 11.1) and E_{00} s a characteristic energy defined by

$$E_{00} = \left\{\frac{qh^-}{2}\right\} \cdot \sqrt{\frac{N_d}{m^* \varepsilon}} \tag{11.2}$$

where $h^- = h/2\pi$, h is Plank's constant, m^* is the effective mass, and ε is the dielectric constant. The tunneling current increases exponentially as the square root of N_d and inversely with ϕ_b.

For normal metal–GaAs contacts, the barrier is ~0.8 eV for most metals. This contact exhibits Schottky behavior. For an ideal ohmic contact, the barrier height needs to be reduced to zero (see Fig. 11.1) [13].

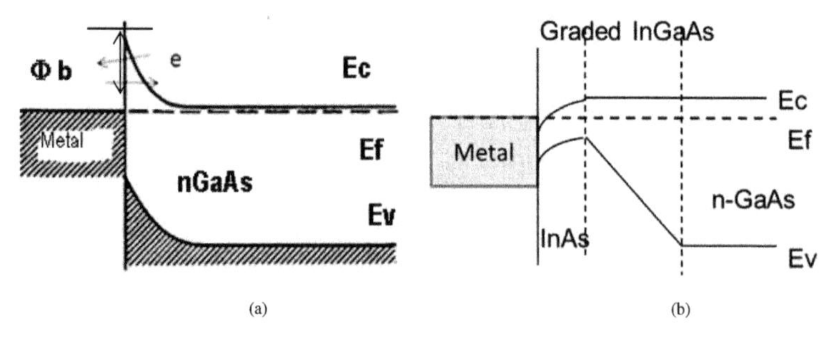

Figure 11.1 (a) Energy band diagram for metal contact to n-type semiconductor and (b) metal contact to InAs on graded InGaAs on GaAs.

In practice, contact schemes are developed on the basis of the creation of highly doped thin layers between the semiconductor contact and the metal or reduction of barrier height, thus reducing either the width or the height of the barrier or both. Figure 11.2 shows the I–V curves for a tunnel junction–type contact and a barrier reduction–type contact in comparison to an ideal ohmic contact.

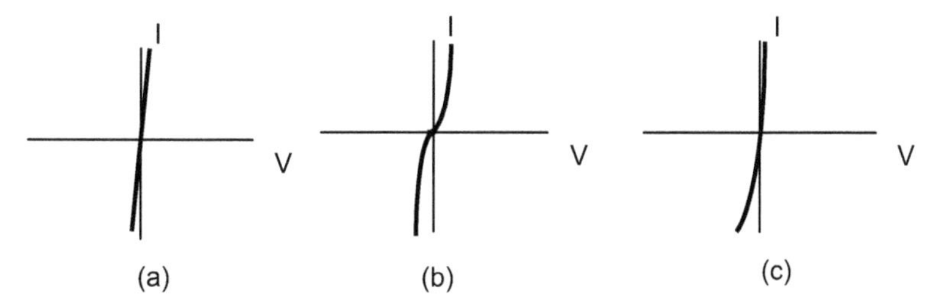

Figure 11.2 I–V curves for an ideal ohmic contact and those based on tunnel junction or high–low (barrier reduction type) junction.

11.3.1 Contact Resistance

If a rectangular bar of semiconducting material is contacted by ohmic contacts on two sides, the total resistance will be the sum of the semiconductor bulk resistance and the resistance of the two contacts. See Fig. 11.3 for details of the geometry.

$$R = 2R_c + R_s \tag{11.3}$$

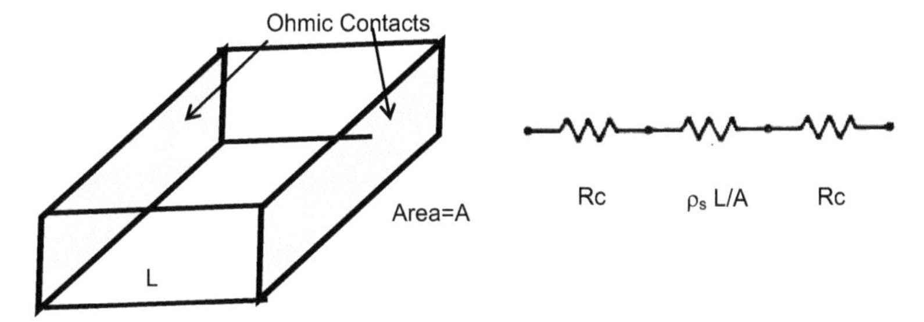

Figure 11.3 Resistance of a rectangular bar of semiconductor showing the resistance of the bar and the two contacts.

The resistance of the semiconductor, R_s, is equal to $\rho_s.L/A$. The contact resistance, R_c, is related to the specific contact resistance, r_c, of the contact by the relation

$$R_c = \frac{r_c}{A} \tag{11.4}$$

If the contact area is small, the resistance will be large. Most semiconductor devices are made with planar, horizontal contacts, not vertical contacts, as in this example, as shown in Fig. 11.4a, even if the device is vertical like an HBT. Therefore the contact resistance is defined to be the resistance between the metal contact and an imaginary line at the edge of the metal. This contact resistance is a function of both the specific contact resistance of the contact and the characteristics of the semiconducting layer under it. As the current flowing into this region is collected by the contact, its magnitude drops off as the distance increases (see Fig. 11.4b). The transfer length is defined as the length at which the current drops to 1/e and is a measure of the quality of the contact. For devices of micron dimensions, transfer lengths of the same order are required.

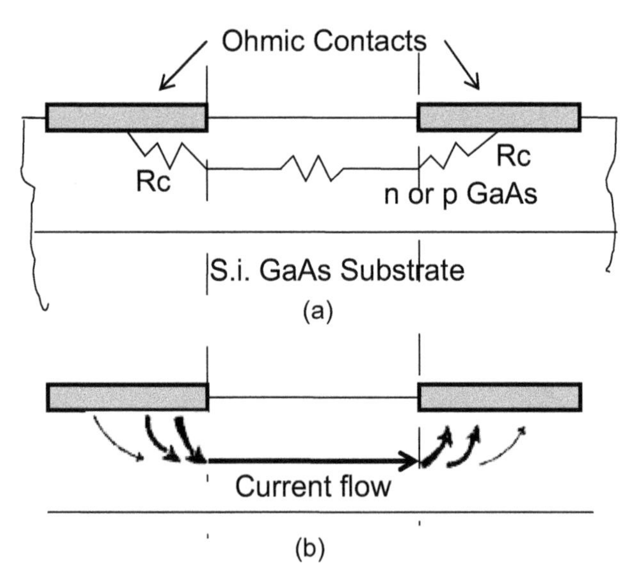

Figure 11.4 Schematic diagram of ohmic contacts to a sheet of semiconducting material showing the resistances (a) and the current flow (b).

11.4 Contact Resistance Measurement by TLM

Contact resistance is measured using a transmission line method (TLM) pattern for process development and production monitoring

[4, 14]. Although process control monitor (PCM) structures that take less space and processing steps can be used, these are not utilized commonly because of associated errors (e.g., isolation can be avoided with circular contacts).

In this model, the conductive layer is assumed to be thin with respect to the width of the contact. Various network models can be used to calculate the contact resistance as a function of sheet resistance R_{sh} and r_c, and all give the same result.

Figure 11.5 shows the representation of the contact as distributed resistors R, conductance G, and capacitors C like a transmission line. The transmission line parameters R and G are given by

$$R = \frac{R_{sh}}{W} \tag{11.5}$$

$$G = W\frac{1}{r_c}(j\omega C) \tag{11.6}$$

where W is the width of the contact, C is the capacitance per unit area, and ω is the frequency. The voltage and current along the contact are given by well-known transmission line equations:

$$V(x) = V_1 \cosh(\gamma x) - I_1 Z \sinh(\gamma x) \tag{11.7}$$

$$I(x) = I_1 \cosh(\gamma x) - V_1 Z \sinh(\gamma x) \tag{11.8}$$

where Z is the characteristic impedance of the transmission line and the γ is the propagation constant.

$$Z = \sqrt{\left(\frac{R}{G}\right)} \tag{11.9}$$

and

$$\gamma = \sqrt{RG} \tag{11.10}$$

If the analysis is restricted to $\omega = 0$ and the contact is assumed to be infinite in length, then

$$Z = R_c = \sqrt{\frac{r_c . R_{sh}}{W}} \tag{11.11}$$

and

$$\gamma = \sqrt{\frac{R_{sh}}{r_c}} \tag{11.12}$$

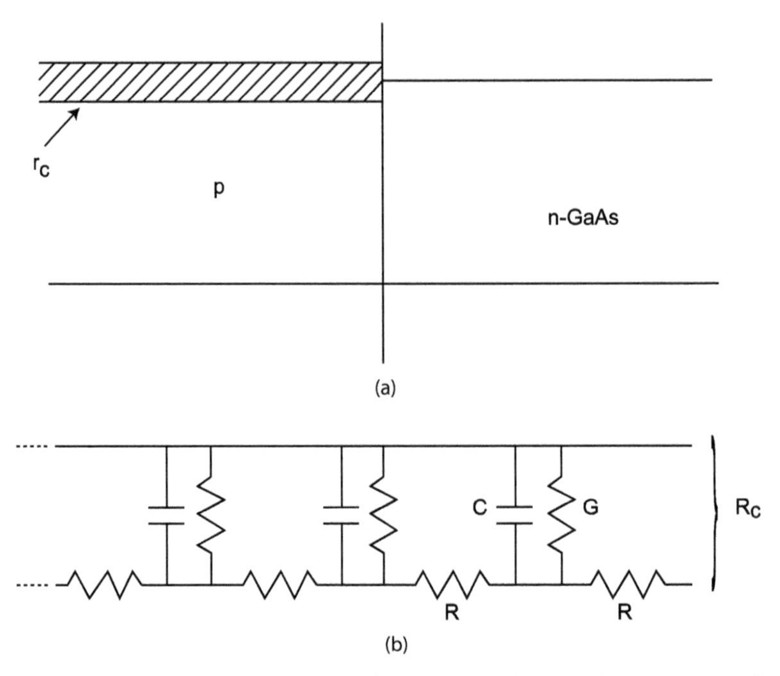

Figure 11.5 (a) Physical model for ohmic contact showing the contact metal with specific contact resistance of r_c over bulk semiconductor of resistivity ρ. (b) Transmission line model for the contact.

The transfer length L_t is the inverse of γ.

$$L_t = \sqrt{\frac{r_c}{R_{sh}}} \quad (11.13)$$

The contact resistance can be derived for the case of finite length:

$$R_c = \sqrt{\frac{r_c.R_{sh}}{W}}\,.\coth\left(\frac{d}{L_t}\right) \quad (11.14)$$

Using Eq. 11.13 in 11.14

$$R_c = \left\{\frac{R_{sh}L_t}{W}\right\}.\coth\left(\frac{d}{L_t}\right) \quad (11.15)$$

For $L_t << d$, the above equation can be simplified as

$$R_c = \frac{R_{sh}L_t}{W} \quad (11.16)$$

This approaches the infinite case when d becomes several times L_t. If the specific contact resistance is not low and small contacts must be used for space reasons, the more complex formula, Eq. 11.15, must be used.

The PCM pattern used to measure contact resistance is shown in Fig. 11.6. Large-area ohmic contacts are formed on a rectangular strip of semiconducting material, defined by implant isolation or mesa etching. The contact area is kept constant and typically the contact spacing (L) is varied from a few microns to tens of microns. The total resistance measured, R, should be given by

$$R = 2R_c + R_{sh}.\frac{L}{W} \qquad (11.17)$$

When the resistance is plotted as a function of length, L, a straight line should result (Fig. 11.6). The intercept on the R axis is $2R_c$. Sheet resistance of the semiconductor, R_{sh}, can be determined from the slope of the line. The intercept on the length axis is related to the transfer length. An error may be introduced into this analysis if the semiconductor material under the contact changes substantially, so its sheet resistance is different from R_{sh}. In this case the transfer length calculation will be in error.

In production, the PCM structure may have fewer number of length variations, for example, only three lengths of constant width. The structure may include width variation for fixed length. When data from the second set is plotted as $1/R$ versus width, ΔW, the lithography process bias can be calculated. (Fig 11.6c).

In summary, the common equations for ohmic contacts are

$$R_c = \frac{r_c.R_{sh}}{W} \text{ for } L_t << d \qquad (11.18)$$

and

$$r_c\, R_{sh}.L_t^2 \qquad (11.19)$$

(From Eq. 11.13)

It is obvious that the transfer length of a good ohmic contact should be small. In practice this should be of the order of tenths of a micron for micron-size contacts.

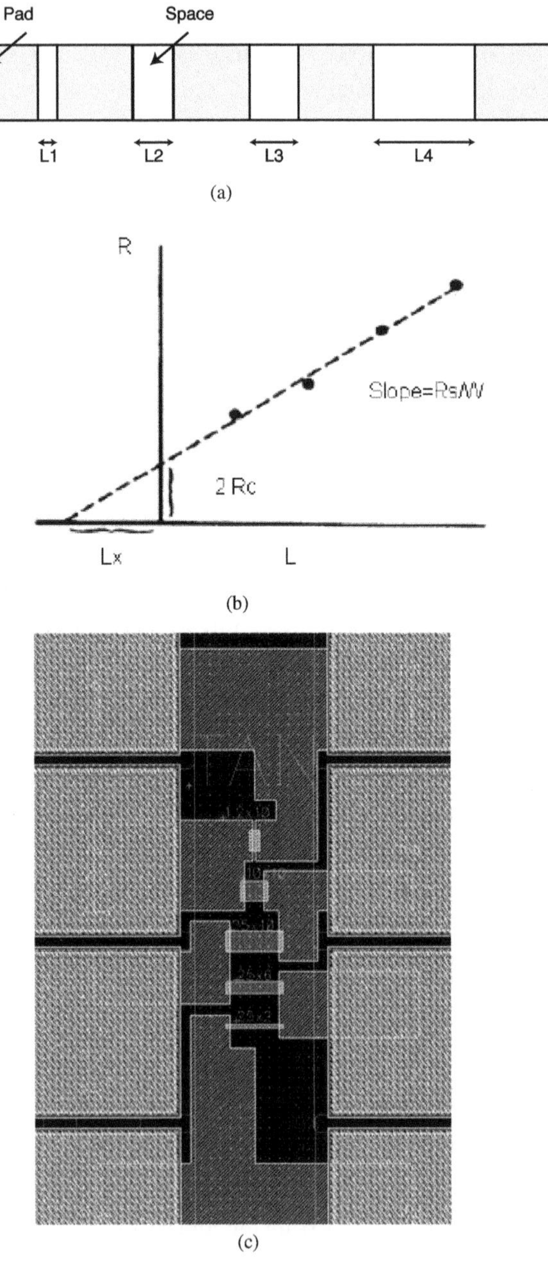

Figure 11.6 TLM measurement of (a) contact layout and (b) straight-line fit to the resistance vs. length data, and layout of PCM structure used to measure TLM and ΔW (c).

11.5 Ohmic Contact Technology for n-Type Contacts

The most common ohmic contacts that are made in the III–V semiconductor integrated circuit (IC) industry are the n-type contacts to FETs and HBTs and p-type contacts to HBT base layers. In FET-type devices, contacts may be needed for heavily doped cap layers for HEMT-type devices or medium-doped layers for ion-implanted FETs or HBT collectors. Contacts to n-type layers are more prevalent and are discussed first. The contacts to p-type layers, which are easier to dope heavily, are discussed later in this chapter.

11.5.1 Epigrown Contacts

11.5.1.1 Contacts with heavy donor doping

In GaAs, n-type dopants cannot be diffused or ion-implanted at high concentration levels needed for field emission–dominated contacts (see Chapter 10). In FET-type devices, fabricated by ion implantation, n+ source and drain regions are created by ion implantation with doping levels up to 4E18/cm^3. Metal–organic chemical vapor deposition (MOCVD) and molecular beam epitaxy (MBE) can be used to grow heavily doped contact layers that can be removed from the circuit area, where not needed. Among the common dopants used, Si and Te have been reviewed under epigrowth (Chapter 4). Doping levels over 6E18/cm^3 (close to 1E19/cm^3) have been achieved in MBE. Ge is found to have stronger amphoteric behavior than Si and offers no advantages. In MOCVD, higher levels of doping can be achieved for Te, but it is easier to grow another contact layer of InGaAs on top and dope it with Te to even higher levels, thus exploiting high doping and low barrier height. For doping levels achievable with a few semiconducting materials, the reader is referred to Table 11.2 [15].

Table 11.2 Maximum doping level

Material	n-Type doping /cm^3	p-Type doping /cm^3
Si	1×10^{20}	1×10^{20}
GaAs	5×10^{18}	2×10^{20}
InGaAs	2×10^{19}	1×10^{20}
InGaP	6×10^{18}	1×10^{19}

11.5.1.2 Contacts with lower barrier height

The second approach to reducing contact resistance is to reduce the barrier height at the metal–semiconductor interface. However, reduction of barrier height is not easy for contacts to GaAs because of Fermi level pinning. Therefore, another lower-bandgap material is deposited first. Ge, InAs, and InGaAs have been used for this purpose. An in situ ohmic contact with a heavily doped Ge layer was first reported by Stall et al. [16]. Ge has a low bandgap and can be doped heavily with arsenic. When gold is deposited on this Ge layer, contact resistance of 1E-7 ohm.cm^2 can be achieved. This type of contact scheme may have advantages in circuits needing patterned deposition for different device types. In HBTs and HEMT-type devices, use of InGaAs-based layers has become common. The Fermi level of InAs is pinned in the conduction band. To reduce the high barrier between InAs and GaAs, Woodall et al. used a compositionally graded $In_xGa_{(1-x)}$ As layer [13]. The energy band diagram for this type of contact scheme is shown in Fig. 11.1b. These contacts can be produced by MBE and MOCVD and routinely produce a contact resistance below 1×10^{-7} ohm.cm^2.

11.5.2 Alloyed Ohmic Contacts

11.5.2.1 Gold:germanium contacts

Nonepitaxial techniques for making ohmic contacts are needed for situations where epigrowth is not cost effective or practical. Ion-implanted FET and HBT collector contacts are made using such conventional techniques of metal deposition followed by high-temperature processing. Au:Ge ohmic contacts with an eutectic composition of 12 wt% Ge were first used in Gunn diodes in 1964, as mentioned earlier. The melting point of this alloy is near 365°C (see Fig. 3.3 for the phase diagram of the Au–Ge metallurgical system). The contact can be alloyed in by heating at 360°C to 380°C. These contacts are rough in texture, vertical diffusion into the substrate can be deep, and the stability is not good. The addition of Ni (discovered accidentally when a contaminated filament was used in evaporation) improved morphology and reliability [17]. By changing the sequence of evaporation of the layers, the amount of Ni, and the addition of gold and other capping layers, further improvements have been achieved. Figure 11.7 shows the layer structure of layered gold and germanium and Au:Ge alloy–based structures. By limiting the amount of gold available at the surface, a more high-temperature stable contact can be achieved [18]. These contacts are stable above 350°C to about 400°C. Several nongold, Ge-based contacts have been tried to improve stability with mixed success, for example, with aluminum. Pd:Ge is the other nongold contact first developed by Sinha et al. [19] and extensively pursued by Lau [12, 20]. The main advantages are shallow depth and smooth morphology for micron-size contacts used for digital, high-density circuits, but the contact resistance is generally high (3×10^{-5} ohm.cm^2).

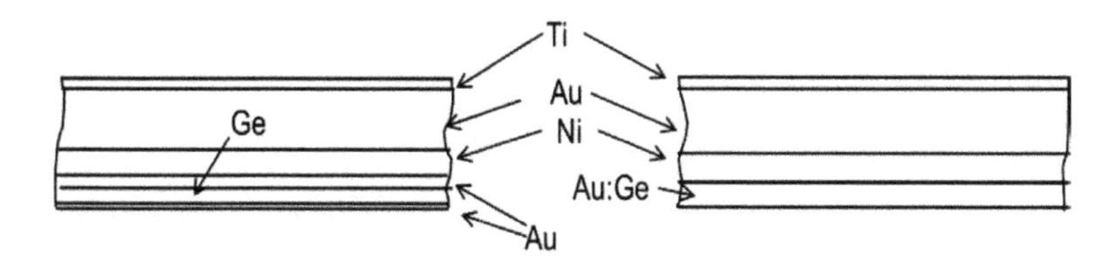

Figure 11.7 n-Type contact layer structures using Au/Ge layered and Au:Ge alloy systems.

11.5.2.2 Silicon:tin contacts

High levels of silicon doping are possible in epigrowth, so it appears to be a good candidate for alloy systems. Silicon alloys have been tried, for example, Au:Si or Ag:Si. It may be noted that even with the addition of Ni, lower resistance or better stability has not been achieved. Tin-based contacts like Ag:Sn or Au:Sn provide low contact resistance, but Sn diffusion creates stability problems.

15.2.2.3 Indium-based contacts

Indium-based contacts attempt to reproduce metal/InAs/GaAs contacts without the need for epigrowth. These contacts have a long history, starting first with the use of pure In and the addition of Ag and Ge. These contacts need a high alloying temperature of about 600°C and the contact resistance values are not very low. The Au:Ge:In contact can be alloyed at a lower temperature of 520°C, but the contact resistance is not any better than Ag:Ge:In.

Significant improvements have been made by using Ni and W with In and using rapid thermal annealing (RTA). These contacts have been used for self-aligned MESFETs [21]. Although the contact resistance is higher, MESFET performance, similar to that with Au:Ge:Ni contacts, has been achieved. Since evaporation of tungsten (W) is not easy, NiInW contacts can be deposited by sputter deposition. Kolawa et al. [22] have prepared NiGe contacts by sputter deposition. They deposited Ni/Ge/WN/Au on GaAs and annealed it at 500°C for 10 minutes to achieve an r_c value of 2×10^{-6} ohm.cm^2 [22]. If compatibility with an aluminum interconnect is desired, gold must be avoided, as in the NiInW system.

11.5.3 Ohmic Contact Deposition

Gold:germanium–nickel contacts can be evaporated and patterned on wafers using a lift-off process. A Au:Ge premixed alloy or individual layers of Ge and Au can be evaporated. Nickel can be evaporated next or a thin layer of nickel can be applied as the first layer for wetting reasons. The total nickel needs to be about 18% by volume of the AuGe for optimum morphology and lowest contact resistance [23, 24]. The total thickness of AuGe and Ni can be about 1000 Å. A layer of gold is deposited on top for lower sheet resistance of the ohmic layer. A thin top layer of Ti may be used for promoting adhesion of interlevel dielectrics without any negative consequences. This also minimizes galvanic corrosion. Wafers with ohmic metallization should be processed through alloying within a reasonable time (of a day or two) in order to avoid corrosion due to humidity.

11.5.4 Alloy Process and Alloying Systems

After the patterning process is complete, the ohmic metallization must be alloyed or sintered to complete the contact formation, as discussed in Section 11.5.2. In GaAs and other III–V compound processing, low-temperature processes are used for alloying contacts to the semiconductor material. Typical alloying temperatures are in the 300°C to 400°C range. For refractory contact schemes, the temperature may be considerably higher. The temperature profile of a rapid thermal processing (RTP) alloy cycle is shown in Fig. 11.8. The optimum temperature and time are determined for each metallization system using design of experiment (DOE) techniques with contact resistance and morphology as the responses. The reliability of the contact can be checked by high-temperature operating life (HTOL). The contact resistance versus temperature data shows a typical bathtub-type response (Fig. 11.9) [25]. This is true for most metal schemes. Figure 11.9 shows the bath-type behavior for AuGe- and Pd-based contacts. The smooth texture of the surface is shown in Fig. 11.10. Spots due to too much Ni are shown in Fig. 11.10b. The alloy process must be optimized for all three responses: contact resistance, morphology, and reliability.

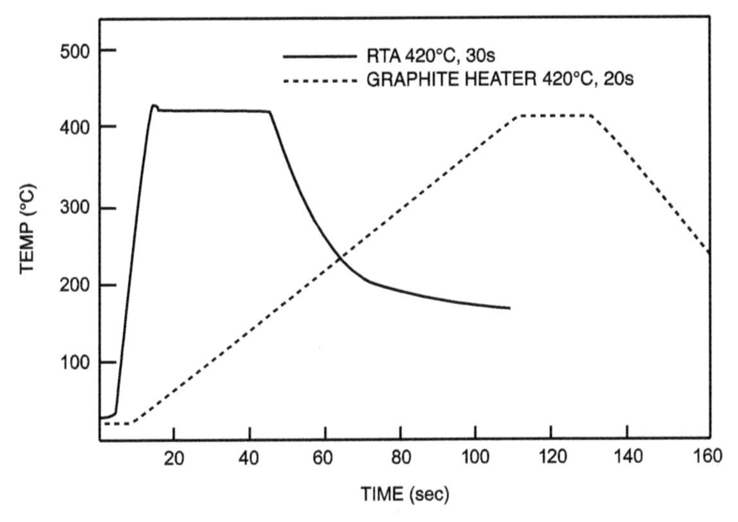

Figure 11.8 RTP thermal cycle for alloying ohmic contacts on III–V wafers (compared to a graphite heater).

11.5.4.1 Alloying systems

The contact can be alloyed in by a furnace, by RTA, or on a hot-plate track system under a proper inert ambient gas. The alloy operation is generally done in the 360°C–400°C temperature range. The steady-state temperature must be controlled to within a few degrees. Good temperature control at as low a temperature as 250°C is desired. Historically alloying was done in a furnace in about 20 to 30 minutes. As soon as RTP became available, it was adopted for GaAs alloying. Details of RTP systems are discussed in Chapter 9. RTP has the advantages of no alloy spiking or balling and good wafer-to-wafer reproducibility. A typical temperature cycle for alloying an n-type Au:Ge/Ni/Au contact to GaAs is shown in Fig. 11.8, which shows a short, precise cycle compared to a graphite strip heater. The heating cycle in a furnace takes a long time and the temperature uniformity from wafer to wafer may be poor. In RTP, a correct temperature profile is not easy to achieve because of errors in sensing low temperature and emissivity differences between wafers. Low-temperature pyrometers are therefore needed for this application. Also, the use of graphite susceptors, which are used to hold the wafers, is recommended for achieving temperature uniformity. Temperature and time should be optimized to achieve the right morphology of the metal alloy, ensure that the contact resistance is as low as it can be, and ensure the device reliability is intact. Device performance as measured by source resistance of FETs or pHEMTs or beta, emitter resistance, etc., of HBTs should not be compromised . The optimum temperature thus determined may be lower than what would be used on the basis of contact resistance alone.

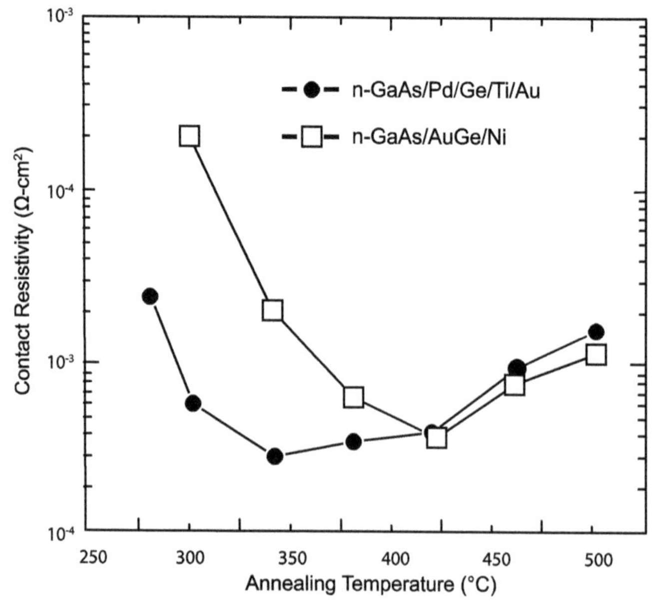

Figure 11.9 Contact resistivity as a function of annealing temperature showing an optimum alloying temperature for lowest contact resistance. © [1998] IEEE. Reprinted, with permission, from Ref. [25].

Figure 11.10 Morphology of gold–germanium/nickel ohmic contacts; the contact on the right has above-optimum nickel.

RTP alloying can be done on hot-plate track-type systems, where a temperature of up to 400°C and alloy times of one to a few minutes can be well controlled. An inert atmosphere should be used during alloying and the rate of lowering the wafers on the hot-plate surface should be well controlled.

11.5.5 Mechanism of Contact Formation

In the Au:Ge/Ni-on-GaAs system, GaAs decomposes, arsenic reacts with Ni and Ge to form Ni_2GeAs pebbles, and Ga diffuses out to alloy with gold. The Ni_2GeAs phase in contact with heavily Ge-doped GaAs is believed to be responsible for the ohmic contact. The heavily Ge doped GaAs is produced by local regrowth of GaAs from AuGe melt. See Fig. 11.11 for a micrograph of the cross section of an ohmic contact and a schematic representation of the same contact region. Phases forming ohmic contacts contain Ni, Ge, and As. The ratio of Ni_2GeAs to the AuGa phase at the interface determines the contact resistance. Ni–Ge regions may show up in low-contact-resistance samples. AuGa is associated with high contact resistance. This mechanism has been verified by extensive transmission electron microscopy (TEM), scanning electron microscopy (SEM), and X-ray diffraction analysis by several workers [26–28].

11.5.6 Refractory Contacts

Thermally stable, low-resistance NiInW contacts to n-type GaAs were first developed by Murakami [29]. The role of In for forming contacts to GaAs is exploited, with the addition of Ni, which eliminates free leftover In on the surface. These contacts were evaporated and annealed by RTA at high temperatures over 700°C. Evaporation of W is difficult; therefore a sputtered contact of Ni:In/W (~90% Ni) annealed at up to 800°C for 10 seconds is more production compatible for a self-aligned MESFET process where RTP must be used for implant annealing anyway [30]. These contacts are stable up to 400°C and are compatible with aluminum-based, silicon-type interconnect processing. For analog circuits the specific contact resistance of this contact may not be acceptable. For an HBT collector, collector layers that have a doping concentration over $1E18/cm^3$, a GeMo contact scheme has been developed [6]. Use of As-doped Ge and As overpressure results in a contact resistance of 2×10^{-7} ohm. cm^2.

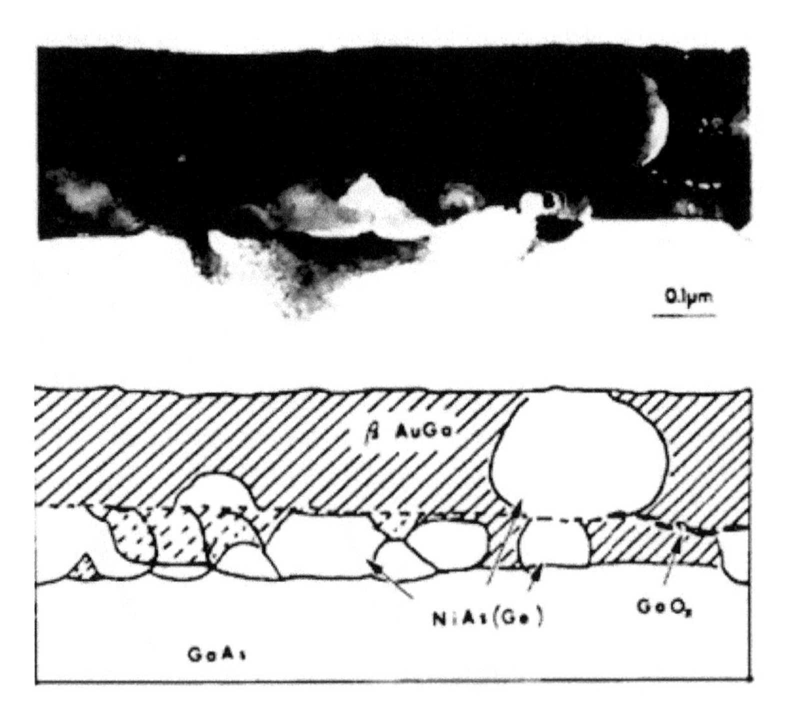

Figure 11.11 Cross-sectional view of Au:Ge/Ni ohmic contact after alloying [26].

For FET- and HBT-type devices, refractory contacts or contacts with a barrier are needed for introduction of a copper interconnect. A Pd:Ge-based contact may see a revival as interest grows in copper as an interconnect metal. Pd:Ge:Cu-based contacts have been demonstrated [8]. It may be noted that copper may diffuse in too deep to increase contact resistance. Barriers like TiN may have to be employed with nonrefractory ohmic contacts.

11.6 Ohmic Contacts to p-Type GaAs

Both alloyed and nonalloyed ohmic contacts are used for contacts to p-type layers. For layers with moderate levels of doping, alloyed contacts using a gold–zinc (Au–Zn) metallization system are commonly used for alloyed ohmic contacts to p-type GaAs and other III–V semiconductor layers. In this scheme, Zn acts as the p-type dopant and gold fulfils the role similar to the Au:Ge/Ni system for an n-type material. Ag:Zn and Au:Be systems have also been used for p-type GaAs for semiconductor doping levels in the E17/cm^3 range. These contacts are alloyed at 380°C to 400°C, similar to n-type contacts. Beryllium (Be) content can be 1% to 2% in the gold alloy. Au:Be alloyed contacts can spike into the p-layer and therefore cannot be used for contacts to thin layers, such as those used for base layers in HBTs. Zinc exhibits anomalous diffusion in GaAs and the diffusion front can penetrate deep. Rapid thermal alloying can be used to minimize penetration of Zn and refractory barrier metals can be used for improving reliability. The contact resistance with this process can be in the low E-6 ohms/cm^2 range and shows bathtub-type behavior as a function of alloying temperature.

With heavily doped p+ layers, for example, carbon-doped ones used for HBT devices, ohmic contacts can be made with any metal. Therefore commonly used Ti/Pt/Au metallization can be used. Figure 11.12 shows the energy band diagram for such a contact. Contact resistance in the $1E^{-6}$ ohm.cm^2 range can be achieved with or without any heat treatment. Deposition of a thin layer of Pt as the first layer has been shown to drop the contact resistance to the E-7 ohm.cm^2 range after short heat treatment [9]. If the Pt layer is too thick, excessive metal sinking due to reaction with the underlayer can take place during subsequent high-temperature processing. Metal spiking can also occur at the edges of the metallization due to gold foot spilling over the Pt and Ti barrier, as shown in Fig. 11.13. Both these effects can lead to device degradation, especially for p-layers less than 1000 Å in thickness. The second layer of Pt also needs to meet the minimum thickness requirements based on the barrier needed; typically 250 Å may be used. Gold thickness needs to be enough to keep the line resistance of contact lines low. For example, for a rectangular HBT, using 1 μm thin base lines, 600 to 1000 Å gold may be necessary. Ni/Ti/Pt contacts with 50 Å Ni as the first layer have shown lower contact resistance with less interdiffusion [31].

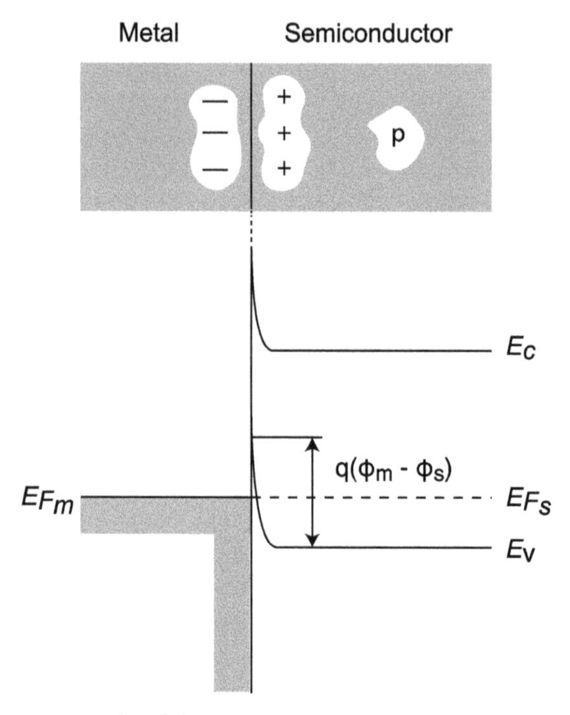

Figure 11.12 Energy band diagram of metal contact to heavily doped p-type semiconductor.

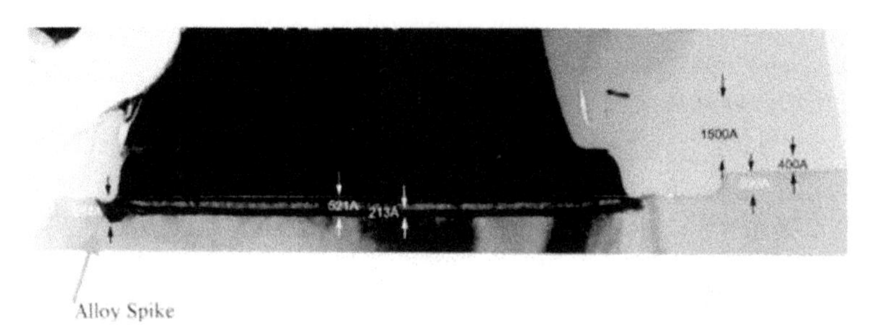

Figure 11.13 Alloy spiking at the edge of HBT base contact ohmic contact.

A further discussion of device performance and reliability issues related to ohmic contacts is covered under those topics. In summary, a barrier layer must be provided to keep the overlay layer of gold from intermixing with the contact layers and the p+ GaAs layer [32].

11.7 Ohmic Contacts to InP Devices

In InP devices, both HEMT and HBT types, transition is made from the InP layer to lattice-matched $In_{0.53}Ga_{0.47}As$. InGaAs can be doped to very high levels, of the order of $3 \times 10^{19}/cm^3$. Alloyed or annealed contacts can be easily made on this layer. If an InAlAs layer is used in the device, it also can be capped by an InGaAs layer. Doping levels can be so high that nonalloyed contacts can be made with standard Ti/Pt/Au metallization or refractory metals like Ti:W and molybdenum. Extremely low contact resistance, in the 10^{-8} ohm.cm^2 range has been achieved [33].

For nonalloyed contacts with metals likeTi/Pt/Au or TiW, contact resistance can be lowered for modest doping levels and ex situ metal contact deposition by using graded InGaAs epitaxial cap layers. The Ga mole fraction is dropped from 0.47 to 0.3 on the surface in the graded layer [34].

11.8 Ohmic Contacts to GaN

GaN is a large-bandgap semiconductor and is becoming popular for high-power electronics and optoelectronics. Due to the ionic nature of GaN, it is necessary to unpin the Fermi level at the surface and metal–GaN interface. It is necessary to identify metals that have work functions smaller than that of n-type GaN. The Ti:Al/NiAu metal system is commonly used for ohmic contacts to GaN/AlN semiconductor devices. The typical layer thickness is 100 Ti/2000 Al/500 Ni/200 Au Å. Ohmic contact resistance in the range of low 10^{-5} ohm.cm^2 is achieved in this system after alloying at a temperature of around 850°C. Ti is needed to form ohmic contacts by the formation of TiN at the surface. Ti/Al metal layers are similar to Ge/Au used for GaAs-type materials; however, the Ti/Al system does not have a low melting point like the Au:Ge eutectic. The Ti/Al ratio is optimized for low contact resistance and may be around 0.15

[35]. Gold is needed to prevent oxidation of this contact; however, Ni is needed in between as a barrier against gold diffusion. Pt or Ti could be used also. The Ni/Au ratio is also optimized in practice. $TiAl_3$ has also been used as a cap, resulting in low contact resistance of 1×10^{-5} ohm.cm^2 after annealing at 700°C–900°C for one minute [36]. On the contrary, the strategy to form ohmic contacts on p-type GaN is to identify metals having a large work function that will result in $\sim 10^{-4}$ ohm.cm^2 of specific contact resistance. It is compounded by a double problem of the constraint of being able to grow heavily doped p-type by addressing the high activation energy (~170 meV) of deep Mg acceptors and the formation of Mg–G complexes, in addition to the difficulty in finding a metal or conducting oxide having a work function greater than p-type GaN (bandgap of 3.4 eV and electron affinity of about 4.1 eV). Apart from such a solution, there has been research into solving this technology problem through surface superlattice and strained layers, use of interlayers, hydrogen extraction, and surface treatments. Ni/Au contacts have been used (work function of 4.1 eV), which required oxygen annealing–induced ohmic formation at 500°C, with specific contact resistances improving from $\sim 10^{-4}$ to 10^{-6} ohm.cm^2 from unannealed values of 10^{-2} ohm.cm^2.

11.8.1 Mechanism

GaN has a wide bandgap (3.4 eV), and most metals form a Schottky contact when deposited on the surface. Either the contacts need to have a low Schottky barrier or the GaN needs to be heavily doped so that tunneling of carriers can occur through the barrier. Therefore, the Schottky barrier has to be reduced or thinned (to form a tunneling junction) before an ohmic contact can form. TiN has a shallow work function and forms a conducting layer on top. Nitrogen outdiffuses from GaN lattice, leading to the formation of TiN and resulting in residual nitrogen vacancies that act as donors to GaN and form shallow n-type centers in GaN. The area in the interface becomes heavily doped, leading to high carrier concentration and shallow barrier creation simultaneously. Deposition of TiN directly on GaN does not give an ohmic contact, and the interreaction of GaN and Ti/Al is necessary for creation of TiN , Ti_2N, $AlTi_2N$, etc., as well as N vacancies [37]. For the p-type contact, it is arising from the formation of an intermediate semiconducting layer with high hole concentration in the Ti/Au system on GaN. This is probably a p-type NiO-based semiconductor in contact with p-GaN, because

Au-rich islands, amorphous Ni–Ga–O, and voids at the interface have combined with NiO at the interface. Additionally, formation of interfacial compounds such as Ga_4Ni_3, Ga_3Ni_2, AuGa, and $GaAu_2$ leads to the formation of Ga vacancies at the interface serving as the acceptor [38, 39].

11.9 Ohmic Contact Corrosion [40]

It was mentioned earlier that III–V wafers must be protected from moisture exposure after contacts are made. When GaAs comes in contact with an electrolyte, electrons are transferred from the semiconductor to the electrolyte until the Fermi levels match. In other words, holes are added to the GaAs.

$$GaAs + 6h^+ \rightarrow Ga^{3+} + As^{3+}$$

Figure 11.14 shows the band diagram of GaAs in contact with an electrolyte. Electrons in the GaAs valence band can get excited to the conduction band (particularly in the presence of light). The holes left in the valence band recombine with Ga or As to generate Ga^{3+} and As^{3+} ions. If the surface of the GaAs has previously been damaged (e.g., by plasma damage), the surface states will make this process easier. The ions are dissolved away and result in pits near the metal edges. Ohmic contacts provide a path for the electron circuit to close. As mentioned earlier, after the ohmic metal is deposited, exposure to moisture or chemicals should be minimized until the surface is sealed up.

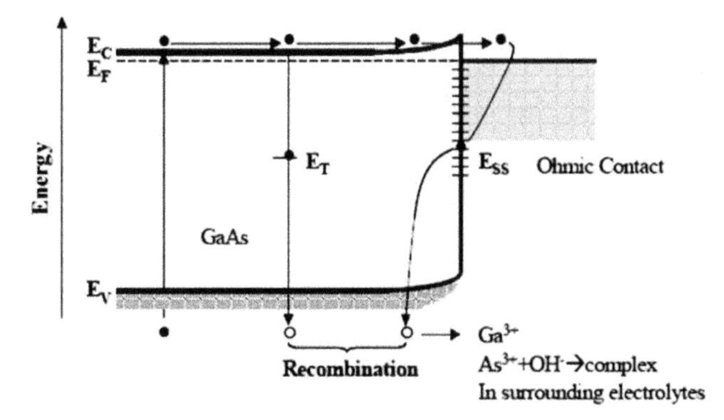

Figure 11.14 Bandgap theory for GaAs corrosion near ohmic contact. Reprinted from Ref. [40] with permission. Copyright © 2004 CS MANTECH.

References

1. V. L. Rideout, A review of the theory and technology for ohmic contacts to group III-V compound semiconductors, *Solid State Electron.*, **18**, 541 (1975).
2. M. N. Yoder, Ohmic contacts in GaAs, *Solid State Electron.*, **23**, 117 (1980).
3. N. Braslau, Ohmic contacts to GaAs, *Thin Solid Films*, **104**, 391 (1983).
4. M. Murakami, Development of ohmic contact materials for GaAs integrated circuits, *IBM Res. Rep. Mater. Sci.*, **5**, 273 (1990).
5. H. H. Berger, *Solid State Electron.*, **15**, 145 (1979).
6. M. Ketaka, *J. Phys. D Appl. Phys.*, **26**, 2075 (1993).
7. A. Y. C. Yu, *Solid State Electron.*, **13**, 239 (1970).
8. S. P. Wang et al., *Copper Metallized Power InGaP HBT*, CS MANTECH (2009).
9. A. Katz, C. R. Abernathy and S. J. Pearton, *Appl. Phys. Lett.*, **56**, 1028 (1990).
10. A. G. Baca, F. Ren, J. C. Zolper, R. D. Briggs and S. J. Pearton, A survey of ohmic contacts to III-V compound semiconductors, *Thin Solid Films*, **308–309**, 599 (1997).
11. S. Sze, *Physics of Semiconductor Devices*, Wiley Interscience, New York (1969).
12. E. D Marshall, S. S. Lau et al., *Appl. Phys. Lett.*, **47**, 298 (1985).
13. J. M. Woodall, *J. Vac. Sci. Technol.*, **19**, 626 (1981).
14. R. Williams, *Modern GaAs Processing Methods*, Artech House, Boston (1990).
15. M. Sun et al., Si implanted subcollector heterojunction bipolar transistors, *Electron. Lett.*, **34**, 1155–1156 (1998).
16. R. Stall et al., *Electron. Lett.*, **15**, 800 (1979).
17. J. B. Gunn, *IEEE Trans. Electron. Devices*, **ED-23**, 705 (1976).
18. S. Tiku, unpublished data.
19. A. K. Sinha, T. E. Smith and H. J. Levinstein, *IEEE Trans. Electron. Devices*, **22**, 218 (1975).
20. L. C. Wang, S. S. Lau et al., Pd:Ge:In ohmic contacts to GaAs, *J. Appl. Phys.*, **69**, 4364 (1991).
21. M. Murukami et al., *J. Appl. Phys.*, **65**, 3546 (1989).
22. K. Kolawa et al., *IEEE Trans. Electron. Devices*, **36**, 1223 (1989).

23. H. J. Buhlmann and M. Legems, *J. Electrochem. Soc.*, **138**, 2795 (1991).
24. N. E. Lumpkin and G. Lumpkin, Investigation of the Role of Ni ..., *J. Mater. Res.*, **11** 1244 (1996).
25. J. S. Kwak, J. L. Lee and H. K. Baik, Improved uniformity of contact resistance in GaAs MESFET using Pd/Ge/Ti/Au ohmic contacts, *IEEE Electron. Device Lett.*, **19**, 481 (1998).
26. M. Murakami, H. J. Kim and W. H. Price, Interfaces between ohmic contacts and GaAs, *IBM Res. Rep. Mater. Sci.*, 269 (1989).
27. H. Goronkin, Ohmic contact penetration at contacts in GaAs/AlGaAs and GaAs FETS, *IEEE Trans. Electron. Devices*, **36,** 281 (1989).
28. T. S. Kuan et al., *J. Appl. Phys.*, **54**, 6952 (1983).
29. M. Murakami et al., *J. Appl. Phys.*, **64**, 1974 (1988).
30. M. C. Hugon et al., *J. Appl. Phys.*, **72**, 3570 (1992).
31. M. Yamagihara and A. Tamura, Ni/Ti/Pt ohmic contacts to p-GaAs for HBT process, *Electron. Lett.*, **32**, 1238 (1996).
32. Q. Xu and Li-Wu Yang, *IEEE Trans. Electron. Devices*, **58**, 2582 (2011).
33. A. Alt and C. R. Bolognesi, *IEEE Trans. Electron. Devices*, **60**, 787 (2013).
34. A. Crespo et al., *Ti/Al/Ni/Au Ohmic Contact on AlGaN/GaN HEMTs*, CS MANTECH (2003).
35. U. Singisetti et al., DRC, 65th Annual, 147 (2007).
36. C. Pelto, Ohmic contacts to n-GaN, *Solid State Electron.*, **45** 1597 (2001).
37. M. Masakatsu, Y. Takao and T. Yasuo, *Trans. JWRI*, **41** (2012).
38. Y.-J. Lin, Z.-D. Li, C.-W. Hsu, F.-T. Chien, C.-T. Lee, S.-T. Shao and H.-C. Chang, Investigation of degradation for ohmic performance of oxidized Au/Ni/Mg-doped GaN, *Appl. Phys. Lett.*, **82**, 2817–2819, (2003).
39. Z. Z. Chen, Z. X. Qin, Y. Z. Tong, X. D. Hu, T. J. Yu, Z. J. Yang,L. S. Yu, G. Y. Zhang, W. L. Zheng, Q. J. Jia and X. M. Jiang, Effects of oxidation by O_2 plasma on formation of Ni/Au ohmic contact top-GaN, *J. Appl. Phys.*, **96**, 2091–2094 (2004).
40. H. Shen et al., *GaAs Corrosion under Ohmic Contacts by Electrochemical Oxidation in HBT Device Fabrication*, CS MANTECH (2004).

Chapter 12

Schottky Diodes and FET Processing

Schottky diodes and FET gates share common processing challenges. Schottky contacts will be discussed first because the basic discussion is applicable to both. FET gate processing will be discussed in more detail later.

12.1 Schottky Diodes

Schottky diodes perform functions in silicon that cannot be achieved by p-n junction devices. The devices have a low turn-on voltage and are fast, being majority carrier (electrons for n-type Si) devices with a short recovery time. In the case of GaAs and other III–V compounds, Schottky diodes and field-effect transistor (FET) gates fulfill the needs for functions that otherwise could not be accomplished. Diffusion processes are not needed and p–n junctions are avoided and lack of a good gate oxide technology is circumvented. The process associated with Schottky diodes and gates is very simple and amenable to submicron gate formation, thus making high-frequency operation easier to achieve.

The basic physics of metal–semiconductor junctions was discussed briefly in Chapter 2. A few results of importance are repeated here. The *I–V* curve of a metal–semiconductor junction showing rectifying or diode behavior, with current density in the forward direction, has the exponential form given by

III–V Integrated Circuit Fabrication Technology
Shiban Tiku and Dhrubes Biswas

ISBN 978-981-4669-30-6 (Hardcover), 978-981-4669-31-3 (eBook)
www.panstanford.com

$$J \approx \exp\left\{\frac{qV}{nkT}\right\} \tag{12.1}$$

The reverse current and the breakdown behavior are discussed later.

12.1.1 Depletion Width

In an n-type semiconductor, when a voltage is applied to a Schottky contact, electrons are depleted out of the depletion region and the applied voltage controls the depletion width (the depth under the metal). The negative charge of the electrons is balanced by the positive charge of the ionized dopant atoms. The depletion width can be derived by Poisson's equation:

$$\frac{d^2V}{dx^2} = -\frac{\rho(x)}{\varepsilon_s} \tag{12.2}$$

where ε_s is related to the dielectric constant k and permittivity of free space by $\varepsilon_s = k\varepsilon_0$.

If the charge density is a constant N

$$\frac{d^2V}{dx^2} = \frac{qN}{\varepsilon_s} \tag{12.3}$$

Integrating twice and using proper boundary conditions

$$V(x) = qNw\frac{\left[x - \left(\frac{x^2}{2w}\right)\right]}{\varepsilon_s} \tag{12.4}$$

$$V(w) = V = \frac{qNw^2}{2\varepsilon_s} \tag{12.5}$$

or

$$w^2 = 2\frac{\varepsilon_s V}{qN} \tag{12.6}$$

Therefore depletion width $w = \sqrt{\{2.(\varepsilon_s V)/qN\}}$. The applied voltage V is related to the gate voltage V_g by $V = V_{bi} - V_g$, where V_{bi} is the built-in voltage with no external voltage applied (V_{bi} values for some common metals on GaAs are listed in Table 2.1). The width of the depletion region goes down with increasing doping and is very narrow for doping levels over 1E18/cm^3.

If the doping profile is not uniform, a more general equation for integration may have to be solved:

$$V = \frac{q}{\varepsilon_s} \cdot \int_0^w x.N(x)dx \tag{12.7}$$

For more exact treatment, the reader should refer to device physics textbooks [1, 2].

The depletion region represents a capacitor formed between the metal electrode and the semiconductor (the depletion width being the insulator in between):

$$C = \frac{\varepsilon_s A}{w} \tag{12.8}$$

The change of the depletion region capacitance as a function of applied voltage is used for determining the doping profile, as discussed in Chapter 21.

12.1.2 Schottky Diode Metallization [3]

Numerous metallic elements and refractory alloys have been studied for Schottky diode/gate metallization for GaAs. As discussed in Chapter 2, the barrier height does not vary much. In a large study with metal–organic chemical vapor deposition (MOCVD)-grown epilayers, with an identical cleaning and deposition procedure, Schottky barrier height (ϕ_b) determined by the I–V method ranged from 0.6 eV to 1.0 eV [3]. The correlation of barrier height to metal function was weak. On the low end metals like Ca and Mg (ϕ_m of ~3 eV) have a low barrier height, around 0.6 eV. On the high end, Pt with a work function near 6 eV has a barrier height of 0.95 eV, a high ϕ_b being preferred for ideal low-leakage diodes and FET gates. Materials such as gold, aluminum, and silver make good Schottky contacts to GaAs. However, heat treatment close to 300°C, required for subsequent processing, results in interreaction, as gold diffuses into the semiconductor and gallium diffuses out. Metal spiking, change of barrier height, and deviation from ideal diode behavior are the result. Metals such as titanium and tungsten have more inert interface properties. These metals can be used as part of a multilayer Schottky diode or FET gate contact. Titanium does form TiAs over 550°C and does not prevent gold diffusion into GaAs over extended periods of time, even at lower temperatures. Pt/Au Schottky gates are not stable either, because Pt intermixes with GaAs at 300°C within minutes and at contact alloying temperatures (around 380°C in rapid thermal processing [RTP]) within seconds. Therefore more complicated three-layer metal schemes, such as Ti/Pt/Au or Ti/W/Au, have been adopted in practice [4]. Sputtered Ti/W (10%–20% Ti) can also be used to separate a top gold layer from the Schottky contact layer. This may be possible only where a contact via can be placed on top of the Schottky contact. WSi_x- and WN_x-type sputtered gates can be used for digital circuits, where the sheet resistance of the gate layer can be high. For submicron gate layers used in analog or mixed-signal circuits, the line sheet resistance cannot be too high, so a complex multilayer gate metallization scheme, which can be patterned by lift-off, is necessary. Therefore a triple-layer, evaporated Ti/Pt/Au metallization scheme has become the dominant Schottky and gate metallization.

12.1.3 Reverse Breakdown

In the reverse direction, a small current flows as the barrier height is lowered with applied voltage. A high barrier height is obviously good for low reverse current. The nature of the *I–V* curve is similar to the forward current due to thermionic emission (see Fig. 2.9). In practice, for integrated circuit (IC) devices, the dominant reverse current may be due to an edge leakage current.

A high reverse breakdown voltage is desired for switching diodes and FET Schottky gate diodes. Reverse breakdown limits the voltage swing on FET gates. The main mechanism for breakdown is avalanche multiplication or impact ionization. At a high reverse electric field, electrons released thermally can gain enough energy to create more electron–hole pairs, which, in turn, gain more energy and create more pairs. This process leads to avalanche and breakdown. The breakdown is a function of the electric field, which can be high at some points near the gate edges. Breakdown is also a function of the doping concentration and surface effects. In GaAs the Fermi level is pinned at the surface; nevertheless it is still sensitive to the details of the gate metal edges. Details of recess etches, cleaning, and surface passivation are important for optimizing gate process for high breakdown. Breakdown behavior can be modeled, but in practice it is optimized using empirical techniques. Field plates can be used to shift peak fields and improve breakdown, like guard rings are used in Si Schottky diodes.

12.2 FET Gate Fabrication

12.2.1 Gate Metallization and Fabrication

All the metal systems discussed under Schottky diodes can be used for gate metallization. Additional constraints may be placed by the requirements of the patterning process, especially needed for submicron gate processes. Good adhesion, thermal stability, and reasonably low resistance are essential. For digital FETs, gate metal resistance can be higher, but for analog FETs that have large widths (finger length), low resistance is important. As discussed under Schottky diodes, multiple-layer metallization with gold overlayers have become standard.

12.2.2 Gate Recess Process

In spite of tremendous improvements in the control of starting materials, that is, wafers, threshold voltage, and Idss current control are difficult to achieve in ion-implanted and even in epigrown channel layers. Therefore, a recess etch process is used to control the actual threshold voltage and source–drain current (Idss). Also for meeting high voltage breakdown and other performance requirements of FETs and pseudomorphic high-electron-mobility transistors (PHEMTs), recess gate geometry is still dominant. Gate recess and deposition processes in these device types are done after ohmic contact formation so that channel currents can be monitored.

The recessed gate process consists of three process steps: photolithography, etching, and metal deposition. Ordinary optical lithography processes are used for gates over 0.5 μm in size; electron beam lithography is used for smaller gate lengths. After the gate area is patterned with a resist, if there is a dielectric layer present on the surface, such as silicon nitride, it is removed by plasma etching. See Fig. 12.1a. Wet etching is used to remove a predetermined thickness of the semiconductor to form the channel. The channel saturated current is measured for this ungated channel using the ohmic source and drain contacts. See Fig. 12.1b. This step is carried out in the fabrication area because the photopattern is still present on the wafer. If the target current is not reached, the wafer is etched further until the current target is achieved. Slow and well-controlled etch chemistry must be used for good control (See Chapter 6). The target current at the recess etch is higher than the final value because the current drops when the metal gate is deposited. The etch times needed can be calculated using algorithms based on prior data for the recess etch process. Current monitoring is done using special test FETs on the wafers. The test FET has to be connected to pads for probing and is wired using ohmic or gate metal and not an interconnect layer metal like the FETs in the circuit. Its etch behavior may depend upon the layout and area of the metal exposed to the chemicals at the recess etch. Recess etch current targeting must take all these factors into account. Dry etch processing has been tried for recess etching; however, it is difficult to minimize damage and maintain threshold voltage control.

Dual recess processes are used to improve breakdown and performance of power FETs. A second photolithography and etch step must be used. A dual recess is also needed for high-electron-mobility transistor (HEMT) processing and will be discussed in Chapter 13.

12.2.3 Gate Formation

Gate formation is the most demanding of the process steps used for fabrication of circuits based on FET-type devices. The gate size, being the smallest, determines the limits for lithography and etch tools. The placement of the gate between the source and the drain and gate size control determine the performance of the circuits. A simple, recess gate geometry is shown in Fig. 12.1c. Different photoresist schemes are used to define the gates. Optical photolithography using a single layer of resist can be used for a gate size over 0.5 μm. Electron beam lithography is used for sub-half-micron gates. To make the lift-off process more production compatible, multilevel resist techniques can be used.

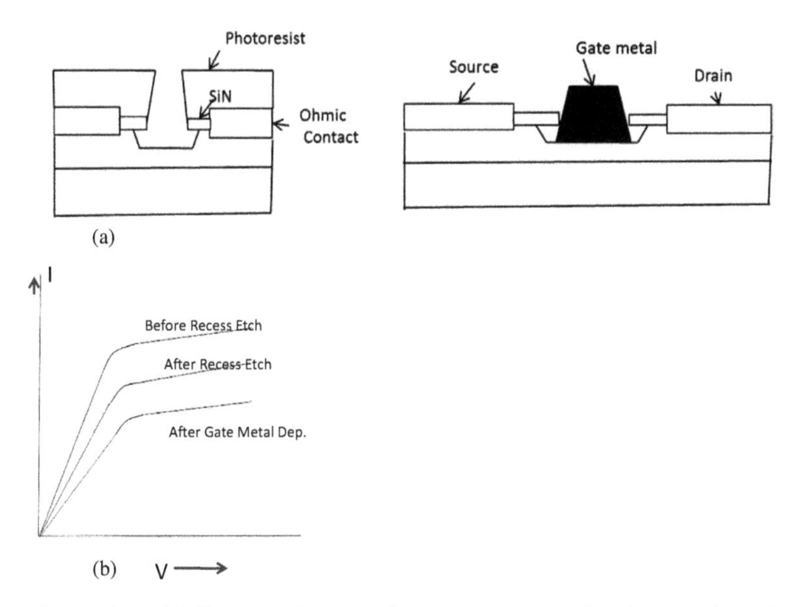

Figure 12.1 (a) Gate metal patterning process steps showing opening of nitride layer and gate metal deposition. (b) *I–V* curves show the change of source–drain current with gate metal deposition.

The cleaning procedure prior to Schottky diode or gate deposition is critical to device performance. The cleaning sequence that removes unwanted photoresist scum, the built-up oxide layer, and prepares the surface prior to going into the deposition system without introducing ion damage must be developed for the GaAs channel layer. If the layer on which the gate metal is deposited is a different III–V material, a proper surface state must be created for optimum performance. Such treatments have been developed over the years on the basis of an understanding of surfaces and through trial and error. Details of these are discussed in this chapter and in Chapter 22.

12.2.3.1 T-gate

GaAs FETs with submicron gates are used in low-noise, high-gain devices at high frequencies. As gate length is reduced, the resistance increases (cross-sectional area of the gate line reduces). Different techniques are employed to retain a short gate length and decrease the resistance of the gate line. T- or mushroom gates are one example.

A simple procedure, used to produce a T-gate of the type shown in Fig. 12.2a, uses two photomask steps. One step is used to define an opening in the dielectric and to do the recess etch and a second photostep is used to pattern the metal. Another possibility is to deposit a dummy photoresist gate and use it to lift off an evaporable dielectric material like SiO, thus leaving an opening for the gate. See Fig. 12.2b. Then an etch process can be used to remove the Si_3N_4 passivation layer and do the recess etch. A second photostep is used to pattern a wider gate metal by ordinary photolithography. Another approach is to deposit a double-layer metal gate with a dry etchable lower part like WN_x so that undercut etching can be used to reduce the gate length. See Fig. 12.2c.

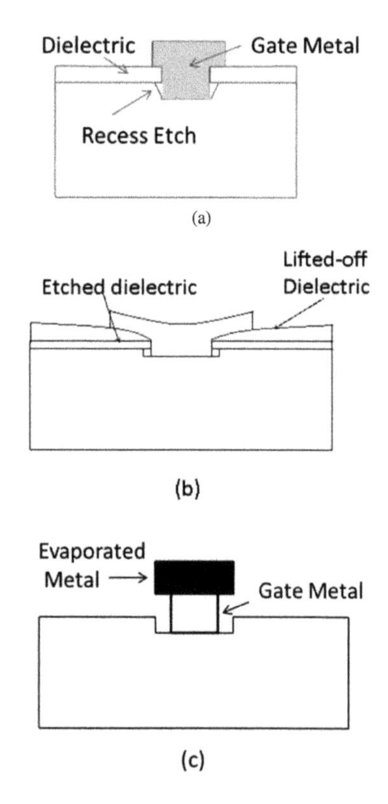

Figure 12.2 T-gate formation by (a) dielectric opening and metal lift-off. (b) Use of dummy gate process to pattern evaporable dielectric. (c) Use of evaporable metal as a mask for dry etching gate metal like WN_x.

12.3 Digital FETs

12.3.1 Gate Fabrication

Digital circuits that utilize FETs as switches and as load devices use a large number of transistors, in thousands, and occupy a large chip area. The performance requirements are different (discussed in Chapter 2). The most important specification is the threshold voltage of FETs, particularly the enhancement mode type, which must be controlled within 50 mV. Ion-implanted FETs and epigrown FETs and heterojunction FETs (HFETs) are capable of meeting this requirement. Digital circuits need complex interconnect processes that are similar to or compatible with silicon processing. Use of mesas is avoided. Digital FETs are generally planar and do not have thick source and drain regions [5]. The resistance of these source and drain regions can have high resistance due to low doping and deep depletion regions. The extrinsic transconductance depends upon the source resistance (see Fig. 2.6).

$$g_{\mathrm{m}}\,(\mathrm{ext}) = \frac{1}{\sqrt{1+g_{\mathrm{m}}Rs}} \tag{12.8}$$

To maintain high g_{m}, the R_{s} source resistance must be minimized. This means that the resistance of the semiconducting region between the gate and the source contact must be minimized. Historically, TiPtAu gates have been used, but refractory gates suitable for self-alignment to reduce the source gate resistance are more common now.

12.3.2 Self-Aligned n+-Technique

Different refractory metal Schottky gate metals have been developed that can withstand annealing required for ion implant activation. These refractory metals are usually deposited by sputtering and patterned by dry etching. WSi_x and WN are good examples [6–8]. These metals maintain good Schottky barrier properties after annealing at high temperatures needed for Si implant ion activation, 850°C, or rapid thermal annealing even at higher temperature. Two examples of basic process steps used for n+-self-alignment are shown in Fig. 12.3. In one scheme, shown in Fig. 12.3a, a two-layer metal gate is shown, where the bottom layer is a metal etchable by plasma etching and the top layer is patterned by lift-off. For example, a nickel mask can be used for the top layer [9] and TiWN as the refractory gate. The nickel mask is defined by optical lithography, and the bottom gate is etched to give an undercut and a submicron gate dimension. The desired ion implantation profile is achieved, with light doping near the gate and higher doping in most of the source gate region. After ion implantation, the nickel metal can be removed before the structure is annealed at high temperature. In another scheme, a silicon nitride or oxide layer is deposited by plasma-enhanced chemical vapor deposition (PECVD) to give a conformal dielectric coverage over the gate metal and then etched back by reactive ion etching (RIE) to give sidewall nitrides around the gate metal. As in the first scheme, the ion implantation results in lightly doped region near the gate edges, thus maintaining reasonable breakdown voltage and lower gate capacitance (see Fig. 12.3c). These geometries are called a lightly doped drain (LDD). The distance of the n+ from the gate is controlled by the thickness of the nitride layer. Low source resistance, R_s, high g_m, and consequently high f_t values have been achieved [10].

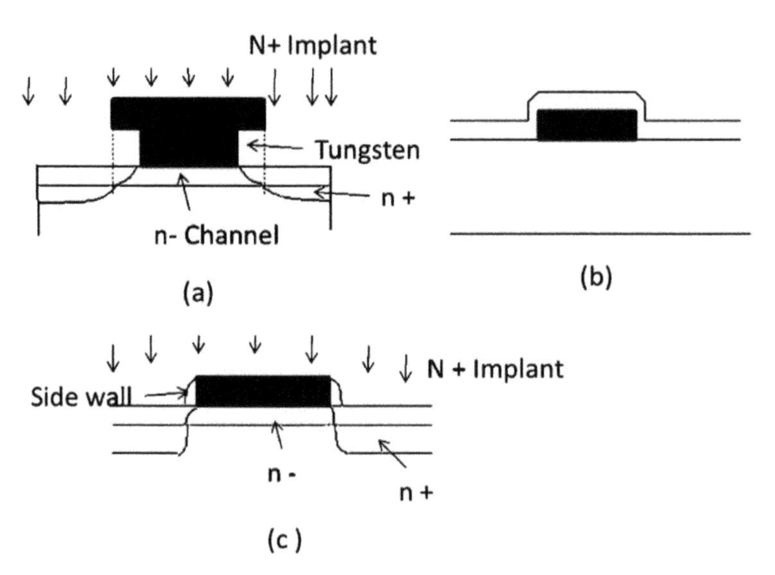

Figure 12.3 Two ways of making a self-aligned n+-implant Using a T-gate (a) and using sidewall dielectric deposition (b, c).

12.3.3 Substitutional Gate Processes

In these processes, a dummy gate or temporary gate is patterned to be used as a mask during ion implantation. This gate is then removed and the wafers are capped and annealed and the real gate is patterned next. In this scheme, refractory gate materials do not have to be used; low-resistance gate materials good for analog devices can be easily used. One of the first published examples of this process was the self-aligned implantation for N+ layer technology (SAINT) process [10]. A short version showing the major process steps is shown in Fig. 12.4. The wafer is implanted to form the channel regions. A thin layer of silicon nitride is deposited all over. A multilevel gate consisting of a resist and SiO_2 is deposited next, using a second layer of the photoresist for pattern definition. The lower resist is undercut in the etch process. After n+ ion implantation, another cap layer is deposited before removal of the substitutional gate. Now the wafer can be annealed. Silicon nitride is opened with the gate critical dimension (CD). The real gate material is deposited with a larger dimension to form a T-shaped gate. Different variations of this process are being used to form low-resistance gates with submicron electrical gate length.

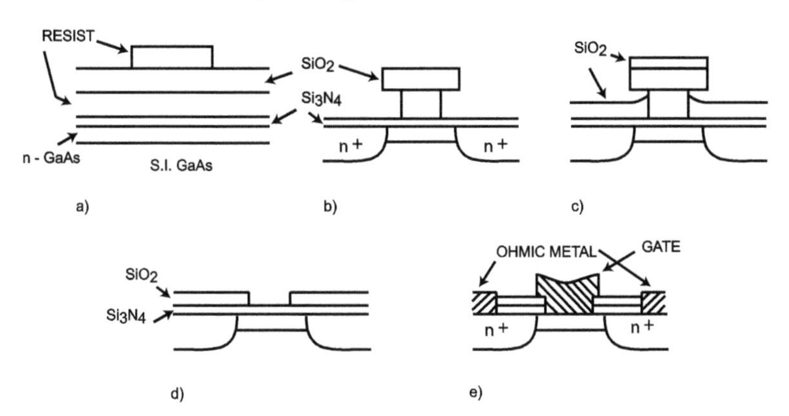

Figure 12.4 Major steps in the SAINT process used to fabricate digital FET gates. © [1922] IEEE. Reprinted, with permission, from Ref. [10].

12.3.4 Mixed-Signal Process

These processes are needed for fabrication of front-end receivers where both analog and digital FETs must be fabricated on the same die. These processes are generally planar, although recessed gate processes have been used for decades at many companies. These processes utilize enhancement-mode (E-mode) and depletion-mode (D-mode) and multiple voltage implants, high-conductivity gate layers, and some scheme for achieving higher breakdown voltage for the analog FETs. This could be placement of a gate offset in the source–drain channel or the utilization of an extra masking step before the n+ implantation to reduce the doping on the drain side of analog FETs (see Fig. 12.5) [11].

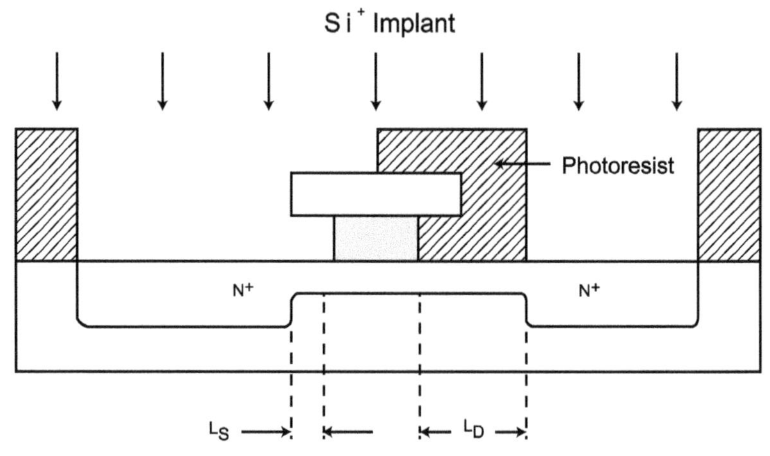

Figure 12.5 Self-aligned n+-implant with a photoresist on the drain side used to offset the n+-doped region for increased gate drain breakdown voltage.

12.4 Heterojunction and Insulated Gate FETs

In circuits where FETs need to operate under forward gate bias conditions, Schottky gate FETs have limitations because of the limitation of the barrier height to about 0.7 V. Under such limitations, the voltage swings are limited and lead to a low noise margin. The search for a metal–oxide–semiconductor field-effect transistor (MOSFET)-type device will be discussed separately. Other attempts have been made to solve this problem by adding an insulating semiconductor layer as in metal–insulator–semiconductor FETs (MISFETs) [12] or adding a higher-bandgap semiconductor on top as in HFETs or heterostructure isolated gate FETs (HIGFETs) [13–15]. AlGaAs or AlInAs layers can be used for the large-bandgap material and the GaAs channel can be replaced by InGaAs for higher mobility in the channel.

For very-large-scale integration (VLSI) digital circuits, device performance as well as production yields are important considerations. Above all, if circuit function can be accomplished in existing silicon fabrication lines, the compound semiconductors are not good candidates. To achieve high system clock rates, high logic density, and low power dissipation, many compound semiconductor devices have been tried, for example, junction FET (JFET), PHEMT, HIGFET, and heterostructure metal–semiconductor field-effect transistor (HMESFET). Most of these have used molecular beam epitaxy (MBE) materials and have not been successful in market penetration because of high defect density and cost.

12.4.1 HMESFET

High production yields of VLSI devices require stringent control of threshold voltage in order to keep the logic swing low and retain good device margin, while for robust circuits a noise margin of 200 mV is required. Voltage swings greater than 1 V are desirable. Conventional MESFET devices have a clamp voltage of about 0.65 V, which limits the logic swing. Vclamp voltage can be increased by using approaches mentioned above; the drawback of using MISFETs or HIGFETs is that these have a high threshold voltage. HFETs have a high clamping voltage, 1.2 V for AlGaAs/GaAs, without any drawbacks and are a good candidate for digital large-scale integration (LSI) circuits. One example is described below [13, 14].

It may be noted that MOCVD-grown epiwafers have been used with AlGaAs as the heterojunction layer, typically with 50% aluminum grown on a p-channel layer. A GaAs cap layer is required for fabrication purposes because of the chemical sensitivity of high Al fraction AlGaAs. All the layers were grown nominally undoped. Figure 12.6 shows a schematic diagram for an HMESFET. Device processing is similar to an ion-implanted FET processed for a self-aligned gate process. Shallow Si ion implants were used for E-mode FETs. Gate material can be WSi_x or WN_x. Optical photolithography was used for achieving 0.5 um gates. SRAM (16 k) was designed and fabricated to demonstrate the capability of this process for achieving high-complexity circuits at with good yield.

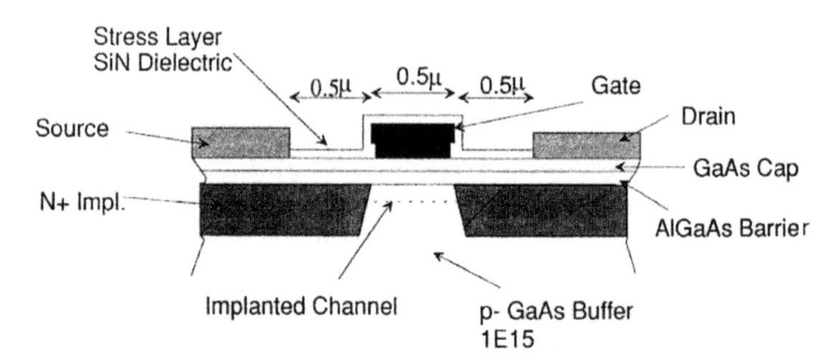

Figure 12.6 HMESFET (HFET) device cross section.

12.4.2 SAG FET Technology: Remarks

In recent years advances in direct ion-implanted FETs have brought the performance of digital FETs, as measured by the *K* factor, V_t control, close to those of HEMT devices for VLSI applications. Manufacturability and yield remain as the main issues for achieving commercial success. Submicron FETs are required for high *K* factors, but threshold voltage becomes an issue because of the short channel effect. Figure 12.7 shows the dependence of threshold voltage for E-mode and D-mode devices on a GaAs channel [16]. As the gate length shrinks, V_t drops. As the device shrinks, the aspect ratio must be maintained to optimize performance. Ion-implanted submicron FETs require shallow implants, which deviate considerably from theoretical predictions due to implant spreading. A high degree of implant tail clipping is necessary, which can be done by buried p implants [17] and use of dielectric stress over the FET to induce piezoelectric charge. For a 0.5 um gate length device, a *K* factor of 445 has been demonstrated with the use of a buried p layer. With just piezoelectric charge a value of 450 is also possible with correct FET orientation. These values are close to those achieved with MBE-grown FETs.

The effect of a dielectric stress–induced piezoelectric charge on V_t of FETs along (011) and (011^-) has been discussed in the early literature on GaAs. The piezoelectric charge can be induced for either FET orientation on GaAs as long as dielectric layers of the proper stress type are employed (compressive or tensile). Equal K factor improvement is observed for both directions down to 0.4 μm gate length. Gate length dependence of V_t for these orientations and 45 degrees is shown in Fig. 12.8. If performance can be relaxed, threshold voltage control is made easier with a 45-degree FET orientation, as the dependence on controlling passivation nitride and the interlayer dielectric does not become critical.

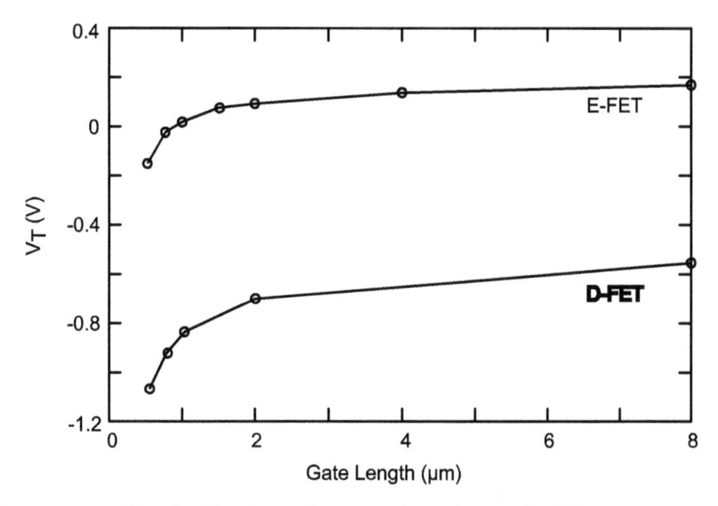

Figure 12.7 Threshold voltage for E-mode and D-mode FETs as a function of gate length.

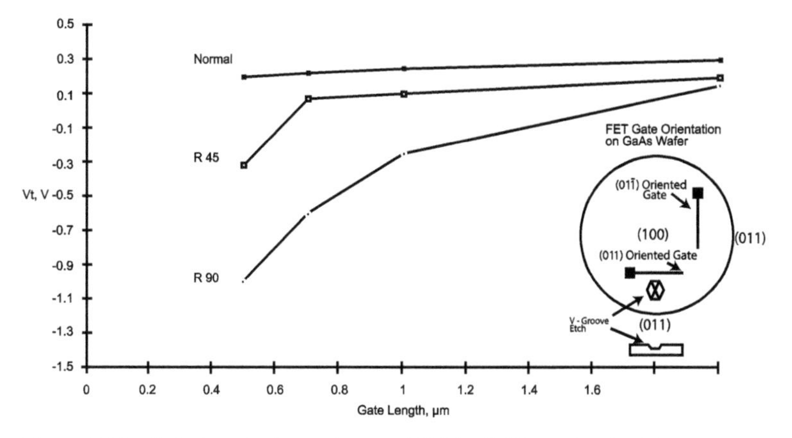

Figure 12.8 Gate length dependence of FETs of different orientations.

Performance numbers for digital MESFETs are listed in Table 12.1.

Table 12.1 Digital MESFET performance

L_g (μm)	g_m	K	V_t	Mode	Ref.
1.0	270	240	-0.5	D	[18]
	380		0.16	E	
0.5	336	445	0.1	E	[17]
	362	420		D	
0.5	140		-2.3	D	Qorvo Foundry
0.5		450	0.22	E	HMESFET [14]
0.5	400	580	0.29	E	MBE grown [19, 20]
0.4	330	400	0.15	E	[8]
	210	260	-0.45	D	
0.3	440		-0.15	E	[21]
0.2	480		-0.45	D	
0.1	370		-1.5	D	

12.5 Pregate Surface Preparation and Passivation

The state of the surface under the gate metal and the ledge nearby is extremely important for achieving reproducible FET performance. In particular, threshold voltage control and breakdown voltage are affected by a lack of surface control. To make sure the gate contact is as close as possible to an ideal Schottky diode, the surface must be free of native oxide or contaminants. So contaminants from the photoprocess must be removed and then the surface cleaned of any native oxide formed on the surface. The metal deposition step must be close-coupled, within 15–20 minutes of the cleaning step. After the gate lift-off or patterning process, the area between the gate and source or drain must be passivated by silicon nitride; again a prenitride clean procedure must be strictly followed to maintain good on-resistance and breakdown voltage.

Details of cleaning and passivation are discussed in Chapter 6.

12.6 Current Developments [22–26]

Large-volume production of FETs on GaAs for VLSI is no longer needed. Numerous attempts have been made to improve the performance of digital FET technology in III–V compounds to make them attractive for commercial circuits. MOSFET devices may provide a performance advantage, but yield issues due to defects (like DX centers; see Chapter 1) and limitations to a large gate voltage persist. Even complementary FET technology has been reported for FETs [22] and MOSFET devices, and commercialization has been started.

The InGaAs channel layer is used to replace the GaAs channel for higher speeds, particularly with an InP substrate, because of perfect lattice matching. Details are discussed in Chapters 4 and 13.

The InGaP layer has been used in place of AlGaAs for a better breakdown field, lower surface recombination, and avoidance of deep-level traps. InGaP also has high etch selectivity with GaAs. This material is also resistant to oxidation. This material has been used for HFET- [23] as well as HEMT-type devices [24]. A low-doped (n-) layer, about 10 nm thick, is used as an insulating layer under the Schottky metal. Schottky diodes on InGaP are known to degrade with thermal processing [25], so other layers are introduced on top or a thin oxide layer is grown or deposited. Ti metal gives a low barrier height (0.6 eV); platinum gives a higher value (0.7 eV) but continues to diffuse in at high temperature steps that follow. This platinum sinking can also be utilized to adjust the threshold voltage.

References

1. S. M. Sze, *Semiconductor Device Physics,* Wiley Interscience, New York (1969).
2. D. Biswas and D. A. Neamen, *Semiconductor Physics and Devices,* 4th Ed., Mc Graw Hill, Special Indian Edition (2012).
3. G. Myburgh et al., *Thin Solid Films,* **325**, 181 (1998).
4. S. K. Ghandhi, *VLSI Fabrication Principles, Silicon and Gallium Arsenide,* Wiley Interscience, New York (1983).
5. R. C. Eden, B. M. Welch and R. Zucca, *IEEE J. Solid State Circuits,* **SC-13**, 419 (1978).
6. A. Callegari et al., *J. Appl. Phys.,* **62**, 4812 (1987).
7. T. Ohnishi et al., *Appl. Phys. Lett.,* **43**, 600 (1983).

8. K. T. Alavi, *IEEE Trans. Electron. Devices*, **42**, 1205 (1995).
9. R. A. Sadler et al., A manufacturable 0.4 μm process for high performance LSI circuits, *GaAs IC Symp.* (1990).
10. K. Yamasaki, et al., GaAs LSI-directed MESFETs with self-aligned implantation for n+-layer technology (SAINT), *Electron. Lett.*, **18**, 119 (1982).
11. R. Williams, *Modern GaAs Processing Methods*, Artech House, Boston (1990).
12. R.A. Kiehl et al., High-speed, low-voltage complementary heterostructure FET circuit technology, *GaAs IC Symp.*, 101 (1991).
13. T. Tsen et al., A low power GaAs 16k SRAM with built-in redundancy, *GaAs IC Symp.* (1990).
14. K. R. Elliott et al., Heterostructure MESFET device for VLSI applications, *GaAs IC Symp.* (1991).
15. M. Kamada and M. Feng, *IEEE Trans. Electron. Devices*, **40**, 1358 (1993).
16. C. J. Anderson et al., MESFET MMIC for high volume applications, *GaAs IC Symp.* (1991).
17. M. Wilson et al., Extending the performance envelope of 0.5um implanted SAG-MESFET's for super-computer applications, *GaAs IC Symp.* (1993).
18. P. O'Neil et al., GaAs integrated circuit fabrication at Motorola, *GaAs IC Symp.* (1993).
19. M. Abe et al., HEMT LSI technology, *GaAs IC Symp.* (1983).
20. N. C. Cirillo, J. K. Abrokwah and M. S. Shur, *IEEE Electron. Device Lett.*, **EDL-5** (1984).
21. T. Shimura et al., High performance and highly uniform sub-quarter micron BPLDD SAGFET with reduced source to gate spacing, *GaAs IC Symp.* (1992).
22. S. M. Baier et al., *IEEE Electron. Device Lett.*, **8**, 260 (1987).
23. H. M. Chuang et al., *Semicond. Sci. Technol.*, **18**, 319 (2003).
24. A. Mahajan et al., *IEEE Trans. Electron. Device*, **45**, 2422 (1998).
25. E. Y. Chang et al., *J. Appl. Phys.*, **74**, 5622 (1993).
26. H. Jaeckel et al., *IEEE Electron. Device Lett.*, **EDL-7**, 522 (1986).

Chapter 13

HEMT Process

13.1 Introduction

High-electron-mobility transistor (HEMT) and pseudomorphic high-electron-mobility transistor (PHEMT) fabrication processes are similar to metal–semiconductor field-effect transistor (MESFET) processes but have some differences because of differences in the structure of the channel and other layers. These layers are epigrown, as discussed in Chapters 2 and 3. During the early years in the development of III–V integrated circuits (ICs), there was a lot of interest in HEMT processes for digital devices. This interest continues but the progress has slowed due to advances in silicon processes that made the need for higher-speed III–V digital devices irrelevant, at least for now, until silicon runs into scaling issues. These devices are used in production for applications in power amplifiers, low-noise amplifiers, and switches for mobile devices. As silicon complementary metal–oxide semiconductor (CMOS) circuits reach current technological limits, interest in low-bandgap III–V materials for digital circuits has grown again. In the interim, work has been continuing on improving the operating frequency limit (300 GHz and above) and performance of devices for the growing analog, wireless, and fast switch market. The processing of field-effect transistor (FET)-type heterojunction devices on GaAs

III–V Integrated Circuit Fabrication Technology
Shiban Tiku and Dhrubes Biswas

ISBN 978-981-4669-30-6 (Hardcover), 978-981-4669-31-3 (eBook)
www.panstanford.com

substrates and other III–V materials grown on InP, for both analog and digital applications, will be discussed in this chapter.

The key to success of CMOS technology has been the reduction of power. For high speed and low power, low-bandgap III–V materials have advantages over silicon. Reduction of direct current (DC) power consumption of a FET is possible by reduction of drain voltage. Narrow-bandgap-based FETs are good candidates for low-bias, low-power circuits because of high electron mobility, as long as gates of length of 0.5 μm and below can be fabricated.

Requirements for pseudomorphic HEMT (PHEMT) devices for switch or power amplifier applications were discussed in Chapter 2. In summary, low on-resistance and insertion loss must be achieved in switch devices and high power, high gain, and linearity are needed for power amplifier devices.

13.2 Device Fabrication

General epitaxial layer structures of HEMT and PHEMT devices were discussed in Chapter 4. The device cross section of a basic HEMT is shown in Fig. 13.1. These device structures are grown by metal–organic chemical vapor deposition (MOCVD) or molecular beam epitaxy (MBE) on (100) semi-insulating GaAs. Surface quality of the starting wafers must be well controlled better than that for the heterojunction bipolar transistor (HBT) epigrowth process, because a conducting interface layer between the substrate and epilayers may interfere with device performance. Low-surface-contaminant wafers (low silicon) ready for epigrowth are used. Thicker buffer layers or superlattice buffers are used in practice in epigrowth. The basic process steps needed to make the device and circuits based on it, from the epilayer structure, are outlined here. As in the case of FET processing, device isolation and ohmic contacts are completed first. This is followed by removal of the heavily conducting contact or cap layers and recess etching down to the layer on which a Schottky gate can be formed. Gate preclean and deposition as well as the passivation of the semiconducting layers near the gate are supercritical for process control. After the devices are formed, the interconnect steps similar to other processes are used.

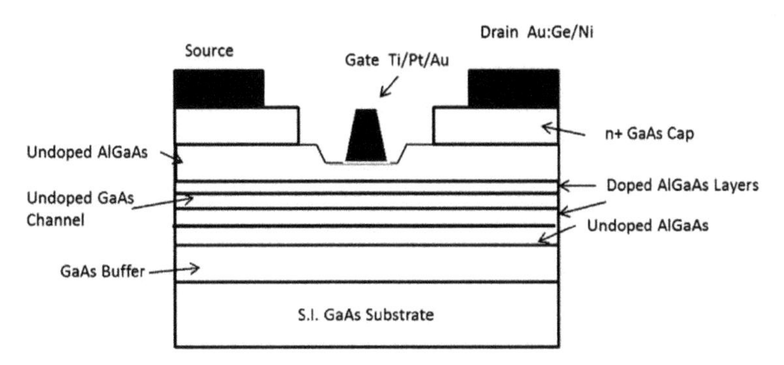

Figure 13.1 Cross section of a basic HEMT device.

For large-scale integration (LSI)-type applications, ion-implanted source and drain regions similar to FETs were developed and used first [1, 2]; these could be self-aligned to a refractory gate that may be a pure refractory metal or a silicide-like WSi_x or a nitride like WN_x, which are used as in the case of self-aligned FET processing. Annealing is needed for ion implant activation. However, annealing of epigrown structures with very thin layers and sharp profiles needed for good performance is not desirable. Therefore other schemes of reducing source and drain resistance have been devised, which mostly rely on use of epigrown, low-bandgap, heavily doped layers. The processing of such structures that do not employ ion implantation for n+ regions or for isolation is described first.

Processing starts with alignment mark formation. These marks may be made by wet etching using standard peroxide-based etchants. The active areas where devices are to be formed must be isolated from each other by removal of conducting layers, thus forming mesa structures, as shown in Fig. 13.2. The requirements of this process step are not as critical as the ones for etching into the device channel area. Therefore dry etching could be used without any consequences of ion damage. Cap layers are next removed by patterning and etching to form source and drain regions and define the channel area in between. The next step is to form ohmic contacts. A photoresist mask is used to define the desired contacts and bond pads using a lift-off processing sequence. Au:Ge/Ni/Au-based metallization is deposited and lifted off to leave behind source–drain contacts and bond pads, etc. For a single recess, only one recess etch step is used. Photoresist patterning and wet etching steps are used to define and etch off the GaAs cap layer and part of the AlGaAs layer. The etch depth is controlled to control the breakdown characteristics of the device. For a double-recessed device, a channel etch may be done first followed by a second recess. Next, the photolithography step is used to pattern the gate. Electron beam lithography may be needed for subhalf-micron gates. Wet etching is again used to etch off AlGaAs (or the GaAs layer for InGaAs channel devices). This etch step is the most critical for control of device parameters and must have good uniformity over the whole wafer. Spray etching systems may be needed for large-size wafers. Generally saturation current is measured to monitor the extent of the recess. Etching may be done in two or three steps, depending upon the degree of threshold voltage/Idss control that is desired. Etch stops may be grown in the epistructure to stop the recess etch for better Idss control and to avoid long process cycle time needed for multiple-recess etches. InGaP has excellent etch selectivity with respect to GaAs [3] and can be used as the etch stop layer and as a Schottky layer. Schottky gate metal is deposited using evaporation. A standard Ti/Pt/Au or Ti/Pd/Au metal sequence is used, the layers being optimized for desired reliability. The preclean prior to metal deposition is close-coupled to keep the surface clean and free of oxide and ensure good adhesion and good Schottky barrier characteristics. The gate metal may be used as the bottom plate of metal–insulator–metal (MIM) capacitors; therefore the deposition process must be optimized for low particles or nodules. The lift-off process must not leave flags, dog-ears, etc., over the gate fingers. Cross-sectional details of the gate and channel area are shown in Fig. 13.3. A transmission electron microscopy (TEM) cross section that shows the details of the channel region and the epilayers is shown in Fig. 13.4.

Passivation silicon nitride is deposited next, again preceded by preclean. The details of the surface preparation before nitride deposition are critical to the performance, breakdown characteristics, and long-term stability of the devices.

All contact openings must be patterned at this time for interconnect processing. If a plated metal layer is used for interconnect, a seed layer is deposited all over the wafer. This can be evaporated Ti/Au or sputtered TiW/Au. The wafers are patterned and can be electroplated in a bath or in an automated plating system. The thickness of the plating is generally several microns. After the plating is complete, the resist is removed and the seed layer in between the metal lines is etched off.

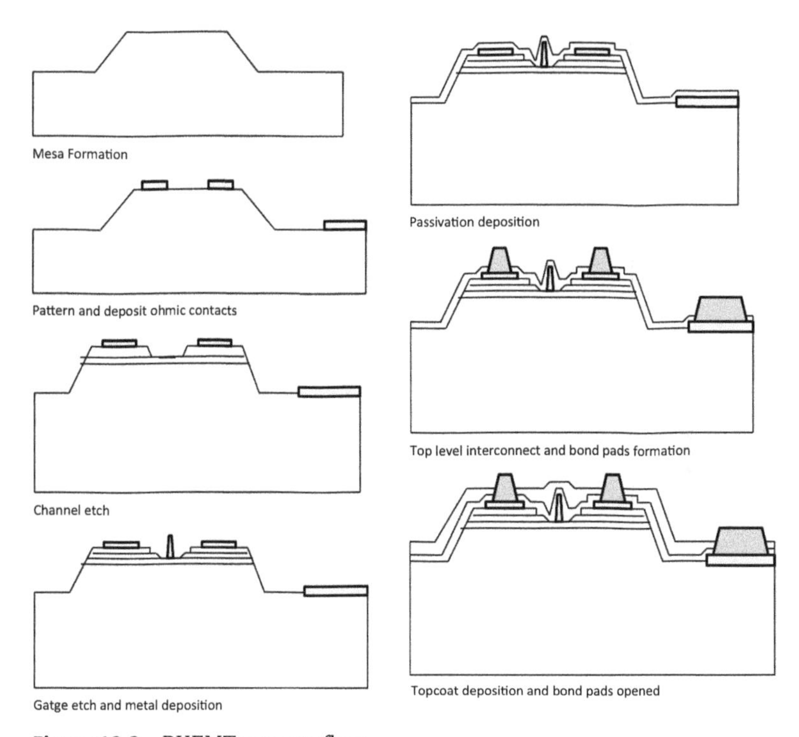

Figure 13.2 PHEMT process flow.

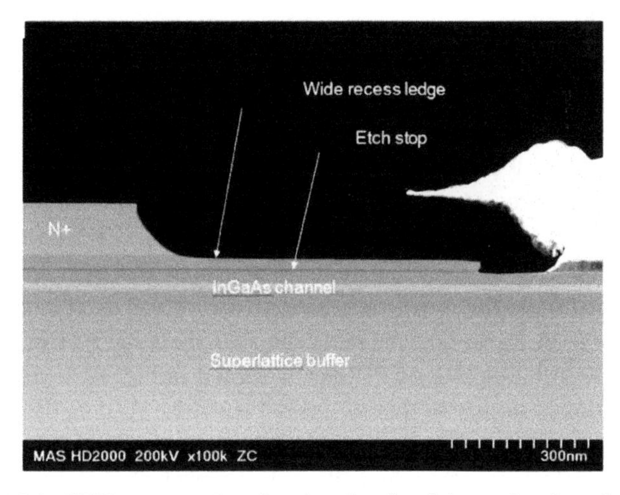

Figure 13.3 TEM cross section showing details of channel and cap layers.

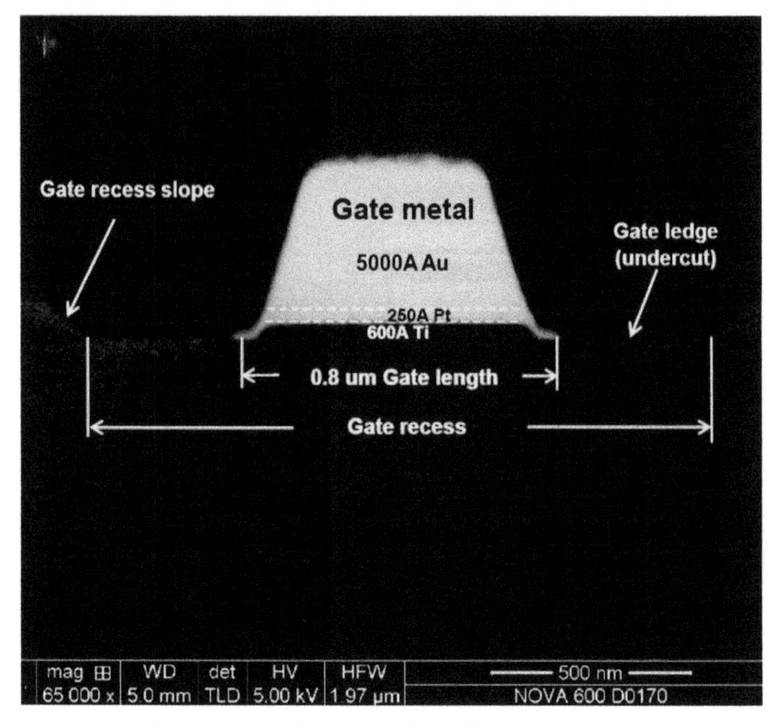

Figure 13.4 Cross section of gate and channel areas.

13.2.1 InGaP HEMT

Low insertion loss is important for PHEMT switch devices. For low insertion loss, device contact resistances must be minimized. Large energy band offsets between AlGaAs and GaAs can cause a discontinuity for electron transport in the contact path. Smaller band offsets were reported by several groups for InGaP/GaAs heterostructures [4, 5]. $In_{0.48}Ga_{0.52}P$ has almost the same energy band gap as $Al_{0.35}Ga_{0.65}As$ (~1.9 eV). Its advantages are fewer traps, less sensitivity to surface oxidation, and high wet etch selectivity. With the improved wet etch selectivity, recess etch uniformity below +/−10 Å can be achieved. The schematic layer structure for an InGaP/AlGaAs/InGaAs PHEMT is shown in Fig. 13.5 [6]. Device fabrication is similar to that described above for AlGaAs HEMTs, the four major steps being device isolation, ohmic contacts, recess etch, and Schottky metal gate deposition. For mesa isolation, wet etching can be used to selectively remove the GaAs cap over the InGaP layer by using the H_3PO_4:H_2O_2:H_2O (3:1:100) chemistry because this has excellent selectivity. HCl:H_2O can be used to remove the InGaP layer and H_3PO_4-based chemistry to remove other epilayers. After the ohmic contacts are formed, these can be alloyed by rapid thermal processing (RTP) at a temperature near 400°C. For a gate recess, H_3PO_4:H_2O_2:H_2O chemistry can again be used for selective etching. Special preclean steps are needed before the Schottky gate deposition. Figure 13.6 shows the layer structure and HEMT device structure in more detail. Good gate Schottky behavior, excellent drain–source *I*–*V* characteristics, high transconductance, and low insertion loss for switches have been demonstrated [3].

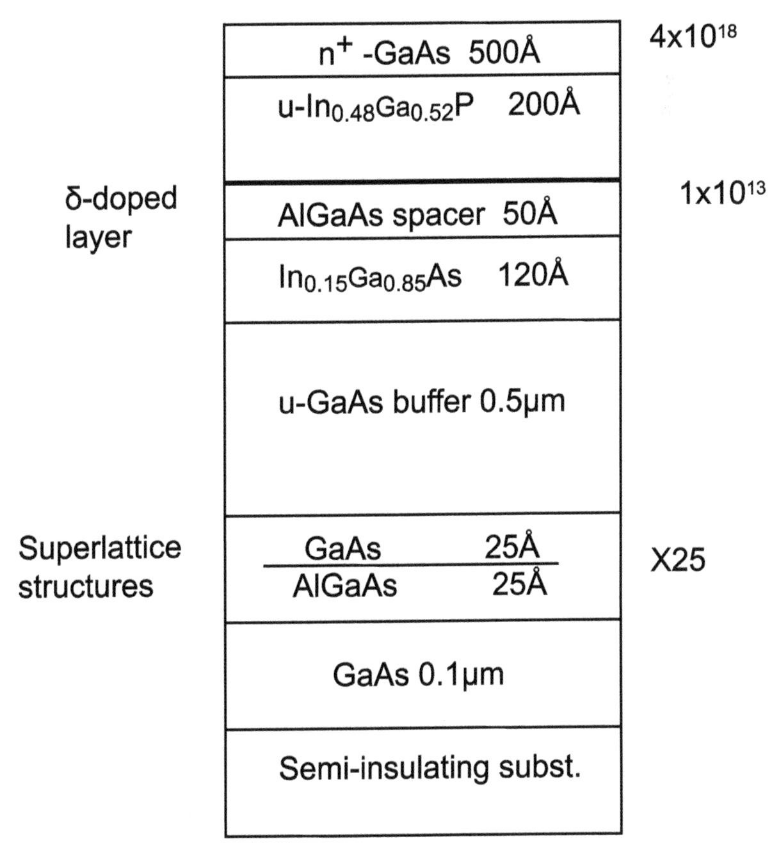

Figure 13.5 Schematic of InGaP/InGaAs/GaAs PHEMT grown by solid-source MBE. © [1996] IEEE. Reprinted, with permission, from Ref. [6].

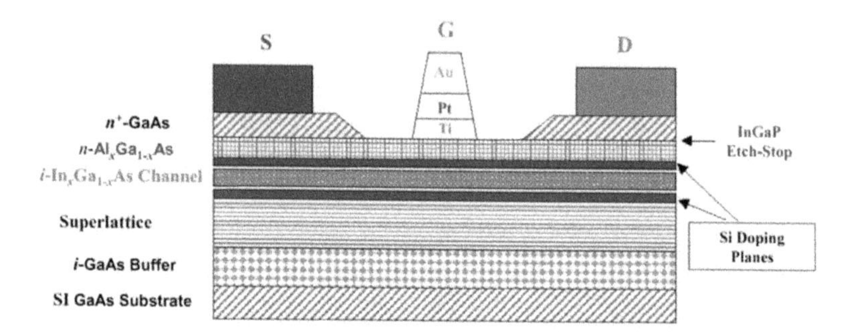

Figure 13.6 Schematic cross section of InGaP/AlGaAs/InGaAs PHEMT.

13.2.2 Low-Noise Process

The potential for low-noise devices was recognized very early in the development of HEMTs. The development of the PHEMT epistructure was a breakthrough in the development of low-noise HEMT devices. In the growth of pseudomorphic layers, a much larger fraction of In can be used to achieve very high electron mobility. Thus a very thin channel layer can be used. The distance of the channel from the gate is well controlled and fixed, resulting in reproducible high transconductance (see Eq. 2.59 in Section 2.3.1).

In the design of the epistructure and the device, there are always trade-offs. For low-noise devices, noise figures (NFs) below 1 dB are generally specified. A trade-off between linearity and return loss must be made while achieving a low NF and ensuring stability. Commercially viable processes have been developed with a 0.15 um gate length to achieve a low NF using a metamorphic epilayer structure (not lattice matched) on a GaAs substrate [7]. In this structure, the InGaAs channel layer is sandwiched between InAlAs layers. The composition of In$_x$Ga$_{1-x}$As can be with x between 0.45 and 0.53. Electron mobility in excess of 10,000 cm^2/V-sec can be achieved in this channel. Figure 15.13 (Chapter 15) shows the schematic cross section of the Triquint (now Qorvo) process. The active layer of the HEMT is made by mesa etching. The 0.15 um T-gate is patterned using electron beam lithography. A similar process has been reported by WIN SEMI for E/D HEMT foundry applications [8].

Enhancement-mode (E-mode) HEMT devices are always preferred by designers because of their low power consumption (normally off) and the fact that a negative power supply is not needed. Using a similar epistructure (except for a channel to achieve E-mode behavior), a 0.35 μm gate length process has been reported with an NF of 0.2 dB. The device with an f_t of 45 GHz and an I_{max} of 325 mA/mm is well suited for low-noise amplifier (LNA) applications in the 1.5 to 2.7 GHz handset market [9].

13.2.3 Power Amplifier Process

For high-power applications, a large gate width is needed, which means that the periphery of the device is high. Devices are laid out with a large number of gate fingers in between source and drain contacts. The device must have good scalability and high breakdown voltage. Historically, the challenges to manufacturing power ICs have been achieving good yield with a large periphery and large gate fingers. Lithography and lift-off patterning of submicron gates has been the main issue. These can be resolved by careful design of the resist process, metallization, and the lift-off process. Control of gate critical dimensions (CDs) needs to be monitored in production.

Multiple gates are not used for power processes because of the increase of gate capacitance. As the number of gates goes up (not gate fingers), f_t, the power-added efficiency (PAE) goes down.

As an example of a high-performance power amplifier process developed by Northrup-Grumman on InP substrates is briefly described below.

The epilayer structure is shown in Fig. 13.7 [10]. The composition of the InGaAs channel is similar to that used for low-noise applications. The $In_xAl_{1-x}As$ has a composition with $x = 0.52$. A cap layer with an In concentration is grown on the top for low contact resistance. The In content and the doping are high enough so that the ohmic contact does not have be of the alloyed type. The contact resistance is below 0.04 ohm.mm. The cap layers are removed before channel and gate processing. The gate length is 50 nm and has a long periphery. Electron beam lithography using a multilayer photoresist is employed. The gate process is generally based on polymethylmethacrylate (PMMA)/resist T-gate processes discussed under gate formation in FET processing. The gate is self-aligned to the gate recess. After the gate is patterned, the devices are passivated with silicon nitride deposited by plasma-enhanced chemical vapor deposition (PECVD), taking the usual precautions in surface preparation, as described above in Section 13.2. A g_m of 2400 mS/mm and an f_{max} of ~1000 GHz have been demonstrated.

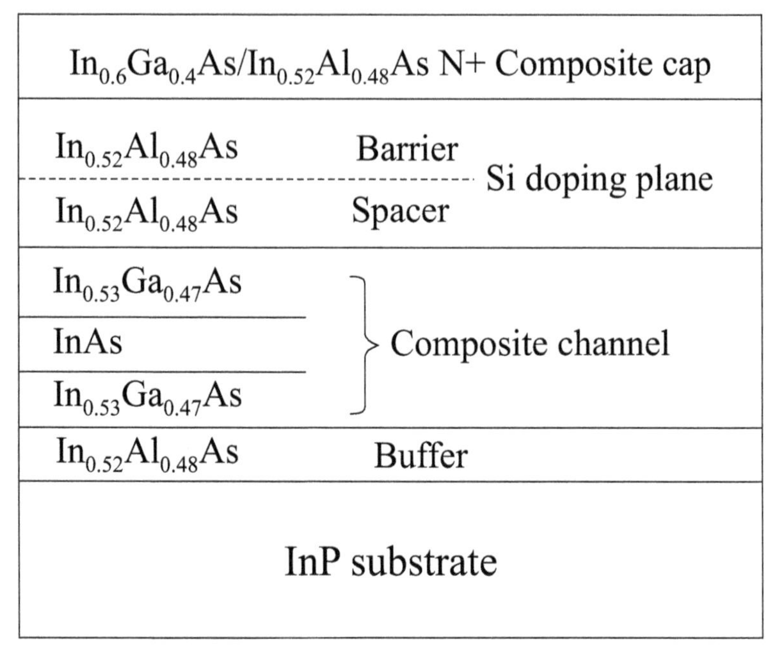

Figure 13.7 Layer structure of InAsAs/InGaAs HEMT grown on InP for power applications.

13.2.4 Switch Process

Requirements for the switch process were mentioned earlier in Section 13.1. These are low insertion loss, low R_{on}, good isolation, and high breakdown voltage. As mentioned above under Section 13.2.1, ohmic contact resistance can be reduced by avoiding the large energy band discontinuities by using InGaP-/GaAs-type heterojunctions rather than AlGaAs/GaAs interfaces. Delta doping is used to keep the overall channel thin. Delta doping, or double delta doping from below and above, improves transconductance and maintains high breakdown voltage. Large gate widths are used to reduce insertion loss and achieve high current.

To avoid harmonics, devices are designed to be as linear as possible, because nonlinearity generates harmonics. Common gate resistors are used in series with the gate to maintain high voltage on the gate in the off state.

Dual or multiple gates have been used to achieve higher radio frequency (RF) gain, higher output impedance, and reduced feedback capacitance. The gates are connected in a cascade configuration [11] (see Fig. 13.8). Dual gates are being utilized for HEMT switch processes to achieve a higher f_t and breakdown voltage at the same time, both for III–V and for III–N type devices [11, 12]. To improve off-state linearity and improve leakage, the region between the two gates can be connected to reverse bias. Therefore, an ohmic contact can be deposited in between the two gate fingers, as shown in Fig. 13.9, showing the IGCC HEMT process layout and cross section (Hitachi) [13]. The process using 0.5 um gates has been used for a single-pole, double-throw antenna switch. Lower second and third harmonics were demonstrated compared to a single-gate PHEMT.

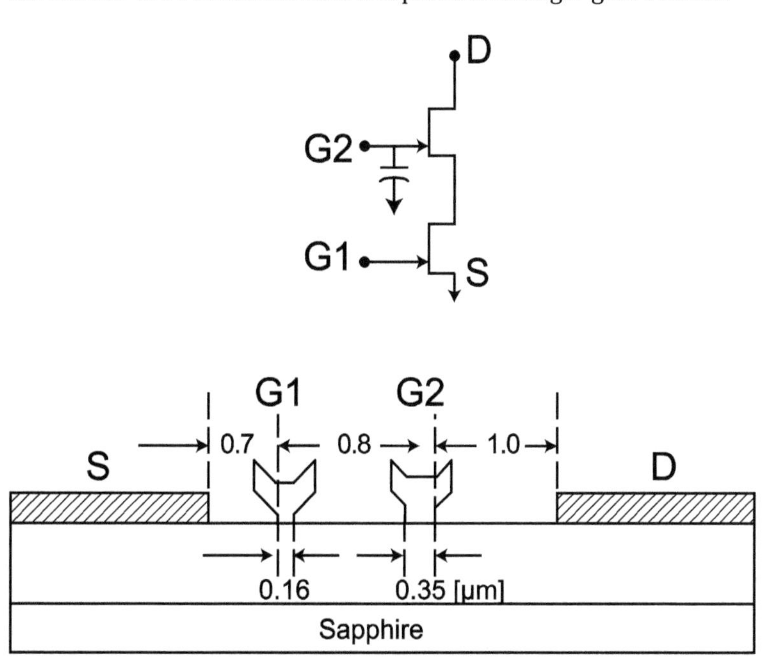

Figure 13.8 Schematic of GaN HEMT dual gates, cascode connected.

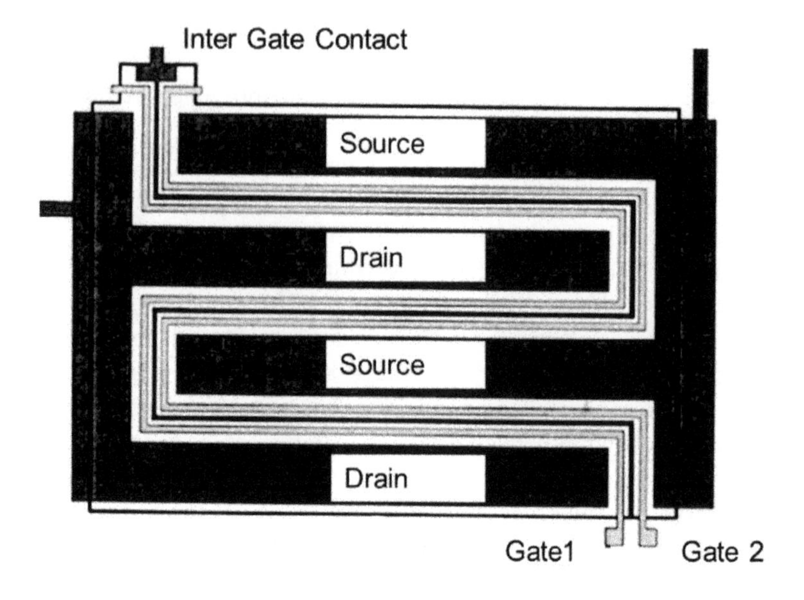

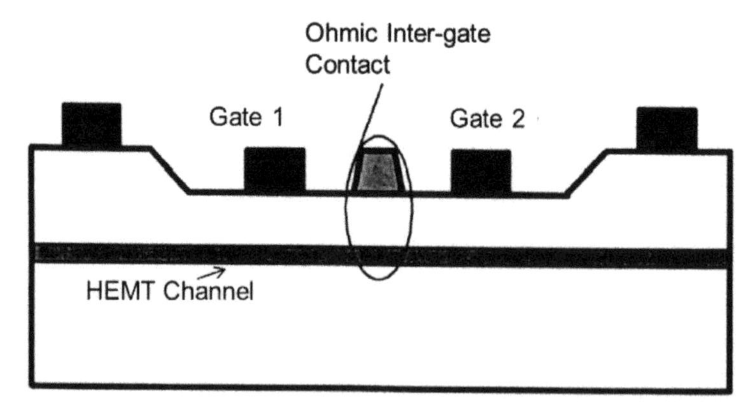

Figure 13.9 IGCC multigate HEMT switch layout and schematic cross section. © [2008] IEEE. Reprinted, with permission, from Ref. [13].

13.3 InP HEMT

The advantages of InP substrates over GaAs have been discussed elsewhere (in Chapter 2). In review, very high electron mobilities can be realized in low-bandgap materials like InGaAs or InAs, and these materials have a better lattice match to InP. These channel layers can be sandwiched between higher-bandgap materials like InAlAs or AlSb to form two-dimensional electron gas (2DEG) heterojunction devices. Cutoff frequencies of over 300 GHz have been reported with sub-0.1 μm gate devices on these type of structures [14]. These devices are promising candidates for analog circuits and also digital ICs, because E-mode FET-type devices can be fabricated. As an example, a cross section of a 0.1 μm gate device is shown in Fig. 13.10 [7, 14]. The device shown uses processing similar to other HEMT processes except for the critical gate step—nonalloyed source and drain contacts and nonsymmetrical gate placement. The recess etches utilize etch chemistries with good selectivity, so the Schottky gate can be formed on InP. The gate is formed by electron beam lithography. WSiN gate metal is sputtered all over, and goes into the openings in the dielectric. Ti/Pt/Au metal is evaporated to act as the top cladding and lifted off on top, and the extra WSiN is etched off.

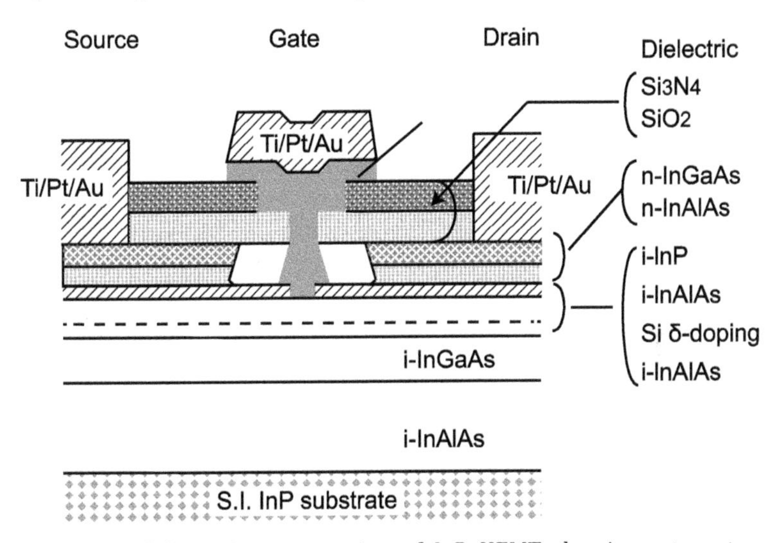

Figure 13.10 Schematic cross section of InP HEMT showing a two-step recessed gate. © [1999] IEEE. Reprinted, with permission, from Ref. [14].

InGaAs HEMTs on InP substrates are being studied with the goal of achieving CMOS-type circuits for low-power, high-speed logic. For high density and a good I_{on}/I_{off} ratio, short gate lengths (below 0.1 μm) and small gate–source spacing are pursued. By using self-alignment, source–drain spacing can be reduced, from the current state of the art of 1.5–2.0 μm by almost a factor of 10. In one example of a novel approach, refractory tungsten contacts are made first [15] and then the gate opening is made using a layer of SiO as a mask to get a controlled spacing between W source/drain and the gate metal that is patterned and lifted off in between. A short description and process details are shown in Fig. 13.11.

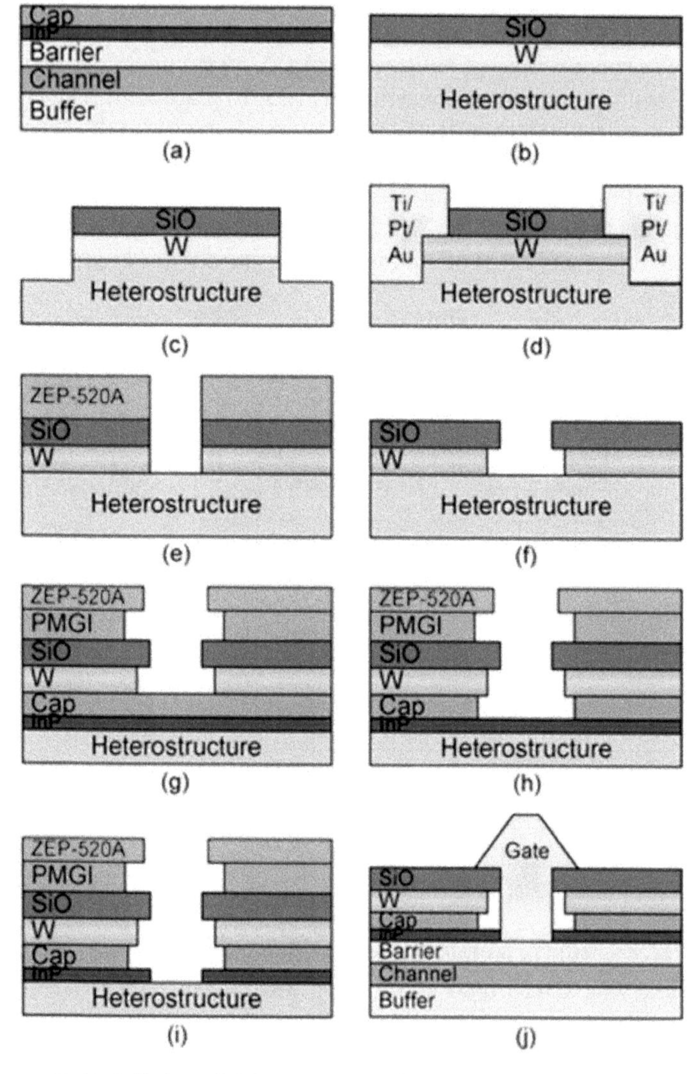

Figure 13.11 Self-aligned HEMT process flow. © [2010] IEEE. Reprinted, with permission, from Ref. [15].

At such short gate-to-source distances leakages become an issue. Development of an oxide or a high-*K* dielectric layer over InP would be a welcome relief.

13.4 Processing Issues

Good control of Idss and threshold voltage (and pinch-off voltage) depends upon control of all process steps involved with gate processing. Control of the process for gate opening in the nitride, if nitride passivation is deposited before the gate opening is made, as well as pregate surface preparation steps are crucial. Plasma bombardment during nitride etching and excessive damage during ashing or de-scum steps must be minimized. Too much exposure to aqueous solutions or lack of control of the close coupling time to the next step must be avoided. The details of the surface treatments before gate deposition as well as nitride passivation are generally proprietary to each company.

Threshold voltage can shift due to gate metal sinking that changes the effective gate-to-channel spacing, so it must be carefully minimized and monitored for the process. Threshold voltage can be controlled to some extent by gate metal deposition. A depletion-mode (D-mode) HEMT may be converted to an E-mode HEMT by use of gate metallization that can be controllably sunk in to create an E-mode FET [8, 16]. However, the metallization may continue sinking in during the life of the device and cause reliability issue.

13.4.1 Gate Walk

Gate placement between the source and drain contacts depends on the alignment capabilities of the photolithography process. In the lift-off process, perpendicular arrival of metal atoms in the deposition process is assumed. In practice the metal arrives at an angle at the edges of wafers. Geometrical details are shown in Fig. 13.12a. As wafer size increases, this problem becomes worse, unless the throw distance in the evaporator is increased to maintain the angle below a few degrees. Otherwise the gate is displaced to one side, as seen in Fig. 13.12b. This phenomenon is sometimes called *gate walk*. If the foot of the gold extends beyond the barrier layers, onto the higher conducting layers (is not symmetrically inside the second recess in a double-recess process), leakage and low breakdown may result.

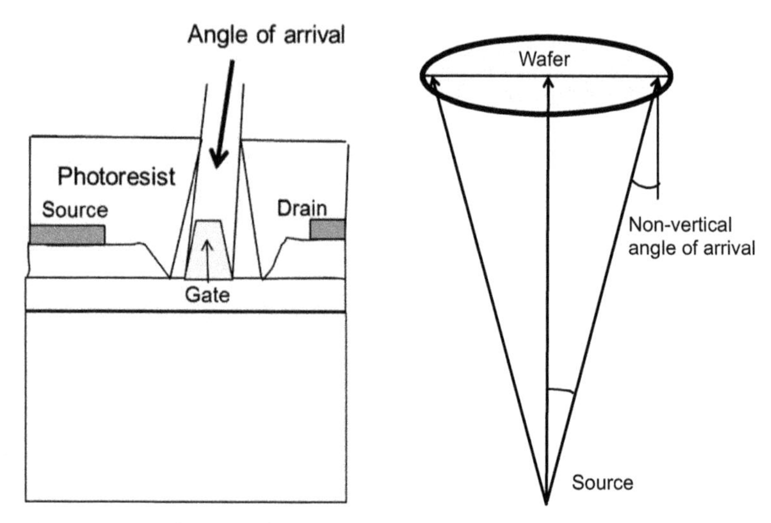

Figure 13.12 Schematic diagram showing the geometrical details of gate patterning and deposition (by evaporation) at wafer edges.

13.4.2 Gate Sinking

As in the case of FETs, HEMTs suffer from process issues related to the gate process. As mentioned earlier in Section 13.2, pregate and postgate formation the surface treatments are crucial to maintaining performance and reliability. In spite of the precautions, problems may occur. Gate sinking is a common problem, as discussed in Chapter 12. These devices generally have long gate periphery, and the chances of a metal barrier not being present are high. Also, electrochemical reasons may enhance the diffusion and corrosion. Figure 13.13 shows a common problem of gate sinking due to an inadequate barrier. Sufficient thickness of Pt or other refractory metals like Mo or TaN is required to suppress this phenomenon. Some groups are trying to use gate sinking to adjust threshold voltage, for example, to convert a D-mode HEMT to an E-mode HEMT; however, this is not desirable from a reliability point of view.

If the metal evaporation process results in gold spilling over the barrier, or if the Ti and Pt underlayers are not as wide as gold, diffusion at the gate edge can be seen, as shown in Fig. 13.14. This gets worse with completion of the interconnect process that demands 300°C or the device packaging process requires a temperature near 260°C.

Figure 13.13 TEM cross section of a gate showing gate sinking.

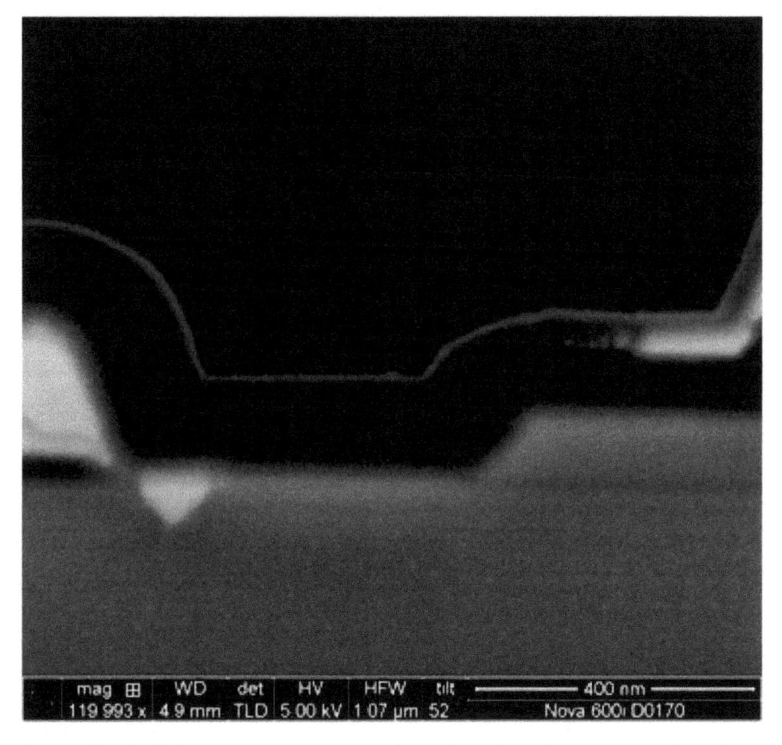

Figure 13.14 FIB cross section near the gate edge showing metal sinking near the gold foot.

13.4.3 Breakdown Voltage Improvement

Gate-to-drain breakdown voltage or off-state source–drain leakage is affected by processing and can vary in production. Breakdown is dominated by surface states near the gate edges and therefore may depend on the details of the recess etch and surface preparation before the passivation nitride is deposited. Generally a compromise must be made between breakdown voltage and high-frequency response in deciding on the prenitride surface preparation that sets the surface states. And careful attention must be paid to these critical process steps.

The drain can be offset with respect to the gate to improve breakdown. A better approach is to use double-recess etching, as shown in Fig. 13.15 [17]. For high-power devices this is a necessity.

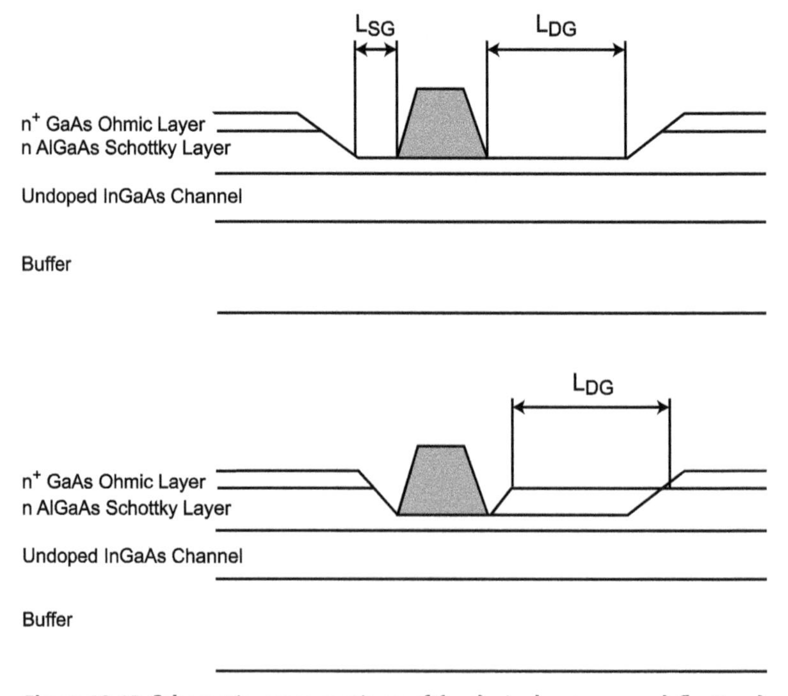

Figure 13.15 Schematic cross sections of (top) single-recess and (bottom) double-recess PHEMT.

References

1. J. H. Huang, J. K. Abrokwah and W. J. Ooms, *Appl. Phys. Lett.*, **61**, 2455 (1992).
2. K. Nishii et al., A novel high-performance WSi-gate self-aligned N-AlGaAs/InGaAs/N-AlGaAs pseudomorphic double heterojunction MODFET by ion implantation, *Proc. IEEE Int. Symp. Compd. Semicond.*, 478 (1998).
3. H.C. Chiu et al., Low inserting loss switch technology Using 6-inch InGaP/AlGaAs/InGaAs pHEMT production process, *IEEE CSIC Digest*, **119** (2004).
4. S. Froyen, A. Zunger and A. Mascarenhas, *Appl. Phys. Lett.*, **68**, 2852 (1996).
5. C. Cai, M. I. Nathan and T. H. Lim, *Appl. Phys. Lett.*, **74**, 720 (1999).
6. A. A. Aziz and M. Missous, High performance electron devices for microwave and optoelectronic applications, *IEEE EDMO Conf. Proc.*, 145 (1996).
7. T. Henderson et al., *High Performance BiHEMT HBT E/D pHEMT Integration*, CS MANTECH (2007).
8. C. K. Lin et al., *Monolithic Integration of E/D-Mode pHEMT and InGaP HBT Technology on 150mm GaAs*, CS MANTECH (2007).
9. T. J. Yao, X. Sun and B. Lin, Ultra low-noise highly linear integrated 1.5 to 2.7 GHz LNA, *IEEE Int. Wireless Symp.* (2013).
10. X. B. Mei et al., Sub-50 nm InGaAs/InAlAs/InP pHEMT for sub-millimeter wave power amplifier applications, *IEEE Conf. InP Relat. Mater.* (2010).
11. M. Rodell and U. Mishra, High power broad band amplifiers based on GaN, *UCSB Rep.* (2002).
12. J. Wenger et al., Dual gate pHEMT with high gain, *IEEE Microwave Guided Wave Lett.*, **2**, 46 (1992).
13. S. Koya et al., Intergate-channel-connected multi-gate PHEMT devices for antenna switch applications, *IEEE MTT Int.* (2008).
14. Suemitsu et al., High-performance 0.1-μm gate enhancement-mode InAlAs/InGaAs HEMT's using two-step recessed gate technology, *IEEE Trans. Electron. Devices*, **46**, 1074 (1999).
15. N. Waldron, D-H. Kim and J. del Alamo, A self-aligned InGaAs HEMT architecture for logic applications, *Trans. IEEE Electron. Device*, **57**, 297 (2010).
16. J.-H. Du et al., *WINSEMI Foundry*, CS MANTECH (2011).
17. R. Menozzi, Off-state breakdown of GaAs pHEMTs, *IEEE Trans. Device Mater. Rel.*, **4**, 54 (2004).

Chapter 14

HBT Processing

14.1 Introduction

The underlying concepts of the heterojunction bipolar transistor (HBT), which were discussed in Chapter 2, have been developed over decades since the 1950s. However, h1h-performance HBTs became possible only after molecular beam epitaxy (MBE) and metal–organic chemical vapor deposition (MOCVD) techniques were available in the 1980s. Details of epitaxial growth were discussed in a separate chapter. A layer structure for the HBT is shown in Fig. 14.1 for an InGaP emitter HBT; the structure is very similar to the AlGaAs/GaAs HBT, which was developed first. Given an epistructure such as the one shown, the fabrication process concentrates on making contact with the various layers, while etching off unwanted portions of the layers, to end up with a device as that shown in Fig. 14.2. The device structure of the HBT and base–collector (BC) diodes are carved out of the epilayer structure without damaging the HBT junctions or the surfaces. Processes have been developed over the years to minimize topology, while making the contacts and isolating adjacent devices from each other. With a combination of etching and isolation by ion implantation this can be achieved. Care must be taken during processing to not degrade device performance or introduce damage that may result in loss of long-term reliability.

III–V Integrated Circuit Fabrication Technology
Shiban Tiku and Dhrubes Biswas

ISBN 978-981-4669-30-6 (Hardcover), 978-981-4669-31-3 (eBook)
www.panstanford.com

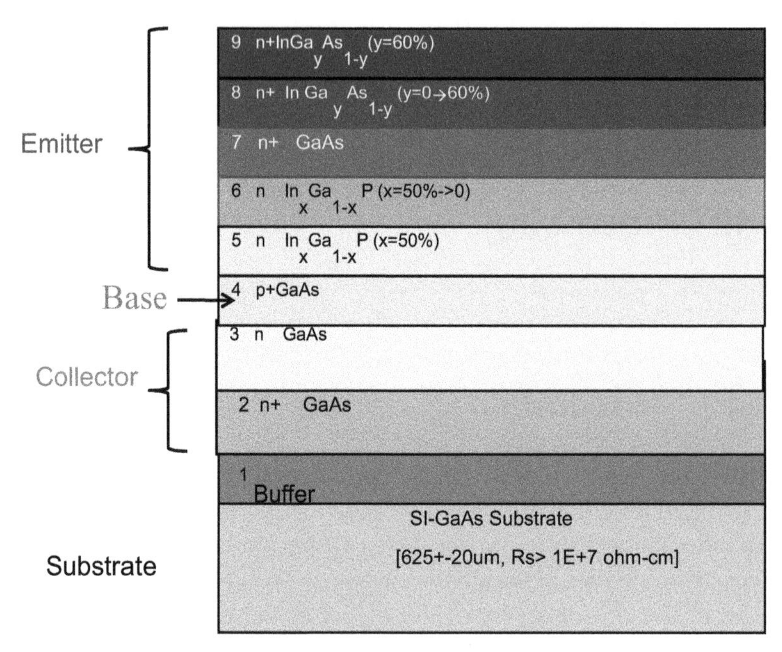

Figure 14.1 Epilayer structure of an InGaP/GaAs HBT.

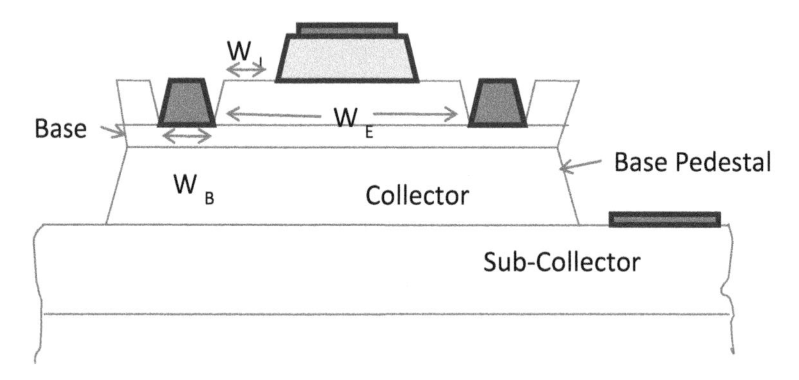

Figure 14.2 Schematic cross section of an HBT. The effect of emitter and base contact sizes on the HBT base–collector junction and the overall HBT can be seen.

14.2 Review of HBT Process Evolution [1–3]

From the very early days of HBT circuit and process development, emphasis was placed on parasitic reduction. This means reduction of the BC junction area and base resistance for common emitter HBT circuits (discussed in the chapter on devices). Self-alignment was already in use in silicon technology, so attempts were made to utilize different self-alignment schemes. A few of these are shown in Fig. 14.3 [2]. Details of the evolution of the HBT process are discussed in more detail in the literature [2, 3]. Non-self-aligned, mesa-etched HBT performance has been adequate for products that have been in production so far. A brief description follows next, and details will be discussed later.

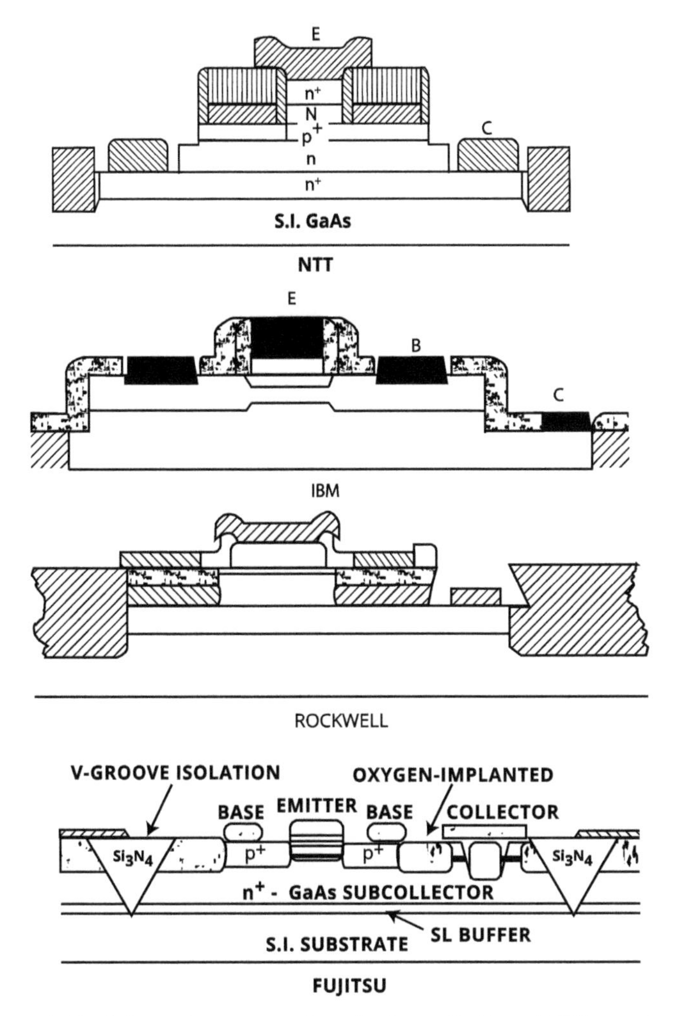

Figure 14.3 Schematic cross section of a few self-aligned HBT structures.

For emitter-up HBTs (see, for example, Fig. 14.2) contact to the top is made first. This contact can be made with any commonly used metal system, such as Ti/Pt/Au or Au:Ge/Ni. Next, an emitter mesa is formed. The top low-resistivity layers are removed to expose the actual emitter layer that is going to form the ledge of the device. The etch depth must be controlled accurately so that the ledge layer is not disturbed. The details of this etch step and other etches used for the base contact are discussed in Chapter 7 in more detail. The second mesa or base pedestal is formed next by etching off the emitter AlGaAs or InGaP layer as well as some of the collector layer. The whole surface of the wafer, including the ledge, is covered by silicon nitride. The base pedestal is made deeper into the collector to make isolation by ion implantation easier, which can be done at this stage. See Fig. 14.2. Base contact areas are opened next in the nitride and then base contact metal is deposited. The base contact etch must also be controlled very precisely because the base layers are the thinnest. To make the collector contact the areas are defined and the collector layer is removed to make contact with the highly doped subcollector layer. The commonly used n-type contact scheme based on Au:Ge/Ni is utilized here.

An alternative, more planar approach to make a base contact is to convert the n-type top layers to p-type layers with ion implantation [4] or diffusion of p-type dopants [5]. A schematic cross section of this approach is shown in Fig. 14.4. The dopant concentrations possible with p-type dopants can be very high, in the 1E20 range for Zn. The p-type dopants can be placed very close to the emitter. The combination of high doping and small spacing can result in low base resistance. The disadvantage of ion implantation is that annealing at high temperatures is required, which is not ideal for junction sharpness even if rapid thermal annealing is used. Also, in this process p–n junctions must not be made in the topmost InGaAs or GaAs layers, because these low-bandgap materials have low turn-on junctions and therefore would form parasitic junctions. So these layers should be removed. The isolation of devices can also be done by implantation damage with a more planar device of the type shown in Fig. 14.2 or 14.3. The mechanism and details of the ion implantation process are discussed in Chapter 3. Historically, boron, oxygen, and hydrogen have been tried. All can convert the semiconductor layers to resistivity of 1E7 ohm.cm, with some thermal treatment after the implant. Boron cannot go too deep and hydrogen has the disadvantage of diffusing around. A very-high-energy ion implanter would be needed with boron or oxygen, so these are good for shallow layer isolation. Helium ion implantation is practical with a 400 keV implanter. A combination of mesa etching and ion implantation can be made to work for a variety of HBT structures.

High power, high frequency and low parasitic requirements make mesa isolation more desirable. The isolation resistance achieved by the mesa can be higher than what is possible with ion implantation. A large number of HBT processes used in different companies for microwave products are based on this basic, emitter-up, mesa-isolated structure. These devices are used in the common emitter mode, where the emitters are connected to the ground plane on the backside of the wafer and the collectors are connected together by a thick metal on the front.

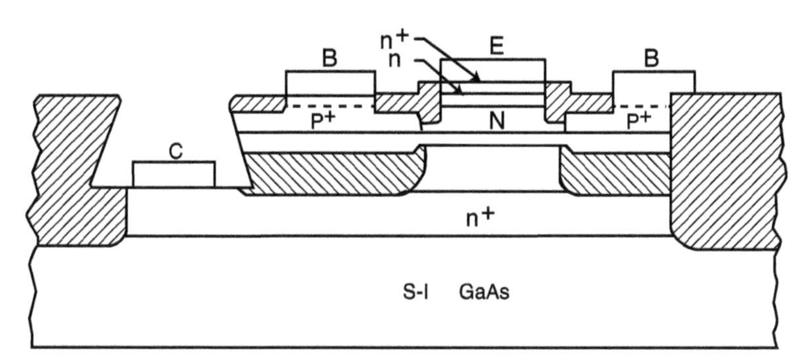

Figure 14.4 Structure of HBT with p+ ion implantation in the extrinsic base region.

14.3 Basic HBT Fabrication Process

Processes that use fewer steps and are simpler give higher production yields. The area around the emitter, particularly the space between emitter and base contacts, is called the ledge. With current lithography tools, critical spacing like the ledge size can be maintained down to 0.5 μm. Therefore, more complicated self-alignment schemes are not needed. Good performance and reliability can be achieved [6–8].

In the self-aligned processing methods mentioned above, the extrinsic base contact area is not covered by the depleted AlGaAs or InGaP emitter layer. Ensuring depletion of this ledge area is beneficial for reducing base surface recombination and thereby promotes good reliability. To leave just the emitter layer on the extrinsic base area, and not the emitter cap layers, a proper wet and dry etch sequence must be used. Process details for AlGaAs/GaAs HBT are described below. Similar process steps are used for InGaP/GaAs HBTs. More details of etching are discussed in Chapter 7.

Process details of a production process are shown in Fig. 14.5. The emitter contact layer is patterned first by the lift-off process, followed by an emitter mesa etch. Details of photoresist steps are not shown in the schematic process steps. The InGaAs emitter cap can be removed by using a $H_3PO_4/H_2O_2/H_2O$-based etch solution. Chlorine-based dry etching, for example, using $SiCl_4/SF_6/He$, can be used to etch GaAs and stop on AlGaAs. At this point, the ledge of the HBT is exposed and needs to be passivated. One of the major advantages of silicon is that native oxide grows on it with a "clean" interface. The transition from semiconductor to insulator is smooth and there are no dangling bonds giving rise to surface states. Numerous attempts have been made and proposed to grow a good oxide or insulator on III-V semiconductors (discussed in Chapter 2). GaAs production processes rely on passivation of the surface by silicon nitride. For emitter-up HBTs, the most crucial region to be passivated is the ledge region between base and emitter contacts. Thin layer of silicon nitride is deposited using plasma-enhanced chemical vapor deposition (PECVD), making sure excessive damage is not introduced.

The base contact is patterned next and an opening etched in the silicon nitride passivation layer. Then a BCl_3-based dry etch can be used to etch AlGaAs for etching down to the GaAs base layer. Selective etching is not available for this step, so an endpoint tool must be used. Development of a production-worthy, high-reliability HBT process needs optimization of all subprocess steps, including de-scums after photolithography, precleans before metal deposition, etc. This optimization is critical for emitter and base contact process steps.

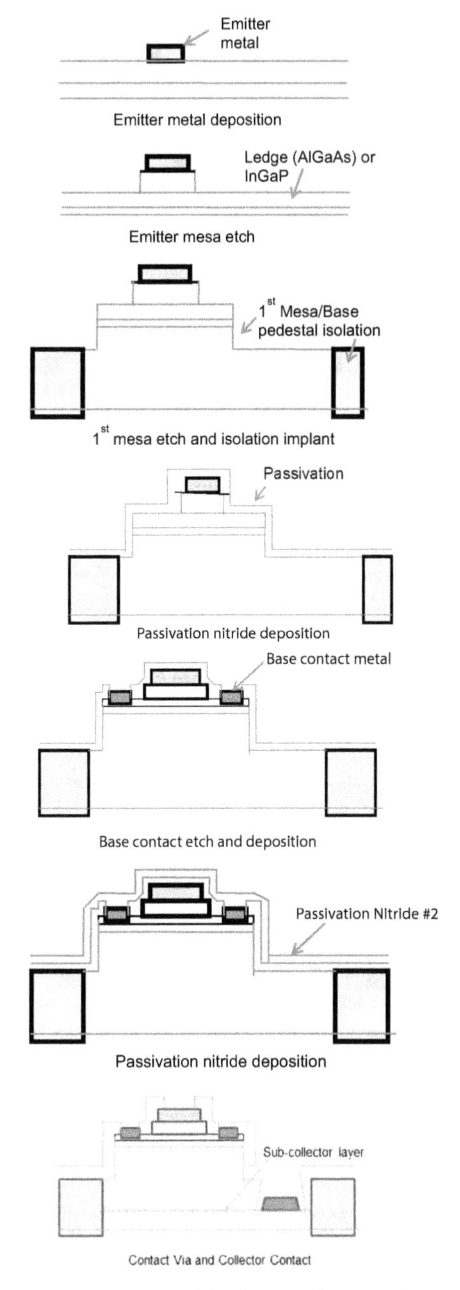

Figure 14.5 Process sequence of the device (front end) process for a non-self-aligned HBT. Reprinted from Ref. [6] with permission. Copyright © 2010 CS MANTECH.

After the base contact is formed, another layer of silicon nitride passivation is deposited all over. The ledge surface near the base contact that might be open is thus passivated. Then via openings over the emitter, base, and collector contacts are formed. Collector contact photo, etch, and metal deposition follows next. The wafers are annealed by rapid thermal annealing (RTA) or alloyed at 360°C to 400°C for short times typically in the 30- to 60-second range to form ohmic contacts. The transistor structure is finished and the wafers can be tested to check the device functionality, if needed. Wafers are moved onto passive component fabrication and interconnect processing. Figure 14.6 shows the cross section of a finished device after contact vias are opened and interconnect metal added. Figure 14.7 shows the typical process sequence for the device and interconnect process.

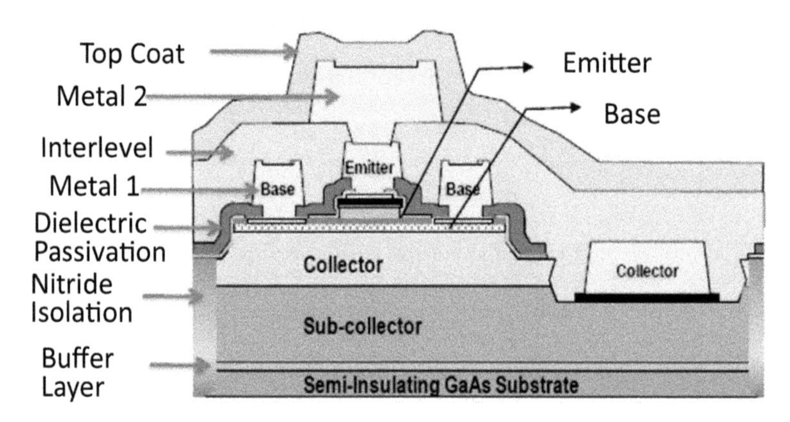

Figure 14.6 Cross section of an HBT after metal interconnect has been added. Reprinted from Ref. [7] with permission. Copyright © 2010 CS MANTECH.

14.4 Self-Alignment

In the emitter-up, common emitter HBT technology, the sizes of the HBT and parasitic are determined mainly by the base pedestal size, which in turn depends upon emitter and base contact size and spacing. Self-alignment in HBTs is mainly the minimization of the base-contact-to-emitter spacing, similar to that of self-alignment of the source contact to gate for field-effect transistors (FETs).

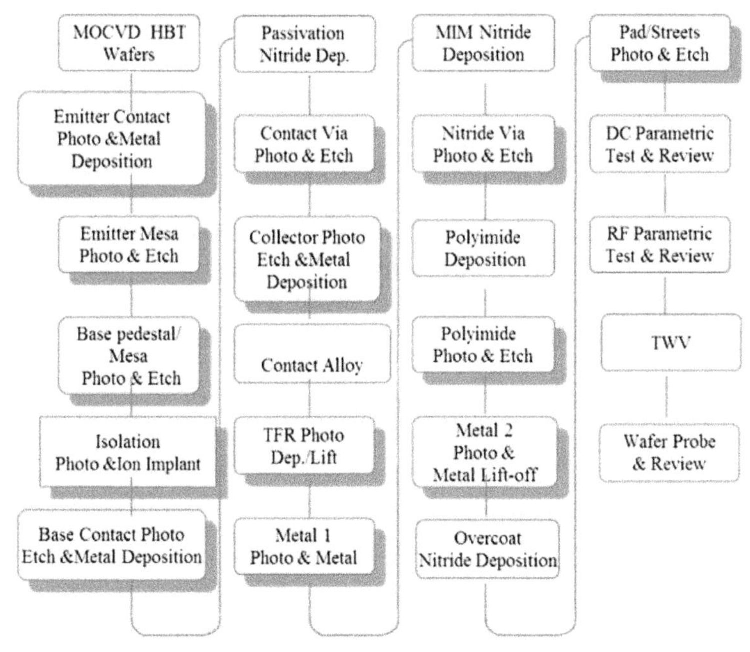

Figure 14.7 Process sequence of a typical HBT process, including a lift-off patterned metal interconnect. Reprinted from Ref. [6] with permission. Copyright © 2010 CS MANTECH.

14.4.1 Base–Emitter Self-Alignment

In an HBT fabrication process in which the base contact is aligned to the emitter mesa or contact metal, enough space must be left to allow for misalignment and critical dimension (CD) variations. With current state-of-the-art projection lithography tools, a spacing of at least 0.5 μm is needed. This spacing adds a significant area to a small-width, micron-size HBT. This results in an increase in the BC capacitance and extrinsic base resistance, R_b. With self-alignment, emitter–base spacing can be reduced to 0.2 μm to keep these parasitics small. Numerous self-alignment schemes have been reported in the literature [1, 3, 8–11]. A few of these schemes that are production worthy are described here.

One of the most common schemes for self-alignment utilizes the emitter metal as a mask for etching the emitter mesa and also for self-aligning the base contact. In this process, illustrated in Fig. 16.8, emitter metal is deposited first. It could be Ti/Pt/Au base metallization or a refractory metal based on tungsten that is dry-etchable [12]. This layer can be used next as the mask for etching emitter fingers. A wet/ dry combination may be used, depending upon the details of the emitter epilayer structure. Care must be taken to ensure the base layer is not etched. A dry etch process with endpoint detection is preferred. The process is designed to result in a controlled undercut. At this point the base contact photo is patterned and the base metal deposited by evaporation. The rest of the processing can be standard involving a second mesa, isolation and collector contact etching, and metal deposition.

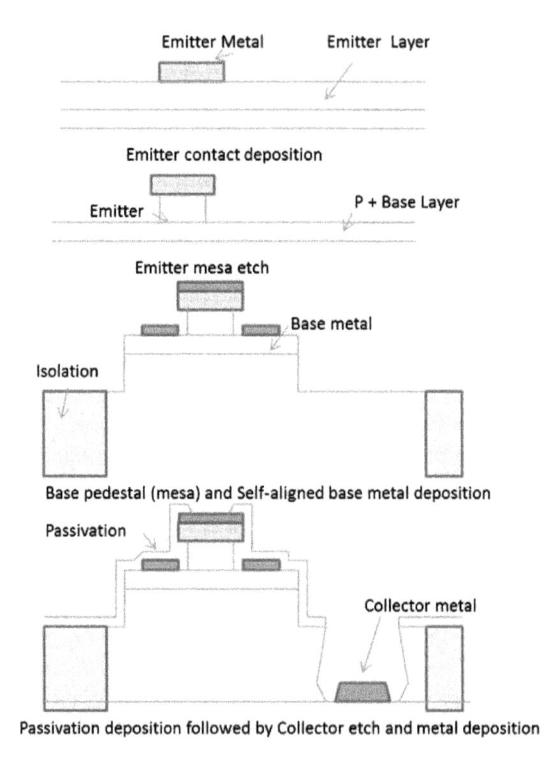

Figure 14.8 Self-aligned HBT process flow based on base metal deposition.

In one version of this self-aligned base metal process, an MBE-grown structure with a Be-doped p+ base layer has been used [13]. AuBe/Pd/Au metallization is used for the base metal.

To make sure that metal shorting does not occur because of off-vertical evaporation, particularly at edges of large wafers, a sidewall process may be utilized. This module is illustrated in Fig. 14.9. In this process, a conformal silicon nitride film is deposited on the wafer after emitter mesa etch using the PECVD process. A reactive ion etching (RIE) step is used next to remove the nitride from the horizontal areas of the wafer, leaving behind only the nitride on the sidewalls of the vertical edges. After this step the process flow can return to the flow described above.

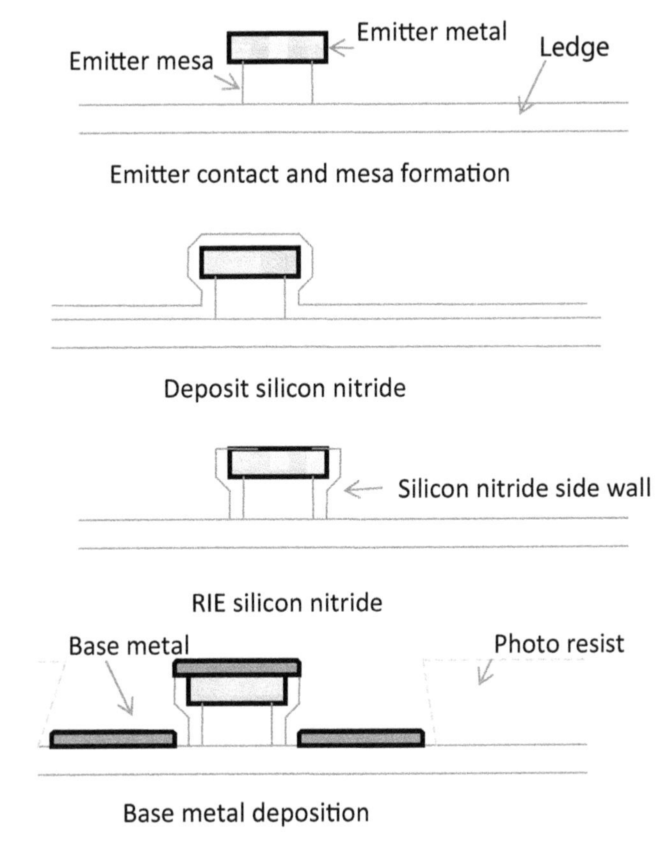

Figure 14.9 Self-aligned HBT process using sidewall formation.

14.4.2 Other Self-Alignment Schemes

It is possible to develop HBT processes that leave a good passivating ledge layer and yet allow self-alignment of the base contact to the emitter. For example, if the InGaP emitter layer is left on the ledge and PdPtAu metallization is deposited self-aligned to the emitter, it can be alloyed through to make contact with the base layer [11]. Another approach is to deposit the base contact metal with Zn in it and then drive the Zn in by rapid thermal processing. However, lateral diffusion of Zn must be avoided by limiting the Zn concentration.

14.5 Collector-Up HBT

The collector-up HBT first proposed by Kroemer [14] and extensively pursued for high-integration-level circuits is shown in Fig. 14.10 [15, 16]. This structure, being planar and allowing higher interconnect packing density, is suitable for digital circuits. Its effective BC area can be smaller than the emitter-up configuration. The base contacts are made by ion implantation and annealing and thus have higher base resistance. Ion implantation needs annealing close to 850°C, where dopant redistribution and junction smearing can take place.

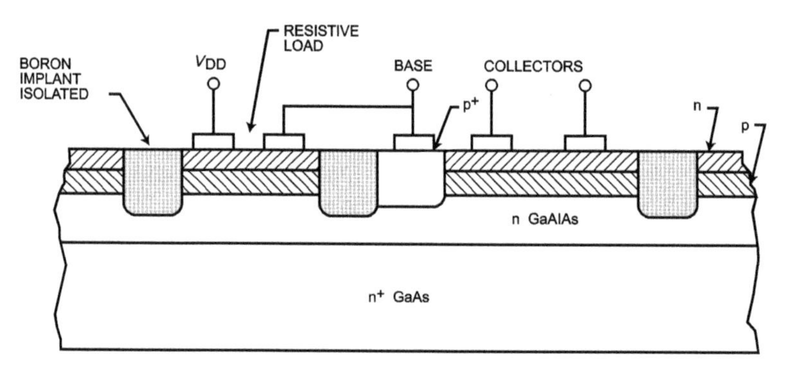

Figure 14.10 Structure developed for the implementation of heterojunction integration injection logic (HI^{2}L).

These limitations can be overcome by using a collector-up mesa structure. In the example shown in Fig. 14.11 [17], the base contact was formed by etching away the collector and depositing Zn alloys for in-diffusion. An oxygen implant was used to convert part of the external emitter into the high-resistivity current-blocking layer. If a p–n junction is used for current blocking, high-temperature annealing is required for activation of the Be$^+$ implant. The mesa structure devices have better performance than planar devices (f_{max}: 128 GHz; Ref. [18]) but not as good as emitter-up ones.

The need for high-speed very-large-scale integration (VLSI) circuits using III–V semiconductors has been pushed out by the progress made by silicon devices. Therefore the process development and production scale-up has helped only mixed-signal processes using the emitter-up configuration. It may be noted that even though the emitter-up configuration is the most used structure, the emitter-down structure has been effectively used by the once popular heterojunction integration injection logic (HI^2L) at Rockwell Science Center (Thousand Oaks, California), leading to greatly enhanced device packing densities.

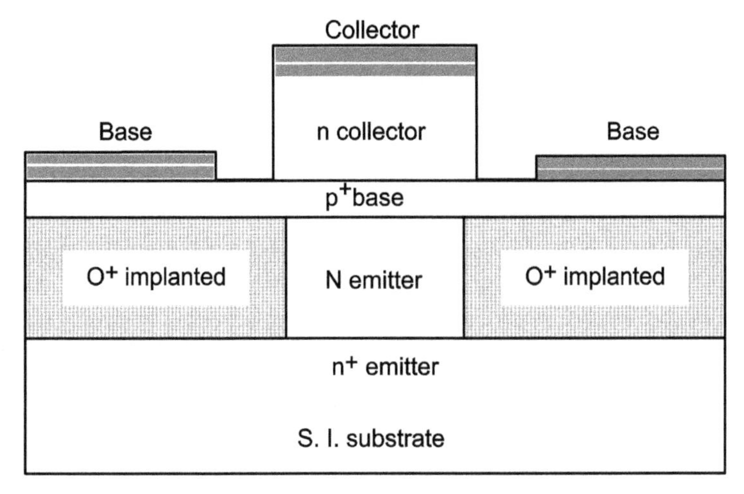

Figure 14.11 Collector-up HBT with oxygen-implanted high-resistivity region as the current-blocking layer.

14.6 Common HBT Epimaterials

The most common epistructures in production today are based on GaAs wafers with InGaP emitters and InP substrates with InGaAs base layers. These will be discussed briefly below. Other epistructures are described in Chapter 3.

14.6.1 InGaP/GaAs HBTs

AlGaAs/GaAs HBTs were the first to be produced in large volumes for power amplifiers in numerous applications, particularly for mobile phones. As the volumes grew and the demands of meeting the specifications of a growing industry grew, it became apparent that a more rugged system with better reliability was needed. The InGaP/GaAs conduction and valence band offsets are better than those for the AlGaAs system. Valence band and conduction band offsets were discussed in Chapter 2 and will be explained further under Section 14.9.2, "Junction Considerations." Wet etch selectivities with the H_3PO_4:H_2O_2 system as well as citric acid–based ones are better for etching GaAs and InGaAs on top of InGaP. The InGaP/GaAs system became of interest and process steps to replace critical etch steps developed for fabricating AlGaAs HBTs were needed. Etch procedures for etching the emitter mesa and stopping on InGaP and etching off InGaP to stop on the GaAs base layer were developed. Wet processes were used first using H_3PO_4:H_2O_2-based solutions. Several etching solutions are available for etching InGaP that can stop on GaAs. HCl-based solutions etch InGaP easily but do not attack GaAs (without the addition of H_2O_2, of course) [19]. With these two critical etch processes in place, a wet etch process for this material system can be easily put together. For better process control, a combination of wet and dry etch processes has been implemented in production. Endpoint controls are needed for stopping on the GaAs base layer during the base contact etch. Details of dry etching will be discussed in Chapter 7.

If the base contact etch is too deep and the combination of etch pits and the base metal alloying into the base layer can give rise to punch-through, a spurious Schottky diode between the base metal and the collector layer will be formed. This diode will be in parallel to the BC junction diode and will increase the V_{ce} offset seen in the I–V characteristics.

14.6.2 InP HBTs [19, 20]

InGaAs ternary compounds exhibit the highest electron mobilities and can be grown lattice-matched to InP substrates (discussed in detail in Chapter 4). InP HBTs are grown with an InGaAs base layer and an InAlAs emitter layer. A process sequence utilizing self-alignment of the base contact to the emitter mesa can be used [19]; just the highlights are described below. See Fig. 14.12. Silicon nitride is deposited all over the wafer. Emitter metal contacts are formed first, etched by an oxidizing selective etch for InGaAs, and used as an etch mask for the emitter mesa. HCl is used to etch the InAlAs emitter layer. As in the AlGaAs HBT process, this etch must stop at the p+ base layer. Base contact metal (Ti/Pt/Au) is deposited self-aligned to the emitter. Wet etching is used to etch down to the subcollector to make contact with the collector. After the contacts are made, SiO_2 is deposited all over and vias are made in it to make the contact holes for the interconnect metal.

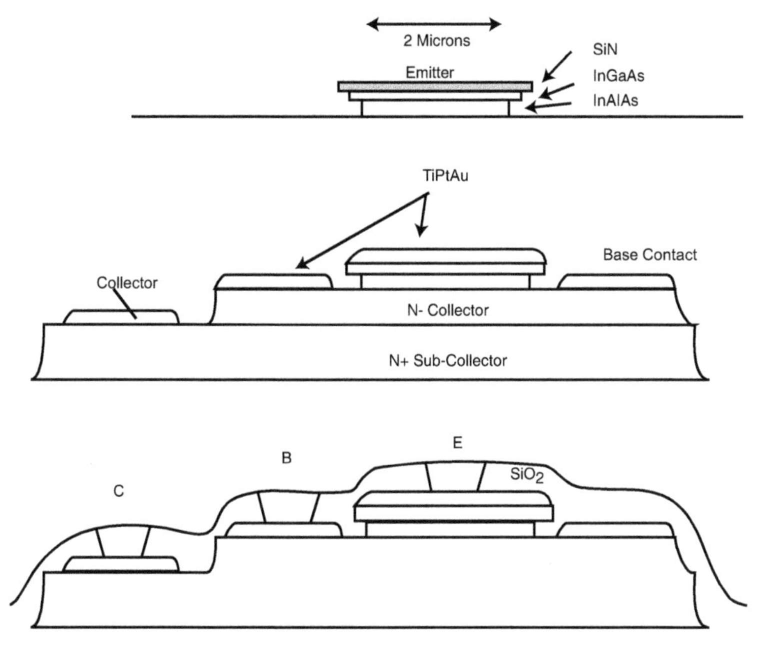

Figure 14.12 Schematic of a wet etch–processed self-aligned InP HBT. From Ref. [19]. Copyright © 1995. Reproduced with permission of John Wiley & Sons, Inc.

14.7 HBT Contacts

Pure metals and alloys are used to make contacts to semiconductor devices. The total contact resistance depends upon the area of the contact and the specific contact resistivity (per unit area). In a nonalloyed system the contact may be close to being an ideal metal–semiconductor junction, two types of which were discussed in Chapter 11. In other situations involving sintered or alloyed contacts, in the contact area there may be regions of sharp ideal junctions and other areas of graded composition or doping. The ideal metal to an n-type semiconductor junction is depicted in Fig. 11.1a. in Chapter 11. Conduction can take place by *thermionic emission* over the barrier ϕ_b when the doping levels are low. When the doping level is high, the depletion layer is small and the electrons can tunnel through the barrier by field emission. In HBT processing, high levels of n-type doping are possible in epigrowth by both MOCVD and MBE. Maximum doping levels achievable for n-type materials are listed in Table 11.1 (in Chapter 11). The highest doping level for GaAs in the subcollector region is close to 5 E18/cm^3. At this level, alloyed contacts are used for the collector. The energy band diagram and *I–V* curve for a realistic contact, which may be a combination of the above-mentioned "ideal" contacts, are shown in Fig. 14.13.

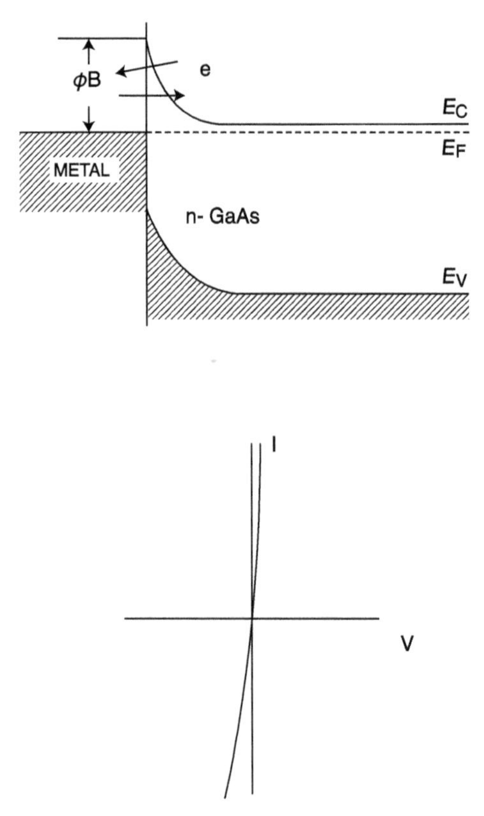

Figure 14.13 Energy band diagram for a collector contact and an almost ohmic *I–V* curve.

For the large-bandgap emitter, it is common to grade the semiconductor bandgap down to lower levels, so a low-resistivity contact can be made with the lower bandgap material on top. Figure 14.14a shows a schematic band diagram for a graded GaAs/InGaAs/InAs metal–semiconductor system. Both InAs and InGaAs can be doped heavily without causing too much surface roughness. Care must be taken in minimizing other defects that might be generated due to excessive dopant incorporation. Such defects can cause stability problems.

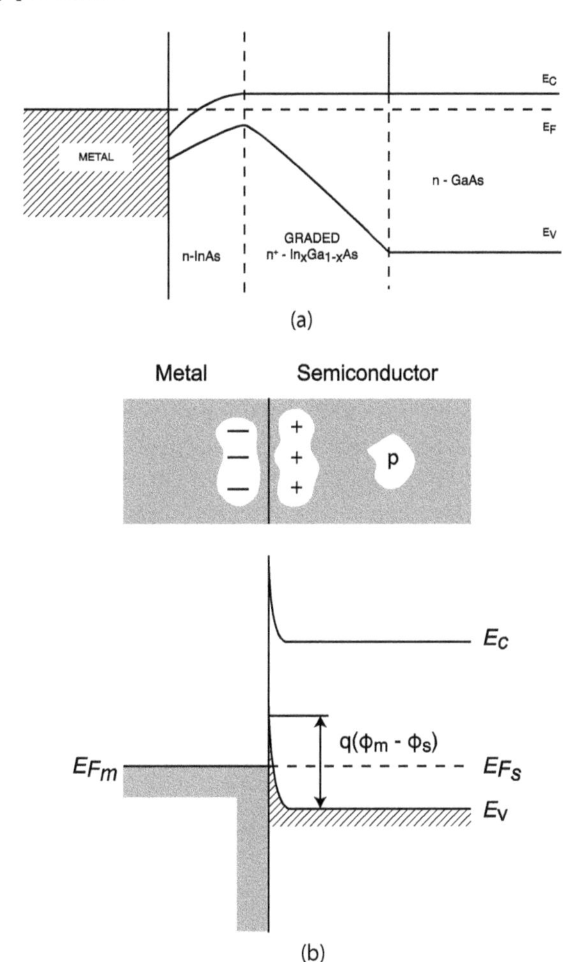

Figure 14.14 Energy band diagram for a (a) graded HBT emitter contact and (b) heavily doped p+ base contact.

For p-type materials, high doping levels are more easily achievable in GaAs, particularly with carbon doping. Details are discussed in Chapter 11. The band diagram for this case is shown in Fig. 14.14b. Zinc is another common dopant for p-type *I–V* semiconductors; the Au:Zn metal alloy can be utilized for contact alloying or a W:Zn deposited layer can be used for self-aligned contacts.

14.8 HBT Geometry

The emitter layout is critical to HBT performance like gates in FETs. Parameters such as R_{ee} (emitter resistance) are directly related to it, and other dimensions of the device are determined by it. Figure 14.15 shows a schematic diagram for a rectangular or stripe device. Both figures of merit, f_t and f_{max} are related to emitter size. f_t is a weak function of emitter width (page 837 in Ref. [8]) and f_{max}, which is parasitic limited, can be related by the following:

$$f_{max} \approx 1/W_e \tag{14.1}$$

f_{max} can be maximized if the width (W_e) is made small. At 2 μm or more, alternating current (AC) crowding effects are negligible, as fingers become small and parasitics start to dominate. Peripheral effects come into play, so devices below 0.5 μm may not be practical. Reliability considerations may also determine the minimum geometries in production. As the emitter length becomes larger for larger-power devices, the base becomes larger in rectangular geometry and R_b can go up. In this situation, the resistance of the base metal must be kept low, and refractory contacts alone, that is, without overlaying metal, may not work.

14.8.1 Base Width

For high f_t base width W_b must be kept small. For high f_{max}, base resistance must be minimized

$$f_{max} \approx \frac{1}{\sqrt{Rb}} \tag{14.2}$$

(See Eq. 2.97, Chapter 2.)

The base width must be large for low R_b. The width of the base contact must be larger than the characteristic length of the contact metal system used. However, as the size increases, BC junction capacitance; C_{bc} increases as the base pedestal size increases. The emitter–collector transit time will also increase with base width.

The collector contact size is more straightforward. The width must be large enough to minimize thermal effects and yet keep the cell size small to keep the overall power amplifier die size small.

14.8.2 Layout Comparison

Circular and rectangular geometries have their advantages and disadvantages. Rectangular stripe and circular geometries are easily scalable (see Fig. 14.15). A dot-geometry HBT shown in Fig. 14.15c is perfectly scalable; however, making contact vias on it is not convenient. This geometry has a good emitter utilization factor and low base resistance. It has higher BC capacitance. The horseshoe emitter shown in Fig. 14.15b is one way to make contact with interconnect metals easier.

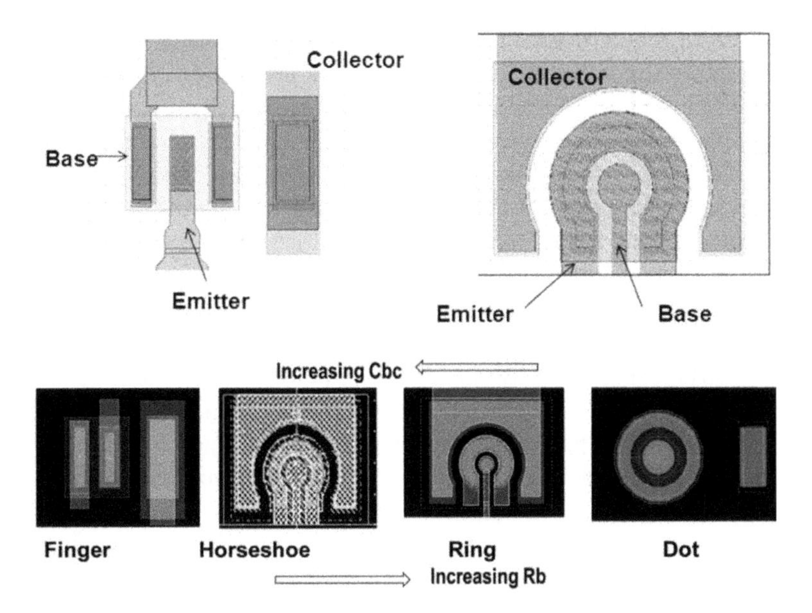

Figure 14.15 Common HBT device layouts: rectangular, horseshoe, and ring. Reprinted from Ref. [7] with permission. Copyright © 2010 CS MANTECH.

14.9 HBT Fabrication Issues

Most of the critical parameters for HBT circuits are determined and controlled during epigrowth. Epilayer design and growth issues are discussed in chapters on those subjects. Unlike threshold control in FET-type circuits, which are process sensitive, the turn-on voltage of HBTs is set by the epiheterojunction. For a definition of the turn-on voltage, V_{be}, and I–V characteristics, refer to Chapter 2.

$$V_{be} = \frac{E_{gb}}{q} - \frac{kT}{q} \ln \frac{QDn\, Nc\, Nv}{Jc\, p_b w_b} \tag{14.3}$$

See the Appendix for HBT equations.

V_{be} depends mainly on the energy bandgap in the base (E_{gb}); it changes slowly with base layer doping and thickness (2 mV change for 10% change in p_b.) Turn-on voltage control is excellent in both MBE- and MOCVD-grown structures. Only if a potential barrier exists at the base–emitter junction (close-to or abrupt junction) or if the acceptor dopants from the base diffuse into the emitter, V_{be} will vary over the wafer. Graded heterojunctions are used in practice to avoid such problems.

During epigrowth of HBTs, multiple layers must be grown in succession in a reasonable time and the properties of the layers must be maintained as if grown by themselves. For example, the doping in the base layer is generally near 2 E20/cm^3 and as high as 4 E20/cm^3. As the base doping level increases, the diffusion of acceptors increases. Be and Zn, which diffuse fast, should be avoided. Carbon doping has led to advancement of the state of the art of HBTs, particularly for MOCVD-grown structures, because high doping levels are possible without danger of diffusion and little memory effect in the reactor during growth.

HBTs, being minority carrier devicesm rely on excellent quality layers, the carrier lifetime in the base should be long. Recombination of electron–hole pairs must be minimized for all small HBT devices, particularly for devices of a few micron dimensions such as those used for digital (~1um emitter) and even power devices where the emitter size may be near 2 μm. Surface recombination velocity of GaAs is high, even if AlGaAs or InGaP is used as the surface passivating layer, so surface recombination is an issue. To minimize out-diffusion of dopants during growth, the emitter growth temperature must be kept low (emitter-up structure). However, the growth temperature needed for high quality and low level of recombination centers is high. A recombination center density of below $10^{15}/cm^3$ is desired. At high growth temperatures, the base dopants can diffuse into the emitter layer and cause a V_{be} shift. In practice, a compromise is needed. Passivation by a depleted semiconductor layer works better than silicon nitride or other dielectric passivations that have been tried. I–V curves of passivated and unpasivated HBTs are shown in Fig. 14.16 [7, 8].

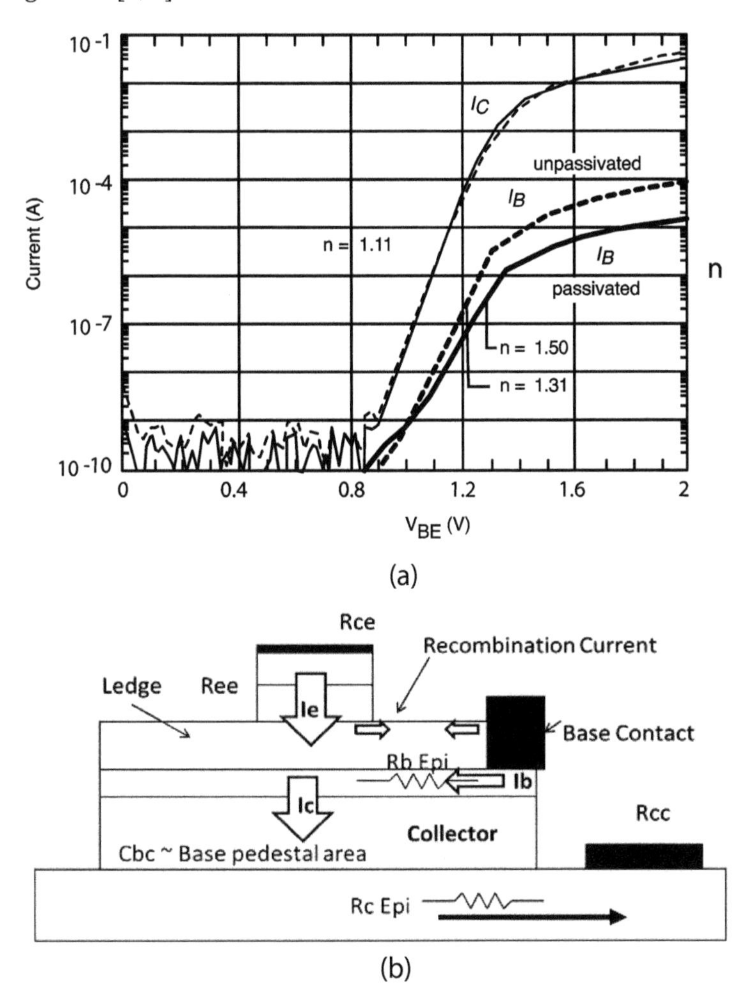

Figure 14.16 (a) Gummel plot of passivated and unpassivated InGaP/GaAs HBTs. The emitter area is 4 × 20 μm². (b) Schematic diagram of an HBT showing recombination current at the ledge. Reprinted from Ref. [7] with permission. Copyright © 2010 CS MANTECH.

To minimize the size of HBTs and thereby reduce C_{bc} (collector–base junction capacitance), the ledge should be reduced. However, there is a limit based on β reduction and reliability. As the ledge size is reduced on layout, the actual size of the passivated ledge goes down. This is due to base contact to emitter alignment tolerances, erosion of the ledge near the base contact, and general feature size control. The effect of reducing the ledge size has been determined by measuring base current increase and the ideality factor. As the size drops n_B increases, ands correspondingly β drops.

The area near the base contact generally gets eroded, so it does not maintain perfect depletion. Also, this region may be exposed to more plasma damage. It may be noted that as this area gets closer to the emitter, the recombination rate increases. See Fig. 14.16 for current loss paths. (In the case of self-aligned HBTs, surface passivation must be maintained and the base contact made through the passivating layer.)

On the basis of such results, the minimum practical limit to date has been about 0.5 μm. In addition to a drop of β at time zero, the increased recombination current leads to generation of more defects with an operating time that results in a β drop and thus reduced reliability.

The collector current and β depend upon the emitter area and not on the BC junction area. To maintain high β, the base recombination current must be minimized.

The wide-bandgap emitter material is optimized for minimization of back-injection of holes from the base to the emitter. The abruptness of the base–emitter junction is determined by the epigrowth equipment (see also Chapter 4). Abrupt junctions result in a discontinuity in the conduction band, as shown in Fig. 14.17a.

14.9.1 Junction Considerations

In an abrupt heterojunction, the total difference in the bandgaps is split between the valence band and the conduction band. The lineup of the bands depends upon the binding energies of the electrons in the different regions. See Fig. 14.17a. For an abrupt junction, the hole barrier is larger than the electron barrier, but a discontinuity exists on the conduction band side. If the junction is graded, the discontinuity is smoothed out, as shown in Fig. 14.17b. In principle,

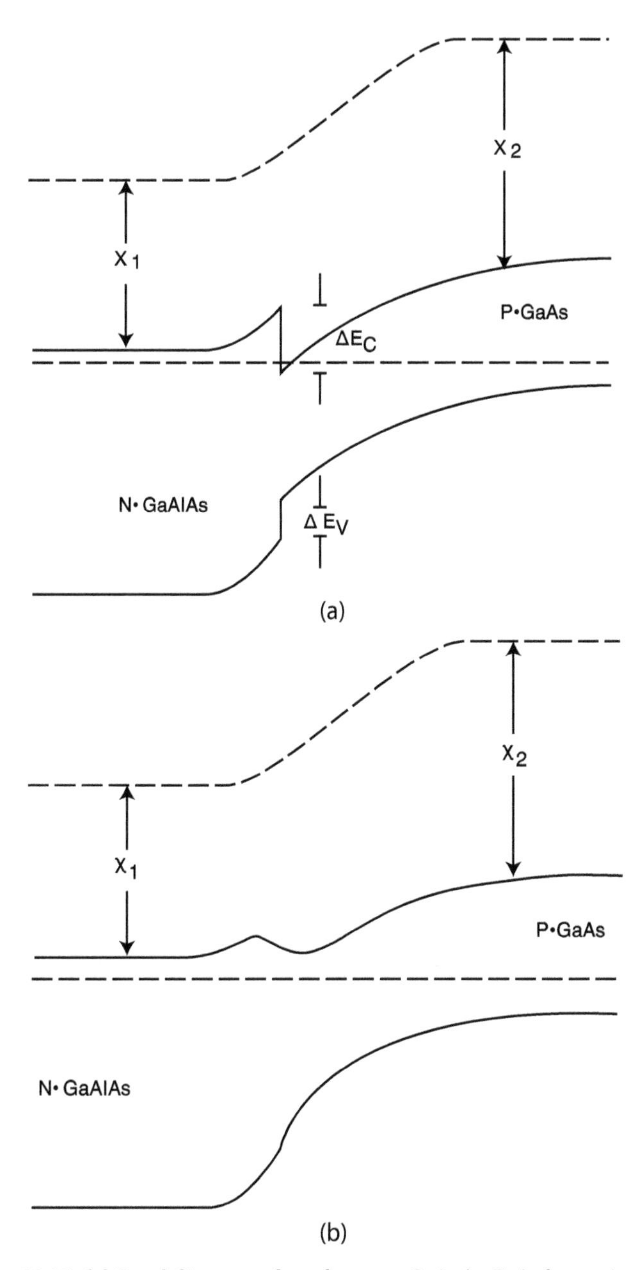

Figure 14.17 (a) Band diagram of an abrupt n-GaAs/p-GaAs heterojunction. X1 and X2 denote electron affinities. (b) Band diagram of a graded n-AlGaAs/GaAs heterojunction.

a discontinuity could help in improving the transit time, as the electrons can be launched with additional kinetic energy ΔE_c. But in practice, abrupt junctions are hard to reproduce. Current gain dependence on junction sharpness is discussed in reviews and books on HBTs [1, 2, 8]. In a graded junction a larger portion of the bandgap difference (ΔE_g) appears on the valence band side, thus giving a larger current gain ($\beta \approx \exp(\Delta E_g/kT)$ (Eq. 2.91, Chapter 2). In MOCVD growth, some amount of grading, close to 10 Å, is unavoidable. By careful grading of the composition over 10–15 Å, epigrowers can eliminate the conduction band spike and achieve reproducible high gain.

14.10 HBT Epilayer Design

Design- and process-related considerations that go into the design of the main HBT layers are discussed in this section.

14.10.1 Emitter Layer Design

For AlGaAs/GaAs HBTs, an aluminum fraction of 25% is commonly used. As the aluminum fraction increases, HBT processing becomes more difficult because of the sensitivity of AlAs to oxidation and chemicals used in processing. Emitter thickness should be at least 500 Å, otherwise current gain drops at high currents. A thickness of 700–1000 Å is commonly used. Emitter doping needs to be optimized for low R_{ee}, while keeping the level low enough so that the base–emitter junction capacitance is not too high. The AlGaAs or InGaP layer that remains on the ledge surface must be thin enough to be fully depleted by the free-surface Fermi level pinning from the top.

On the emitter side, it is difficult to make ohmic contact to wide-bandgap materials like AlGaAs or InGaP. Contacts are therefore made to low-bandgap materials, with graded composition layers in between. Therefore, a layer of doped GaAs is added first and then a heavily doped InGaAs layer is added on top. The epidesigner has to keep the HBT fabrication process in mind, so layers are not too thick for etching or add extra resistance to the emitter contact in the vertical direction.

14.10.2 Collector Layer

Collector doping and thickness are chosen to optimize high-frequency performance, depending upon the application, keeping in mind ease of device processing, particularly collector contact etch and implant isolation. The thickness of the depletion layer in the collector depends upon the doping level. As the doping level is reduced, the depletion width increases, causing an increase in collector collection time and a decrease in f_t. Collector junction capacitance decreases, resulting in an increase in f_{max}.

$$f_{max} = \sqrt{\frac{f_t}{8\pi R_b C_{bc}}} \quad \text{(from Eq. 2.97)}$$

For power amplifier applications, breakdown voltage must be high, over 20 V for some applications. Figure 14.18 shows an example of variation of breakdown voltage in the doping thickness space. See also Ref. [12].

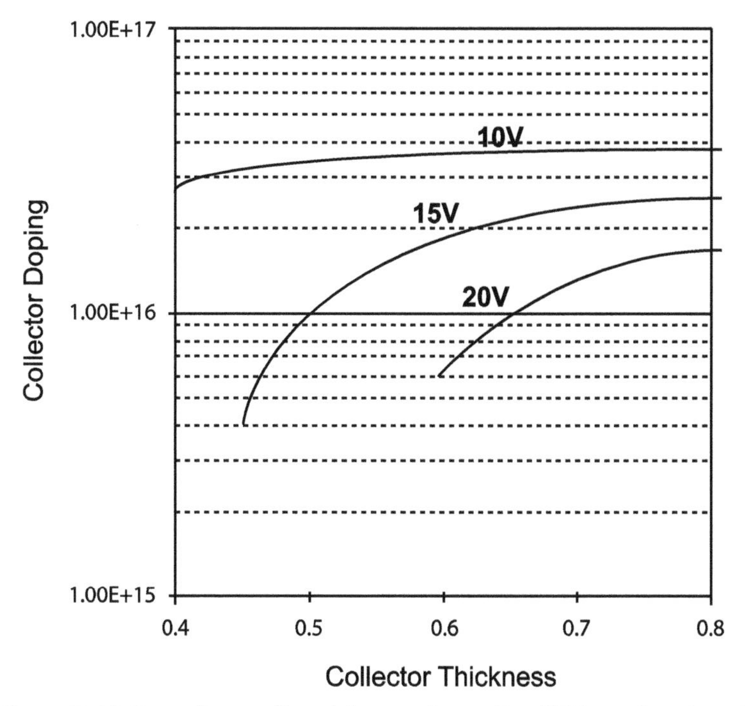

Figure 14.18 Dependence of breakdown voltage of an HBT as a function of collector thickness and doping.

14.10.2.1 Subcollector layer

Emphasis is placed on reducing sheet resistance so that the collector extrinsic resistance is kept low. The highest doping level possible for GaAs is limited to 5E18/cm^3. R sheet values of 10 to 15 ohms/sq. are possible for 5000–7000 Å thick layers. Combined together, the collector and subcollector layers cannot be too thick because the devices need to be isolated by a combination of etching and ion implantation. With even the lightest ions like helium, the thickness is limited to less than 2 μm.

14.10.3 Base Layer Design

The base layer is the most important layer of the HBT and the most critical from the epigrowth point of view. The cutoff frequency depends upon the base transit time and f_{max} is affected by the base resistance, R_b. Dependence of R_b and β on the base thickness is shown in Fig. 14.19. For high f_t the base needs to be thin, and for high f_{max} it needs to be thick. In practice the combination of base thickness and doping is chosen to maximize f_{max}. For MBE-grown HBTs using Be as the dopant, there are practical limits based on reliability (discussed elsewhere). For carbon doping, very high doping levels can be utilized. For power amplifier applications, a base thickness of ~1500 Å is used, and for digital applications, the thickness can be halved. As mentioned earlier, punch-through can occur for very thin base layers, so excellent process control is needed for layers below 500 Å thickness.

14.11 Other HBT Structure Improvements

Figures of merit for an AlGaAs/GaAs HBT are shown in Fig. 14.20. With better structure and better-quality epilayers as well as device geometry, better performance, namely higher f_t and f_{max}, has been achieved. With the use of higher-mobility, low-bandgap materials, HBT performance can be improved dramatically, as shown in Fig. 14.21 [21] for an InGaAs/InAlAs HBT, where an f_{max} of 280 GHz has been demonstrated.

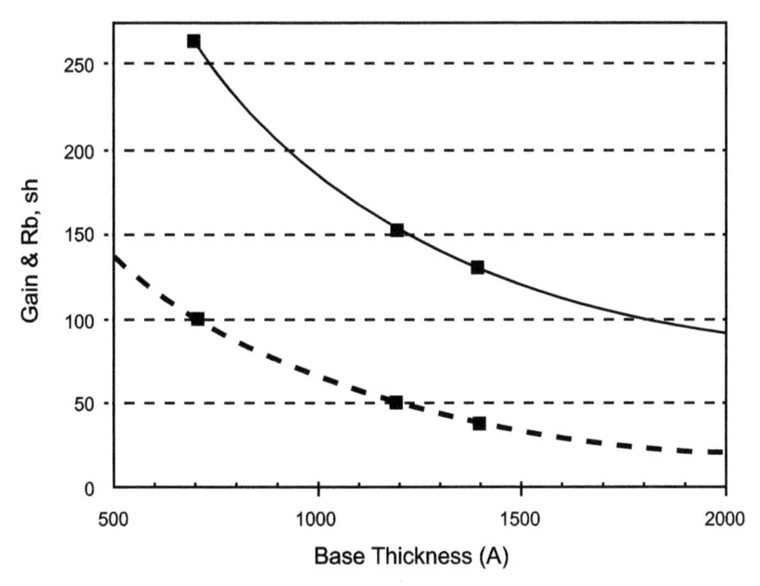

Figure 14.19 Gain/beta (dotted line) and base resistance (solid line) dependence on base layer thickness for an HBT. Reprinted from Ref. [6] with permission. Copyright © 2010 CS MANTECH.

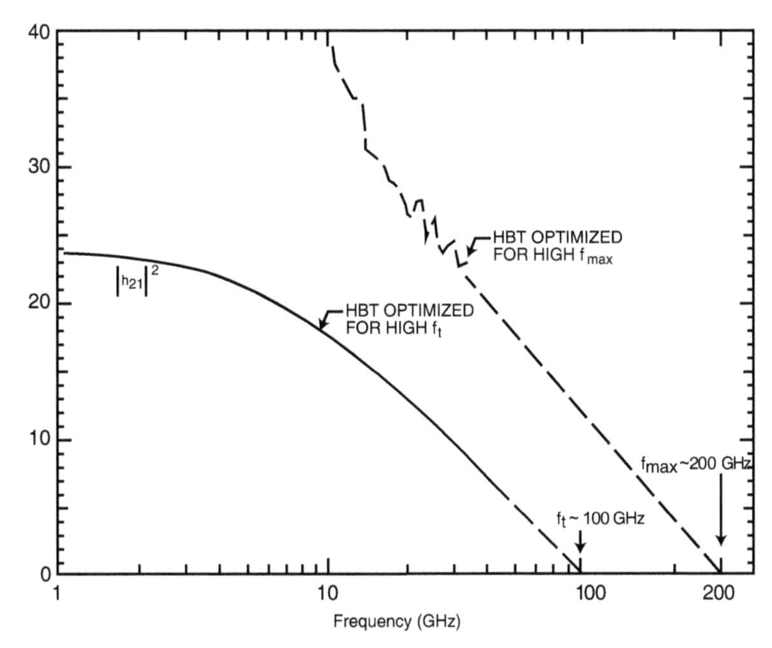

Figure 14.20 f_t and f_{max} for an AlGaAs/GaAs HBT. Courtesy of Rockwell (Peter Asbeck).

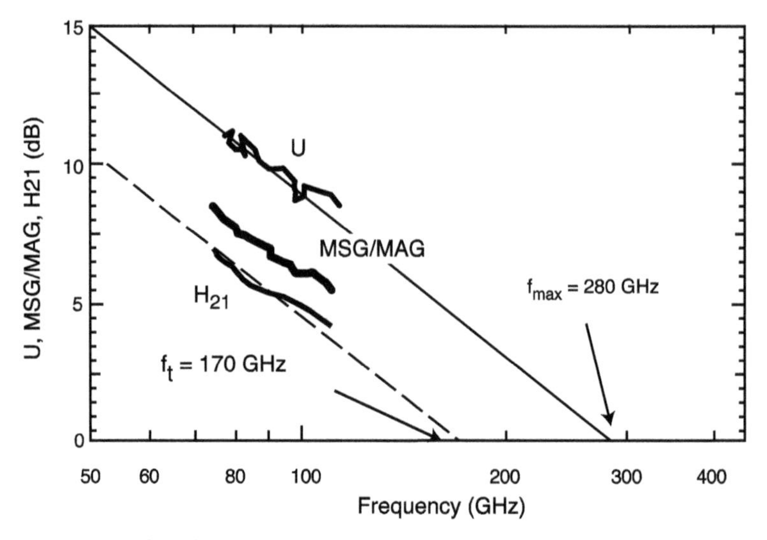

Figure 14.21 f_t and f_{max} (projected using Mason's gail U) for an InGaAs/InAlAs HBT. © [2000] IEEE. Reprinted, with permission, from Ref. [21].

14.11.1 Graded-Base HBTs

The concept of grading the base is an old strategy [14]. Since the hole quasi-Fermi level in the base is flat, very close to the valence band, due to the high doping level, if the bandgap is varied across the base the change appears as a gradient in electron energy. The gradient gives rise to a quasi-electric field that drives carriers toward the collector. A large increase in current gain and cutoff frequency has been reported. See, for example, 66% current gain increase reported by De Luca et al. for a compositionally grade InGaAs base layer [22].

14.11.2 Double-Heterojunction Bipolar Transistor

There are some advantages to insertion of a heterojunction on the collector side also. Charge storage during transistor saturation can be reduced by eliminating the minority carrier hole injection into the collector. The BC breakdown voltage can be increased. The V_{be} offset seen in HBTs disappears due to the equal turn-on voltage of the two junctions. Therefore the knee voltage of the I–V characteristics can be reduced. This can be useful for increasing the efficiency of power amplifiers.

A double-heterojunction bipolar transistor (DHBT) has its disadvantages, though. Epigrowth has to ensure that there are no barriers for electrons due to conduction band discontinuity; this can be done again by grading the junction. Epilayer growth is more complicated because composition grading has to be used to change the composition back to GaAs.

References

1. M. F. Chang and P. M. Asbeck, III-V heterojunction bipolar technology, *Int. J. High Speed Electron. Syst.*, **1**, 245 (1990).
2. P. M. Asbeck, M. F. Chang, K. C. Wang and D. Miller, Heterojunction bipolar transistor technology, *IEEE Trans. Microwave Theory Technol.*, **35**, 1462 (1987) and other publications by Rockwell Science Center authors.
3. B. Bayraktaroglu, GaAs HBTs for microwave integrated circuits, *Proc. IEEE*, **81**, 1762 (1993).
4. P. M. Asbeck et al., *IEEE Electron. Device Lett.*, **EDL-4**, 81 (1983).
5. P. M. Asbeck et al, GaAs-based heterojunction bipolar transistors for very high performance electronic circuits, *Proc. IEEE*, **81**, 1709 (1993).
6. S. Tiku, *Device Processing in III-V Manufacturing*, CS MANTECH (2010).
7. R. Ramanathan, *Compound Semiconductors: Process Flow, Process Integration, Devices and Testing*, CS MANTECH (2010).
8. W. Liu, *III-V Heterojunction Bipolar Transistors*, Wiley Interscience, New York (1998).
9. T. Ohshima et al., A self aligned AlGaAs/GaAs heterojunction bipolar transistor, *Tech Dig., IEEE Gallium Arsenide IC Symp.*, 53 (1985).
10. Mike Sun et al., *Self Aligned InP HBT*, GaAS MANTECH (2002).
11. M. L. Hattendorf, Incorporation of an alloy-though passivating-ledge process into a fully self-aligned InGaP/GaAs HBT process, GaAs MANTECH (2001).
12. P. Kurpas et al., 10 W GalnP/GaAs power HBTs for base station applications, *IEDM*, 681 (2002).
13. F. M. Yamada et al., Reliability of a high performance monolithic IC fabricated using a production GaAsA1GaA.s HBT process, *GaAs IC Symp.*, 271 (1994).
14. H. Kroemer, Heterostructure bipolar transistors and integrated circuits, *Proc. IEEE*, **70**, 13 (1982).

15. H. T. Yuan et al., A 4-k GaAs bipolar gate array, *Tech. Dig., ISSCC*, 74 (1987).
16. H. T. Yuan et al., The development of heterojunction integrated logic, *IEEE Trans. Electron. Devices*, **ED-36**, 2083 (1989).
17. S. Adachi and T. Ishibashi, Collector–up HBTs fabricated by Be+ and O+ ion implantations, *IEEE Electron. Device Lett.*, **EDL-7**, 32 (1986).
18. S. Yamahata, Y. Matsuoka and T. Ishibashi, *IEEE Electron. Device Lett.*, **14**, 173 (1992).
19. B. Jalali and S. J. Pearton (Eds.), In*P HBTs: Growth, Processing and Applications*, Artech House, Boston (1995).
20. S. D. Mukherjee and D. W. Woodard in *Gallium Arsenide Materials Devices and Circuits*, John Wiley and Sons, p. 119 (1985).
21. J. R. Guthrie et al., HBT MMIC 75 GHz abd 78 GHz power amlifiers, *Proc. Int. Conf. DOI*, p. 246, University of California, Santa Barbara (2000).
22. P. M. DeLuca, P. M. Asbeck et al., *IEEE Electron. Device Lett.*, **23**, 582 (2002).

Chapter 15

BiFET and BiHEMT Processing

The integration of GaAs HBTs and FET-type devices offers the advantages of both device types to designers for design of high-performance and mixed-signal integrated circuits. The advantages are similar to the ones offered by a BiCMOS in silicon. This is the latest family of devices in the III–V stable, with all the leading companies and research laboratories being very active in this area. This integration offers competitive combined performance advantages of the individual device families without getting compromised by the drawbacks of either solo device. In particular for power amplifiers, advanced bias circuit design, stage bypassing, adaptive gate switching and analog controls, etc., have been possible with the combination of HBT and FET processes in the BiFET process [1, 2]. Advantages of HBT are high speed, high drive capability, and excellent threshold voltage matching, as well as low $1/f$ noise. Advantages of FETs are high input impedance and low power dissipation. All these can be exploited using the BIFET process. For example, for buffered bias circuits, the reference voltage can be made less sensitive to temperature variations of the HBT threshold (V_{BE}) by use of FETs. Similarly HBT and PHEMT devices have been combined in the BiHEMT process to provide bias control, logic, low-noise amplifiers, as well as RF switches [3–5].

III–V Integrated Circuit Fabrication Technology
Shiban Tiku and Dhrubes Biswas

ISBN 978-981-4669-30-6 (Hardcover), 978-981-4669-31-3 (eBook)
www.panstanford.com

15.1 BiFET Process

Integration attempts have been reported in the literature since the 1990s [6]. Early attempts at integration of heterojunction bipolar transistor (HBT) and field-effect transistor (FET) processes involved selective area regrowth [7–9]. These techniques result in high performance but are not suitable for high-volume, short-cycle-time processing, even if the epigrowth can be done in the same fabrication facility. Formation of metal–semiconductor field-effect transistors (MESFETs) by ion implantation in an HBT process has been demonstrated [10], but it has the obvious problem of interdiffusion of HBT layers due to high-temperature annealing that is needed for implant activation.

15.1.1 Stacked Devices

In stacked structures the HBT and FET device layers are grown on top of each other such that no layers are shared [11]. The advantage is that each structure can be optimized separately. The number of layers is large and the overall yield for epigrowth and process combined is expected to be low. Prescreening of material with a quick large-area device is tougher in this approach.

15.1.2 Merged Devices

The n-type layers already exist in the n–p–n HBT structure and can be used for the channel layers. In a merged structure where all the epilayers are common to both devices, the challenge is to design the material stack in such a way that the trade-off in the performance of the two devices is minimized. The merger of an n-type FET into the emitter or collector layers has been analyzed [1, 2]. Researchers from Rockwell Science Center (Thousand Oaks, California) and UCSD (San Diego) developed the first process to utilize the emitter of an HBT in the AlGaAs/GaAs HBT structure. They also used the p-type base layer as the back-gate layer. This process requires only one additional mask for gate definition [12, 13].

The other n-type layer in the HBT structure that is a candidate for the channel layer is the collector. In the commonly used emitter-up HBT, the collector layer is very deep and the n-channel layer could be placed below the heavily doped subcollector. In this case, the subcollector layer can be used for source and drain contacts. However, in this structure [14] the topology is very steep because the typical collector layers are over 1 μm thick and the depth of the gate could be nearly 2 μm below the top of the HBT device. Figure 15.1 shows the difference in topology. With the FET in the emitter, the topology is well suited to the use of optical lithography for FET gate formation. FET device isolation is also easy with the combination of the formation of the emitter mesas and the use of ion implantation used for HBT isolation. With the FET-below approach, optimizing the isolation dose (e.g., helium) is difficult for the n layers. Overdosing is not a problem in HBT structures, but overdosing can cause leakage for channel like low dose implants.

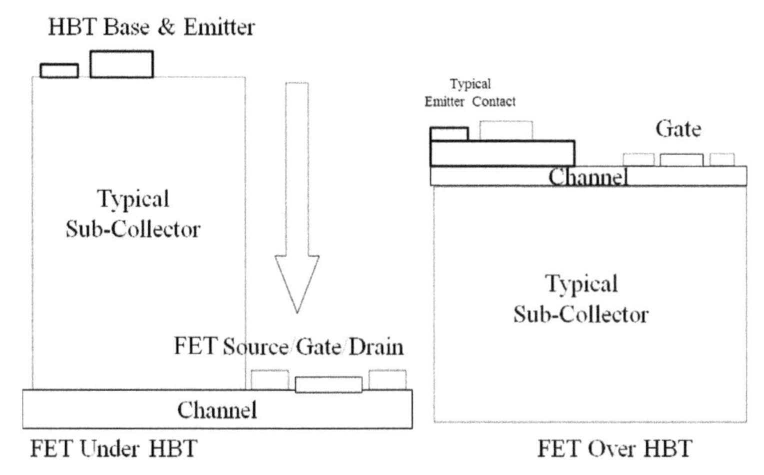

Figure 15.1 Choice of insertion of FET epilayers into the HBT structure. Courtesy of Pete Zampardi, Skyworks Solutions.

The original merged epi-FET process developed at Rockwell and UCSD [1] used AlGaAs/GaAs commercial metal–organic chemical vapor deposition (MOCVD) wafers, with the difference that a highly doped channel layer was inserted between the AlGaAs emitter and the top heavily doped cap layers. See Fig. 15.2 [15]. The FET uses the top cap layers for source and drain. Unique features of the FET are that the AlGaAs emitter layer is fully depleted, thus confining the channel and the p+ GaAs base layer is used as the back gate. In this original form there was no etch stop for channel etching; instead the channel current was monitored during the etch step. After gate deposition, the rest of the processing was all common to HBTs. Good

FET (0.5 μm gate) and HBT characteristics were achieved, and circuit functionality of complex circuits (like static random access memory [SRAM]) was demonstrated.

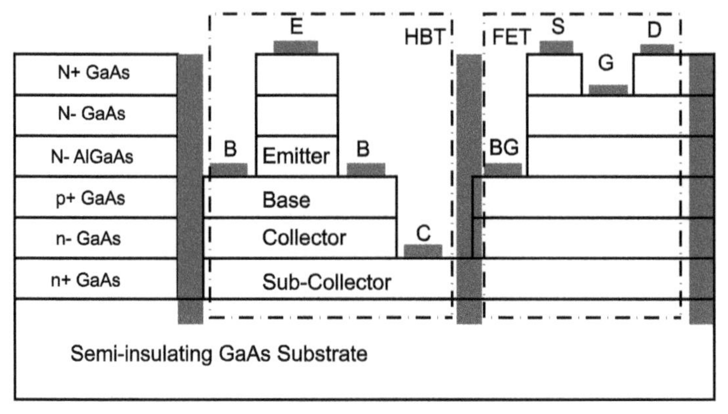

Figure 15.2 Schematic cross section of an AlGaAs/GaAs HBT and a MESFET using common epilayers in the merged structure. Reprinted from Ref. [15] with permission. Copyright © 2010 CS MANTECH.

For better manufacturability an InGaP etch stop layer over the channel layer has been added to the standard InGaP/GaAs HBT structure (see Fig. 15.3) [14, 16]. Details of this high-volume manufacturing process are discussed in brief below.

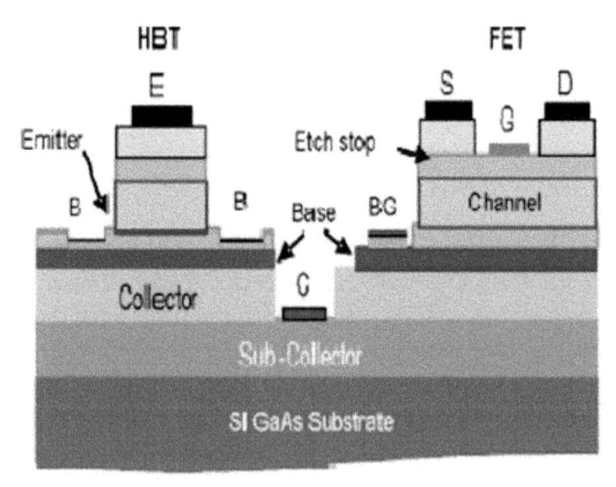

Figure 15.3 Improved merged BiFET structure showing an InGaP/GaAs BiFET schematic cross section with InGaP etch stop. Reprinted from Ref. [16] with permission. Copyright © 2010 CS MANTECH.

15.1.3 Guidelines for Extra Layers

The addition of the channel layer is relatively easy; however, the addition of the InGaP etch stop must be carefully considered. An undoped layer or one that introduces discontinuities in the energy bandgap would be unacceptable as it would raise the emitter resistance (R_{ee}) of the HBT. Thickness must be chosen appropriately for optimum processing and minimum increase of R_{ee}. The threshold voltage and therefore drain current, Idss, depend upon the InGaP thickness. It must be well controlled during growth and its loss due to erosion through processing steps must be minimized and controlled.

15.1.4 Epitaxial Layer Screening

The threshold voltage of the FET depends on the position of the channel charge with respect to the gate and the amount of charge in the channel layer (see Eq. 2.44, Chapter 2). The channel doping and thickness can be monitored by taking the capacitance–voltage (C–V) profile of the emitter–base junction. Figure 15.4a shows the carrier concentration versus the depth plot obtained from a C–V measurement (discussed in Chapter 21). Consistent thickness and doping information can be obtained. Figure 15.4b shows the carrier concentration plot for two batches of wafers that show good control of doping level and channel depth. Also, a large-area device can be fabricated by the epivendor to monitor the threshold voltage of a batch of wafers.

15.1.5 Fabrication Process

The process flow for an HBT has been discussed in Chapter 14. Figure 15.5 shows the process flow with the extra steps added for FET insertion [17]. Emitter contacts that also serve as source and drain contacts for FETs are formed first. Since a highly doped InGaAs layer is used on the top, as part of the cap layers, any refractory metal can be used as the emitter contact. After this, channel etch can be performed by etching off the cap layers and an etch stop layer can be helpful in controlling this etch. After the channel etch, mesas for the emitter and the FET can be defined and etched by dry plasma etching or a combination of wet and dry etch processes. This is followed by

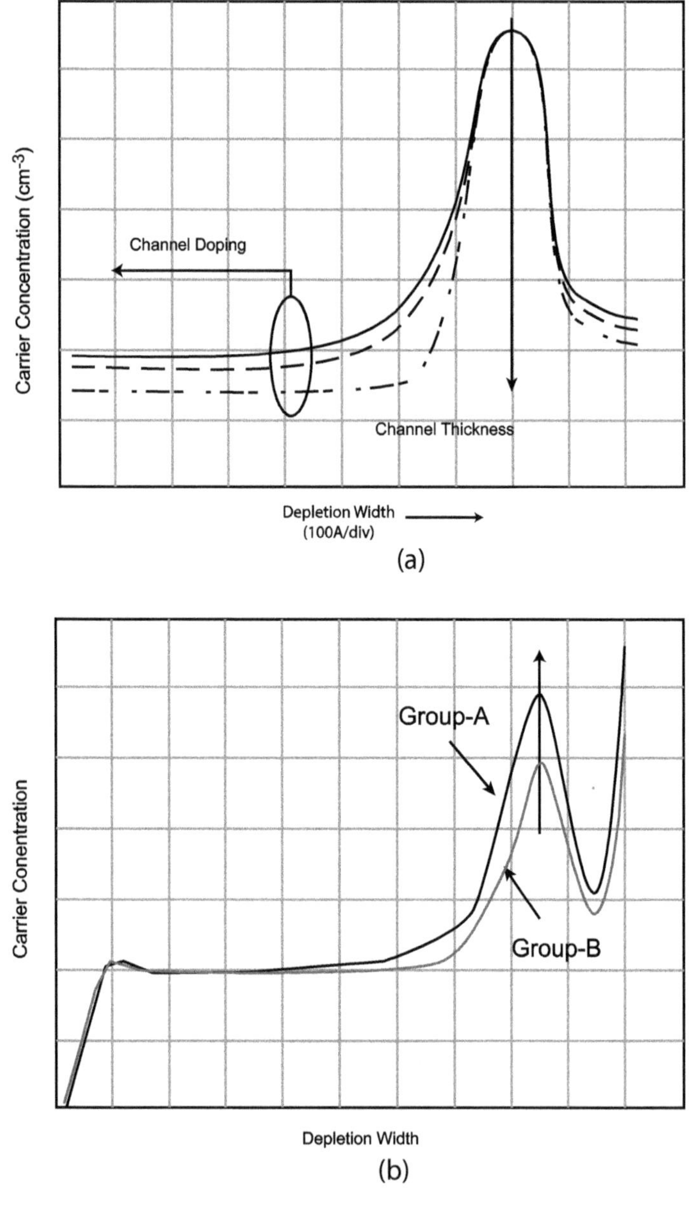

Figure 15.4 (a) Carrier concentration profile obtained from *C–V* measurement. Channel doping differences from sample to sample can be seen. (b) Carrier concentration plot for two groups of epigrowth samples showing identical channel thickness. Courtesy of Cristian Cismaru, Skyworks Solutions.

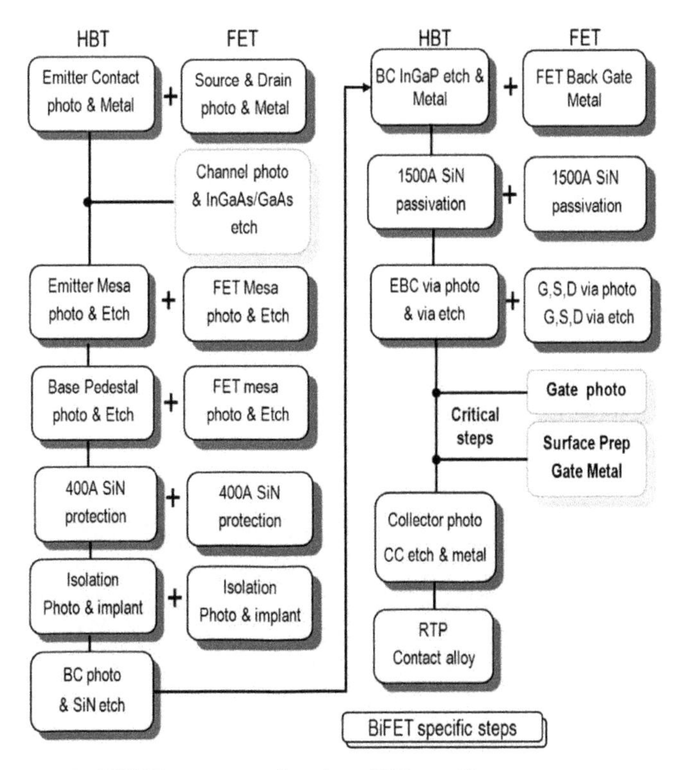

Figure 15.5 BiFET process showing FET-specific extra process steps. Reprinted from Ref. [17] with permission. Copyright © 2007 CS MANTECH.

standard HBT processing steps of the base pedestal or mesa and nitride deposition to protect the surfaces during processing. The base contact is also used for contact to the p+ backgate layer. Device isolation is achieved by combination of the base pedestal (second) mesa and helium ion implantation, just like the HBT process.

15.1.6 BiFET Gate Process

The opening in the nitride for the gate is made as part of the general contact via process for all contacts to HBTs and FETs. The actual gate size can be larger than the opening in the nitride, as depicted in Fig. 15.6. In this particular design the nitride opening forms the gate, and the actual gate metal lithography does not involve submicron line definition. If the InGaP layer is not etched after the channel etch, and the gate is designed to be on InGaP, pregate processing steps have to be well controlled. A few metals have been studied as Schottky contacts to InGaP. Although Pt gives a larger barrier height, it is known to sink into InGaP [18]. Refractory metals, like molybdenum, are more stable. Using Ti as the bottom layer for a Ti/Pt/Au evaporated gate, it is important to prepare the gate surface and control the process to ensure proper surface condition before the gate metal is deposited. Channel thickness, doping, InGaP thickness, and surface states all control the resulting threshold voltage. All process steps involved from opening the nitride (with low plasma damage process), plasma de-scum, and pregate cleans to metal deposition, including close-coupling time between steps, must be carefully controlled.

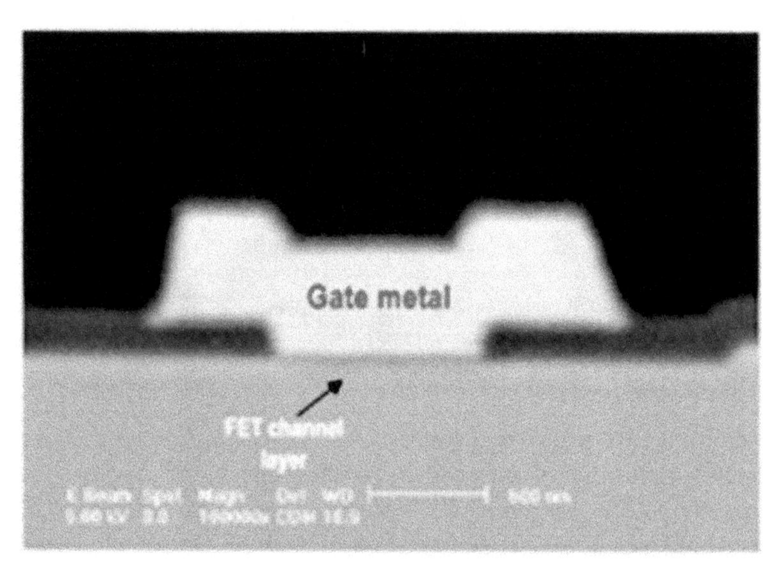

Figure 15.6 SEM crosssection showing details of the FET gate.

If the pregate metal surface preparation is not done properly, the gate metal can sink into the etch stop toward the channel layer and cause poor Schottky characteristics, high ideality factor, and low saturation current. Without a good barrier nonrefractory metals sink in. Figure 15.7 shows a cross section of a gate metal that has sunk in. Also, Ga out-diffusion and formation of voids can result in poor performance and reliability issues. Too much oxide layer on top can cause too high (more negative) threshold voltage [19].

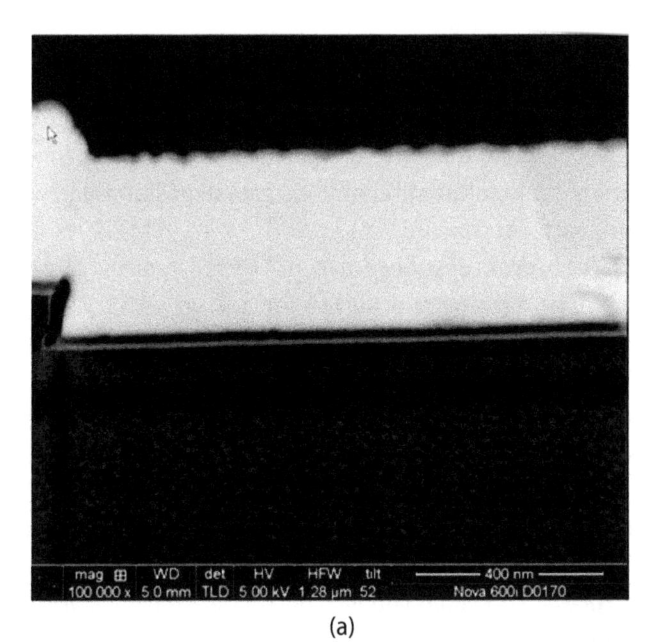

(a)

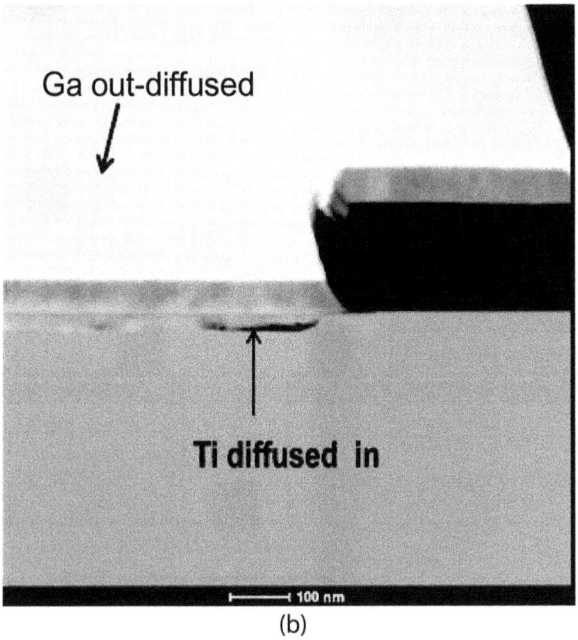

(b)

Figure 15.7 TEM cross sections showing gate sinking (b) in comparison with good gate interface, as shown by FIB cross section (a). Reprinted from Ref. [14] with permission. Copyright © 2006 CS MANTECH.

In the BiFET process, the HBT device performance is not compromised much, but the FET performance cannot be as good as the independent FET process. The control of threshold voltage, although over wider than normal specifications, involves controlling epilayers, pregate surface preparation, gate deposition, and postgate thermal treatment control.

With the process described, a g_m of 165 mS/mm and low leakage (<10 nA/mm) have been achieved for a V_t of −0.35 V [17]. Also, addition of the FET process has been shown to have no effect on the radio frequency (RF) performance of the HBT.

An InGaP/GaAs HBT/junction field-effect transistor (JFET) process has been reported that does not involve any metal gate from above but utilizes only the junction from the p+ base below [20]. In this case, also called voltage-variable resistor by the authors, there is very little change to the epistructure and only one additional mask in the process. Threshold voltage control depends primarily on the control of the channel layer added to the emitter. This technology enables power amplifier bias circuit designs with lower reference voltage (Fig. 15.8).

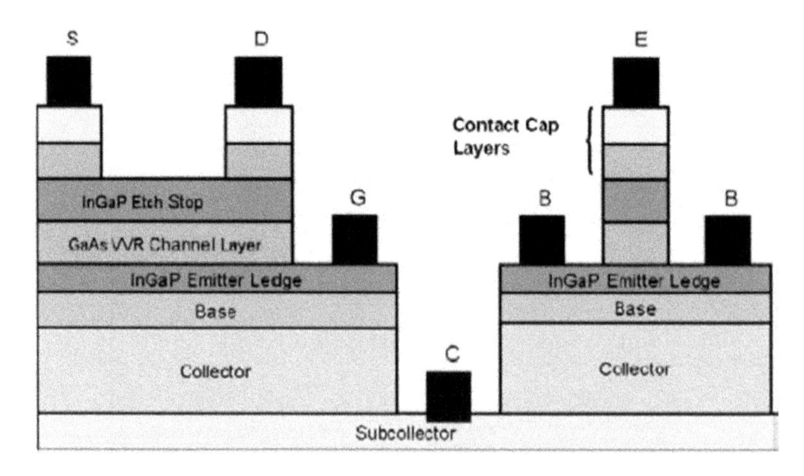

Figure 15.8 HBT/JFET process cross section showing the JFET (VVR) on the left. Reprinted from Ref. [20] with permission. Copyright © 2008 CS MANTECH.

15.2 BiHEMT Process

In this technology epistructures used in the production of HBTs and pseudomorphic high-electron-mobility transistors (PHEMTs) are combined with minimum change, as reported in a production process by Anadigics [4]. The basic concept is to stack the epistructures on top of each other (see Fig. 15.9). The PHEMT structure is grown first, followed by the HBT structure in the same epigrowth run. The growth temperatures for the PHEMT are generally higher than the growth temperature of the HBT. Therefore growing it first avoids the degradation of HBT performance during PHEMT growth. The drawback, of course, is the huge topology in the finished integrated circuit. In this particular structure, the n+ subcollector layer is shared with the HEMT structure for use as the cap layer, which is used for source and drain contacts. The thickness of this layer is a compromise between the collector resistance of the HBT and acceptable topography at the gate recess step. Figure 15.10a,b shows that good high-electron-mobility transistor (HEMT) device performance can be achieved, as shown by the *I–V* characteristics of the PHEMT device.

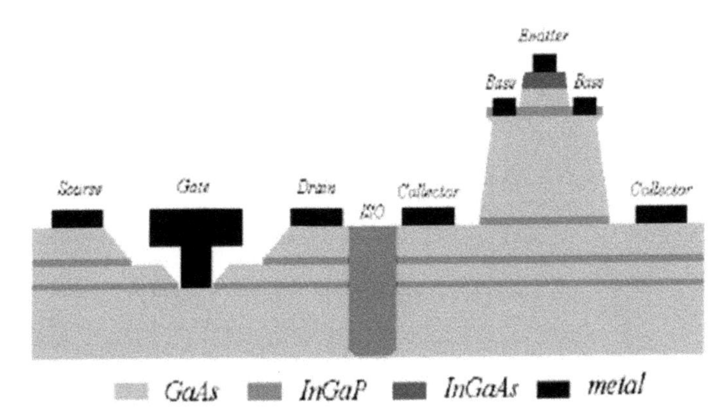

Figure 15.9 Schematic of the Anadigics BiHEMT structure. Reprinted from Ref. [4] with permission. Copyright © 2006 CS MANTECH.

The degradation of PHEMT junctions can also be avoided by the addition of spacers in the epigrowth process, as shown in Fig. 15.11.

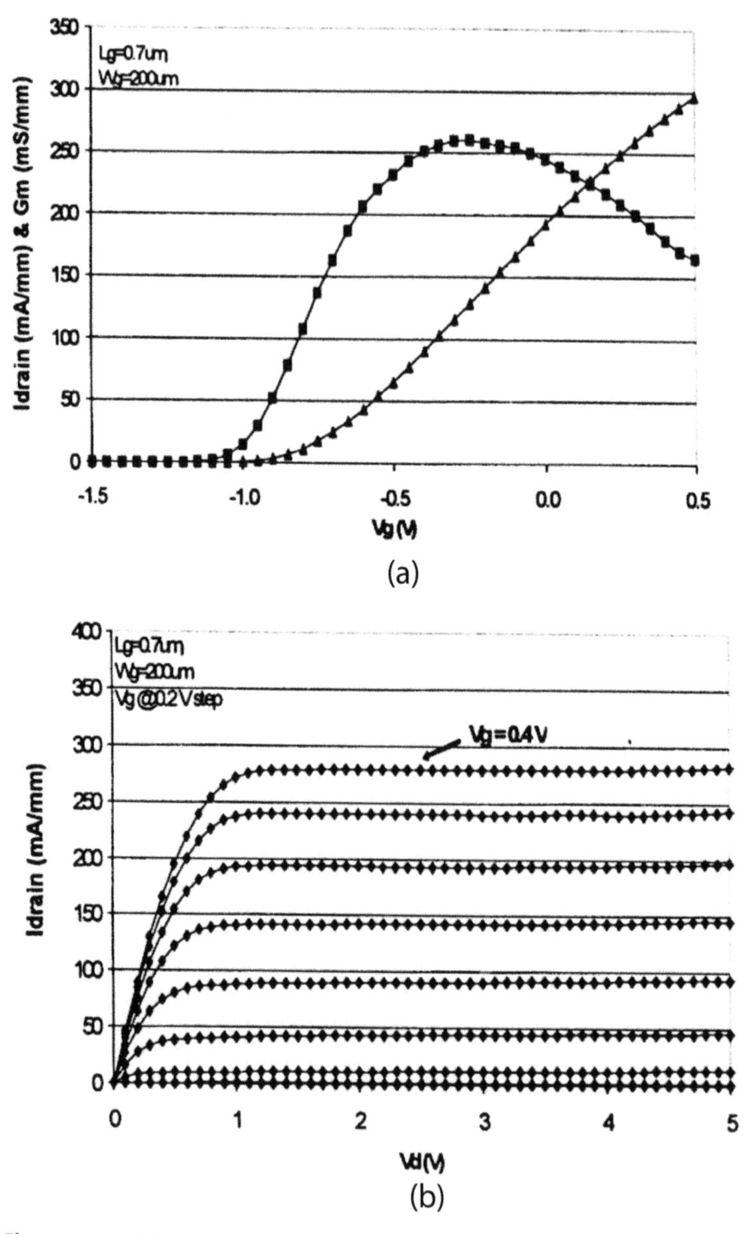

Figure 15.10 (a) Transfer curve of a PHEMT with a 0.7 μm gate (b) *I–V* curves of a PHEMT with 0.7 μm gate length (V_g = 0.2 V/step). Reprinted from Ref. [4] with permission. Copyright © 2006 CS MANTECH.

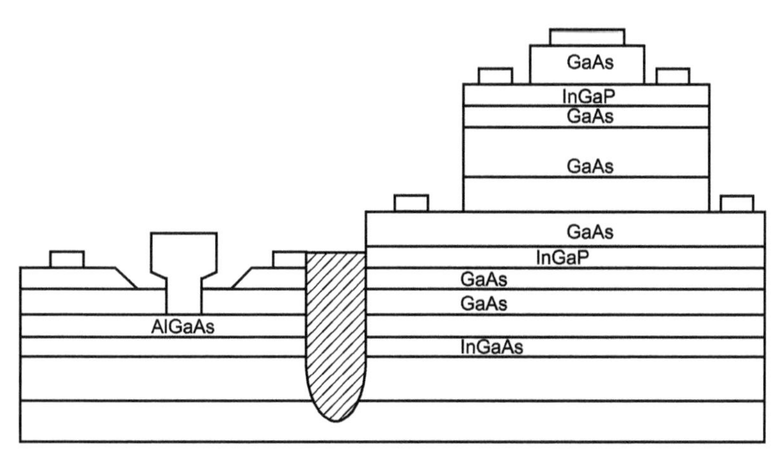

Figure 15.11 Schematic cross section of the WIN Semi BiHEMT. Reprinted from Ref. [5] with permission. Copyright © 2007 CS MANTECH.

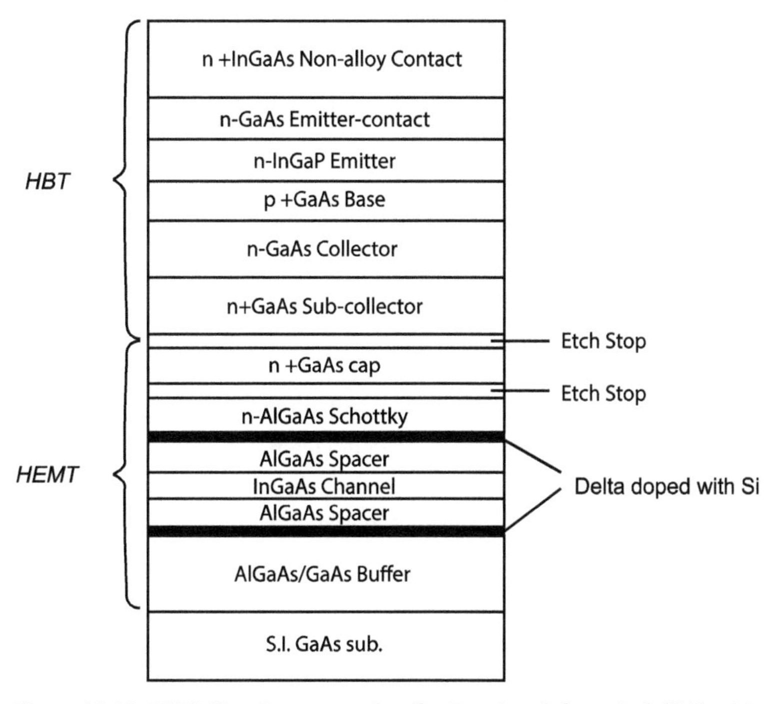

Figure 15.12 BiHEMT epistructure details. Reprinted from Ref. [21] with permission. Copyright © 2010 CS MANTECH.

Or it can be minimized by not using sharp delta doping. Instead the doping can be spread over the thickness of the large-bandgap AlGaAs layer [21].

In the process developed by WIN Semi [5] none of the layers are common, all the layers used for the PHEMT and HBT are simply stacked on each other, as shown in Fig. 15.12. In this case no trade-off of performance is needed. In this process no layers are shared. Common metallization is used for the HBT collector and source/drain contacts for the PHEMT. The use of InGaP etch stops for HBT emitter etch and AlGaAs as the Schottky layer under the gate are also processes that are used in HBT and HEMT processes by themselves. The only difference is that the gate metallization has platinum at the bottom that can be thermally driven in to adjust the threshold voltage. Because of the topology, submicron gate processing (0.5 μm) is the greatest process challenge. Satisfactory performance has been achieved for PHEMTs as well as HBTs.

The process for a bipolar high-electron-mobility transistor (BiHEMT) is obviously a combination of the process steps needed for HBTs and PHEMTs. The process starts with emitter contact fabrication, followed by emitter mesa etch. Then the base contacts are defined and etched, and then metal is evaporated and lifted off. The second mesa for the base is defined and etched next. A nitride passivation layer is deposited and implant isolation done to create insulating regions between devices by helium ion implantation. Collector and PHEMT source/drain contacts are formed next using a standard Au:Ge/Ni metallization and alloy process. Once the HBT device process is complete and the device protected by nitride, the PHEMT process steps of channel recess etch and gate formation are completed. The height difference between the top of the HBT and the gate layer for the HEMT may be close to 2 μm. Because of this extreme topology, the gate process must be developed properly to make gates below 0.7 μm in length. After the front-end process, the usual back-end interconnect process is carried out.

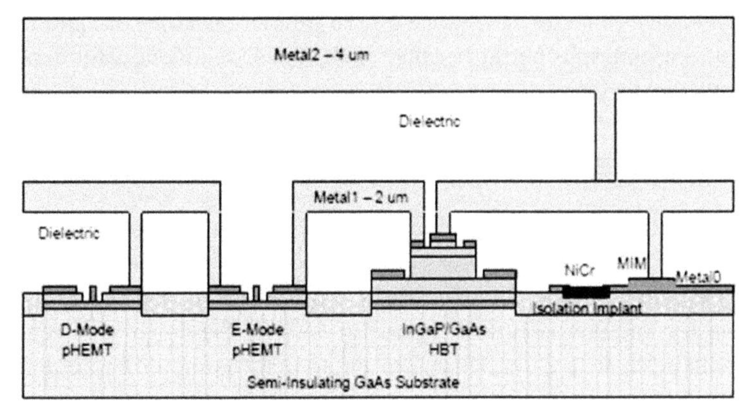

Figure 15.13 Cross section of the Triquint BiHEMT structure showing p- and n-mode HEMT and InGaP/GaAs HBT, along with passive devices and interconnect. Reprinted from Ref. [3] with permission. Copyright © 2007 CS MANTECH.

Triquint has developed a process based on growing an HBT device structure on top of the (enhancement and depletion mode) E/D PHEMT epitaxial structure [3]. Figure 15.13 shows a schematic cross section of the device and interconnect process. They also do not have any common layers in the epistructure. Thin etch stop layers are added for process ease. The PHEMT layers are kept as thin as possible to reduce the thermal resistance below the HBT. Epigrowth of the HBT is done with minimum impact on the PHEMT R_{ON} and maximum current. HBT processing follows the usual sequence of emitter, base, and collector. The HBT subcollector is removed to uncover the PHEMT epistructure. PHEMT processing is done with the usual recess etch, implant isolation, etc. The collector process is combined with the source and drain contacts of the HEMTs. The PHEMT gate deposition is the last device process module. Devices gate lengths of 0.5 μm or 0.7 μm are formed. After the device processing is complete, the resistor, interconnect, and metal–insulator–metal (MIM) capacitor processing is done. Figure 15.12 shows a schematic cross section of a device with the interconnect finished. In this process the HBT parameters are the same as the HBT process by itself (β = 80, collector-base breakdown voltage BVCBo = 24 V, f_t =30 GHz, f_{max} = 55 GHz). The only significant differences are higher collector resistance due to thinner subcollector, and the device isolation is better because of shallower semiconductor depth between devices.

15.2.1 BiHEMT Process and Yield Improvement

The BiHEMT process provides challenges for photolithography because small features must be resolved at different levels on the wafer. For example, for opening of contact vias on top of the HBT emitter and source/drain regions of HEMTs, the height difference could be close to 2 μm. Therefore, depth of focus should be large for the photo process. For patterning of resistors, the resist thickness should be large so that the tops of HBTs are well covered by the resist and hence the resistor material is not left on the HBT emitter.

The problem of steep topology can be solved by making the HBT base pedestal mesa in two or more substeps. This way, metal lines can go up and down the mesas without seams and breaks. This, of course, adds masks and processing steps. Also, the area taken by the HBT increases, although the active device area (base–collector junction remains the same).

It may also be noted that the lift-off process used for defining metal lines must be optimized so that fine geometry lines are not blown away in the process.

References

1. M. F. Chang, Heterojunction BiFET technology for high speed electronic systems, *WOFE Proc. Adv. Workshop*, 15–20 (1997).
2. A. Metzger et al., An InGaP/GaAs merged HBT-FET technology and application to the design of power amplifiers, *CSICS, Paper J1* (2006).
3. T. Henderson et al., *High Performance BiHEMT/E-D pHEMT Integration*, CS MANTECH (2007).
4. M. Shokrani et al., *InGaP-Plus: A Low Cost Manufacturable GaAs BiFET Process Technology*, CS MANTECH (2006).
5. C. K. Lin et al., *Monolithic Integration of E/D-Mode pHEMT and InGaP HBT Technology on 150-mm GaAs Wafers*, CS MANTECH (2007).
6. W. J. Ho et al., A GaAs BiFET LSI technology, *GaAs IC Symp. Tech. Digest*, **47** (1994).

7. D. Streit et al., Monolithic HEMT-HBT integration for novel microwave circuit applications, *Digest GaAs IC Symp.*, 329–324 (1994).
8. K. Kiziloglu et al., *Electron. Lett.*, **33**, 2065 (1997).
9. D. K. Umemoto, D. C. Streit, K. W. Koboyashi and A. K. Oki, 35 GHz HEMT amplifiers fabricated using integrated HEMT-HBT material grown by selective MBE, *IEEE Microwave Guided Wave Lett.*, **4**, 361–363 (1994).
10. D. L. Plumpton et al., *Method to Integrate HBTs and FETs*, US Patent No. 5077231 (1989).
11. Y. F. Yang, C. C. Hsu and E. S. Yang, Integration of GaInP/GaAs HBTs and HEMTs, *Electron. Device Lett.*, **17**, 363 (1996).
12. D. Cheskis et al., Co-integration of GaAlAs/GaAs HBTs and GaAs FETs with a simple, manufacturable process, *IEEE Electron. Device Meeting*, **91** (1992).
13. M. F. Chang, P. M. Asbeck and R. L. Pierson, *Planar HBT-FET Device*, US Patent No. 5250826.
14. Mike Sun et al., *A High Yield Manufacturable BiFET Epitaxial Profile and Process for High Volume Production*, CS MANTECH (2006).
15. S. Tiku, *Device Processing in III-V Manufacturing*, CS MANTECH (2010).
16. R. Ramanathan, *Compound Semiconductors: Process Flow, Process Integration, Devices and Testing*, CS MANTECH (2010).
17. R. Ramanathan et al., *Commercial Viability of a Merged HBT-FET (BiFET) Technology for Power Amplifiers*, CS MANTECH (2007).
18. P. Fay et al., *IEEE Electron. Device Lett.*, **20**, 554 (1999).
19. J.-H. Tsai, *Mater. Chem. Phys.*, **133**, 328 (2012).
20. B. Moser et al., *An InGaP/GaAs HBT/JFET BiFET Technology for PA Bias Circuit Applications*, CS MANTECH (2008).
21. J. Takeda et al., *Degradation of pHEMT Performance in BiHEMTs Caused by Thermal History during HBT Growth and Suggestions for Improvement*, CS MANTECH (2010).

Chapter 16

MOSFET Processing

16.1 Introduction

As device geometries reach a limit in silicon, attention has shifted to exploiting the high electron mobility of III–V compound semiconductors. Peak electron mobility in GaAs is about 8800 cm^2/V-sec, and in very low-energy-bandgap semiconductors like InSb, it can reach 77,000 cm^2/V-sec. Ternaries like low-bandgap InGaAs have been actively pursued for the channel layer and other ternaries studied for the growth of oxides or other insulators under the gate. Metal–oxide–semiconductor field-effect transistor (MOSFET)-type devices on III–V compounds have the potential of devices capable of higher gate voltage swing, higher drive current, and low gate leakage not possible with high-electron-mobility transistor (HEMT)-type devices. In particular, E-mode devices are more desirable because a negative voltage source is not needed and a single power supply can be used. This is a definite advantage for mobile devices.

A III–V metal–oxide semiconductor (MOS) is seen as one of the ways to address the problems in an 11–15 nm technology node, as strained silicon runs out of steam and InGaAs-based nMOS is seen as a solution. Values of saturation velocity in excess of 3×10^7 cm/s have been achieved for 30–40 nm gate length devices for $V_{DS} = 0.5$ V by teams at MIT and Intel [1]. Further increases of injection velocity

III–V Integrated Circuit Fabrication Technology
Shiban Tiku and Dhrubes Biswas

ISBN 978-981-4669-30-6 (Hardcover), 978-981-4669-31-3 (eBook)
www.panstanford.com

are achievable at higher InAs mole fractions, which can be twice as high as those obtained for strained silicon at half the operating voltage. But such III–V-based exotic MOSs must solve the problem of a realistic and cost-effective high-*K* dielectric with good III–V/dielectric interface properties, junction technology down-selection, as well as practical and technologically cost-effective ways of integrating III–V devices on a silicon substrate.

The problem of growing or depositing an electronically clean oxide or a dielectric layer in general has been pursued by researchers, as briefly discussed under Section 6.6 in Chapter 6. In summary, a high interface state density and a pinned Fermi level are the main issues. Many decades have passed with numerous announcements of success. However, no production MOSFET process has emerged. In the meantime, the disadvantage of not having an oxide and the fact that any metal forms a Schottky device were put to good use in the production processes for GaAs integrated circuits (ICs). In recent years (2000 onward) finally there are indications that the goal of making a III–V MOSFET may be achievable [1, 2].

The idea of growing an oxide may be dropped and replaced by growing of a good dielectric. However, proper surface preparation before growth still remains relevant. The main challenge is to grow a dielectric free of trapped charge and other defects.

16.2 Oxidation

There is a consensus in the literature (see, for example, Refs. [3, 4]) based on X-ray photoelectron spectroscopy (XPS) and electrical data that As and As oxides are not desirable on GaAs. These leave dangling bonds; only pure Ga_2O_3 gives the lowest surface state density. Gentle processes like ozone oxidation, low-energy plasma, photoexcitation by ultraviolet (UV), sulfur passivation, and numerous other approaches have been tried. However, the goal of unpinning GaAs or other III–V compound semiconductors and creating surface inversion that is needed for a MOSFET-like operation has not been achieved. Ternary compounds are more useful for growing the top oxide layer, particularly in devices of the type that use high-mobility InGaAs as the channel layer and a larger-bandgap material on top under the gate.

InGaP has a large bandgap and excellent etch selectivity. But growing a clean oxide on it has been a huge challenge. An oxide consisting of a mixture of $InPO_4$ and Ga oxide has been demonstrated [5]. However, on ternary InGaP, deposited on GaAs, there is another problem—the heterointerface has an intermixing layer, which has a huge interface charge, so getting a clean, low-surface-state device is not possible [6] (Skyworks data). When InGaP is grown on an InGaAs channel, and an oxide layer is grown by dry oxidation, a larger turn-on voltage is seen; however, the drain current could not be pinched off [7].

16.2.1 Wet Oxidation

On ternary systems, like InAlP, different groups have had better success by creating native oxides using a wet-oxidation process [8, 9].

MOSFET-based epistructures, shown in Fig. 16.1a,b, were grown by molecular beam epitaxy (MBE) [8]. The structure consists of a pseudomorphic undoped $In_{0.22}Ga_{0.78}As$ channel, an undoped $In_{0.5}Ga_{0.5}P$ layer that serves as both a spacer and an oxidation stop layer, a 20 Å n+-InAlP layer lattice-matched to GaAs, and a 250 Å heavily doped GaAs cap [9]. The choice of which ternary to use for the oxidation layer is critical. An oxide grown on InGaP tends to have poor quality. Use of a ternary based on Al is important. The thickness of this layer is only 20 Å. Solubility of In is low in the ternary oxide, so indium may form clusters before it gets a chance to form stable $InPO_4$. Indium cluster formation on InAlP with a thickness of over 170 Å has been seen before [10]. The thickness of the grown oxide is seen to be 35 Å.

Wet oxidation is used at 440°C for ~10 min. Water is brought into the oxidation furnace by bubbling ultrapure N_2 through a water bubbler held at 95°C. Issues related to In cluster formation have been seen with dry oxidation. A clean layer of oxide is seen in transmission electron microscopy (TEM) (see Fig. 16.1b). The oxide has been analyzed by various techniques, including XPS. The composition of the oxide is a mixture of Ga oxide and $InPO_4$-type phases. The oxide that grows on the source drain cap layers can be removed later.

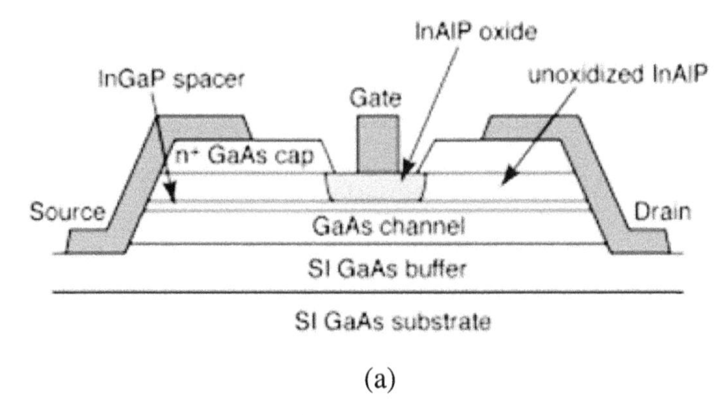

(a)

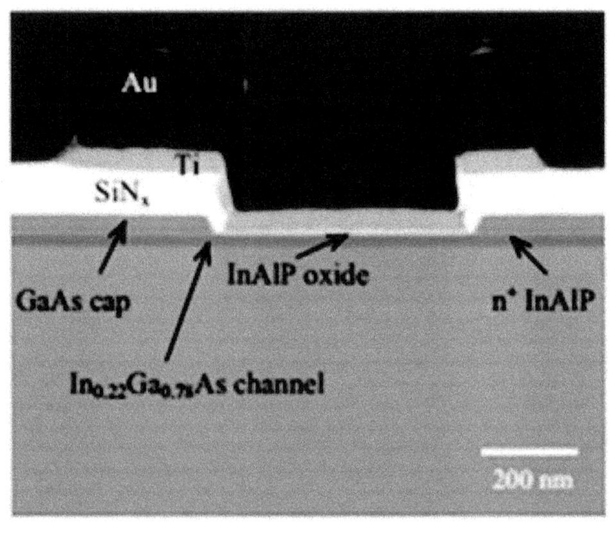

(b)

Figure 16.1 Schematic diagram of metal-defined gate and TEM cross section of a MOSFET with a gate defined by nitride opening, both with native oxide grown on InAlP in the gate opening. Reprinted from Ref. [8] with permission. Copyright © 2007 CS MANTECH. Reprinted from Ref. [9] with permission. Copyright © 2010 CS MANTECH.

The gate length, defined by the opening in the nitride, was 0.5 um. The threshold voltage of the device was 1.05 V. Transconductance g_m of 146 mS/mm and Idss of 95 mA/mm were achieved. The F_t and f_{max} of 34 GHz and 49 GHz, respectively, were reported. These are the best results for a native oxide–grown MOSFET. The gate leakage is reduced by 5 orders of magnitude compared to a heterojunction field-effect transistor (HFET), as shown in Fig. 16.2.

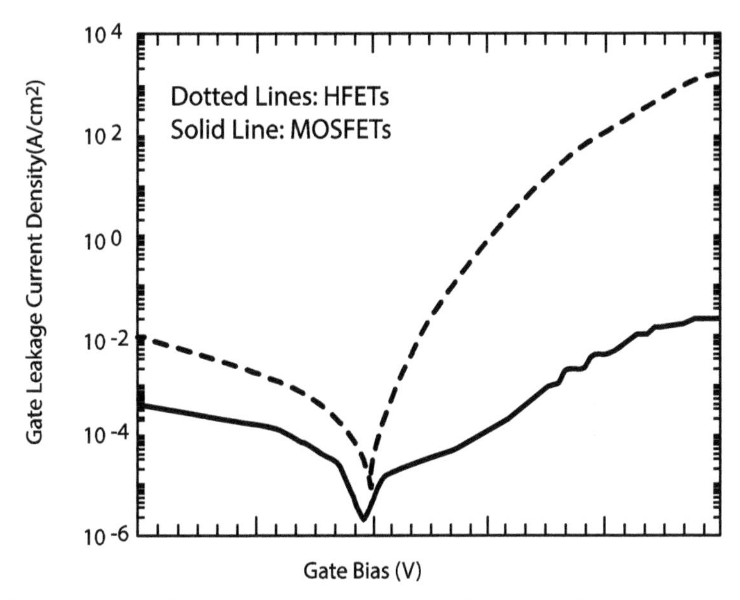

Figure 16.2 Typical gate leakage current of an $In_{0.22}Ga_{0.78}As$ channel MOSFET compared to an HFET on the same epitaxial heterostructure. Reprinted from Ref. [9] with permission. Copyright © 2010 CS MANTECH.

16.2.2 Liquid-Phase Oxidation

This is another low-cost approach (compared to MBE or atomic layer deposition [ALD]). In this method GaAs wafers are immersed in a temperature- and pH-controlled nitric acid solution with Ga in it. The device is MBE-grown with a GaAs channel on top and then oxidized and a Ti/Au gate deposited [11].

To make the oxidation medium, Ga metal is dissolved in hot nitric acid, and then ammonium hydroxide is added. Figure 16.3 shows the important details of making the bath for liquid-phase oxidation (LPO) [12]. The bath is held at a temperature between 30°C and 70°C. The product wafer is immersed for 60 min for a growth of ~100 Å. The process is photoresist compatible. The typical device cross section is shown in Fig. 16.4. Note that the layer being oxidized is InGaP.

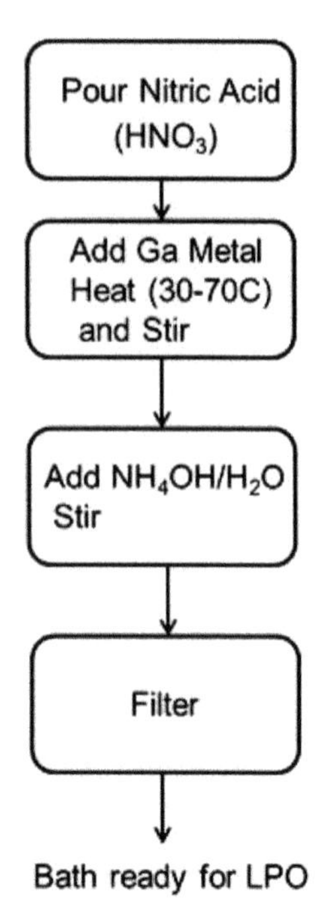

Figure 16.3 Procedure for making an oxidation bath for liquid-phase oxidation (LPO).

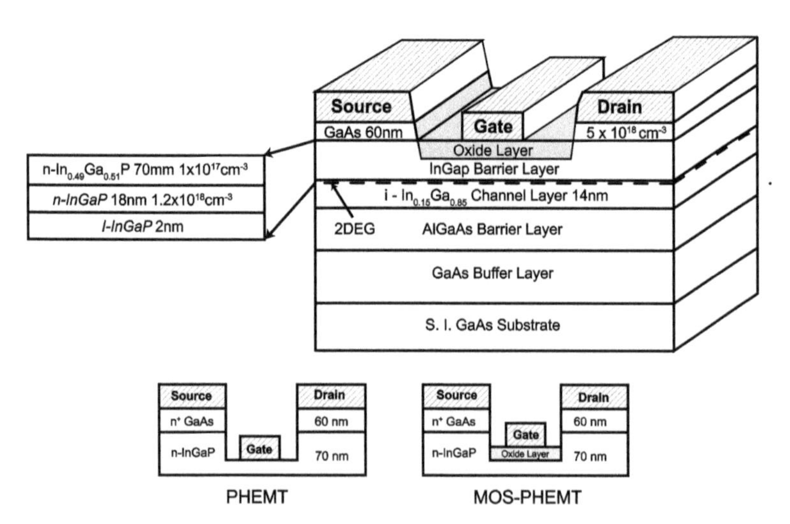

Figure 16.4 Schematic diagram of an InGaP/InGaAs MOS-pHEMT fabricated using liquid-phase oxidation. © [2005] IEEE. Reprinted, with permission, from Ref. [12].

The electrical results from this process are not as good as wet oxidation at high temperature in a furnace. For a V_t of near −2 V, g_m was only about 80 mS/mm for an oxide thickness near 100 Å and only 30 mS/mm for a very thick oxide layer of 1000 Å [13].

16.3 Dielectric Passivation

A gate stack free of trapped charge and other defects is needed to create a MOSFET-type device. Surface state density must be reduced below 10^{12}/cm^2-eV levels for good device performance and lifetime. Such lower values can be achieved and retained through the life of the device by surface passivation using dielectric deposition. Dielectric layers are commonly deposited by plasma-enhanced chemical vapor deposition. Low-energy depositions activated by UV do not produce films of good enough quality. Therefore in practice, the surface of the semiconductor is exposed to plasma before the deposition starts. Energetic ions, chemical radicals, UV light, etc., all cause damage. Ion bombardment must be minimized. Displacement energy for creation of defects by ions is about 40 eV. Careful design of the plasma system is needed to reduce bias voltage, for example, by making the powered electrode area large, and optimize radio frequency (RF) power, pressure, etc. Deposition temperature and rate also need to be optimized to maintain a high-quality dielectric without introducing excessive damage. Electron cyclotron resonance (ECR) systems can be used to reduce damage. Details will be discussed in Chapter 8.

In 1995 Ga_2O_3 deposited in situ came close to the quality of a GaAs/AlGaAs interface [14]. However, the problem of Fermi level pinning at the surface due to surface states cannot be fully resolved, even under low-temperature conditions or with gentle ozone oxidation.

16.4 Atomic Layer Deposition

The difficulty of growing a native oxide on III–V materials is obvious from the above discussion. Therefore, attempts have been made to deposit oxides over the last few decades. Ga_2O_3 has been grown by MBE. Results are better but not good enough. Oxides of Al, Gd, and Hf have also received a lot of attention. These are high-K dielectrics and grow epitaxially on III–V semiconductors in the MBE process, although the lattice constant and crystal structures are very different. These exhibit low dielectric loss and electrical leakage. However, these must be protected from moisture absorption during processing.

More recently, attention has turned to dielectric materials deposited using ALD. In 2003 Al_2O_3 deposited by ALD [3, 15] was reported. In this process, the surface oxide is cleaned up at the beginning of ALD by the heat in the vacuum system. All dangling bonds due to As and As oxide evaporate due to low pressure and high substrate temperature (the temperature is high enough to remove loose oxides, As oxide, and As). Interface state density is reduced before deposition by proper pretreatment. Surface state densities in the range of 10^{12}–10^{13} /cm^2-eV are possible for Ga-rich surfaces. Interface density is normally measured as a number per unit area per unit energy in the semiconductor bandgap. Ammonia-treated surfaces are best for growing Al_2O_3 on GaAs (by the ALD process) to achieve the lowest surface state density [4, 16]. For an Al_2O_3 thickness of 8 nm, capacitance–voltage (*C–V*) measurements from 10 kHz to 30 Hz confirm inversion, as seen in Fig. 16.5, and therefore show that the Fermi level is unpinned. Better MOSFET current densities have also been achieved with ammonium sulfide passivation before atomic layer epitaxy (ALE) (see Ref. [46] in Chapter 4).

InGaAs has the best mobility and has the best potential for going beyond a silicon complementary metal–oxide semiconductor (CMOS). More recent attempts have been made to make a MOSFET directly on it. ALD has the potential of becoming popular because of its applicability to high-*K* dielectrics in silicon technology. Unpinning of the Fermi level shows that the dielectric is of high enough quality. Interface state density around 1.4×10^{12} /cm^2-eV was achieved (as determined by the *C–V* method). Gate leakage was 10^{-4} A/cm^2 at a 3 V gate bias with an 80 Å oxide.Device performance figure of merit, g_m, for a depletion-mode device with 160 Å Al_2O_3 grown on an $In_{0.2}Ga_{0.8}As$ layer with 1 μm gate length, was 105 mS/mm. A strong accumulation current is seen for a positive gate voltage (V_{gs}).

A true accumulation-type device, an enhancement-mode MOSFET is of immense interest to get high currents. This is difficult to realize because the interface state density must be reduced further. Also, to reduce leakage, the oxide thickness cannot be too thin. Therefore the device cannot be scaled down to gate sizes below 100 nm needed for CMOS applications. A possible application in traditional III–V products could be as shown in Fig. 16.6. Here a large 1 μm gate length MOS-HEMT using 3 nm ALD-deposited Al_2O_3 is shown. Figure 16.7 shows the cross section and *I–V* characteristics of an InGaAs channel device made on an InP substrate (lattice matched). This device has a gate length of 0.4 μm on top of 10 nm of Al_2O_3 grown on an $In_{0.65}Ga_{0.35}As$ channel. *I–V* curves show inversion-type E-mode behavior.

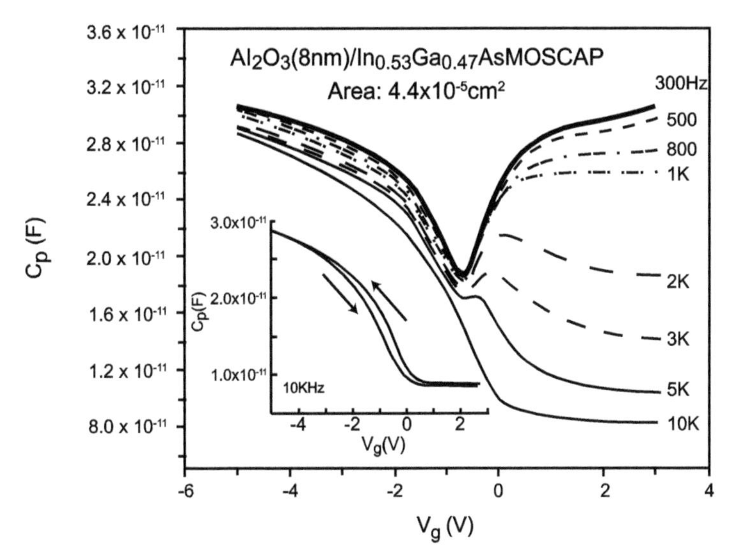

Figure 16.5 *C–V* characteristics of an ALD-grown MOS structure with 8 nm Al_2O_3 on InGaAs, showing inversion [16]. Compare this to the ideal *C–V* for a MOSFET, as shown in Fig. 2.28. © [2007] IEEE. Reprinted, with permission, from Ref. [16].

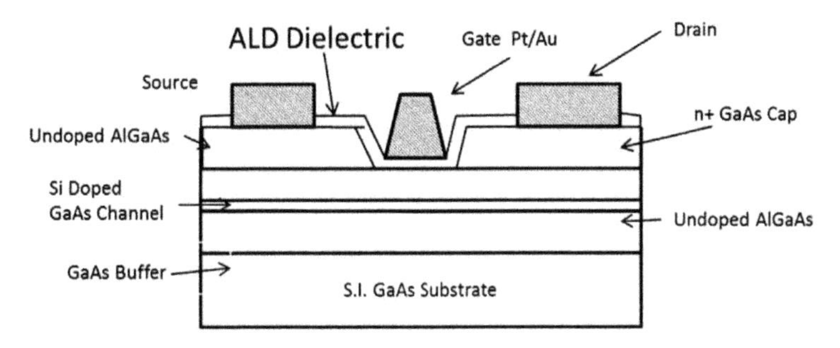

Figure 16.6 GaAs MOSFET with an ALD-grown gate dielectric on a GaAs substrate.

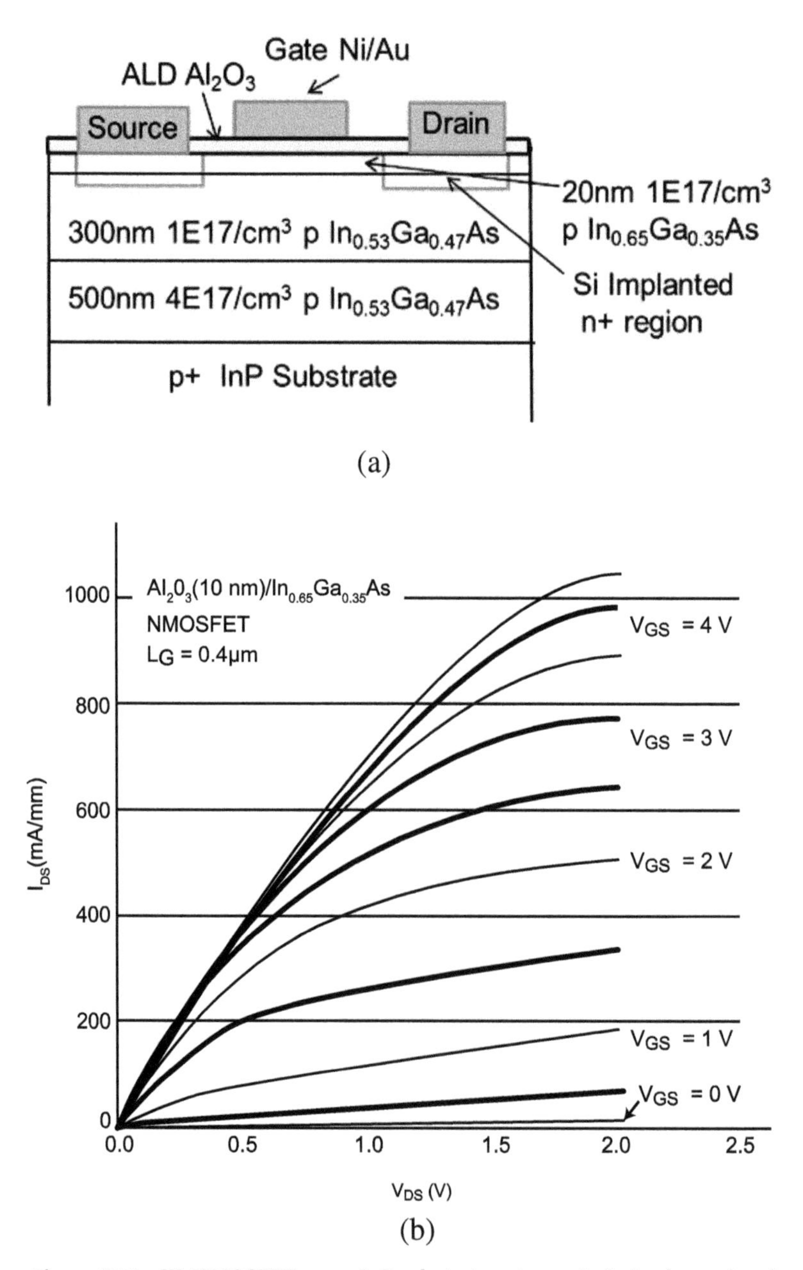

Figure 16.7 III–V MOSFET on an InP substrate using an InGaAs channel and an ALD Al_2O_3 gate dielectric schematic cross section (a) and *I–V* characteristics showing inversion-type E-mode behavior (b). © [2008] IEEE. Reprinted, with permission, from Ref. [15].

Recently Al_2O_3 was deposited on top of $(La{:}Y)_2O_3$ by ALE on a GaAs substrate to achieve an NMOSFET with a record-high drain current for the GaAs channel (330 mA/mm). Figure 16.8 shows a high-resolution TEM cross section of the interface, which remains sharp and perfect even after high-temperature activation (rapid thermal annealing [RTA]) at 750°C [17].

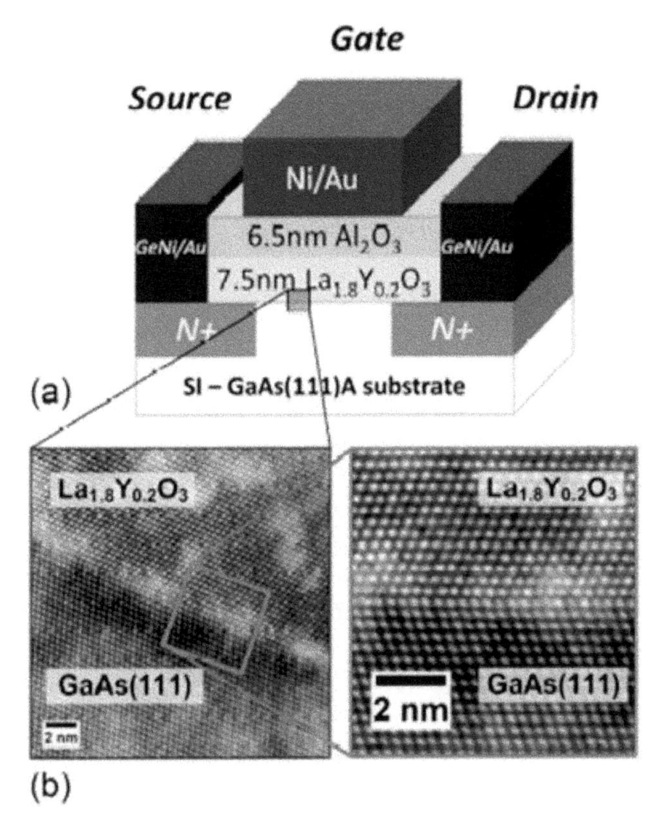

Figure 16.8 TEM cross section of an $Al_2O_3/In_{0.75}Ga_{0.25}As$ interface of an InP MOSFET using an InGaAs channel. © [2013] IEEE. Reprinted, with permission, from Ref. [17].

16.5 p-Type Devices [15]

Achieving a p-type device on III–V materials has been a challenge due to low hole mobility. In Si strain is used to enhance mobility. After decades of effort, ALD is again enabling fabrication of such a device, so CMOS-type ICs can be made. MBE-grown p-type $In_{0.53}Ga_{0.47}As$ on an InP substrate has been demonstrated by Xuan et al. [16] again using ALD Al_2O_3 or HFO_2. Record-high values of p-mobility and transconductance have been shown ($\mu_p = 1550$ cm^2/V-sec at 0.3 MV/cm higher than mobility in silicon).

Finding an excellent dielectric for III–V materials still remains a challenge, but ALD provides the best hope in decades. The material system, an InGaAs channel combined with an ALE- or ALD-deposited dielectric, shows the best potential and may have a revolutionary impact if somehow integrated with large silicon substrates.

16.6 Concluding Remarks

To date deposition of very thin layers of Ga_2O_3/GdO_3 or HfO_2 by MBE or the deposition of Al_2O_3 by ALD on a low-bandgap InGaAs layer remains the main contender for achieving surface inversion and development of MOSFET devices. Oxidation of other ternaries like InAlP could lead to MOSFET-like E-mode devices [18].

References

1. J. A. del Alamo, Nanometer scale III-V compound semiconductors, *Nature*, **317** (2011) and *IEDM*, 696–699 (2010).
2. I. G. Thayne, et al., Review of current status of III-V MOSFETs, *ECS Trans.*, **19**(5), 275–286 (2009).
3. S. Oktyabryski and P. D. Ye, *Fundamentals of III-V Semiconductor MOSFETs*, Springer Science and Business Media (2010).
4. J. M. Woodall, Non-silicon MOSFET technology: a long time coming, in book *Fundamentals of III-V Semiconductor MOSFETs*, Springer Science and Business Media (2010).
5. K.-W. Lee et al., *IEEE Electron. Device Lett.*, **26**, 864 (2005).
6. M. J. Hwu et al., *A Chemistry for Etching Quarternary Interface Layers on InGaAsP Formed between GaAs and InGaP Layers*, Patent PCT US 02/39954 (1995).
7. J.-H. Tsai et al., *Mater. Chem. Phys.*, **133**, 328 (2012).
8. P. Fay et al., *III-V MOSFET with Native Oxide Gate Dielectric*, CS MANTECH (2007).

9. X. Xu and P. Fay et al., *E Mode MOSFET with InAlP Native Oxide*, CS MANTECH (2010).
10. X. Li and P. Fay, *J. Appl. Phys.*, **95**, 4209 (2004).
11. J.-Y. Wu et al., *IEEE Electron. Device Lett.*, **20**, 18 (1999).
12. Y.-H. Wang et al., Liquid phase oxidation of GaAs based materials and their applications, *Proc. 7th Int. Conf. Solid State Int. Circuit Technol.*, 2291 (2005).
13. Y.-H. Wang and K.-W. Lee, Liquid phase oxidation of InGaP and applications, in *Solid State Circuits and Technology*, Ed., P. K. Chu, INTECH, Croatia (2010).
14. M. Passlack, *J. Appl. Phys.*, **77**, 686 (1995).
15. P. D. Ye, High performance III-V MOSFETs enabled by atomic layer deposition, *9th Int. Conf. Solid State IC Technol.*, 1429 (2008).
16. Y. Xuan et al., Sub-micron inversion-type enhancement-mode InGaAs MOSFET with atomic-layer-deposited Al_2O_3 as gate dielectric, *IEEE Electron. Device Lett.*, **28**, 935 (2007).
17. L. Dong, P. D Ye et al., GaAs enhancement-mode MOSFETs enabled by atomic layer epitaxial La1.8 Y0.2 O3 as dielectric, *IEEE Electron. Device Lett.*, **34**, 487 (2013).
18. P. D. Ye, Main determinants for III-V MOSFETs, Birck Nanotechnology Center and School of Electrical and Computer Engineering, Purdue University, yep@purdue.edu.

Chapter 17

Passive Components

Passive components used in electronic circuits are resistors, inductors, and capacitors. In an integrated circuit these and other components like couplers or strip lines need to be made as part of the wafer fabrication process. The various process steps are discussed in this chapter.

17.1 Resistors

Resistors are used in integrated circuits (ICs) for numerous functions that require a range of values over many decades and precision from very crude to less than 1%. Resistors are made in the semiconductor itself or deposited as a thin film. Semiconductor resistors can be made by utilizing any of the techniques used for making n- and p-type layers. Diffused resistors are possible in silicon but for III–V compounds, epilayers or ion implantation must be used.

17.1.1 Semiconductor/GaAs Resistors [1]

These resistors involve the least processing because the resistor layer is already present in the device structure and the contacting and isolation processes are the same that are needed for active

III–V Integrated Circuit Fabrication Technology
Shiban Tiku and Dhrubes Biswas

ISBN 978-981-4669-30-6 (Hardcover), 978-981-4669-31-3 (eBook)
www.panstanford.com

devices anyway. The sheet resistances of the semiconductors that are available depend upon the technology or device type. In the case of field-effect transistor (FET)-type ICs, sheet resistance of the n+ source/drain and n-type channel layers can be utilized. These can be from 100 ohms/sq. to 1000 ohms/sq. In the case of heterojunction bipolar transistors (HBTs), lower values, those of emitter and collector layers, are available. The base layer may be used and could be in the range of one hundred to a few hundred ohms/sq.

The resistance of the semiconductor resistor is the sum of the semiconductor sheet and the two contacts:

$$R = Rs.\frac{L}{W} + 2\frac{Rc}{W} \tag{17.1}$$

where R_c is the contact resistance per unit length, generally measured in ohm-mm. Resistors should be designed with a large L/W aspect ratio so that contact resistance is negligible. For n resistors, with sheet resistivity of 400 to 700 ohms/sq., the specific contact resistance can be high, even with n+ contact regions. Therefore these resistors may require large contact regions to ensure that the resistors scale with dimensions. Figure 17.1 shows a top view and a cross section of a semiconductor resistor in an implanted or epigrown FET-type substrate. The resistor region is defined by an ion-implanted isolation region. The minimum width of the resistor depends upon the minimum dimensions allowed for the semiconductor layer due to the constraints of the isolation process.

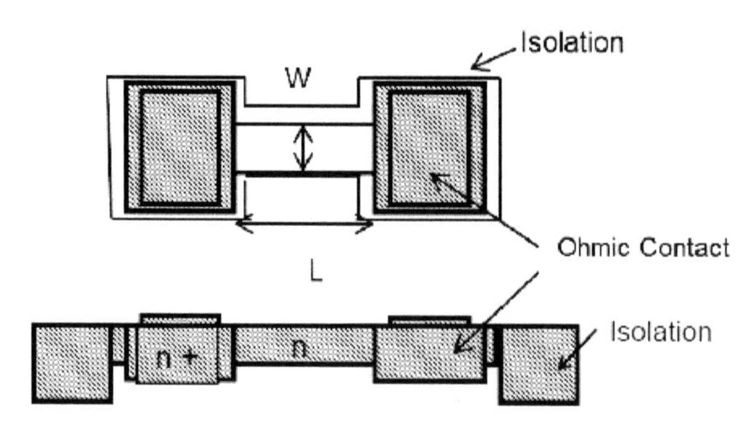

Figure 17.1 Ion-implanted resistor in GaAs defined by selective implantation.

GaAs resistors can be made on an HBT structure using the n+ emitter or the p+ base layer. An example of a p+ resistor is shown in Fig. 17.2. This resistor area is defined by the mesa structure. The lower n-type collector layer is isolated by the p–n junction.

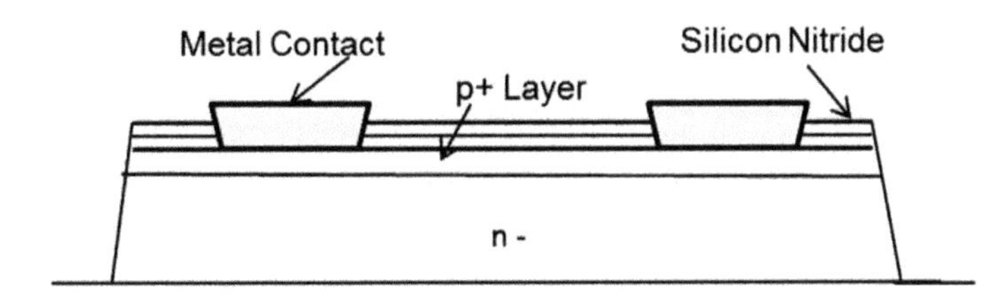

Figure 17.2 A p+ epigrown resistor defined by a mesa structure.

I–V curves for semiconductor resistors show saturation over a certain electric field (see Fig. 17.3), just like an ungated FET channel; therefore the resistors must be operated in the a linear low-field region. Semiconductor resistors have a positive, high temperature coefficient of resistance (TCR), over a few hundred ppm. These resistors cannot be used in situations where a high TCR is unacceptable. These resistors can be used if the design is tolerant, for example, in a voltage divider where only the ratio of the resistors is important. The current-carrying capability of semiconductor resistors is not high because of the poor thermal conductivity of GaAs. Resistor sizes cannot be made larger for heat dissipation due to general area limitations and increased shunt capacitance.

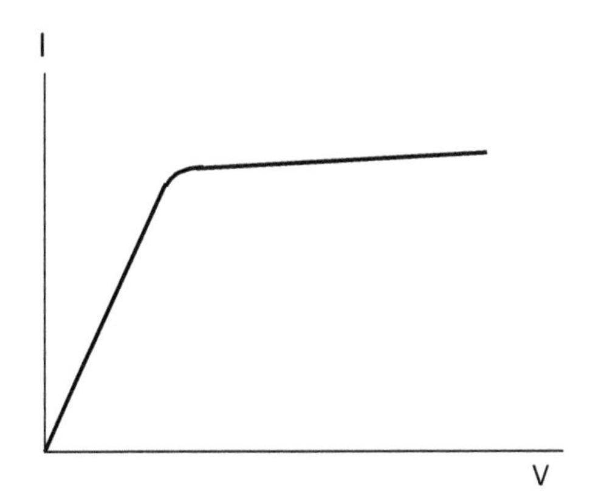

Figure 17.3 *I–V* curve for a semiconductor resistor showing saturation at high voltage.

17.1.2 Thin-Film Resistors [2]

A high TCR and low current-carrying capability of semiconductor resistors are unacceptable for many circuit functions; therefore thin-film resistors (TFRs) are provided as part of the process in GaAs and other III–V semiconductor ICs. TFRs originally developed for thin-film circuits as well as for use as components have been used extensively in monolithic microwave integrated circuit (MMIC) processing. These resistor layers are generally deposited after the active device fabrication is complete, so the deposition, patterning, and contact process must be compatible with the rest of the IC process. Figure 17.4 shows a schematic diagram for a TFR deposited on semi-insulating GaAs with contacts made by an interconnect metal like Ti/Pt/Au. These films must adhere well to the dielectric layer underneath, not get affected by usual lithography processing, and not shift in value through the rest of the interconnect processing, like ashing and thermal cycles needed for the interlayer dielectrics. If there is a small shift, it should be reproducible so that it can be adjusted for. Table 17.1 lists commonly used TFR materials. The materials that are chosen fulfill the basic requirements described above. The thickness range used is in the range of a few hundred to a few thousand angstrom.

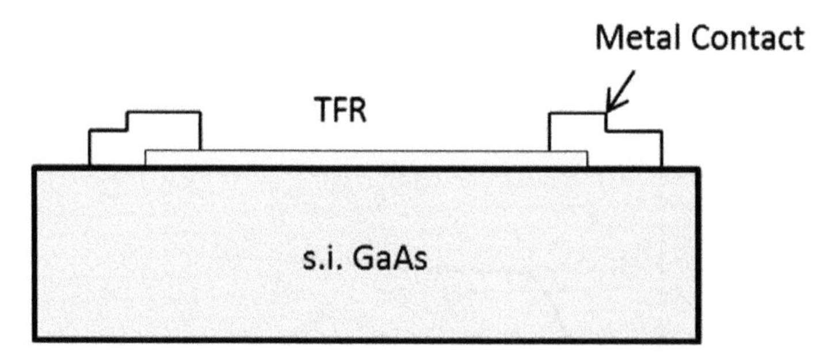

Figure 17.4 Thin-film resistor deposited on semi-insulating GaAs with contacts made by an interconnect metal.

For GaAs ICs two or three ranges of sheet resistivity values are needed to cover circuit applications. Very-low-value resistors, less than 5 ohms/sq., are needed for ballast purposes, 20 to 60 ohms/sq. are needed for precision resistors, high-value resistors are needed for voltage dividers, and superhigh-value resistors are needed for low current load elements. These materials should exhibit good stability, but in practice, the TCR, current-carrying capability, and tolerance values vary a lot.

17.1.3 Common TFR Materials

Numerous materials have been tried and use for TFRs. Some common ones are listed in Table 17.1 and a few are discussed in some detail below.

- **Nichrome:** Nichrome resistors with Cr varying from 20 wt% to 50 wt% are used for resistance range from 35 ohms/sq. to 70 ohms/sq. The thickness needed is in the few-hundred-angstrom range. These films are easy to deposit either by vacuum evaporation or by sputtering. Patterning by lift-off, common in III–V processing, is easy for evaporated films. Achieving low run-to-run variability by evaporation is difficult even with the use of an in situ resistance monitor. Deposition by sputtering gives better control, but lift-off patterning is more difficult. The main advantage of nichrome is that a very low TCR can be achieved, especially by addition of impurities like aluminum, which is done by the material vendor. Nichrome films are susceptible to change through processing and must be protected by dielectric encapsulation. Widths can scale down to 1 μm when deposited by evaporation and patterned using the lift-off process. Nichrome thin films of the order of a few hundred angstroms have remarkably high current-carrying capability, in the range 0.25 to 1 mA/μm width.
- **TaN:** Tantalum nitride–based resistors are extremely stable but must be deposited by sputtering. The films are deposited by direct current (DC) magnetron reactive sputtering of a tantalum target in an argon–nitrogen gas mixture. Figure 17.5 shows the resistivity and TCR as a function of nitrogen partial pressure [3]. Small variations in nitrogen partial pressure can result in large variation in resistivity. The phases formed and resistivity are also sensitive to substrate temperature. For lift-off, a process must be developed for deposition at room temperature. Resistance changes through subsequent

processing, but the change is small and predictable. Film stress can be minimized by optimizing deposition pressure. The TCR is determined by fitting a line to resistance values measured between 25°C and 150°C. TCR values close to zero (<50 ppm) can be achieved. The crystalline phase needed for achieving a low TCR has been identified to be $TaN_{0.1}$ with a body-centered cubic (bcc) structure [4, 5]. Small amounts of hexagonal Ta_2N may be present.

TaN resistors have a current-carrying capability similar to nichrome but are more stable and less prone to compositional changes. The resistors are stable under temperature tests up to 350°C [6], and under high humidity (highly accelerated stress test [HAST]), when protected properly by a silicon nitride overcoat. TaN resistors are dry-etchable, so wherever possible, a dry etch process should be utilized to pattern these. For achieving good selectivity, it is better to deposit a resistor layer on SiO_2 so that CF_4:O_2 chemistry can be used. If that is not possible special chlorine chemistry must be used to etch TaN_x on Si_3N_4. If dry etching cannot be used, a lift-process must be developed. The dielectric-assisted lift-off process works well with thin layers of TaN down to a 2 μm width. The effective electrical width of narrow resistors may be different from the actual width of narrow resistors, and a correction factor may have to be applied to the photomask. These resistors are compatible with Al, Cu, and Au interconnect systems.

- **Ta–W:** This material gives film characteristics that are similar to TaN but is easier to deposit by cosputtering with good composition control, because there is no need to control gas composition, like nitrogen partial pressure for TaN_x. These films may be used in environments where nichrome may not survive.
- **Cr–SiO:** Films are deposited by DC magnetron sputtering from a composite target or by flash evaporation. Film thicknesses used are in the range of a few hundred to a few thousand angstrom; combined with a composition change, a large range of sheet resistivity can be covered. Sheet resistivity of the

order of hundreds of kilo-ohms per square is possible. TCR values go up with resistivity. Sputtering targets are available commercially for 50 wt% Cr to 70 wt% Cr. Aluminum and gold metallization can be used for contacts.

- **WSiN:** These films can be used for deposition of resistors in the few hundred ohms/sq. range. Films can be deposited by reactive cosputtering from W and Si targets or from a composite target in a N_2:Ar gas mixture. The process can be controlled better in cosputtering, because high-density and high-purity targets are available for W and Si. Resistance range and TCR control is also better. These films are stable at high temperatures, but the current-carrying capability is lower than TaN.
- **Ni-Si:** These films have been prepared by flash evaporation with a range of composition and sheet resistance. TCR values go up (more negative) as Si content goes up. Heat treatment is needed for stabilization.

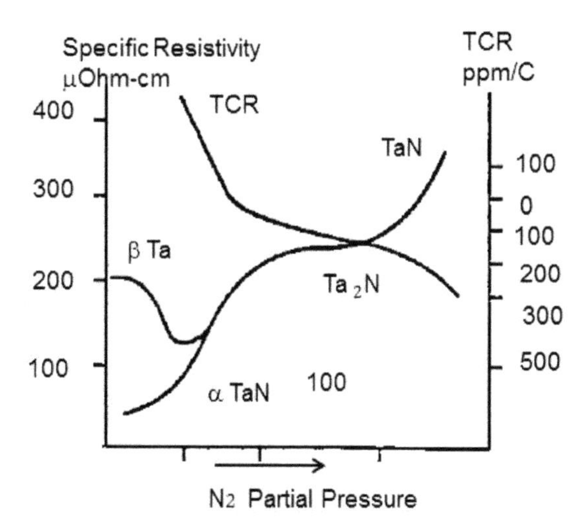

Figure 17.5 Resistivity and temperature coefficient of resistance (TCR) for TaN_x as a function of nitrogen partial pressure during deposition.

Table 17.1 Thin-film resistor materials

Material	Resistivity (ohms-cm)	Ohms/sq.	TCR (ppm/°C)	Ref.
GaAs		100–700	~3000	
NiCr 80%Ni	110 μ	35	100	
NiCr 45%Ni	250 μ	65		
TaN	250 μ	<50		
TaW	200 m		40	
Cr–Si				
Cr–SiO Cermet		10k–500k	>400	[7]
		50–100	<50	[8]
TaSi		20	~100	[9]
$WSiN_x$	1800 m	1500	500	[10]
		300	250	[11]
Ni–P (electroless plated)			5–50	[12]
Ni–Si 20 wt%Si		200	80	[13]
Ni–Si 80 wt%Si			500	

17.2 Capacitors

Capacitors are used in radio frequency (RF) ICs for interstage coupling, tuning, and filtering functions. Tuning capacitors are used as part of inductor-capacitor circuits (LC circuits) for interstate matching. Capacitors are used to isolate DC bias voltages between different amplifier stages and bypass RF voltages to ground. Capacitors fabricated in the same step for different functions must pass the requirements for the most stringent function. Therefore high precision, low loss, and low defect density are needed for these capacitors. Capacitor loss generally goes up with frequency. Capacitors must have a high cutoff frequency and the capacitance value must be independent of the amplitude of the signal. Thin-film capacitors of the metal–insulator–metal (MIM) type are most suitable in this regard. Other types of capacitors are used in silicon circuits, for example, metal–oxide–semiconductor (MOS) devices. In GaAs ICs, Schottky diodes are used as variable capacitors called varactors (discussed in Chapter 2). As with semiconductor resistors, the temperature coefficient of capacitance of varactors is high. Interdigitated lines, stubs in micro-strip transmission lines can act as capacitors. These devices can take up large areas at lower frequencies and are more suitable over 10 GHz. Only MIM capacitors will be discussed in detail here.

17.2.1 MIM Capacitors

MIM capacitors are used in III–V semiconductor circuits with very high capacitance densities, requiring thicknesses down to 500 Å. Figure 17.6a shows a schematic diagram of the cross section of a MIM capacitor. The bottom plate (M1) can be deposited by evaporation or sputtering, as long as the roughness and particle count are low. Silicon nitride is the most commonly used insulator material in the III–V industry. Dielectric layers are deposited by plasma-enhanced chemical vapor deposition (PECVD). Generally a second insulator layer is used on top of the silicon nitride to provide better insulation and planarity to the second metal layer (Fig. 17.6a). Polyimide or other polymer-type films are used for this purpose. This layer is removed from the capacitor area in the via etch step. The top plate can be deposited by any physical vapor deposition (PVD) or plating process. The layout of the capacitor should be such that M2 crossing over the sharp M1 edge is avoided (see Fig. 17.6b). How large the area of the capacitor can be made depends upon the defect levels possible in the process.

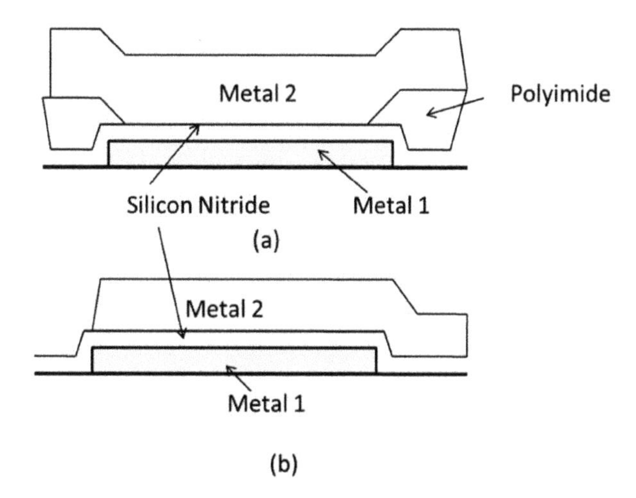

Figure 17.6 Cross section of an MIM capacitor.

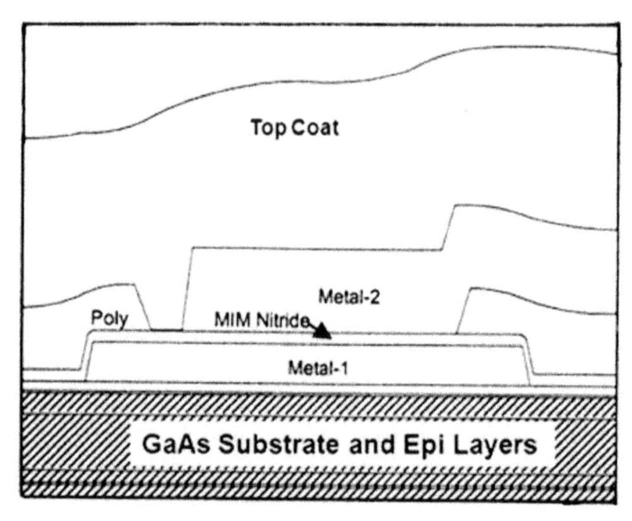

Figure 17.7 Schematic diagram of an MIM capacitor protected by a top coat layer on a GaAs substrate.

Capacitance of a parallel-plate capacitor is given by

$$C = \varepsilon_0 \varepsilon_r . \frac{A}{d} \tag{17.2}$$

where A is the area and d is the thickness of the dielectric. ε_r is the relative dielectric constant of the dielectric material.

MIM capacitors do scale with area for all practical purposes. Fringing fields increase the capacitance at the edges.

$$C = A.C_0 + P.C_f \tag{17.3}$$

where P is the perimeter.

The extra perimeter capacitance, $P.C_f$, can be neglected for large-area capacitors, in practice capacitors with a size over 100 μm. Obviously large and narrow capacitors should be avoided, as these may not scale and may be prone to breakdown.

The quality factor Q is important for MIM capacitors used in high-frequency RF circuits. The total quality factor is a function of losses in the dielectric and the conducting plates:

$$\frac{1}{Q} = \frac{1}{Q_d} + \frac{1}{Q_c} \tag{17.4}$$

where $Q_d = 1/\tan\,\delta$ ($\tan\,\delta$ is the dielectric loss tangent) and $Q_c \approx 1/(\omega R_s C)$ (where R_s is the surface skin resistivity of the conducting plates). The bottom and top plates used in MIM capacitors are generally gold, over 1 μm in thickness. Therefore *Qc* is large, so overall *Q* is dominated by the dielectric quality factor. Thin-film dielectric materials have more losses than bulk materials because it is difficult to achieve bulk properties in thin films. Films should be deposited under conditions that minimize point defects, leading to conduction, and give low defect and pinhole density and nodules or particles.

To increase yield and reduce die size, it is obvious that the MIM capacitor area must be minimized. In a typical MMIC or power amplifier (PA)-type circuit, the MIM capacitors may consume 15% of the die area, but in some cases it may be as high as 30%. The capacitor area can be reduced by increasing the capacitance per unit area, which can be done by either increasing the dielectric constant or reducing the thickness of the layer. Table 17.2 lists a few common dielectric materials used in MIM fabrication. At first sight it appears tempting to utilize the high dielectric constant of tantalum pentoxide (Ta_2O_5) or even the titanates. However, in practice, dielectric losses and long-term reliability make these materials more difficult to integrate into III–V circuits. Ta_2O_5 can be deposited by reactive sputtering with fairly low loss; insertion into production has been delayed by the advances made in deposition of very thin silicon nitride layers [14, 15].

Table 17.2 Properties of some common dielectric materials used in III–V circuit fabrication

Dielectric	Dielectric constant		TCR (ppm)	Formation
	Range	Typical		
SiO_2	4–5	5	50–100	PECVD, sputter deposition
Si_3N_4	5.5–7.5	7.5	25–35	PECVD, sputter deposition
Ta_2O_5	20–25	21	200–400	Sputter deposition, anodize
Al_2O_3	6–11	10	100–500	PECVD, sputter deposition, anodize
Polyimide	3–4.5	3.5	-500	Spin-on and cure
BCB	2.65	2.65		Spin-on and cure
TiO_2	14–110	27		

Source: Reprinted from Ref. [14] with permission. Copyright © 2003 GaAs MANTECH.

17.2.2 Silicon Nitride for MIM

Evaporated gold is used as the bottom plate of the MIM capacitor. Evaporated, sputtered, or plated gold is used as the top plate. Both upper and lower electrodes have thin titanium layers inserted for achieving good adhesion. PECVD is commonly used to deposit silicon nitride films. Parallel-plate systems with batch loading of wafers or automatic loading systems where the deposition takes place in multiple steps are being used in production. The details will be discussed in Chapter 8. The deposition temperature is limited by the temperature that the devices, which are already on the wafer, can tolerate. The typical temperature is 300°C. This low deposition temperature leads to low-density films as well as high hydrogen content. By optimizing film growth conditions, the density of the films (as measured by the wet etch rate) and stress can be optimized, so a very thin layer, down to 500 Å, can be made under production conditions with excellent yield and no long-term breakdown issues. As the thickness is reduced, dielectric quality becomes an issue. At lower thicknesses, below 1000 Å, the properties of the intrinsic film as well as the extrinsic defects become important. The metal bottom plate is deposited by electron beam evaporation under conditions that minimize particles or nodules. High-purity starting materials with very low carbon and other impurities are used. Deposition parameters, power, and ramp rates are optimized for low spitting. Deposition parameters of silicon nitride are optimized for high density, moderate level of stress, and low hydrogen content. The film density can be monitored by measuring the wet etch rate in buffered hydroflouric acid (BHF). Film stress and refractive index are monitored in production. In addition to the usual characterization schemes used in production, Fourier transform infrared spectroscopy (FTIR), ESD, and autoclave testing have been utilized to reduce hydrogen content in these films and ensure minimal moisture ingression [14, 16].

Typical properties of 600 Å thin silicon nitride films are listed in Table 17.3 [14]. The deposition rate of the films is low, and films have compressive stress and low wet etch rate, indicating high density and film quality. Time-dependent dielectric breakdown (TDDB) characteristics of this film, along with thicker films deposited by prior process (less dense film), has been studied [14]. In this method, time to breakdown is measured as a function of voltage, the range being chosen depending upon the thickness. The data is plotted on a semilog scale to project a lifetime at the operating voltage. Figure 17.8 shows lifetime projections for a 600 Å dense film in comparison to films deposited by prior processes for 1000 Å and 2000 Å. A lifetime of over 20 years is projected for this film at an operating voltage of 20 V. These films are also resilient toward moisture absorption. As shown in Fig. 17.9, there is no change in FTIR spectra after the pressure cooker test.

Table 17.3 Typical properties of properly deposited Si_3N_4

Properties	
Deposition rate (Å/sec)	8.84
Refraction index	1.875
Stress (dyne/cm^2)	–2.20E9 (compressive)
Wet etch rate in 10:1 BOE (Å/sec)	13.3
Capacitance density (fF/μm^2)	0.930
Breakdown voltage (V)	70
Breakdown E-field (V/cm)	11.7

Source: Reprinted from Ref. [14] with permission. Copyright © 2003 GaAs MANTECH.

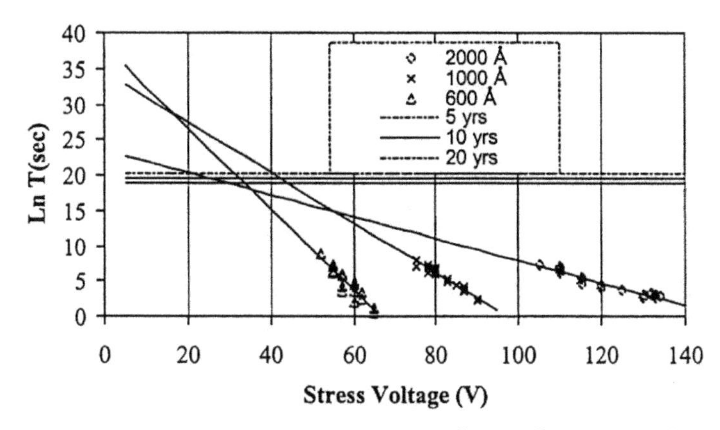

Figure 17.8 TDDB lifetime projection of 600 Å,1000 Å, and 2000 Å Si_3N_4 films deposited at different conditions, measured at various stress voltages at 125°C. Reprinted from Ref. [14] with permission. Copyright © 2003 GaAs MANTECH.

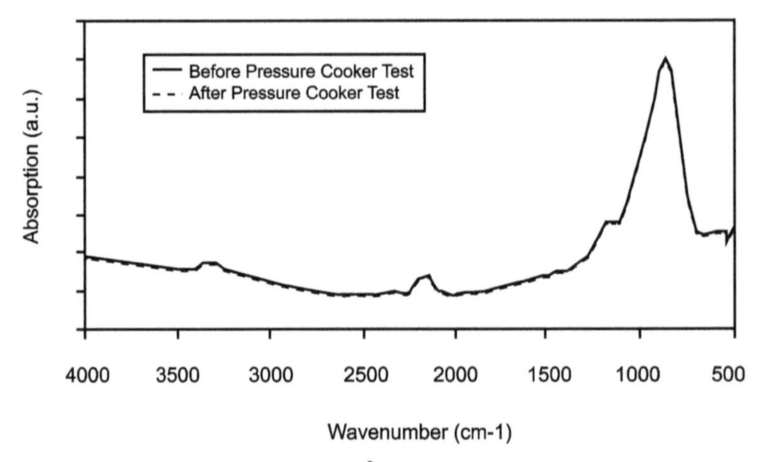

Figure 17.9 FTIR spectra of a 600 Å Si_3N_4 film deposited on a GaAs wafer, showing no degradation or water absorption detected after the pressure cooker test. Reprinted from Ref. [14] with permission. Copyright © 2003 GaAs MANTECH.

Capacitor burnout is determined by a combination of intrinsic and extrinsic defects. The effect of particles and other defects (extrinsic factors) becomes dominant at very low dielectric thicknesses, below 1000 Å. With the recent developments in the deposition of PECVD silicon nitride, the bottom electrode has become the most dominant factor because the process of dielectric deposition has been advanced to the point where pinholes and particles generated in the deposition process are almost nonexistent. The bottom plate is generally deposited by evaporation and patterned by lift-off, so metal nodules and edge defects dominate breakdown. Moisture ingression and mechanical damage are other factors for MIM breakdown that become important.

TDDB characteristics are used often and have been studied in detail in order to evaluate the reliability of MIM capacitors [14, 17, 18]. However, the advantage of a simple ramped breakdown test is that it shows the extrinsic versus intrinsic breakdown. Since the breakdown of MIM capacitors used in MIM circuits occurs at distinct point-like locations, the breakdown cannot be described by distributed dielectric material parameters. Under these conditions, a simple ramped voltage test can be performed [15]. The cumulative distribution of the burnout voltages can be plotted, as shown in Fig 17.10, for wafers with a different particle/nodule count. The area under the curve is small for a sample that approaches intrinsic behavior.

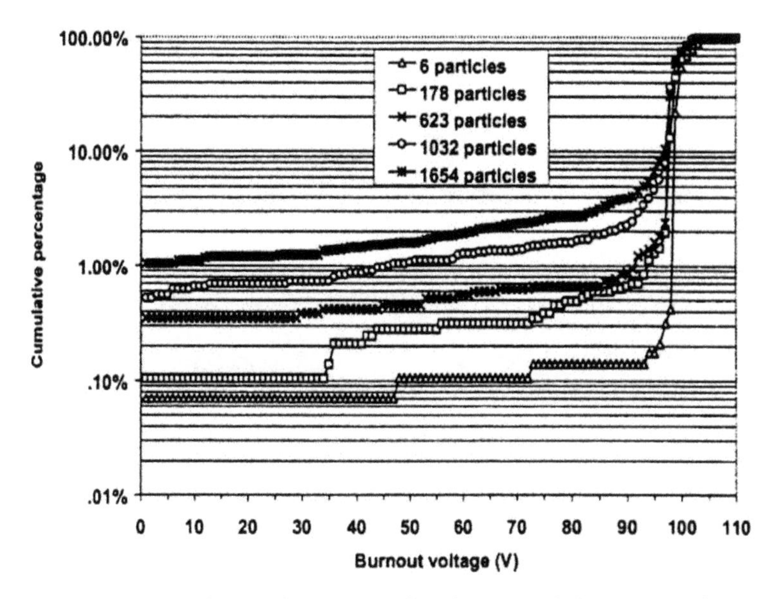

Figure 17.10 Cumulative histogram distribution of burnout voltages measured on wafers with different metal 1 particle counts. Reprinted from Ref. [15] with permission. Copyright © 2003 GaAs MANTECH.

The constant voltage TDDB method [18, 19] is good for characterization of failures due to wear-out, an intrinsic property of the dielectric. In this method, capacitors are stressed at different constant voltage near the film breakdown voltage, until breakdown occurs and the time to breakdown is recorded. The data is plotted on a semilog plot, and the lifetime can be determined by extrapolation [14].

ESD has been shown to be a major cause of failure in ICs. Since MIM capacitors in MMICs may be unprotected, these should be tested for ESD robustness.

As mentioned before, M1 particles and defects dominate the breakdown in MIM capacitors. PECVD nitride gives excellent step coverage over bumps, thereby making it possible to use very thin SiN. However, if the particle is large, breakdown may occur, as shown in Fig. 17.11.

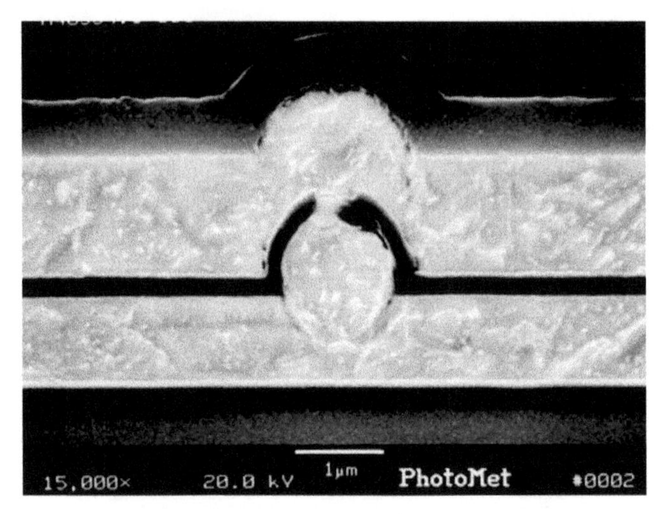

Figure 17.11 Scanning electron micrograph of a shorted region of a capacitor after burnout at low voltage, showing the effect of metal nodules. Courtesy of Skyworks Solutions.

MIM capacitors must be laid out to avoid crossing of the top plate over the sharp edges of the bottom plate, as shown in Fig. 17.6b. Long-term reliability and resistance to moisture have also been related to moisture ingression due to the damage of the overcoat protection nitride during assembly. Moisture can come in through mechanical damage and reach the capacitor edges through the polyimide interlevel dielectric, which acts like a wick for water. If a second poly-overcoat is provided on top of the overcoat protection, this problem of mechanical damage can be avoided.

To improve HAST performance even further, MIM capacitors must be shielded from moisture penetration. This can be done by avoiding contact of polymer-type layers to capacitor edges by use of an additional thin metal layer as the capacitor top plate and a thin silicon nitride dielectric layer, as shown in Fig. 17.12.

17.2.2.1 Stacked capacitors

Capacitors can be stacked to increase the specific capacitance, as long as smooth bottom layers are used and particle and nodule control is exercised. These can be gate layers of FETs or base contact layers in HBT circuits or an additional interconnect layer (M0). Passivation layers may be used as the dielectric for the bottom capacitor.

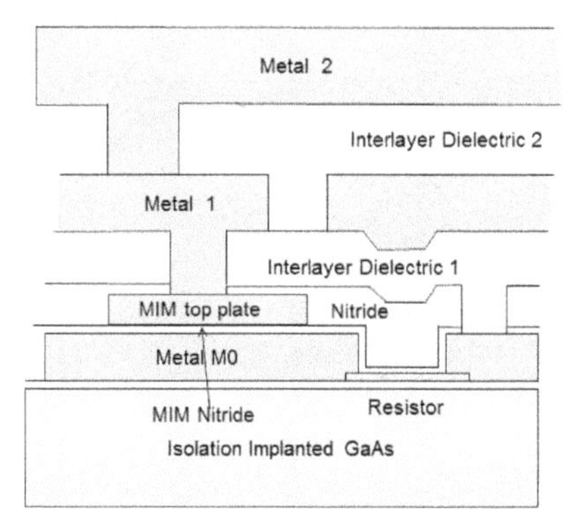

Figure 17.12 MIMC capacitor schematic showing additional metal and dielectric layers to improve HAST performance by keeping the moisture-absorbing layers away from the MIM edge at the cost of added processing (adapted from Fig. 15.13).

17.3 Inductors [20–22]

Inductors are needed in MMICs for tuning purposes. In the frequency range below 10 GHz, lumped inductors are designed to be on-chip. At higher frequencies, distributed inductors may be used in the form of strip or transmission lines.

Metal lines of any length have considerable self-inductance, and at high frequencies mutual inductance between segments increases because of electromagnetic coupling. Since all metal lines have a capacitance to ground, inductor layouts are designed to increase inductance and fabrication focuses on achieving a high Q factor. Spiral inductors that use thick metal layers are most commonly used and can be modeled by designers. A typical layout of spiral inductors is shown in Fig. 17.13. In Fig. 17.13a, metal 1 is used to contact the center of the inductor and is isolated from metal 2 by the interlevel dielectric (nitride plus polyimide). In Fig.17.13b, the spiral loops of metal 1 and metal 2 are added in series to increase the inductance. The two-level inductor is connected in such a way that the current flows in the same direction in both traces that overlap, so the magnetic flux lines add in phase to result in high mutual inductance. However, if a high Q value is desired, the two metals can be overlaid on each other for most of the length (except where the two metals have to cross for the connection to the center of the spiral) (see Figs. 17.13 and 17.14). The center point can also be connected by an air

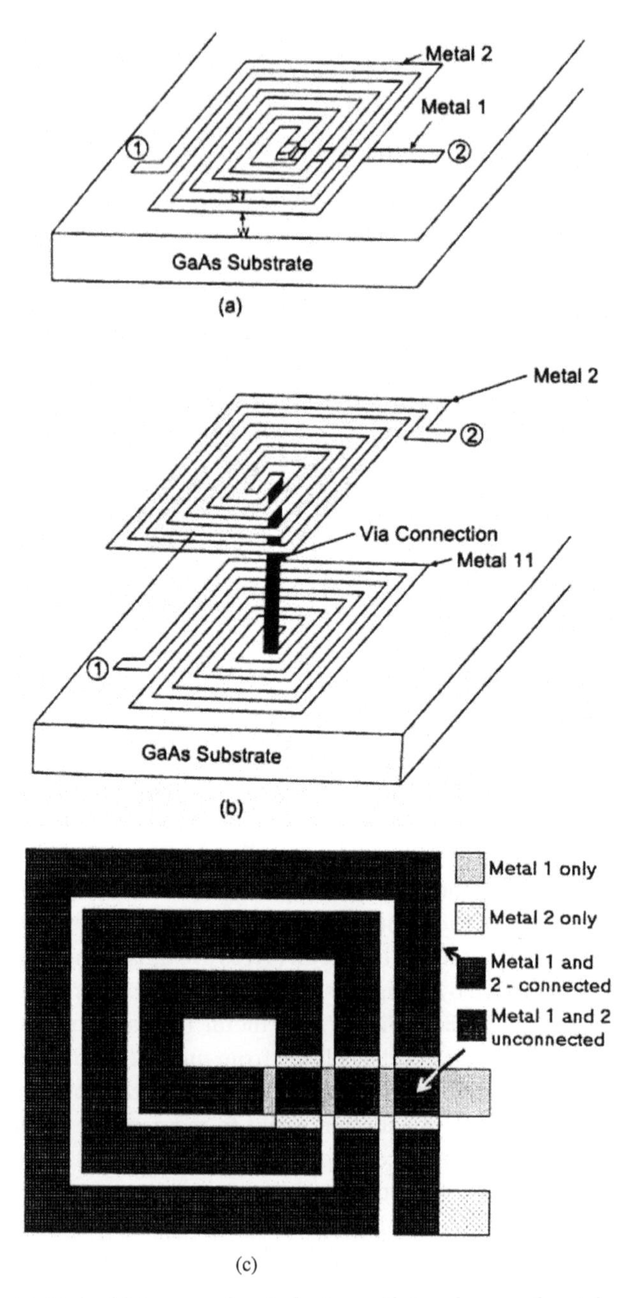

(c)

Figure 17.13 (a , b) Layout of an inductor with two layers of metal overlaid on each other. (c) Typical multimetal-layer inductor layouts used in III–V layouts.

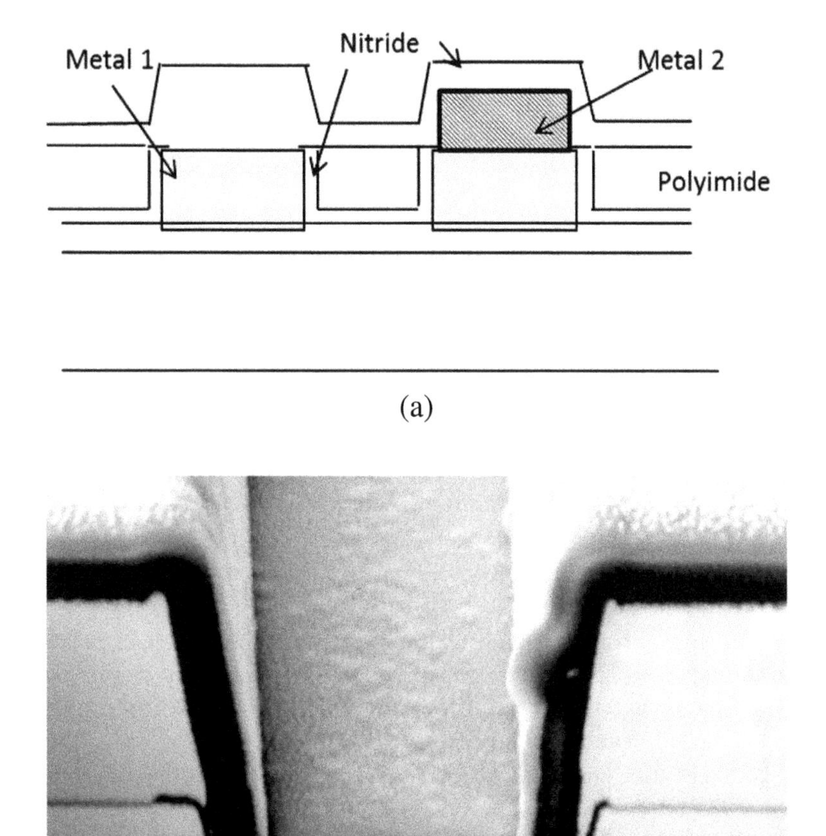

(a)

(b)

Figure 17.14 Cross-sectional schematic of an inductor with two layers of metal overlaid on each other (a). SEM cross section of a two-metal layer inductor (b).

bridge from above or a through-wafer via (TWV) connection through the wafer. Inductors should be placed over a semi-insulating GaAs or ion-implanted isolated epigrown layer. Values of 0.1 nH to a few nH can be realized for loop inductors. For achieving high current-carrying capability and a higher Q factor, two to three metal layers can be overlaid on top of each other. High topology must be avoided in these cases by using proper planarizing schemes so that sharp vertical geometries are avoided. Poor nitride coverage over these can lead to poor HAST performance.

References

1. R. Williams, *Modern GaAs Processing Methods*, Artech House, Boston (1990).
2. A. B. Glasser and G. E. Subak-Sharpe, *Integrated Circuit Engineering*, Addison-Wesley (1979).
3. J. D. Madden, Tantalum-tungsten thin resistor film as an all-environment complement to nichrome film, *Hybrid Circuit Technol.*, p. 27 (1985).
4. K. L. Coates, Multichip modules and high density packaging, *Int. Conf. Multichip Modules*, 490 (1998).
5. H. B. Nie et al., Structural and electrical properties of tantalum nitride thin films fabricated by using reactive radio frequency magnetron sputtering, *Appl. Phys. A: Mater. Sci. Proc.*, **73**, 229 (2001), http://arxiv.org/pdf/cond-mat/0305683.
6. T. Lee et al., Characterization and reliability of TaN thin film resistors, *42nd Int. Reliability Phys. Symp.*, **502** (2004).
7. J. Songand and E. R. Fossum, *IEEE Trans. Electron. Devices*, **36**, 1575 (1989).
8. F. Wu and J. Morris, The dependence of the resistivity of on- chip SiOxCr1-x thin film resistors on process parameters, *Proc. 9th Int. Symp. Adv. Packaging Mater.: Proc., Properties Interfaces*, **84** (2004).
9. A. Trigg et al., Thin film resistors and capacitors for multichip modules, *Electron. Comp. Technol. Conf.*, **228** (1998).
10. H. Sorimachi and T. Hosoya, Properties of W-Si-N thin film resistors, *IEEE ECC Conf.*, **596** (1988).
11. S. Tiku, Rockwell International data.
12. P. Chahal and R. Tummala, A novel integrated decoupling capacitor for MCM-L technology, *Electron. Comp. Technol. Conf.* (1998).
13. J. Kodama , *J. Electron. Mater.*, **24**, 1997 (1995).
14. J. Yota et al., *Development and Characterization of a 600 Å PECVD Si_3N_4 High-Density MIM Capacitor for InGaP/GaAs HBT Applications,*

GaAs MANTECH, gaasmantech.org/Digests/2003/2003PDF/4-6.pdf (2003).

15. M. Rao, S. O'Neil and S. Tiku, *Metal Particle Effects on Thin Film Capacitors in High Volume Manufacturing*, GaAs MANTECH, gaasmantech.org/Digests/2003/2003PDF/4-5.pdf (2003).
16. Y. C. Chou et al., On the development of high density nitrides for MIMCs, GaAs MANTECH (2003).
17. J. Scarpula et al., A TDDB model of Si_3N_4 based capacitors in GaAs MMICs, *Proc. 37th Int. Reliability Phys. Symp.*, 128 (1999).
18. B. Yates, *IEEE Trans. Electron. Devices*, **45**, 939 (1998).
19. M. Brophy, *MIM's the Word*, GaAs MANTECH (2003).
20. R. A. Pucel, *IEEE Trans. Microwave Theory Technol.*, **29**, 513 (1981).
21. B. Piernas et al., Improved three dimensional GaAs inductors, *IEEE Microwave Symp. Digest, MIT-S Int.*, **1**, 189 (2001).
22. I. J. Bahl, *IEEE Microwave Mag.*, **1**, 64 (2000).

Chapter 18

Interconnect Technology

18.1 Introduction

Interconnect technology for silicon very-large-scale integration (VLSI) has grown in complexity due to the demands of the numerous products that require higher and higher functional density. The requirements for interconnect densities have remained very modest for III–V compound circuits because of the lack of demand for high-density digital products. Interconnect development has been driven by the mixed-signal industry. Apart from the gate layer, there has been no need for submicron line/space geometries, and contact via sizes needed are also about 1 micron or a bit smaller in size. In future, if and when submicron III–V devices are integrated at high levels, for example, for devices grown on large silicon wafers or by demands of the optoelectronics industry, interconnect technologies developed for silicon will converge with the III–V process technology.

The drive toward ultra-large-scale integration (ULSI) in the silicon world has led to scaling of interconnects so that circuit speed is not hampered. To minimize interconnect signal delay, reduction in line resistance and reduction of both line length and capacitance are needed; this has resulted in the use of upto nine layers of metal and use of low-K dielectrics [1]. In III–V circuits, one to three layers of metal suffice and only rarely over three layers are used. Emphasis has been placed on high current-carrying capability, low capacitance,

III–V Integrated Circuit Fabrication Technology
Shiban Tiku and Dhrubes Biswas

ISBN 978-981-4669-30-6 (Hardcover), 978-981-4669-31-3 (eBook)
www.panstanford.com

and use of backside metallization, all optimized for electrical and thermal management. GaAs and other III–V circuits need special contacts and Schottky gate materials that are generally gold based (discussed in Chapters 11 and 12). Resistors, metal–insulator–metal (MIM) capacitors, and inductors are used extensively in radio frequency (RF) circuits. Interconnect layers are used to contact all these together. Barrier metals are used to ensure that the top-level metals do not migrate into the devices during processing or during use.

18.2 Interconnect Requirements

18.2.1 Electrical Requirements

The most commonly used metals for electronic circuits are aluminum, gold, and copper. Table 18.1 lists the resistivities of a few common candidate metals. The actual sheet resistivities of films are higher due to the poor texture of evaporated or sputtered films. Although silver has the lowest resistivity it is not used in electronic integrated circuits (ICs) because it is prone to oxidation due to its reactive nature and it creates deep trap levels in the energy bandgap of silicon. Among the nonprecious metals, aluminum and copper, aluminum is preferred because of its use for contacts in silicon and technical challenges presented by copper, lack of etch chemistries, corrosion, etc. Copper is a rapid diffuser in GaAs and as a dopant acts as a trap for electrons and is well known to kill device performance at even parts-per-million (ppm) levels. This basically left a choice of two, gold and aluminum, for consideration in early GaAs circuits. Aluminum was considered widely for digital circuits but has fallen out of favor because of the dominance of analog circuits. Aluminum has the advantages of low cost, availability of process tools, and good adhesion to most other materials. It does need a barrier before it can be put on top of ohmic contacts because of the interreaction of gold and aluminum causing "purple plague" (formation of a brittle intermetallic alloy).

Gold is the most commonly used metallization for the front side as well as the backside for GaAs and other III–V semiconductor circuits. The main advantage is the high current density that can be supported by it due to its electromigration resistance. Other advantages are good bondability and corrosion resistance. High electromigration-limited current densities are needed for high-frequency circuits because skin depth at high frequencies is small. As current increases, current is confined to the outer layer, the "skin" of the conductor. Skin depth is the depth at which current falls to 1/e of the surface value. If the interconnect layer is much thicker than skin depth (0.8 μm for gold at 10 GHz), current is confined to the skin and the extra thickness does not contribute much. The effective current density is much higher than the calculated value. Therefore, the electromigration limit for the metal layer must be high. Electromigration current limits are listed in Table 18.2 for a few common metal systems.

Table 18.1 Sheet resistivity of metal films

Metal	Bulk resistivity (μ ohm-cm)	Film sheet resistivity (measured) (ohms/sq. for 1000 Å)
Al	2.8	0.33
Cu	1.7	0.2
Au	2.4	0.27
Ti	55	10
Pd	11	1.33
Pt	10	1.0
Ag	1.59	0.22

18.2.2 Adhesion and Barrier Requirements

Gold is well known to have poor adhesion because of its noble nature. Therefore an adhesion layer must be provided. Titanium and chromium have been used because of their tendency to bond well to oxide surfaces. Although two component systems solve the problem of adhesion, there is still an issue of gold migration into the GaAs during high-temperature process steps. And titanium itself spikes into GaAs if the amount of titanium is not limited. The intermixing creates reliability problems in the device. Platinum, palladium, and TiW have been used to provide a barrier. A few hundred angstroms of Pt are adequate for processing temperatures up to 400°C and operating temperatures for which the junction temperature is up to 250°C. Sputtered TiW is particularly suited to the use of an aluminum interconnect over gold-based contacts. Sputtered alloys like TiW and Ni–V work as adhesion promoters as well as diffusion barriers.

18.2.3 Diffusion and Electromigration Effects [2, 3]

Interdiffusion of metals and semiconductors is a primary factor in determining the choice of the metal interconnect system. For analog circuits like power amplifiers, electromigration is equally important. Electromigration is the movement of the conductor metal due to high electric current. In a multilayer interconnect metal, or a single-layer metal that is not planar, metal lines go over steps and may have narrow portions, thin regions, and seams. Under high-current conditions these regions have a higher current density and metal ions are pushed, causing void formation. This metal transport is proportional to electron momentum and flux. Figure 18.1 shows a schematic diagram of the forces on a metal ion. The force F1 due to the electric field is small under these conditions compared to F2, the force due to electron wind. Movement due to ionic conduction under electric field is negligible. The electromigration rate is proportional to power, and the equation for the rate has the usual exponential form [2] (see also Chapter 10).

$$\text{Electromigration rate, } R \approx J^2 \exp\left(-\frac{E_\mathrm{d}}{kT}\right) \tag{18.1}$$

where J is the current density and E_d is the activation energy for self-diffusion and is low for low-melting-point materials like aluminum. E_d is 0.5 eV for Al and 1 eV for gold—hence the higher electromigration limit for gold (see Table 18.2). Addition of alloying elements reduces electromigration. Addition of refractory metal interlayers can improve current-carrying capacity even more.

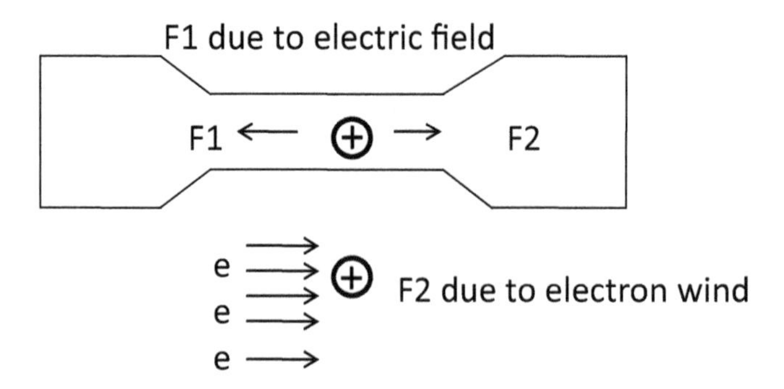

Figure 18.1 Electromigration forces on a metal ion.

Table 18.2 Electromigration limits of common interconnects

Material /Layer	Current limit (A/cm^2)	Ref.
Al	1×10^5	[4]
Al:Cu	2×10^5	
Al/Ti layered	2×10^5	
Au	3×10^5	[5]
	6×10^5	[4]

18.2.4 Interlevel Dielectric Layer Requirements

Silicon nitride is used extensively in GaAs and other III–V semiconductor fabrication for passivation and as an interlevel insulator. In analog and mixed-signal circuits, MIM capacitors are needed and silicon nitride is a good candidate for these. Silicon nitride deposited by plasma-enhanced chemical vapor deposition (PECVD) has a moderately high dielectric constant (7.0) and has high breakdown strength and moisture resistance. It also has good step coverage and can provide protection against a highly accelerated stress test (HAST)-type environment where moisture could penetrate. It is used as a device passivation layer anyway, so it is a good candidate for MIM layers and for stacked capacitors. These layers are deposited at around 300°C; temperatures over this are not desirable for III–V epilayer structures, contact metallurgies, etc. As an interlayer dielectric, it is not a good candidate because of a high dielectric constant and a lack of planarizing or gap fill properties. Polyimide-type materials have a dielectric constant near 3.0 and have excellent planarizing properties. Gaps and crevices created in the thick metal processes are also easily filled by these. Benzocyclobutane (BCB) and polybenzoxazole (PBO) are other polymers used for this purpose.

18.3 Production Interconnect Processes

18.3.1 Baseline Gold Interconnect Process

A general-purpose process for mixed-signal circuits may need two or three layers of metals, with thickness of 1 to 2 microns or more.

These are generally separated by an interlayer of silicon nitride and polyimide-type dielectric materials. Process steps for the interconnect portion of the process are described next.

After the contact layers are formed in metal–semiconductor field-effect transistor (MESFET) or heterojunction bipolar transistor (HBT) processes, thin-film resistors are deposited. At this point the first interconnect metal is deposited to connect active devices and the thin-film resistors. This metal contacts the ohmic or Schottky layers, which are typically thin. An example of M1 contacts to these layers is shown in Fig. 18.2. The importance of having good barrier properties is obvious for maintaining good contact resistance and reliability. This metal also forms the bottom plate of MIM capacitors, so care is taken in minimizing particles (to be discussed in detail under metal deposition) and other lift-off or process defects. For gold layers, an electron beam–evaporated film patterned by lift-off is common. MIM silicon nitride is deposited next. This is followed by via contact photolithography. Silicon nitride is etched using the dry etching process. Since this dielectric layer forms the capacitor it is typically very thin; therefore a second layer of thicker insulator is needed. To minimize the interlayer capacitance and planarize the topology, polymer-type dielectrics are used. Polyimide layers are spun on using track type systems and cured using ovens at around 300°C. Via photopatterning is done next and contact vias and MIM capacitor areas are etched off by oxygen plasma. Polyimide vias should have sloping side walls so that the next layer of metal can go over the steps without breaks or seams. Figure 18.3 shows a schematic diagram and cross section of a typical via. Planarization by use of polyimide is shown in Fig. 18.3c. The wafers are patterned for second-level metal (M2). For lift-off-patterned, evaporated gold-based metallization, an image reverse photoprocess may be used or a negative lift-off resist may be patterned directly. M2 layers are 1.5 to 2 μm thick, although thicker layers are sometimes used. Adhesion of metals to polyimide can be poor if proper surface preparation is not performed before metal deposition. This may involve plasma treatment to roughen the surface chemical treatment to modify the surface chemically, for example, by NH_4- containing photoresist developers or KOH. After lift-off of the patterned metal the wafers are covered by a silicon nitride top coat or final passivation. This is a thick layer that must give good step coverage over thick metal steps and cover any topology created by the polyimide and M2 (or M3) process. Dense, stress-free nitride with good step coverage is desired for avoiding any HAST (or other high-temperature and moisture) failure issues. Bond pads and the street photostep follow. Dry etching is used to etch silicon nitride. If a layer of another polymer is needed on top for prevention of mechanical damage that can be deposited and patterned next. Photodefinable PBO dielectric material can be used for this purpose [1]. Shin-Etsu has optimized a family of photosensitive siloxane materials known as SINR photoresists with a low cure temperature in the range of 220°C. These materials can be used for interlevel dielectrics as well as passivation. This material meets all properties expected of photoresist and is an excellent planarizing material and is ideal for maintaining a low thermal budget.

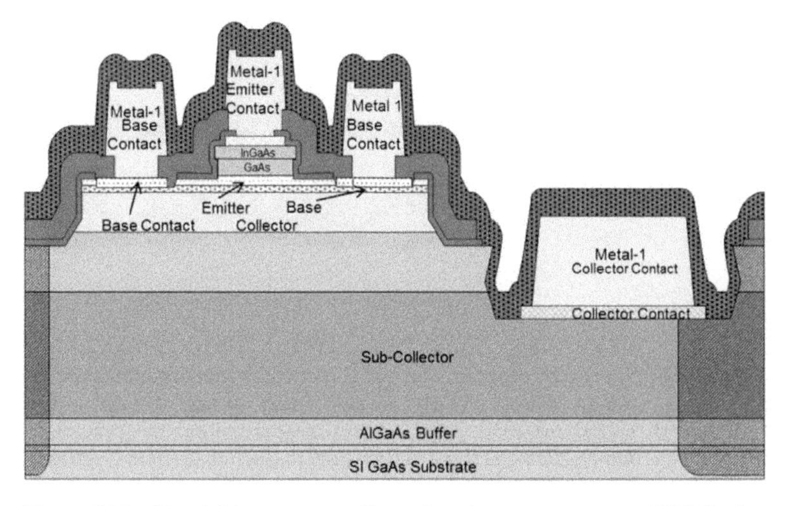

Figure 18.2 Metal 1 interconnect lines showing contacts to an HBT device. Reprinted from Ref. [6] with permission. Copyright © 2010 CS MANTECH.

The wafers are sent for process control monitor (PCM) testing at this point and any backside processing of vias and electroplating is done later. Figure 18.4 shows a schematic crosssection of an interconnect for a two-layer metal interconnect process used by Triquint (now Qorvo) (this example is for a pseudomorphic high-electron-mobility transistor [PHEMT] circuit).

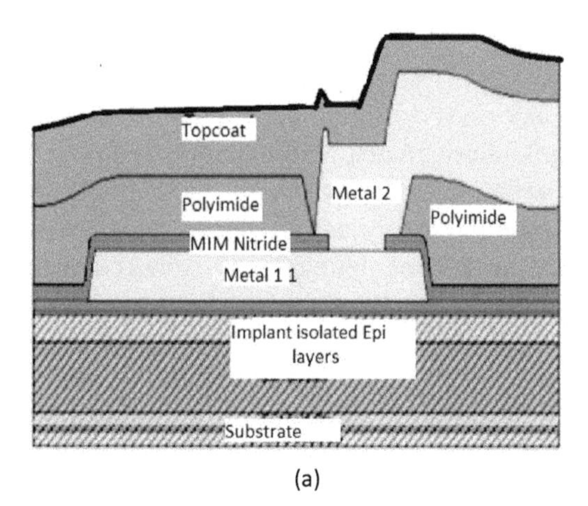

(a)

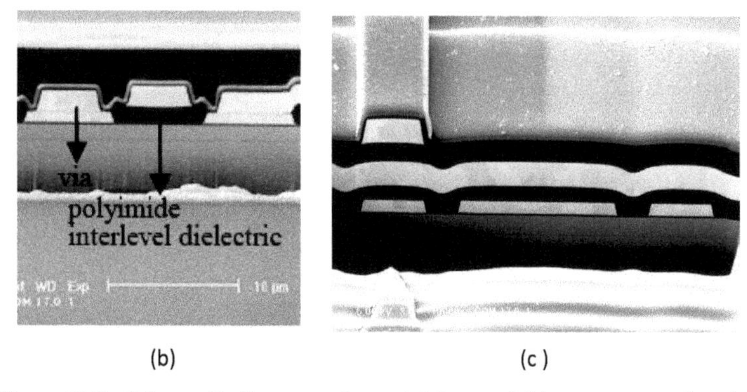

(b) (c)

Figure 18.3 Schematic diagram of a metal 2–metal 1 interconnect showing (a) an M1–M2 via, (b) an SEM cross section of a three-level metal interconnect, and (c) planarization by polyimide. Reprinted from Ref. [7] with permission. Copyright © 2008 CS MANTECH.

18.3.2 Plated Metal Interconnect Process

Two levels of thick plated metal have been used in monolithic microwave integrated circuit (MMIC) fabrication for high-power amplifiers [8]. Evaporation of thick gold layers is not economically feasible and plated gold processes have been developed for circuits that can benefit from thick layers. Low-loss transmission lines, low resistance, high-current power lines, and low-inductance spiral inductors can be made by plated thick gold process. Thick photoresist patterns are needed. Polyimide or BCB is used as the interlayer insulato. A very thick layer of photoresist (6 μm) is used to pattern vias in the polyimide. A second layer of plated gold, 4.5 μm thick, is deposited on top, again using a thick layer of resist for the patterned plating. This process has been used for high-power, 10 W power amplifier fabrication.

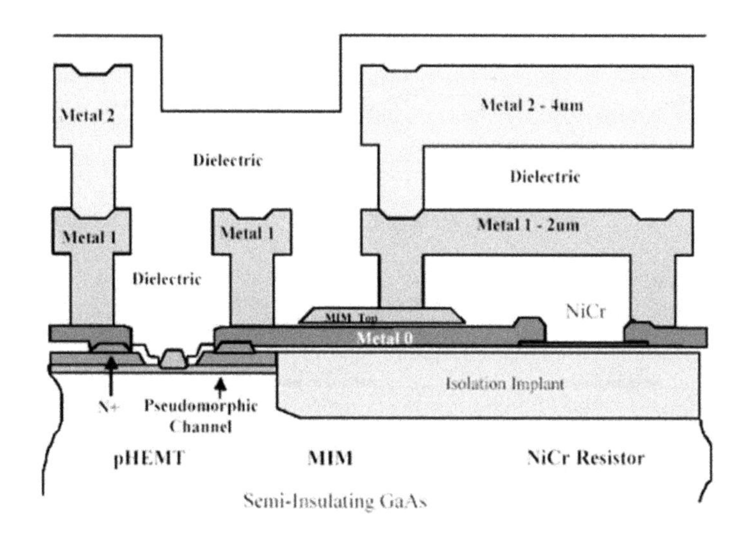

Figure 18.4 Triquint (now Qorvo) schematic for showing a PHEMT, a MIM capacitor, and nichrome resistors interconnected by a three-metal process. Reprinted from Ref. [21] with permission. Copyright © 2004 CS MANTECH.

18.3.3 Air Bridge Process

Long before low-K dielectric materials were available for use at gigahertz frequencies, RF engineers had figured out ways to replace interlayer dielectrics by air, the ultimate low-dielectric-constant material. Air bridge processes thus circumvent the problems of developing low-K, thick metal interconnects. In a typical process sequence [9], first a post photolithography pattern is used to form posts on top of the device contacts (see Fig. 18.5). After the resist is baked at a moderate temperature, a thin layer of seed layer is deposited by sputtering metal, for example, TiW/Au. A second photolithography step is then used to pattern the interconnect layer. Gold metal is then plated on the wafer. After this step, the resist and seed layers are removed, the latter by partial deplating, TiW etch, etc. Finally, a combination of solvent and oxygen plasma processing is used to remove the post photoresist from under the metal, leaving air bridges behind. These circuits are mechanically fragile, may have low yield compared to low-K dielectrics, but are reliable. MMIC circuits with a thick top plate as the ground plane are an alternative to circuits built with wafer thinning, through-wafer vias (TWVs), and thick backside plating.

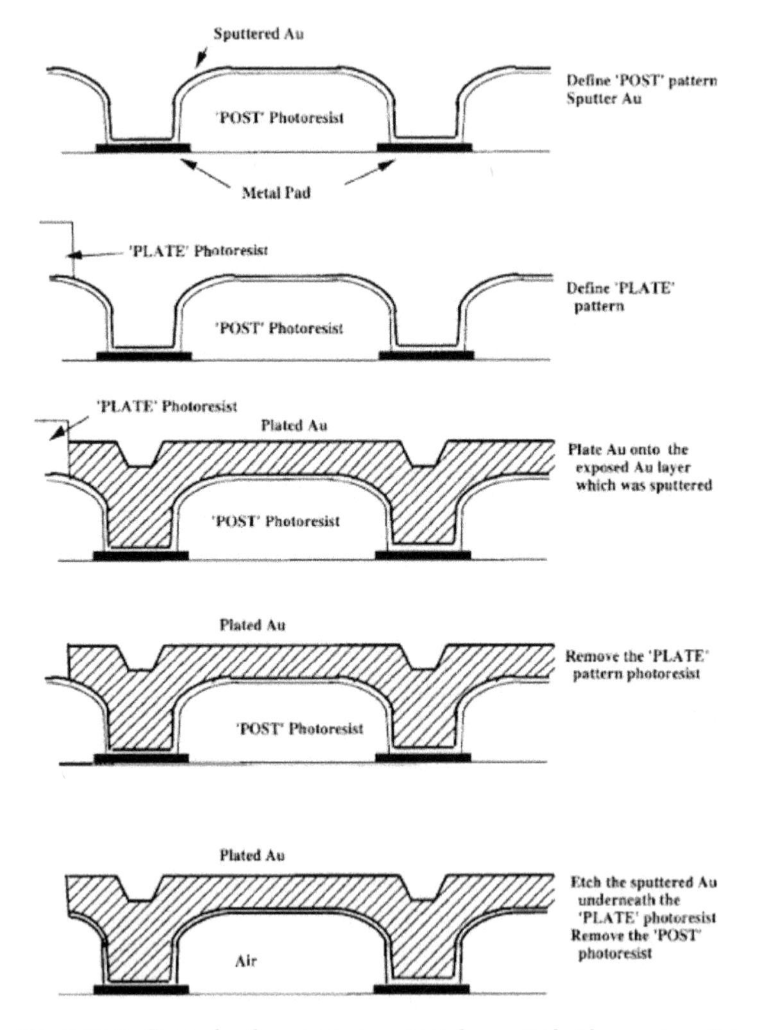

Figure 18.5 Example of a process sequence for an air-bridge process. From Ref. [9]. Copyright © 1998. Reproduced with permission of John Wiley & Sons, Inc.

18.3.4 Digital GaAs Interconnect Process

The choice of interconnect metals and insulators is critical to the performance and cost of GaAs ICs for use in analog or digital circuits. Since the circuit requirements of digital circuits are different, and are driven mostly by high levels of integration, it is advantageous to use the same interconnect technology as the one used in well-developed silicon VLSI. To adapt aluminum-based interconnect technology for GaAs, where gold-based device-level contacts are used, barrier layers must be used to prevent interdiffusion. Multilevel aluminum metallization and SiO_2 interlevel dielectric processes have been used for commercial GaAs digital VLSI [10]. The metallization is sputter-deposited and dry-etched using standard silicon fabrication tools. A schematic diagram of the metallization system used by Vittesse and others is shown in Fig. 18.6. All the advances made in silicon VLSI interconnects can be applied to this interconnect system, for example, spin-on glass for planarization. Lower-level metals, layers 1 and 2, can be used for signal routing. Higher-level metals, layers 3 and 4, can be used for high-current power buses. Planarization with polyimide-type films is utilized at higher layers.

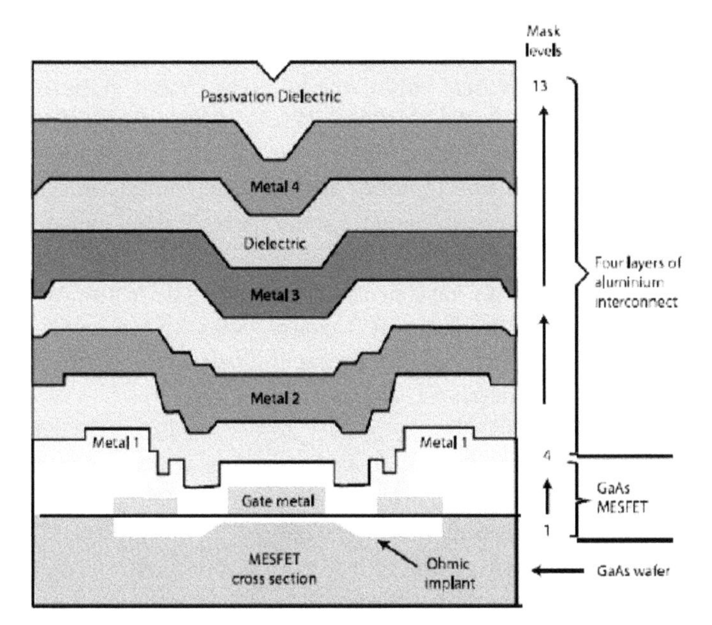

Figure 18.6 Digital interconnect process showing four layers of aluminum (Vittesse Semiconductor). © [1991] IEEE. Reprinted, with permission, from Ref. [10].

18.4 Future of Interconnect Technology

Interconnect development in silicon VLSI has been driven by shrinkage of device size and interconnect line width. The III–V interconnect technologies used to date are there to fill the needs of the analog market. The number of metal layers is going up for mixed-signal circuits and also the need for thicker metal for power and thermal management. Polyimide-type processes have kept up with the demands of planarization, so there have been no developments in via plug-type processes. If and when low-bandgap semiconductor materials find use in ultrafast circuits in highly complex digital circuits, ways to utilize the damascene processes that depend upon via plugs (W, TaN) and chemical mechanical polish (CMP) will be found [1]. This will be particularly true for copper metallization.

18.4.1 Copper Interconnects

Copper is being used for backside plating (will be discussed in Chapter 19), and some early reports have been published for use on the front side. As the price of precious metals has gone up, the interest in copper has, too. Copper has the lowest resistivity (apart from silver) of potential candidates for VLSI wiring. With the use of copper, both pitch and thickness can be reduced, thus reducing capacitance and resistance and hence signal delay. Copper has been used in the silicon industry since IBM announced success by the use of a tantalum diffusion barrier [11]. Major challenges to implementation of copper have been solved since the nineties. Many deposition techniques have been tried, but electroplating was the winner because of the fact that it works well with the damascene-type patterning process. The problem of avoiding diffusion by providing barrier metals has also been resolved.

Copper is a rapid diffuser in GaAs as it is in silicon. Tungsten or TaN plugs have to be provided to act as barriers. To avoid dry etching of copper, damascene-type processes are used in the silicon industry. In damascene patterning (see, for example, Ref. [12]), the via and interconnect lines are etched first in the dielectric material and then inlaid with a barrier metal deposited by PECVD and electroplated metal. CMP is used to planarize the structure and remove the metal in between. In silicon technology, magnetron sputtering, CVD, and atomic layer deposition (ALD) may be used for deposition; in III–V these may not be convenient because of incompatibility to dry etching chemicals.

Simpler copper metal processes for the front side, using evaporated barrier and seed layers [13–15], may be commercialized for analog and other wireless circuits before the need for this damascene process arises. These processes are more compatible with lift-off patterning. Evaporated diffusion barriers are also preferred because these are already in use in III–V fabrication. Using titanium as the adhesion layer, a layer of 600 Å of platinum can be used for a barrier. This thickness can withstand the 300°C cure process normally used for polyimide-type interlayer dielectric processes. However, Pt and Cu do form intermetallics that have high resistance. Therefore, use of refractory barriers may be warranted. For preventing corrosion of copper, a layer of organic protectant or thin silicon nitride may be used on top. If the copper layer is the very top layer, used for wire bonding, then a layer of gold may have to used on top, again needing barriers in between.

Convergence of silicon and III–V interconnect technologies will continue at the present pace until silicon circuits are replaced by a new III–V low-bandgap material system, preferably grown on large silicon wafers. Integration of III–V optoelectronic devices with silicon will accelerate this pace even further.

References

1. K. Buchanan, *The Evolution of Interconnect Technology for Silicon Integrated Circuitry*, GaAs MANTECH (2002).
2. S. K. Ghandhi, *VLSI Fabrication Principles, Silicon and Gallium Arsenide*, Wiley Interscience, New York (1983).
3. F. d'Heurle, Electromigration and failure in electronics: an introduction, *Proc. IEEE* **59**, 1409 (1971).
4. R. Williams, *Modern GaAs Processing Methods*, Artech House (1990).
5. S. Tiku, Rockwell International, unpublished data.
6. S. Tiku, *Device Processing in III-V Manufacturing*, CS MANTECH (2012).
7. J. Yota, Hoa Ly, D. Barone, M. Sun, and R. Ramanathan, *Photodefinable Polybenzoxazole Interlevel Dielectric for GaAs HBT Applications*, CS MANTECH (2008).

8. J. W. L. Dilley and S. K. Hall, *A Manufacturable Multilevel Interconnect Process Using Two Layers of 4.5 μm Thick Gold*, MANTECH (2000).
9. W. Liu, *Handbook of III-V Heterojunction Bipolar Transistors*, Wiley Interscience, New York (1998).
10. J. Mickelson, GaAs digital VLSI device and circuit technology, *IEDM*, **231** (1991).
11. K. Holloway and P. M. Fryer, *Appl. Phys. Lett.*, **57**, 1736 (1990).
12. B. Zhao et al., Dual damascene interconnect of copper and low permittivity dielectric for high performance integrated circuits, *Electrochem. Solid State Lett.*, **1**, 276 (1998).
13. S. W. Chang et al., A gold free fully copper metalized InGaP/GaAs HBT, in *12th GaAs Symposium*, Amsterdam (2004).
14. Y. C. Wu et al., Copper metallization for pHEMT switches, *IEEE Microwave Wireless Compon. Lett.*, **17**, 133 (2007).
15. K. Cheng, *Copper Interconnect on GaAs pHEMT by Evaporation Process*, CS MANTECH (2009).
16. R. N. Hall et al., Electromigration reliability studies of intermetallic contacts having CVD tungsten via plugs, *Solid State Technol.* (1982).
17. I. Deyhimy, *IEEE Spect.*, 33 (1995).
18. D. Barone et al., *Development and Characterization of Photo-Definable Polybenzooxazole Buffer*, GaAs MANTECH (2008).
19. *CRC Handbook of Chemistry and Physics*, Weast, Boca Raton, Florida (1984).
20. A. B. Glasser and G. E. Subak-Sharpe, *Integrated Circuit Engineering*, Addison-Wesley (1979).
21. W. A. Wohlmuth et al., *A 0.5mm InGaP Etch Stop Power pHEMT Process 21 Utilizing Multi-Level High Density Interconnect*, CS MANTECH (2004).

Chapter 19

Backend Processing and Through-Wafer Vias

Section I: Through-Wafer Via Process

19.1 Introduction

Backside processing is an integral part of III–V wafer processing. In its simplest form, it may involve only thinning of wafers, and in most applications it can be very complex, requiring thinning, through-wafer vias (TWVs), backside metallization, etc. In analog and mixed-signal circuit fabrication, large power densities are needed, and thinner die and good electrical ground are essential for thermal and power management. Thin die allow reduction of wire bonds and reduce thermal resistance, even without the use of TWVs. A TWV is a connection from devices on the front side of the wafer to a metallized backside. TWVs reduce both electrical resistance and inductance to ground and allow smaller die sizes. Almost all GaAs analog devices such as power amplifiers and monolithic microwave integrated circuits (MMICs) being manufactured currently require the use of TWVs so as to obtain high power efficiency and good high-frequency performance.

Backside processes tend to be very dirty compared to front-side processes, are labor intensive, and historically had lower yields. GaAs, InP, and SiC substrates are all very expensive, but after front-

III–V Integrated Circuit Fabrication Technology
Shiban Tiku and Dhrubes Biswas

ISBN 978-981-4669-30-6 (Hardcover), 978-981-4669-31-3 (eBook)
www.panstanford.com

side processing is complete, they are even more expensive due to the value addition of the processed device. So, the yield of the backside processing must be high. As the demand for GaAs circuits has grown, so has the need for high-yield backside processing. Consequently a lot of process development effort has gone into improvements for automation, throughput increase, and cost reduction. The process steps used in TWV processing, bonding, wafer thinning, via etching, metallization, and debonding will be discussed in this chapter. The process of die separation by scribing will also be discussed.

19.2 Wafer Bonding

GaAs wafers are thinned to a thickness of 100 μm (4 mils) or less for MMIC circuits. Wafers are typically mounted to a sapphire substrate for support during backside processing. Thinning of wafers and processing of thinned wafers are nearly impossible without some support. Use of a temporary wafer-bonding process allows use of carriers that can be used (and reused) for providing needed support. The choice of substrates for this purpose is limited because the substrate material must have a good thermal expansion match to GaAs and be available in a large size. Other substrate materials, like quartz, borosilicate glasses, and silicon wafers, are used for backside processing. If backside photolithography is needed, the support wafer must be transparent to infrared (IR) radiation because the backside of the circuit needs to be aligned to the front side through the substrate that is transparent to IR. Other substrates, whose composition can be adjusted to match the thermal expansion to that of GaAs have been tried. However, substrate and wafer cracking are still an issue. Sapphire substrates can be used hundreds of times and can even be refurbished for a second life. Therefore expensive sapphire substrates are still in use.

The sapphire substrates are slightly larger than the substrate by about 5 to 6 mm to minimize the edge exclusion zone and to provide support during different operations. In particular, this provides space for clamping in different process tools. Perforated sapphires are used in some processes if the debonding is done by soaking in a solvent bath. The perforations allow the chemical to dissolve the polymer adhesive faster. Perforated sapphire does have a cost disadvantage. Also, chemicals used for backside processing can attack the adhesive and subsequently the front-side circuit if care is not exercised.

Wax was one of the first adhesives used to bond GaAs wafers to sapphire for backside processing. Different types of wax are available having a range of melting or softening temperatures. And different techniques for applying wax evenly were developed. The most sophisticated method was to evaporate wax using a vacuum evaporator. Even this method was not suitable for large-volume production, because of frequent downtime needed for cleaning the evaporator hardware. Due to the increase in volume of the III–V market, different adhesives have been developed and are being used. To ensure robust bond and ease of processing, an adhesive must be carefully chosen. Two important factors are the melting point and ease of debonding. If the melting point is too low, the wafer will debond in process steps in which the wafer gets hot, for example, the very high-energy via etch. The bonding compound should withstand all the different wet clean, electroplating, and dry etch process steps. Different polymer, thermoplastic materials are available commercially as bonding adhesives in spin-on liquid form. Crystal Bond (a trademark of SPI) is one of the common solid bonding materials that are dissolved in acetone to form a liquid with controlled viscosity that then applied on a spin-coating system.

Wafer bonding is done by placing the GaAs wafer on the sapphire substrate with the adhesive, and applying heat and pressure. See Fig. 19.1 for a schematic diagram of a processed wafer bonded to a sapphire substrate. This process needs to be carried out in a relatively clean space so that hard particles are not embedded in the adhesive. Also care is taken to make sure there are no high protrusions on the surface of the wafer thicker than the thickness of the dried adhesive, otherwise the wafer may develop small cracks. Epidefects can be present on the wafer from the start, even with a large height due to pile-up of deposited materials. Also, particles due to coagulation of polymers like polyimide and polymers used for scratch protection must be avoided. The process parameters, mainly pressure and temperature, are optimized for achieving a strong, uniform bond. Adhesive thickness depends upon the heights of thick metal features and nodules and defects, as mentioned above, typical values being in the few to 10 μm range. Uniform pressure across the whole wafer is critical to achieving low total thickness variation (TTV) specification. Commercial bonders are availablefor automating this operation [2].

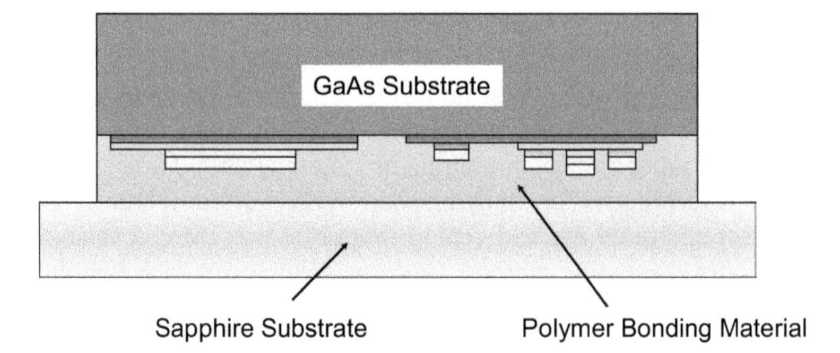

Figure 19.1 Schematic diagram of a GaAs wafer bonded to a sapphire substrate with a polymer bonding material (Crystal Bond adhesive). Reprinted from Ref. [1] with permission. Copyright © 2010 CS MANTECH.

19.3 Wafer Thinning

After the wafer is bonded to the sapphire, it is ready for thinning, which is generally done by grinding and in some cases followed by polishing. The ground wafer can be polished mechanically or by chemical mechanical polish (CMP) to give a finer surface that may make the subsequent processing easier. Typical grind thicknesses are 80 to 100 μm. Grinders can be fitted with real- time thickness monitoring to control thickness to +/−10 μm. Good thickness control is necessary for reproducible packaging process and control of TWV impedance. Thinned wafer thicknesses of below 60 μm have been reported, but in these cases a thick layer of gold is needed to provide mechanical support so that the wafer can be handled.

Historically grinding used to be followed by lapping, but spray etching or polishing are better suited to high-volume production. A certain degree of roughness may be needed for good adhesion of metals, so smooth chemical polish is not necessary. Mechanical debris and polishing contaminants are removed at this step. Since fast grind speeds are utilized in production, there is a considerable amount of damage and stress in the surface layer. This is reduced by a chemical etch in an etch bath that removes damages layer and smoothens the surface (see Chapter 6). Few to 10 microns of material is removed to ensure a minimum required die strength of the finished die.

19.4 TWV Photolithography

TWVs are aligned to metal landing pads on the front side, which in turn are connected to grounding nodes, emitter contacts of heterojunction bipolar transistors (HBTs), etc. Backside photolithography tools are needed for alignment of marks on the front side with marks on the photomasks for the backside. This alignment can be done by using IR light to go through the GaAs and bounce back to the optical system for registration with the photomask (see Fig. 19.2). Split views are shown on the screen and the alignment is done manually by superimposing the two images. Other approach is to use charge-coupled device (CCD) cameras to store the image of the mask alignment marks with respect to the wafer stage and then load the wafer and move the stage to automatically align to the stored marks. TWV sizes are getting smaller, toward 20 μm, and their spacing is also being reduced as via process technology is progressing. Although TWV features are large, in tens of microns, alignment accuracy needs to be in the few-micron range because the TWV via size control may not be better than a few microns, so the via may not land on the metal landing pad. The resist used for this process needs to be very thick to withstand the long time exposure to the high-energy plasma. If the resist gets eroded in the etch process, the backside surface may be etched.

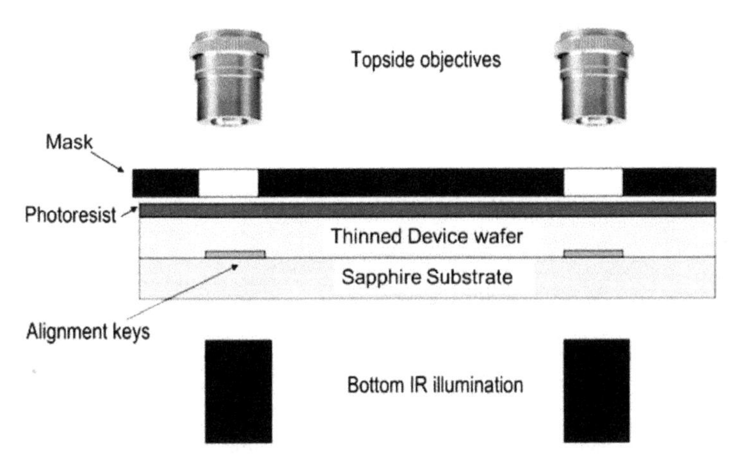

Figure 19.2 Alignment of wafer vias with infrared illumination.

19.5 TWV Etch

Figure 19.3 shows a schematic diagram of a wafer via, with photoresist still present, that has been aligned to land on a metal grounding pad formed during front-side processing. To etch deep vias up to 100 μm deep, reactive ion etching (RIE) using chlorine chemistry was used for years. This process had a slow etch rate and the process times were over a few hours per wafer [3]. The use of inductively coupled plasma (ICP) with RIE allowed the increase in the etch rate by nearly a factor of 10, this reducing etch times to a reasonable time for high-volume production [4, 5]. These dual-frequency systems provide an additional degree of freedom. Plasma ion density and bias voltage can be controlled independently (see Chapter 7 for details). BCl_3/Cl_2 chemistry is generally employed. Inert gases may be added for via profile and smoothness control. By controlling the reactive gas ratios, the degree of polymer formation on the sidewalls can be controlled. Through a combination of gas and power ratio of the ICP to RIE, the shape of the via can be controlled all the way from vertical to isotropic [6, 7]. A via of any taper angle can thus be produced. Multiple process steps can be used to further control the shape so that the bottom portion is vertical. See Fig. 19.4 for different via shapes. This shape can make it easy to coat the bottom corners of the via at seed deposition. The power level can be lowered at the very bottom, where metal back-sputtering may have to be avoided. These systems are available in cluster tool configurations and are suited to very high throughput.

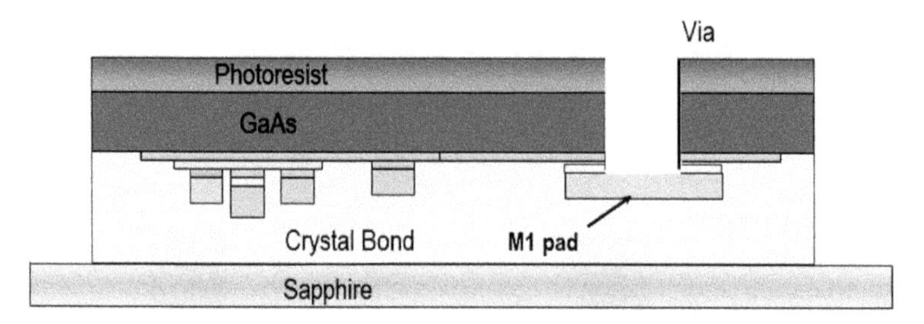

Figure 19.3 Schematic diagram showing a through-wafer via etched to land on a metal pad on the front. Reprinted from Ref. [1] with permission. Copyright © 2010 CS MANTECH.

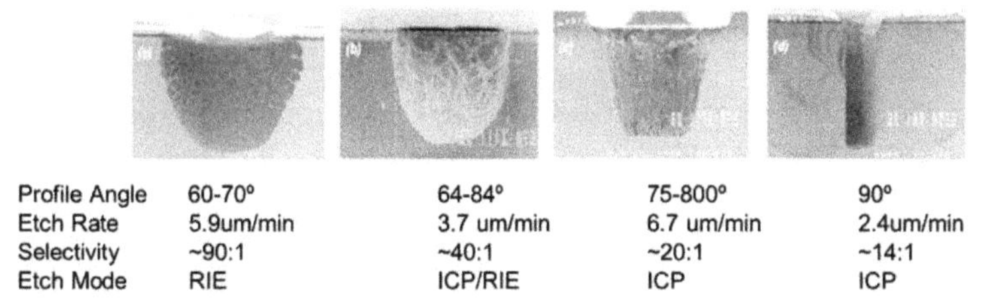

	(a)	(b)	(c)	(d)
Profile Angle	60-70°	64-84°	75-800°	90°
Etch Rate	5.9um/min	3.7 um/min	6.7 um/min	2.4um/min
Selectivity	~90:1	~40:1	~20:1	~14:1
Etch Mode	RIE	ICP/RIE	ICP	ICP

Figure 19.4 Range of TWV shapes from isotropic with pure RIE (a) to vertical with pure ICP. Selectivity is to the photoresist mask [6]. Courtesy of Plasma-Therm LLC.

A common problem seen in a TWV is the formation of pillars. Figure 19.5 shows an example of pillar formation. These are believed to be related to substrate defects [7]. In practice these can be avoided by using a high power (high bias) level in the beginning to avoid the formation of the defect that leads to shadowing of the etch. Pillars can also be caused by insufficient exhaust of the etch system.

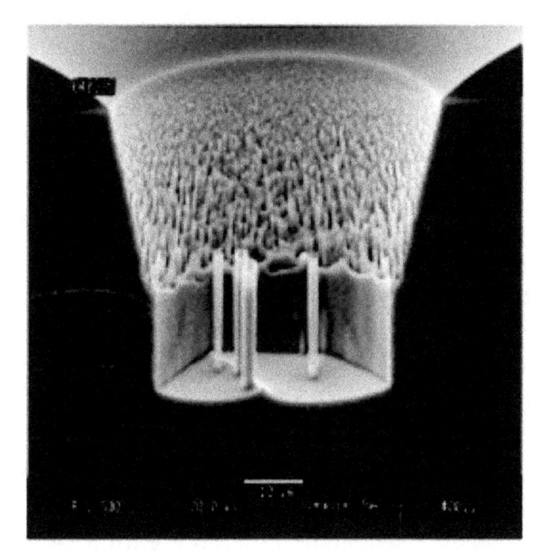

Figure 19.5 Pillar formation in TWV etching [6]. Courtesy of Plasma-Therm LLC.

Following the TWV etch wafers must be stripped of resist completely, and different metal–organic complexes formed during the long extended etch must be removed. An aggressive clean is needed to avoid metal peeling.

19.6 Backside Metallization

Backside metallization needed for good low impedance is generally a thick film of over a few microns thickness. Therefore physical vapor deposition is not practical. Electroplating has been the preferred method of deposition. Choice of metals and details of plating process are discussed in Chapter 20. In summary, gold has been the material of choice and is now being replaced by copper because of the high cost of gold. Gold is easy to bond and is inert, so it does not need an extra protection layer like copper. Regardless of the metal choice, a seed layer is needed. This layer could be electroless-plated nickel, but because of demands of reliability, good adhesion, and coverage in small vias, sputtered seed layers are preferred. TiW/Au or Ni–V/Au bilayers are needed to achieve good adhesion and avoid interdiffusion. The aspect ratio of the vias needs to be large for the sputtered deposit to reach the bottom corners of the via. Tapered vias also help in this regard. If small vertical vias are used for small die size, the sidewall coverage of the seed layer must be improved. Using collimated sputtering is an option, but it is not very efficient [8], so it is not too interesting for precious metals. However, sputter systems that combine sputtering with ion plating are a good option.

As with any metal deposition, a proper precleaning procedure must be used before seed deposition. In practice, a multiple-step process involving, oxygen plasma ashing, and standard GaAs surface cleaning is utilized. Good adhesion of the metal barrier and seed layers must be verified with proper adhesion testing.

After the seed deposition, the wafers are ready for electroplating. The details of this process are covered in Chapter 20. A few highlights are mentioned here.

19.6.1 Backside Plating

Figure 19.6 shows a schematic of a TWV with a seed and plated metal. Historically, rectangular baths were used. Now, automatic electroplaters that are designed with a showerhead that circulates the bath over the wafer are used. These systems typically include a prewet station and several heads to plate a few wafers at a time in a continuous cassette-to-cassette fashion.

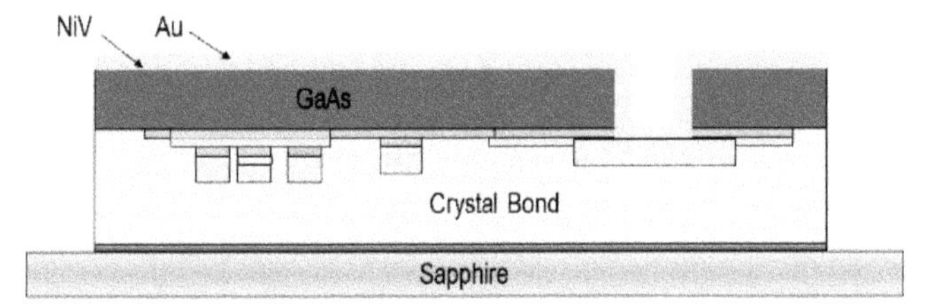

Figure 19.6 Schematic cross section showing a wafer after seed deposition and electroplating. Reprinted from Ref. [1] with permission. Copyright © 2010 CS MANTECH.

The primary costs of the electroplating process come from capital expenditures and from the cost of gold. Reducing the amount of gold deposited has often been viewed as the primary cost-saving technique and has been accomplished by reducing the current density used during deposition. However, with lower current density comes increased cycle time. To avoid additional capital expenditure during a production ramp, it became necessary to significantly decrease the plating time in order to increase wafer throughput. Pulsed plating can be used to increase coverage at the bottom corners. Uniformity can be improved by adding additional electrolyte jets or improving the flow of electrolyte to the center of the wafers, the details being dependent on the design of the plating system. By proper design of the flow and process parameters, low via resistances and high across-wafer uniformity can be achieved, while maintaining low plating cycle time [2]. Actual focused ion beam (FIB) cross sections of wafer vias are shown in Fig. 19.7.

Care must be taken to avoid dissolution of the seed layer and barrier during the start of the plating process, otherwise metal peeling and reliability issues may be seen. A close up cross section of the bottom of a via is shown in Fig. 19.7c, showing good adhesion of the backside metal to the front-side metal as well as good metal coverage in the corners.

Gold as a plating metal is being replaced by copper because of the high cost of gold. Details of the copper plating process are discussed in Chapter 20.

Copper has low electrical resistance and higher thermal conductivity than gold. Its major disadvantage is that it can diffuse fast from the backside TWV into the front side and degrade the performance of field-effect transistors (FETs), HBTs, etc. This problem can be solved by adding a diffusion barrier under the seed layer. Refractory TaN was used early on along with sputtered copper to show good FET device reliability [10] and later on good HBT and pseudomorphic high-electron-mobility transistor (PHEMT) reliability. Sputtered TaN may not cover rough TWV sidewalls and may leave holes through which diffusion can occur. Electroless plating can be used to deposit a seed and a barrier layer to solve this problem. A Ni–P layer has been used as the seed for copper. However, the Ni–P/Cu metal system can still lead to interdiffusion, easily observed by appearance of Kirkendall voids. This problem has been solved by adding electroless nickel in between [11] (Skyworks internal data). An oxidation prevention layer is added on top of copper. This can be a layer of TaN or an organic solderability preservative (OSP) film.This metallization scheme has been shown to withstand temperatures over 260°C needed for packaging and reflow during circuit assembly. Good reliability can also be achieved by ensuring enough coverage of sputter-deposited adhesion and barrier layers like Ni–V and TaN.

Figure 19.7 Cross sections of TWV shapes produced by a combination of RIE and ICP, tapered top and vertical bottom (a) and SEM cross section showing a vertical via (b). The plated metal adhesion and coverage at the wafer backside and inside the via can be verified by cross sections like these. (c) Close-up cross section of the bottom of a TWV showing good plating metal coverage and no gaps between front-side gold and backside metallization. Reprinted from Ref. [9] with permission. Copyright © 2003 CS MANTECH.

Copper films deposited by electroplating give excellent coverage of TWVs down to sizes below 20 μm. These films do tend to have mechanical stress and may warp the thinned wafer, thus making handling difficult during debonding and dicing. Therefore, stress must be reduced by controlling the deposition rate and plating process parameters to achieve proper film texture and density. Stress can also be relieved by proper thermal annealing. Excess stress may lead to reliability issues in the field by causing delamination of metal in or around the TWV.

19.7 Backside Street Etching

Backside streets are needed for separation of individual die if the scribing process is such that the die cannot be separated. If sawing is used (as it was for years when die sizes were large) street etching is not necessary. If the diamond scribe and break process is used, the die will not separate easily, because the gold is not brittle. A photolithography step is needed to pattern the streets. This resist also needs to be thick in order to make sure the resist fills the vias and adequately covers the top tapered edges of the vias. Potassium iodide/iodine-based gold etches are used for etching. Gold can be recovered from the etch chemistry waste. If copper is used as the backside metal, it may be cheaper to avoid the process step of street etching by laser-dicing through the metal (to be discussed in Section 19.9.2).

19.8 Wafer Demounting/Debonding

Demounting of thinned wafers is an extremely tedious process. If perforated sapphires are used, the mounted wafers are soaked in a hot solvent (e.g., GenSolve 500) under agitation. When the wafers are released, they must be handled with great care until mounted on tape for the scribe and break operation. EVG and BLE have introduced automatic debonders for use with nonperforated sapphire [12]. Debonding tools use two hot vacuum chucks to grab the wafer and the carrier as they are separated by heating. A robot is used to load the stack into the cavity between the two chucks, placing the stack sapphire side down on the chuck. The top chuck is lowered to contact the GaAs wafer. Both chucks have vacuum grooves for holding the wafer. As the temperature crosses the melting (transformation) point, the top chuck is tilted gently to debond the wafer from wafer rim. The circuit wafers must be cleaned with solvents to remove wax or adhesive.

19.9 Wafer Dicing

19.9.1 Scribe and Break

The operation of cutting out individual die from a thinned wafer is called dicing. Thinned wafers are cleaned and mounted on UV tape for temporary bonding necessary for the dicing and separation process. This is a stretchable tape that has UV-releasable glue on one side that is held taut in a frame. The separation operation can be done by sawing, diamond scribing, or laser scribing. In sawing, used in the early days of GaAs production, high-speed circular saws are used to cut the wafer with a very thin blade at very high blade speeds. The table, on which the wafer is placed, is a high-precision table that can be moved at controlled speed. Thick gold from the backside street may have to be removed because it may get stuck in between the diamond particles. The blade life depends upon the material being cut. The minimum street width has to be over 40 μm because the kerf (the material removed by the sawing operation) cannot be small due to the blade cutting edge thickness, the vibration during sawing, stage accuracy, etc. Disco Corp. has designed blades that have a small diamond grit bonded by electroplated nickel, which has a longer life. Large 6-inch wafers with very small die, which are being produced now, require thousands of cuts, so the blade may not last for more than a few wafers.

The scribe and break process is a more popular approach used in the GaAs industry. A diamond stylus is used to score lines on the wafer at a controlled speed and angle. The scribe line goes partially through the wafer thickness. Breaking is done by flexing the wafer by using a roller and anvil. The scribe and break process has several advantages: it is faster than sawing and the kerf can be reduced to increase number of die on a wafer. There are a few disadvantages of the scribe and break process: It works only along crystal axes;

generally circuit die are placed along these but not always. It does not work at odd angles, as expected for crystalline wafers. Excessive chipping or bruising can still occur, so scribe force must be reduced. To maintain alignment accuracy, the diamond stylus may have to be realigned after a few cuts are made, otherwise the cut may run off the street.

19.9.2 Laser Dicing

As the demand for semiconductor chips in general and III–V compound semiconductors in particular has gone up and the wafer thickness and die size have gone down, there has been a market drive towards a high-yield, -speed scribe process. Laser dicing has come to fulfill this need. Laser dicing has numerous advantages. The kerf can be smaller than 30 μm, and speed can be an order of magnitude higher. Yields are higher due to low chipping and cracking and cuts can be made at any angle, not just along crystal axes. Constant change of saw blades or diamonds is not required. The process sequence for laser dicing is similar to the scribe and break process. Wafers are mounted on UV tape, coated with a protective water-soluble polymer to prevent damage and contamination, laser-diced, stretched, and cleaned. Laser ablation leaves the molten slag on the edges and nearby. An additional step of recast removal by wet etching is added at the end.

In the ASM Laser Separation International Corp. (ALSI) system [13], pulsed laser radiation at a wavelength of 1064 nm is focused on the wafer with a beam size of about 10 μm. This beam can be scanned at 500 mm/sec. At critical laser intensity the material vaporizes, thus creating a cut through the thickness of the wafer. To optimize the quality of the cut, multiple passes of the beam are utilized, rather than a high-power single cut. After the wafer has been scribed, it is stretched as usual and goes through the cleaning and polymer coating removal process. Molten material resolidifies on the sides of the die to form a recast layer. Microcracks were observed early in the process development, between the bulk material and the recast. These microcracks reduce die strength. Therefore a die strength recovery step, which is similar to stress relief etch, based on standard GaAs peroxide/acid etching is used to remove the damage. After this step, the wafer can be cleaned and is ready for pick and place.

Synova [14] has developed a laser-dicing system that utilizes a water jet to guide the laser beam by total internal reflection at the water–air interface, like a glass optical fiber. Water also cools the wafer and removes the molten material generated by laser ablation.

In the laser-dicing process, etching of metal in the streets can be avoided by developing a scribe process that cuts through the metal layer, thus saving a mask step and processing steps.

After the die are separated, the tape is stretched to pull the die apart slightly. The wafers must be exposed to UV before the die can be picked by automatic pick-and-place machines. This operation must be completed before the life of the UV tape expires, and after UV exposure the die must be picked within a short time, otherwise the residue may be left behind.

Laser dicing can also be done even with backside metallization on the wafer. This saves one photolithography and etching step. Careful optimization of dicing parameters is needed to achieve a clean break, cut through the metal (which may be a few microns thick), and avoid chipping. The process requires a lower cutting speed [15]. Laser dicing has evolved at a fast pace and good progress and innovations are expected.

Dicing processes, as well as many of the TWV processing steps have a large impact on the performance of circuits and reliability of front-side devices. Therefore, these must be evaluated by final RF testing as well as reliability qualification.

Section II: Wafer-Bumping Process [16–18]

19.10 Introduction

The TWV process has limitations for RF circuits for mobile devices. Typically low-resistance and low-inductance ground connections are provided by TWVs. All other input/output connections are made using wire bonds on bond pads. For higher frequencies, wire bond connections to the package (multilayer laminate board) may not be acceptable. Also, it is difficult to make TWVs in InP, which will be needed for the next-generation mobile technology.

Bond pads take a lot of space and have a strong effect on chip size. The height of wire bonds makes the package size large. Use of solder balls with a flip chip solves most of these issues. However, flip chip processing of die is cumbersome and not cost effective. Therefore wafer-level bumping is seen to be the preferred approach. Figure 19.8 [18] shows the difference between GaAs HBT die with TWVs and the same die with copper bumps on the front side. The difference between the two approaches for a finished, packaged module is shown in Fig. 19.9.

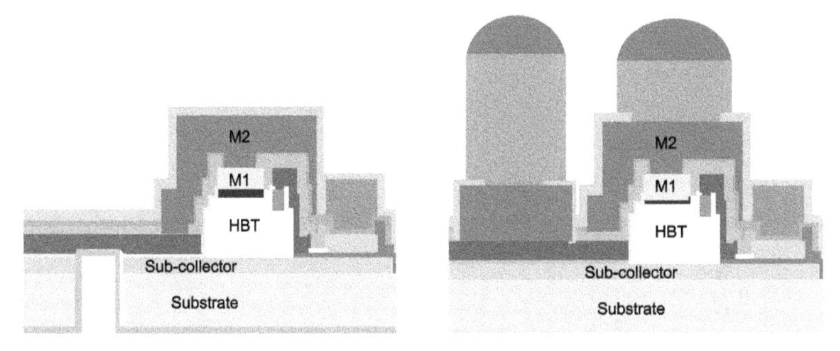

Figure 19.8 Schematic cross sections showing difference between TWV and copper bump. Reprinted from Ref. [18] with permission. Copyright © 2013 CS MANTECH.

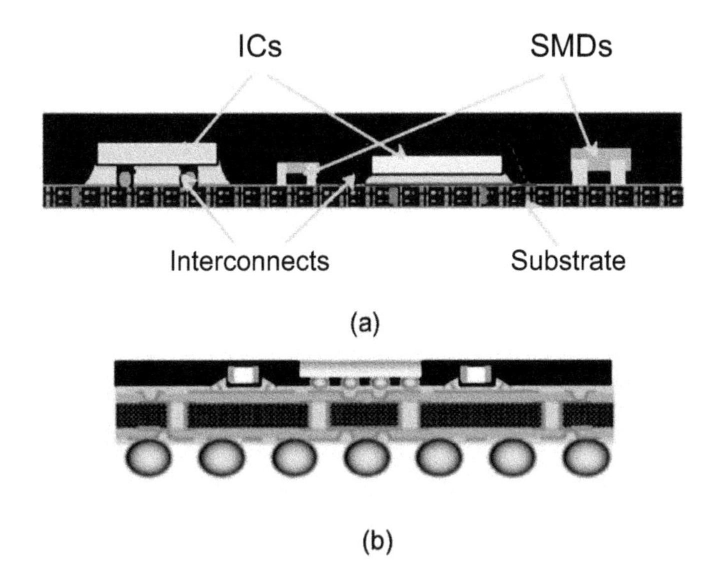

Figure 19.9 Comparison between wire-bonded die on laminate and flipped solder ball die on BGA package. Illustration courtesy of Amkor Technology, Inc.

19.10.1 Advantages of the Wafer-Level Bump Process

Processing is done with thick wafers, thus avoiding breakage of brittle GaAs or InP thinned wafers. Since wafers are not bonded to a sapphire susbstrate, the bonding material attack by a harsh photoresist stripping process is avoided.

RF performance is better because of low impedance provided by the solder joint, and silver epoxy used for attaching the die with a backside metal has higher impedance, even at high silver concentrations.

The overall footprint of the finished package is smaller, and the height is smaller, too.

Larger-size balls can be used

There is no need for underfilling material that is used in flip chip packaging.

Die probe testing can be done at the wafer level.

The process has a lower overall cost, for example, if gold wires are eliminated, and a lower assembly cost.

There is compatibility with future 3D packaging.

19.11 Requirements of Components of the Solder Ball or Pillar Process

Wafer-scale bumping involves two main process components, underbump metallurgy (UBM) and the pillar or solder ball itself. The desired properties of these two are discussed below, before a description of the process, which will follow.

19.11.1 Underbump Metallurgy

UBM provides a stable structure for solder balls or a barrier in the case of pillars. It has to have good adhesion and must act as a good diffusion barrier for copper, tin, etc. The top of the metallurgy must provide good wettability for the solder for it to flow properly (not important for copper pillars). It should not add stress to the wafer or die. If there is some minor probe damage due to testing, it should not affect adhesion and UBM should cover it up.

In silicon IC fabrication, bond pads are aluminum metal, which oxidizes easily and does not bond easily to other metals. Therefore

UBM makes it possible to make the transition from Al to solder or other pillar materials. In the case of III–V materials, pads generally have gold on top (even for copper interconnects), so UBM is designed to make the transition from gold to solder or copper pillars. Figure 19.10 shows the cross section of a solder bump on top of a front-side metal pad with UBM in between. Because of the multiple requirements placed on choice of UBM, one metal layer cannot do the job; therefore multilayer stacks of metals are used. For silicon IC adhesion, barrier and wetting layers need to be provided, so a stack like Cr/Ti/Ni–V is needed. For III–V semiconductors, a TiW/Cu or TiW/Au layer structure fulfils the need of adhesion, barrier, and wetting layers. Ti/Cu and TiN/electroless plated nickel can also be used in the UBM stack.

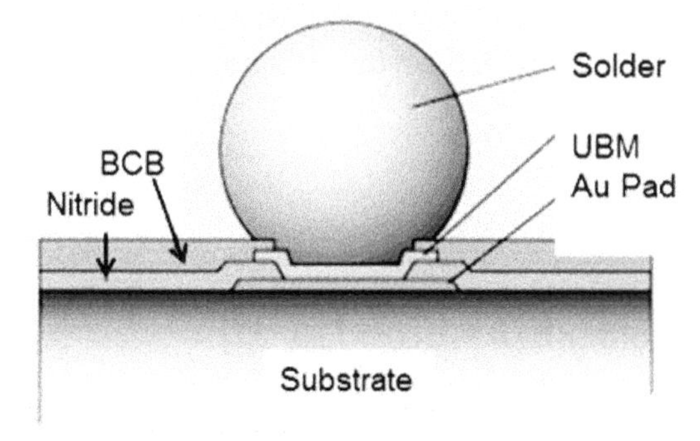

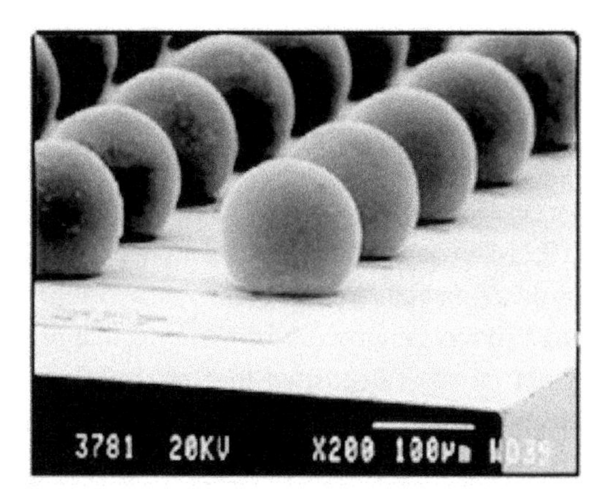

Figure 19.10 Cross section of solder bump and SEM image of an array of bumps after reflow. Reproduced from Ref. [16] with permission from *Solid State Technology*.

19.11.2 Solder Ball and Pillar [16]

Solder bump ball: Solder bumps eliminate the need for solder on the circuit board; the balls are reflowable. The ball is designed to self-center and collapse during assembly.

The composition of the solder ball is chosen on the basis of temperature that can be used to reflow it. This is limited primarily by the temporary bonding material but could be limited by the reflow temperature allowed at the final assembly on the circuit board or package. Generally, lower temperatures are preferred, so historically eutectic solder compositions (Pb:Sn, ~38% Pb) were chosen. Due to the recent requirement of eliminating lead (Pb), Sn/3.5Ag or 99% Sn lead-free solders are used.

Use of copper pillars is gaining acceptance in GaAs packaging, as shown by introduction of this technology by foundry service providers [17–19].

19.12 Fabrication Process

19.12.1 Solder Ball Process [20, 21]

The fabrication process starts after the front-side circuit fabrication is complete. An additional layer of dielectric, like benzocyclobutane (BCB), is applied first, sometimes called repassivation. It planarizes better and acts as a stress buffer. Redistribution and rerouting can be done in the UBM metal and this repassivation used as the interlayer dielectric. Small size pads, 60 μm in size on the die, can be rerouted to larger UBM pads for the pillars or bumps 300–500 μm in size. Figure 19.11 shows the details of a simple process for creating solder bumps on a finished circuit wafer.

Figure 19.12 shows the process for bumping in more detail from UBM up. A UBM metal is deposited first over the whole wafer to provide an electrical path for plating current. The TiW layer can be sputter-deposited on the whole wafer, followed by copper or gold and then another layer of TiW is deposited. Layer thicknesses can be 200 Å TiW/4000 Å Cu/75 Å TiW. Photolithography is used to pattern the pads and create exposed areas that need to be plated. See Fig. 19.12 for details. A very thick resist is needed. For example, Shipley AZ4620 can be spun up to a thickness of 26 μm and Clariant TFP-V can be deposited with a thickness of 65 μm in a single spin operation [20]. After resist patterning, the TiW is removed, with H_2O_2 etching, and the wafer loaded for plating. Copper is plated (~6 μm) followed by solder, which originally used to be a Pb:Sn eutectic but now can be lead-free solder. After plating the resist is stripped. Now the field metal can be removed. TiW is etched off by H_2O_2 and copper or gold removed in commercial cyanide etch baths. Wafers are cleaned and reflowed at 183°C for eutectic alloy (62 wt% Sn) or higher temperatures (200°C to 245°C) for lead-free Sn:Ag or An:Pd solders.

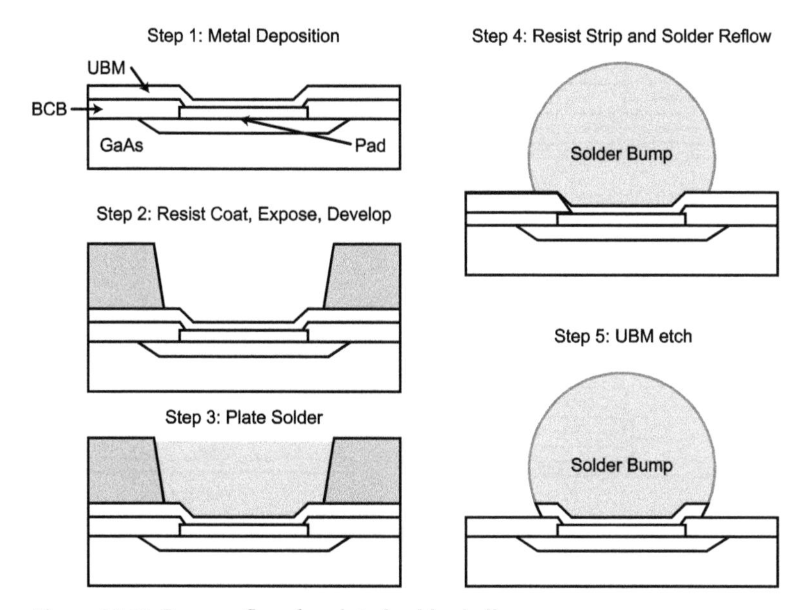

Figure 19.11 Process flow for plated solder ball process.

A cross section showing details of a solder ball on top of UBM is shown in Fig. 19.13. There should be no path for the solder metals to diffuse to the front-side circuit [22].

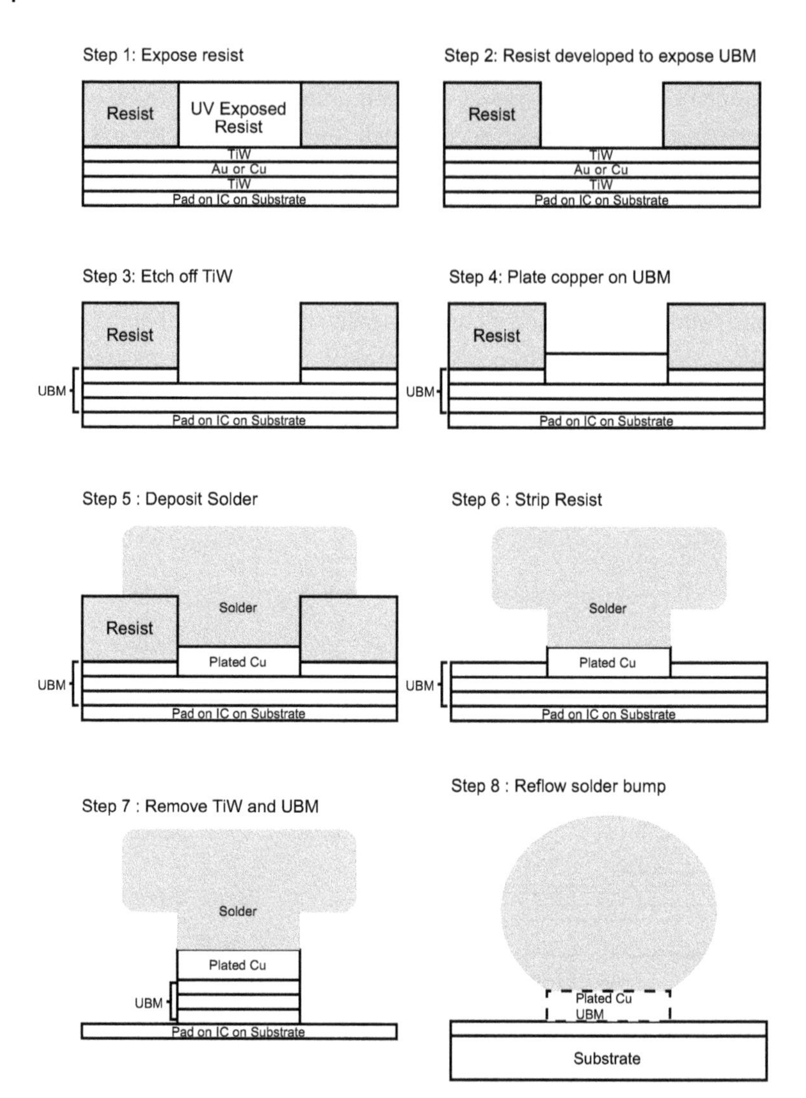

Figure 19.12 Schematic diagram of solder bump process using UBM and plated copper. Reprinted from Ref. [20] with permission. Copyright © 2003 CS MANTECH.

19.12.2 Copper Pillar Process [19]

In this process copper pillars are used instead of solder balls. The pitch can be smaller than round solder balls because the pillar size control is better than solder balls. So a thin copper layer and a thick solder layer are replaced by a thick copper layer (40 μm) and a thin solder layer on top (25 μm). The process flow is shown in Fig. 19.14. The difference from the processes described above is that the thickness of copper is high, so a high deposition rate for copper plating is desired. Win Semiconductor has tried sulfuric acid-based baths and achieved a deposition rate of 2 μm/min, so it takes a reasonable amount of time to deposit 60 μm.

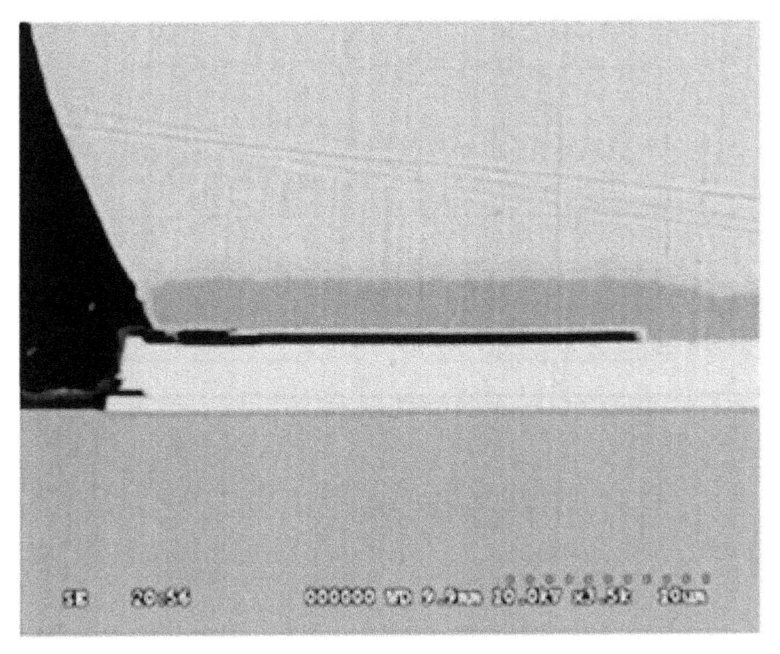

Figure 19.13 Cross section of a solder ball on top of UBM, showing barrier layer. Reprinted from Ref. [22] with permission. Copyright © 2007 CS MANTECH.

Round pillars as well as rectangular bars for thermal cooling with the same height have been demonstrated for this process. Figure 19.15 shows the same pillar height for round and rectangular, bar-shaped pillars [19]. Better electrical and thermal performance has been shown. For HBT power amplifier devices, a higher maximum available gain (MAG) at a higher collector current was reported. Thermal bars were shown to reduce the temperature significantly by 35°C .

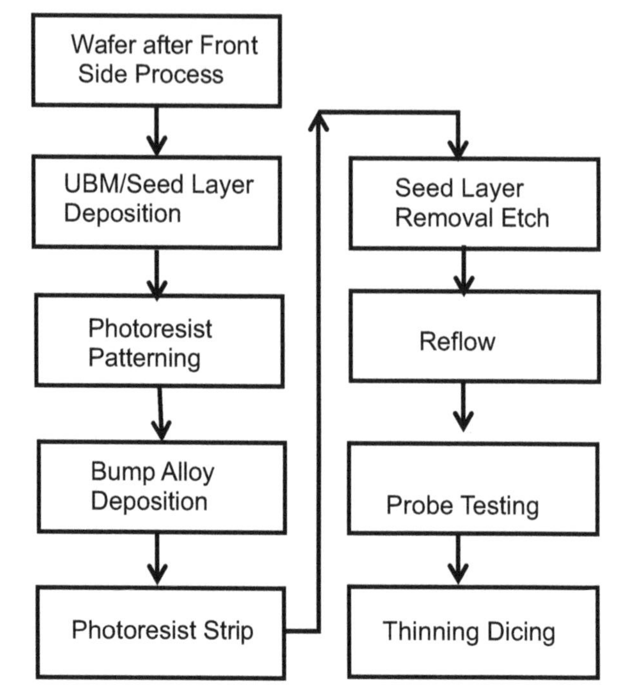

Figure 19.14 Process flow diagram for copper pillar bumping process.

(a) (b) (c)

Figure 19.15 Same height round- and bar-shaped copper pillars with solder cap (a). Round pillar before (b) and after (c) reflow. The pillar height is about 60 mm. Reprinted from Ref. [19] with permission. Copyright © 2012 CS MANTECH.

References

1. S. Tiku, *Device Processing in III-V Manufacturing*, CS MANTECH (2010).
2. T. Glinsner et al., *Reversible and Permanent Wafer Bonding for GaAs Processing*, GaAs MANTECH (2001).

3. H. Hendriks et al., *Benefits and Challenges in Decreasing GaAs through Substrate via Size and Die Thickness*, GaAs MANTECH (2002).
4. P. S. Nam et al., *J. Vac. Sci. Technol. B*, **18**, 2780 (2000).
5. R. Szweda, ICP cluster tool for etching GaAs wafers, *III-V Review*, **13-5**, 48 (2000).
6. R. J. Westerman, D. J. Johnson and Y. S. Lee, Characterizationof 10 μm/min. chlorine based ICP etch process for GaAs vias, Plasmatherm Application notes.
7. R. Westerman, D. Johnson and F. Clayton, GaAs MANTECH (2003).
8. R. A. Powell and S. M. Rossnagel, *Thin Film PVD for Microelectrronics*, Acadamic Press, Boston, MA (1999).
9. D. Anderson, H. Knoedler and S. Tiku, *Cycle Time Reduction during Electroplating of Through Wafer Vias for Backside Metallization of III-V Semiconductor Circuits*, CS MANTECH (2003).
10. C.-Y. Chen et al., Backside copper metallization of GaAs MESFETs using TaN as the diffusion barrier, *IEEE Trans. Electron. Devices*, **48**, 1033 (2011).
11. D. Tsunami et al., *Palladium Diffusion Barrier Grown by Electroplating for Backside Cu Metallization of GaAs Devices*, CS MANTECH (2013).
12. C. Schaefer et al., *High Yield Lithography and Wafer Handling Method for Reliable Back Side Processing*, GaAs MANTECH (2003).
13. M. C. Muller, R. Hendriks and H. P. Chall, *Significant Step in Wafer Yield Optimization and Operation Cost Reduction due to Dicing Innovation*, GaAs MANTECH (2006).
14. D. Perrottet et al., *GaAs Wafer Dicing Using the Water Jet Guided Laser*, GaAs MANTECH (2005).
15. N. M. Dushkina and B. Richerzhager, *Dicing GaAs Wafers with Microjet*, www.Synova.ch.
16. D. S. Patterson, The back-end process: step 7; solder bumping step by step, *Solid State Technol.*, http://electroiq.com/blog/2001/07/the-back-end-process-step-7-solder-bumping-step-by-step/.
17. F. Juskey, *TriQuint RF Innovation for Peak Performance*, CS MANTECH (2010).
18. H.-C. Chang et al., *Device Characteristics Analysis of GaAs/InGaP HBT Power Cells Using Conventional Through Wafer Via Process and Copper Pillar Bump Process*, CS MANTECH (2013).
19. T. Hsiao, G. Chen, S. Chou and H. Liao, *Manufacturing of Cu-Pillar for III-V MMIC Thermal Management*, CS MANTECH (2012).

20. P. Sricharoenchaikit, *Building Solder Bumps on GaAs Flip Chip Schottky Devices*, CS MANTECH, gaasmantech.org/Digests/2003/2003PDF/8-19.pdf (2003).
21. X. Zeng et al. (Northrup Grumman), *Wafer Level Bump Technology for III-V MMIC Manufacturing*, CS MANTECH (2010).
22. S. Combe et al., *Development of a Lead Free Solder Bumped RFIC Switch Process*, CS MANTECH (2007).

Chapter 20

Electroplating

20.1 Electroplating History [1]

Electroplating started in the early 1800s, when Luigi V. Brubnatelli used Volta's voltaic pile to plate a thin layer of gold onto two large silver medals. Thirty-five years later, John Wright found that both gold and silver could be electroplated using potassium cyanide as the electrolyte. Shortly thereafter Henry and George Elkington obtained the first patents for gold and silver electroplating processes on the basis of Wright's electrolytes. The use of gold plating in electrical equipment is as old as the electrical industry itself. Gold creates a metal contact surface that does not corrode and oxidize when used for electrical contacts and bonding components. It has been used in the electronics industry since the 1940s; it has been used for corrosion resistance, bonding of surfaces for soldering, and more recently low-resistance interconnects in monolithic microwave integrated circuit (MMIC) circuits.

20.2 Electroplating Fundamentals

Electroplating is the process of depositing a coating of a desired metal on top of a base metal by means of electrolysis. Electrolysis

III–V Integrated Circuit Fabrication Technology
Shiban Tiku and Dhrubes Biswas

ISBN 978-981-4669-30-6 (Hardcover), 978-981-4669-31-3 (eBook)
www.panstanford.com

is carried out in a solution or bath, which generally is an aqueous solution of the material to be deposited. The electrodes, a cathode and an anode, are immersed in the solution and connected to a battery or power source. The laws of electrolysis were formulated by Faraday in 1833 and remain as the basis for electroplating. Faraday's laws specify that 1 Faraday (96,400A-sec) of charge yields 1 g equivalent of substance on the cathode. The gram equivalent is obtained by dividing the atomic weight of the metal by the number of electrons required per atom. If an alloy is deposited, then Faraday's law applies to the sum of the constituents and also to impurities or hydrogen that evolves at the cathode.

The substrate to be coated must be conductive or it must be covered by a conductive layer called the seed layer so that it can act as the cathode when connected to the negative voltage source. The anode historically consisted of the sacrificial material and could be gold for gold deposition but now is invariably stainless steel or platinized titanium for the semiconductor industry. When the current is turned on, the positively charged metal ions in the bath move toward the cathode and plate out as metal atoms as they regain their electrons.

20.3 Electroplating Bath Types

Electroplating has been historically done in rectangular baths, filled with an electrolyte, into which a cathode to be plated and an anode are placed (see Fig. 20.1). A simple direct current (DC) power supply can be used where the current can be set to a desired value. In production systems with pulse plating options, pulse width and current can be controlled. The solution is agitated during plating. Modern production systems can be more complicated as systems are designed for higher throughput, better thickness uniformity, and automatic wafer handling.

As far as chemistry is concerned, four types of baths have been used historically for plating metals. The first three are sodium- or potassium cyanide–based solutions: (i) unbuffered alkaline (pH 8.5 to 13), (ii) buffered acidic (pH 3 to 6), or (iii) neutral (pH 6 to 8). The fourth bath uses solutions of complex ions without cyanide. Cyanide baths were chosen very early over other solutions, such as copper sulfate, because the copper potential is less negative and the deposition process is slower, so the deposit is not spongy or porous. Lower-pH baths give higher-density deposits but may contain unwanted impurities and result in nonuniform colored deposits. Most cyano-complexes decompose and release poisonous hydrogen cyanide gas when exposed to acids; therefore baths must be alkaline. Alkaline baths give deposits that are columnar and contain measurable carbon. These baths contain $KAu(CN)_2$, KCN, K_2HPO_4, etc. Phosphate is added to ensure stable pH, otherwise a drop in pH at the anode will result in the formation of unwanted colored deposits.

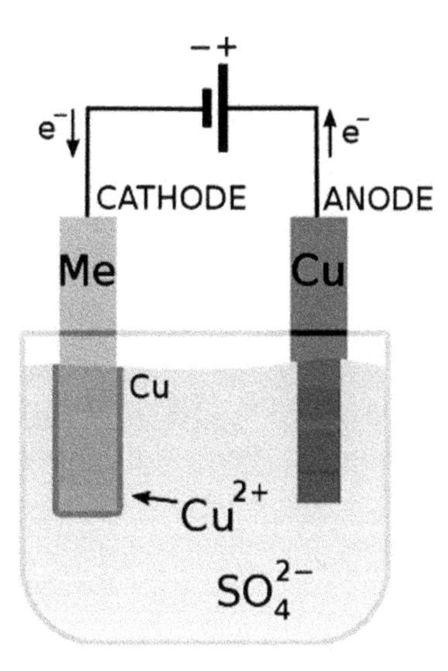

Figure 20.1 Basic plating apparatus for copper plating.

Unbuffered alkaline baths that contain high concentrations of $Au(CN)_2$ give hard, fine-grained deposits. Buffered baths with citrate or phosphate added also give hard deposits and brighter plated films. Precise formulations of plating baths are proprietary because other agents added to control finish and brightness are kept as trade secrets. But general guidelines for bath parameters that need to be controlled in production are supplied by the bath vendors. These may include pH, cell voltage, specific gravity or Baumé, gold content, temperature, total electrolyte content, conducting salt concentrations, etc.

The main advantage of cyanide-based baths is the inherent stability of the baths, as measured by the equilibrium or stability constant. The equilibrium constant is defined as the ratio of the concentrations of products of the chemical reaction divided by that of the reactants. A high value of the stability constant (usually greater than 10^2) indicates that at equilibrium the solution products are not likely to dissociate into reactants.

Cyanide baths have a stability constant of 10^{39} [1, 2]. The stability constant being related to the equilibrium constant for the reaction $Au(CN)_2^- = Au^+ + 4(CN)^-$

However, there are two main drawbacks to cyanide baths. A cyanide bath is highly toxic and needs extreme care in preventing personnel from getting exposure. The second disadvantage is that it can attack the positive photoresist, preventing its use in patterned plating. Gold sulfite baths are being used as a more environmentally friendly alternative. The stability factor for gold sulfite is much lower, 10^{10}. Therefore stabilizing agents must be used to prevent precipitation of gold out of the solution. The gold–thiosulfate complex, $[Au(S_2O_3)_2]^{3-}$ has a stability coefficient of 10^{26}. The electrodeposition takes place with the following reaction:

$$[Au(S_2O_3)_2]^{3-} + e^- \rightarrow Au + 2(S_2O_3)^{2-}$$

The stability of the thiosulfate ion itself is low, which causes a build-up of thiosulfate in the bath [3]. However, a stable bath can be created by combining thiosulfate and sulfite without adding any stabilizing agents, such bath chemistries are commercially available (Table 20.1).

Table 20.1 Composition of a mixed sulfite–thiosulfate gold bath

$NaAuCl_4$_2H2O	0.06 M
Na_2SO_3	0.42 M
$Na_2S_2O_3$_$5H_2O$	0.42 M
Na_2HPO_4	0.03 M
Tl^+ (added as Tl_2SO_4)	0.03 M
pH	6
Temperature	60°C
Current density	5 mA/cm^2

Note that thallium is added as Tl_2SO_4 for brightening, discussed later in this chapter. Other chemistries have been tried for achieving good deposits for microelectronic applications, with varying degrees of success.

20.4 Electroplating Deposition Process

Plating processes may not be 100% efficient—part of the charge transported may be wasted for hydrogen evolution or reduction of other ions and not deposition of desired metal on the cathode. In most electronic applications, a uniform deposit is desired (except for filling via holes, to be discussed later and in Chapter 19). The deposit should meet the requirements of resistivity or density, surface smoothness, hardness, and stress. Metal ions are not deposited in atomic layers over the whole wafer, but like in other chemical deposition processes, ions stick to favored nucleation sites. The adions diffuse over the surface until they are incorporated into the plated lattice at steps, edges, or other irregularities. The growth of these metal nuclei advances until they contact other growing nuclei or crystallite sites and bunch up to form crystallites or grains, which are bounded by grain boundaries that may have impurities in them. The growth in general is not epitaxial. The growth pattern depends on the bath, impurities, and details of the deposition process, such as the growth rate. In cyanide baths, adsorbed cyanide ions cause formation of fine grain deposits. Impurities in the grain boundaries increase film resistance and stress. Stresses in deposits can also result from incorporation of foreign material such as oxides or hydroxides, sulfur, carbon, hydrogen, and metallic impurities (possibly from the anode or other hardware). Electroplated deposits are bonded by the first layer deposits, and the strength of the bond depends upon lattice forces. The plated film adhesion is therefore as good as the substrate metal unless there is a huge mismatch in atomic size or lattice constant between the metals. Adhesion therefore primarily depends upon substrate preparation.

20.4.1 Pulse Plating

Plating with current excitation by pulses rather than steady DC has been used to improve plating film quality as well as higher throwing power for coverage of deep holes and geometries. Pulse plating results in a higher nucleation rate and thus finer grains. Impurities are believed to be desorbed between pulses. Diffusion paths are blocked due to high anion concentration, surface mobility decreases, and the structure becomes more columnar, orientated to the seed layer crystals.

20.5 Metal Deposition Mechanisms

An electrode normally carries an electric charge, so water molecules, which are dipoles, and other complex ions are attracted to the electrode. The dipoles of different species surround the electrode and form an electrical double layer that has a measurable capacitance. Metal ions to be deposited have to make their way through this double layer for deposition.

20.5.1 Polarization

When the electrode potentials are at their equilibrium value, the anodic and cathodic currents are equal and there is no change. To produce a change, an external potential is applied; this is called polarization. Due to impurities, oxides, or contaminants at the surface of the electrode (wafer to be plated), extra voltage is needed to initiate plating. This voltage is higher than the equilibrium voltage. The total polarization or applied potential needed to maintain a steady current is the sum of the overvoltages needed for activation polarization (minimum energy for the reaction to start), concentration polarization (due to mass transport not keeping up with cathodic deposition), and resistance polarization (overvoltage needed for ohmic resistance of the bath).

20.5.2 Diffusion and Mass Transport

Diffusion is the nonrandom movement of ions or neutrals in response to the concentration gradient. The region near the electrode where the concentration of species drops from the bulk value is called the diffusion layer. This boundary is not sharp, so an arbitrary drop of concentration of 1% is defined as the diffusion layer [1]. Figure 20.2 illustrates the concept of the diffusion layer.

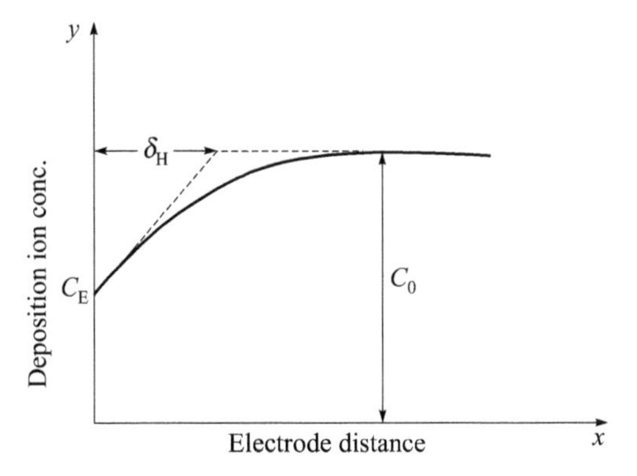

Figure 20.2 Metal ion concentration near the cathode during electroplating deposition. C_0 is the bulk concentration and δ_H is the effective diffusion layer or Nernst thickness.

When an external potential is applied, and the overvoltage starts a current, ionic migration starts; however, metal ion movement can only supply a fraction of the metal ions needed to satisfy the rate of deposition process at the cathode. This fraction is called the transport number. The remainder must be supplied by diffusion or convection (called agitation in practice). In the presence of complex ions and salts the transport number can be nearly zero. Therefore, plating systems and processes are optimized to improve the overall transport of depositing cations through the bath and across the double layer. The agitation effectively reduces diffusion layer thickness, and the overall transport of ions is increased.

Pulse plating, where a steady DC current is replaced by pulses of a different duty cycle, can be used to improve plating coverage into deep vias because it allows ions more time to diffuse into the growing deposit and thus give denser coverage at the bottom of the vias.

20.5.3 Microthrowing Power

The ability of a plating process to produce deposits of uniform thickness over irregular surfaces is called throwing power. Throwing power over large-size features is called macrothrowing power; for micron-size vias, it is called microthrowing power. Throwing power can be measured by comparing thicknesses in valleys compared to peaks after adjusting for current density differences. In alkaline baths or in general baths with complex ions, a high concentration polarization causes current efficiency to fall off rapidly as current density is increased. This results in more uniform deposits on uneven surfaces. Chemical additives to the plating bath can help improve microthrowing power and general plating uniformity. Most plating baths contain brighteners and some contain levelers.

Electroplating process accentuates roughness by depositing faster on peaks than in the valleys, because the current density is highest at the peaks, where the electric field is highest. Leveling in plating is similar to planarization in dielectric deposition. Leveling agents are added to baths to improve throwing power [1, 4]. These agents tend to accumulate on high points reducing the deposition rate there and allowing the valleys to be filled. See Fig. 20.3 for the mechanism of leveling.

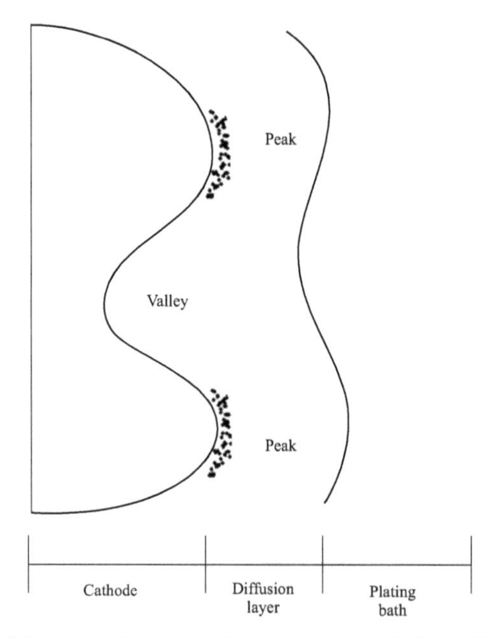

Figure 20.3 Schematic showing micro-roughness of the cathode, with the leveling agent accumulated at the peaks, thus allowing the valleys to fill up to improve microthrowing power.

20.5.4 Brightening

Plated films may not stay smooth on highly polished surfaces causing a matte finish. Bright surfaces result when crystallites in the deposit are smaller than the wavelength of light (0.4 μm). Also, oriented grains are better for high brightness. Brighteners are generally organic molecules that suppress the tendency of plating to give a matte finish. Thallium compounds are also used for brightening by a mechanism of grain refining. Thallium ions also promote the deposition of organic additives. Brighteners appear to suppress fast growth at preferred points, like crystallographic differences, lattice kinks, or growth steps. For other details and mechanisms, the reader is referred to standard texts [1].

20.6 Process Monitoring

Resistivity of gold deposits is supersensitive to metal impurities. Therefore, resistivity monitoring is sufficient for impurity detection. A 1% Fe impurity can raise resistivity from th eideal 2.2 μ·ohm-cm by an order of magnitude. Other metallic impurities have similar (but not so large an effect. Color changes are seen even at lower impurity levels, before the changes in electrical resistivity can be detected by statistical process control (SPC) monitoring of film resistance.

Other properties that are monitored in production are film stress and via resistance. For thin substrates below 100 μm thickness, even a small amount of stress can cause bowing and handling or process issues in subsequent process steps. A high stress level may be caused by a rise of total carbon in the bath. Stress is a function of bath temperature, as seen in Fig. 20.4a. It drops and then rises sharply with an increase of temperature. Current density has a similar response, shown in Fig. 20.4b [5, 6].

20.6.1 Hull Cell

A Hull cell is a basic tool to check plating baths over a tenfold change in current density (see Fig. 20.5). It is a trapezoidal container of nonconducting material with one side at a 37.5° angle to the other. The anode is placed against the right angle and the cathode is placed on the angled side. Current is passed through the cell for a few minutes and the cathode film examined for texture. The current density varies over the cathode because of the variation of the distance from the anode. The cells are made in a small known volume, so additions of additives in grams can be converted to ounces per gallon of a real plating system. Hull cell tests are quick and easy. If problems are detected, bath chemical analysis is performed for more quantitative analysis of bath constituents and additives.

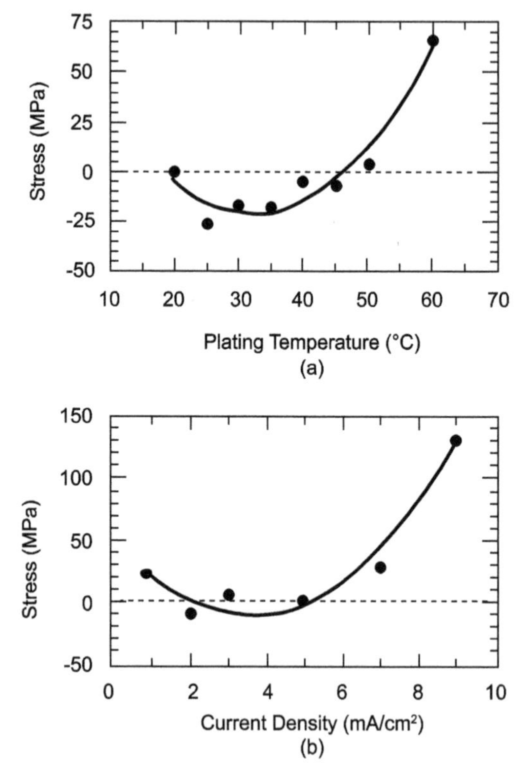

Figure 20.4 Stress dependence on bath temperature and current density, plated at 50°C, from a bath containing 75 ppm thallium. From Ref. [6]. Copyright © 2010. Reproduced with permission of John Wiley & Sons, Inc.

20.7 Electroless Plating

In microelectronic fabrication, sometimes, it is difficult to provide an electric conduction path to the feature to be plated. In such cases, electrons must be provided to the ions to be plated out from the bath itself. This is done by furnishing reducing agents in the bath chemical formulation. Deposition thicknesses can be very uniform, and insulating substrates can be plated. Electroless plating is useful for plating into through-wafer vias, because seed layer coverage by other physical deposition methods inside the vias can be poor. Nickel is the most common metal deposited by the electroless technique for electronic applications.

20.8 Copper Electroplating [1, 7]

Copper plating has been used for a long time in electrical circuits. Copper is becoming increasingly important for interconnects in the III–V semiconductor industry due to the high cost of gold. In many high- volume fabrication processes, copper has replaced gold for backside deposition. Cyanide baths have been used in automotive and for electrical applications like printed circuit boards because of the excellent throwing power. Copper cyanide is insoluble in water, but baths are made by using alkali metal cyanide and other additives.

Acid copper baths have been used for numerous applications since the 1800s. These baths contain copper sulfate and sulfuric acid. Other acids such as hydrochloric acid may be added to improve the poorer throwing power of acid baths. Just like gold baths, additional agents for brightening, surface finish, and hardening are added to commercial baths.

Copper itself is used as the anode. Oxygen-free high-conductivity (OFHC) copper is recommended. The basic film structure is columnar compared to the more fine-grained structure for cyanide baths. Periodic current reversal is used for leveling.

Small amounts of additives cause drastic changes in the deposit, just like in gold electrodeposition. Bath chemistries are generally proprietary; the number of additives can be very large. So, the plating engineer needs to work with the plating bath provider to control the process and avoid yield loss due to poor adhesion, off-color and high-resistivity deposits, film stress, and nonuniformity. A few general guidelines should be kept in mind.

Good uniformity, throwing power, and film quality can be achieved by lowering ohmic resistance rather than mass transport, so the plating solution conductivity must be kept high. A high sulfuric-acid-to-copper-ratio bath must be used, with a ratio of 10 and above recommended for high-aspect-ratio geometries [7]. Levelers and brighteners should be incorporated; with their addition the plating rate can be increased, while maintaining good uniformity. The plating system must be designed to ensure adequate flow everywhere on the wafer to avoid mass transport problems.

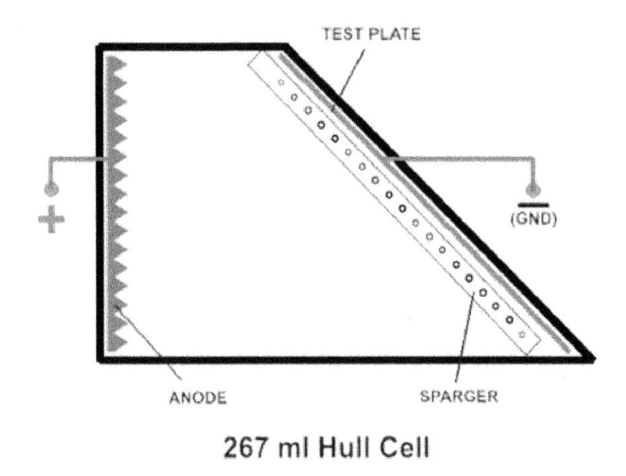

Figure 20.5 Schematic for a 267 mL Hull cell, a commonly used small size.

20.8.1 Large-Volume Production [8]

For large-volume production continuous multihead systems are used. A plating system may consist of four heads for plating and cleaning and prewetting stations. Details of one plating bath are shown in Fig. 20.6. The electroplating solution is held in a tank and circulated through the plating bath. Wafers are held with the plating side down over an anode that is also used for supplying an electrolyte jet. The anode design should ensure adequate flow of the electrolyte over the whole wafer [9]. The wafer seed layer is connected to the cathode connection of the power supply by spring-loaded hooks. The sequence of cleaning, prewetting, immersion in the bath, and application of plating power is controlled carefully to ensure good adhesion and make sure the seed layer is not deplated before plating starts. Good corner coverage at the bottom of the via is shown in Fig. 20.7.

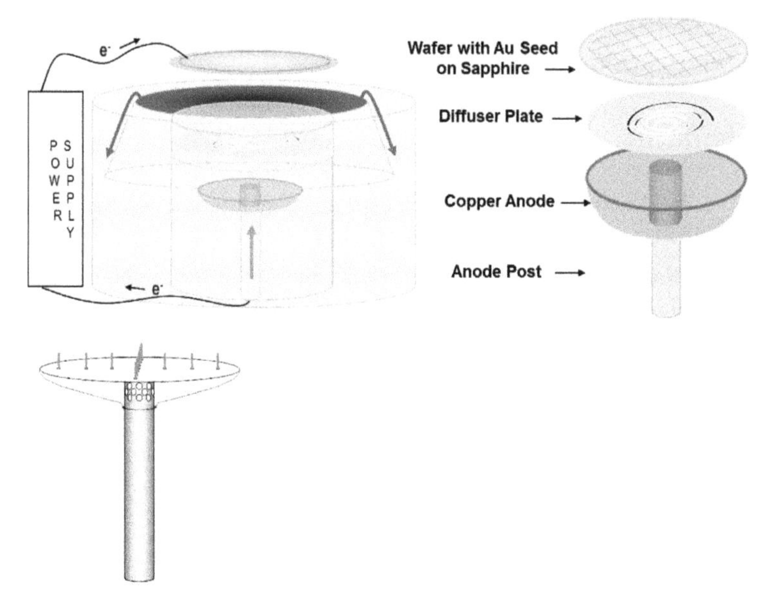

Figure 20.6 Details of one plating bath for a large-volume plating system for copper plating on top of a gold seed and details of showerhead cup with a central hole. Reprinted from Ref. [9] with permission. Copyright © 2011 CS MANTECH.

Figure 20.7 Gold plating coverage improvement (courtesy Skyworks Solutions).

More details of copper plating relevant to III–V processing are discussed in Chapter 19.

References

1. F. A. Lowenheim, *Modern Electroplating*, Wiley Interscience, New York (1974).
2. M. Kato and Y. Okinaka, Some recent developments in non-cyanide gold plating for electronics applications, in *Gold 2003: New Industrial Applications for Gold*, Springer International, Vancouver, Canada (2003).
3. A. Sullivan and P. Kohl, *J. Electrochem. Soc.*, **144**, 1686 (1997).
4. L. Oniciu and L. Mureşan, Some fundamental aspects of levelling and brightening in metal electrodeposition, *J. Appl. Electrochem.*, **21**, 565 (1991).
5. W. J. Dauksher et al., *Microelectrron. Eng.*, **23**, 235 (1994).
6. P. A. Kohl, Electrodeposition of gold, in *Modern Electroplating*, Eds., M. Schlesinger and M. Paunovic, John Wiley and Sons (2010).
7. P. Stransky, A review of copper plating high aspect ratio plated through holes, Finishing.com online resource.
8. D. Anderson, S. Tiku et al., *Cycle Time Reduction during Electroplating of Through Wafer Vias for Backside Metallization of III-V Semiconductor Circuits*, CS MANTECH (2003).
9. J. Riege et al., *Plating Showerhead System for Improved Backside Wafer Plating*, CS MANTECH (2011).

Chapter 21

Measurements and Characterization

21.1 Introduction

Measurements of physical and electrical properties of films and layers that are used in the fabrication of electronic devices are essential for the process development of these materials. Monitoring and statistical process control (SPC) of these are essential for maintaining process control and device quality during production. Electronic properties of bulk semiconductors and layers grown or deposited on wafers must be monitored. It is necessary to evaluate parameters such as carrier concentration, carrier mobility, and doping profiles of implanted or epigrown layers. Electrical properties of various layers deposited for contacts and interconnects are measured routinely for process control. Different physical and electrical techniques are discussed in this chapter. Some specific methods such as ohmic contact measurement are discussed in the chapters on those specific topics. Most common device parameters for field-effect transistors (FETs) and heterojunction bipolar transistors (HBTs) are also described.

21.2 Sheet Resistance

Sheet resistance of semiconducting and conducting layers is generally measured by the four-point probe method or contactless methods.

III–V Integrated Circuit Fabrication Technology
Shiban Tiku and Dhrubes Biswas

ISBN 978-981-4669-30-6 (Hardcover), 978-981-4669-31-3 (eBook)
www.panstanford.com

21.2.1 Four-Point Probe Method [1, 2]

For a rectangular layer, the resistance R is given by

$$R = R_S.\frac{l}{w} \tag{21.1}$$

where R_S is the sheet resistance of the layer, l is the length, and w is the width of the rectangle.

A four-point probe is the most commonly used method for measuring resistivity, R_S, of conducting metal or heavily doped semiconducting layers. In this method four probes, equally spaced along a line, are placed on the layers. A current (I) is passed through the outer two and the voltage across the inner two (V) is measured, as shown in Fig. 21.1a. To calculate the sheet resistivity from the measured voltage between two points, consider the geometry of Fig. 21.1b. If the spacing, s, between the points is small with respect to the size of the sheet, it can be considered infinite. The potential at any point P is given by [3]

$$\psi_p = \frac{I.Rs}{2\pi}.\ln\left(\frac{r_2}{r_1}\right) + A \tag{21.2}$$

where r_1 and r_2 are the distances of point P from the positive and negative voltage points and A is is a constant of integration. For the geometry of Fig. 21.1c

$$\psi_1 = \frac{I.Rs}{2\pi}.\ln(2) + \mathrm{A} \tag{21.3}$$

$$\psi_2 = -\frac{I.Rs}{2\pi}.\ln\left(\frac{r_2}{r_1}\right) + \mathrm{A} \tag{21.4}$$

Therefore $\psi_1 - \psi_2 = \frac{I.Rs}{\pi}.\ln(2)$ (21.5)

From the above and $V = \psi_1 - \psi_2$

$$R_S = \left\{\frac{\pi}{\ln 2}\right\}.\frac{V}{I}$$

$$R_S \approx 4.53\ \frac{V}{I}. \tag{21.6}$$

Equation 21.6 is used in practice. The formula can be corrected for a finite sheet, but for all practical purposes, where the wafer diameter is 100 times the probe spacing, the factor 4.53 is close. This formula is valid for sheet resistance near 10 ohm-cm but is used over a wider range of resistivity. More accurate formulas have been derived for other cases, and correction factors have been derived for wafers of smaller diameter.

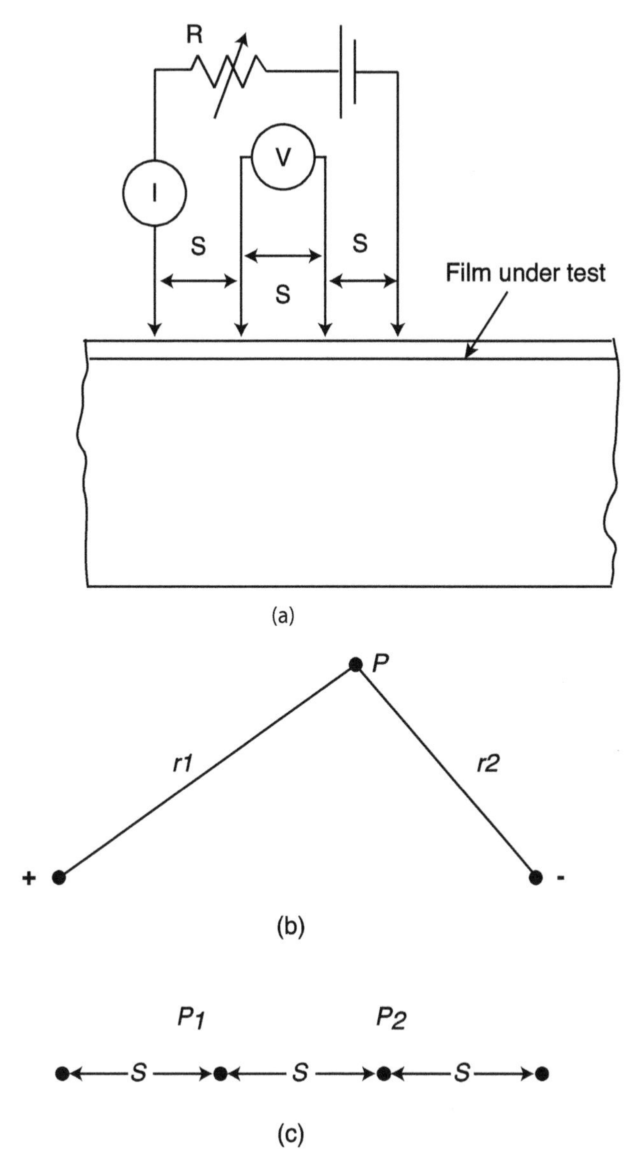

Figure 21.1 Schematic diagram for the four-point probe method (a). Potential at point P due to field applied between two points (b).

This is a nondestructive measurement method; however, the pins may damage the layer being measured if excessive pressure is applied, and therefore the method is used generally on test wafers.

21.2.2 Van der Pauw Method [1–3]

This technique developed by Van der Pauw is applicable to arbitrary-shaped layers of uniform thickness and composition. Typical samples are made as a square with contacts placed at the corners. Results are sensitive to the placement of these contacts, unless the active area is defined by isolating active regions between the contacts to make a clover leaf shape (see Fig. 21.2). The specific resistivity can be determined by passing a current through adjacent terminals and measuring the voltage developed across the other two.

$$\rho = \left\{\frac{\pi}{\ln 2}\right\} .t.\,R \tag{21.7}$$

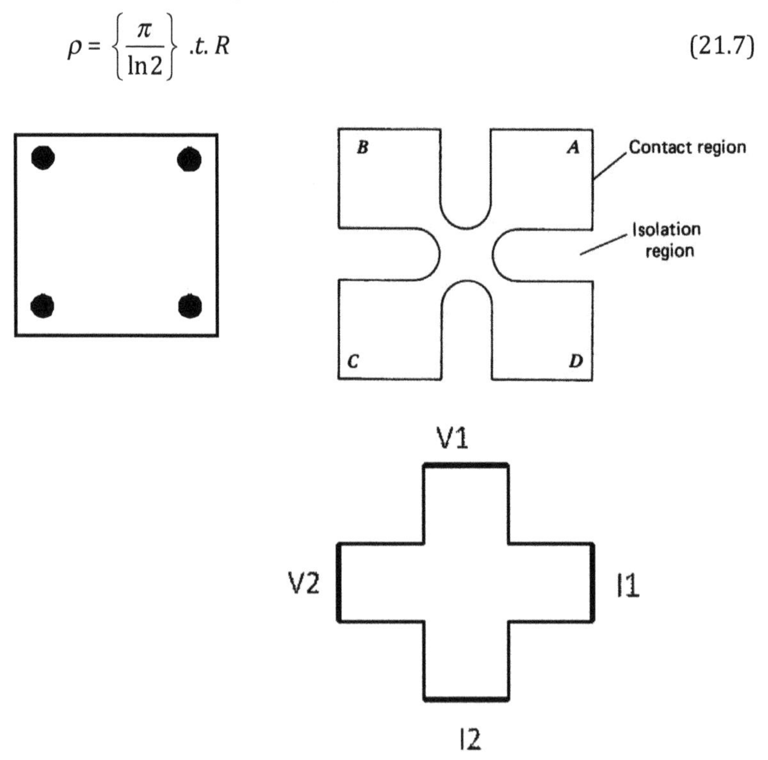

Figure 21.2 Sample geometry showing simple four ohmic contacts on a square sample and layouts more suitable for measurements on wafers by Van der Pauw clover leaf or cross.

21.3 Contactless Resistivity Measurement [4]

The above-described methods need sample preparation or damage the film under test. Contactless methods are fast and nondestructive. Numbers of methods have been tried, but the most common ones in production are based on eddy current or dielectric loss measurement.

In the eddy current method, two coils, as shown in Fig. 21.3, are used to measure the effect of introduction of a conducting sample between the two coils at radio frequency (RF). Voltages are measured before and after the sample is moved in between. The lower head is small, but the upper head for measurement is large (14 mm for the Lehighton system) and determines the resolution. NIST-traceable standards are used for calibration. Resistivity in the range of 0.1 ohm /sq. to 10 kilo-ohms/sq. can be measured by this system.

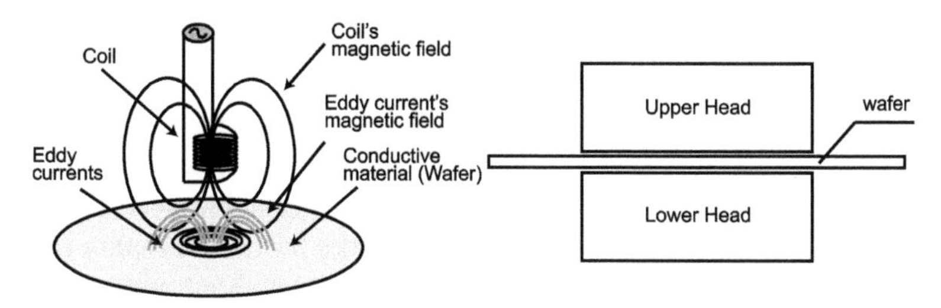

Figure 21.3 Contactless measurement system, Lehighton type.

Systems based on dielectric loss measurement use microwave frequency. The permittivity of a conducting layer can be given as

$$\varepsilon = \varepsilon_0\varepsilon_r\{1 - \tan\delta\} \quad (21.8)$$

where $\tan\delta$ is the effective loss tangent of the layer, two components, and dielectric and conductor portion

$$\tan\delta = \tan\delta_d + \tan\delta_c = \tan\delta_d + \frac{\sigma}{\omega\varepsilon_0\varepsilon_r} \quad (21.9)$$

Conductivity σ (and resistivity) is computed from the conductor portion of the loss tangent. RF and microwave contactless methods give similar results.

21.4 Carrier Mobility

21.4.1 Hall Mobility

In GaAs and other III–V semiconductors, mobility is not related to carrier concentration (as in silicon), because it varies as a function of the degree of compensation, the net carrier concentration being the difference between the n- and p-type dopant sites. Mobility thus varies with defects acting as compensating sites and scattering centers. In such cases mobility must be determined by making measurements with the fat-FET (a FET with a large gate length) method or with the Hall effect method. The fat-FET method gives the drift mobility and the Hall effect method measures the so-called Hall mobility. Hall mobility will be described first in this section.

Advantages of this method are that a small sample can be used or the test can be done with a test structure isolated on a wafer. This technique is not useful if the doping concentration is not constant.

The layer to be evaluated is grown on a semi-insulating substrate and a bar-shaped piece is cleaved out for Hall measurement (see Fig. 21.4). The piece is held in a magnetic field while current is forced through two contacts along the length of the bar, and the electric field is measured across the other two faces of the bar. Hall mobility can be determined by measuring voltage with and without a magnetic field applied. The additional voltage in the presence of a magnetic field is called Hall voltage, V_H.

Figure 21.4 shows the sample dimensions and Hall effect geometry [3]. According to the Hall effect, an electric field is generated mutually perpendicular to the current and the magnetic field and is given by

$$\mathbf{E}_H = \frac{R_H.I.\mathbf{B}}{A} \tag{21.10}$$

where R_H is the Hall coefficient, I is the current, A is the cross-sectional area ($w.t$), and $\mathbf{B}$ is the magnetic field.

$V_H = \mathbf{E}_H.w$ (21.11)

$$R_H = \frac{A.\mathbf{E}_H}{I.\mathbf{B}} = \left\{\frac{t}{I.\mathbf{B}}\right\}.V_\mathbf{B} \tag{21.12}$$

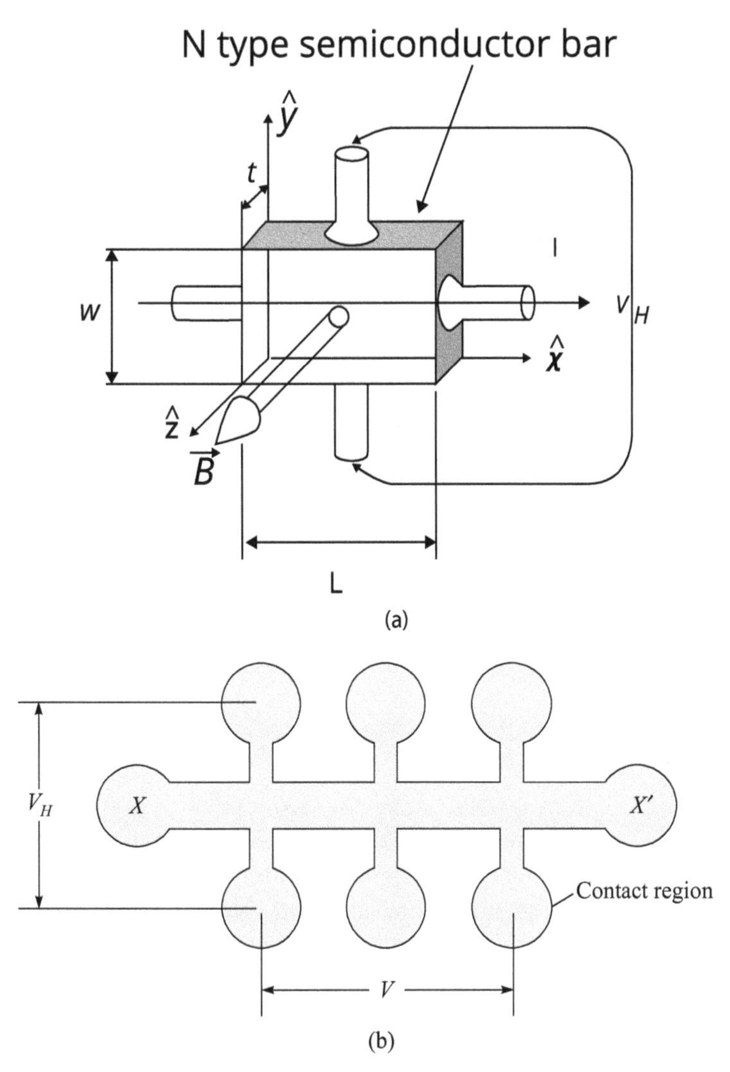

Figure 21.4 (a) Hall effect; the diagram shows field orientations. The magnetic field (**B**) is along the z direction. (b) Contact layout geometry of the sample for Hall measurement.

Hall mobility is given by

$$\mu_H = \frac{R_H}{\rho} \tag{21.13}$$

where ρ is the resistivity of the bar.

Samples can also be made using a photopattern on a wafer. Resistivity can be measured on the same sample by the Van der Pauw method or the transmission line method (TLM) to characterize the sample and estimate the carrier concentration.

21.4.2 Drift Mobility [5, 6]

Drift mobility is based on the response of electrons to the electric field as opposed to the magnetic field in the case of Hall mobility, and the two may differ but are related. Drift mobility can be measured with process control monitor (PCM) structures that can be placed on wafers, a large-length device with ohmic contacts on each side, and a FET-like gate in between. Figure 21.5 shows the geometry of a fat-FET used for such measurements. The current for a channel of width W and depth or thickness a is given by

$$I = q\mathbf{v}NWa = q\mu_\mathrm{d}ENWa \tag{21.14}$$

where $\mathbf{v}$ is the electron velocity = $\mu_\mathrm{d}\mathbf{E}$, $\mathbf{E}$ is the electric field = V/L , and a is the channel thickness.

Therefore

$$\mu_\mathrm{d} = \left\{\frac{L}{qNWa}\right\}\left(\frac{I}{V}\right) \tag{21.15}$$

This equation is valid under the assumption of a low electric field, below that required for saturation velocity, negligible contact resistance, and same surface depletion over the whole length L.

If the doping level N is not known, it can be measured by the capacitance–voltage (C–V) method on a separate structure nearby or by gathering I–V and C–V data on a fat-FET (Fig. 21.5). From C–V measurements, the $N(x)$ versus depth x can be determined.

$$I = q\,\mu_\mathrm{d}\,\mathbf{E}\,N\,W\,a \tag{21.16}$$

$$dI = q\,\mu_\mathrm{d}\,(x)\,\mathbf{E}\mathrm{N}(x)\,W\,dx \tag{21.17}$$

$$\frac{dI}{dV_g} = q.\mu_d.\mathbf{E}NW\frac{dx}{dV_g} \tag{21.18}$$

where V_g is the gate voltage applied to the fat-FET.

Using $C = \varepsilon\frac{A}{x}$ and $N(x) = \frac{C^3}{q\varepsilon_s A^2}\left(\frac{dV}{dC}\right)$ and using Eq. 21.25 (from the C–V method in Section 21.5 below)

$$\frac{dx}{dV_g} = \left\{-\frac{eA}{C^2}\right\}\left(\frac{dC}{dV_g}\right) \tag{21.19}$$

can be derived from the above equations.

$$\frac{dx}{dVg} = \left\{-\frac{eA}{C^2}\right\}\frac{C^3}{q\varepsilon_s A^2}.Nx = \frac{C}{qAN} \tag{21.20}$$

Combining Eqs. 21.16 and 21.20

$$\frac{dI}{dV_g} = \mu_d \frac{\mathbf{E}C}{L} = \mu_d \frac{V}{L}.\frac{C}{L} \tag{21.21}$$

or $$\mu_d = \left\{\frac{dI}{dV_g}\right\}.\left\{\frac{L^2}{CV}\right\} = g_m \frac{L^2}{CV}. \tag{21.22}$$

where $g_m = \dfrac{dI}{dVg}$ is the transconductance of the fat-FET.

Transconductance and the capacitance data can be measured using the fat-FET geometry or can be measured on two structures, a FET-type device and a capacitance dot placed nearby.

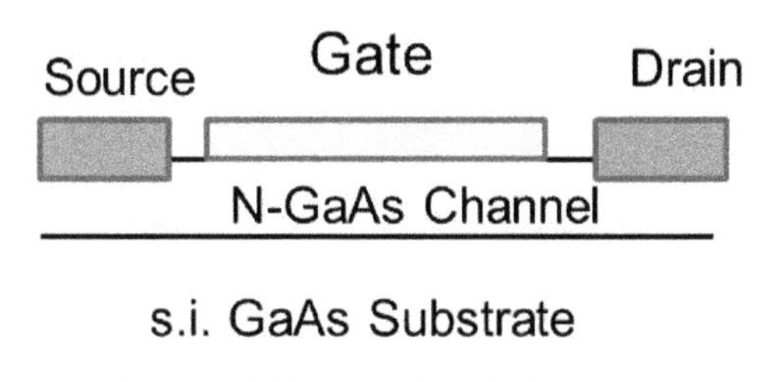

Figure 21.5 The fat-FET structure for *I–V* measurement for drift mobility determination.

21.5 Doping Profile by *C–V* Method [7]

Secondary ion mass spectrometry (SIMS) is a direct measurement technique to measure profiles of dopants, impurities, or the composition of epilayers. This technique is slow and expensive and does not give the effective carrier concentration profile but the total dopant concentration. The *C–V* technique is faster and more relevant. In this technique a diode structure is used, as shown in Fig. 21.6. For GaAs a metal layer is deposited in the form of small circles. The second contact can be a large-area contact of the same metal, which acts as an ohmic contact. Or the layer can be grown on an n+ substrate and the ohmic contact made on the backside. For a PCM monitoring the ohmic contact on product wafers, the PCM *C–V* structure can be made like a large-area Schottky diode. Figure 21.6a shows the depletion layer formed under the metal, which forms a capacitor. Figures 21.6b and 21.6c show the schematic diagram of the geometry and the doping profile. If the voltage on the Schottky contact is changed by ΔV, the depletion layer width increases by Δx; this uncovers an additional charge $qN(x)\Delta x$.

$$\Delta \mathbf{E} = \frac{\Delta V}{x} = \left(\frac{q}{\varepsilon_s}\right) N(x) \tag{21.23}$$

Incremental small-signal capacitance

$$C = \varepsilon_s \cdot \frac{A}{x} \tag{21.24}$$

Eliminating x and Δx from these two equations, $N(x)$ is found to be [3, 5]

$$N(x) = -\frac{C^3}{q\varepsilon_s A^2}\left(\frac{\Delta V}{\Delta C}\right) \tag{21.25}$$

By measuring the capacitance versus voltage for a known-area device, a point-by-point doping profile can be determined.

21.6 Schottky Diode Parameter Measurement

Schottky barriers are used extensively in III–V circuits as diodes and as gates in FET*type devices. The behavior of these and the performance of the devices depend on the characteristics of these contacts. These contacts are sensitive to processing parameters and thus monitoring of the Schottky diode parameters is essential for good process control. The two basic parameters that are measured are the barrier height and the ideality factor. Leakages are also monitored in practice, but those are straightforward to measure.

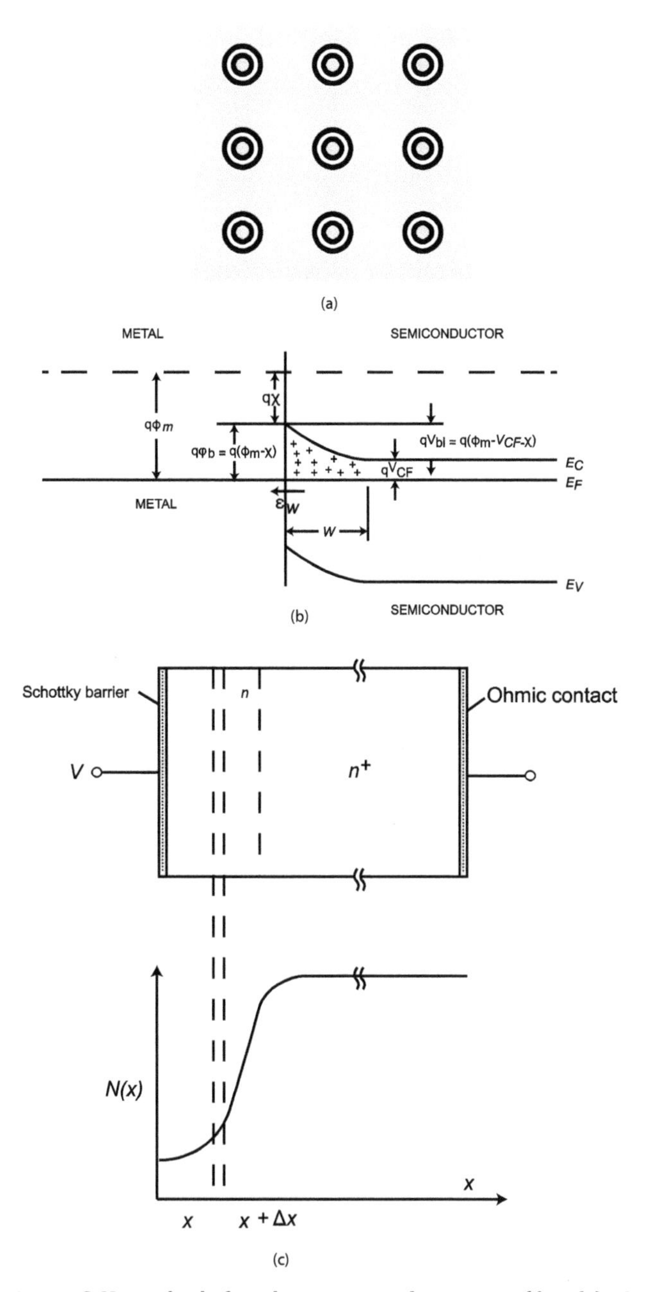

Figure 21.6 *C–V* method for determining doping profile. (a) Sample geometry showing metal dots in a sea of metal, (b) energy band diagram of Schottky contact, and (c) doping profile.

21.6.1 Current–Voltage Method

Schottky diodes were discussed in Chapter 2. The *I–V* relationship can be written as

$$J = \frac{I}{A} = A^{**}\, T^2 \exp\left\{\frac{qV}{nkT}\right\}. \tag{21.26}$$

for $V >> kT/q$. This equation can be written as

$$\ln J = \ln (J_s) + \frac{qV}{nkT} \tag{21.27}$$

where $J_s = A^{**} \exp\left\{-\frac{q\phi_b}{kT}\right\}$.

Therefore a plot of ln (J) versus V should be linear with a slope of q/nkT (see Fig. 21.7). The relationship does not hold at low currents because of leakages and at high currents due to series resistance, etc. The ideality factor n can be calculated from the slope. The intercept on the I axis, J_s, can be used to calculate the barrier height ϕ_b, knowing the value of the Richardson constant A^{**} (120 A/cm^2/K^2). Due to the logarithmic nature of the dependence, the error due to the value of A^{**} and area measurement is small. At room temperature kT/q is about 0.026 V, so the measurement voltages should be over 1 V (for the condition $V >> kT/q$).

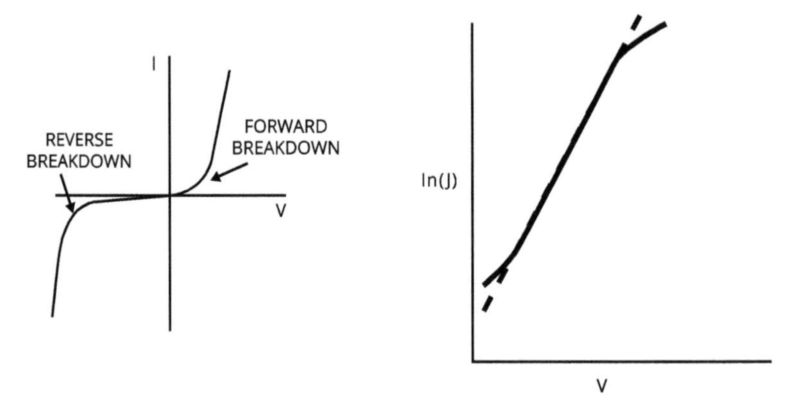

Figure 21.7 Schottky diode *I–V* measurement for determination of barrier height and ideality factor.

21.6.2 Activation Energy Method

Refer to Section 2.1.2 in Chapter 2 (see Eqs. 2.37 and 2.40). As in other activation energy methods, measurements must be made at several temperatures. This method does not need area measurement, so it can avoid errors due to size differences between designed versus actual area. Current I for a device with area A is measured. From Eq. 21.26

$$I = A.\,A^{**}\,T^2 \exp\left\{\frac{qV}{nkT}\right\}. \tag{21.28}$$

For V >> kT/q (fixed $V = V_f$)

$$\frac{I}{T^2} = \exp\left\{\frac{qV_f}{nkT}\right\}. \tag{21.29}$$

$$\ln\left(\frac{I}{T^2}\right) = \ln\,(A.A^{**}) - \frac{q}{kT}\,(V_{bi}\;\frac{V_f}{n} \tag{21.30}$$

I $/T^2$ (current/temperature^2) is plotted against $1/T$ on a semilog plot, The slope of the line is measured.

$$\text{Slope} = \frac{q}{kT}\left(V_{bi} - \frac{V_f}{n}\right) \tag{21.31}$$

There are two unknowns in this equation, V_{bi} and n. If V_{bi} is known, n can be computed. Initially n may be determined by the I–V method to determine the value of V_{bi} for a metal–semiconductor system.

21.6.3 Capacitance–Voltage Method for Schottky Diode Barrier Height Measurement

This method can be used for epigrown or implanted samples where uniform doping can be assumed. As has been discussed elsewhere (Section 2.1.2 in Chapter 2) the fundamental equation applicable here is Poisson's equation, which can be used to derive the capacitance of the depleted region of depth w

$$\frac{d^2V}{dx^2} = -\frac{\rho(x)}{\varepsilon} \tag{21.32}$$

where $V(x)$ is the potential, ρ is the charge density, and ε is the permittivity of the semiconductor ($=k.\varepsilon_0$, k being the dielectric constant). For the case of uniform doping, N, the equation becomes

$$\frac{d^2V}{dx^2} = -\frac{qN}{\varepsilon} \tag{21.33}$$

Integrating this equation and applying the boundary condition $dV/dx = 0$ at $x = w$, the depletion width

$$\frac{dV}{dx} = q\frac{N(w-x)}{\varepsilon} \tag{21.34}$$

Integrating this equation once more and using the boundary condition $V(0) = 0$

$$V(x) = = \frac{qNw\left(x - \dfrac{x^2}{2w}\right)}{\varepsilon} \tag{21.35}$$

and $V(w) = \dfrac{qNw^2}{2\varepsilon}$ (21.36)

From this the Schottky barrier depletion width w is derived as

$$w = \sqrt{\frac{2\varepsilon(V_{\mathrm{bi}} - V_g)}{qN}} \tag{21.37}$$

The capacitance of the depletion region is given by

$$C = \frac{\varepsilon A}{w} \tag{21.38}$$

where A is the area of the Schottky metal (or gate of the fat-FET). Using Eqs. 21.37 and 21.38, we can derive

$$\frac{1}{C^2} = \frac{2V}{q\varepsilon NA^2} \tag{21.39}$$

If $1/C^2$ is plotted as a function of V, the slope of the line should be equal to $2/q\varepsilon N$ and the intercept should be V_{bi}. The ideality factor cannot be determined by this method. A big advantage is that the area of the Schottky diode does not have to be known, which can have errors because of lithography differences. The area of the Schottky contact must be large enough to have a large enough capacitance that can be measured by the available measuring instrument.

21.7 FET Characteristics

21.7.1 FET Transconductance

The most common measurements of the FET are the threshold voltage and transconductance measurements shown in Fig. 21.8a. With the source as ground, drain current is measured as a function of the gate voltage for a fixed drain voltage (voltage high enough for FET current to be in the saturated region). The threshold voltage is determined as the slope of the line fit to the square root of I_d versus V_{gs}. The pinch of voltage is measured as the gate voltage at which the current drops to a low predetermined value (e.g., 2.5% of Idss). The transconductance g_m is determined (and can be plotted) as the slope of the I_d versus V_g curve, $\partial I_{ds}/\partial V_g$ for $V_g = 0$ for depletion-mode (D-mode) and $V_g = 0.6$ V for enhancement-mode (E-mode) devices. *I–V* characteristics of a typical FET are shown in Fig. 21.8b. Idss is the saturated drain–source current defined as the current when V_{dsis} is set at a voltage (say $V_{ds} = 2.5$V).

The on resistance of the FET is defined as the slope of V_{ds} versus I_{ds} calculated at $I_{ds} = 0$

$$R_{on} = \frac{\partial V_{ds}}{\partial I_{ds}} \tag{21.40}$$

For digital FETs, the proportionality factor, k, is reported, where $I_{ds} = k(V_{gs} - V_p)^2$, (see Eq. 2.55, Chapter 2), where V_p is the pinch-off voltage. Slope m is calculated for the plot of I_{ds} versus V_{gs}, and then $k = m^2$.

21.7.2 FET Source Resistance Measurement

If the gate is forward-biased and I_{gs} is measured as a function of V_{gs}, the slope of the *I–V* curve will be the sum of the source and gate resistance. To separate the source resistance a different approach is used. In the common source mode, voltage stimulation is provided on the gate(see Fig. 21.9). The drain is essentially floating, and under a very small current flow, V_{ds} is measured. The source resistance is calculated as

$$\frac{1}{Rs} = \frac{\partial I_g}{\partial V_{ds}} \tag{21.41}$$

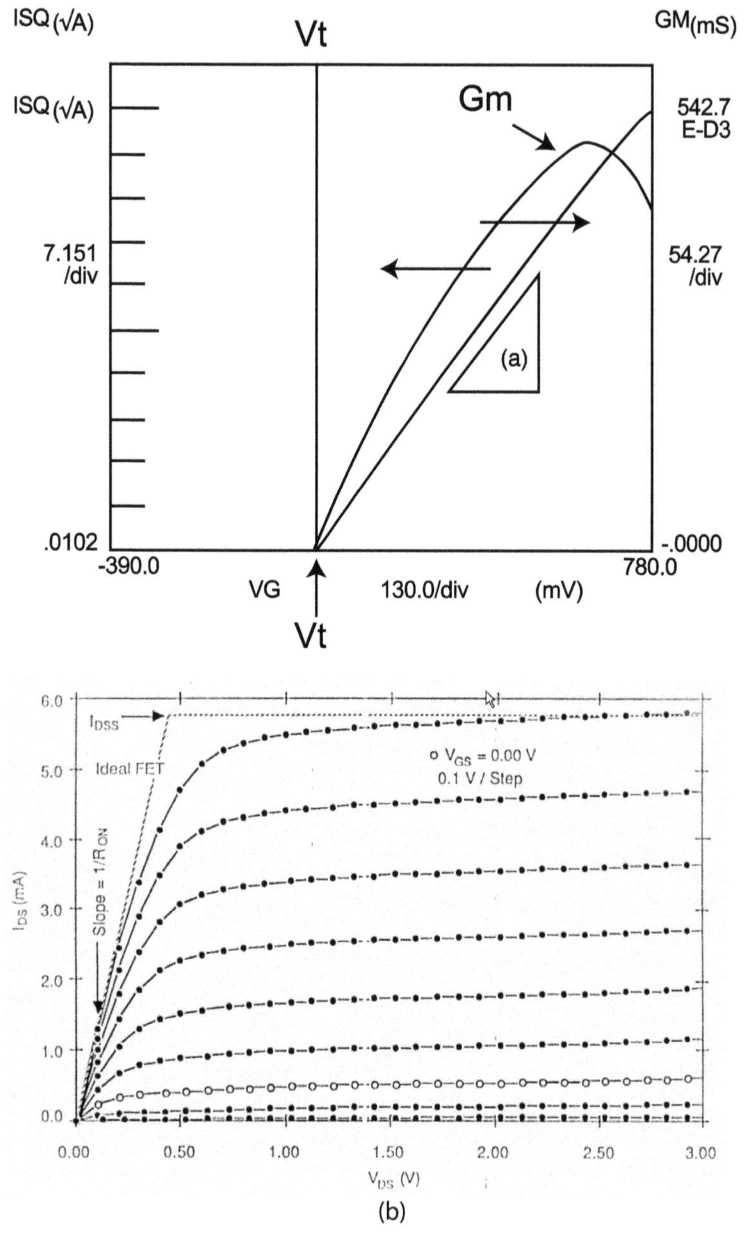

(b)

Figure 21.8 (a) Noise figure measurement setup when the noise source is off and input termination is 50 ohms; (b) output power without and with the noise source on, showing degradation of the signal-to-noise ratio.

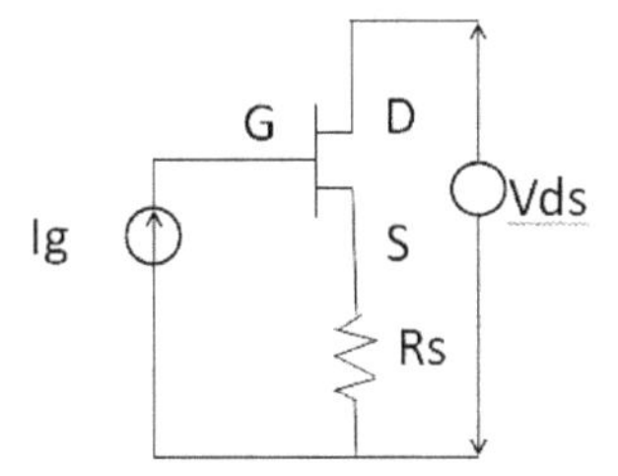

Figure 21.9 FET source resistance measurement.

21.8 HBT Parameter Extraction

21.8.1 Output *I–V* characteristics

This is a series of plots of the collector current I_C as a function of V_{CE}, the collector–emitter voltage, as I_B the base current is stepped up. For a bipolar junction transistor (BJT), these characteristics intercept the voltage axis at zero V_{CE}. However, for an HBT, there is an offset ΔV_{CE} (see Fig. 2.21b in Chapter 2) due to the difference in emitter–base and base–collector junction turn-on voltages.

21.8.2 Gummel Plot

HBT gain (β) is commonly measured by the Gummel method. In this method, the collector–emitter voltage V_{BE} is kept constant (generally V_{BC} = 0) and V_{BE} is varied. The base current I_B and the collector current I_C are measured as a function of V_{BE}. The plot is made on a semilog scale. Direct current (DC) gain (β) is calculated as I_C/I_B at different collector current levels. Other parameters for base–emitter and base–collector junctions, for example, ideality factors, can be derived from the measurements. See Fig. 2.16 in Chapter 2 for a Gummel plot. β is low at low current and reaches an optimum value at high current. At very high currents, β drops due to high injection and the Kirk effect [8].

21.8.3 Emitter Resistance

R_{EE} is measured by a method similar to R_S measurement in FETs. The transistor is connected in a common emitter configuration and I_C is set to zero (see Fig. 21.10). The voltage drop across R_C is nearly zero because $I_C \approx 0$, and the V_{CE} drop appears across R_{EE}. The slope of the plot of I_B versus V_{CE} gives R_{EE}.

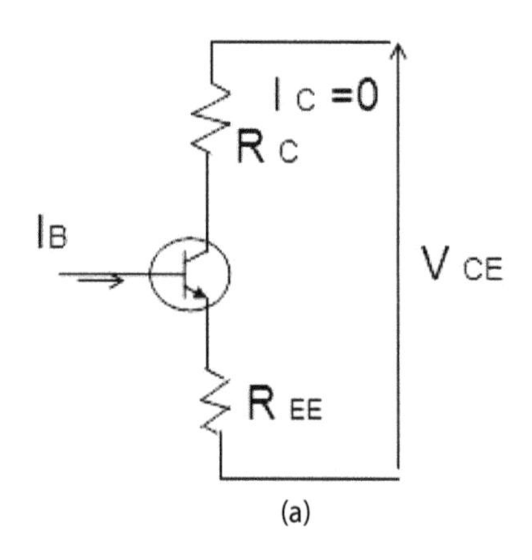

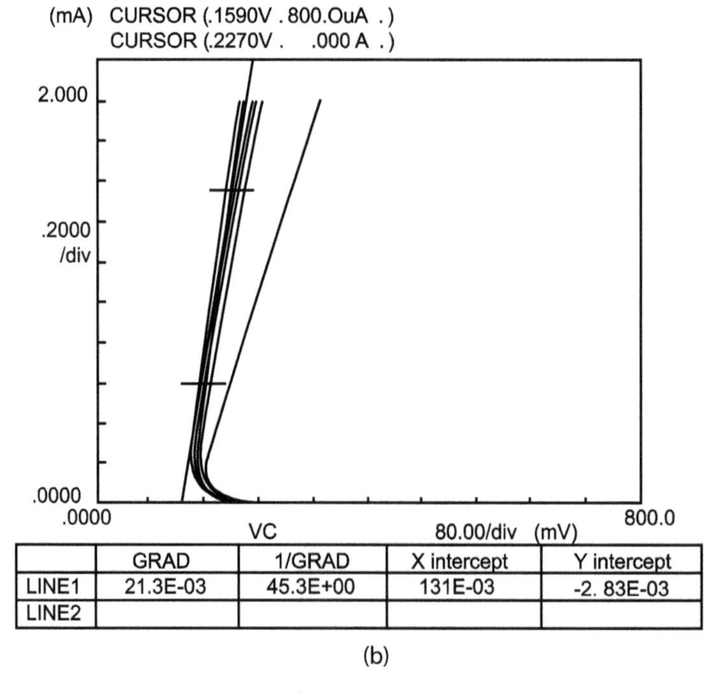

	GRAD	1/GRAD	X intercept	Y intercept
LINE1	21.3E-03	45.3E+00	131E-03	-2. 83E-03
LINE2				

(b)

Figure 21.10 R_{EE} measurement for an HBT using the flyback method. (a) The cicuit and (b) the I–V curve. The slope R_{EE} is 45.3 ohms for this example.

Breakdown voltages of collector–base and emitter–base junctions and between emitter and collector nodes can be measured as voltages at predefined small currents.

21.8.4 V_{CE} Offset

This is an important figure of merit for power at idle, because the operating voltage shifts to higher values and the quiescent current shifts up a bit with a high V_{CE} offset. Reduction of knee voltage, as shown in Figs. 2.21 and 2.22 in Chapter 2, is key to achieving high power efficiency.

21.8.5 R_{on}

As in the case of FET-type devices, the on resistance, R_{on} is minimized, while maintaining other parameters like breakdown voltages (see Fig. 2.15, Chapter 2).

21.9 RF Characterization [8–11]

21.9.1 Introduction

In Chapter 2, the main figures of merit for FET and HBT devices were introduced. Most of these can be measured using DC or low-frequency measurements. However, for high-frequency figures of merit, which are relevant at the operating frequency, RF measurements must be made to measure or extract these figures of merit. Therefore, a simplified version of these methods is described here.

RF measurements are made on RF PCMs, laid out specifically to reduce additional parasitic impedances. An example of an RF PCM is shown in Fig. 21.11, where an HBT is connected by low resistance and inductance traces to the pads. For the common emitter configuration, the emitter contact is connected to large ground pads. Using special probes that can function at gigahertz frequencies, these structures are measured on network analyzers.

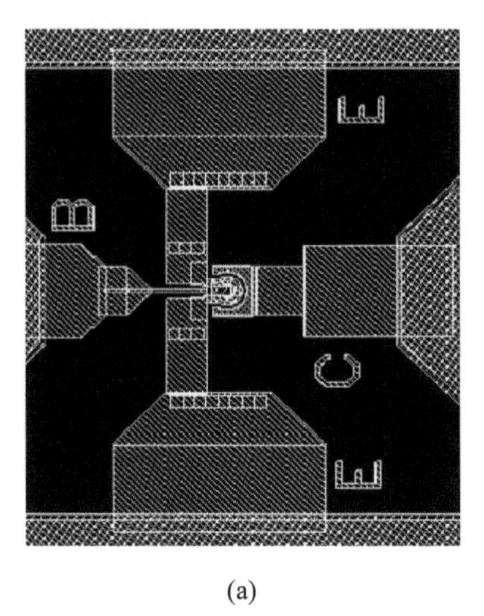

(a)

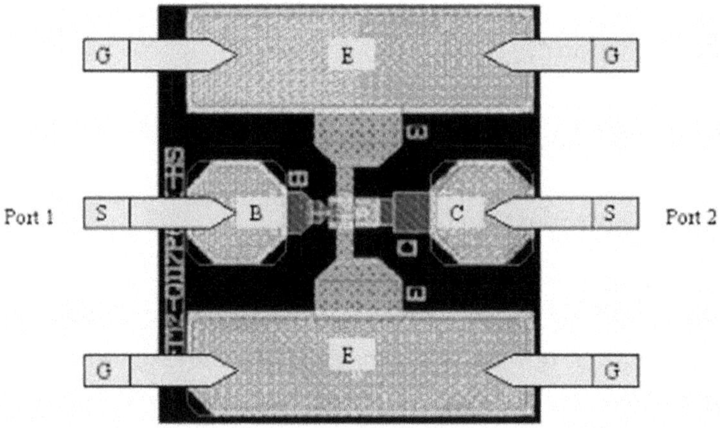

(b)

Figure 21.11 Layout of an HBT showing details of the connections to pads (a) and the RF probable structure (b), with low resistance traces from emitter to ground for the common emitter configuration.

21.9.2 S-Parameter Measurements [9]

Scattering, or S, parameters, as these are commonly called, are routinely used in RF device and circuit characterization. A brief introduction is given here for the understanding of the critical parameters and figures of merit of RF circuits.

Any RF electrical circuit can be viewed as a two-port network, with an input and an output port, as shown in the simple schematic in Fig. 21.12. DC power or bias inputs are internal to the network. V_1 and I_1 are the inputs and V_2 and I_2 are the outputs. The network can be a simple device like a diode or a complicated monolithic microwave integrated circuit (MMIC) module. At low frequencies, impedance Z-parameters can be used to describe the relationship of the output to input parameters.

$$V_1 = Z_{11}I_1 + Z_{12}I_2 \tag{21.42}$$

$$V_2 = Z_{21}I_1 + Z_{22}I_2 \tag{21.43}$$

The relationship can also be written in terms of hybrid, or H, parameters:

$$V_1 = h_{11}I_1 + h_{12}V_2 \tag{21.44}$$

$$V_2 = h_{21}I_1 + h_{22}V_2 \tag{21.45}$$

Simple measurements are made to obtain these parameters. For example

$$Z_{11} = V_1/I_1 \text{ when } I_2 = 0 \tag{21.46}$$

Therefore if output is left open, V_1 and V_2 are used to calculate Z_{11}.

Similarly for H-parameters

$$h_{21} = \frac{I_2}{I_1} \text{ when } V_2 = 0 \tag{21.47}$$

So if the output is shorted and I_2 and I_1 are measured, h_{21} can be calculated. This parameter has been previously introduced as short-circuit current gain.

At microwave frequencies, high-gain devices and circuits are not stable if shorted. Also, it is not easy to short two terminals at high frequency. Perfect zero impedance is hard to achieve at RFs. Also, an open circuit may not be truly open at microwave frequencies. An appreciable amount of capacitance and leakage may be present.

S-parameter measurements circumvent the problems associated with making ideal terminations by using impedance matching at the input and output. Transmission lines with characteristic impedance, generally 50 ohms-sec, are used to connect source and load. This is shown in Fig. 21.12. When the circuit and input power is turned on, there are incident as well as reflected or scattered waves at the input as well as output. These waves, like in any transmission line, result in standing waves. Voltages and currents are given by

$$V_1 = \mathbf{E}_{i1} + \mathbf{E}_{r1} \tag{21.48}$$

and $I_1 = (\mathbf{E}_{i1} - \mathbf{E}_{r1})/Z_0$ (21.49)

where Z_0 is the characteristic impedance of the transmission line. A new set of variables, representing complex voltage waves, is now defined:

$$a_2 = \frac{\mathbf{E}_{i1}}{\sqrt{Z_0}}\, b_1 = \frac{\mathbf{E}_{r1}}{\sqrt{Z_0}}\quad a_2 = \frac{\mathbf{E}_{i2}}{\sqrt{Z_0}}\, b_2 = \frac{\mathbf{E}_{r2}}{\sqrt{Z_0}}$$

where a_1, a_2 represent normalized complex incident voltage waves, and b_1, b_2 represent reflected voltage waves. The square of the magnitude of these represent power (V^2/Z_0) incident or reflected at the corresponding ports.

S-parameters are defined on the basis of these as follows:

$$b_1 = S_{11}a_1 + S_{12}a_2 \tag{21.50}$$

$$b_2 = S_{21}a_1 + S_{22}a_2 \tag{21.51}$$

S_{11} can be measured by making a_2 =0, and $S_{11} = b_1/a_1$. This can be achieved simply by terminating the transmission line with its characteristic impedance, which is not hard to do, like making shorts or opens. Similarly the other scattering parameters can be easily measured. These parameters can be written in matrix form:

$$\begin{bmatrix} b_1 \\ b_2 \end{bmatrix} = \begin{bmatrix} S_{11} & S_{12} \\ S_{21} & S_{22} \end{bmatrix} \begin{bmatrix} a_1 \\ a_2 \end{bmatrix}$$

The definitions of the S-parameters are given below:

S_{11} = Input reflection coefficient with output matched

S_{12} = Reverse reflection coefficient with input matched

S_{21} = Forward transmission coefficient with output matched

S_{22} = Output reflection coefficient with input matched

The power gain is the square of the magnitude of S_{21} and is given by $|S_{21}|^2$.

S-parameters can be used for a complete understanding and characterization of two-port RF networks. For power amplifiers in the linear region, all equivalent circuit parameters can be computed from these.

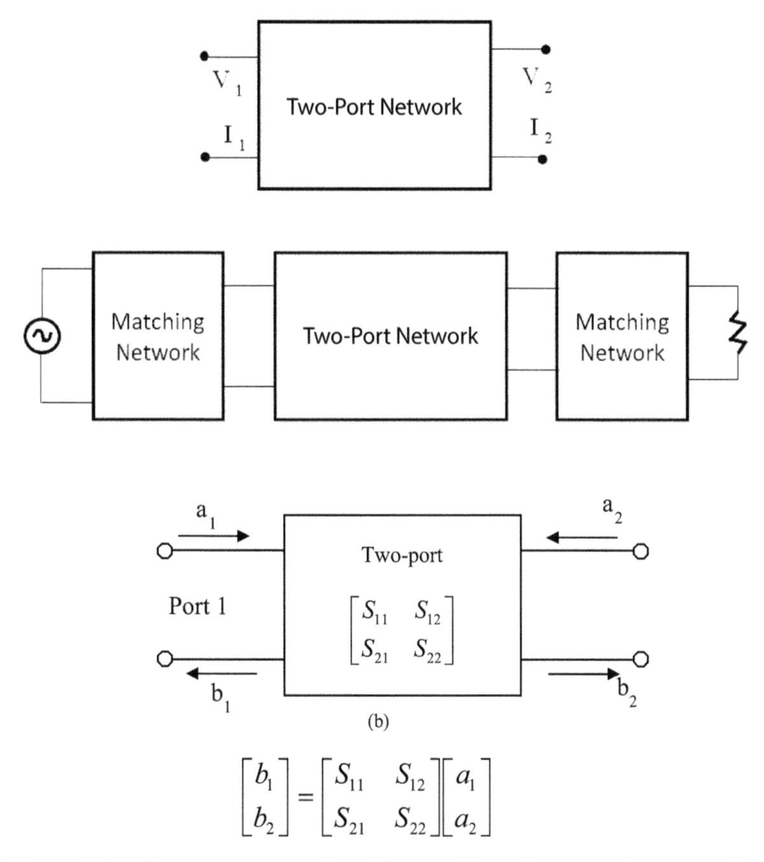

$$\begin{bmatrix} b_1 \\ b_2 \end{bmatrix} = \begin{bmatrix} S_{11} & S_{12} \\ S_{21} & S_{22} \end{bmatrix} \begin{bmatrix} a_1 \\ a_2 \end{bmatrix}$$

Figure 21.12 Two-port network with matching input and output for S-parameter measurements.

21.9.3 RF Figures of Merit [9–11]

These two figures of merit are used in the development of new devices and technologies and in the monitoring of production processes. These were introduced in Chapter 2. Figure 21.13 shows these definitions in schematic form. f_t is projected by plotting h_{21}, short-circuit current gain, which can be derived from S-parameters [12].

$$h_{21} = -\frac{-S_{21}}{\{(1-S_{11})(1+S_{22})+S_{12}S_{21}\}} \tag{21.52}$$

h_{21} can be computed and plotted against frequency to project f_t. For a FET-type device (see Eq. 2.57 in Chapter 2)

$$f_t = \frac{g_m}{2\pi C_{gs}} \tag{21.53}$$

C_{gs} can be calculated by using this expression using the measured value of f_t.

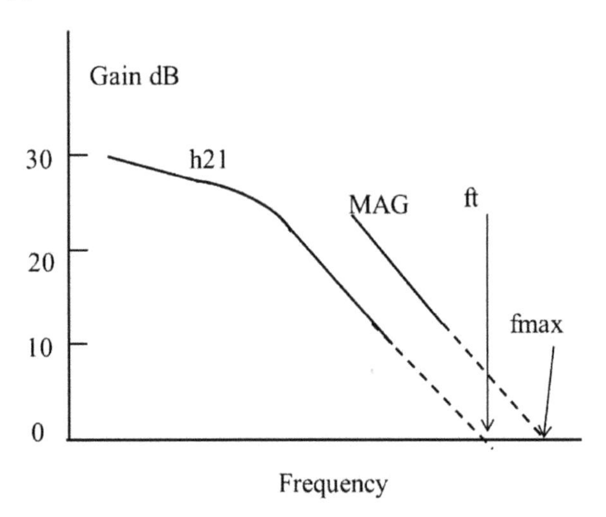

Figure 21.13 f_t and f_{max} determined from projections of measured h_{21} and maximum stable gain (MSG) data to unity gain (0 dB).

In the case of HBTs, f_t is measured again by plotting h_{21}. f_{max} is calculated from the projection of maximum stable gain (MSG) to unity. MSG is calculated from the measurement of S-parameters S_{21} and S_{12}.

$$\text{MSG} = \left|\frac{S_{21}}{S_{12}}\right| \tag{21.54a}$$

MAG, defined above, is the maximum power gain that can be obtained using matching on input and output.

$$\left|\frac{S_{21}}{S_{12}}\right| k - \sqrt{k^2 - 1} \tag{21.54b}$$

However, there is always a small amount of feedback in RF circuits, so additional gain can be obtained by using external feedback to neutralize feedback. The gain under these conditions is called unilateral power gain, U, and is given by

$$U = \frac{|S_{21} - S_{12}|}{2k|S_{21}S_{12} - \text{Re}(S_{21}S_{12})} \tag{21.55}$$

where $\underline{k}$ is the stability factor. This condition is not the same as the reverse reflection coefficient $S_{12} = 0$. The presence of feedback neutralization changes other S-parameters. f_{max} is often calculated as the projected frequency at which U reaches zero (instead of MAG).

21.9.3.1 VSWR

Impedance matching is never perfect in RF circuits, so there is always a reflected wave present. The combination of the incident and reflected waves gives rise to a standing wave, as shown in Fig. 21.14. The extent of reflection can be measured as the voltage standing wave ratio (VSWR), which is defined as

$$\text{VSWR} = \frac{V_{max}}{V_{min}} \tag{21.56}$$

$$\text{VSWR is also equal to } \frac{1+\Gamma}{1-\Gamma} \tag{21.57}$$

In an ideal circuit, $V_{max} = V_{min}$ and 0; therefore VSWR = 1. High values of VSWR can cause high voltages in an RF circuit and may cause breakdown.

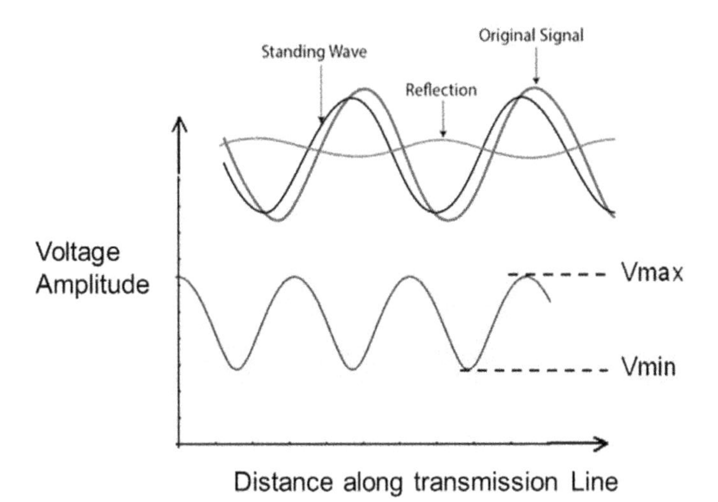

Figure 21.14 Standing waves created as a result of reflection of an incident wave. The reflection coefficient is defined as V_{max}/V_{min}.

21.9.3.2 Load pull test

Microwave and RF circuits are tested under load conditions to ensure proper functionality over frequency, power, etc. This is done using automatic test equipment (ATE) but, in principle, can be a simple setup, as shown in Fig. 21.15, that includes a signal source, an amplifier, an impedance tuner, the device under test (DUT), an output tuner, a spectrum analyzer, a power meter, etc. During the test impedance is varied to determine the best matching conditions.

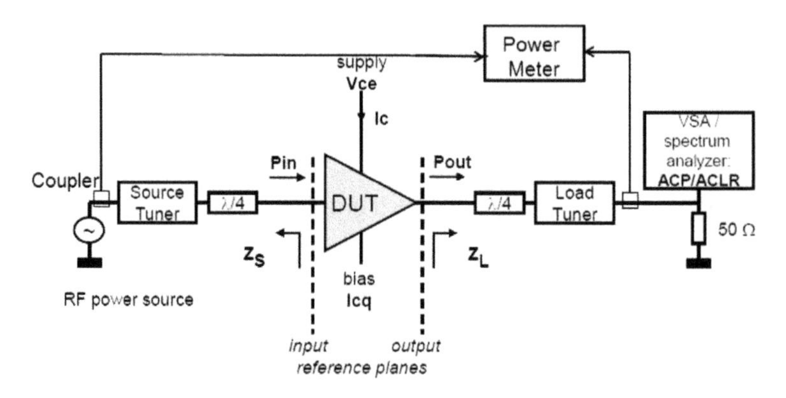

Figure 21.15 Basic load pull setup for RF circuit for gain, power-added efficiency (PAE), adjacent channel power (ACP), adjacent channel leakage ratio (ACLR), etc.

An example of the results from a load pull test are shown in Fig. 21.16 for P_{out}. Circuits are designed to work under these tuning conditions in the region of the Smith chart with the highest P_{out}. Tuning can be optimized quickly to optimize P_{out}, etc.

21.9.3.3 PAE

Power-added efficiency (PAE) is measured as

$$\eta = \frac{P_{out} - P_{in}}{PDC} \tag{21.58}$$

21.9.3.4 Linearity

Linearity is measured by measuring output power as a function of input power. A figure of merit for linearity is the 1 dB compression point and is defined as the input power at which the output power drops by 1 dB. Figure 21.17 illustrates this figure of merit.

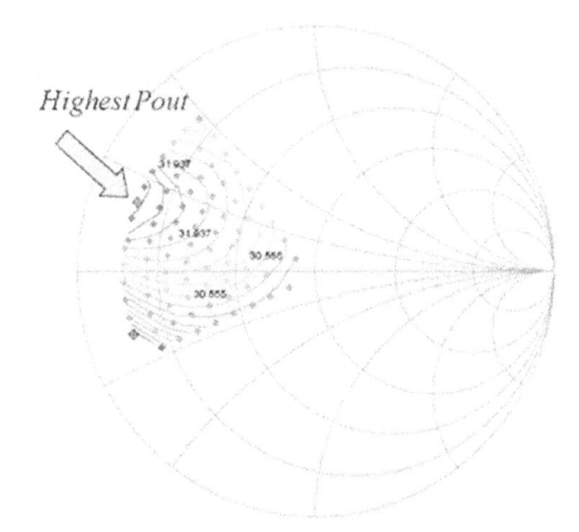

Figure 21.16 Load pull data displayed on a Smith chart, showing region of highest P_{out}.

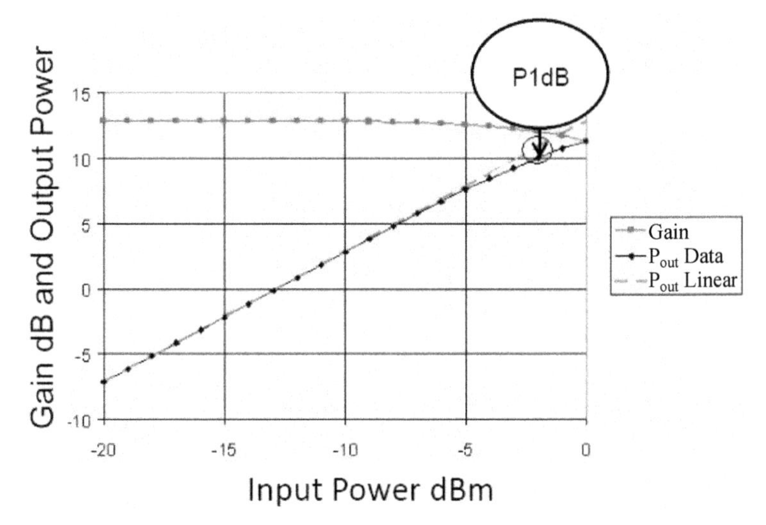

Figure 21.17 Power amplifier linearity and deviation from it at high power levels. The point at which gain drops by 1 dB from the linear is shown—1 dB compression power.

21.9.3.5 Noise figure

All circuits amplify incoming noise (N_i) and add their own noise to the output signal (N_o). The noise figure (NF) of a device or circuit measures the noise contribution of the device under test. It is defined as the signal-to-noise ratio at input to that at output (see Fig. 21.18).

$$F = \frac{\frac{S_i}{N_i}}{\frac{S_o}{N_o}} \tag{21.59}$$

F can also be defined as

$$F = \frac{N_o}{N_i.G} = \frac{N_a + N_i.G}{N_i.G} \tag{21.60}$$

where N_a is the noise added by the DUT.

The commonly used NF is given by

$$\text{NF} = 10 \log F \tag{21.61}$$

Measurements are made with the noise source off and the noise source on, as shown in Fig. 21.18, often referred to as cold and hot, respectively.

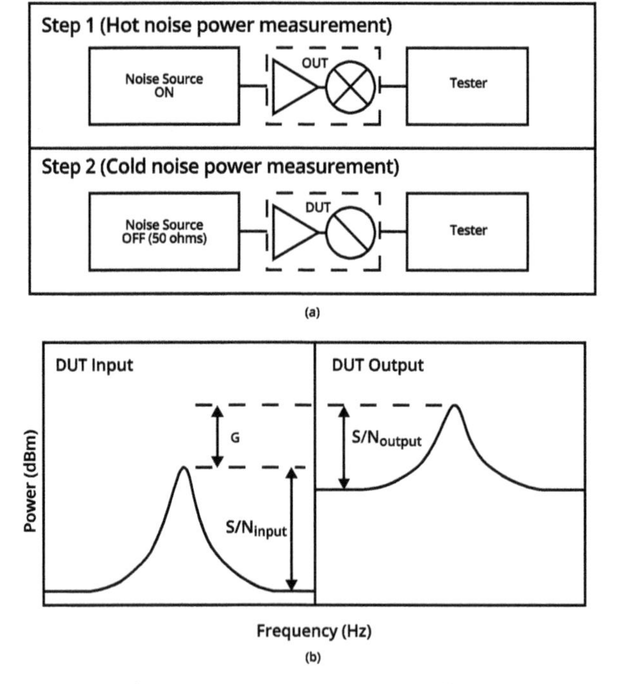

Figure 21.18 Noise figure measurement, setup, and output power without and with the noise source on.

21.9.4 Smith Chart [10, 11]

A Smith chart is a graphical aid designed for RF engineers to solve and visualize solutions to complex problems related to transmission lines, waveguides, amplifiers, etc. These are universally used in all high-frequency device and circuit simulation, design, and characterization.

A Smith chart is plotted as a complex reflection coefficient plane and is scaled to normalized impedance, generally 50 ohms. The transformation from the impedance plane (unbounded) to the reflection coefficient plane (Γ) (bounded for positive values) is shown in Fig. 21.19. The reflection coefficient, Γ, is the ratio of reflected power to input power and can be shown to be

$$\Gamma = S_{11} = \frac{Z - Z_0}{Z + Z_0} \tag{21.61}$$

On a Smith chart constant resistance and reactance lines appear as circles.

A Smith chart is used to plot all parameters of interest in RF, including, complex voltage and current transmission coefficients, power reflection and transmission coefficients, reflection loss, return loss, maximum and minimum of voltage and current, and VSWR, P_{out}, gain, NF, etc.

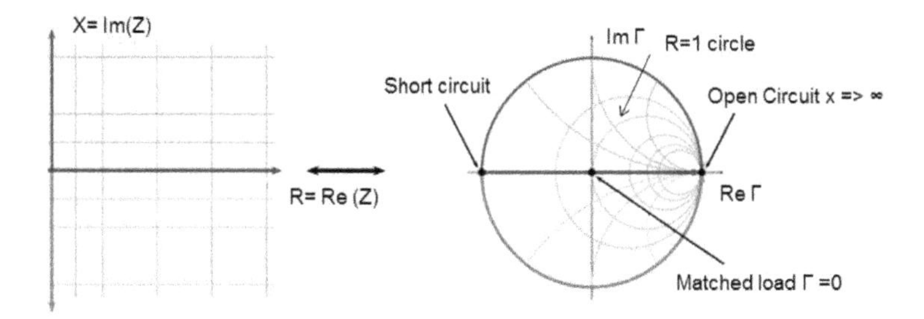

Figure 21.19 Illustration of complex impedance plane transformation into a reflection coefficient (Γ) plane to form a Smith chart. Reproduced from Ref. [11] with kind permission of Fritz Caspers (CERN).

21.10 Film Thickness and Refractive Index

Film thickness can be measured often by measuring an etch step in the film by stylus-type instruments. For faster measurements, especially on product wafers, nondestructive ways of measuring film thickness are preferred. Ellipsometry and interferometry are commonly used in semiconductor processing.

21.10.1 Ellipsometry [12]

Ellipsometry is used to measure thin-film thickness and refractive index of all transparent materials used in semiconductor IC fabrication. In GaAs processing, it is used for thickness measurement of dielectric films, photoresists, and polymer films. It can also be used for determination of film quality, composition, crystallinity, etc., that are associated with change of optical properties. Figure 21.20 shows the main optical components of an ellipsometer, a light source, a polarizer, an analyzer, and a detector. In ellipsometry, light from a source, such as a laser, is sent through a polarizer. The linearly polarized light is reflected from the sample surface and becomes elliptically polarized. This light is sent through a rotating polarizer, called the analyzer, to a detector, which converts the light to an electronic signal.

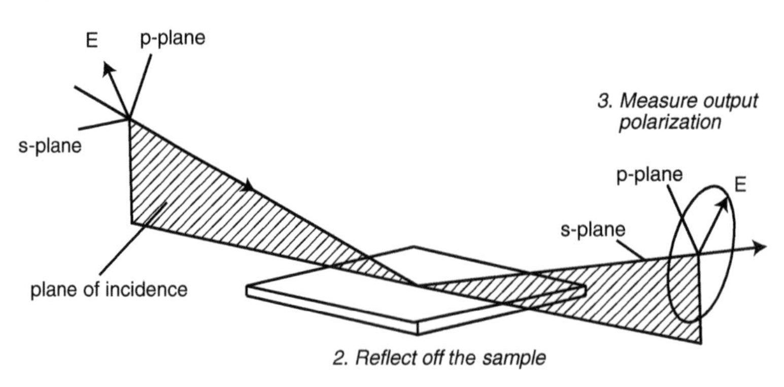

Figure 21.20 Schematic diagram for an ellipsometer used to measure film thickness and refractive index.

Plane polarized light can be broken down into two components: the p component is parallel to the plane of incidence, and the s component is perpendicular to it. The change of polarization is measured in terms of the magnitude and phase changes of these two components. See Fig. 21.20 (2) [12]. The change of polarization is commonly described as

$$\rho = \tan(\psi) . e^{i\Delta} \tag{21.62}$$

where $\psi = \tan -1 \{R_p/R_S\}$, R_p is the reflection coefficient of the p component, and R_S is the reflection coefficient of the s component.

$$\Delta = \Delta p - \Delta s$$

ψ(Psi) and Δ(Del) are the outputs of an ellipsometer measurement and can be used for the calculation of thickness and refractive index of the film.

When light reflected from the surface interferes with the light traveling through the film, both the amplitude and the phase change. The phase information D is very sensitive to the film thickness down to the nanometer (10 Å) range. Figure 21.21 shows the sensitivity of Del and Psi to the thickness and refractive index of oxide, oxynitride, and nitride films. In commercially available systems, a model is constructed for the substrate/film stack. Unknown material properties are varied to get the best fit of experimental data to model for the sample. The best fit is achieved through regression.

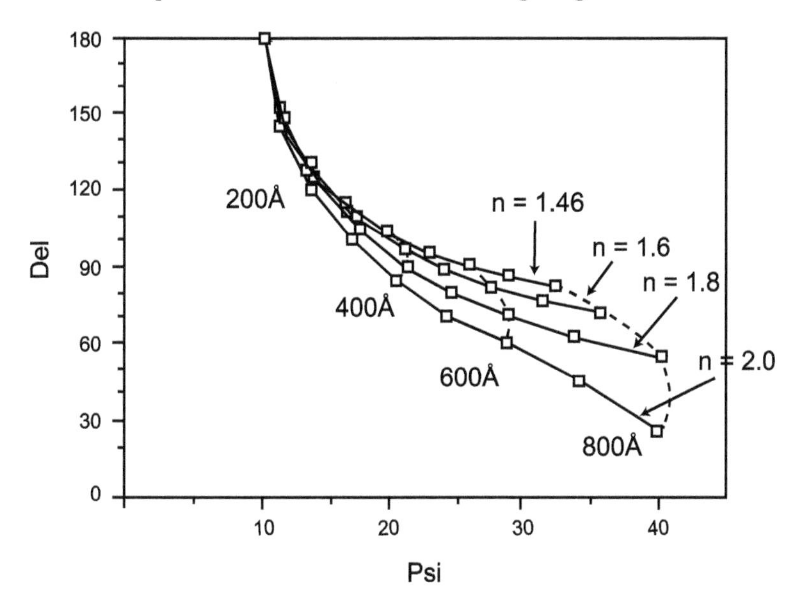

Figure 21.21 Del and psi variation with thickness and refractive index of oxide.

21.10.2 Interferometry

Interference of light reflected from two different planes, surface and film substrate interface, can give rise to maxima and minima in intensity [1].

$$t = \frac{m\lambda}{(2n\cos\Phi')} \tag{21.63}$$

where Φ is the arrival angle at the film (refractive index n) and Φ' is the arrival angle at the film–substrate interface. From the measurements of the maxima and minima, film thickness can be determined.

Interferometry can be used for measurement of film thickness in growth or deposition systems in real time by using transparent windows and properly positioning the light source and detectors.

21.11 Film Stress Measurement [13, 14]

Film stress control is an issue in all thin-film processing for silicon IC fabrication or optical coatings. Excessive stress causes film peeling. Stress control is of particular importance in GaAs processing because of the piezoelectric nature of GaAs. Charges induced by the piezoelectric effect can cause a parametric shift of devices, such as the threshold voltage shift in FETs and junction leakage for HBTs. Stress measurement can also be used as an indicator of film quality like density, which may be harder to measure in production. It is easier to monitor film quality by stress measurement than the wet etch rate.

Film stress in metal and dielectric films can be intrinsic, caused by nucleation and growth processes (discussed in Chapter 8), or extrinsic, caused by differences in thermal expansion coefficient of the film and the substrate. Film stress control is possible by control of deposition parameters in deposition of metals and dielectrics. The most effective way is to control ion bombardment in sputtering and plasma-enhanced chemical vapor deposition (PECVD).

Film stress on semiconductor wafers is calculated from curvature measurements. The curvature is a result of the film trying to return to its original state when stretched or compressed on to the wafer. See Fig. 21.22 for a better understanding. A laser beam is directed at the surface and the reflected beam is detected using a position-sensitive photodiode. In this manner, the curvature of the sample surface is measured. Figure 21.23 shows a schematic diagram for such a system [13], a similar concept being used by most equipment vendors. Measurements are made before and after deposition of the film to be measured. The stress is calculated from the curvature. The curvature can be measured locally and in different orientations of the wafer to get a map of the whole wafer. Figure 21.24 shows the output from a stress measurement tool. Stress is calculated from the radius of curvature, R, using the following equation [14]:

$$\sigma = \left(\frac{1}{R}\right)\left\{\frac{E}{6(1-\nu)}\right\}\left(\frac{T^2}{t}\right) \tag{21.64}$$

where E is Young's modulus (ratio between direct tensile stress and strain) and ν is Poisson's ratio for the substrate (negative ratio of lateral strain to direct tensile strain), T is substrate thickness, and t is film thickness.

The above equation is true for the no-bow-starting condition. In practice, the wafer always has some bow with radius of curvature R_1. Under these conditions, the calculation can be simplified to the following equation:

$$\sigma = \left(\frac{R_1 - R_2}{R_1 R_2}\right)\left\{\frac{E}{6(1-\nu)}\right\}\left(\frac{T^2}{t}\right) \tag{21.65}$$

where R_1 and R_2 are pre deposition and postdeposition radii of curvature, respectively.

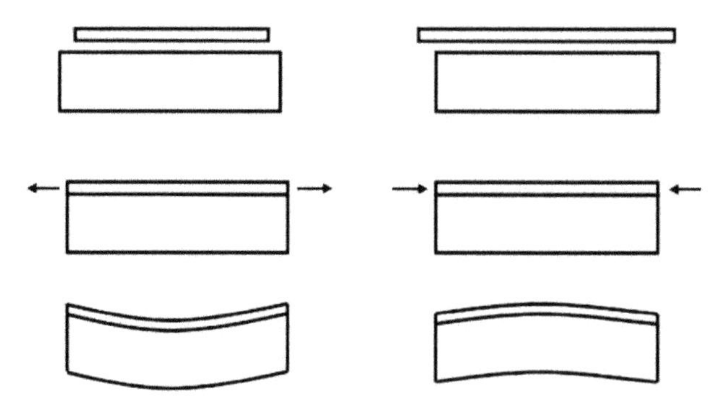

Figure 21.22 Stress in thin films causes bowing of the substrate, as shown above for tensile (left) and compressive (right) stress. Imagine a free film stretched or compressed to wafer size. The films try to return to their original shape, curving the wafer as a consequence.

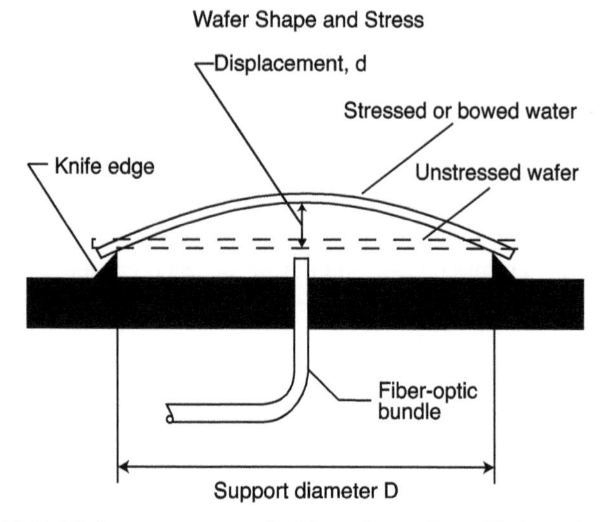

Figure 21.23 Wafer curvature method based on reflected light technique.

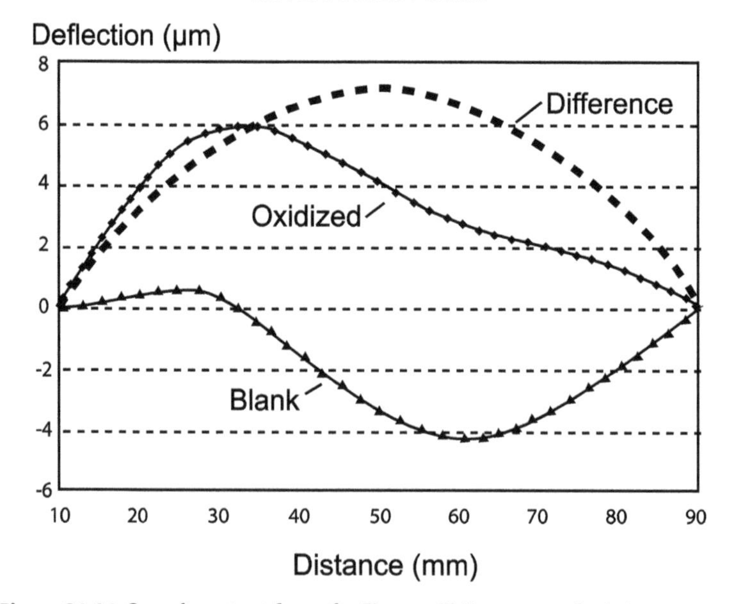

Figure 21.24 Sample output from the Tencor FLX stress analysis instrument, showing how stress is calculated from a change in wafer curvature.

References

1. W. R. Runyan, *Semiconductor Measurements and Instrumentation, Texas Instruments Electronics Series,* McGraw Hill Book (1975).
2. L. J. van der Pauw, A method of measuring resistivity and Hall effect of discs of arbitrary shape, *Philips Res. Rep.,* **13**, 1 (1958).
3. S. K. Ghandhi, *VLSI Fabrication Principles, Silicon and Gallium Arsenide,* Wiley Interscience, New York (1983).
4. J. Krupka, D. Nguyen and J. Mazierska, *Microwave and RF methods of contactless mapping of the sheet resistance and the complex permittivity of conductive materials and semiconductors, Meas. Sci. Technol.,* **22**, 085703 (2011).
5. J. Hillibrand and R. D. Gold, Determination of the impurity distribution in junction diodes from capacitance-voltage measurements, *RCA Rev.,* **21**, 245 (1960).
6. R. Williams, *Modern GaAs Processing Methods,* Artech House, Boston, London (1990).
7. R. A. Pucel and C. F. Krumm, *Electron. Lett.,* **12**, 242 (1976).
8. W. Liu, *Handbook of III-V Heterojunction Bipolar Transistors,* Wiley Interscience, New York (1998).
9. Agilent Technologies, *S Parameter Techniques for Faster, More Accurate Network Design,* www.Hp.com/go/tmaappnotes.
10. Smith Chart, *Noise Figure,* http://www.ieee.li/pdf/viewgraphs.
11. F. Caspers, *R. F. engineering basic concepts, CERN,* http://arxiv.org/pdf/1201.4068.
12. Woolam online resource on ellipsometry.
13. Flexus brochure and website.
14. P. Singer, Film stress and how to measure it, *Semicond. Int.,* **54** (1992).

Chapter 22

Reliability

22.1 Introduction

Semiconductor circuits are expected to perform under adverse environmental conditions, and III–V semiconductor devices are no exception. Reliability is a fundamental concern in the development of semiconductor devices and processes, and each new process step or module must be verified to not degrade expected reliability. Reliability can be defined as the probability that a part will perform a required function under defined conditions for a specified period of time. Different users have different expectations; some commercial devices may be expected to work for hours, whereas devices for satellite systems may require 20 years. A large number of variables influence reliability; these include design, manufacturing processes, and the application. As a product matures, the design is optimized for each application. Reliability issues from processing can be minimized by incorporating statistical process control methodology in the fabrication of the device and constant reliability monitoring and process improvement.

The failure rate of semiconductor devices when plotted against time, like most products, shows a bathtub type of shape, as seen in Fig. 22.1. The early period is characterized by infant mortality, dominated by failure of weak parts. After this initial phase is over, the rate drops to a stable value, nearly constant, the value being

III–V Integrated Circuit Fabrication Technology
Shiban Tiku and Dhrubes Biswas

ISBN 978-981-4669-30-6 (Hardcover), 978-981-4669-31-3 (eBook)
www.panstanford.com

dependent upon the random failure rate. At the end of this period, a rapid wear-out phase begins, signifying the end of life. In time all devices fail and the failure rate falls to zero.

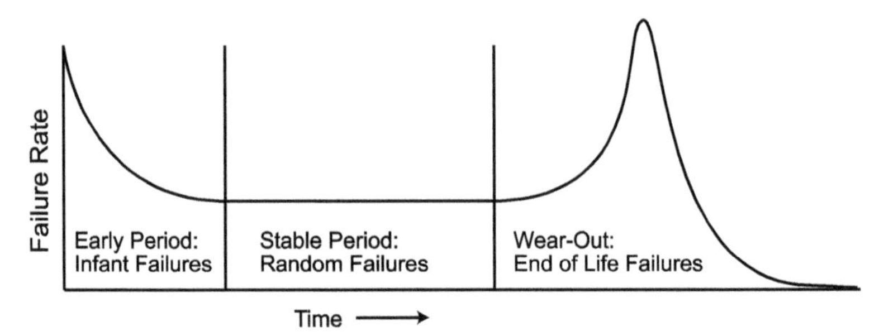

Figure 22.1 Bathtub curve for failure rates.

22.2 Basic Reliability Testing [1, 2]

Reliability theories and testing are based on the fact that most chemical changes that occur in devices are governed by the Arrhenius equation, whereby the rate of change varies exponentially with temperature (T):

$$R = A\exp\left\{-\frac{E_\mathrm{a}}{kT}\right\} \tag{22.1}$$

where R is the reaction rate, A is the proportionality constant, sometimes a slow function of temperature, E_a is the activation energy, and k is the Boltzman constant (8.6×10^{-5} eV/K). Different failure mechanisms have different activation energies; the mechanisms and the value of the activation energy associated with them will be discussed later. The exponential dependence on temperature makes it possible to test semiconductor devices in a reasonable time at elevated temperatures. The data from these tests can be used to predict the lifetime at the operating temperature. If a physical change leads to a failure that can be defined, for example, drop in β by a certain percentage for a heterojunction bipolar transistor (HBT), in time t (assuming a constant reaction rate), at temperature T

$$t = \frac{1}{R} \tag{22.2}$$

Rearranging the Arrhenius Eq. 22.1

$$t = \left(\frac{1}{A}\right)\exp\left\{\frac{E_a}{kT}\right\} \tag{22.3}$$

or $$\ln(t) = C + \frac{E_a}{kT} \tag{22.4}$$

which leads to

$$\ln\left(\frac{t_2}{t_1}\right) = \left(\frac{E_a}{k}\right)\left(\frac{1}{T_2} - \frac{1}{T_2}\right) \tag{22.5}$$

To accurately predict lifetimes at normal temperature, at least three high-temperature tests are conducted. The median life from each of the tests is transferred to an Arrhenius plot, which has log time as the x axis and $1/T$ as the y axis. A straight-line fit to this data is then used to project the lifetime at any temperature.

Most reactions are thermally activated, and activation energy measures how rapidly the mean lifetime changes with temperature. Plotting failure data on a log time versus $1/T$ graph, E_a can be computed. A high activation energy does not necessarily mean a better or more reliable device. Figure 22.2 shows the lifetime as a function of temperature for different activation energies [1]. In this example, lifetimes below 150°C are longer for a high activation energy, but above 150°C, the reverse is true. At very high temperatures, the mechanism of degradation itself may switch, leading to different activation energies. This can be seen as a change of slope of the lifetime versus $1/T$ data.

Accelerated life tests should be done with large sample sizes so that the lifetime can be predicted with a high degree of confidence. In practice, a more practical or economical sample size is chosen. For a large group being subjected to accelerated testing, devices fail at different times. The median life is the time at which 50% of the devices have failed. Failures are going to follow a statistical distribution function, and testing must be continued until 50% failures have accumulated. Semiconductor devices almost always follow a log-normal distribution, among a variety of other distributions, for example, normal, Weibull, exponential, etc. For normal distribution, a plot of the logarithm of time to failure as a function of percentage cumulative failure is a straight line. The cumulative distribution function, $F(t)$ gives the distribution of cumulative failures over time.

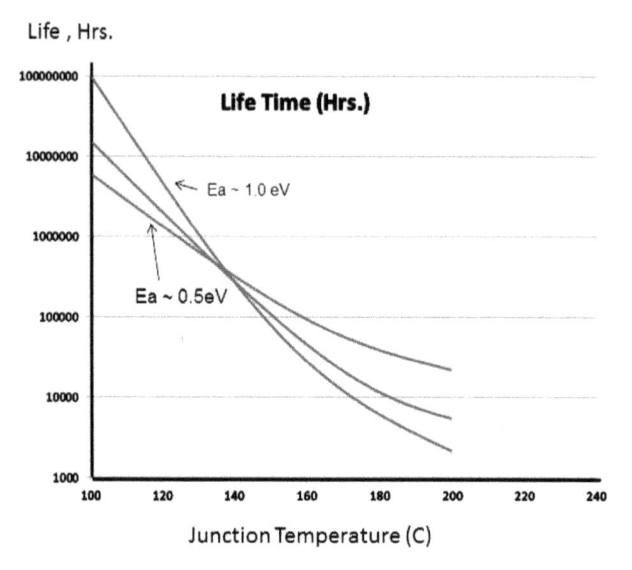

Figure 22.2 Lifetime as a function of temperature for different activation energies.

22.3 Test Procedure

A group of devices is packaged so that they can be powered up inside a high-temperature oven and left on. Data, that is, the number of devices that fail, is collected periodically. Failure is predefined; this may be a 20% drop in source–drain current for a field-effect transistor (FET)-type device or a known drop in current gain β for an HBT. Figure 22.3 shows the β drift plot for a number of devices. This data is plotted on a log-normal graph, as shown in Fig. 22.4. The test is continued until over 50% of the devices have failed to determine the MTTF. For the example shown in Fig. 22.4a, the MTTF is about 400 hours. This procedure is repeated at higher temperatures, preferably over three temperatures (Fig. 22.4b). From this data the median time to failure is determined for each temperature. This data is plotted on a $1/T$ versus log time plot, as shown in Fig. 22.5. The MTTF at the expected operating temperature may be predicted from this data and used as a qualification parameter for the process. The activation energy can be determined by plotting the data for different temperatures on this Arrhenius plot (see Eq. 22.1 in the section above).

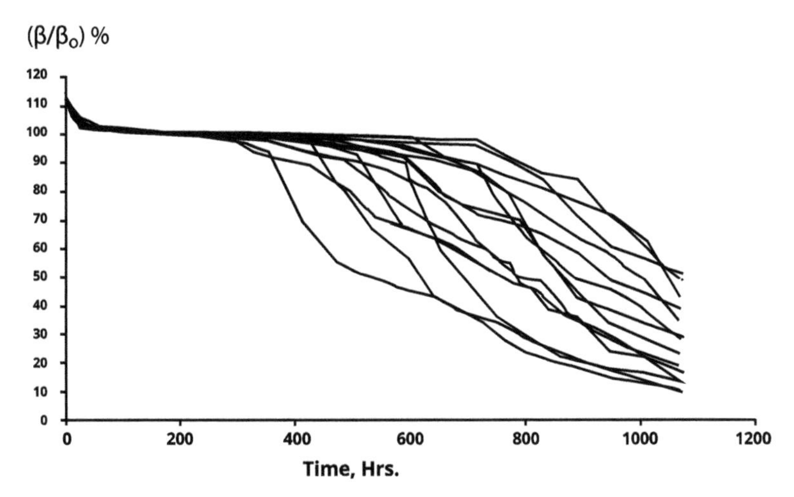

Figure 22.3 Example of β vs. stress time plot for 16 HBT devices.

If infant failures are observed, seen as bimodal distribution on log-normal plots, these should not be included in the data to determine the MTTF or the activation energy. Failure analysis should be done on sample parts to ensure that the failure mechanism is consistent across all parts. If a second mechanism of failure is present, it may also result in a difference in the slope of data for different temperatures on the log-normal plots.

22.3.1 Step Stress Test

Reliability testing is time consuming and if the right temperatures (the three temperatures) are not chosen, test must be repeated at different temperatures. To save time and determine what temperature range to use, a quick step stress test must be performed first. In a step stress test, the same group of devices is subjected to testing at a number of successively higher temperatures for a fixed time, for example, 24 hours. The starting temperature can be low enough where the failure rate is expected to be low, and the highest temperature can be high enough to result in a cumulative failure rate close to 100%. In general, the threshold level for device degradation and the temperature range for standard reliability can be determined from this test.

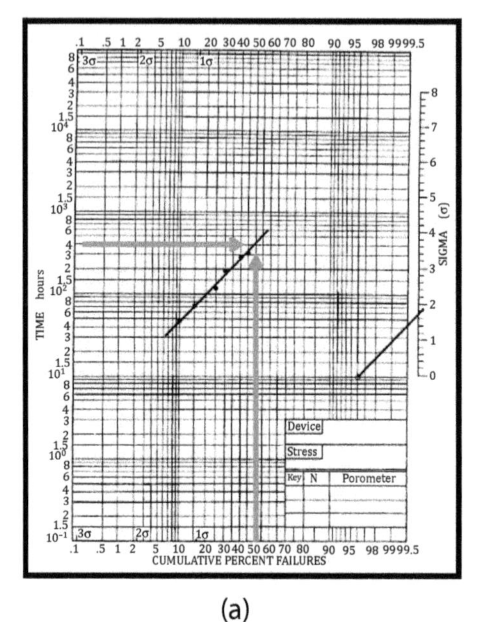

(a)

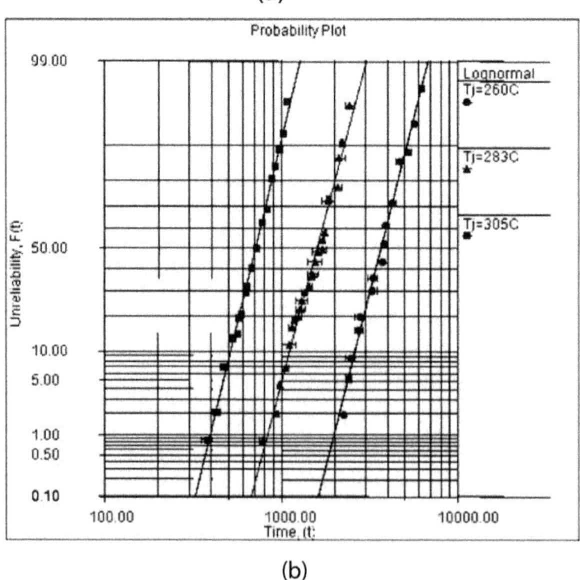

(b)

Figure 22.4 (a) Example of cumulative percent failures vs. time plotted on log-normal paper for a sample of 20 devices [1]. (b) Example of cumulative percent failures vs. time plotted on log-normal paper for three different junction temperatures, 305°C ■, 283°C ▲, and 260°C •. Courtesy of Skyworks Solutions.

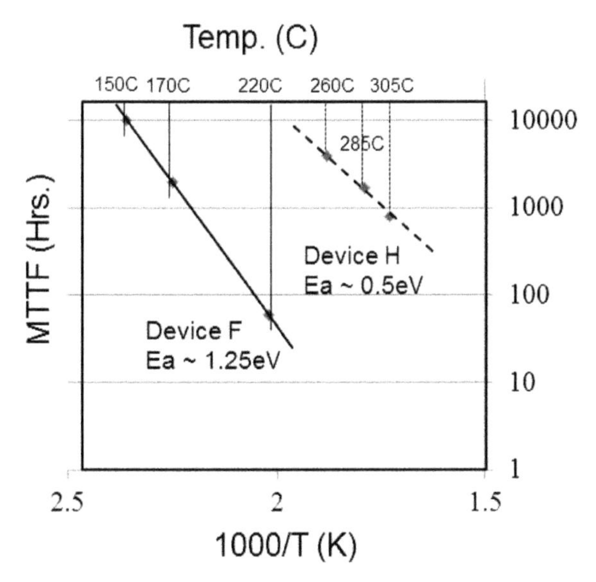

Figure 22.5 Arrhenius plot (time vs. linear 1/temperature) based on three temperature failure data. MTTF can be predicted from this plot for an expected operating temperature.

22.3.1.1 Temperature measurement

Semiconductor devices get hot during life testing, and the effective temperature is higher in the device compared to the base plate that is equilibrated with the oven ambient. The effective temperature of the hottest region must be used for plotting of data and determination of the activation energy. GaAs has high thermal resistance, so the difference in temperature can be large compared to silicon devices. Techniques used for measuring temperature are infrared (IR) microscope, liquid crystals, etc.

22.3.1.2 Simulation of operation

The simplest test to implement would be the storage test with no bias, but it is not recommended for prediction of reliability. Such a test could be used for process development or screening of process or material parameters. The most realistic test would be to use both direct currect (DC) bias and radio frequency (RF) drive, but that may be difficult in practice with a large number of devices needed for statistically valid data. Also, RF fixtures may not stand high temperatures. Therefore use of DC bias is more common.

22.4 FET/HEMT Failure Modes and Mechanisms [3–8]

Some of the common failure modes of FET-type devices are discussed in this section and those for HBT devices are discussed in Section 22.5. Failure often leads to performance degradation and occasionally to catastrophic breakdown due to acceleration by mechanical, electrical, or environmental stress. Failure can be attributed to poor device design, poor choice of fabrication process, abnormal application, or a combination of several factors. Some failure mechanisms that lead to these failures are common to the different families of III–V devices. These mechanisms can be deduced from careful failure analysis of the failed devices.

Most of the failure modes of FET-type devices can be detected by DC parameter testing of devices. Some of the common modes are shown in Fig. 22.6. FET- and high-electron-mobility transistor (HEMT)-based circuits fail in the field due to Idss current degradation; threshold voltage, V_t, change; on resistance (R_{on}); or R_{ds} change. P_{out} may drop due to all or some of the above parameter failures. The most common modes of failure for these will be discussed below in some detail.

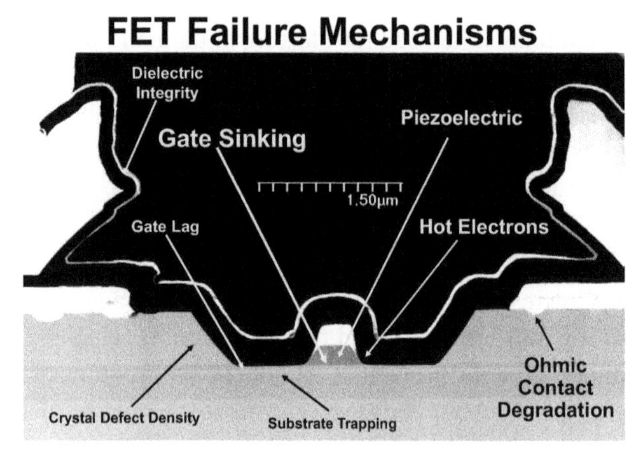

Figure 22.6 FET failure mechanisms. Reproduced from Ref. [7], provided with permission of author.

Table 22.1 Common III–V IC failure modes

Failure mode	Failure mechanism
FET/HEMT	
Idss degradation	Gate sinking, contact degradation, surface effects
V_t change	Gate sinking, hydrogen effects
R_{ds}, R_{on} increase	Ohmic contact degradation, Gate edge pitting
P_{out} drop	All of the above
HBT	
Power drop	
β degradation	I_b increase:
	Ledge defect accumulation
	Junction degradation
	Void formation under emitter
	Base contact sinking at edges
	Collector current drop:
	Collector contact degradation
V_{be} increase	Junction degradation, R_b increase

22.4.1 Gate Sinking

In gate sinking, one of the components of the gate metal diffuses into the channel region, causing the channel to narrow down. This causes a threshold voltage shift and reduction of Idss. The threshold voltage can also change if the barrier properties of the gate–channel interface change. For gold-based gates, if a path exists for gold to migrate, it will intermix with GaAs. The diffusion mechanism is thermally driven, as expected. Titanium, which is the most common gate adhesion layer metal, has an activation energy of diffusion (2.56 eV) close to the value measured for gate sinking (2.65 eV) [7]. In FET-type devices on GaAs or other III–V semiconductors, the thickness of Ti must be minimized and titanium must be followed by a barrier layer. Use of refractory gate metals avoids this. If gold-based gates are needed for low gate resistance, such as for mixed-signal and analog circuits, a refractory barrier like Pt must be provided. For this reason the industry standard for GaAs FET-type devices is triple-layer structures consisting of Ti/Pt/Au or Ti/Pd/Au. The first

Ti layer is used for good adhesion; Pt and Pd are used as the barrier. Diffusion through Pt or Pd is faster than through alloys like Ti:W, particularly refractory barrier layers that have oxygen or nitrogen inclusions. Pure metals tend to have large grain boundaries, which are good paths for diffusion. Poor surfaces, for example, those with ion damage or nondense oxide layers, can also provide diffusion paths.

22.4.2 Ion-Induced Failure Mechanisms

In AlGaAs/GaAs heterojunction field-effect transistors (HFETs) a change in the two-dimensional electron gas (2DEG) concentration can cause a shift of threshold voltage and current. The degradation may take place by creation of deep levels in the barrier by field-aided diffusion at the heterointerfaces. Diffusion of hydrogen into the doped channel can also drop the active dopant concentration.

Diffusion of dopants and impurities from the substrate below the surrounding area can cause a drop in the level of active dopants. Generation of traps can have a similar effect; these traps can be generated by electron impact in the passivation–semiconductor interface.

In HEMT devices, additional mechanisms related to heterojunctions can be present. Diffusion can cause softening of the AlGaAs/GaAs interface by intermixing. Also, the n-type dopants, for example, Si, can diffuse out of AlGaAs into the channel layer, causing an increase in scattering, a drop in electron mobility, and a drop of transconductance, g_m.

GaAs circuits, particularly metal–insulator–metal (MIM) capacitors in them, show high sensitivity to mechanical damage because of the effect of moisture that gets in. Failure is accelerated if moisture can find a path to the MIM capacitors and create ionic conduction paths near the edges. Gold corrosion and nickel migration have also been seen under moisture conditions, accelerated by chlorine from processing.

22.4.3 Effect of Hydrogen [9]

The effect of hydrogen on the reliability of GaAs devices has been discussed in numerous publications [9, 10]. For FET-type devices, degradation of Idss and shifts in parametric data have been observed. Source of hydrogen may be packaging materials. Hydrogen diffuses into the channel region and compensates the silicon donors by forming Si–H. In the case of HBTs, excess hydrogen may be present in epilayers (hydrogen is used as the carrier gas) or be introduced through ion implantation during implant isolation. In this case, hydrogen combines with carbon to form C–H, thus changing the effective doping level. Reactivation takes place by decomposition of the C–H complexes via minority carrier injection. Use of He and boron for ion implant isolation can minimize hydrogen-related issues. Hydrogen effects in HBTs are discussed in more detail later in Section 22.5.

22.4.4 Mobile Ion Contamination

Mobile ion contamination at very low levels is feared in semiconductor processing due to known device parameter failure and reliability problems. In FET-type devices, threshold voltage shifts and loss of current are common. In HBTs, increase in contact resistance, β degradation, and leakage increase can occur. Mobile ions like Na^+, K^+, and Cl^- from processing are commonly avoided. Copper, chrome, and other deep impurities kill mobility in III–V compounds and compensate dopants. A barrier must be used if such metals are used in interconnect processing.

Ionic contamination can result from processing of the device front end, interconnect, and backside processing. High-temperature baking can be used for screening suspected contaminated lots.

22.4.5 Hot Electron Trapping

Hot electrons are known to be trapped in gate oxides of metal–oxide–semiconductor field-effect transistors (MOSFETs), causing threshold voltage shifts. In GaAs metal–semiconductor field-effect transistor (MESFET) devices, a similar effect has been seen under high-power conditions, causing a *power slump*. This problem is different from burn-in-type shifts due to ion migration. Under RF overdrive, hot electrons are generated near the drain region of FETs where the electric field is the highest (see Fig. 22.7). Some of the hot electrons can tunnel into the silicon nitride passivation to form traps [4, 11]. The Schottky barrier height and gate ideality factor remain unchanged; therefore under small signal conditions the device behavior is normal. Improved device design and passivation nitride can limit the hot electron effect.

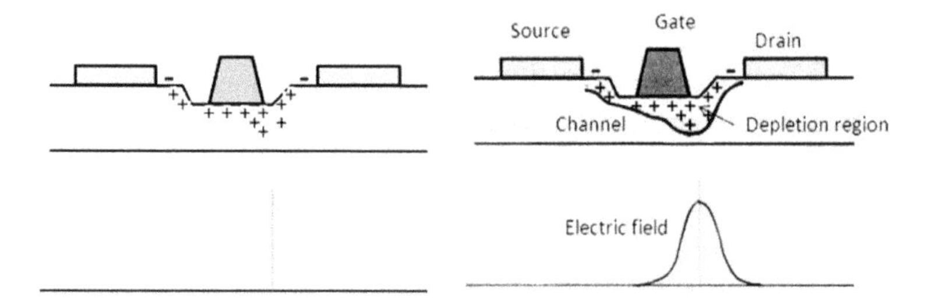

Figure 22.7 Hot electron effects in the high-electric-field region of the channel degrade FET reliability.

22.4.6 Surface State Effects [12]

The performance and reliability of electron devices depend on the nature of the active device surface, which critically depends upon the passivation layer. In GaAs devices silicon nitride is used for this purpose, and the nature of the quality of the surface depends on the deposition process and the precleaning. If the surface is not well passivated, the density of the surface states goes up, which enlarges the depletion region and changes the electric field and device performance. Presence of surface states gives rise to hopping conduction, which lowers the breakdown voltage.

22.5 HBT Degradation and Reliability [13–15]

HBTs should be tested at a junction temperature that is low enough for the degradation mechanism to be the same as that under normal operating conditions. This temperature is in the range of 230°C to 280°C. The operating current density must be nearly 15 kA/cm^2 (V_{ce} less than 10 V). Three failure modes are observed, as illustrated in Fig. 22.8.

- Initial β degradation: For carbon-doped HBTs, this is mostly due to hydrogen dissociation.
- Sudden β degradation: Sudden degradation, particularly infant mortality, is caused by space charge recombination damage on the ledge near the emitter. Carbon and metallic

precipitates can also form near the base, as found by Ueda 14], of failed InGaP HBTs.

- General β drift: This is similar to drift under normal operation but not exactly the same [12].

Some common HBT failure modes are pictorially depicted in Fig. 22.9 (Triquint, now Qorvo). Some of the mechanisms mentioned will be discussed briefly here.

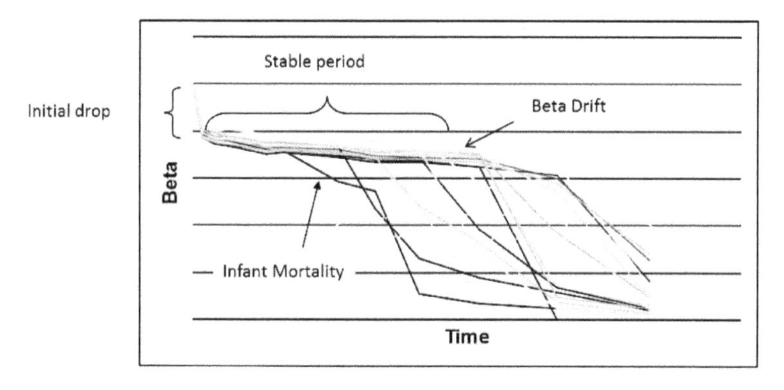

Figure 22.8 HBT failures due to β degradation showing initial β drop, infant mortality, and drift.

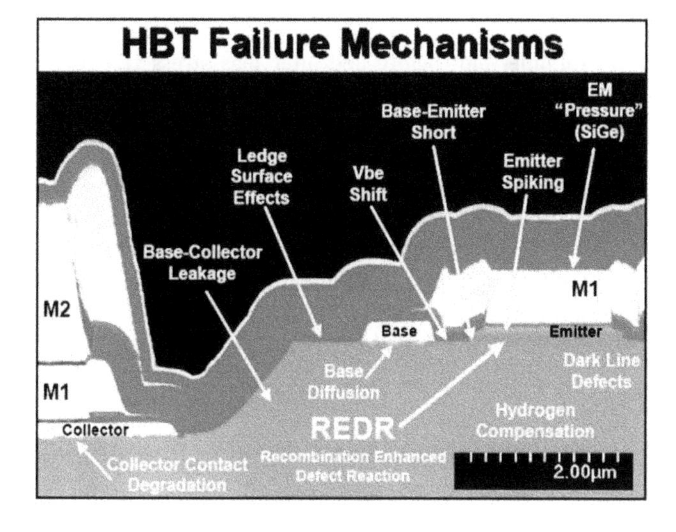

Figure 22.9 Common HBT failure modes. Reproduced from Ref. [7], provided with permission of author.

For HBTs, Gummel plots taken at different stress times generally show an increase in the base current; the collector current is seen to be stable. The base current (I_b) increase results in low β. Figure 22.10 shows this for an AlGaAs/GaAs HBT. At high V_{be}, I_b increase is due to recombination in the base region itself, due to diffusion of dopants or creation of defects under a high field (see Fig. 22.11a). I_b increase at low V_{be} may be related to space charge recombination–enhanced defect generation in the ledge near the emitter (Fig. 22.11b). If I_c changes, the emitter resistance may be changing due to an unstable emitter epilayer or contact (changes in the collector epi- or ohmic contact are less likely).

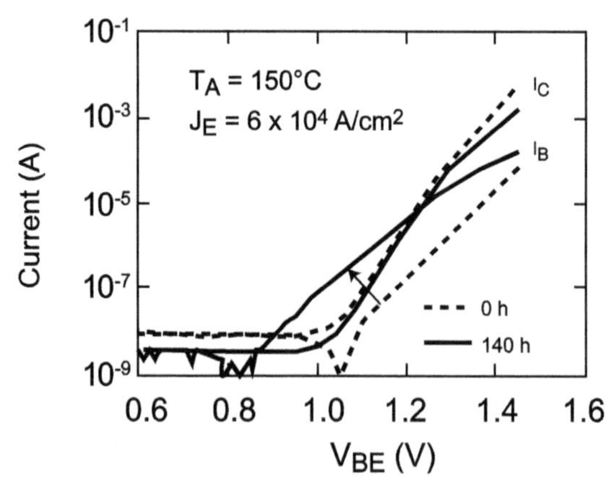

Figure 22.10 Gummel plot before and after bias stress test for AlGaAs/GaAs HBT. Reprinted from Ref. [10], Copyright (1999), with permission from Elsevier.

At higher operating temperatures other mechanisms of degradation come into play. Figure 22.12 shows collector contact degradation and base contact spiking at higher temperatures. Reliability can be improved by changing epimaterials and structure as well as metallization used for contacts, such as refractory contacts. Improvement due to change from the AlGaAs emitter to InGaP is shown in Fig. 22.13.

22.5.1 HBT Reliability Issues Related to Base Doping

During the development of HBTs, epistructures were grown by both molecular beam epitaxy (MBE) and metal–organic chemical vapor deposition (MOCVD). The MBE-grown HBTs with Be doping in the base layer were seen to have reliability issues. Although easier to incorporate, Be proved to be prone to outdiffusion due to a high diffusion coefficient. The sharpness of the heterojunction was lost and the device behaved more like an AlGaAs/AlGaAs homojunction device. The stability issues were later resolved by improving Be incorporation using improved arsenic (As) sources in the growth process [15].

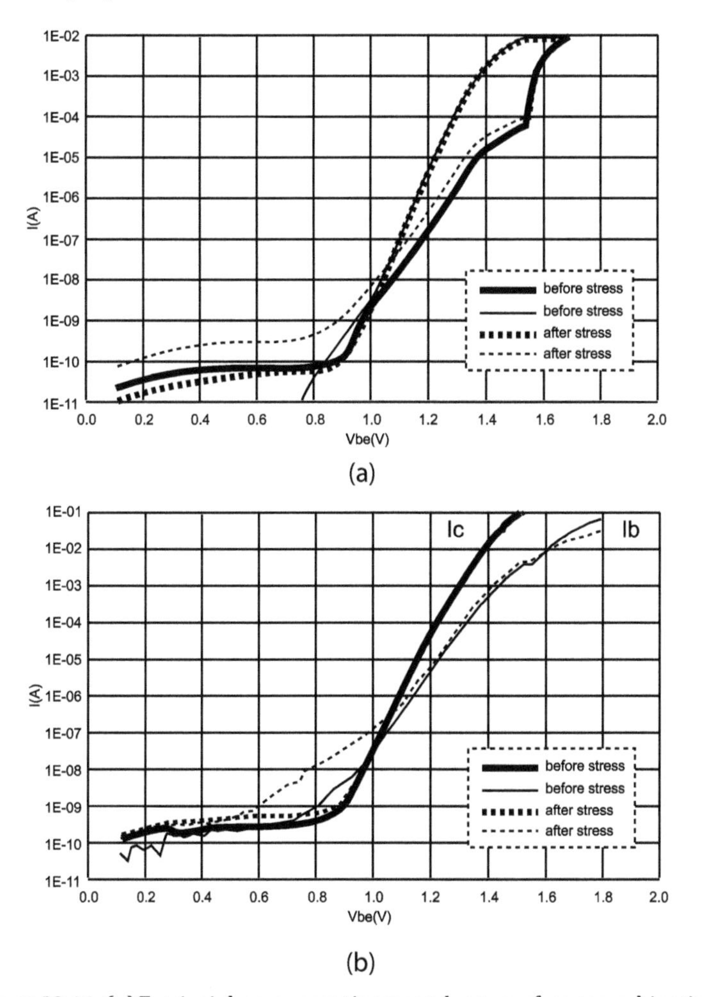

Figure 22.11 (a) Extrinsic base current increase due to surface recombination caused by ledge degradation (pink to green curve). Collector current is unchanged (blue and red curves). (b) Base current increase due to tunnel recombination caused by bulk defects in the emitter–base junction (pink to green curve).

Carbon doping of the base solves the Be diffusion problem and can be easily achieved using MOCVD.

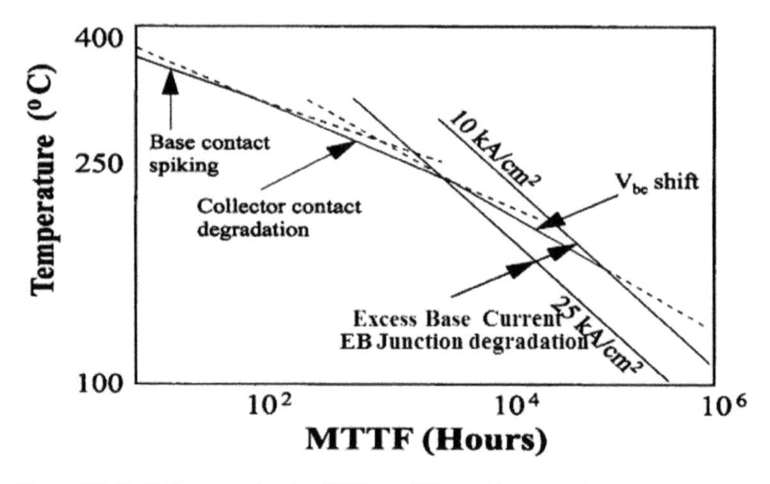

Figure 22.12 Failure modes for HBTs at different junction temperatures.

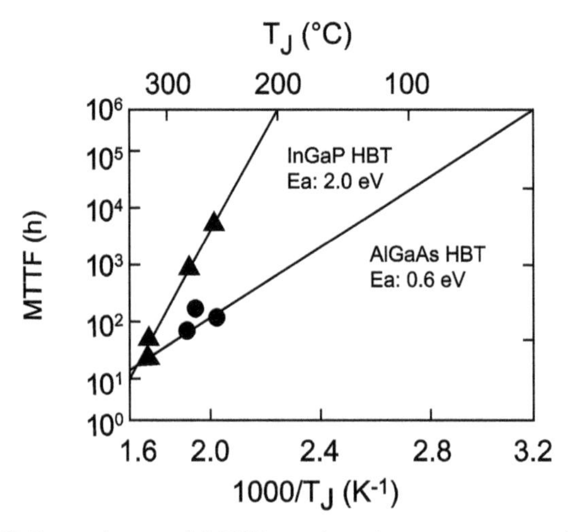

Figure 22.13 Dependence of MTTF on junction temperature for AlGaAs and InGaP emitter HBTs, showing higher activation energy for InGaP. Reprinted from Ref. [10], Copyright (1999), with permission from Elsevier.

22.5.2 Hydrogen and Ion Implant Isolation [15–17]

MOCVD epigrowth utilizes hydrogen carrier gas and metal-organic compounds contain hydrogen. Therefore, concerns about hydrogen migration have been raised and in some cases verified. HBT epilayers are thick, nearly 2.5 μm, and contain highly doped n+ layers. Hence, for device isolation by ion implantation, multiple high-energy H+ implants are required to achieve relatively flat damage profiles and avoid high concentration regions. Pearton et al. [16] have investigated the diffusion of H+ implants and the intermixing of epilayers to explain gain degradation in HBTs. They concluded that annealing over 500°C is needed to optimize isolation resistance. However, at such high temperatures redistribution of implanted H+ takes place. They verified by secondary ion mass spectrometry (SIMS) analysis that H migrates and piles up at the base–emitter junction because of strain there.

Hydrogen can come from many sources in an HBT device. First, it is present in MOCVD layers. As much as $10^{19}/cm^3$ is expected in the base layer (with a doping of 4 E19) [14]. The silicon nitride material used for passivation also contains hydrogen on the order of $10^{21}/cm^3$. But since devices implanted with He do not show the same gain degradation as those implanted with H+, it is evident that the hydrogen of concern migrates from the isolation implanted regions. Hydrogen gets closer to active device regions due to lateral straggle, which can be of the order of 1 μm. This hydrogen eventually reaches the base layer. Effective base doping is reduced due to the formation of C–H complexes. This causes V_{be} shift and β change.

The variation of turn-on voltage, V_{be} and β change has been reported by Ueda (1999) [14] for InGaP/GaAs HBT devices using either He or H implants. When an HBT without H isolation is operated, V_{be} increases exponentially with time as C–H complexes present in the base are broken up and the effective doping level goes up. With H+ isolation, a high level of H+ migrates from isolation regions, first causing a drop in V_{be} and then reversing the process as C–H complexes break up. Due to this behavior, the use of He for isolation of HBTs is recommended.

22.6 Ohmic Contact Degradation

The most common contact scheme used in III–V integrated circuit (IC) fabrication is the Au:Ge:Ni metallurgical system (discussed in Chapter 11). This system is used for contacts to FET- and HEMT-type devices, as well as to HBTs. Over the years numerous attempts have been made to replace this contact with more stable systems that can withstand higher-temperature processing and last longer in operation. However, a satisfactory solution has not been found yet. Therefore degradation mechanisms of this system are described here. The main degradation is the outdiffusion of gallium from GaAs into the gold layer. The nickel layer in this contact scheme serves multiple purposes: It is intended to act as an adhesion layer and a barrier for gold and gallium diffusion and for the formation of ternary phases like Ni:Ge:As, which are essential for good electrical contact formation. This layer cannot be replaced by other barrier metals like Ti, Pt, or W without affecting the contact resistance. The other mechanism for degradation is diffusion of gold and nickel into the GaAs, which can compensate the doping due to Ge and thus change the carrier concentration in the contact area and nearby semiconductor. The ohmic contact can also degrade as a result of formation of unwanted phases because of interdiffusion. Proper cleaning of the surface before deposition and deposition of layered contacts rather than premixed Au:Ge alloy and rapid thermal processing (RTP) alloying of these at higher temperatures close to 400°C have been seen to improve ohmic contact stability [18].

22.7 Other III–V IC Failure Mechanisms

General failure mechanisms related to devices as well as circuit components other than active devices, including interconnects, are discussed in this section.

22.7.1 Electromigration

Electromigration is the movement of atoms due to momentum transfer from electrons under heavy current density conditions (discussed in Chapter 18). As the current density is increased in a conducting line, semiconductor or metallic, it can experience failure due to formation of voids and narrow necks in hot spots and accumulation of metal atoms at other spots that can form hillocks. Eventually the necks lead to opens and the hillocks may cause shorts by crossing metal lines. This kind of failure occurs in narrow lines that carry heavy current or very thin lines used for resistors.

Gold-based metal lines are therefore used in mixed-signal circuits where the current densities are over 1×10^5 A/cm^2. (Electromigration-based current limits are discussed in Chapter 18). Line widths are designed to meet the expected current density requirement. For other metal systems like Al, use of Ti cladding can increase electromigration resistance. Electromigration can also take place under contacts if the current density limits are exceeded.

22.7.2 Moisture Ingression and Corrosion

FET and HBT ICs are packaged in nonhermetic modules. The ICs and modules are designed to pass highly accelerated stress test (HAST), THB, and pressure cooker–type tests. In general, active devices are less prone to failure than MIM capacitors that are part of the circuit. Processing of top metallization and nitride coat are optimized to prevent moisture ingress. Moisture eventually causes MIM failure, which may lead to active device failure. Active devices also fail if moisture finds a path to the ohmic contacts, channels of FETs, or junctions of HBTs. Presence of leftover chlorine from HCl treatments must also be avoided to prevent failure due to corrosion and migration of water into the circuit from pads and die edges.

22.7.3 Stress-Induced Burnout

Semiconductor devices often show a burn mark on a large portion of the circuit, indicating that the device had a catastrophic failure or burnout. The burnout can be a result of any failure mechanism that causes a runaway situation due to localized power dissipation and thermal runaway. Burnout that is instantaneous may not leave any evidence of the mechanism behind. ESD, electrical overstress (EOS), RF voltage, or current spikes can cause burnout of this nature. EOS, in particular, refers to improper application or use beyond that intended by design. This kind of burnout can be prevented by robust design and by providing ESD protection.

In FET-type devices, burnout due to gate–drain breakdown can be prevented by device layout, structure, or topology, as well as process improvements that improve surface quality. Source–drain burnout is related more to thermal activation and current crowding. Metal migration through defects can also lead to shorts and subsequent instantaneous burnout. This type of failure can also occur in HBTs and can also be mitigated by better device design, more robust epistructure, and ESD protection.

Long-term burnout is the result of device parametric drift with aging, which eventually leads to localized current and thermal runaway. Use of better passivation, thermal management, defect-free substrates, etc., can lead to improved lifetime. Schottky contacts, resistors, and MIM capacitors can also be susceptible to ESD or EOS.

Awareness of ESD and its effect on reliability and implementation of protective measures in the fabrication process can reduce ESD issues.

22.8 GaN Device Reliability [19]

GaN device reliability suffers from frequency dispersion, current collapse, and gate and drain lag transients. Reliability issues due to material include epitaxial layer quality, strain in the epitaxial layer stack, and the level of Al% in the AlGaN barrier. The degradation mechanisms related to the surface are traps and metal–semiconductor diffusion. Defects can be created by processing, such as dry plasma etching or deposition. Defects can be reduced by proper surface treatments and dielectric passivation. AlN deposition has been shown to improve mobility; however, reliability was not improved [20]. Bulk defects that degrade reliability are deep in the energy bandgap and have been detected by photoionization.

Idss drop can be caused by gate Schottky metal instability due to high operating temperature and RF stress [21] and also gate–drain edge pit formation, possibly due to the inverse piezoelectric effect. The inverse piezoelectric effect, which is the creation of mechanical stress as a result of a high field, combined with the presence of moisture, has been shown to cause pits under high drain bias in the off condition [22].

The choice of gate metals used and the shape of the gate profile in order to tailor electric field profiles can also be a factor.

Ti/Al ohmic contacts are stable against oxidation and cracking when sealed with Ni/Al but do wear out after long operation. Contacts to n+ AlGaN may be more stable.

References

1. R. Williams, *Modern GaAs Processing Methods*, Artech House (1990).
2. R. Shaw, Reliability overview, JPL Publication, 96-25 (1996).
3. A. Christou and W. M. Webb, *Reliability of Compound Semiconductor Analog Integrated Circuits,* RIAC, University of Maryland, Google Books (2006).
4. S. Kayali, G. Ponchak and R. Shaw, GaAs MMIC reliability assurance guideline for space applications, JPL Publication, 96-25 (1996).
5. C. Canali et al., Gate metallization sinking into the active channel in Ti/W/Au metallized power MESFETs, *IEEE Electron. Device Lett.*, **EDL-7**, 185 (1986).
6. T. Hashinaga, M. Nishiguchi and H. Nishizawa, Active layer thinning due to metal-GaAs reactions IN GaAs ICs with Ti/Pt/Au gate metallization, *Proc. GaAs Rel. Workshop* (1991).
7. W. Roesch, *Outstanding Issues in Compound Semiconductor Reliability,* CS MANTECH (2005).
8. F. Magistrali et al., Reliability issues of discreet FETs and HEMTs, in *Reliability of MMICs,* Ed., A. Christou, John Wiley and Sons (1992).
9. S. Kayali, Hydrogen effects on GaAs, status and progress, *Proc. GaAs Rel. Workshop* (1995).
10. O. Ueda et al., Current status of reliability of InGaP/GaAs HBTs, *Solid State Electron.,* **41**(10), 1605 (1997).
11. Y. A. Tkachenko et al., Hot electron induced degradation of MESFET, *Proc. IEEE GaAs IC Symp.,* 259 (1994).
12. B. Yeats et al., Reliability of InGaP emitter HBTs at high collector voltage, *GaAs IC Symp. Tech. Digest,* **73** (2002).
13. K. Feng et al., Reliability of commercial InGaP/GaAs HBTs under high voltage operation, *IEEE GaAs IC Digest,* **71** (2003).
14. O. Ueda, Reliability issues in III-V compound semiconductor devices: optical devices and GaAs-based HBTs, *Microelectron. Rel.,* **39**, 1839 (1999).

15. W. Liu, *Handbook of III-V Heterojunction Bipolar Transistors*, Wiley Interscience, New York (1998).
16. S. J. Pearton et al., *J. Vac. Sci. Technol.*, **B13**, 15 (1995).
17. N. M. Johnson, Hydrogen in compound semiconductors, *Mat. Res. Soc. Symp. Proc.*, **262**, 369 (1992).
18. S. Tiku, Rockwell International, unpublished data.
19. Y. Otoki, CS MANTECH (2013).
20. A. Christou, CS MANTECH (2013).
21. A. R. Barnes and F. Vitobello, CS MANTECH (2012).
22. F. Gao et al., *Role of Electrochemical Reactions in the Degradation of AlGaN/GaN HEMTs*, CS MANTECH (2014).

Chapter 23

GaN Devices

Section I: GaN Electronic Devices

23.1 Introduction [1]

Wide-bandgap (WBG) semiconductors have gained popularity in the last decade due to their potential in electronic applications that require high voltage, high power, high temperature, and high switching speed. Devices based on WBG materials are already used in commercial photonics, power conditioning, light-emitting diodes (LEDs), and high-power microwave devices. AlN, GaN, and InN have a direct energy bandgap that ranges from 6.2 eV to 0.78 eV, form a continuous alloy system, and thus cover the red-to-ultraviolet wavelength. Therefore III–N materials have immense potential for optical devices. Schottky power diodes made with GaN on sapphire have shown a breakdown voltage near 10 kV. Devices with an operating voltage of 600 V are possible on GaN. The unique properties of GaN make it suitable for high-electron-mobility transistor (HEMT) and heterojunction field-effect transistor (HFET) device applications. AlGaN/GaN HEMTs are already being produced in large quantities for high-power amplifiers such as for cellular base stations. These also hold promise for more applications in mixed-signal, millimeter-wave, and sensing applications

III–V Integrated Circuit Fabrication Technology
Shiban Tiku and Dhrubes Biswas

ISBN 978-981-4669-30-6 (Hardcover), 978-981-4669-31-3 (eBook)
www.panstanford.com

For power amplifiers, different technologies are being used or being considered: Si-lateral-diffused metal-oxide semiconductor (LDMOS) and bipolar transistors, GaAs heterojunction bipolar transistors (HBTs), SiC (silicon carbide) metal-semiconductor field-effect transistors (MESFETs), and GaN HEMTs. The relevant materials properties of competing materials, including GaN, are shown in Table 23.1. Because of high bandgap, high breakdown voltage, and low on resistance, the power density can be very high. The high power per unit width translates into smaller devices that not only are easier tofabricate but also offer much higher impedance.

High breakdown voltage allows these to be used at high voltage. Use of high drain voltages eliminates or at least reduces the need for voltage conversion. High output impedance enables easier low loss matching.

Commercial systems (e.g., wireless base stations) operate at 28 V and a low-voltage technology would need a voltage stepdown from 28 V to the required voltage. GaN devices can easily operate at 28 V. The higher efficiency that results from this high operating voltage reduces power requirements. High power and gain can allow replacement of multiple stages of amplifiers in a module by just one power amplifier die. Higher thermal conductivity (see Table 23.1) of the substrate allows high-temperature operation and simplifies cooling, an important advantage for reducing the cost and weight of cooling systems. WBG materials offer a rugged and reliable technology capable of high voltage-high temperature operation and industrial, automotive, and aircraft and defense applications.

Crystal and epilayer growth, device types, and fabrication processes, as well as reliability of GaN devices and circuits, are discussed in this chapter.

23.2 Bulk Crystal Growth [2]

Gallium nitride substrates are available only in a very small size. Phosphorus- or arsenic-based III–V semiconductors can be grown from melt with relative ease due to a melting point of 1238°C with an arsenic dissociation pressure of about 1 atm. At temperatures needed for GaN growth, nitrogen has very low solubility in gallium; therefore

Table 23.1 Properties of candidate materials for high-power electronics and figures of merit

Material	Mobility, μ (cm^2/V.sec)	Dielectric constant, ε	Bandgap E_g (eV)	Breakdown field E_b (MV/cm)	Thermal conductivity (mW/cm-K)	T_{max} (°C)	Lattice constant (Å)
Si	1300	11.9	1.12	0.3	1500	300	5.431
GaAs	5000	12.5	1.42	0.4	460	300	5.653
4H–SiC	260	10	3.2	3.5	4900	600	4.360
GaN	1500	9.5	3.4	2	2200	700	3.185

a very high vapor pressure of N_2 is needed to grow GaN. GaN has a melting point of 2500°C with a dissociation pressure of nitrogen estimated at around 45,000 atm., making it virtually unattractive for liquid-phase growth. Heteroepitaxyon substrates like sapphire and silicon was tried historically. Growth on foreign substrates has disadvantages. These are presence of defects, strain, and thus lower performance and reliability. Bulk-grown boules are a necessity for large-volume production of high-quality devices. Even epigrowth can be simplified if a native substrate is used, because elaborate and expensive buffer layers would not be needed. Potential advantages of bulk crystal availability would be low defect density and smooth morphology of the epidevice layers. Thermal conductivity of GaN is higher than that of sapphire. And electrically conductive substrates could be used for devices that can be fabricated with ohmic contacts on the backside, for example, Schottky diodes or LEDs.

Since large-volume boule growth, like that used for GaAs or InP cannot be used, pseudobulk growth methods are described below.

23.2.1 Hydride Vapor-Phase Epitaxy [2, 3]

Hydride vapor-phase epitaxy (HVPE) is a practical method that can be used to grow layers thick enough to be separated from the substrate. The earliest GaN substrate was grown by halogen transport vapor-phase epitaxy, which for historical reasons is more commonly called HVPE from reactions of GaCl and NH_3. The main advantages of this process are user-friendly growth conditions, low pressure, relatively low growth temperature, and high growth rate. The halide epigrowth process was discussed in Chapter 4, where the source and the substrate are held at different temperatures. Figure 23.1 shows a schematic diagram of such a system for GaN. Two temperatures are shown, but more controlled temperature zones are used in practice. HCl is passed over Ga metal held at 860°C to form GaCl. GaCl flows over to the substrate zone and is directed by a showerhead to the substrate held at a higher temperature of 950°C–1050°C. Ammonia is fed through separately, and H_2 is used as the carrier gas.

$$GaCl + NH_3 \rightarrow GaN + HCl + H_2$$

A growth rate of greater than 100 μm/hr can be achieved. Layers thicker than 200 μm can be grown, which could be separated to make a freestanding substrate for epigrowth. Semi-insulating, n- or p-type layers have been grown. Unintentional doping by silicon and oxygen from the quartz-ware is unavoidable, and that results in n-type material. Intentional Si doping can be done by using SiH_4 or Si_2H_6. For p-type doping, the standard III–V dopants can be used, namely Zn, Cd, Be, and Mg.

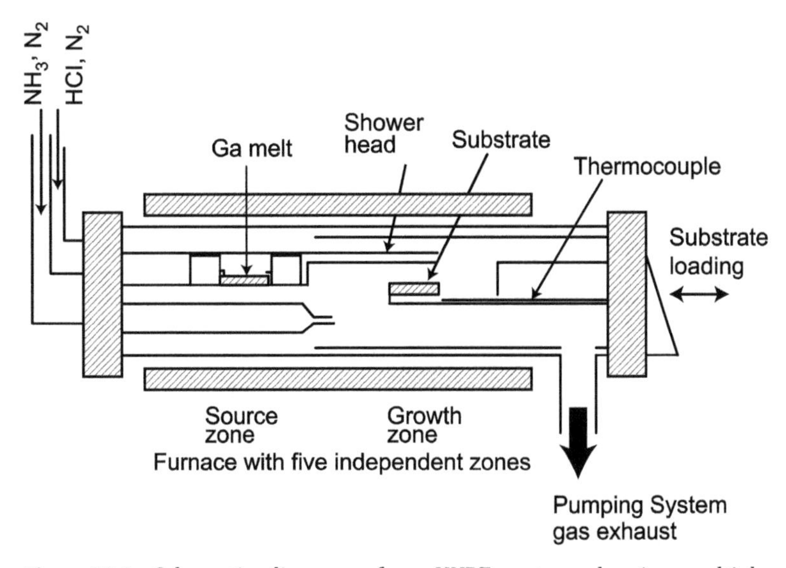

Figure 23.1 Schematic diagram of an HVPE system showing multiple temperature zones. Reproduced from Ref. [3] with permission from JMOe.

HVPE systems with vertical orientation have been used in high-volume production. In the configuration shown in Fig. 23.2 [4] from Aixtron, the source gases are fed through an annular tube system, with Ga in the central tube. Reactant gases are kept apart in these systems until these reach the substrate. Otherwise unwanted reactions will take place, including formation of NCl_3, a dangerous explosive. In principle, InN and AlN layers can be grown in such a system but are difficult in practice. Dopant gases are fed to achieve n- and p-type layers. As mentioned before, the contamination from the walls tends to make the material n type.

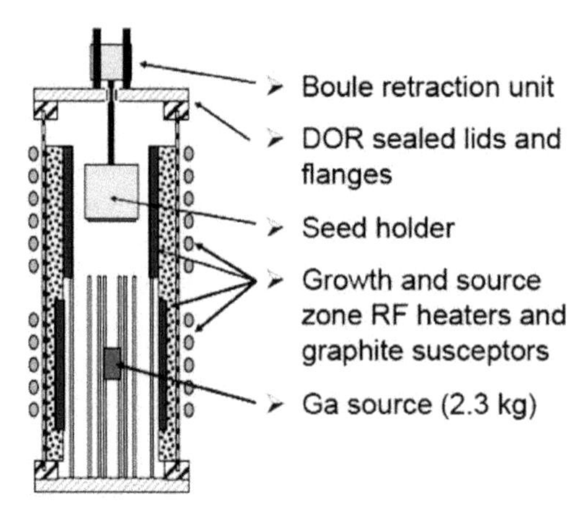

Figure 23.2 Simplified schematic drawing of a vertical HVPE reactor from Aixtron for growing free-standing GaN. Reprinted from Ref. [4] with permission. Copyright © 2007 CS MANTECH.

23.2.1.1 MOCVD templates [5]

The use of metal–organic chemical vapor deposition (MOCVD) templates for GaN growth on sapphire was first shown in molecular beam epitaxy (MBE) growth of thin GaN layers. Later, it was successfully employed in HVPE GaN growth [5]. The authors showed that MOCVD GaN templates can provide a good buffer for growth of thick HVPE GaN layers with very good structural and optical characteristics, although with cracks at the interface. Two-inch boule growth ranging from 2.5 and 3.5 mm up to 10 mm has been reported, with 4 inches being the maximum-reported current size, with threading dislocation densities of the order of less than $5 \times 10^6/\text{cm}^2$. Several vendors such as Sumitomo, Furukuwa, Hitachi, Mitsubishi, Samsung, Saint-Gobain, and Kyma are in the market. The resistivity of n-type substrates required for optoelectronic applications is 0.02 ohm-cm or lower (doped with Si), while that of semi-insulating substrates (doped with Fe) for microwave applications is greater than 10^6 ohm-cm (source: Kyma). The doped substrates have higher crystallinity than the semi-insulating ones. It may be noted that the sizes of such substrates are 5×20 mm^2; 2-inch circular substrates are difficult to form compared to rectangular ones.

23.2.2 High-Pressure Solution Growth

High-pressure solution growth (HPSG) is achieved by a direct reaction between NH_3 and Ga at 1500°C under 16,000 atm., achieving growth rates of 1–10 μm/hr (by Karpinsky et al. [6]). Figure 23.3 shows a basic schematic diagram. Earlier HPSG attempts at lower pressures of 8000 atm. had abysmally low growth rates of 1 μm/day. Even though this method has lower growth rates, it yields high-quality GaN with about half of the theoretical limit of full width at half maximum (FWHM) of X-ray diffraction (XRD) in omega mode at 20 arc seconds and a very low dislocation density of the order of $10^2/cm^2$.

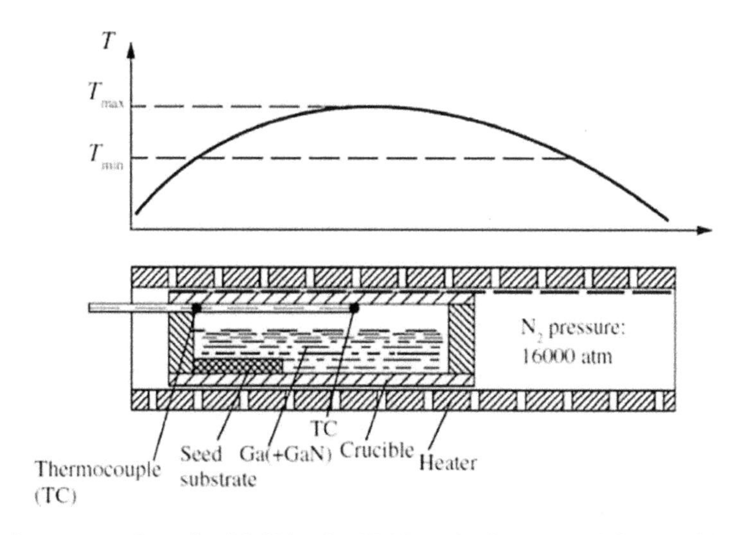

Figure 23.3 Growth of GaN by the HPSG method. Reprinted from Ref. [6], Copyright (1984), with permission from Elsevier.

23.2.3 Ammonothermal Growth

This method is similar to hydrothermal growth. Unlike the HPSG method, the pressures required is about 10%–20% to about 2000–3000 atm. at temperatures of about 500°C and achieving growth rates of 1–2 μm/hr. GaN is dissolved in ammonia under high pressure using a solubilizing agent also called a mineralizer. The dissolved nutrient is transported to the growth region for crystallization on a seed crystal. The mineralizer can be basic, based on NH^{2-}, acidic NH^{4+}, or neutral. Potential advantages of this process could be good quality, because the growth takes place under conditions close to equilibrium. In practice, the quality may depend upon the seed. The expected growth rate and hence the capacity of the systems is low. Crystals of 50 mm diameter with thickness of a few millimeters have been grown using this process. The crystalline quality is excellent at a theoretical limit of about 10 arc seconds, with dislocation densities around $5 \times 10^3/cm^2$ or lower.

There is a fourth growth type under investigation. This process, called *solution growth*, is similar to liquid-phase epitaxy (LPE) discussed in Chapter 4. For GaN, sodium flux can be used to make possible growth at low temperature and low pressure. These techniques are still under development.

At this time, GaN boule growth is still in the early stages of development, and no mature technology is available due to issues described above.

23.3 Epitaxial Growth

Figure 23.4 shows the structure of a basic HEMT. The devices are grown as epistructures. Sapphire, SiC, and AlN have been tried as substrates, and growth on silicon is an ongoing goal of research and development.

The lattice constant of GaN is 3.185 Å. Sapphire has a lattice constant of 3.76 Å; hence there is a huge mismatch. The SiC (4H) lattice constant is 4.36 Å, and it is 3.08 Å for 6H–SiC. MBE and MOCVD have been tried on sapphire and SiC. Lattice mismatch, cracking on cooling due to thermal mismatch, stacking faults, and dislocations have been studied for all substrates. An AlN interface layer is deposited first on <111> silicon. Silicon carbide has become the preferred substrate because of high performance and easy thermal management in the high-power niche market. In addition to thermal conductivity, the electron mobility in this material is pretty decent; therefore high power can be achieved at high frequency. Power levels of 800 W at 2.9 GHz and 500 W at 3.5 GHz have been demonstrated [7].

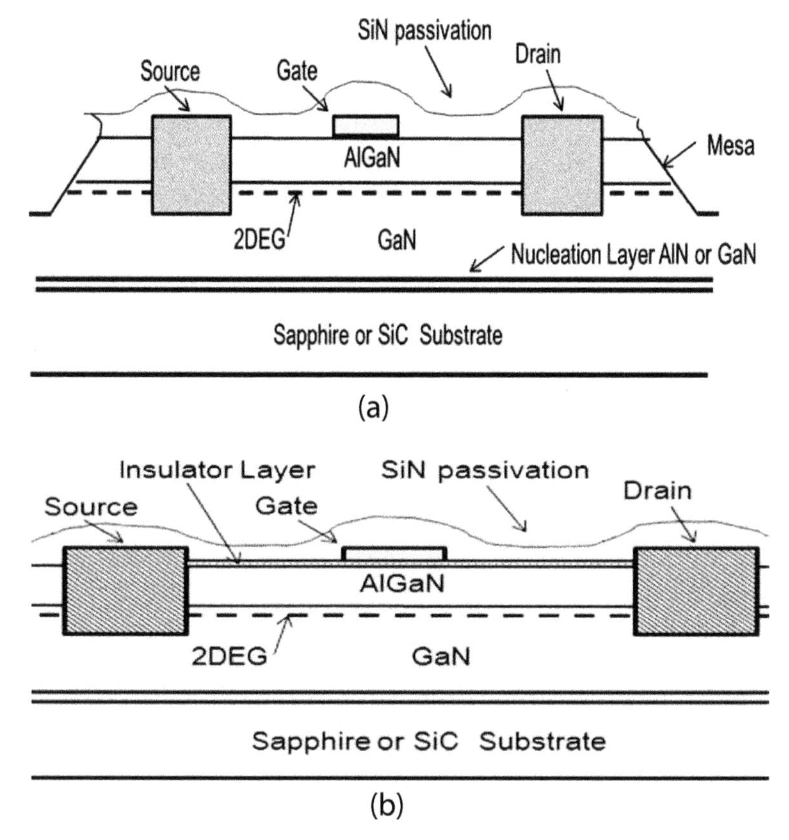

Figure 23.4 A basic AlGaN/GaN HEMT device (a) and the same device with an insulator under the gate (b).

The epitaxial layers may be either grown entirely by MBE or MOCVD. Achieving perfect stoichiometry is always a challenge because of the difficulty of N incorporation.

23.3.1 MBE

MBE has the usual advantages discussed in Chapter 4. Conventional MBE cannot be used for III–N compounds. N_2 does not dissociate on the hot substrate in vacuum. The reactivity of N_2 can be increased by plasma-assisted molecular beam epitaxy (PAMBE) (Fig. 23.5) [8]. In radio frequency molecular beam epitaxy (RFMBE), low-energy species can be inserted without damage by high-energy ions (see Fig. 3.58 in Ref. [8]). In reactive MBE, N_2 is replaced by an NH_3 source, and NH_3 decomposes on the growing surface by pyrolysis. At the High Performance Devices Lab (IIT Kharagpur, India) Biswas et al. (coauthor of this book) typically initiated a nucleation layer of 50 nm of AlN required before the GaN was grown either on a sapphire or (111) on a Si substrate at a temperature above 900°C with an RF plasma power of 600 W at a growth pressure of the order of 10^{-5} Torr, with a N_2 flow of 3 sccm. This usually will ensure a Ga face than the deleterious N face, which leads to spontaneous polarization-related two-dimensional electron gas (2DEG) charges of the order of $10^{13}/cm^2$. With an improper nucleation layer, the N face will lead to poor spontaneous polarization and is not desirable for GaN HEMT growth. Subsequent to AlN growth, GaN growth is performed at an RF plasma power of 650 W at a growth pressure of the order of 10^{-5} Torr, with a N_2 flow of 4 sccm, usually leading to a growth rate of around 0.6–0.7 μm/hr. It may be noted that the growth initiation by nucleating layers of AlN, subsequent maintenance of Ga-rich conditions, and proper streaky surface reconstructions (1 × 1) are necessary for devices to utilize the spontaneous polarizations and absence of harmful surface states above the terminating layer that leads to devastating current slumps in power devices.

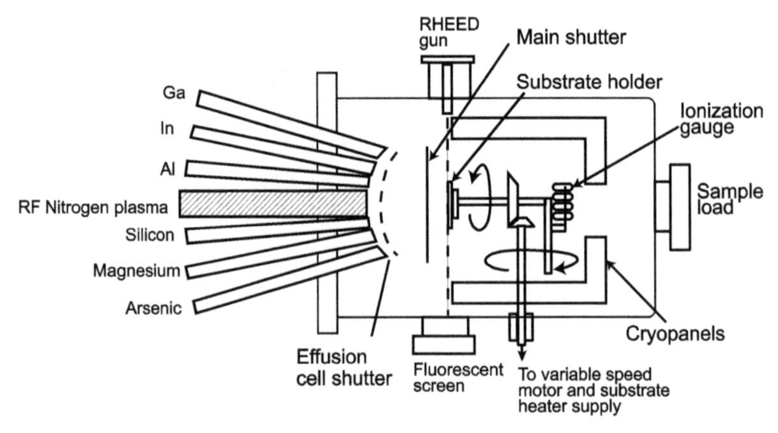

Figure 23.5 Schematic diagram of PAMBE. Reproduced from Ref. [3] with permission from JMOe.

23.3.2 OMVPE

Typically there are two well-established technologies, planar MOVPE and close-coupled showerhead MOCVD. From Noble laureate Nakamura to well-known manufacturer Aixtron-based planetary reactors, planar MOVPE technology has been the mainstream, where the technology varies from a simple horizontal tube reactor to complicated planetary systems. The depletion profile of the reactant gases from inlet to outlet is compensated by substrate rotation where the growth conditions are averaged into a radial symmetry. While the early systems had infrared (IR) lamps for heating the susceptor carrying the multiwafer configured substrates, the modern ones are equipped with an inductive RF heater. While the group III metalorganics thermally decomposed at 350°C to 400°C, only initial cracking of group V NH_3 occurs at 800°C to 900°C, at typical growth temperatures of GaN epitaxial layers at 1000°C to 1100°C, while the majority of NH_3 is decomposed catalytically on the growing surface through dissociative adsorption occurring in conjunction with adsorbed group III compounds. In situ determination of growth rates and surface morphology is estimated through measurements of Fabry–Perot (FP) interferences as well as optical pyrometers.

While GaN growth is done routinely, special care is required for the growth of InGaN as it suffers from the miscibility gap, while AlGaN does not, making it less susceptible to temperature variations. Growth of AlGaN also has the possibility of pre-reactions of TMAl and NH_3, leading to the formation of adducts in the gas phase, thereby decreasing the growth rate, which needs to be addressed by process optimization and reactor design.

Close-coupled showerhead MOCVD has the potential for scalability. Vertical-axially symmetric, fast-rotating disc reactors can intrinsically achieve very uniform boundary layers, deposition rates, and higher alloy compositions. The growth precursors diffuse through the boundary layer and are defined by a classical mass transfer–limited regime in MOCVD, which fixes the growth rate and alloy composition, which is an invariance with the position on the susceptor, leading to reactors with high growth uniformity. Ideally, it assumes that the susceptor at a fixed temperature is immersed in an infinite volume of gas, which is realistically not the case. The flow of gas bending over the edge of the susceptor leads to distortion of the boundary layer. Additionally, *wall drag effects* arising from the stationary gas in contact with the reactor chamber wall get worse for tall reactors, leading to a short bell-jar design necessary for superior growth of higher alloy compositions of AlGaN. The shorter design also reduces the susceptor edge-driven boundary layer distortion. Furthermore, thermal buoyancy–driven recirculation cells further degrade interfacial quality and uniformity, unless careful thermal designs are implemented. Many commercial computational fluid dynamics–driven software programs are available for optimization of reactor geometries for exacting requirement of nitride semiconductors. In addition to standard in situ measurement techniques mentioned earlier, intensive reflectometry techniques are used in such close-coupled showerhead reactors, which allows for accurate automatic control of reactant flows on the basis of the live situation of growth progress in the reactor.

MOCVD or organometallic vapor-phase epitaxy (OMVPE) of nitrides was started by Manasevit [9] and work on improving crystal quality has been going on for decades. As expected, NH_3 is used for nitrogen and TMG, TMAl, and TMIn are used as sources for group III elements.

As in the case of other III–V compounds, the reaction is as below:

$$(CH_3)_3Ga + NH_3 \rightarrow GaN + 3(CH_4)$$

23.3.2.1 Doping

Generally, SiH_4 is used for n-type Si doping and TMMg is used for p-type doping. Other n-type dopants are Si, Ge, S, Se, and Sn. Mg and Zn can be used for p-type doping. Cyclopentadienyle Mg (Cp2Mg), which is $(C_3H_5)_2$ Mg, can be used for Mg doping. Mg forms Mg–H, which is inert and must be activated by low-energy electron beam irradiation (LEEBI) to actuate Mg [10].

GaN and AlGaN layers are typically grown at 1000°C at growth rates of ▯1 μm/hr. AlN has less than 1% mismatch on 6H and 4H–SiC (3.49% and 3.53% for GaN). In SiC, a polar crystal, the N bond to Si is stronger on the Si face of SiC than that of Al or Ga. Growth on well-prepared SiC produces Al-polarity AlN, which leads to Ga-polarity GaN. GaN nucleation on SiC is typically performed using AlN grown at high temperatures near 1080°C for better electrical and optical properties. Good devices can be made with nucleation layers grown at a lower temperature of 900°C. HEMT structures are grown at Cree by MOCVD on 100 mm semi-insulating 4H–SiC. AlN is used as the nucleation layer. Insulating Fe-doped GaN with a mobility as high as 2000 cm^2/V-sec can be grown.

23.3.3 HVPE

This process described under bulk growth can be used for epilayers. GaN tends to grow with a large-background n-type carrier concentration because of native defects (N deficiency as expected). Therefore, HVPE produces layers with the best transport and structural properties. This technique has been successfully used to grow templates (from 0.5 to 100 microns), which allow growth of subsequent epitaxial layers by MBE and MOCVD without the need of an elaborate buffer layer. The various kinds of template substrates that are available, apart from GaN on silicon for microwave applications, also include low-defect 5–20 μm thick n-GaN, p-GaN, ultraviolet (UV) transparent AlGaN on sapphire templates for UV LEDs, and AlN on sapphire/silicon/SiC for high-power field-effect transistor (FET) applications. The challenges of such strategy are to counter the drawbacks of high defect density, cracking, and substantial substrate bowing due to thick layers. Stress control techniques have been utilized to counter such severe bowing [11].

23.4 Device Physics and Device Types [11–16]

III–N materials have a wurtzite crystal structure, different from the zinc-blende structure of the III–V semiconductors (Fig. 23.6). In this structure the dipoles do not cancel because of nonsymmetry, so there is spontaneous polarization and fixed charge sheets at interfaces and surfaces. The piezoelectric effect is also present like the zinc-blende structure but is an order of magnitude stronger. Figure 23.7 shows the energy band diagram for the undoped AlGaN/GaN interface. The usual conduction and valence band discontinuities are present at the heterointerface like other III–V heterojunction semiconductors. In this case, however, there is polarization and an electric field present. Figure 23.8 shows a comparison of the AlGaN/GaN interface with that of zinc-blende materials, like AlGaAs/GaAs [12]. Both band discontinuity and the presence of a polarization field contribute to the formation of a 2DEG. The polarization charge dominates. At the AlN/GaN interface, $6.4e^{13}/cm^2$ charge is present. Similarly at the AlGaN/GaN interface in an HEMT device, the 2DEG is entirely due to polarization charge ($>10^{13}$) and doping is not needed. As a result, the carrier concentration in not affected by temperature. The mobility of electrons is in the 1200–2000 cm^2/V-sec range.

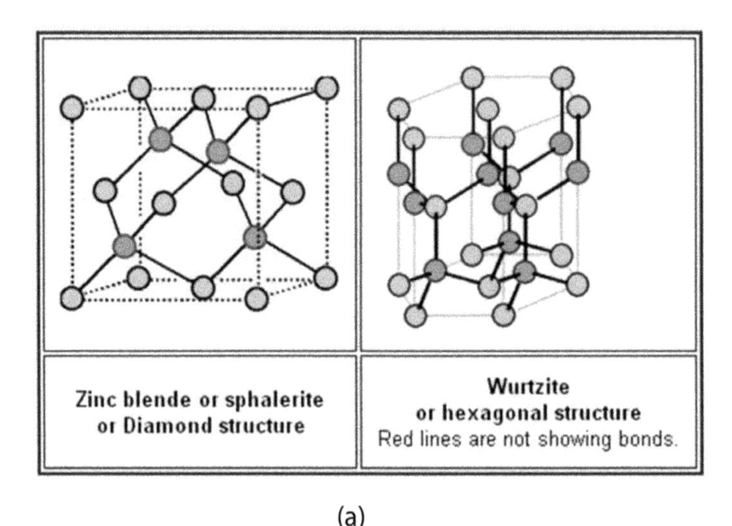

(a)

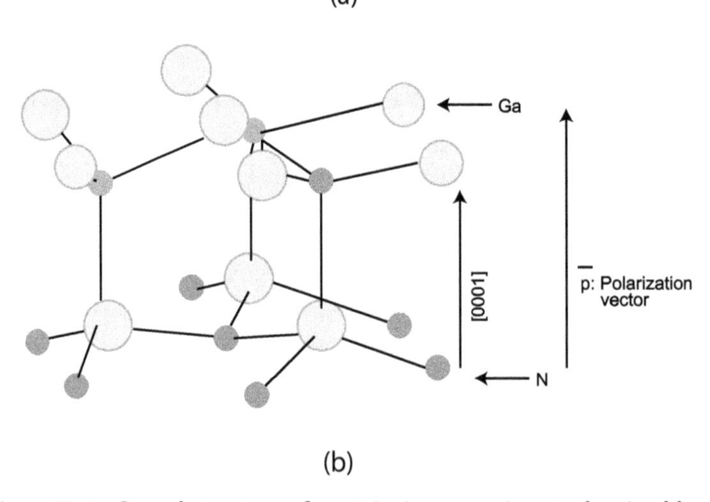

(b)

Figure 23.6 Crystal structure of wurtzite in comparison to the zinc-blende structure of GaAs (a). Details of bonding and polarization (b).

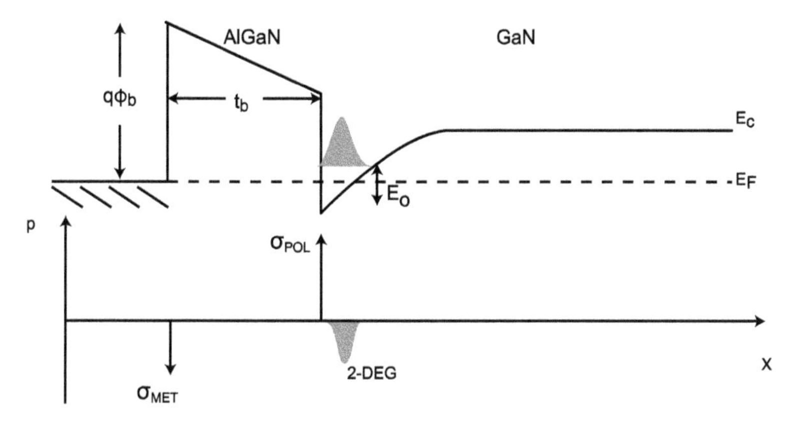

Figure 23.7 Energy band diagram of the metal/AlGaN/GaN structure.

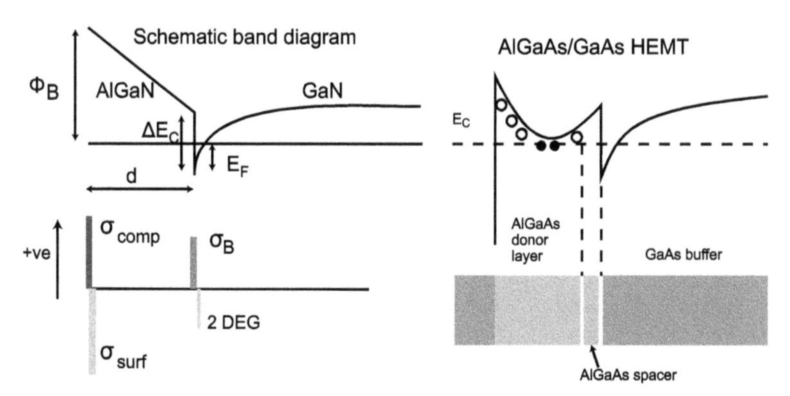

Figure 23.8 Comparison of an AlGaN/GaN HEMT with that of AlGaAs/GaAs, showing 2DEG created without doping in the AlGaN. © [2008] IEEE. Reprinted, with permission, from Ref. [12].

The electron velocity comparison for Si, GaAs, and GaN is shown in Fig. 23.9. At high electric fields both SiC and GaN have an advantage of higher electron mobility [14].

High breakdown voltage and low R_{ON} are easily possible in GaN devices. The figure of merit (FM) for high-voltage, high-power devices (power-handling capability)

$$\mathrm{FM} = \frac{V_{\mathrm{BR}}^2}{R_{\mathrm{ON}}} \tag{23.1}$$

is 100 times better than that of silicon devices.

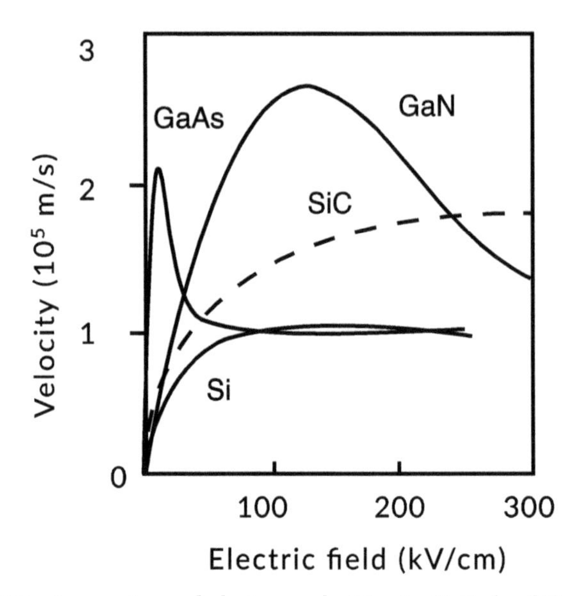

Figure 23.9 Comparison of electron velocities in Si, GaAs, SiC, and GaN, showing the highest numbers for GaN at a high field. © [2002] IEEE. Reprinted, with permission, from Ref. [14].

A wide range of device types is possible with GaN, as mentioned earlier. Only device types used in monolithic microwave integrated circuit (MMIC) and power electronics are discussed briefly here.

An HEMT is the obvious device of choice for high-power MMICs because of favorable properties and ease of fabrication. AlGaN/GaN FET-type devices grown on SiC or sapphire will be discussed in detail in this chapter. A metal–insulator–semiconductor heterostructure field-effect transistor (MISHFET) with the addition of insulators or a metal–oxide–semiconductor heterostructure field-effect transistor (MOSHFET) with the addition of oxide under the gate can be fabricated with improved performance at higher power levels and voltages. Figure 23.4b shows a schematic diagram of FET- (HEMT-) type devices with an optional insulator layer under the gate metal. Figures 23.10 and 23.11 show the transfer curves and gate leakage currents for three types of devices: HFET, MISHFET, and MOSHFET. Higher drain currents and very low gate leakage can be achieved by the addition of insulators or oxides. As expected, deposition or growth of good insulators is not trivial.

GaN devices have tremendous potential in power electronics as diodes, switching transistors, etc. HBT-type devices are not practical at the moment and will be mentioned briefly at the end of this chapter.

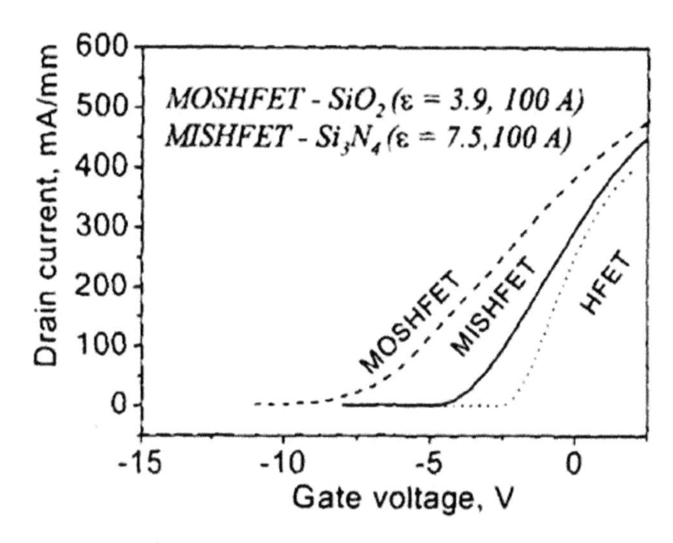

Figure 23.10 Transfer characteristics of MOSHFET, MISHFET, and HFET. © [2002] IEEE. Reprinted, with permission, from Ref. [14].

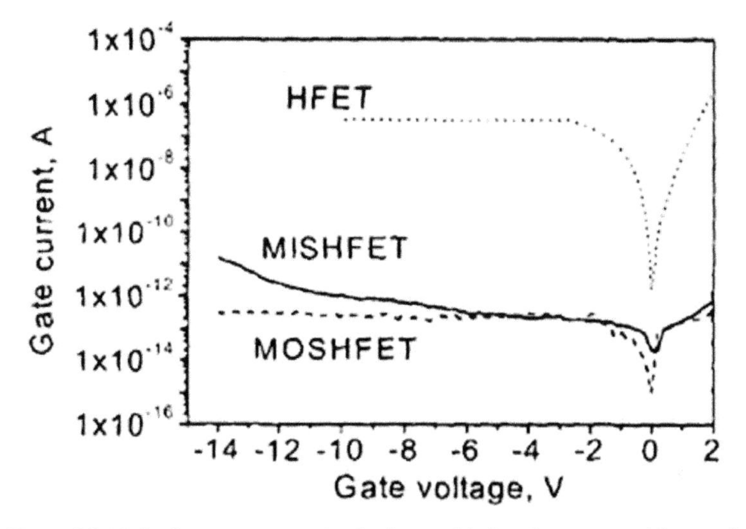

Figure 23.11 Leakage currents in devices with insulator or oxide can be orders of magnitude lower. © [2002] IEEE. Reprinted, with permission, from Ref. [14].

23.5 Process Technology

Process techniques needed for fabrication of diode- and FET-type devices are discussed in this section.

23.5.1 Etching and Surface Passivation of GaN

There are no useful wet etches available for GaN. Sodium hydroxide, NaOH, at 125°C or molten KOH can be used, but these are not practical in integrated circuit (IC) processing where masks, like photoresists, have to be used. Therefore dry plasma etching is used in device fabrication.

Dry plasma etches are used in mesa formation and removal of layers. Surfaces need to be precleaned before deposition of passivation layers or gates. Low-damage dry etching processes are needed prior to metal deposition. Wet cleaning and oxygen plasma treatment unpin the surface to some extent.

For mesa etching or selective etching of layers, chlorine-based plasmas are used for etching GaN and AlGaN. Flourine and oxygen addition reduces the etch rate at the AlGaN interface (same mechanisms as the AlGaAs/GaAs interface, due to low volatility of Al flourides). Higher selectivity has been achieved with $Ar/Cl_2/CH_4/O_2$ chemistry [8, 17]. Good mesa isolation (200 nm deep) has been shown by dry etching by Cl_2:CH_4:He: Ar [18].

Silicon nitride deposition by plasma-enhanced chemical vapor deposition (PECVD) is used for surface passivation. In addition to PECVD silicon nitride, atomic layer deposition (ALD) AlN has been tried.

23.5.2 Ohmic Contacts

Making a low-resistance ohmic contact is challenging for high-bandgap materials. Ohmic contacts are made by alloying in Ti/Al metallization into GaN, but contacts can be made directly through AlGaN. Ohmic contact formation relies on the regrowth of TiN-conducting regions as the Ti diffuses in the alloying process in the presence of nitrogen gas. Temperatures close to 900°C are needed, typically for 30 seconds. Specific ohmic contact resistance less than

10^{-5} ohm-cm^2 can be achieved for semiconductor doping levels in the 10^{19}/cm^3 range [19]. If TiN is deposited, the contact exhibits thermionic emission, whereas a Ti contact annealed in N_2 shows tunneling behavior. A second layer of Ti/Al was added originally to reduce Ga outdiffusion and improve surface morphology [8]. However, Ni/Au or Pt/Au can be used to achieve the same result and avoid oxidation. Typical ohmic contacts may have a layer structure like Ti/Al/Ni/Au of 100 Å/2000 Å/500 Å/200 Å [20]. As in the case of other III–V compounds, surface preparation before ohmic contact deposition is important. Very aggressive wet etches like hot aqua regia or boiling KOH are needed for GaN [8]. Reactive ion etching (RIE) surface preparation that promotes interdiffusion may be preferred in practice.

Arsenic contamination must be avoided, as it has an adverse effect on morphology and FET leakage current [21].

Low R_c (3×10^{-7} ohm-cm^2) has been achieved by Ti/Al/Mo/Au alloyed at 850°C and also by Ta/Al/Mo/Au (1×10^{-6} ohm-cm^2) [22]. Figure 23.12 shows the ohmic contact resistance as a function of the anneal temperature. Ti completely mixed with AlGaN and GaN, and no interface was left after alloying. In the case of Ta/Al/Mo/Au, the interface was still intact after alloying, as expected for a refractory metal. W and WSi metal systems form more stable ohmic contacts but need to be alloyed at a higher temperature, near 1000°C [23, 24], to give R_c of 8e^{-5} ohm-cm^2.

Nonalloyed ohmic contacts can be fabricated by implanting source and drain regions heavily by Si ion implantation. Contact morphology can be smoother [11].

In III–V materials, grading of ternary WBG semiconductors to low-bandgap materials like InAs is used for achieving low contact resistance. In the III–N case, InN low-bandgap material is less stable thermally, so grading has not been utilized. Ohmic contacts with lower contact resistance can be achieved by using regrown n+ epilayers. In this case nonalloyed ohmic contacts can be formed with Ti or Mo/Au. Ohmic contacts to p-type GaN are much harder to make. Ni/Au metal system produces high R_c ($\sim 10^{-2}$ ohm-cm^2 range). Contacts to P+ are much easier, and these are needed for contacts to the HBT base (to be discussed in Section 23.8.1).

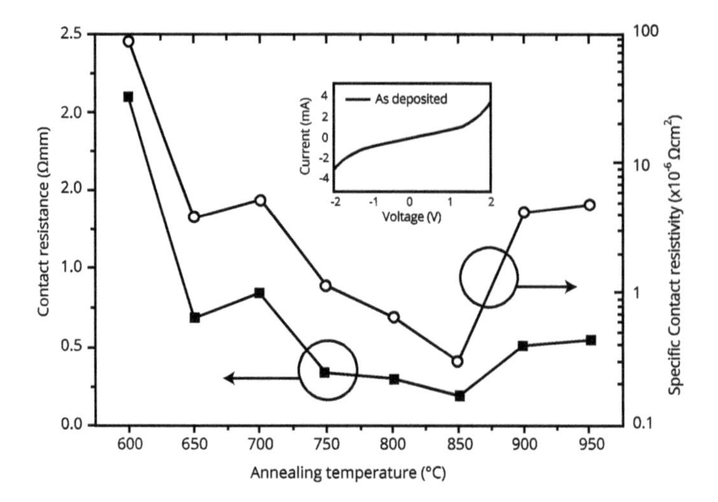

Figure 23.12 Ohmic contact resistance as a function of RTP anneal temperature. Optimum temperature is seen to be in the 750°C to 850°C range. Reprinted with permission from Ref. [22]. Copyright © 2005, American Vacuum Society.

23.5.3 Schottky Contacts

The barrier height of various metals shows a lot more variation on GaN than on GaAs because the Fermi level is pinned at the surface to a less extent compared to GaAs. Figure 23.13 shows a wide range of barrier heights [23]. The surface state density is an order of magnitude lower, $\sim 1.8 \times 10^{13}/\text{cm}^2$ [17]. Pt has the highest barrier height of 1.0–1.1 eV; Ti shows the lowest. Pt is thermally stable up to 400°C, but Ni is even more stable up to 600°C and is preferred, and Ni/Au is commonly used in practice . Also, PtSi has promise. Ideality factors vary from better than 1.1 for Pt to greater than 1.3.

TaN gate metal is very stable at high temperatures, as expected, but the barrier height is only 0.75 eV compared to 1.2 eV for Ni, which is stable up to 300°C and is therefore preferred [20]. The barrier height of Ni is ~1 eV according to other recent reports [25].

Schottky contacts with metals like Ti, Nb, Cr, W, and Mo are poor (for obvious reason that these form conducting nitrides at the interface).

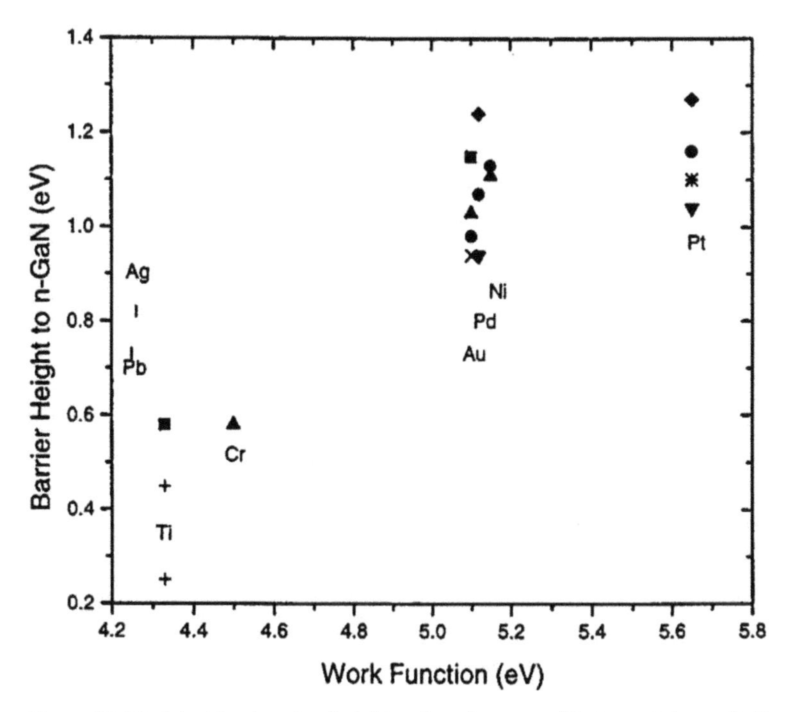

Figure 23.13 Schottky barrier heights of various candidate metals on GaN. Reprinted with permission from Ref. [23]. Copyright © 1999, AIP Publishing LLC.

Precleaning treatments with HCl, HF, and NH_4OH reduce the surface oxide. However, controlled oxidation is necessary to avoid diffusion of gate metal with high-temperature postgate processing.

23.5.4 Implant Isolation

Mesa etching for isolation creates a step for interconnects and is avoided in some processes. Implant isolation is suitable for planar processes more suitable for submicron gates. P/He implantation has been shown to produce very high sheet resistivity ($>10^{11}$ ohms/sq.) [26]. This double implant relies on compensation of n-type carriers and damage by He. On the other hand, N implants rely on damage only. From temperature dependence of resistivity, activation energy can be determined, which is 0.71 eV. These point to deep levels in the energy bandgap, created by damage. As in the case of III–V implant isolation, annealing is needed for optimization of resistivity, which goes up with annealing temperature and then drops. For GaN, annealing temperatures are much higher, ~750°C.

23.6 Device Fabrication

Device fabrication of the basic AlGaN/GaN HEMT (shown in Fig. 23.4) starts with the definition of the active device area. This can be either done by Cl_2 mesa etching or by ion implant isolation. The ohmic contacts are formed by partially etching the AlGaN in the source and drain regions, depositing the ohmic metals and rapid thermal annealing at ~900°C. For recessed gates like those used in GaAs FET/HEMT processing, recess etching in the gate region is performed. Next, the gate is defined by lift-off processing using Ni/Au metallization. Device processing is completed with deposition of a Si_3N_4 passivation layer. This layer serves a critical purpose in avoiding current collapse and eliminating dispersion between the large-signal alternating current (AC) and direct current (DC) characteristics of the HEMT.

23.6.1 AlN/GaN HEMT

AlN has a larger bandgap difference with GaN and can produce higher polarization-induced charge density. But AlN is prone to chemical attack and must be protected during processing. If Al_2O_3 is deposited on top, it must be removed by RIE before an ohmic contact can be made. Otherwise the alloy process must make the contact through it; rapid thermal processing (RTP) at 850°C for 30 seconds is needed [27].

23.6.2 Device Performance Optimization

Figure 23.14 shows a basic HFET AlGaN/GaN device and three ways to reduce the contact resistance. Figure 23.14b shows a selectively regrown n-type GaN layer over which the contacts are made. Figure 23.14c shows the use of a highly conducting cap layer that is removed from the channel area. Figure 23.14d shows an ion-implanted device where the n+ implant is self-aligned to the gate as in standard GaAs self-aligned gate processing. The gate is formed first before Si ion implantation is used to form source and drain. High-temperature annealing, at a temperature much higher than that needed for ohmic contact formation, is needed to activate the silicon implant.

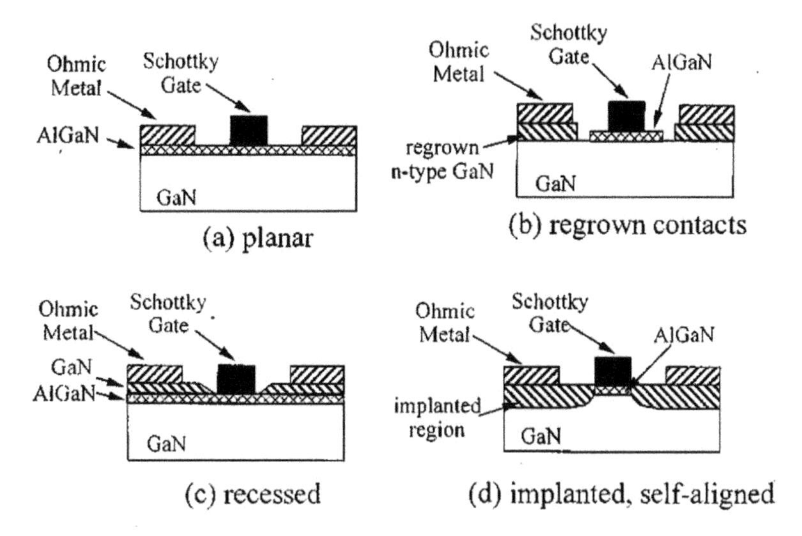

Figure 23.14 Basic GaN HFET types showing a planar contact (a), a regrown contact (b), a recessed structure (c) and a self-aligned implanted structure (d). Reprinted with permission from Ref. [26]. Copyright © 1995, AIP Publishing LLC.

Historically, the main issue with GaN-based HEMT-type devices has been "current collapse," that is, drop of Idss after short operating time.

Approaches to fix current collapse are [1, 28]:

- N GaN cap structures, mentioned above
- Recessed gate
- Field-modulating plates or plates
- Surface passivation

Figure 23.15 shows the use of a basic field plate. Since very high electric fields are present on the drain side, field plates are added to reduce the field. The field plate is electrically connected to the gate and is placed on an insulator layer. With field plates, a power density of 40 W/mm and a breakdown voltage of 10 kV can be achieved [7]. To keep the gate-to-channel capacitance low, the field plate cannot be made too large. A second field plate is added to shape the field even further. This field plate is connected to the source, as shown in Fig. 23.15. To keep the feedback capacitance low, a thicker insulator layer is used between the second field plate and the channel (Fig. 23.16). Device simulations are used to calculate the fields and optimize the field plates (to be discussed further in Section 23.7).

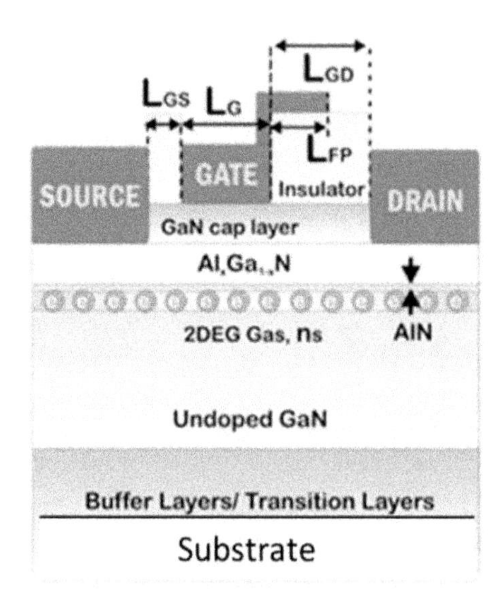

Figure 23.15 GaN device with field plate connected to the gate with insulator in between. © [2014] IEEE. Reprinted, with permission, from Ref. [1].

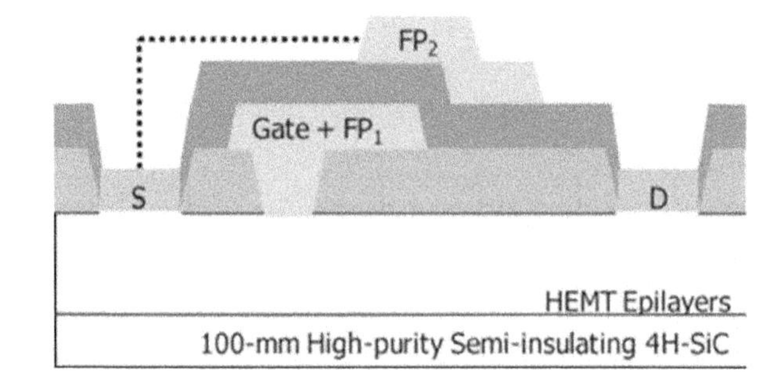

Figure 23.16 Use of a second field plate to reduce the schematic cross section of the AlGaN/AlN/GaN HEMT RF structure showing integrated first field plate and source-connected second field plate. © [2012] IEEE. Reprinted, with permission, from Ref. [7].

23.6.2.1 Surface passivation

Surface passivation by deposition of insulators, for example, silicon nitride, reduces the creation of surface states and thus reduces current collapse.

23.6.3 Normally Off GaN Devices

Normally off devices are always preferred by designers, so attempts have been made to fabricate an enhancement-mode (E-mode) HEMT device using III–N materials.

The following approaches have been tried [29]

- Use a recessed gate.
- Incorporate F into GaN. The damage is annealed out, but semiconductor carriers are compensated by F acceptors.
- Use a p–n junction gate instead of Schottky metal. The 2DEG is depleted.

  Use a recessed gate to make the AlGaN layer thinner (illustrated in Fig. 23.17) [1]. In this approach the recess etch is use to thin the AlGaN layer so that the 2DEG is reduced. The dimensions of the device are shown in Fig. 23.17. The epilayer structure has a 40 nm AlN buffer, 3 μm thick undoped GaN, and a 30 nm Al0.25 Ga 0.75 N barrier layer. The recess etch reduces the AlGaN layer thickness under the gate. Figure 23.18 [29] shows the V_t change with AlGaN thickness. Standard Ni/Au gate metallization is used. Source drain regions are also standard Ti:Al. The device uses chemical vapor deposition (CVD)-deposited 360 nm SiN and 600 nm SiO_2 as passivation.

As the threshold voltage shifts, no increase in R_{ON} is seen, even when V_t approaches 0 V. On-state and off-state I–V characteristics are shown in Figs. 23.19a and 23.19b, respectively.

Devices are available from EPC (40 V, 33 A demonstrated).

23.6.4 MMIC Fabrication [7]

A number of GaN MMIC foundries have developed processes for MMIC fabrication. In the process provided by Cree, after the basic transistor device is completed, standard passive components such as metal–insulator–metal (MIM) capacitors, thin-film resistors, and through-wafer slot vias are fabricated. Figure 23.20 shows a schematic of an MMIC fabricated on a GaN on SiC substrate.

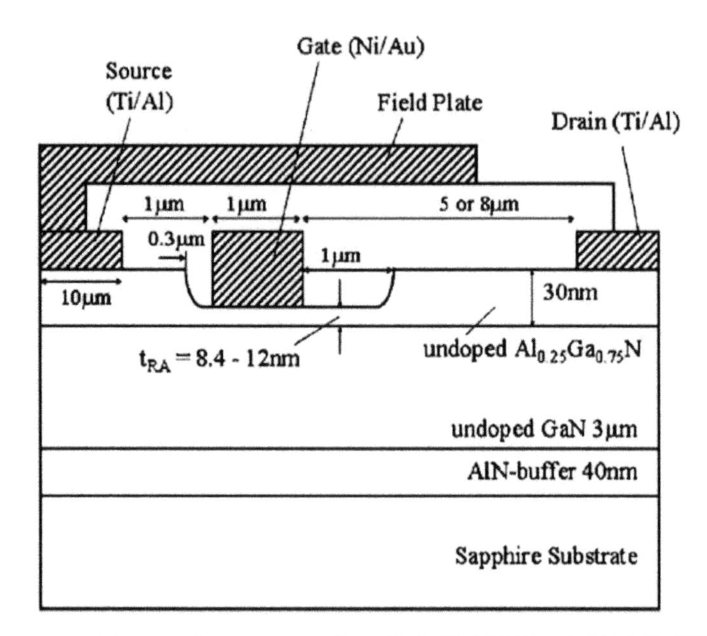

Figure 23.17 Recessed-gate normally off GaN HEMT. © [2014] IEEE. Reprinted, with permission, from Ref. [1].

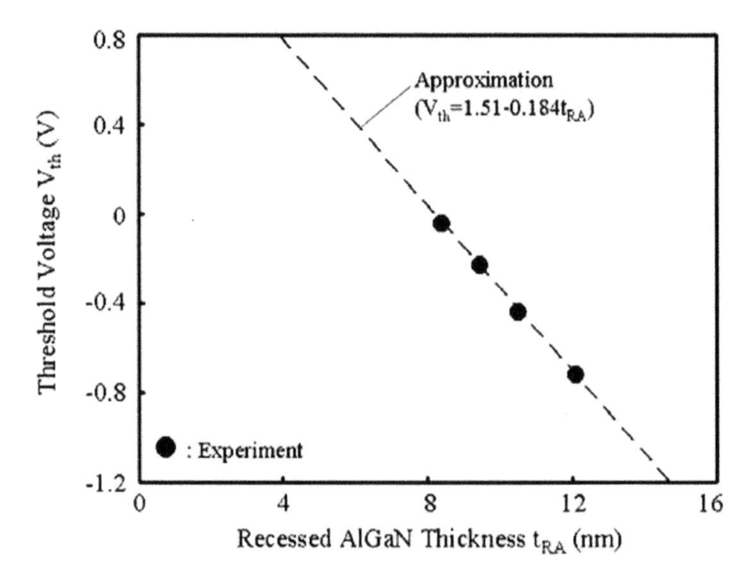

Figure 23.18 Threshold voltage dependence on AlGaN layer thickness. E-mode operation below 8 nm. © [2006] IEEE. Reprinted, with permission, from Ref. [29].

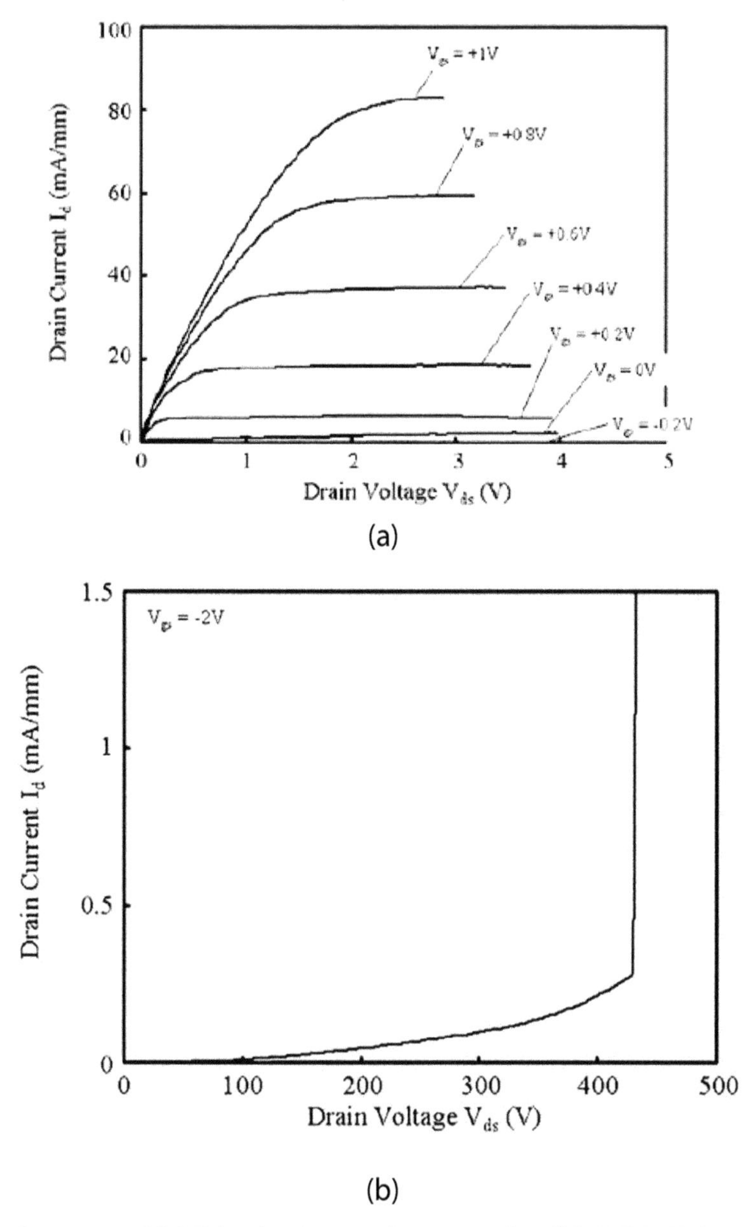

Figure 23.19 (a) *I–V* family of curves for a normally off GaN HEMT shown in Fig. 23.17; threshold voltage of −0.14 V. (b) Off-state *I–V* characteristics of device shown in Fig. 23.17, with V_t = 0.14 V, showing extremely low leakage current. © [2006] IEEE. Reprinted, with permission, from Ref. [29].

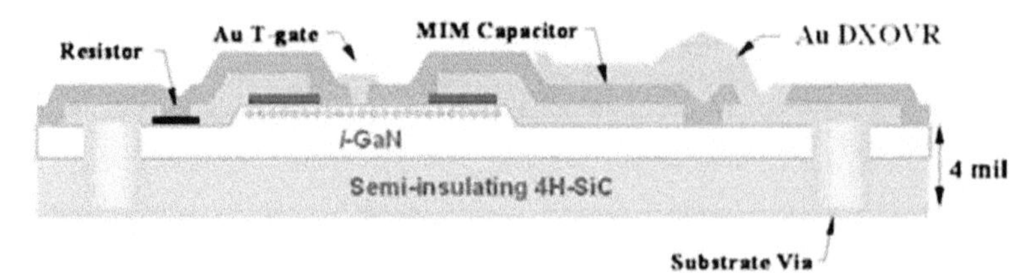

Figure 23.20 A portion of an MMIC device completed with an interconnect and through-wafer via. © [2012] IEEE. Reprinted, with permission, from Ref. [7].

The MIM capacitors were developed to support peak voltages greater than 100 V. Through-wafer slot vias are implemented in the 100 m thick SiC substrates to simplify layout and increase gain. In addition to thin-film resistors, two GaN semiconductor resistors with 70 and 400 ohms/sq. resistance are offered. Bulk GaN resistor layers are covered by thick dielectric insulators, enabling metal crossovers. A 0.4 μm gate length 28 V process provides 4.5 W/mm of gate periphery for circuits between DC to 8 GHz, while a 0.25 μm gate length 40 V process provides 7 W/mm of gate periphery between DC and 18 GHz.

23.7 Reliability [30]

23.7.1 General Reliability Concerns

The failure mechanisms of GaN devices are similar to other III–V devices; however, due to high electric fields and higher operating temperature, the effects of surface traps, intermetallic diffusion, and electromigration become more important. Trap-related effects that have been reported are g_m frequency dispersion, current collapse, and light sensitivity. On the drain side, there is a higher chance of electromigration due to the high field. A lot of reliability improvement has been focused on this. Figure 23.21 shows the common reliability concerns. Some of the common failure mechanisms of III–V FET-type devices are briefly described below.

23.7.1.1 Gate sinking

Ni/Au- and Pt/Au-based metallizations are seen to be very stable on GaN, as are Mo:Au, as expected from III–V gate studies. The Schottky barrier may change in the burn-in process but stabilizes to 0.95 eV [25].

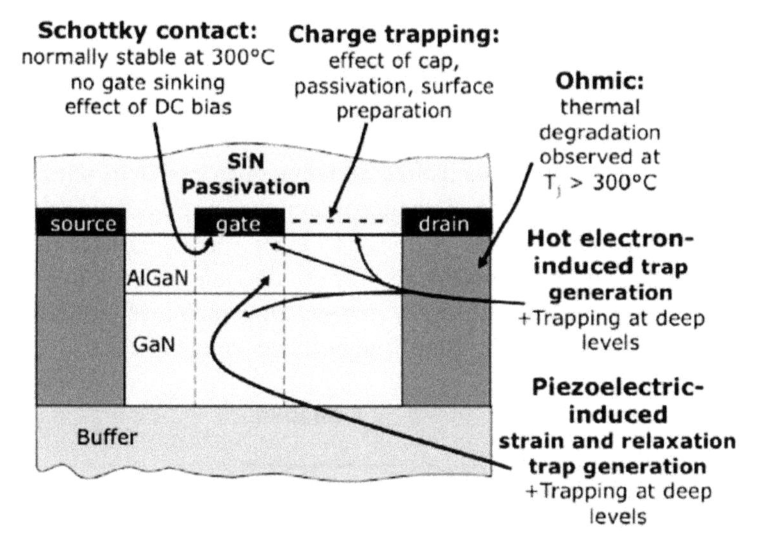

Figure 23.21 Cross section of a typical GaN device showing observed degradation mechanisms. © [2008] IEEE. Reprinted, with permission, from Ref. [25].

23.7.1.2 Ohmic contacts

Ohmic contacts based on metallizations described above are very stable on GaN. *I*–*V* curves remain unchanged at junction temperatures of 300°C–390°C. No interdiffusion has been observed by scanning transmission electron microscopy (STEM) analysis. Although Ni/Au layers are added to ohmic contacts to prevent oxidation, degradation can occur over a long time. Adding a GaN cap layer on top of AlGaN can result in a more reliable ohmic contact.

23.7.2 Current Collapse

A lot of work has been done on understanding the phenomenon of current collapse. The dynamic on resistance increases in continuous switching operation mode. The change of *I*–*V* characteristics is shown in Fig. 23.22 [31]. The change starts immediately upon start of operation and can be seen after a few minutes and continues over a long operating time. Two mechanisms have been suggested. The first degradation mechanism is believed to be the generation of traps by hot electrons. Many groups have verified the mechanism and suggested device modifications to mitigate this problem [32]. Trap generation and ways to limit it will be discussed in more detail later. The second mechanism suggested is the inverse piezoelectric effect. Mechanical stress generated by the application of a high field can cause strain, point, and line defects and thus accumulate to result in physical crystal damage. Limiting of the maximum electric field is therefore required. Material quality is a big factor for this effect because if pre-existing strain is present, devices can degrade faster under the applied field. Critical gate–drain voltages for avoiding generation of damage at the gate–drain edge have been determined [33].

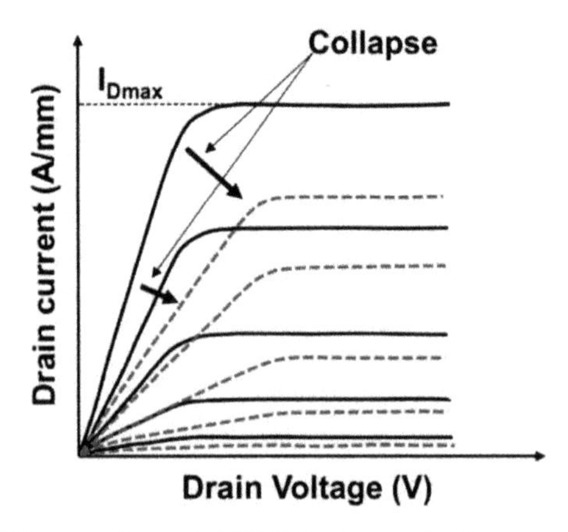

Figure 23.22 *I–V* curve for GaN HEMT before and after stress. Reproduced from Ref. [31] with kind permission of the author.

Electroluminescence (EL) has been used to evaluate hot electron effects [25], the intensity of EL being elated to degradation. Results of this study show that:

- Degradation involves accumulation of negative charge on the AlGaN surface. Figure 23.23 schematically shows the effect of surface charge on the AlGaN surface, which causes reduction of channel charge and thus current collapse [11].
- The state of the surface can be improved by silicon nitride passivation, and surface traps can be reduced by NH_3 plasma treatment before the silicon nitride deposition is started.

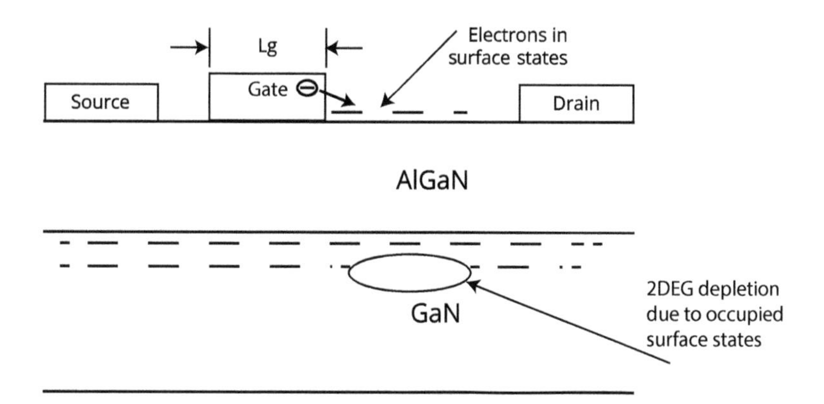

Figure 23.23 Depletion of 2DEG due to surface state formation.

- The main solution to this reliability issue has been the use of field plates discussed above. Figure 23.24 shows the single- and dual-field plate designs by Saito et al. [32]. The effect of the length of the field plates and the thickness of the SiN were analyzed to optimize device geometry. The FM used was the percentage change in R_{ON}. For a single-field plate increasing the SiN thickness from 100 nm to 150 nm improved device reliability. For the dual-field plate version, a 4 um ledge field plate (LGFP in Fig. 23.24) works better than a 2.5 μm ledge.

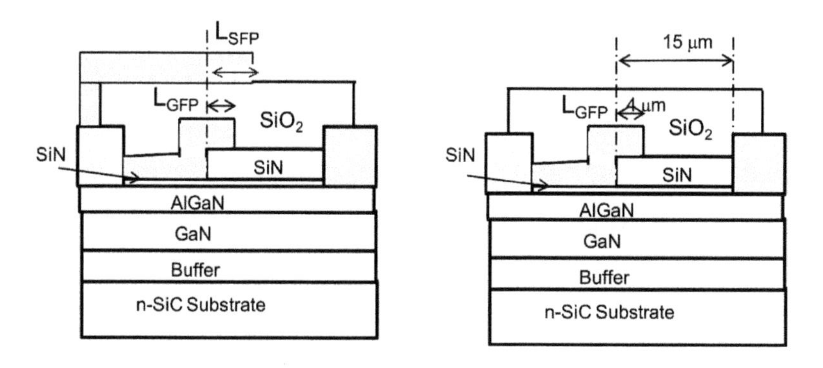

Figure 23.24 Use of dual-filed plates for minimization of current collapse.

- Degradation has been also seen to be related to the recess etch. The larger the recess, the greater the degradation. So, the use of a field plate is very important for E-mode devices using recess etching.
- A third, though minor, degradation mechanism is electrochemical corrosion. Figure 23.25 shows pitting seen at the gate–drain edge under the off condition, with drain voltage on [34]. The pitting does not occur if the device is operated under dry argon, confirming the role of moisture.

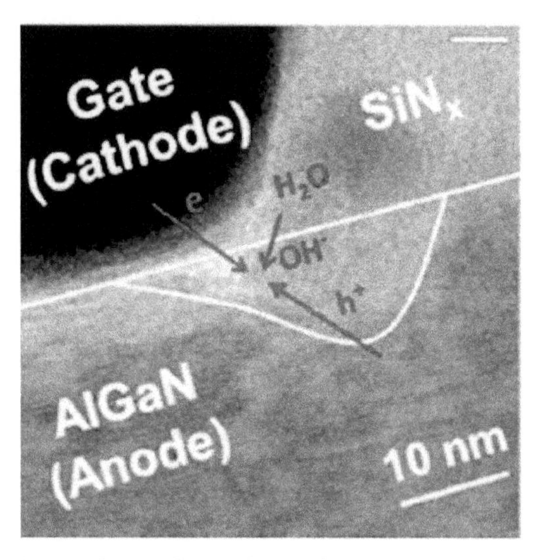

Figure 23.25 Pitting due to electrochemical corrosion at the drain edge due to formation of an electrochemical cell. Reprinted from Ref. [34] with permission. Copyright © 2014 CS MANTECH.

23.8 III–N HBT Devices [35, 36]

FET-type GaN devices suffer from surface states; therefore it appears that junction devices like HBTs should offer more promise to exploit the high-power, high-temperature capabilities of GaN. However, HBT devices in a III–N material system are difficult to make because of the lack of good p+ layers. There are no good p-type dopants available for GaN. Mg is a deep acceptor (180 mV deep). The activation efficiency is extremely small, close to 1%.

For optical devices like LEDs, where the p layer is used to inject holes, the quality of the layer is not critical. In HBTs, which are minority carrier devices, little recombination leads to low β, and thus very-high-quality, highly doped base layers are needed, which is not possible. Therefore, HBTs are not attractive yet for commercialization.

23.8.1 GaN HBT Device Challenges

In spite of the immense difficulties a few research groups have succeeded in demonstrating reasonable HBT performance [35–38].

A common approach to making HBTs with nitrides has been to grow an n–p–n epistructure with AlGaN as the large-bandgap emitter on GaN grown on sapphire. This structure benefits from the hole blocking at the AlGaN/GaN emitter–base junction. There are several challenges to this approach.

First of all, at the epitaxial growth stage, it is very hard to grow a p-type GaN layer. Mg is the best-known acceptor (see Chapter 23), and its energy level is located at 170 mV above the valence band, which is not very shallow. Its activation efficiency is only 1% or so. GaN has a tendency to grow with N vacancies, V_N^+, which act as donors. So, the layer has to be doped very heavily to get p+ behavior. Magnesium has a memory effect in MOCVD systems. So the combination of very heavy doping and the memory effect results in degradation of junction sharpness. The epilayer quality overall is not very good, so dislocations and other defects penetrate through the thick collector all the way through the junctions, leading to leakage. Minority carrier lifetime is poor in GaN. Due to all these reasons, achieving high β is a challenge [11].

In the device-processing phase, again a lack of wet etches is an issue, as discussed in Chapter 23. So, plasma etching has to be used. There are no etch stops, so stopping at the base layer during the emitter mesa etch is a problem. These etches are highly energetic and cause damage to the underlying layers. Therefore the base layer is damaged in the process of revealing the base for base contact. The sheet resistance of the base epilayer is already high, and the damaged semiconductor under the base contact causes high base resistance. So, R_b is very high. The FM f_{max} goes down (see Section 2.5.3 in Chapter 2). The damage to the base layer also causes the memory effect in electrical characteristics.

To circumvent the problem of ion damage, selective regrowth of the emitter layer has been pursued [11]. This adds additional complexity and requires in-house epigrowth capability. Otherwise, wafers have to be sent out for regrowth.

In particular, Dupuis and Shen [35] have shown an HBT β of greater than 100 with the emitter-up structure. The schematic for a GaN HBT is shown in Fig. 23.26. For high-power applications, a power density > 3 MW/cm^2 was achieved. A double-heterojunction structure is used; the layer structure is described in Table 23.2. The composition of the base layer is $In_{0.03}Ga_{0.97}N$. Above In content of 0.05, the base layer has higher defects. The Mg doping level in this layer was 1E18/cm3. For In incorporation a lower growth temperature, 850°C, is used. GaN is grown at 1050°C. To avoid polarization charge and lattice strain, the base layer is graded on both the collector and the emitter side. (Polarization charge that plays a key role in FET-type devices is not desirable for HBTs.)

For device fabrication, inductively coupled plasma (ICP) etching, using Cl_2 chemistry, was used to reveal the base layer. ICP etching is used to etch down to the collector layer (see Fig. 23.27). To remove damage, the surface is treated with $KOH/K_2S_2O_8$ solution under UV. Ni/Ag/Pt is used as the base contact metal and Ti/Al/Ti/Au is used for making contacts to the n-type emitter and collector.

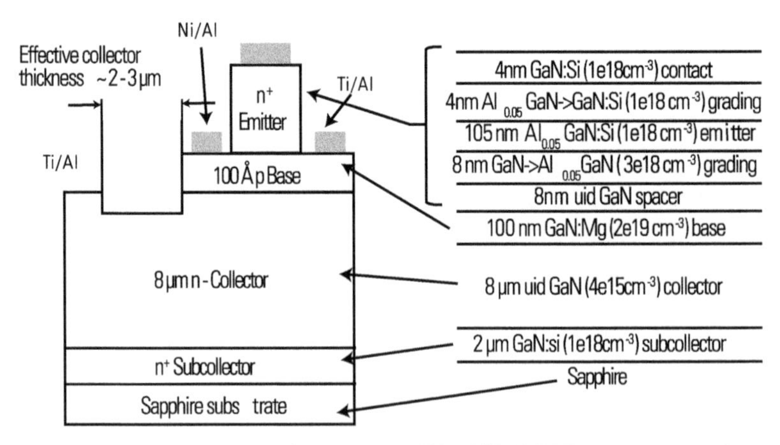

Figure 23.26 Cross-sectional emitter-up AlGaN/GaN HBT grown on sapphire [37]. Courtesy of Patrick Fay, University of Notre Dame.

Table 23.2 Summary of layer structure variations of n–p–n GaN/InGaN DHBTs on a sapphire substrate

Layer	Material		Thickness	Free carrier concentration
	Structure A	Structure B		
Emitter cap	GaN	GaN	70 nm	$n = 1 \times 10^{19}$ cm^{-3}
Emitter grading	$In_xGa_{1-x}N$ (x = 0–0.03)	$In_xGa_{1-x}N$ (x = 0–0.05)	30 nm	$n = 1 \times 10^{18}$ cm^{-3}
Base	$In_xGa_{1-x}N$ (x = 0.03)	$In_xGa_{1-x}N$ (x = 0.05)	100 nm	$p = 2 \times 10^{18}$ cm^{-3}
Collector grading	$In_xGa_{1-x}N$ (x = 0.03–0)	$In_xGa_{1-x}N$ (x = 0.05–0)	30 nm	$n = 1 \times 10^{18}$ cm^{-3}
Collector	GaN	GaN	500 nm	$n = 1 \times 10^{17}$ cm^{-3}
Subcollector	GaN	GaN	1000 nm	$n = 3 \times 10^{18}$ cm^{-3}
Buffer layer	GaN	GaN	2500 nm	UID
Sapphire Substitute				

Source: Reproduced from Ref. [35] with kind permission of Prof. Shen.

23.8.2 Current State-of-the-Art Performance

The performance achieved by the device described above is briefly listed below:

$\beta > 100$, $f_t > 5$ GHz, $f_{max} = 1.2$ GHz, and V_{BD} (breakdown voltage) = 90–160 V.

When the same structure was grown on a native GaN substrate, β was only a little better at 110.

The HBT devices fabricated with Ni-based contacts to p-type bases exhibit a poor lifetime. The V_{be} and collector voltage offset of the I–V characteristics increase from 2 V to 5 V in a matter of minutes when the device is operated. Recently, this problem has been addressed by making the p+ contact more stable by using a Pd-based contact. A current gain (β) of 80 at a current density of 95 kA/cm^2 was achieved.

The potential of GaN HBTs for high-power, high-temperature applications seems achievable, but a breakthrough is needed for high-frequency applications because of the memory effect and f_{max} limitation.

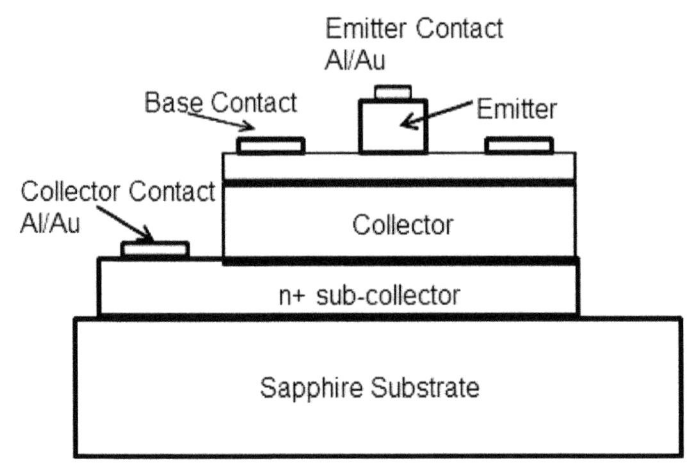

Figure 23.27 Emitter-regrown AlGaN/GaN HBT cross section [38].

23.9 Other Devices

MOSHEMT-type devices have better potential in an AlGaN/GaN material system. Gate dielectrics deposited by ALD or simply sputter deposition are being investigated [39].

Section II: GaN Optical Devices

23.10 Introduction to LEDs [40]

Light-emitting diodes (LEDs) have a long history starting with observation of emission from SiC and II–VI compound semiconductor diodes in the beginning of the last century. Work on III–V semiconductors started later with the development of IR LEDs using a GaAs/AlGaAs material system. Interest in the visible part of the spectrum led to work on GaP, GaAsP, and AlGaAs. Recently, the desire to cover the whole visible spectrum and generate white light has been responsible for an immense worldwide effort on III–N (nitride) materials. LEDs based on InGaN and AlInGaP have been developed in the last two decades. The advantages of LEDs over conventional light sources are obvious. They exhibit very high efficiency, a long lifetime (over 50,000 hr), small size, dimming capability, and a range of colors. The main challenge has been achieving blue color and reducing production cost by lowering the cost of substrate crystals, epitaxial layer growth, and mass production.

LEDs consist of a conducting semiconductor substrate with n and p epilayers grown on it to make a p–n junction and conducting contacts to the two layers and contacts to the package. The most common structures may have an active n layer between two larger-bandgap p-type layers. For higher efficiency quantum well (QW) structures may be used.

The p–n junctions being common to all LED and laser structures are discussed briefly below. For detailed device physics, readers are referred to the literature on LEDs [40–42] and semiconductor physics [43, 44].

23.11 Junction Luminescence

LEDs are p–n junction diodes where large numbers of holes are injected under forward bias and the recombination of these with electrons in the junction regions leads to light emission. The excitation happens at very low voltages compared to electroluminescent devices, which work at high voltages. In a light-emitting p–n junction, excess carriers, beyond equilibrium, are created, as electrons are constantly being excited from the valence band. Figure 23.28 shows a schematic diagram of a cross section of an LED and the energy band diagram under forward bias. As these excess carriers recombine in the junction region, photons are emitted, the wavelength being dependent on the bandgap of the semiconductor. II–VI compounds have bandgaps that are matched to visible light, but these are hard to fabricate. III–V and III–N compound semiconductors are easier to make to cover the whole visible range. Figure 23.29a shows the variation of wavelength with the energy bandgap. Since III–V compounds have good miscibility (see Chapter 3), a range of wavelengths can be produced by using ternaries and quaternaries. Figure 23.29b shows that the energy bandgap varies smoothly over the composition and wavelength can be varied continuously between the two values for the two compounds (Fig. 23.29c).

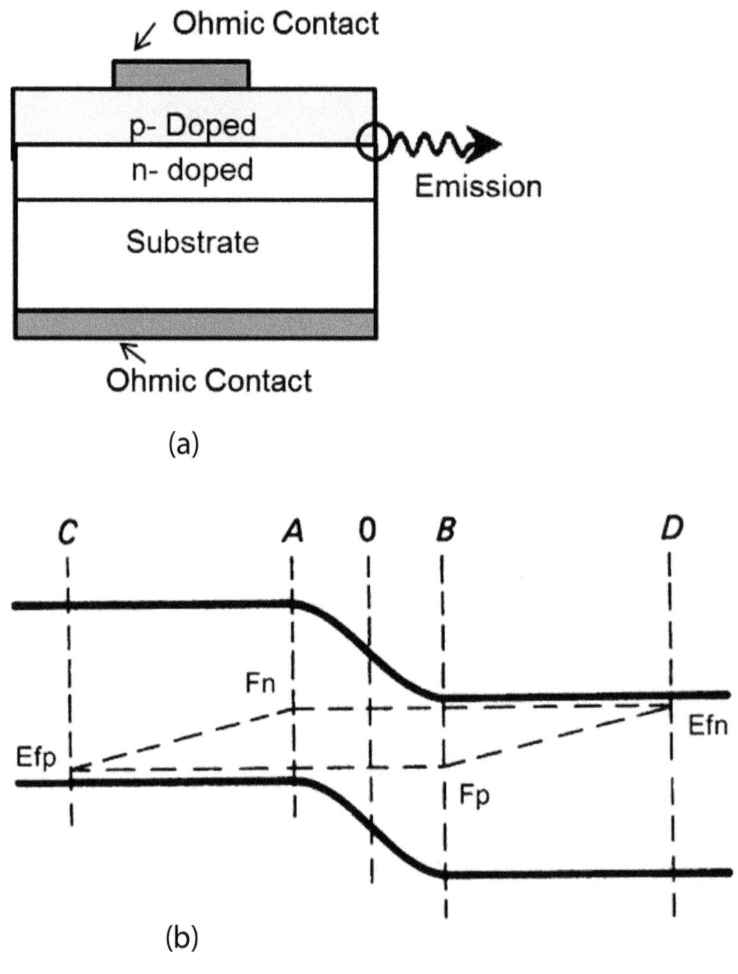

Figure 23.28 Schematic cross section of a p–n junction (a) and energy band diagram under forward bias showing quasi-Fermi levels E_{fp} for holes and E_{fn} for electrons (b). From Wang, *Fundamentals of Semiconductor Theory & Device Physics*, 1st Ed. ©1989 Printed and reproduced by permission of Pearson Education, Inc., New York.

LEDs are used commonly as light sources and for short-distance optical communication. Two common types of LED geometries are shown in Fig. 23.30. These can be fabricated on GaAs or InP substrates. The cladding and the ohmic contact layers depend on the active layer, to be discussed later in Section 23.15.

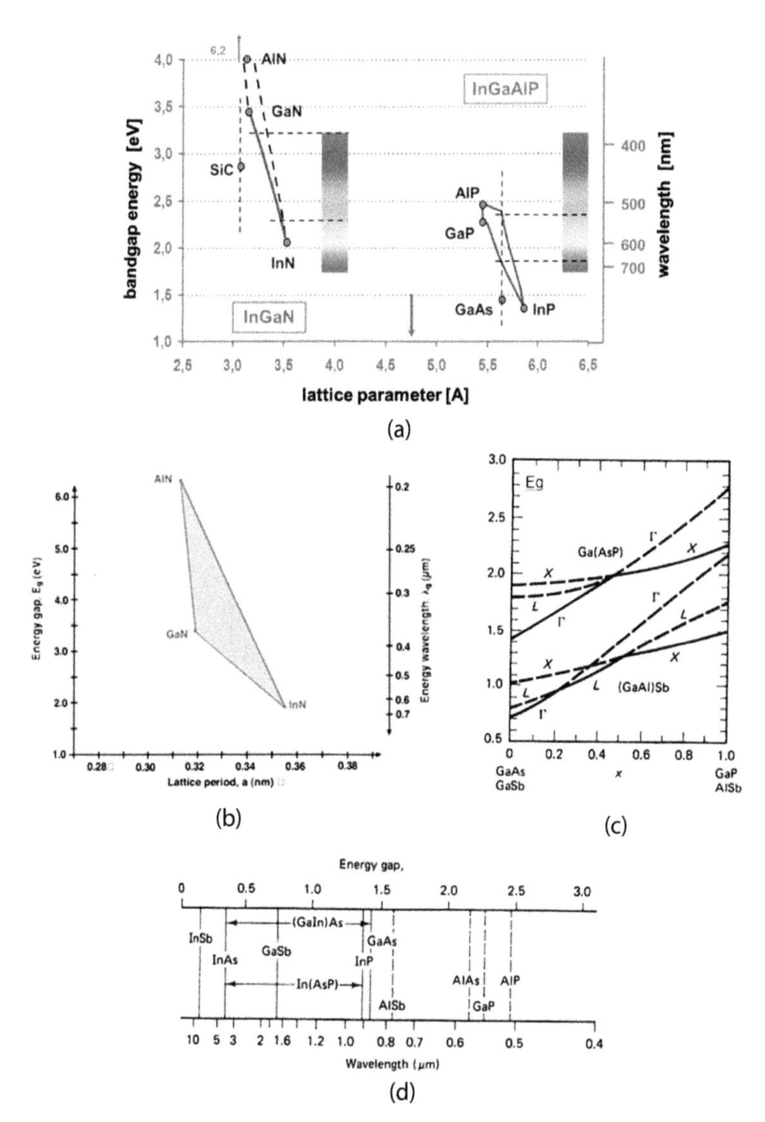

Figure 23.29 (a) Energy gap vs. lattice parameter chart for nitride and phosphide material systems. Corresponding color charts are shown. Lattice parameter for sapphire: 4.76 Å. Courtesy of Heribert Zull, OSRAM Optical and OSRAM online resource. (b) Details of the AlN–GaN–InN system. (c) Energy bandgap change with composition for GaAs–GaP and GaSb–AlSb systems [47]. (d) Change of bandgap with composition for material systems (From Wang, *Fundamentals of Semiconductor Theory & Device Physics*, 1st Ed. ©1989 Printed and reproduced by permission of Pearson Education, Inc., New York).

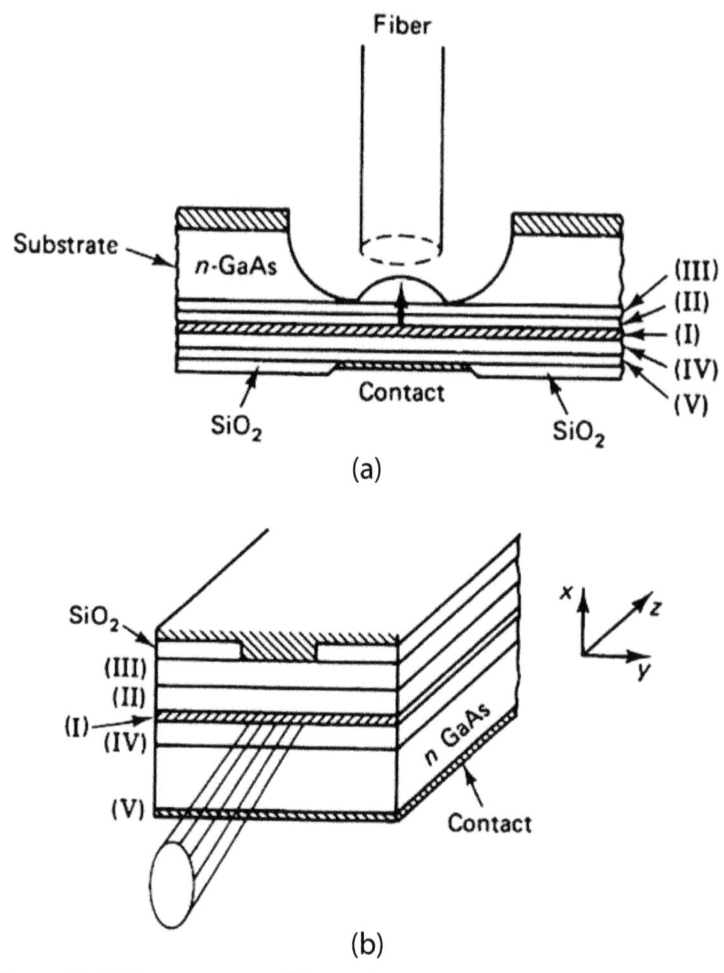

Figure 23.30 Two common LED configurations, surface emitting (a) and edge emitting (b). I: active layer; II and IV: cladding layers; III and V: contact layers. Active layer can be InGaAsP, cladding layer can be InP, and contact layers can be n-InP and p-InGaAsP. From Wang, *Fundamentals of Semiconductor Theory & Device Physics*, 1st Ed. ©1989 Printed and reproduced by permission of Pearson Education, Inc., New York.

23.11.1 Device Behavior: Electrical

Simple abrupt junctions are used to make LED devices. In Chapter 2, p–n junction physics was discussed briefly. The built-in voltage at the junction under equilibrium was derived as (Eq. 2.12)

$$V_{\text{bi}} = \frac{kT}{e}.\ln\frac{N_{\text{A}}N_{\text{D}}}{n_i^2} \tag{23.2}$$

The depletion layer width is given by (see Eq. 2.31)

$$Wd = \sqrt{\frac{2\varepsilon(V_{\text{bi}} - V)}{e}\left(\frac{1}{N_{\text{A}}} + \frac{1}{N_{\text{B}}}\right)} \tag{23.3}$$

Current–voltage characteristics, derived in Chapter 2, are of the form

$$I = I_{\text{s}}\exp\frac{eV}{kT}\ \{1\} \tag{23.4}$$

Under forward-bias conditions, {exp (eV/kT) – 1} ≈ exp (eV/kT), and the I–V characteristics of an LED are given by

$$I = I_{\text{s}}\exp\left\{\frac{e(V - V_{\text{bi}})}{kT}\right\} \tag{23.5}$$

The current–voltage characteristics are show in Fig. 23.31 for semiconductors with different bandgaps, showing increasing threshold voltage, V_{th}, with increasing E_{g}.

In practice, there is a series resistance present with the diode, so the external applied voltage is given by

$$V = V_{\text{th}} + IR_{\text{s}} \tag{23.6}$$

The I–V characteristics with series and shunt resistance are shown in Fig. 23.32.

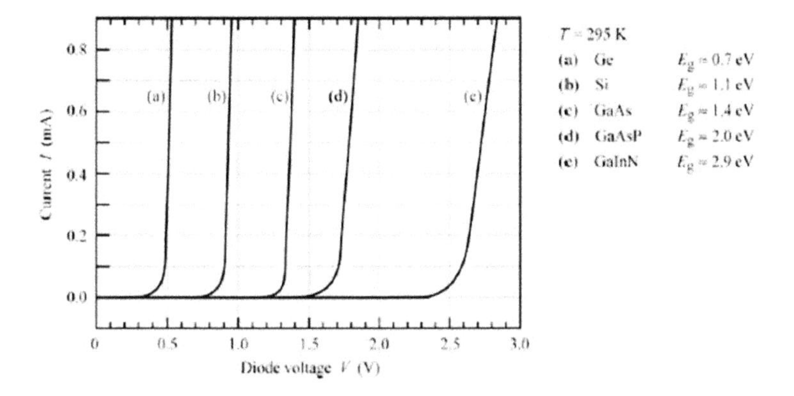

Figure 23.31 I–V characteristics of p–n junctions made from semiconductors with energy bandgap varying from 0.7 eV for Ge to 2.9 V for GaInN [40]. Reprinted with permission from Cambridge University Press.

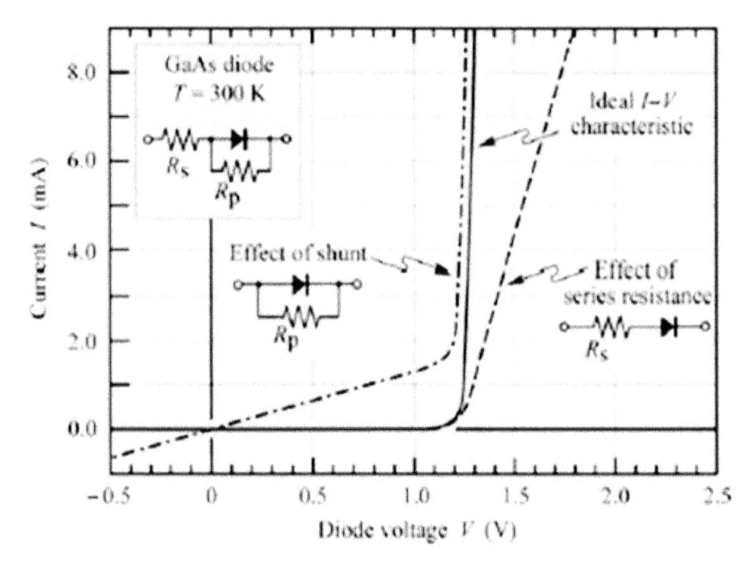

Figure 23.32 *I–V* curves for a GaAs p–n junction diode showing the effect of series and shunt resistance [40]. Reprinted with permission from Cambridge University Press.

23.11.2 Optical Characteristics

Figure 23.37 shows the valence and conduction n bands of a semiconductor that is doped to degeneration. The electron and hole energies are given by

$$E_e = E_c + \frac{h^2k^2}{2m_e*} \tag{23.7}$$

$$E_e = E_v - \frac{h^2k^2}{2m_h*} \tag{23.8}$$

The emitted energy at radiative recombination is equal to

$$E = E_e - E_h \approx E_g = hv \tag{23.9}$$

The emitted photon energy is nearly equal to E_g.

Ideally one injected electron leads to one photon. The electrical efficiency is given by

$$\eta_{int} = P(hv)/(IV) \tag{23.10}$$

The emitted energy is emitted into free space in a wide beam, and only a portion of it is collected. The optical efficiency is called the extraction efficiency, $\eta_{\text{extraction}}$.

$$\eta_{\text{ext}} = \eta_{\text{int}} \cdot \eta_{\text{extraction}} \tag{23.11}$$

23.12 LED Processing

23.12.1 Visible LEDs

A major breakthrough in the development of blue LEDs was the achievement of p-type doping in III–N compounds [45]. This led to the rapid development of a host of III–N layers for covering the whole visible spectrum. The sequence of LED processing involves (i) epitaxy, (ii) device fabrication, and (iii) packaging. For epitaxy, MOCVD is mostly used. The standard metal–organic sources for Al, Ga, and In are used, as described in Chapter 4. Hydride sources are used for N and P. Dopant atoms are Mg, Si, Te, etc., again using common sources used in MOCVD. The substrate temperature for III–N epilayers is around 1000°C. Quarternary layers of AlInGaN are used for violet, blue, and green [46], with AlInGaP layers being used for yellow, orange, and red, as shown in Fig. 23.29 [47].

As discussed in Chapter 13, sapphire and SiC are used for nitrides. GaAs is used for AlInGaP growth. Simple structures shown in Fig. 23.30 were utilized first.

In a simple double-heterostructure LED using a GaAs active layer, shown in Fig. 23.30, the epigrowth starts with an n-type contact layer, followed by a large-bandgap AlGaAs cladding layer, and then the active layer. On the top side the growth continues with larger-bandgap AlGaAs, followed by a p+ GaAs contact layer. For a structure grown on n-type InP wafers, an n-type InP layer is grown, followed by the active layer InGaAsP. On the top side of the active layer, p-InP is grown to make the heterojunction, followed by the p-InGaAsP contact layer.

For GaN-based devices, thick layers of n-GaN can be grown by using Si as the dopant (from the SiH_4 source). The p-type GaN has the usual challenges, as discussed in GaN epigrowth. Mg is the commonly used dopant and is activated by low-energy electron beam irradiation (LEEBI). Layers are annealed at 800°C–1000°C under nitrogen. For thin n-type active layers, or layers in a QW structure, n-type doping is not necessary because of induced polarization and piezoelectric charges (discussed in the "GaN HEMT" section). The epiwafers are patterned into mesas (a few hundred microns in size) using photolithography and RIE etching. See Fig. 23.33 for typical device geometry. Ti/Au layers (300 Å/1500 Å) are used as the n-type contact. For current spreading, a transparent indium tin oxide (ITO) layer can be sputter-deposited on top, below the p-type contact. SiO_2 is deposited on top of the whole structure for encapsulation. A window is opened in the oxide to make the ohmic contact to the top side. The Ni/Au thin film (100 Å/1500 Å) is deposited on the p-electrode. This contact is annealed at 515°C in an O_2/N_2 atmosphere for 5 minutes.

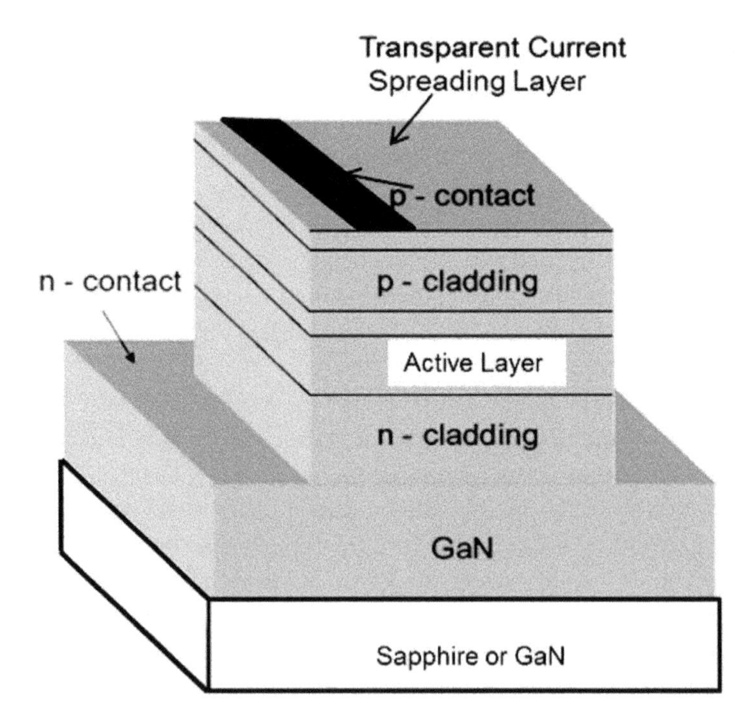

Figure 23.33 A simple mesa LED shown with MQW active layer. Simpler structures use double heterojunctions for the active layer. The whole structure can be on top of a sapphire substrate.

Now, more efficient structures needing precise layer control using QWs are used. An example is shown in Fig. 23.34. In this example, the process starts with a 2-inch sapphire substrate. A low-temperature nucleation layer of GaN is deposited first. This is followed by a 2000 nm GaN buffer and then a 1500 nm Si-doped (1e19/cm^3) contact layer. The active layer consists of InGaN/GaN multiple QWs. A thin AlGaN:Mg electron-blocking layer follows. The contact is made on the top by p-GaN, doped with Mg to 1.5E 20/cm^3.

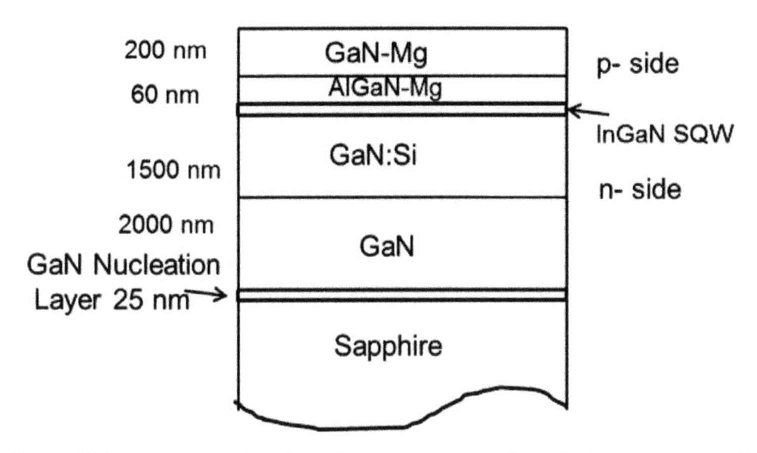

Figure 23.34 An example of epilayer structure for single-quantum-well LED grown on sapphire. Reproduced from Ref. [52] with kind permission of the author.

To improve the efficiency even further, more complicated structures have been developed. The active layer is replaced by QWs. Figure 23.35 shows a multiple-QW AlGaN/GaN/GaInN laser epilayer structure without (a) and with (b) doping. Note that the electron-blocking layer has a higher Al concentration than the confinement layer [40]. Both optical confinement and carrier confinement improve as the number of QWs goes up, as seen in Fig. 23.36 [40].

23.12.2 UV LEDs

Blue-violet LEDs are being produced commercially, and interest has shifted toward even smaller wavelengths. In this case the In content must be reduced in double heterostructures or QWs. Low-wavelength devices are of interest in generating white light by the use of phosphor materials. In principle, white light can be generated by the RGB combination method. But, the excitation of RGB phophors by UV light is seen as a better approach. UV LEDs have other applications, for example, high-density data storage, water purification, etc.

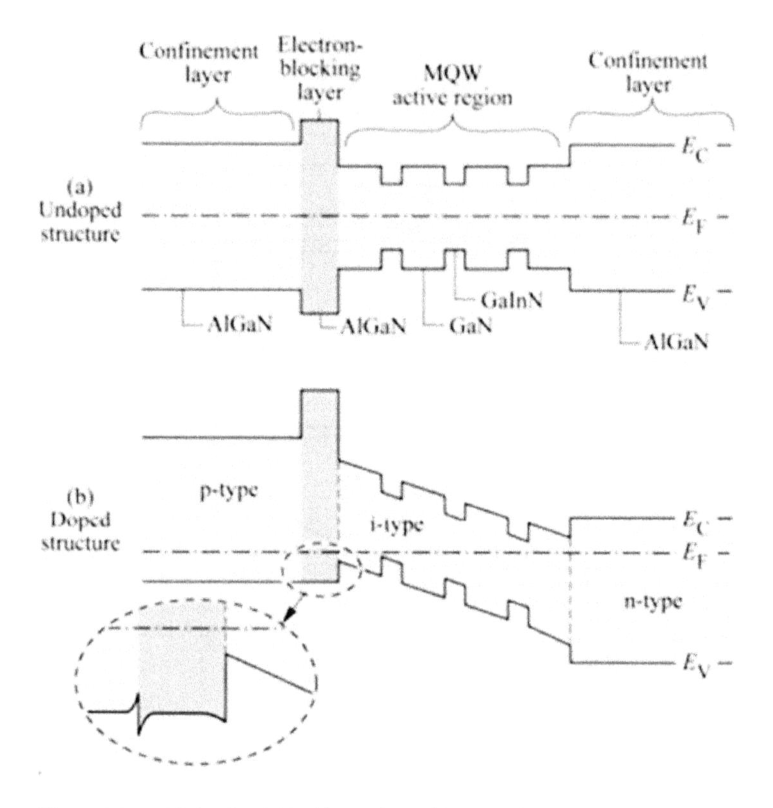

Figure 23.35 Multiple-QW AlGaN/GaN/GaInN laser epilayer structure, without (a) and with (b) doping. Note that the electron-blocking layer has a higher Al concentration than the confinement layer [40]. Reprinted with permission from Cambridge University Press.

23.12.3 Epitaxial Growth

GaN has a large lattice mismatch with sapphire, which leads to defect generation and low-quality crystal structure of the epilayer. The defects cause increase of nonradiative recombination. For UV radiation, very low In content is needed (see Fig. 23.29a). At low In concentration, phase separation does not occur, so the localized states that are needed for radiative recombination do not form. Therefore the emission is weak. The potential barrier at low In content is also lower, so an AlGaN optical and electronic barrier is needed. For shorter wavelengths, Al content should be high in the AlGaN. A higher-Al-concentration layer generally has poorer crystal quality. The quality of the epilayer can be improved by growing a thick layer of GaN first to form a pseudosubstrate. Introduction of a low-temperature (LT)-grown layer on top of the GaN buffer, before the growth of high-temperature (HT) AlGaN, improves the quality even further [48, 49].

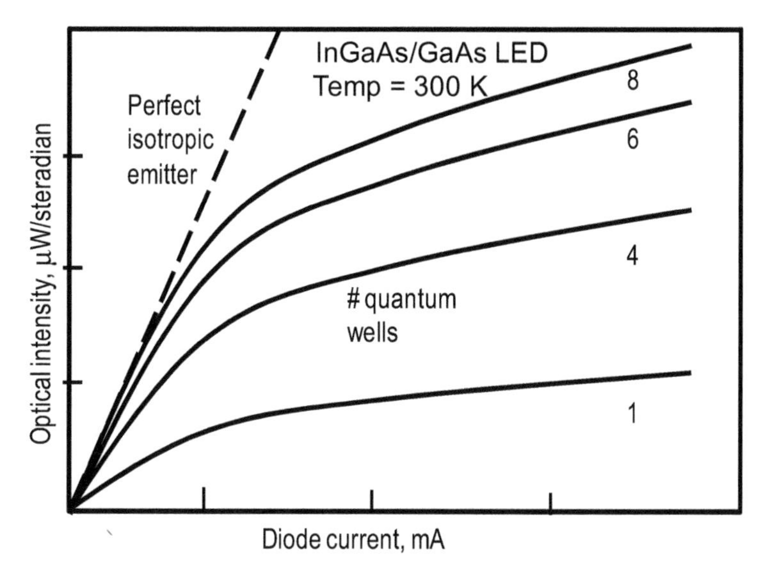

Figure 23.36 Output intensity vs. number of quantum wells.

Another approach to reducing defects is strain balancing by adding underlayers with stress of the opposite sign under the buffer layer, for example, by adding a compressive stress layer of AlGaN on sapphire.

One method to improve the overall quality is to deposit an AlN nucleation layer by sputtering before the wafer is loaded into the epigrowth chamber [50].

23.13 Current Challenges [51]

23.13.1 Current Performance

Efficiency of up to 60% has been achieved for blue LEDs. With good heat sinking, distributed contacts, and good package design, currents of 50 A/cm^2 can be used achieving 100–140 lumens/W overall efficiency. Great progress has been made in the last decade,; however, some challenges remain.

- **Green gap**: After the challenge of developing blue color, a few challenges remain. One of these is the green gap being filled by the development of InGaN active layers of proper composition. At a high In concentration needed for green at 525 nm, good material quality is hard to achieve. GaN and other ternaries are piezoelectric materials. Polarization charge is used to advantage in HEMT-type devices (see Chapter 23); however, in optical heterojunction devices, spontaneous and piezoelectric polarization charge separation leads to significant reduction of electron and hole wavefunction overlap, which results in reduction of radiative recombination. Therefore in InGaN/GaN QWs use of nonpolar InGaN is being investigated.
- **Efficiency droop**: Efficiency of LEDs can drop due to gradual reduction of the heterojunction barrier that can cause carrier leakage. Also, nonradiative recombination can increase with time of operation, as defects are generated. Extra barriers, for example, AlInGaN quaternaries, can be used to provide a more stable barrier.
- **Light extraction**: Improvement of light extraction is an ongoing effort. Use of surface layers and microlenses is being investigated.
- **High-quality GaN substrate**: The thermal conductivity of GaN is five times that of sapphire. Availability of a high-quality large area could benefit optical devices like it would HEMT-type devices, because of the obvious improvement of epilayer quality, discussed in the "GaN HEMT Devices" section.

23.14 Introduction to III–N Lasers

23.14.1 Basic Principles

LASER is an acronym for light amplification by stimulated emission of radiation. For a solid-state laser an optical amplifier and a resonator are needed. In the p–n junction semiconductor laser, the amplifying element is a p–n junction, which is strongly forward biased. The resonator is made by confining the active region electronically and optically to the equivalent of a laser cavity. Figure 23.37 shows the band diagram of a p–n junction, showing a degenerate semiconductor region with a quasi-Fermi level for holes in the valence band and a quasi-Fermi level for electrons in the conduction band. Electrons from the n side and holes from the p side are confined in a lower-bandgap region. These recombine with emission of photons.

$$h\nu = E_1 - E_2 << E_{\mathrm{fn}} - E_{\mathrm{fp}} \tag{23.12}$$

See Fig. 23.37 for the terminology, *E*–*k* diagram, and simplified energy band diagram for a single junction.

In a double-heterostructure laser, two junctions are used, thus creating carrier confinement from both sides and an optical resonator. Electrons are confined to the lower-bandgap region in the middle. The pumping process ensures that the electron concentration due to injection into the laser cavity is much higher than thermal equilibrium. Population inversion is easier to achieve in a double-heterojunction structure. Optically the region is defined by the step in the refractive index. The photons are reflected back at the edges of the heterostructures, the fraction depending upon the magnitude of the refractive index step. Carriers combine spontaneously in this region via stimulated emission when excited by existing photons. There are always some losses due to nonradiative recombination ion the active region, which leads to loss. The reflected photons from the edges stimulate emission and cause amplification of optical modes. Laser action occurs when the current density is high enough and net gain overcomes the losses and one mode is sustained by proper geometry and design.

The color and operating voltage depend upon the energy bandgap just as in the case of an LED.

The threshold current density of a laser is given by [43]

$$J_{\text{th}} = e \cdot \frac{N_1 d}{\tau} \approx \frac{N_1}{\mu} \tag{23.13}$$

where N_1 is the carrier density required to produce gain, d is the thickness of the active layer, and τ is the effective lifetime, with radiative and nonradiative recombinations occurring in parallel. For high efficiency, the threshold current should be low, so the lifetime, τ, should be large and N_1 should be small. Therefore, crystal quality is the primary factor for achieving high efficiency.

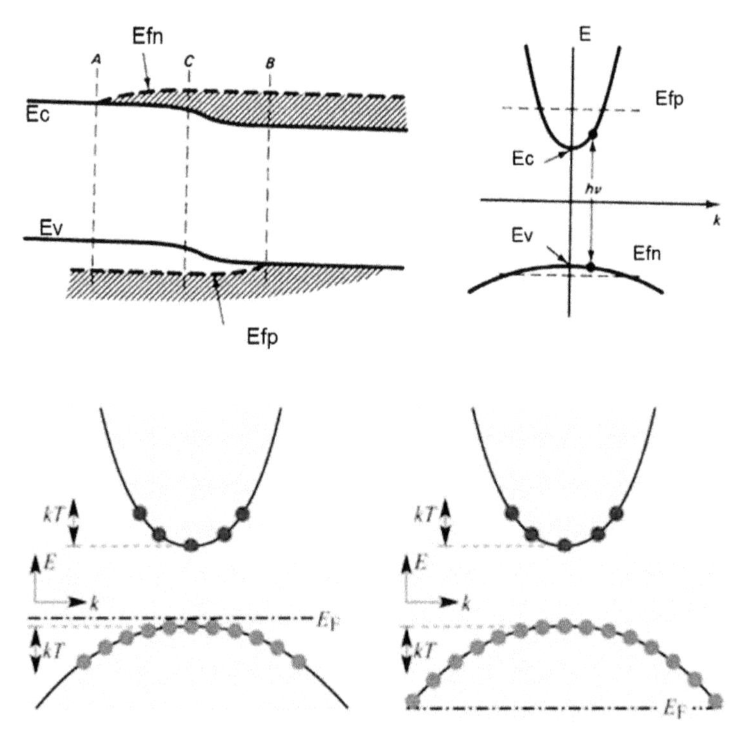

Figure 23.37 p–n junction diode energy band and E–k diagrams under degenerate conditions showing population inversion (From Wang, *Fundamentals of Semiconductor Theory & Device Physics*, 1st Ed. ©1989 Printed and reproduced by permission of Pearson Education, Inc., New York). Nondegenerate vs. degenerate doping diagram [40](reprinted with permission from Cambridge University Press).

23.15 Diode Laser Fabrication

The coherence of laser emission is the result of confinement and stimulated emission. So the spectrum is narrow. In ordinary optically pumped lasers, mirrors are used for optical confinement; however, in solid-state lasers, optical reflection must occur at crystal facets that act as mirrors. The simplest and most commonly used type is the Fabry–Perot (FP) laser.

23.15.1 Fabry–Perot Semiconductor Diode Laser

A resonator that has two parallel mirrors is called an FP resonator. To understand the operation of the FP laser it is first necessary to understand the FP filter. The principle of the FP filter is illustrated in Fig. 23.38. When two mirrors are put opposite to one another they form a resonant cavity. Light bounces between the two mirrors. When the distance between the mirrors is an integral multiple of half wavelengths, the light will reinforce itself. Wavelengths that are not resonant undergo destructive interference with them and are reflected away.

This principle also applies to the FP laser although the light is emitted within the cavity itself rather than arriving from outside. In some sense every laser cavity is an FP cavity. We consider a laser to be FP when it has a relatively short cavity (in relation to the wavelength of the light produced). Multiple wavelengths that are produced are related to the distance between the mirrors by the following formula:

$$\lambda = \frac{2L.n}{N} \tag{23.14}$$

where n is the refractive index, L is the length, and N is an integer. In practice, the geometry is made narrow, only a few microns in width, and the length supports only the single optical mode, not multiple modes. Figure 23.38 shows the layer structure and geometry of a simple III–V FP laser made on a GaAs substrate. Notice that it uses a double heterostructure, in this case a GaAs active layer bounded by WBG AlGaAs. The corresponding energy band diagram is shown in Fig. 23.39. In a single junction, the current required for lasing is very high. With the double heterostructure, the lasing current drops.

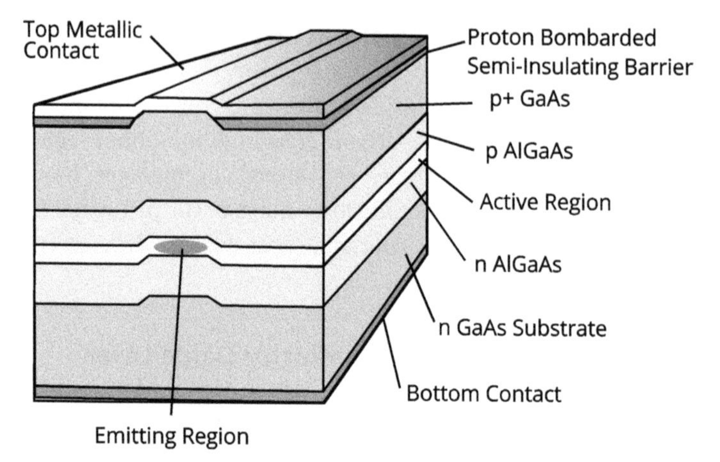

Figure 23.38 Fabry–Perot laser–based simple III–V semiconductor double heterostructure. Figure reproduced from *Britney Spears' Guide to Semiconductor Physics* with kind permission of Carl Hepburn.

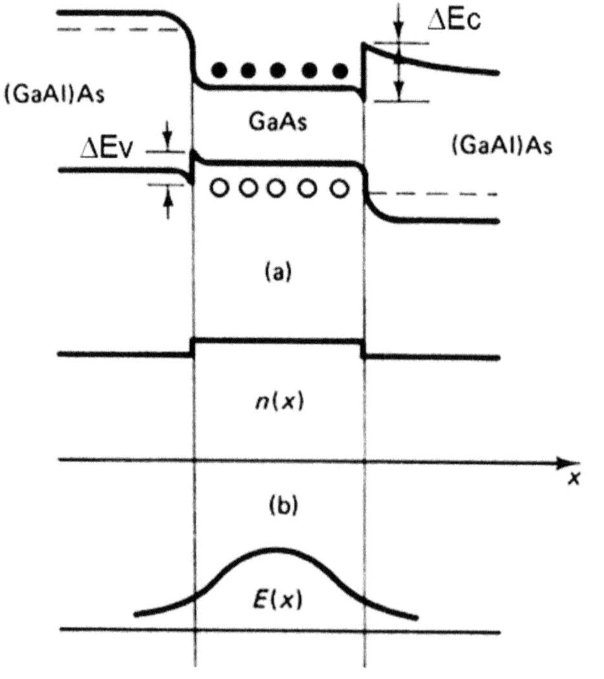

Figure 23.39 Double-heterostructure laser structure showing optical (refractive index) and electronic (electric field) confinement.

Figure 23.40 shows a schematic diagram of a nitride-based QW blue LED laser [52], which consists of the following:

- Active region AlGaN/GaN multiple or single QW. AlGaN is used as the electron barrier.
- Confinement layers: n- and p-type GaN
- Cladding layers: n- and p-type AlGaN

This structure is grown on sapphire, so it cannot be cleaved. Therefore a photopattern is depositedm and chemically assisted ion beam etching is used to make the facets.

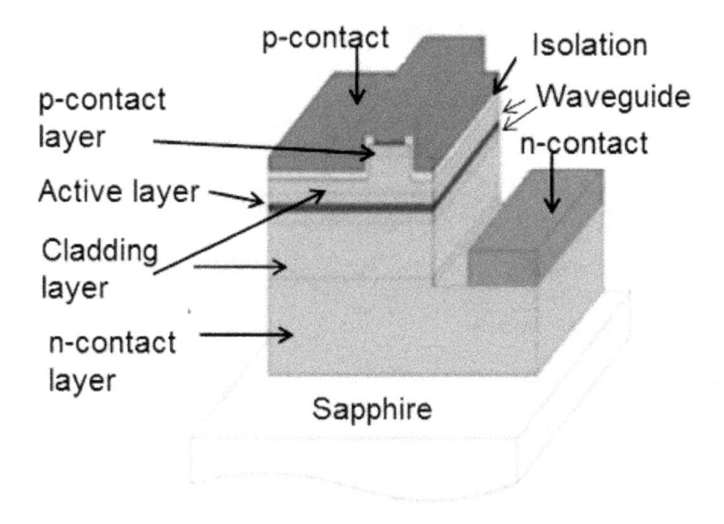

Figure 23.40 Schematic view of an AlGaInN QW blue emitting laser with etched facets. Reproduced from Ref. [52] with kind permission of the author.

23.15.2 VCSEL

In contrast to transverse emitting devices, vertical cavity surface emitting lasers (VCSELs) emit narrow beams from the surface. Figure 23.41 shows the difference between the two types. VCSELs are more difficult to fabricate than FP lasers described earlier. There are more layers and more precise thickness control is needed. However, these offer many advantages.

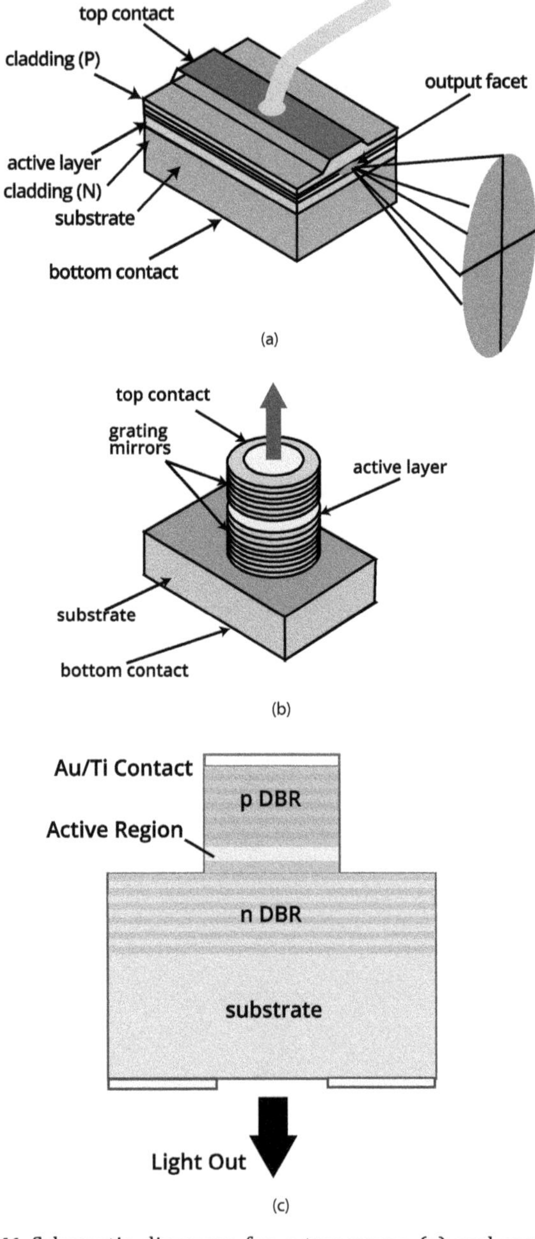

Figure 23.41 Schematic diagrams for a transverse (a) and vertical cavity surface emitting lasers (VCSELs) (b) and (c) airpost VCSEL using distributed Bragg reflectors (DBRs). Figure reproduced from *Britney Spears' Guide to Semiconductor Physics* with kind permission of Carl Hepburn.

VCSELs do not need cleaved facets or parallel etched faces. These can be fabricated and tested at the wafer level. These require lower power and deliver a circular beam with low divergence and are thus suitable for optical communication.

In a VCSEL, both photons and current flow are confined to a small volume, as shown in Fig. 23.42 [53]. Bragg reflectors are used for optical confinement. These consist of alternating layers of high- and low-refractive-index materials, each layer being quarter wavelength thick. Typically 20 Bragg pairs are needed on each side. MOCVD and MBE are both suitable for growth of this structure. The active layer is placed inside higher-refractive-index layers and may be a few QWs thick.

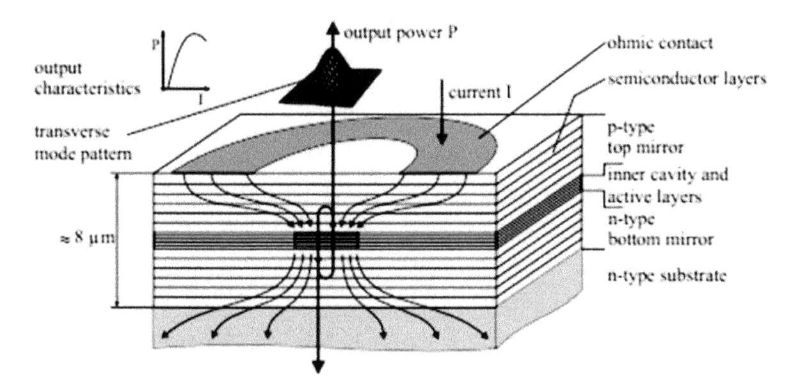

Figure 23.42 Schematic cross section and operating principle of VCSEL. With kind permission from Springer Science+Business Media: Ref. [53], Fig. 3.1 (p. 54).

For electrical confinement, either a mesa structure can be used or proton bombardment can be used to turn the surrounding semiconductor to an insulator just like proton ion implant isolation. An oxide layer can be used for surface passivation and optical reflection. For mesa etching of VCSELs, the reader is referred to Chapter 7.

References

1. J. Millan, A survey of WBG semiconductors, *IEEE Trans. Electron. Devices*, **29**, 2155 (2014).

2. T. Paskova, D. A. Hanser and K. R. Evans, GaN substrates for III-V devices, *Proc. IEEE*, **98**, 1324 (2010).

3. N. M. Nasser et al., GaN hetero epitaxial techniques, *J. Microwave Opto Electron.*, **2**, 22 (2001).
4. B. Schineller, K .Kaeppeler and M. Hueken, *Vertical-HVPE as a Production Method for Free-Standing GaN-Substrates*, CS MANTECH (2007).
5. T. Paskova et al., Hydride vapor phase homoepitaxial growth on GaN on MOCVD-grown "templates," *MRS Proc.*, **595** (1999).
6. J. Karpinski, J. Jun and S. Porowski, Equilibrium pressure of N2 over GaN and high pressure growth of GaN, *J. Cryst. Growth*, **66**, 1 (1984).
7. R. S. Pengelly et al., Review of GaN on SiC for high electron mobility power transistors and MMICs, *IEEE Trans. Microwave Theory Tech.*, **60**, 1764 (2012).
8. H. Morkoc, *Handbook of Nitride Semiconductors and Devices*, Wiley (2008).
9. H. M. Manasevit, F. M. Erdmann and W. I. Simpson, The use of metalorganics in the preparation of semiconductor materials, *J. Electrochem. Soc.*, **118**, 1864–1868 (1971).
10. H. Amano et al., P type conduction in Mg doped GaN treated with low energy electron beam irradiation, *Jpn. J. Appl. Phys.*, **28**, L2112 (1989).
11. O. Kovalenkov, V. Soukhoveev, V. Ivantsov, A. Usikov and V. Dmitriev, Thick AlN layers grown by HVPE, *J. Cryst. Growth*, **281**, 87–92 (2005).
12. U. Mishra, L. Shen and T. E. Kazior, GaN-based RF power devices and amplifiers, *Proc. IEEE*, **96**, 287 (2008).
13. C. Y. Chang et al., *Very Low Sheet Resistance AlGaN/GaN HEMT*, CS MANTECH (2009).
14. M. Shur et al., Wide band gap electronic devices, *IEEE Conf. Proc. 4th Conf. Dev. Circuits Syst.*, D051-1 (2002).
15. P. Fay, Lecture on GaN devices given at Skyworks Solutions.
16. R. Gaska et al., *App. Phys. Lett.*, **79**, 2832–2834 (2001).
17. A. C. Schmitz et al., *J. Electron. Mater.*, **27**, 255 (1998).
18. H. Sun et al., *IEEE Electron. Device Lett.*, **32**, 1056 (2011).
19. M. E. Lin et al., *Appl. Phys. Lett.*, **64**, 1003 (1994).
20. Y. Knafo, *Investigation of Metal Contact Stacks for Sub-Micron GaN HEMTs*, CS MANTECH (2005); see also A. Crespo et al., *Ti/Al/Ni/Au Ohmic Contact on GaN*, CS MANTECH (2003).
21. J. Bell et al., *GaN on Si HEMT Process Transfer and Qualification*, CS MANTECH (2012).

22. F. Mohammed et al., Ohmic contact formation mechanism of Ta/Al/Mo/Au and Ti/Al/Mo/Au metallizations on AlGaN/GaN HEMTs, *J. Vac. Sci. Technol. B*, **23**(6), 2330–2335, Nanoscale.mntl.illinois.edu/Publications/publications/264.pdf? (2005)

23. J. Pearton et al., GaN electronics, *Appl. Phys. Rev.*, 86 (1999).

24. M. W. Cole, F. Ren and S. J. Pearton, *J. Electrochem. Soc.*, **144**, L275 (1997).

25. G. Meneghesso et al., Reliability of GaN high electron mobility transistors: state of the art and perspectives, *IEEE Trans. Device Rel.*, **8**, 332 (2008).

26. S. J. Pearton et al., Ion implantation doping and isolation of GaN, *Appl. Phys. Lett.*, **67**, 1435 (1995).

27. K. Chabak et al., *Processing Methods for Low Ohmic Contact Resistance in AlN/GaN MOSHEMTs*, CS MANTECH (2009).

28. T. Oku, Y. Kamo and M. Totsuka, AlGaN/GaN HEMTs passivated by a Cat-CVD SiN Film, *Thin Solid Films*, **516**, 545–547 (2008).

29. W. Saito et al., Recessed-gate structure approach toward normally off high-voltage AlGaN/GaN HEMT for power electronics applications, *IEEE Trans. Electron. Devices*, **53**, 356 (2006).

30. A. Christou, *Reliability of GaN Devices*, CS MANTECH (2013).

31. M. T. Hassan, *Mechanism and Suppression of Current Collapse in AlGaN/GaN HEMT*, PhD dissertation, University of Fukui, Fukui, Japan (2013).

32. W. Saito, Reliability of GaN HEMT high voltage switching applications, *IEEE Int. Rel. Phys. Symp. IPRS* (2011).

33. J. Joh and J. A. del Alamo, Mechanisms for electrical degradation of GaN high-electron mobility transistors, *IEDM Tech. Digest*, **385** (2006).

34. F. Gao, C. V Thompson, Jesús del Alamo and T. Palacios, *Role of Electrochemical Reactions in the Degradation Mechanisms of AlGaN/GaN HEMTs*, CS MANTECH (2014), gaasmantech.com/Digests/2014/papers/009.pdf.

35. R. D. Dupuis, S. C. Shen, J. H. Ryou and P. D. Yoder, *Development of High Power NPN GaN/InGaN Double-Heterojunction Bipolar Transistor*, https://smartech.gatech.edu/handle/1853/46602.

36. Y.-C. Lee et al., *NpnGaN/InGaN HBT with Pd Base Contact*, CS MANTECH (2014).

37. J. Simon, University of Notre Dame, https://www3.nd.edu/~gsnider/EE666/666_05/JSimon_HBT.

38. U. Mishra, P. Parikh and Y. F. Wu, GaN HEMTs for commercial and DoD applications, http://my.ece.ucsb.edu/Mishra/classfiles/overview.pdf.
39. L. Pang et al., *GaN MOSHEMT Using Sputtered Gate SiO_2 and Post Annealing Treatment*, CS MANTECH (2013).
40. E. F. Schubert, *Light Emitting Diodes*, Cambridge Press (2006).
41. N. E. J. Hunt and E. F. Schubert, *Electron. Lett.*, **28**, 2169 (1992).
42. M. Razeghi and M. Henini, *Optoelectronic Devices*, http://kqd.eecs.northwestern.edu
43. S. Wang, *Fundamentals of Semiconductor Theory and Device Physics*, Prentice-Hall International (1989).
44. S. M. Sze, *Physics of Semiconductor Devices*, Wiley Interscience (1974).
45. H. Amano et al., *J. Lumin.*, **40–41**, 121 (1988).
46. Heribert Zull, OSRAM online resource.
47. F. Fonstadt, MIT open courseware.
48. S. Nakamura et al., *Appl. Phys. Lett.*, **62**, 2390 (1993).
49. H. Amano et al., *Phys. Status Solidi B*, **216**, 683 (1999).
50. W. C. Lai et al., *J. Displ. Technol.*, **9**, 895 (2013).
51. N. Tansu et al., III-Nitride photonics, *IEEE Photon. J.*, **2**, 241 (2010).
52. S. Müller, R. Quay, F. Sommer, F. Vollrath, R. Kiefer, K. Köhler and J. Wagner, Epitaxial growth and device fabrication of GaN based electronic and optoelectronic structures, *10th European Workshop on MOVPE*, Lecce (Italy), June 8–11, 2003.
53. R. Micalzik and K. J. Ebeling, Operating principles of VCSELs, in *Vertical Cavity Surface-Emitting Laser Devices*, Eds., K. Iga and H. E. Li, Springer (2003).

Chapter 24

RF MEMS

24.1 Introduction

Microelectromechanical systems (MEMS) based on silicon micromachining are finding numerous applications in a host of industries and revolutionizing the next generation of sensors in automobile, chemical, and biomedical industries. MEMS are also used in accelerometers and micro-robots and have immense potential in nanotechnology. Silicon-based MEMS have been in use for over two decades, for example, for deformable mirrors.

The interest in III–V MEMS is driven by demonstrated excellent radio frequency (RF) performance of such devices and their compatibility with existing production monolithic microwave integrated circuit (MMIC) processes. As an example, a MEMS switch shows good isolation better than 30 dB and an insertion loss less than 0.3 dB at 2 GHz [1], thus making it a candidate for switching in cell phone applications. Applications could include switching of antennas, transmission/receiving, bands, filters, and power amplifiers (PAs).

Other applications include tunable capacitors and suspended spiral inductors, which have less parasitics, Fabry–Perot filters, reconfigurable or steerable antennas and integrated circuits (ICs), reconfigurable satellite communication systems, and optical switching.

III–V Integrated Circuit Fabrication Technology
Shiban Tiku and Dhrubes Biswas

ISBN 978-981-4669-30-6 (Hardcover), 978-981-4669-31-3 (eBook)
www.panstanford.com

24.1.1 Differences with Silicon [2]

III–V semiconductors are useful for RF and optical applications because of precise control possible with epilayers and stress control through lattice mismatch. Compound semiconductors are direct-energy-bandgap materials, so MEMS fabrication on these can be combined with optical devices like lasers. For silicon SiO_2 and borophosphate silicon glass (BPSG) are used as sacrificial layers, so HF and buffered oxide etch (BOE) are used as chemical etches. For III–V compounds there are more choices of sacrificial compounds and therefore etchants. Polyimide-type polymers are already in use in MMIC processing, so these can be easily incorporated as sacrificial layers, too. Compound semiconductors are piezoelectric materials; therefore this property can be used for actuation.

24.2 Basics of MEMS

RF MEMS devices have relatively simple geometries compared to complicated silicon MEMS. Simple beams or patterned cantilevers are used as the moving element, and movement is limited to short distances. The basic actuation mechanisms are listed below.

- Electrostatic: This is the most common mechanism used and the simplest to fabricate. These consume almost no control power. However, it is difficult to get good separation with simple designs, so structure modifications are used in practice [3].
- Piezoelectric: Two materials of dissimilar piezoelectric constants are used like a bimetallic strip to make the moving element. When an electric field is applied, the differential contraction causes bending.
- Magnetic and electrothermal actuation: These actuation types are slow for RF applications and the electrothermal activation must, in fact, be avoided as a parasitic in practice.

24.3 Ohmic Contact Switches

Conceptually, this is the simplest MEMS switching device. A schematic diagram of a metal/air/metal system is shown in Fig.

24.1. Conducting metallic bumps are used for making and breaking contact in this simple cantilever geometry. A wedge is used as the fulcrum. In practice these devices have problems of corrosion and stiction, which is the static friction that needs to be overcome to enable relative motion of stationary objects in contact. The lifetime of these devices is limited.

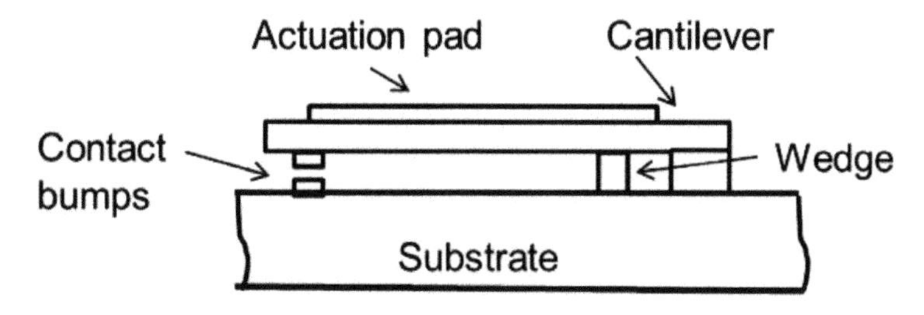

Figure 24.1 Schematic of an ohmic contact MEMS cantilever device.

24.4 Capacitive Switches

24.4.1 Electrostatic or Capacitive Excitation

For RF applications, the problems of a simple ohmic switch can be solved by adding a dielectric layer on top of the metal, thus forming a metal–insulator–metal (MIM)-type switch. This increases the switch lifetime to make these devices practical. The area of the contacts must be kept low and the distance at zero excitation reduced to minimize the off-state capacitance and improve isolation.

Let us consider the simplest case of electrostatic actuation [4]. Figure 24.2 shows a schematic of a capacitive switch with a fixed bottom plate and a movable top plate on a beam. The plates are separated by air. When a voltage is applied, the Lorentz force will be given by

$$F = qE \tag{24.1}$$

The actual force on a conducting beam may be half because some charge is shielded by the surface charge.

The force per unit area, $P_e = \rho s(\mathbf{E_0}/2)$
where $\mathbf{E_0}$ is the electric field on the surface.

and

$$\rho s = \varepsilon_0 \mathbf{E_0}$$

where ε_0 is the permittivity of free space (here air):

$$P_e = \varepsilon_0 \frac{\mathbf{E}_0^2}{2} \text{ (N/m}^2\text{ units)} \tag{24.2}$$

The cantilever acts like a spring with a constant K.

$$F = Kx = PA \tag{24.3}$$

where A is the area and x is the deflection.

$$x = \frac{PA}{K} \; \varepsilon_0 \frac{\mathbf{E}_0^2}{2} \cdot \frac{A}{K} = \varepsilon_0 \frac{A.\mathbf{E}_0^2}{2K} (m) \tag{24.4}$$

The deflection depends upon the composition of the cantilever and the area, as expected. $\mathbf{E_0}$ is equal to V/g, where g is the gap between the moving beam and the fixed plate. The force generated, P_eA, must overcome the stiction force to break and make the contact. The cantilever beams are designed to reduce K (for a certain A) and contacts are designed to reduce the stiction forces.

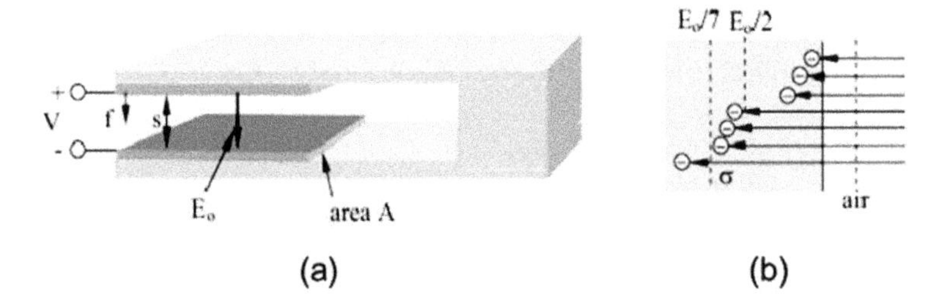

Figure 24.2 Electrostatic MEMS switch (a) and forces on a charged conductor (b) [4].

The structure designed for lateral movement or partially overlapping capacitor plates are good examples of these (Fig. 24.3).

The figure of merit of RF MEMS has similarity to other RF circuits. The cutoff frequency is related to capacitance and on-resistance.

$$f_c = \frac{1}{2\pi R_{ON} C_{OFF}} \tag{24.5}$$

So, reducing the on-resistance and keeping the capacitance low are the goals of layout design and the fabrication process.

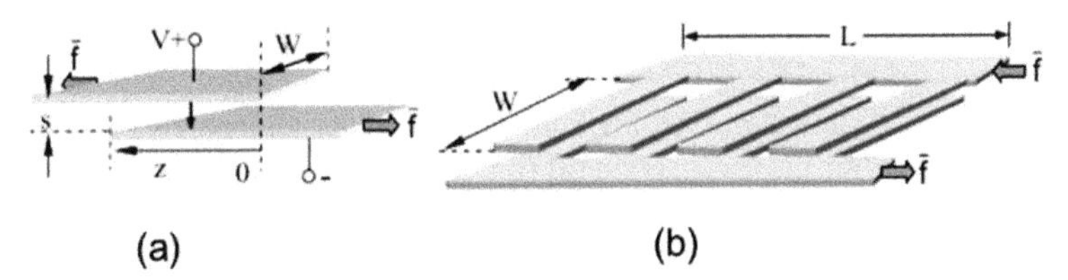

(a) (b)

Figure 24.3 Electrostatic actuator with partially overlapping plates [4].

24.4.2 Actuation Voltage [5]

The actuation voltage for electrostatic excitation can be derived by equating the electrostatic force with the force exerted by the beam due to its stiffness. The capacitance of the contact system is given by

$$C = \varepsilon_0 \frac{A}{g_0} \tag{24.6}$$

where g_0 is the gap.

The electrostatic force is given by

$$F_e = ½(V^2)\,\frac{dC(g)}{dg} = -(½)\varepsilon_0 AV^2/g^2 \tag{24.7}$$

This force must equal the force exerted by the beam stiffness:

$$F_s = K(g_0 - g) \tag{24.8}$$

Using the above two equations ($F_s = F_e$)

$$\varepsilon_0 AV^2/2g^2 = K(g_0 - g) \tag{24.8}$$

Therefore

$$V = \sqrt{2\frac{K}{\varepsilon_0}\frac{g2}{A}(g_0 - g)} \tag{24.9}$$

As the beam bends due to the force, at $g = 2g_0/3$, the force exceeds a critical limit and the beam collapses, and the actuation voltage is given by

$$V_p = \sqrt{\frac{8K}{27\varepsilon_0}\frac{g3}{A}} \tag{24.10}$$

The actuation voltage depends upon g and K. The value of g, the gap, depends on the capability of the MEMS process technology. Therefore, the actuation voltage depends mainly on the value of the spring constant K.

Historically, the main drawback of MEMS devices was their high actuation voltage. In the capacitive mode, voltages over 25 V were needed. As mentioned above, to reduce the voltage, the K factor of the cantilever needs to be low. The *K* factor depends upon the material of the metal; however, the choices of the metal are limited to the ones generally used for MMIC fabrication. Therefore, research and development has focused on the geometry of the beam.

Hinged structures have been shown to reduce activation voltage to a reasonable level of below 10 V. Low actuation voltage is also desirable for good reliability. Figure 24.4 shows the concept of a hinged cantilever beam that has been fabricated on a GaAs substrate. The RF signal travels over the coplanar waveguide on the bottom, over which the metal pad is suspended using the hinged structure [6]. In this case the voltage was applied to the bottom plate, and the RF is pulled to ground when the switch is closed.

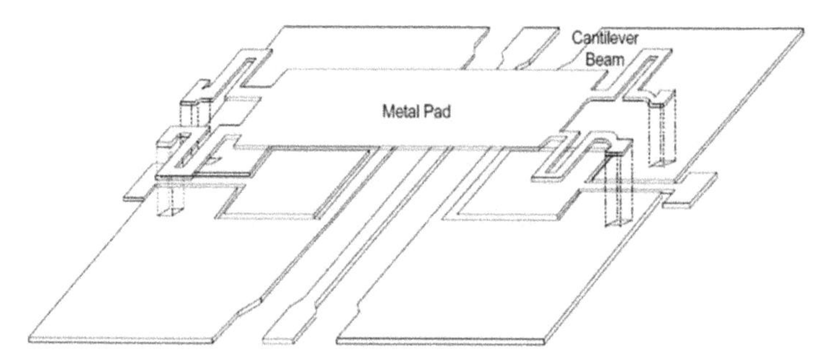

Figure 24.4 Schematic diagram of a hinged MEMS switch. Reprinted from Ref. [6] with permission. Copyright © 2001 GaAs MANTECH.

The *K* factor can be reduced also by making the hinged structure even more flexible by using serpentine structures. With meander structures, actuation voltage drops as the number of meanders increases [3, 5]. An extreme case of a meander structure, a serpentine, is shown in Fig. 24.5, with simulated displacement represented by color [5].

Figure 24.5 Serpentine cantilever structure as an extension of a meander structure. Reprinted from Ref. [5] with permission from IJAREEIE.

24.4.3 Piezoelectric Excitation [7]

Figure 24.6a shows a piezoelectric material film under applied voltage. The material has been poled (domains aligned) to result in this mode of expansion and contraction (as opposed to shear, etc.). When the bilayer is under an electric field, the structure bends, as shown in Fig. 24.6b.

In the case of piezoelectric excitation of MEMS devices, an inverse piezoelectric effect is utilized, which means that the application of the electric field results in stress and consequent strain or deflection.

$$\lambda = d/L \propto \sigma = F/A \tag{24.11}$$

$$d = g_P.E \tag{24.12}$$

where g_P is the piezoelectric constant and d is the deflection of a bar of length, L. The piezoelectric constant needs to be high for high deflection. In the case of polycrystalline materials, layers need to be dense and strongly poled to result in high polarization. For III–V compounds, crystalline quality is better due to use of epilayers.

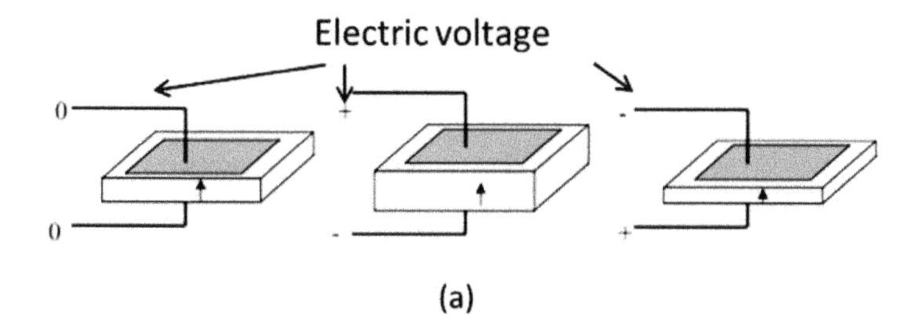

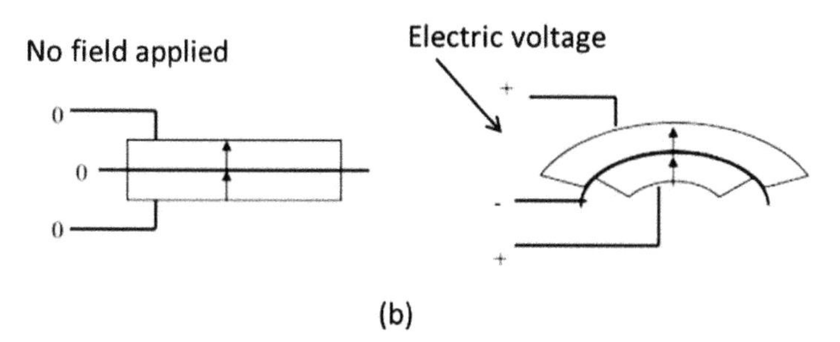

Figure 24.6 Piezoelectric film in thickness expansion mode, single layer (a). Deflection is double layer (b).

24.5 Process Technology

A couple of simple processes for making a MEMS device using metal beams and polymer sacrificial layer are described first.

24.5.1 OMMIC Process

In the first process described recently by OMMIC Foundry [8, 9], a MEMS switch is added to an MMIC circuit. Process steps are shown in Fig. 24.7.

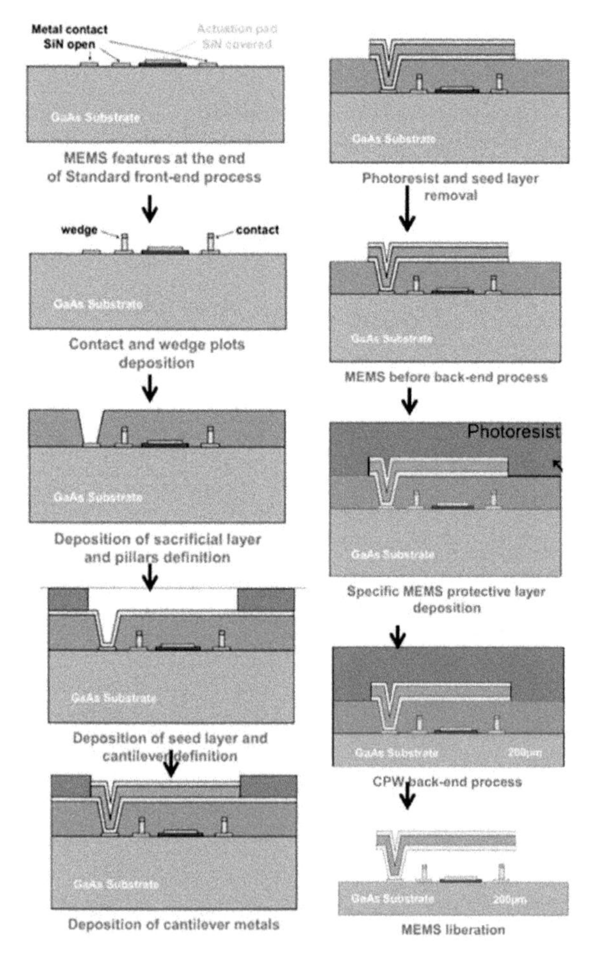

Figure 24.7 MEMS front-end process (OMMIC process). Reprinted from Ref. [8] with kind permission of OMMIC.

The MEMS process starts after the front-side processing of the MMIC device is complete. Contact and wedge posts are added first. A sacrificial layer is deposited next. Then pillars are defined and a seed layer is deposited all over. Now the cantilever is defined by photolithography and the metal deposited and photoresist lifted off. At this point, the redundant seed layer is etched off. If the wafer needs to be thinned, the devices are protected by a thick photoresist layer during the grinding process. The MEMS device cantilever is released. If through-wafer vias (TWVs) are needed, the MEMS device is not released until the TWV process is complete. During the TWV process, the MEMS device is protected by polymer-type layers and silicon nitride, as shown in Fig. 24.8.

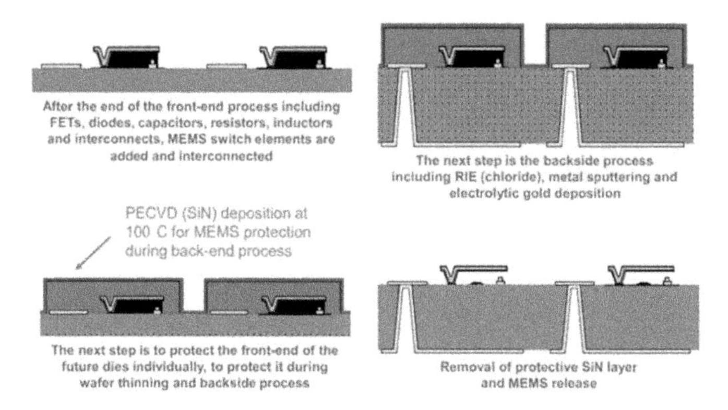

Figure 24.8 Back-end process for MEMS processing showing through-wafer via fabrication. Reprinted from Ref. [8] with kind permission of OMMIC.

Packaging of MEMS devices is carried out using packaging processes needed for devices that must be placed inside a cavity (acoustic filters, etc.). A schematic of a packaged MEMS device is shown in Fig. 24.9.

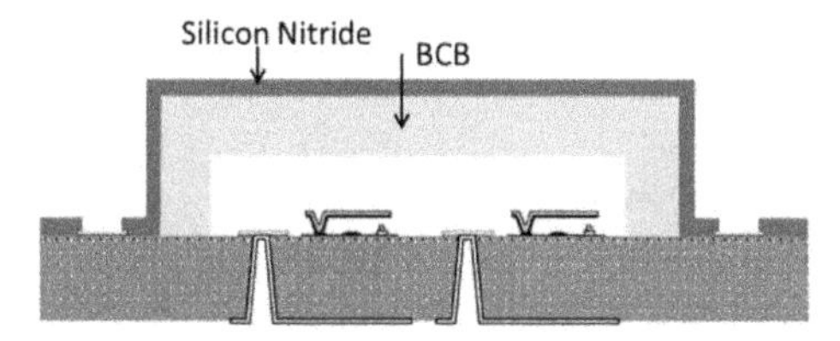

Figure 24.9 Packaged MEMS device schematic. Reprinted from Ref. [8] with kind permission of OMMIC.

24.5.2 University of Illinois Process [6]

This process module can be adopted for Si or III–V wafers. The process involves five mask layers. Process steps are illustrated in Fig. 24.10a. The coplanar waveguide uses 1 μm standard gold-based films. Plasma-enhanced chemical vapor deposition (PECVD) silicon is used as the dielectric. A via hole layer and a metal bump layer complete the lower plate processing. A layer of thick polyimide, a few microns generally, is deposited next. Vias are made in this layer to form anchors, dimples, etc. A thick gold layer is deposited using a standard air bridge methodology (see also Chapter 18). The polyimide is then removed by wet etch and the cantilever released using CO_2 supercritical drying technology [10].

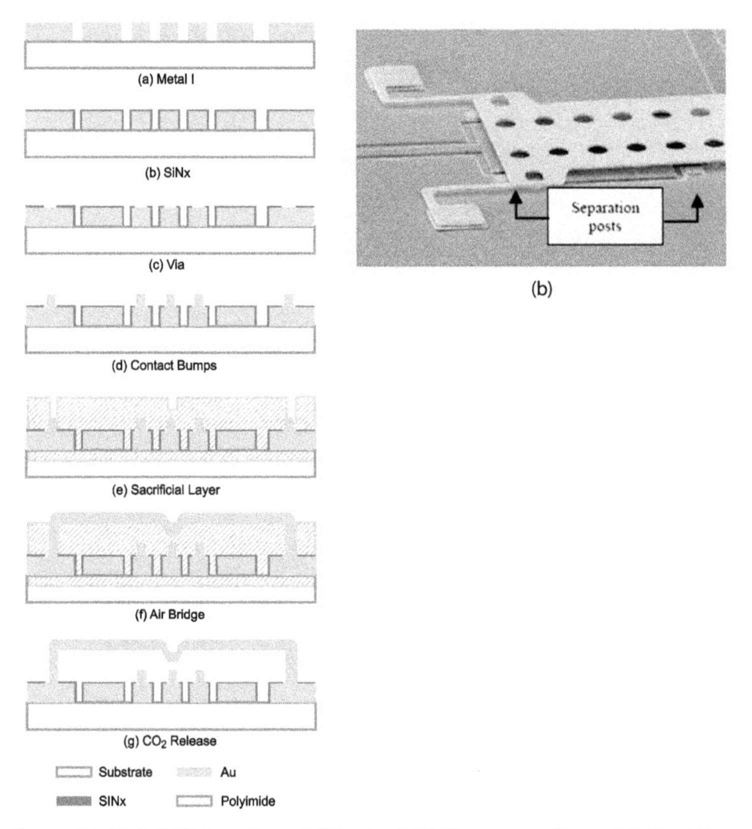

Figure 24.10 (a) University of Illinois MEMS process flow. (b) Cantilever with holes for higher deflection and separation posts for low stiction. Reprinted from Ref. [10] with permission. Copyright © 2003 GaAs MANTECH.

The design uses a hinged cantilever with holes. The voltage is applied to the bottom electrode and the cantilever pulled down to short the RF signal. Metal bumps acting as separation posts are added to reduce stiction issues (see Fig. 24.10b), thus improving lifetime.

24.6 Examples of Applications

24.6.1 Waveguide Switch [2]

Figure 24.11 shows an example of a waveguide switch. Here a 4 μm thick waveguide structure is grown on top of a 2 μm AlGaAs layer with a high Al fraction of 0.7. Once the beam structure is patterned, the structure is released by using HF-based wet etch. This structure needed a high voltage of 25 V to actuate. Switching speed was 40 μsec; however, with changes in design , 10 μsec is possible.

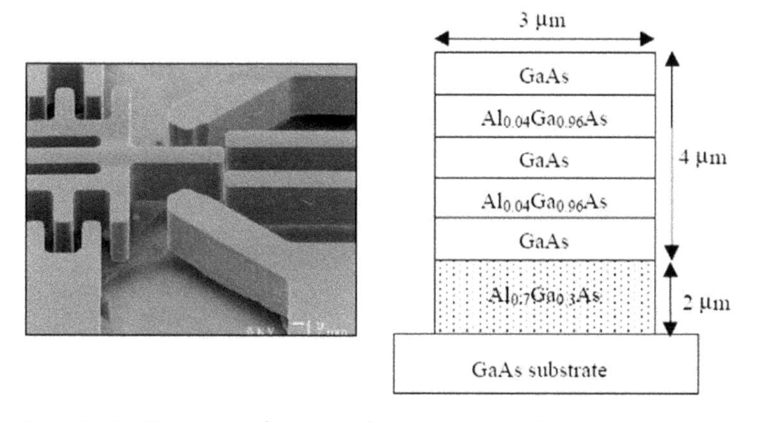

Figure 24.11 Top view of waveguide switch and schematic cross section of the layer structure. Shaded layer is the sacrificial layer. Reprinted from Ref. [2] with permission. Copyright © 2001 GaAs MANTECH.

24.6.2 Fabry–Perot Filter

An early example of a Fabry–Perot filter designed and fabricated with micromachining technology is shown in Fig. 24.12 [11]. The structure is formed on an InP substrate. The cavity is made between two distributed Bragg reflectors (DBRs) (described in Chapter 23), in an approach similar to tunable lasers. An InGaAs/InP DBR is used at the bottom. The bridge (or beam) is made of InP, InGaAs being etched away to form the cavity. The top DBR is made of Si–SiO_2–Si–SiN*x*, $\lambda/4$ layers on top of the InP bridge. The bridge is actuated using electrostatic means [11].

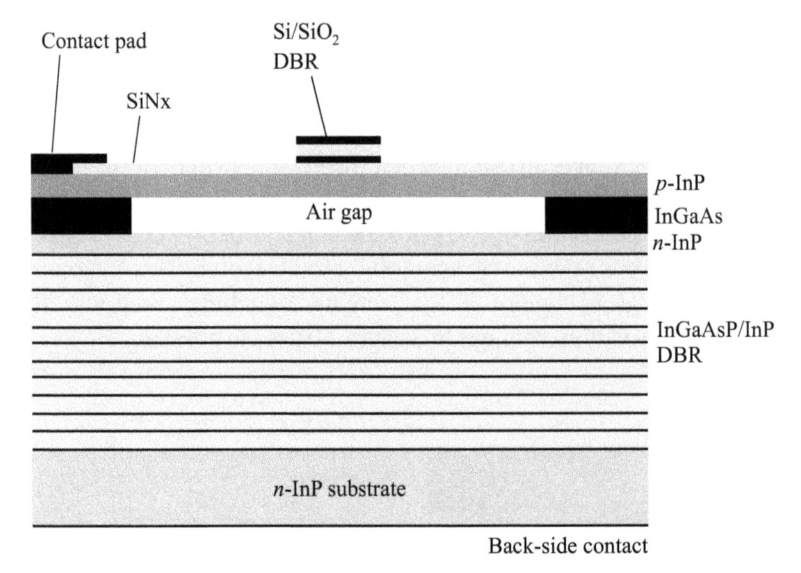

Figure 24.12 Fabry–Perot MEMS device based on InP using a tunable cavity between distributed Bragg reflectors (DBRs).

24.6.3 MEMS on MMIC

Using the process described in Section 24.5.1, an RF sensing device shown in Fig. 24.13 has been fabricated by Grandchamp et al. [9]. The design uses a 150 × 280 μm active area flexible cantilever supported by two anchors. The device has four contact points, two anchors, and a wedge for a clean on-off function. The cantilever is divided into four flexible arms. The contact resistance was 1 ohm. The switching voltage was high, 60 to 80 V. Switching times were 1 μsec for open to close and 20 μsec for close to open.

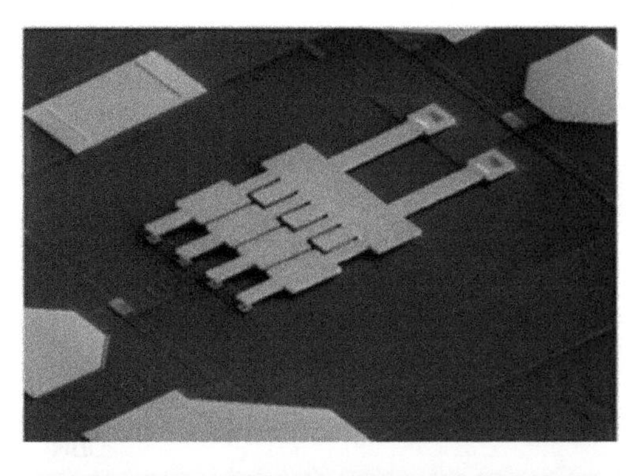

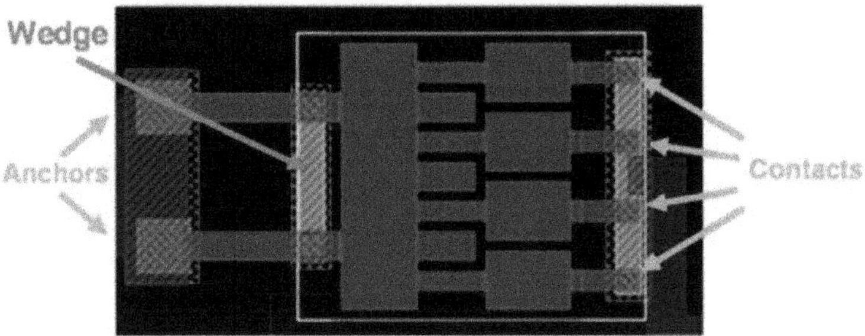

Figure 24.13 MEMS on MMIC example showing SEM view and schematic. Design is optimized for low activation voltage and limiting double transitions. Reprinted from Ref. [9] with kind permission of OMMIC.

24.6.4 Hybrid Circuits

Examples are tunable optical lasers and filters [12]. The resonant frequency or mode is fixed for optical III–N devices. To make these tunable to different frequencies, MEMS-type structures can be utilized. A multilayer mirror or reflector is fabricated and then removed from the substrate by etching away a sacrificial layer underneath. The separated reflector is then attached to a vertical cavity surface emitting laser (VCSEL) device by placing it on posts and the interconnect is completed. As the suspended element is deflected by excitation, the optical path changes, thus changing the fundamental resonant frequency. See Fig. 24.14. The process to make the VCSEL and reflector on the same substrate would be a lot more complicated.

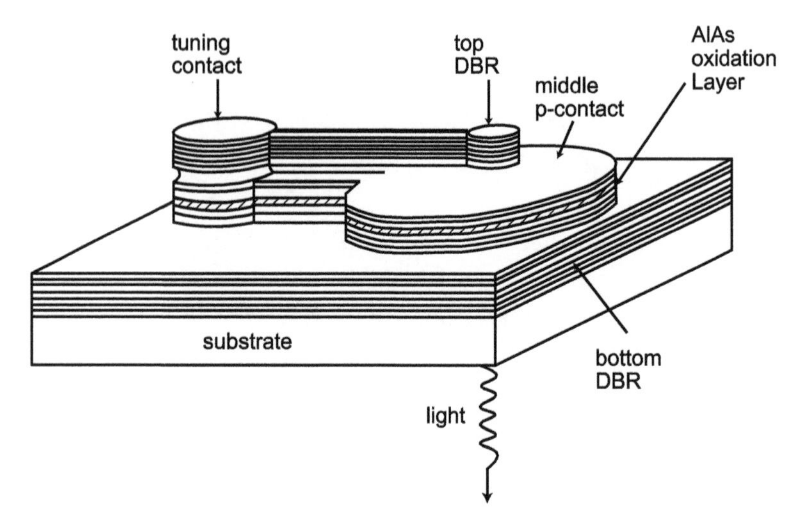

Figure 24.14 Schematic diagram of a micromechanical tunable VCSEL. Reprinted from Ref. [2] with permission. Copyright © 2001 GaAs MANTECH.

The DBR element can be made by depositing a thick AlAS buffer layer on the substrate, followed by a GaAs/AlGaAs multilayer stack, each layer being a quarter-wavelength thick. The whole structure is grown by molecular beam epitaxy (MBE). After patterning and completion of the reflector, the AlAs sacrificial layer is etched. The researchers removed the 250 × 250 μm mirrors and flip chip attached to the VCSEL. For details of VCSEL operation, see Chapter 23.

References

1. P. Gammel, G. Fischer and J. Bouchaud, *RF MEMS and NEMS technology, devices and applications, Bell Labs Tech. J.*, **10**, 29 (2005).
2. O. Blum Spahn et al., *Promise and Progress of GaAs MEMS and MOEMS*, GaAs MANTECH (2001).
3. S. Lucyszyn, *Review of radio frequency microelectromechanical systems technology, IEE Proc.-Sci. Meas. Technol.*, **151**, 93 (2004).
4. D. H. Staelin, *Electromagnetics and Applications*, http://ocw.raf.edu.rs/courses/electrical-engineering-and-computer-science/6-013-electromagnetics-and-applications-spring-2009/readings/MIT6_013S09_notes.pdf.

5. S. Sarkar and A. V. Juliet, Design of a low voltage RF MEMS capacitive switch with low spring constant, *Int. J. Adv. Res. Electron. Instr. Eng.*, **3**, 8966 (2014).
6. S. C. Shen et al., *Development of Broadband Low Voltage RF MEMS Switches*, GaAs MANTECH (2001).
7. J. R. Philips, Piezoelectric technology primer, CTS Wireless Corp.
8. B. Grandchamp (OMMIC), Process notes, *RF MEMS Workshop*, Paris (2013), http://www.microwave-rf.com/docs/11h15-OMMIC.pdf.
9. B. Grandchamp et al., Monolithic integration of RF MEMS switch and GaAs MMIC for RF sensing applications (2012), http://www.microwave-rf.com/docs/11h15-OMMIC.pdf. See also http://www.armms.org/media/uploads/08_armms_apr12_bgrandchamp.pdf.
10. R. Chan et al., *Ultra Broadband MEMS in Si and GaAs Substrates*, GaAs MANTECH (2003).
11. N. Chitica et al., Monolithic InP-biased tunable filter with 10-nm bandwidth for optical data interconnects in the 1550-nm band, *IEEE Photon. Technol. Lett.*, **11**, 584 (1999).
12. M. C. Harvey, et al., *Towards Hybrid MEMS, Tunable Optical Filters and Lasers*, MANTECH (2004).

Appendix

Useful Physical Constants

Angstrom unit	Å	10^{-4} mm $=10^{-10}$ m
Avogadro's number	N_{AVO}	6.023×10^{23} molecules/g-mole
Bohr radius	a_{B}	0.53 Å
Boltzman's constant	k	8.62×10^{-5} eV/K
Electron charge	q	1.602×10^{-19} coulombs
Electron rest mass	m_0	9.11×10^{-28} g
Electron volt	eV	1.6×10^{-19} joules
Micron	μm	10^{-6} m
Permittivity of free space	ε_0	8.854×10^{-14} farad/cm
Permeability of free space	μ_0	1.257×10^{-8} henry/cm
Planck's constant	h	6.625×10^{-34} joule-sec
Standard atmosphere		1.033×10^{4} kg/cm^2 = 760 mm Hg (Torr)
Thermal voltage (300 K)	kT/q	0.0259 volts
Velocity of light in free space	**c**	2.998×10^{10} cm/sec
Wavelength associated with 1 eV	λ	1.239 μm
Base of natural logarithm	e	2.71828

Table A.1 Properties of selected semiconductors at room temperature, 300 K

Material	Atomic molecular weight	Bandgap E (eV) at 300 K	Lattice parameters a,c (Å)	Melting point (°C)	Thermal conductivity ($Wcm^{-1}K^{-1}$)	Mobility μ ($cm^2V^{-1}s^{-1}$)		Dielectric constant ε_r
						Electron	Hole	
Group IV								
C (Diamond)[Dd]	12.01	5.47[l]	3.567	3577	30,000	1800	1200	5.7
4H–SiC [W]	40.1	3.2[l]	4.360	1800	4900	800	115	9.7
6H–SiC [W]		3.0[l]	3.08, 15.12	1800	4900	370	90	9.7
3C–SiC [Z]		2.2[l]	4.36	1800	3,500–4,500	750	40	9.7
Si[Dd]	28.09	1.12[l]	5.431	1415	1500	1500	450	11.9
Ge[Dd]	72.6	0.66[l]	5.646	937	600	3900	1900	16.2
α-Sn[Dd]	118.69	0.08	6.4912	505.2	640	2500	2400	
III–V								
BN[Z]	24.82	4.6	3.615	3300	200			
BP[Z]	41.78	2.1	4.538	2800		500	70	
BAs[Z]	85.73	1.5	4.777	2300				
AlN[Z]	40.99	5.11[D] (theory)	4.38	2400	823	–	–	8.5
W		6.2	3.112, 4.982		2000			
AlP[Z]	57.95	2.46[l]	5.451	1870	900	60	–	9.8
AlAs[Z]	101.90	2.16[l]	5.665	1740	910	294	420	10.06

Material	Atomic molecular weight	Bandgap E (eV) at 300 K	Lattice parameters a,c (Å)	Melting point (°C)	Thermal conductivity ($Wcm^{-1}K^{-1}$)	Mobility μ ($cm^2V^{-1}s^{-1}$)		Dielectric constant ε_r
						Electron	Hole	
$AlSb^Z$	148.73	1.58^I	6.136	1057	600	200	550	
GaN^W	83.73	3.39^D	3.189, 5.185	1500	2200	1500	30	8.9
^{Z}		3.2-3.3	4.52					5.3
GaP^Z	100.69	2.26^I	5.451	1480	1100	110	75	11.11
$GaAs^Z$	144.63	1.43^D	5.653	1238	460	8500	400	13.18
$GaSb^Z$	191.47	0.72^D	6.096	706	350	5000	850	15.69
InN^W	128.83	1.89	0.7/0.8	156				
InP^Z	145.79	1.35^D	5.869	1062	700	4600	150	12.56
$InAs^Z$	189.74	0.36^D	6.058	943	260	33,000	460	15.15
$InSb^Z$	236.57	0.17^D	6.479	525	150	80,000	1250	
II–VI								
ZnO^R	81.38	3.35^D	4.580	1975	234	200	180	
ZnS^Z	97.44	3.54^Z	5.420	1827	270	165	5	7.45
ZnS^W		3.91^W						
$ZnSe^Z$	144.34	2.7^D	5.667	1500	180	540	28	7.6
$ZnTe^Z$	192.99	2.26	6.1					9.7
CdS^Z	144.47	2.42^D	5.832	1397	270	340	50	8.4

(Continued)

Table A.1 (*Continued*)

Material	Atomic molecular weight	Bandgap E (eV) at 300 K	Lattice parameters a,c (Å)	Melting point (°C)	Thermal conductivity ($Wcm^{-1}K^{-1}$)	Mobility μ ($cm^2V^{-1}s^{-1}$)		Dielectric constant ε_r
						Electron	Hole	
$CdSe^Z$	191.37	1.70^D	6.050	1258	–	800	–	9.7
$CdTe^Z$	240.01	1.56^D	6.482	1097	60	1050	100	10.6
HgS **(α-HgS)**	232.656	-0.17^S	5.85			250		18
$HgSe^Z$	279.55	-0.1^S	6.07			20,000		25.6
$HgTe^Z$	328.19	-0.15^S	6.46			25,000	350	2.1
PbS^R		0.37				800	1000	190
$PbSe^R$		0.26				1500	1500	230
$PbTe^R$		0.25				1600	750	500
Ternary								
$Al_{0.48}In_{0.52}As$		1.46			–	900	180	12.9
$In_{0.53}Ga_{0.47}As$		0.75			50	7000	300	12.5

Crystal structure: [W] Wurtzite (hexagonal) [Z] Zinc blende (cubic)
[Dd] Diamond
[C] Cinnabar(hexagonal) [R] Rock salt (NaCl) [D] Direct bandgap
[S] Semimetal [i] Indirect bandgap

Source: Reprinted from EMF Limited (2005). III-Vs Review, 2002(Advanced Semiconductor Buyers' Guide), pp. 76–78, Copyright (2005), with permission from Elsevier.

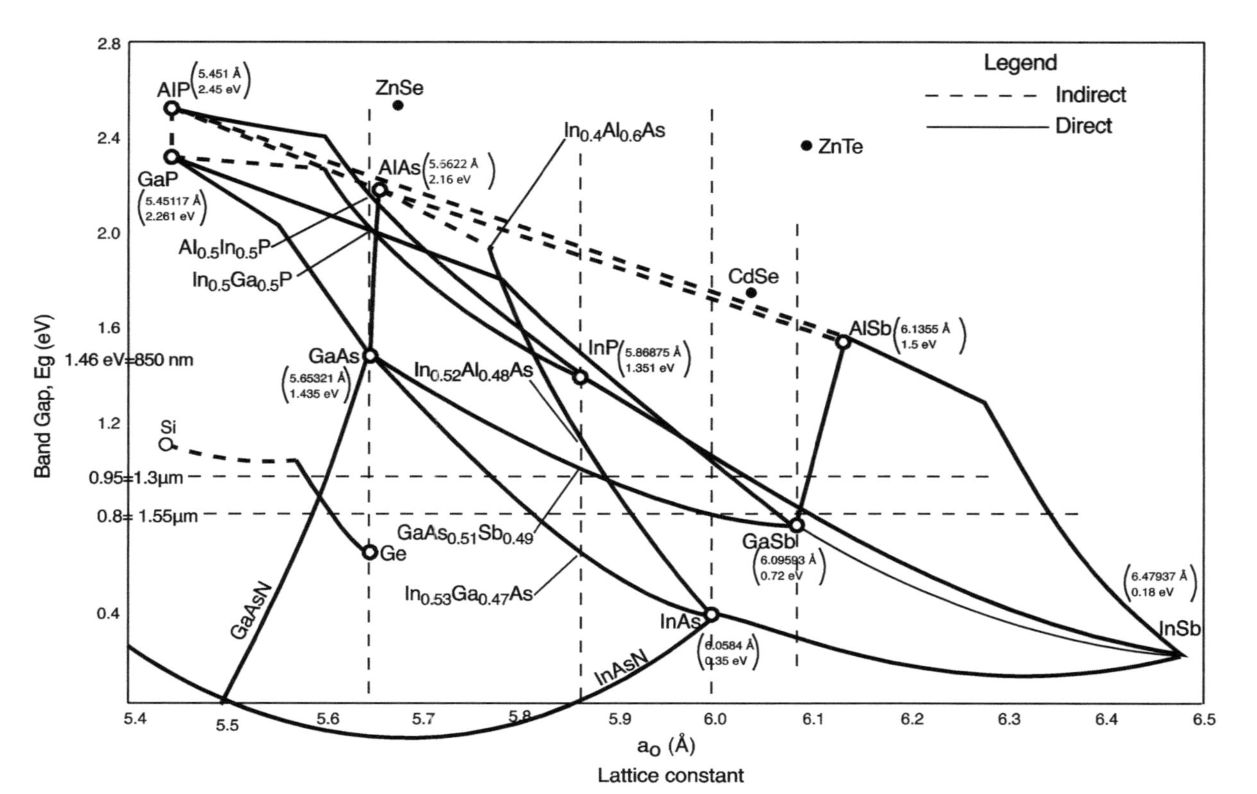

Figure A.1 Bandgap vs. lattice constant. Reprinted from EMF Limited (2005). III-Vs Review, 2002(Advanced Semiconductor Buyers Guide), pp. 76–78, Copyright (2005), with permission from Elsevier.

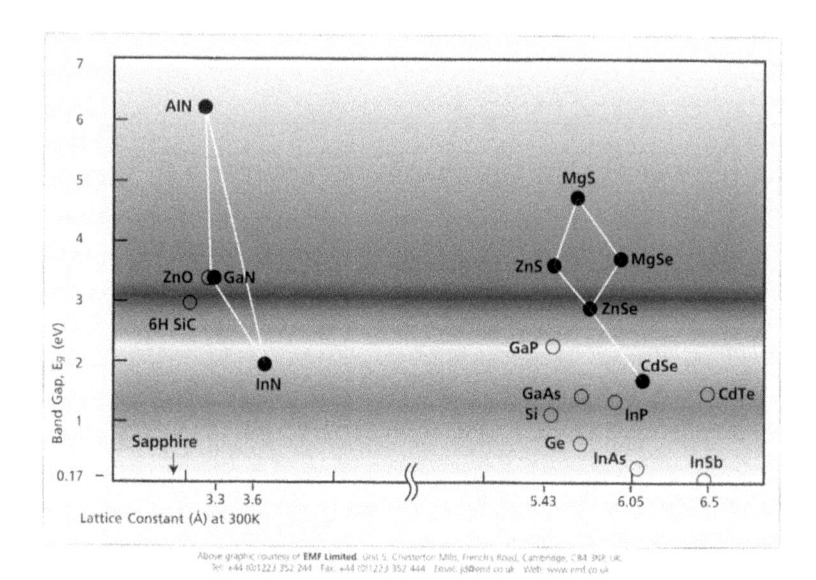

Figure A.2 Bandgap vs. lattice constant. Reprinted from EMF Limited (2005). III-Vs Review, 2002 (Advanced Semiconductor Buyers' Guide), pp. 76–78, Copyright (2005), with permission from Elsevier.

Acronyms

2DEG	two-dimensional electron gas
ACP	adjacent channel power
ALD	atomic layer deposition
ALE	atomic layer epitaxy
ARC	antireflective coating
BCB	benzocyclobutene
BEP	beam-equivalent pressure
BGA	ball grid array
BHF	buffered hydroflouric acid
BJT	bipolar junction transistor
BOE	buffered oxide etch
BPSG	borophosphate spin on glass
BVCBO	breakdown voltage collector to base
CCD	charge-coupled device
CMOS	complementary metal–oxide semiconductor
CMP	chemical mechanical polish
DBR	distributed Bragg reflector
DUT	device under test
DUV	deep ultraviolet
DX	deep localized energy states seen in AlGaAs and other III–V semiconductors
E/D	enhancement/depletion mode
ECR	electron cyclotron resonance

EFR	early failure rate
EOS	electrical overstress
ESD	electrostatic discharge
FER	front-end receiver
FET	field-effect transistor
FTIR	fourier transform infrared spectroscopy
HAST	highly accelerated stress test
HBT	heterojunction bipolar transistor
HEMT	high-electron-mobility transistor
HFET	heterojunction field-effect transistor
HMDS	hexamethyl di-silazane
HTOL	high-temperature operating life
HVPE	hydride vapor-phase epitaxy
ICP	inductively coupled plasma
Idss	drain current under saturation
LDMOS	lateral-diffused metal–oxide semiconductor
LEC	liquid-encapsulated Czochralski
LED	light-emitting diode
LN2	liquid nitrogen
LNA	low-noise amplifier
LPO	liquid-phase oxidation
LSI	large-scale integration
LSP	low-silicon process
MAG	maximum available gain
MBE	molecular beam epitaxy
MEM	microetched machine
MFC	mass flow controller
MIS	metal–insulator semiconductor
MMIC	monolithic microwave integrated circuit
MOCVD	metal–organic chemical vapor deposition
MOMBE	metal–organic molecular beam epitaxy

MOSFET	metal–oxide semiconductor field-effect transistor
MQW	multiple quantum well
NF	noise figure
NMP	*N*-methylpyrrolidone
OSP	organic solderability preservative
PA	power amplifier
PAC	photoactive compound
PAE	power-added efficiency
PBN	pyroletic boron nitride
PCM	process control monitor
pHEMT	pseudomorphic high-electron-mobility transistor
PL	photoluminescence
PM	periodic maintenance
PMMA	polymethylmethacrylate
PZT	lead zirconatetitanate
RGA	residual gas analyzer
RHEED	reflection high-energy electron diffraction
RTA	rapid thermal annealing
RTP	rapid thermal processing
SAINT	self-aligned implantation for n+ layer technology
SIMS	scanning ion mass spectroscopy
SMD	surface-mount devices
SOG	spin on glass
SPC	statistical process control
SQW	single quantum well
SRAM	static random access memory
THB	temperature humidity bias
TTV	total thickness variation
UBM	underbump metallurgy
UHV	ultrahigh vacuum
ULSI	ultra-large-scale integration

VCSEL	vertical cavity surface emitting laser
VLSI	very-large-scale integration
WBG	wide bandgap
XPS	X-ray photoelectron spectroscopy

Index

www.ingramcontent.com/pod-product-compliance
Ingram Content Group UK Ltd.
Pitfield, Milton Keynes, MK11 3LW, UK
UKHW021114180425
457613UK00005B/80

* 9 7 8 9 8 1 4 6 6 9 3 0 6 *